XI International Scientific and Technical Conference "Applied Mechanics and Dynamics Systems"

Journal of Physics: Conference Series Volume 944

Omsk, Russia
14 – 16 November 2017

ISBN: 978-1-5108-5751-3
ISSN: 1742-6588

Printed from e-media with permission by:

Curran Associates, Inc.
57 Morehouse Lane
Red Hook, NY 12571

proceedings.com

Some format issues inherent in the e-media version may also appear in this print version.

Printed by Curran Associates, Inc. (2018)

For permission requests, please contact the Institute of Physics at the address below.

Institute of Physics
Dirac House, Temple Back
Bristol BS1 6BE UK

Phone: 44 1 17 929 7481
Fax: 44 1 17 920 0979

techtracking@iop.org

Additional copies of this publication are available from:

Curran Associates, Inc.
57 Morehouse Lane
Red Hook, NY 12571 USA
Phone: 845-758-0400
Fax: 845-758-2633
Email: curran@proceedings.com
Web: www.proceedings.com

TABLE OF CONTENTS

IOP Conf. Series: Journal of Physics: Conf. Series **944** (2017) 011001 doi:10.1088/1742-6596/944/1/011001

PREFACE

The object of the scientific and technical conference "Applied Mechanics and System Dynamics" is the analysis of the current state, trends and prospects for the development of applied mechanics and system dynamics with the aim of improving technology and production and creating new-generation equipment.

The conference addresses topical issues of development and research on the problems of mechanical engineering, strength and dynamics of mechanical systems.

IOP Conf. Series: Journal of Physics: Conf. Series **944** (2017) 011002 doi:10.1088/1742-6596/944/1/011002

Peer review statement

All papers published in this volume of *Journal of Physics: Conference Series* have been peer reviewed through processes administered by the proceedings Editors. Reviews were conducted by expert referees to the professional and scientific standards expected of a proceedings journal published by IOP Publishing.

Complex method to calculate objective assessments of information systems protection to improve expert assessments reliability

A Zh Abdenov[1], V A Trushin[2] and G A Abdenova [3]

[1] Novosibirsk State Technical University, Novosibirsk, Russia
[2] Novosibirsk State Technical University, Novosibirsk, Russia
[3] L.N. Gumilyov Eurasian National University, Astana, Kazakhstan

[1] e-mail: amirlan21@gmail.com
[2] e-mail: rastr89@mail.ru

Abstract. The paper considers the questions of filling the relevant SIEM nodes based on calculations of objective assessments in order to improve the reliability of subjective expert assessments. The proposed methodology is necessary for the most accurate security risk assessment of information systems. This technique is also intended for the purpose of establishing real-time operational information protection in the enterprise information systems. Risk calculations are based on objective estimates of the adverse events implementation probabilities, predictions of the damage magnitude from information security violations. Calculations of objective assessments are necessary to increase the reliability of the proposed expert assessments.

Keywords: risk assessment, information security, information system, the objective assessment of the damage, SIEM nodes, information security violations.

1. Introduction

Development of a system with intelligent services for the protection of confidential information (CI) requires the successful implementation of various activities to protect information resources (IR) in computer systems (CS). These activities involve the solution of a number of tasks, in particular, the creation of a system for monitoring security threats. Monitoring systems implement the current and a posteriori approaches to information protection, and the main goal of its creation is to reduce the number of incidents affecting the IR in CS to a minimum level of risk and to minimize the resulting damage. In this paper, we consider an incident to be any illegal, unresolved, unfavorable events (UE) that occur in the company's information systems (IS).

Today, many enterprises widely use various computer systems to search for customers, through employment of software products with intellectual properties, to conclude contracts and sell goods and services. Each of these companies is engaged in the collection, processing, storage and exchange of business information. Company's business processes generate metadata that is stored in various flexible and effective forms in order to ensure rapid data retrieval, use and movement between various enterprise entities and external partners.

The convenience of using the Internet for information exchange makes not only the information that the company sends via the Network, but also the information resources (IR) in any information system (IS) that the company has at its disposal and which are connected to the global public Network, vulnerable. At the same time, the importance of ensuring the safety of IR is understood by the management of every enterprise, organization, as well as private persons, notebook owners. Therefore, the issues of building an information security (IS) system in terms of protection from the most likely attacker, who has sufficient means for penetrating the information structure of the enterprise, are very relevant. Designed IS systems should be reliable with respect to the level of importance, secrecy and criticality of the protected information. At the same time, the cost of reliable protection of IR in the IS

should not exceed the possible damage from the security violation of the protected information.
One of the most promising and effective areas in the development of a security threat monitoring system is currently considered to be SIEM-systems (Security Information and Event Management), which allows for management of information and security events. The main purpose of the SIEM-system construction and operation is to significantly increase the level of information security (IFS) in the information infrastructure by providing the ability to manipulate security information in real-time and to actively manage security events [1, 2]. It is assumed that the active management of statistical information of the past, present and forecasting nature with respect to security incidents and events is based on automatic mechanisms using accumulated information on the history of the analyzed events and the forecast of future events, as well as on automatic adjustment of the monitoring parameters of events to the current state of the protected system [2, 3].
Discussion of issues arising in the development of SIEM-systems for service information infrastructures is carried out in the MASSIF project (Management of Security Information and Events in Service Infrastructure) of European Union's Seventh Framework Program for Research and Technological Development [1]. We assume that one of the issues discussed within the framework of the MASSIF project should be the questions of filling and continuously improving the individual SIEM system nodes with the ability to solve problems of incident classification, forecasting, filtering, and optimizing the use of IR protection in CS [4-8]. The main initial data used by the SIEM system to solve these tasks are records of various audit logs (logs), logging events in the information infrastructure, called "security events". Data on events reflect such actions by users and programs that may have an impact on the IS.
Next generation SIEM-systems aside from the standard node functions should include extensions such as, the analysis of events, incidents and their consequences, decision-making and visualization of information. We will reveal some mechanisms of functional nodes filling according to the hierarchy levels of the SIEM-system: normalization means the reduction of the formats of log entries collected from various sources to a single internal format, which will then be used for their storage and subsequent processing [2, 3]; filtration of the security events is to adjust the current assessments of the security state of IR in the CS and calculate these assessments for a future time interval [5, 7]; prioritization determines the significance and criticality of security events based on the rules defined in the system [1, 8, 9]; analysis of events, incidents and their consequences include procedures for modeling events, attacks and their consequences, vulnerability analysis, risk assessments, forecasting and filtering of NS assessments and incidents [10, 11]; The Decision Support System (DSS) determines the development of measures to optimize and reconfigure security tools to prevent attacks on the CS [2, 3]; Generation of reports and warnings means the formation, transmission, display and (or) printing of the results of operation [1–4].
In this paper, we will propose and disclose some elements of filling of individual SIEM system nodes with algorithms for solving risk calculation problems on the basis of objective assessments, as well as estimating the predictions and filtering of the number of UE, the amount of damage to IR in the CS [5, 6, 10, 11].
At the same time, there arises the problem of estimating the state of possible channels of information leakage and the corresponding assessment of the damage that may occur in the event of information leakage or any other security breach, and likelihood of causing such damage. To determine the adequacy of the cost of the protection system, it is necessary to compare the extent of damage and the likelihood of its application with the costs of providing protection. Unfortunately, it is very difficult and troublesome to obtain an objective real value of each IR in the IS due to a lot of uncertainties, as well as adequate automated and automatic control and accounting of UE in computer systems. Therefore, only expert assessments are often used.
In solving practical problems of information protection, the quantitative assessment of its vulnerability (inability to deter an attack) is of paramount importance. Unauthorized access (UA) to information in the automated and automatic system (AAS) is possible not only through direct access to databases (object), but also in many other ways that do not require such access. In this case, the intentional actions of intruders are the main danger. The impact of random factors in itself does not lead to unauthorized access, it only contributes to the emergence of channels of unauthorized information

retrieval (CUIR), which can be exploited by an attacker.
In this paper, for the purposes of protecting IR in enterprise IS, we will review various approaches to risk management. When calculating risks in IS, we will base not only on well-known expert [3], but also objective assessments of the probabilities of the number of NS implementations. We will also perform prediction assessments of the damage magnitude from IR security violations.
A risk-based approach to assessing damage from attacks by violators and selecting measures to minimize it has been termed risk management. Risk management refers to a full range of successive processes that comply with existing international standards and practices for enterprise risk management [4]. In existing risk management techniques, their identification is carried out by various methods, such as team "brainstorming", drawing up sequences of processes and event trees, analyzing the system architecture, operational modeling, scenario analysis, HAZOP studies [5, 6]. At the present stage, one of the methods of risk management is well tested on expert assessments [6]. Following the approach of Kumamoto and Henley [9], as well as the ideas reflected in [5], we adopt a narrower practical direction of the formal definition of parameters that characterizes the risks of IR security breach. First, it is necessary to classify the types of security infringement attacks. Secondly, it is necessary to accumulate observational data on each type of attack that characterizes the number of violations of each type of attack and the cost estimate of damage from IR security violations.
Methodical assessment of damage to IS caused by UE depends on the risk assessment of UE probability in IS. Evaluation of the objective probabilities of the UE occurrence is one of the important tasks in the algorithm for calculating risk assessments. Note also that the use of these objective assessments to improve the effectiveness of the calculation of risk assessments in enterprise IS is an important task in the methodology for calculating assessments of potential damage from attacks by violators.

2. The analysis and classification of possible UAs to company's information in different zones based on territory and probability

The following methods and methods of committing computer crimes are known: the introduction of unauthorized data; manipulation of unauthorized data; illegal use of files; creation of unauthorized files; overriding of internal control mechanisms; unauthorized destruction and modification of data on exit; unauthorized manipulation of computer programs or documentation; unauthorized manipulation of data processing; manipulation of errors, system failures; unauthorized use of passwords and codes; unauthorized transmission and interception of communications; theft of computer equipment, software; intentional destruction or damage of equipment, software or data, etc.
Territorially potentially possible UAs to company information can take place in different zones [11]:
– external uncontrolled zone of the company;
– zone within the company's controlled territory;
– AAS zone;
– AAS resource zone;
– the database zone.

Furthermore, UAs require following events to take place simultaneously in order to receive information:

– the attacker must gain access to the corresponding zone;
– the necessary CUIR should be available in the accessed zone;
– the manifested CUIR should be accessible to an attacker of the appropriate category (here, it means that the attacker must have the appropriate level of education, skills and equipment to be able to take advantage of CUIR);
– the CUIR at the time of access by an attacker should have protected information.

2.1. External uncontrolled zone.

2.1.1. Calculation of the UE objective probability occurrences

Consider first zone: external uncontrolled zone. All computer crimes can be divided into three classes:

the interception of information; unauthorized access; «data manipulation».
Consider the first class. It:
A) electromagnetic interception, for example, registration of radiation generated by a processor, printer, monitor;
B) direct interception, for example, direct connection to data transmission channels.
Consider the second class:
A) illegal access to the line of the legitimate user;
B) The kind of crime that is called «boarding» is when «computer pirates» get into other people's information systems by guessing their password.
Consider the third class («data manipulation»):
A) type of crime – code substitution, for example, with different code variations;
B) type «Trojan horse» – a secret introduction to someone else's program of such commands that gives attacker ability to execute new, unplanned by the user, software functions;
C) computer viruses – act on the principle, for example, erase all the data of this program, go to the next one and do the same. They have the property of moving through communication networks from one system to another, spreading as a viral disease;
D) the kind of «logical and temporary bomb» – the effect of falsely entered commands on programs under certain conditions and time, etc.
Now we will make a list of the significant types of UEs arising in the IS, leading to a deterioration in the system performance of the computer system (CS). Let this list be a set of various UEs $\{O_1, O_2, \dots, O_m\}$. Let us distinguish from this set some significant subset of UEs, leading to a perceptible violation of the safety of IR in the CS. We denote this subset as $O = \{O_{i_1}, O_{i_2}, \dots, O_{i_m}\}$, for example O_{i_1} - the number of UA with respect to the launch failures of individual CS nodes; O_{i_2} - the number of UEs with respect to the incorrect information applied to a specific information process and data processing, etc.
After building a subset O we proceed to the subset's elements properties analysis on the basis of the quantitative indicators of the UEs and the magnitude of the damage that has occurred in the past. Let $O = \{O_{i_1}, O_{i_2}, \dots, O_{i_m}\}$ – the set of all significant UE, leading to a decrease in the system efficiency of the CS. The mathematical expectation of the damage caused by i's UE in time ΔT (For example, 1 month, 1 half year, 1 year, etc.) can be represented by the formula:

$$e(O_i, \Delta T) = M[e(O_i) \cdot f_i], i = \overline{1, m}, \tag{1}$$

where $e(O_i)$ – a random amount of already occurred UA damage in case of a single UA event; f_i – random amount of UE of type i in time ΔT; m – the total number of all types of already occurred UEs.
If UEs don't have consequences, damage from each UE is independent, then

$$e(O_i, \Delta T) = M[e(O_i)] \cdot M[f_i], \quad i = \overline{1, m}, \tag{2}$$

the damage for the whole set of significant UE will be determined by using the formula

$$E(O, \Delta T) = \sum_{i=1}^{m} M[e(O_i)] \cdot M[f_i]. \tag{3}$$

Algorithm 1

Step 1.1. For simplicity, we consider only one type of UE, for example, we set a specific value i = 1. Further, the quantitative indicators of the UE, for example, μ years we will compile in the table 1:

Table 1. Quantitative indicators of UA that occurred during the last μ years by months (or several quarters, half-year or more rare occurrences of UA).

t	1	2	3	4	5	6	7	8	9	10	11	12
1	$f_1^{(1)}$	$f_2^{(1)}$	$f_3^{(1)}$	$f_4^{(1)}$	$f_5^{(1)}$	$f_6^{(1)}$	$f_7^{(1)}$	$f_{18}^{(1)}$	$f_9^{(1)}$	$f_{10}^{(1)}$	$f_{11}^{(1)}$	$f_{12}^{(1)}$
2	$f_1^{(2)}$	$f_2^{(2)}$	$f_3^{(2)}$	$f_4^{(2)}$	$f_5^{(2)}$	$f_6^{(2)}$	$f_7^{(2)}$	$f_{18}^{(2)}$	$f_9^{(2)}$	$f_{10}^{(2)}$	$f_{11}^{(2)}$	$f_{12}^{(2)}$
$\vdots$	...	...	...	...	...	...	...	...	...	...	...	...
μ	$f_1^{(\mu)}$	$f_2^{(\mu)}$	$f_3^{(\mu)}$	$f_4^{(\mu)}$	$f_5^{(\mu)}$	$f_6^{(\mu)}$	$f_7^{(\mu)}$	$f_8^{(\mu)}$	$f_9^{(\mu)}$	$f_{10}^{(\mu)}$	$f_{11}^{(\mu)}$	$f_{12}^{(\mu)}$

Step 1.2. From the data in Table 1, using formula (4) we can calculate one line of columns averages for monthly (quarterly, semiannual, annual relatively rare occurrences of UA) quantitative values of UA (see Table 2):

$$f_t^{ycp} = \sum_{i=1}^{\mu} f_t^{(i)} / \mu, \ t = \overline{1,12}. \tag{4}$$

Table 2. The average line quantitative occurred UE.

t	1	2	3	4	5	6	7	8	9	10	11	12
f_t^{ycp}	f_1	f_2	f_3	f_4	f_5	f_6	f_7	f_8	f_9	f_{10}	f_{11}	f_{12}

Similar tables desirable to construct in relation to other types of UE that have already occurred within certain period of time, for example one month.

Step 1.3. With respect to the averaged data (see Table 2) it is possible to construct a discrete linear stationary stochastic model in the form of state-space (SS) [9, 10] (according to the procedure described in [9]) of the form:

$$x(t+1) = a \cdot x(t) + b \cdot u(t) + w(t), \quad x(0) = \bar{x}_0, \tag{5}$$

$$f^{ycp}(t+1) = x(t+1) + v(t+1), \quad t = \overline{0, N-1}. \tag{6}$$

where $x(t)$ is the true number of the UA that took place during the month t (could be also a few quarters, a few semesters, as long as a few years for the rare type of UA); $u(t)$ is the externally observable control action in the form of fixed type of UE at the time t ; $w(t)$ is the white Gaussian unobserved effects at a time t with mathematical expectation of zero and unknown variance Q; $x(0)$ is the number of UE at the beginning time $t = 0$ with mathematical expectation $\bar{x}_0$ and unknown variance $P(0)$; a, b are the unknown coefficients in the model dynamics (5); t is the month of the year (or number of quarters, semesters or years); $N = 12$ is the the number of months in a year; $f^{ycp}(t)$ is the observed random number of UE in a month (Data from the company's record book); $v(t)$ is the white Gaussian sequence of observation errors on the number UE within defined period of time, for example, a month with zero mean and unknown variance R .

In this step, it is necessary to evaluate all of the variances associated with noise of the dynamics model $\hat{Q}$, initial state value of the noise $\hat{P}(0)$, with the noise of the measuring system $\hat{R}$ according to the

formulas set forth in [12].
Step 1.4. Assessments for the coefficients in the model dynamics (5) can be calculated based on the method of least squares (OLS) [9, 10].
Step 1.5. The constructed model (5), (6) will provide the most reliable estimates of the number UE (in real time, using Kalman filter) with respect to each month as a filter estimates [10] for a subsequent, for example, (μ +1) year. The resulting filter estimates should be rounded to the nearest whole number.
Step 1.6. Filtering assessments will calculate the objective probability assessment of UE occurrences. For example, we propose the following procedure for calculating the probability for a particular type of UE.
Suppose we are interested in the probability of the UA, for example, in each of the previous months of the μ -th year. In order to calculate this number we need to count total sum (for the average amount of) of UE filtering assessments during the whole μ of the year ($F^{(\mu)}$), then filtration assessment of UE quantity during each month ($f^{(\mu)}(t)$) is divided by the total sum of UE filtration values of the quantity UE ($F^{(\mu)}$) for one μ is year determined by the formula:

$$p^{(\mu)}(t) = f^{(\mu)}(t)/F^{(\mu)}, \quad t = \overline{1,12}, \tag{7}$$

where $p_t^{(\mu)} = p^{(\mu)}(t)$ is the objective probability of a particular type of UE for each month μ – year and of all 12 months (see. Table 3).

Table 3. Objective probabilities of particular type of UE occurrences within the μ – th year

t	1	2	3	4	5	6	7	8	9	10	11	12
$p_t^{(\mu)}$	$p_1^{(\mu)}$	$p_2^{(\mu)}$	$p_3^{(\mu)}$	$p_4^{(\mu)}$	$p_5^{(\mu)}$	$p_6^{(\mu)}$	$p_7^{(\mu)}$	$p_8^{(\mu)}$	$p_9^{(\mu)}$	$p_{10}^{(\mu)}$	$p_{11}^{(\mu)}$	$p_{12}^{(\mu)}$

In addition, for μ –th year we will have:

$$\sum_{t=1}^{12} p^{(\mu)}(t) = 1, \quad \mu = 4, 5, \dots.$$

The efficiency of the algorithm 1 is illustrated in the test example in [12].

2.1.2. An objective valuation of the damage prediction of security breaches IR

Algorithm 2

Step 2.1. Let $O = \{O_i, i = \overline{1,m}\}$ – many types of UE, leading to the security breaches of IR. In section 2.1.1. this work has been proposed assessment procedure for calculating monthly (or quarterly or semi-annual, etc.) objective probability of the number of certain type of attacks on IR in the company's IS. Suppose that company's information security (IFS) department possesses statistics regarding the monthly damage assessment, which corresponds to a monthly amount of particular i is the type of information security breaches, i.e. data Table 1 values correspond to the values of Table 4.

Table 4. Quantitative monthly (or quarterly or half-yearly, etc.) indicators damage from IFS violations depending on i –th type attack

t	1	2	3	4	5	6	7	8	9	10	11	12
1	$S_1^{(1)}$	$S_2^{(1)}$	$S_3^{(1)}$	$S_4^{(1)}$	$S_5^{(1)}$	$S_6^{(1)}$	$S_7^{(1)}$	$S_8^{(1)}$	$S_9^{(1)}$	$S_{10}^{(1)}$	$S_{11}^{(1)}$	$S_{12}^{(1)}$
$\vdots$	…	…	…	…	…	…	…	…	…	…	…	…

μ	$S_1^{(\mu)}$	$S_2^{(\mu)}$	$S_3^{(\mu)}$	$S_4^{(\mu)}$	$S_5^{(\mu)}$	$S_6^{(\mu)}$	$S_7^{(\mu)}$	$S_8^{(\mu)}$	$S_9^{(\mu)}$	$S_{10}^{(\mu)}$	$S_{11}^{(\mu)}$	$S_{12}^{(\mu)}$

Table 4 $\{S_t^{(i)},\ t=\overline{1,12}, i=\overline{1,\mu}\}$ – outputs that characterize the monthly (quarterly, semi-annual) amount of damage caused by the UE in the IS of the company during i –th sampling period.

Step 2.2. Note that company's performance losses not always directly proportional to the number of UE which took place. Nevertheless, based on the data of Table 4 we can construct a linear discrete model in the form of SS, which will correspond to the averages of data columns in Table 4. The elements of the averaged lines are calculated using equation (8) (see Table 5):

$$s_t^{(y)} = (\sum_{i=1}^{\mu} S_t^{(i)})/\mu,\ t=\overline{1,12}. \tag{8}$$

Table 5. The average monthly line (quarterly, semiannual, etc. with respect to a time interval) quantitative value of damage caused to the company's IR.

t	1	2	3	4	5	6	7	8	9	10	11	12
$s_t^{(y)}$	$s_1^{(y)}$	$s_2^{(y)}$	$s_3^{(y)}$	$s_4^{(y)}$	$s_5^{(y)}$	$s_6^{(y)}$	$s_7^{(y)}$	$s_8^{(y)}$	$s_9^{(y)}$	$s_{10}^{(y)}$	$s_{11}^{(y)}$	$s_{12}^{(y)}$

Step 2.3. Based on the data line $\{s_t^{(y)},\ t=\overline{1,12}\}$ the algorithm described in [10, 11], we can construct a linear stochastic model in the SS form of a stationary type

$$s(t+1)=\hat{c}\cdot s(t)+\hat{d}\cdot+w(t),\quad s(0)=s_0,\quad t=\overline{0,11}, \tag{9}$$

$$s^y(t+1)=s(t+1)+v(t+1),\quad t=\overline{0,11}. \tag{10}$$

In this case, first on the basis of data in Table 5 let's calculate line noise variance estimation model view (9), (10), in particular assessment of variances evaluation Q, R, $P(0)$ [12].

Step 2.4. Furthermore, let's calculate dynamics model coefficients (9) using OLS [9, 10].

Step 2.5. Suppose we have data of quantitative observations of damage caused to the company's IR in $(\mu+1)$ year (see. Table 6) and on the basis of the Kalman filter equations [10] we obtain line filter estimates (see. Table 7).

Table 6. Monthly (quarterly, half-yearly on several years) quantitative measures of damage from IFS violations depending on i -th type of attack in $(\mu+1)$ year.

t	1	2	3	4	5	6	7	8	9	10	11	12
$S_t^{(\mu)}$	$S_1^{(\mu)}$	$S_2^{(\mu)}$	$S_3^{(\mu)}$	$S_4^{(\mu)}$	$S_5^{(\mu)}$	$S_6^{(\mu)}$	$S_7^{(\mu)}$	$S_8^{(\mu)}$	$S_9^{(\mu)}$	$S_{10}^{(\mu)}$	$S_{11}^{(\mu)}$	$S_{12}^{(\mu)}$

Using the equations of the Kalman filter and the data in Table 6, we obtain the sequence estimates of filtering $\{\hat{s}(t\,|\,t), t=\overline{1,12}\}$, relatively to more reliable data of monthly quantitative damage (see. Table 7).

Table 7. Monthly (or quarterly or half-yearly, etc. relative to a few years) quantitative filtration damage assessments in $(\mu+1)$ -th year.

t	1	2	3	4	5	6	7	8	9	10	11	12
$\hat{s}(t\,|\,t)$	s_1	s_2	s_3	s_4	s_5	s_6	s_7	s_8	s_9	s_{10}	s_{11}	s_{12}

Step 2.6. Using rounded to the nearest whole number data filtering estimates on quantitative indicators, UE instances during μ year monthly (or quarterly or half-yearly, etc. relative to a few years), and the data in Table 7 with respect to the filtering estimates of quantitative indicators of the damage inflicted on the company's IR in μ year, you can get the average damage caused by a single UE $\{e(t),\ t=\overline{1,12}\}$. To do this, let's divide the data table row 7 divided into corresponding, rounded to the nearest whole number of UE instances. Calculated data can be summarized in Table 8.

Table 8. The average damage from a single UE instance.

t	1	2	3	4	5	6	7	8	9	10	11	12
$e(t)$	e_1	e_2	e_3	e_4	e_5	e_6	e_7	e_8	e_9	e_{10}	e_{11}	e_{12}

Step 2.7. Predicting the UE quantity of the (f_i^{pre}) i-th type by using appropriate model in the form of SS and corresponding averaged damage from the single UE according to Table 8, we can estimate the prediction of the damage to the company in t - the month (or quarter or half-year, and so on a few years) $(\mu+1)$ (current) year.
Operability of the 2 algorithm is illustrated in the test example in [12].

2.1.3. Controls the security group

Security measures (security controls), used in the company for IR and IS protection, can be divided into three main groups: technical, operational and management [6]. Groups, in turn, are divided into families. Listed these security measures in the standard are listed in [13].
The group management control includes security measures for IS, which focus on risk management and information security. The group consists of five families of controls [6, 13]. The group operating controls include security measures for IS, which is primarily implemented and executed by people. The group consists of nine families of controls [6, 13]. The group of technical controls include security measures for the IS, which are primarily implemented and executed through the steps in hardware, software, or firmware components of the system. The group consists of four sets of controls [6, 13].
Experts who conduct assessment may come from different professional training, such as: technical, financial, engineering, and management, with their own individual perceptions, attitudes and motivations in determining the amount of the realized loss from UE. Therefore, before calculating the amount of risk in the company's IS it is vital for all experts to get familiar with all objective assessments and predictions regarding the filtering of monthly (or quarterly or semi-annual, etc. relative to a few years) UE quantities and the damage caused by UE which allow experts to offer the most realistic expert estimates.
Procedure risk assessment based on expert judgment is comprised of six steps [6].
Now consider the risk calculation methodology according to the procedure set forth above.

2.2. Methods of calculating

2.2.1. Example 1. Methods of calculating the objective quantity prediction evaluations UE for external uncontrolled zone

Presumably, in the case of uncontrolled outer zone relative to the UE quantity of UA in IR CS, even in six months, in the range 0-5 of UE. In order to construct a discrete linear stochastic model in the form of SS with two unknown coefficients, you must have 4–5 times more observations than the number of unknown coefficients. In this regard, we must stretch the whole time quantization interval with respect to the collection of a sufficient number of UE, for four or five years, provided that the sample size of UE is no less than eight or ten (N = 10). Under these assumptions, the time interval between the observations will take at least six months.

Algorithm 1.

1. Suppose that one implementation of a time series (TS) observation for 5 years is $N = 10$. In order to improve the data quality with respect to the UE quantitative analysis, we take the number of bands equal to instances, for example, $k = 6$ and compile these data in Table 9 [14]: All data in Table 9 are modeled through the use of a uniform distribution law using the restriction on the number UE no more than 5, and using the procedure of rounding to a whole number.

Table 9. Modeled 6 TS values with sample volume $N = 10$ in relation to quantitative indicators of UE.

No	1st year		2nd year		3rd year		4th year		5th year	
	1	2	3	4	5	6	7	8	9	10
$y_1(i)$	3	2	4	2	2	4	2	2	5	0
$y_2(i)$	2	2	0	2	5	2	4	2	2	2
$y_3(i)$	2	2	4	3	4	5	2	4	4	4
$y_4(i)$	4	2	1	2	2	4	2	0	0	2
$y_5(i)$	2	2	1	4	1	2	1	1	2	1
$y_6(i)$	2	2	2	2	4	2	1	3	2	2

2. For $k = 6$ instances calculate the average quantities of UE quantities in the columns. Then we get average TS that corresponds to the number of UE for every six months during 5-year time period [14]. The data are summarized in Table 10.

Table 10. Averaged TS, with sample volume $N = 10$, characterizing quantitative indicators of UE.

No	1st year		2nd year		3rd year		4th year		5th year	
	1	2	3	4	5	6	7	8	9	10
$\bar{y}(i)$	2.500	2.0000	2.0000	2.5000	3.0000	3.1667	2.0000	2.0000	2.5000	1.8333

3. Building a linear discrete stationary stochastic model in the SS form based on the averaged data. Model coefficients are calculated by OLS acquired the following values: a =0.2233; b =1.7958 [10].

4. On the basis of the recurrence formulas given in the paper [12], we calculate dynamics model noise variance Q, the initial state $P(1)$ and the noise of the measurement system R based on an average TS data $\{\bar{y}(i), \overline{1, N}\}$. After calculation, variances took the following values: Q = 0.0982; P (1) = 0.0982; R = 0.1013.

5. Calculated variables in the preceding two paragraphs have allowed based on the Kalman filter equations to calculate the estimates and predictions, respectively, with respect to evaluation of the filtration, the amount of UE with anticipation intervals for half a year, during the preceding year,

Table 11. The calculated assessment predictions and the corresponding filtering estimates from the previous year.

No.	1st year		2nd year		3rd year		4th year		5th year	
	1	2	3	4	5	6th	7th	8	9	10
$\hat{y}(i\mid i-1)$	2.50	2.354	2.282	2.274	2.329	2.391	2.417	2.289	2.274	2.329
$\hat{y}(i\mid i)$	2.50	2.176	2.14	2,388	2.667	2.783	2.207	2.143	2,388	2.079
$z(i)$	2.50	2.000	2.000	2.500	3,000	3.167	2.000	2.000	2.500	1.833

Figure 1 shows plots reflecting changes in the number of UE based on TS values in relation to real (z), predicted (XP) and filtration ratings (XF)

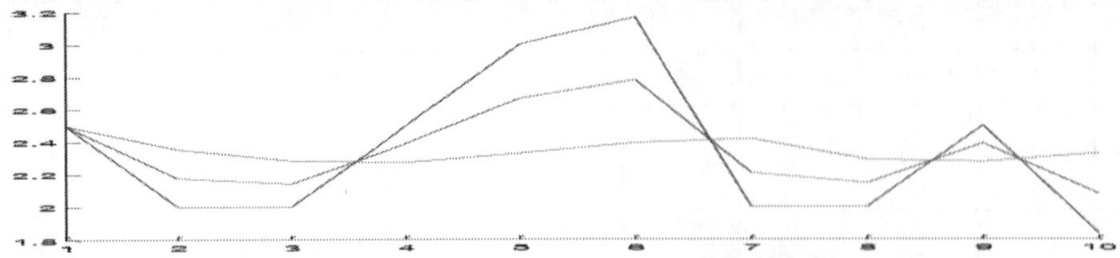

Figure 1. Graphs constructed based on real (z), the predicted (XP) and filtration (XF) estimates.

6. Calculate filtration assessments rounded to the nearest whole number. The calculation results are summarized in Table 12

Table 12. The calculated assessment predictions and the corresponding filtering estimates from the previous year.

No.	1st year		2nd year		3rd year		4th year		5th year	
	1	2	3	4	5	6th	7th	8	9	10
$\hat{y}(i\mid i)$	2.50	2.176	2.14	2,388	2.667	2.783	2.207	2.143	2,388	2.079
xfc	3	2	2	2	3	3	2	2	2	2
pf	0.13	0.087	0.087	0.087	0.13	0.13	0.087	0.087	0.087	0.087

where xfc - n – vector of filtering estimates rounded to the nearest whole number (see. Table 12).

7. Calculate the objective probability estimates regarding UE instances in the previous year based on the algorithm 1, step 1.6 pf - n – vector of objective semi-annual probabilities (see the last line of Table 12).

8. We use filtering estimates to predict the number of incidents occurred in the first half of the $(\mu+1)$ -th year. For this purpose we calculate the equation coefficients of linear approximation μ -th occurence on the basis of filtration estimates in the form of TS with the sample size $N=10$: XF = [2.5000 2.1755 2.1396 2.3878 2.6674 2.7825 2.2067 2.1430 2.3882 2.0790]. As a result of the calculations, we obtain the following assessments: a=0.2385; b = 1.7631.

9. Based on the Kalman filter equations, we obtain the following prediction estimate 2.3541. Further on the basis of the next observation $z^{(i+1)}(1)=5$ it is possible to calculate

the filtration estimate $XF^{(j+1)}(1) = 3.6887$. Next, we can calculate the prediction estimate for the next six months: $XP^{(j+1)}(2) = 2.6195$. Based on the corresponding observation $z^{(i+1)}(2) = 2$ we can calculate the filtration rating $XF^{(i+1)}(2) = 2.307$ and etc.

2.2.2. Example 2. The method of calculating the objective prediction valuation of damage magnitude from security breaches IR

Algorithm 2

1. Let $O = \{O_i, i = \overline{1, m}\}$ - be many types of UE, leading to a violation of the security of the IR, of which we choose one significant type. In clause 2.1.1. This work proposed a procedure for calculating the estimation of the semi-annual objective probability of the number of violations of a certain type of attack on the IR in the enterprise's IS. Suppose that the company's IFS department has statistics on a semi-annual average damage estimate, depending on the occurrence of a single incident, i.e. the values of the data in Table 1 correspond to the values of the data in Table 4, provided that the average loss from the implementation of a single damage is $\bar{s} = 21$ c.u. (conventional units). Then table 4 will correspond to the table with the following values.

Table 13. Modeled $k = 6$ TS with sample size $N = 10$ in the form of data related to quantitative indicators of damage corresponding to the number of UE from Table 9.

No	1st year		2nd year		3rd year		4th year		5th year	
	1	2	3	4	5	6	7	8	9	10
$S_1^{(i)}$	63	42	84	42	42	84	42	42	105	0
$S_2^{(i)}$	42	42	0	42	105	42	84	42	42	42
$S_3^{(i)}$	42	42	84	63	84	105	42	84	84	84
$S_4^{(i)}$	84	42	21	42	42	84	42	0	0	42
$S_5^{(i)}$	42	42	21	84	21	42	21	21	42	21
$S_6^{(i)}$	42	42	42	42	84	42	21	63	42	42

Table 14

The averaged 6-month line relative to 5 years and 6 occurrences of the quantitative indicators of damage inflicted by the enterprise's IR based on the data in Table 13

t	1	2	3	4	5	6	7	8	9	10
$s_i^{(y)}$	70,83	42	42	70,83	63	66,5	42	42	70,83	42

2. Based on the data string $\{s_i^{(y)}, i = \overline{1, 10}\}$ (see Table 14), using the algorithms described in [7, 12], we can construct a discrete linear stochastic stationary model in the form of SS view (9), (10).

We calculate the coefficients of the dynamics model (9): $\hat{c} = 0.1736, \hat{d} = 45.0051$.

3. Based on the data line of Table 14, with the sample size $N = 10$, estimates of the noise variances of the model of the form (9), (10) are calculated. The calculations gave the following variance estimates: Q = 88.4390; R = 56.7539; P (0) = 88.4390.

4. Suppose that we have the initial data of observations of quantitative indicators of damage inflicted by the company's IR in $(\mu + 1)$ year consistently for two semesters with the initial condition in the

form of a filtration estimate XF (1 | 1) = XF (1) = z (1) = 42 and observations for the two half-year z (2), z (3) and on the basis of the Kalman filter equations. Based on the Kalman filter equations, we first obtain a semi-annual estimate of the damage prediction XP (2|1) = 52.2963 and, taking into account the actual damage monitoring z (2) = 45, we calculate the corresponding damage filtration estimate XF (2 | 2) = 47.8006.
For the second half of the year, we take the initial condition XF (2 | 2) = 47.8006. Using this initial condition, we can calculate the prediction estimate for the second half of the year XP (3|2) = 53.3033. Next, having received at the time $t=3$, observation z (3) = 54, we can calculate the filtration estimate XF (3|3) = 53.7326, which, as we note, corrects the prediction estimate based on objective observation.

Table 15. Half-yearly quantitative indicators of damage from violations of the IFS, depending on i -th type of attack in $(\mu+1)$ year.

t	1	2	3	4	5	6	7	8	9	10	11	12
$S_t^{(\mu)}$	$S_1^{(\mu)}$	$S_2^{(\mu)}$	$S_3^{(\mu)}$	$S_4^{(\mu)}$	$S_5^{(\mu)}$	$S_6^{(\mu)}$	$S_7^{(\mu)}$	$S_8^{(\mu)}$	$S_9^{(\mu)}$	$S_{10}^{(\mu)}$	$S_{11}^{(\mu)}$	$S_{12}^{(\mu)}$

5. Using the Kalman filter equations and the data in Table 5, we obtain a sequence of filter estimates $\{\hat{s}(t\,|\,t), t=\overline{1,12}\}$, with respect to semiannual more reliable quantitative indices of the damage (Table 15).

Table 16. Semiannual quantitative indicators of filtering assessment of damage in $(\mu+1)$ year.

t	1	2	3	4	5	6	7	8	9	10	11	12
$\hat{s}(t\,|\,t)$	s_1	s_2	s_3	s_4	s_5	s_6	s_7	s_8	s_9	s_{10}	s_{11}	s_{12}

6. Using rounded to the nearest whole number filtering data relatively to the quantitative indicators of occurred UE within $(\mu+1)$ year by month, and the data of Table 7 regarding the filtration estimates as quantitative indicators of damage inflicted on the enterprise's IR in $(\mu+1)$ year, it is possible to obtain the average damage caused by a single instance of UE $\{e(t),\ t=\overline{1,12}\}$. To do this, it is necessary to divide lines of Table 7 by corresponding rounded to the whole numbers of UE occurrences. The calculated data can be summarized in Table 17.

Table 17. The averaged damage inflicted by a single incident of UE.

t	1	2	3	4	5	6	7	8	9	10	11	12
$e(t)$	e_1	e_2	e_3	e_4	e_5	e_6	e_7	e_8	e_9	e_{10}	e_{11}	e_{12}

7. Predicting the number of UE (f_i^{pre}) i -th type with the help of the corresponding model in the form of a PS and the corresponding average damage according to the data of Table 17, one can obtain an estimate of the prediction of the amount of damage that will be done to the enterprise in t -th half year in relation to $(\mu+2)$ implementation.
The efficiency of the methodology according to Algorithm 2 is illustrated in the test example, as well as in [12].

2.3. Zone-controlled territory AAS zone, AAS resource zone, the database zone.

In this section, we will consider all the above-mentioned zones of the controlled territory, AAS zones, AAS resource zones, and the database zones. As already mentioned above computer crimes can be divided into three classes: the interception of information; unauthorized access; "data manipulation". All classes can be considered separately.

With respect to all the last four zones, using the methodology described in paragraphs 2.1.1. – 2.1.3., it is possible, based on the relevant observational data, to calculate estimates of predictions and assessments of UE quantity filterings, as well as appropriate estimates of damage that may be incurred, for example, by a particular company.

Now, let's assume that there are three ISs of the conditional company, on which we assess the risks: IS-1, IS-2, IS-3. After reviewing the results of objective assessments, the experts discovered certain vulnerabilities and threats related to different control families. (The initial data and numerical calculations are shown in [15]). It is required to carry out expert evaluations to assimilate the above-described methodology.

After ranking the IS according to the level of risk in descending order, the following calculated data were obtained (see Table 18).

Table 18. Ranks of information systems.

Information System	IS-2	IS-3	IS-1
Level of Risk	0,224	0,217	0,213

Table 18 shows that the IS-2 receives the highest level of risk. This means that the probability of implementing threats and damage to this system is greater than for other systems. Therefore, the leaders of the organization, first of all, need to pay attention to the security of IS-2.

2.4. Assessment of the vulnerability of information processed in AAS

In view of the above methodology for calculating the number of UEs and the corresponding assessments of the damage inflicted to IR in the IS, it is possible to derive a formula for assessing the vulnerability of information processed in AAS. To do this, you can enter, for example [11], the following indicators: $P^{(d)}{}_{ikl}$ - probability of access of the criminal of the k-th category to the l-th zone; $P^{(k)}{}_{ijl}$ - probability of the presence (manifestation) of the j-th CUIR in the l-th zone of the i-th component of the AAS; $P^{(n)}{}_{ijkL}$ - probability of access of the violator of the k-th category to the j-th CUIR in the i-th zone of the i-th component of the AAS, provided that the intruder accesses the l-zone; $P^{(n)}{}_{ijl}$ - probability of access of the violator of the k-th category to the j-th CUIR in the i-th zone of the i-th component of the AAC, provided that the intruder accesses the l-th zone;

$$P_{ijkl} = P^{(d)}{}_{ikl} * P^{(k)}{}_{ijl} * P^{(n)}{}_{ijkl} * P^{(n)}{}_{ijl}, \tag{11}$$

- The probability of UA in one component of AAS by a single intruder of one category via one type of CUIR, let's call the basic indicator of the vulnerability of information (from the point of view of the UA).

Taking into account (1) the expression for the basic indicator will look like:

$$P^{(k)}{}_{ijk} = 1 - \prod_{l=1}^{5}[1 - P_{ijkl}] = 1 - \prod_{l=1}^{5}[P^{(d)}{}_{ikl} P^{(k)}{}_{ijl} P^{(n)}{}_{ijkl} P^{(n)}{}_{ijl}]. \tag{12}$$

The basic vulnerability indicators calculated in this way are of limited practical importance. To solve the problems associated with the development and operation of information security systems, use generalized vulnerability indicators, generalized by an index (I, j, k) or by their combination. For example, {K *} is a subset of interest to us from the set of potential violators. Or {1 *}, {J *} - is a subset of the components of AAS and CUIR. Then in general, the overall vulnerability index will be defined as:

$$P_{(I^*)(J^*)(K^*)} = 1 - \prod_{\forall i}[1 - P^{(b)}_{ijkl}]\prod_{\forall j}[1 - P^{(b)}_{ijkl}]\prod_{\forall k}[1 - P^{(b)}_{ijkl}]. \tag{13}$$

It is also necessary to consider the method of calculating the vulnerability index taking into account the time interval at which the vulnerability is assessed. In this case, it should be borne in mind that the longer the time interval, the more opportunities for the intruder for malicious actions and the greater the probability of a change in the state of the AAS. It is possible to determine such time intervals (not reducible at a point) at which the processes associated with a breach of security are homogeneous. Let's call these intervals small. Such a small interval, in turn, can be divided into very small intervals, the vulnerability of information on each of which is determined independently of the others. The dependence on each of the allocated intervals will be the same because of the homogeneity of the ongoing vulnerability processes. Then, through P_t^m we denote the vulnerability index of interest to us at a point (on a very small interval), and through the same P indicator on a small interval. We obtain the following formula:

$$P = 1 - \prod_{i=1}^{n_t}[1 - P_t^m] \tag{14}$$

where t - is a variable index of very small intervals, which represent a small interval; n_t -the total number of very small intervals. This approach is valid if the conditions for violation of information security remain unchanged throughout the interval under consideration. In real AAS, these conditions can vary.

3. Summary and conclusions

The current practice of applied technical or economic research indicates that in order to achieve success with regard to the safety of IR in the company's IS, the researcher should be well versed in three areas: 1) technical and economic theory; 2) mathematical modeling, i.e. the art of formalizing the formulation of the problem, which consists in the ability to translate a problem from the language of problem-oriented to the language of abstract mathematical models; 3) the corresponding software. Therefore, in this work, a systematic presentation of mathematical methods and models of analysis of safety measures was given and was aimed at studying theoretical approaches and was aimed at developing practical skills in their development and application to the calculation of risks with respect to the information systems under investigation.

Risk management is based on data that must be captured, accumulated, analyzed, stored, processed for the purpose of assessing the potential damage from user errors and attacks on IRs in the company's IS, the choice of measures to minimize it, the calculation of prediction estimates and the filtering of all possible parameters and indicators associated with the IFS. In this paper, in particular, methods have been proposed that allow obtaining estimates of the objective probability in the possibility of UE occurrence, estimating the objective cost of damage from IR security breaches in the company's IS prediction and filtering of the damage value estimation corresponding to the quantitative indicators of the UE. All the basic calculations of the IFS indicators in the company's IS used the capabilities of a discrete linear stochastic stationary model in the form of a SS and the Kalman filter equations to obtain more reliable estimates of the state of the object under study.

The work concentrates those methods and models that are most often used for analyzing and developing solutions for the rational choice and distribution of security measures relative to IR, taking into account the influence of safety measures on other controls, etc. In the process of risk assessment, the interdependencies between security measures are taken into account, which makes it possible to prioritize implementation of security measures and develop an adequate risk management strategy.

References

[1] Kotenko I V Sayenko I B 2012 SIEM-systems for management of information and security events *Information protection. INSIDE* 5 pp 54–65
[2] Miller D R Harris Sh Harper A A Van-Dyke S 2011 Black Ch. Security Information and Event Management (SIEM) Implementation McGrawHill Companies p 430
[3] ISO / IEC 27005: 2008. Information technology. Security techniques. Information security risk management
[4] 2006 *Risk management: Implementation principles and inventories for risk management* (ENISA)

[5] Zapechnikov S V 2010 *Information Technology Security* 1 pp 21–27
[6] Chi-Chun Lo Wan-Jia Chen 2011 *Expert Systems with Applications* 39 pp 248–257
[7] Abdenov A Zh Zarkumova-Reichel R N 2015 *Issues of Information Protection* 1 pp 64–70
[8] Kumamoto H Henley E 1996 *Probabilistic risk assessment and management for engineers and scientists* (IEEE, New York) p 620
[9] Zarkumova-Reichel R N Abdenov A Zh 2012 *Forecasting the number of incidents in the enterprise's information security system using a dynamic model* Fundamental Research 6(2) pp 429–434
[10] Sinitsyn I N 2006 *Filters of Kalman and Pugachev: Textbook* (Moscow: Logos) p 640
[11] Meshcheryakov R V Praskurin G A 2004 *Theoretical basis of computer security Razdel 2* (Tomsk: Tomsk State University of Control Systems and Radioelectronics) pp 97–148
[12] Abdenov A Zh Trushin V A Abdenova G A Inozemtseva Yu A 2016 *Artificial intelligence and decision-making* 3 pp 87–99
[12] NIST SP 800-30 2012 *Guide for conducting Risk* (Broadway: Assessments National Institute of Standards and Technology)
[13] Klimova E G Platov G A Kilanova R V 2014 *Computational technologies* 19 pp 27–37
[14] Abdenov A Zh Belkin S A Zarkumova-Reichel R N 2014 *Methodics of risk assessment for information systems on the basis of expert assessments* (Novosibirsk: NSTU) p 71

The algorithm of numerical calculation of constraints reactions in a dynamic system of transport machine

A L Akhtulov
Omsk State Technical University, 11, Mira ave., Omsk, 644050, Russia

e-mail: ahtulov-al1949@yandex.ru

Abstract. The questions of construction and practical application of the automation system for the design of components and aggregates for the construction of transport vehicles are considered, taking into account their dynamic characteristics. Based on the results of the studies, a unified method for determining the reactions of bonds of a complex spatial structure is proposed. The technique, based on the method of substructures, allows us to determine the values of the transfer functions taking into account the reactions of the bonds. After the carried out researches it is necessary to note, that such approach gives the most satisfactory results and can be used for calculations of complex mechanical systems of machines and units of different purposes. The directions of increasing the degree of validity of technical decisions are shown, especially in the early stages of design, when the cost of errors is high, with careful thorough working out of all the elements of the design, which is really feasible only on the basis of automation of design and technological work.

Keywords: transport machine, dynamic system, complex structure, communication reaction, algorithm, design automation system

1. Introduction

In recent years, problems have arisen associated with the need to take into account the elastic properties of structures that perform spatial motion, which causes the emergence of a number of new situations that have a significant effect on the dynamic characteristics of the object.

The complexity of the objects created caused the expediency of using new computational models based on the development of modern scientifically grounded methods for studying structures in order to create efficient systems and programs for automating the design of new equipment objects that represent complex structures.

The motion of a medium of simple structure, used in the mechanics of a deformed body, is described by three functions - the projections of the displacement vector. The motion of the medium of a complex structure is characterized by the collection of a large number of functions. These functions are generalized coordinates of the individual elements of which the medium consists. Thus, the concept of a complex structure generalizes the classical concept used in the theory of elasticity. Historically, one of the first models of an elastic medium that can not be described within the framework of the classical theory of elasticity is the continuum introduced for the first time in 1909 by E.F. By Cosser [1]. Another paper [2] describes the vibration fields in the zone of the boundary of the contact of two bodies of a complex structure by the methods of the theory of vibration conductivity.

The purpose of this work is the development of models for determining the dynamic characteristics of transport machine designs as complex technical systems [3], taking into account the reactions of bonds and the creation on their basis of design techniques.

2. Formulation of the problem

The design of transport vehicles as a complex object from the point of view of its design, production and preparation of operation can be considered as a multi-level, hierarchical structure [4] The problem is [5] derivation of equations of reactions of the elements of the transport machine structure, with unrelated interaction with each other (Fig. 1).

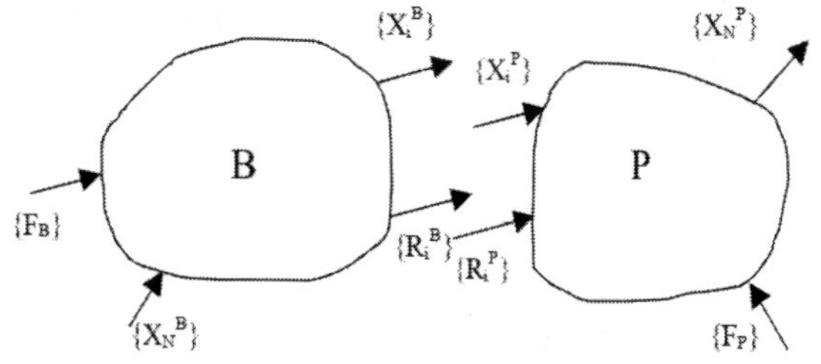

Figure 1. Scheme of interaction of unrelated components of the transport machine

Physically, this interaction occurs in a number of design points that have an indirect effect on both elements. Mathematically, this means that the vector of generalized displacements $\{x^B_i\}$ one element of the transport machine at these points should be equal to the displacement vector $\{x^P_i\}$ another, which can be written in the form

$$\{x^B_i\} = \{x^P_i\} \tag{1}$$

for any time t.. A similar expression can be written for the vectors of generalized reaction forces at common points of interaction

$$\{R^B_i\} = - \{R^P_i\} \tag{2}$$

for any time t.

Using the scheme of unbound bodies [6] (Figure 1), it is easy to derive the equations of motion for both elements

$$\begin{bmatrix} M_B & \\ & M_P \end{bmatrix}\begin{Bmatrix} \ddot{x}_B \\ \ddot{x}_P \end{Bmatrix} + \begin{bmatrix} K_B & \\ & K_P \end{bmatrix}\begin{Bmatrix} x_B \\ x_P \end{Bmatrix} = \begin{Bmatrix} F_B \\ F_P \end{Bmatrix} + \begin{Bmatrix} 0 \\ R_i^B \\ 0 \\ R_i^P \end{Bmatrix}, \tag{3}$$

where $\{x_B\}$ is the vector of generalized displacements, $[M_B]$ is the mass matrix, $[K_B]$ is the stiffness matrix, and $\{F_B\}$ is the vector of external forces for one of the elements of the transport machine. By analogy, we also introduce the notation for another element. Equations (3) are unrelated equations of motion for non-damped elements, i.e. at the moment, almost everywhere, damping is excluded from the equations. It is assumed that the damping can be taken into account by the generally accepted method of damping the vibration modes.

3. Theory

To obtain the equations of motion of the transport machine system, it is necessary to eliminate the previously unknown reaction-force vectors $\{R^B_i\}$ and $\{R^P_i\}$. Two most commonly used transformations are known to solve this problem. In the first one, [7] a non-free and free element of the transport machine is used, and in the second [8] both elements of the transport machine are not accepted when the interaction surface is free. In both approaches, the following separation of the vector $\{x_P\}$

$$\{x_P\} = \begin{Bmatrix} x_N^P \\ x_i^P \end{Bmatrix}. \tag{4}$$

where $\{x_N^P\}$ reflects the unlimited movement of the element, which can be represented as a sum

$$\{x^P_N\} = [S_P]\{x^P_i\} + \{\overline{x}_N^P\}, \tag{5}$$

IOP Conf. Series: Journal of Physics: Conf. Series **944** (2017) 012002 doi:10.1088/1742-6596/944/1/012002

where $[S_P] = -[K^P_{NN}]^{-1}[K^P_{Ni}]$ (6)

and the vector $\{\bar{x}^P_N\}$ is the vector of unbounded displacements of the element with respect to the interaction surface.

The equations of motion in the case of a free element [9] are derived by means of expressions (1) and (5) in the form of a coordinate transformation

$$\begin{Bmatrix} x_B \\ x_P \end{Bmatrix} = \begin{Bmatrix} x^B_N \\ x^B_i \\ x^P_N \\ x^P_i \end{Bmatrix} = \begin{bmatrix} I & 0 & 0 \\ 0 & I & 0 \\ 0 & S_P & I \\ 0 & I & 0 \end{bmatrix} \begin{Bmatrix} x^B_N \\ x^B_i \\ \bar{x}^P_N \end{Bmatrix}. \tag{7}$$

Using the notation, we can write in the form

$$\begin{Bmatrix} x_B \\ x_P \end{Bmatrix} = \begin{bmatrix} I_B & L_B & 0 \\ 0 & T_P & I_P \end{bmatrix} \begin{Bmatrix} x^B_N \\ x^B_i \\ \bar{x}_N \end{Bmatrix}, \tag{8}$$

As well as equations (2), (3) and (7), we obtain the following equations of motion for the case of a free element of the construction of a transport machine

$$\begin{bmatrix} I^T_B M_B I_B & I^T_B M_B L_B & 0 \\ L^T_B M_B I_B & L^T_B M_B L_B + T^T_P M_P T_P & T^T_P M_P I_P \\ 0 & I^T_P M_P T_P & I^T_P M_P I_P \end{bmatrix} \begin{Bmatrix} \ddot{x}^B_N \\ \ddot{x}^B_i \\ \ddot{\bar{x}}^P_N \end{Bmatrix} + \\ + \begin{bmatrix} I^T_B K_B I_B & 0 & 0 \\ L^T_B K_B I_B & L^T_B K_B L_B + T^T_P K_P T_P & 0 \\ 0 & 0 & I^T_P K_P I_P \end{bmatrix} \begin{Bmatrix} x^B_N \\ x^B_i \\ \bar{x}^P_N \end{Bmatrix} = \begin{Bmatrix} I^T_B F_B \\ L^T_B F_B \\ 0 \end{Bmatrix}. \tag{9}$$

If both elements are formally described identically, i.e.

$$\{x^B_N\} = [S_B]\{x^B_i\} + \{\bar{x}^B_N\}. \tag{10}$$

Then a suitable coordinate transformation is a transformation of the form

$$\begin{Bmatrix} x^B \\ x^P \end{Bmatrix} = \begin{Bmatrix} x^B_N \\ x^B_i \\ x^P_N \\ x^P_i \end{Bmatrix} = \begin{bmatrix} I & S_B & 0 \\ 0 & I & 0 \\ 0 & S_P & I \\ 0 & I & 0 \end{bmatrix} \begin{Bmatrix} \bar{x}^B_N \\ x^B_i \\ \bar{x}^P_N \end{Bmatrix}. \tag{11}$$

The resulting equations of motion for non-free elements of the transport machine structure with a free interaction surface (boundary) can be written in the form

$$\begin{bmatrix} I^T_B M_B I_B & I^T_B M_B T_B & 0 \\ T^T_B M_B I_B & T^T_B M_B T_B + T^T_P M_P T_P & T^T_P M_P I_P \\ 0 & I^T_P M_P T_P & I^T_P M_P I_P \end{bmatrix} \cdot \begin{Bmatrix} \ddot{\bar{x}}^B_N \\ \ddot{x}^B_i \\ \ddot{\bar{x}}^P_N \end{Bmatrix} + \\ + \begin{bmatrix} I^T_B K_B I_B & 0 & 0 \\ 0 & T^T_B K_B T_B + T^T_P K_P T_P & 0 \\ 0 & 0 & I^T_P K_P I_P \end{bmatrix} \cdot \begin{Bmatrix} \bar{x}^B_N \\ x^B_i \\ \bar{x}^P_N \end{Bmatrix} = \begin{Bmatrix} I^T_B F_B \\ T^T_B F_B \\ 0 \end{Bmatrix}. \tag{12}$$

Matrices $[S_B]$ and $[T_B]$ are defined similarly to $[S_p]$ and $[T_p]$. The main difference between equations (9) and (12) is the presence in the last additional term due to the rigidity of the unbound system.

The columns of the matrices $[T_P]$ and $[T_B]$ will be referred to as coupling modes, which allows us to represent the motion of structural elements using a combination of coupling modes, rigid-body motion modes, and modes of a fixed boundary. The introduction of such a simplification allows the modes of motion of a rigid body not to be considered separately from the bond modes when using the transformation (5). This is precisely the transformation of the coordinates, which leads to the appearance in (9) and (12) of a term due to the rigidity of the unbound system.

The forces acting on the structural element can be assigned to two categories [8]: the forces arising in the bonds and the forces applied by means of sources external to the system.

The first category includes forces acting through links to other elements, as well as reactions acting on the system with fixed connections. The forces arising in the bonds work on the movements of the element, while the reactions in the fixed bonds do not work.

The forces acting in the bonds between the elements are further subdivided into two categories: R_s^R is the force applied to the statically determined connection and R_r^C is the force applied to the redundant link.

External forces with respect to the system can be distributed or concentrated. Let f be the intensity of the distributed load at an arbitrary point, and let it be the concentrated force at some point with displacement.

The virtual work [8], produced by these forces on the virtual displacement of the element, is given by the expression

$$\delta W = \sum_s R_s^R \delta\tilde{u}_s^R + \sum_r R_r^C (\delta\tilde{u}_r^R + \delta\tilde{u}_r^C) + \int \bar{f} \cdot \delta\bar{u}\, dV + \sum_i \bar{F}_i \cdot \delta\bar{u}_i \,. \tag{13}$$

Where $\delta\tilde{u}_s^R$ is the virtual displacement of the s-th statically determinable connection during a hard transfer; $\delta\tilde{u}_r^R$ is the virtual movement of r-th redundant communication during hard transfer; $\delta\tilde{u}_r^C$ is the virtual movement of the r-th redundant communication at offset.

Note that the movements of the links are distinguished by signs in the form of a wavy line above the letter. These displacements are assumed to be scalar quantities, since the bonds themselves determine the direction of displacements [6], which can be determined through typical constraints that limit the movement of elements and their reactions

$$\left.\begin{aligned} \tilde{u}_s^R &= \sum_j \tilde{\phi}_{sj}^{RR} \rho_j^R \\ \tilde{u}_r^R &= \sum_j \tilde{\phi}_{rj}^{CR} \rho_j^R \\ \tilde{u}_r^C &= \sum_j \tilde{\phi}_{rj}^{CC} \rho_j^C \end{aligned}\right\}, \tag{14}$$

where $\tilde{\phi}_{sj}^{RR}$ is the typical displacement of the s-th statically determined connection for the j-th rigid form; $\tilde{\phi}_{rj}^{CR}$ is the typical movement of the rth redundant link at the j-th rigid form; $\tilde{\phi}_{rj}^{CC}$ is the typical movement of the r-th redundant link at the j-th form of displacement due to the coupling.

Substituting expression (14) into (13), we obtain the following equation for virtual work:

$$\begin{aligned} \delta W = {} & \sum_j \delta\rho_j^R \left[\sum_s R_s^R \tilde{\phi}_{sj}^{RR} + \sum_r R_r^C \tilde{\phi}_{rj}^{CR} + \int \bar{f} \cdot \tilde{\phi}_j^R dV + \sum_i \bar{F}_i \cdot \tilde{\phi}_{ij}^R \right] + \sum_j \delta\rho_j^C \times \\ & \times \left[\sum_r R_r^C \tilde{\phi}_{rj}^{CC} + \int \bar{f} \cdot \tilde{\phi}_j^C dV + \sum_i \bar{F}_i \cdot \tilde{\phi}_{ij}^C \right] + \sum_j \delta\rho_j^N \times \left[\int \bar{f} \cdot \tilde{\phi}_j^N dV + \sum_i \bar{F}_i \cdot \tilde{\phi}_{ij}^N \right]. \end{aligned} \tag{15}$$

Virtual work can also be written through generalized forces and movements

$$\delta W = \sum_j P_j^R \delta\rho_j^R + \sum_j P_j^C \delta\rho_j^C + \sum_j P_j^N \delta\rho_j^N . \tag{16}$$

IOP Conf. Series: Journal of Physics: Conf. Series **944** (2017) 012002 doi:10.1088/1742-6596/944/1/012002

Comparing relations (15) and (16) and taking into account that the generalized virtual displacements $\delta\rho_j (j = 1,2,3,...)$ are independent, we obtain the following equations for generalized forces:

$$\left.\begin{aligned} P_j^R &= \sum_s R_s^R \tilde{\phi}_{sj}^{RR} + \sum_r R_r^C \tilde{\phi}_{rj}^{CR} + \int \bar{f} \cdot \bar{\phi}_j^R dV + \sum_i \bar{F}_i \cdot \bar{\phi}_{ij}^R \\ P_j^C &= \sum_r R_r^C \tilde{\phi}_{rj}^{CC} + \int \bar{f} \cdot \bar{\phi}_j^C dV + \sum_i \bar{F}_i \cdot \bar{\phi}_{ij}^C \\ P_j^N &= \int \bar{f} \cdot \bar{\phi}_i^N dV + \sum_i \bar{F}_i \cdot \bar{\phi}_{ij}^N \end{aligned}\right\}. \tag{17}$$

It is convenient to write the system of generalized external forces separately in this form:

$$\left.\begin{aligned} F_j^R &= \int \bar{f} \cdot \bar{\phi}_j^R dV + \sum_i \bar{F}_i \cdot \bar{\phi}_{ij}^R \\ F_j^C &= \int \bar{f} \cdot \bar{\phi}_j^C dV + \sum_i \bar{F}_i \cdot \bar{\phi}_{ij}^C \\ F_j^N &= \int \bar{f} \cdot \bar{\phi}_j^N dV + \sum_i \bar{F}_i \cdot \bar{\phi}_{ij}^N \end{aligned}\right\}. \tag{18}$$

Taking expression (18) into account for generalized forces, we can rewrite equation (17) in the matrix form

$$\{ P \} = [\tilde{\phi}]^T \{ R \} + \{ F \}, \tag{19}$$

Where $\{P\} = \begin{Bmatrix} P^R \\ P^C \\ P^N \end{Bmatrix}$, $\{R\} = \begin{Bmatrix} R^R \\ R^C \end{Bmatrix}$, $\{F\} = \begin{Bmatrix} F^R \\ F^C \\ F^N \end{Bmatrix}$, $[\tilde{\phi}] = \begin{bmatrix} \tilde{\phi}^{RR} & 0 & 0 \\ \tilde{\phi}^{CR} & \tilde{\phi}^{CC} & 0 \end{bmatrix}$.

Here it should be noted that the translations of the links are transformed into generalized displacements by means of a matrix $[\tilde{\phi}]$ in this way:

$$\begin{Bmatrix} \tilde{u}_R \\ \tilde{u}_C \end{Bmatrix} = \begin{bmatrix} \tilde{\phi}^{RR} & 0 & 0 \\ \tilde{\phi}^{CR} & \tilde{\phi}^{CC} & 0 \end{bmatrix} \begin{Bmatrix} \rho^R \\ \rho^C \\ \rho^N \end{Bmatrix}, \tag{20}$$

where $\tilde{u}_R$ is the complete movement of the statically determinate connection; $\tilde{u}_C$ is the complete redundancy transfer.

Using the continuity condition [10] in the relations of the elements, it can be shown that the equilibrium of the interacting forces in these bonds is automatically satisfied. In order to show the validity of this statement, it is necessary to clearly formulate and track the steps leading to the definition of a system of generalized forces. Substitution of relation (19) into (2) shows that the force vector for the system has the form

$$\{Q(t)\} = [\beta]T [\tilde{\phi}]^T \{R\} + [\beta]T \{F\}. \tag{21}$$

In this equality, the matrix is $[\tilde{\phi}]$ composed of matrix components arranged as follows:

$$[\tilde{\phi}] = \begin{bmatrix} [\tilde{\phi}]_1 & & & & & & \\ & [\tilde{\phi}]_2 & & & & & \\ & & \cdot & & & & \\ & & & \cdot & & & \\ \cdot & & & & [\tilde{\phi}]_i & & \\ & & & & & [\tilde{\phi}]_s & \\ & & & & & & \cdot \\ & & & & & & & \cdot \end{bmatrix}. \tag{22}$$

Similarly, the components of the vectors $\{R\}$ and $\{F\}$ are the corresponding vectors for the individual elements of the construction [11], arranged in this order:

$$\{R\}=\begin{Bmatrix}\{R\}_1\\ \{R\}_2\\ \cdot\\ \cdot\\ \cdot\\ \{R\}_r\\ \{R\}_s\\ \cdot\\ \cdot\\ \cdot\end{Bmatrix}\quad \{F\}=\begin{Bmatrix}\{F\}_1\\ \{F\}_2\\ \cdot\\ \cdot\\ \cdot\\ \{F\}_r\\ \{F\}_s\\ \cdot\\ \cdot\\ \cdot\end{Bmatrix}. \tag{23}$$

Since the components of the vector of the system {R} are the forces acting in the bonds of the elements, i.e. Are internal with respect to the entire structure under consideration, they do not affect the generalized forces of the system {Q (t)}. Therefore, the first term on the right-hand side of (21) disappears, and the system is simplified.

$$\{Q\ (t)\} = [\beta]T\{F\}. \tag{24}$$

Thus, the assertion concerning the self-equilibrium of internal forces can be carried out by showing that the first term on the right-hand side of (21) actually vanishes, that is

$$[\beta]^T[\tilde{\phi}]^T\{R\}=\{0\}. \tag{25}$$

The proof can be carried out by applying the principle of a minimum of virtual work [3] to two interrelated elements of the system. It is not difficult to show that two reactions in the general connection are equal in magnitude and opposite in direction with continuity of displacements, which is the proof of the validity of (24).

Consequently, the coupling reactions of a system of several bodies consisting of p rigid bodies with q bonds can be described by a linear overdetermined system of equations:

$$\overline{Q}\cdot g=\overline{q}^z \tag{26}$$

with $6_{p\times q}$ – distributed matrix $\overline{Q};q$ – a vector of the generalized coupling reaction; q and $6p$ are the vectors denoting the coupling reactions $\overline{q}^z$ at the center of mass.

A numerical solution can be a numerical standard approach, for example, the Gauss elimination method [5]. In this paper it is shown that it is possible to use a block matrix structure $\overline{Q}$ as the basis for the algorithm.

Thus, the reactions of the positions of L_k, $k = 1\ (1)\ l$ of the bodies K_i and K_j under consideration; $i, j = 1$ (1) p are connected (Fig. 2) with the help of the radius vector r_{Gk} and the rotation matrix G_k determine the stability of the coordinate system [4].

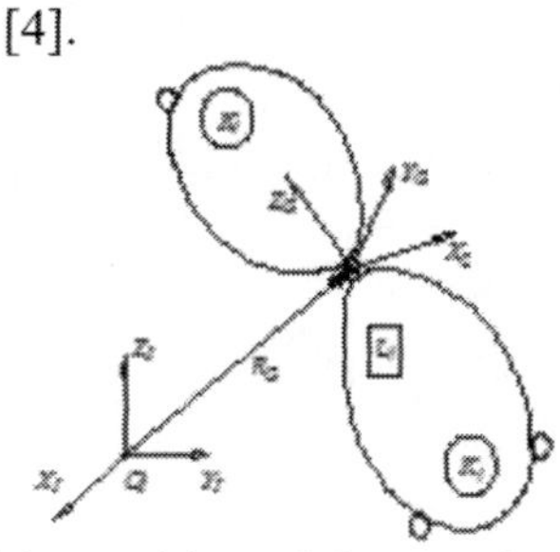

Figure 2. Geometrical description of the position of the reaction of bonds

Next, the scalar factor gm, $m = 1\,(1)\ q$ is carried out for each direction of the position, which corresponds to the magnitude of the generalized coupling reaction in this direction and still retains a visual significance.

If we describe the law of conservation of momentum and the law of angular momentum for bodies with identical vector starts, we obtain the equation of the bond reaction [12].

IOP Conf. Series: Journal of Physics: Conf. Series **944** (2017) 012002 doi:10.1088/1742-6596/944/1/012002

$$\begin{bmatrix} I_E|0 \\ \ldots\ldots \\ \underbrace{0|I_E}_{\bar{I}_E} \end{bmatrix} \mathrm{RGTg} = \mathrm{qz0} \qquad (27)$$

With a block matrix $\overline{T} = Diag\{T_{T1} \ldots T_{Ti}; T_{R1} \ldots T_{Ri}\}, \overline{G} = Diag\{G_1 \ldots G_i; G_1 \ldots G_i\}$,

$$\overline{R} = \begin{bmatrix} E & & & 0 \\ \ldots & \ddots E & \ldots & \ldots \\ \tilde{r}_{G1} & & \ddots E & \\ & \ddots r_{Gi} & & \ddots E \end{bmatrix},$$

while the symbol is a cross product. The constant matrix IE is the matrix of the incidence of a system of several bodies, where, as necessary, the unit elements are replaced 3×3 by the unit matrix and the zero elements by means 3×3 of the zero matrix.

To solve equation (27), first calculate the *G*-vector $y = \overline{R}\overline{G}\overline{T}g$ from (27). Then, for systems with a complex structure, it is possible to find a solution, for example, by direct use of the Gauss algorithm [5].

In systems with S kinematic nodes, a similar solution for y is not possible, since the matrix IE reflects the slope 6s [9]. Therefore, we obtain a 6s-parametric solution $y = \hat{y} + Y\lambda$ with a particular solution vector $\hat{y}$, a constant $6l \times 6s$-matrix of a homogeneous solution of Y and a 6_s vector of an arbitrary parameter λ.

For the vector g, only equation:

$$\overline{T}g = \overline{G}^T R^{-1}(\hat{y} + Y\lambda) \qquad (28)$$

$$\overline{G}^T = \overline{G}^{-1}$$

And

$$\overline{R}^{-1} = \begin{bmatrix} E & & & 0 \\ & \ddots E & & \\ -\tilde{r}_{G1} & & E & \\ & \ddots -\tilde{r}_{Gl} & & \ddots E \end{bmatrix}.$$

What corresponds to the vector component for all positions of the multiple-body system.

Each q component, which corresponds to the closed position of the direction, contains, depending on the circumstances, the components of the vector q. For each node, 6 equations are obtained for the calculation of the vector. The further course of action corresponds to a system with a building structure.

4. Results

On the basis of the obtained algorithm, a computer program is developed that leads the numerical calculation of the communication reactions, according to which the calculation time is much less than required for the Gauss algorithm.

In Fig. 3 represents the object and its typical element e.

$$\{M_p\}\{x_p\} + \{K_p\}\{x_p\} = \{F_p\}. \qquad (29)$$

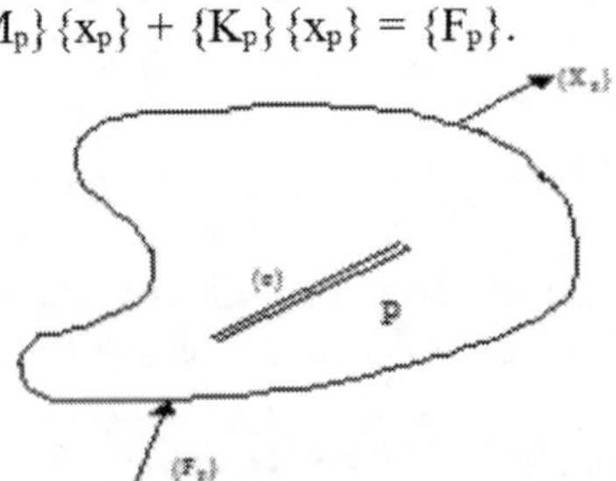

Figure 3. The object and the definition of its element e

It is required to determine the load or the vector of internal forces acting on this element. The force vector can be represented in the form

$$\{F_e^P\} = [k_e][T_e]\{x_P\}, \tag{30}$$

where $[x_p]$ is the displacement vector of the payload, $[k_e]$ is the stiffness matrix of the element e and $[T_e]$ is the transformation matrix reflecting the displacements of the element e through the displacements of the payload, which are determined from the equation of the form (28).

We assume that the approach based on the formulation of Newton's equations for each mode [7] makes it possible to more accurately determine the vector $\{X_p\}$ and, consequently, the internal force vector $\{F_e^P\}$. The use of such an approach leads to the need to solve the equation relative to $\{x_p\}$ and the modal decomposition $\{x_p\}$, that is,

$$\{x_p\} = \{K_p\}^{-1}[\{F_p\} - [M_p]\{x_p\}]. \tag{31}$$

Assuming that M the mod of the free element and the N modes of the cantilevered second element are stored in $[\phi_B]$.

First N + M related equations in the system are replaced by N M systems of two equations, each of which reflects the interaction of one mode of one element with one mode of another, namely:

$$\left[\begin{array}{c|c} I + \{\phi_i^B\}_i^T T_P^T M_P T_P \{\phi_i^B\}_i & \{\phi_i^B\}_i^T T_P^T M_P I_P \{\overline{\phi}_N^P\}_j \\ \hline \{\overline{\phi}_N^P\}_j^T I_P^T M_P T_P \{\phi_i^B\}_i & 1 \end{array}\right] \cdot \left\{\begin{array}{c} \ddot{q}_{Bi} \\ \ddot{\overline{q}}_{Nj} \end{array}\right\} + \tag{32}$$

$$+\left[\begin{array}{c|c} \omega_{Bi}^2 & 0 \\ \hline 0 & \overline{\omega}_{Pj}^2 \end{array}\right] \cdot \left\{\begin{array}{c} q_{Bi} \\ \hline \overline{q}_{Nj}^P \end{array}\right\} = \left\{\begin{array}{c} \{\phi_N^B\}_i^T F_N^B \\ \hline 0 \end{array}\right\} \begin{array}{l} i = 1,2,\ldots,M \\ j = 1,2,\ldots,N \end{array}$$

In equation (30), it was assumed that the interaction surface was statically determined, ie. $[T_P^T K_P T_P] = [0]$.

Then, for each of the $N \times M$-modal reactions of the element, the limiting value q_{BP} is introduced. This is realized by compiling a new model for the input action (forcing function [8]) in equation (30). A rather complex forcing function is replaced by a much simpler (for example, a force pulse), which generates the same maximum reaction value as the original force. Thus, from the equation with some subsequent simplifications it turns out to be possible to obtain an analytical expression of the reaction, which makes it possible to determine its maximum value. Then the limit value of the payload reaction qP is determined by summing the modal limit values q_{BP} ((in absolute units or in the rms sense, which allows the introduction of weighting coefficients)). The responses of the payload element are determined by summing the contributions of all the modal components.

As noted above, the forcing function in equation (30) is replaced by a modal δ-function of a given amplitude FB. This amplitude is determined by analyzing the transient processes of the transport machine in the presence or absence of an artificial load. Advantages of the method of analyzing the spectrum of maximum loads are the small amount of computation and the brevity of each computational cycle.

It can be noted that this method can be effectively used in cases where the mass of the structure is not the main factor or when preliminary estimates of the dynamic responses at the initial stages of the payload design process are preliminary calculated.

Therefore, in principle, it is necessary to consider only methods that take into account only free vibrations. The advantage of this approach is to simplify the process of obtaining and analyzing data. In this case, for free vibrations, the equation of motion, neglecting damping, is written in a simplified form:

$$[M]\{\ddot{q}\} + [K]\{q\} = 0\,. \tag{33}$$

The solution of this equation, using the Euler substitution, can be represented in the following form:

$$\{q\} = \{q_1, q_2, \ldots, q_N\} = \{D_1, D_2, \ldots, D_N\} \cdot e^{\omega_C \cdot t} = \{D\} \cdot e^{\omega_C \cdot t}\,, \tag{34}$$

Where {D} is a column vector of the oscillation amplitudes; ω_C is the circular frequency of its own undamped oscillations.

Substituting (32) into the equation of free oscillations, we obtain a system of linear homogeneous algebraic equations with respect to the coefficient Dj $[G]\{D\} = 0$,

$$\{D\}=\begin{Vmatrix} D_1 \\ \vdots \\ D_N \end{Vmatrix}; \qquad [G]=\begin{Vmatrix} G_{11} & \dots & G_{1N} \\ \vdots & \dots & \vdots \\ G_{N1} & \dots & G_{NN} \end{Vmatrix} \tag{35}$$

where $G_{ij}=-m_{ij}\cdot\omega_C^2+K_{ij}$.

From this we obtain the condition for oscillations of the construction at natural frequencies

$$\begin{vmatrix} (-m_{11}\omega_C^2+K_{11}) & \dots & (-m_{1N}\omega_C^2+K_{1N}) \\ \dots & \dots & \dots \\ (-m_{N1}\omega_C^2+K_{N1}) & \dots & (-m_{NN}\omega_C^2+K_{NN}) \end{vmatrix}. \tag{36}$$

From (33) we find N roots $(\omega_C^2)_1\le(\omega_C^2)_2\le\dots\le(\omega_C^2)_N$.

(37)

After finding the roots, we find the coefficients D_1, ..., D_N. Each root of equation (36) has its own system of coefficients. Most often, not the coefficients themselves, reduced to a given natural frequency of oscillations, are used, but their ratio:

$$\Phi_i=\frac{D_i}{D_1},\ (i=1, 2, 3, \dots, N), \tag{38}$$

Called the eigenmodes of the system's vibrations (modes).

When solving the characteristic equation (36) with allowance for damping, we obtain N roots,

$$(\lambda_C^2)_1\le(\lambda_C^2)_2\le\dots\le(\lambda_C^2)_N, \tag{39}$$

Where $\lambda_C^2=\omega_C^2(1-\xi_c^2)$ is the circular frequency of the self-damped oscillations; ξ_C – damping parameter of natural oscillations.

Thus, the equation has 2N roots, which determine 2N values of λ, from which, on the basis of the Laplace transform [10] with zero initial position and velocity {D}={Q*} $[B(\lambda)=(\lambda^2[M]+\lambda[C]+[K])^{-1}$, where $[B(\lambda)=(\lambda^2[M]+\lambda[C]+[K])^{-1}$ is the matrix of transfer functions.

And then, the transfer function $B_{ij}^J(\lambda)$ for the characteristic points i and j of the subsystem J can be determined analytically [4] or experimentally [12]. Due to the presence of damping, there is a lag in phase between the disturbance and the reaction. Consequently, in the general case, they are complex numbers with real and imaginary part, or modulus and phase angle, as a function of frequency λ.

In the analytic determination of the transfer function, it is necessary to know the values of the normal forms V_{ik}, representing the reaction of the k-th tone at the point i, as well as the frequencies of the k-th tone ω_k and the generalized masses m_k. For example, the transfer function transfer, or conductivity, with small or proportional damping is defined as

$$B_{ij}^J(\lambda)=\sum\frac{V_{ik}\cdot V_{jk}}{m_k\left[(\lambda_k^2-\lambda^2)+i2\eta_k\lambda_k\lambda\right]}, \tag{40}$$

where η_k is the calculated or measured value of the damping coefficient (generally depending on the frequency).

For the movements and rotations of the transport machine as a solid, the corresponding eigenfrequencies in equation (38) are zero. From equation (38) it is possible to obtain other expressions for the transfer functions. For example, the transfer function in terms of speed or acceleration is obtained by multiplying equation (38) by $i\omega$ and $-\omega^2$, respectively.

5. Discussion of results

In the experimental determination of the transfer function (Fig. 4), the structure is excited by a vibrator creating a perturbing force that is close in shape to a sinusoid with a slowly varying frequency; The reaction of the system is recorded. After appropriate filtering, the analog data processing system allocates a phase shift between them as a function of frequency; Then these data are digitized. In the analytical determination of transfer functions, the calculation of vibration modes is performed taking into account the boundary conditions imposed by the characteristics of the chosen model. However, in experimental studies it is expedient to determine the transfer functions for free, non-attached subsystems. For this purpose, the "soft suspension" method is often used, when the frequencies of the solid body on the suspension are removed from the fundamental disturbance frequency.

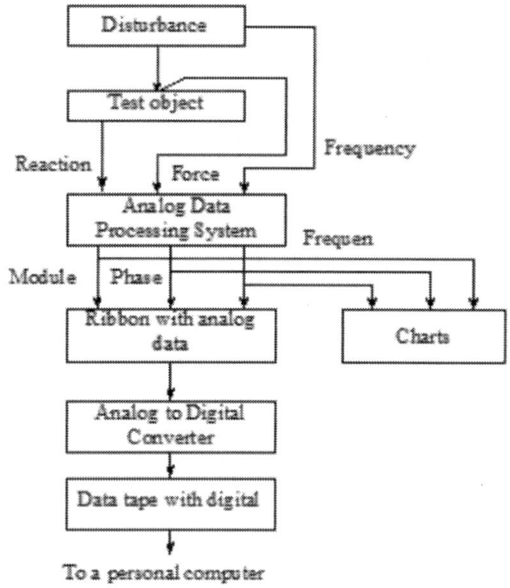

Figure 4. Block diagram of the transfer function determination system

On the other hand, it is often possible to use the symmetry of subsystems with respect to axes or planes, when the support devices can be installed so that they have no effect on the reaction being investigated. The most widely used is the elastic fixing of the subsystem at the points under study with measurements for each level of disturbance of all forces and displacements of the support (Fig. 5), the main components of the signal,

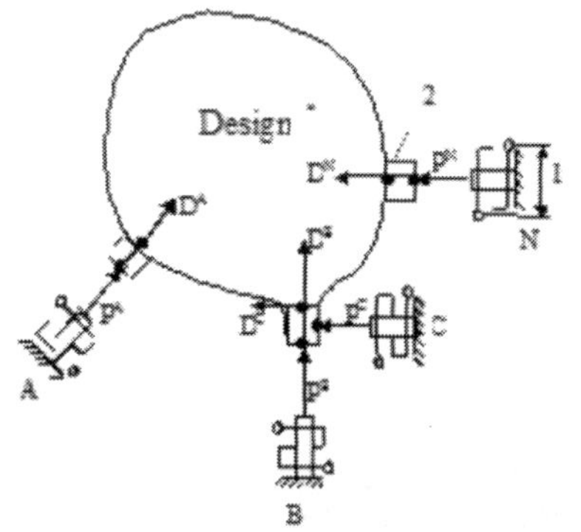

Figure 5. Excitation of structures by several vibrators with simultaneous measurement of all supporting reactions: DB – typical movement; PB is a typical support force; 1 is the excitation voltage; 2 – displacement sensor and dynamometer

those. Module of the ratio of the reaction to the disturbance and, as will be shown below, the influence of the test equipment can be eliminated by computation, resulting in the transfer functions of an unattached system.

As will be shown below, the influence of the test equipment can be eliminated by computation, resulting in the transfer functions of an unattached system.

We represent through the real [H] and imaginary [J] matrices in the form:

$$([M]+j[J])\cdot[B(\lambda)]=[E], \quad (41)$$

Where [E] is the identity matrix.

On the basis of the solutions obtained, the vector-column of the accelerations of the system can be represented in the form

$$\{\ddot{q}\} = \sum_{j=1}^{2N} \lambda_{cj}^2 \{\Phi\}_j \cdot e^{\lambda_j t}, \tag{42}$$

Where N is the number of measurement points.
Displacements at any point in the system are obtained from their own forms by modal overlap

$$q_i = \sum_{k=1}^{L} \Phi_{ik} \cdot q_k \cdot \tag{43}$$

Then, taking into account (37) and (38) from expression (36), we can write:

$$B(\omega)_{ik} = \frac{\ddot{q}_{ik}}{\ddot{x}_c \cdot M \cdot \omega_c} = \frac{k_{gik}}{\bar{k}_k} \omega_C, \tag{44}$$

Where i = 1, 2, 3, ..., N is the number of measurement points; K = 1, 2, 3, ..., L is the number of proper forms.
The ratio of the acceleration at the output to the input frequency is usually denoted by the acceleration transfer coefficient of the construction [11]

$$k_{gik} = \frac{\ddot{q}_{ik}}{\ddot{x}_C}.$$

Then we get

$$k_{gik} = \frac{\bar{k}_k \cdot B(\omega)_{ik}}{\omega_C}. \tag{45}$$

6. Conclusion

The dynamic properties of linear systems in solving they described problem are usually characterized by the ratio of the reaction to the external action in the frequency function, that is, the transfer functions. It is also established that the nature of the signal passing through a complex system determined by its structure (stiffness and inertial mass characteristics), as well as the type of the input action, have, in the main, nonlinear dependences on the magnitude of the input action.
Thus, with the linearization of the construction of the transport machine as a complex mechanical system [11], the proposed method of substructures allows us to determine both its own forms and frequencies of oscillations, and its individual parts, using solutions for individual elements and aggregates. In this case, the dynamic model of the individual elements and aggregates of the transport machine can be reduced to a spatial system of solids modeled by finite element methods [14, 15], connected together and the base by elastic-inertial elements representing the bonds.

References

[1] Koshelev A I Chelnak S I 2000 *Regularity of solutions of certain boundary value problems for quasilinear elliptic and parabolic systems* (St. Petersburg: St. Petersburg University) p 335
[2] Belyaev A K Pal'mov V A 1980 *Theory of vibratory conductivity Problems of Dynamics and Strength* (Riga: Zinatne) 36 pp 93–102
[3] Devyatkov V V 2017 *Methodology and technology of simulation studies of complex systems: current state and development prospects: monograph* (Moscow: University textbook) pp 448
[4] Akhtulov A L 2005 Methodology of construction and practical application of the automation system for the design of transport vehicles *Bulletin of the Siberian State Automobile and Highway Academy (SibADI). Omsk: Publishing House* «LEO» Issue 3 pp 14–29
[5] Kron G 1963 *Diakoptics; the piecewise solution of large-scale systems* (London: MacDonald) pp 166
[6] Benfie1d W A Hruda R F 1971 *Vibration analysis of structures by component mode substitution* AIAA J pp 9
[7] Hurty W C 1965 *Dynamic analysis of structural systems using component modes* AIAA Journal 3 pp 678–685

[8] Craig R.R., Bampton M.C.C. "Coupling of structures for dynamic analysis". AIAA J., 1968, 6.
[9] Brohn D., Brown D., Henderson R., Rathbone A. "Modeling of Steel Structures for Computer Analysis". The Steel Construction Institute, 1995. 74s.
[10] Przemieniecki J.S. "Matrix of structural analysis of substructures". AIAAJ., 1963, 1, pp. 138-147.
[11] Akhtulov, A.JI. Calculation of oscillations of systems of large order by the method of substructures / A.L. Akhtulov. // Problems of the mechanics of modern machines: Mater. Intern. Conf. (Ulan-Ude, June 20-25, 2000). Ulan-Ude: VSGTU, 2000
[12] Sohor, Yu.N. Tensor analysis of networks and diakoptics in engineering calculations / Yu.N. Sohor. M.: Lambert Academic Publishing (LAP), 2012. 200 p.
[13] Akhtulov A.L. "Building The Automation System For Designing Load-Lifting Cranes Of Bridge Tyoe". European Science and Technology: materials of the XVII international research and practice conference, Munich, June 7th – 8th, 2017. publishing office Vela Verlag Waldkraiburg – Munich – Germany, 2017. pp. 29-35
[14] Bathe, Klaus-Jürgen "Finite element procedures". New Jersey: Prentice Hall, 1996. 1052 s.
[15] Serpik, I.N. Finite element method in solving problems of the mechanics of load-bearing systems / I.N. Serpic. M .: Publishing house of the Association of Construction Universities, 2015. 200 p.

IOP Conf. Series: Journal of Physics: Conf. Series **944** (2017) 012003 doi:10.1088/1742-6596/944/1/012003

Primitive polynomials selection method for pseudo-random number generator

I V Anikin, Kh Alnajjar

Kazan National Research Technical University named after A.N.Tupolev-KAI, Kazan, 420111, Russia.

e-mail: anikinigor777@mail.ru, khkazan@mail.ru

Abstract. In this paper we suggested the method for primitive polynomials selection of special type. This kind of polynomials can be efficiently used as a characteristic polynomials for linear feedback shift registers in pseudo-random number generators. The proposed method consists of two basic steps: finding minimum-cost irreducible polynomials of the desired degree and applying primitivity tests to get the primitive ones. Finally two primitive polynomials, which was found by the proposed method, used in pseudorandom number generator based on fuzzy logic (FRNG) which had been suggested before by the authors. The sequences generated by new version of FRNG have low correlation magnitude, high linear complexity, less power consumption, is more balanced and have better statistical properties.

1. Introduction

Nowadays high quality pseudorandom numbers have a critical importance to many scientific applications. Pseudorandom number generators based on linear feedback shift registers (LFSRs) [1] are among the fastest long-period generators currently available. They require less hardware for implementation, have more higher speed of operations and produce pseudorandom sequences with good statistical properties. In papers [2,3] we suggested new type of pseudorandom number generator based on fuzzy logic - FRNG. The structure of this generator is presented on Figure 1.

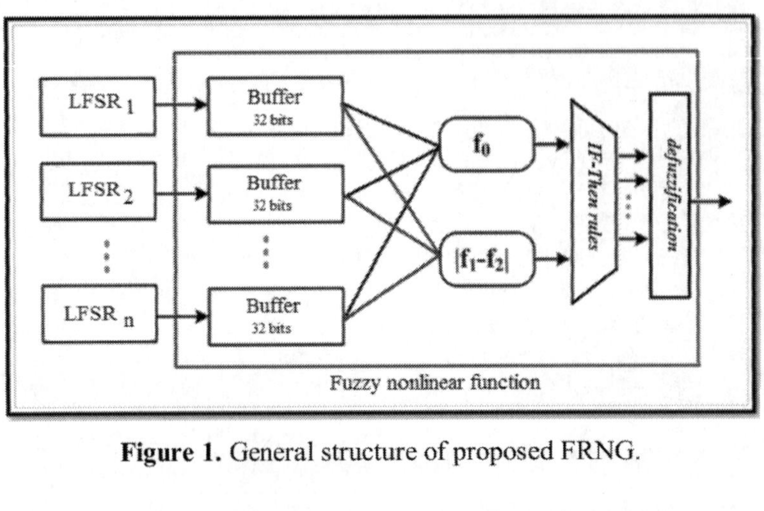

Figure 1. General structure of proposed FRNG.

This generator involves several LSFRs, buffers and fuzzy non-linear function with two fuzzy variables f_0 (number of ones) and $|f_1 - f_2|$ (the difference between number of two consecutive ones (0110) and number of two consecutive zeros (1001)), and a group of fuzzy if-then rules.

In [3] we evaluated the quality of suggested generator and showed that it passed basic statistical tests from NIST and Diehard bundles. Moreover, this generator showed better quality than some other well-known generators. Nevertheless, it is extremely important to investigate crucial parameters of the suggested generator and choose the best ones to improve the quality and the efficiency of generated pseudo-random numbers sequences. One of the basic parameters are characteristic polynomials of LFSRs which are used in FRNG.

It's well known that a binary sequence generated by LFSR possesses good statistical properties if its characteristic polynomial is primitive. For example, the period T of such sequence (is called M-sequence) is 2^n-1, where n is the degree of the polynomial. M-sequences are used for obtaining uniformly distributed random numbers [4] and they widely used in practice, for example in modeling or cryptography. The quality of M-sequences grows with increasing the degree n of the characteristic polynomial of LFSR, so it is very important in practice to obtain primitive polynomials with big degrees.

There are several published tables with primitive binary polynomials. Watson in [5] gives one primitive of degree n for every $n<100$. Stahnke in [6] gives a primitive with a minimum number of nonzero coefficients (trinomial or pentanomial) for every $n \leq 168$. One primitive pentanomial for every degree M_j, $8 \leq j \leq 27$, is also presented in [7] (here M_j denotes the jth Mersenne exponent, the prime for which $(2^{M_j}-1)$ is also prime). 5,9,17-weight primitive polynomials of degree up to 800 over GF(2) are presented in [8].

Aside from uniform distribution, good F_2-linear generators are also required to have characteristic polynomials $P(x)$ whose number of nonzero coefficients is not too far from the half of degree, i.e., approximately $n/2$ [9]. In particular, generators for which $P(x)$ is a trinomial or a pentanomial, which have been widely used in the past, should be avoided. It is known, that these generators have different statistical weaknesses [10].

In this paper we propose a method of finding good primitive binary polynomials. These polynomials have good statistical properties, high diffusion capacity (due to relatively high number of non-zero coefficients) and low power consumption.

The proposed method includes two main stages:

- Finding minimum-cost irreducible polynomials of the desired degree.
- Applying primitivity tests to find primitive polynomials.

This paper is organized as follows: in the section 2 we describe the step related with finding minimum-cost irreducible polynomials of the desired degree, in the section 3 we describe the step related with applying primitivity tests to get only primitive polynomials, in the section 4 we describe some experimental results, in Conclusion we give the basic results of the research.

2. Finding minimum-cost irreducible polynomials of the desired degree

It was proved in [8], that characteristic polynomial $f(x)$ of special type can be used to construct an n-stage standard LFSR with minimum-cost (concerning the number of 2-input XOR gates needed to implement the LFSR). These polynomials can be described by the general expression (1):

$$f(x) = (1+x^{b_1})(1+x^{b_2})\ldots(1+x^{b_m})+x^n \tag{1}$$

Parameters of these polynomials should satisfy the following conditions (2):

$$b_1 \geq 1, b_1 < b_2, (b_1+b_2) < b_3, \cdots, (b_1+b_2+\ldots+b_{m-1}) < b_m, (b_1+b_2+\ldots+b_m) < n \tag{2}$$

We will use this type of polynomials to fit the requirements of high efficiency regarding power consumption which directly depends on the number of 2-input XOR gates needed to implement it in hardware (In our case it's equal to $m+1$). These types of polynomials have also a high diffusion capacity which depends on the number of non-zero coefficients, which should be not too far from the half of degree $n/2$. This number is defined as $t = 2^m + 1$. For example if $n = 11$ then the most suitable choice of m is 2 where we can get a polynomial with $t = 2^2 + 1 = 5$ non-zero coefficients. We can take $b_1 = 1, b_2 = 5$ to get the following irreducible polynomial as example:

$$f(x) = (1+x^1)(1+x^5)+x^{11}$$

We can select all the irreducible polynomials of degree $n = 11$ and satisfy the conditions defined by (2) in this case ($b_1 \geq 1, b_2 > b_1, b_1 + b_2 < 11$). Table 1 contains all possible variants that fit these requirements:

Table 1. List of all 5-wights (t=5) irreducible polynomials of the degree n=11.

b_1	b_2	$f(x)$
1	5	$(1+x^1)(1+x^5)+x^{11}$
1	6	$(1+x^1)(1+x^6)+x^{11}$
1	7	$(1+x^1)(1+x^7)+x^{11}$
1	8	$(1+x^1)(1+x^8)+x^{11}$
2	3	$(1+x^2)(1+x^3)+x^{11}$
2	6	$(1+x^2)(1+x^6)+x^{11}$
2	7	$(1+x^2)(1+x^7)+x^{11}$
3	5	$(1+x^3)(1+x^5)+x^{11}$
3	6	$(1+x^3)(1+x^6)+x^{11}$
3	7	$(1+x^3)(1+x^7)+x^{11}$

In the previous example (n=11, m=2) the complete list of irreducible polynomials is not too long, which is not the general case. The number of polynomials will grow very fast with increasing the parameters n and m. For example if m=5 and the desired degree is 67 we will get a list of 147 irreducible polynomials with $t=33$ non-zero coefficients.

In most practical applications we need a polynomials with high quality. Therefore there is a need to find primitive polynomials with the degree as high as possible. Table 2 contains some values of m, the corresponding number of non-zero coefficients t, and the range of degrees which should be used with them.

Table 2. Values of m with corresponding number of non-zero coefficients t and the suitable ranges of degrees n of polynomials.

m	t	n
4	17	35-66
5	33	67-130
6	65	131-258
7	129	259-514
8	257	515-1026

We used *Mathematica* software to write a program which find all irreducible polynomials of type (1) with 2^m+1 non-zero coefficients for desired degree. This program enumerates all the possible values of the parameters (b_1, b_2,...,b_m) which satisfy all the previous mentioned conditions (2). The result is a list of numbers (b_1, b_2,...,b_m) ending with the degree of the polynomials. For example if m=5, n=67, one of the resulting element will be (2, 3, 7, 16, 33, 67) that means:

$$f(x) = (1+x^2)(1+x^3)(1+x^7)(1+x^{16})(1+x^{33}) + x^{67}$$

We can expand the expression and get the irreducible polynomial with t=33 terms:

$$f(x) = 1 + x^2 + x^3 + x^5 + x^7 + x^9 + x^{10} + x^{12} + x^{16} + x^{18} + x^{19} + x^{21} + x^{23} + x^{25} + x^{26} + x^{28} + x^{33} + x^{35} + x^{36} + x^{38} + x^{40} + x^{42} + x^{43} + x^{45} + x^{49} + x^{51} + x^{52} + x^{54} + x^{56} + x^{58} + x^{59} + x^{61} + x^{67}$$

Finally, at the end of this step we obtain the full list of irreducible polynomials which satisfy the conditions (2). Then we need to select only the primitive ones from them, and that will be done in the next step.

3. Applying primitivity tests and finding primitive polynomials from the list

After generating the lists of irreducible polynomials over GF(2) we should test them to get only the primitive ones [11]. The primitivity test of a given polynomial f is effectively performed with using the following set of conditions [12]:

- $f(0) = f(1) = 1$
- $min\{k : f | x^{2^k} - x\} = n$
- for all primes $p | 2^n - 1$, $x^{(2^k - 1)/p} \neq 1 \; mod \; p$.

The first condition eliminates polynomials which is divisible by x and x+1. In our case the polynomial f has the type defined by (1) and this condition is fulfilled automatically.

To check if the polynomial f satisfies the second condition, it is necessary to calculate n residues $x^{2^k} - x \; mod \; f$. The total number of elementary operations over GF(2) which should be done to test the condition (2) is bounded by O(tn^2), which is big, taking into account the number of polynomials that is needed to be checked. The problem is solved in [14] where condition (2) is modified by checking two subconditions:

- $gcd(f, f') = 1$
- $gcd(f, x^{2^k} - x) = 1$, $2 \leq k \leq 12$.

Here $\gcd(f, g) = 1$ is the greatest common divisor of the polynomials f and g [13].

The computing complexity of factorization of $2^n - 1$, which is needed in the third condition, is very high. This makes it inefficient and unreasonable to include this checking in the primitivity test and it makes the testing process extremely slow. Fortunately we can avoid it according to the famous Cunningham project [14]. We build a much faster method by storing the factors in database depending on [14]. Furthermore, if we choose the degree n to be a prime, then the number of factors of $2^n - 1$ will be very small, so the speed of the primitivity test will be very high. For example, when $n=63$ the factors of $2^{63} - 1$ is {{7,2},{73,1},{127,1},{337,1}, {92737,1},{649657,1}} we have to check six cases, but with n=67 the factors of $2^{67} - 1$ will be {{193707721,1},{761838257287,1}}. So we have to check only two polynomials. The complexity of this primitivity check is $O(F \cdot t \cdot n^2)$, where F is the number of distinct prime factors of $2^n - 1$.

In our algorithm we listed the prime factors of ($2^n - 1$) for all primes n between 35 and 1026 in a table according to [15]. Then we created the program in Mathematica software [16] to generate the lists of high efficient irreducible polynomials of the desired degree with the suitable m (see Table II). Finally we wrote the algorithm and program for primitivity test to pick up only primitive polynomials from the created list of irreducible polynomials.

4. Experimental results

Applying the suggested method we found all primitive polynomials with prime degree *n* between 35 and 1026. Some examples are represented in the Table 3.

Table 3. Some primitive polynomials which had been gotten by the suggested method.

n	*t*	*f(x)*
37	17	(3, 4, 9, 17, 37)
61	17	(2, 3, 15, 39, 61)
67	33	(1, 2, 4, 9, 41, 67)
89	33	(1, 5, 10, 17, 39, 89)
149	65	(3, 5, 9, 19, 37, 75, 149)
251	65	(1, 2, 4, 14, 23, 141, 251)
313	129	(2, 3, 6, 12, 24, 65, 169, 313)
509	129	(3, 6, 13, 27, 54, 119, 266, 509)
983	257	(2, 5, 12, 28, 58, 118, 243, 512, 983)

In order to increase the efficiency of the suggested in [2,3] FRNG, we replaced two primitive characteristic polynomials (3) currently used in FRNG by two primitive ones (4) selected from the obtained lists with using the proposed method.

$$P_1(x)=1+x^{83}+x^{84}+x^{86}+x^{89}$$
$$P_2(x)=1+x^{91}+x^{97} \tag{3}$$

$$P_1(x)=(1+x\ )(1+x^5)(1+x^{10})(1+x^{17})(1+x^{39})+x^{89}$$
$$P_2(x)=(1+x\ )(1+x^4)(1+x^7)(1+x^{20})(1+x^{53})+x^{97} \tag{4}$$

Finally we evaluated the quality of our FRNG (with two selected primitive polynomials (4)) with using NIST [17] and Diehard [18] bundles. The new FRNG successfully passed all randomness tests. Furthermore the new version of FRNG became more secure against algebraic attacks due to using special type of polynomials which have high diffusion capacity and high linear complexity, which increase the immunity of the proposed generator against correlation attacks

5. Conclusion

Finally, we can conclude that the proposed method of selection primitive polynomials is very useful for the field of pseudo-random numbers generators. Selected primitive polynomials increase the efficiency of pseudorandom numbers generators. As a result the sequences generated by new version of FRNG will have low correlation magnitude, high linear complexity, less power consumption, will be more balanced, they will have good statistical properties. It makes the resulting FRNG more suitable for various applications such as modeling, telecommunication, cryptography, authentication, etc.

6. References

[1] Lewis T G and Payne W H 1973 Generalized Feedback Shift Register Pseudorandom Number Algoritm *Journal of the ACM (JACM)* **20** 456-468.

[2] Anikin I V and Alnajjar K 2015 Fuzzy stream cipher system *Proc. Int. Siberian Conf. on Control and Communications* (Omsk).

[3] Anikin I V and Alnajjar K 2016 pseudo-random number generator based on fuzzy logic *Proc. Int. Siberian Conf. on Control and Communications* (Moscow).

IOP Conf. Series: Journal of Physics: Conf. Series **944** (2017) 012003 doi:10.1088/1742-6596/944/1/012003

[4] Tausworthe R C 1965 Random numbers generated by linear recurrence modulo two *Math. Comp.* **19** 201-209.
[5] Watson J 1962 Primitive polynomials (mod 2) *Math. Comp.* **16** 368-369.
[6] Stahnke W 1973 Primitive binary polynomials *Math. Comp.* **27** 977-980.
[7] Kurita Y and Matsumoto M 1991 Primitive t-nomials (t = 3; 5) over GF(2) whose degree is a Mersenne exponent ≤ 44497 *Math. Comp.* **56** 817-821.
[8] Wang Lang-Terng, Touba Nur A, Brent Richard P, and Wang Hui 2011 On Designing Transformed Linear Feedback Shift Registers with Minimum Hardware Cost *UT-CERC-12-03*.
[9] Wang D and Compagner A 1993 On the use of reducible polynomials as random number generators *Mathematics of Computation* **60** 363–374.
[10] Lindholm J H 1968 An analysis of the pseudo-randomness properties of subsequences of long m-sequence *IEEE Transactions on Information Theory* **IT-14 (4)** 569–576.
[11] Watson E J Primitive polynomials (mod 2) *Mathematics of Computation* **16(79)** 368-368.
[12] Lidl R and Niederreiter H 1983 Finite Fields, Encyclopedia *Math. Appl* **20**.
[13] Cohen E Arithmetical functions of a greatest common divisor *Proceedings of the American Mathematical Society* **11(2)** 164-171.
[14] Heringa J R, Blote H W and Compagner 1992 A New Primitive trinomials of Mersenne-exponent degrees for random number generation *International Journal of Modern Physics* **3** 561-564.
[15] Brillhart J, Lehmer D H, Selfridge J L, Tuckerman B, and Wagstaf S S 1988 Factorization of $b^n \pm 1$, b = 2, 3, 5, 6, 7, 10, 11, 12 up to high powers **22** *Amer.Math. Soc.*
[16] https://www.wolfram.com/mathematica/
[17] Rukhin A, Soto J, Nechvatal J, Smid M, Barker E, Leigh S, Levenson M, Vangel M, Banks D, Heckert A, Dray J and Vo S 2010 A statistical test suite for random and pseudorandom number generators for cryptographic applications *NIST Special Publication 800-22, Revision 1a*.

Synthesis and analysis of discriminators under influence of broadband non-Gaussian noise

V M Artyushenko[1] **and V I Volovach**[2]

[1] Technological University, 42, Gagarin str., Korolev, 141070, Russia
[2] Volga Region State University of Service, 4, Gagarin str., Togliatty, 445017, Russia

E-mail: volovach.vi@mail.ru

Abstract. We considered the problems of the synthesis and analysis of discriminators, when the useful signal is exposed to non-Gaussian additive broadband noise. It is shown that in this case, the discriminator of the tracking meter should contain the nonlinear transformation unit, the characteristics of which are determined by the Fisher information relative to the probability density function of the mixture of non-Gaussian broadband noise and mismatch errors. The parameters of the discriminatory and phase characteristics of the discriminators working under the above conditions are obtained. It is shown that the efficiency of non-linear processing depends on the ratio of power of FM noise to the power of Gaussian noise. The analysis of the information loss of signal transformation caused by the linear section of discriminatory characteristics of the unit of nonlinear transformations of the discriminator is carried out. It is shown that the average slope of the nonlinear transformation characteristic is determined by the Fisher information relative to the probability density function of the mixture of non-Gaussian noise and mismatch errors.

1. Introduction

The most important element of any tracking meter is a discriminator. The issues of synthesis and analysis of discriminators are described in a great number of works [1, 2, etc.]. Studies show that nonlinear inertial units of transformation of instantaneous values or envelopes of processed signals play the most significant role in signal processing algorithms under the influence of non-Gaussian noise. Discriminators are the most important among such devices. Theory and practice of implementation of discriminators is fairly well developed. However, when considering these issues primarily a form of signal was taken into account, but hardly ever the probability density function (PDF) of the influencing noise $W_{п}(n)$ and mismatch error $W(\varepsilon)$ were counted for.

2. The synthesis of the discriminators of tracking meters when exposed to non-Gaussian broadband noise

First, confirm that you have the correct template for your Let us consider the synthesis and analysis of the discriminators of tracking meters under the influence of additive broadband noise with the non-Gaussian distribution [3–9].

Given the input of the receiving device receives samples $\{y_h\}$ of an additive mixture of carrier signal $s(\lambda_h, t_h)$ and non-Gaussian broadband noise n_h

$$y_h = s\left(\lambda_h, t_h\right) + n_h, \quad h = \overline{1,H},$$

where λ_h is the information parameter of the signal.

Provided that the demodulation is accurate, using the results in [10], we write equations for estimation that meet the criterion of maximum a posteriori probability:

$$\hat{\lambda}_h = \hat{\lambda}_{e.h} + \hat{\sigma}^2_{\varepsilon.h}\left[\hat{\lambda}_{e.h}, y_h - s\left(\hat{\lambda}_{e.h}\right)\right]B'_{\lambda.h};$$

$$\hat{\sigma}^2_{\varepsilon.h}\left[\hat{\lambda}_{e.h}, y_h - s\left(\hat{\lambda}_{e.h}\right)\right] = \left\{\frac{\partial^2}{\partial\lambda_h^2}\ln W\left(\lambda_h|\lambda^{h-1}\right) - B''_{\lambda.h}\right\}^{-1}\Bigg|_{\lambda_h=\hat{\lambda}_{e.h}},$$

where $\hat{\sigma}^2_{\varepsilon.h}\left[\hat{\lambda}_{e.h}, y_h - s\left(\hat{\lambda}_{e.h}\right)\right]$ is the variance of the posterior distribution of the estimation; $\hat{\lambda}_{e.h}$ is the extrapolated value λ for the h-th step;

$$B'_{\lambda.h} = s'_{\lambda}\left(\hat{\lambda}_{e.h}\right)z\left(n_{\Sigma.h}\right) \tag{1}$$

is the output effect of the discriminator; $s'_{\lambda}\left(\hat{\lambda}_{e.h}\right)$ is the derivative of the signal with respect to a measured parameter λ; $z\left(n_{\Sigma.h}\right)$ is a characteristic of the nonlinear transformation unit (NLTU), depending on the type of the PDF of noise $W_n(n)$ and mismatch error $W(\varepsilon)$

$$z\left(n_{\Sigma.h}\right) = -\frac{\mathrm{d}}{\mathrm{d}n_{\Sigma.h}}\ln W_{\mathrm{n}}\left[y_h - s\left(\hat{\lambda}_{e.h}, t_h\right)\right] = -\frac{\mathrm{d}}{\mathrm{d}n_{\Sigma.h}}\ln W_{\mathrm{n}}\left[s\left(\lambda_{e.h}, t_h\right) + n_{\mathrm{n}.h}s\left(\hat{\lambda}_{e.h}, t_h\right)\right] =$$
$$= -\frac{\mathrm{d}}{\mathrm{d}n_{\Sigma.h}}\ln W_{\mathrm{n}}\left(n_{\varepsilon.h} + n_h\right) = -\frac{\mathrm{d}}{\mathrm{d}n_{\Sigma.h}}\ln W_{\mathrm{n}}\left(n_{\Sigma.h}\right);$$

where $n_{\varepsilon.h} = s\left(\lambda_h, t_h\right) - s\left(\hat{\lambda}_h, t_h\right)$ is the discrepancy between the received signal and reference signal, which can be both deterministic or random; $B''_{\lambda.h}$ is the derivative $B'_{\lambda.h}$ with respect to the information parameter λ.

Note that this difference also carries the information on the information parameter of the signal.

A device that computes $B'_{\lambda.h}\left[y_h - s\left(\hat{\lambda}_{e.h}\right)\right]$ is called the discriminator. Its block diagram is shown in Figure 1.

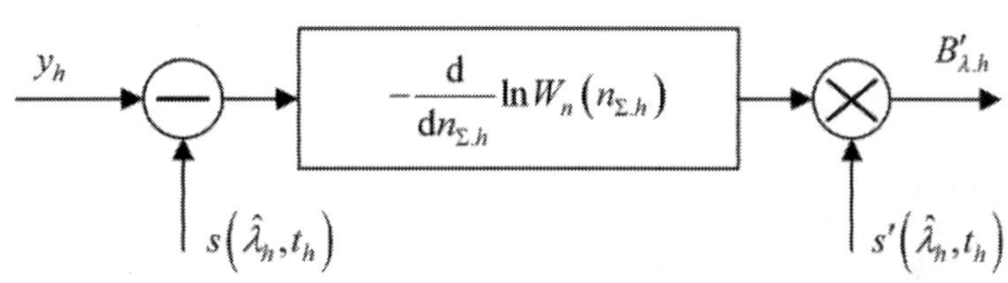

Figure 1. The block diagram of the discriminator of a tracking meter exposed to non-Gaussian broadband noise.

It should be noted, that in contrast to the known results, the characteristic of the NLTU which is included in the block diagram of the discriminator is different.

When exposed to non-Gaussian noise the discriminator realization was only considered for the asymptotic case when precision of the demodulator was high, it was assumed that $z\left(n_{\Sigma.h}\right) \approx z\left(n_{\mathrm{n}.h}\right)$.

As a rule, this condition is not met, therefore we analyze the characteristics of the discriminator in the presence of non-Gaussian noise and with permanent and accidental discrepancies between the reference $s(\lambda_h, t_h)$ and the input signal values with an extrapolated estimate of the information parameter $s\left(\hat{\lambda}_{e.h}, t_h\right)$, that is, when mismatch errors are non-zero.

It should be noted that the optimum discriminator can fairly accurately operate only when tracking precision is high. Initially we will analyze this particular case which is practically important though, and then we consider a more general case, when detuning is arbitrary.

Expanding the function $z(n_{h.\Sigma})$ in a Taylor series at the point $\hat{\lambda}_h$ and taking only two first components, which is justified with high precision tracking, we will get:

$$z(n_{\Sigma.h}) \approx z(n_{\text{n}.h}) + \varepsilon_h s'_\lambda(\hat{\lambda}_{\text{e}.h}) z'_\text{n}(n_{\text{n}.h}),$$

where $\varepsilon_h = \lambda_h - \hat{\lambda}_{\text{e}.h}$.

Substituting this ratio into the expression determining the output effect of the discriminator (1), we get

$$B'_\lambda(n_{\Sigma.h}) \approx s'_\lambda(\hat{\lambda}_{\text{e}.h}) z'_\text{n}(n_{\text{n}.h}) + \varepsilon_h \left[s'_\lambda(\hat{\lambda}_{\text{e}.h})\right]^2 z'_\text{n}(n_{\text{n}.h}) = \alpha(t_h) + \beta(\varepsilon).$$

The obtained expression shows that the output effect of the discriminator consists of two random components $\alpha(t_h)$ and $\beta(\varepsilon)$.

The component $\beta(\varepsilon) = \varepsilon_h \left[s'_\lambda(\hat{\lambda}_{\text{e}.h})\right]^2 z'_\text{n}(n_{\text{n}.h})$ describes the signal part of the output effect and is called the discriminatory characteristic (DC) of the discriminator.

The component $\alpha(t_h) = s'_\lambda(\hat{\lambda}_{\text{e}.h}) z'_\text{n}(n_{\text{n}.h})$ describes the noise part of the output effect and is called the fluctuation characteristics (FC) of the discriminator.

Let us consider and analyze these characteristics in detail.

Take the mathematical expectation $\beta(\varepsilon)$ obtained by set averaging and time averaging as the useful signal at the output of the discriminator. In this case provided the precision is high we get

$$\bar{\tilde{\beta}}(\varepsilon) = \bar{I}_{\text{F.n}} \left[\tilde{\bar{s}}'_\lambda(\hat{\lambda}_{\text{e}.h})\right]^2 \varepsilon = \bar{\tilde{K}}_\text{d}^2 \varepsilon, \tag{2}$$

where $\bar{\tilde{K}}_\text{d}^2 = \bar{\tilde{I}}_{\text{F.n}} \left[\tilde{\bar{s}}'_\lambda(\hat{\lambda}_{\text{e}.h})\right]^2 \varepsilon = I_{\text{F.n}} W_{\text{s}'}$ is the slope of the DC obtained by set and time averaging of the signal component; $I_{\text{F.n}} = \int z'_\text{n}(n_{\text{n}.h}) W_\text{n}(n_{\text{n}.h}) dn_{\text{n}.h}$ is the amount of Fisher information relative to the noise with the density of probability distribution $W_\text{n}(n_{\text{n}.h})$; $W_{\text{s}'} = \lim_{H\to\infty} \frac{1}{H} \sum_{h=1}^{H} \left[s'_\lambda(\hat{\lambda}_{\text{e}.h})\right]^2$.

Note that hereinafter, wavy line above a variable means time averaging, while straight line above a variable means set averaging.

When noise is Gaussian,

$$I_{\text{F.n}} = \sigma_\text{n}^{-2},$$

where σ_n^2 is the variance of affecting noise, which coincides with the known results [11].

As it can be seen from the expression (2), DC has a linear section.

In contrast to the case of Gaussian noise, the slope of DC is determined not only by characteristics (form) of a processed signal and variance of noise, but also by the amount of Fisher information $I_{\text{F.n}}$ relative to the noise $W_\text{n}(n_{\text{n}.h})$.

We define the variance of the fluctuation component at the output of the discriminator by averaging $\alpha(t_h)$ over the set

$$\sigma_{\varepsilon.\text{F}}^2 = \alpha^2(t_h) = \left[\tilde{s}'_\lambda(\hat{\lambda}_{\text{e}.h})\right]^2 \bar{z}_\text{n}^2(n_{\text{n}.h}) = W_{\text{s}'} I_{\text{F.n}} = \bar{\tilde{K}}_\text{d}^2. \tag{3}$$

The obtained relation confirms the well-known result that it $\tilde{\bar{K}}_{\mathrm{d}}^{2}$ also characterizes the variance of the fluctuation error at the output of the discriminator [12].

The obtained relation confirms the well-known result that it also characterizes the variance of the fluctuation error at the output of the discriminator

Dividing the noise component by $\tilde{\bar{K}}_{\mathrm{d}}^{2}$ we determine its equivalent value $\vartheta_h = \alpha_h \tilde{\bar{K}}_{\mathrm{d}}^{-2}$. Accordingly, the variance is equivalent to the noise component $\sigma_{\alpha}^{2} = \tilde{\bar{K}}_{\mathrm{d}}^{2}$.

With non-Gaussian broadband noise and arbitrary discrepancies, the process at the output of the discriminator is the result of a complex nonlinear transformation of the sum of two random processes $n_{\varepsilon.h}$ and $n_{\mathrm{n}.h}$.

In accordance with the above, we determine the slope of DC by averaging the derivative of the output effect of a measured parameter of the discriminator over the set and the time

$$\tilde{\bar{K}}_{\mathrm{d.n.}\Sigma}^{2} = \tilde{\bar{B}}_{\lambda}'' = \left[\tilde{\bar{z}}_{\mathrm{n}}'(n_{\Sigma.h})\tilde{\bar{s}}_{\lambda}'(\hat{\lambda}_{\mathrm{e}.h})\right]_{\lambda}' = \int_{-\infty}^{\infty}\left[\bar{s}_{\lambda}'(\hat{\lambda}_{\mathrm{e}.h})\right]^{2} z_{\mathrm{n}}'(n_{\Sigma.h})W_{\mathrm{n}}(n_{\Sigma.h})\mathrm{d}n_{\Sigma.h} = I_{\mathrm{F.n.}\Sigma}W_{\mathrm{s}'},$$

where $W_{\mathrm{n}}(n_{\Sigma.h}) = \int_{-\infty}^{\infty} W_{\varepsilon}(n_{\varepsilon.h})W_{\mathrm{n}}(n_{\Sigma.h} - n_{\varepsilon.h})\mathrm{d}n_{\varepsilon.h}$ of total error defined by resultant.

In contrast to the previous case, when detuning is arbitrary, the slope of DC is determined by the amount of Fisher information relative to the PDF or the error $W_{\mathrm{n}}(n_{\Sigma.h})$ of the mixture of noise and mismatch errors.

It is known that if the PDF $W_{\mathrm{n}}(n_{\Sigma.h})$ is a resultant $W_{\mathrm{n}}(n_{\mathrm{N}})$ with an arbitrary density, then [11]

$$I_{\mathrm{F.n.}\varepsilon} \le I_{\mathrm{F.n}}. \tag{4}$$

Consequently, the slope of the DC in the case when mismatch error is taken into account, satisfies the inequality

$$\tilde{\bar{K}}_{\mathrm{d.n.}\Sigma}^{2} \le \tilde{\bar{K}}_{\mathrm{d}}^{2}.$$

Time averaging is performed on the processing interval T. When processing random processes, this interval exceeds the interval of correlation of the information process as a rule.

As $z(n_{\Sigma.h})$ is an odd function of the argument, then $\tilde{\bar{\beta}}(\varepsilon) = 0$. This means that the DC is not biased.

Noise component at the output of the discriminator can be written as

$$\alpha(t_h) = B_{\lambda}' - \tilde{\bar{\beta}}(\varepsilon) = B_{\lambda}'.$$

As $\tilde{\bar{\beta}}(\varepsilon) = 0$, the variance can be defined as follows:

$$\sigma_{\varepsilon}^{2} = \left(\tilde{\bar{B}}_{\lambda}'\right)^{2}.$$

Taking into account that

$$\overline{\left[\frac{\partial \ln W(x)}{\partial x}\right]^{2}} = \overline{\left[\frac{\partial^{2} \ln W(x)}{\partial x^{2}}\right]} = I_{\mathrm{F.x}},$$

we can finally get

$$\sigma_{\varepsilon}^{2} = I_{\mathrm{F.n.}\varepsilon}W_{\mathrm{s}'}.$$

IOP Conf. Series: Journal of Physics: Conf. Series **944** (2017) 012004 doi:10.1088/1742-6596/944/1/012004

Taking into account the expression (4) and mismatch errors $\sigma_{\varepsilon}^{2} < \sigma_{\varepsilon.F}^{2}$, the equation (3) can serve as an upper estimate of the variance of the fluctuation component at the output of the discriminator.

To get a quantitative estimate when comparing relative efficiency of non-linear discriminators in non-Gaussian noise with their efficiency in Gaussian noise, it is convenient to use the ratio

$$\mu_{e.d}^{2} = \frac{\beta_{d.nG}^{2}(\varepsilon)\sigma_{\varepsilon.nG}^{-2}}{\beta_{d.G}^{2}(\varepsilon)\sigma_{\varepsilon.G}^{-2}} = -I_{F.n.\Sigma}\sigma_{n}^{2}, \tag{5}$$

where $\beta_{d.nG}^{2}$, $\beta_{d.G}^{2}(\varepsilon)$, $\sigma_{\varepsilon.nG}^{2}$, $\sigma_{\varepsilon.G}^{2}$ are respectively, average values and variances of the output signal of the discriminator with non-Gaussian and Gaussian noise.

This ratio shows that if absolute values of the average signal of mismatch error are equal, when we compare two discriminators, more effective will be the one which provides the maximum signal-to-noise ratio or the smaller value of σ_{ε}^{2}, which is the same.

For a particular case when the characteristic of the discriminator is described by the linearized ratio

$$\mu_{e.d}^{2} = I_{F.n}\sigma_{n}^{2} = \mu_{0a}^{2},$$

the efficiency of non-linear processing in the discriminator (5) coincides with the efficiency of the nonlinear transformation units $z(_{n.h})$ obtained in the theory of asymptotically optimal reception [11]. (PDF) of the influencing noise $W_{n}(n)$ and mismatch error $W(\varepsilon)$ were counted for.

3. Analysis of information loss caused by linear section of nonlinear transformation unit

As known, there are three sections in the characteristic of the discriminator. If the mismatch error $\varepsilon_{h} = \lambda_{h} - \hat{\lambda}_{e.h}$ (where λ_{h} is the information parameter of the signal at the h-th step; $\hat{\lambda}_{e.h}$ is extrapolated estimation value λ at the h-th step) is described by the Gaussian PDF, the middle section of discrimination characteristic (DC) of the discriminator is linear. Steepness and length of this section is determined by the variance of the mismatch error σ_{ε}^{2} and it changes in the process of operation of the demodulator. The nature of the end sections of the DC is determined by the form and parameters of the PDF of additive noise $W_{n}(n)$ and steepness of these sections is determined by the variance of the noise σ_{n}^{2}.

Figure 2 shows an example of the characteristic $z(n_{\Sigma.h})$ of the discriminator for the case when the PDF of the total error $(n_{\Sigma.h})$ can be described by bimodal distribution $W_{n}(n_{\Sigma.h}) = \sum_{i=1}^{2} p_{i}N(m_{i}, \sigma_{2n}^{2} + \sigma_{\varepsilon}^{2})$.

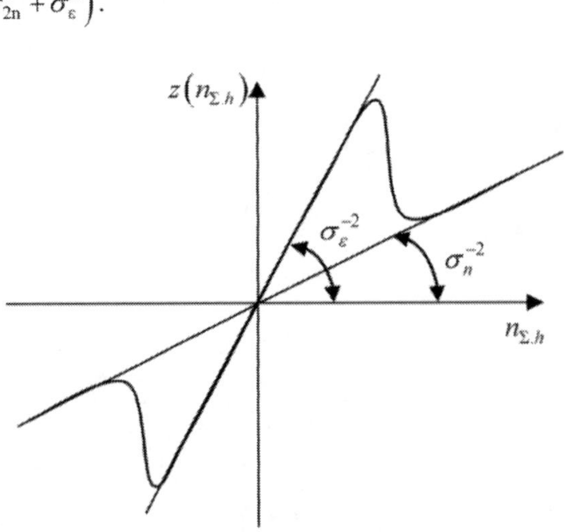

Figure 2. The characteristic of the unit of nonlinear transformations of the discriminator.

Let us consider and analyze the information loss connected with the linear section of the characteristic of the NLTU of the discriminator.

Taking into account mismatch errors leads to the necessity of introducing the linear section to the characteristics of the NLTU, which therefore reduces the efficiency of nonlinear processing.

The boundaries and slope of the linear section change in the process of operation and are determined by the value of current a posteriori variance $\hat{\sigma}^2_{\varepsilon.h}$.

To estimate the information loss caused by the presence of the linear section of the characteristic of the NLTU, the stationary mode of operation of the meter has to be considered when

$$\hat{\sigma}^2_{\varepsilon.h} = \hat{\sigma}^2_{\varepsilon.h-1} = \sigma^2_{\varepsilon}.$$

In this case, the amplitude of the NLTU can be represented as:

$$z_n(n) \approx \begin{cases} \dfrac{n}{\sigma^2_{\varepsilon}}, & |n| \le \Delta; \\ -\dfrac{W'_{\text{n}.\Sigma}(n)}{W_{\text{n}.\Sigma}(n)}, & |n| > \Delta, \end{cases}$$

where Δ is the boundary of linear section, which can be estimated by the value $k_\Delta \sigma^2_\varepsilon$; k_Δ is a constant coefficient.

We estimate the average slope of the characteristic of NLTU before the introduction of the linear section

$$K_{\text{d}} = -\int_{-\infty}^{\infty} \frac{\text{d}}{\text{d}n} z_{\text{n}}(n) W_{\text{n}}(n)\text{d}n = \int_{-\infty}^{\infty} z_{\text{n}}^2(n) W_{\text{n}}(n)\text{d}n = I_{\text{F.n}},$$

where $I_{\text{F.n}}$ is Fisher information relative to noise with the density of probability distribution $W_{\text{n}}(n)$.

The increment in the signal/noise ratio (SNR) at the output of such NLTU is estimated by

$$\mu^2_{0\text{a}} = I_{\text{F.n}} \sigma^2_{\text{n}},$$

which is called the ratio of the amplitude noise suppression [12].

With the introduction of the linear section the average slope of the NLTU changes and can be estimated by the expression

$$K_{\text{d.l}} = -\int_{-\infty}^{\Delta} z'_{\text{n}}(n) W_{\text{n}}(n)\text{d}n + \sigma_\varepsilon^{-2} \int_{-\Delta}^{\Delta} W_{\text{n}}(n)\text{d}n + \int_{\Delta}^{\infty} z'_{\text{n}}(n) W_{\text{n}}(n)\text{d}n = I_{\text{F.n}} - I_{\text{F.l}},$$

where

$$I_{\text{F.l}} = \int_{-\Delta}^{\Delta} z'_{\text{n}}(n) W_{\text{n}}(n)\text{d}n - \sigma_\varepsilon^{-2} \int_{-\Delta}^{\Delta} W_{\text{n}}(n)\text{d}n, \tag{6}$$

characterizes Fisher information loss relative to noise due to the introduction of linear section in the characteristic of the NLTU.

In general, the ultimate loss of Fisher information due to errors can be estimated by the expression [13]

$$I_{\text{F.l}} = \int_{-\infty}^{\infty} z'_{\text{n}}(n) W_{\text{n}}(n)\text{d}n - \int_{-\infty}^{\infty} z'_{\text{n}}(n_\Sigma) W_{\text{n}.\Sigma}(n_\Sigma)\text{d}n_\Sigma = \int_{-\infty}^{\infty} z'_{\text{n}}(n) W_{\text{n}}(n)\text{d}n -$$
$$- \iint_{-\infty}^{\infty} z'_{\text{n}}(n) W_{\text{n}}(u-n) W_\varepsilon(u)\text{d}u\text{d}n.$$

The slope of the characteristic of the NLTU with taking into account mismatch errors will be determined by the expression

$$K_{\text{d.}\varepsilon} = \iint_{-\infty}^{\infty} \frac{\text{d}}{\text{d}n_\Sigma} z_\text{n}(n_\Sigma) W_\varepsilon(u) W_\text{n}(u - n_\Sigma) \text{d}u \text{d}n_\Sigma .$$

The losses mentioned above lead to the reduction of SNR at the output of the NLTU by

$$\mu_1^2 = I_{\text{F.1}} \sigma_\text{n}^2 .$$

As a result the $\mu_{\text{0.a}}^2$ value should be considered as ultimate, which is achieved only in the absence of errors of tracking.

We estimate the value $I_{\text{F.1}}$ for rather general case when the PDF can be described by poly-Gaussian distribution [14]

$$W_\text{n}(n) = \sum_{m=1}^{M} p_m N\left(m_{\text{n.}m}, \sigma_{\text{n.}m}^2\right); \Sigma p_m = 1 . \tag{7}$$

In this case

$$z_\text{n}(n) = -W_\text{n}'(n) / W_\text{n}(n);\; z_n'(n) = -\left[W_n''(n) W_n(n) - \left(W_n'(n)\right)^2\right] W_n^{-2}(n). \tag{8}$$

Substituting (7) and (8) in expression (6), after simple but cumbersome transformations, we get

$$I_{\text{F.1}} = \int_{-\Delta}^{\Delta} \frac{\left(W_\text{n}'(n)\right)^2}{W_\text{n}(n)} \text{d}n - W_\text{n}(n) \Big|_{-\Delta}^{\Delta} - \int_{-\Delta}^{\Delta} W_\text{n}(n) \text{d}n ,$$

or

$$I_{\text{F.1}} = I_1 - \sum_{m=1}^{M} \frac{p_m}{\sqrt{2\pi\sigma_{\text{n.}m}^2}} \left\{ (\Delta - m_{\text{n.}m}) \exp\left(-\frac{(\Delta - m_{\text{n.}m})^2}{2\sigma_{\text{n.}m}^2} \right) + (\Delta - m_{\text{n.}m}) \exp\left(-\frac{(\Delta - m_{\text{n.}m})^2}{2\sigma_{\text{n.}m}^2} \right) - \right.$$
$$\left. -0{,}5 \left[\text{erf} \frac{(\Delta - m_{\text{n.}m})}{\sqrt{2}\sigma_{\text{n.}m}} + \text{erf} \frac{(\Delta + m_{\text{n.}m})}{\sqrt{2}\sigma_{\text{n.}m}} \right] \right\};$$

$$I_1 = \int_{-\Delta}^{\Delta} \frac{\left(W_\text{n}'(n)\right)^2}{W_\text{n}(n)} \text{d}n ;\; \text{erf}(x) = \frac{2}{\sqrt{\pi}} \int_0^{\pi} \exp\left(-\xi^2\right) \text{d}\xi , \tag{9}$$

where erf(x) is the error function (probability integral).

4. Conclusion

Thus, the synthesis and analysis of the discriminator of the tracking meter operating in conditions of exposure to broadband non-Gaussian noise is carried out. It is shown that in case of exposure to broadband non-Gaussian noise the discriminator the tracking meter contains the nonlinear transformation unit, the characteristic $z(n_{\Sigma.h})$ of which is determined by the form of the PDF $W_\text{n}(n_\text{n})$ of noise and mismatch errors $W(\varepsilon)$.

Characteristics of the discriminators $\tilde{\bar{K}}_\text{d}^2$ and σ_ε^2 are determined by the amount of Fisher information $I_{\text{F.n.}\Sigma}$ relative to the PDF of the mixture of non-Gaussian noise $n_{\text{n.}h}$ and mismatch errors $n_{\varepsilon.h}$.

The main calculated ratios are given, which allows us to estimate the efficiency of signal processing in the discriminators in presence of non-Gaussian broadband noise, taking into account mismatch errors. The parameters of DC and FC of discriminators operating under these conditions are found.

It is shown that the efficiency of non-linear processing increases with decreasing mismatch error and when $n_{\varepsilon.h} \to 0$, it is estimated by the amount of Fisher information relative to influencing noise $I_{\text{F.n}}$.

Numerical analysis of the expression (9) shows that the expansion of the boundaries of the linear section of the DC leads to information loss increase and when $\Delta \to 0$, $I_{\text{F.l}} \to I_{\text{F.n}}$. It means that the efficiency of nonlinear transformation is zero and DC degenerates into linear one.

Note that to estimate the efficiency of some real NLTUs, characteristics of which differ from the optimal ones, the above ratio of real amplitude suppression of noise can be used

References

[1] Ellingson S W 2016 *Radio System Engineering* (Cambridge: Cambridge University Press)

[2] McClaning K and Vito T 2001 *Radio Receiver Design* (London: Noble Publishing Corporation)

[3] Kassam S A 1989 *Signal detection in non-Gaussian noise* (New York: Springer Verlag)

[4] Park J, Shevlyakov G and Kim K 2012 Distributed Detection and Fusion of Weak Signals in Fading Channels with Non-Gaussian Noises *IEEE Communications Letters* **16**(2) pp 220–3

[5] Bandiera F Dodde V and Ricci G 2015 Radar detection and range estimation of a point-like target in non-Gaussian noise *Proc. IEEE Radar Conf. (RadarCon)* DOI: 10.1109/RADAR.2015.7131051

[6] Yang J Cheng Y Wang H Li Y and Hua X 2015 Unknown stochastic signal detection via non-Gaussian noise modelling *2015 IEEE International Conference on Signal Processing, Communications and Computing (ICSPCC)*

[7] Palahina E and Palahin V 2016 Signal detection in additive-multiplicative non-Gaussian noise using higher order statistics *Proc. 26th Int. Conf. Radioelektonika (RADIOELEKTRONIKA)*

[8] Artyushenko V M, Volovach V I and Shakurskiy M V Analysis of influence of uncorrelated additive non-Gaussian noise on accuracy of motion parameters measurement in short-range radio systems 2015 *Proc. EEE Int. Siberian Conf. Control and Communications (SIBCON-2015)* 7147279

[9] Artyushenko V M and Volovach V I Measuring information signal parameters under additive non-Gaussian correlated noise 2016 *Optoelectronics, Instrumentation and Data Processing* **59**(6) pp 22–8

[10] Artyushenko V M and Volovach V I 2016 The demodulation signal under the influence of additive and multiplicative non-Gaussian noise *Proc. IEEE East-West Design & Test Symp. (EWDTS'2016)*

[11] Novoselov O N and Fomin A F 1991 *Bases of the theory and calculation of information and measuring systems* (Moscow: Mechanical engineering) (in Russian)

[12] Kuznetsov P I, Stratonovich R L and Tikhonov V I 1965 *Non-Linear Transformations of Stochastic Processes* (Oxford: Pergamon Press)

[13] Valeyev V T and Sosulin Y G Detection of weak, coherent signals in correlated non-Gaussian noise 1969 *Radio technology and electronics* **14**(2) (in Russian)

[14] Tihonov V I and Harisov V N 1991 *Statistical analysis and synthesis of radio engineering devices and systems* (Moscow: Radio and communication) (in Russian)

Development of the method of aggregation to determine the current storage area using computer vision and radiofrequency identification

A Astafiev, A Orlov and D Privezencev

Vladimir State University, Vladimir, Russia

e-mail: Alexandr.Astafiev@mail.ru

Abstract. The article is devoted to the development of technology and software for the construction of positioning and control systems in industrial plants based on aggregation to determine the current storage area using computer vision and radiofrequency identification. It describes the developed of the project of hardware for industrial products positioning system in the territory of a plant on the basis of radio-frequency grid. It describes the development of the project of hardware for industrial products positioning system in the plant on the basis of computer vision methods. It describes the development of the method of aggregation to determine the current storage area using computer vision and radiofrequency identification. Experimental studies in laboratory and production conditions have been conducted and described in the article.

1. Introduction

Currently, due to the need of industrial development and the implementation of international quality standards the introduction of new technologies to control product movement is required. Implementation of such control is possible through introduction of automatic identification and recognition systems. The degree of perfection of the product movement control system depends on the efficiency of production in general. Therefore, the solution of control problems in industrial plants, consisting in the development of new and improvement of existing methods and means of information processing, in order to increase the efficiency of the production of motion control systems is essential.

A large number of companies, both domestic and foreign markets hardware and software solutions have long been engaged in the development of such systems. However, the currently existing solutions allow you to automate industrial processes in which a product is transported next to a stationary sensor (eg, a conveyor line with permanently mounted image sensor), or sensor control is performed manually to identify stored products (eg, hand-held portable barcode readers). Use existing systems for more complex tasks, where the transportation of goods is carried out using several routs (eg transport truck), or using the manual method of identification is technically impossible or contrary to the rules of safety, do not allow to automate the process of monitoring the movement of products due to low reliability of the results.

One of these tasks is to develop a system of automatic identification of markings on metal pipes on the example of pipe electric welding shop № 3 (Welding Shop №3) of OJSC «Vyksa Steel Works» (VSW). Storage of pipes at Welding Shop № 3 made by the organization «heaps» of the same type of products in the two types of warehouses: indoor (inside the plant premises) and open (outdoor storage). Transportation of pipes is performed in 5 different ways: conveyor, overhead crane with hooks, a bridge crane with electromagnets, vacuum loader and street autoloader with hooks. Pipes come to the shop on the conveyor belt, and leave it in rail cars.

The main difficulties of implementing automatic control are:

1. The presence of 5 different ways to transport products, which involves the development of a universal approach for all types of transportation.
2. The presence of both indoor and outdoor spaces, which involves the development and use of algorithms to artificially shade and lighten the image, by using methods of digital image processing.
3. Inability to use manual methods of data collection due to safety rules (need to stand clear of the crane and the loader boom).
4. Use the visual markings on the tubes is complicated due to their distortion from friction between the pipes.
5. The use of radio-frequency labels is not possible due to lack of credibility, since the products are made of metal, which badly distorts radio waves.
6. The hardware is to be designed as a stand-alone installation, placed on a transport device without making any changes in their structure (as this is against the Rosstandart rules) and remote units (as during the rise / descent of traverse the pipes are beating against the sides of the trucks and each other).

The main application purpose is to develop a method for automatic identification of pipes based on the technology of mathematical data and software aggregation of the digital processing of heterogeneous features a unified point of images and statistical analysis of radio frequency identifiers, which will automate the process of motion control products on the Welding Shop №3 OJSC "VSW". Automation of pipe movement control will quickly and accurately obtain information on the status of storage, reduce the risk of regrading and pipe loss, as well as increase the efficiency of the search and transporting a given product to the place of storage or shipment.

2. Development of the project of hardware for industrial products positioning system in the plant on the basis of computer vision methods

Based on the functional requirements of the system, it can be concluded that the developed system should work equally well on various transport devices and in different weather and lighting conditions. Therefore, it is advisable to develop a custom universal video device to obtain homogeneous data for use in a variety of production environments.

As the practice shows the most "narrow" place of such systems is the communication channel between the video sensor and video data processing unit. Therefore, it is advisable to place a video processing device in the vicinity of the video sensor and transfer already processed and recognized identifiers of transported products to the warehouse server.

Implementing such systems sharply raises questions about the presence of power supply and communication channel. To address the question of power it was decided to supply the device with independent replaceable batteries. Data transfer is made using Wi-Fi wireless networks, which makes the inclusion of Wi-Fi device transmitter mandatory.

Given all of this video device of automatic detection and marking identification should include:

1. The vision sensor.
2. Microcontroller.
3. Power supply.
4. Wi-Fi transmitter.

General view of the device can be represented in a pattern (Figure 1):

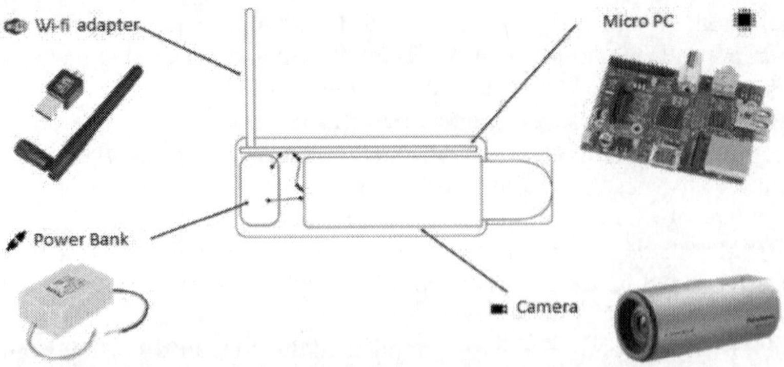

Figure 1. General view of the video device automatic detection and marking identification.

The hardware implementation of the video device for automatic detection and marking identification may consist of a single device, which includes all necessary elements of the system (for example, smartphone or terminal), or consist of individual functional units.

Necessary conditions of work of automatic marking identification system:

1. The presence of sustained and stable Wi-Fi coverage in the territory of operating of identifying equipment.
2. The regular replacement of power supply elements and equipment maintenance.
3. Bar coding markings stamped onto the tube circumference (at least 4 places).

Detection, localization and identification of markings on industrial products is performed using the algorithms described in the publications [1–3].

3. Development of the project of hardware for industrial products positioning system in the territory of a plant on the basis of radio-frequency grid

As the basic data for the implementation of tracking of industrial products emerge in the course of its movement through the territory of the plant, it is expedient to develop a hardware project to receive and process all the data. The main types of traffic information is the information about who moves the products, how and its route. Production and processing of this information will allow to organize a permanent automatic traceability of industrial products on the plant.

The paper proposes the development of a stand-alone device, consisting of a reading, processing and transmission of information equipment. The developed device is mounted on the transport device, and to ensure the positioning, plant territory is marked with RFID tags, thus creating a radio frequency grid.

Thus, the hardware part of the system can be divided into 5 levels:

1. The RF tag for labeling storage areas.
2. Equipment to read RFID tags.
3. Equipment for collecting and processing statistical data.
4. The equipment for the transmission of data to the enterprise server.
5. Software and hardware that is used by the company.

Laboratory prototype was developed for testing the project hardware industrial products positioning system on the territory of the enterprise on the basis of radio-frequency grid for experimental studies. Laboratory prototype consists of a microcontroller, a manual RF reader, power supply unit and a laptop (Figure 2).

IOP Conf. Series: Journal of Physics: Conf. Series **944** (2017) 012005 doi:10.1088/1742-6596/944/1/012005

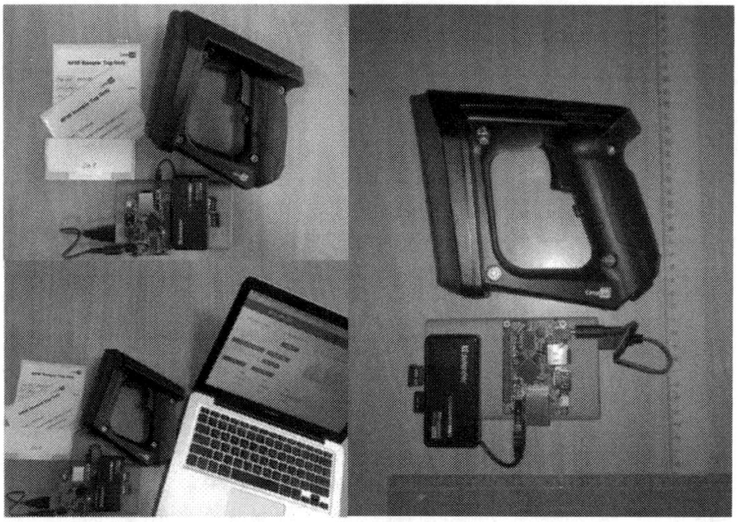

Figure 2. View of a laboratory prototype.

The work of the laboratory setup is based on the algorithms described in the article [1–3].

4. Development of the method of aggregation to determine the current storage area using computer vision and radiofrequency identification

The work of algorithms to determine the current storage area in a production environment is highly dependent on the presence of various kinds of interference, introducing distortions in the raw data set for analysis. In view of this, aggregation algorithms based on different technologies and approaches will improve the accuracy of their results.

If the algorithm for determining the current storage area based on methods of computer vision returns the result of the current position Ctv and evaluation, showing the accuracy of the calculations ptv, and the algorithm to determine the current storage area on the basis of methods of RFID returns the current position Crfid and evaluation, showing the accuracy of the calculations prfid, then the methodology of aggregation algorithms that determine the current storage area can be represented by the following set of steps:

1. Obtaining data from algorithms to determine the current storage area on the basis of methods of computer vision (Ctv, ptv) and radio frequency identification (Crfid, prfid).
2. Comparison of the results of Ctv and Crfid.
3. If Ctv = Crfid, then one of the values of Ctv or Crfid is transferred to PCS warehouse database as an identifier of the current storage area.
4. If Ctv ≠ Crfid, then ID value of storage area with the higher accuracy of the calculations is transferred to PCS warehouse database. In this case, the field is set in the database which shows that the results of determination of the storage area should be refined.
5. Data on the current storage area are sent to warehouse database.

The work of this method can be represented by the following formula:

$$(C_{db}, p) = \begin{cases} (C_{tv}, 0), IF\ C_{tv} = C_{RFID} \\ (C_{tv}, 1), IF\ C_{tv} \neq C_{RFID}, p_{tv} > p_{RFID} \\ (C_{RFID}, 1), \quad IF\ C_{tv} \neq C_{RFID}, p_{tv} < p_{RFID} \end{cases}$$

where Cdb – the storage area identifier, recorded in the warehouse database; p – the need for additional control of the recorded information flag, 1 – is necessary to monitor, 0 – data is correct.

The generalized scheme of the technique for algorithm aggregation to determine the current storage area based on methods of computer vision and RFID is represented in a pattern (Figure 3):

IOP Conf. Series: Journal of Physics: Conf. Series **944** (2017) 012005 doi:10.1088/1742-6596/944/1/012005

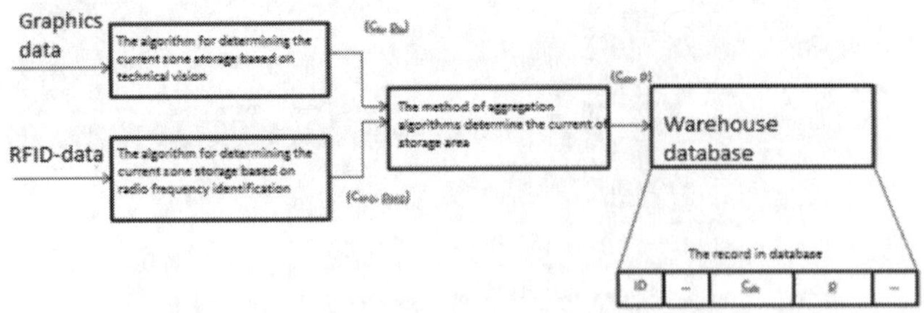

Figure 3. A generalized scheme of this method.

Refinement of the result for current storage area can be done in two ways:

1. Manual mode by doing inventory in a warehouse or in a specific storage area.
2. Automatic mode by checking the correctness of the specified data in the subsequent transportation of industrial products, whose position has been determined inaccurately.

The choice of method to clarify the current storage area is done by enterprise staff through establishing the rules of work in the storage area.

5. Conclusion

The article is devoted to the development of technology and software for the construction of positioning and control systems in industrial plants based on aggregation to determine the current storage area using computer vision and radiofrequency identification. It describes the developed of the project of hardware for industrial products positioning system in the territory of a plant on the basis of radio-frequency grid. It describes the development of the project of hardware for industrial products positioning system in the plant on the basis of computer vision methods. It describes the development of the method of aggregation to determine the current storage area using computer vision and radiofrequency identification. Experimental studies in laboratory and production conditions have been conducted and described in the article.

The work is executed at financial support of the grant of the President of the Russian Federation № MK-991.2017.9.

References

[1] Provotorov A D Privezentsev D G Astafiev A 2015 Development of Methods for Determining the Locations of Large Industrial Goods During Transportation on the Basis of RFID *Procedia Engineering Volume* 129 pp 1005–1009

[2] Astafiev A V Orlov A A Provotorov A V Astafiev A V 2015 The localization algorithm of symbolic and bar-code labels on industrial products for the control of product movements *Stability and Control Processes in Memory of V.I. Zubov* (SCP) pp 615–616 DOI: 10.1109/SCP.2015.7342230

[3] Astafiev AV Orlov A A Privezentsev D G 2016 Method of controlling the movement of large metal products with the use of algorithms for localization and recognition of bar code markings *Dynamics of Systems, Mechanisms and Machines, Dynamics 2016 (Omsk)* DOI: 10.1109/Dynamics.2016.7818969

[4] Zhiznyakov A L Privezentsev D G Zakharov A A Zhiznyakov A L 2015 Using fractal features of digital images for the detection of surface defects *Pattern Recognition and Image Analysis* Vol 25 Is 1 pp 122–131

[5] Bondarevsky A C Zolotov V R 2009 Design of radio frequency identification (RFID) – a problem situation *Modern high technologies* 9 pp 19–23.

[6] To Huy 2015 Gnat research and application of the technology rfid (radiofrequencyidentification) *Duy. Neat. International research journal* 6(37) Part 1 pp 34–37

[7] Kamozin Denis Y 2013 Comparison of the effectiveness of the use of bar code technology and

RFID technology in logistics processes *Proceedings of ISEA* 3
[8] Ke-Sheng Wang 2014 Intelligent and integrated RFID (II-RFID) system for improving traceability in manufacturing *Advances in Manufacturing* Vol 2 Is 2 pp 106–120
[9] Worapot Jakkhupan, Somjit Arch-int Yuefeng Li 2015 Erratum to: An RFID-based traceability system *Telecommunication Systems* Vol 58 Is 3 pp 273–273
[10] Mahir Oner Alp Ustundag Aysenur Budak 2016 An RFID-based tracking system for denim production processes *The International Journal of Advanced Manufacturing Technology* pp 591–604

The development of small-scale mechanization means positioning algorithm using radio frequency identification technology in industrial plants

A Astafiev, A Orlov and D Privezencev
Vladimir State University, Vladimir, Russia

e-mail: Alexandr.Astafiev@mail.ru

Abstract. The article is devoted to the development of technology and software for the construction of positioning and control systems for small mechanization in industrial plants based on radio frequency identification methods, which will be the basis for creating highly efficient intelligent systems for controlling the product movement in industrial enterprises. The main standards that are applied in the field of product movement control automation and radio frequency identification are considered. The article reviews modern publications and automation systems for the control of product movement developed by domestic and foreign manufacturers. It describes the developed algorithm for positioning of small-scale mechanization means in an industrial enterprise. Experimental studies in laboratory and production conditions have been conducted and described in the article.

1. Introduction

Currently, because of the need to move to new intellectual digital production technologies and implement international quality standards, it is necessary to introduce new science-based approaches to controlling the movement of products and small-scale mechanization of warehouses. This is due to the fact that the warehouses of large industrial organizations, at the current level of hardware and software, cannot fully comply with domestic and foreign quality standards in the field of product tracking, regulated by GOST and ISO. A number of domestic and foreign standards include strict requirements for organizing the process of product traceability. Any unit produced in the enterprise must be accompanied by a detailed history of its manufacturing, starting at the beginning of the production process. Registration processes of materials used in the production of a product unit and the main stages of its movement, at the current level of development of software and hardware automation of industrial enterprises, are documented in electronic document management systems, but the movement of products within the warehouse territories requires greater detail. The lack of detailed product movement information can lead to product re-sorting and theft. Automation of product movement control in the last decade is mainly done with the use of technical vision, however, in the last few years, systems based on radio frequency identification methods are rapidly developing in the market of industrial automation technologies. However, current solutions allow automation of industrial processes in which either the products are transported past a stationary sensor (for example, a conveyor line with a stationary video sensor or radio frequency reader) or manual control of the sensor to identify the stored products (for example, manual portable bar code readers and RFID tags). Using the existing systems to solve more complex problems, where the transportation of products is carried out by more than one route (for example, transportation by a forklift truck) or using manual identification method is technically impossible or contradicts safety rules. Implementation of these systems won't help automate the process of controlling the product movement in due to the low reliability of the received results. Based on this, the development of new algorithms, methods and systems for monitoring the movement of industrial products is a substantial scientific and technical task [1–3].

All such developments should be based on the already implemented software and hardware solutions for documenting information on traceability and automated means of transportation and storage. If documenting of information is implemented in almost all industrial enterprises in the form of separate electronic document management systems, the automation of small-scale mechanization often requires a complete replacement of the fleet of transport devices, due to the lack of effective ways to organize their positioning and status tracking, which entails enormous financial and time costs for the enterprise. Therefore, the development of technology and software for the construction of positioning and control systems for small mechanization in industrial plants based on radio frequency identification methods is a substantial scientific and technical task [4, 5].

The purpose of the work is to consider and use a fundamentally new approach to automating small mechanization facilities for industrial enterprises with a complex transport infrastructure. The use of this approach will allow solving fundamentally new automation tasks in which the use of existing solutions does not yield the required result and will provide improvement of known scientific results by revising them with the proposed approach.

2. Subject overview

A great contribution to the development of radio frequency identification technology and product movement control systems in various spheres of life was made by Bondarevsky A.S., Zolotov R.V., Do Zuy Nyat, Kamozin D.U., Manish B., Shahram M., Ke-Sheng Wang , Worapot Jakkhupan, Somjit Arch-int, Yuefeng Li, Mahir Oner, Alp Ustundag, Aysenur Budak and many others [5–10].

Application of these knowledge-intensive technologies makes it possible to automate the processes of controlling the product movement at industrial plants, ultimately, to increase the efficiency and reliability of transportation control and warehouse inventory control of manufactured products.

However, they are not without flaws. The use of existing software and hardware solutions is more aimed at organizing automated warehouse inventory control and less suitable for automating product movement control, in the absence of universal methods and algorithms. In support of this, at a number of industrial enterprises, developers of RFID systems attempted to organize traceability of products by automatic movement control based on radio frequency identification. As a result, it became clear that automatic control of the product movement is possible only in certain areas of the production process. Such areas are conveyor lines and transport tunnels, where the transportation of products is carried out along the permanently installed radio frequency identification equipment (RFID tunnels). In other production and warehouse areas, automatic control over the movement of products is impossible. This is due to the lack of universal methods and algorithms for product identification in the process of its transportation along unmapped routes.

Positioning of objects and people using information technology is quite a substantial task. These technologies can be used to solve social, industrial and other types of tasks. Currently, there are a large number of approaches to positioning using a large number of technologies, among them:

– Satellite navigation technologies (GPS, GLONASS);
– Local positioning technology (infrared and ultrasonic);
– Technology of technical vision;
– Radio-frequency technologies.

The use of satellite navigation technology and positioning are tightly integrated into our daily lives. They are used for navigation and transport tracking, monitoring and coordination of various kinds of events. The accuracy of positioning is 10–15 meters outdoors. Unfortunately, the application of this technology inside production facilities is almost impossible. An exception is the installation of expensive equipment for organizing GPS-positioning indoors, the unit of which can cover no more than 10 square meters, which is unacceptable for most industrial plants, whose sizes can be tens of kilometers.

Local positioning technologies are highly accurate – about 2 centimeters, but with a short range of 5–10 meters. With these attributes, they are used to achieve local accurate results and, in general, are used for flaw detection (analysis of welds, detection of chips, dents, etc.). The use of local positioning technology for small-scale mechanization is not economically effective and will lead to huge financial costs.

The use of vision technology for solving positioning problems is a relatively young concept. Currently, there are a huge number of methods and algorithms for solving localization and positioning problems, but their effectiveness depends a lot on meeting a large number of requirements, which include the quality of materials used for production of visual labels, cleanliness and lighting of premises, staff attentiveness, etc. Failure to comply with even one of the requirements can lead to a significant reduction in positioning accuracy or make it completely inoperative.

Radio-frequency technologies have found wide application in sales (organization of security in stores). Positioning based on radio frequency technologies can be divided into two categories: positioning on passive RFID tags (distance up to 5 meters) and active RFID beacons (distance up to 80 meters), but all of them are based on the principle that the moved object is marked with an RFID tag and the reading equipment is stationary. This approach allows to effectively automate production processes, where the product movement routes are strictly limited, for example, conveyor lines. However, for the positioning of chaotically moving small mechanization means, this approach will lead to a significant increase in the cost of the positioning system. Instead of a few readers they would need ten times as many.

Considering all the information stated above it is possible to draw a conclusion, that development of technology and software for the construction of positioning and control systems for small mechanization in industrial plants based on radio frequency identification methods is a substantial scientific and technical task.

The development of software and hardware for movement control systems is carried out by: PCT-Invent (Russia, Saint-Petersburg), AiTiProject (Russia, Moscow), Impinj (USA, Seattle), Motorola (USA, Morrisville), Nordic ID (Finland, Salo), FEIG (Germany, Weilburg).

Development of positioning systems based on radio frequency identification is carried out by the following scientific organizations:

– Human positioning systems, in particular patients in medical institutions: Shonan Institute of Technology (Japan, Fujisawa), Institute of Medicine (Kathmandu, Nepal), National Patient Safety Foundation (USA, Boston) and others.

– Systems for positioning moving non-metallic objects: East China Jiaotong University (China, Nanchang), Universiti Sains Malaysia (Malaysia, Nibong Tebal), University of Adelaide (Australia, Adelaide), Wellness Convergence Research Center (Korea, Daegu) and many others.

However, the tasks of developing and implementing automatic systems for tracking products in production are still unresolved. Currently, industrial enterprises still have a number of problems, the solution of which is not realized with the help of modern product movement control systems.

Other paragraphs are indented (BodytextIndented style).

3. The development of small-scale mechanization means positioning algorithm using radio frequency identification technology

In order to determine the current position of the transporting device for continuous automatic control of the transported products, an algorithm for determining the current storage area was developed.

Let M be the set of tags on the racks:

$$M=\{m_i\}$$

The data coming from the reader is accumulated in the sequence F, consisting of the data sets received from the reader.

$$F=(f),$$

$$f=\{s,m,t\},$$

m is the identifier of the tag, s is the level of the signal from the tag at time t.

To calculate the average signal level from the tag for a time τ, use the following formula:

$$\overline{s_i} = \frac{1}{n}\sum_k \begin{cases} s_k, if\ m_k = m_i, t-\tau < t_k < t \\ 0, else \end{cases},$$

Where n is the number of readings over a period of time τ, defined by the formula:

$$n = \frac{\tau}{t_c},$$

Where tc is the time interval that regulates when the reading takes place.

To determine the current rack, you need to select the label identifier of the rack, as mi, where i is the index of the maximum value $\overline{s}_i$.

4. Experimental research

During the pilot studies, a large number of different typical close to production situations were modeled (Figure 1). Among them:

– movement between two storage areas;
– movement between three or more storage areas;
– movement between storage areas with the presence of "noise" (other radio frequency tags that are not tags of storage areas)
– movement between storage areas with partial overlapping of non-metallic and metal barriers.

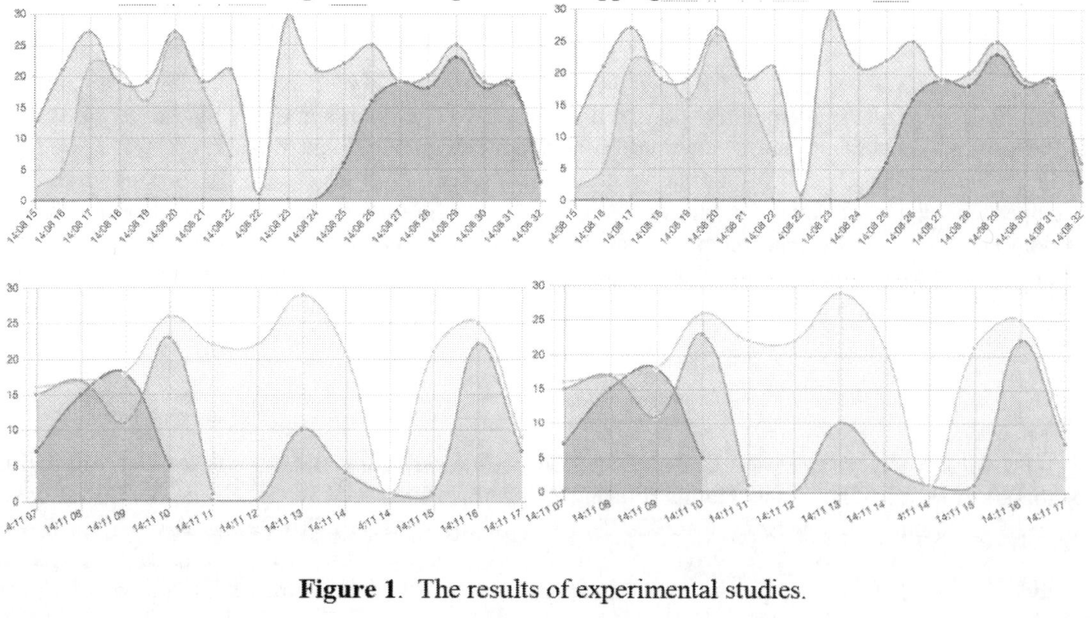

Figure 1. The results of experimental studies.

IOP Conf. Series: Journal of Physics: Conf. Series **944** (2017) 012006 doi:10.1088/1742-6596/944/1/012006

Experimental research was carried out at the industrial enterprise of JSC Vyksa Steel Works. During the research, the labeled metal products were moved between the shelves by means of small-scale mechanization, in particular a bridge crane with a load-carrying beam. The technological map of the product movement is shown in Figure 2.

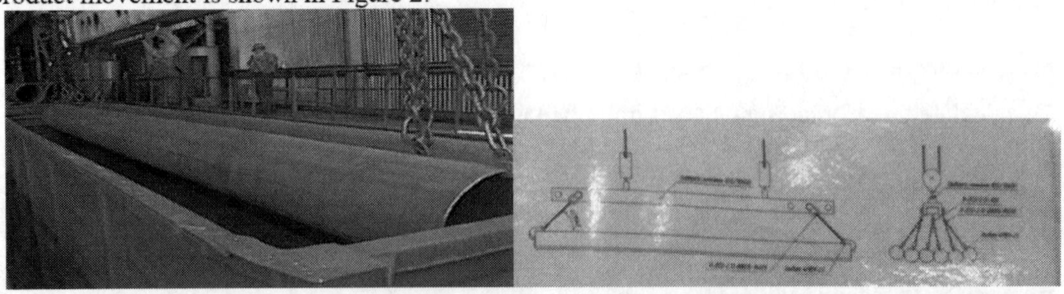

Figure 2. A technological map for the transport of products by overhead cranes with a load-carrying cross-beam.

Figure 3 shows the interpreted data on four experiments on beam movement between racks.

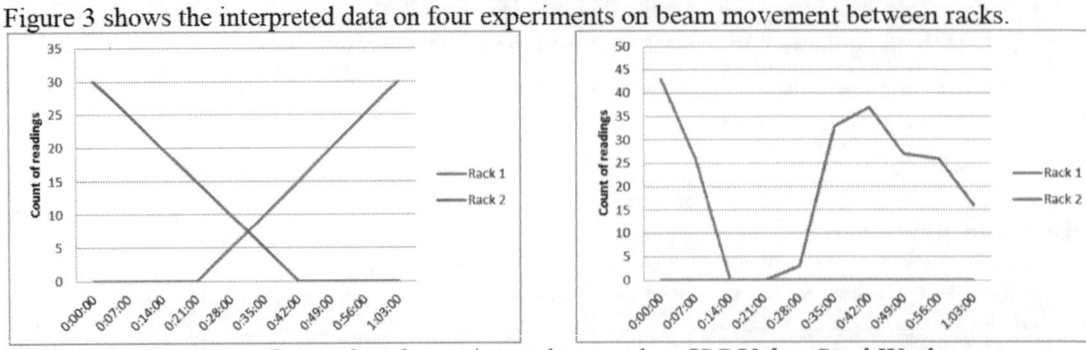

Figure 3. The results of experimental research at JSC Vyksa Steel Works.

Experimental studies have shown the correctness of the algorithm for determining the current storage zone in laboratory and production conditions.

5. Conclusion

The article is devoted to the development of technology and software for the construction of positioning and control systems for small mechanization in industrial plants based on radio frequency identification methods, which will be the basis for creating highly efficient intelligent systems for controlling the product movement in industrial enterprises. The main standards that are applied in the field of product movement control automation and radio frequency identification are considered. The article reviews modern publications and automation systems for the control of product movement developed by domestic and foreign manufacturers. It describes the developed algorithm for positioning of small-scale mechanization means in an industrial enterprise. Experimental studies in laboratory and production conditions have been conducted and described in the article.

The work is executed at financial support of the grant of the President of the Russian Federation № MK-991.2017.9.

References

[1] Provotorov A, Privezentsev D, Astafiev A 2015 Development of Methods for Determining the Locations of Large Industrial Goods During Transportation on the Basis of RFID *Procedia Engineering* Volume 129 pp 1005–1009 (2015) DOI: 10.1016/j.proeng.2015.12.163

[2] Astafiev A V. Orlov A A, Provotorov A V 2015 The localization algorithm of symbolic and bar-code labels on industrial products for the control of product movements Stability and Control Processes in Memory of V.I. Zubov *(SCP) International Conference, St. Petersburg* 5–9 Oct. 2015 pp.

615–616 DOI: 10.1109/SCP.2015.7342230

[3] Astafiev A V, Orlov A A, Privezentsev D G 2016 Method of controlling the movement of large metal products with the use of algorithms for localization and recognition of bar code markings *2016 Dynamics of Systems, Mechanisms and Machines, Dynamics 2016,* Omsk DOI: 10.1109/Dynamics.2016.7818969

[4] Zhiznyakov A L, Privezentsev D G, Zakharov A A 2015 Using fractal features of digital images for the detection of surface defects *Pattern Recognition and Image Analysis*Volume 25 Issue 1 pp 122–131

[5] Bondarevskaya A C, Zolotov V R 2009 Design of radio frequency identification (rfid) - a problem situation *Modern high technologie* № 9 pp 19–23

[6] To Huy 2015 Gnat Research and application of the technology rfid (radiofrequencyidentification) Duy. Neat. *International research journal* № 6(37) Part 1 pp 34–37

[7] Kamozin Denis Y 2013 Comparison of the effectiveness of the use of bar code technology and RFID technology in logistics processes *Proceedings of ISEA* No 3

[8] Ke-Sheng Wang Intelligent and integrated RFID (II-RFID) system for improving traceability in manufacturing *Advances in Manufacturing* June 2014 Volume 2 Issue 2 pp 106–120

[9] Worapot Jakkhupan, Somjit Arch-int, Yuefeng Li 2015 Erratum to: An RFID-based traceability system *Telecommunication Systems* March 2015 Volume 58 Issue 3 pp 273–273.

[10] Mahir Oner, Alp Ustundag, Aysenur Budak 2016 An RFID-based tracking system for denim production processes *The International Journal of Advanced Manufacturing Technology* pp 1–14

Modeling of Autovariator Operation as Power Components Adjuster in Adaptive Machine Drives

P D Balakin, V N Belkov and L O Shtripling

Omsk State Technical University, 11, Mira ave., Omsk, 644050, Russia

e-mail: tmm@omgtu.ru

Abstract. Full application of the available power and stationary mode preservation for the power station (engine) operation of the transport machine under the conditions of variable external loading, are topical issues. The issues solution is possible by means of mechanical drives with the autovaried rate transfer function and nonholonomic constraint of the main driving mediums. Additional to the main motion, controlled motion of the driving mediums is formed by a variable part of the transformed power flow and is implemented by the integrated control loop, functioning only on the basis of the laws of motion. The mathematical model of the mechanical autovariator operation is developed using Gibbs function, acceleration energy; the study results are presented; on their basis, the design calculations of the autovariator driving mediums and constraints, including its automatic control loop, are possible.

1. Introduction

As it is known [1], one of the significant and achievable objectives for adapting actual mechanical machines systems is the automatic control of the power components of the energy flow from the engine to the machine operating element. This is achieved by means of a targeted variation in the transfer function of the mechanical system rate, which provides the necessary power operation mode or engine stationary operation at an economy mode under conditions of variable external force loading of the operating element.

In machines with an operator (manual) control, the variation of the transfer function is made by an operator, performing a step switching motion of the transmitted power flow through various chains of the kinematic circuit. It is significantly less often when the mechanical system includes a motion variator converter which allows the operator to vary the kinematic dimensions of the driving mediums smoothly, without breaking the force flow, and, as a consequence, the transfer function of the mechanical part of the machine drive.

Recently, especially in the transport engineering, automatic transmissions are serially produced and widely used, their availability improving the machines performance, simplifying the control under multimode operation typical for these machines.

Automatic transmissions, as a rule, include a torque converter, a multi-stage gearbox of an planetary circuit, multi-disk shift units, power and command element of hydraulics and electronics. In general, this is a complex and expensive technology, requiring considerable power produced by the power plant for its operation.

The variator circuits of mechanical drives are structurally simpler, despite the fact that in order to control the drive transfer function and normal forces level in frictional contacts, hydraulics is used allowing to produce considerable forces in control loops by means of the variator transfer function, i.e. a special power source integrated into the control loop is used in the designs [2–6].

Successful implementations of V-belt variators schemes with sliding pulleys and hydraulic control in the running transmissions of domestic combines, in drives of threshing drums, in reel drives are known. Despite the fact that the fundamental issues of creating V-belt variators have been solved, the problem of providing the operational life for the V-belt remains an issue, since the conditions of its operation are extremely unfavorable, so the synthetic or split-type belt is the resource-determining element of the variator design and the variators with such a belt are realized mainly in low-power machinery only: motor scooters, snowmobiles, mini-cars for inter-quarter trips, etc.

Recently, Honda, Nissan and Audi automakers have developed an equal-resource split-type V-belt, with the main aggregates, containing trapezoidal steel elements connected into a loop by means of a multilayer steel strip. The belt allows the level of normal forces up to 104 N and is able to transmit torque up to 300 Nm, which makes the variator drive promising for a wide range of transport machines.

2. Research objective

The development of mechanical variator and autovariator drives is prevented by the absence of models for the dynamic operation of the reduced masses on either sides of the two-degree-of-freedom nonholonomic constraint of the main driving mediums.

The problem of variator operation modeling is set and solved, the operation being based solely on the use of the laws of motion. The controlling means for transfer function is the targeted driving mediums motion, additional to the main one, by means of an integrated control loop that uses the variable power component of the main transformable power flow, i.e., the proposed design does not have a special source of energy. Such a variator is able to make a significant evolution to achieve the energy perfection of the drive and auto-control by means of the transformed power components.

Similar structures of motion transducers are called autovariators, numerous technical solutions of autovariators, for example [7–8], are obtained on the basis of the mechanical systems designing principle based on the property of the designed system to adapt to the systems real parameters, to operating modes.

The modeling of the autovariator operation is of current concern. It allows to get information on the driving mediums and constraints load at an early stage of the preliminary design, which is necessary for the design calculations of all the autovariator drive components, and the integrated control loop in particular.

3. Theory

The main and current task for transport machines is the full utilization of the available power by maintaining the stationary mode of operation of the power plant (engine) under conditions of variable external loading.

M_1 and M_2 are the power moments and shafts rates on the engine and the operating elements of the machine (the driving wheel of the transport machine), respectively; η being mechanical efficiency of the drive, the variation ratio equals:

$$M_1\omega_1\eta = M_2\omega_2, \qquad (1)$$

IOP Conf. Series: Journal of Physics: Conf. Series **944** (2017) 012007 doi:10.1088/1742-6596/944/1/012007

from

$$\omega_2 = \frac{M_1 \omega_1 \eta}{M_2}. \tag{2}$$

Assuming that M_1=*const* and ω_1 =*const* and, in the first approximation, η = const, a hyperbolic constraint ω_2 is obtained with the variable M_2, but the transfer function Π^{ω} of the autovariator

$$\Pi^{\omega} = U_{1,2} = \frac{\omega_1}{\omega_2}$$

is linear with respect to M_2; that is, in maintaining the task conditions, a multiple variation M_2 requires a multiple variation Π , and this circumstance is decisive in the synthesis of the parameters of the integrated control loop of the autovariator.

We compose the autovariator dynamic model, taking the differential equation of the coordinates φ1 and φ2 derivatives ratio of the main driving mediums in its basis. To be definite, r and ρ are assumed the kinematic dimensions for the main driving mediums, and the result is

$$r\dot{\varphi}_1 = \rho\dot{\varphi}_2 . \tag{3}$$

Despite the formal division of variables, equation (3) is not integrated due to the time uncertainty of the variable *ρ*.

A proper consideration of the mutual effect on the motion of the kinematic chains located on different sides of the nonholonomic constraint is given by a model constructed on the Appell equations using the Gibbs function, the accelerations energy:

$$s = \sum_{i=1}^{n} \frac{Jr_i \ddot{\varphi}_i^2}{2} , \tag{4}$$

Where Jr_1 is the inertial characteristic of the kinematic chain reduced to φ_1, and the specific partial derivative of the Gibbs function

$$\frac{ds}{d\ddot{\varphi}_1} = Jr\ddot{\varphi}_1 = M_1 \tag{5}$$

Where M_1 is the force parameter, reduced to the generalized coordinate, from external forces. The acceleration energy for a two-shaft variator, referring to the two reduction links associated with shafts 1 and 2, equals:

$$s = \sum_{i=1}^{n} \frac{J_1 \ddot{\varphi}_1^2}{2} + \sum_{i=1}^{n} \frac{J_2 \ddot{\varphi}_2^2}{2} \tag{6}$$

Where J_1 and J_2 are reduced to shafts 1 and 2, the inertial characteristics of the kinematic chains, divided by a nonholonomic constraint. After the transformations of [1], the differential equation of motion of the shaft 1 is obtained:

$$(J_1 + J_2 U_{2,1}^2)\ddot{\varphi}_1 + J_2 U_{2,1} \dot{U}_{2,1} \dot{\varphi}_1 = M_1 + M_2 U_{2,1} \tag{7}$$

for shaft 2

$$(J_2 + J_1 U_{1,2}^2)\ddot{\varphi}_2 + J_1 U_{1,2} \dot{U}_{1,2} \dot{\varphi}_2 = M_1 U_{1,2} + M_2 \tag{8}$$

Referring to the features of the technical specification for the autovariator design for transport machine drive, namely, when M_1=const and $\dot{\varphi}_1 = \omega_1 = const$ on coordinate φ_1 is preserved, equation (7) is simplified to the following form:

$$J_2 U_{2,1} \dot{U}_{2,1} \dot{\varphi}_1 = M_1 + M_2 U_{2,1} \tag{9}$$

or

$$\dot{U}_{2,1} = \frac{M_1 / U_{2,1} + M_2}{J_2 \dot{\varphi}_1} \tag{10}$$

As the transfer function $U_{2,1}$ is linear with respect to M_2, it can be represented as $U_{2,1}=kM_2$ and M_2 remains the only variable in (10), and if the variation law of M_2 is known in time, then (10) is integrated, thus the problem with of the output autovariator chain motion can be solved in quadratures.

4. Discussion

The ratio of the variation $U_{2,1}=U_{2,1}$ (M_2) is the initial one for designing of the integrated autovariator control loop. «+» in the numerator (10) means the algebraic addition of multidirected reduced moments M_1 and M_2 and the numerator's vanishing signifies the end of the transient process, the motion becomes steady with the transfer function $U_{2,1}$=*const*. The study of the motion model on (7) and (8) shows that the operation of the main driving mediums is affected not only by the transfer function, but also by the rate of its variation. Moreover, this effect has a damping character for the transmission of transport machines, i.e. the more sharply the transfer function changes, the less controllable the rate of the chain system is.

As in (10), the numerator is the difference in the power characteristic before and after the nonholonomic constraint, i.e., actually, this difference is an excessive moment, at M_1=*const*, being completely determined by the variables $M_2(t)$ or

$$M_1 / U_{2,1} + M_2(t) = Mexc \tag{11}$$

The excess moment can be represented by the time integrable function, and then $\dot{U}_{2,1}$ in (10) has a finite analytic definition.

Several variants *Mexc=Mexc(t)* are considered.

$$\frac{dU_{2,1}}{dt} = \frac{Mecx(t)}{J_2 \dot{\varphi}_1}$$

or

$$dU_{2,1} = \frac{1}{J_2 \dot{\varphi}_1} Mexc(t) dt \tag{12}$$

The result is

$$U_{2,1} = \frac{1}{J_2 \dot{\varphi}_1} \int_{t_0}^{t_1} Mexc(t) dt = \frac{Mexc(t)}{J_2 \dot{\varphi}_1} (t_1 - t_0) \,. \tag{13}$$

At $t_0=0$, $U_{2,1} = \frac{Mexc(t)}{J_2 \dot{\varphi}_1}$, the latter means a linear variation $U_{2,1}$ from time at constant *Mexc(t)*, and rate variation $U_{2,1}$ significantly depends on inertia J_2 of the driven chain.

The variant of linear increase or decrease of *Mexc(t)*, determined by the value *k*, is quite general.

This variant has an independent value and also it can be used as a linearized representation of *Mexc(t)* on the studied interval for any function *Mexc(t)*. If *Mexc=kt*, where *k=const*, then: $dU_{2,2} = \frac{1}{J_2\dot{\varphi}_1} ktdt$ and $U_{2,1} = \frac{k}{J_2\dot{\varphi}_1}\frac{t^2}{2}$, i.e. the variation $U_{2,1}$ at a linear variation *Mexc(t)* depends on time in the second power and depends additionally on the variation rate *Mexc(t)*, determined by the value *k*.

The most general case is the analytic representation of *Mexc(t)* by a polynom of any power or by a trigonometric function; in any case, under the given conditions, it is possible to obtain the analytic expression $U_{2,1}$ in the function *Mexc(t)*.

The results of mathematical modeling in the Excel program are given below. The input motion is carried out with a constant angular rate $\omega_1 = 10\frac{1}{c}$ and the nominal value of the inertia moment $J_1=0{,}5kgm^2$ applied to this shaft. The ratio of the autovariator's correlation to the inertia of the driven chain J_2 is studied for two values $J_2=1kgm^2$ and $J_2=4kgm^2$.

Several ratio variants of the variation in the excess force moment *Mexc(t)* are examined.

Mexc(t)=const=$5H_M$ is defined. The variation $U_{2,1}(t)$ is presented in **Figure 1.**

When *Mexc(t)* varies according to a linear law, namely *Mexc(t)=kt* with the accepted values J_2, the variation in $U_{2,1}(t)$ occurs, as shown in **Figure 2.**

The effect of the slope *k* of the linear characteristic Mexc (t) = kt at $k = 1$ and $k=4$ at constant $J_2=2kgm^2$ is shown in **Figure 3.**

5. Summary and conclusion

The results analysis shows that an increase in inertia of the driven chain results in a decrease in the variation rate of the transferred ratio $U_{2,1}$, and an increase in the steepness *k* of the linear characteristic *Mexc(t)=kt* increases the variation rate $U_{2,1}$, which is physically reasonable. The obtained ratio can be used as the initial basis for constructing driving mediums and constraints of a mechanical autovariator, especially this refers to the synthesis of the integrated control loop with the transferred ratio of this adaptive regulator by power components in the drive of transport machines.

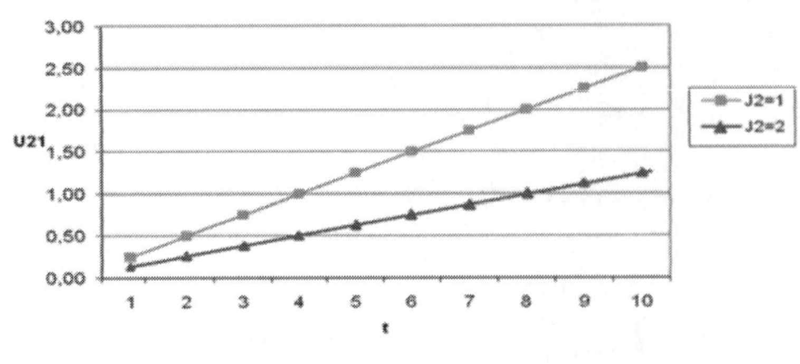

Figure 1. Variation $U_{2,1}(t)$ at *Mexc(t)=const* depending on J_2.

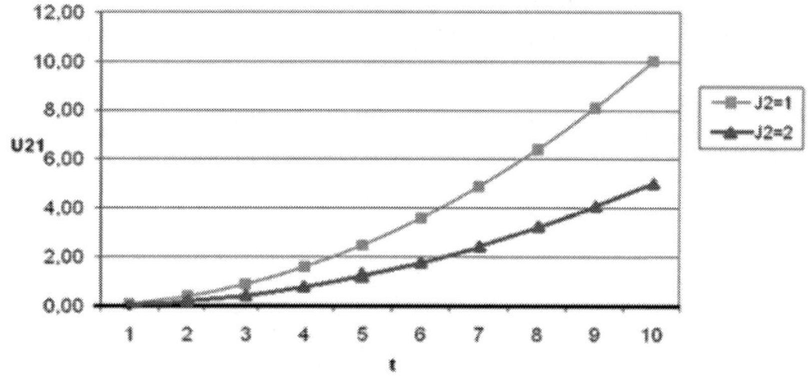

Figure 2. Variation $U_{2,1}(t)$ at $Mexc(t)= kt$ depending on J_2.

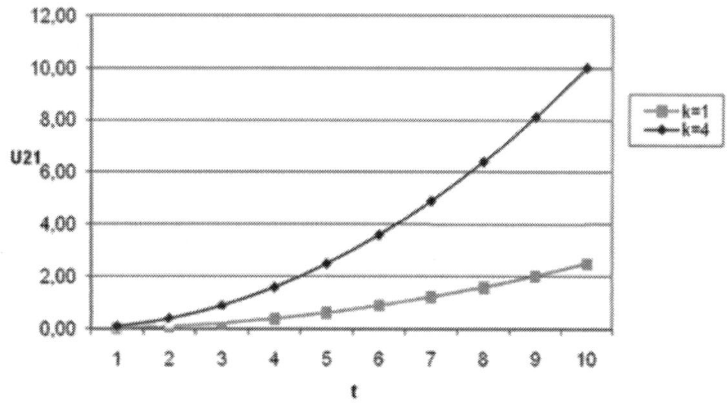

Figure 3. Variation $U_{2,1}(t)$ at $Mexc(t)= kt$ depending on k.

References

[1] Balakin P D 2016 Elements of real mechanical systems theory (Omsk: Published by Omsk State Technical University) p 272

[2] Yildiz A Bottiglione F and Carbone G 2015 An experimental study on the shifting dynamics of the chain CVT *Proc of The Third Intl. Conf. on Advances in Mechanical and Robotics Engineering AMRE-2015*

[3] Brown A W, van Rooij J and Frank A A 2004 The Design of an Inline GCI Chain CVT For Large Vehicles, *International Continuously Variable and Hybrid Transmission Congress* (UC Davis) 54

[4] Srivastava N and Haque I 2009 A review on belt and chain continuously variable transmissions (CVT): dynamics and control *Mechanism and Machine Theory* 44(1) pp 19–41

[5] Bottiglione F, De Pinto S, Mantriota G and Sorniotti A 2014 Energy Comsumption of a Battery Electric Vehicle with Infinitely Variable Transmission *Energies* 7 (12) 8317–8337

[6] Mangialardi L and Mantriota G 1992 The Advantages of USING Continuously Variable Transmission in Wind Power Systems *Renewable Energy* 2(3) pp 201–209

[7] Pat RU 2122670 MKU 6 F 16 H 9/18 Avtomaticheskij klinoremennyj variator / Balakin P D and Bienko V V Bjul № 33

[8] Pat RU 2127841 MKU 6 F 16 H 9/00 Shkiv / Balakin P D and Bienko V V Bjul № 8

IOP Conf. Series: Journal of Physics: Conf. Series **944** (2017) 012008 doi:10.1088/1742-6596/944/1/012008

Supersonic plasma outflow in a plasmochemical method of amorphous silicon thin films formation

L V Baranova, V I Strunin, G Zh Khudaibergenov

Dostoevsky Omsk State University, pr. Mira 55a, Omsk, 644077, Russia

Abstract. As a result of the numerical modeling of gasdynamic functions of a nozzle of Laval there obtained its parameters which form supersonic plasma jet outflow in a process of amorphous silicon thin films deposition. According to the nozzle design parameters, there obtained amorphous silicon thin films and studied uniformity of the thickness of the synthesized coatings. It was also performed that due to a low translational temperature at the nozzle exit the relaxation losses reduce significantly, "freezing" the vibrational degrees of freedom and the degrees of freedom of the transverse motion of the particles, and increasing the energy efficiency of the film formation process. All this is caused by the fact that on the surface of a growing film only the products of primary interaction of electrons with molecules of a silicon-containing gas in the plasmatron do interact.

1. Introduction

Among other methods of hydrogenous amorphous silicon thin films (a-Si:H) synthesis the most perspective ones are supposed to be jet methods nowadays. They differ according to the way of gaseous reagents activation [1, 2]. Traditional methods of a-Si:H and mc-Si thin films deposition based on the use of plasma flows of various electrical discharges limit the development of electronics of large areas and have a relatively low growth rate of the films. Any attempts to increase growth rates lead to a decrease in the quality of the produced films. It is suggested to overcome the disadvantages of traditional methods using the method of producing amorphous silicon films based on silane decomposition in plasma of a radio-frequency capacitive discharge (RFC-discharge) outside the disposition chamber with the following formation of supersonic jets from the decomposition products that outflow into the vacuum deposition chamber through a system of supersonic nozzles.

The aim of the study is to increase the effectiveness of technological deposition processes of amorphous silicon thin-film coatings basing on the jet plasmochemical method with a preliminary activation of a gas in the area remoted from the growth chamber [3].

2. Problem Definition

To create supersonic outflow in order to deliver the decomposition products of the silane-containing gas mixture from the plasmatron to the substrate and to freeze recombination processes during the delivery there were made calculations of parameters of a nozzle of Laval (figure1) and the ones of gasdynamic functions of the pressure, temperature and speed of gas in the jet (figures 2–5).

Calculations of pressure and temperature distributions along the nozzle length are necessary for determination of the initial conditions at which the outflow of the silane decomposition products in the plasmatron to the film deposition substrate is carried out by a supersonic jet, which provides the «freezing» of the vibrational and transverse degrees of freedom according to the direction of motion. The products of primary interaction of electrons with molecules of a silicon-containing gas in the jet do not interact with each other.

A nozzle of Laval, which forms supersonic outflow of silane dissociation products to deliver a «frozen» composition of the dissociation products in the plasmatron to the film deposition substrate, is profiled by means of certain gasdynamic calculations [4].

To analyze the gas flow in a nozzle of Laval the following assumptions were made:

1. The gas is considered to be ideal.
2. The gas flow is isentropic and adiabatic.
3. The gas flow is stationary and one-dimensional, all flow parameters at any point of the nozzle vary only along the axis of the nozzle.
4. Mass flow rate of the gas in all cross sections of the flow is the same.
5. The influence of all external fields (including gravitational) and forces is neglected.
6. The nozzle symmetry axis is a spatial coordinate.

The following initial parameters were used for calculations:

– P_1=26.66 Pa – pressure in the plasmatron chamber;
– a_1 =40^0 – convergence angle;
– a_2 =15^0 – angle of expansion;
– T_1=500 K – stagnation temperature;
– R=207.85 J/(kg·K) – universal gas constant;
– G=3·10^{-7} kg/s – mass flow of the gas;
– P_2=10^{-2} Pa – pressure in the environment;
– k = 1.67 – adiabatic index.

3. Results and discussion

Stagnation parameters are such gas parameters which correspond to the zero-flow velocity. The relationship between the stagnation parameters and the flow ones can be determined by means of gasdynamic functions: P – pressure, T – temperature, V – velocity.

One of the gasdynamic function arguments will be the Mach's number which is also supposed to be local and depends on the coordinate: x: $M=V/a$, where V is a local flow velocity, a is a speed of a sound in the gas. Another argument is a velocity coefficient which is also known as reduced velocity λ (the ratio of the speed of gas in the given section of the nozzle to the speed of a sound in the critical section):

$$\lambda=V/a_{cr},$$

where a_{cr} is the speed of a sound in the critical section.

The mass flow rate G is related to the critical section area S_{cr}, the pressure P_1 and the gas temperature T_1 as follows (1) [4]:

$$G=\frac{P_1}{\sqrt{T_1}}S_{cr}\left(\frac{2}{k+1}\right)^{\frac{k+1}{2(k-1)}}\sqrt{\frac{k}{R}}\,. \tag{1}$$

The critical section S_{cr} and the length of the subsonic part of the nozzle l_u are expressed from the flow rate equation.

Parameters in any section of the nozzle can be found if the parameters in the plasmatron chamber are known. They can be calculated from the following relationships:

pressure:

$$\pi=\frac{P}{P_*}=\left(1-\frac{k-1}{k+1}\lambda^2\right)^{\frac{k}{k-1}}. \tag{2}$$

temperature:

$$\frac{T_*}{T}=1+\frac{k-1}{2}M^2. \tag{3}$$

IOP Conf. Series: Journal of Physics: Conf. Series **944** (2017) 012008 doi:10.1088/1742-6596/944/1/012008

One more gasdynamic function used in the calculations is specific time flow $q(\lambda)$ which is expressed by means of the ratio of the areas occupied by a non-viscous core in the critical and analyzed sections,

$$q(\lambda) = \left[\frac{d_{cr}-2\delta^*_{cr}}{d-2\delta^*}\right]^2, \quad (4)$$

where d_{cr}, and d are the diameters of the critical and analyzed sections (in mm), δ^*_{cr} and δ^* are the boundary layer displacement thicknesses for the critical and analyzed sections (in mm) respectively.

As a first approximation the values of the boundary layer displacement thickness δ^*_{cr} and δ in the calculations of gasdynamic functions are assumed to be zero. In the following approximations their values are obtained due to boundary layer calculation.

The gasdynamic function takes its maximum value $q(\lambda)=1$ in the critical section. In this section the velocity coefficient has the value of $\lambda=1$. The relation between the velocity coefficient λ and the function $q(\lambda)$ is expressed as follows:

$$q(\lambda) = \lambda\left[\left(1-\frac{k-1}{k+1}\lambda^2\right)\left(\frac{k+1}{2}\right)\right]^{\frac{1}{k-1}}, \quad (5)$$

For the diverging section the expressions for the supersonic section S_d and the length of the sound part of the nozzle l_d are valid:

$$d_d = 2\sqrt{\frac{S_d}{\pi}}, \quad (6)$$

$$l_d = \frac{(d_u-d_d)}{\left(2\,tn\frac{\alpha_2}{2}\right)}, \quad (7)$$

Other parameters can be calculated according to the section length. The flow in the converging section of a nozzle of Laval moves with a subsonic velocity where the velocity coefficient has the value of $\lambda<1$. The convergence of the result is provided by the scheme:

$$\lambda = \frac{q(\lambda)}{\left(\frac{k+1}{2}\right)^{\frac{1}{k-1}}\left(1-\frac{k-1}{k+1}\lambda^2\right)^{\frac{1}{k-1}}} \quad (8)$$

the Mach's number:

$$M = \sqrt{\frac{\frac{2}{k+1}\lambda^2}{1-\frac{k-1}{k+1}\lambda^2}} \quad (9)$$

The critical velocity is:

$$a_{cr} = \sqrt{\frac{k}{k+1}RT^*} \quad (10)$$

T^* is the temperature of the stagnated flow.

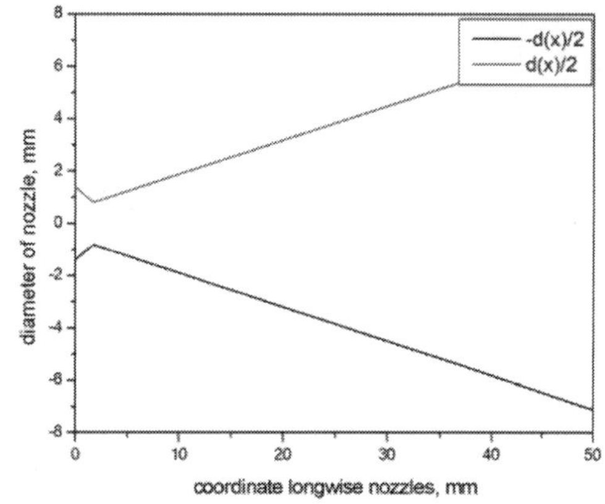

Figure 1. Profile of a nozzle of Laval.

As a result of the numerical modeling of the supersonic nozzle the following values of its parameter are obtained (figure 1):

The length of the subsonic part of the nozzle constituted $l_u=1.6$ mm, the general length was $l=50$ *mm*, the areas of the subsonic part, the critical and the diverging sections of the nozzle had the values of $S_u=4.1$, $S_{cr}=1.5$, $S_d=140$ mm^2 respectively.

Figures 2-5 presents the charts of the dependence of the speed of gas and the speed of a sound, the Mach's number, the pressure and the temperature on nozzle length (where the initial point is 0):

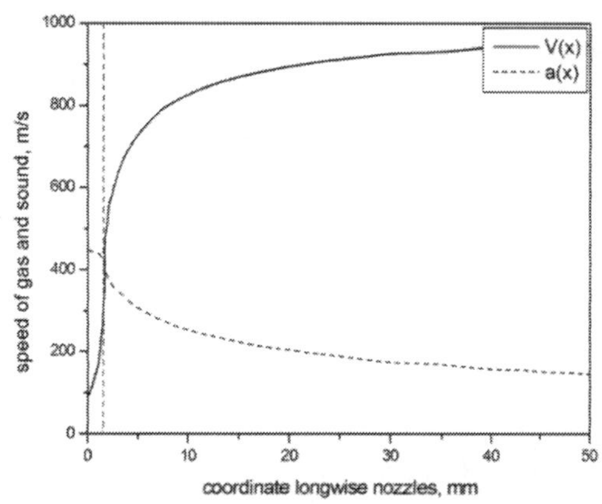

Figure 2. Dependence of speed of gas and speed of a sound on nozzle length.

The speed of gas increases in the area of the critical section (vertical dashed line) (see figure 2). It reaches the value of 900 m/s along the length of the nozzle. The Mach's number (figure 3) reaches the value M=6 which is not something unusual for such kinds of pressures [5].

Figure 4 shows that the condition of the supersonic outflow is being fulfilled:

$$P_{cr} = P_* \left(\frac{2}{k+1}\right)^{\frac{k}{k-1}} > P_e, \tag{11}$$

where P_* is the pressure in the chamber (stagnation pressure); P_{cr} is the pressure in the critical section of the nozzle; P_e is the pressure in the environment; k is the heat capacity ratio.

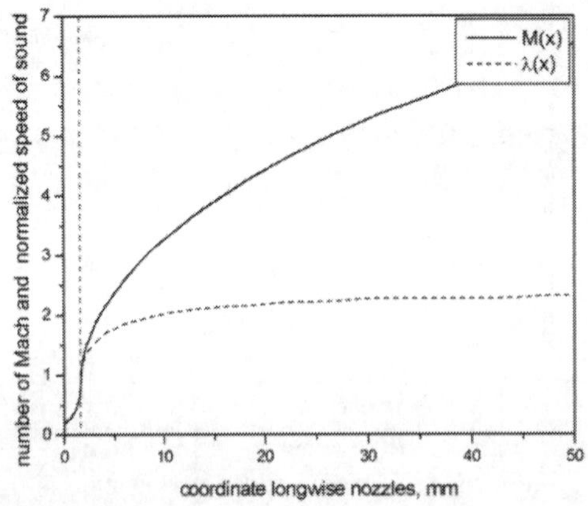

Figure 3. Dependence of Mach`s number and speed of a sound on nozzle length.

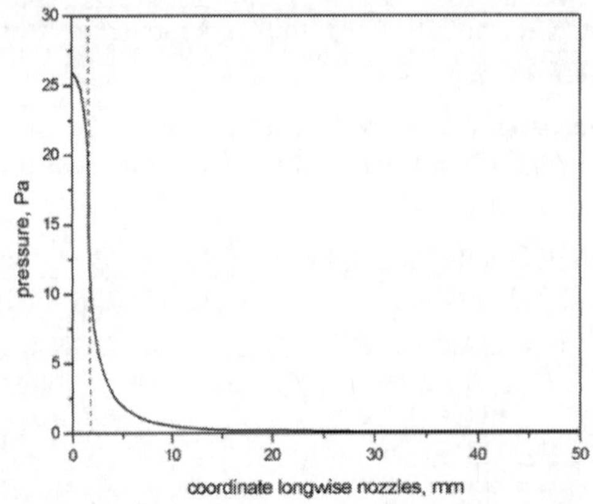

Figure 4. Gasdynamic function of pressure on nozzle length.

The analysis of gasdynamic temperature function on nozzle length (figure 5) shows that due to a low value of translational temperature at the nozzle exit the relaxation losses reduce significantly, «freezing» the vibrational degrees of freedom and the degrees of freedom of the transverse motion of the particles, and increasing the energy efficiency of the process.

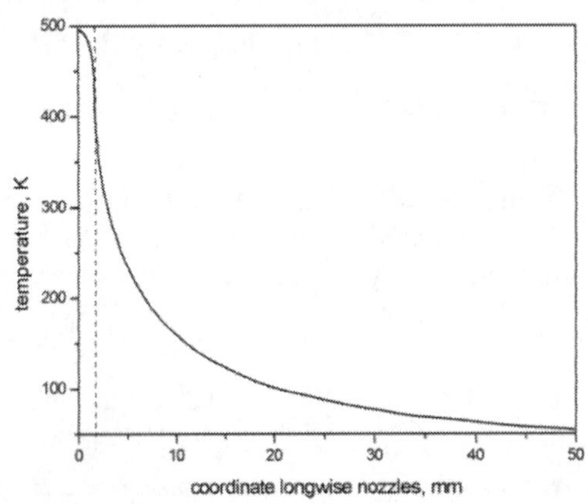

Figure 5. Chart of gasdynamic function of temperature on nozzle length.

To measure the dynamic pressure in the free stream of the flowing plasma of the argon a Prandtl tube and thermistor pressure sensors were used [6]. The results of the measurements are presented in figure 6. The measurements were made without a sample (substrate). The chart shows that the pressure decreases while it spreads from the nozzle which is demonstrated by the modeling calculations as well.

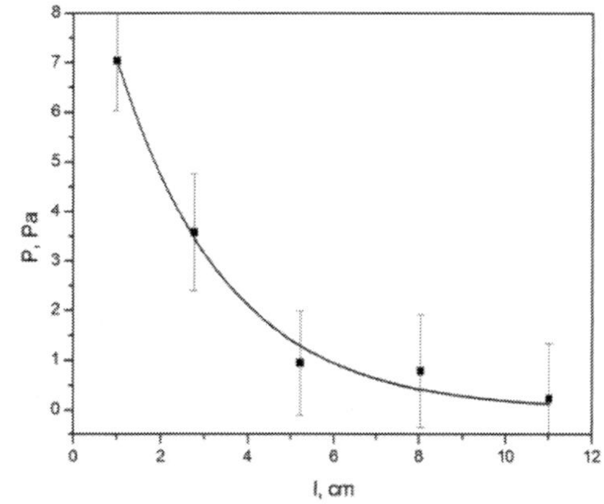

Figure 6. Dependence of the dynamic pressure in the free stream of the flowing plasma of the argon from the distance from the nozzle.

One should remember that thermistor pressure sensors have a relatively low accuracy and their measurement error constitutes ± (10%–60%). Thus, the obtained results are rather of a qualitative nature.

The deposition of amorphous silicon thin films was carried out in a vacuum unit NNV6.1. The glow-discharge plasma was excited in a plasmatron with a diameter of 70 mm, and a length of 30 mm. Parameters of nozzle: $S_u=4.1$, $S_{cr}=1.5$, $S_d=140\ mm^2$. The pressure in the chamber varied within 26.66 Pa and was controlled by the capacitive vacuum sensor SETRA 760. Radio frequency glow discharge was excited by the generator with the oscillating power 600 W (field intensity constituted E = 50 V/cm, frequency f =13.56 MHz). Two sliding vane rotary pumps and a diffusion pump evacuated the vacuum chamber till the ultimate vacuum pressure of $6\cdot10^{-4}$ Pa. Mass flow controller allowed to control the flow of the working gas within the range of 4-60 sccm. After the pressure in the plasmatron chamber was increased to the performance range, the glow discharge was excited.

Figure 7 demonstrates the results of thickness measurements of the amorphous silicon films from distance from the center. The deposition was carried out for different values of the distance from the nozzle to the substrate (30 and 50 mm), the deposition time was 20 min. It is clearly seen that for the case of 30 mm the films have a highly uneven thickness due to the uneven plasma flow density. There is also a large pressure drop in the center and along the edges of the jet. If we put the substrate in the distance of 50 mm from the nozzle edge, the drop in pressure will constitute approximately 30 % and on the substrate there will be an area with a radius of 60 mm. However, in this case the deposition velocity decreases sharply which can be explained by a decrease of the local gas density near the substrate. To get uniform thickness coatings with a non-uniform coating thickness of less than 5 %, it is suggested to install more than one nozzle and even a nozzle array arranged so that gas jets intersect 60% of the distance from the nozzle to the substrate [7].

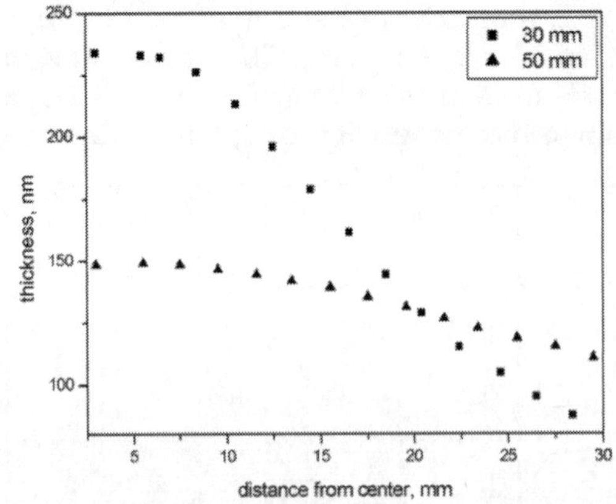

Figure 7. Film thickness from distance from the center.

Conclusions

Calculations of pressure and temperature distributions along the length of the nozzle prove that the delivery of silane dissociation products in the plasmatron to the substrate is accomplished by a supersonic jet. It was also performed that due to a low translational temperature at the nozzle exit the relaxation losses reduce significantly, "freezing" the vibrational degrees of freedom and the degrees of freedom of the transverse motion of the particles, and increasing the energy efficiency of the film formation process. All this is caused by the fact that on the surface of a growing film only the products of primary interaction of electrons with molecules of a silicon-containing gas in the plasmatron do interact. When using the nozzle array, the supersonic outflow is maintained by the increase of the mass flow rate of the gas.

References

[1] Ikoma Y Endo T et al. Growth of Ultrathin Epitaxial 3C-SiC Films on Si(100) by Pulsed Supersonic Free Jets of CH_3SiH_3 1999 *Japan. J. Appl. Phys.* Vol 38 L301-L303

[2] Skovorodko P Sharafutdinov R Shchukin V and Konstantinov V Gas dynamic aspects of silicon thin layers deposition using excitation of a free jet of the working gas mixture by an electron beam 2012 *AIP Conf. Proc.* **1501** p 1437 DOI: 10.1063/1.4769708

[3] Strunin V Baranova L Lyahov A Popov V Hudaybergenov G Jacob S Optical properties of amorphous silicon thin films deposited by plasma chemical jet method *2009 International Conference and Seminar on Micro/Nanotechnologies and Electron Devices* **DOI:** 10.1109/EDM.2009.5173929

[4] Daych M Zyryankin A 1984 *Hydro and gas dynamics* (Moscow: Energoatomizdat) p 384

[5] Abramovich G N 1976 *Applied gas dynamics* (Moscow: Nauka) p 824

[6] Strunin V Khudaibergenov G Shemakhin A Zheltukhin V Modeling of gas-dynamic expiration freely expanding gas jet in space with low pressure background 2014 *Herald of Omsk University* **72** pp 53–59

[7] Pat. 2165476 RU IPC 7 C 23 C 16/24 C 16/50. *Method for depositing amorphous silicon films and a device for its implementation* Baranova L Strunin V Khudaibergenov G

Acknowledgment

The reported study was funded by Russian Foundation for Basic Research according to the research project № **16-42-550936 p_a**»

Fire-tube boiler optimization criteria and efficiency indicators rational values defining

P A Batrakov[1], A G Mikhailov[1] and V Yu Ignatov[2]

[1]Omsk State Technical University, 11, Mira Pr., Omsk 64405, Russia
[2]Saratov State University named after Y.A. Gagarin, 77, Politekhnicheskay, Saratov 410054, Russia

e-mail: peter_1cool@mail.ru

Abstract. Technical and economic calculations problems solving with the aim of identifying the opportunity to recommend the project for industrial implementation are represented in the paper. One of the main determining factors impacting boiler energy efficiency is the exhaust gases temperature, as well as the furnace volume thermal stress. Fire-tube boilers with different types of furnaces are considered in the study. The fullest analysis of the boiler performance thermal and technical indicators for the following engineering problem: Q=idem, M=idem and evaluation according to η, B is presented. The furnace with the finned ellipse profile application results in the fuel consumption decrease due to a more efficient heat exchange surface of the furnace compared to other examined ones.

1. Introduction

Energy efficiency is the efficient use of energy resources. The purposes of energy efficiency are the resources saving, efficient energy consumption, fuel costs reduction, performance improving [1]. The Russian Federation is the world's third largest energy consumer and its economy has a high level of energy intensity [2]. Therefore, energy efficiency and energy saving are included in the five strategic areas of the Russian Federation prioritized technological development. Taking this into account, when using the efficient heat exchange surfaces, in improving the low and medium power fire-tube boiler furnace design it is necessary to obtain the technical and economic calculations results making it possible to recommend the given project for industrial implementation.

2. Problem statement

Optimization problem formulation includes several stages: optimality criterion selecting, optimized parameters selecting, necessary limits applying and objective optimization function recording.
The most common optimization criteria are the economic and technological performance indicators.
The fire-tube boiler mass and size and performance characteristics optimization problem is possible to be recorded as follows:

$$f(T_{ex}, q_v) \rightarrow \max;$$
$$390\ \text{K} \leq T_{ex} \leq 400\ \text{K};$$
$$250\ \text{kW/m}^3 \leq q_v \leq 1100\ \text{kW/m}^3$$

where T_{ex} is the exhaust gases temperature; q_v is the furnace volume thermal stress.
One of the main factors impacting the heat loss with exhaust gases is the temperature T_{ex}. For decreasing T_{ex} the heating surface area is increased. T_{ex} value influences not only the unit efficiency but also the capital costs required for additional heat exchange equipment installation. Tex reduction

leads to the efficiency increase, fuel consumption and costs decrease. However, in this case the heating surface areas are increased (at the low temperature difference the heat exchange surface area is necessary to be increased), consequently, the installation and operating costs are increased. Therefore, for the newly projected boiler units or other heat consuming systems T_{ex} value is defined according to the technical and economic calculation in which Tex influence not only over the efficiency but also over the capital and operating costs value is taken into account. State-of-the-art boiler units using gaseous fuel have Tex from 390 to 400 K.

The value $q_v = Q/VT$ represents the heat amount released at the fuel certain quantity combustion per time unit and accounted for 1 m^3 of the furnace volume. If the value qv does not range the limits defined practically, fuel is not cleanly burnt for the period it is in the furnace. The boiler units operating experience has shown that for the different kinds of fuel, combustion methods and furnaces designs, the accepted value qv varies over a wide range. For the low and medium power fire-tube boiler using the gaseous fuel, qv varies from 250 to 1100 kW/m^3.

Further the gradient projection method as the descent method variation is considered. The algorithm described below is intended to solve the following problems:

$$\text{Min } f(x),$$
$$\varphi_i(x) \le 0.\ I = 1. \ldots, m,$$
$$\varphi_i(x) = 0.\ I = m + 1. \ldots, l,$$

where f, φ_i are the continuously differentiable functions and the given algorithm is the direct generalization of the steepest descent method. The operation principle of both methods is the same – to trend towards the minimized function speediest decrease. Only in the steepest descent method searching for the absolute minimum this direction is antigradient, while in the gradient projection method solving the conditional minimization problems it is defined taking into account the restrictions and results from the orthogonal antigradient projecting for the particular linear variety. The last approximates the acceptable area boundary part in parallel to which the ordinary iteration step will be taken. Generally, as the boundary is nonlinear this step will get out of the feasible set even if the initial point belongs to it. Thus, in the gradient projection method the movement along infeasible points is possible. However, the restrictions violation level is strictly controlled and keeps low by means of the steps lengths correcting and limiting.

3. Theory

For comparing the heat exchange surfaces, the following values as: heat quantity Q transferred through the surface, fuel consumption B, mass M are pointed out as the main characteristics.

To solve the problem where fire-tube boilers with different types of furnaces and varying mass are considered [3], the heating surface cannot absolutely be a criterion to evaluate the compared variants. Accordingly, the following formulation of the technical problems main types is proposed:

1. Q = idem, M = idem – evaluation by η, B.
2. Q = idem, B = idem – evaluation by M.
3. B = idem, M = idem – evaluation by η, Q.
4. Q = idem, F = idem – evaluation by η, B, M.
5. B = idem, F = idem – evaluation by η, Q, M,

Where Q is the heat quantity transferred through the surface, M is the fire-tube boiler mass, η is the fire-tube gross efficiency, B is the fuel consumption, V is all the boiler design volume, F is the cross-sectional area.

The fullest analysis of the fire-tube boiler performance thermal and technical indicators is possible in solving the following engineering problem:

By maintaining the fixed heat exchange surface area of the furnace section and boiler as a whole (Q = idem, M = idem – evaluation by η, B).

The transfer from the circle profile to the ellipse and finned ellipse ones at the exhaust gases recommended temperature constant interval corresponds to the fuel consumption reduction approximately by 1.1 m^3/hr due to the more efficient heat exchange surface of the finned ellipse profile furnace compared to other considered variants (Fig. 1) where points indicate the boilers operation nominal characteristics.

The transfer from the circle profile to the ellipse and finned ellipse ones results in the furnace volume reduction and at the furnace recommended thermal stresses constant interval corresponds to the fuel consumption decrease approximately by 15–17 Nov. 20161.1 m^3/h (Fig. 2) in the exhaust gases specified temperature range.

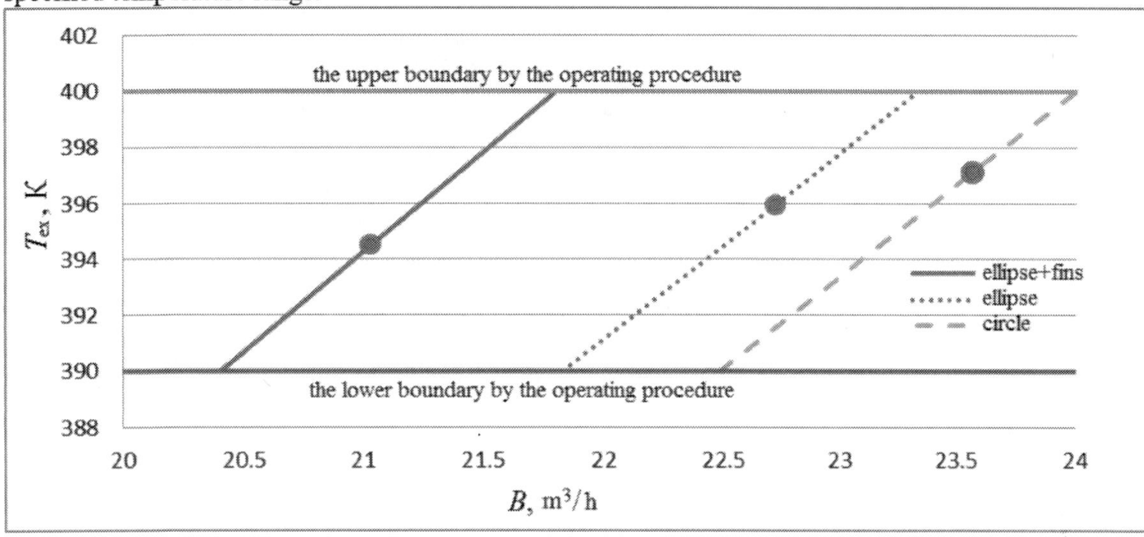

Figure 1. The exhaust gases temperature dependence on the fuel consumption.

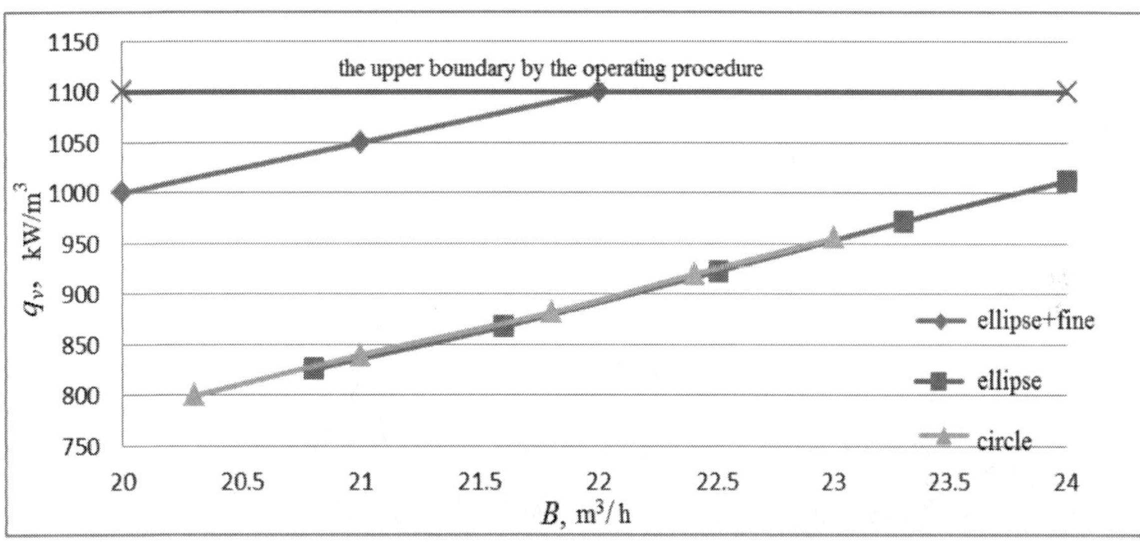

Figure 2. The furnace volume thermal stress dependence on the fuel consumption.

According to Fig. 3 one can define the area which includes the function maximum $\eta = f(a/b, \delta)$, where a/b is the minor and major semiaxes lengths ratio, δ is the finning coefficient that corresponds to $\eta = 0.95$, $a/b \approx 1.3$ and $\delta \approx 1.32$ when meeting the conditions: $1.0 \leq a/b \leq 1.5$; $1.0 \leq \delta \leq 1.4$; $390 \leq Tex \leq 400$ K; $250 \leq qv \leq 1100$ kW/m^3.

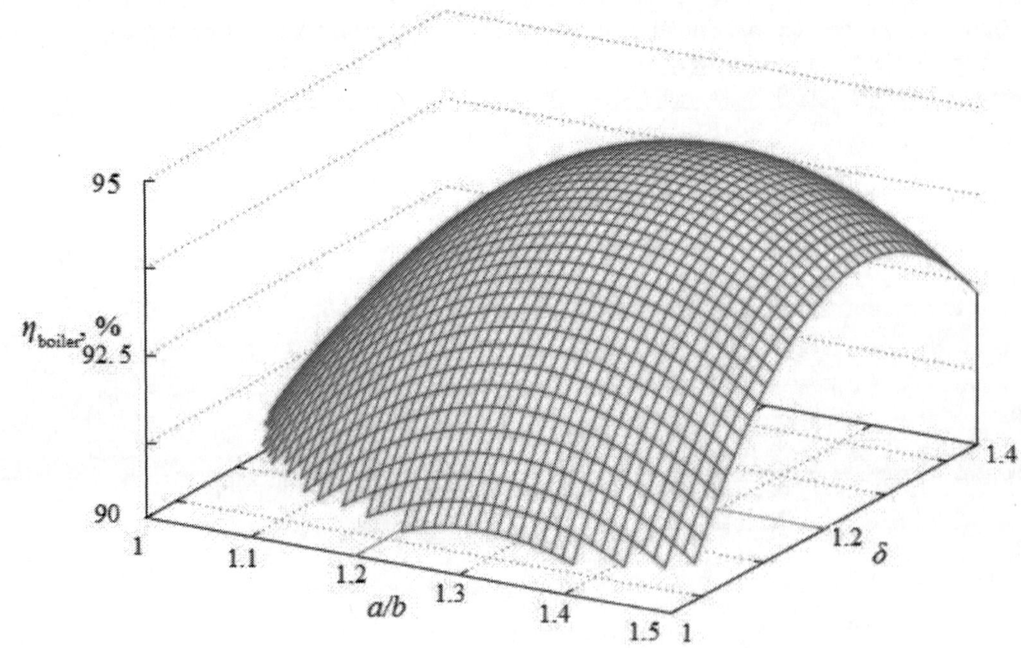

Figure 3. The calculation dependences for η = f (a/b, δ).

According to Fig. 4 one can define the area which includes the function maximum B = f (a/b, δ), that corresponds to B = 21.6 m^3/h, a/b ≈ 1.3 and δ ≈ 1.32 when meeting the conditions: 1.0 ≤ a/b ≤ 1.5; 1.0 ≤ δ ≤ 1.4; 390 ≤ T_{ex} ≤ 400 K; 250 ≤ qv ≤ 1100 kW/m^3.

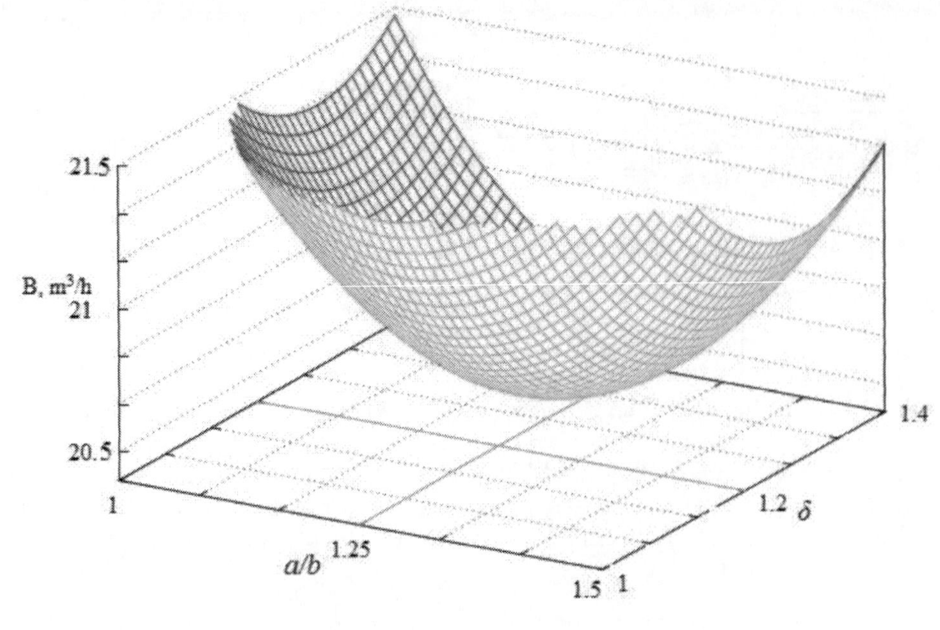

Figure 4. The calculation dependences for B = f (a/b, δ).

IOP Conf. Series: Journal of Physics: Conf. Series **944** (2017) 012009 doi:10.1088/1742-6596/944/1/012009

4. Conclusion

Thus, the general range of maximum values for gross efficiency η calculated by the descent method is possible to be pointed out. The range of minimum values for gas consumption B was defined according to the same method. Moreover, in both cases maximum, minimum values of these thermal and technical characteristics correspond to $a/b \approx 1.3$ and $\delta \approx 1.32$.

References

[1] Vilenskij P L Livshic V N Smoljak S A 2001 *The Investment Projects Efficiency Evaluation: Theory And Practice* (Moscow: Delo)

[2] Gencler I V Petrova E F Sivaev S B 2009 *Energy Saving in an Apartment Complex* (Tver: Nauchnaja kniga)

[3] Batrakov P A Mrakin A N Selivanov A A 2016 Numerical Study of the Nitrogen Oxide Formation in Non-Circular Profile Furnaces of Gas-Tube Boilers *Dynamics of Systems, Mechanisms and Machines (Dynamics)* (Russia: Omsk) DOI: 10.1109/Dynamics.2016.7818974

Prospects of using medium-wave band for radio communication with rescue mobile teams of EMERCOM of Russia

I F Bazhukov[1], I V Dulkejt[2], S A Zavyalov[2], Yu V Lvova[3], A N Lyashuk[2], P I Puzyrev[2], S G Rekunov[3] and E A Chaschin[2], S V Sharapov[3]

[1] Vorkuta Arctic Integrated Emergency and Rescue Center of EMERCOM of Russia, Vorkuta, Russia
[2] Omsk State Technical University, 11, Mira Ave., Omsk 64405, Russia
[3] St. Petersburg University of NFFS of EMERCOM of Russia, St. Peterburg, Russia

e-mail: dulkeytiv@yandex.ru

Abstract. The results of tests in-situ of the prototype of medium-wave mobile radio station «Noema-SV» in Western Siberia, Omsk region and Vorkuta Arctic Integrated Emergency and Rescue Center of EMERCOM of Russia are presented. Radio paths tests in-situ in the Far North show the possibility of radio communication with rescue mobile teams of EMERCOM of Russia in the medium-wave band within distances of several tens of kilometers of rugged topography. The radio range on a flat terrain increases to several hundreds of kilometers. Shortened medium-wave band antennas developed at OmSTU and employed by rescue mobile teams of EMERCOM of Russia were used in.

1. Introduction

The short-wave band (SW), used traditionally for ionospheric long distance (up to a thousand km) and super-long distance (several thousand km) radio communication, is limited in its application by complicated conditions of radio-wave propagation caused by the non-stationary state of the ionosphere. It is challenging and often impossible to use short-wave band in the Arctic, since the ionosphere in polar areas is more perturbed. Moreover, its closest to the Earth layer D absorbs much of the radio wave energy passing through this layer. The most dangerous types of abnormal absorption at high latitudes are the absorption of polar cap and auroral absorption.

The degree of absorption depends on a solar elevation angle, site latitude, solar activity, and radio paths geography [1, 2], which can be divided into three groups.

1. Short-range paths, being entirely inside the polar cap and crossing no auroral oval. Reliability of communications is basically determined by absorption of the polar cap on them.

2. Mid-range paths with at least one end point being located in an auroral zone. Fade-outs are the most probable on such paths, and signal attenuation can reach 30–60 dB depending on the emission frequency.

3. Long-range paths crossing an auroral oval. In this case radio conditions are more favorable as opposed to the second case.

Thus, in waters of the Northern Sea Route the issue is to organize radio on the mid-range radio paths where VHF could not work any longer, but SW has not worked yet. This task can be solved by means of the medium-wave band (MW) of the surface wave [3]. The longer wave length allows you to connect by the surface wave over distances far exceeding the line of sight due to diffraction and refraction. This ensures communication in this band over distances from several tens and even hundreds of kilometers providing transhorizon radio communication.

2. Problem statement

In order to determine the potential range of signal transmission between mobile objects in the medium-wave (MW) frequency bands on the real radio path, tests in-situ of the prototype of marine medium-wave mobile radio station «Noema-SV» (Figure 1) were conducted, this station comprising of: «Noema-SV» transceiver of 8 W, off-chip power amplifier «Noema-SV-U1» of 100 W and whip antenna of 6 m with the inductive antenna tuning unit with the capacity of 100 W [4].

Figure 1. General view of the prototype of marine medium-wave mobile radio station «Noema-SV».

The two prototypes of radio station «Noema-SV» were mounted on motor cars (Figure 2) and KamAZ 4803 and TRECOL-39294 based accident rescue vehicles (Figure 3). Shortened medium-wave band antennas developed in OmSTU were used in tests and proved to be of high efficiency, despite their small size compared to the wavelength used for radio communication.

Figure 2. Antenna location of medium-wave mobile radio station «Noema-SV» on motor cars.

Figure 3. Antennas location of radio station «Noema-SV» on KamAZ 4803 and TRECOL-39294 based accident rescue vehicles.

3. Testing procedure

Tests were performed in two stages on real radio paths. The first stage of the tests took place on 12 April 2017 in Omsk region, the second stage was conducted from 19 to 21 April, in the Komi Republic at Vorkuta Arctic Integrated Emergency and Rescue Center of EMERCOM of Russia - a branch of the North-West Regional Rescue Center of EMERCOM of Russia (Vorkuta AIERC of EMERCOM of Russia – branch of the NWRSRT of EMERCOM of Russia). The radio stations worked in the frequency range of 440–450 kHz during the tests.

North-West Regional Rescue Center of EMERCOM of Russia (Vorkuta AIERC of EMERCOM of Russia – branch of NWRSRT of EMERCOM of Russia). The radio stations worked in the frequency range of 440–450 kHz during the tests.

During the tests, the range of the following types of signals was tested:

– analog telephony, emission classJ3E, on the lower sideband;

– digital selective calling messages (DSC), emission class F1B, frequency shift of 170 Hz, transfer rate of 100 b/s recommended in the ITU-R M. 493-12 guidelines, except that transmitted information was arbitrary;

– digital data at the rate of 1200 b/s, emission class - fast frequency shift keying (FFSK.).

In determination of the radio range for analog voice telephonyvoice message communication sessions of satisfactory quality were taken into consideration (3 points on a five-point scale). While receiving digital messages, sessions complying with a standard were those without errors.

4. Results of experiments and discussion

The choice of method to clarify the current storage area is done by enterprise staff through establishing the rules of work in the storage area.

Radio paths tests in-situ in the Far North including showed the possibility of radio communication with rescue mobile teams of EMERCOM of Russia in the range of medium-wavelength over distances of several tens of kilometers of rugged topography. The radio range on a flat terrain increases to several hundreds of kilometers.

Figures 4 and 5 show the location of radio paths in the field tests (Figure 4 is the first stage and Figure 5 is the second stage of the tests) [5].

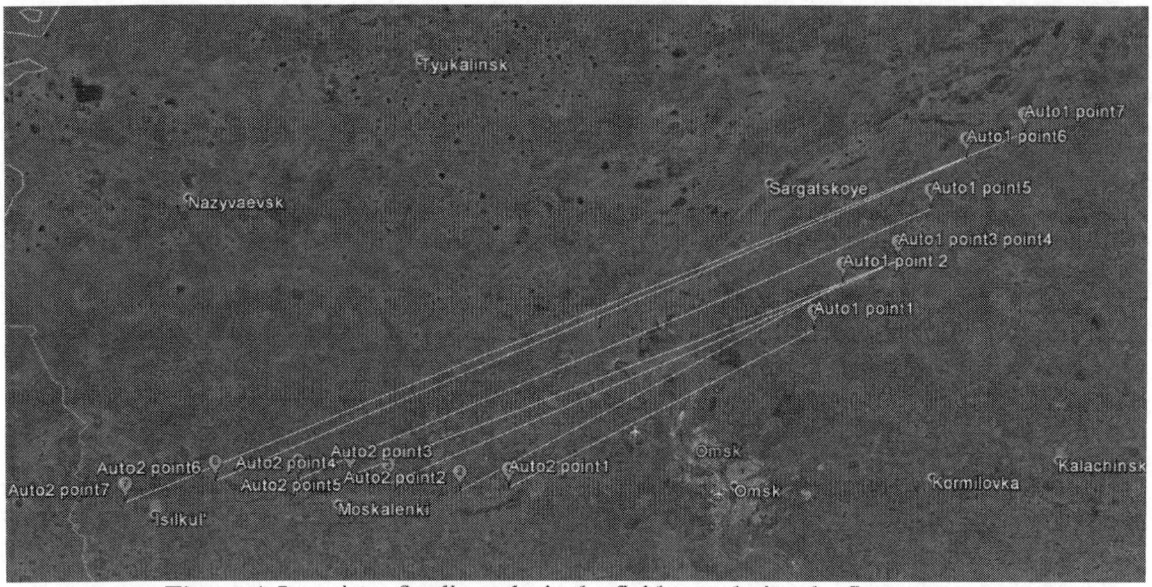

Figure 4. Location of radio paths in the field tests during the first stage.

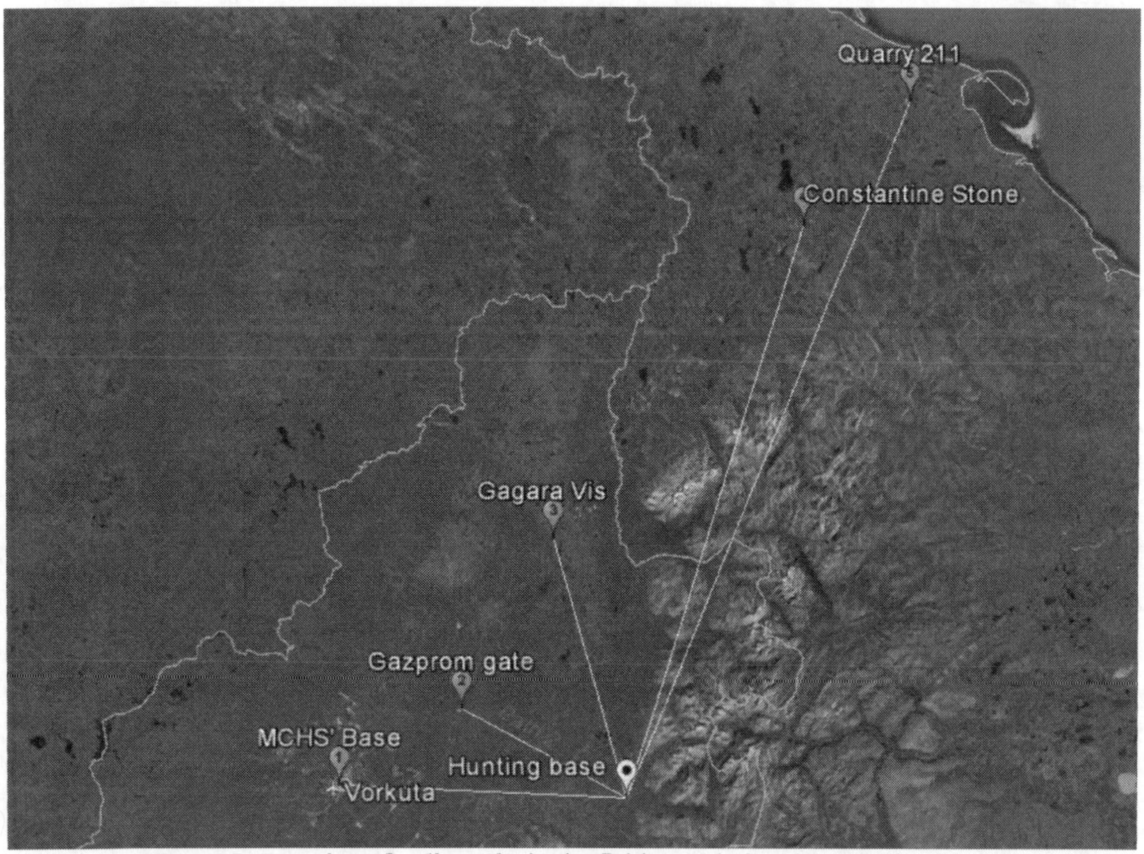

Figure 5. Location of radio paths in the field tests during the second stage.

At the first stage of the tests with the transmitter power of 100 W, the voice message radio range reached up to 130 km (radio path 3 in Figure 4), and the receiving range of digital data with the transfer rate of 1200 b/s was 190 km (radio path 6 in Figure 4). The same radio range of 190 km was provided with DSC signals, but with a transmitter capacity of 8 W. With 100 W transmitter capacity, the consistentsignal reception of DSC was provided to ranges more than 220 km (radio path 7 in Figure 4).

The second stage of the test showed the possibility of a sustained voice signal radio communication and message communication of a digital selective call in the MF band with the transmitter capacity of 100 W up to the distance of 60 km (radio path 1-3 in Figure 5), in motion including (radio path 1 in Figure 5). There was no radio communication on radio paths of 120–155 km (paths 4 and 5 in Figure 5).

The radio communication loss on radio paths 4 and 5 in Figure 5 is due to a significant change in topography. At the first stage of the tests and on the entire radio paths 1, 2 and 3 of the second stage of the tests (Figure 5), the altitude differences of local terrain did not exceed 100 m and were much smaller than wavelengths for radio communications. Cross-section path profiles 1, 2 and 3 are shown in Figures 6, 7 and 8, respectively [5]. On radio paths 4 and 5 (Figure 5), the altitude differences of local terrain were of several hundred meters and exceeded the wavelength for radio communications. Cross-section path profiles 4 and 5 are shown in Figures 9 and 10, respectively [5].

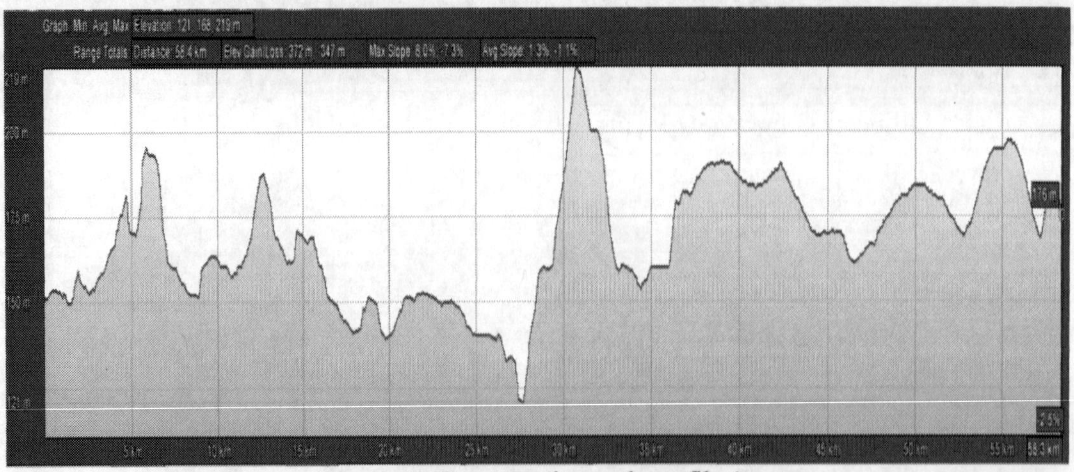

Figure 6. Cross-section path profile 1.

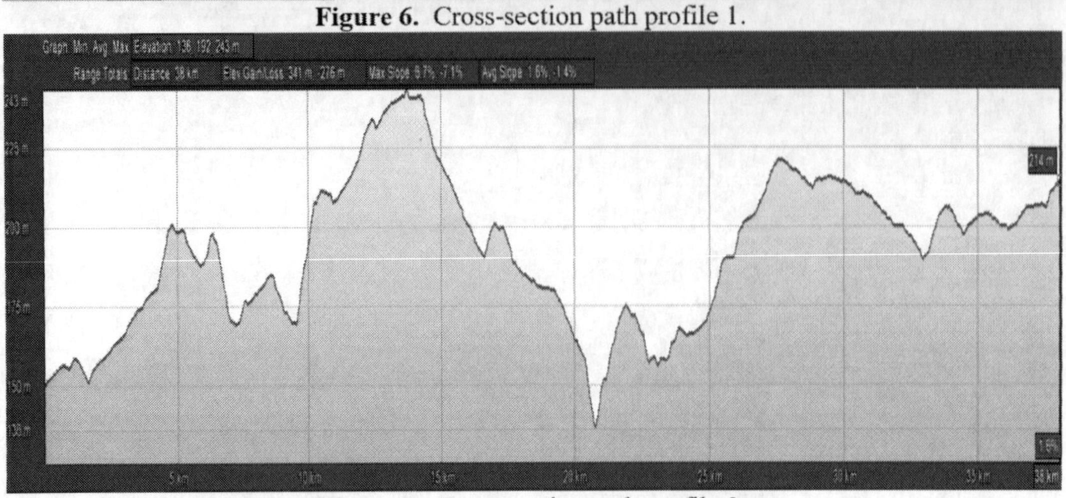

Figure 7. Cross-section path profile 2.

IOP Conf. Series: Journal of Physics: Conf. Series **944** (2017) 012010 doi:10.1088/1742-6596/944/1/012010

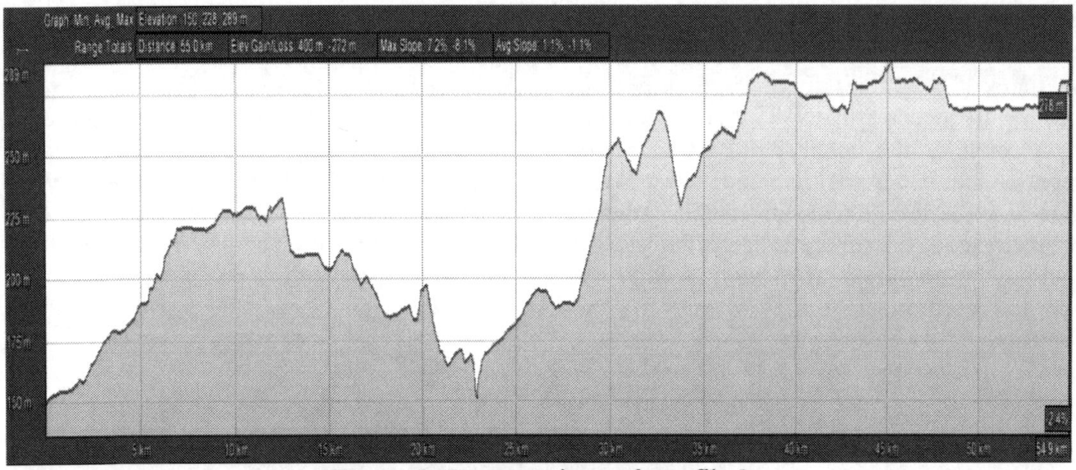

Figure 8. Cross-section path profile 3.

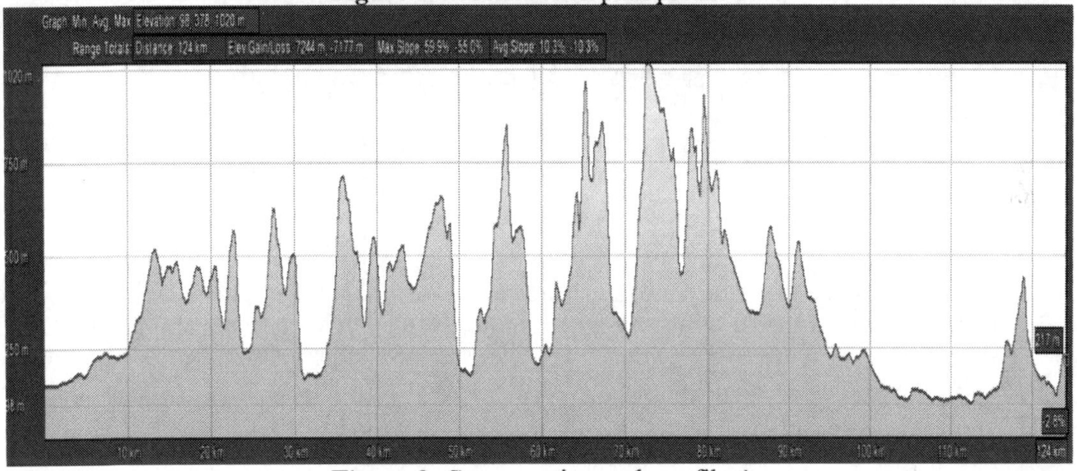

Figure 9. Cross-section path profile 4.

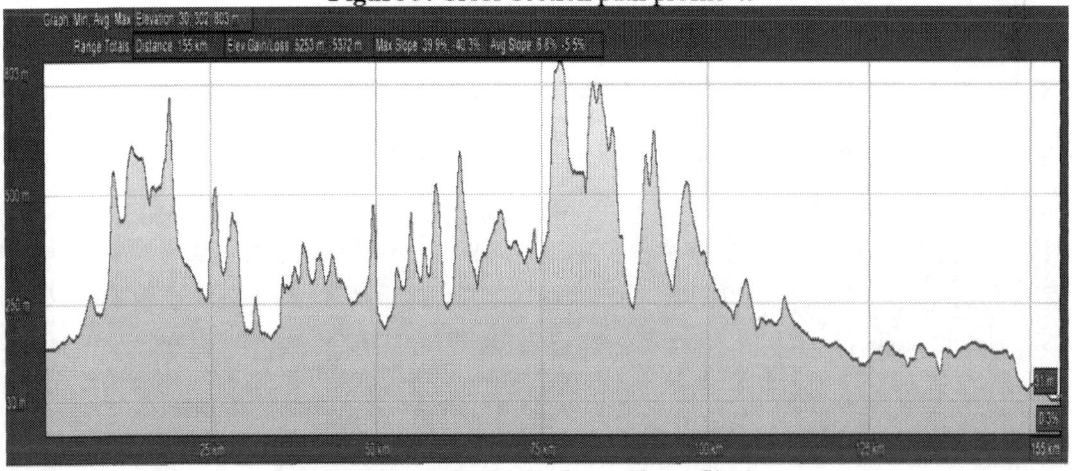

Figure 10. Cross-section path profile 5.

To solve the problem of communications with «dead» zone objects outside radio visibility on mid- and short - range radio paths, the short-wave communication method can be applied with near vertical ionospheric exposure, known as NVIS (Near Vertical Incidence Skywave Propagation) [6]. The key points for such type of radio communications are the application of NVIS antennas and choice of an operating frequency.

In applying NVIS radio communication, radio waves reflecting from layer E at the height of 90...120 kilometers are absorbed by layer D at the height of 50 kilometers. Layer D does not have the maximum of height density and smoothly transacts to layer E, passing it twice [6]. Therefore, while selecting frequencies for NVIS radio communications, the conflicting requirements should be taken into account: as the signal frequency increases, the reflection factor of a vertical down going wave decreases, but in case of frequency decrease, its ionospheric absorption increases. So, a fairly narrow range of applicable frequencies remains suitable for radio communication, this range being dependent on time of day, season and solar activity determining electron density. The antenna used in the test provides the possibility to work in nearly vertical incidence sky wave mode (Figure 11) [7], but abnormal absorption of radio waves in the Arctic remain this type of radio communication problematic.

Figure 11. Whip antenna located on a motor car as NVIS antenna.

For the Arctic Zone of the Russian Federation, installation of MW radio communication network with automatic relaying of signals by remote repeaters can be preferable due to the possibility to operate in an autonomous, unattended mode. "Dead" zones arising from a local terrain and optimum repeater locations can be calculated using the recommendations of International Telecommunications Union ITU-R P.526-10 "Propagation by Diffraction" [8].

5. Conclusion

Radio paths tests in-situ in the Far North including showed the possibility of radio communication with rescue mobile teams of EMERCOM of Russia in the range of medium-wavelength over distances of several tens of kilometers and in the plain terrain up to several hundred kilometers, in motion including.

Radio communication was provided in the vicinity of the town of Vorkuta. Much narrow-band interference indicates in itself the intensive use of medium-wavelength range in the Far North. This range is proved to be a very promising and sometimes a single frequency resource for the operation of various radio engineering systems.

Shortened medium-wave band antennas developed at OmSTU can be located on motor cars used by rescue mobile teams of EMERCOM of Russia, these antennas providing good efficiency.

It is necessary to continue the tests with a special emphasis on «dead» zone radio communications due to topographic obstacles in highlands.

References

[1] Dulkeyt I, Zavyalov S, Chaschin E, Shigabutdinov A 2015 Some aspects of providing information security of hydrocarbons extraction and transportation in the Arctic Zone of the Russian Federation *Procedia Engineering* vol 113 pp 344–348
[2] Bryksenkov A Dulkejt I Zavyalov S Kosykh A Khazan V Shigabutdinov A 2014 Creation of a single information space for navigation in the North Sea route through the integration of the

equipment and use of information technology *Maritime information management systems* 3 (6) pp 290–297

[3] Dulkejt I Zavyalov S Kosykh A Ljashuk A Chaschin E 2016 The results of the tests in-situ of the mobile medium-wave mobile radio station on medium range radio path *Omsk Science Bulletin: Series Instruments, machines and technology* № 3 (147) pp 82–86

[4] Liashuk A, Zavyalov S, Chashin E 2016 Small Monopole Transceiver Antenna for Medium Frequencies *17th International Conference of Young Specialists on Micro: Nanotechnologies and Electron Devices (EDM)* pp 140–144

[5] Google Planet Earth URL: //https://www.google.com/intl/ru/earth/ (accessed date: 10.05.2017).

[6] Polyakov V 2009 NVIS near field communication technology on SW *Special-purpose machinery and radio* № 1 pp 59–63

[7] ICOM IC-AH-5NV NVIS Fiberglass Antenna Element (14.8 Feet Long) for HF Short Range NVIS Communications URL: //http://www.hamradio.com/detail.cfm?pid=H0-012951 (accessed date: 10.05.2017)

[8] Recommendation ITU-R P.526-6 Propagation by Diffraction URL: //https://www.itu.int/dms_pubrec/itu-r/rec/p/R-REC-P.526-6-199910-S!!PDF-E.pdf (accessed date: 12.05.2017)

IOP Conf. Series: Journal of Physics: Conf. Series **944** (2017) 012011 doi:10.1088/1742-6596/944/1/012011

Computer Modeling of Phase Transitions of Semibounded Antiferromagnets

S V Belim and E V Trushnikova
Dostoevsky Omsk State University, pr. Mira 55a, Omsk, 644077, Russia

Abstract. The article gives a machine modeling of surface and bulk phase transitions in semibounded antiferromagnetic Ising model. Distinct values of exchange integrals proportion were considered. Dependency of temperature of phase transitions from surface energy value was calculated. System phase diagram was constructed. Availability of tetracritical point was shown. Conditions of surface phase transition were obtained. Temperature and properties of tetracritical point was determined.

1. Introduction

The phenomenon of surface magnetism consists of distinction between ordering temperature of the surface spins and ordering temperature of other system spins. The reason for this phenomenon is difference between values of exchange integral on surface and exchange integral in depth of sample. Surface ordering for antiferromagnetic systems was observed in a series of contributions [1, 2, 3]. In the abstract, within the framework of the mean-field theory surface magnetism was reported in paper [4], in which it was shown that the phase transition temperature on the surface may vary from the corresponding Néel temperature. System phase diagram was obtained in this and in other papers [5, 6, 7]. It consist of three phases: disordered phase (SD/BD), surface-ordered bulk-dis-ordered phase (SO/BD), surface-ordered bulk-ordered phase (SO/BO) Figure 1.

In addition, three transition lines crossing at a tetracritical point are observed on phase diagram. SD/BD to SO/BD transition is referred to surface phase transition, SO/BD to SO/BO transition is referred to extraordinary phase transition, SD/BD to SO/BO transition is referred to ordinary or bulk phase transition. Crossing of these three transition lines form a multicritical point. Phase transition in this point named special phase transition. Transition names was first introduced in the paper [8]. Critical behavior of semibounded Ising model was considered in the papers [9, 10] as part of field-theoretical approach.

The purpose of our article is research of phase transitions in semibounded antiferromagnetic Ising model by machine modeling at distinct values of surface energy and construction of phase diagram.

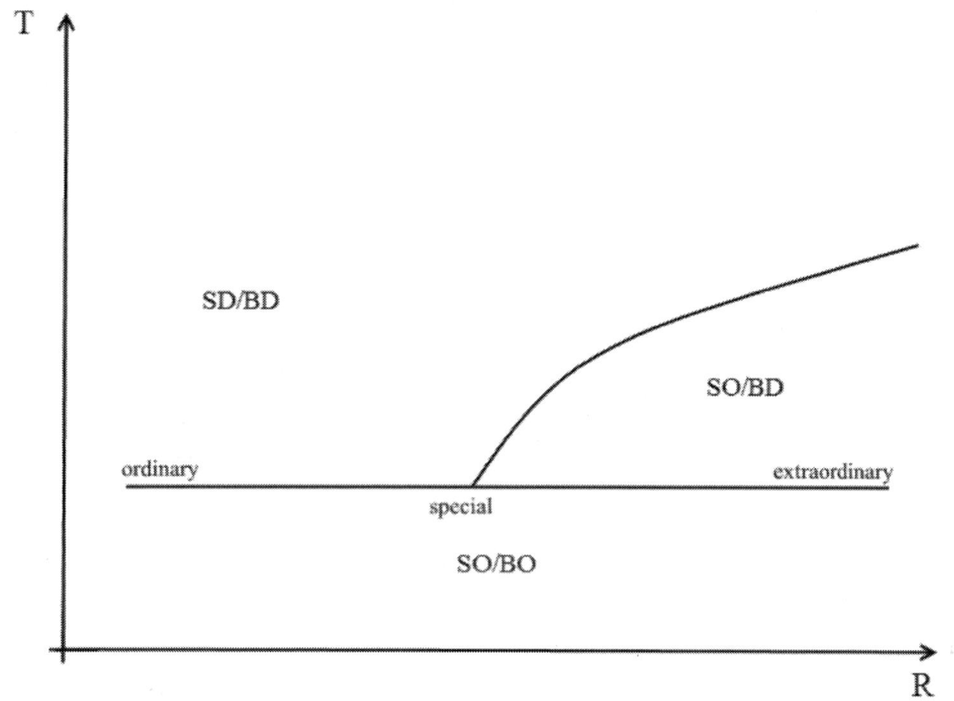

Figure 1. Phase diagram with tricritical point.

2. Description of the system

Hamiltonian of semibounded antiferromagnetic Ising model may be written as:

$$H = -J_B \sum_B S_i S_j - J_S \sum_S S_i S_j, \tag{1}$$

where S_i is value of spin i point (+ ½ or – ½). Addition is carried out over nearest neighbours. The second sum include surface spins only, the first sum include the others. Surface (J_S) and bulk (J_B) exchange integrals have different values in most real systems. Besides J_S may have values bigger then J_B as well as smaller then J_B.

In this paper machine modeling of 3D systems was carried out with cubical lattice having linear dimensions $L \times L \times 2L$ by Monte Carlo using the Metropolis algorithm. Free boundary plane was given by equation $z = 0$. System was found semi-space $z \geq 0$. Periodic boundary conditions were used. Spins from plane $z = L$ were considered neighbours for spins from plane $z = 2L$. Theory of finite-dimensional scaling was used for determination phase transition temperature.

Let us introduce two order parameter – m and m_S to describe the antiferromagnetic ordering. Order parameter m is calculated as staggered magnetization of all system. It is equal to difference between magnetic moments of two sublattice. Surface order parameter m_S was calculated as staggered magnetization of free-surface located spins.

Fluctuation relations were used to monitor the behaviour of heat capacity and susceptibility in bulk of system and on its surface:

$$C = NK^2\left(\langle E^2\rangle - \langle E\rangle^2\right), \tag{2}$$

$$\chi = NK\left(\langle m^2 \rangle - \langle m \rangle^2\right), \tag{3}$$

$$C_S = SK^2\left(\langle E_S^2 \rangle - \langle E_S \rangle^2\right), \tag{4}$$

$$\chi_S = SK\left(\langle m_S^2 \rangle - \langle n_S \rangle^2\right), \tag{5}$$

$$K = |J_B| / k_B T, \tag{6}$$

$$N = 2L^3, \tag{7}$$

$$S = L^2. \tag{8}$$

(7) – number of system points, (8) – number of surface points, E – internal energy, E_S – surface energy. Angle brackets signify thermodynamic averaging.
Critical temperature of transition was defined by bulk and surface Binder's quartic cumulants:

$$U = 1 - \frac{\langle m^4 \rangle}{3\langle m^2 \rangle^2}, \quad U_S = 1 - \frac{\langle m_S^4 \rangle}{3\langle m_S^2 \rangle^2}. \tag{9}$$

Temperature of phase transition can be defined by the position of the intersection point of cumulants for systems with different sizes of L. Temperature of bulk phase transition T_N was defined by intersection point of bulk cumulants U, temperature of surface phase transition T_S was defined by intersection point of surface cumulants U_S.

3. Results of computer modeling

Computer experiment was carried out for systems with linear sizes from L = 12 to L = 32 in increments of L = 4. The number of steps of Monte Carlo for the spin was equal to 100 000. The number of steps of Monte Carlo for the spin was equal to 100 000. The value of the ratio of the exchange integrals $R = J_S / J_B$ ran from R = 1.0 to R = 2.0 in increments of ΔR = 0.1. The modeling was carried out in the interval from R = 1.30 to R = 1.45 with step ΔR = 0.01. The critical temperature of the phase transition on the surface of the system and in the bulk of the system was determined for each value of R. Phase diagram of the system shown in figure 2.

IOP Conf. Series: Journal of Physics: Conf. Series **944** (2017) 012011 doi:10.1088/1742-6596/944/1/012011

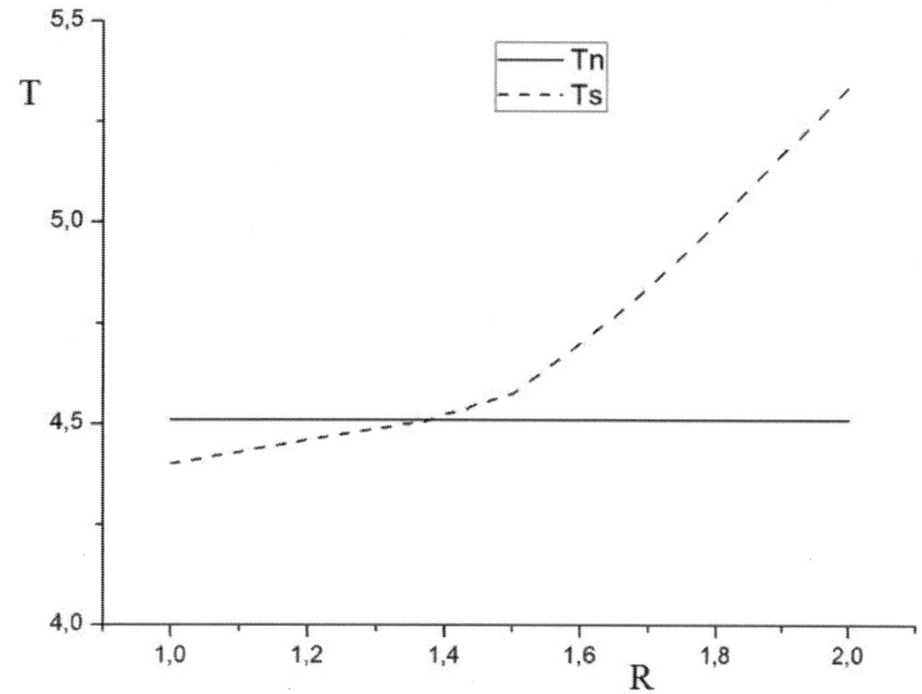

Figure 2. Phase diagram of the system.
The temperature of bulk phase transition T_n dependence on relations of exchange integrals R shown as a solid graph. The temperature of surface phase transition T_S dependence on R shown as a dotted graph

One can see from the phase diagram, the presence of the surface-disordered bulk-ordered phase (SD/BO) was observed in the system. The possible existence of this condition was experimentally detected for Fe_3BO_6 [11, 12] and $FeBO_3$ [13]. Surface energy lower than bulk energy is a criterion for the persistence of this phase.

Also it is worth noting that tetracritical point is present on the phase diagram of antiferromagnets that observed at $R = 1.38$, in contrast to the theoretical predictions, and similar results for the ferromagnetic systems.

The distribution of magnetization in the layers close to the surface is significantly different on the left and right of tetracritical point. Figure 3 shows the magnetization dependence on the layer number with the Neel temperature $T_n = 4.51$ at $R = 1.00$ for the system with linear size $L = 32$. It is seen from the figure the magnetization increases deep into the sample.

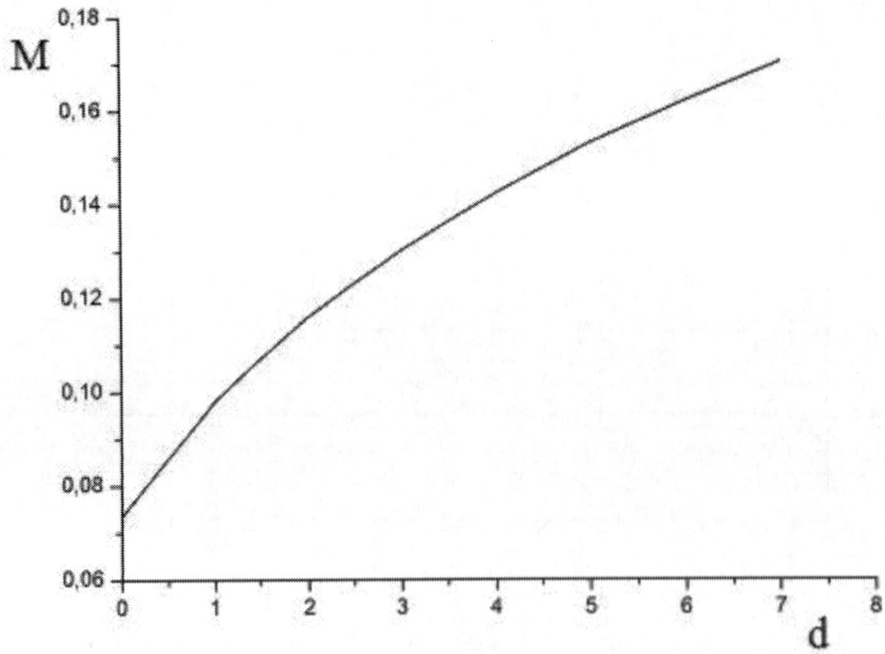

Figure 3. The dependence of the magnetization on the distance to the free surface d at $R = 1.00$.

Figure 4 shows the magnetization dependence on the layer number with the Neel temperature $T_n = 4.51$ at $R = 1.50$ for the system with linear size $L = 32$.

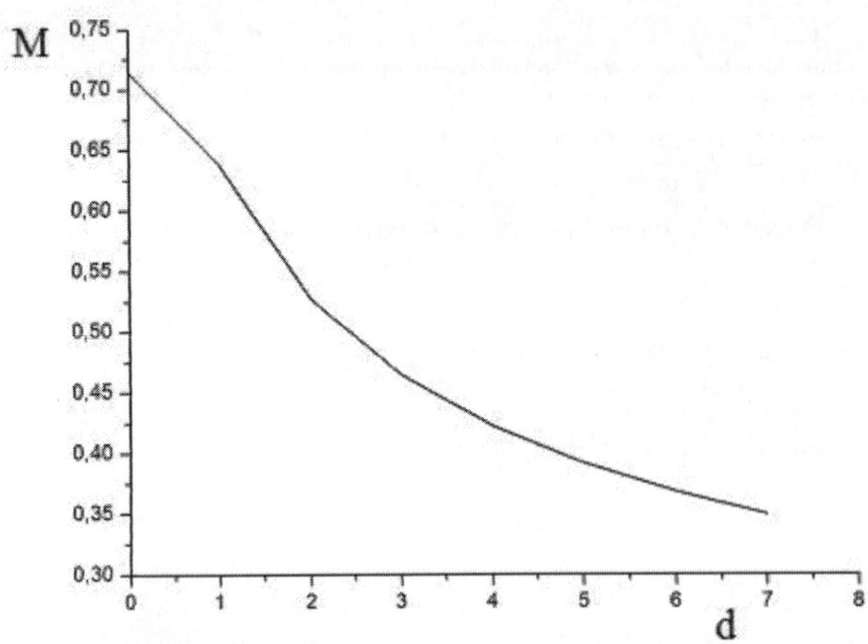

Figure 4. The dependence of the magnetization on the distance to the free surface d at $R = 1.50$.

Figure 5 shows the magnetization dependence on the layer number with the Neel temperature $T_n = 4.51$ at $R = 1.38$ for the system with linear size $L = 32$.

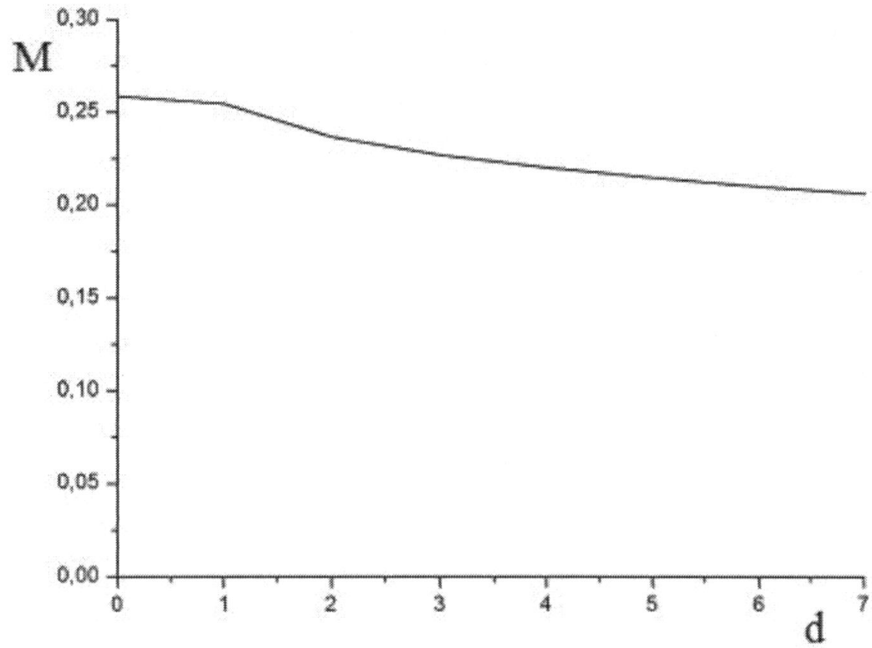

Figure 5. The dependence of the magnetization on the distance to the free surface d at $R = 1.38$.

From the comparison of the last three figures it is evident that the character of distribution of magnetization changes with distance from the surface of the system when passing through tetracritical point.

References

[1] Aderhold J Davydov V Yu Fedler F Klausing H Mistele D Rotter T Semchinova O Stemmer J and Graul J 2001 InN thin films grown by metalorganic molecular beam epitaxy on sapphire substrates *Journal of Crystal Growth* **222** pp 701–705 DOI: 10.1016/S0022-0248(00)00986-6

[2] Campagna M 1985 Surface Magnetism: Recent Progress and Opportunities *Journal of Vacuum Science & Technology A: Vacuum, Surfaces, and Films* 3(3) pp 1491–1495

[3] Kamzin A S Grigor'ev L A 1993 Mössbauer investigation of the surface and volume spin-reorientation phase transition in Fe3BO6*JETP Lett* 57(9) pp 552–556

[4] Kamzin A S., Grigor'ev L A 1994 Simultaneous triple-radiation Mossbauer spectroscopy investigation of surface and bulk magnetic properties of Fe-jBoo near Neel temperature. *JETP* **78** pp 200–207

[5] Mills D L 1971 Surface Effects in Magnetic Crystals near the Ordering Temperature *Phys. Rev.*, **3**(ll) pp 3887–3894

[6] Binder K 1983 Magnetic surface phenomena. Phase transition and critical phenomena. Ed. by Domb C. And Lebowitz *J.L. NY Academ.* Press **3** pp 325–331

[7] Kaganov M L 1972 Surface Magnetism. *JETP* **35**(3) pp 631–633

[8] Diehl H W 1997 The Theory of Boundary Critical Phenomena. *J. Mod. Phys.* B 11 pp 3503–3523

[9] Lubensky T C, Rubin H 1975 Critical phenomena in semi-infinite systems. II. Mean-field theory *Phys. Rev.* B. **12** pp 3885–3901

[10] Belim S V 2006 Critical behavior of disordered systems with a free surface. *J. Exp. Theor. Phys.* **103** pp 611–623

[11] Belim S V 2008 Multicritical behavior of systems with a free surface *J. Exp. Theor. Phys.*, **106** pp 773–780

IOP Conf. Series: Journal of Physics: Conf. Series **944** (2017) 012011 doi:10.1088/1742-6596/944/1/012011

[12] Kamzin A S, Grigor'ev L A 1990 Mossbauer Study Of The Surface And Bulk Properties Of Fe3B06 In The Critical-Temperature Region. *Fizika tverdogo tela* **32**(11) pp 3278–3281

[13] Kamzin A S, Grigor'ev L A 1993 Mossbauer Study Of Spin-Flip Phase-Transition On The Fe350e Surface. *Jetp letters* **57**(9) pp 557–561

[14] Kamzin A S, Grigor'ev L A, Kamzin S A 1994 Mossbauer Study Of The Surface And Bulk Reorientational Phase-Transition In Ga-doped Fe3B06 Macrocrystal *Fizika tverdogo tela* **36**(5) pp 399–1415

Algorithm for detection of steganographic inserts type LSB-substitution on the basis of an analysis of the zero layer

S V Belim, D E Vilkhovskiy

Dostoevsky Omsk State University, pr. Mira 55a, Omsk, 644077, Russia

e-mail: sbelim@mail.ru, vilkhovskiy@gmail.com

Abstract. All articles *must* contain an abstract. The abstract text should be formatted using 10 point Times or Times New Roman and indented 25 mm from the left margin. Leave 10 mm space after the abstract before you begin the main text of your article, starting on the same page as the abstract. The abstract should give readers concise information about the content of the article and indicate the main results obtained and conclusions drawn. The abstract is not part of the text and should be complete in itself; no table numbers, figure numbers, references or displayed mathematical expressions should be included. It should be suitable for direct inclusion in abstracting services and should not normally exceed 200 words in a single paragraph. Since contemporary information-retrieval systems rely heavily on the content of titles and abstracts to identify relevant articles in literature searches, great care should be taken in constructing both. Keywords – search for LSB-inserts, analysis of steganography container, revealing of steganography inserts.

1. Introduction

The most common method of embedding steganographic inserts to date is to replace the least significant bits (LSB- substitution) [1]. This method is based on the fact that the replacement of one to four least significant bits of the color representation of the image pixels remains almost invisible to the human eye. The greatest way for hiding information is the blue component, which is due to the structure of the retina. To date, many algorithms have been developed for embedding information in images, audio and video streams, but the method of LSB-replacement, historically the first one, is widely used. All methods of hiding information are focused on such an image transformation, which is not visible to the human eye. In this connection, the actual task is to develop image analysis algorithms for the presence detection of steganographic inserts.

To date, all existing algorithms are focused on the definition of the fact of the presence or absence of the steganographic insert in the image. In articles [2, 3] a method of statistical analysis Chi-square based on the assumption of random distribution of the least significant bits of color in image. This method gives good results when the container is evenly filled and is weakly applicable when randomly selecting pixels to replace the lower bits. In the article [4], steganoanalysis is performed on the basis of comparing the least significant bits in neighboring bytes using the Markov chain formalism. In [5], a method for detecting a steganographic insert based on the use of artificial neural networks. It is shown that for a sufficiently large volume of the training sample, the neural network is able to determine the presence of an insert with an error not exceeding 15%. All known to date methods of stegoanalysis of the LSB-substitution method are effective when filling the steganographic container by at least 50% [6]. In [7], a method for detecting embedded information based on information compression algorithms. The main idea of the method is that the random data is compressed more weakly than the ordered ones. This approach makes it possible to determine with high accuracy the presence of a steganographic insert when filling a container from 40%. This method was further developed in [8] based on the usage of pre-processing for image, allowing to use it at a much lower container coverage.

It should be noted that to date, there are no algorithms that determine the bytes in which the least significant bit was done. This problem is close to the problem of detecting pixels damaged by impulse noise. For impulse noise, a color change of an arbitrarily selected byte is randomly selected. However, the task of searching for an embedded message is more complex, since the change amount is only one bit.

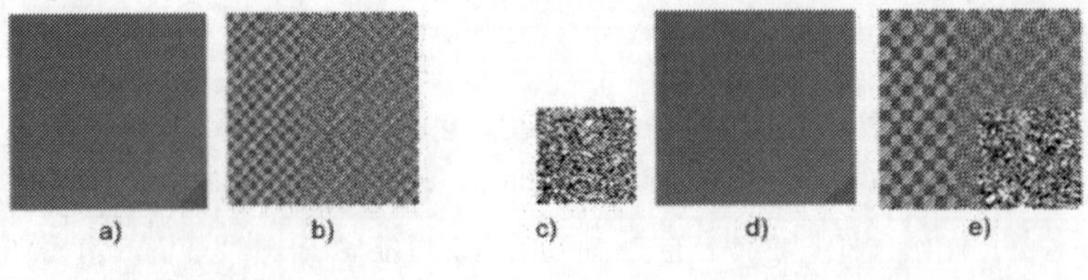

Figure 1. Comparison of the zero layers: a) the original image, b) the zero layer of the original image, c) the embedded pixel map, d) the image with the embedded message, e) the zero layer of the image with the embedded message.

To date, there are several methods for detecting corrupted pixels. First of all, it is necessary to single out the method SDROM [9], which, being historically the first, served as the basis for algorithms [10, 11]. There are also other approaches based on the search for associative rules [12, 13], the clustering method [14], and the hierarchy analysis method [15].
In this article an algorithm for detecting pixels in the image in which substitution is made at the least significant bit steganography embedding messages based on automatic analysis of the zero layer.
Formatting the title, authors and affiliations
Please follow these instructions as carefully as possible so all articles within a conference have the same style to the title page. This paragraph follows a section title so it should not be indented.

2. Formulation of the problem

We will analyze images in which information can be embedded in the form of steganographic inserts in the lower bit of the blue component. The analysis of the blue component is due to the fact that embedding in it is the least noticeable visually, and therefore it is recommended to use it for making message embedding. Similarly, other components can be analyzed without loss of generality of the proposed method. We start with two assumptions. First, it is not known whether there is a steganographic insert or not. Secondly, we will assume that the steganographic insert fills a certain rectangular area whose dimensions and position are unknown. The task is not only to determine the presence of the steganographic insert, but also the area in which the embedding. The second assumption substantially complicates the problem, since a situation is possible in which all the lower pixels of the blue component are replaced. We solve the problem, based on the assumption of incomplete substitution of the zero layer.

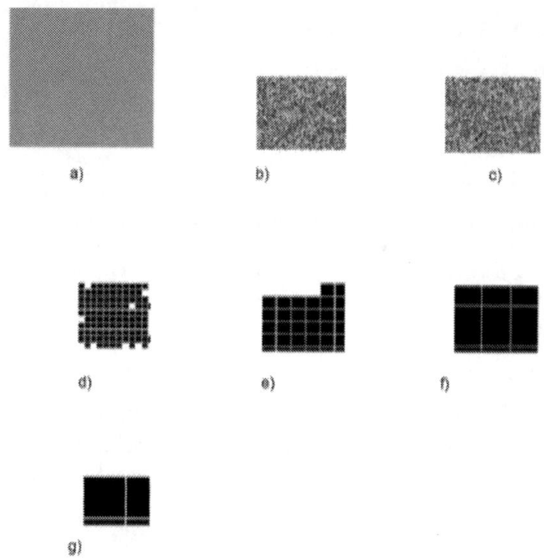

Figure 2. The results of the algorithm for automatic allocation of the embedding region for a uniformly filled image: a) Image with embedded insert, b) embedded pixels map, c) zero layer, d) Automatically selected area of embedding when R_0=15, e) Automatically selected area of embedding when R_0=30, f) Automatically selected area of embedding when R_0=50, g) Automatically selected area of embedding when R_0=100.

The zero layer is a matrix of zeros and ones. The patterns of the distribution of zero and unit values of the zero layer without embedding are due to the structure of the image. The embedding of a message makes a change to the zero layer, changing the density of the distribution of unit values. Figure 1 shows an image, its zero-layer without embedding, and a zero layer with embedded message.

As can be seen from Figure 1, the presence of an embedded message can be detected visually from the analysis of the zero layer. We set ourselves the goal of automatically determining the embedding area.

3. Algorithm for finding the embedding area

To select the embedding area, we use an algorithm based on the FOREL taxonomy algorithm [16]. In its classical form, FOREL combines points into taxa that lie inside the circle.

In our case, we will build rectangular taxa. We introduce the density of unit values$p$. If in some area of the image contains N pixels and N_1 of them has a value one, Unit density index p = N1/N. We will look for rectangular regions having a density of unit values of a given value p_0. Also, as an input parameter of the algorithm, we set the parameter R_0, which determines the initial size of the taxon. The algorithm consists of the following steps:

Step 1. Choose the initial value of the taxon size $R = R_0$.

Step 2. We randomly choose a point with coordinates(x_1, y_1), that will be the center of the taxon. We construct a square whose upper-left corner has coordinates $(x_1 - R,\ y_1 - R)$, and the lower-right corner of the coordinate $(x_1 + R,\ y_1 + R)$.

Step 3. We seek the center of mass coordinates of points lying inside a square-built(x_2, y_2).

Step 4. If the points $(x_1,\ y_1)$ and (x_2, y_2) is the same, then go to Step 5, otherwise $x_1 = x_2, y_1 = y_2$ and go to Step 2.

Step 5. Calculate the density of unit values p.

Step 6. If $p > p_0$, then $R := 1.1R$ and go to Step3.

Step 7. If $p < p_0$, then $R := 0.9R$ and go to Step 3.

Step 8. If $p = p_0$, then go to Step 2.

The algorithm is performed until all the points of the zero layer are combined into some taxa.

As areas in which a message can be embedded, we select taxa whose size is not less than 10% of the size of the original image. The results of this algorithm for a uniformly filled image are shown in Figure 2.

As can be seen from Figure 2, as a result of the operation of the algorithm on auto-generated image with a uniform fill, the embedding area is determined quite accurately. However, even for artifacts with gradient fills, complications arise, since a continuous change in the color of the image as a whole is manifested as bands of identical values on the zero layer (Figure 1, b). These bands can be eliminated by pre-processing the image.

4. Image preprocessing algorithm

As was shown above, the zero layer of a gradient-filled image represents bands of zeros and ones. We use the linear transformation:

$$d(x, y) = ax + by - e$$

Where $a = c(x+1, y) - c(x, y)$, $b = c(x, y+1) - c(x, y)$ – the color value of the pixel located at the point with coordinates (x, y). Let's define e, as minimal value of $c(x, y)$ on the set of all pixels of the image. In the case where the color of the image is a fill with a constant gradient, the function $d(x, y)$ will have a constant value $(d(x, y) = const)$. An algorithm for making decisions about changing a pixel can be applied to a function $d(x, y)$, rather than to a function (x, y).

To apply this linear transformation to photographic images, it is necessary to define the gradient fill areas. We will calculate the second derivatives of the function (x, y) and identify the regions in which they have a zero value. Due to the fact that the areas of ideal gradient filling on photographic images are extremely rare we will require the fulfillment of three more "soft" conditions:

$$\left|\frac{\partial^2 c(x,y)}{\partial x^2}\right| \leq 2, \left|\frac{\partial^2 c(x,y)}{\partial y^2}\right| \leq 2, \left|\frac{\partial^2 c(x,y)}{\partial x \partial y}\right| \leq 2.$$

Not a strict inequality instead of zero is introduced in order to take into account small deviations from the gradient fill and not to lose the embedded bits.

After identifying connected regions which satisfy the conditions for the second derivatives necessary to determine the coefficients of the function $d(x, y)$. To find them, the least squares method was used. After that, the value of the function $d(x, y)$, to which the algorithm for searching the message embedding areas was applied was calculated.

5. Discussion of results and conclusions

A computer experiment was conducted for various color images, both auto-generated and photographic. The embedded message was a text string which is represented as a sequence of bits. Embedding was made in the blue component, as the least visible to the human eye. Embedded bits filled the rectangular area. The task of stegoanalysis was to determine the area in which the substituted bits.

Initially, the algorithm was tested on a rectangular auto-generated image with a gradient fill. The results are shown in Figure 3.

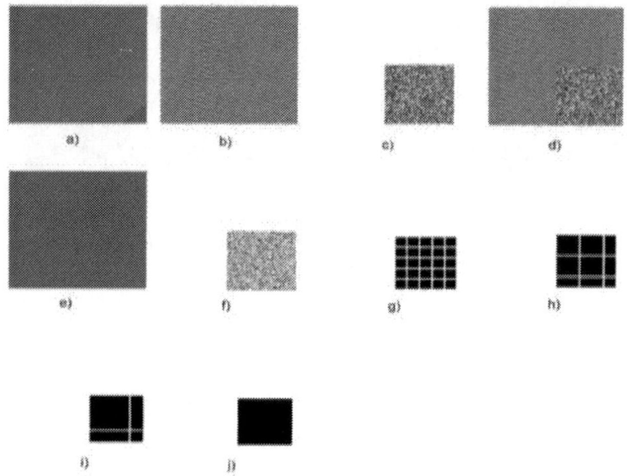

Figure 3. The results of the algorithm for automatic allocation of the embedding region for a gradient-filled image: a) Original image, b) Zero layer of the original image, c) Embedded pixel map, d) Zero layer with embedded message, e) Image after pre-processing, f) Zero layer of image after pre-processing step, g) Automatically selected area of embedding when R_0=15, h) Automatically selected area of embedding when R_0=30, i) Automatically selected area of embedding when R_0=50, j) Automatically selected area of embedding when R_0=100.

As can be seen from Figure 3, the image preprocessing algorithm allows processing gradient-filled images as efficiently as single-color images. Both the uniform and the gradient fill on the results of the algorithm work is affected by the choice of the initial size of the taxon R_0.

References

[1] Adelson E 1990 Digital Signal Encoding and Decoding Apparatus. U.S. Patent. No. 4,939515
[2] Provos N Honeyman P 2001 Detecting steganographic content on the internet *CITI Technical Report 01-11* (University of Michigan)
[3] Westfeld A Pfitzmann A 2000 Attacks on Steganographic Systems: Breaking the Steganographic Utilities EzStego, Jsteg, Steganos and STools-and Some Lessons Learned *International Workshop on Information Hiding* pp 61–76
[4] Golub V A Dryuchenko M A 2009 Steganographic information detection in JPEG files with the help of complex usage of several stego-attackes *Infocommunication Technologies* 7(1) pp 44–50
[5] Abdenov A Zh Leonov L S 2010 Ispolzovaniye neyronnykh setey v slepykh metodakh obnaruzheniya vstroyennoy steganograficheskoy informatsii v tsifrovykh izobrazheniyakh *Polzunovsky vestnik* 2 pp 221–225
[6] A. Westfeld, A. Pfitzmann. 2000 Attacks on Steganographic Systems. Breaking the Steganographic Utilities EzStego, Jsteg, Steganos and S-Tools - and Some Lessons Learned. Lecture Notes in Computer Science, 1768 pp 61–75
[7] M.Yu. Zhilkin 2008 Stegoanaliz graficheskikh dannykh v razlichnykh formatakh. Doklady TUSURa **2**(18) pp 63–64
[8] Monarev V A 2012 Sdvigovyy metod obnaruzheniya skrytoy informatsii. *Vestnik SibGUTI* **4** pp 62–68

IOP Conf. Series: Journal of Physics: Conf. Series **944** (2017) 012012 doi:10.1088/1742-6596/944/1/012012

[9] Abreu E, Lightstone M, Mitra S K, Arakawa S K 1996 A new efficient approach for the removal of impulse noise from highly corrupted images. *IEEE Transactions on Image Processing, IEEE Transactions* on 5 pp 1012–1025.
[10] Garnett R, Huegerich T, Chui C, He W 2012 A Universal Noise Removal Algorithm with an Impulse Detector. *IEEE Trans Image Proccess **14***(11) pp 1747–1754

A graph-graph approach to the analysis of the set of associative rules

S V Belim, T B Smirnova, and A N Mironenko

Dostoevsky Omsk State University, 55A, Mira ave., Omsk, 644077, Russia

Abstract. The article proposes a method for processing a set of associative rules, which makes it possible to identify additional relationships between the set of objects under study. The proposed approach consists of three stages. At the first stage, associative rules are revealed from the statistical data. At the second stage, constructed weighted oriented graph of relationships between the objects of the system. The third stage analyses the graph and identifies the community (community), which allows you to determine the groups of the most related objects. As an example, is given an analysis of the activities of public organizations. The result of the work is a method that allows to identify patterns from the analysis of a set of associative rules, and not just from a separate associative rule.

1. Introduction

The traditional use of the method of forming of associative rules from dataset consists of the study of peculiarities of consumers demand for goods in trade networks [1, 2]. Information about the correlation interest to buying of various goods, is used later for their positioning in trading rooms. In recent times, observed an expansion of the use of this method. For example for processing images [3, 4], analysis dissemination of biological species [5], search by photo of the surface points of interest in mining [6] and others.

Quite a large number of works devoted to researching social networks with help of searching association rules. So in article [7] on the basis of association rules are detected mutual influence users of social networks. Then stand out the most influential users, which can be used for effective dissemination of information. In works [8, 9], based on the associative rules formed from Facebook entries, a connection is established between the student's gender and the selected training courses. Similar researching's [10], based on Facebook entries, and allowed to study the influence of the social network on the educational process of students in Turkey. In [11], the authors on the basis of the analysis of social networks identify factors that affect the hobby of users. The possibilities of using associative rules in criminology are investigated in the work [12]. The authors of this work investigated the correlation between motives of crime and the type of crime. In article [13], with the help of associative rules, constructed a social network is based on the database of terrorist attacks. The paper [14] is devoted to the study of the influence of the nature of people on the addiction to drug addiction through the construction of associative rules based on data from narcological clinics. The application of the method of searching for associative rules to population census data within the framework of the SPIN project is presented in [15].

A set of associative rules are formed automatically based on statistical data. After the associative rules that satisfy certain statistical constraints are obtained, is carried out the search for «interesting» relationships. To interesting, usually include non-obvious correlations between objects. At this stage, we have to face a number of difficulties. First, the selection of associative rules is carried out manually, which can create significant difficulties with a large database of transactions and a large number of identified associative rules. Secondly, we have to limit ourselves to a relatively small amount of generated transactions with a high degree of support. In this case, the information contained in the remaining associative rules is lost. Thirdly, each associative rule is considered separately, which can lead to the loss of regularities, manifested by transitivity. Thus, the problem of constructing algorithms for automatic processing of a large set of associative rules is actual.

This article proposes a method for processing a set of associative rules formed on the basis of statistical data based on the construction of a graph model, taking into account information contained in associative rules with low support, as well as interrelations between objects, manifested by transitivity.

2. Statement of the problem and methods of solution

Consider the set of associative rules A, which are formed on the basis of the set of transactions T.

Let I be the set of all answers that can be present in the transaction. Each transaction t is a collection of elements from I $(t \sqsubseteq I)$. D is the set of all transactions. It is said that transaction T contains a set of elements of X if $X \sqsubseteq t$ and $X \sqsubseteq I$. An associative rule is the implication $X \Rightarrow Y$, where $X \sqsubseteq I$, $Y \sqsubseteq I$, and $X \cap Y = \varnothing$.

Every associative rule characterized by a certain set of parameters. The first parameter, called support, shows the frequency detector of this rule in an existing set of transactions. Support rule's $X \Rightarrow Y$ is calculated as a percentage of transactions containing lots of $X \cup Y$:

$$supp(X \Rightarrow Y) = \left(\left(N \times (X \cup Y)\right)/|D|\right) \times 100\% \qquad (1)$$

where $N \times (X \cup Y)$ is the number of transactions containing the set $X \cup Y$.

The validity of the rule shows the probability with which Y follows from X. The reliability of the associative rule $X \Rightarrow Y$ is calculated as the percentage of transactions containing both X and Y in the set of transactions containing X:

$$conf(X \Rightarrow Y) = supp(X \Rightarrow Y)/supp(X) \qquad (2)$$

The task of finding associative rules is to find sets of elements whose support is not lower than minsupport. From the found sets, rules are distinguished with the reliability not less than minconfidence.

Build a graph G of correlations based on the identified Association rules. As the set of vertices of the graph, we select the set of objects encountered in the constructed associative rules. The edges will correspond to associative rules, so the graph will be oriented. As the weight of the edges we will use the validity of the corresponding associative rules.

Apply the search community's algorithm (community) [16] to this graph. To do this, you must first obtain reduce form the weights matrix $e=E/m$, where

$$m = \sum_{i,j=1}^{N} E_{ij} \qquad (3)$$

N is the number of vertices. In the reduced the weights matrix, the element e_{ij} shows the weight fraction of the given arc in the total weight of the graph. In the future, under the weights matrix, the reduced form will be understood. Easy to see, that

$$\sum_{i,j=1}^{N} e_{ij} = 1 \qquad (4)$$

To identify communities uses modularity function (modularity), showing the optimality of dividing a graph into sub-graphs:

$$Q(e) = \sum_{i=1}^{N} e_{ii} - \sum_{i=1}^{N} p_{ii} \quad (5)$$

Where p_{ii} is expected connectivity. In the canonical model [16] p_{ii} is defined through the outgoing degree vertices in a_i, and incoming degree vertices b_i:

$$p_{ii} = a_i b_i \quad (6)$$

In this case, modularity is written as:

$$Q(e) = \sum_{i=1}^{N} e_{ii} - \sum_{i=1}^{N} a_i b_i \quad (7)$$

where

$$a_i = \sum_{j=1, j \neq i}^{N} e_{ij}, \qquad b_i = \sum_{j=1, j \neq i}^{N} e_{ij}, \quad (8)$$

The modularity value shows the difference between the fraction of all edges lying within the communities for a given graph and the expected value computed for a graph in which all vertices have the same degree and the edges are randomly located. The modularity function will have a null value for a graph with a uniform random distribution of edges. Deviation from randomness leads to a positive value for the modularity. Deviation from randomness in the location of edges indicates the presence of regularities in the connection of objects.

To search for communities on graphs algorithm searching couplers is used. Select the sub-graph *G'* in the graph *G* and replace all its vertices with one vertex, while the vertices of the sub-graph *G/G'* remain unchanged. The formed vertex is connected by arcs to those vertices of the graph *G/G'* with which the vertices included in the coupler. The sum of the weights of the vertices and arcs that are included in the coupler is equal to the weight of the vertex coupler. The community will call sub-graph source graph, which during the formation of the coupler maximizes the function modularity of the graph *Q(e)*.

At the output of the algorithm we get a graph consisting of couplers. And the formation of new ties reduces the value of the modularity function. Each screed corresponds to a subgraph in the original graph. Pixels corresponding to the vertices of a single subgraph belong to the same segment. Our task is to identify communities on the graph of associative rules.

3. An example of analysis of a set of associative rules

Consider the results of applying the proposed algorithm to the analysis of the activities of public organizations of russian germans. Initial information has been extracted from the questionnaires filled the heads of organizations. As the transaction selected one questionnaire. Each questionnaire questions in different formats. The first type of questions suggested two variants of the answer «Yes» or «No». The second type of question allowed the choice of one of four or five options, and allowed the additional response «other», which was not coded, because it contained uncertainty. Also in the questionnaire there were questions with a nondeterministic answer, but they were not taken into account in the formation of transactions, as all the answers were different and could not lead to the identification of associative rules with any appreciable support. Each variant of the answer was coded by its record. The question was identified by an identifier of one or two Latin letters. For example, the question designated «A» had two response variants, coded «A1» and «A2». In questions containing a choice of two alternatives, one cannot limit oneself to coding only one of them, despite the possibility of an unambiguous recovery of the second. Such a restriction may lead to the loss of associative rules.

A total of 107 were processed questionnaires, each of which was viewed as an independent transaction. After coding the transaction were obtained with different numbers of entries from 24 to 50. On the basis of transaction data, associative rules were searched with the support of not less than 60 %, and accuracy not less than 80 %. For finding Association Rules was used algorithm APriori

IOP Conf. Series: Journal of Physics: Conf. Series **944** (2017) 012013 doi:10.1088/1742-6596/944/1/012013

[17]. The resulting set of associative rules contains the expected interactions. At the same time, it should be noted that in the formation of associative rules, there are only 9 statements from 173 possible. Between the remaining statements, associative rules with sufficiently high support and reliability are absent.

Table 1 lists the notation for claims occurring associative rules. The text of your paper should be formatted as follows:

Table 1. Notation for claims occurring in associative rules simple table.

v_1	more than 50 % of the visitors center are russian germans
v_2	the Internet is used in the work of the centre several times a day
v_3	organization uses language courses for adults
v_4	German language skills have improved over the past 10 years
v_5	organizations in Russia are partners in the work of the
v_6	Cooperation with the MCHK is carried out continuously with high degree of efficiency
v_7	In the organization know, that Germany is implementing special programs
v_8	For the study of the German language children's and youth language clubs are used

Figure 1 show a graph based on association rules with one precondition.

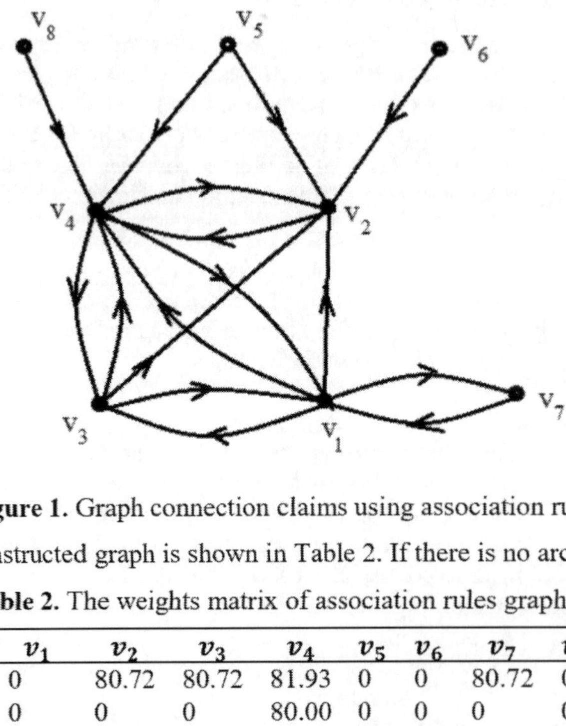

Figure 1. Graph connection claims using association rules.

The weights matrix constructed graph is shown in Table 2. If there is no arc weight is zero.

Table 2. The weights matrix of association rules graph.

	v_1	v_2	v_3	v_4	v_5	v_6	v_7	v_8
v_1	0	80.72	80.72	81.93	0	0	80.72	0
v_2	0	0	0	80.00	0	0	0	0
v_3	82.72	82.72	0	85.19	0	0	0	0
v_4	80.95	80.95	82.14	0	0	0	0	0
v_5	0	83.54	0	82.28	0	0	0	0
v_6	0	88.16	0	0	0	0	0	0
v_7	80.72	0	0	0	0	0	0	0
v_8	0	0	0	90.00	0	0	0	0

Because the original graph has a small number of vertices, the task can be solved by brute force. Function modularity source graph is equal to $Q=-0.1497$. The combining of vertices v_1, v_3, v_4 in a community leads to a value of the function modularity $Q_{1,3,4}=-0.0361$, that is such consolidation is profitable and shows the close communication of these vertices. Combining into one community of vertices v_1, v_3, v_4 and v_2 resulted in the value of modularity $Q_{1,2,3,4}=-0.0233$. Other variants of combining do not increase the modularity feature, that is are not profitable.

From this analysis we can conclude about the close association of such aspects of the activity of public organizations of Russian Germans: "more than 50% of the visitors center are Russian Germans", "the Internet is used in the work of the centre several times a day", "organizati". These four areas most closely interrelated and need to be considered together. It should be noted that the relationship identified by the proposed graph-theoretic approach does not follow directly from the formed associative rules. The corresponding set has little support.

4. Conclusion

The article proposes a new method for automatic analysis of a set of associative rules. The representation of connections between objects in the form of an oriented graph allows us to reveal regularities that do not manifest themselves in the traditional «manual» analysis of associative rules. Analysis of the relationship graph with the help of the search for vertex communities makes it possible to determine the most closely related objects of the original set. The increase of informativeness in revealing the regularity in comparison with the traditional approach is based on consideration of more associative rules and taking into account the transitivity relationship. The application of the proposed method to the analysis of the activities of public organizations made it possible to reveal new regularities that do not directly follow from one associative rule.

References

[1] Agrawal R, Imielinski T and Swami A 1993 *Proc. ACM SIGMOD International Conference on Management of Data Washington* (DC, USA) pp 207–216
[2] Shaheen M, Shahbaz M and Guergachi A 2013 *Elsevier Knowledge-Based Systems* pp 261–273
[3] Belim S V, Mayorov-Zilbernagel A O 2014 *Herald of computer and information technologies* vol 12 pp 18–23
[4] Belim S V, Mayorov-Zilbernagel A O 2014 *Science and Education of the Bauman MSTU* vol 12 pp 716–737
[5] Atepalikhin M S, Kassal B Y and Belim S V 2014 *Herald of Omsk university* vol 2(72) pp 25–29
[6] Lee I, Cai G and Lee K 2013 *46th Hawaii International Conference on System Sciences* pp 1580–1588
[7] Erlandsson F. Brodka P, Borg A and Johnson H 2016 *Entropy* vol 18 p 164
[8] Erlandsson F, Borg A, Johnson H and Bródka P 2016 *Advances in Network Science* (Springer International Publishing: Cham, Switserland) pp 126–135.
[9] Nancy P, Geetha Ramani R and Jacob S 2013 Mining of Association Patterns in Social Network Data (Face Book 100 Universities) through Data Mining Techniques and Methods *Proceedings of the Second International Conference on Advances in Computing and Information Technology (ACITY)* vol 178 (Springer: Berlin/Heidelberg, Germany) pp 107–117
[10] Selman Bozkır A, Güzin Mazman S and Ebru Akçapınar Sezer 2010 Technological Convergence and Social Networks in Information Management IMCW 2010 *Communications in Computer and Information Science* vol 96 (Springer: Berlin/Heidelberg, Germany) pp 145–153
[11] Yu X, Liu H, Shi J, Hwang J N, Wan W and Lu J 2014 *IEEE International Congress on Big Data* (Anchorage, USA) pp 310–314
[12] Pereira B L, Brandão W C 2014 *IADIS International Conference Applied Computing* pp 159–166
[13] Górecki J, Slaninová K 2013 *Acta academica karviniensia* pp 56–68

[14] Zahedi F, Zare-Mirakabad M R 2014 *Journal of AI and Data Mining* vol 2 pp 135–139
[15] Malerba D, Esposito F and Lisi F A 2001 *Proceedings of the Joint Conferences on "New Techniques and Technologies for Statistcs" and "Exchange of Technology and Know-how" (ETK-NTTS'01)* pp 541–550
[16] Newman M E J 2003 Mixing patterns in networks *Phys. Rev. E* 67 026126
[17] Agrawal R, Srikant R 1994 *Proc. of the 20th International Conference on VLDB* (Santiago, Chile) pp 487–499

Non-destructive testing method for determining the solvent diffusion coefficient in the porous materials products

V P Belyaev, S V Mishchenko and P S Belyaev

Technological Institute, Tambov State Technical University, 106 Sovetskaya St, Tambov, 392000, Russia

Abstract. Ensuring non-destructive testing of products in industry is an urgent task. Most of the modern methods for determining the diffusion coefficient in porous materials have been developed for bodies of a given configuration and size. This leads to the need for finished products destruction to make experimental samples from them. The purpose of this study is the development of a dynamic method that allows operatively determine the diffusion coefficient in finished products from porous materials without destroying them. The method is designed to investigate the solvents diffusion coefficient in building constructions from materials having a porous structure: brick, concrete and aerated concrete, gypsum, cement, gypsum or silicate solutions, gas silicate blocks, heat insulators, etc. A mathematical model of the method is constructed. The influence of the design and measuring device operating parameters on the method accuracy is studied. The application results of the developed method for structural porous products are presented.

1. Introduction

Most studies of the diffusion in the products made from porous building materials is focused on the moisture diffusion coefficient determination. In the applied measurement devices unidirectional isothermal mass transfer is usually used in specially prepared samples for the stationary or non-stationary measurement processes [1 - 3]. The methods are characterized by substantial costs and time for experiment preparation. Applied instruments are also cumbersome and have significant cost. For the determination of spatial moisture distributions inside the specimens, made from building materials, traditional weight methods with the cutting of the test specimen into pieces are often used. Now there are modern methods of nondestructive testing: nuclear magnetic resonance [4], game scope [2], neutron radiography, computed tomography and x-rays [3]. However, the methods are not able to carry out non-destructive testing of finished products as it is necessary to have samples of a given configuration, and the measurement process is characterized by a significant duration, which does not allow to obtain a rapid information of the products quality. A significant increase in efficiency control may be obtained through the use of dynamic and pulse methods. However, the known pulse techniques (methods instantaneous source mass) are also associated with the need to manufacture samples [5].

2. Formulation of the problem

The aim is to develop a new impulse method for measuring the diffusion coefficients of solvents distributed in the solid phase of porous materials (PM). The method should ensure the possibility to investigate finished products without destroying them and significantly reduce the time spent on monitoring. To achieve the goal it is necessary to solve the following tasks:

- develop a physical and mathematical model of the method that provides a fast flowing mass transfer process in the controlled product and get without the product destruction the necessary measurement;
- to develop and investigate a mathematical model of errors in measuring the diffusion coefficient;
- to investigate the influence of the measuring device design parameters and the regime variables for the accuracy measurement process of determining the diffusion coefficient, give recommendations on rational parameters of measuring devices;
- to test experimentally the working capacity of the developed method on products for construction purposes.

3. Theory

The research results are based on the analytical theory of heat and mass transfer, mathematical physics, mathematical modeling, the classical metrology theory and mathematical statistics. The main idea of the developed method is to use the impulse action on a PM product in the form of a solvent dose. Then, the response to this effect is analyzed as a the solvent concentration change in the PM solid phase. From the point of the mathematical mass transfer process description, the sample is viewed as an unbounded body of the canonical form - the ball. For this purpose, the checked sample must have a flat surface of specified dimensions. This allows one-sided access to the product and replacement the measuring device on this part of the surface. In such case non-destructive control of the diffusion coefficient of solvents is possible. Figure 1 represents the physical model of the method.

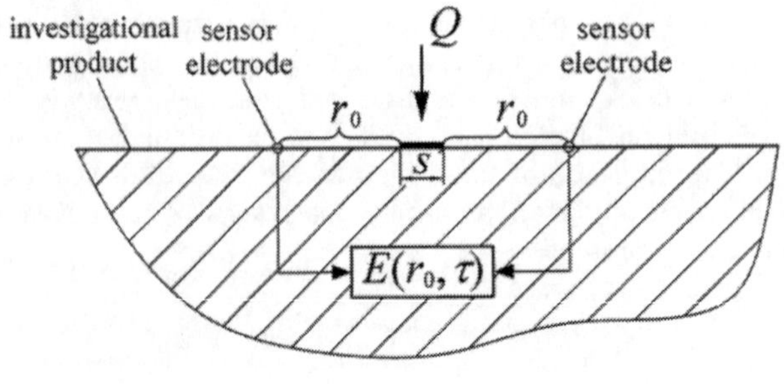

Figure 1. Physical model of the method.

When implementing the method, product is used with a uniform initial distribution of the solvent. At the beginning of the experiment, a pulse solvent dose Q is applied within a «spot» of diameter s on the flat sample surface. After the pulse is applied, the flat product surface is waterproofed. With the help of electrodes, located on a concentric circle with radius r_0 in respect to the point of impulse action on the sample, the electromotive force (EMF) change $E(r_0,\tau)$ of the electrochemical converter (ECC) is measured in time[6]. The process of spreading a solvent in a solid PM product is described by the following boundary-value mass transfer problem in an unbounded medium:

$$\frac{\partial C(r,\tau)}{\partial \tau} = D\frac{1}{r^2}\frac{\partial}{\partial r}\left[r^2\frac{\partial C(r,\tau)}{\partial r}\right] + \frac{Q}{\rho_0}\delta(r,\tau),$$

$$\tau > 0,\ 0 \le r < \infty,\ C(r,0) = C_0;\ \frac{\partial C(0,\tau)}{\partial r} = 0;\ C(\infty,\tau) = C_0,$$

where $C(r_0,\tau)$ - the solvent concentration on the sphere surface of radius r_0 with respect to the solvent dose impulse delivery point at time τ; D - the solvent diffusion coefficient; $\delta(r,\tau)$ - Dirac delta function; ρ_0 - the density of the absolutely dry test material; Q - the amount of liquid phase supplied from the dispenser to the flat sample surface; C_0 - initial concentration of solvent in the material at the time $\tau = 0$.
The diffusion coefficient can be found from the formula [7]:

$$D = r_0^2/(6\tau_{max}), \tag{1}$$

where τ_{max} is the time corresponding to the maximum on the $C(r_0,\tau)$ curve of the concentration change at a distance r_0 from the source.

Due to the monotonicity of static ECC characteristic [8], the moments of reaching the maximum concentration $C(r_0,\tau)$ and the converter maximum EMF $E(r_0, \tau)$ is coincide, it is possible to realize the measurement method without preliminary calibration of the ECC. The experience with the application of electrochemical converters [8, 9] shows that it is expedient to use the range of the static characteristic near the value $0.5C_e$, characterized by high sensitivity and stable signal. Here C_e is the diffusing substance concentration in the PM solid phase at equilibrium with saturated solvent vapors at a given temperature. Therefore, in carrying out the experiment it is expedient to obtain the value C_{max} in the vicinity of the given value:

$$C_{max} \approx 0.5C_e \tag{2}$$

The values of the ECC location coordinate r_0 and the solvent dose Q supplied at the beginning of the experiment active part exert a significant influence on the C_{max} value. In this case, the solvent concentration change in the zone of source action is described by the function:

$$C(r,\tau) - C_0 = Q/\left(8\rho_0(\pi D\tau)^{3/2}\exp\left[r^2/4D\tau\right]\right). \tag{3}$$

From (3), taking (1) into account, assuming for simplicity $C_0 = 0$, we can obtain the dependence of the attained maximum C_{max} at $\tau = \tau_{max}$ from the solvent dose Q:

$$C_{max}(r_0,\tau_{max}) = Q/\left(8\rho_0(\pi/6)^{3/2}\left(\sqrt{e}\right)^3 r_0^3\right). \tag{4}$$

We will carry out a study of the proposed method as the design measuring device parameters and the technological conditions of the measurement process affect on the diffusion coefficient accuracy. The main design parameters are: the minimum permissible flat section dimensions and the product thickness in the area of action of the pulsed source; allowable «spot» size s of substance dosing; the coordinate r_0 of the ECC electrodes location relative to the location of the pulse application. The main technological variables are the parameters of the applied mass pulse: the duration t and the power Q. The basis for finding the design measuring device parameters is the determination of coordinate r_0 for monitoring the EMF change after pulse application.

The mathematical model of the relative error root-mean-square estimate in determining the sought diffusion coefficient has the form [8, 10]:

$$\delta D = \sqrt{4(\delta r_0)^2 + (\delta\tau_{max})^2 + (\delta_m)^2}, \tag{5}$$

where $\delta r_0 = \Delta r_0 / r_0$ and $\delta\tau_{max} = \Delta\tau_{max}/\tau_{max}$ are, respectively, the relative error in determining the coordinate of the calculated cross-section and the time to reach the maximum on the curve of the concentration change in the calculated section; δ_m is the total value of the methodical error due to the incomplete correspondence of the mathematical model used to real physical processes.

It was shown in [8] that when ECC electrodes are located on the equipotential samples surfaces with respect to a pulsed source of solvent dose, a value $\pm a/r_0$ can be used as an estimate δr_0, where a is the ECC electrodes radius. The value $\delta\tau_{max}$ essentially depends on the nature of the change in the concentration $C(r_0,\tau)$ and the error in its determination. In figure 2, 3 it is shown as an example, the calculated dependences in the relative C_{max}/C_e concentrations change and the error $\delta C_{max} = \Delta C/C_{max}$ on r_0 obtained in the study of the diffusion coefficient of ethanol in a dry state gas silicate plate (530 kg/m^3) for different dose values of the solvent pulse Q. The dependences of figure 3 are obtained on the basis of (4), assuming the ΔC constancy with the relative error δE (2) of measuring the EMF (δE = 0.5% in the chosen rational ethanol concentration range). Studies show that for a given material, for example, with an impulse of 8.0×10^{-6} kg, the measuring device must be equipped with a sensor located at a distance $r_0 \leq 2.5$ mm.

IOP Conf. Series: Journal of Physics: Conf. Series **944** (2017) 012014 doi:10.1088/1742-6596/944/1/012014

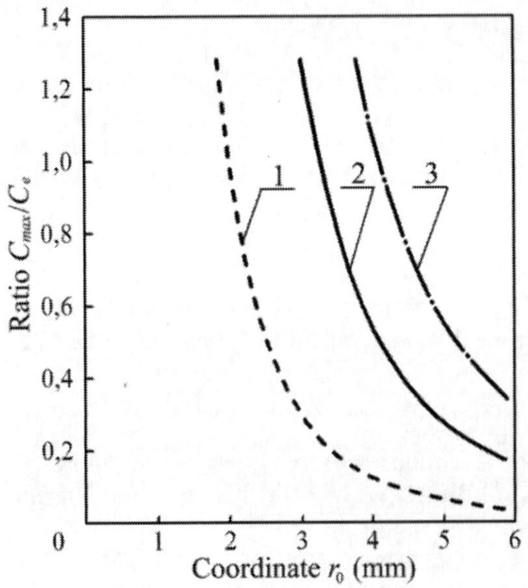

Figure 2. The influence of the sensor location on the maximum concentrations reached at different source powers Q (kg):
1 – 8.0 × 10^{-6}; 2 – 3.5 × 10^{-5}; 3 – 7.0 × 10^{-5}.

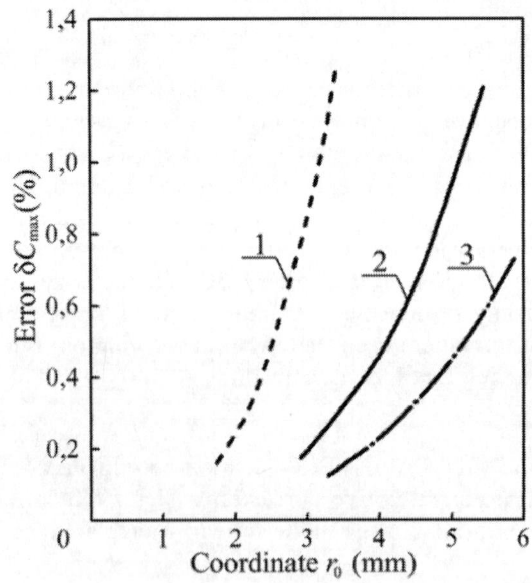

Figure 3. The influence of the sensor location on the error in their measurements at different source powers Q (kg):
1 – 8.0 × 10^{-6}; 2 – 3.5 × 10^{-5}; 3 – 7.0 × 10^{-5}.

At r_0> 3 mm, the maxima on the change curves $C(r_0, \tau)$ and, respectively $E(r_0, \tau)$, will be inadequate. This will lead to a significant increase in the relative error δC_{max} and δE_{max}, respectively, the concentration and EMF of the ECC and, consequently, the relative error in determining the time to reach the maximum on the curve of the concentration change in the calculated section. At a pulse of 7.0 × 10^{-5} kg, the sensors information with coordinates r_0<4 mm does not allow to calculate the diffusion coefficient at all. The value of C_{max} exceeds C_e, and the ECC static characteristic is on the saturation plateau [8].

Figure 4 presents the results of numerical experiments when we determine the dependence $\delta\tau_{max}$ of the ethanol diffusion coefficient in a gas silicate plate on r_0 obtained in the study for the same values of solvent doses. In the same place, the error curve is shown for the ECC electrode diameters to 0.2 mm.

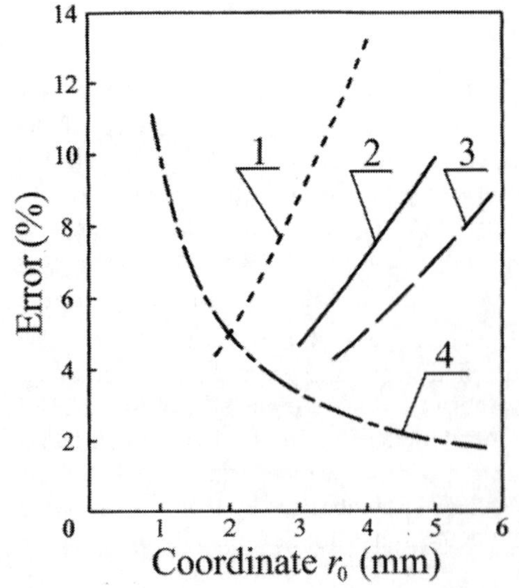

Figure 4. The influence of the sensor position on the error δr_0 (4) and $\delta\tau_{max}$ at different source powers Q (kg):
1 – 8.0 × 10^{-6}; 2 – 3.5 × 10^{-5}; 3 – 7.0 × 10^{-5}.

The analysis of the curves (figures 2, 4) shows that if the condition (2) is satisfied when setting the experiment with different source power, then the error $\delta\tau_{max}$ has approximately the same values. At the same time, on the graphs of the dependence $\delta\tau_{max}$ on r_0 (figure 4), it is necessary to take the sections, corresponding to the range of the best ECC sensitivity near the value 0.5 C_e (figure 2). Therefore, the distance choice r_0, according (5) and δ_m - const, is achieved by a compromise solution based on the error δr_0 and duration of the experiment. As the analysis result (figure 4), it was decided to place the main electrodes of the electrochemical converter at a distance r_0= 4 mm from the point of impulse action and additional electrodes at a distance of 3 and 5 mm. Removing the convertor from the source more than 5 mm causes an increase in the duration of the experiment, in the dimensions of the products flat sections and in the overall dimensions of the measuring device with a slight increase in the control accuracy. Using (4), and taking into account the calculated constants, it is possible to calculate in advance the required impulse action solvent dose Q which provides the fulfillment of condition (2):

$$Q \approx 3.41\rho_0 C_e r_0^3 \, . \tag{6}$$

The rationale for using the remaining design parameters and regime variables of the diffusion coefficient measurement process was obtained by analyzing the methodical error δ_m. They are chosen from the condition of reaching negligibly small values δ_m in comparison with the remaining resulting error δD components according to [8, 11].The minimum dimensions d_∞ of the flat section diameter for controlled massive PM product and a measuring probe are obtained from the condition: d_∞ >20 r_0. The product thickness h in the area of action of the source must be at least 10 r_0. Dimensions of the "spot" diameter s, within which the solvent dose is applied to the test material surface, should not exceed 0.05 × r_0. The pulse duration t of the mass must not exceed 0.05 × τ_{max}. Based on the maximum location value (5 mm) of the ECC electrodes we determined the minimum permissible diameter of the products flat piece (100 mm) on which the measuring device is placed. The minimum sizes of controlled massive products in the area of the point source action are also determined. The dimensions of the products must exceed the hemisphere dimensions of a radius 50 mm centered at the point of pulse application. This allowed us to take the upper estimate of 5% for the total methodical error δ_m.Taking this into account, the root-mean-square estimate of the relative error δD in determining the unknown diffusion coefficient according to (5) and taking into account the data in figure 2 for distances 3, 4 and 5 mm up to a point source, respectively, is 12; 11 and 10% (for δE = 0.5% and the condition for using the optimal doses, calculated by equation (6) for each r_0).

4. Experimental results

To carry out experimental studies, we used the information-measuring system developed by the "National Instruments" technology [12] with a modernized measuring device [8]. Before carrying out the active experiment, the required value of the pulse mass was preliminarily determined by formula (6). In figure 5, as an example, the results of measuring the diffusion coefficient of ethanol in gas-silicate dry slabs 50 mm thick with a 530 kg / m^3 density are presented which obtained using the proposed method. The total number of studies i for each PM-solvent system was at least 20 runs, the ethanol pulse value for this example was 3.2×10^{-5} kg, r_0 = 4 mm. Experimental studies have shown that the mathematical expectation of the ethanol diffusion coefficient in the tested gas-silicate plate with a confidence probability $\alpha = 0.95$ is 5.4×10^{-9} m^2/s. The random error of the diffusion coefficient determining is less than 9% the duration of the experiment does not exceed 12 minutes.

5. Discussion of the results

This method allows us to measure quickly the diffusion coefficient of polar solvents in porous materials products without destroying them. Products must have at least one flat surface with a diameter more than 100 mm. A significant increase in control performance is ensured by:

- excluding the need of making samples for research;
- using more dynamic mass transfer process under impulse action;

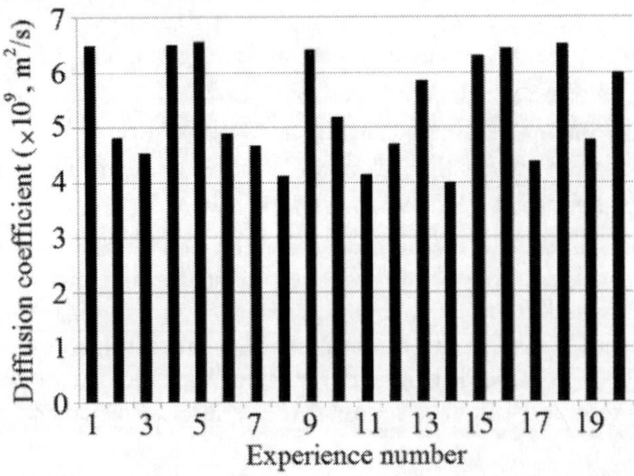

Figure 5. Histogram of the ethanol diffusion coefficient values in a gas silicate plate.

- elimination the need for preliminary calibration of the electrochemical converter used for each new material and diffusing solvent (tens of hours).

The efficiency of this method was checked by comparing the obtained data with the results of determining the diffusion coefficients of moisture and ethanol in samples from massive porous materials by the zonal method [13] and the sorption method [14]. The discrepancy in the numerical values of the coefficients did not exceed 8 ... 10%.

6. Conclusion

The dynamic method has been developed that makes it possible to significantly improve the efficiency of controlling the polar solvents diffusion coefficients in finished articles from porous materials without destroying them. The method can be used for the operational diffusion coefficient control in portable devices for finished construction products made from porous materials. As a studies result, the minimum permissible dimensions of the product flat surface on which a measuring portable device can be placed are determined. The rational value of the electrochemical converter electrodes location coordinate relative to the point of impulse action application is determined from the permissible accuracy and duration of the measurement process. The equation is obtained for calculating the required solvent dose for impulse action. The statistical processing the obtained experimental data and its comparison with the known results indicate the operability of the proposed nondestructive testing method.

Acknowledgments
Applied scientific research was carried out with the financial support of the Russian Federation Education and Science Ministry (the unique identifier of the project RFMEFI57716X0214).

References

[1] Remond R and Almeida G 2011 *Wood Mater. Sci. Eng.* **1** 23
[2] Nizovtsev M I, Stankus S V, Sterlyagov A N, Terekhov V I and Khairulin R A 2008 *Int. J. Heat Mass Transf.* **51** 4161
[3] Roels S and Carmeliet J 2006 *Int. J. Heat Mass Transf.* **49** 4762
[4] Koptyug I V and Sagdeev R Z 2002 *Russian Chem. Rev.* **71** 789
[5] García-Gutiérrez M J L C, Cormenzana J L, Missana T, Mingarro M and Molinero J 2006 *J. Iberian Geol.* **32** 37
[6] Mishchenko S V, Belyaev P S, Gladkikh V A, and Safronova E N 1999 *Drying Technol..* **17** 2151
[7] Carslaw H S and Jaeger J C 1959 *Conduction of Heat in Solids* (Oxford: Clarendon Press) p 487

[8] Belyaev V P, Belyaev M P, Mishchenko S V and Belyaev P S 2014 *Meas. Tech.* **56** 1190
[9] Belyaev V P, Mishchenko S V and Belyaev P S 2015 *Meas. Tech.* **58** 574
[10] Gurov A V, Sosedov G A, Rodina A E and Ponomarev S V 2013 *Meas. Tech.* **55** 1187
[11] Shashkov A G, Volokhov G M, Abramenko T N and Kozlov V P 1973 *Methods to determine thermal conductivity and thermal diffusivity [in Russian]* (Moscow: Énergiya) p 336
[12] Belyaev M P, Belyaev V P and Divin A G 2010 *Vestn. Tambovsk. Gos. Tekh. Univ.* **16** 797
[13] Kartashov É M and Rudobashta S P 2010 *Diffusion in chemical-technology processes [in Russian]* (Moscow: KolosS) p 478
[14] Scheffler G A and Plagge R 2011 *Proc. 5th Int. Autoclaved Aerated Concrete Conf.* September 2011, Bydgoszcz (Poland) 337

IOP Conf. Series: Journal of Physics: Conf. Series **944** (2017) 012015 doi:10.1088/1742-6596/944/1/012015

Pulsation-based method for reduction of nitrogen oxides content in torch combustion products

I A Berg[1], S V Porshnev[1], V Y Oshchepkova[1], and M Kit[2]

[1]Ural Federal University, Ekaterinburg, Russia

[2]Language Interface Inc., Seattle, USA

e-mail: berg77777@gmail.com

Abstract. Out of all ways to fuel burn the torch combustion systems is used most often. Even though the processes in the steam boiler are stochastic, the system can be controlled rather easily by changing the flowrate of the air pumped into it and – in case of balanced flue units – exhausters load. Advantages offered by torch-based combustion systems are offset by a disadvantage resulted in oxidation of nitrogen contained in the air. This paper provides rationale for an NOx content reduction method that employs pulsation mode of fuel combustion; it also describes combustion control and monitoring system employed for implementation of this method. Described methodology can be used not only for pulsation combustion studies but also for studies of torches formed by conventional burning systems. The outcome of the experimental study supports the assumption that it is possible to create conditions for NOx content reduction in flue gases by means of cycling the fuel supply on/off valve at the rate of 6 Hz.

1. Introduction

Fuel combustion processes have been thorough studied and are widely used in industry. Out of all ways to fuel burn the torch combustion method used most often. In this method, fuel and oxidizer (usually air) are supplied to the burner to be mixed and produce a burning torch on the burner outlet. The degree of air-fuel mixing directly affects the ratio of burned fuel to the total amount of fuel supplied to the burner and this degree depends on the burner design. The phenomenon of incomplete combustion of fuel supplied to the burner is called chemical underburning of fuel. Fuels conventionally used for combustion in torch include pulverized coal, fuel oil or gas.

The main purpose of the burner is to achieve high degree of air-fuel mixing in order to ensure maximum rate of fuel combustion as compared with fuel-bed or fluidized bed furnaces. This translates into higher temperatures in the torch core and high thermal stresses in the furnace. Since the combustion process comes to completion at a short zone (the torch length), combustion products produced by all burners further mix with each other forming a hot homogeneous mixture that comes into the heat-exchanging units of the plant. Even though the processes in the boiler are stochastic, the system can be controlled rather easily by changing the flowrate of the air pumped into it and – in case of balanced flue units – exhausters load. A typical automatic torch-combustion control system includes a negative feedback loop. It is worth noting that the load control is performed for the overall unit rather than individual burners, which makes the control system considerably simpler. The load on each burner changes in proportion to the total load.

Advantages offered by torch-based combustion systems are offset by a disadvantage resulted from the fact that a great amount of heat is released in a short path where combustion takes place and that results in oxidation of nitrogen contained in the air. In effect, a number of environmentally noxious substances belonging to nitrogen oxide group (NOx) are released to the atmosphere.

This is why it is so important to reduce NOx content in combustion products generated in industrial and power-generation units. To solve this problem designers usually create heterogeneous lower-temperature zones where combustion process slows down [1]. However, this approach suffers from incomplete burning of fuel (i.e. chemical underburning) and the underburned fuel is carried out from the boiler along with combustion products, which does not help maintain clean environmental conditions near the heat plant. Therefore, for improvement of environmental safety of fuel combustion processes heat and power plants (HPP) call for fundamentally new control systems whose operation would be based not only on data on the overall combustion process outcome, but also on processes that take place in the combustion zone.

This paper provides rationale for an NOx content reduction method that employs pulsation mode of fuel combustion; it also describes combustion control and monitoring system employed for implementation of this method.

2. Rationale for the method for nox content reduction through the introduction of pulsation mode of fuel combustion

A critical parameter to take into account when burning fuels in power generation units, industrial furnaces, etc., is the excess air factor α, which is the ratio between actual volume of air supplied to the furnace and the minimum amount of air necessary to completely burn the fuel fed to the furnace

$$\alpha = \frac{V^{в}}{V^{0}}, \tag{1}$$

where $V^{в}$ – actual volume of the air supplied to the furnace, V^{0} – analytically found volume of air needed to completely burn the supplied fuel. This parameter affects both the combustion process and cost-efficiency of the unit. The volume of the needed air can be found from the content and flowrate of the air fed into the furnace. The actual volume of air fed to the furnace in practice is found from air flowrate per boiler, e.g. through the readings of orifice meters.

When using the torch method for fuel combustion efforts are made to ensure optimal value of αopt, that is calculated based on the overall fuel supply and air supply flowrates per boiler. In addition, the following basic technical and cost parameters should be taken into account: losses of heat carried away with residual gases q2 and heat losses due to chemical underburning q3. These parameters are used in the analysis of the power unit thermal balance. Since all flows are distributed nearly evenly between the burners, this assumed value of α is optimal for each burner.

The loss of heat carried out with residual gases q2 takes place because enthalpy of the combustion products leaving the unit is greater than that of the air and fuel coming into the boiler, so that part of the produced heat is released into the atmosphere. This parameter depends on the volume of combustion products and, therefore, on the excess air factor α. The greater factor α is, the more air is fed to the furnace and more combustion products are created. Supplying air in the volumes greater that minimum needed to burn the fuel causes some heat to be spent for heating the air that is not involved in the combustion process and releasing that heat into the atmosphere along with flue gases.

Chemical underburning takes place when the excess air factor falls below 1 since in that case there amount of air fed to the boiler is insufficient to completely burn the fuel. The loss due to underburning is the amount of heat q3 is the amount of heat that would be generated at complete oxidation of the chemical underburning products.

To cope with this, the optimal α is thought to have a value that excludes chemical underburning and, at the same time, minimizes heat losses caused by carryover with residual gases. This value is calculated in the course of boiler design and then it needs to be adjusted at the start-up testing phase because it depends on fuel type and content, furnace or combustion chamber geometry, the unit operating loads range, etc. As a rule, modern gas-fired steam boilers of high capacity are designed to operate at the excess air factor of $\alpha = 1,05 \div 1,1$. When using solid fossil fuels the factor is assumed to be $\alpha \approx 1,2$ in order to allow for more complex mixing and ignition processes resulting in poorer interaction between the air and fuel particles.

As noted above, proper operation of torch systems causes extensive formation of nitrogen oxides that can be broken down into fuel, prompt and thermal NOx due to different nature of their formation. Fuel oxides are made of nitrogen contained in the fuel and they usually emerge when burning solid or liquid fuels. The mechanism of formation of prompt NOx was suggested by Fenimore [2]: CHm radicals formed as interim components only in the front of the flame react with the air nitrogen, causing the formation of nitrogen cyanide that undergoes further oxidation to NO [3]. Prompt NOx in the front of the flame are formed during the entire combustion process and presently it is practically impossible to reduce their content. Thermal nitrogen oxides emerge because the heat energy generated in the combustion chamber when burning fuel is sufficient for the nitrogen contained in the air to undergo direct endothermal oxidation reaction [4]

$$\tfrac{1}{2}N_2 + \tfrac{1}{2}O_2 + 21,5\,{}^{ккал}\!/_{моль} = NO. \qquad (2)$$

Most nitrogen oxides are formed through this thermal mechanism and the NOx formation rate dependency from the air excess factor is known. Typical curves showing dependency between NOx content in combustion products and factor α ($C_{NO_x} = f(\alpha)$) for two different boilers operating on gas fuel are shown on Figure 1 [5] (data was obtained from tests conducted in the course of start-up operations).

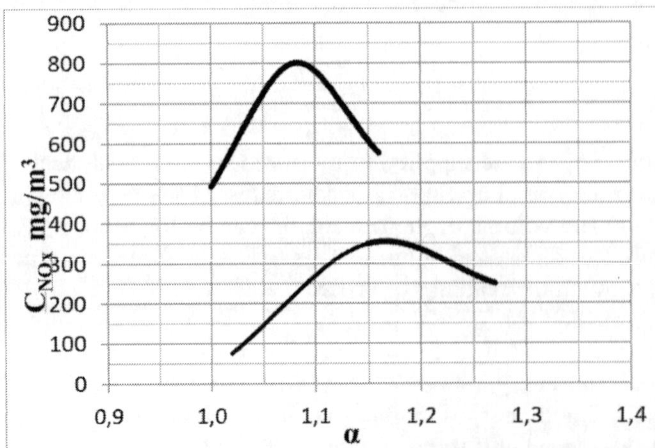

Figure 1. Dependency between NO_x content in combustion products and factor α , data was obtained from tests (upper curve – of TGM-94 steam boiler of Tashkent HPP, lower curve – of BKZ-320-140 of Engelsk HPP).

Figure 1 shows that the graphs are similar: αopt = agr max(f (α)). Where $\alpha < \alpha_{opt}$ (overrich fuel mixture and insufficient oxidizer) or $\alpha > \alpha_{opt}$ (weak fuel mixture), NOx formation is considerably slower. Therefore, NOx content can be reduced if, instead of burning fuel in a single stream at the optimal air access factor αopt (from the heat losses standpoint), it will be burned in several streams αi ≠ αopt, $i = \overline{1,I}$ is the stream number, so that $\bar{\alpha} = \frac{1}{I}\sum_{i=1}^{I}\alpha_i = \alpha_{opt}$, i.e. the overall air excess factor at the furnace inlet stays optimal. To implement this requirement the air or fuel can be redistributed between tiers of burners. In this solution, the boiler furnace is apparently broken into zones where fuel first is burned at the excess air volume much lower than that required and then the fuel undergoes additional oxidation with air streams supplied through the burners of the top tier. However, this configuration is difficult to control since regulation of the thermal load on the unit requires separately change fuel or oxidizer flowrates for each burner. It also requires agitation of combustion products in each burners by the time they are leaving the chamber in order to prevent chemical underburning and abrupt increase of the temperature in the downstream heat exchangers that would take place due to afterburning of combustible gases.

We suggest that the shortcomings described above can be removed if, instead of stationary combustion process, combustion will change in time through changes in the excess air factor $\alpha(t)$, where $\alpha(t)$ is a time-dependent function, e.g. a periodic one. To control air flowrate a regulation device should be installed in the fuel supply line connected to the burner. That device will change fuel flowrate periodically while the air supply flowrate will be kept unchanged. Under these conditions the excess air factor will change following a similar periodic pattern. This will make the torch consisting of alternating zones where $\alpha < \alpha_{opt}$ and $\alpha > \alpha_{opt}$ while the average excess air factor per cycle $\bar{\alpha}$ will be equal to the optimal factor for this unit α_{opt}. In the course of longitudinal heat and mass exchange in the torch and combustion products in the furnace these zones will mix with each other and the interim combustion products will undergo further oxidation thus ensuring complete extraction of heat and reduced formation of NOx.

Estimates made by the authors show that the optimal pattern $\alpha(t)$ for excess air factor changes is a sequence of square pulses described as

$$\alpha(t) = \begin{cases} \alpha_1, t - T_1 \cdot \mathrm{round}(t / T_1) < T_1, \\ \alpha_2\ t - T_2 \cdot \mathrm{round}(t / T_2) < T_2, \end{cases} \tag{3}$$

where $\alpha_1 < \alpha_{opt}$, $\alpha_2 > \alpha_{opt}$, $T_1 = T_2 = T$ (pulsation period) where NOx content can be reduced by 65%. This mode will be called pulsation mode hereafter as suggested in (3). As shown in (3), NOx content in combustion products depends on parameters α_1, α_2, T. Right choice of these parameters will ensure creation of torch zones with different excess air factors, as well as achieve such a gas-dynamic mode in the furnace that the zones will have enough time to mix with each other within the furnace and the underburned products will be oxidized completely.

Since the known theoretical combustion models depend on a number of parameters which values are either unknown or greatly differ in estimates made by different researchers, it was not feasible to obtain reliable values for α_1, α_2, T. This necessitated targeted experimental studies of pulsation combustion mode. The outcome of these studies is discussed below.

3. Methodology developed for the experimental studies of pulsation combustion

The experimental study was conducted using an automated unit for pulsation combustion studies specifically developed for that purpose. [6]. The unit includes a direct-flow jet burner with the outlet diameter of 20 mm that creates a turbulent diffusion torch without premixing.

This torch is similar to that conventional torch combustion system create (Figure 2). Two vales are installed into the gaseous fuel supply line. The first valve regulates the total gas pressure and flowrate, the second operates in on/off mode creating pulsations of fuel flowrate in order to arrange pulsation combustion.

Figure 2. Photograph (taken in the visible spectrum) of the work zone with the burner of the automated measuring unit for pulsation combustion studies.

The experiments were run following the procedure shown below:

1. Burner is ignited and upstream valve adjusted to set gas pressure and, therefore, the average flowrate of gas supplied to the burner.
2. Combustion pulsation mode maintained through opening/closing the on/off valve that changes the excess air in the burner at a preset rate. When studying steady combustion process without pulsations this on/off valve stays open.
3. Sequences of torch images in IR range recorded with thermal imager FLIR 7700M with the resolution of 320 x 240 pixels for the period of 10 seconds at the frame rate of 412 Hz. Each sequence saved as an matrix T(i,j,k) of images T(i,j) with the matrix size of 320×240×4120. Figure 2 shows an example of a single frame T(i,j).
4. Time series compiled based on the obtained frame sequences. These time series (TS) represent instantaneous sizes of the areas where ignition begins $N_{low}(k)$ and the torch core $N_{core}(k)$, $k = \overline{1,4120}$.
5. Power spectral density (PSD) of the time series $N_{low}(k)$, $N_{core}(k)$ normalized to the maximum value found through Fast Fourier Transform (FFT).
6. Primary periodic components of the time series $N_{low}(k), N_{core}(k)$ found through the singular spectral analysis (7) and their PSD computed.

It should be noted that this methodology can be used not only for pulsation combustion studies but also for studies of torches formed by conventional burning systems, e.g. when precommissioning and tuning existing units.

4. The experiments outcome and discussion

We now will discuss normalized power spectral densities of time series $N_{low}(k), N_{core}(k)$ shown on Figures 3 through 5.Other paragraphs are indented (BodytextIndented style).

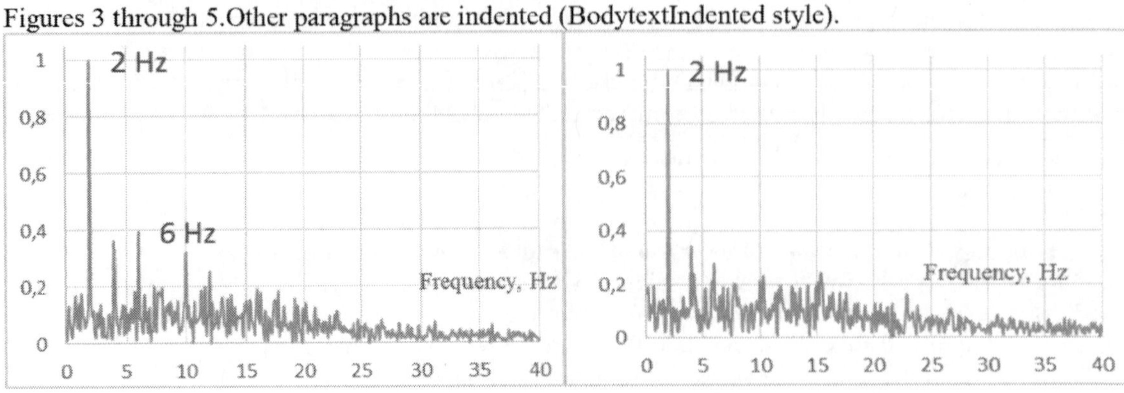

Figure 3. Normalized PSD for time series $N_{low}(k)$ (below) and $N_{core}(k)$ (above). Pulsation combustion mode with fuel supply cycling rate of 2 Hz.

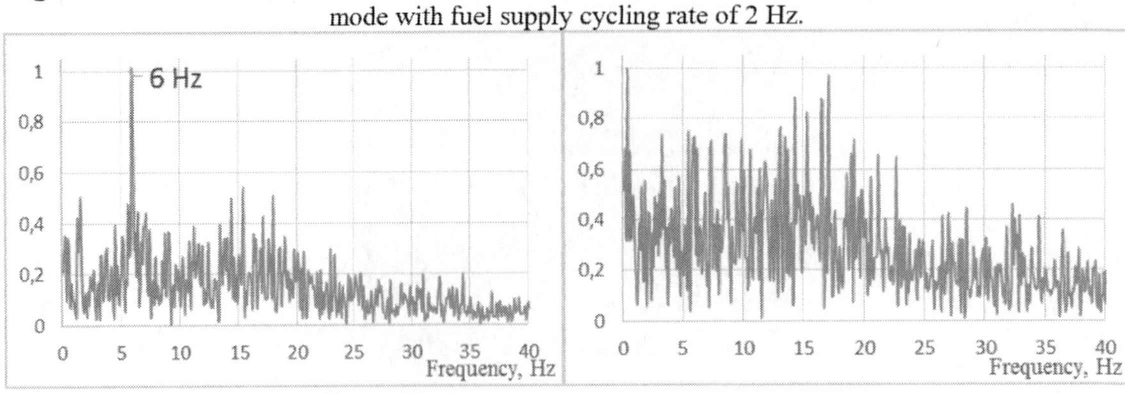

Figure 4. Normalized PSD for time series $N_{low}(k)$ (below) and $N_{core}(k)$ (above). Pulsation combustion mode with fuel supply cycling rate of 6 Hz.

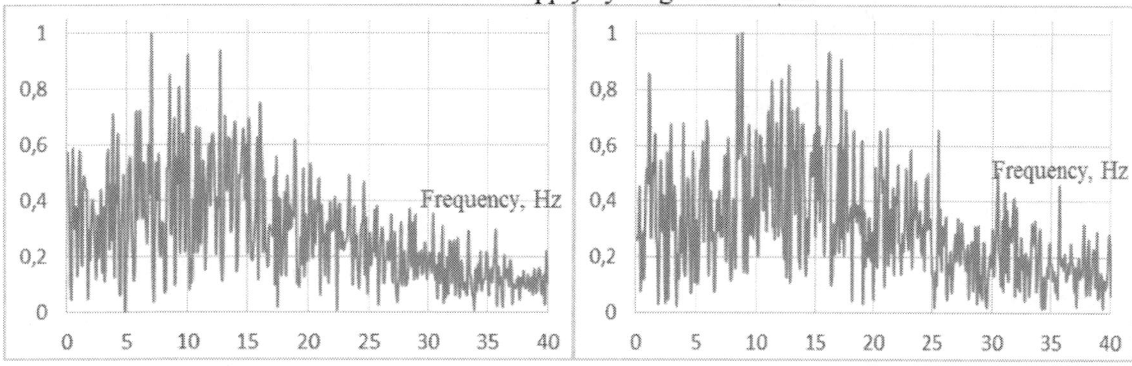

Figure 5. Normalized PSD for time series $N_{low}(k)$ (below) and $N_{core}(k)$ (above). Pulsation combustion mode with fuel supply cycling rate of 10 Hz.

Figure 3 shows that the maximal energy belongs to the PSD spectral component with the frequency equal to that of the fuel supply on/off valve cycling rate. On comparison of the absolute energies of these spectral components for on/off valve cycling rate of 1 Hz, 2 Hz, 3 Hz and 4 Hz, the conclusion was drawn that increase of the cycling rate causes the energy of these spectral component to decrease while the energy of other spectral components increases. This findings are supported by the results shown on Figures 4 and 5 demonstrating that maximum energy falls into spectral components of the time series $N_{low}(k)$, with frequencies not divisible to the on/off valve cycling rate. One can see on Figure 5 that at the cycling rate of 10 Hz the PSD components of both time series $N_{low}(k)$, and $N_{core}(k)$ being discussed are similar to PSD components of time series obtained at steady supply of fuel. These combustion modes can be viewed as quasi-continuous processes where, as well as in the steady mode, no heterogeneous zones exist in the torch and, subsequently, reduction of NOx content by means of pulsation combustion is not feasible.

It was also found that at the cycling rate of 3 Hz and below the torch is unstable while the quasi-continuous torch exhibit steady burning. Therefore, even though cycling at the rates of 3 Hz and below creates conditions for NOx reduction in the combustion products, it is not recommended to use this mode in practice due to unstable combustion.

Yet, it is feasible to create such conditions that the fuel burns quasi-continuously at the root of the torch while maintaining pulsation mode in the torch core where the primary volume of fuel is burned (Figure 4). This ensures necessary conditions for reduction of NOx content in flue gases.

5. Conclusion

The outcome of the experimental study supports the assumption that it is possible to create conditions for NO_x content reduction in flue gases by means of cycling the fuel supply on/off valve at the rate of 6 Hz. It was found that in this case the fuel supplied to the burner burns quasi-continuously in the torch root while in the torch core, where the main volume of the fuel is burned, the pulsation combustion process is maintained.

References

[1] Ganapathy V. 2003 *Industrial Boilers and Heat Recovery Steam Generators: Design, Applications, and Calculations* (Taylor & Francis) p 648
[2] Fenimore C P 1971 Formation of Nitric Oxide in Premixed Hydrocarbon Flames *13th Symp. (Int'l.) on Combustion The Combustion Institute* p 373
[3] Warnatz J, Maas U and Dibble R. W 2006 *Combustion. Physical and chemical fundamentals, modeling and simulation, experiments, pollutant formation* (Springer)

[4] Zeldovich Y B, Branblatt G, Librovich V B and Makhviladze G M 1985 *The mathematical theory of combustion and explosions* (Consultants Bureau)
[5] Shulman V L 1989 Application of the industry's technological methods of reducing of nitrogen oxides emissions Guidelines (Sverdlovsk: Uraltechenergo)
[6] Berg I A, Khudyakov P Yu and Oschepkova V Yu 2016 Automation of analytic complex for the investigation of pulsating combustion *Fundamental research* vol 6(1) 2016 pp 24–28
[7] Porshnev S V and Rabaia F 2016 *The study of features of using the singular spectrum analysis method for the analysis and forecasting the time series* (Ulyanovsk: Zebra)

IOP Conf. Series: Journal of Physics: Conf. Series **944** (2017) 012016 doi:10.1088/1742-6596/944/1/012016

Mode and climatic factors effect on energy losses in transient heat modes of transmission lines

A Ya Bigun[1], O A Sidorov[2], D S Osipov[1], S S Girshin[1], V N Goryunov[1] and E V Petrova[1]

[1]Omsk State Technical University, 11, Mira ave., Omsk, 644050, Russia
[2]Omsk State Transport University, Omsk, Russia

e-mail: barsbigun@list.ru

Abstract. Electrical energy losses increase in modern grids. The losses are connected with an increase in consumption. Existing models of electric power losses estimation considering climatic factors do not allow estimating the cable temperature in real time. Considering weather and mode factors in real time allows to meet effectively and safely the consumer's needs to minimize energy losses during transmission, to use electric power equipment effectively. These factors increase an interest in the evaluation of the dynamic thermal mode of overhead transmission lines conductors. The article discusses an approximate analytic solution of the heat balance equation in the transient operation mode of overhead lines based on the least squares method. The accuracy of the results obtained is comparable with the results of solving the heat balance equation of transient thermal mode with the Runge–Kutt method. The analysis of mode and climatic factors effect on the cable temperature in a dynamic thermal mode is presented. The calculation of the maximum permissible current for variation of weather conditions is made. The average electric energy losses during the transient process are calculated with the change of wind, air temperature and solar radiation. The parameters having the greatest effect on the transmission capacity are identified. Keywords – transient thermal modes; heat balance equation; overhead power lines; the conductor temperature; temperature dependence of resistance; power losses.

1. Introduction

Electricity consumption is constantly growing in the world [1, 2]. There is made systematic introduction of new generating capacity to reduce electricity deficit. There is a problem of insufficient capacity of transmission lines in economically developed areas in the maximum load consumption hours. With current increasing there is an increase of energy losses. To increase the transmission capacity and to reduce the losses made the construction of new lines is developed. The event is financially costly and time-consuming to be implemented. One solution to the problem is the usage of transmission lines dynamic thermal mode assessment. This method makes possible to increase the capacity of electric power systems equipment. [3–6].

When operating the power system to prevent malfunction associated with the overheating of the power lines when there is an increase in transmission power, static thermal models were developed [7–9]. The models are used to estimate the conductors' temperature and maximum transmission capacity of power lines when designing and operation. The heat absorbed with a conductor includes: the heat from the flow of current and heat from the sun-light conductor surface (Figure 1). Heat transfer from the conductor into the environment is by convection (Q_c) and the thermal radiation (Q_r) [9, 10]. Static models of the thermal mode of overhead lines conductors produce evaluation of conductor heating and maximum current based on the worst cooling conditions [11–12].

When calculating the permissible load of lines in a stationary mode, the permissible temperature of an aluminum conductor with a steel core is assumed equal to 70 °C. Ambient temperatures can vary from 20°C to 40°C. As a general rule, the average temperature for the hottest month is taken for the calculations. The wind speed is 0.2 m/s (calm). Solar radiation power is considered 1000 W/m^2. This power corresponds the most to radiation at noon. Real weather conditions when there is operation of power lines differ significantly from the worst cooling conditions the most time of the year. The fact consideration makes possible to increase the line capacity. [1, 13, 14].

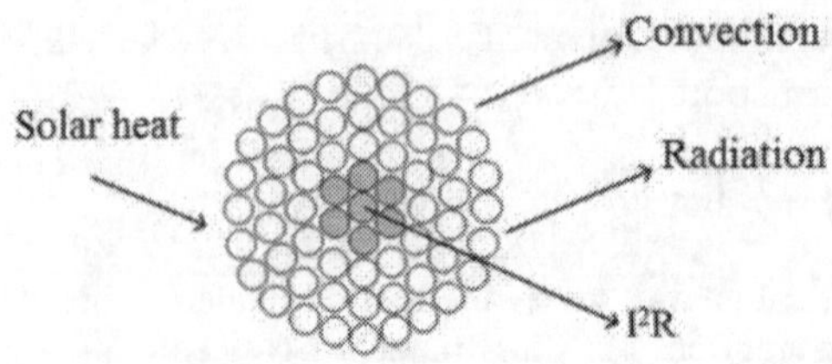

Figure 1. Links between incoming and outgoing heat energy in the conductor.

According to the documents in the electricity a consumer should receive a high quality electricity and in full. This requirement shall not be violated, despite the emerging emergency situations in the system. In case of accidents, some of the lines are disconnected. Line disconnection occurs due to the conductor excessive heating. This fact makes necessary to use thermal dynamical state of the conductors of overhead power lines. This method will allow increasing the capacity of existing lines without modernization in compliance with all safety requirements. [15, 16]. The conductor temperature is monitored with direct or indirect temperature control devices. Indirect control devices are ambient temperature recording devices in real time, wind speed and direction devices, solar heating, and the degree of slack arrow tension of the conductors [16]. The findings are transferred to the calculation systems.

2. Mathematical Simulation

When there is no possibility of installing devices of the conductor temperature direct control one must use mathematical models that consider the mode-climatic factors and calculate the conductor temperature in real time. The mathematical model should be based on a heat balance equation of the conductor in the transient temperature conditions [9]:

$$Q_c + Q_r = A_c(T - T_{amb})^k + A_r(T^4 - T^4_{amb}) \quad (1)$$

$$\Delta P_0(1+\alpha\Theta) = C\frac{d\Theta}{dt} + d_{cn}\left[\pi\alpha_{forc}(\Theta - \Theta_{amb}) + \pi\varepsilon_n C_0\left(T^4 - T^4_{amb}\right) - A_s q_{sol}\right] \quad (2)$$

where T and T_{amb} are absolute temperature of the conductor and ambient temperature; A_c and A_L are constant coefficients; k is exponent that depends on convection conditions; α_{forc} is heat transfer coefficient of forced convection; ε_n is conductor surface emissivity for the infrared radiation; C_0 = 5.67·10^{-8} W/(m^2·K^4) is constant blackbody radiation; Θ and Θ_{amb} are temperature of conductors, respectively, and ambient temperature in °C; T and T_{amb} are the same but in K (absolute temperatures); A_s is absorptivity of the conductor surface for the solar radiation; q_{sol} is solar radiation flux density on the conductor; d_{cn} is the conductor diameter; $\Delta P_0 = I^2 r_0$ are active power losses in the conductor per unit length at $\Theta = 0$ °C; I is current in the cable; r_0 is per unit length resistance of the cable at $\Theta = 0$ °C; α is temperature coefficient of resistance.

The heat capacity per unit length C and convection heat transfer coefficient α_{forc} are defined with [9, 15]:

$$C = C_{sp,Al} M_{Al} + C_{sp,st} M_{st} \tag{3}$$

$$\alpha_{forc} = 0.044 \frac{k_V \left(P_{atm} V\right)^{0.6}}{\left(T_{amb} d_{cn}\right)^{0.4}} \tag{4}$$

where k_v is coefficient characterizing the wind attack angle; P is atmospheric pressure; v is wind speed; $C_{sp.Al}$, $C_{sp.st}$, are specific heat capacity mass of aluminum and steel; M_{Al}, M_{cm} is weight of aluminum and the steel portion of the conductor per unit length.

On the basis of the method of least squares, equation (2) can be transformed to the form:

$$\frac{d\Theta}{dt} = A_1\Theta^2 + A_2\Theta + A_3 \tag{5}$$

Coefficients A_1, A_2, A_3 are given in [9, 15]. Equation (5) can have different solutions depending on the type of equation roots

$$A_1\Theta^2 + A_2\Theta + A_3 = 0 \tag{6}$$

The case of real roots of equation (6) is of practical interest

$$\Theta_{1,2} = \frac{-A_2 \mp \sqrt{A_2^2 - 4A_1A_3}}{2A_1} \tag{7}$$

On the basis of (7) solution (5) can be written as:

$$\Theta(t) = \Theta_2 + \frac{\Theta_1 - \Theta_2}{1 - \Theta' e^{-t/T_n}} \tag{8}$$

$$T_n = -\frac{1}{A_1\left(\Theta_1 - \Theta_2\right)\cdot 60} \tag{9}$$

$$\Theta' = \frac{\Theta_0 - \Theta_1}{\Theta_0 - \Theta_2} \tag{10}$$

where Θ_0 is conductor temperature at time $t=0$ (initial condition).

Real roots (7) of equation (6) take the values $\Theta_1 > \Theta_2$. Solution (8) is valid only when $\Theta_0 > \Theta_2$. Calculations show that the temperature Θ_2 has strongly negative values not exceeding the ambient temperature.

Parameter T_n, determines the time scale (inertia) of the process. This parameter is equivalent to the time constant in standard exponential function. However, the quantitative meaning of this parameter is more complicated.

The average temperature Θ_{av} and power losses ΔW in the three-phase line of length l during the time T_p is determined with the equations

$$\Theta_{av} = \Theta_1 + \left(\Theta_1 - \Theta_2\right)\frac{T_n}{T_p}\ln\left(\frac{1 - \Theta' e^{-T_p/T_n}}{1 - \Theta'}\right) \tag{11}$$

$$\Delta W = 3\Delta P_0 (1 + \alpha\Theta_{av}) T_p l \quad (12)$$

3. Results of numerical simulation

The developed approach for the analysis of transient thermal modes of overhead transmission lines is implemented in the form of an algorithm and a calculation program. Numerical simulating was conducted for the conductor mark ACSR Lynx 175 mm^2. The conductor parameters and conditions of the numerical experiment are presented in Table 1.

Table 1. Conditions adopted in the calculation of the thermal process in the conductor acsr lynx 175 mm^2.

The parameter name and designation	Numeric value
The conductor diameter d_{cn}	0.01953 m
Chase resistance at 0°C r_0	0.0001440 Ohm/m
Temperature coefficient of resistance α	0.0043 °C^{-1}
Specific heat capacity of aluminum $C_{sp,Al}$	922 J/(kg·°C)
Specific heat capacity of steel $C_{sp,st}$	452 J/(kg·°C)
Aluminum mass per unit length M_{Al}	0.497 kg/m
Steel mass per unit length M_{st}	0.3276 kg/m
The emissivity of the conductor surface ε_{cn}	0.6
Absorption capacity of the conductor surface for solar radiation A_s	0.6
Atmospheric pressure P_{atm}	100000 Pa
Coefficient of the wind attack angle k_V	1
Permissible conductor temperature Θ_{cn}	75 °C
The calculation period T_n	60 min

A model of transient temperature mode of conductors is developed. The average losses of electrical energy depending on the mode and climatic factors are calculated. The maximum permissible currents are obtained with the variation of ambient conditions.

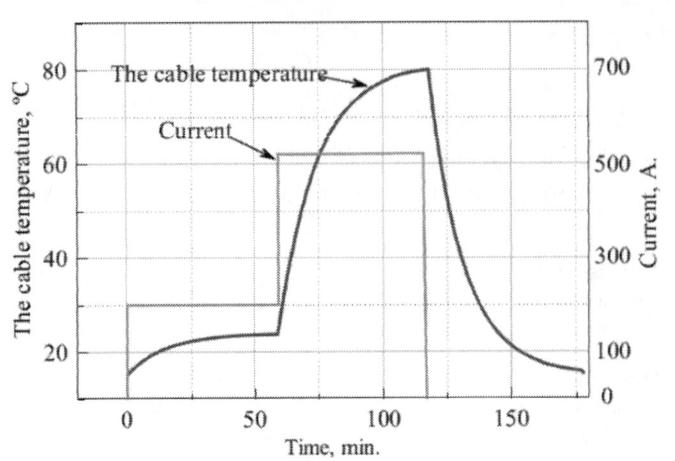

Figure 2. Change the temperature of ACSR Lynx cable on load current.

Data given in Table 2 were obtained by carrying out a numerical experiment. The most interesting results are the graphs of the achievement of a stationary mode with a change in any parameter. The above numerical experiment shows that the parameter T_n when changing one of the parameters that affects the conductor temperature remains constant provided that the remaining parameters are not changed. Besides the possibility of determining the final temperature of transition temperature mode a proposed model allows to derive average temperature during the conductor cooling or heating. Figure2 shows the curve of the conductor heating and cooling for a time interval 60 min and the wind speed 0.2 m/s. Current changes instantaneously from 0A till 200A, from 200A till 500A and from 500A till 0A. In practice, the current changes constantly. This change can occur in a wide range. The last fact demonstrates the need for using a dynamic thermal model of conductors.

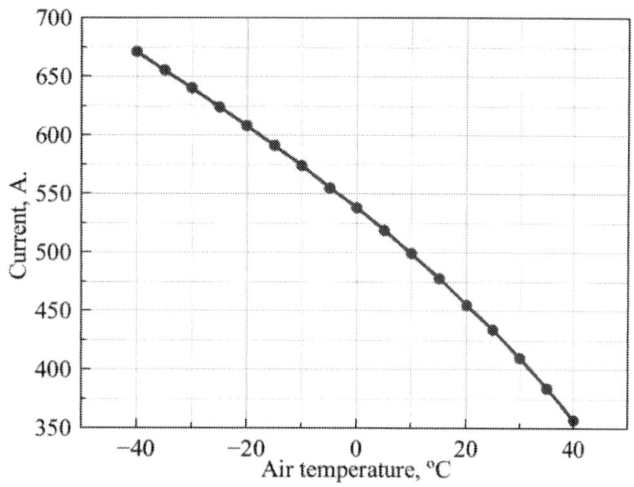

Figure 3. Links between the permissible current load and the ambient temperature.

Table 2. A slightly more complex table with a narrow caption.

The wind speed	Solar radiation		Current power	Parameter T_n (min)	Temperature of stationary mode	Initial temperature of the conductor, °C	Average temperature during transient process
	Direct	Dissipated					
0.2	0		0→200	14.233	23.642	15	21.701
0.2	0		200→519	14.233	80	23.642	70.109
0.2	0		519→0	14.233	15	80	29.371
5	0		0→200	2.799	16.64	15	16.566
5	0		200→519	2.799	26.478	16.64	26.477
5	0		519→0	2.799	15	26.478	15.509
15	0		0→200	1.462	15.868	15	15.847
15	0		200→519	1.462	20.978	15.868	20.853
15	0		519→0	1.462	15	20.978	15.137
0.2→5	0		519	2.799	26.478	80	26.255
5→15	0		519	1.462	20.978	26.478	21.112
15→0.2	0		519	14.233	80	20.978	66.117
0.2→15	0		519	1.462	20.978	80	22.408
0.2	500	100	0	13.737	25.09	15	22.891
0.2	250	50	0	13.979	20.065	15	18.954

Figure3 - Figure8 show the effect of weather conditions on the maximum line current and average energy losses during the transition process. One considers the effect of climatic factors on the permissible current in detail. An analysis is started with the ambient temperature. The parameter during the day can be varied over a wide range.

Effect of ambient temperature (Figure3) on the load current of the conductors shall be considered provided that the conductor temperature does not exceed 70 °C, and wind speed is 0.2 m/s. Solar light intensity is set equal to 0 W/m^2.

Figure3 shows the necessity to reduce the current if other parameters remain unchanged. Based on the obtained dependence of the current on the air temperature, it can be seen that at a temperature -40°C the current exceeds 1.5 times the current at 20 °C. The temperature of 20°C is accepted as the base. When the temperature is 40 °C the current is less than the basic one to 95A. Thus, the improvement in the capacity of transmission lines and its reliability is possible when considering the actual ambient temperature.

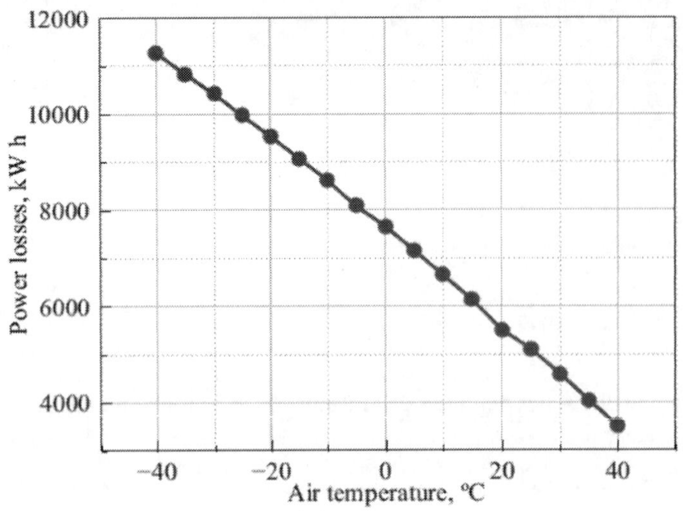

Figure 4. Dependence of average energy losses on the ambient temperature.

Figure4 shows the dependence of electric energy losses on the ambient temperature. At the initial time period the conductor temperature was equal to ambient temperature of 15 °C. Later the current was changed to the maximum value. The maximum is to be taken in such a way and the conductor temperature does not exceed 70 °C. The estimated time was 60 min, and the length of the line was assumed to be 50 km. Wind speed is 0.2 m/s, the solar radiation is not considered. Figure4 shows that losses are reduced with an increase in the ambient temperature. It is due to the fact that when the ambient temperature increases the current flowing through the conductor must be reduced, so that the conductor temperature would not exceed 70 °C. When the temperature is -40 °C the current is significantly higher than the current is at +40 °C. When calculating the average energy losses, the square of the current is taken into account. Therefore, the graph has the same form as the maximum permissible current.

To calculate the maximum current load in the Electrical Installation Code (EIC), it is proposed to use coefficients that correspond to a certain temperature. Temperatures indicated in the EIC range from -5°C till +50°C (Table 3). Time sampling is in increments of 5°C. The disadvantage of this change is that it is not possible to assess the permissible current in real time due to the fact that in reality the ambient temperature is not an integer. Minimum EIC temperature is limited to -5°C, although the Siberian, Ural and Far Eastern federal regions have average temperature of the winter months is -9 and below. This fact is also not in favor of using the proposed EIC factors when calculating the permissible current and temperature of the conductor. When temperature is -40°C difference between the currents is 103.333 A, and it is 17.76%, i.e. the lines are under loaded. It is due to the fact that as the temperature decreases, the cooling conditions improve. Within the temperature range from -5°C till +40°C differences are no more than 4%. When temperature is +50 differences are 25A or 9.59%. The analysis showed that the proposed method for calculating is preferable when calculating the permissible current in real mode. As in the initial data temperatures may take any value. Also the fact is in favor of this method that it is possible to produce a solution at low temperatures over a wide range.

Table 3. A slightly more complex table with a narrow caption.

Ambient temperature, °C	Correction coefficients for currents according to EIC	The current calculated with the coefficients of EIC, A	The current calculated on the basis of the proposed model, A	Differenc e of currents, A	Difference, %
-40	1.24	478.64	581.973	103.333	17.75563
-35	1.24	478.64	569.346	90.706	15.93161
-30	1.24	478.64	556.456	77.816	13.98421
-25	1.24	478.64	543.267	64.627	11.89599
-20	1.24	478.64	529.742	51.102	9646583
-15	1.24	478.64	515.837	37.197	7.210999
-10	1.24	478.64	501.505	22.865	4.559277
-5	1.24	478.64	486.691	8.051	1.654232
0	1.2	463.2	471.334	8.134	1.72574
5	1.17	451.62	455.361	3.741	0.821546
10	1.13	436.18	438.687	2.507	0.571478
15	1.09	420.74	421.213	0.473	0.112295
20	104	401.44	401.813	0.373	0.092829
25	1	386	386	0	0
30	0.95	366.7	362.591	-4.109	-1.13323
35	0.9	347.4	340.319	-7.081	-2.08069
40	0.85	328.1	316.179	-11.921	-3.77033
45	0.8	308.8	289.677	-19123	-6.60149
50	0.74	285.64	260.64	-25	-9.59177

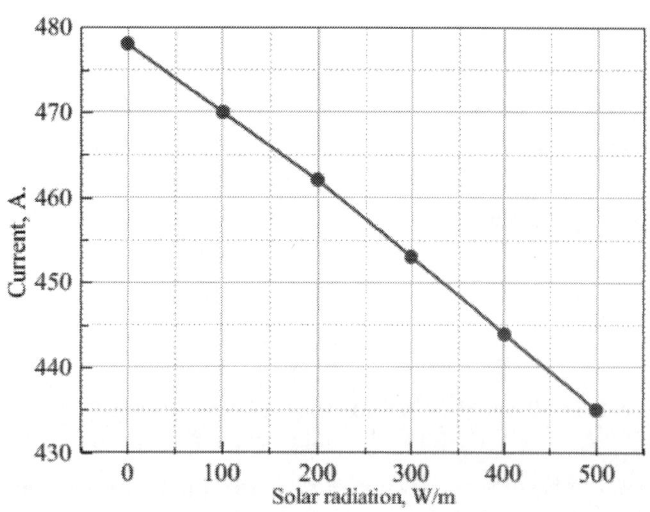

Figure 5. Links between permissible current load and solar radiation.

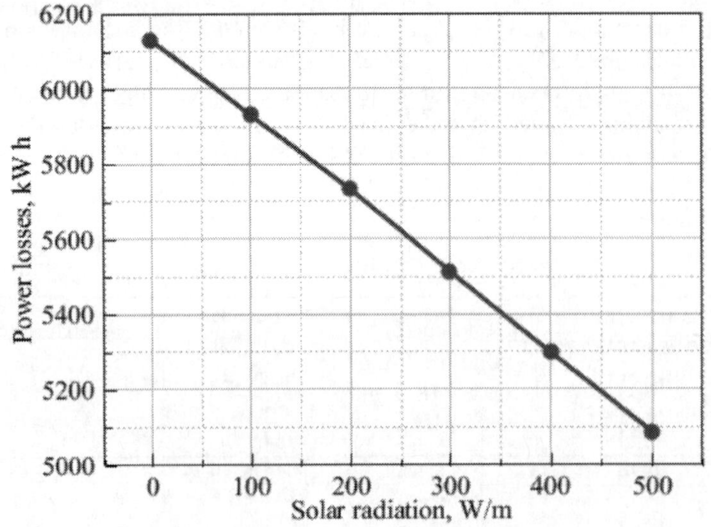

Figure 6. Dependence of average energy losses on solar radiation.

An important factor affecting the capacity of lines is solar radiation. When absorbing solar radiation there is additional conductor heating. Limited allowable current is determined under the condition that the conductor temperature does not exceed temperature of 70 °C, and wind speed is not more than 0.2 m/s, ambient temperature is 15°C. The ratio of the light intensity and current carrying capacity of the conductors is shown in Figure 5. When the solar radiation is changed by 500 W/m the current change is 43 A. This feature makes possible to increase the load on the line in winter with the evening maximum of consumption. When the conductor heating is increased with solar radiation, the maximum current is reduced. This fact leads to reduction in both energy losses (Figure 6) proportional to the square of the current.

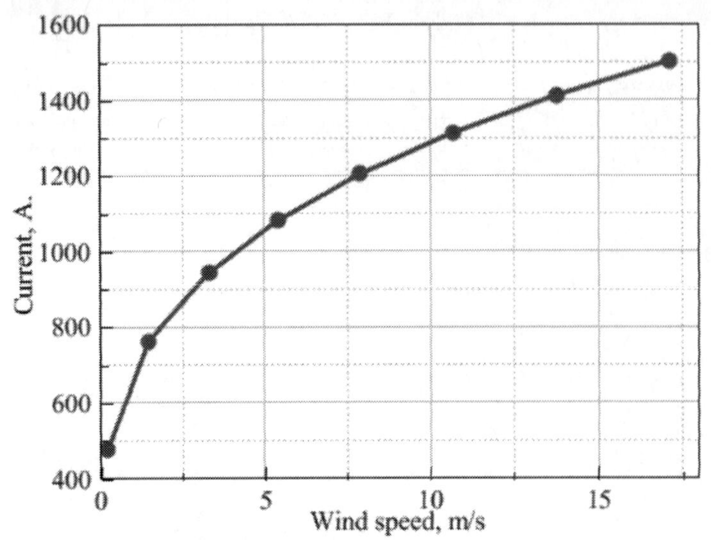

Figure 7. Links between wind speed and transmission capacity.

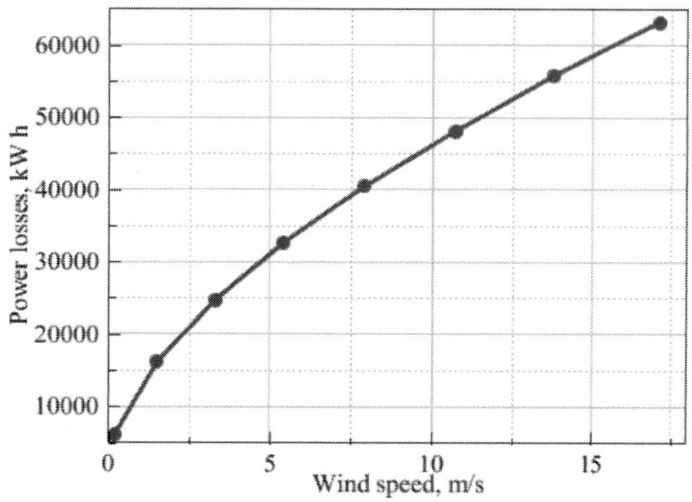

Figure 8. Dependence of the average losses on the wind speed.

One important parameter of the environment when calculating the permissible current and the conductor temperature is wind speed, as wind equal to ambient temperature significantly contributes to the conductor cooling. The dependence of the change in the current load on wind speed is shown in Figure 7. The minimum wind speed at calculation was assumed equal to 0.2 m/s, since there is always movement of air masses near the conductors. The wind speed value is classified as calm. The above dependence shows that an increase in wind speed leads to an increase in the permissible current. This fact is explained with the fact that increasing wind speed improves cooling conditions, and it in turn makes possible to load additionally the overhead transmission line. Figure7 shows the wind changes from 0.2 m/s (calm) till 17.1 m/s (strong wind), as in most cases the wind speed changes in the range during the year. With an increase in wind speed from 0.2 m/s till 17.1 m/s the current is increased in 3.14 times. Current increase leads to an increase in electric power loss (Figure 8).

4. Conclusions

Thermal power lines calculations are important. Electric power losses, transmission capacity of electrical networks and sag depend on conductor temperature. In its turn the conductor temperature depends on the current load, ambient temperature, wind speed and solar radiation. Due to the fact that the active resistance of the conductor has temperature dependence, it is a nonlinear element. So it is necessary to calculate resistance based on the heat balance equations.

Mode and climatic factors continuously change over time. This fact makes relevant to the calculation of the dynamic thermal mode of the lines. Considering transient thermal modes is necessary for a fair determination of the maximum temperature of conductors. It makes possible to predict more accurately the possible maximum current load. The proposed method is an analytical solution of the heat balance equation of conductors in a transient mode allows determining both as current as electric power losses during the transit thermal process.

The difference between the maximum current value when there is no solar radiation and at 500 W/m^2 is 8.99%, electrical energy average losses are 17.02%. When changing conductor temperature from –40°C till +40° C difference between the currents is 46.8%, average losses are 68.9%. When the wind changes from calm to strong current difference is 68.2%, average losses are 90.3%. The presented data shows that solar radiation affects the transmission capacity in lesser extent. The next most important parameter is the ambient temperature. The wind has the biggest effect. Thus, in determining the ability to transmit more power through existing lines ambient temperature and wind should be considered.

References

[1] Yan Y, Bao W, Xin J, Lin H, Li Z and Zhong H 2015 A thermal model based dynamic rating system for overhead transmission lines *5th Int. Conf. on Electric Utility Deregulation and Restructuring and Power Technologies (DRPT)* (Changsha) pp. 2758–2763

[2] Oleinikova I, Mutule A, Grebesh E and Lvovs A 2015 Line parameter estimation based on PMU application in the power grid *IEEE 5th Int. Conf. Power Engineering, Energy and Electrical Drives (POWERENG)* (Riga, Latvia) pp 453–457

[3] Yang Y, Harley R G, Divan D and Habetler T G 2009 Thermal modeling and real time overload capacity prediction of overhead power lines *Proc. IEEE SDEMPED* (Cargese, France) pp 1–7

[4] Beryozkina S, Sauhats A, Banga A and Jakusevics I 2013 Testing thermal rating methods for the overhead high voltage line *12th Int. Conf. on Environment and Electrical Engineering*, (Wroclaw) pp 215–220

[5] Roberts D, Taylor P and Michiorri A 2008 Dynamic thermal rating for increasing network capacity and delaying network reinforcements *CIRED Seminar-2008: SmartGrids for Distribution* (Frankfurt, Germany) pp 1–4

[6] Girshin S S, Bigun A Y and Petrova E V 2016 Analysis of dynamic thermal rating of overhead power lines in the conditions of forced convection considering non-linearity of heat transfer processes 2nd *Int. Conf. on Industrial Engineering, Applications and Manufacturing (ICIEAM)* (Chelyabinsk) pp 1-6

[7] Goryunov V N, Girshin S S, Kuznetsov E A, Petrova E V and Bigun A Y 2016 A mathematical model of steady-state thermal regime of insulated overhead line conductors *IEEE 16th Int. Conf. on Environment and Electrical Engineering (EEEIC)* (Florence) pp 1–5

[8] Cigr´e Working Group 22.12, *Thermal behaviour of overhead conductors*, Cigr´e Brochure 207, August 2002

[9] Girshin S S, Kuznetsov E A and Petrova E V 2016 Application of least square method for heat balance equation solving of overhead line conductors in case of natural convection *2nd Int/ Conf. on Industrial Engineering, Applications and Manufacturing (ICIEAM)* (Chelyabinsk) pp 1–5

[10] Mutule A, Grebesh E, Oleinikova I and Obushevs A 2016 Methodology for transmission line capacity assessement based on PMU data *Power Systems Computation Conference (PSCC)* (Genoa) pp 1–5.

[11] Girshin S S, Bubenchikov A A, Bubenchikova T V, Goryunov V N and Osipov D S 2016 Mathematical model of electric energy losses calculating in crosslinked four-wire polyethylene insulated (XLPE) aerial bundled cables *ELEKTRO* (Strbske Pleso) pp 294–298

[12] IEEE Standard for Calculating the Current-Temperature of Bare Overhead Conductors *IEEE Std* 738-2006 (Revision of IEEE Std 738-1993)

[13] Holbert K E and Heydt G T 2001 Prospects for dynamic transmission circuit ratings *Proc. IEEE Int. Symp. Circuits and Systems (ISCAS 2001)* vol. 2 (Sydney, NSW, Australia) pp 205–208.

[14] Fu J, Abbott S, Fox B, Morrow D J and Abdelkader S 2010 Wind cooling effect on dynamic overhead line ratings *Proc. 2010 UPEC Conf.* (Cardiff, Wales, UK) pp 1–6.

[15] Girshin S S, Gorjunov V N, Bigun A Y, Petrova E V and Kuznetsov E A 2016 Overhead power line heating dynamic processes calculation based on the heat transfer quadratic model *Dynamics of Systems, Mechanisms and Machines (Dynamics)* pp 1–5.

[16] Albizu E F, Mazon A J, Bedialauneta M and Sagastabeitia K 2011 Overhead conductor monitoring system for the evaluation of the low sag behaviour *Proc. 2011 IEEE PowerTech Conf.* (Trondheim, Norway) pp 1–6

IOP Conf. Series: Journal of Physics: Conf. Series **944** (2017) 012017 doi:10.1088/1742-6596/944/1/012017

Electroinduction disk sensor of electric field strength

S V Biryukov, M A Korolyova
Omsk State Technical University, 11, Mira ave., Omsk, 64450, Russia

e-mail: sbiryukov154@mail.ru

Abstract. Measurement of the level of electric fields exposure to the technical and biological objects for a long time is an urgent task. To solve this problem, the required electric field sensors with specified metrological characteristics. The aim of the study is the establishment of theoretical assumptions for the calculation of the flat electric field sensors. It is proved that the accuracy of the sensor does not exceed *3%* in the spatial range $0<a< 0.2$ and in the entire spatial range of measurement is negative. Using the methods of calculation of electrostatic fields, this article examines single-axis electromotive disk sensor radius R, and the estimation errors caused by the inhomogeneity of the field. The maximum of this error is *3%* in the spatial range from *0* to *5R* to the source field that allows you to design better quality sensors used in different measuring systems of wide application.

1. Introduction

Around us the electric field causes an adverse impact both on technical and biological objects. To study the effect of electric fields on these objects the sensors and measuring instruments on their basis are necessary. The development of such measuring instruments began with the 70-ies of the last century. Analyzing manufactured and designed devices with a flat sensitive elements such as [1], [2] and [3] we see that the shape of the sensing elements is a form of a petal [1] and [2], or the shape of a flat square [3], which is made of conductive material. According to the information about the approved types of measurements of Federal information fund by providing a single measurement of the Russian Federation [4] referred to such measurements as IESP-01, which are sensitive elements made in the form of a flat electrode. However, the error of such measurements is 20%, which is unacceptable in accordance with the new regulations on electromagnetic safety of production and life companies.

In this labor, the study is subjected to single-axis flat panel detector of the electric field made in the form of a conductive disc. The reason for this study was that currently on the market there are a large number of devices for measuring the strength of the electric field [4], that are not supported by research, articles, patents for utility model and invention. In this regard, on the market of instruments for measuring electric field strength than does not confirm its metrological characteristics. Therefore, they ascribe errors reaching 20 %, which is not always acceptable.

2. Formulatin of the problem

In the physical basis of most sensors is the phenomenon of electrostatic induction, that is, the appearance of electric charges on the conductor surface under the action of an electric field. Such charges are called induction. In a constructive-based sensor can lie as dielectric and conductive substrate. The difference in these sensors is that the sensors with a dielectric substrate have floating potential of the midpoint, and the sensors of the conductive substrate have the potential mid-point equal to the potential of a point in space place of the sensor. Our study will be single-axis electromotive sensor with a conductive substrate in the form of a disc. The aim of the study is the creation of electric field intensity sensor with an accuracy not more than 5 %

3. Theory

Theory of the sensor operation is based on the consideration of a conducting disk of thickness h and radius R ($h << R$) is placed in a uniform and non-uniform quasi-static electric field $E_0 = E \cdot \sin\omega t$. Hereinafter simply E_0.

Consider the interaction of the sensor tension with electric fields of different discontinuity. As these fields will choose a uniform electric field and a point source. A uniform field will be considered as the reference model field. In relation to it will be evaluated the accuracy of the sensor working in real conditions, where the fields are often uneven. The non-uniform of the fields to be considered as one part in assessing the error limits of the sensor. Further, we assume that the sensor accuracy in the other uniform fields less than the field of the point source. Field of the point source is chosen from the condition of maximum that can be modeled in the analysis of the behavior of the sensor in a non-uniformly field. It should be noted that there are other sources of electric fields with greater heterogeneity than the point source. These include the electric field of the dipole and quadruple [5]. However, the electric field of these sources rapidly decreases inversely proportional to the third and fourth degree, respectively, of the distance from the investigated point of the field. At distances where all these fields exist, measurement techniques, based on the phenomenon of electrostatic induction, unsuitable.

The sensor is a conductive circular plate 1, of radius R and thickness h ($h << R$). On two opposite surfaces of the plate are conductive sensitive elements 2 and 3, as shown in Figure 1.

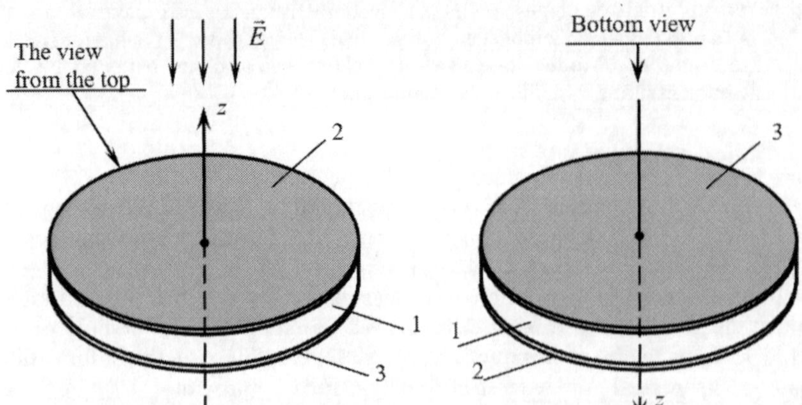

Figure 1. The sensor of the electric field with the conductive substrate in the form of a circular plate.

The sensitive elements 2 and 3, a thin conductive layer with thickness δ, are the same size and shape, and located at a distance l from the surface of conductive plate 1 (Figure 2).

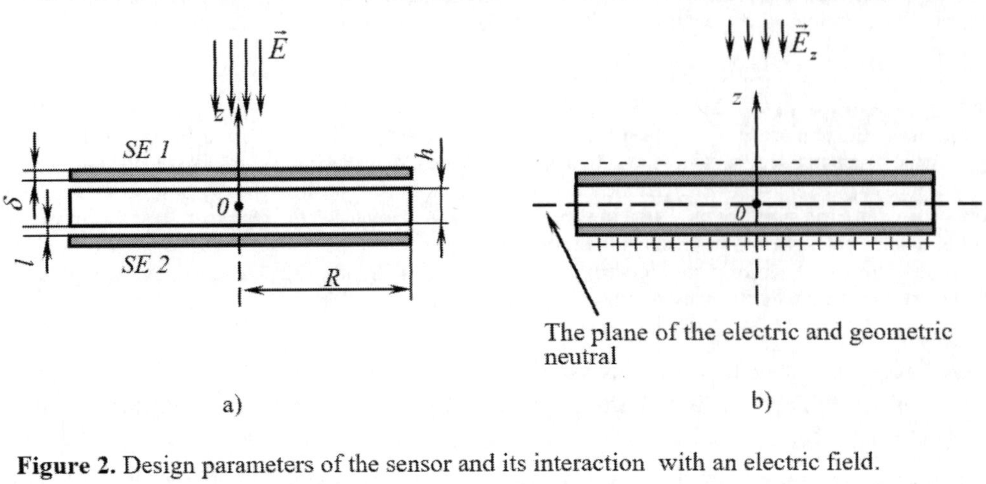

Figure 2. Design parameters of the sensor and its interaction with an electric field.

Adopted assumptions $\delta << h$ and $l << R$ give reason to believe that the potential of the sensitive elements 2 and 3 is equal to the potential of the conductive plate 1, and the sensing element is something other than the surface of the plate. Further other measures will be taken, which allows considering conductive sensing element surface to a conducting plate. Thus, the sensor in General is something else, like a flat conductive plate in the form of a disk.

A. Sensor in a homogeneous field

Place a conductive circular plate in a uniform electric field perpendicular to its force lines. Then, according to the theory, the plate will occur the separation of electric charges: positive charges will move in the direction of the field, the negative against of this direction. As a result, on one surface of the plate induced by a negative and the other positive charges, separated by a plane of the electrical and geometrical neutral (see Fig.2, b). Induced on the surfaces of the plates of the external electric field electric charges create their own electric field directed opposite to the external field. As a result, within plate resultant field will be zero. In this case all points of the plate will have the same potential equal to the potential of a point in space the place of the plate.

Value of these charges can be determined from the interaction between a conducting plate with a uniform electric field. The solution of this problem shows that the surface of the conducting plane, there is only the normal part E_n of the electric field intensity equal to the intensity of the external uniform field E_0. This electric field according to the Gauss theorem, and determines the surface density of electric charge

$$\sigma = -2\varepsilon\varepsilon_0 E_n = -2\varepsilon\varepsilon_0 E_0, \quad (1)$$

where ε – dielectric permeability of the medium surrounding the conductive disk; ε_0 – dielectric constant. As can be seen from the formula (1) surface charge density on a conducting disk in a uniform field constant.

Therefore, in this case the charge is distributed on the surface of the disc evenly.

The total charge on the plane of the conductive disk can be determined from the formula

$$Q = \iint_s \sigma \cdot dS \quad (2)$$

Then the size of the charges according to the formula (2), induced on the two planes of the conductive disk can be calculated by introducing polar coordinates

$$Q_O = \pm \iint_s \sigma \cdot dS = \pm 2\varepsilon\varepsilon_0 E_O \int_0^{2\pi} \int_0^R \rho d\rho d\varphi = \pm 2\pi\varepsilon\varepsilon_0 R^2 E_O, \quad (3)$$

where Q_O – the charge induced by a homogeneous field; the sign "-" – refers to the upper plane of the disk, and the sign "+" – to the lower plane of the disk (see Fig. 2, b).

It follows from formula (3) that the charges on the surfaces of the conducting disk are proportional to the electric field strength. Therefore, they can act as a measure of a strength. If you remove them from the surfaces of the plate and measure, you can get an electric field strength sensor. Such a sensor was described at the beginning of the article. Since the sensor has two sensing elements, a dual sensor is obtained. Therefore, when the sensor is switched on in differential mode, it is total, that is, the differential charge will be doubled, according to expression

$$Q_O^{dif} = (+Q_O) - (-Q_O) = 2Q_O = 4\pi\varepsilon\varepsilon_0 R^2 E_O. \quad (4)$$

The analysis of formulas (3) and (4) for the values of charges induced by an electric field on the surfaces of a conducting disk shows their dependence on the radius R of the disk plate. For further investigations, it is more convenient to have a certain value that does not depend on the dimensions of the plate and the value of the electric field strength, and such a value can be the normalized value of the charge

$$Q_O^{nor} = \frac{Q_O}{\pi\varepsilon\varepsilon_0 R^2 E_O} = 2 \cdot \tag{5}$$

B. A sensor in an inhomogeneous field of a point source

We place the conducting disk plate in the field of the point source. As a point source, we will consider a positive point charge q at a distance d from the conductive plate. And in this case negative plates are induced on one side of the conducting plate, and positive charges on the other. Field lines and induced on the conducting plate charges is presented in figure 3,and the equipotential surface of a point source in the presence of the flat electrode obtained by simulation in the Elcut program, shown in Fig. 3,b.

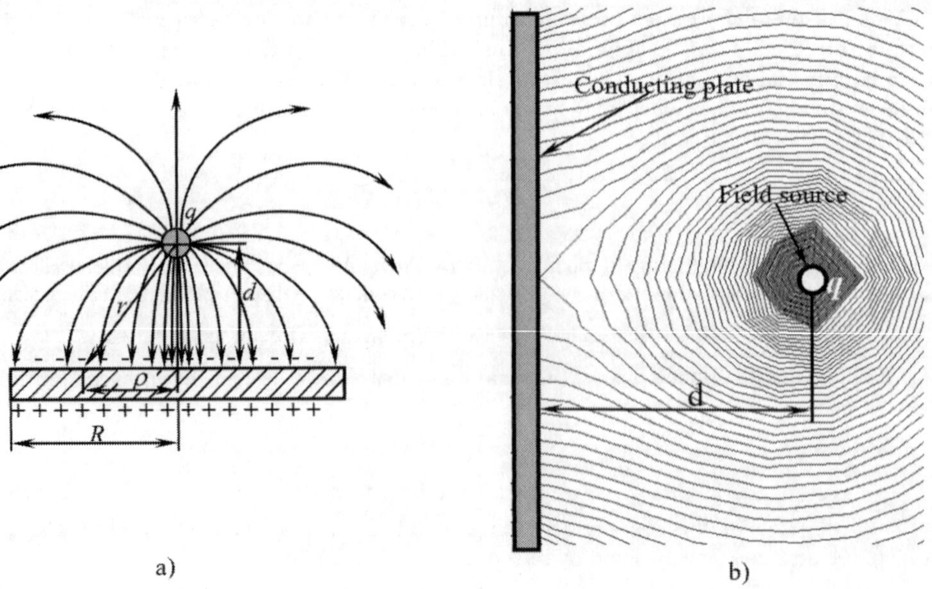

Figure 3. View of power lines a), and equipotential surfaces, b) in the system of a point charge - conducting disk plate.

Using the method of images in the plane it is possible to get the potential near the surface of the conducting plane, the normal part of the electric field and surface charge density on the surface of the conductive disk

$$\sigma = -\frac{2\varepsilon\varepsilon_0 d^3}{r^3} E_H , \tag{6}$$

where r - the distance from the charge q to the observation point; E_H - the intensity of the non-uniform electric field at the observation point.

Using formula (6), one can know the surface charge density at each point of the surface of the conducting plate. Consider the point of the surface of the conducting plate at a distance ρ from its center (see Fig. 3), provided that the point charge q is located above the center of the plate. Since r in expression (6) is the distance from the point charge to the observation point, we express it in terms of the characteristic ρ

$$r = (d^2 + \rho^2)^{\frac{1}{2}}. \tag{7}$$

Then the expression for the surface charge density as a function of the distance ρ from the center of the conducting plate will have the form

$$\sigma(\rho) = -\frac{2\varepsilon\varepsilon_0 d^3}{\left(d^2+\rho^2\right)^{\frac{3}{2}}} E_н = -\frac{2\varepsilon\varepsilon_0}{\left[1+\left(\frac{\rho}{d}\right)^2\right]^{\frac{3}{2}}} E_н \tag{8}$$

As can be seen from formula (8), the surface density of charges on a conducting disk, unlike a homogeneous field, is not constant over the entire area of the conducting plate, and depends not only on the distance ρ from the center of the plate, but also on the distance of the plate to the source of the field d. Consequently, in this case the charge is distributed unevenly over the surface of the plate.

The values of the charges induced on the two planes of the plate are formed by an external inhomogeneous field and are determined by the expression (2). Using the tables of integrals [6] we find these charges

$$\begin{aligned} Q_H &= \pm\iint_S \sigma_н \cdot dS = \int_0^{2\pi}\int_0^R \frac{2\varepsilon\varepsilon_0 d^3 E_н}{\left(d^2+\rho^2\right)^{\frac{3}{2}}}\rho d\rho \cdot d\varphi = \\ &= \pm 2\pi\varepsilon\varepsilon_0 R^2 E_H \frac{2}{\left(\frac{R}{d}\right)^2}\left[1-\frac{1}{\sqrt{1+\left(\frac{R}{d}\right)^2}}\right], \end{aligned} \tag{9}$$

where the sign "-" – refers to the upper plane of the disk, and the sign "+" – to the lower plane of the disk (see Fig. 2, b).

When the sensor is switched on in differential mode, its total differential charge will be equal to

$$\begin{aligned} Q_H^{диф} &= (-Q_H) - (-Q_H) = 2Q_O = \\ &= 4\pi\varepsilon\varepsilon_0 R^2 \frac{2}{\left(\frac{R}{d}\right)^2}\left[1-\frac{1}{\sqrt{1+\left(\frac{R}{d}\right)^2}}\right] E_H \end{aligned} \tag{10}$$

For the convenience of further research, we introduce the normalization of the size of the charge, similarly, as in the investigation of a homogeneous electric field

IOP Conf. Series: Journal of Physics: Conf. Series **944** (2017) 012017 doi:10.1088/1742-6596/944/1/012017

$$Q_H^{нор} = \frac{Q_H}{\pi\varepsilon\varepsilon_0 R^2 E_H} = \frac{4}{\left(\frac{R}{d}\right)^2}\cdot\left[1-\frac{1}{\sqrt{1+\left(\frac{R}{d}\right)^2}}\right]. \qquad (11)$$

Analyze the differences in the behavior of the electric field strength sensor in extreme cases - in a homogeneous (reference) field and in the field of a point source with a significant heterogeneity.

First of all, let us consider how electric charges are distributed on the surfaces of a conducting plate-sensor. We use the mathematical editor MathCAD and construct graph of the electric charge density on the surface of sensor, that in a homogeneous and inhomogeneous field, depending on the relative distance to the source of the field $a = R/d$, where R – the radius of the sensor plate; d – the distance

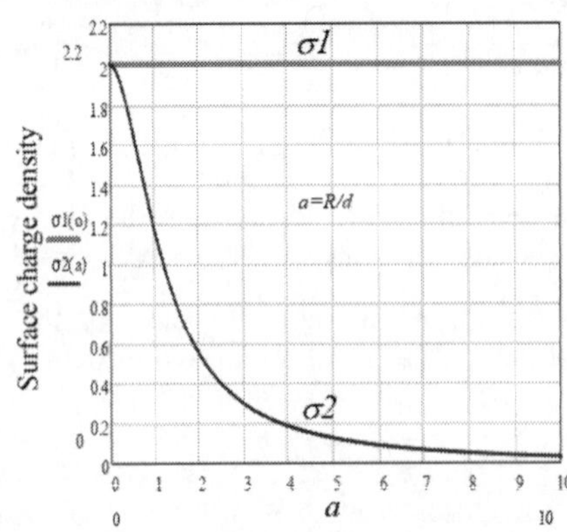

Figure 4. Dependences of the electric charge density on the surface of sensor sensitive elements, located in a homogeneous σ1 and in an inhomogeneous σ2 electric field

from the center of the sensor to the source of the field (Fig. 4).

It follows from Fig. 4 that the electric charge density on the sensor surface in a uniform field is constant across the entire sensitive element, and in an inhomogeneous field it decreases to the edge of the sensitive element. And this decrease is the stronger the higher the heterogeneity of the field, that is, the closer the sensor to the source of the field.

We estimate the infelicity of the sensor from the inhomogeneity of the electric field. To this end, we use formulas (4) and (10) and a normalization $a = R/d$, where R – the radius of the sensor plate; d – the distance from the center of the sensor to the source of the field. The normalizing measure a characterizes the proximity of the sensor to the source of the field. Thus, the smaller a, further the sensor is from the source of the field, and the field becomes more homogeneous. The most acceptable range for changing parameter a is from 0 to 1.

Taking into account this normalization, the formula for the error from the inhomogeneity of the electric field takes the form

$$\delta = \frac{Q_H^{диф} - Q_O^{диф}}{Q_O^{диф}} = \left\{\frac{2}{a^2}\left[1-\frac{1}{\sqrt{1+a^2}}\right]-1\right\}\times 100\,\%. \qquad (12)$$

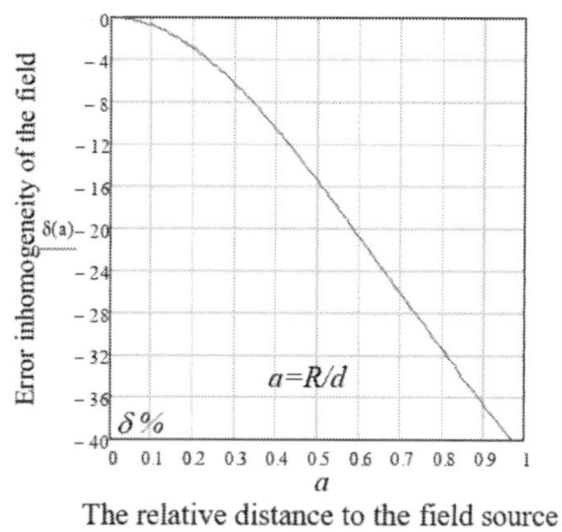

Figure 5. Dependence of the field inhomogeneity infelicity on the relative distance to the source of the field

We use the mathematical editor MathCAD and build graphs of the infelicity from the inhomogeneity of the electric field as a function of the characteristic a. The infelicity graph is shown in Fig. 5.

4. The results of the surveys

Surveys have shown (see Fig. 5) that the accuracy of the sensor in all spatial measurement range of negative and already at $a > 0.2$ beyond 3 %. In this regard, the sensor is suitable for measurements at distances from the source d field that is five times the radius of the conductive disk of the sensor ($d=5R$).

5. Conclusion

From the survey results it follows that the considered tension sensor has an infelicity from the inhomogeneity of the field up to *3 %* in the spatial range from *0* to *5R* from the field source, where R - the radius of the disk plates of the sensor. The sensor gives lower values of charges in an inhomogeneous field, this may lead to a biased assessment of the influence of the electric field on the technical and biological objects. The advantage of sensor is its simplicity, allowing to make its methods of deposition of conductive and dielectric layers on a thin conducting substrate. In addition, because of its small dimensions, the sensor allows you to place it inside the device.

In conclusion, we assume that the solution of the problem of minimizing the error from the inhomogeneity of the field lies in the optimization of the sensor sensitive elements size. Further research will be conducted in this direction.

References

[1] Michael F. Steward. Electric field measurement system. US patent US005315232A/№5315232; filed 03.01.1991; published 24.05.1994.

[2] William H. Beasley, Leon G. Byerley, Jody F/ Swenson, Ivan G. Bogoev. Low power, low maintenance. Electric field meter. US patent US006984971B1/№6984971B1; filed 14.03.2002; published 10.01.2006.

[3] Mathieu Baicry, Matthieu Le Prado. . Device for measuring an electric field in a conducting medium and method of calibrating such a device. US patent № US20160238646 A1, USA, / №US 15/045,445; filed 17.02.2016; published 18.08.2017.
[4] Federal information fund by providing a single measurement. URL http://www.fundmetrology.ru/ (date 26.06.17)
[5] 2002 *Handbook of Physics* ed Walter Benenson, John W. Harris, Horst Stöcker and Holger Lutz (Springer: Frankfurt am Main, Germany)
[6] Dwight H B 1961 *Tables of integrals and other mathematical date* (New York: Macmillan company)

Spline wavelets use for output processes analysis of multi-dimensional non-stationary linear control systems

Y I Bityukov[1], Y I Deniskin[1] and G Y Deniskina[1]

[1] Moscow Aviation Institute (national Research University) MAI, Moscow, Russia

Abstract. Methods for solving systems of differential equations based on the use of wavelets have become very popular recently. Debeshy wavelets, coiflets were used among others. The disadvantage of such wavelets is that they do not have an analytical expression. Therefore, there are great difficulties in integrating and differentiating expressions containing these wavelets. The article presents an algorithm for the numerical solution of a system of differential equations with variable coefficients based on spline-wavelets on a interval. The algorithm presented generalizes the well-known method based on Haar wavelets which are a special case of spline-wavelets. The results of the article are used to analyze the output processes of multi-dimensional non-stationary linear control systems.

1. Introduction

Many mathematical applications require numerical approximation of solutions of differential equations. One of the methods for constructing such an approximation is based on Haar wavelets. A vast number of works have been devoted to the various applications of these wavelets. An efficient numerical method for solution of nonlinear evolution equations based on the Haar wavelets approach is proposed in the article [1]. Haar wavelets techniques for the solution of ordinary differential equations is discussed in articles [2], [4] and in the book [5]. Integral equations are considered in the article [3].

Since Haar wavelets are orthogonal the system of linear equations for finding wavelet coefficients of approximant becomes sparse. Unfortunately the approximants are piecewise constant in solving integral equations, and the smoothness class of the approximant will be one less than the order of the differential equation in solving differential equations. Applying a spline wavelet, the special case of which are Haar wavelets, can remove such shortcomings. One can construct approximants of any smoothness class using a spline wavelet. It should be noted that spline wavelets in the general case are a semiorthogonal system, but as it is shown in the article [6] the system of linear equations for determining the wavelet coefficients of approximation is pseudo-sparse i.e. contains a large number of elements close to zero. This article is devoted to the use of a spline wavelet to the construction of approximants of any smoothness class of solutions to linear systems of differential equations.

2. Problem formulation

Let there be given a linear system of differential equations

$$\begin{cases} \mathrm{d}x_1/\mathrm{d}t = a_{11}(t)x_1 + a_{12}(t)x_2 + \ldots + a_{1m}(t)x_m + f_1(t); \\ \mathrm{d}x_2/\mathrm{d}t = a_{21}(t)x_1 + a_{22}(t)x_2 + \ldots + a_{2m}(t)x_m + f_2(t); \\ \ldots \\ \mathrm{d}x_m/\mathrm{d}t = a_{m1}(t)x_1 + a_{m2}(t)x_2 + \ldots + a_{mm}(t)x_m + f_m(t), \end{cases} \tag{1}$$

and the initial conditions are given

$$x_1(t_0) = x_{10}, x_2(t_0) = x_{20}, \ldots, x_m(t_0) = x_{m0}. \tag{2}$$

We assume that $a_{ij}, f_i \in C[t_0;t_1]$ for all $i, j = 1,2,\ldots,m$. The aapproximants of any class of smoothness are required to be constructed $C^n[t_0;t_1]$ system solution (1), satisfying the initial condition (2).

3. Theoretical description

The approach to the construction of wavelet systems on an interval proposed in the present paper will be briefly considered in this section [7]. Let the net function φ belongs to the real space $L_2(\mathbf{R})$ satisfying the equation

$$\varphi(x) = \sqrt{2}\sum_{k\in\mathbf{Z}} u_k \varphi(2x-k), \quad u_k \in \mathbf{R} \tag{3}$$

and has a compact support $[a;b]$. Let $\varphi_{jk}(x) = 2^{j/2}\varphi(2^j x - k), x \in [a;b]$ be defined. Function φ is called scaling in the theory fo wavelets, and equation (3) is called scaling relation [8], [9]. It is clear that only a finite number of such functions will be different from zero on an interval $[a;b]$. Let these be functions be $\varphi_{j,0}, \varphi_{j,1}, \ldots, \varphi_{j,n_j-1}$ for definiteness. If linear spaces $V_j = \mathrm{lin}\{\varphi_{j,0}, \varphi_{j,1}, \ldots, \varphi_{j,n_j-1}\}, \dim V_j = n_j$ are considered then by virtue of equation (3) $V_0 \subset V_1 \subset \ldots \subset L_2[a;b]$ will be satisfied. So $\varphi_{j-1,k} = \sum_{s=0}^{n_j-1} p_{s,k}\varphi_{j,s}$. Let us introduce the notation as in [7]

$$\mathbf{\Phi}_j(x) = \left(\varphi_{j,0}(x), \varphi_{j,1}(x), \ldots, \varphi_{j,n_j-1}(x)\right),$$

$$\mathbf{P}_j = \left(p_{s,k}\right)_{s=0,k=0}^{n_j-1,n_{j-1}-1}.$$

In this case $\mathbf{\Phi}_{j-1} = \mathbf{\Phi}_j \mathbf{P}_j$. Let us denote the orthogonal complement to the space V_{j-1} in the space V_j by W_{j-1}. By virtue of the fact that $V_j = V_{j-1} \oplus W_{j-1}$ and $W_{j-1} \subset V_j$, then W_{j-1} is finite-dimensional space

$$W_j = \mathrm{lin}\{\psi_{j,0}, \psi_{j,1}, \ldots, \psi_{j,m_j-1}\}, \dim W_j = m_j$$

and $\psi_{j-1,k} = \sum_{s=0}^{n_j-1} q^j_{s,k}\varphi_{j,s}$. Functions $\psi_{j,k}$ are called wavelats, and spaces W_j are called wavelat spaces [8], [9]. We again introduce matrices [7]

$$\mathbf{\Psi}_j(x) = \left(\psi_{j,0}(x), \psi_{j,1}(x), \ldots, \psi_{j,m_j-1}(x)\right),$$

$$\mathbf{Q}_j = \left(q^j_{s,k}\right)_{s=0,k=0}^{n_j-1,m_{j-1}-1}.$$

In this case $\mathbf{\Psi}_{j-1} = \mathbf{\Phi}_j \mathbf{Q}_j$. It is necessary to mention that $n_j + m_j = n_{j+1}$. Let us assume that $f \in L_2(X)$ and $\Pi_j : L_2(X) \to V_j$. In this case

$$\Pi_j f = \sum_{k=0}^{n_j-1} c_{jk}\varphi_{jk} = \Pi_{j-1}f + \Pi^W_{j-1}f = \sum_{k=0}^{n_{j-1}-1} c_{j-1,k}\varphi_{j-1,k} + \sum_{k=0}^{m_{j-1}-1} d_{j-1,k}\psi_{j-1,k}.$$

This equation can be rewritten in a matrix form if we introduce vectors

$$\mathbf{C}_j = \left(c_{j,0},\dots,c_{j,n_j-1}\right)^T, \quad \mathbf{D}_j = \left(d_{j,0},\dots,d_{j,m_j-1}\right)^T.$$

The first vector governs the approximant of the function f, and the second vector repsresents wavelet coefficients characterizing excursion $\Pi_{j-1}f$ from $\Pi_j f$. As it is shown in [7] there is an equation found $\mathbf{C}_j = \mathbf{P}_j\mathbf{C}_{j-1} + \mathbf{Q}_j\mathbf{D}_{j-1}$. This equation allows to restore the approximant $\Pi_j f$ by more rude approximation $\Pi_{j-1}f$ and wavelet coefficients.

Since linear operators (projectors) $V_j \to V_{j-1}, V_j \to W_{j-1}$ are defined by some matrices $\mathbf{A}_j, \mathbf{B}_j$, then $\mathbf{C}_{j-1} = \mathbf{A}_j\mathbf{C}_j$, $\mathbf{D}_{j-1} = \mathbf{B}_j\mathbf{C}_j$.

By the wavelet decomposiotion of f function we mean the finding of vectors $\mathbf{C}_0, \mathbf{D}_0, \mathbf{D}_1, \dots, \mathbf{D}_{j-1}$. Matrices $\mathbf{Q}_j$ and $\mathbf{P}_j$ are known as synthesis filters. Matrices $\mathbf{A}_j$ and $\mathbf{B}_j$ are known as analysis filters. The multitude $\{\mathbf{A}_j, \mathbf{B}_j, \mathbf{P}_j, \mathbf{Q}_j\}$ is known as filterbank.

As it is shown in paper [7] there is a followng link between matrices $\mathbf{A}_j, \mathbf{B}_j$ and $\mathbf{P}_j, \mathbf{Q}_j$

$$\begin{pmatrix}\mathbf{A}_j \\ \mathbf{B}_j\end{pmatrix} = \begin{pmatrix}\mathbf{P}_j & \mathbf{Q}_j\end{pmatrix}^{-1}.$$

We shall consider now how to determine the matrix $\mathbf{Q}_j$. We introduce the following notation. If $\mathbf{f} = (f_1,\dots,f_r)$, $\mathbf{g} = (g_1,\dots,g_r)$ are some vectors then $[(\mathbf{f},\mathbf{g})] = \left((f_i,g_j)\right)_{i,j=1}^r$ is a scalar product matrix. As it is shown in paper [7] matrix $\mathbf{Q}_j$ satisfies the following equation $\mathbf{P}_j^T\left[(\mathbf{\Phi}_j,\mathbf{\Phi}_j)\right]\mathbf{Q}_j = 0$.

We turn to spline wavelets on a segment now. We define B-splines of order n as convolution [10]

$$N_n = N_{n-1} * N_0, \; N_0(x) = \begin{cases} 1, x \in [0;1), \\ 0, x \notin [0;1). \end{cases}$$

As it is shown in [10] if we define the function $\varphi(x) = N_n(x)$ it satisfies the equation $\varphi(x) = \sum_{k=0}^{n+1} \frac{C_{n+1}^k}{2^n}\varphi(2x-k)$, where $C_{n+1}^k = \frac{(n+1)!}{k!(n+1-k)!}$. The article [11] represents the filterbank that corresponds to the function $\varphi(x) = N_n(x)$, that is the following results are valid.

Lemma 1. *The function* $\varphi(x) = N_n(x)$ *defines a sequence of subspaces*

$$V_{\alpha,0} \subset V_{\alpha,1} \subset \dots, V_{\alpha,j} = \text{lin}\left\{\varphi_{j,-n}, \varphi_{j,-n+1}, \dots, \varphi_{j,2^j\alpha(n+1)-1}\right\}$$

of the space $L_2[0;\alpha(n+1)], \alpha = 1,2,\dots$ *such that* $\overline{\bigcup_{j=0}^{+\infty} V_{\alpha,j}} = L_2[0;\alpha(n+1)]$.

IOP Conf. Series: Journal of Physics: Conf. Series **944** (2017) 012018 doi:10.1088/1742-6596/944/1/012018

Lemma 2. *We have the par*

$$\sum_{k=-n}^{2^j\alpha(n+1)-1} \varphi_{j,k}(x) \equiv 2^{\frac{j}{2}}, x \in [0;\alpha(n+1)].$$

If $V_{\alpha,j} = V_{\alpha,j-1} \oplus W_{\alpha,j-1}$, *then* $\dim W_{\alpha,j-1} = 2^{j-1}\alpha(n+1)$.
Let us assume that

$$\lambda_{m,k} = \int_k^{k+1} N_n(z)N_n(z-m)\,dz\,,\ m=-n,\ldots,n,\ k=0,1,\ldots,n$$

and $\omega_{i,k} = \omega_{k,i} = \sum_{s=n-i+1}^{n} \lambda_{k-i,s}$, $\theta_{i,k} = \theta_{k,i} = \sum_{s=0}^{n-k}\lambda_{i-k,s}$, $1 \le i \le k \le n$. we introduce the vector $\mathbf{p} \in \mathbf{R}^{2^j\alpha(n+1)+n}$ which we define by the par

$$\mathbf{p} = \begin{cases} \left(C_{n+1}^0\ C_{n+1}^1 \ldots C_{n+1}^k\ C_{n+1}^k \ldots C_{n+1}^1\ C_{n+1}^0 0\ldots 0\right)^T, & n=2k; \\ \left(C_{n+1}^0\ C_{n+1}^1 \ldots C_{n+1}^k\ C_{n+1}^{k+1}\ C_{n+1}^k \ldots C_{n+1}^1\ C_{n+1}^0 0\ldots 0\right)^T, & n=2k+1. \end{cases}$$

We define shift operator $R_s : \mathbf{R}^m \to \mathbf{R}^m$ by the following rule

$$R_s\mathbf{a} = \begin{cases} \left(0,\ldots,0,a_1,\ldots,a_{m-s}\right)^T, & m > s \ge 0; \\ \left(a_{|s|+1},\ldots,a_m,0,\ldots,0\right)^T, & -m < s < 0, \end{cases} \qquad \mathbf{a} = (a_1,\ldots,a_m)^T.$$

If $|s| \ge m$ then $R_s\mathbf{a} = 0$.

Lemma 3. *Matrices* $\mathbf{P}_j$ *and* $[(\mathbf{\Phi}_j,\mathbf{\Phi}_j)]$ *for the sequence of subspaces* $V_{\alpha,0} \subset V_{\alpha,1} \subset \ldots$ *are as follows*

$$\mathbf{P}_j = \frac{1}{2^{n+\frac{1}{2}}}\left(R_{-n}\mathbf{p}\quad R_{-n+2}\mathbf{p}\quad \ldots\quad R_{n-2+2^j\alpha(n+1)}\mathbf{p}\right);$$

$$[(\mathbf{\Phi}_j,\mathbf{\Phi}_j)] = \left(\mathbf{d}_1,\ldots,\mathbf{d}_n,\mathbf{q},R_1\mathbf{q},\ldots,R_{2^j\alpha(n+1)-n-1}\mathbf{q},\ \mathbf{u}_1,\ldots,\mathbf{u}_n\right)^T,$$

where

$$\mathbf{d}_s = \left(\omega_{1,s},\omega_{2,s},\ldots,\omega_{n,s},q_{n-s+1},\ldots,q_n,0,\ldots,0\right)^T,\ \mathbf{u}_s = \left(0,\ldots,0,q_n,\ldots,q_s,\theta_{1,s},\ldots,\theta_{n,s}\right)^T,$$

$$\mathbf{q} = \left(q_n,q_{n-1},\ldots,q_1,q_0,q_1,\ldots,q_{n-1},q_n,0,\ldots,0\right)^T \in \mathbf{R}^{2^j\alpha(n+1)+n},\quad q_k = \left(N_n(\cdot),N_n(\cdot-k)\right).$$

The matrix transposed to $\mathrm{T}_j = \mathrm{P}_j^T[(\Phi_j,\Phi_j)] = 2^{-n-\frac{1}{2}}(t_{i,s})_{i=1,s=1}^{2^{j-1}\alpha(n+1)+n,\,2^j\alpha(n+1)+n}$ *is as follows*

$$2^{n+\frac{1}{2}}\cdot \mathrm{T}_j^T = \left(\mathbf{L}_1,\ldots,\mathbf{L}_n,\mathbf{w},R_2\mathbf{w},\ldots,R_{2^j\alpha(n+1)-2n-2}\mathbf{w},\ \mathbf{L}_{2^{j-1}\alpha(n+1)+1},\ldots,\ \mathbf{L}_{2^{j-1}\alpha(n+1)+n}\right),$$

where $\mathbf{w} = \left(\mathbf{p}^T R_{-2n}\mathbf{q},\mathbf{p}^T R_{-2n+1}\mathbf{q},\ldots,\mathbf{p}^T R_{n+1}\mathbf{q},0,\ldots,0\right)^T \in \mathbf{R}^{2^j\alpha(n+1)+n}$,

$$\mathbf{L}_i = \left((R_{-n+2i-2}\ \mathbf{p})^T\mathbf{d}_1,\ldots,(R_{-n+2i-2}\ \mathbf{p})^T\mathbf{d}_n,0,\ldots,0\right)^T + (R_n \circ R_{-3n+2i-2})\mathbf{w},\quad i=1,\ldots,n,$$

$$\mathbf{L}_{i+1} = \left(0,\ldots,0,(R_{-n+2i}\ \mathbf{p})^T\mathbf{u}_1,\ldots,(R_{-n+2i}\ \mathbf{p})^T\mathbf{u}_n\right)^T + (R_{-n}\circ R_{-n+2i})\mathbf{w},\quad i = 2^{j-1}\alpha(n+1),\ldots,n-1+2^{j-1}\alpha(n+1).$$

Using the lemma 3 in the article [11] are found $2^{j-1}\alpha(n+1)$ of linearly independent solutions $\mathbf{h}_s=(h_{1,s},h_{2,s},\ldots,h_{2^j\alpha(n+1)+n,s})^T$ of system of linear equation $\mathbf{T}_j\mathbf{h}_s=0$. These solution represent the columns of the matrix $\mathbf{Q}_j=\left(\mathbf{h}_1,\ldots,\mathbf{h}_{2^{j-1}\alpha(n+1)}\right)$. Columns $\mathbf{h}_s$ are chosen in such a way that the functions

$$\psi_{j-1,s}(x)=\mathbf{\Phi}_j(x)\mathbf{h}_s=\sum_{i=1}^{2^j\alpha(n+1)+n} h_{i,s}\cdot\varphi_{j,-n+(i-1)}(x),$$

as possible woyuld represent shifted versions of one function i.e. would have one form (exept for, of course, the boundary wavelets). We introduce the abbreviated notation for matrices composed of matrix elements $\mathbf{T}_j$:

$$\mathbf{T}_j\begin{pmatrix} i_1,\ldots,i_k \\ j_1,\ldots,j_m\end{pmatrix}=\begin{pmatrix} t_{i_1,j_1} & \ldots & t_{i_1,j_m} \\ \ldots & \ldots & \ldots \\ t_{i_k,j_1} & \ldots & t_{i_k,j_m}\end{pmatrix}.$$

For internal wavelets (the carrier is contained in a segment $[0;\alpha(n+1)]$)

$$\mathbf{h}_s=(0,\ldots,0,h_{2s-n-1,s},\ldots,h_{2s+2n,s},0,\ldots,0)^T,\quad s=n+1,\ldots,2^{j-1}\alpha(n+1)-n,$$

where $\mathbf{T}_j\begin{pmatrix} s-n,\ldots,s+2n \\ 2s-n-1,\ldots,2n+2s\end{pmatrix}(h_{2s-n-1,s},\ldots,h_{2s+2n,s})^T=0.$

The solutions corresponding to the boundary wavelets were chosen as follows. For $s=1,2,\ldots,n$ we assume $\mathbf{h}_s=(0,\ldots,0,h_{s,s},\ldots,h_{2n+2s,s},0,\ldots,0)^T$, where

$$\mathbf{T}_j\begin{pmatrix} 1,\ldots,s+2n \\ s,\ldots,2s+2n\end{pmatrix}(h_{s,s},\ldots,h_{2s+2n,s})^T=0.$$

For $s=2^{j-1}\alpha(n+1)-n+1,\ldots,2^{j-1}\alpha(n+1)$ we assume

$$\mathbf{h}_s=(0,\ldots,0,h_{2s-n-1,s},\ldots,h_{2^{j-1}\alpha(n+1)+n+s,s},0,\ldots,0)^T,$$

where

$$\mathbf{T}_j\begin{pmatrix} s-n,\ldots,n+2^{j-1}\alpha(n+1) \\ 2s-n-1,\ldots,n+2^{j-1}\alpha(n+1)+s\end{pmatrix}(h_{2s-n-1,s},\ldots,h_{2^{j-1}\alpha(n+1)+n+s,s})^T=0$$

The following lemma is obvious.

Lemma 4. *Let us assume that* $f\in L_2[0;n+1]$, *then* $\Pi_j f=\mathbf{\Phi}_j\mathbf{C}_j^*$, *where* $\mathbf{C}_j^*=[(\mathbf{\Phi}_j,\mathbf{\Phi}_j)]^{-1}[(f,\mathbf{\Phi}_j)]$.

3.1. Integrals from spline wavelets

Let us assume that $\mathbf{Q}_j=\left(\mathbf{h}_1^j,\ldots,\mathbf{h}_{2^{j-1}(n+1)}^j\right)$, where $\mathbf{h}_s^j=\left(h_{1,s}^j,h_{2,s}^j,\ldots,h_{2^j(n+1)+n,s}^j\right)^T$. According to the results of the previous section

$$\psi_{j-1,s}(x)=\sum_{i=s}^{2s+2n} h_{i,s}^j\,\varphi_{j,-n+i-1}(x),\quad s=1,\ldots,n, \tag{4}$$

$$\psi_{j-1,s}(x)=\sum_{i=2s-n-1}^{2s+2n} h^{j}_{i,s}\varphi_{j,-n+i-1}(x),\quad s=n+1,\dots,2^{j-1}(n+1)-n, \tag{5}$$

$$\psi_{j-1,s}(x)=\sum_{i=2s-n-1}^{2^{j-1}(n+1)+n+s} h^{j}_{i,s}\varphi_{j,-n+i-1}(x),\quad s=2^{j-1}(n+1)-n+1,\dots,2^{j-1}(n+1). \tag{6}$$

As well as in the papers [1]-[5] for the ease of convenience we introduce the following notation

$$w_l(x)=\varphi_{0,l-n-1},\, l=1,2,\dots,2n+1,$$

$$w_l(x)=\psi_{j,s}(x),\ l=2^j(n+1)+n+s,\ j=0,1,\dots,s=1,\dots,2^j(n+1).$$

Figure 1 represents graphs of some functions w_l for the case $n=5$.

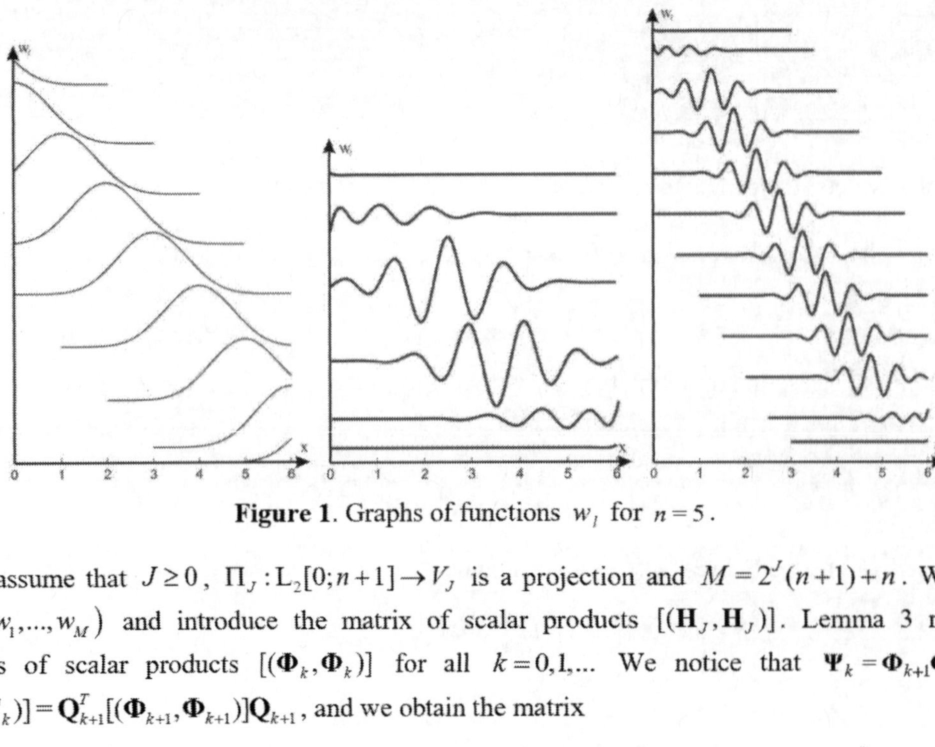

Figure 1. Graphs of functions w_l for $n=5$.

Let us assume that $J\ge 0$, $\Pi_J:\mathrm{L}_2[0;n+1]\to V_J$ is a projection and $M=2^J(n+1)+n$. We denote $\mathbf{H}_J=(w_1,\dots,w_M)$ and introduce the matrix of scalar products $[(\mathbf{H}_J,\mathbf{H}_J)]$. Lemma 3 represents matrices of scalar products $[(\mathbf{\Phi}_k,\mathbf{\Phi}_k)]$ for all $k=0,1,\dots$ We notice that $\mathbf{\Psi}_k=\mathbf{\Phi}_{k+1}\mathbf{Q}_{k+1}$ and $[(\mathbf{\Psi}_k,\mathbf{\Psi}_k)]=\mathbf{Q}^T_{k+1}[(\mathbf{\Phi}_{k+1},\mathbf{\Phi}_{k+1})]\mathbf{Q}_{k+1}$, and we obtain the matrix

$$\left[(\mathbf{H}_J,\mathbf{H}_J)\right]=\begin{pmatrix} \left[(\mathbf{\Phi}_0,\mathbf{\Phi}_0)\right] & 0 & 0\dots & 0 \\ 0 & \mathbf{Q}_1^T\left[(\mathbf{\Phi}_1,\mathbf{\Phi}_1)\right]\mathbf{Q}_1 & 0\dots & 0 \\ \dots & \dots & \dots & \dots \\ 0 & 0 & 0\dots & \mathbf{Q}_J^T\left[(\mathbf{\Phi}_J,\mathbf{\Phi}_J)\right]\mathbf{Q}_J \end{pmatrix}.$$

As $V_J=V_0\oplus W_0\oplus V_1\oplus\ldots\oplus W_{J-1}$, then for $f\in\mathrm{L}_2[0;n+1]$ we have $\Pi_J f=\sum_{l=1}^{M}c_l w_l=\mathbf{H}_J\mathbf{C}_J$ where $\mathbf{C}_J=(c_1,\dots,c_M)^T$. As it is done in papers [1]-[5] we define functions

$$\xi_{1,l}(x)=\int_0^x w_l(t)dt,\quad \xi_{\nu+1,l}(x)=\int_0^x \xi_{\nu,l}(t)dt=\frac{1}{\nu!}\int_0^x (x-t)^{\nu} w_l(t)dt,\nu=1,2,\dots \tag{7}$$

According to defining functions w_l and pars (4)-(6) function $\xi_{\nu+1,l}(x)$ is a linear combination of functions $\eta_{n,\nu}^{j,s}(x)=\int_0^x (x-t)^\nu N_n(2^j t-s)dt$.

Lemma 5. *There is the following recurrence equation found*

$$\eta_{n,\nu}^{j,s}(x)=\frac{x^{\nu+1}}{\nu+1}N_n(-s)+\frac{2^j}{\nu+1}\left(\eta_{n-1,\nu+1}^{j,s}(x)-\eta_{n-1,\nu+1}^{j,s+1}(x)\right), \tag{8}$$

where

$$\eta_{0,\nu}^{j,s}(x)=\begin{cases}\dfrac{(x-a)^{\nu+1}-(x-b)^{\nu+1}}{\nu+1}, & [a;b]=[0;x]\cap\left[\dfrac{s}{2^j};\dfrac{s+1}{2^j}\right]\neq\varnothing;\\ 0, & [a;b]=[0;x]\cap\left[\dfrac{s}{2^j};\dfrac{s+1}{2^j}\right]=\varnothing.\end{cases} \tag{9}$$

Proof. By the characteristic B-splines [10] there is a par found $N_n'(x)=N_{n-1}(x)-N_{n-1}(x-1)$. It follows that by the integrating by parts formula we obtain

$$\eta_{0,\nu}^{j,s}(x)=-\frac{(x-t)^{\nu+1}}{\nu+1}N_n(2^j t-s)\Big|_0^x+2^j\int_0^x\frac{(x-t)^{\nu+1}}{\nu+1}(N_{n-1}(2^j t-s)-N_{n-1}(2^j t-s-1))dt=$$

$$=\frac{x^{\nu+1}}{\nu+1}N_n(-s)+\frac{2^j}{\nu+1}\left(\eta_{n-1,\nu+1}^{j,s}(x)-\eta_{n-1,\nu+1}^{j,s+1}(x)\right).$$

Par (9) is obvious. The lemma is convinced.

Formulas (8), (9) allow us to find the value of a function $\eta_{n,\nu}^{j,s}(x)$ at any point without integration. So for $l=1,2,\dots,2n+1$ we obtain

$$\xi_{\nu+1,l}(x)=\frac{1}{\nu!}\eta_{n,\nu}^{0,l-n-1}(x),\quad l=1,2,\dots,2n+1.$$

For $l=2^j(n+1)+n+s,\quad j=0,1,\dots,\quad s=1,\dots,2^j(n+1)$ we obtain

$$\xi_{\nu+1,l}(x)=\frac{2^{\frac{j+1}{2}}}{\nu!}\sum_{i=s}^{2s+2n}h_{i,s}^{j+1}\eta_{n,\nu}^{j+1,-n+i-1}(x),\ s=1,\dots,n;$$

$$\xi_{\nu+1,l}(x)=\frac{2^{\frac{j+1}{2}}}{\nu!}\sum_{i=2s-n-1}^{2s+2n}h_{i,s}^{j+1}\eta_{n,\nu}^{j+1,-n+i-1}(x),\quad s=n+1,\dots,2^j(n+1)-n;$$

$$\xi_{\nu+1,l}(x)=\frac{2^{\frac{j+1}{2}}}{\nu!}\sum_{i=2s-n-1}^{2^j(n+1)+n+s}h_{i,s}^{j+1}\eta_{n,\nu}^{j+1,-n+i-1}(x),\quad s=2^j(n+1)-n+1,\dots,2^j(n+1).$$

The pars obtained are valid for all $\nu=0,1,\dots$

3.2. Application of a spline wavelets to the solution of linear systems of differential equations

Two equations are considered in the projection methods for solving linear equations; the first in the complete normed space X

$$Kx\equiv x-\lambda Hx=f, \tag{10}$$

And the second one in its complete subspace X_j

$$K_j x_j \equiv x_j - \lambda H_j x_j = \Pi_j f, \tag{11}$$

where H is a continuous linear operator in X, H_j - is a continuous linear operator in X_j and Π_j is a continuous linear operator projecting X to X_j, i. e. $\Pi_j(X) = X_j$, $\Pi_j \circ \Pi_j = \Pi_j$. Equation (10) is called an exact equation, and equation (11) is called an approximate equation. It is assumed that the following conditions are satisfied:

1. (Condition of aboutness of operators H and H_j.) $\|\Pi_j Hx_j - H_j x_j\| \le \rho_j \|x_j\|$ is satisfied for each $x_j \in X_j$.
2. (The condition of a good approximation of elements of the form Hx by elements from X_j.) There is an element $x_j \in X_j$ for each $x \in X$ such that $\|Hx - x_j\| \le \rho_{1,j} \|x\|$.
3. (The condition of a good approximation of the free term of the exact equation.) There is an element $f_j \in X_j$ such that $\|f - f_j\| \le \rho_{2,j} \|f\|$. Unlike the previous conditions $\rho_{2,j}$ here generally speaking, depends on f.

If operator K has a continuous inverse, the equation (10) has a solution and if $\lim\limits_{j\to+\infty} \rho_j = 0$, $\lim\limits_{j\to+\infty} \rho_{1,j} = 0$, $\lim\limits_{j\to+\infty} \rho_{2,j} = 0$ then $\lim\limits_{j\to+\infty} \|x - x_j\| = 0$, where x_j is the solution for equation (11), and x is the solution for equation (10).

We consider a linear system of differential equations (1)-(2). Without loss of generality we can assume that $t_0 = 0$, $t_1 = n+1$, where $n \in \mathrm{N}$. We introduce the notation $y_i(t) = \mathrm{d}x_i(t)/\mathrm{d}t$. Then subject to conditions (2) we obtain $x_i(t) = x_{i0} + \int_0^t y_i(\tau)d\tau$. System (1) can be rewritten in the form

$$\begin{cases} y_1 = \int_0^t \left(a_{11}(t)y_1(\tau) + a_{12}(t)y_2(\tau) + \ldots + a_{1m}(t)y_m(\tau)\right)d\tau + g_1(t); \\ y_2 = \int_0^t \left(a_{21}(t)y_1(\tau) + a_{22}(t)y_2(\tau) + \ldots + a_{2m}(t)y_m(\tau)\right)d\tau + g_2(t); \\ \ldots \\ y_m = \int_0^t \left(a_{m1}(t)y_1(\tau) + a_{m2}(t)y_2(\tau) + \ldots + a_{mm}(t)y_m(\tau)\right)d\tau + g_m(t) \end{cases} \tag{12}$$

where $g_i(t) = f_i(t) + \sum_{j=1}^m a_{ij}(t)x_{j0}$. We rewritten the system obtained in the matrix form $\mathbf{Y}(t) = \int_0^t \mathbf{A}(t)\mathbf{Y}(\tau)d\tau + \mathbf{G}(t)$, where $\mathbf{Y} = (y_1, \ldots, y_m)^T$, $\mathbf{G} = (g_1, \ldots, g_m)^T$, $\mathbf{A} = (a_{ij})_{i,j=1}^{m,m}$. As a complete normed space we consider the space $X = (\mathrm{L}_2([0; n+1]))^m = \mathrm{L}_2([0; n+1]) \times \ldots \times \mathrm{L}_2([0; n+1])$, with the norm $\|\mathbf{Y}\| = \sqrt{\sum_{i=1}^m \|y_i\|_2^2}$, And as its full subspace we consider $X_j = (V_j)^m$. Let us assume that $\mathbf{\Pi}_j^m : X \to X_j$ is defined by a par

$$\mathbf{\Pi}_j^m(\mathbf{Y}) = \left(\Pi_j y_1, \ldots, \Pi_j y_m\right)^T.$$

We denote $\mathbf{A}_i = (a_{i,1},...,a_{i,m})^T \in X$ as i row of matrix $\mathbf{A}$.
We define operators $K: X \to X$, $H: X \to X$ and $H_j: X_j \to X_j$ by pars

$$(K\mathbf{Y})(t) = \mathbf{Y}(t) - \int_0^t \mathbf{A}(t)\mathbf{Y}(\tau)d\tau, \quad (H\mathbf{Y})(t) = \int_0^t \mathbf{A}(t)\mathbf{Y}(\tau)d\tau, \quad H_j = \mathbf{\Pi}_j^m \circ H.$$

Condition of aboutness of operators H and H_j is satisfied with $\rho_j = 0$. Let us assume that $\mathbf{Y} \in X$ and

$$\mathbf{Y}_j(t) = \left(\int_0^t (\mathbf{\Pi}_j^m \mathrm{A}_1(t))^T \mathbf{Y}(\tau)d\tau, ..., \int_0^t (\mathbf{\Pi}_j^m \mathrm{A}_m(t))^T \mathbf{Y}(\tau)d\tau \right)^T \in (V_j)^m.$$

Then $\rho_{1,j} = (n+1)\sqrt{\sum_{i=1}^m \left\| \mathbf{A}_i - \mathbf{\Pi}_j^m \mathbf{A}_i \right\|^2}$. It is obvious that $\lim_{j\to+\infty} \rho_{1,j} = 0$. It follows that the condition of a good approximation of elements of the form $H\mathbf{Y}$ by elements from $(V_j)^m$ is also satisfied. Finally for the certain $\mathbf{G} \in X, \mathbf{G} \neq 0$ we obtain $\mathbf{G}_j = \mathbf{\Pi}_j^m \mathbf{G}$, and $\rho_{2,j} = \dfrac{\left\| \mathbf{G} - \mathbf{G}_j \right\|}{\left\| \mathbf{G} \right\|}$. Then $\lim_{j\to+\infty} \rho_{2,j} = 0$.

We shall seek the solution of the approximate equation

$$\mathbf{Y}_J - \mathbf{\Pi}_J^m \circ H\mathbf{Y}_J = \mathbf{\Pi}_J^m \mathbf{G}, \tag{13}$$

in the for

$$\mathbf{Y}_J = \left(\sum_{i=1}^M c_i w_i(t) \quad \sum_{i=1}^M c_{M+i} w_i(t) \quad \dots \quad \sum_{i=1}^M c_{(m-1)M+i} w_i(t) \right)^T,$$

where $M = 2^J(n+1)+n$. Then we can rewrite the equation (13) in the form of the systems of linear equations for the determination of unknown wavelet coefficients

$$\sum_{i=1}^M c_{(s-1)M+i}(w_i, w_l) = \sum_{i=1}^M \sum_{k=1}^m c_{(k-1)M+i}(a_{sk}\xi_{1,i}, w_l) + (g_s, w_l), \qquad s = 1,\dots,m, l = 1,\dots,M.$$

The approximate solution of system (1) has the form

$$x_{s,J}(t) = \sum_{i=1}^M c_{(s-1)M+i}\xi_{1,i}(t) + x_{s,0}, \ s = 1,2,\dots,m.$$

4. Results of experiments

Multi-dimensional non-stationary linear systems are described by the equations of state and output

$$\mathrm{d}\mathbf{x}(t)/\mathrm{d}t = \mathbf{A}(t)\mathbf{x}(t) + \mathbf{B}(t)\mathbf{g}(t), \quad \mathbf{x}(0) = \mathbf{x}_0 \tag{14}$$

$$\mathbf{y}(t) = \mathbf{L}(t)\mathbf{x}(t), \tag{15}$$

where $\mathbf{x} = (x_1,...,x_m)^T$ is a m-dimensional state vector; $\mathbf{g} = (g_1,...,g_r)^T$ is a r-dimensional input stimulus vector; $\mathbf{y} = (y_1,...,y_k)^T$ is a k-dimensional output vector; t is time; $\mathbf{A}(t) = (a_{i,j}(t))_{i,j=1}^{m,m}$, $\mathbf{B}(t) = (b_{i,j}(t))_{i,j=1}^{m,r}$, $\mathbf{L}(t) = (l_{i,j}(t))_{i,j=1}^{k,m}$ are m×m, m×r, k×m dimention matrices respectively. The search for the law of change of the state vector and the output vector according to the given input signal shall be understood to mean the analysis of the output process.

We shall seek an approximate solution of the system (14) in the form

$$x_{s,J}(t)=\sum_{i=1}^{M}c_{(s-1)M+i}\xi_{1,i}(t)+x_{s,0},\; s=1,2,\ldots,m,$$

where $\mathbf{x}_0=\left(x_{1,0},\ldots,x_{m,0}\right)^T$ and coefficients c_s are found from the linear equation systems

$$\sum_{s=1,s\neq k}^{n}\sum_{i=1}^{M}(a_{k,s}\xi_{1,i},w_\alpha)c_{(s-1)M+i}+\sum_{i=1}^{M}(a_{k,k}\xi_{1,i}-w_i,w_\alpha)c_{(k-1)M+i}=-\sum_{s=1}^{n}(a_{k,s},w_\alpha)x_{s,0}-\sum_{s=1}^{r}(b_{k,s}g_s,w_\alpha),$$
$$k=1,\ldots,n,\;\alpha=1,\ldots,M.$$

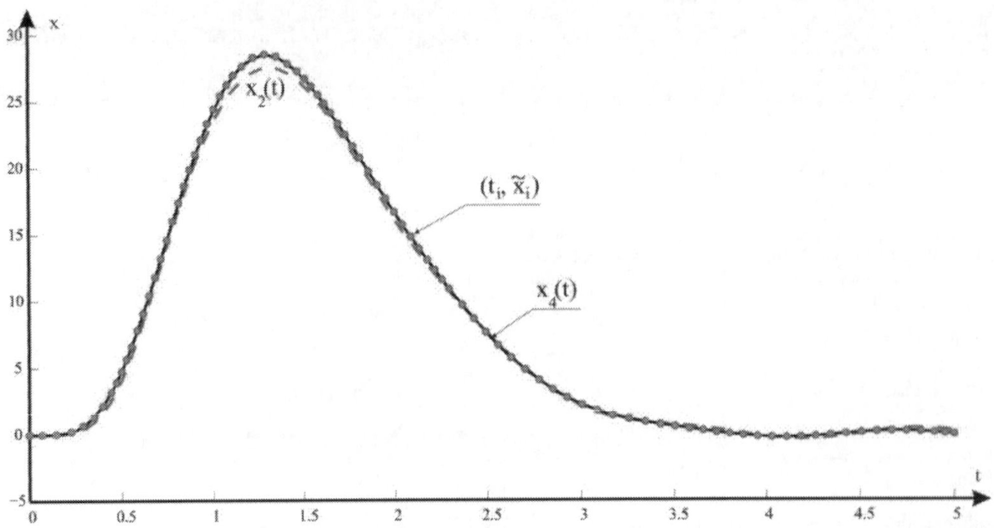

Figure 2. Approximation graphs $x_2(t)$ (dashed line), $x_4(t)$ (continuous line) and a grid function graph $\{(t_i,\tilde{x}_i)\}$ (points), obtained using the Runge-Kutta method.

Example 1. *Let us consider a nonstationary automatic control system whose behavior is described by a differential equation*

$$\sum_{k=0}^{5}a_k(t)x^{(k)}(t)=g(t),$$

Where coefficients $a_k(t)$ *are determined from the following expression:*

$$\begin{pmatrix}a_0(t)\\a_1(t)\\a_2(t)\\a_3(t)\\a_4(t)\\a_5(t)\end{pmatrix}=\begin{pmatrix}0{,}5596 & 1{,}8918 & 2{,}5825 & 1{,}7855 & 0{,}6277 & 0{,}0909\\0{,}7113 & 2{,}3843 & 3{,}222 & 2{,}1975 & 0{,}7588 & 0{,}1065\\0{,}3717 & 1{,}2333 & 1{,}6449 & 1{,}1038 & 0{,}3728 & 0{,}0507\\0{,}1002 & 0{,}3278 & 0{,}43 & 0{,}2827 & 0{,}093 & 0{,}0122\\0{,}014 & 0{,}0449 & 0{,}0576 & 0{,}0369 & 0{,}0118 & 0{,}0015\\0{,}0008 & 0{,}0025 & 0{,}0031 & 0{,}0019 & 0{,}006 & 0{,}00007\end{pmatrix}\begin{pmatrix}1\\t\\t^2\\t^3\\t^4\\t^5\end{pmatrix}.$$

Find the response of the system to the input effect

$$g(t) = (85,7661 + 338,5984t + 497,0437t^2 + 406,9496t^3 +$$
$$+186,9354t^4 + 46,7809t^5 + 4,8258t^6)e^{-4t}.$$

Initial conditions are zero. Research interval is $[0;5]c$.

Solution. Since the initial conditions are zero, an approximate solution of this problem will be sought in the form $x_J(t) = \sum_{s=1}^{M} c_s \xi_{5,s}(t)$, where coefficients $c_1,\dots,c_M$ are defined from the linear equation system

$$\sum_{s=1}^{2^J(n+1)+n} c_s \left(a_5 w_s + \sum_{k=0}^{4} a_k \xi_{5-k,s}, w_l \right) = (g, w_l), \quad l = 1,\dots,2^J(n+1)+n.$$

Figure 2 shows graphs of the third and fifth approximations $x_2(t)$ (dashed line), $x_4(t)$ (continuous line) and a grid function graph $\{(t_i, \tilde{x}_i)\}$ (points) obtained using the Runge-Kutta method.

Example 2. *The behavior of a linear nonstationary system is described by the following system of differential equations*

$$\begin{pmatrix} dx(t)/dt \\ dy(t)/dt \end{pmatrix} = \begin{pmatrix} t^2 & 1-t \\ 1+t & t-t^2 \end{pmatrix} \begin{pmatrix} x(t) \\ y(t) \end{pmatrix} + \begin{pmatrix} t^2 & 0 \\ 1 & t \end{pmatrix} \begin{pmatrix} g_1(t) \\ g_2(t) \end{pmatrix}$$

Find the response of the system to the input effect

$$g_1(t) = 0,23315158t^9 - 3,89665t^8 + 26,4309725t^7 - 93,4794t^6 + 183,95t^5 - 200,83t^4 +$$
$$+122,255277t^3 - 50,135386t^2 + 13,095959t - 2,8237;$$
$$g_2(t) = -0,071962459t^{13} + 1,3465024t^{12} - 10,98105044t^{11} + 51,1908385t^{10} - 150,5098287t^9 +$$
$$+291,295256t^8 - 378,61242t^7 + 336,683591t^6 - 213,9681871t^5 + 106,48891t^4 - 47,3676t^3 +$$
$$+19,56997t^2 - 3,863587t - 0,0004283.$$

for initial conditions $x(0) = -1, y(0) = 2$ *over a time slot of* $[0;2]$ *c*.

Solution. An approximate solution will be sought in the form $x_J(t) = -1 + \sum_{s=1}^{M} c_s \xi_{1,s}(t)$, $y_J(t) = 2 + \sum_{s=1}^{M} c_{M+s} \xi_{1,s}(t)$, where $M = 2^J(n+1)+n$, and coefficients $c_s, s = 1,2,\dots,2M$ are defined from the linear equation system

$$\sum_{s=1}^{M} c_s \left((w_s, w_l) - \int_0^{n+1} t^2 \xi_{1,s}(t) w_l(t) dt \right) - \sum_{s=1}^{M} c_{M+s} \int_0^{n+1} (1-t) \xi_{1,s}(t) w_l(t) dt =$$
$$= \int_0^{n+1} (t^2 g_1(t) - t^2 + 2(1-t)) w_l(t) dt, l = 1,2,\dots,M;$$
$$\sum_{s=1}^{M} c_s \int_0^{n+1} (1+t) \xi_{1,s}(t) w_l(t) dt + \sum_{s=1}^{M} c_{M+s} \left(\int_0^{n+1} (t-t^2) \xi_{1,s}(t) w_l(t) dt - (w_s, w_l) \right) =$$
$$= \int_0^{n+1} (2t^2 - t + 1 - g_1(t) - t g_2(t)) w_l(t) dt, l = 1,2,\dots,M.$$

Figure 3 shows graphs of the first and thrid approximations $x_0(t), y_0(t)$ (dashed line), $x_2(t), y_2(t)$ (continuous line) and a grid function graph $\{(t_i, \tilde{x}_i)\}$, $\{(t_i, \tilde{y}_i)\}$ (points) obtained using the Runge-Kutta method.

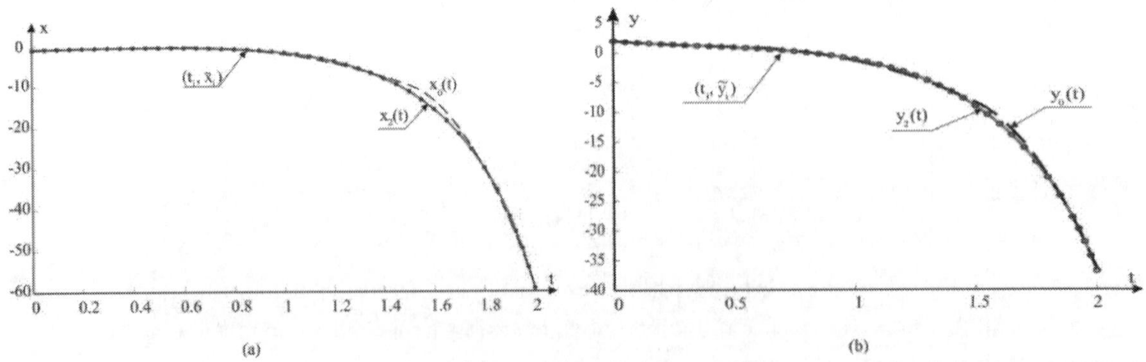

Figure 3. Approximation graphs: (a) $x_0(t)$ (dashed line), $x_2(t)$ (continuous line) and a grid function graph $\{(t_i, \tilde{x}_i)\}$ (points) obtained using the Runge-Kutta method; (b) $y_0(t)$ (dashed line), $y_2(t)$ (continuous line) and a grid function graph $\{(t_i, \tilde{y}_i)\}$ (points) obtained using the Runge-Kutta method.

5. Summary and conclusion

In this paper we have considered the use of spline wavelets for the approximate solution of linear systems of differential equations and as an application for analyzing the output processes of multidimensional linear nonstationary control systems. As a novelty of the present paper one can note the obtainment of finite formulas for approximations of any C^n smoothness class of solutions of linear differential equations systems by means of application of spline-wavelets constructed on the basis of a B-spline of arbitrary order n. Known methods based on Haar wavelets are obtained from the methods presented in this article when $n = 0$.

References

[1] Lepik Ü 2007 Numerical solution of evolution equations by the Haar wavelet method *Appl. Math. Comput* **185** pp 695–704
[2] Lepik Ü 2008 Haar wavelet method for solving higher order differential equations *Int. J. Math. Comput* **1** pp 84–94
[3] Lepik Ü 2007 Application of the Haar wavelet transform to solving integral and differential equations *Proc. Estonian Acad. Sci. Phys. Math* **56** pp 28–46
[4] Lepik Ü 2005 Numerical solution of differential equations using Haar wavelets *Math. Comput. Simul* **68** pp 127–43
[5] Lepik Ü and Hein H 2014 Haar wavelets with applications (Berlin: Springer) p 207
[6] Blatov I A and Rogova N V 2013 Application of semiorthogonal spline wavelets and the Galerkin method to the numerical simulation of thin wire antennas *Computational Mathematics and Mathematical Physics* **53** pp 727–36
[7] Finkelstein A and Salesin D H 1994 Multiresolution curves *ACM SIGGRAPH* pp 261–268
[8] Frazier M. W 1999 *An introduction to wavelets through linear algebra* (New York: Springer) p 503
[9] Daubechies I 1992 *Ten lectures on wavelets* (Philadelphia: SIAM) p 358
[10] Chui Ch K 1991 *An introduction to wavelets* (Boston: Academic press) p 412
[11] Bityukov Yu I and Akmaeva V N 2016 The use of wavelets in the mathematical and computer modelling of manufacture of the complex-shaped shells made of composite materials *Bulletin of the South Ural State University. Ser. Mathematical Modelling, Programming and Computer Software* **9** pp 5–16

Numerical modelling of cylindrical composite materials heating in dynamic modes

K Blinov, S Galunin and Yu Blinov

St. Petersburg Electrotechnical University,
Prof. Popova St. 5, 197373 St. Petersburg, Russia

E-mail: kyblinov@gmail.com

Abstract. Presented paper describes the dynamic processes, which are happens during the process of creating the steel pipes with protection layer. This protection layer saves the pipes from chemical reactions on inner or outer side during their using in industry processes. The main aim of presented work is receiving of requested temperature profile in the place of connection of the steel and protection material.

1. Introduction

The receiving of composite materials is quite difficult process. Many parameter parameters must be taken into consideration during the designing phase of creating the installation. The easiest way to make some finance and time economy is using the numerical modelling of the process as a part of designing phase. It is necessary to receive requested temperatures in the workpiece, because under- and overheating of the materials will decreased the quality of the final product.

2. Numerical models and simulation algorithm

In the frame of this work it is need to receive the results of the heating process for the case, when the edge of the pipe is coming over the induction system and comes to “quasi”-steady state mode. The pipe is coming over the inductor and the heating temperature on the connection place between steel and protection material must be close to 1000 °C. The starting temperature of the process is equals to 20 °C. Underheating will not give the required connection between both materials which reduce the quality of the workpiece and overheating can damage final multilayer pipe. As it is shown in figure 1 a and 1 b the numerical models include: inductor (1), steel pipe (2), protection material (3) which is located inside the pipe in one case and outside in the second one, and thermal isolation (4) to save the inductor from the heating processes. Figure 1 shows the numerical model on 60th calculation step. Numerical simulation has been done in 2D and numerical models have cycling symmetry over Y-axis. The investigated installation has the same inductor. The difference between both numerical models consists only in the workpiece. The dimensions of the steel pipes are the same, but the protection material locates inside or outside from it. The frequency of the current in both models is equals to 1000 Hz and the current value has been chosen to receive 1000 °C ± 1% in the protection material on the time, when the process starts to be in “quasi”-steady state mode. The numerical models have been done in ANSYS® Software [1] which gives a possibility to make structural analysis on the next design steps.

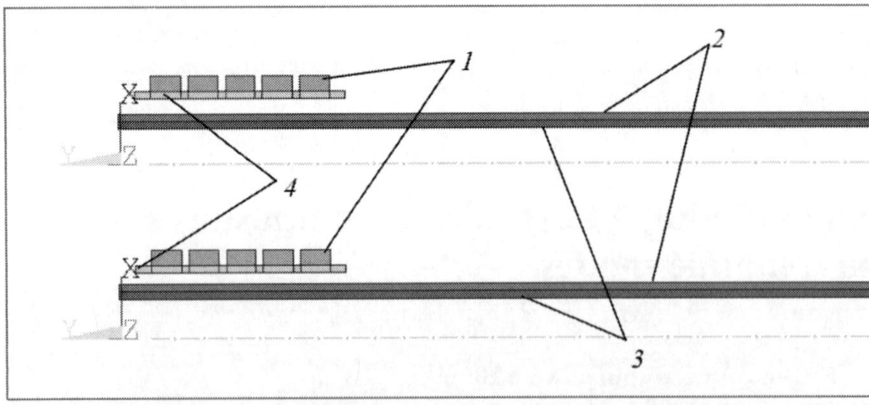

Figure 1. Numerical models of both induction systems.

The numerical models include the movement of the workpiece over the inductor and figure 2 shows the location of the edge of the workpiece relatively the inductor on different calculation steps (40th – 2-a, 50th – 2-b, 60th – 2-c and 100th – 2-d). On 100th calculation step the numerical model comes to "quasi"-steady state mode in electromagnetic and thermal calculation.

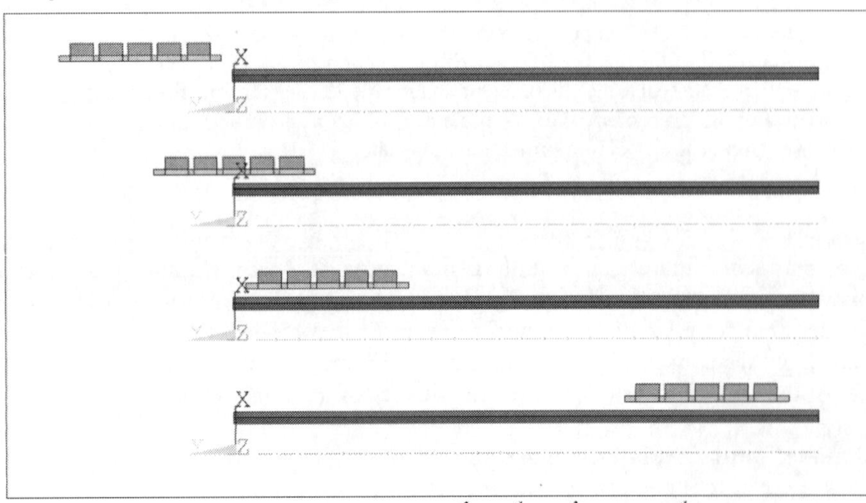

Figure 2. Location of the pipes edge on 40th, 50th, 60th and 100th calculation steps.

The numerical algorithm for numerical models is presented in figure 3. The presented algorithm already has been used for other tasks [2], therefore this algorithm does not need to been verified. On the first part of simulation algorithm comes the creating of electromagnetic and thermal databases for each location of the workpiece relatively the inductor. Then comes the second part of the algorithm: the electromagnetic calculation for the first calculation step. The main result of the electromagnetic calculation is Joule heat distribution in the workpiece, which is taking into account in the third part of the algorithm: thermal analysis. Besides the Joule heat generation, thermal analysis also includes temperature depended thermal properties like thermal conductivity and specific heat, and also it includes convection and radiation of induction system. Then the algorithm comes back to the second part – electromagnetic analysis. The received on previous step temperature distribution is includes to second electromagnetic calculation step, electromagnetic properties of the system have a correction to received temperature profile and starts the second electromagnetic calculation. The cycle of the algorithm is looped until last calculation step, when the system starts to be in "quasi"-steady state mode.

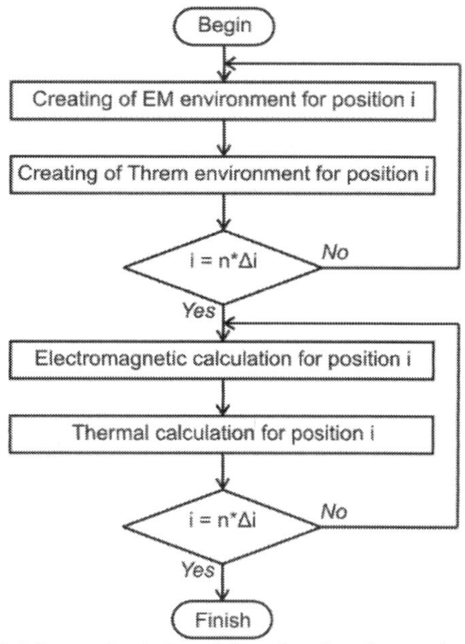

Figure 3. Numerical algorithm for the dynamic models.

3. Electromagnetic analysis

As it was described earlier, the main result of electromagnetic analysis is a Joule heat distribution in the workpiece because of initial conditions for thermal calculation. The figure 4 shows the Joule heat distribution on. 40th (4-a), 50th (4-b) and 60th (4-c) calculation step for the numerical model, where protection material locates inside the steel part of pipe. The view of Joule heat on 100th calculation step looks completely like on 60th calculation step and will not be presented here. On 100th calculation step only the values of Joule heat are a little bit different because of temperature depended electromagnetic properties. The maximal values of the Joule heat are presented in figure 4-b because the electromagnetic field of "free" part of the inductor has an influence to the part of the workpiece which is already located in the inductor. In figure 4 are shown only the steel parts of the workpiece because the protection material of the workpiece is sturdy against to induction frequencies.

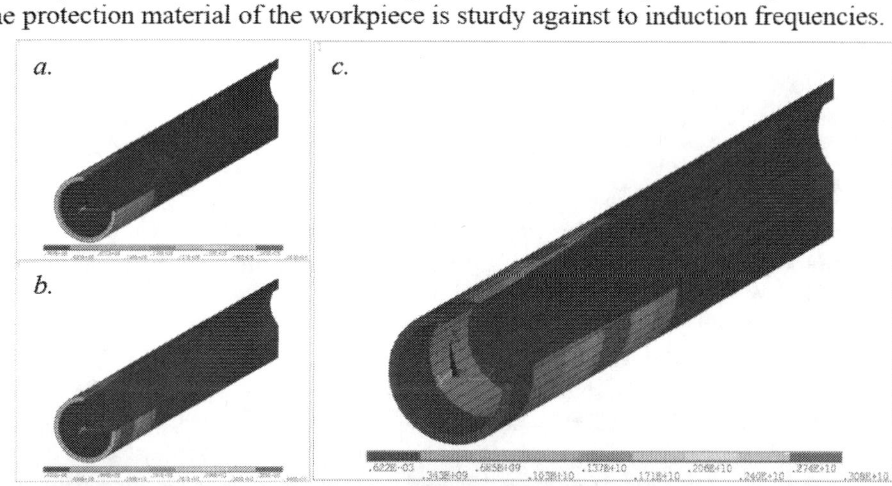

Figure 4. Joule heat distribution in the steel during the simulation process.

In other numerical model, where the protection material locates outside from the steel, the Joule heat distribution is repeats the Joule heat distribution of the first model because of the same dimensions of the steel part of the workpiece and induction system. Only the values of Joule heat are a little bit different because of the temperature depended electromagnetic properties of the induction system.

4. Thermal analysis

The temperature distribution in the protection material and inductor insulation of the numerical model, where protection material locates inside the steel pipe on 40th (5-a), 50th (5-b), 60th (5-c) and 100th (5-d) calculation step is presented in figure 5. It is possible to see, that on 40th calculation step the temperature is increased only for 29 °C from the staring temperature, but on the 50th step, when the edge of the workpiece is located in the middle of induction coil, the increasing is already equals to 348 °C. On 60th calculation step the value of the maximal temperature is equals to 930 °C and on 100th calculation step, when the system is already came to "quasi"-steady state mode, the value of the maximal temperature is equals to 1006 °C. The rings of cold temperature in thermal insulation on last calculation steps are coming from boundary conditions. Induction coils are cooling down by water and have temperature 20 °C, but the thermal insulation is heating up from convection and radiation processes, therefore maximal temperature of thermal insulation is equals to 775 °C. It is possible to see in figure 6-a. The temperature distribution in this system on last calculation step is shown in figure 6-b. The maximal temperature in complete workpiece is 1032 °C what is on 26 °C more than in the protection material, but maximal temperature 1000 °C ± 1% must be only in the protection material.

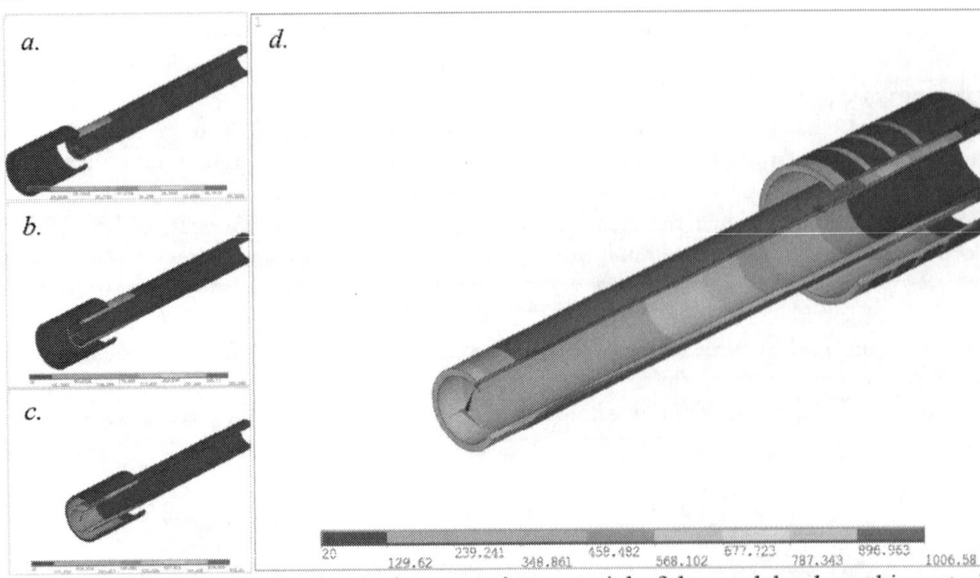

Figure 5. Temperature distribution in the protection material of the model, where this material is located inside the steel pipe, during the heating process.

Figure 7 accordingly shows the temperature distribution in the protection material and in inductor insulation on 40th, 50th, 60th and 100th calculation step in the model, where protection material locates outside the steel part of the workpiece. The maximal values of temperatures are higher than in previous case. On 40th calculation step the maximal temperature is equals to 66 °C, what is for 17 °C more than in previous case. The maximal temperature of 50th calculation step is 578 °C, what is for 210 °C more than in previous case. On 60th calculation step the maximal temperature in the workpiece is for 24 °C more than in previous case and equals to 954 °C. But on the last calculation step the maximal value of the temperature in protection material is equals to 1007 °C, what coincide to initial task of the simulation.

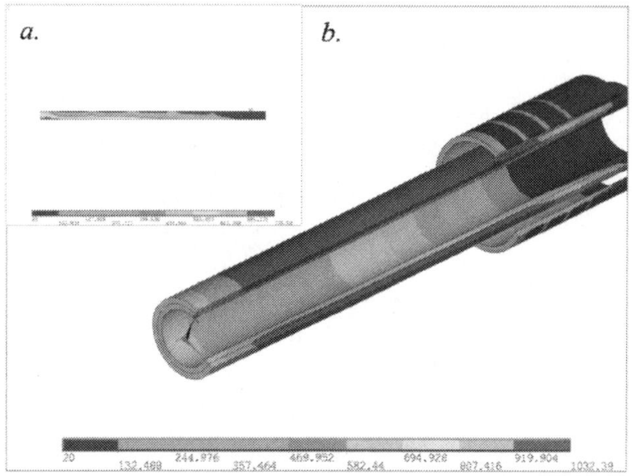

Figure 6. Final temperature distribution in the numerical model, where protection material is located inside the steel pipe.

Thermal distribution in thermal insulation (8-a) and in all system of induction composite pipe heating system in "quasi"-steady state mode (8-b) with outer protection material are shown in figure 8. The thermal insulation is not heated too much like in previous case because the outer protection material is heated most by thermal conductivity from steel pipe and has other thermal properties. The thermal isolation has been heated only for 93 °C. The maximal temperature in the steel pipe is equals to 1033 °C but the protection material has been heated to required temperature.

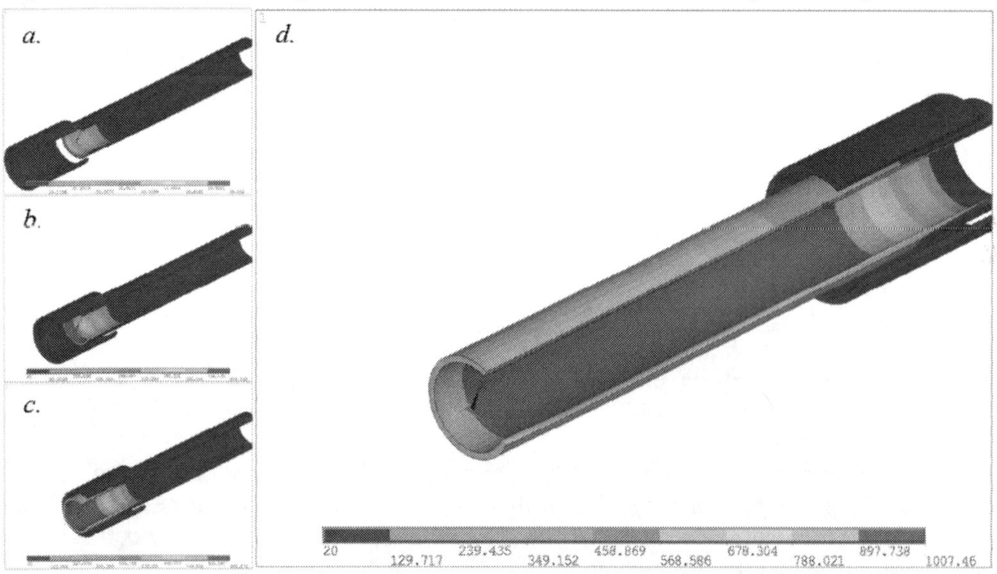

Figure 7. Temperature distribution in the protection material of the model, where this material is located outside the steel pipe, during the heating process.

The dynamics of maximal temperature in the protection material (dash curves) and in the steel pipe (permanent curves) for the numerical models where protection material locates inside (grey curves) and outside (black curves) during the induction heating simulation is locates in the figure 9. In case, where protection material locates outside the steel pipe, the heating of both materials is happens more similar till the moment, when edge of the workpiece is coming out from the inductor. In other case, the heating of steel pipe comes faster.

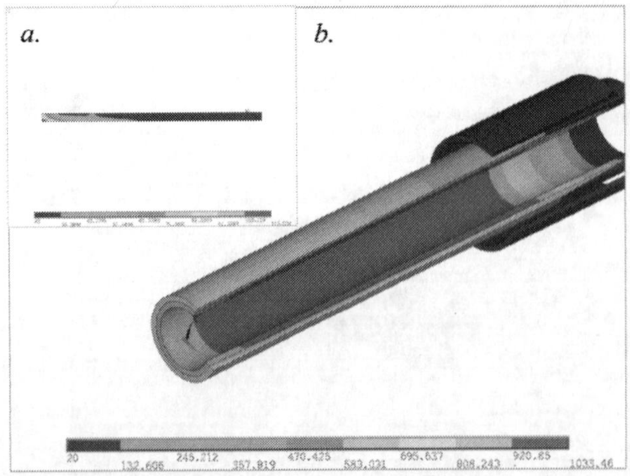

Figure 8. Final temperature distribution in the numerical model, where protection material is located outside the steel pipe.

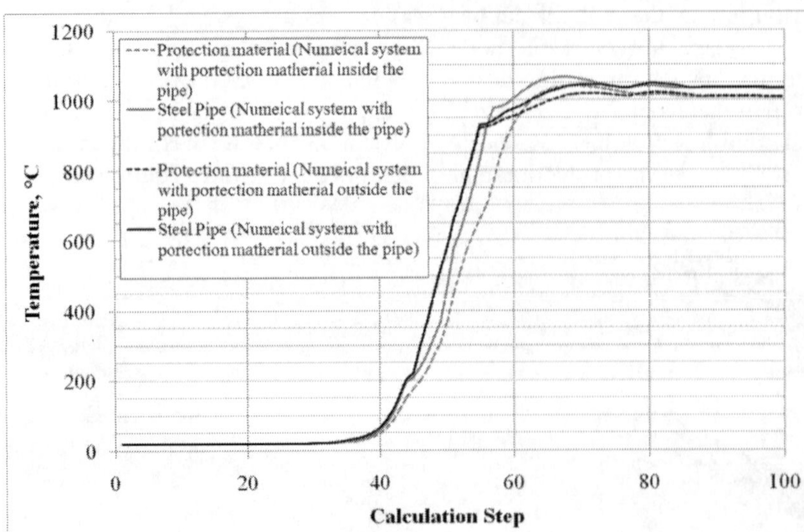

Figure 9. Dynamics of the maximal temperature in both models during all heating process.

5. Conclusions and outlooks

The developed dynamic numerical models have been done. This gives more information about the dynamics of the composite tube heating. The next work is directed to simulate the numerical systems with variable value of the current to make overheating of the protection material much smaller in the time, when edge of the workpiece is coming during the induction coils. But received results are already extending to using them in design phase of induction heating installation for receiving the composite pipes.

References

[1] http://www.ansys.com/

[2] Blinov K, Nikanorov A, Nacke B and Klöpzig M 2011 Numerical simulation and investigation of induction through-heaters in dynamic operation mode *COMPEL - The international journal for computation and mathematics in electrical and electronic engineering* ISSN: 0332-1649 pp 1539 – 1549

Study of power-to-weight ratio of the electrothermal propulsion system of nanosatellite maneuvering satellite platform

V N Blinov, I S Vavilov, V V Kositsin, A I Lukyanchik, V I Ruban and V V Shalay

Omsk State Technical University (OmSTU), 11, Mira ave., Omsk, 644050, Russian

E-mail: lukyanchik1991@mail.ru

Abstract. The direction of the solution of the actual task of maneuvering satellite platforms (MSP) design for nanosatellite weighing up to 10 kg, power-to-weight ratio of PS up to 8 W(electrothermal micro engine (ETME) 5 W, vaporizer 2 W, electrovalve up to 1 W) and with characteristic velocity up to 60 m/s were considered on the basis of studies of the propulsion system(PS) with ETME. The aim of study is the confirmation of technical possibility of nanosatellites design with mass up to 10 kg, power-to-weight ratio up to 8 W and with characteristic velocity up to 60 m/s on the basis of PS prototype experimental studies. In the course of the research tasks were solved to determine the design of PS and ETME of nanosatellit's MSP, determine the electric parameters of PS depending on power consumption that determining specific impulse of ETME, and estimate the implemented characteristic velocity of the nanosatellite. The PS constructive scheme of nanosatellite mass of 10 kg was design, PS experimental prototype was produced and PS experimental research on ammonia were conducted. The 200°C was reached per 900 s at 5 W ETME power consumption with nitrogen, that equivalent to specific impulse of ammonia ETME 124/136 s when entering the stationary mode. 2 W energy consumption of a two-thread liquid ammonia vaporizer is experimentally substantiated. The using of electrovelve stepped control cyclogram allowed to reduce the average power consumption to 1 W.

1. Introduction

At present, the relevance of MSP with ammonia PS design is becoming increasingly important, as modern nanosattelites are designed for solving of a wide range of task, including the tasks of orbital maneuvering: Elimination of orbiting errors of nanosatellite by rocket launch vehicles, maintenance of orbital parameters, interorbital maneuvering, construction of orbital groups of nanosatellites and MSC, retraction of the nanosatellite into the orbit of utilization. There are also a number of specific application and scientific tasks: inspection of other MSC and orbital objects, monitoring of Earth outer space, space objects removal into the orbit of burial, etc. [1-3].

The using of ammonia PS in nanosatellites is due to the posibility of obtaining high values of the specific impulse of ETME with minimal energy consumption. This leads to a reduction in the mass of fuel expended on the implementation of a given characteristic velocity of the nanosatellite as part of the orbital manoeuvres as well as the mass of the PS and its means of adaptation to nanosatellite.

The recently created ammonia PS with ETME were part of the "Ugatusat" MSC (Russia, 2008) weighing 35 kg and "BX-2" MSC (China, 2016) weighing 47 kg. These ammoniac PS can be considered as analogues of PS for nanosatellites.

In the MSC "Ugatusat", the weight of the fueling ammonia was 0.4 kg, the power consumption of the ETME was 30 W. The automatic control system include: electrovalve, vaporizer, reducer, fueling and drainage clutch.

Vaporization of ammonia Is accomplished by preliminary gasification of ammonia in the vaporizer and final heating in the ETME. PS thrust is 30 mN, specific impulse - up to 200 s [1].

In PS of MSC "BX-2" vaporization of ammonia was realized due to the heating of ammonia by solar radiation. Vaporizer, separator and reducer were introduced in PS design. PS thrust is 85 mN, specific impulse is 100 s.

In Ukraing the ammonia PS were designed for MSC "Sich - 2M" (2011-2012) and MSC "Microsat".

At the stage of maneuvering nanosatellite designing, while limiting its mass with the realization of a adjusted characteristic velocity, the task is to redistribute a adjusted mass to the maximum possible mass of PS and the minimum possible mass of electronic components and the design of a nanosatellite.

In turn, the mass of the PS is redistributed to the maximum possible mass of fueling ammonia and the minimum possible mass of the design of the PS.

This approach to the design of ammonia PS leads to a failure to reserve elements of the pneumohydraulic system of PS and to the use of a highly reliable "cold" ETME switching scheme.

A mass-related problem is the choice of optimum nanosatellite power supply from the condition of obtaining a adjusted characteristic velocity.

The costs of the characteristic velocity for the implementation of typical orbital maneuvers of a nanosatellite weighing 10 kg can be:

- Elimination of launch errors (inclination of the orbit, period of rotation and altitude) - 12 m/s;
- sustaining a lifetime of 1 year in a circular orbit 600 km, 700 km - 3 m/s, 1.1 m/s;
- a maneuver from a circular orbit 600 km to 650 km - 27 m/s;
- move into orbit of utilization with the lifetime of 25 years from a circular orbit of 700 km - 31 m/s.

Analysis of the costs of the characteristic velocity indicates that the choice of the tasks of the orbital maneuvering of the nanosatellite should be carried out based on the capabilities of the nanosatellite in terms of the available reserves of the characteristic velocity, taking into account the power supply limitation.

For example, the CAN-X4 nanosatellite (Canada) has a characteristic velocity margin of only 14 m/s at a mass of 7 kg. The power supply system of the nanosatellite provides a "peak" power input of up to 10 W.

The nanosatellite is built on Sel's Generic NanosateLlite Bus platform. The platform has a cubic shape with a side of 20 cm and is equipped with CNAPS PS.

The CNAPS PS force module has a size of 18×12.5×7 cm. As a working fluid, liquid sulfur hexafluoride is fed from two tanks with a capacity of 300 ml each.

The working body is voporized and discharged through four nozzles, controlled independently. The specific impulse of the nozzle is 35 s.

When the nanosatellite's dimensions are limited, the fuel reserves are determined by the layout of the MSP with PS and ETME.

In accordance with figure 1 shows the configuration of PS with ETME according to the classical scheme, in which the fuel tank of PS is cylindrical with torosferic bottoms, and the ETME is located near the tank. This layout increases the size of the PS by the length of the ETME.

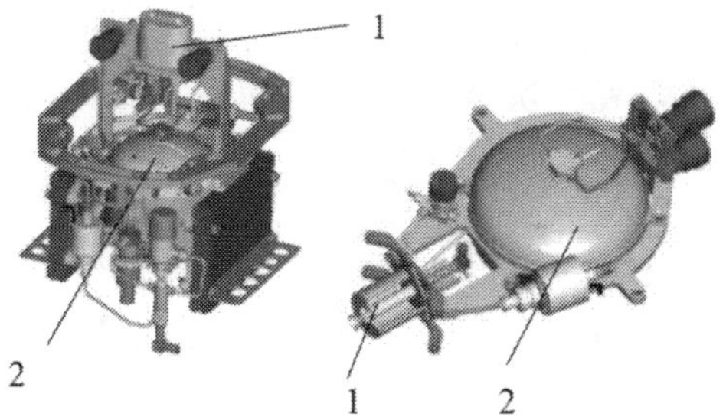

Figure 1. The PS layouts of the classical scheme:
1 – fuel tank; 2 – ETME.

The length of the ETME is determined by the layout of its heating zone and the current outputs zone. With their co-axial arrangement, the length of the ETME is maximized (Figure 2). The reduction of the length of ETME is obtained in the T-shaped layout of the heating zone and the current outputs zone (Figure 3). The greatest length has an ETME with a circuit matched with the vaporizer. In this ETME, the functions of the ETME and the vaporizer of a working flued are implemented in one structure (Figure 4) [1, 4, 5].

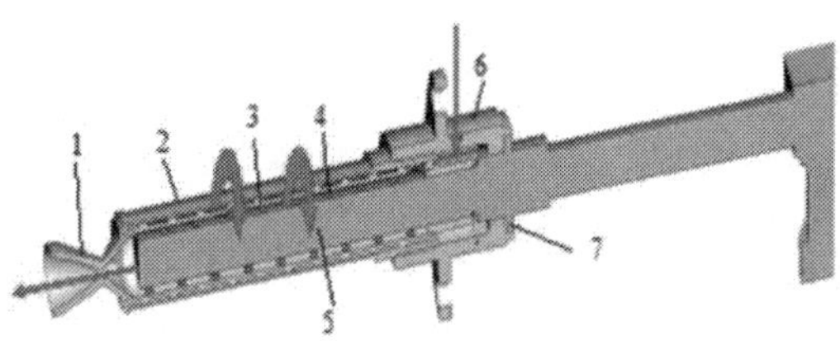

Figure 2. ETME scheme with coaxial arrangement of the heating zone and the zone of the current collectors: 1 – the proffiling nozzle; 2 – Hull; 3 – internal hull with gas lines; 4 – gas flow generator; 5 – heater; 6 – attachment point of the microengine; 7 – heater flange.

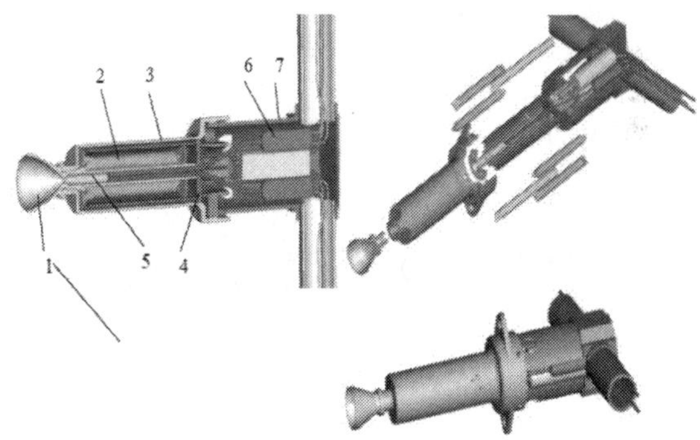

Figure 3. ETME with a T-shaped arrangement of the heating zone and the zone of the current collectors: 1 – the profiling nozzle; 2 – tubular heating element (main and reserve); 3 – hull; 4 – a barrel with a collar; 5 – gas pipeline; 6 – thermocouple; 7 – the body of the current collectors.

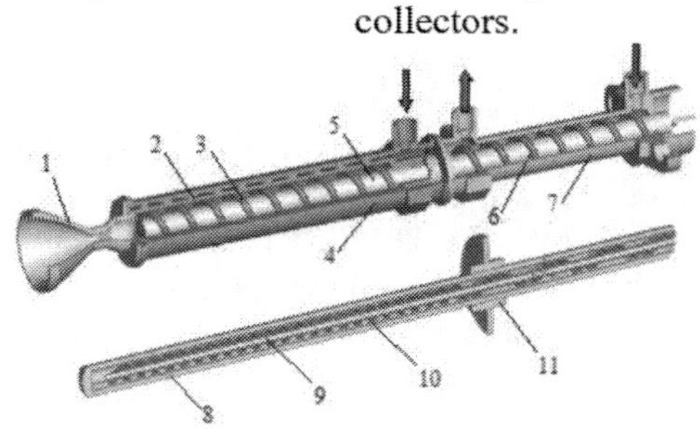

Figure 4. General view of the ETME scheme combined with the vaporizer: 1 – is the nozzle; 2 – gas pipeline; 3 – swirler of ETME; 4 – ETME hull; 5 – heater; 6 – vaporizer swirler; 7 – vaporizer hull, 8 – heater hull; 9 – single-channel metal tube with thermocouple; 10 – heater; 11 – flange.

Designed ammonia ETME with a thrust of 30 mN are characterized by a thrust price of up to 2 W/mN and are used in a number of MSC with a mass of 30-120 kg. The specific impulse of such

ETME at a power consumption of 60 W does not exceed 250 s. Their circuit design with appropriate modifications allows them to be used as part of a PS for a nanosatellite [4, 5].

The developed ETME for small spacecrafts differ in their constructive diversity and in the type of fuels used [5 ± 9]. Often in ETME can use as a working fluid compressed nitrogen, which has environmental safety. A nitrogen-based working flued is used in the practice of ground-based experimental testing of ammonia ETME, as well as in some flight CPS. So the firm SSTL (Great Britain) for MSC "UoSAT 12" designed an ETMD of the type "Resistojet" with a working fluid of nitric oxide.

In the ETME of the "Resistojet" type, nitrogen oxide is vaporizing by means of an electric heating element receiving power from the batteries of the MSC. With a specific impulse of 127 s, a thrust of 93 mN (9.5 gs) and an energy consumption of 90 W, the ETME provides a total increase in the velocity of the MSC "UoSAT 12" at 10.4 m/s.

In general, energy-consuming elements of the remote control are an electrovalve, an evaporator and an ETME. When designing a PS for a nanosatellite, the task is to reduce the energy consumption of an ETME, an electrovalve and a vaporizer.

One of the important characteristics of remote control is the time of one-time activation. An increase in the duration of one-time activation of the remote control makes it possible to reduce the number of remote control inputs during the development of the working fluid. A critical element in this case is an electrovalve due to possible overheating during operation.

In a number of works, the possibility of creating maneuvering nanosatellites is based on the results of autonomous studies of ETME while the energy consumption ranges of 5 W to 30 W. However, the relevance of the research ETME in the composition of the PS.

2. Formulation of the problem

The task is to determine the ways of designing and improving the design and construction of ammoniac PS with ETME for MSP of nanosatellites.

The methodological basis for the formation of the design and structural appearance of the PS is the multi-purpose method of structural design [5].

At the heart of the accepted methodology for creating ammonium PS for nanosatellite MPS is experimental testing of new technical solutions in conditions meeting safety requirements. Therefore, compressed nitrogen is used instead of toxic ammonia as a modeling working body.

In this connection, the following experimental studies are of practical interest for the purpose of creating a DV for the nanosatellite SMEs:

- electric tests of PS in a vacuum chamber when working on nitrogen;
- autonomous electrical tests of the electrovalve with control of the power consumption at the open and hold modes.

3. Formatting the text

PS was designed by a multi-purpose structural design method, in which the fuel tank is the completing structure of the PS and is developed optimally for a specific target task [1, 5].

The design of the remote control for the nanosatellite ICP is considered under the limitations of the nanosatellite dimensions 200×200×200 mm. When designing the PS, the layout of the increased density was used due to the execution of a toric-shaped fuel tank with the placement of ETME in the internal cavity of the tank (Figure 5).

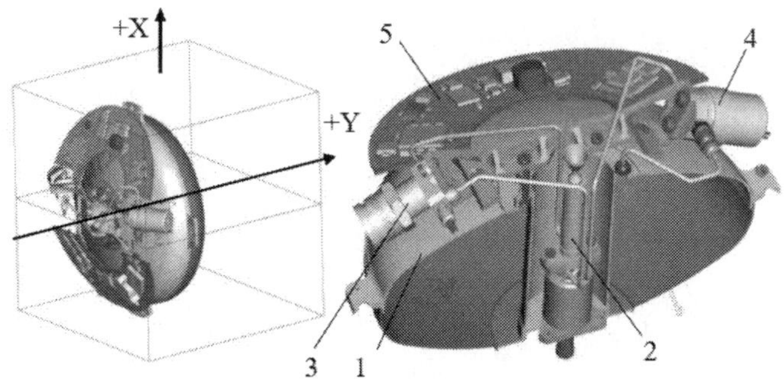

Figure 5. Scheme of the PS with increased density of the layout for the nanosatellite MSP with ETME with vaporizer: 1 – fuel tank; 2 – ETME with the vaporizer; 3 – throttle assembly; 4 – electro valve; 5 – board with electronic components.

To carry out experimental studies, a PS scheme with a two-pass vaporizer, an electrovalve, a pressure regulator, and an experimental PS sample was designed(Figure 6 – 8).

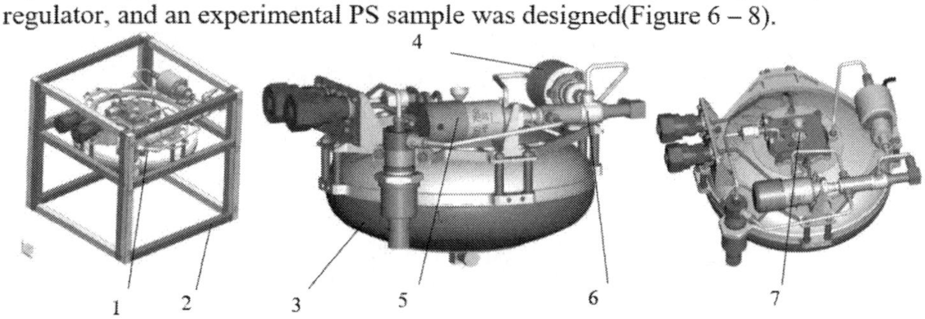

Figure 6. Scheme of the MSP of a nanosatellite with an increased density of the layout for carrying out experimental studies: 1 – PS; 2 – the hull of a nanosatellite; 3 – a fuel tank; 4 – electro valve; 5 – pressure regulator; 6 – vaprorizer; 7 – ETME.

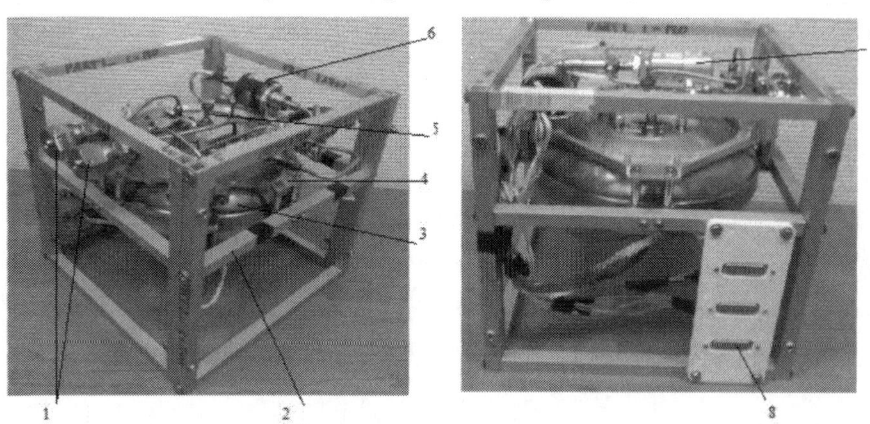

Figure 7. Experimental PS of a MSP of nanosatellite with an increased density of configuration for conducting experimental studies: 1 – filling and drainage coupling; 2 – the hull of a nanosatellite; 3 – a fuel tank; 4 – the rack of the PS; 5 – ETME; 6 – electro valve; 7 – pressure regulator; 8 – electrical connectors.

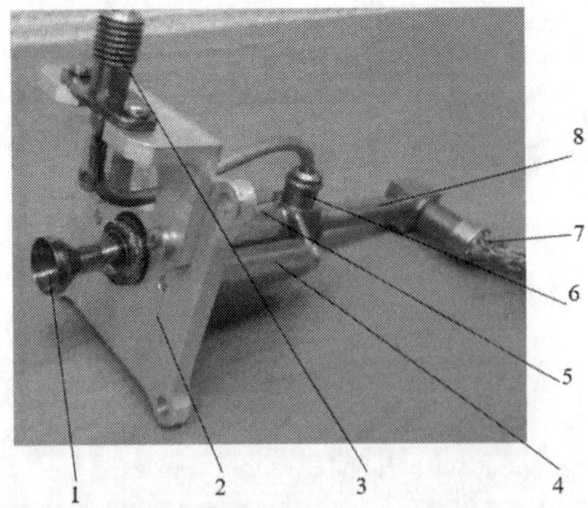

Figure 8. Experimental ETME: 1 – nozzle; 2 – titanium platinum; 3 – connecting the working fluid; 4 – racks; 5 – thermal insulation;6 – ETME gas inlet; 7 – current leads (heating element and thermocouple); 8 – self-contained heating element with built-in thermocouple.

The pneumohydraulic circuit is shown in accordance with Figure 9.

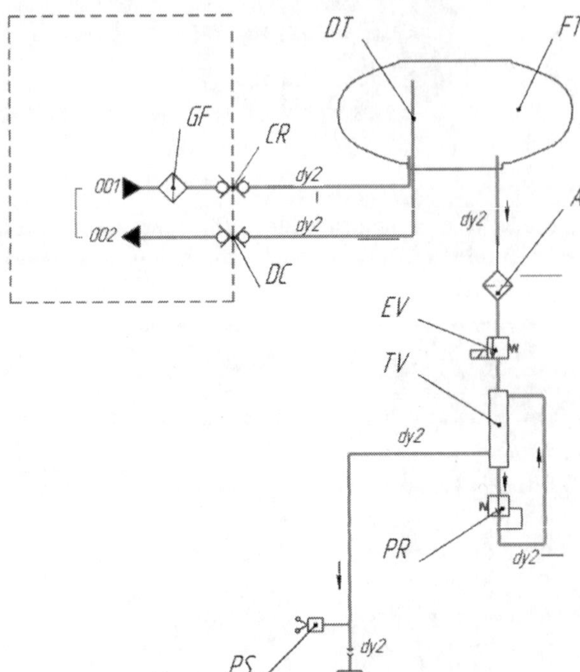

Figure 9. Pneumohydraulic circuit of PS: GF – ground filter, DC – drainage coupling; CR – clutch refilling; DT – drainage tube; FT – fuel tank; AF – airborne filter; EV – electro valve; TV – toway vaporizer; PR – pressure regulator; PS – pressure sensor; ETME – electrothermal microengine.

The characteristics of the PS are given in Table 1.

IOP Conf. Series: Journal of Physics: Conf. Series **944** (2017) 012020 doi:10.1088/1742-6596/944/1/012020

Table 1. PS Characteristics

№	Parameter	Value
1	PS thrust, mN	up to 30
2	ETME type	with self-contained heater
3	Construction mass, kg	2.28
4	Pressure in fuel tank, MP	1.9
5	Fuel tank capacity, l	0.917
6	Volume of filled working fluid, l	0.78
7	Volume of gas cushion, l	0.137
8	Average value of ammonia filled, kg	0.484
9	Energy consumption range, W	5-30

The range of power consumption is assumed equal to 10-30 W and was carried out in order to determine the amount of heat loss of the working fluid (loss of characteristic velocity) at elevated powers of 20-30 W, which are difficult to realize in the nanosatellite power supply system.

The electrical tests of the PS have been carried out for the purpose of research:

– dynamics of heating of ETME and vaporizer without supplying the working fluid under normal atmospheric conditions;

– dynamics of heating of the ETME and the vaporizer without supplying the working medium in the vacuum chamber;

– dynamics of heating of the ETME and the vaporizer with the supply of a working fluid in the vacuum chamber (the "cold" start-up method);

– functioning of the PS in accordance with the cyclogram ("hot" start-up method).

During the research, the bench base of the research laboratory "Propulsion systems of microthrust for small spacecrafts" of OmSTU (Figure 10, 11) was used.

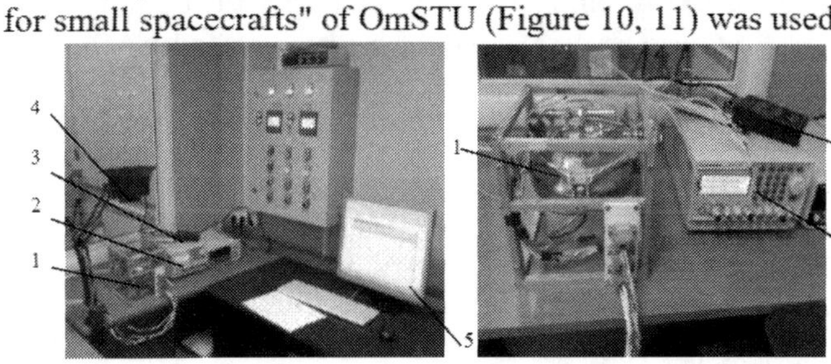

Figure 10. Experimental stand for PS on air without supply of a working fluid: 1 – experimental sample of PS; 2 – multi-channel power supply APS-7205L; 3 – multichannel temperature meter MIT-8; 4 – cooling system; 5 – personal computer.

Figure 11. Experimental bench for PS study in a vacuum chamber with a working fluid feed: 1 – hose of the pipeline of supply of working fluid to CPS; 2 – experimental sample of CPS; 3 – vacuum chamber; 4 – balloon with a working fluid (nitrogen).

To reduce the charging cycles of the PS with nitrogen, a test circuit with a balloon with nitrogen connected to the PS is implemented in accordance with Figure 12.

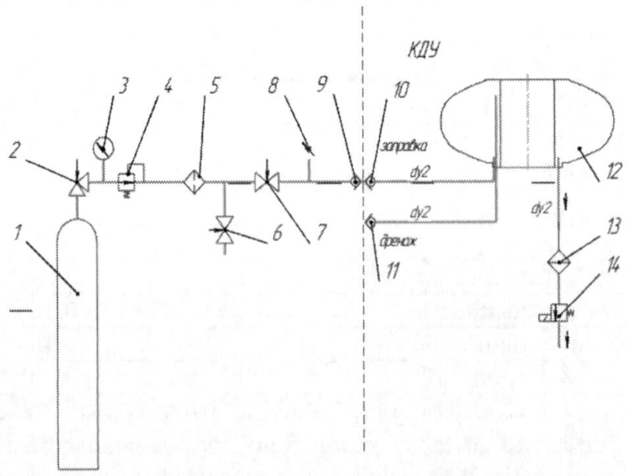

Figure 12. Scheme of connecting a baloon with nitrogen to the PS: 1 – a baloon with nitrogen; 2, 6, 7 – the valves; 3 – manometer; 4 – reducer; 5 – the filter; 8 – manometer; 9,10 – filling couplings; 11 – drain coupling; 12 – fuel tank of the PS;13 – the filter; 14 – electrovalve.

A normally closed solenoid valve is used as the shut-off valve for supplying the working fluid of the PS. The resistance of the coil of the solenoid R = 160 Om, with the nominal voltage corresponding to the opening of the valve with full stem lift is 27 V, and the power consumed by the valve is 3.5 W.

Holding the valve in the open state requires a deliberately lower supply voltage, and correspondingly, less power consumption, which reduces the total power consumption of the PS. For this purpose, a step-by-step cycle of switching on the solenoid valve was used, with a short-time supply of a guaranteed valve opening voltage, followed by a step-by-step decrease thereof to the guaranteed confinement in the open state.

The control of the voltage applied to the valve winding was carried out according to the scheme shown in Figure 13.

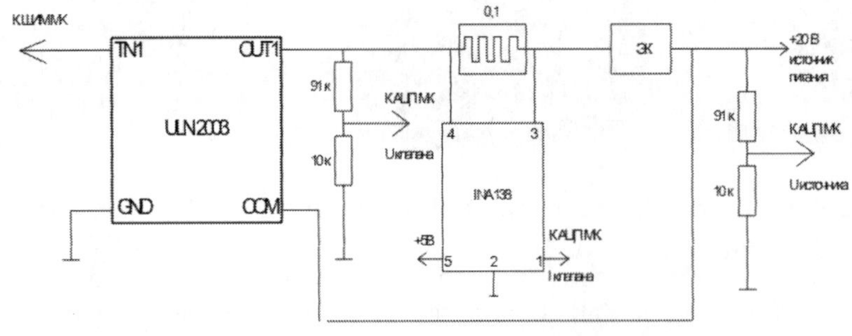

Figure 13. Principal electrical diagram of the solenoid valve control.

At the inlet of the solenoid valve, a pressure of 1.9 MPa was applied, corresponding to the maximum pressure of the working fluid in the tank. Opening was recorded using a float rotameter, installed after the solenoid valve.

The increase in the operating time of the solenoid valve in the PS can be reached by cooling it with ammonia by contact removal of heat with a screw flow line with ammonia (Figure 14).

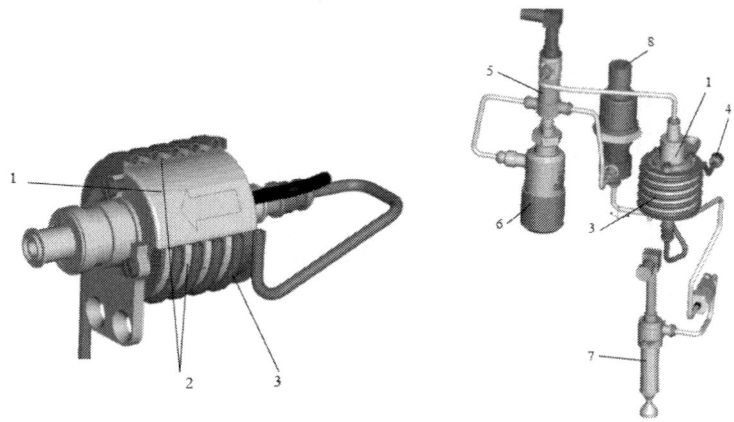

Figure 14. Start valve with cooling system: 1 – electro valve; 2 – cooling jacket; 3 – a tube with ammonia; 4 – ammonia inlet from the tank; 5 – vaporizer; 6 – pressure regulator; 7 – ETME; 8 – pressure sensor.

The interaction of the mass characteristics of the nanosatellite, the thrust of the ETME, the specific impulse, the time of operation and the number of ETME(PS) inclusions is described by the expression [1]:

$$V=-g_0\cdot P_{UD}^{V}\ln\left(\frac{m_{NS}-\frac{P_{ETME}}{P_{UD}^{R}}T^{R}(N-1)-\frac{P_{ETME}}{P_{UD}^{V}}T^{V}N}{m_{NS}-\frac{P_{ETME}}{P_{UD}^{R}}T^{R}\cdot(N-1)}\right)-g_0P_{UD}^{R}\ln\left(\frac{m_{NS}-\frac{P_{ETME}}{P_{UD}^{V}}T^{V}N-\frac{P_{ETME}}{P_{UD}^{R}}T^{R}N}{m_{NS}-\frac{P_{ETME}}{P_{UD}^{V}}T^{V}\cdot N}\right)$$

where, V is the characteristic velocity realized by the PS in the nanosatellite (NS);
P_{UD}^{V} – the average specific impulse of the ETME when entering the mode;
P_{UD}^{R} – the average specific impulse of the ETME after entering the mode;
m_{NS} – start mass of nanosatellite;
P_{ETME} – ETME thrust;
T^{V} – time of ETME release to the mode;
T^{R} – ECMD operating time on the mode;
N – the number of ETME(PS) inclusions.

4. Results of experiments

Below are the results of studies of the vaporizer and ETME in the composition of the PS, as well as the results of autonomous tests of the PS solenoid valve. The results of the experiments on the temperature of the vaporizer are given in Table 2.

Table 2. Results of experimental studies of evaporator heating dynamics

№	Characteristic	Value			
1	Vaporizer power consumption, W	2	3	4	5
2	Maximum vaporizer temperature when operating under normal conditions without gas, °C	76.36	87.79	98.55	105.68
3	Maximum vaporizer temperature when operating under normal conditions without gas in a vacuum, °C	120.74	142.23	150	156.17

4	The maximum temperature of the evaporator when operating in vacuum with the supply of a working fluid, °C	89.88	101.73	110	112.3
5	The set time for the maximum temperature of the evaporator, s	1200	1200	1200	1200
6	Dialing time up to 90% of the maximum temperature of the evaporator, s	600-800			

The results on the investigation of the temperature of the ETME under a cold startup scheme are shown in Figure 15 – 17.

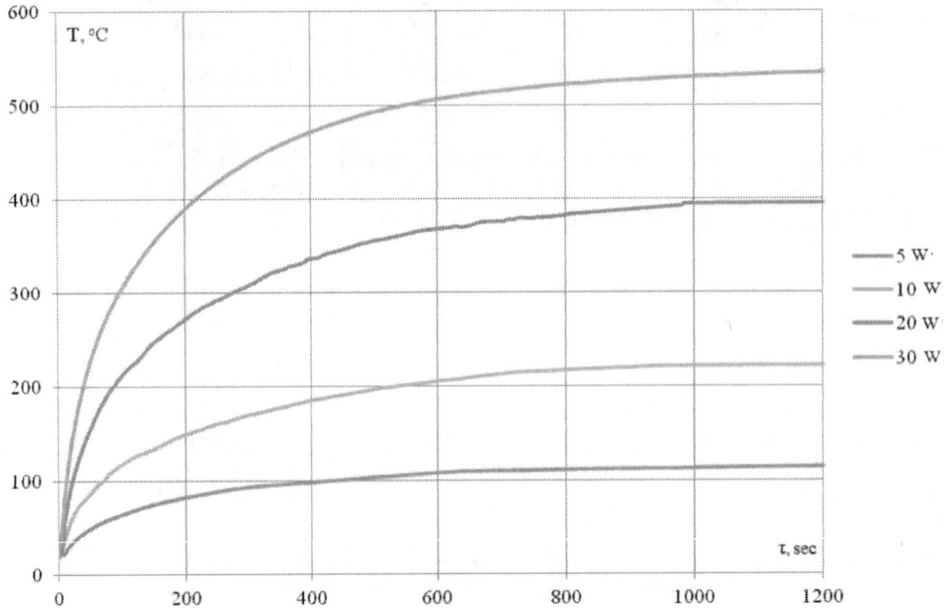

Figure 15. Dependence of the temperature of the ETME on the heating time at different power supply without gas under normal atmospheric conditions.

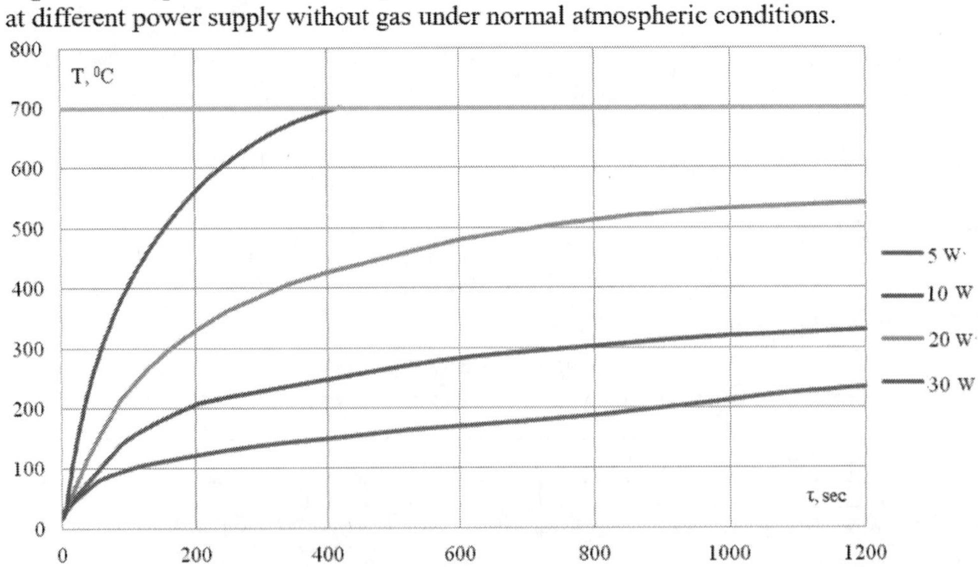

Figure 16. Dependence of the temperature of the ETME on the heating time at different power supply without gas in a vacuum with a heating temperature limitation of 700 °C.

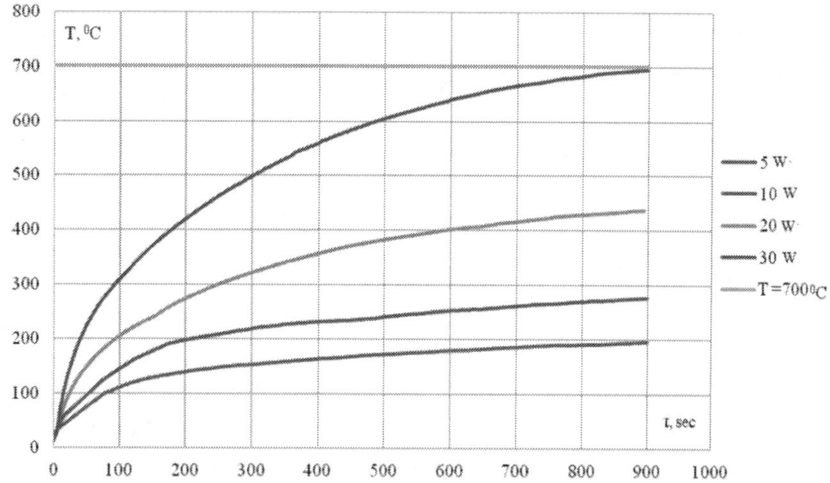

Figure 17. Dependence of the temperature of the ETME on the heating time at different power supply with gas supply at a cold start-up method in vacuum conditions with a temperature limit of 700 °C.

The resistance of the coil of the solenoid valve used is 160 Ω, while the nominal voltage corresponding to the opening of the electrovalve with a full stem lift is 27 V, and the power consumed by the valve is 3.5 W.

Studies have shown that holding the solenoid valve in the open state requires a deliberately lower supply voltage and lower power consumption, which reduces the total power consumption of the PS. For this, a step-by-step cyclogram of switching on the solenoid valve is used. A short-term supply of a guaranteed valve opening voltage was carried out, followed by a stepwise reduction of it to a guaranteed confinement in the open state.

At the inlet of the solenoid valve, a pressure of 1.9 MPa was applied, corresponding to the maximum pressure of the working fluid in the tank of PS. Opening was recorded using a float rotameter, installed after the solenoid valve.

S a result of the experiment, a cyclogram including the following stages (Figure 18) was obtained:

– the opening voltage of the solenoid valve for a duration of 3 s, with a power consumption of 1.85 W;

– smooth decrease of voltage applied to the solenoid valve, within 7 s, with a decrease in power to the value of the holding of the solenoid valve;

– holding the solenoid valve, with the required duration of the PS, with a power consumption of 0.12 W.

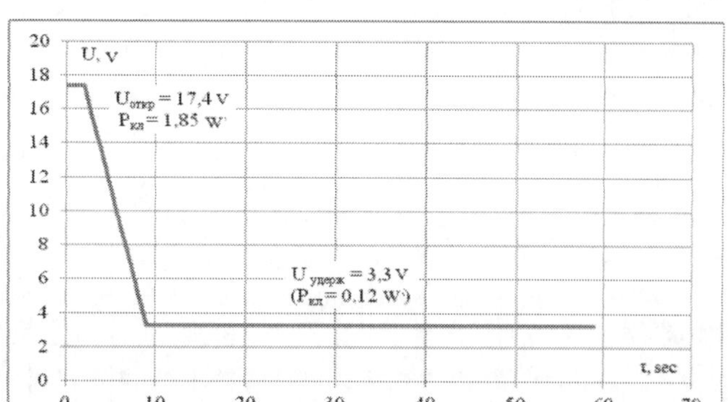

Figure 18. Cyclogram of the solenoid valve

5. The discussion of the results

The considered power capacity of the nanosatellite consists of the energy consumption of the vaporizer, ETME and electrovalve. The main purpose of the vaporizer is gasification of liquid ammonia supplied from the fuel tank. In accordance with Figure 9 the vaporizer is made by a two-pass, which allows gasification of liquid ammonia with minimal energy consumption (Figure 19) [10].

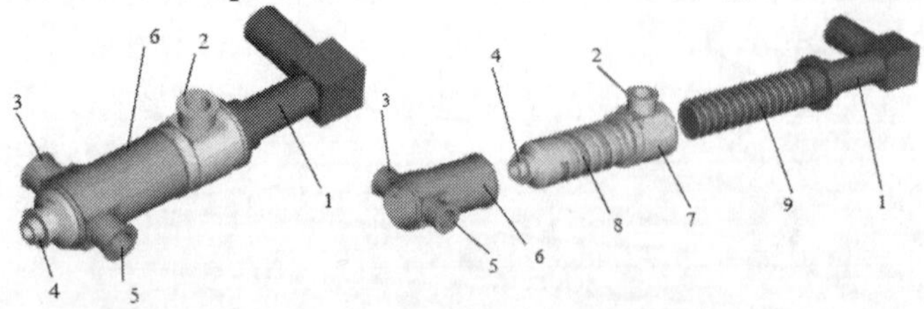

Figure 19. Two-threaded vaporizer of the liquid ammonia: 1 – heating element; 2, 5 – ammonia inlet / outlet fittings; 6,7 – external and internal hull; 8 – two-threaded ammonia feed channel; 9 – swirler.

Researches show:

– at the power consumption of the vaporizer is changed from 2 W to 5 W, the maximum temperature of the enclosure is 90 °C - 112 °C;

– The dialing time of 90% of the maximum temperature of the vaporizer is 600 s - 800 s.

In view of the fact that the autonomous heating element of the vaporizer allows the use of a hot start-up circuit (starting before the working fluid is supplied), the vaporizer must be switched on before the ETME is switched on for 600 s - 800 s for its heating. At the same time, the energy consumption of the vaporizer can be assumed equal to 2 W.

Another measure of reducing the power consumption of the PS is the use of ETME combined with the vaporizer scheme(Figure 4). The main considered energy-consuming element of the PS is ETME.

The obtained experimental values of the ETME temperatures for energy consumption of 5 W, 10 W, 20 W, and 30 W made it possible to estimate the specific impulse of the ETME when entering the mode(P_{UD}^{V}) and mode (P_{UD}^{R}) when operating on nitrogen (table 3).

Table 3. Results of ETME experimental researches

№	The specific impulse with Nitrogen, s	ETME energy consumption, W			
		5	10	20	30
1	P_{UD}^{V}	100	105	120	144
2	P_{UD}^{R}	105	114	133	164

Table 4 shows the calculated specific impulse values of ETME with autonomous heating element and a profiled nozzle with using nitrogen and ammonia.

Table 4. The calculated specific impulse values at different power consumption

№	Temperature, °C	47	127	207	287	367	447	527	607
1	The specific impulse with Nitrogen, s	85	95	105	114	123	131	139	147
2	The specific impulse with ammonia, S	107	122	136	149	161	173	183	194

From the table it follows that the use of ammonia is preferable to the specific impulse of the ETME.

The increase in the specific impulse of the ETME is achieved due to an increase in power consumption to 10 W - 30 W. This can be achieved only by increasing the mass of the nanosatellite (by increasing the size and mass of the solar battery - SB). For stationary SB, the specific mass of the SB can reach up to 4.93 kg/m^2. Let's estimate the expediency of using the energy consumption in a nanosatellite PS 10W - 30W.

As a criterion, we take the realizable characteristic velocity of the PS in the nanosatellite composition of the reduced mass, taking into account the mass of the SB of the considered power (Table 5).

Table 5. Dependence of the characteristic velocity on the specific impulse and energy consumption

№	ETME energy consumption, W	Nanosatellite characteristic				
		Mass of nanosatellite, kg	ETME specific impulse with ammonia, P^V_{UD}/P^R_{UD}, s	Working time, TV/TP, s	Mass of ammonia, kg	Nanosatellite characteristic velocity, m/s
1	5	10.0	124/136	900/300	0.484	63
2	10	10.5	136/149			65
		11.0				62
		12.0				57
3	20	12.0	159/196			68
		13.0				63
		14.0				58
4	30	14.0	187/220			68
		15.0				63

The analysis showed that the desire to realize on the nanosatellite energy consumption for the PS of 10 W - 30 W does not give significant advantages for the realized characteristic velocity in comparison with the nano-satellite with the mass of 10 kg and the power consumption of the PS of 5 W.

In this regard, in the future, it is necessary to use the power consumption of an ETME of 5 watts. Improving the design of ETME in terms of reducing its mass to increase temperature while limiting energy consumption in conjunction with the producing of ETME by additive technology. Investigations of the power consumption of the solenoid valve showed that when it is held in the open state, the power consumption can be 0.12 W. A step-by-step cyclogram of power control of the solenoid valve can be realized by the onboard control system of the nanosatellite with minimal mass costs.

The use of the obtained cyclogram will reduce the overall power consumption of the PS.

In addition, together with the application of the electrovalve cooling system (Figure 14), the duration of a single switching can be increased by reducing the heat dissipation of the solenoid valve with a reduced power consumption. In practice, to ensure the guaranteed opening and retention of the solenoid valve in the open state, when designing the onboard control system, the magnitude of the control voltages should be increased by 10-20% of those determined during the experiment.

6. Summary and conclusion

1. As a result of the experimental studies carried out by the PS, the possibility of designing a nanosatellite maneuvering platform with ammonium PS and ETME was confirmed.

2. The realized characteristic velocity of a nanosatellite with an ammonia PS with an energy consumption of an ETME of 5 W, an vaporizer of 2 W and an electrovalve of less than 1 W is estimated at 63 m / s.

3. The modes of operation of the PS are defined:ETME

- cold start with simultaneous supply of voltage and working fluid (hot start is also possible);
- -vaporizer- start by hot scheme with preheating for 600 s - 800 s;
- -electrovalve is a step-by-step cyclogram of the voltage supply.

4. Defined constructive ways to upgraiding PS:

- the implementation of a toric-shaped fuel tank with the placement of ETME in the internal cavity of the tank;
- the implementation of ETME combined with the vaproizer circuit;
- equipping the solenoid valve with an ammonia cooling system;
- implementation of the vaporizer by two-way scheme;
- production of ECME by additive technology for weight reduction.

References

[1] Blinov V N, Schalay V V, Zubarev S V, Kositsin V V, Ruban V I and Khodoreva E V 2014 *Studies of electrothermal microengines of corrective propulsion systems for maneuvering small spacecrafts: monograph* (Omsk: OmSTU Publishing House) p 264

[2] Blinov V N, Sechenov I N and Shalai V V 2015 *Small spacecrafts: reference allowance* (Omsk: OmSTU Publishing House) p 184

[3] Volotseyev V V, Tkachenko I S and Safronov S L 2012 Selection of design parameters for universal platforms of small spacecrafts *Bulletin of the Samara State Aerospace University behalf S.P. Korolev* (Samara: SSAU Publishing House) pp 35 – 47

[4] Blinov V N, Vavilov I S, Kositsin V V, Ruban V I, Khodoreva E V and Schalay V V 2015 Design features and research of Electrothermal Microthrusters with Autonomous Heating Elements for the Purposes of Small Space Vehicle Orbital Manoeuvring *Indian Journal of Science and Technology* vol. 8 (27) p 10

[5] Blinov V N, Vavilov I S, Kositsin V V, Ruban V I, Khodoreva E V and Shala V V 2015 The studies of small space vehicles ammoniac electrothermal engine units design and structural layout *Modern Applied Science* vol. 9, no. 5 (Published be canadian Center of Science and Education) pp 337 – 357

[6] Coxhill I, Gibbon D and Drube M 2008 The Evolution of Xenon Resistojet Propulsion Systems at SSTL Surrey Satellite *5th International Spacecraft Propulsion Conf.*(Crete) p 6

[7] Gibbon D, Ward J and Ray N 2000 The Design, Development and Testing of a Propulsion System for the SNAP-1 Nanosatellite *14th Annual AIAA/USU Conference on Small Satellites* (Logan: Utah) p 7

[8] Sweetin M N, Lawrence T, Leduc J 1999 Low-cost maneuvers for minisatellites using novel resistojet thrusters *Proceedings Institution of Mechanical Engineers* vol. 213 . part G. pp 223 – 231

[9] Hodnenko V P and Chromov A V 2009 Correctional propulsion system for a small spacecraft vol. 109 *Electrical mechanic tasks* (Moscow) pp 27 – 32

[10] Zubarev S V, Nazarenko L N and Piyukov C A 2011 Ways to optimize the design of the heater of an electrothermal micro engine *Problems of development, manufacture and operation of rocket-space and aviation equipment: materials of the sixth All-Russia. sci. tech. Conf.* (Omsk: OmSTU Publishing House) pp 86 – 89

Technical vision system for analysing the mechanical characteristics of bulk materials

A V Boikov, V A Payor and R V Savelev
Saint-Petersburg mining university, 21 line V.O, Saint-Petersburg, 2, 199106, Russia

e-mail: boikov_av@mail.ru

Abstract. In this article actual topics concerned with mechanical properties of bulk materials, usage of computer vision and artificial neural networks in this research are discussed. The main principles of the system for analysis of bulk materials mechanical characteristics are described. Bulk material outflow behaviour with predefined parameters (particles shapes and radius, coefficients of friction, etc.) was modelled. The outflow was modelled from the calibrated conical funnel. Obtained dependencies between mechanical characteristics and pile geometrical properties are represented as diagrams and graphs.

1. Introduction

Bulk materials are widely spread in nature and are also actively used in different industries. In spite of their wide application, there is no united theory that could predict the behaviour of flowing medium or classify the mechanical characteristics of particle interaction based on simple and fast manipulations. Accurate mathematical description of flowing medium behaviour is becoming impossible because of huge amount of contemporaneous affecting factors. In the majority of cases, table values of friction parameters for individual materials are used to define the mechanical characteristics of bulk material, but such method doesn't consider all aspects of particle interaction, such as particle shape, grain-size composition, cohesion/adhesion, etc. Defining the friction parameters of flowing medium has huge applied significance in technical sciences, particularly while designing technological facilities for concentrating and mining industries. Those parameters are affecting the whole technological process, the performance of facilities and their durability.

Modern devices for defining the properties of bulk materials represent mechanical aggregates that are way too complex and expensive in exploitation [1]. They don't presume determination of flowing medium status parameters in dynamic and are able to define only one or two parameters into one device, thus, automation of this process within one universal machine are unsolved problems.

Proposed system with technical vision, created in LabVIEW, allows simplify the construction of devices using for bulk material analysis and also is able to contemporaneously define not single, but variety of parameters [2]. The advantages of technical vision for defining the mechanical characteristics of flowing medium are obvious – digital image can be processed different ways and unlimited times. Moreover, hardware GPU acceleration based on NVIDIA CUDA technology (NVIDIA Corporation) provides desirable soft real-time system.

Developed system was tested on bulk materials that were modelled using DEM (Discrete Element Method) in Rocky DEM software (Engineering Simulation and Scientific Software). Received modelling results, exported as video files, were pre-processed using NI Vision Toolkit (National Instruments) and then were passed to artificial neural network. Using of neural network provides structured data and allows reveal more accurate trends and dependences (Figure 1).

IOP Conf. Series: Journal of Physics: Conf. Series **944** (2017) 012021 doi:10.1088/1742-6596/944/1/012021

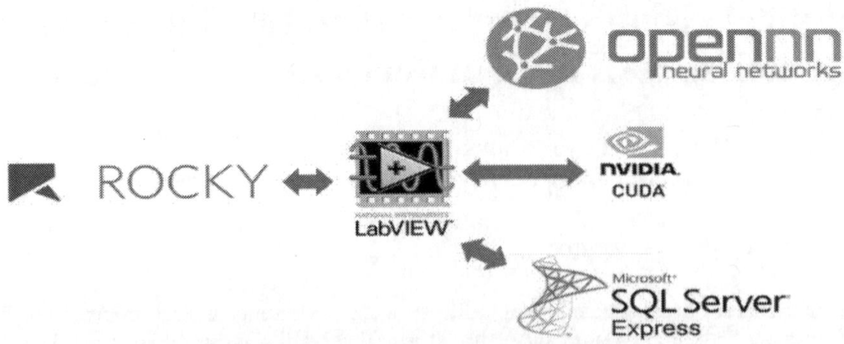

Figure 1. Common concept of the system.

2. Problem description and modelling

Static or dynamic particle-particle and particle-boundary friction analysis is a complex technical problem. The triboprocess study is complicated by mechanical, electrical, magnetic and other nature of the friction process. Physical mechanical properties of bulk materials are directly associated with triboprocess, but nowadays there are lack of theoretical or experimental dependences to consider specific flowing medium as uniform object with precisely known parameters. On the other hand, lots of industries daily face the lack of information problem about mechanical parameters of bulk materials therefore decreasing efficiency of technological processes and frequently occurs accidents [3].

One of the way to resolve this problem is experiential dependency detection between parameters of bulk materials and properties that are approachable to measure in static or dynamic. At the moment, there are huge amount of different approaches to define such properties but all of them are based on real devices with low level of automation and require huge time expenditures and high qualification. Another way is discrete element modelling which allows model the behaviour of flowing medium with desired parameters, such as particle shape, grain-size composition, coefficients of dynamic and static friction, boundary condition, percentage of moisture, etc. In this case, modelling was held in Rocky DEM software with partnership of CADFEM CIS specialists from Saint-Petersburg [4]. It was decided to model outflow of bulk material from the calibrated conical funnel. The funnel has simple geometry and usually is used to define such important characteristic as fluidity, angle of repose and other parameters (Figure 2).

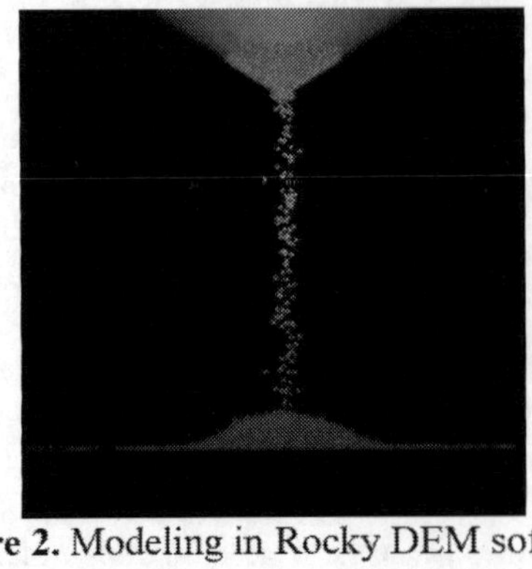

Figure 2. Modeling in Rocky DEM software.

3. Data processing

The computer vision system was developed with NI LabVIEW, including NI Vision Toolkit (National Instruments). Virtual Instrument captures a video stream from file or camera. In this research we used only videos of bulk material simulations modeled with Rocky DEM software. Video processes frame-by-frame, each frame processing consists of following stages (Figure 3):

IOP Conf. Series: Journal of Physics: Conf. Series **944** (2017) 012021 doi:10.1088/1742-6596/944/1/012021

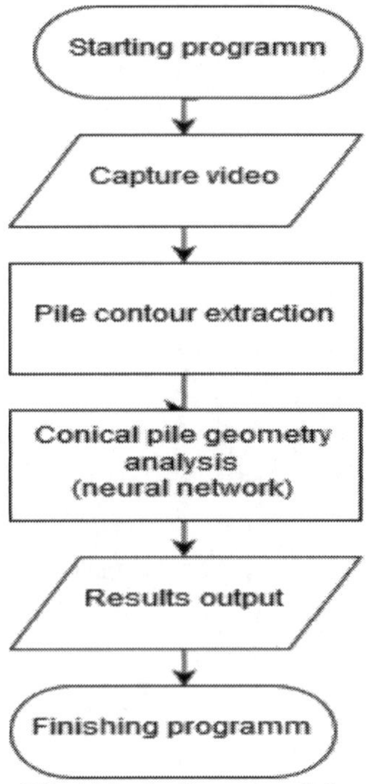

Figure 3. Simplified block diagram of developed system.

- Preparing. Image binarization for further processing and contour analysis
- Filtering. Removing waste particles, construction elements, etc.
- Contour capture. Pile surface detection and converting it to array of points coordinates
- Geometry analysis. Taking contour points

Video processing results are displayed on screen (Figure 4), computed pile geometry parameters, such as height, width, angels of repose, surface radius – are highlighted directly on the video recorded to the MS SQL Express database (Microsoft Corporation) for further analysis and approximation. Communication between database and virtual instrument is implemented by LabVIEW Database Connectivity Toolkit (National Instruments).

Figure 4. Example of data processing.

Frame-by-frame video processing requires valuable computational resources, thus, suggested system uses a hardware computing acceleration on GPU. LabVIEW provides this function with GPU Analysis Toolkit (National Instruments) for NVIDIA CUDA. The application of this approach allows significantly increase the program's performance and provide the necessary processing speed of the video, which is important in a real experiment data processing. GUI of developed vision system is presented in Figure 5.

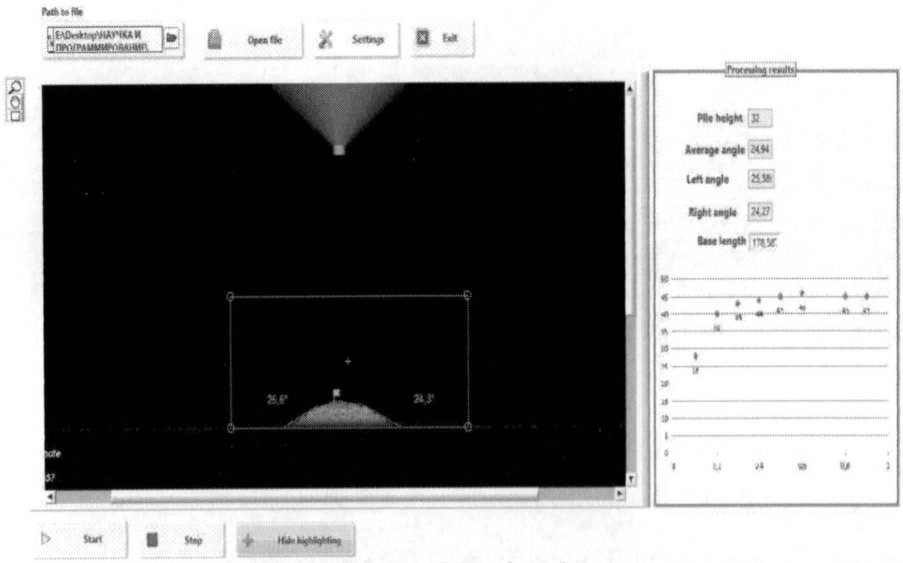

Figure 5. GUI of developed system

Experimental data analysis is carried out by neural network, configured and trained to trace dependency between conical pile geometry parameters and matter friction coefficients. Neural network was realized on C++ with OpenNN library (Artelnics).

4. Results

The neural network was trained on the basis of the parameters of the flowing medium specified in the Rocky models. The neural network analyzed the static and dynamic parameters obtained during the processing of video files. Table 1 shows an example of the result of processing one of the models.

Table 1. Example of model processing.

Friction coefficient	Parameter			
	tan(angle of repose)	*Height, mm*	*Base length, mm*	*Radius, mm*
0.1	0.174	118	810.5	75
0.2	0.312	136	751.6	40
0.3	0.338	141	735.2	45
0.4	0.384	147	726.3	90
0.5	0.391	148	716.2	30
0.6	0.387	148	709.4	40

In virtue of received parameters neural network automatically computed coefficients of friction of given bulk materials. Results of computing are presented in Table 2.

IOP Conf. Series: Journal of Physics: Conf. Series **944** (2017) 012021 doi:10.1088/1742-6596/944/1/012021

Table 2. Post-execution comparison.

Given coefficient	0.1	0.2	0.3	0.4	0.5	0.6
Computed coefficient	0.11	0.193	0.313	0.391	0.519	0.58

Calculated relative error is 1.3% what is quite precise result. Moreover, neural network formed the dependencies between coefficient of friction and parameters from video processing (Figure 6).

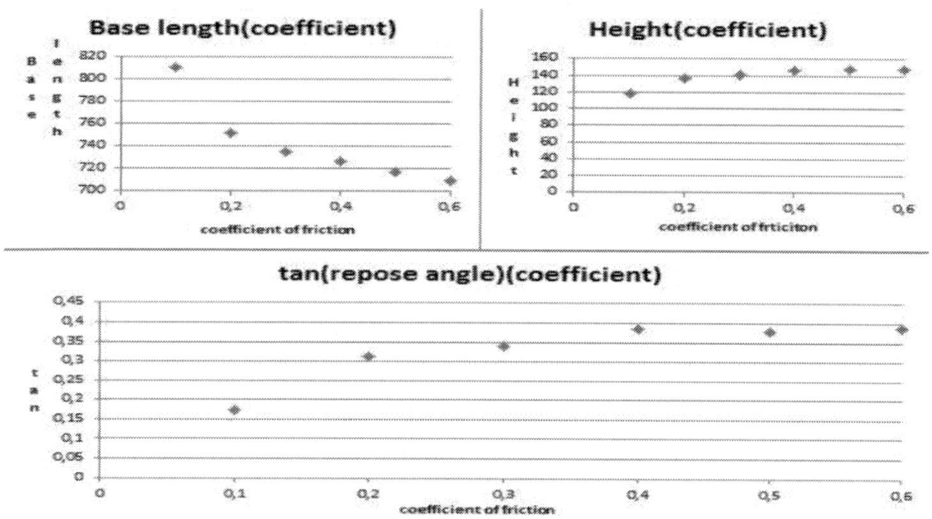

Figure 6. Experimental dependencies.

Conclusion

At present, system is still under development, the laboratory stand is being designed (Figure 7) for carrying out experiments with real bulk materials, primary using conical funnel.

Figure 7. Basic concept of the future real stand.

Neural network is supposed to be trained on several virtual models created with Rocky DEM and applied to real experiment. This approach will make system developing faster. Based on obtained dependencies it is possible to predict bulk materials behaviour in certain conditions, this knowledge in needed at many technical applications. It is planned to conduct experiments using other types of stand with movable elements (Figure 8). On the other hand, the system proposed above might be useful in industrial engineering and mining machine construction and the design of process equipment with the required parameters, in order to increase economic efficiency and reduce the costs of their design and production.

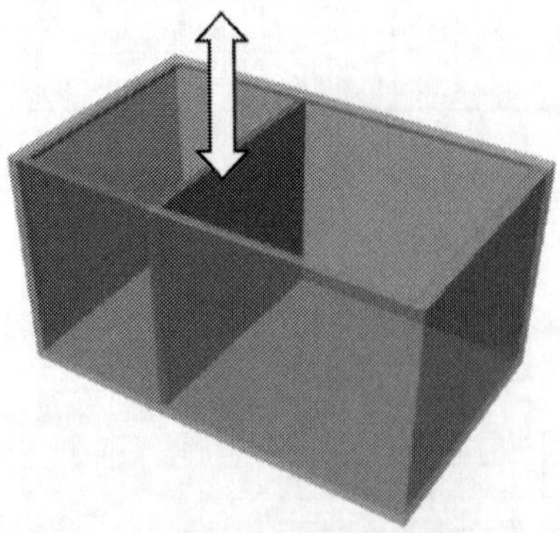

Figure 8. Prototype of new experiment stand with movable elements.

References

[1] Hu G, Liu M, Liu Y, Wan H and Liu L A DEM simulation system for equipment of powder processing and bulk materials transportation 2010 *International Conference on Mechanic Automation and Control Engineering* pp 923–926 DOI: 10.1109/MACE.2010.5536021

[2] Wouters I M F and Geldart D Characterising sem-cohesive powders using angle of repose *Particle and Particle Systems Characterization* vol 13 is 4 pp 254-259 August 1996

[3] South China Morning Post Accessed data: 05.06.2017 http://www.scmp.com/news/china/policies-politics/article/1894877/shenzhen-landslide-declared-industrial-accident-not

[4] Oficial Rocky DEM software website in Russian Accessed data: 05.06.2017 http://www.rocky-dem.ru/

Improving generalized inverted index lock wait times

A Borodin, S Mirvoda, S Porshnev and O Ponomareva

IRIT-RTF, Ural Federal University, Yekaterinburg, Russia

Abstract. Concurrent operations on tree like data structures is a cornerstone of any database system. Concurrent operations intended for improving read\write performance and usually implemented via some way of locking. Deadlock-free methods of concurrency control are known as tree locking protocols. These protocols provide basic operations(verbs) and algorithm (ways of operation invocations) for applying it to any tree-like data structure. These algorithms operate on data, managed by storage engine which are very different among RDBMS implementations. In this paper, we discuss tree locking protocol implementation for General inverted index (Gin) applied to multiversion concurrency control (MVCC) storage engine inside PostgreSQL RDBMS. After that we introduce improvements to locking protocol and provide usage statistics about evaluation of our improvement in very high load environment in one of the world's largest IT company.
Keywords: Relational databases; Tree data structures; Concurrency control

1. Introduction

Generalized inverted index (Gin) is special tree-like data structure used in PostgreSQL for improving performance on set data types, such as arrays. It abstracts well known inverted index data type in a way that index doesn't know which operation it improves. Specific data type can implement its own index method strategy [1].

In this concept gin is different form ordinary B-tree, which support only predefined comparison operators and similar to GiST [GIST].

Gin define 4 different built-in strategies for arrays it can accelerate [2]:

- overlap – two arrays overlaps in any greater than zero number of values,
- contains – every element of the right array exist in the left array (order ignored),
- is contained by – every element of the left array exist in the right array (order ignored),
- equal – array on the left equals array on the right (order matters).

A Gin index consists of a B-tree index constructed over key values, where each key is an element of some indexed items i.e. element of array, and where each tuple in a leaf page contains either a pointer to a B-tree over item pointers (posting tree), or a simple list of item pointers (posting list) if the list is small enough.

The primary aim for Gin in PostgreSQL is high performance full text search querying. But it can be used in any task which can be presented as an array of tokens. It worth noting that Gin stores and looks for keys, not item values, that's why it heavily used for high performance querying of text array or numeric array for example article keywords, email recipients etc.

A. Index Layout

The index consists of pages, each containing as many index tuples as space allows. Index tuple is a structure containing: value of the indexed attribute (key datum) and tuple identifier (TID). TID is a structure pointing to physical location of data on disk. It tells on which page within a table or index the tuple is located and where on that page. For every table row, corresponding index tuple should exist, even if another row with the same key value already has been indexed. This duplication can cause page split, for one or more pages of the index.

TID duplication on posting lists affecting page size can be mitigated with compression similar to bitmap packing, which allows them to fit in much smaller. Compression is used for both the lists stored in-line in entry tree items, and in posting tree leaf pages.

Index layout composed of internal (entry tree) and leaf pages (posting tree or posting list). Internal page layout is looks exactly like B-tree and has no important differences from B-tree. But leaf pages have different layout and, as mentioned above, depends on number of elements.

When number of elements is small enough, they are stored as simple sorted list of tuple identifiers (TID), called posting list (Figure 1).

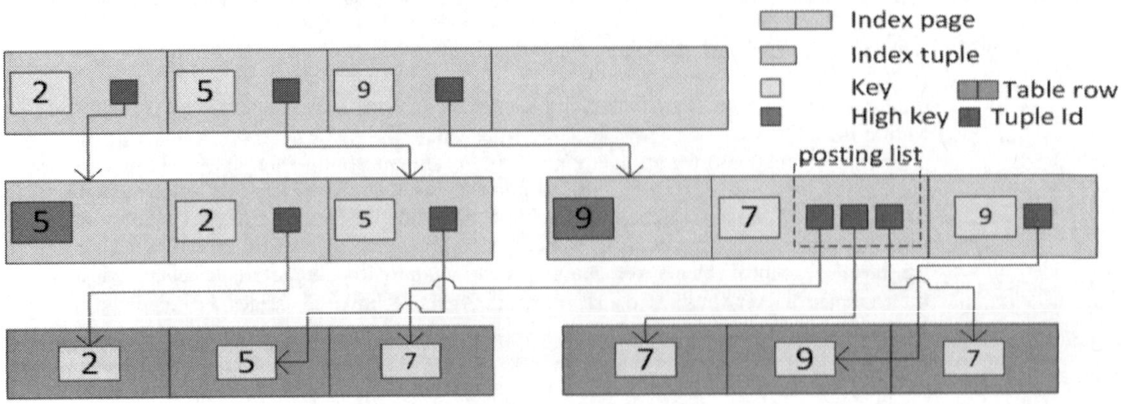

Figure 1. Posting list data structure.

When it is not, leaf pages contain a simplified B-tree called posting tree (Figure 2). High key is a special case of index tuple. It contains key value such that all tuples on the page are lower than or equal that value.

Layout features heavily used for managing concurrency issues, for example, we always certain about values of interest stored exactly righter and (or) deeper in the three hierarchy.

Another important aspect of design for concurrency shown in next section.

Posting list resides on the leaf page of entry tree. It contains ordered identifiers of heap tuples of data. If the posting list exceeds size of inline attribute (also called TOAST size, usually big attributes are stored in the oversized attribute storage - TOAST), then it is converted into posting tree.

IOP Conf. Series: Journal of Physics: Conf. Series **944** (2017) 012022 doi:10.1088/1742-6596/944/1/012022

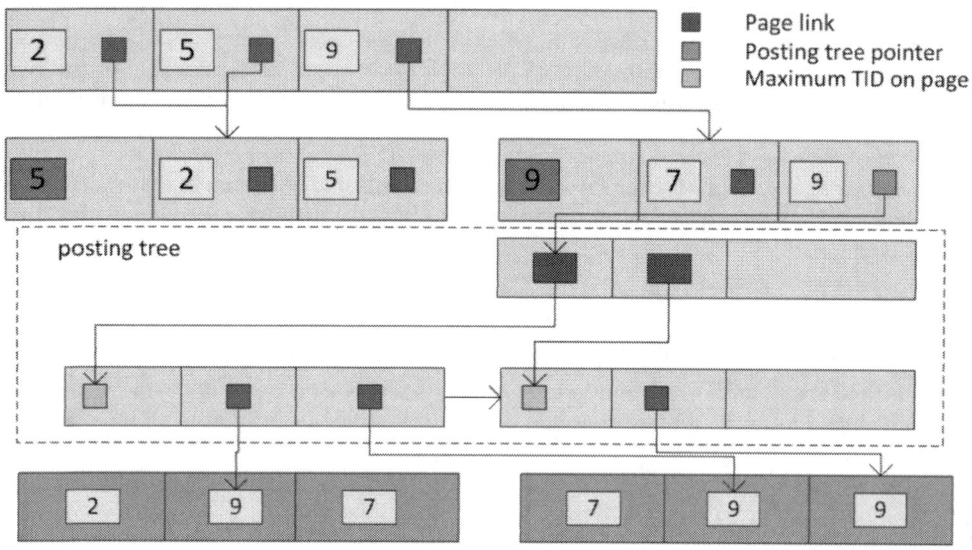

Figure 2. Posting tree data structure.

Key space for the entry tree is values of subelements of indexed attribute. Keys pace for posting list is identifiers of tuples in heap (TIDs). Keys pace for posting tree is, essentially, the same as the key space of posting lists. But posting tree is stored as B-tree.

Both posting trees and posting lists are ordered to make operation of intersection more efficient. E.g. if we scan an index for the query «green apple», we scan posting list of «apple» and posting tree of «green», and find every TID of tuple mentioning both «green» and «apple». This operation is performed for O(A+B) in worst case where A and B is size of posting trees or lists. In average complexity of the search in posting tree is $O(R+H \cdot \log B)$, where R is the size of resulting tuples stream, H is height of B-tree and B is number of tuples in a block (page).

2. Concurrent architecture

As noted earlier the entry tree and each posting tree is a B-tree with right-links connecting sibling pages at the same level known also as B-link Tree [3].

To guarantee tree properties preservation during concurrent operations data structure have to obey so-called tree locking protocol, see [4, 5].

Numerous efficient locking protocols are described in [6]. It is worth noting that there are two kinds of locking objects: locks and latches. The first one is heavyweight operating system feature, the second one is lightweight language level construct or programming technique.

In PostgreSQL Gin tree locking protocol implemented with number of peculiarities which we show further.

a. Deadlock freedom

To avoid deadlocks, pages must always be locked in the same order: left to right, and bottom to top.

b. Searching

When searching, the tree is traversed from top to bottom, so the lock on the parent page must be released before descending to the next level. Concurrent page splits move the key space to right, so after following a downlink, the page actually containing the key we're looking for might be somewhere to the right of the page we landed on. In that case, we follow the right-links until we find the page we're looking for.

c. Deletion

To delete a page, the page's left sibling, the target page, and its parent, are locked in that order, and the page is marked as deleted. However, a concurrent search might already have read a pointer to the page, and might be just about to follow it. A page can be reached via the right-link of its left sibling, or via its downlink in the parent.

d. Insertion/Page split

A search descending the tree must release the lock on the parent page before locking the child, or it could deadlock with a concurrent split of the child page; a page split locks the parent, while already holding a lock on the child page. However, posting trees are only fully searched from left to right, starting from the leftmost leaf.
According to the design of Postgres [1] concurrent data editing implemented as snapshot isolation known as MVCC (Multiversion Concurrency Control) which guarantee that readers never block writers and writers never block readers [7].
Snapshot isolation lead to "dead" (not visible to any living transaction) tuples multiplication over time and degradation of index performance (when Postgres resolves TID provided by index, it can be "dead").
To overcome this problem from time to time Postgres execute special VACUUM service aimed to delete "dead" tuples and rearrange index pages.
Naïve implementation of can be as simple as:

```
foreach data_tuple
    check visibility
    if false
    drop data_tuple
foreach index
    drop index
    create index
```

At first glance, it looks like ordinate write activity, but actually it can lead to whole index and data rewrite, which is not viable for current volumes of data.
That's why vacuuming implemented differently for each one of the native PostgreSQL index (btree, gist, gin, hash) to provide best characteristics, achievable for specific data structure.
In the following section, we cover problem and solution of vacuuming in case of the general inverted index.

3. Problem statement

Vacuum of posting tree is doing two passes through posting tree:
1. Function ginVacuumPostingTreeLeaves() takes LockBufferForCleanup, effectively excluding all inserts. Then it traverses down trough tree taking exclusive lock, effectively excluding all reads. On leaf level it calls ginVacuumPostingTreeLeaf() function, which deletes all dead tuples. If there are any empty page, root lock is not released, it passes to stage two.
2. If there are any empty pages, ginScanToDelete() function scans through tree, deleting empty lead pages, then deleting empty inner pages, if they emerged after leaf page deletion. Between the above VACUUM invokes vacuum_delay_point(), which can hand for a while, holding CleanupLock on root leaf.

These lead to long lock wait times even for middle sized data load: for 50 GB data and 1 GB Gin index size exclusive write lock holds for 90 seconds or more, preventing any reads and writes from data.
Proposed and implemented solution changes function ginVacuumPostingTreeLeaves() such a way that now this it doing same depth first search as pass 1, but without cleanup lock, acquiring only read locks on inner pages and exclusive lock on leafs before eliminating dead tuples. If it finds empty leafs, it computes minimal subtree, containing only empty pages and start scan for empty pages from parent page pointing to found subtree. This scan acquires cleanup lock on root of scan (not necessarily root of posting tree).
Cleanup lock ensures no inserts are inside subtree. Scan traverse subtree depth first taking exclusive locks from left to right. For any page being deleted all leftmost pages were locked and unlocked before. New reads cannot enter subtree, all old read scans were excluded by lock\unlock.
Thus there shall not be deadlocks with ginStepRight() function. We get lock on page being deleted, then on a left page. ginStepRight() takes lock on left page, than on right page. But it's presence is excluded by cleanup lock and DFS scan with locks of upper and left parts of tree.
These modifications completely solve problem with vacuuming middle sized data, but can be improved even more.
The previous paragraph's reasoning only applies to searches, and only to posting trees.

4. Analisys

To protect from inserters following a downlink to a deleted page, vacuum simply locks out all concurrent insertions to the posting tree, by holding a super-exclusive lock on the parent page of subtree with deletable pages.

Inserters hold a pin on the root page, but searches do not, so while new searches cannot begin while root page is locked, any already-in-progress scans can continue concurrently with vacuum in corresponding subtree of posting tree. To exclude interference with readers, vacuum takes exclusive locks in a depth-first scan in left-to-right order of page tuples.

Leftmost page is never deleted. Thus, before deleting any page we obtain exclusive lock on any left page, effectively excluding deadlock with any reader, despite taking parent lock before current and left lock after current.

We take left lock not for a concurrency reasons, but rather in need to mark page dirty. In the entry tree, we never delete pages.

This is quite different from the mechanism the B-tree implementation uses to make page-deletions safe; it stamps the deleted pages with an XID and keeps the deleted pages around with the right-link intact until all concurrent scans have finished.

5. Future work

There are numerous ways to improve current approach. It's possible to rebuild locking protocol and B-tree algorithm in GIN to provide possibility of merging almost-empty pages as described in [4]. But the simplest yet not very clear technique to apply is a posting tree truncation. If the whole posting tree is empty, then we could mark root page as leaf and remove all other pages in tree without any locking.

Currently, during first posting tree scan we are detecting situation when root page shall be deleted, and prevent it, leaving leftmost pages of vacuumed tree. Posting tree height is never reduced. But in this situation, we could just replace the whole posting tree with the new empty root page. The concurrency protocol of this approach is not obvious: there may be still searches scanning through the tree, it's not possible to safely wipe pages.

This trick was suggested by PostgreSQL committers, but, after discussion we left it for future patches. From the performance point of view, this can be beneficial idea. Both, performance of VACUUM and performance of scans. But doing so, we risk to leave some garbage pages in case of a crash. And it is not clear how to avoid these without unlinking pages one by one.

To implement, debug and evaluate described changes of the indexing, we successfully used techniques covered in our earlier works on this subject [8, 9].

6. Acknowledgment

The results were obtained with the financial support of the Russian Scientific Fund, Agreement N17-00-00000.

References

[1] Stonebraker M and Rowe L A (1986) The design of Postgres *Proceeding SIGMOD '86 Proceedings of the 1986 ACM SIGMOD international conference on Management of data* Vol 15 No 2 pp 340–355

[2] Interfacing Extensions To Indexes https://www.postgresql.org/docs/current/static/xindex.html

[3] Lehman P L and Yao S B 1981 Efficient Locking for Concurrent Operations on B-Trees *ACM Transactions on Database Systems* 6(4) pp 650–670

[4] Lanin V and Shasha D 1990 *Tree Locking on Changing Trees* (New York: New York University)

[5] Silberschatz A, and Z Kedem 1980 Consistency in hierarchical database systems *Journal of the ACM* Vol 27 pp 72–80.

[6] Graefe G 2010 A survey of B-Tree Locking Techniques *ACM Transactions on Database Systems* vol 35 16.

[7] Momjian B 2013 *Mvcc unmasked, unpublis hed* https://momjian.us/main/writings/pgsql/mvcc.pdf

[8] Borodin A Mirvoda S Kulikov I and Porshnev S 2017 Optimization of Memory Operations in Generalized Search Trees of PostgreSQL *International Conference: Beyond Databases, Architectures and Structures* (Springer, Cham) pp 224–232

[9] Borodin A Mirvoda S and Porshnev S 2015 Database Index Debug Techniques: A Case Study *International Conference: Beyond Databases, Architectures and Structures* (Springer International Publishing) pp 648–658

Design of Inspection Method for Control of Gap Dimension Between Reinforcing Ring and Vessel's Surface

R R Chernyateva

Ufa State Petroleum Technological University, 1, Kosmonavtov st., Ufa, 450062, Russia

Abstract. Discrepancy between real connecting pipe's unit's characteristics and unit's quality rating on base vessel's nozzle's reinforcing method analysis is shown. The most common misfit is the dimension of gap between reinforcing ring and vessel's surface. Nowadays gap's inspection methods, which are based on taking measurement of gap in one point via signal's nozzle does not allow to estimate the quality of producing by all the contact surface. For reasoning the currency of new connecting pipe's gap's inspection method development finite-element method's modeling of unit's stress mode was done. Authors suggested as an alternating inspection method to produce reinforcing rings with two signal nozzles, which allows to estimate gap's dimension in several points and to increase the objectivity of inspection. Stress mode modeling of reinforcing ring with two nozzles was done. Analysis of modeling results showed, that for inspection gap's dimension between connecting pipe and vessel inspection method with two nozzles could be used. That allows to take measurement of gap's dimension in the most loaded areas of vessel.

1. Introduction

Nowadays the main way to reinforce the holes in vessels and facilities used in technological processes in oil refineries and petrochemical enterprises is the installation of a reinforcement (overhead) ring, in accordance with the GOST R 52857.3-2007 and the GOST 24755-89 (ST SEV 1639-88).

The existing technology of assembling and welding of reinforcing rings and the design of this unit are well developed and have extensive experience of use. However, there are questions in the manufacture, which require additional consideration [1, 2]. One of the main questions is the assembly technology and control of the hole reinforcement assembly. The requirements of GOST R 52630-2006 on the limitation of the gap size to 3 mm are not always fulfilled in practice. The gap is controlled with a probe according to the outer diameter of the reinforcing ring [3].

Authors suggested as an alternating inspection method to produce reinforcing rings with two signal nozzles, which allows to estimate gap's dimension in several points and to increase the objectivity of inspection. Stress mode modeling of reinforcing ring with two nozzles was done. Analysis of modeling results showed, that for inspection gap's dimension between connecting pipe and vessel inspection method with two nozzles could be used. That allows to take measurement of gap's dimension in the most loaded areas of vessel [4].

2. Statement of the problem and its solution

Figure 1 shows that the gap between the body and the ring reaches 5 mm. At the same time, monitoring of this gap for today near the installation of the fitting is practically impossible, and, as a rule, it is not produced. Reducing the gap around the perimeter of the ring is made by clamps, attached to the body of the facility, or, as a rule, by a striking tool, which is also undesirable, and in some cases simply unacceptable in the manufacture of vessels and apparatuses [5].

Avoid incomplete fitting of the overlay ring to the shell in the manufacture of the "fitting with a clamping ring" assembly is rather difficult, due to a number of reasons: deviations during rolling of the overhead ring, especially at large thicknesses, also deflections of rolled shells; impossibility of individually roll one ring under repair conditions at the place of operation of the facility, due to the complex geometry of the body parts, in particular the bottom shape and other reasons [6].

Figure 1. Photo of the reinforcing ring attachment.

Thus, there is a need to improve the technology of installing the reinforcing ring, as well as the technique of monitoring the gap between the reinforcing ring and the surface of the facility to be reinforced.

To solve these issues, taking into account the studies carried out to assess the stresses of the nozzle arising in the welding node, as well as well-known information on the stress-strain state of the choke assemblies, the following technical solutions were adopted [7–10]:

To reduce the gap between the reinforcing ring and the reinforced surface of the body of the facility, use П-shaped screw clamp mounted on the flange of the fitting;

Control of the gap between the reinforcing ring and the surface to be reinforced is carried out in the most dangerous areas of the nozzle welding unit - in small diameter nozzles, this zone is near the edge of the hole in the diametrical cross-section along the generatrix – in the meridian section (Figure 2). In fittings with a diameter of more than 150 mm, the zone of maximum stress moves to the intersection of the union and the shell in the circumferential section;

The control is carried out through the hole in the ring with the simplest measuring instruments - depth gages, in this case the caliper is examined – with a ruler for measuring depths, the measurement error is 0,05 mm and is acceptable for these purposes.

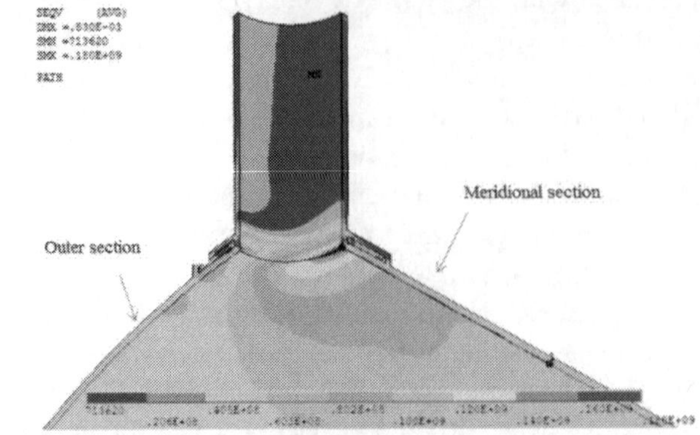

Figure 2. The field of distribution of equivalent stresses in a chimney unit, reinforced with overhead rings, Pa.

In view of the fact that one hole is already provided in the reinforcing ring near the perimeter, it is necessary to drill an additional hole in the reinforcing ring, placing it near the fitting, with the condition that it does not enter the welded joint zone.

In this case, it becomes necessary to evaluate the influence of stresses occurring in the body of the facility and the reinforcing ring in the case of a ring weakening by additional holes. Previously, the hole diameter is assumed to be equal to the existing one in the normative documentation, i.e. equal to

IOP Conf. Series: Journal of Physics: Conf. Series **944** (2017) 012023 doi:10.1088/1742-6596/944/1/012023

10 mm, if necessary, the diameter of the hole can be reduced to 5 mm – the condition under which the measurement can be made using a depth ruler of the caliper SC-1.

Additional holes in the reinforcing ring, arranged to control the size of the gap, cause stress concentration. Given the absence of a standard procedure for calculating the stress state of the device body loaded under such conditions, it was decided to carry out an assessment of the stresses occurring in the body of the apparatus and the reinforcing ring using the ANSYS program.

The design model consists of a shell with a union, a reinforced ring (Figure 3). Geometric parameters of the model:

- Diameter of shell 1 m, thickness 7 mm, length 2.4 m;
- Diameter of the union 100 mm, thickness 5 mm, radius 200 mm;
- Width of reinforcing ring 72 mm, thickness 5 mm;
- The gap between the shell and the reinforcing ring is 3 mm.

Enter material properties. As initial data for solving problems, bilinear properties of the material were specified. In the elastic zone, the modulus of elasticity $e = 2\cdot1011$ Pa, the Poisson's ratio $v = 0.3$, the yield strength 210 MPa, the tangential module 480 MPa were set.

The geometric model is divided into a grid of finite elements of a given size. In the calculation, volumetric eight-node elements are used, the number of elements along the thickness of the shell, the union and the reinforcing ring near the nozzle welding unit is at least 4 to ensure the accuracy of the results.

Boundary conditions and loads:

Symmetry for the nodes of the model lying in the planes XY, YZ, XZ, i.e. the displacement of the nodes lying in planes in the direction normal to the plane is zero;

In the reinforcing ring, a hole with a diameter of 10 mm is modeled, located in this case near the wall of the nozzle, designed to control the gap in the manufacture of the chimney assembly (Figure 3, c).

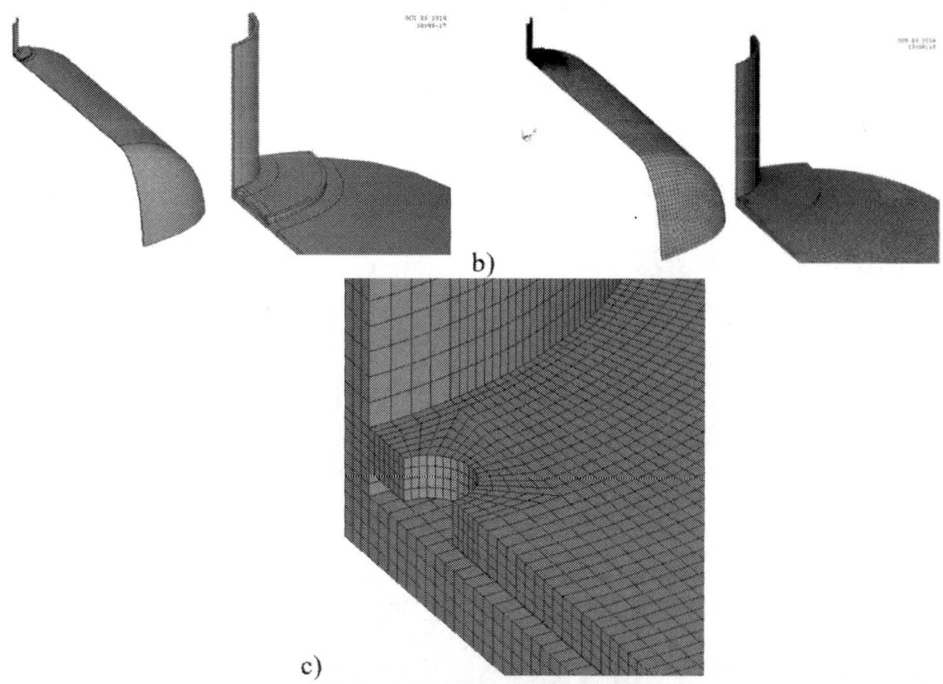

a) b) c)

Figure 3. Calculation model.

The results of the calculations are presented in Figures 4–7.

IOP Conf. Series: Journal of Physics: Conf. Series **944** (2017) 012023 doi:10.1088/1742-6596/944/1/012023

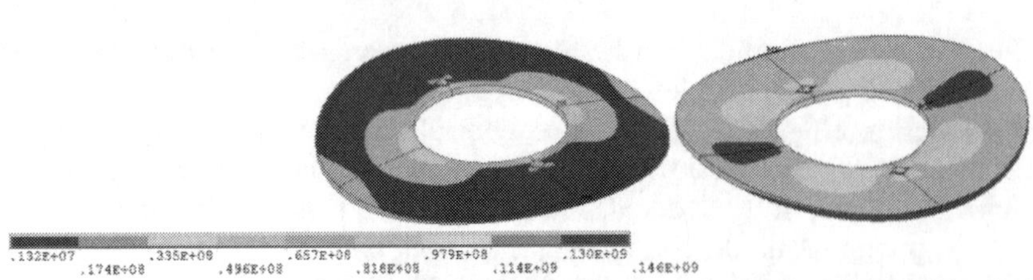

Figure 4. Distribution of equivalent stresses in the reinforcing ring with a hole near the nozzle, Pa

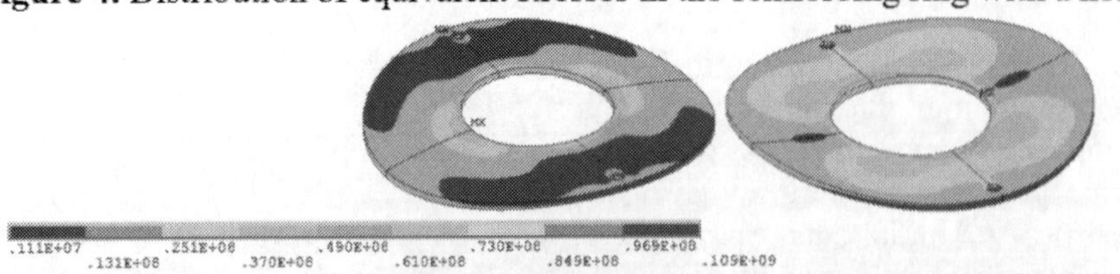

Figure 5. Distribution of equivalent stresses in the reinforcing ring with a hole in the periphery (the existing version), Pa

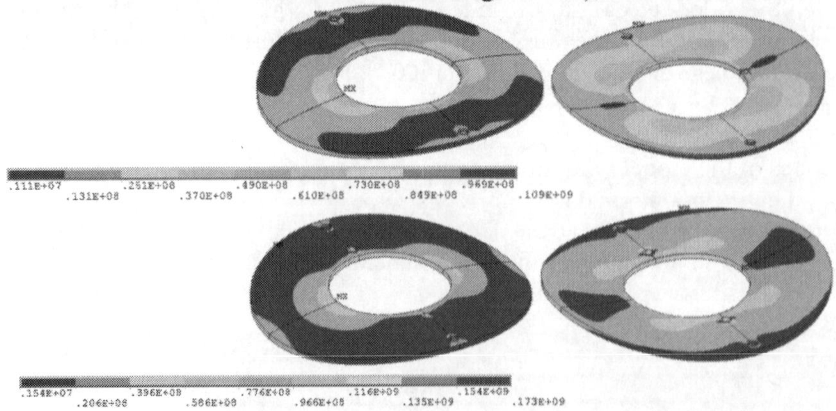

Figure 6. Distribution of equivalent stresses in a reinforcing ring with two holes, Pa

For clarity, the graphs of equivalent stresses in the meridional direction (on the inner surface of the ring) are shown in Fig. 7.

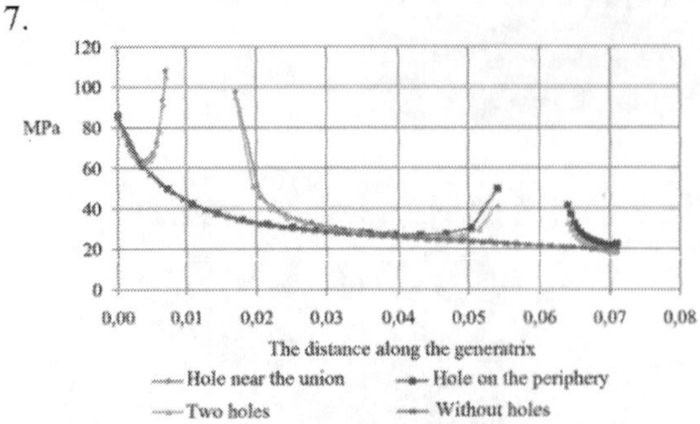

Figure 7. The graph of equivalent stresses in reinforcing rings of different designs in the meridional direction.

Stress concentration factor j as known is determinate as ratio of maximum equivalent stresses in specific area with stress raiser to maximum equivalent stresses in the same area without stress raiser. Next results of stress concentration factor j were received:

For single nozzle near connecting pipe $j = 2.8$;

For single nozzle near outside edge of the reinforcing ring $j = 2.19$;

For two nozzles near connecting pipe $j = 2.73$;

For two nozzles near outside edge of the reinforcing ring $j = 1.89$.

Analysis of received from modeling data showed that the maximum level of stresses in reinforcing ring appears in the area of reinforcing ring and connecting pipe's joint in circular cross-section. Thus the level of stresses in presence of nozzle does not exceed the level of allowable stresses, which equals 177x106 pascal for gost 14249-89 design conditions and there is no impact of 10 mm nozzles at distance 47 mm between them on stress level.

Cross-effect of nozzles will appear only in the case of their disposition in immediate proximity according to the peculiarity of reinforcing ring' loading.

Unit's stress mode's analysis showed that in spite of connecting pipe's unit embodiment regardless of nozzle's disposition the maximum level of stresses is in meridian vessel's section (figure 8). The level of stresses in this area does not change. On the base of modeling results received graph of equivalent stresses in vessel (figure 9).

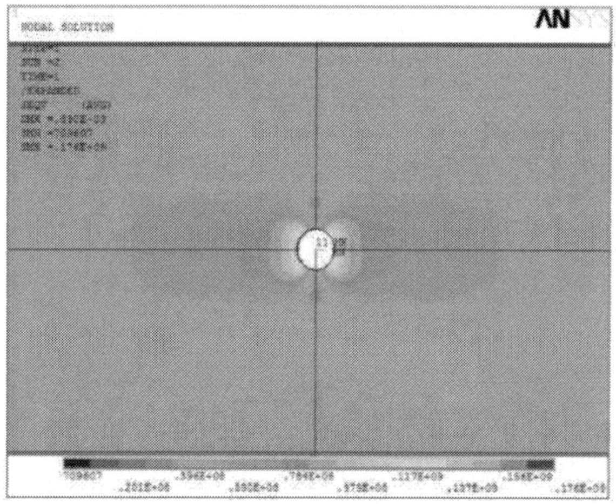

Figure 8. Field of equivalent stresses in vessel's body on inner surface, Pascal.

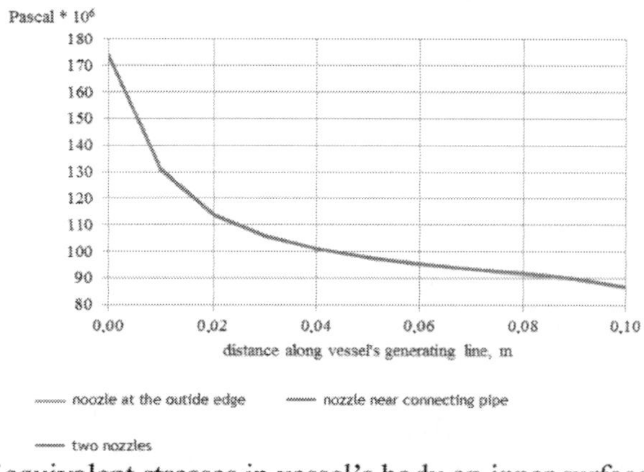

Figure 9. Graph of equivalent stresses in vessel's body on inner surface in meridian cross-section.

3. Conclusions

1 Inspection method on base of using additional nozzle in reinforcing ring could be designed for controlling gap between reinforcing ring and vessel's surface.

2 Modeling results showed that the level of maximum stresses is not in the area of additional joints in reinforcing ring, but in the area of reinforcing ring's and pipe's joint in circular cross-section. Thus in stress concentration zone level of stresses does not exceed the level of allowable stresses.

3 Connecting pipe unit's stress mode analysis showed that in spite of unit embodiment and nozzle's dislocation the level of maximum stresses does not change.

References

[1] Konesev S G, Khazieva R T and Kirillov R V 2016 *X Dynamics of Systems, Mechanisms and Machines* (Dynamics) pp 1–7
[2] Konesev S G, Khazieva R T, Kirillov R V and Konev A A 2017 *Journal of Physics: Conference Series (JPCS)* **1**
[3] Fisher W W, Kelley M E and Lomas J E 2003 *Journal of applied behavior analysis* **3** 387
[4] Usamentiaga R Venegas P Guerediaga J Vega L and López I 2012 *Infrared physics & technology* **6** pp 491–498
[5] Apisov I V 2014 *Electronic scientific journal «Oil and Gas Business»* **5** 223-237
[6] Kuzavkov V M, Rozinov A Y, Shtaits V V 2005 *Russian journal of nondestructive testing* **1** pp 39–44
[7] Pimstein P G, Mordina G M, Barabanova L P 2003 *Chemical and Petroleum Engineering* **4** pp 3–5
[8] Skopinskij V N, Rusanov O A, Nazarov N A 2007 *Machine Building and Engineering Education* **20** pp 23–33
[9] Sabitov M H and Ponikarov S I 2011 *Herald of the Kazan Research university* **20** pp 208–212
[10] Cvik L B, Hramenok M A and Shapova M V 2008 *Herald of Universities* **1** pp 18–24

Effect of baffle spacing and baffle cut on thermal-hydraulic characteristics of the fluid flow

R R Chernyateva
Ufa State Petroleum Technological University, 1, Kosmonavtov st., Ufa, 450062, Russia

Abstract. This article presents the results of investigations of the influence of baffle spacing and baffle cut on the size of dead zone formed near the cross baffles using numerical simulation methods. It is showed the structure of an additional baffle plate which can be used to reduce the dead zone and smoother flow distribution over the cross section.

1. Introduction

This article is devoted to the study of fluid flow in the shell side of shell-and-tube heat exchanger (STHE) and the assessment of the influence of geometric parameters on the thermal-hydraulic characteristics [1, 2]. More intense heat transfer can be achieved by increasing the fluid velocity, but at the same time, it leads to a large increase in pressure drop that reduces the efficiency of heat transfer and increase operating costs [3, 4].

Transfer of certain thermal energy at lower fluid velocity require more heat exchange surface that will increase the cost of the heat exchanger. Thus, it is important to find ways to improve the thermal efficiency of heat exchange equipment by selecting optimal geometric parameters providing the most efficient heat transfer [5, 6].

The efficiency of heat transfer in STHE can be increased by restricting dead zones formed near the cross baffles because of which the pressure drop is increased without a corresponding increase in heat transfer coefficient [7]. The size of the dead zone depends on the baffle spacing and the baffle cut. Thus, the optimal ratio of parameters will reduce pressure drop [8].

This article presents the results of investigations of the influence of baffle spacing and baffle cut on the size of dead zone formed near the cross baffles using numerical simulation methods. It is showed the structure of an additional baffle plate which can be used to reduce the dead zone and smoother flow distribution over the cross section [9].

2. Main part

An important direction of economic development is the economical use of material and energy resources [10, 11]. One aspect of savings is the provision of high thermal efficiency of heat exchange equipment, in particular shell-and-tube heat exchangers (STHE).

STHE is widely used in the oil, gas, oil refining, petrochemical and many other industries. This is due to the reliability of the design and the variety of types and designs for different operating conditions [12]. Common elements for all STHE structures are casing, tubes, distribution chambers, tube plates, dividers and supports. In all designs of the STHE, there are transverse dividers used to increase the velocity of the coolant, to organize the cross current and to maintain the same distance between the tubes.

One of the reasons for the decrease in the efficiency of STHE is the formation of stagnant zones in the flow around the transverse dividers. The location and magnitude of the stagnant zones depend on the geometric parameters of the flow section, in particular the height of the free segment of the divider wall h_w and the distance between the dividers L_b, since these parameters determine the cross-sectional area of the cross and parallel flows. To minimize the size of stagnant zones, it is necessary to choose the optimal ratio of these parameters [13].

At present, such a ratio is not regulated by normative documents. According to the standards of the Tubular Exchanger Manufacturers Association (TEMA), it is recommended to choose a separation distance from 0,2D to D and a free segment height of 0.15D to 0.4D, with the ratio of the height of the window to the distance between the dividers varying widely from 0.15 to 2 [14–16].

A stagnant zone is the region in which the velocity of the fluid is substantially lower than the velocity of the main flow v_B, motions of small velocity are possible. As a rule, stagnant zones are formed behind a streamlined divider and near the casing. Figure 1 shows the longitudinal section of the intertube space of STHE with diameter D = 400 mm, blue is the isosurface $v = 0.8\ v_B$. It can be seen from the figure that behind the streamlined divider a region is formed in which the velocity of the coolant is lower than in the main stream. The size of this area is not the same for different values of the distance between the dividers and the height of the divider window. With a certain ratio of these geometric parameters, a significant increase in the dimensions of the stagnant zone is observed.

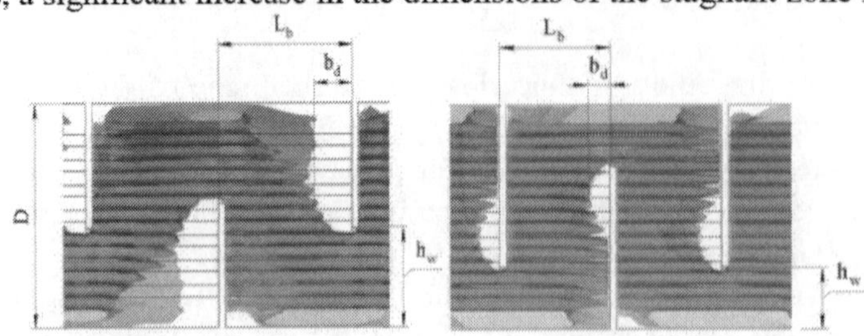

L_b = 250 mm, h_w = 178 mm L_b = 200 mm, h_w = 110 mm

Figure 1. Areas with low velocity of coolant flow.

A stagnant zone is formed when a transverse divider flows around at any ratio of parameters, it is practically impossible to eliminate it. A finite-element modeling of flow past a chess bundle of tubes in a rectangular region is carried out with a non-uniform distribution of the flow across the section. The results of modeling the relative width of the stagnant zone from 0 to 80% have shown that it has a negative effect on the thermohydraulic characteristics: the average value of the heat transfer coefficient for different widths of the stagnant zone varies insignificantly, and the magnitude of the pressure drop significantly increases. It reduces the efficiency of heat transfer. Also in the stagnant zones there is a local increase of temperature, which can cause contamination of the transfer surface, since the polymerization of the deposits depends on the wall temperature of the pipe and can increase at a temperature above the critical temperature. Thus, when designing heat exchangers, it is necessary to select a geometric parameter ratio at which the width of the stagnant zone will be minimal.

Studies of the influence of the distance between the dividers and the height of the one-segment window on the heat transfer in the intertube space of the STHE were carried out by the method of computational hydrodynamics realized in the finite element calculation module ANSYSCFX [17–19].

The finite-element model represents the flowing part of the STHE intertubular space with a diameter of B = 400 mm with a checkerboard of tubes d = 20 mm in diameter located in steps of s_t = 26 mm. The geometric parameters of single-segment transverse dividers changed in accordance with the values of the investigated parameters: the distance between the baffles L_b = 150 ... 325 mm, the height of the cut-out h_w = 87.200 mm. The input of the coolant in this task was not considered, the liquid inlet was set through the cutout of the divider, and the uneven distribution of the flow velocity was taken into account. It is assumed that the flow and heat transfer in the calculated region are stationary, the following uniqueness conditions are given:

physical conditions: liquid – water;

boundary conditions: input with a given velocity distribution and flow temperature; output - with given pressure; the wall is a smooth adiabatic wall with the condition of adherence on the surfaces of the dividers and the cylindrical surface of the casing; the wall is a smooth wall with a predetermined temperature and a sticking condition on the outer surfaces of the heat exchange tubes;

initial conditions: temperature and pressure.

Figure 2 shows the velocity field in the annular space with a separation distance of 250 mm between the dividers and a window height of 133 mm. Figure 3 shows the graphs of the velocity variation in section A-A in the gaps between the tubes at different distances from the axis of the apparatus along the lines indicated in Figure 2. It can be seen from Figures 2 and 3 that the distribution of the flow between the dividers is unequal. In the axial direction, one can distinguish the main flow and the stagnant zone, the velocity in which is much lower. In this case, the width of the stagnant zone is from 50 to 80 mm, which on the average is 29 % of the distance between the dividers. In the transverse direction, the velocity distribution is practically uniform, except for the region located near the inner surface of the casing (Figure 3, line 8) corresponding to the bypass flow path running between the peripheral tubes and the casing.

The bypass flow has a higher speed, which is due to a lower resistance of the bypass path, but it only flushes the outer row of tubes and therefore only slightly participates in the heat exchange process. The proportion of the bypass current, depending on the design of the apparatus, can reach 20–30 % [16].

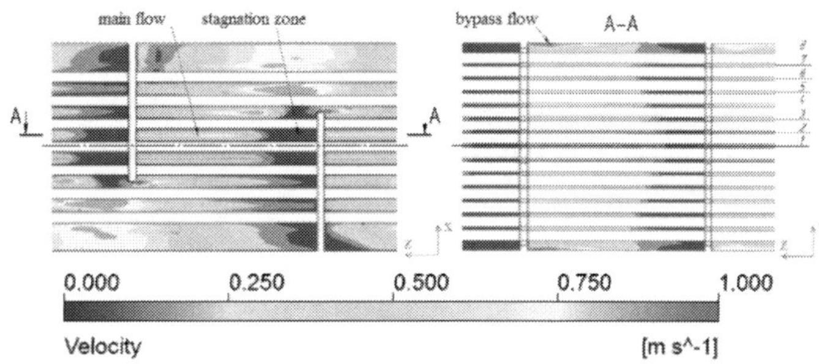

Figure 2. Flow rate distribution

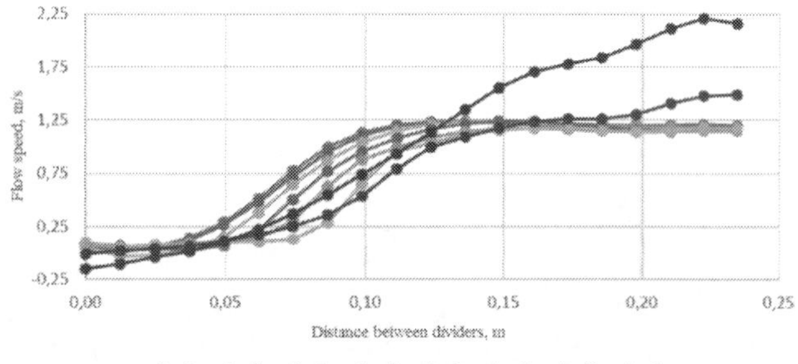

Figure 3. Graphs of velocity change in cross section at different distance from the machine axis along the Y axis.

Thus, in the cross section, three regions can be conventionally distinguished: the region of the main stream, the bypass flow and the stagnant zone.

At a constant window height h_w = 133 mm and the flow area in the divider window S_w = 20397 mm2, the size of the stagnation zone and the degree of unevenness of the flow distribution vary depending on the distance between the dividers (Figure 4).

L_b= 325 mm L_b= 250 mm

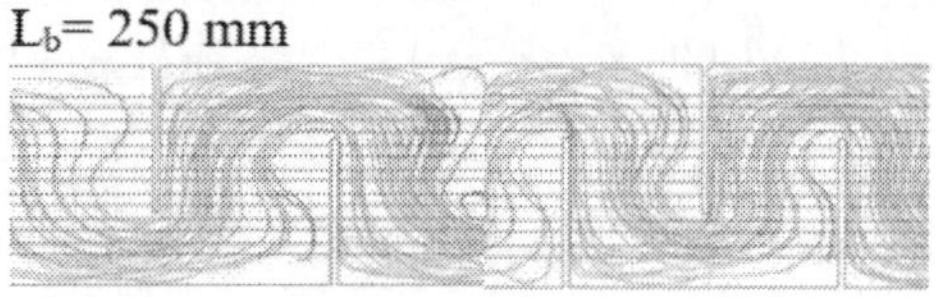

IOP Conf. Series: Journal of Physics: Conf. Series **944** (2017) 012024 doi:10.1088/1742-6596/944/1/012024

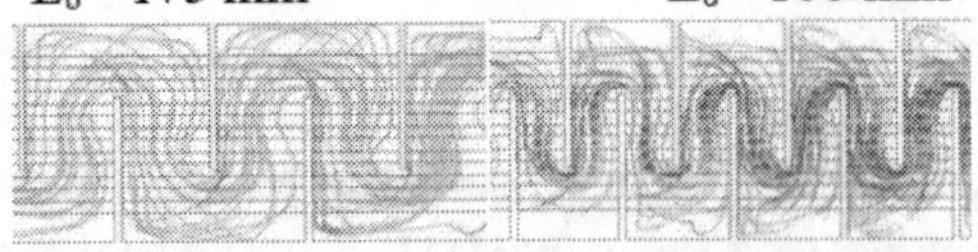

Figure 4. Lines of fluid flow for different locations of dividers.

Figure 5 shows a graph of the flow rate across the flow cross-section between dividers for different distances L_b.

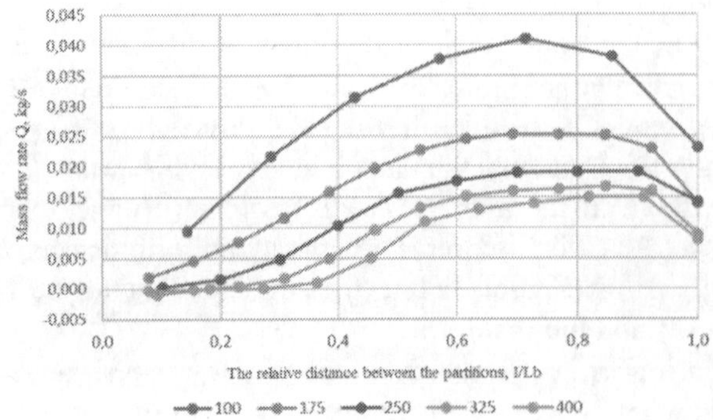

Figure 5. The graph of the flow rate across the flow section.

With a larger distance between the dividers L_b = 250 ... 400 mm and, accordingly, a larger cross-sectional area of the cross flow, S_c = 23750.38750 mm^2, the low-speed range is much wider, and at a distance of 100 and 175 mm between the baffles, S_c = 8750 and 16250 mm^2, respectively, that is, the distribution of the flow over the section is more uniform.

Thus, in cases where $S_c<S_w$, the relative width of the stagnant zone is smaller and the distribution of the flow between the dividers is more uniform.

The tendency to a more equal distribution of the flow with a smaller distance between the baffles is maintained at a fluid velocity of 0.4 to 2 m/s.

With a constant distance between dividers L_b = 250 mm (S_c = 23750 mm^2), the width of the stagnant zone varies depending on the height of the divider window.

In cases where $S_w > S_c$, with a divider window height of 133 to 178 mm, which corresponds to the area of the flow cross-section in the divider window from 20397 to 33397 mm^2, the relative width of the stagnant zone does not change significantly. The smaller height of the divider wall h_w = 87, 110 mm (S_w = 12824, 16615 mm) contributes to the formation of a wider stagnation zone (Figure 6).

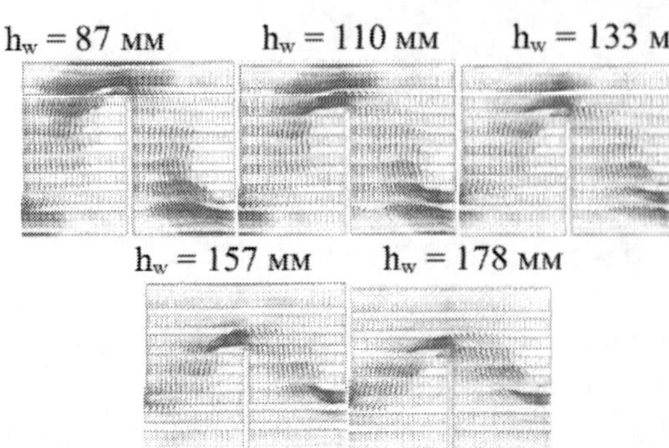

Figure 6. The velocity vector in the longitudinal section of the STHE.

Study of the partial effect of the distance between the dividers and the height of the window on the distribution of the liquid flow showed that both these parameters affect the size of the stagnant zone. An increase in the stagnation zone is observed when the area of the cross-sectional flow cross-section is larger than the cross-sectional area in the divider window.

To assess the joint effect of the height of the window and the distance between the dividers, models with L_b = 150 ... 275 mm and h_w = 87 ... 178 mm are considered. For the considered options, the ratio of the height of the window to the distance between the dividers lies in the range from 0.31 to 1.18.

For each model, the maximum width of the stagnant zone and its fraction in the distance between the dividers are determined. Figure 7 shows the dependence of the relative width of the stagnant zone b_d/L_b on the ratio h_w/L_b.

With a ratio $h_w/L_b > 0,85$, the width of the stagnation zone is 16,18% of the distance between the dividers. With a reduction in the cutout or an increase in the distance between the dividers, the relative width of the stagnant zone widens, reaching more than 40%. In the opposite case, when $h_w/L_b < 0,85$, the width of the stagnant zone behind the divider practically does not change, but with this ratio of parameters a stagnant zone appears near the casing. Thus, the ratio of the height of the window and the distance between the dividers, equal to 0,85, can be considered optimal for ensuring a minimum size of the stagnant zone behind the baffle wall.

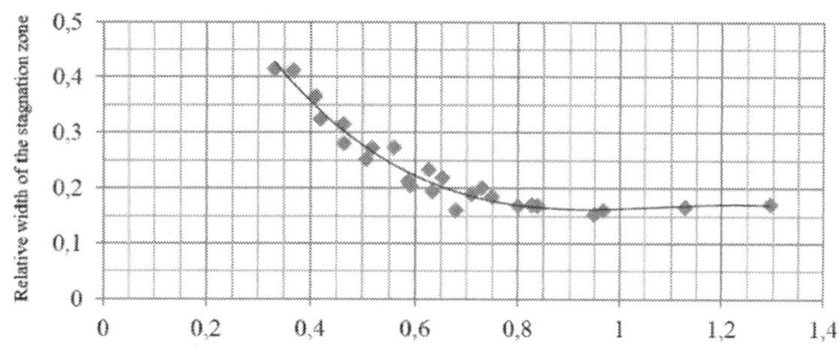

Figure 7. Dependence of the relative width of the stagnant zone on the ratio h_w/L_b.

From the foregoing, it follows that the STHE design, in which the distance between the dividers is smaller than the height of the divider cutout, is optimal insofar as it provides the minimum width of the stagnant zone. The height of the free segment of the divider determines the height of the cross-current region, and an increase in the cut-out of the divider is undesirable. Thus, in order to ensure the highest heat transfer intensity, smaller values of h_w should be taken and, consequently, L_b.

On the other hand, it must be taken into account that the close arrangement of the dividers leads to a more significant increase in the hydraulic resistance than the heat transfer coefficient, i.e. decreasing of the efficiency of heat transfer.

To solve this problem, it is possible by means of additional constructive elements - dividers (figure 8). The additional divider is a rectangular plate with holes for mounting in the tube bundle and is placed between the transverse dividers. Functionally, it acts as a bump, blocking the path of part of the main stream and directing it along the transverse divider to narrow the stagnant zone.

In order to determine the optimal location of the additional divider, several options were considered for placing the bump at different heights and distance from the transverse divider. For STHE diameter of 400 mm with a distance of 325 mm between the dividers, it is optimal to place the bump at ho = 89 mm from the machine axis and at a distance Lo = 120 mm from the transverse divider (Figure 9). The velocity fields in the longitudinal section, shown in Fig. 9, show that the bump block contributes to the narrowing of the stagnant zone. The use of an additional divider allowed reducing the pressure drop by 21.2% and increasing the average heat transfer coefficient by 1.3%. This improves the thermal and hydraulic design perfection, characterized by the ratio of the transferred thermal energy to the mechanical energy expended for pumping the coolant.

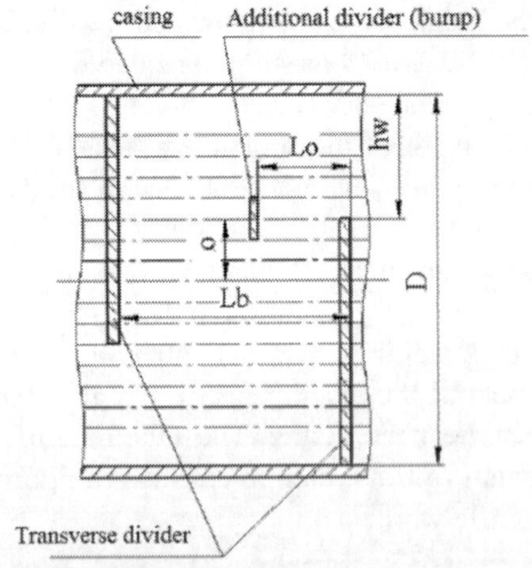

Figure 8. Placing an additional divider.

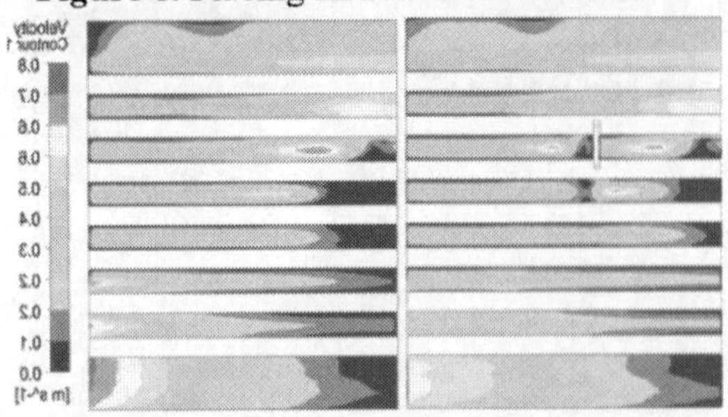

Figure 9. Velocity fields for different location of an additional divider.

Table 1 shows the thermohydraulic characteristics of the section of the annular space STHE, enclosed between adjacent transverse walls.

The coefficient of thermal-hydraulic perfection characterizes the ratio of the amount of heat Q given off by the surface to the power N spent for pumping the coolant relative to the surface [20].

Table 1. Thermohydraulic characteristics.

Characteristics	Without additional Dividers	With additional Dividers	Δ, %
Thermal energy, W	159300	161300	1,3
Mechanical energy, W	2,638	2,083	-21
Coefficient of heat transfer, W/m·K	2648	2678	1,1
Differential pressure, Pa	445,9	351,4	-21,2
Thermohydraulic perfection	60386	77436	28,2

Thus, the use of additional dividers in the STHE construction - bumpers, allows to reduce the hydraulic resistance of the flowing part due to a more even flow distribution and narrowing of the stagnant zones in the cross current region. This contributes to improving the thermal and hydraulic characteristics of the structure and increasing thermal efficiency. The use of bumpers allows to increase the distance between dividers, reducing the cost of the heat exchanger.

3. Conclusions

As a result of the study of the influence of the distance between the transverse dividers L_b and the height of the cutout of the divider wall h_w on the thermal efficiency of the STHE, the dependence of the size of the stagnant zone formed behind the transverse divider on the ratio of the indicated parameters h_w/L_b was revealed.

It is proposed to use additional dividers in the design of the tube bundle STHE, which help reduce the hydraulic resistance due to a more even flow distribution.

References

[1] Luzzi L, Cammi A, Marcello V D and Fiorina C 2010 *Chemical engineering science* **16** pp 487–488
[2] Muzychka Y S, Yovanovich M M 2001 *Journal of enhanced heat transfer* **4** pp 243–259
[3] Chiu Y W, Jang J Y 2009 *Applied thermal engineering* **29** pp 250–258
[4] Yan X, Xiao Z J, Huang Y P, Wang F 2006 *Hedongli gongcheng* **27** 77-82
[5] Hamid A H A, Hussam W K, Sheard G J 2016 *Numerical heat transfer* **8** pp 847–869
[6] Han L, Chang H, Zhang J, Xu T 2015 *Plasma science and technology* **4** pp 347–352
[7] Guo L, Qin F, Chen J, Chen Z 2008 *Applied thermal engineering* **28** pp 907–914
[8] Shulman Z P, Khusid B M, Zaltsgendler E A, Ivashkevich E V, Ryklina I L, Erenburg V B *Experimental thermal and fluid science* **1** pp 9–26
[9] Bakhtizin R N, Pavlova Z H 2015 *Proceedings* **1** pp 35–38
[10] Konesev S G, Khazieva R T, Kirillov R V 2016 *X International IEEE Scientific and Technical Conference* pp 1–7
[11] Konesev S G, Khazieva R T, Kirillov R V, Konev A A 2017 *Journal of Physics: Conference Series (JPCS)* **803**
[12] Konesev S G, Khazieva R T 2015 *Modern problems of science and education. Electronic scientific journal* **1**
[13] Karetnikov D V, Rizvanov R G, Fairushin A M, Kolokhov K S 2013 *Welding international* **7** pp 557–560
[14] GOST R 53677-2009 2011 Standartinform 34
[15] Standards of Tubular Heat Exchanger Manufacturers Association 2007
[16] Spravochnik po teploobmennikam 1987 p 352
[17] Chetvertkova O V, Rizvanov R G 2012 *Neftegazovoe delo* **3** pp 109–112
[18] Chetvertkova O V, Ardashirova E A 2012 *Neft' i gaz* 72
[19] Chetvertkova O V, Mironov A V, Rizvanov R G, Sharafiev R G 2015 *Neftegazovoe delo* **3** p 452–465
[20] Antufev V I 1966 *Effektivnost' razlichnyh form konvektivnyh poverhnostei nagreva* p 183

Upgrading weld quality of a friction stir welded aluminum alloys AMG6

I K Chernykh, E V Vasil'ev, E N Matuzko, and E V Krivonos
Omsk State Technical University, 11, Mira ave., Omsk, 644050, Russia

e-mail: vnchrnkh@gmail.com

Abstract. In the course of introduction of FSW technology into the industry there is a keen interest in this process; there are issues such as how does joining take place, what is the structure of the joint, and where there are dangerous zones. The objective of this research is to obtain information about the structure of the joint, what are the temperatures that arise during the joining, what strength is apply to the tool when joining the material, what tensile strength of joint, and where fracture tended to occur. Specimens were produced at different modes of welding at a tool rotation speed of 315 to 625 rpm and tool travel speed of 40 to 125 mm/min. During the experiment, the strength applied to the tool was measured, which reached 8000–16000 N (Fz) and 400-1400 N (Fx) and the temperature on the surface of the tool, which is in the range 250–400°C. Before the welding process the tool was heated to a temperature in the range of 100–250 degrees, but the tensile strength is not had a tangible impact. The tensile strength is about 80 % of that of the aluminum alloy base metal tensile strength, and fracture tended is occur not at the line of joint but follow the shape of the tool. In the transverse cross section of a FSW material there is a microstructural regions such as weld nugget, thermomechanically affected zone and heat-affected zone with parent material.

1. Introduction

In all branches of mechanical engineering there are many assembly works. Depending on the industry, assembly works can reach large volumes percentage of the total amount of technology process of production of the device or the machine. For example, the body of the big vessels may contain welded seams with a total length up to 1000 km, at the same time the main assembly works at assembly of vessels are welding works. Among other industries where a lot of welding work is using during assembly there are also an aircraft industry, rocketry, car building, automotive industry. The research of this process takes place at the Moscow Institute of Aviation Materials [1], at the Tomsk Polytechnic University [2], in the Southern Ural State University and other institutes in the territory of Russia. Higher education institutions and the enterprises make out patents [3-7]. Foreign firms and research laboratories also conduct researches and works on introduction of this process in the industry, among them there is ESAB, SAPA (Sweden), Boeings Eclipse Aviation Corporation (USA), EADS (France, Germany), Institute Soudure (France), DanStir (Denmark), etc.

Friction stir welding is a solid state welding process used for welding similar and dissimilar materials by the tool, which have tool shaft, shoulder and pin with threads (Figure 1).

Figure 1. Working part of FSW tool. 1 – Pin, 2 – shoulder, 3 – threads.

The perspective directions of researches are joining of titanic alloys and joining of dissimilar materials.[8] At the same time, application of an additional effect on a joint and use the special tools may to achieve the increase of efficiency of welding and tensile strength of a joint. However, researches on aluminum alloy welding remains relevant, since aluminum alloys are using in such industries as rocketry, aircraft industry, shipbuilding; for this reason the interest in joining this material is constantly present. Research of this paper is directed to upgrading of tensile strength and efficiency. The authors suppose that tensile strength of joint of aluminum alloys may reach the tensile strength of parent material, to reach this strength it is necessary to research microstructure of joint and investigate broken specimens for identify dangerous zones.

2. Statement of the problem

For carrying out the experimental works, there is the following inventory: educational and research FSW machine based on a vertically milling machine, and the tool for welding (Figure 2).

Figure 2. Educational and research FSW machine.

The machine is equipped with a research system for filing the parameters of the welding process which allows you to file the strength Fx (the strength acting in the welding direction), Fy (the strength acting perpendicular to the welding direction in the plane parallel to the surfaces of the details being welded – the welding plane) and Fz (the strength acting in the direction perpendicular to the welding plane), and the torque Mz.

To file the temperature, the Fluke Ti400 thermal imager was used, the area of the captured image covered the shoulder of tool and the surface of the details being welded. For the testing tensile strength, the Instron 8802 fatigue testing system was used.

3. Theory

1. Welding process

The physics of process of thermomechanical welding assumes itself thermal and mechanical impact on a weld material. In the course of FSW, heating is provide with a sliding friction between the tool and the welded products and the plastic strain caused by the tool pressure upon the surface of material and introduction of a pin upon joint depth and tool travel.

Research that was made earlier proved that temperature during the welding equally depends on the tool rotation speed and tool travel speed, but the temperature before welding depends only on rotation speed and speed of introduction of a pin. The temperature in a welding zone increase during the tool travel will not reach a particular range of values yet. In order to make the weld uniform throughout its length, we need to conduct a research about arising strengths and temperature in the course of welding and check affect whether preheating of the tool and the welded products up the 250°C on strength and temperature after the start tool travel.

2. Tensile strength of a joint

The peculiarity of the microstructure of friction stir weld is that grains in a weld nugget subjected to strong deformation and their size much smaller than size of grains in a parent material, and the joint structure is divided into zones with different sizes of grains. It is assumed that the heat-affected zone have full strength of a parent material because the heat-affect in this area not enough to recrystallize grains. Dangerous zone in the transverse cross section of a friction stir weld is a weld interface where the microstructure changes. This weld interface shown in Figure 3. Between zones B and C and between weld nugget and zone C.

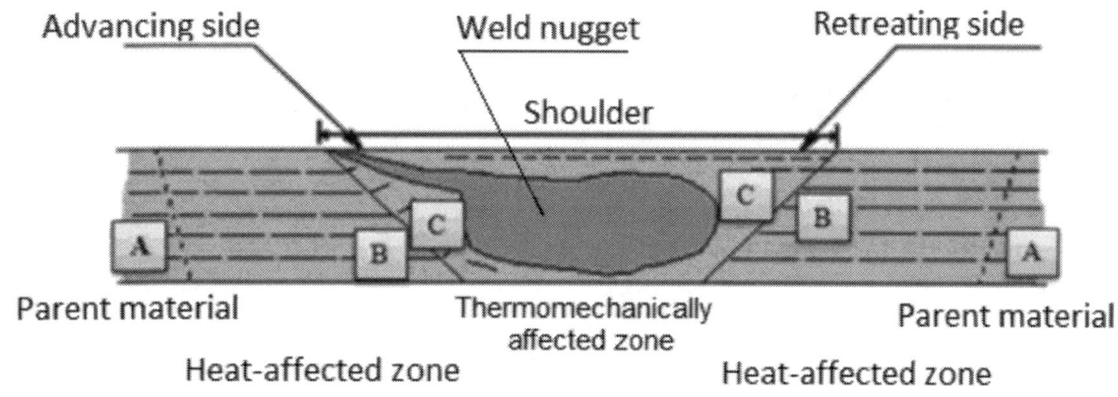

Figure 3. Illustration of different microstructural regions in the transverse cross section of a friction stir welded material. A, parent material; B, heat-affected zone; C, thermomechanically affected zone.

The shape of a pin also affects the strength of a joint strength and microstructure, which is proved in the paper [9]. For identify dangerous zones it is necessary to get a microstructure image, for this reason were made samples for research.

4. Experimental results

Plates of 2-mm-thick AMG6 aluminum alloy were welded with the tool rotating at speeds of 315 to 625 rpm and travel speed, that is, welding speed, was of 40 to 125 mm/min. Herewith several welds were preheated before welding. The vertical strength was applying to the tool shown on the Figure 4, and the horizontal strength shown on the Figure 5.

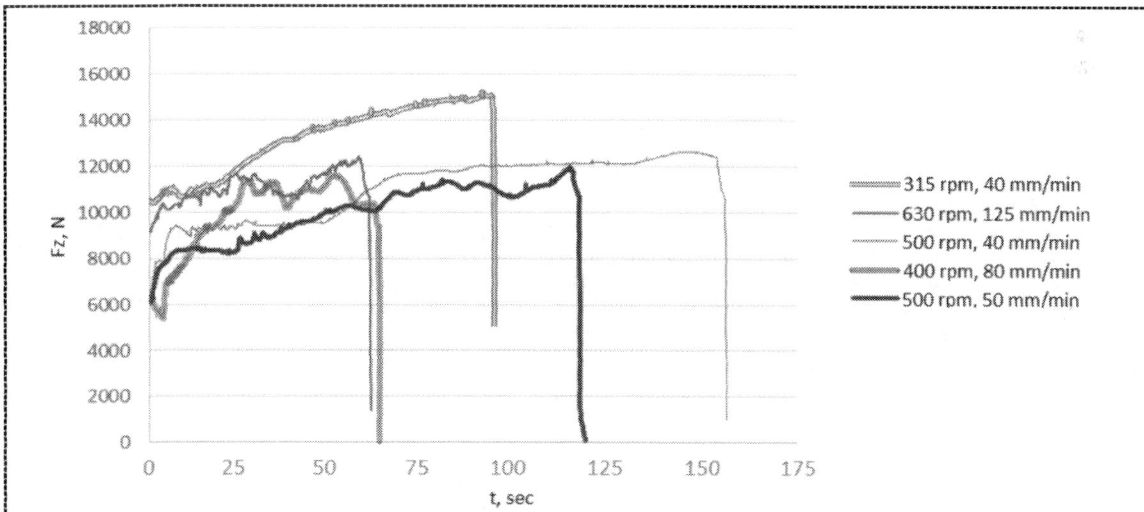

Figure 4. Vertical strength.

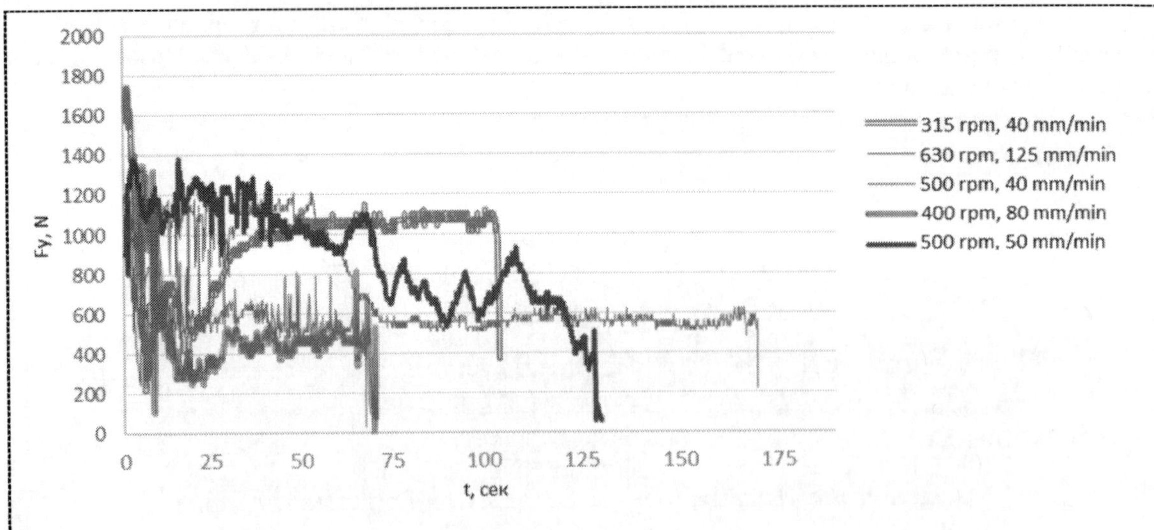

Figure 5. Horizontal strength.

For testing tensile strength depending on the welding conditions 60-mm-width specimens was made and testing with 2 mm/min speed of tension. The results we have it shown on a Figure 6 and Table 1.

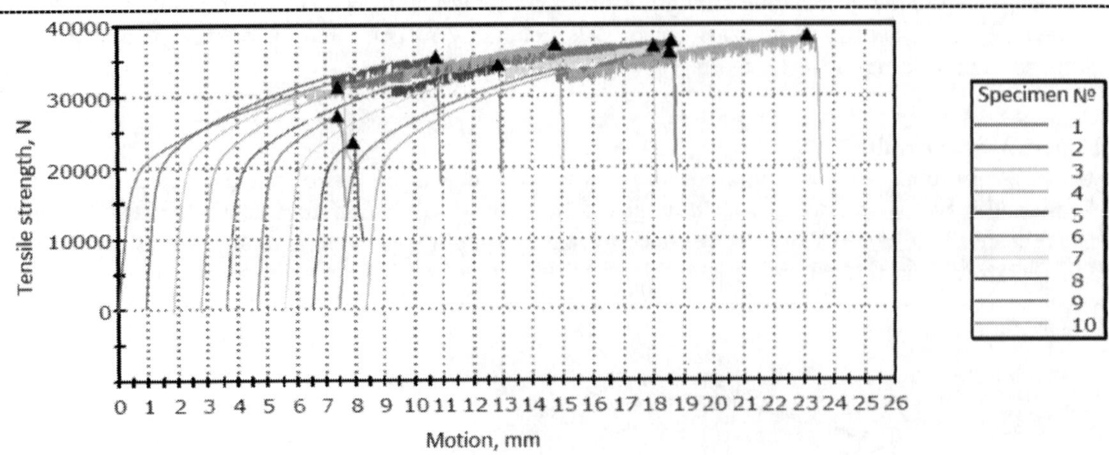

Figure 6. Tensile test of joints. From left to right – 1,7 – 500 rpm, 50 mm/min; 2 – 630 rpm, 125 mm/min; 3 – 315 rpm, 40 mm/min; 4 – 500 rpm, 55 mm/min with preheating, 5 – 400 rpm, 80 mm/min with preheating, 6-9 – 500 rpm, 50 mm/min with preheating, 10 – 400 rpm, 40 mm/min.

Table 1. Tensile strength of specimens.

№ of specimen	Tensile strength, N
1	37793.62
2	35572.62
3	31405.99
4	37291.45
5	34828.50
6	27322.94
7	36964.35

IOP Conf. Series: Journal of Physics: Conf. Series **944** (2017) 012025 doi:10.1088/1742-6596/944/1/012025

Table 1. Tensile strength of specimens.

№ of specimen	Tensile strength, N
8	23670.91
9	36163.15
10	38448.54

For getting a microstructure image were made samples by wire spark erosion, grinding and polishing were manual handling, etching process took place in 20 % caustic soda bath. Samples were analyzed by means of inverted-stage microscope Axio Observer A1m.

Microstructure image we got was connected in full images of the transverse micro-section of a friction stir welded material (Figure 7–10). There is a thin lines that shows the weld interface, the dotted line shows the visible part of a nugget zone, the thick line shows the rupture line after tensile testing. Scale ruler at the bottom corresponds to 1600 microns. 1 – heat-affected zone with parent material, 2 – thermomechanically affected zone, 3 – weld nugget.

Figure 7. Transverse micro-section of friction stir weld at a tool rotation 500 rpm and tool travel 50 mm/min.

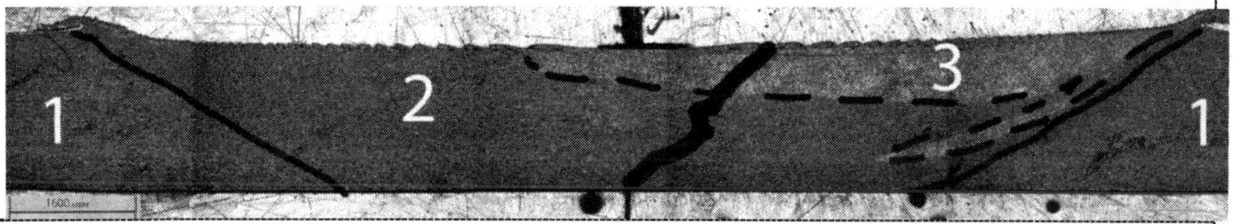

Figure 8. Transverse micro-section of friction stir weld at a tool rotation 315 rpm and tool travel 40 mm/min.

IOP Conf. Series: Journal of Physics: Conf. Series **944** (2017) 012025 doi:10.1088/1742-6596/944/1/012025

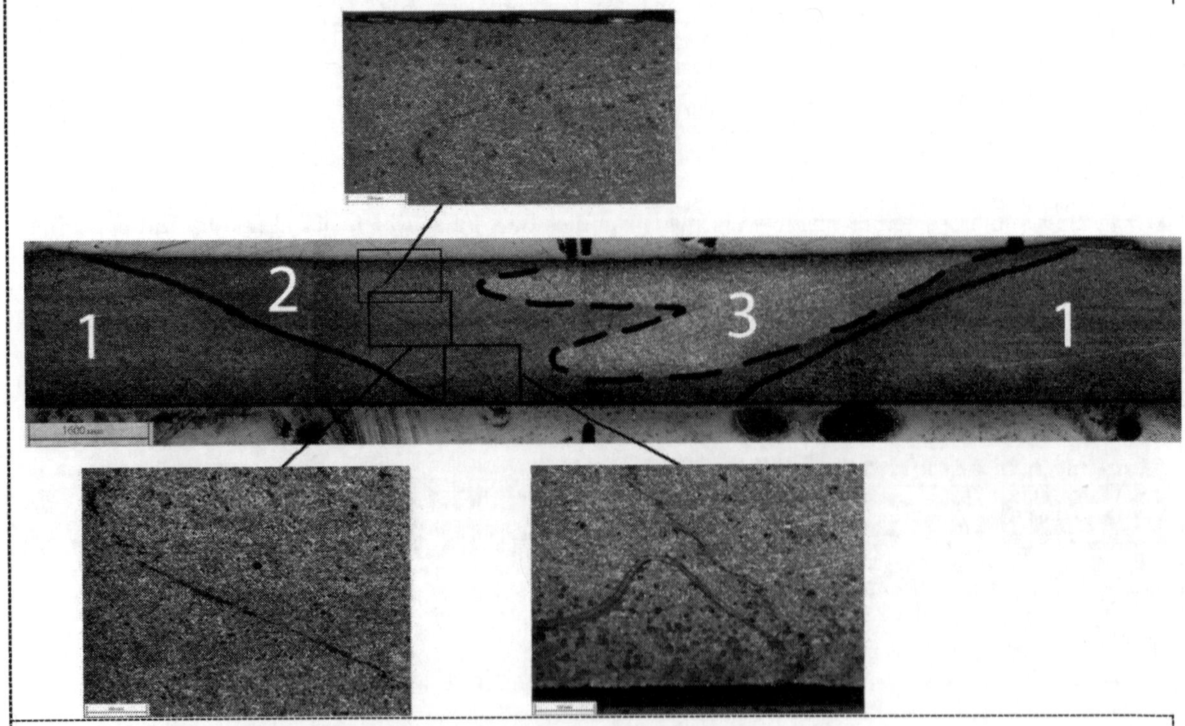

Figure 9. Transverse micro-section of friction stir weld at a tool rotation 500 rpm and tool travel 40 mm/min with enlarged images of the defect «hat crack».

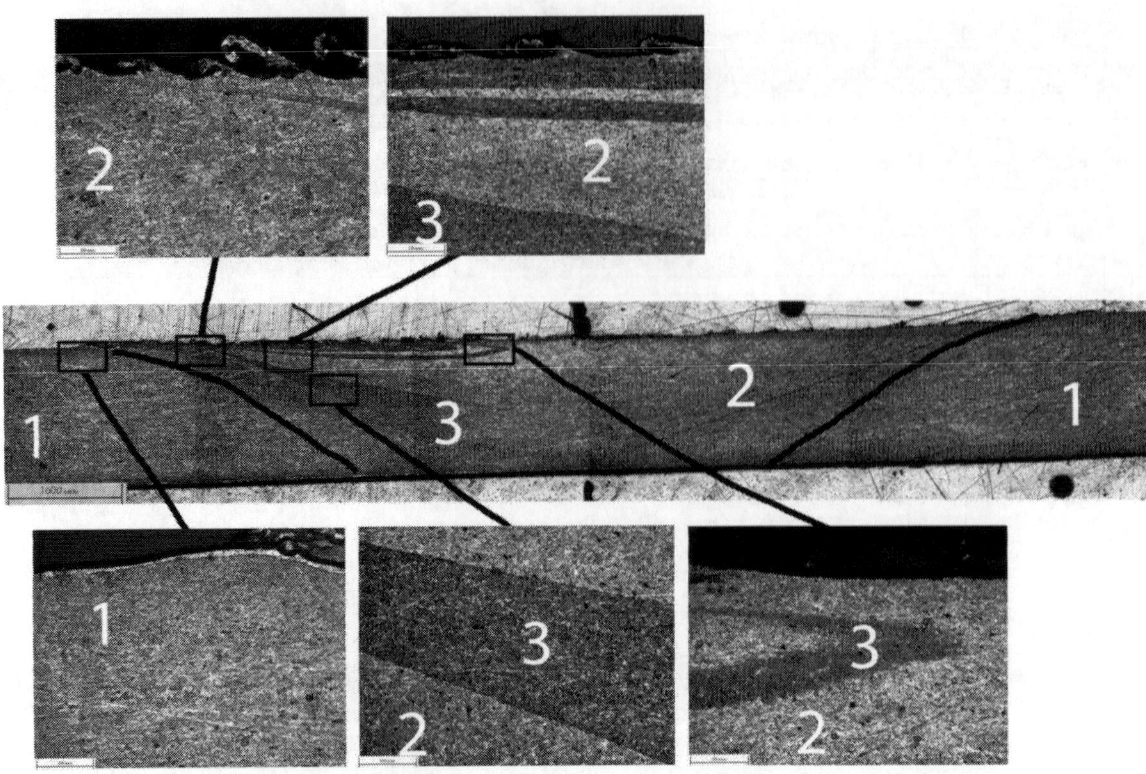

Figure 10. Transverse micro-section of friction stir weld at a tool rotation 630 rpm and tool travel 125

mm/min with enlarged images of the weld zones and weld interfaces.

Temperature were arise in the place of friction between the surface of the details being welded and shoulder of a tool shown in Figure 11.

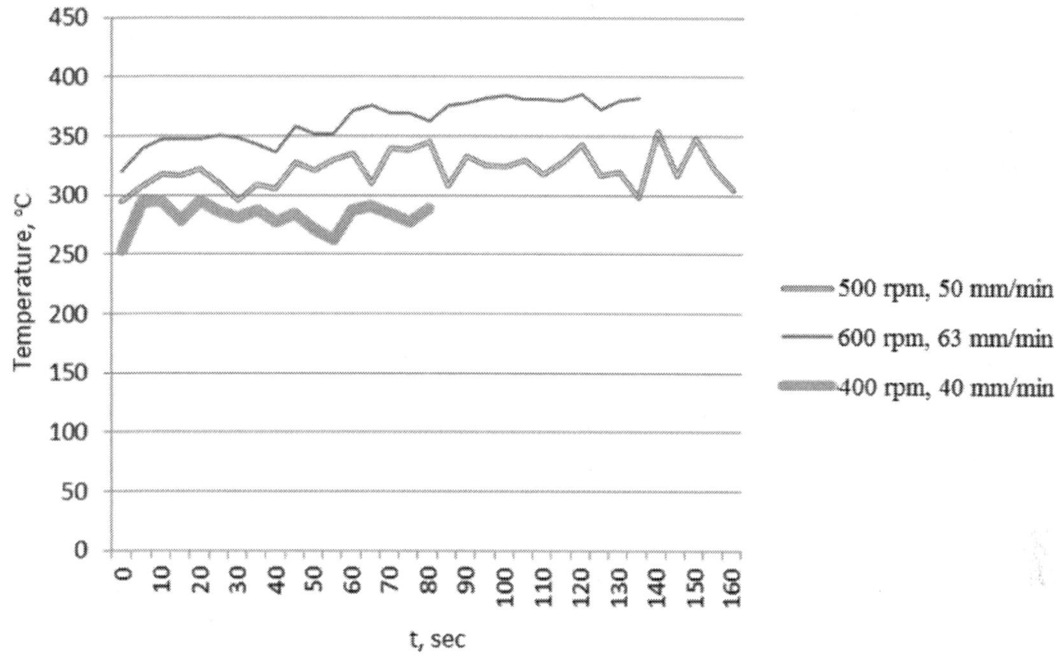

Figure 11. Temperature graph in the place of friction on the surface of the tool for the various welding condition.

5. Discussion

As seen on the graphs (Figure 4, 5) preheating (which was done under the following welding condition: tool rotation 500 rpm and tool travel 40 mm/min; tool rotation 400 rpm and tool travel 80 mm/min, tool rotation 500 rpm and tool travel 50 mm/min) had not a tangible impact on the strength compared with the strength applied to the tool without preheating. The temperature that arise during the joining with preheating more by 40–70°C compared with the joining without preheating, but temperature also had not a tangible impact on the tensile strength (Figure 6, Table 1).

The microstructure images we got (Figure 7–10) show the interface between thermomechanically affected zone and parent material and clearly see the weld nugget. These zones can be identified from parent material by color and microstructure. The joint has an asymmetrical structure where the advancing side and retreating side as seen, their mutual position depends on the direction of tool rotation. The weld nugget generally has a tear-drop shape as seen on the Figure 7. The fracture of the welded samples is occur not at the line of a joint but follow the shape of the tool. Figure 7-8 show that fracture follow the cone shape of the pin and is located close to the weld nugget in the thermomechanically affected zone. The tensile strength is about 80 % of that of the aluminum alloy base metal tensile strength (Table 1) except when specimens has joint defect. One of the possible joint defects is shown in Figure 9. The "hat craks" can reduce the tensile strength up to 40 %.

6. Conclusion

To obtain additional, exhaustive information about the friction stir welding process, it is necessary to carry out multifactorial experiments with varying factors affecting the quality of the weld, such as welding condition, tool shape etc.

To upgrading tensile strength of a friction stir weld must be a deeper research of the microstructure of joint. We presume that tensile strength of joint can be upgrading by changing the microstructure by thermal action. It is plan to carry out experiments using the device for local heating, which the authors are developing. In our opinion, the thermal acting during the joining on welded details will affect the change in microstructure of joint, and it will lead to the fact that sizes of grains from different sides of the interface will not be so various, that it will upgrade the tensile strength of joint.

Aknowledgment
Ability to use inventory of a TsPVIK PO "Polet" is acknowledged. Omsk State Technical University is acknowledged for ability to establish Student Design Bureau the direction of which is aimed at studying this welding process.

References

[1] Lukin B I, Ioda E N, Bazeskin A V, Zhegina I P, Kotel'nikova L V and Ovchinnikov V V 2012 *Welding International (Welding production)* **4** pp 45–48
[2] Kolubaev E A 2011 *Sovremennyye problemy nauki i obrazovaniya (Modern problems of science and education)* **6** pp 1–2
[3] Alifirenko E A, Bakshayev V A , Vasil'yev P A, Dodon R V, Oryshchenko A S, Osokin E P, Pimenov A V, e.a. Ustanovka svarki treniyem s peremeshivaniyem krupnogabaritnykh konstruktsiy (Device for friction stir welding of large-sized structures), Patent RF, no. 156976, 2015
[4] Alifirenko E A, Zarubin V M, Oryshchenko A S, Osokin E P, Pavlova V I 2009 Instrument dlya svarki treniyem s peremeshivaniyem alyuminiyevykh splavov i sposob svarki (Tool for friction stir welding aluminum alloys and welding method) Patent RF no. 2357843
[5] Egorov A V, Shillo G V, Makarov N V 2017 Sposob svarki treniyem s peremeshivaniyem i ustroystvo dlya yego osushchestvleniya (The method of friction stir welding and the device for its implementation) Patent RF no. 2630147
[6] Boytsov A G, Kachko V V 2010 Sposob svarki treniyem s peremeshivaniyem i instrument dlya svarki treniyem s peremeshivaniyem (A friction stir welding method and a friction stir welding tool) Patent RF no. 2393071
[7] Krysanov O N, Kachko V V, Klippenstein A D 2010 Sposob svarki treniyem s peremeshivaniyem (A friction stir welding method) Patent RF no. 2393070
[8] Grigorenko G M, Adeyeva L I, Tunik A Yu, Stepanyuk S N, Poleshchuk M A and Zelenin E V 2014 *The Paton Welding Journal (Automatic welding)* **4** pp.13–23
[9] Palanivel R, Koshy Mathews P 2011 *Materials and technology* **45(6)** pp 623–626

Experimental study of modification mechanism at a wear-resistant surfacing

R R Dema, R N Amirov and O B Kalugina

Nosov Magnitogorsk State Technical University, Magnitigorsk, Russia

e-mail: demarr78@mail.ru, kalugina.olga@bk.ru

Abstract. In the study, a simulation of the crystallization process was carried out for the deposition of the near-eutectic structure alloys with inoculants presence in order to reveal the regularities of the inoculant effect and parameters of the process mode simulating surfacing on the structure of the crystallization front and on the nucleation rate and kinetics of growth of equiaxed crystallites of primary phases occurring in the volume of the melt. The simulation technique of primary crystallization of alloys similar to eutectic alloys in the presence of modifiers is offered. The possibility of fully eutectic structure during surfacing of nominal hypereutectic alloys of type white cast irons in wide range of deviations from the nominal composition is revealed.

Keywords: surfacing; wear-resistant alloys; eutectic; modification; simulation technique.

1. Introduction

The high-carbon wear-resistant iron-based surfacing materials having structure of white cast iron are widely applied to surface and volumetric hardening by an arc surfacing of the parts working in the conditions of an abrasive wear in combination with load impact of moderate intensity[1]. But compared to cast products of white wear-resistant cast iron, in the surfaced and sprayed coating, formation of structure and properties take place in strongly non equilibrium conditions [2, 3]. Therefore coatings have the highest working properties when forming structure different from cast iron product [4, 5].

In surfaced coatings the complex of high working properties (impact value and wear resistance) is reached under following characteristics of structure: [6]:

- the degree of eutectical close to one;
- finely dispersed structure of an eutectic;
- lack of carbides of cementite type etc.;

Existence of dendrites of solid solution in coatings with eutectic structure, considerably reduces the resistibility to abrasing effect. In most cases, presence of fragile primary carbides in the hypereutectic alloys doesn't allow their use even with the minimum load impacts [8].

Features of formation of structure during surfacing (a deviation of a chemical composition of the built-up metal from nominal, rapid mixing, etc.) cause need of receiving a non-equilibrium quasieutectic microstructure of circumeutectic alloys by growth slowdown of primary phases. It is possible under the displaceable crystallization curves [7] and/or under inoculation. The microalloying by elements forming refractory particles in liquid-alloy (in relation to white cast iron – carbides Ti, Y, Nb), has a beneficial effect on creation of the additional centers of crystallization (sort modifiers I) with the following microstructural refinement [9, 10].

The modification by metals with coordinate fusing temperature with the base metal aid changes of the melt properties (viscosity, a surface tension, diffusivity, etc.). Being adsorbed on surfaces of the growing crystallites, they lead to changing of nature of crystallization of separate phases of a eutectic (sort modifiers II).

2. Materials and methods

Modeling of regularities of crystallization under surfacing of the circumeutectic alloys in the presence of modifiers was carried out by a technique of microscopical analysis of processes of hardening of optically transparent materials according to the scheme "motionless bath – the rotating sample" [12] (fig. 1).

As components of model salt system, we have chosen saltpeter K NO_3, ($Melting\ point$ = 334 °C), modeling solid solution, and the KClchloride of potassium ($Melting\ point$ = 778 °C) imitating a refractory phase like carbides and restrictedly dissolved in KN03. The eutectic point corresponds to 4,6% of KCl on the weight and temperature of 312 °C. Chloride sodium NaCl was used as the sort modifier I ($Melting\ point$ = 801 °C), insoluble in KN03, and KOH potassium hydroxide as the sort modifier II, which restrictedly dissolved in KN03 and KS1.

Figure 1. Appearance of a surfacing bath

For modification of a surfacing bath, alloy of salts melted by immersion of a heated spiral of a motionless glow element (see the figure 1). The quantity and the nature of heat sharing entered into a bath are regulated by the heating temperature, form and the sizes of a spiral of an element, and also by rotation speed of the faceplate.

The area of interest of a bath was combined with an optical axis of a microscope and kept its position within sight that provided a continuity of observation process.

During optimisation of methods, crystallization processes researches for the salt system KNO3 + KCl were made. Mass concentration of KCl varied from 0 to 6%. Special attention was paid on detection of regularities of influence of parameters of the process mode imitating a surfacing on a structure of the crystallization front (CF) and on intensity of origin and kinetics of equiaxed crystallites growth (EC) arising in fusion volume.

3. Experimental results

The carried out research made it possible to reveal the regularities of the inoculants effect and process parameters imitating surfacing on the structure of the crystallization front and on the nucleation rate and the growth kinetics of equiaxed crystallites of primary phases occurring in the melt volume. Results of observations of crystallization process were fixed f with video recording by the digital camera. The CF structure obtained by results of data processing, are given in the figure 2.

CF is basically plane for pure saltpeter in the speed range studied (fig. 2 a).EC emergence isn't recorded. Occasionally there are perturbations of plane CF with transition to meshy type of crystallization[13]. There is a step movement of meshy CF, after that the plane CF is recovered. With speed growth of a surfacung the frequency of bounces of CF and their expansion increases[14].

Existence of particles of uncontrollable impurity, accidental mechanical impacts on a fusion and changes of surfacing speed, etc. can be the possible cause of perturbations of FC.

On addition of over 1% of chloride potassium into saltpeter, CF gets strongly marked meshydendritic structure when axes of growth of crystallites of the second order are designated. Before CF emergence EC is observed (the figure 2). After origin EC grow mainly in the direction of the basic EC where it is necessary to expect the increased concentration of impurity in fusion. Around growing EC the movement of the basic CF is slowed down. After the contact of dendrites of the basic CF with EC their joint growth begins, and the CF growing from EC has meshy type with thinner structure.

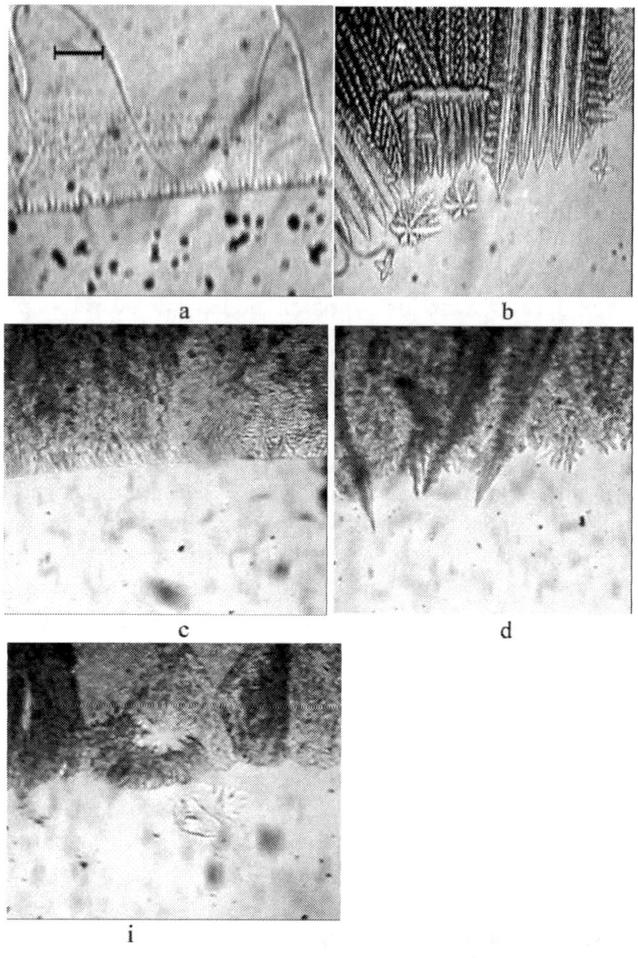

Figure 2. Structure of the crystallization front depending on components of modifiers and surfacing speed:*a) b) c) d)* – $V_{of\ sufacing} = 0{,}12\ mm/s$; *i)* - $V_{of\ sufacing} = 0.28\ mm/s$;

During results of observations of crystallization process were fixed with video recording by the digital camera. Special attention was paid on detection of regularities of influence of parameters of the process mode imitating a surfacing on a structure of the crystallization front (CF) and on intensity of origin and kinetics of equiaxed crystallites growth (EC) arising in fusion volume.

Results of microscopic observations were fixed by video recording. Results of an experiment are presented on the figure 3.

The quasiflat FC typical to crystallization of an eutectic structure alloy under small speed of surfacing of vH is given in fig. 3, and. EC are absent. With increase of vH by 2 – 3 times the morphology of an eutectic (fig. 3,*b*) changes. Formation of the eutectic branched dendrites is observed. EC are also absent that confirms the eutectic nature of crystallization.

Violation of quasiflat FC and emergence of small EC of primary phase is observed on addition of small number of 0,25% sort modifier I to the eutectic alloy (fig. 3, *c*). The effect amplifies under any deviations from the eutectic concentration and with increase of vH. Under surplus of NaCl quantity the EC and speed of their growth promptly increase (overmodification). At the same time the main CF doesn't change practically.

An effect of the sort modifier II is more complex. The additive of 0,5% KOH to the eutectic alloy gives the effect similar to increase of $v_{н}$ i.e. the eutectic morphology changes (fig. 3 *d*). The eutectic dendrites become more thinly branched. Under surplus of KOH they are integrated, also there is a high probability of formation large EC of primary phase (fig. 3 *j*).

In hypoeutectic alloy EC origin becomes active in fusion with their subsequent rapid growth. The CA is almost completely formed from large EC (fig. 3 *i*). In this case KOH acts as the sort modifier I.

The most interesting is the effect of appending of optimum KOH quantity to hypoeutectic alloys. As a result of growth slowdown of a refractory phase, it is observed completely eutectic crystallization for nominally hypoeutectic alloys with concentration till 7% of KC1 (fig. 2, *f*)In practice it means a possibility of receiving the eutectic structure under surfacing in the wide range of deviations from nominal structure of alloys[14,16]. As sort modifiers II for white cast iron, sulfur phosphorus, copper, and other elements can be used. Their search is a separate object of research [12, 13].

4. Results and discussion

The work carried out allows to discover regularities of influence of agents of modifiers and parameters of the mode of the process imitating a surfacing on a structure of crystallization front (CT) and on intensity of origin and kinetics of growth of equiaxed crystallites (EC) of primary phases arising in volume of fusion . Results of microscopic supervision were fixed with video recording.

The quasiflat FC typical to crystallization of an eutectic structure alloy under small speed of surfacing of vH is given in fig. 2a, and. EC are absent. With increase of vH by 2 – 3 times the morphology of an eutectic (fig. 3, b) changes. Formation of the eutectic branched dendrites is observed. EC are also absent that confirms the eutectic nature of crystallization.

Violation of quasiflat FC and emergence of small EC of primary phase is observed on addition of small number of 0,25% sort modifier I to the eutectic alloy (fig. 3, *c*). The effect amplifies under any deviations from the eutectic concentration and with increase of vH. Under surplus of NaCl quantity the EC and speed of their growth promptly increase (overmodification). At the same time the main CF doesn't change practically.

An effect of the sort modifier II is more complex. The additive of 0,5% KOH to the eutectic alloy gives the effect similar to increase of $v_{н}$ i.e. the eutectic morphology changes (fig. 3, *d*). The eutectic dendrites become more thinly branched. Under surplus of KOH they are integrated, also there is a high probability of formation large EC of primary phase (fig. 3, *j*).In hypoeutectic alloy EC origin becomes active in fusion with their subsequent rapid growth. The CA is almost completely formed from large EC (fig. 3, *i*). In this case KOH acts as the sort modifier I.

The most interesting is the effect of appending of optimum KOH quantity to hypoeutectic alloys. As a result of growth slowdown of a refractory phase, it is observed completely eutectic crystallization for nominally hypoeutectic alloys with concentration till 7% of KC1 (fig. 3, *f*). In practice it means a possibility of receiving the eutectic structure under surfacing in the wide range of deviations from nominal structure of alloys.

For simulation verification test, an experiment on surfacing flux cored wire of brand PP-AN125 (PP-200H15S1GRT) was performed. Surfacing mode parameters are: arc current – 350 A; arc volts 30... 32 V; surfacing rate - 0,5 cm / page. Structure of the built-up metal is hypoeutectic with primary crystals of austenite.

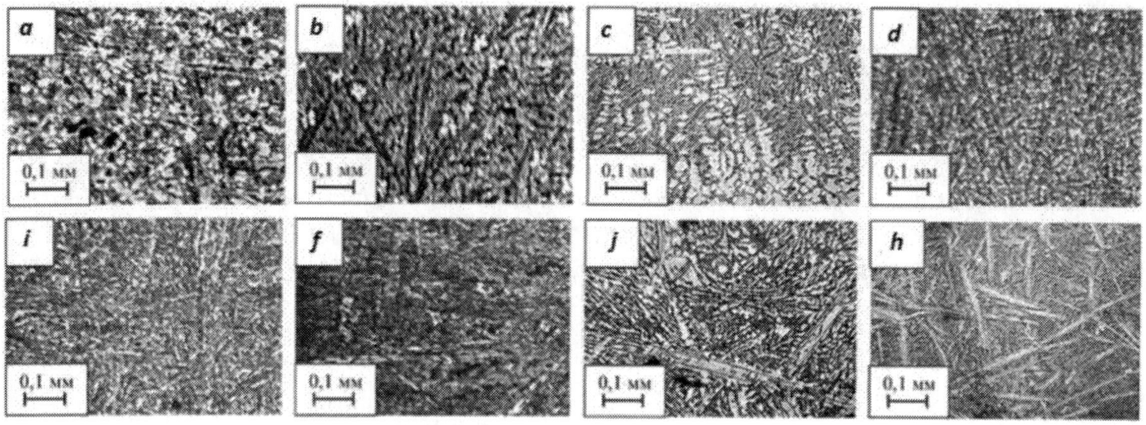

Figure 3. Microstructure of the built-up metal

For obtainment the eutectic and hypoeutectic microstructures, the built-up metal was in addition boron doped by pre-application of a layer of FB20 ferroboron powder to the built-up sample, during surfacing this powder were fixed by zapon lacquer to prevent deflation.

As a sort modifier II silicocalcium was added to ferroboron powder, this choice is based on data of work [18] on efficiency of an silicocalcium additives to the charge flux-cored wire in case of surfacing of high-strength gray cast iron. The quantity of a ferroboron and a silicocalcium per unit area of a sample was being changed.

During the analysis of a microstructure special attention was paid on existence and the size of primary phases and morphology of eutectic. Results of the metallographic analysis are given in the structure chart (fig. 2) and in microstructures (fig. 3).

The chart shows that completely eutectic structure in the absence of the modifier can be obtained in narrow range (2,5...3,0%) of boron concentration austenite dendrites (see fig. 3,*a*) emerged with reduction of boron concentration, primary carboborite (see fig. 3, j) emerged in case of increase of concentration.

5. Conclusions

With appending 0.2% Ca, the area of the eutectic concentration extends to 2.3... 3.7%. The carbide phase is ground (see fig. 4, f). In case of increase of Ca containing over 0,3% the effect of modifying weakens. The area of the eutectic concentration is narrowed again, the sizes of primary phases considerably increase, i.e. we gain effect of overmodification.

In the available samples We determined deformation of crack origin as a crash-worthiness index and also HRC hardness by a technique [5].In nominally hypereutectic built-up metal after addition of the modifier it was succeeded to reach an increase of crash-worthiness by 2 – 3 times as a result of formation of the quasieutectic microstructure. Hardness of the built-up metal practically didn't change.

It is recommended to add silicocalcium to the charge flux-cored wire of PP-AN125 or PP-AN170 types, and also the surfacing electrodes like T-620 for modifying for the purpose of increase of crash-worthiness of the built-up metal.

Acknowledgement

The authors are thankfully acknowledge the financial support by the Ministry of Education and Science of the Russian Federation for the project № 11.2054.2017/Project task within the framework of the state task for 2017-2019(Number for publication: 11.2054.2017 / 4.6) to carry out this investigation.

References

[1] Charchenko M V 2010 *Energy consumption decreasing in wide strip hot rolling on the base of modeling and effective mode of lubrication rolls choice*: diss. of master of techn. sc. Magnitogorsk

[2] Platov S I, Amirov R N, Dema R R and Yaroslavtsev AV 2012 Wearing and working rolls durability predicting mathematical model in cuatro stand at a lubrication supply *Proizvodstvo Prokata (Rolled Products Manufacturing* **9** pp 38–43

[3] Dubovsky S V, Dema R R, Harchenkoand M V and Yaroslavtsev A V 2011 Efficiency of feed technological lubricants in the stand number 7-9 continious broad and strip hot rolling 2000 JSC "MMK" integrated assessment research *Proizvodstvo Prokata (Rolled Products Manufacturing* **12** pp 6–8

[4] Ehlliot R 1987 Upravlenie ehvtekticheskim zatverdevaniem [Eutectic solidification management] (Moscow: Metallurgiya) p 352

[5] Nefedyev S P, Dema R R, Nefedyeva S A and Yaroslavtcev A V 2015 Microstructure of cast iron after plasma bleaching *Journal of Chemical Technology and Metallurgy* **50** 2 pp 213–216

[6] Youn T K, Amanov A, Cho I S and Park I G 2011 Effects of ultrasonic nanocrystalline surface modification on the high-frequency fretting wear of CP titanium and Ti-6Al-4V alloy *Nanotechnology Materials and Devices Conference (NMDC)* (IEEE, South Korea)

[7] Vorozhishchev A N, Dema R R and Kazakova T V 2016 Modeling of a Thermal Massive Body Depending on the Cooling Liquid Volume, as Exemplified by Rolls for a Hot-rolling Mill *Procedia Engineering* **150** pp 1007–1012

[8] Bojan Podgornik and Vojteh Leskovcek 2014 Wear mechanisms and surface engeneering of forming tools *Materials and technology* (Ljubljana: Institute of Metals and Technology) **4** pp 313–324

[9] Vineet Shibe and Vikas Chawla 2014 A Review of Surface Modification Techniques in Enhancing the Erosion Resistance of Engineering Components *International journal of research in mechanical engineering and technology* **4** 2 pp 92-95

[10] Emeljushin A N, Petrochenko E V and Nefedyev S P 2012 Comparison of structure and properties of the cast and built-up wear-resistant materials *Litejnye processy [Foundry processes]* **11** pp 141–145.

[11] Artemenko Yu A, Bartenev D V, Makushenko AV etc. 2008 Research of processes of crystallization under surfacing of wear-resistant alloys on the transparent salt models *Tekhnologiya metallov[Technology of metals]* **8** pp 41–42

[12] Lasitsa A and Churankin V 2014 Dynamics of wear surface layers modified by complex treatment *Dynamics of Systems, Mechanisms and Machines (Dynamics)* Omsk, Russia

[13] Youn T K, Amanov A, Cho I S, Lee C S and Park I G 2011 Wear Resistance Characterization for Plated Connectors *Nanotechnology Materials and Devices Conference (NMDC)* **8** 1

[14] Dema R R Platov S I Amirov R N and Yaroslavtsev A V 2012 Wearing of working rolls continuous mill 2000 mathematical model at a lubrication supply *Latest problem of modern science, technique and education* **2** № 70 pp 165–167

[15] Platov S I, Dema R R, Amirov R N and Gatatulina U H 2013 Estimation of wearing material volume changing at at a lubrication supply *Mechanical equipment of metallurgical plant* **2** pp 16–18

[16] Nefedyev S P, Dema R R and Kotenko D A 2015 Abrasive and impact-abrasive wearing quality of solid surfaced coat *Bulletin of the south Ural State University* **15** № 1 pp 103–106

[17] Lyubich A.I and Pustovgar A.V 2002 Silicocalcium influence on structure and properties of the built-up metal *Svarochnoe proizvodstvo. [Welding production]* **6** pp 46–47

[18] Vorotnikov V Ya, Ivanov S V and Yu A Artemenko 1983 Resistence test technique of the surfacing alloys to shock loads *Avtomaticheskaya svarka [Autowelding]* **9** pp 61–62

Selectivity of the gas sensor based on the 50%In_2O_3–50%Ga_2O_3 thin film in dynamic mode of operation

I E Demin and A G Kozlov

Omsk State Technical University, Mira ave., 11, 644050, Omsk, Russia

e-mail: IvanDemin3467@gmail.com

Abstract. The article considers the gas sensor with the sensitive layer based on the 50%In_2O_3–50%Ga_2O_3 thin film. The temperature and concentration dependencies of gas-induced resistance response of this sensor and the dynamical dependencies of its resistance response on the test gases in air are investigated. The test gases were ethanol, acetone, ammonia and liquefied petroleum gas. The information parameters of the sensor in the dynamical mode of operation were considered to improve its selectivity. The presented results show that the selectivity of the sensor in this mode may be improved by using the following information parameters: gas-induced resistance response in steady state, activation energy of the response and pre-exponential factor of the temperature dependence of the response time constant.

1. Introduction

Semiconductor gas sensors based on oxide materials are widely used in the fields of safety, control and optimization of technological processes. Advantages of this type of sensor are high sensitivity to a large amount of combustible and oxidizing gases, simplicity of design, temperature and chemical stability, low cost [1].

An essential disadvantage of semiconductor sensors is their low selectivity, in other words their low ability to recognize the type of gas. Commonly used approach to overcome this difficulty is to increase the number of information parameters of the sensor. The information parameter is some parameter calculated from output signal of the sensor and used to identify a gas type or to quantify its concentration. In addition to the widely used amplitude of the gas-induced resistance response, following characteristics may be used to calculate information parameters: dynamics of response and recovery under exposure to abruptly changing concentration of sample gas [2], dynamics of response at a changing operating temperature [3–5], dynamics of gas diffusion along the thickness of a sensitive layer [6].

The best technical solution for applying these approaches is to use multi-sensor systems. The following trends can be distinguished in developing multi-sensor systems:

1) systems based on sensor arrays with a differently modified surface;
2) systems based on arrays of identical sensors with different operating temperatures;
3) systems based on sensor arrays with different electrode configurations;
4) systems based on sensor arrays with sensitive layers made of various semiconductor materials [2].

A promising approach from the point of view of the simplicity of design and fabrication of a multi-sensor system is the second approach based on identical sensors, which have different operating temperatures. One of the main trends in research in the field of semiconductor gas sensors is the search

IOP Conf. Series: Journal of Physics: Conf. Series **944** (2017) 012027 doi:10.1088/1742-6596/944/1/012027

for new gas-sensing materials that provide high selectivity. An important approach in this direction is the use of composite oxide materials.

Recently, among semiconductor composite oxide materials, a system based on indium and gallium oxides has attracted interest. The studies [7, 8] have shown that films of 50%In_2O_3–50%Ga_2O_3 have a high sensitivity to a number of gases. However, for the use of these films as sensitive layers of gas sensors, it is necessary to analyse their selectivity to the gases, and to suggest the most suitable operating mode for sensors based on them. As a perspective mode of operation one should consider the dynamic mode, for which there is a number of information parameters of the sensor allowing one to determine concentration of a gaseous medium.

Thus, the aim of the work is to study the selectivity of a semiconductor gas sensor based on the In_2O_3-Ga_2O_3 thin films under the dynamic mode of its operation.

2. Experimental

Semiconductor thin films with 50%In_2O_3–50%Ga_2O_3 composition for gas sensors were obtained in two stages. At the first stage, the target with desired composition was made from the powders of indium and gallium oxides. In the second stage, pulsed laser deposition of these films was performed from the obtained target to a glass ceramic substrates. The modes of the target preparation and the pulsed laser deposition of the films were considered in detail in [9].

The analysed gases were ethanol, acetone, ammonia and a mixture of liquefied petroleum gas (the LPG). As a characteristic of the gas-induced resistance response, the value of S was used

$$S = (R_0 - R_g) / R_g, \quad (1)$$

where R_0 and R_g – resistance of the film in the natural air and in the air with the sample of the analysed gas, respectively.

Temperature and concentration dependencies of gas-induced resistance response of the films and its dynamical resistance response on the analyzed gases in air were investigated experimentally. The research methodology was presented in [9].

3. Formatting the text

A Figure 1 represents the obtained normalized dynamic dependences of the resistance response for the sensor based on the 50%In_2O_3–50%Ga_2O_3 film for 25 ppm of acetone in air, at different operating temperatures. The dynamic dependencies of the normalized resistance response for other analyzed gases have a similar form. All the obtained dependences can be approximated by functions

$$S(t)/S_0 = 1 - \exp(-t/\tau), \quad (2)$$

where $S(t)$ – gas-induced resistance response of the sensor at time t, S_0 – gas-induced resistance response of the sensor in steady state, τ – response time constant.

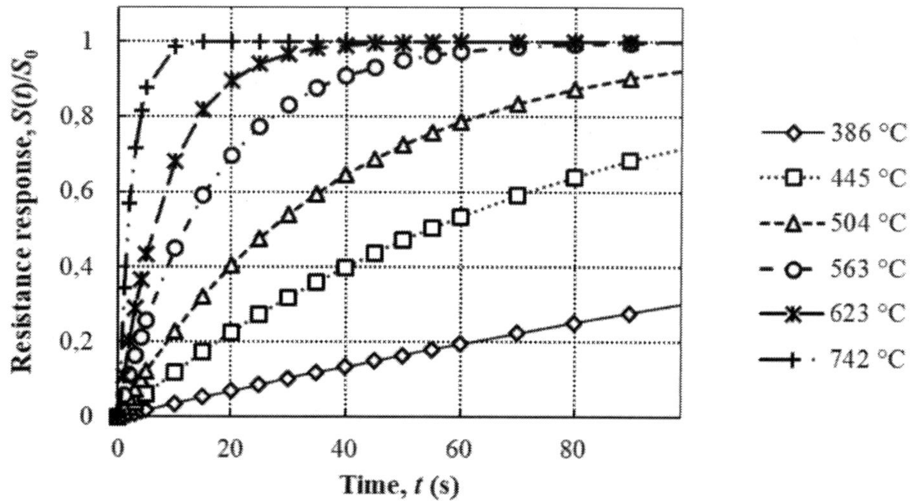

Figure 1. Normalized dynamic dependences of the gas-induced resistance response of the sensor based on the 50%In_2O_3–50%Ga_2O_3 film for 25 ppm of acetone in air.

The response time constants (τ) were calculated from the approximation of the experimental data by the functions (2) using the least squares method. The obtained values of the response time constants for 25 ppm concentration of the analysed gases at the different operating temperatures are presented in Table 1.

Table 1. Sensor response time constants for 25 ppm concentration of analyzed gases at different operating temperatures.

Analyzed gas	Temperature (T), °C						
	386	445	504	563	623	682	742
	Response time constants (τ), s						
Ethanol	139	41.1	22.9	9.00	5.13	2.62	1.62
Acetone	277	78.5	38.7	16.8	8.80	4.20	2.38
Ammonia	342	268	212	21.6	10.8	3.55	1.35
LPG	4655	968	423	143	72.4	24.0	13.1

As noted above, semiconductor gas sensors based on oxide materials have a low selectivity and the recognition of the type of an analysed gas by measuring only one information parameter with their help is nearly impossible. This is confirmed by the experimentally obtained temperature and concentration dependences of the gas-induced resistance response of the 50%In_2O_3–50%Ga_2O_3 film in steady state. For example, Figure 2a shows the temperature dependences of the resistance response of the studied film at 25 ppm ethanol (graph 1) and acetone (graph 2) in air. From the presented data, it can be seen that these dependences are insignificantly different in the whole investigated temperature range. Figure 2b shows the concentration dependences of the resistance response of the 50%In_2O_3–50%Ga_2O_3 film on ethanol and acetone. Analysis of these dependencies shows the ambiguity in determining the type of the gas probe. For example, the response value of the film resistance at 25 ppm of acetone in air is equal to its response at 40 ppm ethanol in air. The given examples clearly demonstrate the impossibility of gas recognizing with the help of the In_2O_3–Ga_2O_3 thin film sensor, basing only on measurements of the resistance response in the steady state.

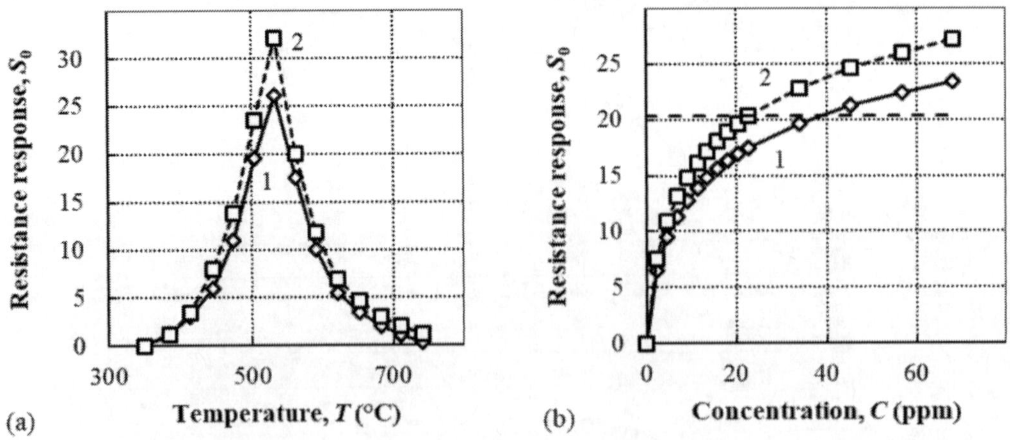

Figure 2. Temperature (a) and concentration (b) dependencies of the gas-induced resistance response for sensor based on the 50%In_2O_3–50%Ga_2O_3 film for ethanol (1) and acetone (2): Fig. 2(a) – gas concentration is 25 ppm; Fig. 2(b) – working temperature is 563 °C.

The solution of this problem is possible when using the parameters of the dynamic characteristics of the sensor behaviour in the medium of the analysed gas. Figure 3 represents dynamic dependencies of the resistance response of the sensor based on the investigated film at 563 °C and at the step change of the test gas concentration (acetone) up to the specified values. The values of the response time constant determined from these dependences are the same for all the concentrations of the analysed gas and depend only on the gas type.

Thus, the value of the sensor response time constant at the certain operating temperature can be an information parameter that does not depend on the gas concentration and allows one to determine the type of gas being analysed. However, for each gas the value of the sensor response time constant depends on the operating temperature, and for some gases these values at the certain operating temperatures may be approximately equal. For example, according to the data presented in Table. 1, in the operating temperature range of 623–682 ° C, the response time constant of the sensor resistance for ammonia and acetone is close, which reduces the selectivity of the sensor operating in dynamic mode. Thus, the complete consideration of the temperature dependence of the response time constant is required in order to unambiguously determine the type of a test gas.

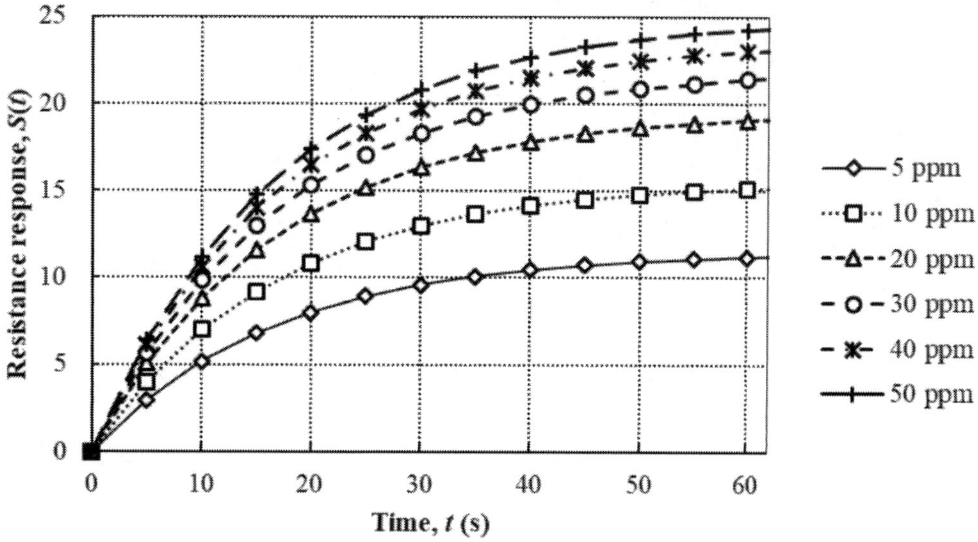

Figure 3. The effect of gas concentration on the dynamics of the gas-induced resistance response: test gas – acetone, working temperature – 563 °C.

Figure 4 presents the temperature dependences of the response time constant for sensor on the 50%In_2O_3–50%Ga_2O_3 film for all the tested gases. These dependences have a linear form in the coordinates of the inverse temperature and the logarithmic coordinates for the response time constant.

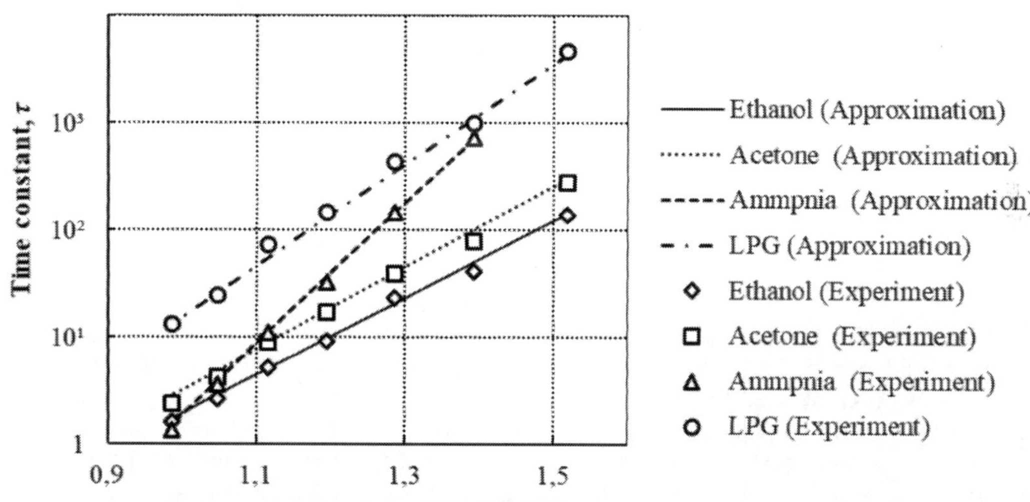

Figure 4. Temperature dependences of the response time constant for sensor on the 50%In_2O_3–50%Ga_2O_3 film for all the tested gases.

The obtained experimental temperature dependences of the response time constant for all the tested gases may be represented in an analytical form as follows

$$\tau = \tau_0 \cdot \exp(-E_a/kT), \quad (3)$$

where E_a – activation energy of the response time constant, τ_0 – pre-exponential factor [9, 10].
The parameters τ_0 and E_a included in (3) may be determined from the experimental dependences in the preliminary calibration of the sensor. These parameters are independent of the concentration of the test

gas and the working temperature of the sensor. They can be used as information parameters for identifying the type of test gas during real operation mode of the sensor.
The values of parameters S_0, τ_0 and E_a for sensor based on the 50%In_2O_3–50%Ga_2O_3 film for the different operating temperatures and for the concentration of the test gases of 25 ppm are given in Table 2. The presented data show that this set of information parameters is unique for each of the test gases. They allow identifying the type of gas being analysed and determining its concentration. For example, as noted above, the response time constant of the sensor resistance in the operating temperature range of 623–682 °C for ammonia and acetone has similar values (Table 1). However, according to the data presented in Table 2, these gases can be identified by the information parameters τ_0 and E_a.

Table 2. The values of parameters S_0, τ_0 and E_a for sensors on the 50%In_2O_3–50%Ga_2O_3 film.

№	Temperature, °C	Parameter	Test gas			
			Ethanol	Acetone	Ammonia	LPG
		S_0	7.00	9.00	0.59	1.03
1	445	τ_0, µs	495	512	0.548	308
		E_a, eV	0.712	0.755	1.297	0.935
		S_0	20.5	24.4	0.8	1.80
2	504	τ_0, µs	495	512	0.548	308
		E_a, eV	0.712	0.755	1.297	0.935
		S_0	18.5	21.00	1.80	4.00
3	563	τ_0, µs	495	512	0.548	308
		E_a, eV	0.712	0.755	1.297	0.935
		S_0	6.47	7.85	2.15	3.78
4	623	τ_0, s	495	512	0.548	308
		E_a, eV	0.712	0.755	1.297	0.935

The considered approach to increase the selectivity of gas sensors based on 50%In_2O_3–50%Ga_2O_3 films requires the use of the more sophisticated measuring circuits that must ensure dynamic measurements at different operating temperatures. There are two options for implementing such a measurement process. In the first variant, one sensor is used the operating temperature of which is periodically varied. The second option is to use a multi-sensor system in which each sensor has its own operating temperature. In both cases, the parameters S_0, τ_0 and E_a are calculated, basing on the dynamic measurements of the gas resistance response. These parameters are used to identify the type of an analysing gas and to determine its concentration.

4. Conclusion

The presented results by the example of the gas sensor on the 50%In_2O_3–50%Ga_2O_3 film show that the selectivity of semiconductor gas sensors can be improved by using the dynamic mode of measurement and determining, from the measurement results, the set of values of the following parameters: gas-induced resistance response in steady state (S_0), activation energy of response time constant (E_a) and pre-exponential factor (τ_0). To determine these parameters, it is necessary to carry out dynamic measurements at different operating temperatures and use more sophisticated measuring schemes. The choice of the optimal temperature operation conditions of the sensors for the dynamic measurements requires the additional studies.

Acknowledgments
This work was carried out within the project «Young scientist»; research effort № 17072B.

References
[1] Mizsei J 2016 *Procedia Engineering* **168** pp 6–221
[2] Korotcenkov G and Cho B K 2013 *Sens. Actuators, B* **188** pp 28–709
[3] Ponzoni A, Depari A, Comini E, Faglia A, Flammini A and Sberveglieri G 2011 *Procedia Engineering* **25** 1173–6
[4] Nakata S, Ozaki E and Ojima N 1998 *Anal. Chim. Acta* **361** pp 93–100
[5] Maziarz W, Potempa P, Sutor A and Pisarkiewicz T 2003 *Thin Solid Films* **436** pp 31–127
[6] Yamazoe N and Shimanoe K 2011 *Sens. Actuators, B* **154** pp 82–277
[7] Korotcenkov G, Boris Iu, Brinzari V, Han S H, Cho B K and Lychkovsky Yu N 2015 *Ceram. Int.* **41** 7478–88
[8] Korotcenkov G and Cho B K 2017 *Sens. Actuators, B* **244** pp 182–210
[9] Demin I E and Kozlov A G 2015 *Proc. 2015 Int. Siberian Conf. on Control and Communications (SIBCON)* (Hoboken: IEEE) 7147037
[10] Korotcenkov G, Brinzari V, Golovanov V and Blinov Y 2004 *Sens. Actuators, B* **98** pp 5–41

Diagnostics of heavy mining equipment during the scheduled preventive maintenance

M Yu Drygin, N P Kuryshkin

Department of information and automated manufacturing systems, T. F. Gorbachev Kuzbass State Technical University, 28 Vesennyaya street, 650000, Kemerovo, Russia

E-mail: kemnik@mail.ru

Abstract. Intensification of production, economic globalization and dramatic downgrade of the workers' professional skills lead to unacceptable technical state of heavy mining equipment. Equipment maintenance outage reaches 84 % of the total downtime, of which emergency maintenance takes up to 36 % of time, that excesses 429 hours per year fr one excavator. It is shown that yearly diagnostics using methods of non-destructive check allows to reduce emergency downtime by 47 %, and 55 % of revealed defects can be eliminated without breaking the technological cycle of the equipment.

1. Introduction

During the USSR existence, breakup and redistribution of the state property to the private the system of scheduled preventive maintenance (SPM) which was one of five-year planning instruments seemed lose its significance. However profound analysis of modern coal mining industry structure shows the same system of management with the only exception that a capitalist country, which Russia definitely is, requires cost saving and profit maximization. Within the last 10 years coal mine staff competence has dramatically decreased, less young and qualified workers as well as middle and top managers are employed [1]. Intensification of production and above mentioned processes caused inappropriate technical state of coal mines' equipment [2].

In 1980–2011, 5 644 electric shovels with bucket capacity from 4.6 to 42 m^3 were delivered to mining enterprises of the Russian Federation and CIS countries, and in 1999–2011, 259 hydraulic power shovels with bucket capacity more 5 m^3 were delivered. Further tendencies are oriented on bringing into service single large capacity power shovels with bucket capacity of 20-55 $м^3$ [3].

At the moment the main equipment for coal extraction and loading are crawler shovels with electromechanical drive EKG. At the same time the main part of excavator fleet (63 %) includes excavators with bucket capacity of 8–15 m^3 manufactured by OMZ (IZ-KARTEKS, former Izhora factories). In 2016, 80 % of excavators of Rissian mining industry had excessive service life.

2. Problem description

Comparative study of work of more than 200 excavators in Kuzbass – the largest coal mining region in Russia – showed that imported equipment meets the wear norm of 82 % and home produced equipment meets that norm of 110 %. At the same time imported equipment has low variability of meeting the norm, that means that all machines work in the same way, and home produced equipment has high variability. Some of home produced machines over fulfill the norm more than twice but some of them do not meet even a half of the norm. Working experience shows that home produced excavators have significant overcapacity which is designed in and confirmed by record high productivity over fulfilling the performance standards.

Thus the optimal way to increase efficiency of imported machines is their running in strict accordance with manufacturing facility requirements and their efficiency meeting the "origin" norm. Increasing efficiency of home produced excavators can be reached by determining a potential efficiency norm based on the technical state of the machine and by optimizing maintenance processes [4].

3. Theory

Analysis of technical state of more than 300 excavators run at Kuzbass coal mines in 2011–2016, showed that the most part of equipment was run with significant number of major defects (often hundreds of them), and was in inappropriate technical state (more than 86 %). This tendency comes from low quality of equipment maintenance and repair system.

Independent performance records of the equipment at the opencast coal mine Vinogradovskiy (PAO «Kuzbasskaya Toplivnaya Kompaniya») within the period of five years (from February 1, 2012 to December 31, 2016) for five excavators ESh-10/70, ESh-11/70, ESh-13/50 and eight excavators EKG-5a showed the following results. Of the total service time of the excavators ESh (178537.5 hours) 75% of time (134587 hours) they offloaded the rock mass and 25 % or 43950.5 hours they stood idle. Excavators EKG of the total service time (299621 hours) offloaded the rock mass 126461.7 hours (42%), and stood idle for 173159.3 hours (58 %).

All downtime periods can be divided into production and non-production delays. Production delays include all periods of downtime caused by the necessity of offloading the rock mass (maintenance outage, climate-induced outage, operational delays, etc.). Non-production downtime periods are directly caused by absence of necessity in equipment operation and are not connected with any operational factors (reserve, stand-by duty, conservation, writing-off, etc.).

For excavators EKG production delays made 41% and non-production delays made 59 % (101925 hours) of the total downtime. For excavators Esh production delays made 98 % and non-production delays made 2 % (816.5 hours) of the total downtime. Therefore the total downtime of non-production delays at the mining enterprise made 102741.5 hours. The obtained result shows low level of machine work load, i.e. within the period of 102741.5 hours the studied excavator fleet was functional but not needed that in turn points at excessive capacity at the mining enterprise.

The graphs given below represent the distribution of production delays for ESh excavator fleet (Figure 1) and EKG excavator fleet (Figure 2) respectively.

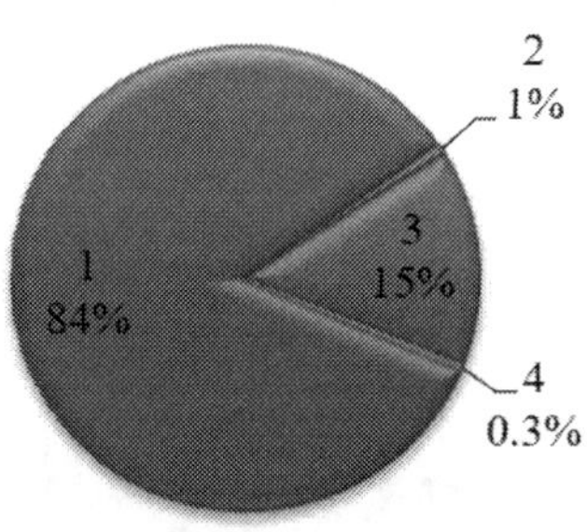

Figure 1. Distribution of ESh Excavators' Production Delay Time: 1 – maintenance outage, 2 – climate-induced outage, 3 – operational delays, 4 – downtime not defined.

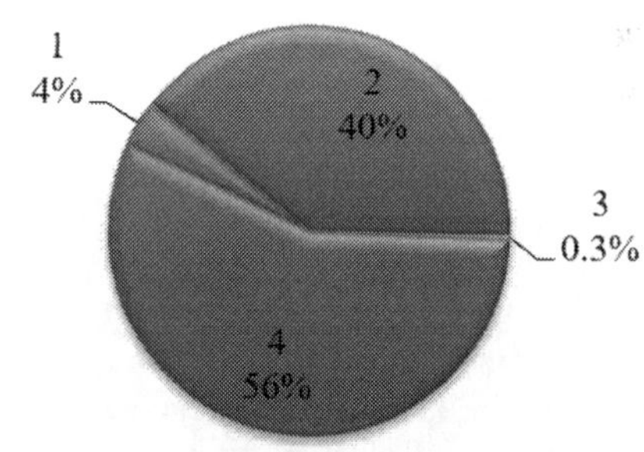

Figure 2. Distribution of EKG Excavators' Production Delay Time: 1 – climate-induced outage, 2 – operational delays, 3 – downtime not defined, 4 – maintenance outage.

Total maintenance outage of the excavator fleet under survey is significant 76338.9 hours. Below, in the Table 1, types and time of maintenance outage (connected with the technical state of the equipment) are listed).

IOP Conf. Series: Journal of Physics: Conf. Series **944** (2017) 012028 doi:10.1088/1742-6596/944/1/012028

Table 1. Maintenance outage distribution.

Downtime type	ESh, hr.	EKG, hr.
Emergency power cutoff	46.3	74.3
Emergency maintenance	10737.5	14505.3
Yearly maintenance	7322.4	6758.9
Yearly maintenance (nightly outage)	6911.5	6624.0
Diagnostics	23.0	52.8
RGTI shutdown	17.5	5.5
Absence of spare parts (delivery)	1038.4	1791.0
Absence of crane	6.0	185.8
Absence of maintenance fitter	7.8	71.3
Absence of welding specialist	26.0	60.4
Scheduled preventive maintenance	5444.9	5344.4
Scheduled preventive maintenance (nightly outage)	4133.7	4255.0
Prestarting procedures	0	30.7
Preparation for repair	69.0	5.0
Power line repair	248.0	354.4
Rectifying the violations	151.0	26.1
Expert review	8.0	3.0
Total	36190.9	40148.0

Significance of maintenance outage is usually determined by the used repair system. Thus Russian mining enterprises are currently functioning in terms of the Regulation on the Equipment Operation System Based on Scheduled Preventive Maintenance (SPM), introduced in 1983. The main items of the Regulation are in line with the standards set by the International Organization for Standards [5]. According to the Regulation scheduled preventive maintenance is the complex of repeating scheduled and technical activities including maintenance, control and operation of equipment.

In terms of the Regulation there are types of turnaround maintenance: on a shift basis, daily, weekly, on a ten days basis, and season maintenance, and repairs: (capital (C), medium (M), current (Cur), and monthly (Tm)). Regular maintenance on the shift basis is stated as basic preventive activity oriented on significant increase of operating life and presupposed to be performed within the period of shift turnover and within operational downtime periods [6]. Thus an obligatory condition is properly organized shift turnover when an engine operator or a shift man accepting the equipment personally checks its technical state and records detected defects in the shift turnover log, but now it is done only to meet formal requirements, as the staff qualification has dramatically decreased.

Turnaround maintenance is performed in accordance with instructions of equipment producer factories with account for the Regulation. Thus, for example, season maintenance should include probing oils, checking sealing materials of vehicle body, doors etc. All detected failures and defects are subject to registration, and their particular reasons must be revealed. Those failures and emergencies that cause shutdown of equipment for the period exceeding three days must be investigated by a commission with drawing up an Act [7], but this is also done only to meet formal requirements.

Repair frequency should be set due to component life measured in equipment running time units. For example, excavators it is measured due to the amount of processed rock mass in thousands of m^3, and coefficients are calculated with regard for actual operating conditions.

Distribution of Esh and EKG excavator fleet of the opencast coal mine Vinogradovskiy of PAO Kuzbasskaya Toplivnaya Kompaniya is presented on the graphs of Figure 3, Figure 4, Figure 5 and Figure 6, and on Figure 4 and Figure 5 categories of other downtime types, making insignificant 1% of total downtime, are given separately.

IOP Conf. Series: Journal of Physics: Conf. Series **944** (2017) 012028 doi:10.1088/1742-6596/944/1/012028

The most significant are: emergency downtime periods (ESh – 30%, EKG – 36%), nightly outage during yearly and scheduled maintenance (ESh – 19% и 11%, EKG – 17% и 11%), absence of spare parts (ESh – 3%, EKG – 4%).

On the assumption that the SPM system allows reduction of duration for all repair types due to unit replacement and multishift repair works with fulfilling the discrepancy list and repair works during the night shift, it is possible to reduce downtime periods of the fleet by 21924.2 hours within five years.

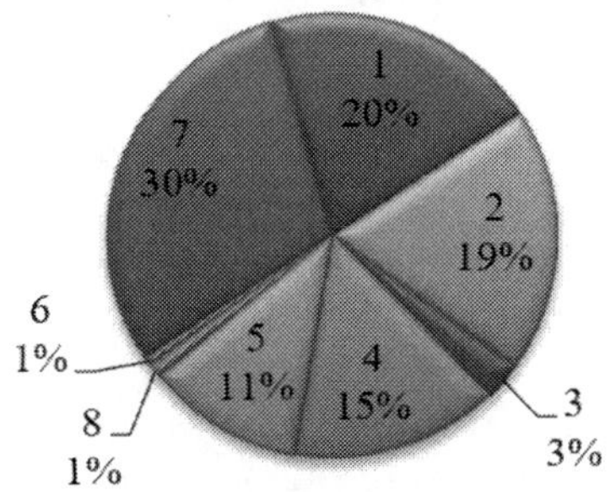

Figure 3. Distribution of ESh Excavators' Main Maintenance Outage Time: 1 – yearly maintenance, 2 – yearly maintenance (nightly outage), 3 – Absence of spare parts (delivery), 4 – Scheduled preventive maintenance, 5 – Scheduled preventive maintenance (nightly outage), 6 – Power line repair, 7 – Emergency maintenance, 8 – Other.

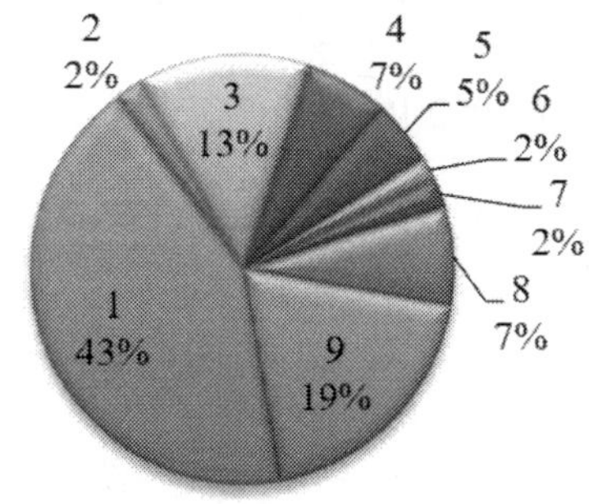

Figure 4. Distribution of ESh Excavators' Other Maintenance Outage Time: 1 – rectifying the violations, 2 – expert review, 3 – emergency power cutoff, 4 – diagnostics, 5 – RGTI shutdown, 6 – absence of crane, 7 – Absence of maintenance fitter, 8 – Absence of welding specialist, 9 – preparation for repair.

Downtimes due to absence of spare parts are caused by inappropriate planning of repair and breaking logistics back. The problem is solved by profound check, repair planning and, for the most part, by bringing the order in the supply service.

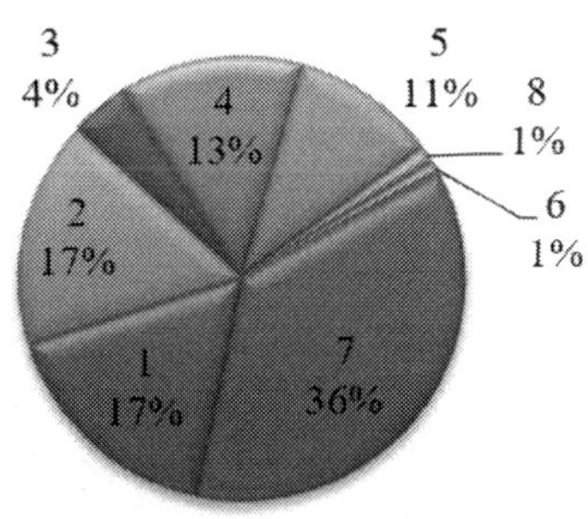

Figure 5. Distribution of EKG Excavators' Main Maintenance Outage Time: 1 – yearly maintenance, 2 – yearly maintenance (nightly outage), 3 – absence of spare parts (delivery), 4 – scheduled preventive maintenance, 5 – scheduled preventive maintenance (nightly outage), 6 – power line repair, 7 – emergency maintenance, 8 – other.

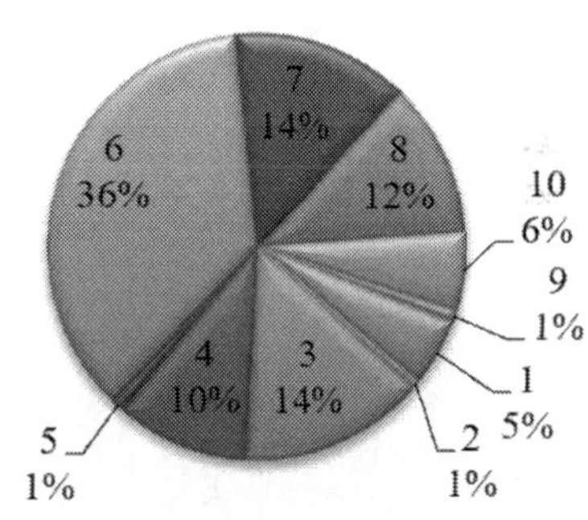

Figure 6. Distribution of EKG Excavators' Other Maintenance Outage Time: 1 – rectifying the violations, 2 – expert review, 3 – emergency power cutoff, 4 – diagnostics, 5 – RGTI shutdown, 6 – absence of crane, 7 – absence of maintenance fitter, 8 – absence of welding specialist, 9 – preparation for repair, 10 – prestarting procedures.

In the current SPM system planning of repair is divided into long term (within the five years' period), yearly and current (on monthly basis). Planning is based on norms set by the Regulations and the data about the equipment technical state collected during its operation. In the yearly planning the SPM system presupposes obligatory setting up the periods and the amounts of works according to the discrepancy list and adjusting the amount of work in the current planning due to the technical state of

the equipment. The result of the planning is a five-year schedule of capital, middle and current repair works, annual schedule of current repair works and a monthly schedule [8].
In the view of the above said it can be affirmed that the SPM system is based not only on scheduled replacement of the spare parts but also on permanent control of the technical state: during the shift turnover – with a record in the shift turnover log, and during the planning of middle, current and annual repair works. Thus the discrepancy list, being an essential part of SPM is one of the most important documents, and the balance of the whole system depends on its proper drafting.

4. Experimental results

In order to reduce emergency downtime since 2012, on the excavator fleet under survey once a year before annual repair works complex diagnostics was performed including the diagnostics of rotating and electrical equipment, and of metal structures, according to maintenance requirement cards worked out due to producer factory technical specifications. The diagnostics was performed in 323 test locations on excavators EKG and in 393 test locations on excavators ESh by using visual and dimensional test (V&DT), vibration-based diagnostics and heat monitoring. Average time of diagnostic works on the equipment was 12 hours, and the report comprises more than 250 pages.
Due to internal organizational problems of the enterprise not all detected discrepancies were fully eliminated. On the one hand that allowed to reduce downtime to acceptable values, on the other hand it allowed to carry out an experiment with monitoring time and conditions of development of different defects to an emergency shutdown.
Figure 7 shows a graph (averaged readings of the whole EKG and ESh excavator fleet under survey) which allows to evaluate the efficiency of such tests, and the difference between the trademarks ESh and EKG did not exceed 2 %.

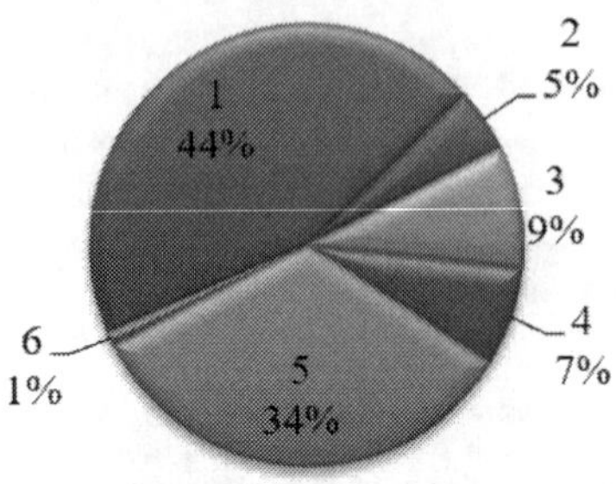

Figure 7. Distribution of the Main Downtime Causes During the Modelling of the system: 1 – detected, 2 – replaced / detected before replacement, 3 – replaced / not detected before replacement, 4 – regular check was not performed, 5 – not detected, 6 – other.

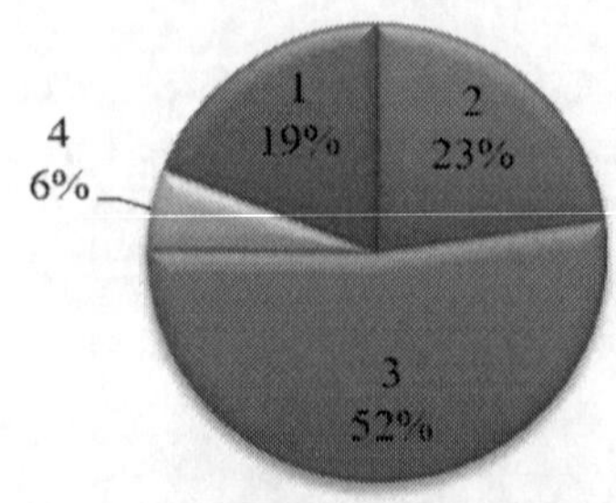

Figure 8. Distribution of Other Downtime Causes During the System Modelling: 1 – absence of access, 2 – detected during the initial check, 3 – power line, 4 – production delays.

On the Figure 8 other insignificant causes of equipment downtime are presented. They make only about 1 % of the total downtime. A modeled situation "What would happen if all major defects were eliminated" allowed to reveal the following categories of defects that had caused emergency equipment shutdowns. The category «Detected» – conclusions comprise defects the early elimination of which would have prevented emergency repair. The category «Regular check was not performed» – annual diagnostics had not yet been performed at the moment of emergency repair. The category «Not detected» – the conclusions do not comprise revealed defects the early elimination of which could have prevented emergency repair.

The category «Power Line» – downtime could not have been detected during the diagnostics due to a voltage slump. The category «Absence of access» – downtime in the nodes unavailable at the moment of diagnostics. The category «Production delays» – downtimes due to the causes that could not have been revealed by way of non-destructive control, like pneumatic system heating. The category «Replaced / detected before replacement» includes downtime periods connected with repeated defects of the spare parts which had been replaced more than one a year. That was why it had not been tested (ropes, teeth, high-voltage cable (often worn out), less often – engine, bucket, etc.). Before their replacement the conclusion comprised defects the early elimination of which could have prevented emergency repair. The category «Replaced / not detected before replacement» is identical to the previous one, but before its replacement the conclusion did not comprise defects the early elimination of which could have prevented emergency repair. The category «Detected during the initial check». The category "Detected by the initial check – defects on the spare part had not been revealed, but emergency repair could have been prevented by the initial check of the spare part.

5. Discussion

The modeled situation showed that 44 % emergency delays could have been prevented only by a diagnostic check performed only once a calendar year. More 3.08 % (43% of the category «"Regular check was not performed») could have been avoided if regular annual check is strictly performed once a year, that will make almost 47 % or 11943 hours provided that this method of diagnostics is used. It is 1038.5 shifts of excavator fleet work within the period under survey or 17.3 shifts per unit of equipment within a year. Downtime of the excavator fleet within the period under survey is presented in Table 2.

Table 2. Modelled time of excavator fleet downtime.

Detectability criteria of defects	Emergency outage, hr.
Detected	11071.5
Detected during the initial check	59.0
Replaced / detected before replacement	1198.4
Replaced / not detected before replacement	2288.6
Check was not performed	1850.5
Power line	135.0
Not detected	8575.7
Production delays	15.0
Absesnce of access	49.1
Total	25242.8

As 93 % of discrepancies detected by diagnostic tests within a period of time cause an emergency shutdown (Figure 9), one of the most important results of diagnostic testing is the reduction of emergency downtime. It allows not only increase productivity rate due to more time available and reduce repair expenses, but also plan the equipment work load more effectively because failures in work performance caused by unplanned downtime are avoided, and that in turn allows to increase workforce productivity.

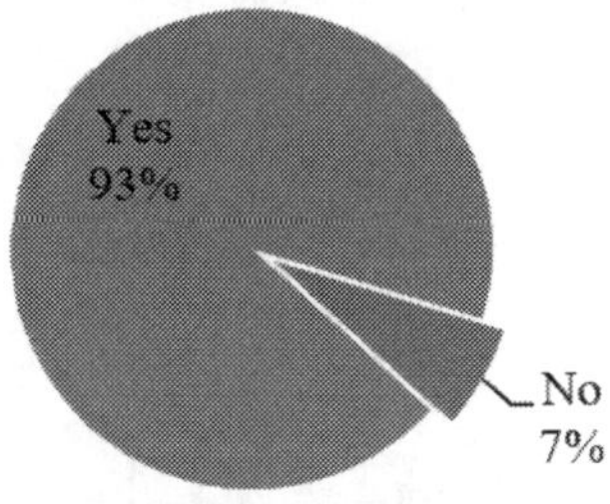

Figure 9. Percentage of Defects Causing Emergency Shutdown of the Excavator.

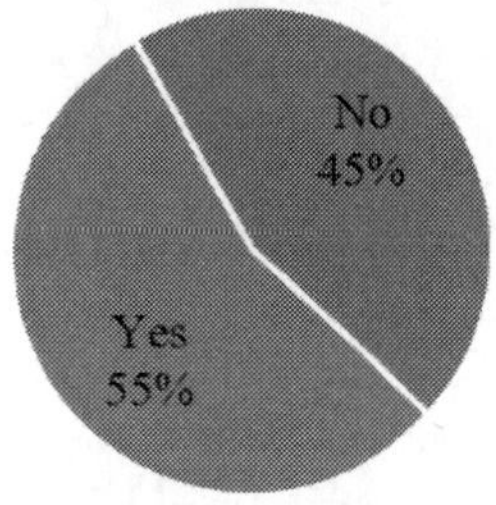

Figure 10. Percentage of Defects Repairable Without Equipment Removal from Service.

The second by significance is the fact that the amount of defects that must be eliminated during forced idle hours by efforts of the brigade without breaking the technological cycle of the equipment makes 55 % (Figure 10).

IOP Conf. Series: Journal of Physics: Conf. Series **944** (2017) 012028 doi:10.1088/1742-6596/944/1/012028

A significant number of defects revealed during annual diagnostic tests are accumulated within a long period of operation. Year after year these defects accumulate due to violation of the shift turnover norms and absence of maintenance on the shift, day, week, ten days' basis and season maintenance. At the best case it is limited only to cleaning and oiling. Capital, current and monthly repairs are performed in order to eliminate only emergency failures evidently being a threat of soon emergency shutdown of the equipment.

Therefore the SPM system requires long-term and short-term planning. It is based on scheduled repair works with regard for the actual technical state of the equipment at every moment of time according to the discrepancy list drafted by a qualified specialist. In the circumstances concerned it becomes clear that significant emergency downtime periods of the equipment of Kuzbass opencast coal mines are nothing less but major violation of the SPM system regulations. But it does not cancel the need of it modernization to the system of repairs due to actual technical state of the equipment. In practice it leads to even more profound diagnostic testing based on modern methods of non-destructive testing for long-term prediction of residual operation life in order to break from one of the most economically ambiguous aspects of SPM – keeping a large repair parts stock based on calculating nodes' non-failure operation time.

6. Summary and conclusion

The performed analysis of the excavator fleet downtime and annual diagnostic testing allowed to come to the following conclusions:

1. Inappropriate technical state of the equipment is caused by incorrect use of the SPM system in what concerns low service standards and incorrect evaluation of actual technical state.
2. Maintenance outage of equipment reach 84 % of the total downtime of which maintenance outage takes up to 36 % exceeding 429 hours per year for one excavator.
3. More than 93% of defects in excavator will cause emergency shutdown in the long-term prospective.
4. Diagnostic tests performed at least once a year allow to reduce emergency downtime by 47 %, and 55 % of revealed defects can be eliminated without breaking the technological cycle of the equipment.

References

[1] Vlasov E A 2007 Qualification of Workers and Consistent Factors of Its Raising *News of Volgograd State Pedagogical University* **1**. pp 78–82 URL: http://cyberleninka.ru/article/n/kvalifikatsiya-rabochih-kadrov-i-zakonomernosti-ee-povysheniya (Assessed: 10.04.2017)

[2] Drygin M, Kurychkin N and Bakanov A 2017 Strategy of Russian Coal Mining Enterprises' Excavator Park Technical State Correction *The 1st Scientific Practical Conference International «Innovative Mining Symposium (in memory of Prof. Vladimir Pronoza»*

[3] Anistratov K Y Analysis of Mining Excavator and Dump Truck Market in Russian Federation and CIS Countries *Gornoye Delo – informational and analytical web portal*. URL: http://www.gornoe-delo.ru/articles/detail.php?ID=7508

[4] Drygin M, Kurychkin N and Bakanov A 2017 Ways of increasing excavator fleet productivity in Russian coal open pits (Kuzbass case study). *The 1st Scientific Practical Conference «International Innovative Mining Symposium (in memory of Prof. Vladimir Pronoza)»* **15**

[5] GOST R ISO 17359-2015 Control and Diagnostics of Equipment State. General Guide. 2016 Introduction 2015-10-20. (Moscow: Standartinform) p 33

[6] Regulation on Scheduled Preventive Maintenance at Opencast Mining Works of the USSR 1983 (Moscow: Ministry of Coal Industry) p 46

[7] Mityushin V S 2004 Myth 2: Maintenance and Repair Works Are Impossible to Plan. *Compass of Industrial Restructurization* **3** pp 36–38

[8] System of Schedule Preventive Maintenance (Electronic source). URL: http://www.sdelaemsami.ru/oborudovanie/kran107.html (Assessed: 10.04.2017).

Raising quality of maintenance and control of metallic structures in large-load technological machines

M Yu Drygin, N P Kuryshkin

Department of information and automated manufacturing systems, T. F. Gorbachev Kuzbass State Technical University, 28 Vesennyaya street, 650000, Kemerovo, Russia

E-mail: kemnik@mail.ru

Abstract. Active growth of coal extraction and underinvestment of coal mining in Russia lead to the fact that technical state of more than 86% of technological machines at opencast coal mines is unacceptable. One of the most significant problems is unacceptable state of supporting metallic structures of excavators and mine dump trucks. The analysis has shown that defects in these metallic structures had been accumulated for a long time. Their removal by the existing method of repair welding was not effective – the flaws reappeared in 2–6 months of technological machines' service. The authors detected the prime causes that did not allow to make a good repair welding joint. A new technology of repair welding had been tested and endorsed, and this allowed to reduce the number of welded joints' flaws by 85% without additional raising welders' qualification. As a result the number of flaws in metallic structures of the equipment had been reduced by 35 % as early as in the first year of using the new technology.

1. Introduction

Coal is one of the most significant resources that stabilize the situation in Russian power industry. The advantages of opencast coal mining are obvious because 70 % of the total amount of coal mined Russia-wide is extracted at opencast mines. Kuznetsk Basin (Kuzbass) is one of the largest in Russia and in the world [1] Within the last five years there is positive increase in exported coal amounts, and Kuzbass places Russia to the third place among the most significant coal exporters after Australia and Indonesia. Kuzbass coal is exported to 85 countries of the world [2]. At the same time within the last twenty-five years investment of coal industry were insignificant. Even in 2010 which was the most stable year they made about 5 % of coal sales revenue [3].

Although despite negative investment environment out-of-date technological equipment and technological devices were replaced. But this replacement was insignificant for excavator fleet and dump truck fleet. That is the reason why opencast coal mines still use life-expired technological machines. This lead to the fact that 70 % of equipment practically every year undergoes safety review performed by experts, and the technical state of more than 86 % of equipment is unacceptable.

Considering the tendencies of early 2000s which were based on total replacement of excavators produced in Russia with hydraulic excavators of foreign origin it may be said that this strategy had failed to prove its value. The advantages demonstrated by the imported equipment come to nothing by the fourth or fifth year of service. Thus introduction of imported equipment required significant current operating costs besides capital expenditures, and tough interrepair maintenance requirements caused the necessity of the whole maintenance system reconsideration. Thus in most of the cases such equipment undergoes post-warranty maintenance and requires permanent presence of a technical-

IOP Conf. Series: Journal of Physics: Conf. Series **944** (2017) 012029 doi:10.1088/1742-6596/944/1/012029

support engineer, and absence of any technical specifications and structural complexity of basic mechanisms makes impossible to repair the equipment by the enterprises' technical support services [4].

2. Problem description

Above mentioned highlights the need to focus the attention on the technical state of domestically produced technological machines. The results of examination of 123 equipment fleet units at AO HK SDS-Ugol (one of the three leading Russian coal mining companies) as part of annual flaw-detection survey of supporting metallic structures (mining booms, frames, dump truck bodies etc.) highlighted unacceptable technical state of the equipment (Table 1). Examinations were performed by methods of visual and dimensional testing (V&DT), ultrasonic testing (UT) and by metal magnetic memory method (MMM).

The nature of the reviewed flaws shows that the flaws had been accumulating within 3-5 years at least. Despite the fact that coal mining enterprises in Russia and particularly in Kuzbass are regulated by Federal Environmental, Industrial and Nuclear Supervision Service, and 70 % of equipment undergo annual safety review, a set of examinations necessary for equipment operational life extension are performed with formal approach to diagnostic works. They are fragmentary and are performed in minimal amount of the safety review, and this is not enough for prevention of such flaws.

Table 1. Results of metallic constructions examination.

Examinations		**Index**
Number of performed examinations including:		123
– primary		59
– removal checks		51
– expert reviews and others		13
Welded joints checked by	number	9885 pcs.
the methods of V&DT,	extension	3999 m
UT, MMM	area	8812 m^2
Total volume of sent conclusions		9118 pages
Flaws		
Flaws detected		5646 pcs. 2126 m
Including those found during primary examinations		4554 pcs. 1600 m
Cracks detected		2679 pcs. 523.3 m

Accumulation of flaws had been induced by the fact of their untimely removal, and thus within six months of the flaws described in Table 1 29% of flaws by number and 28% by extension was removed. The least number of flaws were removed at AO Chernigivets – 13% by number and 18% by extension. The most number – at ZAO Prokopievskiy Ugolnyi Razrez – 51% by number and 45% by extension.

More than 90 % of flaws detected in metallic structures were removed several times but reappeared. Along with this ultrasonic testing showed that in more than 98 % their was a lack of penetration in the welded joint. Figure 1 shows the photo of a typical repair welded joint from backside. As it can be seen a lack of penetration extends through the whole length of the joint.

IOP Conf. Series: Journal of Physics: Conf. Series **944** (2017) 012029 doi:10.1088/1742-6596/944/1/012029

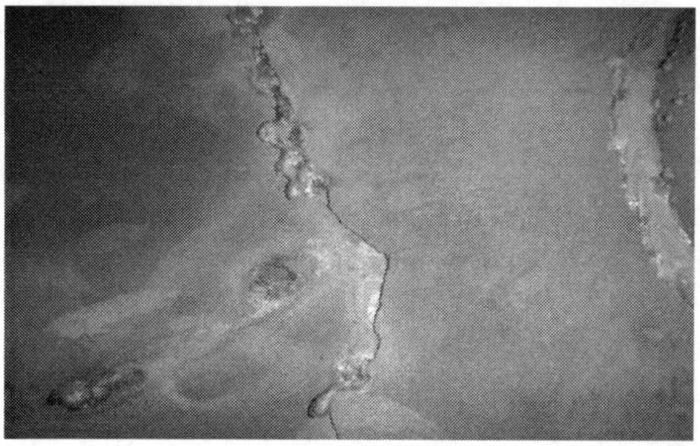

Figure 1. Lack of root penetration in the repair welding joint.

3. Theory

Unlike welding new metallic structures where welders' task is prepared groove of a future welded joint, the process of repair welding is organized in the way that the welders should groove and prepare the metallic structure for welding themselves. The main factors that complicate the process and do not allow to groove the welded joint in accordance with the National State Standard are [5, 6, 7]:

– curvature of the flaw and, as follows, non-linear required groove;

– different position of the flaw;

– meeting of several parallel flaws (Figure 2);

– the necessity to remove totally the welded metal and the heat-affected zone of the fractured welded joint, leading to double or triple extension of the groove area (Figure 3);

– varying thickness of the metal down the line of the flaw extension;

– absence of information about the thickness of the welded metal.

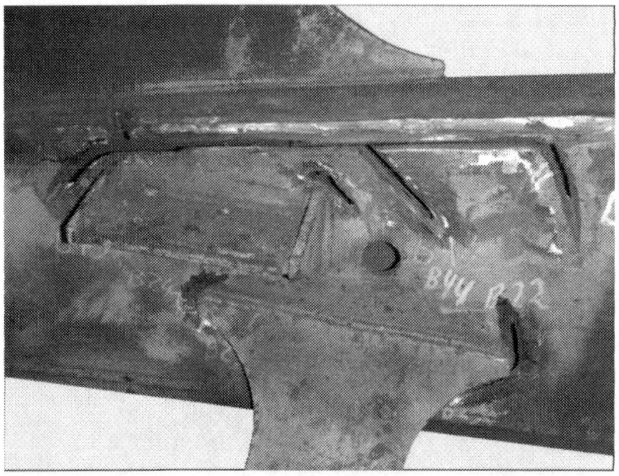

Figure 2. Opening of break on the mine dump truck frame.

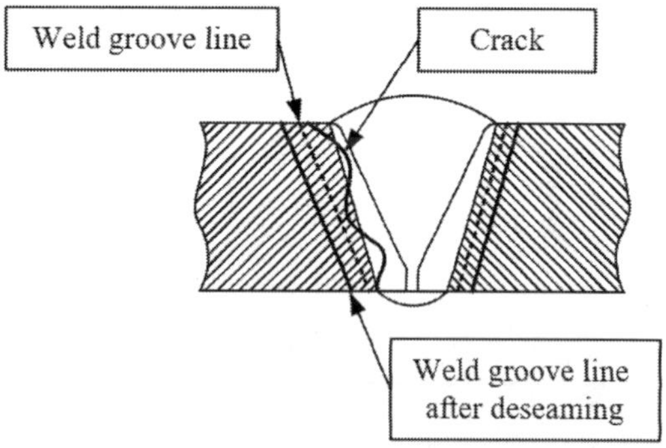

Figure 3. Groove of the fractured welded connection.

All this, complicating the work of the welder who is not familiar with the groove, leads to a typical mistake – narrow groove with leaving some metal for fusion (Figure 4), and it does not allow complete root penetration with a temper bead.

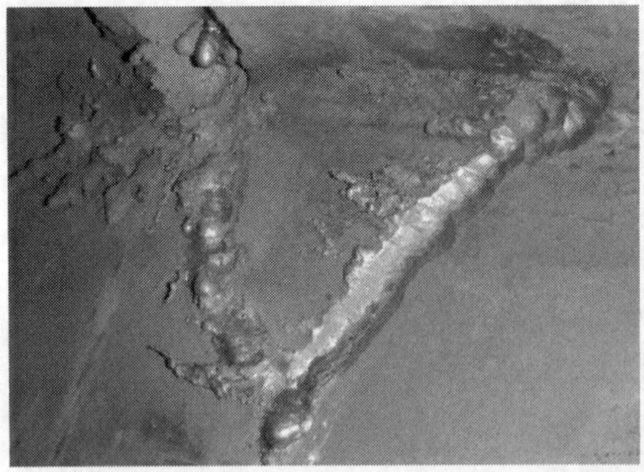

Figure 4. Improper opening of break for making a welding joint.

4. Experimental results

In order to get precise information about the causes of reappearance of the flaws on repair welded joints and to detect the reasons of lack of root penetration a special methodology allowing to model the typical working conditions for the welder making repair welded joints had been developed.

Multipass welding of metallic specimens with different thickness at their different locations was modelled. Welded joints' groove was performed by the welders themselves. Welders of the 5th and 6th categories worked with metallic specimens 16 mm thick and welders of the 3rd and 4th categories worked with specimens 6 mm thick. The criteria of the specimens' assessment met the instructions for capital repair of excavators ESh and EKG and dump trucks Belarusian Auto works 7555 и Belarusian Auto works 75131. The last additional criterion was evaluation of the welded joint for the presence of welding stress using the metal magnetic memory method.

193 welders of AO HK SDS-Ugol enterprises had participated in the experiment. The analysis of more than 80 m of welded joints in 338 welding specimens showed the results given below in Table 2.

From the Table 2 it can be seen that the dependence of the number no-go specimens' number on the welders' qualification is insignificant. On the specimens 16 mm and 6 mm thick respectively were observed: irregular ripples (91 % and 67 %), deviations of the welded joint size form the required by the National State Standard (100 % and 99 %), lack of root penetration through the full length of the welding joint (41 % and 46 %).

Table 2. The results of experimental welding specimen.

№	No-go criteria naming	Presence of flaws in the specimens with thickness of 16 mm, %	Presence of flaws in the specimens with thickness of 6 mm, %
1	Beadings, rough transition to the basic metal	2	28
2	Irregular ripples	91	67
3	Height of the crests excesses 1 mm	70	18
4	Ribbing along the welded joint excessing 3.0 mm	30	39
5	Height of the welded joints' step structure excesses 2.0 mm	17	15
6	Uneven welded joint filling	0	14
	Open craters	15	12
	Pores	15	23
	Air holes	7	11

IOP Conf. Series: Journal of Physics: Conf. Series **944** (2017) 012029 doi:10.1088/1742-6596/944/1/012029

№	No-go criteria naming	Presence of flaws in the specimens with thickness of 16 mm, %	Presence of flaws in the specimens with thickness of 6 mm, %
	Burns-through of the edges	0	0
	Breaks of the welded joint	0	3
	Cracks	0	3
7	Deviations of the weld size from the National State Standard	100	99
8	Flaws with the diameter excessing 1 mm	100	93
	Number of flaws excesses 1 per 100 mm	100	92
	Space between the flaws less than 45 mm	100	91
9	Undercuts without correction	57	62
	Undercuts more than 0.5 mm deep	37	45
10	Specimen's edge displacement	7	18
11	Lack of root penetration through the full length	41	46
12	Bending of the specimen (more than 3 mm)	13	1
13	Welded joint's magnetic field intensity	61	0

The chart presented on Figure 5 shows the results of the assessment of the welded joints quality. The welded joints were done by the workers of AO HK SDS-Ugol enterprises.
The number of specimens of the required quality according to visual and dimensional test done only by the welders of OOO Shakhtoupravleniye Mayskoye was 5.6 %. All other enterprises showed practically zero result. The specimens of ZAO Prokopievskiy Ugolnyi Razrez welders had passed visual and dimensional test and made 38.5% of their total amount. Only 5.6 % of specimens done by the welders of OOO Shakhtoupravleniye Mayskoye met the joint requirements of V&DT and UT.
The results of analysis of flaws' distributions according to the welders' categories shown on Figure 6 show that the quality of the welded joints does not depend on the welders' qualification. Thus on the one hand welders of the 6th category had expectedly more specimens meeting the norms of V&DT (11.2 % more than the welders of the 5th category), on the other hand they had less (2.8 % less) welded joints meeting the norms of UT than welders of the 5th category. And only 1.1% of specimens done by the welders of the 5th category had met the joint norms of V&DT and UT.

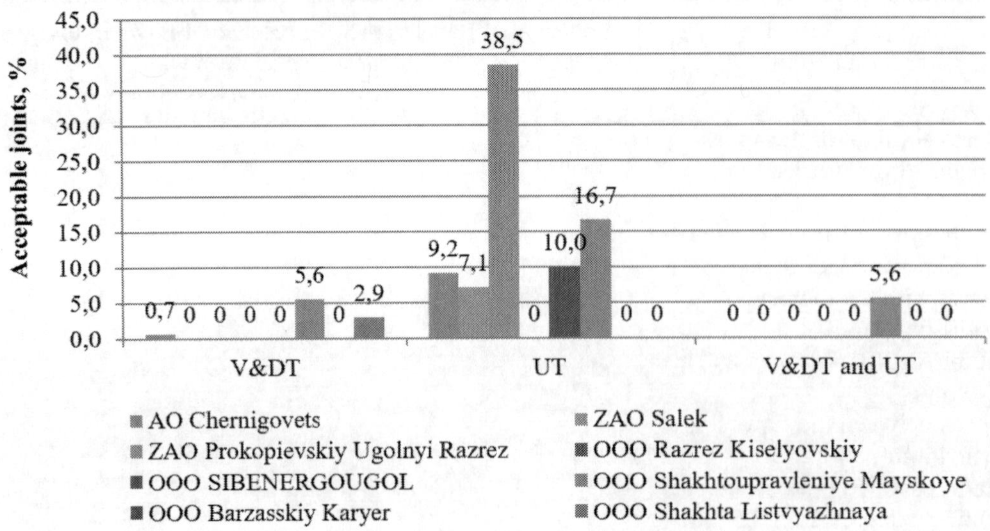

Figure 5. Results of the control of the specimens produced by the welders at the enterprises of AO HK SDS-Ugol.

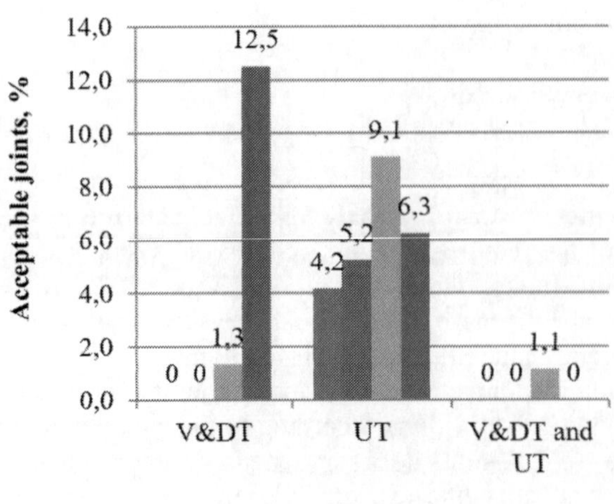

Figure 6. Influence of the welder's job class on the quality of welding joints at AO HK SDS-Ugol.

The results of the experiment with welding specimens evaluation had therefore shown that there were no specimens meeting the norms set by documentary standards. The most occurring flaws were those connected with lack of root penetration (Figure 7, *a*) and form deviations of the welded joint surface (Figure 7, *b*).

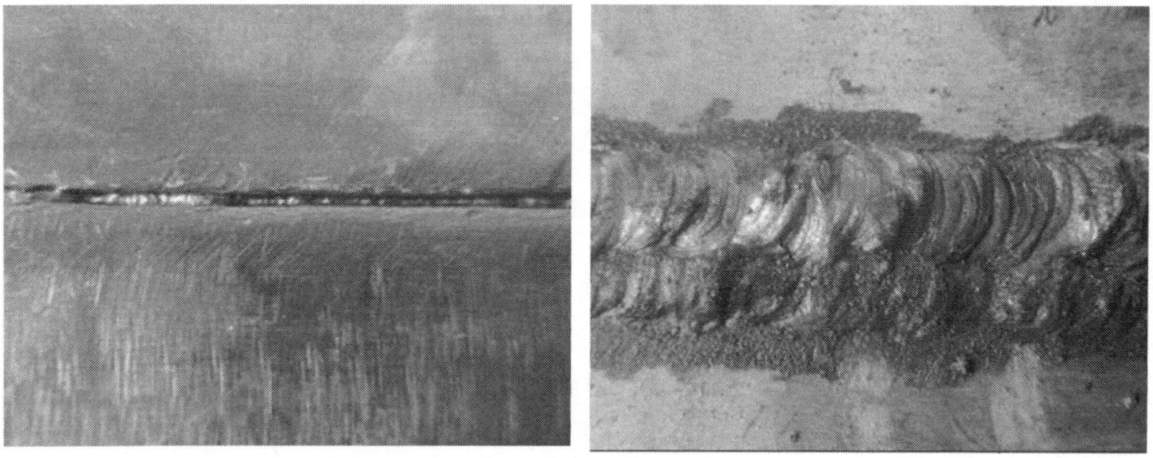

a) *b)*

Figure 7. Typical root defects *a)* and the surface of *b)* repair welding joint.

5. Discussion

More than 91 % of flaws detected by V&DT are removed by deseaming of the weld face (if necessary with backing weld and subsequent deseaming). This method allows to remove such flaws as irregular ripples, beadings tough transition to the base metal, excessive height of the crests step structure of the welded joints and undercuts. It does nt require expensive raising of welders' qualification and increases solidity of welded joints due to removal of macro- and micro acceptable flaws where welding stresses are consentrated. In the conditions of alternating strains (where welded joints usually function) it leads to the loss of their operational life due to appearing and extension of fatigue cracks.

Removal of the flaws detected by UT is no less significant problem. The analysis of welded specimens' ultrasonic testing allowed to conclude that 85 % of flaws are flaws of groove (preparation for welding). They are not directly caused by the qualification of the welder making a joint, but to a greater extent they are flaws of a specialist preparing for the welding works. The solution of this problem lies in the area of repair conducting technology. Thus introduction of the method of slotted groove with installing a background plate allows to solve the problem (Figure 8).

Figure 8. Slotted groove of fractured welded connection with installing a background plate.

For practical approval and introduction of this technology there was an order for the whole enterprise at ZAO Prokopievskiy Ugolnyi Razrez: during the repair works on metallic structures welding joints' groove with installing a background plate is obligatory. This allowed as early as in the first year of functioning to reduce the number of detectable flaws in metallic structures of the technological machines on average by 35 % in comparison with other enterprises of AO HK SDS-Ugol (Figure 9).

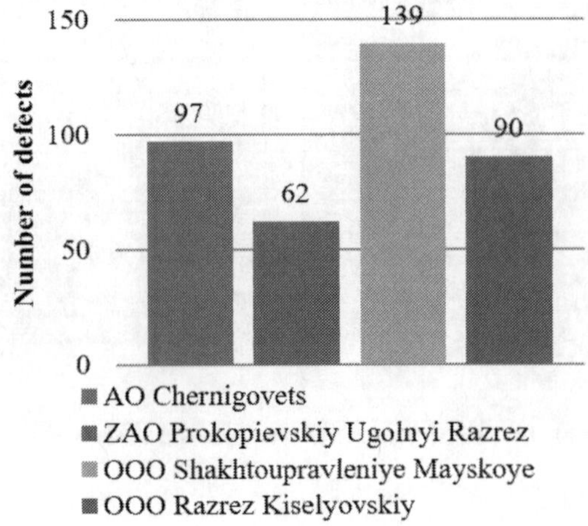

Figure 9. Average number of defects in metallic structures detected during each examination.

According to the results of electric and gas welders' work evaluation at eight enterprises of AO HK SDS-Ugol it was found that specimens done by welders of ZAO Prokopievskiy Ugolnyi Razrez have on average 25 % less flaws in comparison with other enterprises. And the specimens with a background plate there showed no flaws.

The survey proved solidity of repair welded joints made according to the introduced technology. Thus lack of root penetration cases reduced by 85 %, and operational life of the machines raised from 2–6 months to almost unlimited. Within five years of observation 98 % of welded joints remained solid.

6. Summary and conclusions

1. The flaws of technological machines' metallic structures accumulate during a long period of time. Removal of these flaws by repair welding was inefficient – the period of a welded joint's life was 2–6 months. It was caused by low quality of repair and low number of removed flaws (within 6 months only 29 % of detected flaws were removed).
2. According to the results of welding specimens evaluation showing the work of 193 welders of the 3–6th categories at AO HK SDS-Ugol enterprises, no specimen of the total amount of 338 met the requirements of documentary standards.
3. The flaws detected by ultrasonic testing are in 85 % the flaws of the groove (preparation for welding works) and are not directly caused by low qualification of the welder as a specialist making the welded joint.
4. The flaws of repair welded joints (91 %) detected by visual and dimensional testing are removed by the welded joint deseaming.
5. Introduction of the method of slotted grove with installing a background pate allowed to reduce the number of lack of root penetration cases by 85 % without raising welders' qualification and extend the life of the welded joints from 2–6 moths to almost unlimited. This allowed to reduce the number of equipment metallic structures' flaws by 35 % as early as in the first year of new technology introduction.

IOP Conf. Series: Journal of Physics: Conf. Series **944** (2017) 012029 doi:10.1088/1742-6596/944/1/012029

References

[1] Arden H and Tverdov A 2014 Resource and Reserve Valuation Practices in CIS Countries. *Mineral Resource and Ore Reserve Estimation.* The AusIMM Guide to Good Practice

[2] Russian Economy, Figures and Facts. Part 5. Coal Industry. URL: http://utmagazine.ru/posts/10449-ekonomika-rossii-cifry-i-fakty-chast-5-ugolnaya-promyshlennost (accessed: 25.05.2017)

[3] Statistic review of Emerging Markets Information Service on "Coal Mining Sector Russia" (2013) received from the website https://www.emis.com/sites/default/files/Russia%20Coal%20Mining%20Sector%20Report.pdf (accessed: 28.05.2017)

[4] Drygin M, Kurychkin N and Bakanov A 2017 Strategy of Russian Coal Mining Enterprises' Excavator Park Technical State Correction *The 1st Scientific Practical Conference "International Innovative Mining Symposium (in memory of Prof. Vladimir Pronoza)"* **15** p 03011

[5] Alyoshin N, Chernyshov G, and Gladkov E 2004 Welding. Cutting. Control. vol. 1, ed N P Alyoshin and G G Chernyshov (Moscow: Mashinostroyeniye) p 624

[6] Belyanin P, Kolesnikov K and Adamov E 2006 Technology of welding, brazing and cutting. *Mechanical Engineering. Encyclopedia* in 4 vol ed K F Frolov, (Moscow: Mashinostroyeniye) **3** p 767

[7] Vasilyev K, Vill V and Volchenko V 1978 *Welding in Mechanical Engineering* vol 4 ed N A Olshanskiy (Moscow: Mashinostroyeniye) **1** p 501

Increasing the quality of excavators' planetary reduction gearboxes on the basis of dimensional analysis and geometrical characteristics of tooth wheels

M Yu Drygin, N P Kuryshkin

Department of information and automated manufacturing systems, T. F. Gorbachev Kuzbass State Technical University, 28 Vesennyaya street, 650000, Kemerovo, Russia

E-mail: kemnik@mail.ru

Abstract. The article describes the problem of extending operational electrically driven excavators' reducing gear boxes which break down nearly every six months. The main reason of the function loss is the breakdown of the bearing assembly of one of the pinions. The authors performed kinematic, dynamic and geometric calculation of gearing to detect the breakdown reasons. The main reason is that alignment condition is not provided in the differential part and in the power return gear. All toothed gear wheels must be manufactured with the positive bias of the generating rack profile, but in fact all pinions and idle gears are manufactured with negative bias. This lead to an intolerable radial clearance in the gearing and skew of the floating master gears.

1. Introduction

Fuel and energy park has the outmost importance in the world economy. Despite of its turn-down coal is one of the most important among the energy resources of Russia and it stabilizes the situation in the energy industry [1, 2]. Open-cut coal mining has obvious advantages compared with underground mining: higher productivity, low cost-price and safety of mining works. This is the reason why about 70 per cent of coal in Russia are open-cut, and this figure is constantly increasing. Thus in 2015, compared with 2014, it increased by 2 %, and by 41 %, in comparison with 2000 [3].

Working efficiency of opencast coal mines is determined by productive capacity and performance reliability of the in-line equipment and, first of all, excavators. Analysis of Russian mining enterprises' excavator fleet shows that its main part (63%) includes excavators EKG-10 with bucket capacity of 8…15 m^3 produced by OMZ IZ KARTEKS [3, 4].

Mechanical section of an excavator EKG-10 comprises four basic mechanisms: lifting mechanisms (540 kW) and swing mechanisms (200 kW), as well as crowding (125 kW) and traveling (54 kW) gears. Electric motor power is given in brackets. Every mechanism includes electric motors and a reducing gear box. As you can see from the given data bowl lift mechanism has the highest power supply capacity. Operating practice of excavators EKG-10 in opencast coal mines of Kuzbass shows that the reducing gear box of the lifting mechanism breaks down nearly every six months. Given that, the repair, if it is properly organized and necessary spare parts are available, takes not more than three days, that is why it does not influence a lot the breakdown time, but it is always very expensive and excesses the budget for the yearly repair of the whole excavator. The breakdown of the lifting mechanism reducing gear box can hardly be predicted and it is always an emergency.

2. Problem description

The check of broken down reducing gear boxes of lifting mechanisms of excavators EKG-10 at three Kuzbass opencast coal mines showed absolutely the same results. The bearing cage of one of the central part's pinions was fractured, fragments of ball race collars were broken down, and bearing rollers were lost (Figure 1), then they engaged the toothed wheel rim of the drum, and, as a result the teeth were fractured. The character of wearing of the teeth of the master gear that engages the pinions (Figure 2) shows that it worked with an obvious tilt (Figure 2). This tilt caused sideways bending on the axis of the carrier and the planetary bearing. The result of this bending is shown on the Figure 1.

Figure 1. Fracture of the planetary bearing of the lifting mechanism reducing gear box differential part.

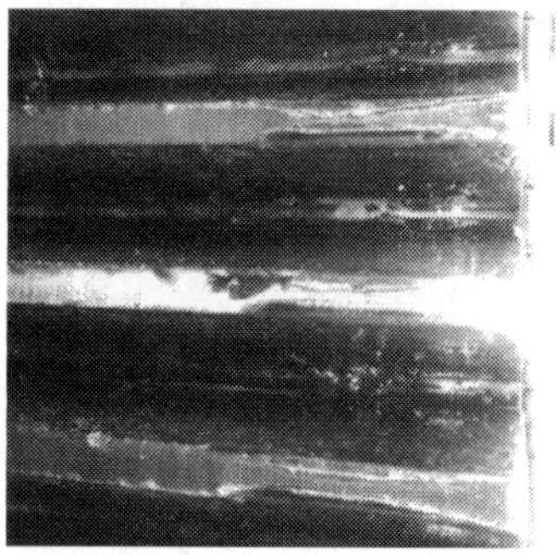

Figure 2. Wearing of the master gear of the lifting mechanism reducing gear box differential part.

In view of the above the authors set a task to study kinematics and dynamics of the reducing gear box, to analyze the breakdown process and to give recommendations for extending operational life of the gear box.

3. Theory

Bowl lifting mechanism gear box of the excavator EKG-10 comprises two identical closed rack-type differentials with one common output element – it is a drum where two strings of the wire rope cable lifting the excavator bucket are spooled simultaneously (Figure 3). Every closed rack-type differential includes a differential part (central wheels z_1 and z_3, three pinions z_2 and a carrier) and a power return gear train (wheels z_5, four idler gears z_4 and a tooth wheel rim z_3). All toothed gear wheels have module m = 10 mm.

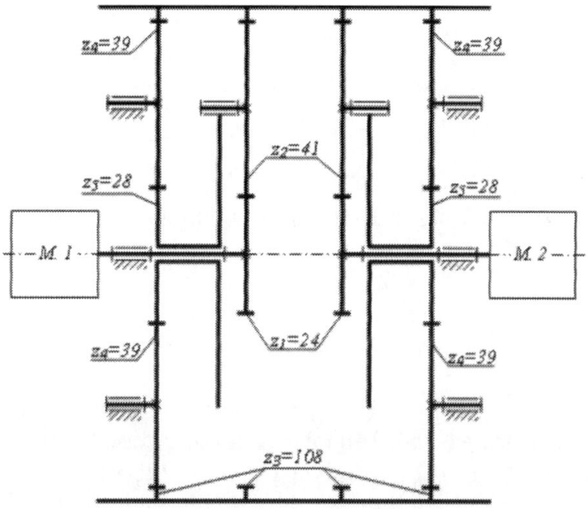

Figure 3. Kinematic diagram of the excavator EKG-10 lifting mechanism reducing gear box.

The central toothed gear wheel z_1 is manufactured as floating without a bearing assembly to provide equal distribution of load on the pinions z_2 of the differential part. Torque from the engine is transmitted to it by the torsion bar via the clutch gear. To prevent axial deflection of the central wheels z_1 on their axis between them a 30 kg distance rod is installed (it is not shown on Figure 3).
Gravity force of the distance rod can produce bending moment causing the work of the master gear z_1 with a tilt. To find the reason of the tilt we calculated the forces of pinions influencing it and compared them with the gravity force of the distance rod. According to the certificate maximum heave is F_{max} = 1000 kN. Calculated heave is assumed $F_n = F_{max}/2$ =500 kN. One string of the wire rope cable circumferential force is $F_{тр}$ = 500 / 2 = 250 kN. Hoist drums torque:

$$M_б = F_{тр} \cdot \frac{d_б}{2} = 250 \cdot \frac{1.32}{2} = 165 \text{ кНм}.$$

Gear ratio of the reducing gear box is calculated using the formula for a closed rack-type differential [5]:

$$u_{13} = \frac{-z_3}{z_5} + \left(1 + \frac{z_3}{z_5}\right) \cdot \left(\frac{-z_3}{z_1}\right).$$

After substitution of the teeth numbers we will get $u_{13} = -25.71$.
Then the master gear z_1 resistive torque, taking into account the reducing gear box efficiency rate, is:

$$M_в = \frac{M_б}{u_{13}\eta} = \frac{165}{25.71 \cdot 0.8} = 8.02 \text{ кNm}.$$

Master gear circumferential force:

$$F_t = \frac{2M_в}{d_1} = \frac{2 \cdot 8.02}{0.24} = 66.83 \text{ кN},$$

where d_1 is the reference diameter ($d_1 = mz_1 = 10 \cdot 24 = 240$ mm = 0.24 m).
Circumferential force on the master gear from one pinion:

$$F_t' = \frac{F_t}{3} = \frac{66.83}{3} = 22.28 \text{ кN}.$$

Then master gear radial stress from one pinion is:

$$F_r = F_t' \cdot tg\alpha = 22.28 \cdot 0.364 = 8.11 \text{кN},$$

where α is the standard pressure angle of a typical generating rack profile ($\alpha = 20^o$).
Thus the master gear is centered by three radial forces 8.11 kN each, and therefore distance rod gravity force equal to 0.3 kN cannot influence its tilt.
The second reason of the master gear work with significant tilt can be excessive spaces in toothed gearing of the reducing gear box differential part. Alignment condition preliminary check showed that toothed gear wheels z_1, z_2, z_4 and z_5 must have been manufactured with generating rack profile positive bias. Indeed alignment conditions with shifting coefficients $x_1 = x_2 = x_4 = x_5 = 0$ for the differential part and power return gear trains are written respectively as:

$$z_3 = z_1 + 2z_2, \quad z_3 = z_5 + 2z_4. \tag{1}$$

Substitution of teeth numbers to the formulae (1) shows that toothed wheel rims must have the number of teeth equal to 106, but not 108 as they have in fact.
To verify this version under the conditions of a servicing station of the open cast coal mine Prokopyevskiy we disassembled a broken down reducing gear box of an excavator EKG-10 and measured the main geometrical parameters of the toothed gear wheels of the differential part.

4. Experimental results

The measurements were performed with reliable measuring tools: gear-tooth micrometer BV-5046, caliper gauges of the types ShTs, ShTsK and ShTsTs, gear-tooth micrometer of the type MZ with special attachments. The measured parameters are: outside diameter d_a, mm; root diameter d_f, mm; base tangent length w_n for n teeth; base tangent length w_{n+1} for $(n+1)$ teeth. The measurements were

performed five times for each wheel on different sections, and then the arithmetic mean value of the parameter was calculated.
The results of master gear geometrical parameters measurements on an unworn section are presented in the Table 1 as an example.

Table 1. The results of calculation the master gear parameters.

Measured Parameter	1 meas.	2 meas.	3 meas.	4 meas.	5 meas.	Mean
Outside diameter d_a, mm	268.1	268.0	268.2	268.0	268.1	268.08
Root diameter d_f, mm	225.75	226.0	225.5	225.25	225.5	225.6
Base tangent length for $n = 4$ teeth, mm	109.9	109.9	109.4	109.8	109.7	109.74
Base tangent length для $n + 1 = 5$, mm	139.4	139.2	138.8	139.4	139.45	139.25

Calculation of the shifting coefficient of the master gear was performed according to the following algorithm [6].
1. Normal pitch to addendum angles:

$$p_n = w_{n+1} - w_n = 139.25 - 109.74 = 29.51 \text{ mm},$$

where w_{n+1} – is the base tangent length for ($n + 1 = 5$) teeth, mm;
w_n – base tangent length for ($n = 4$) teeth.
2. Tooth thickness at the base circle:

$$s_b = w_{n+1} - n \cdot p_n = 139.25 - 4 \cdot 29.51 = 21.21 \text{ mm}.$$

3. Base radius:

$$r_b = (mz/2)\cos\alpha = (10 \cdot 24/2)\cos 20^0 = 112.763 \text{ mm}.$$

4. Shifting coefficient of generating rack profile:

$$x = \frac{z[(s_b/2r_b) - \text{inv}\alpha] - 0.5\pi}{2\text{tg}\alpha} = \frac{24[(21.21/2 \cdot 112.763) - 0.014904] - 0.5 \cdot 3.14159}{2 \cdot 0.364} = 0.4514.$$

5. Root diameter:

$$d_f = mz - 2(h_a^* + c^* - x)m = 10 \cdot 24 - 2(1 + 0.25 - 0.4514) \cdot 10 = 224.029 \text{ mm}, \quad (2)$$

where h_a^* – is the addendum factor ($h_a^* = 1$);
c^* – is the clearance coefficient ($c^* = 0.25$).
The obtained result is coherent with the measured ($d_f = 225.6$ mm). Therefore the shifting coefficient calculated according to base tangent lengths of an *unworn* gear is valid.
Computational error for d_f calculation was:

$$\Delta = \frac{225.6 - 224.029}{225.6} \cdot 100\% = 0.7\%.$$

On this basis we may conclude that generating rack profile shifting coefficient can be calculated using the root diameter according to the formula obtained from (2):

$$x = \frac{d_f - mz + 2(h_a^* + c^*)m}{2m}. \quad (3)$$

With this conclusion and the fact that base tangent lengths on the worn toothed gear wheels are less than design values, shifting coefficients for the rest of the toothed gear wheels were calculated according to the formula (3). The results of the measurements and calculations are presented in the Table 2.

Table 2. The results of measurement and calculation of gears' parameters in the reducing gear box.

Parameters	Pinion 1	Pinion 2	Pinion 3	Central Wheel
Outside diameter d_a, mm	425.14	424.32	424.4	1075.478
Root diameter d_f, mm	380.58	380.08	380.4	1121.378
Shifting coefficient x	-0.221	-0.246	-0.23	–

5. Discussion

Using the results of measurements and calculation of toothed gear wheels' parameters presented in the Tables 1 and 2, we analyzed the real alignment condition of the reducing gear box differential part. On the Figure 4 the calculation model of the reducing gear box differential part shows that:

$$d_{a3} = d_{a1} + 2d_{f2} + 4c^{*}m. \tag{4}$$

And also:

$$d_{f3} = d_{f1} + 2d_{a2} + 4c^{*}m. \tag{5}$$

After substitution of the altered parameters to (4) and (5) we got respectively:

$$d_{a3} = 268.08 + 2 \cdot 380.35 + 4 \cdot 0.25 \cdot 10 = 1038.78 \text{ mm}.$$

$$d_{f3} = 225.6 + 2 \cdot 424.62 + 4 \cdot 0.25 \cdot 10 = 1084.84 \text{ mm}.$$

The measured outside and root diameters of the central wheel with internal teeth: d_{a3} = 1075.478 mm and d_{f3} = 1121.378 mm. The discrepancy is: Δ_{a3} = 1075.478 – 1038.78 = 36.698 mm and Δ_{f3} = 1121.378 – 1084.84 = 36.578 mm.

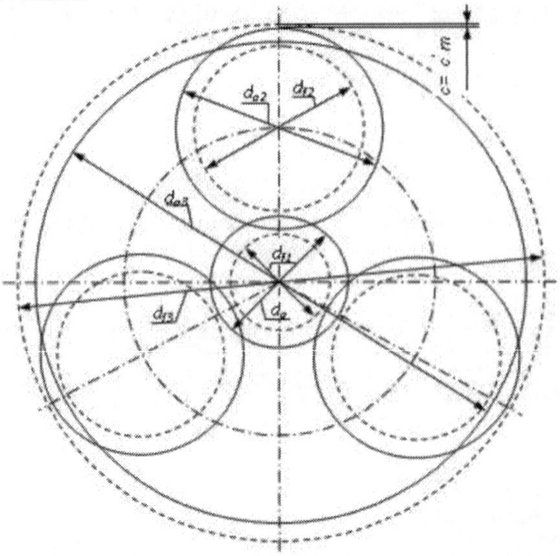

Figure 4. Calculation model of the reducing gear box differential part.

The obtained discrepancy is that very intolerable radial space in the gearing which causes the master gear tilt. As the angular rate of this gear in the reducing gear box is the highest and taking into account that it is free from radial movements its rotation becomes oscillating. In such gearing only the external from the engine half of the gear is put on the job. This is confirmed by specific wearing of the gear (see Figure 2). Besides alternating in sign performance mode of the reducing gear box leads to shock loads in the gearing. The same conclusion can be made according to the character of the rotation motion of the carrier with the pinions. If there is no any equalizing mechanism providing even load distribution between the pinions, the main power will be transmitted by one pinion. This is confirmed by the fact that one pinion had only one bearing block fractured (see Figure 1). Gearing with a tilt caused intolerable axial loading on this bearing block that lead to its fracture.

6. Summary and conclusions

Thus the results of the analysis of broken down bowl lifting mechanism reducing gear boxes' real state, measurement and calculation of geometrical parameters of the toothed gear wheels in their differential part as well as dynamic calculation of the gearing allowed to conclude the following:

1. The main reason of bowl lifting mechanism reducing gear box emergency breakdowns is the fracture of the bearing block of one of the pinions in the differential part.
2. Gravity force of the distance rod between floating master gears of the reducing gear box does not influence significantly their behavior.
3. According to alignment condition in the differential part and the power return gear train all wheels must be manufactured with positive bias of the generating rack profile. However in fact pinions and idler gears are manufactured with negative bias. This lead to intolerable radial space in the gearing about 37 mm.
4. Intolerable radial space in the gearing lead to significant tilt of the floating master gears, their oscillating movement and edge contact with the pinions. This contact caused pinions' bending moment fracturing the bearing block.
5. The radical outcome of the existing situation is eliminating the radial space by manufacturing all toothed gear wheels with the positive bias.

References

[1] Arden H and Tverdov A 2014 Resource and Reserve Valuation Practices in CIS Countries *Mineral Resource and Ore Reserve Estimation* (The AusIMM Guide to Good Practice)

[2] Statistic review of Emerging Markets Information Service on "Coal Mining Sector Russia" (2013) received from the website https://www.emis.com/sites/default/files/Russia%20Coal%20Mining%20Sector%20Report.pdf (Assessed: 25.04.2017)

[3] Anistratov K Y Analysis of Mining Excavator and Dump Truck Market in Russian Federation and CIS Countries. *Gornoye Delo – informational and analytical web portal.* URL: http://www.gornoe-delo.ru/articles/detail.php?ID=7508 (Assessed: 10.04.2017)

[4] Drygin M, Kurychkin N and Bakanov A 2017 Strategy of Russian Coal Mining Enterprises' Excavator Park Technical State Correction. *The 1st Scientific Practical Conference "International Innovative Mining Symposium (in memory of Prof. Vladimir Pronoza"* **15** p 03011

[5] Ivanov K S 2007 Gear Automatic Adaptive Variator with Constant Engagement of Gears. *Proceedings of the 12th World Congress in Mechanism and Machine Science.* (Besancon. France) **2** pp 8–182

[6] Bolotovskiy I A 1986 *A guide to the geometric calculation of serrated involute and worm gears* (Moscow: Mashinostroenie) p 448

Decrease in vibroactivity of gear drives

O S Dyundik, I P Zgonnik

Omsk State Technical University, 11, Mira ave., Omsk, 644050, Russia

E-mail: tmm@omgtu.ru

Abstract. The analysis of technical and technological solutions to reduce the influence of dynamic intermating process to vibroactivity of a gear drive is carried out. The problem of a complete elimination has no solution due to a discrete way of power flow transformation by a gear drive. It is shown that a reduction of mesh stiffness is a controlled parameter in vibroactivity decrease. We have proposed to reduce the stiffness by means of a gear rim radial cut that provides small initial and stepwise variable stiffness of gearing. The technical solutions resulting in a smooth, dependent on a power flow running of stiffness by means of built-in control chain and usage of elastomeric materials are given. As a result, it is noted that the reduction of mesh stiffness at the same time leads to the increase in static deformation δ, and the dynamic load under intermating decreases. Elastomeric parameters' variation t can provide the achievement of calculated mesh stiffness, depending on a variable power flow.

1. Introduction

Gear drives have some advantages such as converting the motion between shafts irrespectively of location in space, compactness, high bearing capacity and efficiency, reliability, manufacturability. These advantages determine a wide spread of these types of transmissions in mechanical engineering objects. However, along with the advantages, gear drives objectively have disadvantages, such as increased vibroactivity caused by the dynamic tooth changeover, under inherent pitch error, explained by primary and power errors, as well as by variable mesh stiffness both in one-pair and in multipair toothing. The problem of vibroactivity decrease of gear drives is urgent and is widely considered in foreign literature, such as in [1, 2].

2. Problem statement

Let us set and solve the problem of analysis for technical and technological solutions to reduce the influence of dynamic intermating process to vibroactivity of a gear drive, as the problem of a complete elimination of this phenomenon has no solution due to the discrete way of power flow transformation by the gear drive.

3. Experiment methodology

It is known that specific dynamic loading P_{dyn} causing vibrations under intermating is defined as follows:

$$P_{dyn} = Av\sqrt{m_{ef}(\delta+\Delta p)\cdot c}\ , \qquad (1)$$

where A is a proportionality factor (its value is not important for a quality process estimate);
v is a peripheral velocity in initial cylinders of transmission wheels (m/s);
m_{ef} is a gear drive mass reduced to a polar point;
δ is a static flexural deformation of a tooth under loading (a power error of a base pitch);
Δp is a primary error of the base pitch;
c is the mesh stiffness.

Analyzing (1) let us note that the mesh stiffness c is the only controlled parameter. To decrease stiffness is necessary to reduce dynamic loading P_{dyn} which generates a forced vibration with intermating frequency

$$f=nz,$$

where n is a revolution number of a gear (or a wheel), z is a number of teeth in the wheel.
This recommendation is realized in many technical solutions of gear drives. Thus, gear drives with increased tooth height are used in high-speed airborne gearboxes. There are well-known methods of reducing teeth stiffness by specific technological cuts, by drilling in rims, etc. However, one should keep in mind that the reduction of mesh stiffness at the same time leads to the increase in static deformation δ.
Let the tooth be a cantilever beam (figure 1), static deflection δ under the force P will be the following:

$$\delta=\frac{P\cdot l^3}{3EJ}, \qquad (2)$$

where P is the force operating in gearing;
l is a tooth height;
E is a material elasticity modulus;
J is an inertia moment of a root section.

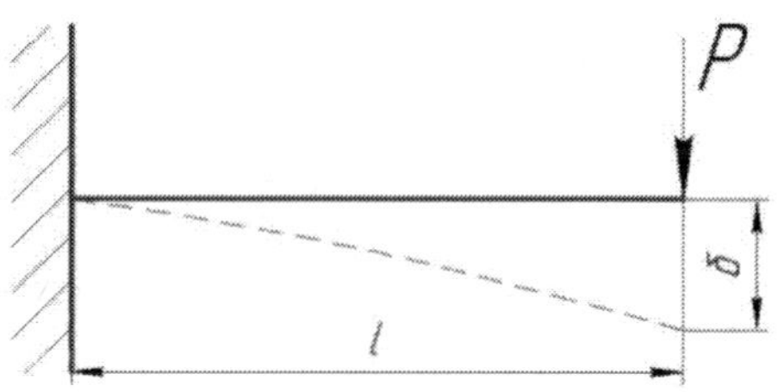

Figure 1. Defining of a tooth deflection as a cantilever beam.

4. Experimental results

As a result we can see that the reduction of mesh stiffness leads simultaneously to the increase in static deformation δ, and the dynamic load under intermating decreases.
Thus following (2) if to increase twice L console length of the tooth with a cut, the static deflection δ will increase eightfold, and stiffness $c=\frac{P}{\delta}$, with the same P will also decrease eightfold. According to (1) this will result in threefold reduction of specific dynamic loading R_{dyn}.

5. Results and discussion

There is a very progressive technical solution [3], which provides the small initial mesh stiffness under a simultaneous deformation restriction of a bearing teeth pair. It is achieved due to teeth hollows in a wheel body are made with rectilinear radial grooves of substantial length l. To limit gearing deformation, the grooves enter hollows via arcform cuts with (0,2÷0,4) mm width.

IOP Conf. Series: Journal of Physics: Conf. Series **944** (2017) 012031 doi:10.1088/1742-6596/944/1/012031

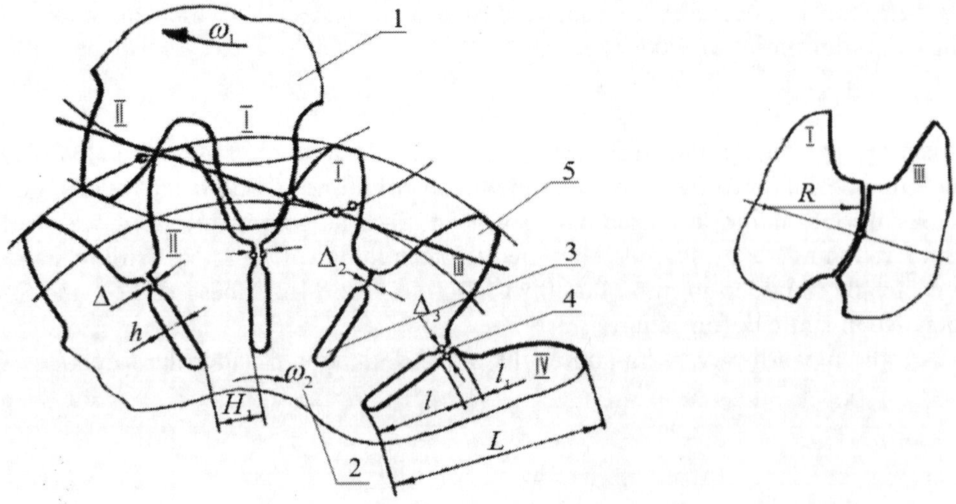

Figure 2. A gear wheel with small initial stepwise variable mesh stiffness [3].

The technical solution [4] (figure. 3) obtained in the development of [3] is more universal. This solution provides an adjustable control of stiffness depending on the level of power flow transferred by the transmission by changing of an active part of radial cut. The gear rim is movably installed on a shaft by means of bearings, thus there is no rim axial displacement towards the shaft, but a relative angular motion is allowed. In wide parts of radial cuts in the proposed gear drive there are special sliders 2 which are connected with a hub 4 by means of flexible rockers 3 made as flat curved springs hingedly fixed on the hub. It should be noted that the hub is rigidly fixed on a shaft. Flexible elastic elements 5 between the shaft and the gear rim are installed to obtain a desired angular shaft displacement to the gear rim. The proposed elastic elements 5 can be designed as an elastomeric ring or a metal spring of a particular form.

The tooth height regulation is provided with the built-in control chain including the elastic element 5 which deformation depends on power flow parameters and this deformation is used for the motion of special sliders 2 in radial cuts. The position of the sliders determines the cut length and, therefore, the mesh stiffness.

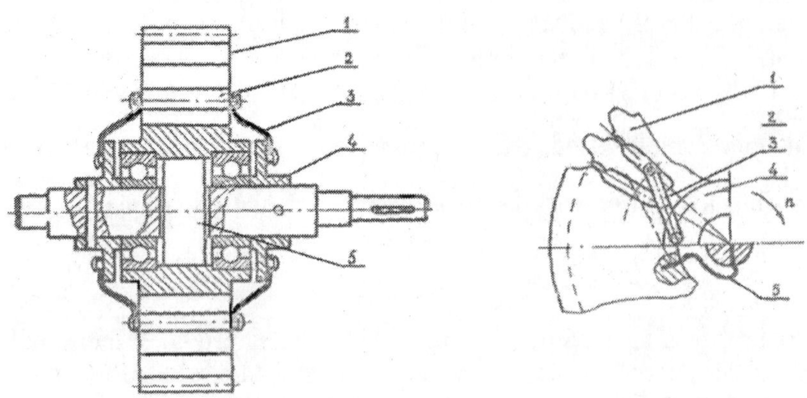

Figure 3. A gear wheel with self-guided mesh stiffness [4].

However, it should be noted that the technical solution [4] is technologically complicated and its implementation is hardly possible for universal mechanical gearboxes. It is much simpler to execute the mesh stiffness control by a gear drive with automatically controlled stiffness on the technical solution [5] where complicated in production sliders with flexible rockers are excluded from the control chain and are replaced with technologically more available elastomeric inserts located in a radial cut in a gear wheel hollow of a modified design according to figure. 4, a, which can also be realized by elastic inserts (figure. 4, b).

6. Conclusion

Since the deformation of such inserts is non-linear and essentially depends on the elastomer shape and its fixing method in a cut, it is possible to provide the achievement of calculated mesh stiffness depending on a variable power flow by the variation of these parameters. This calculation is mandatory, and using of elastomeric inserts is prospective for the decrease in vibroactivity of gear drives. The implementation of technical solution [5] has no restrictions and is applicable in universal gearboxes.

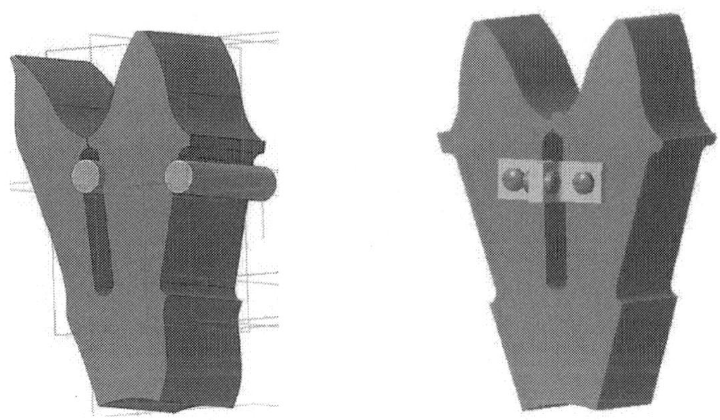

a) b)

Figure 4. A part of gear wheel with variable flexural mesh stiffness [5].

References

[1] Folęga P, Burdzik R and Wojnar G The optimization of the ribbing of gear transmission housing used in transportation machines 2016 *Journal of Vibroengineering* vol 18 issue 4 pp 2372

[2] Jedliński Ł 2016 A new design of gearboxes with reduced vibration and noise levels vol. 17 No 4 Diagnostyka pp 93–98

[3] A. c. 1733777 USSR, ICI F 16 N 55/14. Gear drive / Balakin P D, Ryazantseva I L, Troyan O M № 4750851/28 applic. 16.10.89 publ. 15.05.92 Bulletin No. 18

[4] A. c. 2225552 The Russian Federation IPC F 16 N 55/14 Gear drive / Balakin P D, Fillipov Yu O, Mikhaylik O S, applicant and owner of Public Educational Institution of Higher Professional Training Omsk State Technical University No. 2002119115/11 applic. 15.07.2002 publ. 10.03.2004 Bulletin No. 7

[5] Patent 112968 The Russian Federation, IPC F 16 N 55/14 Gear drive / Balakin P D, Dyundik O S, Dyundik E A No. 2011131847/11 applic. 28.07.2011 publ. 27.01.2012 Bulletin No 3

Single product lot-sizing on unrelated parallel machines with non-decreasing processing times

A Eremeev[1,3], **M Kovalyov**[2] **and P Kuznetsov**[3]

[1]Sobolev Institute of Mathematics, Siberian Branch of Russian Academy of Sciences, Pevtsov 13, 644099, Omsk, Russia
[2]United Institute of Informatics Problems, National Academy of Sciences of Belarus, Surganova 6, 220012, Minsk, Belarus.
[3]Omsk State University n.a. F.M. Dostoevsky, Mira 55a, 644077, Omsk, Russia

Abstract. We consider a problem in which at least a given quantity of a single product has to be partitioned into lots, and lots have to be assigned to unrelated parallel machines for processing. In one version of the problem, the maximum machine completion time should be minimized, in another version of the problem, the sum of machine completion times is to be minimized. Machine-dependent lower and upper bounds on the lot size are given. The product is either assumed to be continuously divisible or discrete. The processing time of each machine is defined by an increasing function of the lot volume, given as an oracle. Setup times and costs are assumed to be negligibly small, and therefore, they are not considered. We derive optimal polynomial time algorithms for several special cases of the problem. An NP-hard case is shown to admit a fully polynomial time approximation scheme. An application of the problem in energy efficient processors scheduling is considered.

1. Introduction

In this paper, we study a problem which arises in scheduling workload on parallel processors or production planning of a single product on a number of parallel machines or reactors. There is a demand for at least A units of a single product to be produced on m, $m\geq 2$, unrelated parallel machines in lots. The size of a lot is the number of units this lot includes. It can take either integer value or real value, in which cases the product is called either discrete or continuous, respectively. Machine i, $i=1,\ldots,m$, requires $p_i(x)$ time units to produce a lot of size x, where $p_i(x)$ is a non-decreasing function with non-negative finite values. If a lot of size x is assigned to machine i, then the lot size must satisfy the relation $l_i\leq x\leq u_i$, where l_i and u_i are given lower and upper bounds, $i=1,\ldots,m$. The problem is to partition at least A units of the product into lots and assign these lots to the machines. In one version of the problem, the maximum machine completion time should be minimized, in another version of the problem the sum of machine completion times is to be minimized. Both criteria are related to the fair distribution of machine workloads. Setup times and costs are assumed to be negligibly small, and therefore, they are not considered. It is assumed that the values A and u_i $i=1,\ldots,m$, are positive rational numbers, and the values l_i, $i=1,\ldots,m$, are non-negative rational numbers.

The assumption that lot sizes must belong to the intervals $[l_i,u_i]$, or otherwise equal to zero, may reflect some technical constraints of the machines. In the computational applications, such constraints may be caused by limited RAM capacities on each of the processors (defines u_i), minimal cost-effective or minimal time-effective work load on each of the processors (defines l_i). In the industrial applications, these constraints may be due to a limited reactor size (defines u_i) and minimal amount of product that can be processed in one lot without damaging the equipment or due to economical reasons (defines l_i), see e.g. [9].

We denote this problem as $R|1,GT,\beta|\gamma$, where $\beta\in\{cntn,dscr\}$, $\gamma\in\{C_\Sigma,C_{\max}\}$. Following the scheduling traditions, notation "*R*" here refers to unrelated parallel machines. Notation "1" in the middle field indicates the case of a single product. Notation "*GT*" is used to to indicate that at most one lot can be assigned on the same machine (*GT* stands for Group Technology). Abbreviations "*cntn*" and "*dscr*" specify continuous and discrete product, respectively. Maximum machine completion time and total machine completion time are denoted as "$C_{\max}$" and "C_Σ", respectively.

1.1. Mathematical programming problem formulation

The following mathematical programming formulation for the problem $R|1,\alpha,\beta|\gamma$ can be given.

$$\min\sum_{i=1}^{m} p_i(x_i), \quad \text{if } \gamma = C_\Sigma \tag{1}$$

$$\min f_{\max}, \quad \text{if } \gamma = C_{\max} \tag{1'}$$

s.t.

$$f_{\max} \geq p_i(x_i), i = 1,\ldots,m, \quad \text{if } \gamma = C_{\max} \tag{2}$$

$$\sum_{i=1}^{m} x_i \geq A \tag{3}$$

$$z_i l_i \leq x_i \leq z_i u_i, i = 1,\ldots,m \tag{4}$$

$$x_i \in \mathbf{Q}_+, \text{ if } \beta = cntn \tag{5}$$

$$x_i \in \mathbf{Z}_+, \text{ if } \beta = dscr \tag{5'}$$

$$z_i \in \{0,1\}. \tag{6}$$

Here $\mathbf{Q}_+$ denotes the set of non-negative rational numbers and $\mathbf{Z}_+$ is the set of non-negative integers. The variables are the maximum machine completion time $f_{\max}$, the production volume x_i on machine i, and the number of lots z_i on machine i, $i=1,\ldots,m$. Relation (3) ensures that the required quantity of the product is assigned to the machines. Relations (4) connect production volume x_i with the number of lots z_i on each machine i. Conditions (5) and (5') address the assumptions that the product is either continuously divisible (5) or discrete (5').

1.2. Related research

The problem formulated above is closely related to the problem studied by Dolgui et al. [4], in which there are several products, machine-dependent lot size lower bounds, machine and sequence dependent setup times, and the objective is to minimize $C_{\max}$. This problem is strongly NP-hard due to a reduction from the well-known strongly NP-hard Traveling Salesman Problem. It was proved to be NP-hard in the ordinary sense even if the number of products $n=2$. Several dynamic programming algorithms for the special cases of the problem were developed.

The so-called Supply Scheduling Problem (SSP) is also closely related to the problem considered in this paper. In SSP, there are m providers that supply a certain product to a manufacturing unit. If provider i is not used, then the corresponding delivered quantity is $x_i=0$. If provider i is used, then the delivered quantity x_i must be between the given lower and upped bounds l_i and u_i. The demand at the manufacturing unit is A. The delivery cost for sending a quantity x_i from provider i to the manufacturing unit is $c_i(x_i)$, where $c_i(\cdot)$ is a cost function which can be linear as e.g. in Chauhan et al. [2], or can be given by an oracle as in Chauhan et al. [3] and Ng et al. [8]. The goal is to minimize the total delivery cost, subject to the condition that the manufacturing demand is satisfied. The SSP is NP-hard in the ordinary sense and several FPTASes are proposed for different versions of this problem in [2,3,8]. A Fully Polynomial Time Approximation Scheme (FPTAS) is a collection of $(1+\varepsilon)$-

approximation algorithms A_ε such that algorithm A_ε guarantees the relative error ε, and it runs in time polynomial in $1/\varepsilon$ and in the problem instance length in binary encoding. A generalization of SSP to the case of concave non-decreasing cost functions and several feasible intervals for the delivered quantity was studied by Eremeev et al. [5] and an FPTAS was developed.

In this paper, we derive optimal polynomial time algorithms for several special cases of the problem $R|1,GT,\beta|\gamma$. For an NP-hard case of the problem we provide a fully polynomial time approximation scheme. The obtained results extend and/or improve the algorithmic results on $R|1,GT,\beta|\gamma$ from [6], where the processing times were assumed to be proportional to the lot sizes and all the numeric parameters A, l_i, u_i and p_i, $i=1,\ldots,m$, were assumed to be integers. Note that under the assumption of processing times being proportional to lot sizes, the problem may be considered as a special case of an applied production scheduling problem in chemical industry [9] or as a special case of fuel supply management problem [1]. In [6], we also considered another version of the problem, denoted $R|1,lot,\beta|\gamma$, where each machine may produce more than one lot, assuming that all processing times are proportional to workload assigned to a machine. The exact and approximate algorithms obtained there for $R|1,lot,\beta|\gamma$ may be extended straightforwardly to the case of non-decreasing continuous processing times, therefore we do not give a detailed analysis of $R|1,lot,\beta|\gamma$ here.

1.3. Application to energy efficient scheduling

The assumption that for each machine i, $p_i(x)$ is a non-decreasing function of the lot volume x allows us to consider an application of $R|1, GT,cntn|\gamma$ in the so-called *energy efficient scheduling* [7]. Suppose that the machines are speed-scalable processors, i.e. the processing time of each machine is x_i/s_i, where s_i is a tunable processing speed (subject to optimization). When a processor runs at a speed s, its energy consumption is s^{a_i} units of energy per time unit. The parameter a_i here is a constant for each processor (practical studies show that $1<a_i\leq 3$). Suppose that the total amount of work (e.g. computation) A needs to be performed on a set of m unrelated parallel processors whose workloads are either zero or between l_i and u_i and all computations should be finished by a given deadline d. The objective is to minimize the total energy consumption. Then allocation of a lot x_i to processor i implies that the minimal energy consumption $E_i(x_i)$ over d time units on this processor is attained if it works at the minimal speed $s_i=x_i/d$, therefore $E_i(x_i)=(x_i/d)^{a_i}d$, which is a convex increasing function of x_i and may substitute $p_i(x)$ in the problem formulation when $R|1, GT,cntn|\gamma$ is applied to energy efficient scheduling as described above. In this case, C_Σ criterion implies minimization of total energy consumption and C_{max} implies minimization of maximal energy consumption over all processors.

The paper is structured as follows. In the next section, we briefly discuss the complexity and approximability of the problem with the criterion of total machine completion time minimization for discrete and continuous products. Section 3 is devoted to the makespan minimization and contains the main contributions of this paper, namely the polynomial-time exact algorithms of different time complexity for several special cases of the problem. Section 4 gives concluding remarks and further research directions.

2. Minimizing total machine completion time

For the problem $R|1,GT,\beta|C_\Sigma$ with continuous non-decreasing non-negative functions $p_i(x_i)$, we note that it is equivalent to the earlier studied Supply Scheduling Problem with continuous non-decreasing cost functions. The latter problem is NP-hard [3] but admits an FPTAS, as shown in [3,8]. In this case, we have to assume that all functions $p_i(x_i)$ are given in the problem input as oracles, the same applies to the supplementary functions $q_i(C):=\sup\{x_i|p_i(x_i)\leq C\}$, and all of these oracles are polynomial-time computable.

3. Minimizing maximum machine completion time

In this section, we describe a polynomial-time algorithm for optimal solving the problem $R|1,GT,\beta|C_{max}$ under the assumption that the processing times are given by piece-wise linear increasing convex functions $p_i(x_i)$, $i=1,\ldots,m$.

3.1. Continuous product

First we describe an algorithm for problem $R|1,GT,cntn|C_{max}$. We begin with computing values $p_i(l_i)$ and $p_i(u_i)$, i=1,…,m, and sorting these values in a non-decreasing order. Let $D_1,\ldots,D_t$, $t\leq 2m$, denote the set of all distinct values $p_i(l_i)$ and $p_i(u_i)$, ordered so that $D_1<\ldots<D_t$. Denote the optimal makespan value by C^*. It is easy to see that there exists an index k^* such that either (a) $C^*=D_{k^*}$, $1\leq k^*\leq t$, or (b) $D_{k^*}<C^*<D_{k^*+1}$, $1\leq k^*\leq t-1$. Assuming for a while that k^* is known, let us partition the set of machines into the following three subsets: subset $Before_{k^*}$ consists of machines i, for which $p_i(u_i)<D_{k^*}$; subset $Between_{k^*}$ consists of machines i, for which $p_i(l_i)\leq D_{k^*}\leq p_i(u_i)$ in the case (a) and $p_i(l_i)\leq D_{k^*}<D_{k^*+1}\leq p_i(u_i)$ in the case (b); and subset $After_{k^*}$ consists of machines i, for which $D_{k^*+1}<p_i(l_i)$. These three definitions account for all possible locations of the interval $[D_k,D_{k+1}]$ with respect to the interval $[p_i(l_i),p_i(u_i)]$. In both cases (a) and (b), if k^* is given, then there exists an optimal solution x^* such that $x_i^*=u_i$ for all $i\in Before_{k^*}$, $x_i^*=0$ for all $i\in After_{k^*}$ and $x_i^*=\max\{x_i|p_i(x_i)\leq C^*\}$ for all $i\in Between_{k^*}$. Furthermore, in the case (b), $\sum_{i=1}^{m} x_i^* = A$.

Index k^* and the corresponding values C^* and x_i^*, $i=1,\ldots,m$, can be found by the enumeration of the index $k=1,\ldots,t$. Consider iteration k of this enumeration. To handle the case (a), calculate $x_i^*=u_i$, $i\in Before_k$, $x_i^*=0$, $i\in After_k$ and $x_i^*=\max\{x_i \mid p_i(x_i)\leq D_k,\ x_i\in[l_i,u_i]\}$, $i\in Between_k$. If $\sum_{i=1}^{m} x_i^* \geq A$, then set $C^{(a,k)}:=D_k$, otherwise set $C^{(a,k)}:=\infty$. To handle the case (b), calculate $x_i^*=u_i$, $i\in Before_k$ and $x_i^*=0$, $i\in After_k$. Values C^* and x_i^*, $i\in Between_k$, are found as optimal solution values of the following mathematical programming problem, denoted as MPP.

$$\begin{gathered}\text{Min } C,\\ \text{s.t.}\\ p_i(x_i)\leq C,\\ \sum_{i\in Before_k} u_i + \sum_{i\in Between_k} x_i = A,\\ x_i\in[l_i,u_i],\ i\in Between_k.\end{gathered}$$

If MPP has a solution, then we set $C^{(b,k)}$ to be equal to its optimal value. Otherwise we set $C^{(b,k)}=\infty$. It is easy to verify that

$$C^* = \min\{C^{(a,k)}, C^{(b,k)} \mid k=1,\ldots,t\}.$$

Note that the subsets $Before_{k+1}$, $Between_{k+1}$ and $After_{k+1}$ may be computed on the basis of subsets $Before_k$, $Between_k$ and $After_k$ so that the total time for finding all subsets $Before_k$, $Between_k$ and $After_k$, k=1,…,t-1 is $O(m)$.

MPP can be converted into a linear programming problem if all $p_i(x_i)$ are convex increasing piece-wise linear functions. Indeed, suppose that

$$p_i(x_i)=\sum_{0\leq j'<j}(\tau_i^{(j'+1)}-\tau_i^{(j')})p_i^{(j')}+(x_i-\tau_i^{(j)})p_i^{(j)}$$

whenever $x_i\in[\tau_i^{(j)},\tau_i^{(j+1)}]$, assuming that $\tau_i^{(1)},\ldots,\tau_i^{(K_i)}$ are the points that separate K_i linear segments of $p_i(x_i)$, where $p_i^{(1)}<\ldots<p_i^{(K_i)}$ are the linear coefficients of the segments. We assume that $\tau_i^{(0)}=0$ and

$\tau_i^{(K_i)} = u_i$. Then the supplementary problem reduces to the following linear programming problem, where each variable $x_i^{(j)}$ measures the usage of j-th segment $[\tau_i^{(j-1)}, \tau_i^{(j)}]$, $j=1,\ldots,K_i$, $i \in Between_k$:

$$\text{Min } C,$$

$$\text{s.t.}$$

$$\sum_{j=1}^{K_i} x_i^{(j)} p_i^{(j)} \le C, \; i \in Between_k,$$

$$\sum_{i \in Before_k} u_i + \sum_{i \in Between_k} \sum_{j=1}^{K_i} x_i^{(j)} \ge A,$$

$$x_i^{(j)} \in [0, \tau_i^{(j+1)} - \tau_i^{(j)}], \; i \in Between_k, \; j=1,\ldots,K_i\text{-}1.$$

Since linear programming problems are polynomially solvable, the following proposition holds.

Proposition 1. *If all processing times $p_i(x_i)$ are convex non-negative increasing piece-wise linear functions, then problem $R|1,GT,cntn|C_{max}$ can be solved in polynomial time.*

If the processing times are proportional to the lot size, i.e. $p_i(x_i)=x_i p_i$ for some given positive scalars $p_1,\ldots,p_m$ then the supplementary linear programming problems are solvable analytically [6] which allows to improve the runtime in this special case as follows.

Proposition 2 [6]. *In the special case where the processing times are proportional to lot sizes, the problem $R|1,GT,cntn|C_{max}$ is solvable in $O(m \log m)$ time.*

3.2. Discrete product

The discrete problem $R|1,GT,dscr|C_{max}$ can be solved by a bisection search w.r.t. C_{max} criterion. The initial lower bound for C_{max} is equal to 0 and the initial upper bound UB_0 for C_{max} is $\max_{i=1,\ldots,m} p_i(u_i)$. In case the processing times are given by polynomial-time computable oracles, the upper bound UB_0 is also upper bounded by $2^{\text{poly}(|I|)}$, where $|I|$ denotes the problem input length and poly($\cdot$) is some polynomial. On each iteration of the bisection algorithm, when a new tentative value of criterion C is chosen by the bisection, in a polynomially bounded time we compute m values

$$q_i(C):=\max\{x_i \in \mathbf{Z}_+ | p_i(x_i) \le C\}, \; i=1,\ldots,m,$$

to find the lot sizes corresponding to C. If some of these lot sizes turn out to be outside the feasible region, they need to be adjusted either by setting to 0 (if $q_i(C) < l_i$) or by setting to u_i (if $q_i(C) > u_i$). After such adjustments, we compute the processing time of an adjusted lot in order to update the bisection search interval. The standard justification of the bisection search algorithms leads to the following proposition.

Proposition 3. *If all processing times $p_i(x_i)$ are non-negative non-decreasing functions given by polynomial-time computable oracles, then problem $R|1,GT,dscr|C_{max}$ can be solved in polynomial time.*

Let us now consider the special case of $R|1,GT,dscr|C_{max}$ where the processing times are proportional to lot sizes.

We will show now that the discrete problem $R|1,GT,dscr|C_{max}$ can be solved in $O(m^2)$ time by an iterative solution of $R|1,GT,cntn|C_{max}$ problem.

Let LB^0 and C^* denote the makespan values of optimal solutions to a pair of identical instances of $R|1,GT,cntn|C_{max}$ and $R|1,GT,dscr|C_{max}$ respectively. Let $\boldsymbol{x}^0$ be an optimal solution to the instance of $R|1,GT,cntn|C_{max}$. Obviously, $LB^0 \le C^*$. At the same time, the vector $\lceil \boldsymbol{x}^0 \rceil := (\lceil x_1^0 \rceil, \ldots, \lceil x_m^0 \rceil)$ is a feasible solution to the discrete instance, therefore the makespan of $\lceil \boldsymbol{x}^0 \rceil$, denoted by UB^0, is not less than C^*.

Without a loss of generality, we assume that in $\lceil \boldsymbol{x}^0 \rceil$ the makespan UB^0 is attained on machine 1. Two alternatives are possible: (A) $C^*=UB^0$ or (B) $LB^0 \le C^* < UB^0$. In the case (A), the vector $\lceil \boldsymbol{x}^0 \rceil$ is an optimal solution.

Consider alternative (B). In an optimal solution to the discrete problem, machine 1 completes at some moment $T_1 \le C^*$. Note that there exists an optimal solution $\boldsymbol{x}^*$ to the instance of $R|1,GT,dscr|C_{max}$, such that $x_1^* \ge \lfloor x_1^0 \rfloor$, since vector $\boldsymbol{x}^0$ yields a lower bound LB^0 to the optimum C^*. In what follows, we will assume that an optimal solution $\boldsymbol{x}^*$ satisfies this condition.

Let us round down x_1^0, fix it, and keep the rest of the components in $\boldsymbol{x}^0$ unchanged for a while. In order to satisfy constraint (3), which may be violated by the modified vector $\boldsymbol{x}^0$, we solve a new supplementary instance of $R|1,GT,cntn|C_{max}$ w.r.t. variables $x_2, \ldots, x_m$, where the right-hand side A is decreased by $\lfloor x_1^0 \rfloor = x_1^*$. Now let LB^1 be an optimal makespan value in the new $R|1,GT,cntn|C_{max}$ subproblem. Note that $LB^1 \le C^*$.

The iterative process continues. On the second iteration we round up the last optimal solution to the supplementary instance of $R|1,GT,cntn|C_{max}$ and denote the rounded solution as $\lceil \boldsymbol{x}^1 \rceil$. Find the corresponding makespan value UB^1. Without loss of generality we assume that UB^1 is attained on machine 2 and continue as we did with machine 1.

The total number of iterations is at most m. Note that in case at some iteration alternative (A) takes place, all subsequent subproblems $R|1,GT,cntn|C_{max}$ may turn out to be infeasible.

An optimal solution to $R|1,GT,dscr|C_{max}$ is now obtained either as a vector consisting of all fixed values of coordinates x_i^* or as one of the vectors $\lceil \boldsymbol{x}^i \rceil$, if at some iteration i the alternative (A) actually took place. It is sufficient to compare the makespan values of all these vectors and find the least of them to identify an optimal solution.

Note that in the algorithm described above, it is sufficient to sort the values D_i just once, which takes $O(m \log m)$ operations. After that, we use the remaining part of the algorithm from Proposition 2 at most m times, which has the overall time complexity $O(m)$. So we have

Proposition 4. *In the special case, where processing times are proportional to lot sizes, the problem* $R|1,GT,dscr|C_{max}$ *is solvable in time* $O(m^2)$.

4. Conclusions and discussion

We have extended the previous results, known for $R|1,GT,\beta|\gamma$ problem with integer input data and processing times proportional to lot sizes, to more general cases where the processing times are given by oracle computable functions and the numeric parameters of the problem are rational numbers. The obtained results are applicable in logistics, production scheduling and scheduling parallel processors, in particular, for the speed-scalable processors.

The further research might be undertaken in analysis of computational complexity and development of efficient exact and approximation algorithms for the case of more than one product and other generalizations of the problem $R|1,GT,\beta|\gamma$. Some technical assumptions made in this paper may be possible to relax. In particular, it is plausible that the exact algorithm proposed for $R|1,GT,cntn|C_{max}$ in Section 3 under the assumption of convex and piece-wise linear processing times may be generalized to a wider class of convex functions, still keeping the runtime polynomially bounded. We also expect that the continuity assumption on functions for processing times, used in Section 2, may be slightly relaxed.

Acknowledgments

The first author is supported by the Program of fundamental scientific research of the State Academies of Sciences, task I.5.1.6. The second author is supported by a joint project of Belarusian Fund for

Fundamental Research and Siberian Branch of Russian Academy of Sciences "Combinatorial models and methods of Scheduling Theory and Their Applications in Decision Support Systems".

References

[1] Austin L M and Hogan W W 1976 Optimizing the procurement of aviation fuels *Management Science* **22** 5 pp 515–527
[2] Chauhan S S, Eremeev A V, Romanova A A and Servakh V V 2007 Approximation solution of the supply management problem *J. of Appl. and Industrial Math.* **1** 4 pp 433–441
[3] Chauhan S S, Eremeev A V, Romanova A A, Servakh V V and Woeginger G J 2005 Approximation of the supply scheduling problem *Oper. Res. Lett.* **33** 3 pp 249–254
[4] Dolgui A, Eremeev A V, Kovalyov M Y and Kuznetsov P M 2010 Multi-product lot sizing and scheduling on unrelated parallel machines *IIE Transactions* **42** 7 pp 514–524
[5] Eremeev A V, Kovalyov M Y and Kuznetsov P M 2008 Approximate solution of the control problem of supplies with many intervals and concave cost functions *Automation and Remote Control* **69** 7 pp 1181–1187
[6] Eremeev A V, Kovalyov M Y and Kuznetsov P M 2016 Scheduling the single-item production on parallel machines with lot size bounds (in Russian) *Proc. 7th Int. Conf. "Tanaev's Readings"* (28–29 March 2016, Minsk) pp 72–76
[7] Gerards M E T, Hurink J L and Holzenspies P K F 2016 A survey of offline algorithms for energy minimization under deadline constraints *J. of Sched.* **19** pp 3–19
[8] Ng C T, Kovalyov M Y and Cheng T C E 2008 An FPTAS for a supply scheduling problem with non-monotone cost functions *Naval Res. Logistics* **55** pp 194–199
[9] Shaik M A, Floudas C A, Kallrath J and Pitz H-J 2009 Production scheduling of a large-scale industrial continuous plant: short-term and medium-term scheduling *Computers and Chem. Engineering* **33** 8 pp 670–686

Structure and characteristics of chromium steel coatings alloyed with boron carbide

E N Eremin[1], A S Losev[1], S A Borodikhin[1], A E Matalasova[1], I A Ponomarev[1], K E Ivlev[2]

[1] Omsk State Technical University, 11, Mira ave., Omsk, 644050, Russia
[2]Omsk Scientific Center of the Siberian Branch of the Russian Academy of Sciences, Omsk, Russia

e-mail: weld_techn@mail.ru

Abstract. This study explores the problems arising from the increase of wear resistance on the coatings of details of a wide range of applications, obtained by surfacing the Fe - Cr system with flux-cored wires. It has shown that insignificant wear resistance of such steel under conditions of metal friction against another metal is due to their relatively low hardness and the absence of strengthening phases. It also shows the effect of boron carbide on the structure and the characteristics of chromium steel obtained by the surfacing process. It was established that the use of high-chromium flux-cored wires alloyed with boron carbide aids the production of a deposited metal of a composite type, with a dispersed hardening based on chromium carboboride. The deposited metal with such structure has a high wear resistance and the hardness of 55 ... 58 HRC and can be used for surfacing cladding the hardening, corrosion-resistant coatings.

1. Introduction

A significant range of parts used in various industries is made from chromium steel, which combines sufficiently high strength with corrosion resistance [1]. Prospectively these can be used as surfacing materials in order to obtain wear-resistance on a wide range of details (parts), thus increasing their service life and providing a significant economic effect. Therefore, the development of these wear-resistant compounds is an urgent task.

2. Statement of a problem

The majority of chromium steel compounds are based on the Fe-Cr system. It serves as a base for the development of solid steel and flux-cored surfacing wires containing 13–17 % chromium [2–4]. At the same time, its wear resistance becomes negligible if friction is present. This happens due to the relatively low steel hardness and the absence of the strengthening phase.

Therefore, the task ahead consists of obtaining the coatings made from chromium steel with a high-hardness strengthening phases.

3. Theory

One of the ways to imrpove the properties of the deposited metal is by adding boride compounds into the dispersion hardening process [5–10]. Alloying the deposited metal with boron carbide opens extra prospects [7, 8, 11, 12]. At the same time, the use of boron carbide in surfacing chromium steel with flux-cored wires has not been sufficiently studied.

Therefore, this study explores the influence of boron carbide on the structure and properties of chromium steel obtained by the surfacing process.

4. Results of the experiments and discussion

As our research object, we chose chromium steel 10X15, obtained by surfacing with a flux-cored wire that contains an additional 2 % of boron carbide.

The surfacing was carried out on St3 steel plates, 200 × 50 × 10 mm in size with experimental flux-cored wires 2,4 mm in diameter in argon in three layers. Surfacing mode: current strength is 230 A; the voltage is 24 V; the speed of surfacing is 20 m/h. The metal was analyzed upon the completion of the surfacing process.

We carried out the metallographic research of the deposited metal on an optical microscope AXIO Observer A1m (Carl Zeiss). The microstructure was detected by chemical etching in a reagent of the composition: $CuSO_4$ – 4 g; HCl – 20 ml; H_2O – 20 ml.

Electron microscopy examination was carried out with a raster-type electron microscope JEOL JSM-6610-LV with add-on device Inca-350 of energy-dispersive analysis (EDA).

Durometric research WAS carried out on the metal samples after surfacing using TK-2 hardness testers by the Rockwell method and Shimadzu HMV-2 using the Vickers method.

The effectiveness of alloying chromium steel with boron carbide is convincingly manifested when comparing the hardness of the deposited metals. The results of the distribution of hardness along the height of the surfaced bead are shown in figure 1.

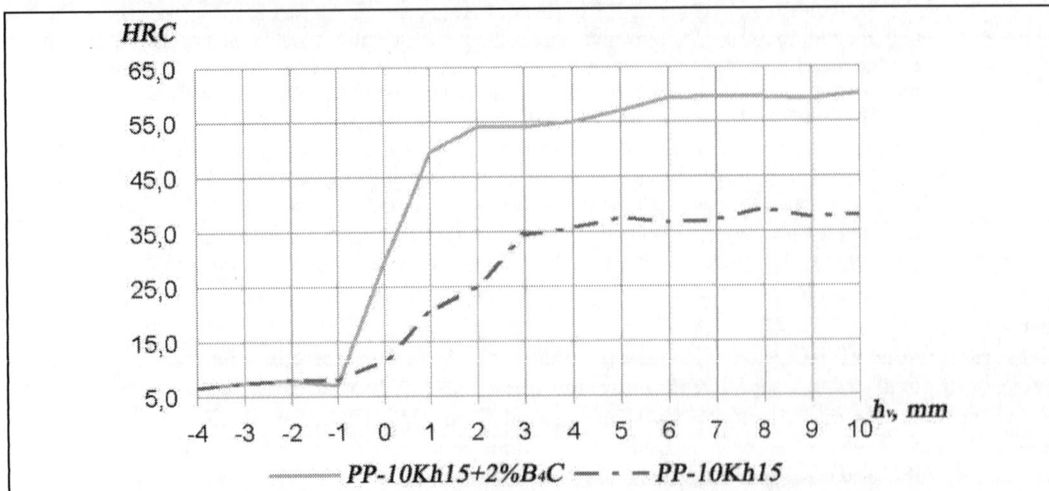

Figure 1. Relationship between the hardness value and the distance from the fusion line (0) during the surfacing process using the compounds in question.

Through the analysis of the collected data, it can be noted that the hardness of the deposited metal is almost equally distributed over the height of the coating. Evidentally, this is due to the invariance of its obtained structure, resulting, in small part, from the main metal. The hardness values of the metal with boron remain stable in the range of 55 ... 58 HRC and significantly exceed the hardness of the metal without boron, which remains in the range of 35 ... 38 HRC.

Metallography research carried out to identify these differences has shown that the deposited metal without boron has a ferritic-martensitic structure (fig. 2, a). In addition, a large amount of δ-ferrite precipitate was observed on the grain boundaries (fig. 2, b).

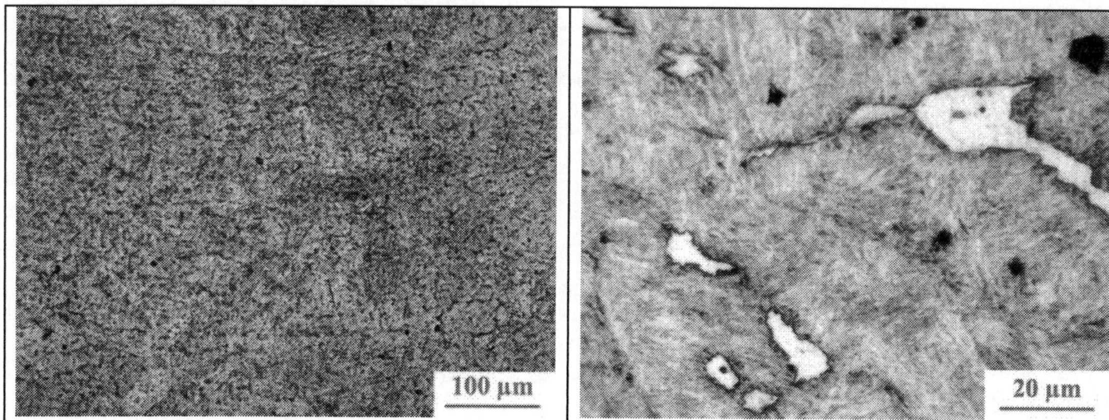

a) b)

Figure 2. The microstructure of the metal surfaced with flux-cored wires type PP-10Kh15.

The analysis of the microhardness structure of the metal without boron is shown in figure 3 and confirms the presence of a ferritic-martensitic matrix, δ-ferrite and carbides. The hardness of the matrix is low and amounts to 408 ... 429 HV. The hardness of δ-ferrite is even less – 358 ... 376 HV, and carbides are 551 ... 609 HV.

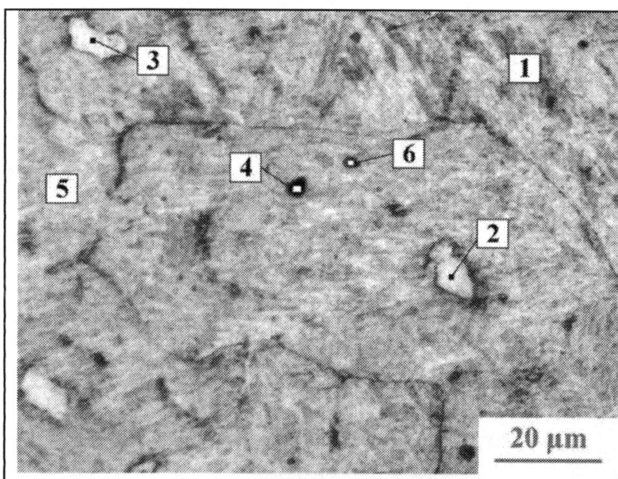

Puncture №	Hardness, $HV_{0,01}^{*}$ and $HV_{0,05}$
1	408
2*	376
3*	358
4*	551
5	429
6*	609

Figure 3. The measurement field and the values of the microhardness of the structural constituents of the metal surfaced with a flux-cored wire type PP-10Kh15.

The results of the qualitative energy dispersive analysis (EDA) of this metal, carried out through the raster-type electron microscopy method, are shown in figure 4.

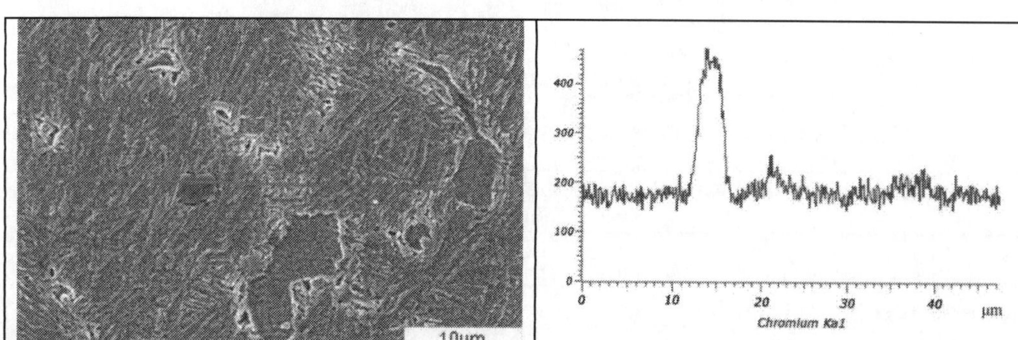

b)

a)

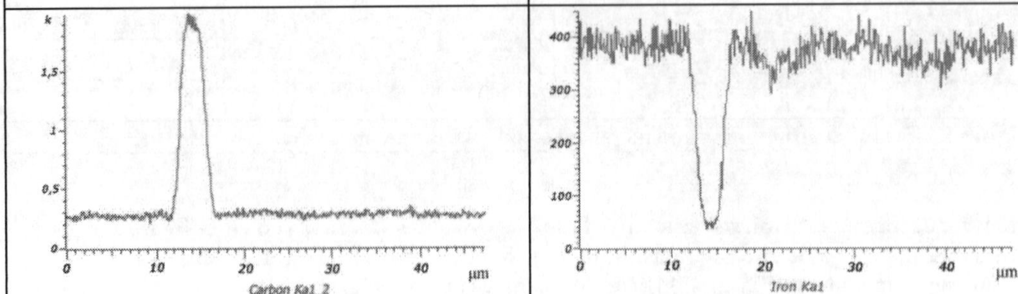

c) d)

Figure 4. The results of scanning the structure of the deposited coating obtained by the flux-cored wire type PP-10Kh15: a) image of a scanned microstructure; b)-d) concentration spectrograms of the distribution of elements along the scanning line.

The obtained data shows that chromium carbides serve as a base for the hardening of this metal.

Adding boron carbide into the deposited metal leads to the formation of a composite structure (fig. 5). It has significant dendritic characteristics. A coarse eutectic is located along the boundary of the dendritic cells, while a large amount of strengthening phases is observed in the martensitic matrix.

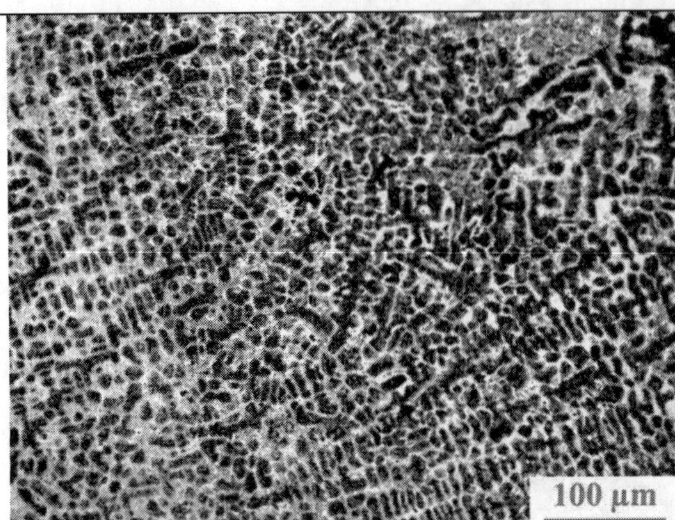

Figure 5. The microstructure of the metal surfaced with flux-cored wire type PP-10Kh15+2%B_4C.

The results of the microhardness research carried out on the structural components of the deposited metal are shown in figure 6.

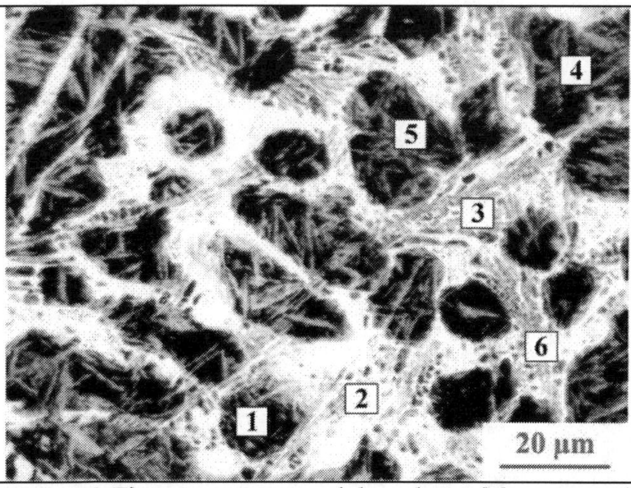

Puncture №	Hardness, $HV_{0,05}$
1	499
2	859
3	924
4	532
5	603
6	891

Figure 6. The measurement and the values of the structural constituents microhardness of the metal surfaced with a flux-cored wire type PP-10Kh15+2%B_4C.

It is established that the hardness of the matrix is in the range of 499…603 HV. A large amount of eutectic is observed. The hardness of the precipitated strengthening phases is significantly increased and is in the range of 859 ... 924 HV. Thus, the microhardness of the deposited metal structure significantly exceeds the microhardness of the constituent metal without borides.

The chemical composition of the areas coated through surfacing with flux-cored wires PP-10Kh15+2%B4C obtained by the EDA method are shown in figure 7.

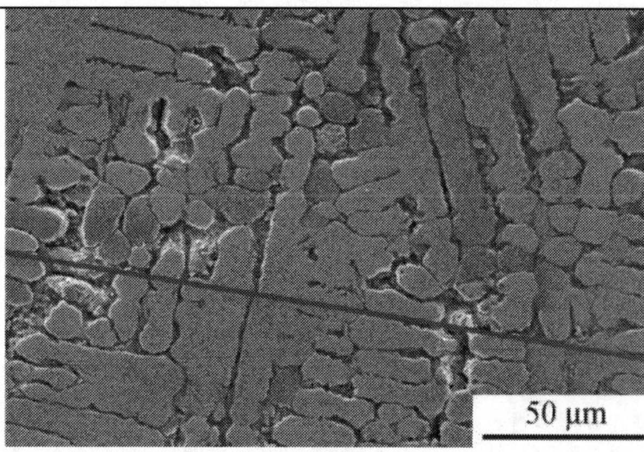

a)

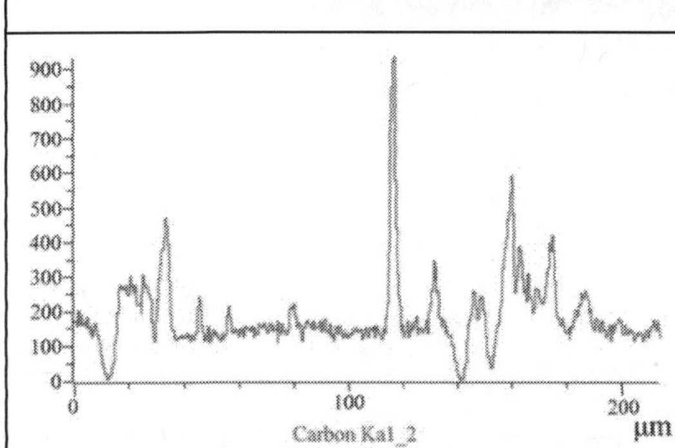

b)

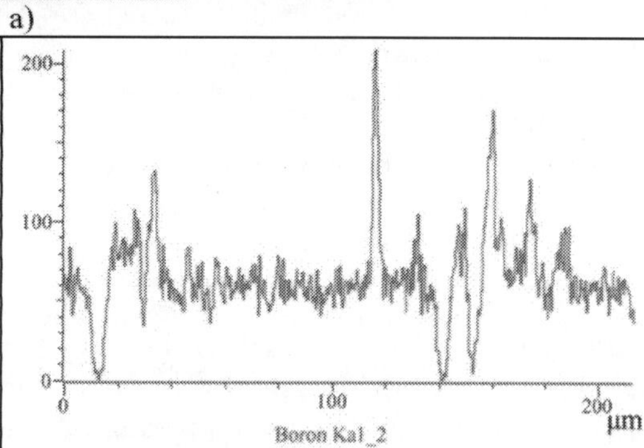

c)

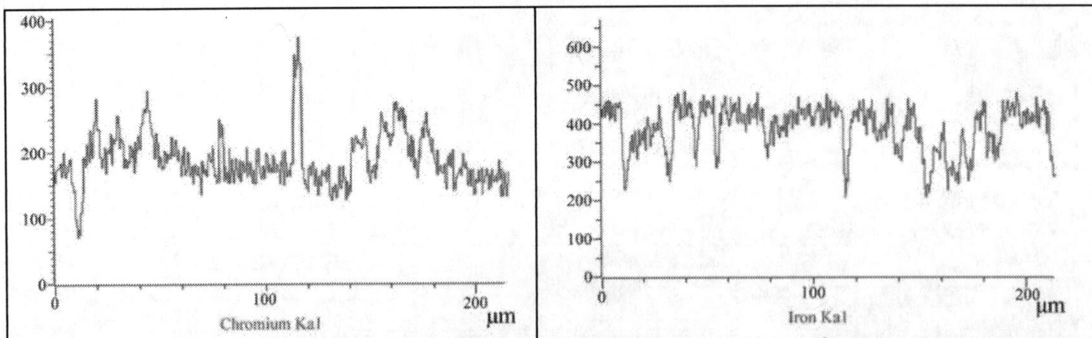

d) e)

Figure 7. The results of scanning the deposited coating structure obtained by the flux-cored wire type PP-10Kh15+2%B_4C: a) image of a scanned microstructure; b)-e) concentration spectrograms of the elements distribution along the scanning line.

As the scanning results show, chromium carboboride acts as the main strengthening phase of this metal. The formation of the hardening phase with high hardness provides the deposited metal with an increased wear resistance.

5. Conclusion

The use of high-chromium flux-cored wires alloyed with boron carbide aids in the production of a composite type deposited metal, with high-hardness chromium carboboride serving as a base for its dispersed hardening. The deposited metal with this structure has a high wearing resistance and can be used to surface strengthening, corrosion-resistant coatings.

Acknowledgments

This research was made possible through the Russian Science Foundation grant (project №17-19-01224).

References

[1] Shlyamnev A P Svistunova T V Lapshina O B Sorokina N A Matorin V I Stolyrov V I Bogolybskii S D Kozlova N N Edneral A F 2008 *Corrosion-resistant, heat-resistant and high-strength steels and alloys* (Moscow: Prommet-splav) p 336

[2] Frumin I I 1979 *The surfacing materials of the CMEA member countries* (Kiev; Moscow: VINITI) p 619

[3] Stepin V S Starchenko E G Volobuev Y S Egorov M Y 2006 *Valve construction* 2 55–56

[4] Yuzvenko Y A, Kirelyuk G A 1975 *Surfacing by flux-cored wire* (Moscow: Mashinostroyeniye) p 45

[5] Lyakishev N P, Pliper Y L, Lappo S I 1986 *Boron-containing steels and alloys* (Moscow: Mashinostroyeniye) p 192

[6] Artemiev A A Sokolov G N DubtsovY N Lysak V I 2011 *Izvestiya vysshikh uchebnikh zavedeniy. Poroshkovaya metallurgiya i funktsionalnyye pokrytiya* 2 pp 44–48

[7] Eremin E N 2013 *Welding International* 27 2 pp 144–146

[8] Eremin E N Losev A S 2014 *Welding International* 28 6 pp 465–468

[9] Zhong L Xiang C Yan-xiang L Kai-hua H 2009 *J. of Iron and Steel Research,International* 16 3 pp 37–42

[10] Raghavan V 2003 *Journal of Phase Equlibria* 24 5 pp 459–460

[11] Dankin A A Svetlopolianskiy V I Caleda V N 1993 *J. Svarochnoe proizvodstvo* 2 pp 8–10

[12] Sheenko I N Gaponov O P 1969 *J. Svarochnoe proizvodstvo* 5 pp 27–28

IOP Conf. Series: Journal of Physics: Conf. Series **944** (2017) 012034 doi:10.1088/1742-6596/944/1/012034

Estimation of resistance to destruction of multi-element coatings

E N Eremin[1], V M Yurov[2], S A Guchenko[2] and V Ch Laurynas[2]

[1]Omsk State Technical University, 11, Mira Pr., Omsk 644050, Russia
[2]Karaganda State University named after E. A. Buketov, 28, Universitetskaya str., Karaganda 100028, Republic of Kazakhstan

e-mail: weld_techn@mail.ru

Abstract. It is proposed to experimentally determine the work of destruction of a coating by their surface tension. The surface tension of the coating is determined from the dimensional dependence of its microhardness. Comparison of the destruction of multi-element coatings with nitride coatings shows that the use of multi-element coatings is very promising for machine building.

1. Introduction

Already in the middle of the last century it became clear that it was necessary to go beyond the path of creating expensive special alloys for the needs of many industries, but to develop coating technologies for already known steels and alloys that had the necessary technological properties. This is due to the fact that 90% of the details of mechanisms and machines fail due to surface wear.

There is a large flow of publications on this subject, of which we mention only the works [1–6], where a more extensive bibliography is given.

In recent years, the concept of highly entropic or multielement alloys and coatings based on them has been developed [7–11].

Stability of structure and composition, as well as high performance characteristics of high-entropic systems, create a very attractive possibility of forming coatings on their basis in order to improve the surface characteristics or to use them as protective films preventing harmful impurities from entering the near-surface layers.

2. Formulation of the problem

Surface wear of machine parts refers to surface phenomena and is the subject of numerous studies. These studies led to the emergence of a new direction in materials science - surface engineering [12].

Most of the existing methods for assessing the destruction and wear of materials, including coatings, are based either on a large number of statistical experimental data [13] or on complex mathematical calculations [14].

Therefore, a simple quantitative assessment of the energy of the destruction of coatings will be very useful for testing the technology of their application.

3. Theory

The work required to destroy the coating is:

$$W = \sigma \cdot S, \tag{1}$$

where σ is the surface tension of the coating, and S is the coating area.
The method of experimental determination of the surface tension of coatings was proposed by us in [15].
The dependence of the microhardness of the deposited coating on its thickness is described by the formula:

$$\mu = \mu_0 \cdot \left(1 - \frac{d}{h}\right), \tag{2}$$

Where μ is the microhardness of the deposited coating; μ_0 is a "thick" sample; H is the thickness of the deposited coating.
The parameter d is related to the surface tension σ by the formula:

$$d = \frac{2\sigma \upsilon}{RT}. \tag{3}$$

Here σ is the surface tension of a massive sample; υ – the volume of one mole; R is the gas constant; T is the temperature.
In the coordinates μ – 1/h (1/h – the inverse thickness of the deposited coating), a straight line is obtained, the slope of the slope which determines d, and the surface tension of the deposited coating (σ) is calculated from (3).
As an example, let us consider the determination of the surface tension of titanium reinforcing coatings nitride on steel.
The results are shown in figures 1. In the coordinates $\mu / \mu_0 \sim 1 / h$, the experimental curve is rectified in accordance with (2), giving a value of h = 1.3 μm. For titanium nitride υ = 11.44 cm^3 / mol and from the ratio (3) for surface tension, we obtained: σ = 0.474 J / m^2.

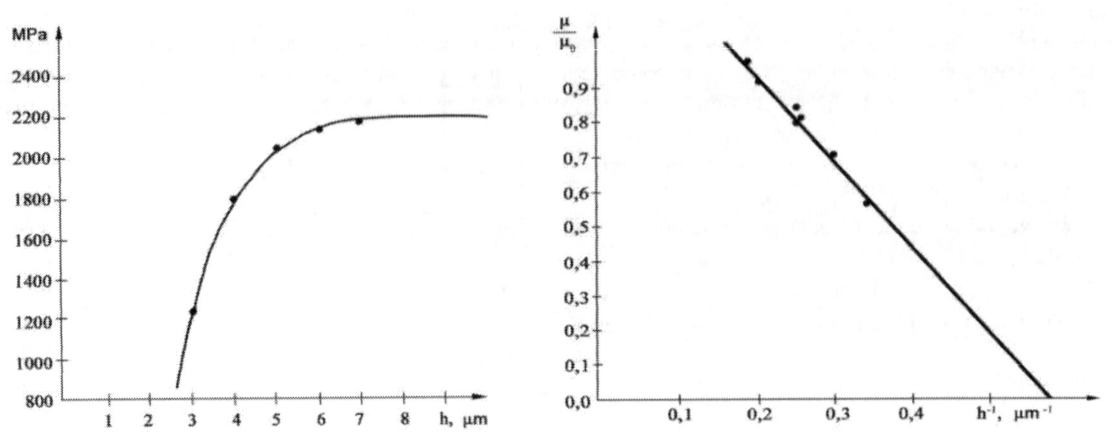

a b

Figure 1. Dependence of microhardness on thickness (a) and inverse thickness (b) titanium nitride coating on steel X12.

4. Experimental results

For the deposition of coatings, composite cathodes and cathodes made of X12 steel were used simultaneously. With the help of these cathodes, coatings were applied on the NNV-6.6П unit on a steel substrate of steel X12 in argon gas for 40 minutes at an arc current Iu = 80 A, a reference voltage U = 200 V, and a gas pressure in the chamber $P = 5 \times 10^{-5}$ Pat.
The surface tension of the coatings was determined according to the procedure described above. The work of destruction of the coating with an area of 1 m^2 was calculated from formula (1). The results are shown

in table 1.

Table 1. Work of destruction of multielement coatings obtained in argon.

Coating	W, J	Coating	W, J
12X18H10T+Zr	0,95	12X18H10T+Zn-Al	1,08
12X18H10T+Zn-Cu-Al	1,07	12X18H10T+Al	1,12
12X18H10T+Fe-Al	1,27	12X18H10T+Cu	1,42

5. Discussion of the results

Let's compare the results with nitride coatings, which are widely used in various fields of engineering. The results are shown in table 2.

Table 2. Work of destruction of nitride coatings.

Nitride	W, J	Nitride	W, J
TiN	0.474	NbN	0.670
ZrN	0.518	TaN	0.735
HfN	0.610		

Comparison of table 1 and table 2 shows that it is much more difficult to destroy a multielement coating than a coating of metal nitrides.

High-entropy alloys and coatings, as a rule, have a single-phase structure [13]. In this case, work on the destruction of coatings can be estimated without using the experimental value of surface tension, according to the formula [16]:

$$\sigma = 0{,}7 \cdot (X_1 T_{m1} + X_2 T_{m2} + \cdots + X_n T_{mn}), \tag{4}$$

where T_m is the melting point of the i-th component of the coating. Xi is the percentage of the i-th component.

6. Conclusion

The results presented in the work show that the work of the coating destruction can be estimated experimentally by its surface tension. Knowing the composition of the coating and the melting point of its components, the work of destruction can also be estimated from formula (4). Summarizing, we can say that high-entropy coatings should have high performance characteristics.

7. References

[1] Polcar T, Parreira N M G and Novak R 2007 *Surface and Coatings Technology* vol 201 pp 5228–5234
[2] Choi E Y, Kang V C, Kwon D H, Shin D W and Kim K H 2007 *J. Mater. Process. Tech. Technol.* vol 187–188 pp 566–572
[3] Silva P N, Dias J P and Cavaleiro A 2005 *Surface and Coatings Technology* vol 200 pp 186–191
[4] Cozza R C 2013 *Surface and Coatings Technology* vol 215 pp 224–233
[5] Sundberg J, Nyberg H, Sarhammar E et al. 2014 *Surface and Coatings Technology* vol 258 pp 86–94
[6] Lemm J D, Warmuth A R, Pearson S R and Shipway P H 2015 *Tribology International* vol 81 pp 258–266
[7] Otto F, Yang Y, Bei H, George E P 2013 *Acta Materialia* vol 61 pp 2628–2638
[8] Tsai M H, Yeh J W 2014 *Mater. Res. Lett.* vol 2 p 107-123
[9] Schuh B, Mendez-Martin F, Vulker B et al. 2015 *Acta Materialia* vol 96 pp 258-268

[10] Pogrebnyak A D, Baghdasaryan A A, Yakushchenko I V and Beresnev V M 2014 *Uspekhi Khimii* vol 83 11 pp 1027–1061
[11] Komarov F F, Pogrebnyak A D and Konstantinov S V 2015 *Journal of Technical Physics* vol 85 pp 106–111
[12] Sidorenko S I, Pashchenko V N and Kuznetsov V D 2001 *Material science basics of surface engineering* (Kiev, Naukova Dumka)
[13] Gong G S and Chukin M V 2006 *Optimization of processes of technological and operational deformation of products with coatings* (Magnitogorsk, MSTU)
[14] Botvina L R 2008 *Destruction: kinetics, mechanisms, general patterns* (Moscow, Nauka)
[15] Yurov V M, Laurinas V Ch, Guchenko S A and Zavatskaya O N 2014 *Strengthening technologies and coatings* 1 pp 33–36
[16] Yurov V M 2011 *Eurasian Physical Technical journal* vol 8 № 1(15) pp 10–14

Acknowledgments
The work was carried out at the expense of a grant from the Russian Science Foundation (№17-19-01224).

IOP Conf. Series: Journal of Physics: Conf. Series **944** (2017) 012035 doi:10.1088/1742-6596/944/1/012035

Research of AISI 321 steel microrelief after ultrasonic impact treatment with marker applying

A A Fedorov, D A Polonyankin, A I Blesman and D V Postnikov

Omsk State Technical University, 11 Mira Avenue, Omsk, 644050, Russia

E–mail: fedot83@gmail.com

Abstract. The paper consideres the influence of AISI 321 steel microstructure formed by ultrasonic impact treatment on the microrelief regularity of preliminary prepared by turning (Ra=2.15 μm) and grinding (Ra=0.36 μm) austenitic stainless steel samples. The developed technique application for exploration the microstructure influence on the microrelief local irregularity of AISI 321 steel made it possible to establish the absence of a significant effect of twins and grain boundaries on the regular microrelief formation. The most probable reasons for irregular microrelief domain's formation could be considered the design features of dynamic technological module and the quality of initial surface preparation before the ultrasonic impact treatment.

1. Introduction

Ultrasonic impact treatment (UIT) is one of the effective method for strength and wear resistance increasing of heavily loaded friction surfaces by the regular microrelief formation on the treated surface (surface's texturing by micro–dimples), nanocrystallization of the surface layer and generation of residual compressive stresses that favorably influence on the fatigue strength.

It is clear from references [1–7, 9, 10] that UIT improves the physical–mechanical and tribological characteristics as well as the fatigue strength of various parts and friction units due to the presence of a regular microrelief consisting of bulges and valleys (micro–dimples). Synonyms for «regular microrelief» definition used in the scientific literature are the «surfaces' textured by micro–dimples» or the «nano–corrugated surfaces».

An essential problem of UIT is obtaining completely regular microrelief such as the dimensions of all dimples are the same both in diameter and in depth and are located at equal distances. Probably such fully regular microrelief makes better tribotechnical characteristics compared to partially regular (not completely regular). The purpose of this exploration is to investigate the influence of AISI 321 stainless steel microstructure on the irregular microrelief formation during ultrasonic impact treatment.

2. Experimental details

Currently researchers [1–9] have complicacies in obtaining completely regular microrelief formed by ultrasonic impact treatment similar to our research team (see figure 1).

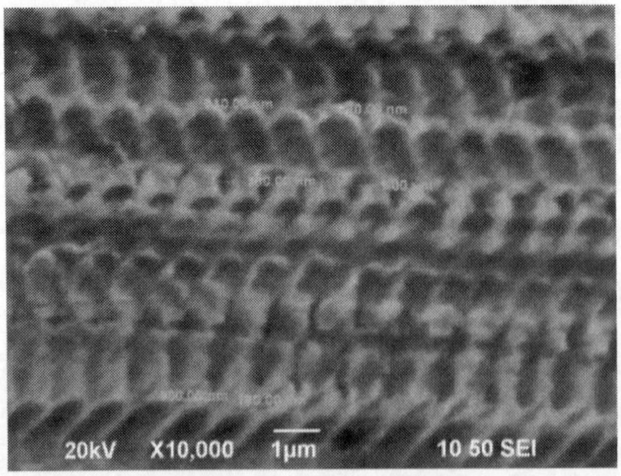

Figure 1. Partially regular microrelief formed by UIT, SEM, × 10000.

It can be seen from figure 1 that the dimples (valleys) have various shapes, sizes and deepness. The bulges which separate the dimples also have different sizes and tilt angles relative to the horizontal line although the treatment of this surface was carried out under invariable modes. In our opinion a range of factors influence the variety of the surface microrelief patterns after UIT the main of which are:

1) The rigidity of the dynamic technological module construction which produces impacts on the surface with ultrasonic frequency.

2) Quality of the initial surface preparation before UIT (turned, ground, polished).

3) Physical–mechanical characteristics of the processed material (microstructure, hardness, stress state).

As it was noted above the influence of AISI 321 stainless steel microstructure on the irregular microrelief formation during ultrasonic impact treatment was investigated. Turned (Ra=2.15 μm) and ground (Ra=0.36 μm) samples in the form of rollers with 25.5 mm of diameter from corrosion–resistant high–alloyed austenitic AISI 321 steel were prepared for exploration. The content of the main AISI 321 steel elements is: C=0.12%; Cr=18.64%; Ni=10.1%; Ti=0.54%.

Ultrasonic impact treatment of the samples was produced by a spherical indenter made of a tungsten carbide (VK8) with a sphere radius of 3 mm. To generate impacts on the samples' surfaces was used the dynamic technological module DTM–07 constructed on the basis of the magnetostrictive transducer PMS15A–18. Processing modes: ultrasonic tool oscillation frequency f=18 kHz, amplitude ξm=50 μm, static load Pst=20 N, relative travel velocity of the ultrasonic tool and sample v=2 m/min, longitudinal feed of the tool s=0.05 mm/rev.

To assess the microstructure influence on the local irregularity of microrelief it occurs a necessity of a special label (marker) applying to the sample surface nearby to explored area for convenience of searching it after removing microrelief by polishing. The most common method of marking is ion milling by a focused ion beam which is often embedded to a scanning electron microscope. Wire electrical discharge machining (WEDM) could be used for marker applying as an alternative of an expensive ion milling method (see figure 2).

The developed technique for exploration the microstructure influence on the local irregularity of microrelief consists of the following procedures:

1. Marker's applying of the by WEDM.

2. Sample's examination by scanning electron microscopy (the stage is set in such way that the edge of the sample is oriented horizontally (see figure 3a). The vertex of the marker semicircle is taken as the origin of the coordinate system).

3. Taking a micrograph of microrelief irregularity near the origin of the coordinate system by SEM.

4. Fixing on the XOY coordinate system with following calculation of the irregular microrelief area location using a scale ruler (see figure 3b).
5. Sample's mounting.
6. Grinding of a plane parallel to the cross section plane on a grinding machine.
7. Manual polishing of the sample plane until the removal of microrelief (herewith the thickness of the removable layer is controlled by a micrometer).
8. Sample's etching.
9. Contemplation of the previously selected domain's microstructure by light microscopy or SEM in order to establish the reasons of microrelief irregularity.

Figure 2. Markers applied by WEDM.

Morphology exploration of the turned and ground samples' subjected to UIT was carried out by JEOL JCM–5700 scanning electron microscope in high vacuum mode. Signal type was secondary electrons (SEI). *SpotSize* parameter was 50–60, accelerating voltage value was 20 kV, magnification ranged from 50 to 10 000×.

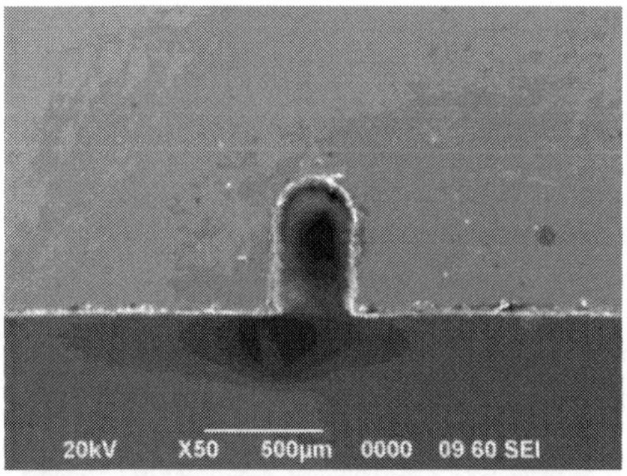

Figure 3a. Marker applied by WEDM, SEM: Alignment of the sample's edge.

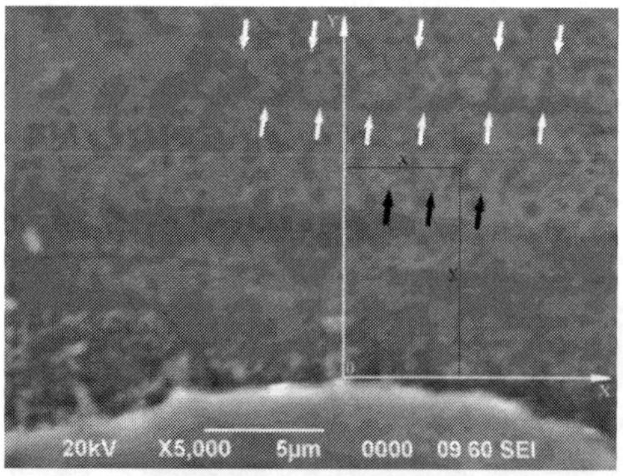

Figure 3b. Position of the coordinate system and areas with regular (indicated by white arrows) and irregular (indicated by black arrows) microrelief.

3. Results and discussion

Analysis of the AISI 321 steel samples' microstructure showed that the grain size varies in the range from 20 to 40 µm. Comparison of the distance between the dimples with the grain size (see figure 4) allows us to conclude that the ultrasonic tool during a single passage of any grain impacts it from 20 to 40 times forming a regular microrelief consisting of bulges and valleys (micro–dimples).

It can be seen from figure 4 that each dimple with serial number from 20 to 40 may occur exactly on the grain boundary or on the twin. The contact of the ball center attached to the ultrasonic tool and impacting the sample can probably lead to deviations from the regularity of the micro–dimples location because of the material deformability in the grain boundary area may differ from the material deformability at the grain center. Usage the developed technique for exploration samples subjected UIT and containing areas with a violation of microrelief regularity (see figure 5) allows us to conclude that grain boundaries and twins do not significantly affect the microrelief regularity of previously turned and ground surfaces.

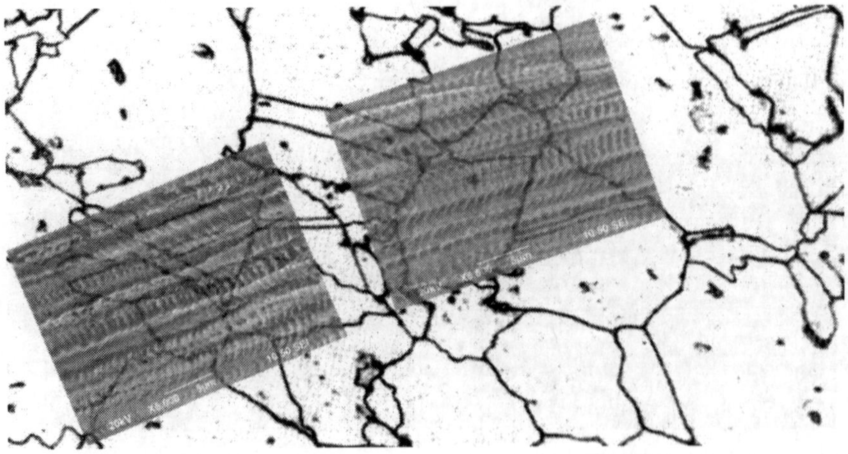

Figure 4. Comparison of the grain size with the distances between the dimples (the figure was obtained by superimposing the SEM micrographs with increased transparency on the microstructure photo in the image editor).

Areas with a significant accumulation of grain boundaries is characterized with a high degree of the microrelief regularity and conversely in some cases series of major grains with size ranges from 35 to 40 μm with a small number of boundaries and twins were located under the areas with an apparent violation of the microrelief regularity (see figures 5a, 5b).

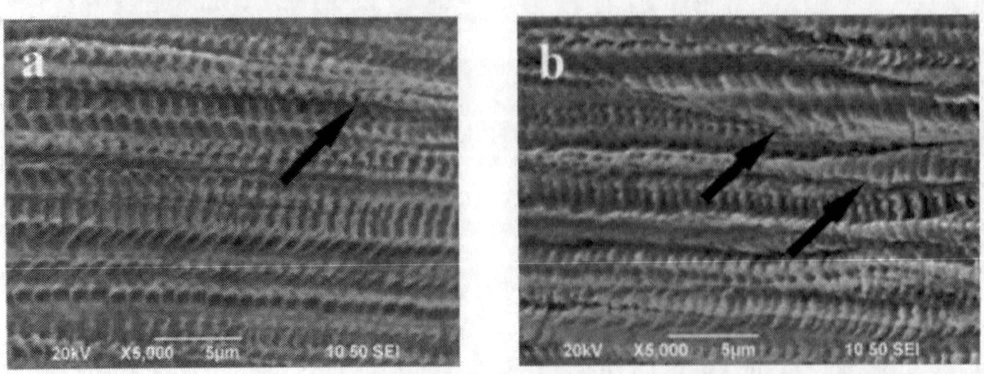

Figure 5. SEM micrograph of sample subjected to UIT. Black arrows indicate areas with an apparent violation of the microrelief regularity.

It should be noted that previously ground samples contain a fewer number of areas with a violation of the microrelief regularity in comparison with samples subjected to preliminary turning. Preliminary preparation of the surface before UIT was earlier considered by us [11].

4. Conclusion

This work has shown that the grain boundaries and twins in austenitic AISI 321 steel do not have a significant effect on the regular microrelief formation. The high surface concentration of grain boundaries and twins does not contribute to the areas formation with an irregular arrangement of micro–dimples. Thus the most probable reasons of the areas formation with irregular microrelief include the design features of the dynamic technological module that impacts on the surface with ultrasonic frequency as well as the quality of the preliminary surface preparation before UIT. Surface grinding as a method of the surface preliminary preparation has a relatively higher influence on the degree of AISI 321 steel microrelief regularity after UIT compared to turning.

5. References

[1] Gujba A K, Ren Z, Dong Y, Ye C and Medraj M 2016 *Surf. Coat. Technol.* **307** p 157
[2] Amanov A, Cho I S, Pyoun Y S, Lee C S and Park I–G 2012 *Wear* **286–287** p 136
[3] Amanov A, Pyoun Y S, Cho I S, Lee C S and Park I G 2011 *J. Nanosci. Nanotechno* **11(1)** p 701
[4] Amanov A, Sasaki S and Young–SikPyun 2013 *Proc. Eng.* **68** p 491
[5] Cherif A, Pyoun Y S and Scholtes B 2009 *J. Mater. Eng. Perform.* **19(2)** p 282
[6] Cho I S, Amanov A, Kwak D H, Jeong B J and Park I G 2015 *Mater. Design* **65** p 401
[7] Lesyk D A, Martinez S, Dzhemelinskyy V V, Lamikiz A, Mordyuk B N and Prokopenko G I 2015 *Surf. Coat. Technol.* **278** p 108
[8] Panin A V, Kazachenok M S, Kozelskaya A I, Hairullin R R and Sinyakova E A 2015 *Mat. Sci. Eng. A. Struct.* **647** p 43
[9] Pyun Y S and Kayumov R 2012 *Int. J. Mod. Phys. Conf. Ser.* **06** p 527
[10] Wu B, Zhang J, Zhang L, Pyoun Y–S and Murakami R. 2014 *Appl. Surf. Sci.* **321** p 318
[11] Fedorov A A, Polonyankin D A, Blesman A I, Postnikov D V, Linovsky A V and Bobkov N V 2017 *AIP Conf. Proc.* **1876** 020065

Determination of Calorific Ability of Fuel Briquettes on the Basis of Oil and Oil Slimes

O A Fedyaeva, E G Poshelyuzhnaya, E M Rakhmatulina, VA Zakharov and T E Fisenko
Omsk state technical university, 11, Mira ave., Omsk, 644050, Russia

kosatine@mail.ru

Abstract. Utilization and neutralization of oil slimes is one of important environmental problems of the oil-extracting, oil-processing and petrochemical industry. The easiest and economic way of utilization of oil slimes is their use as a part of the bricketed boiler fuel. In this work the highest calorific ability of crude oil, the oil slimes and fuel briquettes made on their basis is defined. A research problem was carrying out the technical analysis of oil fuels on the content in them analytical moisture, the cindery rest and volatiles. It is established that in comparison with oil slimes crude oil possesses bigger highest calorific ability, has smaller humidity and an ash-content. The highest calorific abilities of the boiler briquettes made of samples of crude oil, oil slimes and peat made 14 – 26 MJ/kg.

1. Introduction

Environmental pollution by oil and oil products which have negative effect on the soil, a surface water and the geological environment [1 - 4] became one of the most important environmental problems of our country and the whole world. World and domestic experience of development of oil fields showed that modern technologies of oil production don't exclude its flood. It is counted that about 16.5 % of oil are lost at its production, preparation, transportation and processing. Now at the enterprises of the oil-extracting, oil-processing and petrochemical industry about 4.5 million tons of oil slimes [5, 6] are saved up. Oil slimes represent the liquid, pastelike or solid waste consisting of oil or oil products, water and firm particles (sand, clay, oxides of metals) [7]. They are formed during the drilling of wells, preparation of oil for transportation, in system of reverse water supply, at sewage treatment, repair of the equipment and cleaning of tanks. Oil slimes classify on three main groups: soil, benthonic and reservoir. Soil slimes are formed at hit of oil products on the soil, benthonic – at subsidence of oil on a bottom of reservoirs, reservoir – at storage and transportation of oil products in capacities. Crude oil contains soil particles, water, salts and gases. These impurity cause corrosion of the equipment therefore clear oil before transportation. Water after removal from oil is pumped in oil layer, and mechanical impurity form soil oil slime with oil.

Reservoir oil slimes are formed as a result of physical and chemical interaction of oil products with water, air oxygen, mechanical impurity and walls of the tank. As showed researches of many authors [8, 9], the composition of reservoir oil slimes changes over a wide range: hydrocarbons – 5-90 %, water – 1-52 %, solid impurity – 0.8-65 %. Respectively density (830 - 1700 kg/m^3), hardening temperature (270 – 353 K) and flash temperature (308 – 393 K) oil slimes change. Now in Russia there are in operation more than 40 thousand horizontal and vertical cylindrical tanks of various capacity for storage of oil and oil products. At long storage reservoir oil slimes are divided into the separate layers having various properties. The top layer represents the flooded oil product with the content of fine mechanical impurity to 5 %. It belong to the class of emulsions "water in oil". 70-80 %

of oils, 6-25 % of asfalten, 7-20 % of pitches, 1-4 % of paraffin, 5-8 % of water are its part. The organic part of the top layer of oil slime on properties and structure is close to the initial oil product which is stored in tanks. The center contains 70-80 % of water and 1.5-15 % of mechanical impurity. It represents an emulsion like "oil in water". The following layer completely consists of water solution of salts. The benthonic layer represents the firm phase consisting of organic chemistry – to 45 %, water – to 25 %, solid mechanical impurity and a rust – 52-88 %.

There are some directions of use of oil slimes. At dehydration and drying of this waste their return to production for the subsequent processing according to the existing schemes in target products is possible [10]. Perhaps also use of oil slime in drilling as an additive to boring solutions, in production of construction materials (a front tile, concrete, a brick, expanded clay, waterproofing and heat-insulating materials), road construction, the fuel industry (receiving fuel oil, the bricketed solid fuel materials), an igneous metallurgy (extraction of V, Li, B, Sr, Sc, Ge, etc.) [11, 12].

Here is an opinion that burning of oil slimes is ecologically dirty way of their utilization. However, experimental data according to the contents benzo(a)pyrene in emissions of various enterprises of Kurgan showed that in emissions of the boiler rooms working at natural gas benzo(a)pyrene it is more, than in the boiler rooms using oil and mix of coal with gas [13]. So, emissions benzo(a)pyrene of the boiler rooms working at gas make 1341 ng/m^3, working at gas and coal – 594.7 ng/m^3, working at oil – 718.9 ng/m^3, working at fuel oil – 1219.1 ng/m^3, burning coal – 2633.1 ng/m^3. On the basis of these data it is possible to conclude that use of oil slimes as an additive in the bricketed boiler fuel is ecologically proved. Oil slimes can be applied in mixes with peat, coal dust, sawdust or other cheap combustible substances. Their contents in briquettes shouldn't exceed 30-40 % of masses. As binding material it is possible to use a lignin – lignosulfonate, 3,4- polyisoprene, epoxy, etc.

In the real work for production of fuel briquettes peat was used. It well adsorbs the polynuclear, unsaturated and aromatic hydrocarbons containing in oil slimes and possesses high calorific ability.

2. Problem Definition

Determination of the highest calorific ability of the fuel briquettes made of mixes of oil fuels and peat was the purpose of our work. A research problem was carrying out the technical analysis of samples of crude oil, soil and reservoir oil slimes on the contents in them analytical moisture, the cindery rest and volatiles. Samples of oil fuels were taken on the Pioneer field of the Tyumen region.

3. Theory

The highest calorific ability of samples of oil fuels was counted according to the technical analysis with use of a formula of Gutal:

$$Q_B = 82 \cdot K + \alpha \cdot V_G ,$$

where K is the maintenance of the nonvolatile ashless rest, %;
V_G is the content of volatiles in relation to the combustible mass of fuel, %; α is the coefficient depending an exit of volatiles.

The technical analysis of fuel included definition in tests of analytical moisture on a method Dyne and Starck, the content of ashes and an exit of volatiles at long and short-term heating of tests at 1073±25 K respectively [14].

The content of analytical moisture in fuel test (in %) was calculated on a formula:

$$W_a = \frac{V_0}{V} \cdot 100\ \% ,$$

where V is the volume of test of fuel, cm^3; V_0 is water volume in the receiver trap, cm^3.

The content of volatiles in relation to the combustible mass of fuel (in %) was determined by a formula:

$$V_G = V_A \cdot \frac{100}{100-W_a-A_a} ,$$

where V_A is the content of volatiles in relation to analytical test of fuel, %; A_a is the maintenance of the cindery rest in test of fuel, %.

IOP Conf. Series: Journal of Physics: Conf. Series **944** (2017) 012036 doi:10.1088/1742-6596/944/1/012036

V_A values and A_a were counted with use of formulas:

$$V_A = \frac{G_2}{G_1} \cdot 100 - W_A \,;$$

$$A_a = \frac{G_2}{G_1} \cdot 100\,\% \,,$$

where G_1 is the mass of a hinge plate of fuel; G_2 is the mass of ashes.
The maintenance of the nonvolatile ashless rest in fuel test (in %) was found from expression:

$$K = 100 - (V_G + A_a).$$

4. Results of Experiments
Results of the technical analysis of oil fuels are presented in tab. 1

Table 1. Results of the technical analysis of oil fuels.

Oil fuel	Crude oil	Reservoir oil slime	Soil oil slime
Content of analytical moisture W_a, %	0.84	8.40	2.44
The content of volatiles in relation to analytical test of fuel, V_A, %	42.19	58.55	27.56
The content of volatiles in relation to the combustible mass of fuel V_G, %	42.90	70.38	29.90
The maintenance of the nonvolatile ashless rest K, %	56.98	33.04	67.04
Maintenance of cindery rest A_a, %	0.83	8.41	5.40
Coefficient α	76	50	93
The highest calorific ability Q_B, MJ/kg	33.19	26.06	23.27

The highest calorific abilities of the oil fuels and fuel briquettes made of samples of oil fuel and peat are presented in tab. 2.

Table 2. The highest calorific ability of oil fuels and fuel briquettes.

The highest calorific ability Q_B, MJ/kg		
Crude oil	Reservoir oil slime	Soil oil slime
33.19	26.06	23.27
Mix: 30 % of oil fuel + 70 % of peat		
16.96 – 26.76	14.82 – 24.62	13.98 – 23.78

5. Discussion of results
The comparative analysis of oil fuels (tab. 1) showed that reservoir oil slimes contain the greatest number of water (8.40 %) and the cindery rest (8.41 %). These slimes have the highest exit of volatiles (70.38 %). This indicator for soil oil slimes and crude oil makes 29.90 and 42.90 % respectively. Unlike oil slimes, crude oil possesses the greatest calorific ability – 33.19 MJ/kg, a smaller ash-content and humidity. Considering that availability of water in oil slimes and crude oil reduces formation of soot at their burning, the studied oil fuels can be used for production of the bricketed boiler fuel. As a filler of briquettes peat as it well adsorbs organic substances was chosen and possesses high calorific ability (10-24 MJ/kg). In the tab. 2 results of determination of the highest calorific ability of the boiler briquettes made of samples of oil fuel and peat are presented (70 % of mass of peat + 30 % of mass of oil fuel). The received values of the highest calorific ability of briquettes made 14 – 26 MJ/kg.
Thus, sharing of peat and oil fuels in the form of briquettes is a perspective and economic way of utilization of oil slimes of various type as slurry moisture replaces the water vapor added to a burning zone at smokeless burning of fuel.

6. Conclusion

With use of data of the technical analysis on Gutal's formula the highest calorific ability of oil fuels is defined. For crude oil it made 33.19 MJ/kg, for oil slimes – 23-26 MJ/kg. Calorific abilities of the boiler briquettes made of oil fuels and peat made 14 – 26 MJ/kg.

Reference

[1] Ermashova N. A., Ognetova M.P., Lushnikov S.V., Volkov V. M. Research of influence of oil storage on pollution of the geological environment//*Ecology and industry of Russia*, 2004, № 12. pp. 32 - 36.

[2] Soromotin A.V. Oil pollution of lands in a zone of an average taiga of Western Siberia//*Ecology and the industry of Russia*, 2004. № 8. pp. 8 - 11.

[3] Bachurina B. A., Odintsovo T.A. Problems of diagnostics and control of oil pollution of natural geosystems//*Geology, geophysics and development of oil and gas fields*, 2005. № 9-10. pp. 79 - 84.

[4] Haustov A.P., Rehden M. M. Environmental problems of an assessment of formation of oil slimes at accidents//*the Ecological bulletin of Russia*, 2011. № 7. pp. 34 - 39.

[5] Mazlova E.A., Menshikova I.A. Slurry waste of the oil and gas companies// *Environment protection in an oil and gas complex*, 2010. № 1. pp. 20 - 23.

[6] Bokovikova T.N., Shperber E.R., Trukhan E.A., Guseva N. I., Shperber D. R. Classification and system of the address with waste of primary oil refining//*Oil processing and petrochemistry*, 2013. № 11. pp. 36 - 40.

[7] Mazlova E.A., Meshcheryakov S. V. Ecological characteristics of oil slimes//*Chemistry and technology of fuels and oils,* 1999. № 1. pp. 40 - 42.

[8] Weismann Ya.I., Glushankova I.S., Rudakova L.V., M. S Clerks. Researches of physical and chemical properties and thermal destruction of waste of the oil processing enterprises//*Scientific researches and innovations*, 2010. Vol. 4. №. 3. pp. 21 – 27.

[9] Ibatullin R. R., Mutin I.I. Research of properties of oil slimes and ways of their utilization//*Oil economy*, 2006. № 11. pp. 116 – 118.

[10] Fedyaeva O. A. *Industrial ecology*: abstract of lectures. Omsk: Publishing house of OmGTU, 2007. 144 p.

[11] Minigazimov N. S., Rasvetalov V.A., Minigazimov I.N., Tarraf A. *Tekhnika and technology of utilization of oil waste*. Ufa "Gil", 2010. 316 p.

[12] Barakhnina V. B. Use of oil slime as a secondary material and raw resource//*Ecological bulletin of Russia*, 2011. № 9. pp. 16 - 21.

[13] Nevolina Z.A. Pollution of atmospheric air Kurgan benzo(a)pyrene/ In book: *All-Russian conference on problems of resistant organic pollutants* (M., on October 28-29, 2009): Center of the international projects. pp. 109-113.

[14] Belyanin B. V., Erich W. N., Korsakov V. G. *Tekhnichesky analysis of oil products and gas*. L: Chemistry, 1986. 224 p.

IOP Conf. Series: Journal of Physics: Conf. Series **944** (2017) 012037 doi:10.1088/1742-6596/944/1/012037

Control system high-precision laser to obtain the ensemble of ultracold ions Th3+

V V Florentsev[1], V Yu Zhdamirov[1], I I Rodko[1], N A Borodulya[1], A P Biryukov[2]

[1]SPA CADETECH JSC, 3, building 2/4, office 24, Kashirskoe shosse, Moscow, 115230, Russia
[2]National Research Nuclear University MEPhI, 31, Kashirskoe shosse, Moscow, 115409, Russia

E-mail: VVFlorentsev@cadetech.ru

Abstract. One of key problems of nuclear standard frequency development is preparation assembly of ultracold thorium ions in Pauli trap. In this case semiconductive frequency-stabilized lasers with external resonator on frequencies 690 nm, 984 nm, and 1088 nm are used for excitation of corresponding electronic dipole and quadrupole cooling transitions for Th3+ ions. In the paper the results of development and creation of unified laser module, which is able to be used as base for full-featured system designed for laser cooling of Th3+ ions, are presented. The module is able to fine-tune necessary wavelength with accuracy ±5 nm.

1. Introduction

Laser cooling of atoms and ions confined in various traps are by far one of the most common methods to obtain and study the quantum properties of individual ultracold (1 uK) atoms and ions [1]. Significant progress is achieved with the use of such physical systems not only in the traditional ultra high resolution laser spectroscopy but also in new areas such as spectroscopy and quantum logic [2], the optical frequency and time standards [3].

From the standpoint of laser cooling of ions in a trap a number of advantages can be noted. This caused by the fact that the possibility of confining ions in the trap for a long time, up to several days, allows to explore an extremely narrow spectral lines without limitations due to the finite time of particle-field interaction. Spatial localization of ions in a trap in a small volume makes effective the use of ultra deep laser cooling techniques including the method of the sympathetic cooling [4, 5]. It is possible to suppress the Doppler shifts and other effects of decoherence up to the second order which is necessary to perform well protected quantum logic operations. Under the conditions of cooling up to the vibrational ground state of the ion the wave packet is localized in a volume less than a wavelength of light. Therefore, by using laser radiation one can selectively act on certain ions while monitoring their internal state, and translational motion. Since the ions are in an ultrahigh vacuum, it is possible to eliminate the interaction with other particles and significantly reduce the interaction with the environment.

By now it is possible to cool all of the alkali (e.g., Rb, Cs) and alkaline-earth metals (for example, Sr), all the inert gases in the metastable state (except Rn), some other elements (In, Al). This area is booming, and since 2007 were cooled Er [6], Cd [7], Hg [8]. Laser cooling of any new element is a major research challenge because it requires precise knowledge of the system levels, the presence of strong transition for this element, and availability of suitable laser sources.

Currently, the best results in the development of optical clocks on cold atoms and ions achieved at laboratory facilities in the United States (National Institute of Standards, NIST), Germany (PTB), the UK (NPL), France and Japan. Relative accuracy obtained for optical clocks on electronic transitions in single ions Al + in an electromagnetic trap is 7×10-18 [9] and several units of 10-17 for clocks on electronic transitions in neutral atoms of Sr in an optical lattice [10]. This accuracy corresponds to a lag in the fraction of a second for the lifetime of the Universe (13.7 billion years).

Two advanced research groups have started to address the problem by experimental optical spectroscopy of thorium ions: Georgia Institute of Technology (USA) and PTB (Germany). In the United States a system for laser spectroscopy of electronic transitions was set, and thorium ion trap for ion Coulomb crystals of Th3+ ions was constructed; first Coulomb crystals containing up to 10^4 thorium ions Th3+ were received and detected [11]. In Germany, laser system for spectroscopy of electronic transitions was created, and thorium ion trap for Coulomb crystals of Th+ was designed; first Coulomb crystals containing up to 103 ions of Th+ were produced and detected [12]. The estimation of an error budget based on standard nuclear transition in thorium was given. Given into account the quadratic and linear Zeeman effect, Stark effect, linear Doppler effect, blackbody radiation, the influence of the gravitational field and micromotion of the ion in an electromagnetic trap, the total uncertainty of the clock is 10–19, and can achieve a level of ~10–20 [13].

It should be noted that despite the significant progress in the field of direct cooling of thorium ions trapped in a linear Paul trap, the keeping time of the ions at temperatures of the order of 1 mK is limited and does not exceed 30 minutes. This is because thorium ion has ion «uncomfortable» electronic level structure from the viewpoint of the required laser sources, and the cooling transition has a small width. So the task of setting the laser on the corresponding transition is complicated greatly. However, the laser cooling of the ensemble of thorium ions in the trap is the necessary initial phase of the nuclear spectroscopy of the isomeric transition.

The problem of the isomeric transition excitation is even more complicated in terms of experimental realization, as direct excitation of the transition is virtually impossible because of the large uncertainty energy range of the desired transition, and also due to the long lifetime of the excited state. To measure the energy of the isomeric level different mechanisms of excitation of the nuclear transition were suggested. The first one relates to the doping of broadband crystals with 229Th ions followed by scanning the nuclear transition at the synchrotron source in the vacuum ultraviolet region. However, besides the fact that there is currently no synchrotron radiation source of the desired power in Russia, the problem of synthesizing these crystals with the necessary degree of purity is also resource consuming. Another approach appears to be more realistic in the Russian conditions and is based on the implementation of the mechanism of inverse electron bridge (excitation of nuclear states by the resonant excitation of the electron system) proposed by the head of this project [14].

At the current moment in Russian Federation experiments on laser cooling of thorium ions in traps are not being carried out. But they are a fundamental step for nuclear frequency standard.

In the paper the results of development and creation of unified laser module are given.

2. Approach and methodology

Fine tuning of wavelength and high quality factor of lasers are ensured by configurable external resonator and original electronic control block [15, 16]. Comparative analysis of technical specifications of developed laser system show that suggested laser system for cooling thorium ions is not inferior to currently used foreign analogs.

Development and creation of universal platform for high-precision laser system with high quality factor is performed in several steps:

- choosing optical system of laser (with resonator by Littrov or Littman scheme);
- calculating and modelling of optical-mechanical system;
- creating of constructing documentation;
- calculating and modelling of proportional-integral control controller for controlling the system for maintaining temperature of specified laser diode pn-transition;

- production of external system resonator movement control system.
- currently the schemes of external resonators of semiconductor lasers, represented on Figure 1.

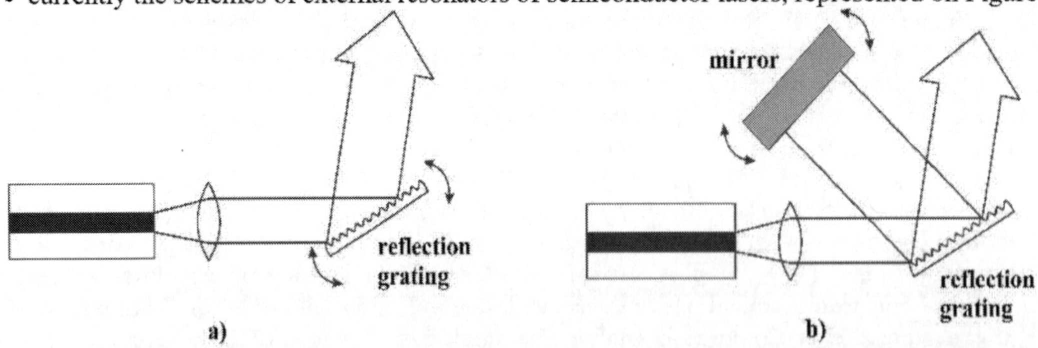

Figure 1. Semiconductor laser with an external cavity of the scheme: a) Littrov's; b) Littman's.

Littrov's resonator feedback is performed by diffraction at non-zero order of lattice, and output of radiation from resonator at zero order of diffraction. However, this arrangement does not allow to work at very high angles of diffraction that are necessary for increasing the dispersion of the resonator. Littman resonator already allows to work at high angles of diffraction grating, and the share of output from the resonator radiation can be easily controlled by selection of exit mirror transparency. However, in Littman's scheme the emitted ions interact with the lattice during full detour in resonator two times, thus the quality factor of resonator decreases significantly. In single-mode lasers with Fabry-Péro interferometer the spectrum contains a large number of single modes. Temperature and electric current influent on frequency on laser emission

3. Experiment setup

For maintaining constant temperature of laser working mode is necessary to use PI-controller with fixed specifications of maintaining temperature with accuracy of 0.01 K in the range 295–330 K [17, 18], with constant of time of controlling not exceeding five seconds and with a control object made on Peltier element. The flowchart for automatic controller is represented on Figure 2.

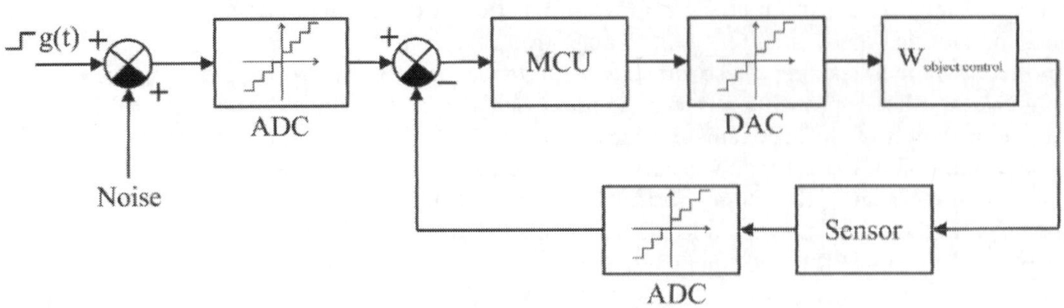

Figure 2. Block diagram of ACS thermos-controller diode laser.

For the laser diode have been obtained curves of the wavelength on temperature, for currents 70–110 mA. The results of an experiment with a laser diode MLD-694-050. The spectral characteristics of the research goniometer-spectrometer GS-5 that has, at the reflection diffraction-grating. Measurement and setting of temperature was made on done to research the thermal-controller. The temperature sensor is a current sensor AD590 [19, 20] produced by Analog Devices Corporation. Heating and cooling the laser diode to upkeep the set temperature, is carried out by a Peltier element.

4. Results and discussion

The curve for the wavelength of the laser radiation on the temperature of the diode crystal is drawn on Figure 3.

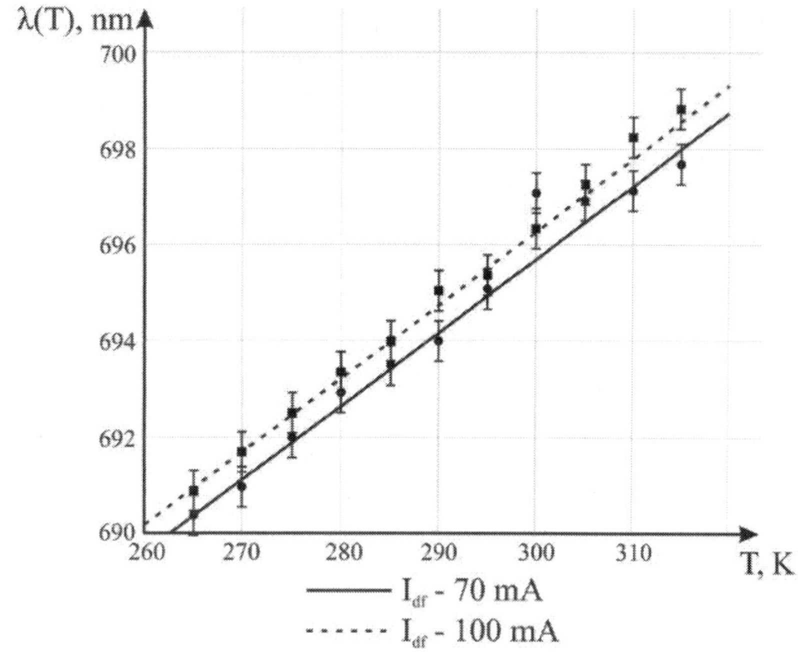

Figure 3. Block diagram of ACS thermos-controller diode laser.

The curve is defined by the ratio ά (I, T) [21, 22] displays the dependence of the wavelength from temperature and current:

$$\lambda(I,T) = \gamma TI + \alpha\gamma \tag{1}$$

The resulting coefficient included in the function of the system temperature control. The equation of dependence of the temperature set values of the wavelength at a constant value of current:

$$T(\lambda) = (\lambda set - \alpha I)/(\gamma I) \tag{2}$$

The equation for regulation of thermal control is:

$$P_{\text{Adj}} = \frac{1}{K_p}\left(\left(\frac{\lambda_{\text{set}} - \alpha I}{\gamma I} + T_{\text{meas}}\right) + \frac{1}{K_I}\int_0^{t_{set}}\left(\frac{\lambda_{\text{set}} - \alpha I}{\gamma I} + T_{\text{meas}}\right)dt\right) \tag{3}$$

- PAdj – adjustment function;
- Kp – the coefficient of proportional component;
- Ki – the coefficient of the integral component;
- λset – set wavelength;
- γ – the coefficient of thermal;
- α –the coefficient of current for wavelength and constant temperature;
- Tmeas – the measured value of the temperature diode now.

Installation shown on Figure 4 was utilized to measure the laser wavelength depending on the current through the pn-junction. The installation is a heavy goniometer spectrometer with a laser system attached to one ocular and an infrared camera attached to the other ocular.

IOP Conf. Series: Journal of Physics: Conf. Series **944** (2017) 012037 doi:10.1088/1742-6596/944/1/012037

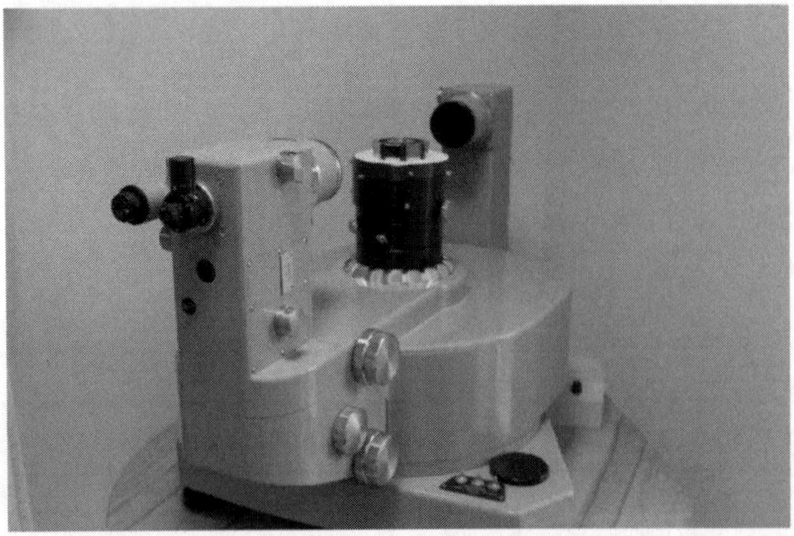

Figure 4. Appearance goniometer spectrometer GS-5 to research the spectral characteristics of the semiconductor laser.

At low current densities through the pn-junction of the radiation intensity is low, it is incoherent and the wavelength is shifted to minimum semiconductor laser operating range . By increasing the current, when the number of photons produced by the recombination exceeds the number of photons absorbed in the substance (current threshold), the emission intensity sharply increases – it becomes coherent and spectral maximum is shifted to the right as seen in Figure 5.

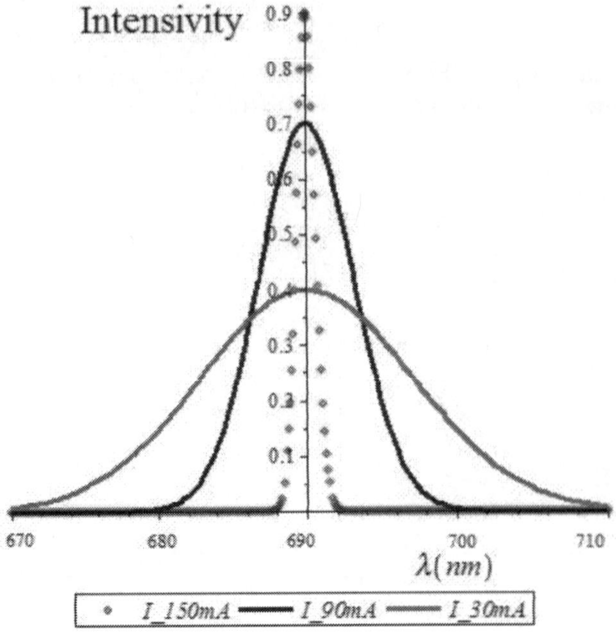

Figure 5. Graph of spectral peaks of the laser radiation on the current density flowing through the laser diode.

Furthermore, when a small spectral width of the pump current is sufficiently large, such behavior is consistent with the spontaneous emission of a conventional semiconductor light emitting diode with increasing current up to the rated values - generation is observed and the emission spectrum is narrowed. With further increase of current spectral characteristic appreciably narrows the quality

IOP Conf. Series: Journal of Physics: Conf. Series **944** (2017) 012037 doi:10.1088/1742-6596/944/1/012037

factor of the laser is increased, in addition significantly increases the stability of the resonator and the laser output power. By increasing the current density flowing through the pn-junction and the corresponding angle tuning external cavity laser diode passes from multimode to single mode lasing mode, with an average width of a single longitudinal mode is equal to 0.5–1.5 nm. Further reduction in the spectral linewidth selected mutual adjustment of the current strength, temperature and the reflection angle of the external cavity.

After successful testing and adjustment works of the laser module on the working wavelength of 690 nm we started to develop a semiconductor laser module for laser diode with an operating range of wavelengths 1080–1090 nm. Given the previous experience in the development of optical-mechanical circuits of the laser module laser system with an external cavity of Littrow model has been completely redesigned concept and approach in the design of the external cavity. The improved design (Figure 6) grating external cavity located on a slotted V-shaped holder, with the possibility of adjusting the vertical and horizontal reflection angle, with mutual compensation of the oscillations of the mechanical system holder.

Figure 6. 3D-model of advanced opto-mechanical system for diode laser LD 1088 nm with a removed housing cover.

In addition, two additional alignment system micrometric screws are provided for balancing the vertical node, with additional piezoceramic nozzles for fine adjustment. Conjugation of the two V-shaped holders split due not only to input additional degrees of freedom, but also removal of excess strain on the holder assembly design that subsequently a positive effect on the mechanical stabilization of the assembly of the external cavity. Since the implementation of this scheme will increase the interval between successive adjustment of the semiconductor laser module of optical-mechanical system.

Additionally, it is worth considering the laser diode collimator block and holder shown in Figure 7. Figure 7 represent 3D-model improved opto-mechanical system for the laser diode LD 1088 nm with a removed side housing cover and the side cover removed collimator-holder.

The collimator of the laser diode holder is formed in a monolithic body of duralumin, and introduced into heat-insulated from the rest of the opto-mechanical system capsule, to maintain a predetermined temperature.
This is necessary in order to reduce the environmental impact directly on the thermal stabilization of the laser diode.

Figure 7. Appearance of modernized design of the laser module.

To reduce the heat-affected thermostabilizing camera-collimator holder to neighboring nodes opto-mechanical systems, heat exchange of the collimator is made not to the base of the optical system, and the radiator grille, derived on the cover of the laser module housing. Figure 7 shows the appearance of modernized design of the laser module, depicting the cooling radiator assembly thermoregulation collimator carrier, as well as basic input-output interfaces the control signals from the aforementioned control ACS modules.

5. Acknowledgments

This work was supported by Russian Foundation for Basic Research (projects No.14-08-00487a) and Ministry of Education of Russia (project No.3.1803.2014/K).

References

[1] Phillips W D 1998 Nobel Lecture: Laser cooling and trapping of neutral atoms, *Rev. Mod. Phys.* 70 721
[2] Schmidt P O et al. 2005 Cold Ions and their Applications for Quantum Computing and Frequency Standards *Science* 309 p 795
[3] Bloom B J et al. 2014 An optical lattice clock with accuracy and stability at the 10^{-18} level *Nature* 506 p 71
[4] Takamoto M and Katori H 2003 Spectroscopy of the $_1S^0 \rightarrow {}_3P^0$ Clock Transition of Sr87 in an Optical Lattice *Phys. Rev. Lett.* **91** 223001
[5] Rosenband T et al. 2007 Observation of the $_1S^0 \rightarrow {}_3P^0$ Clock Transition in $_{27}Al^+$, *Phys. Rev. Lett.* **98** 220801
[6] Berglund A J, Lee S A and McClelland J J 2007 Sub-Doppler laser cooling and magnetic trapping of erbium *Phys. Rev. A.* **76** 053418
[7] Brickman K A, Chang M S, Acton M et al. 2007 Magneto-optical trapping of cadmium *Phys. Rev. A.* **76** 043411

[8] Hachisu H, Miyagishi K, Porsev S G et al. 2008 Trapping of Neutral Mercury Atoms and Prospects for Optical Lattice Clocks *Phys. Rev. Lett.* **100** 053001

[9] Chou et. al. 2010 Frequency Comparison of Two High-Accuracy Al^+ Optical Clocks *Phys. Rev. Lett.* **104** 070802

[10] Matthew D. Swallows et al. 2011 Suppression of Collisional Shifts in a Strongly Interacting Lattice Clock *Science* **331** pp 1043–1046

[11] Chapman et al. 2009 Multiply Charged Thorium Crystals for Nuclear Laser Spectroscopy *Phys. Rev. Lett.* **102** 233004

[12] Peik E, Zimmermann K, Okhapkin M and Tamm Chr 2011 Prospects for a Nuclear Optical Frequency Standard based on Thorium-229, *arXiv.org*, arXiv: 0812.3548

[13] Campbell C J, Radnaev A G, Kuzmich A, Dzuba V A, Flambaum V V and A. Derevianko 2011 A Single-Ion Nuclear Clock for Metrology at the 19th Decimal Place *ArXiv*: 1110.2490v1 [physics.atom-ph] 11

[14] Tkalya E V, Akhrameev E V, Arutyunayn R V, Bol'shov L A, Kondratenko P S 2014 Excitation of atomic nuclei in hot plasma through resonance inverse electron bridge *Physical Review C* vol 90 034 614

[15] Zou J, Zhang C and Bay D.-Y. 2015 Design of fuzzy-PID controller of constant temperature for semiconductor laser diodes *International Journal of Control and Automation* Vol 8 Is. 2 pp 01–08 DOI: 10.14257/ijca.2015.8.2.01

[16] Remmington S 2004 High Performance Temperature Control in Laser Diode Test Applications *ILX Lightwave Corporation United States*

[17] Carl E. Wieman and Leo Hollberg 1991 Using diode lasers for atomic physics *Review of Scientific Instruments Reviews* Vol 62 pp 1–20

[18] Vasil'ev V A and Zhmud' V A 2000 A temperature stabilization system for diode lasers *Instruments and Experimental Techniques* Vol 43 pp 580–581

[19] Han Z H and Feng Z H 2012 SCM based high precision temperature controller design *Applied Mechanics and Materials*, Vol 313–314 pp 549–552 DOI: 10.4028/www.scientific.net/AMM.313-314.549

[20] Maksimkin A I, Baryshev G K, Semenenko A N, Rodko I I, Varyatchenko E P and Ostrovskii V R 2014 System of pulsed solar energy supply of ionizing radiation detectors *Bulletin of the Lebedev Physics Institute* Vol 41 Is 7 pp 205–208. DOI: 10.3103/S1068335614070069

[21] Vetrov A A, Danilov D A, Esipov S S, Komissarov S S and Sergushichev A N 2009 Comparing thermal and electrical methods for controlling the wavelength of semiconductor-laser radiation, *Journal of Optical Technology* Vol 76 Is 8 pp 520–525 DOI: 1070-9762/2009/080520-0615.00

[22] Savelyev A V, Novikov I I, Chunareva A V, Gordeev N Y, Maximov M V, Payusov A S, Arakcheeva E M, Shchukin V. A. and Ledentsov N. N. 2011 A temperature-stable semiconductor laser based on coupled waveguides *Semiconductors* Vol 45 Is 4 pp 550–556 DOI: 10.1134/S1063782611040208

Intelligent control system for continuous technological process of alkylation

E S Gebel[1], R A Hakimov[2]

[1]Omsk State Technical University, Mira Ave., 11, Omsk, 644050, Russia
[2]Ltd. «Avtomatika servis», Neftezavodskaya street, 51, Omsk, Russia

e-mail: gebel_es@mail.ru

Abstract. Relevance of intelligent control for complex dynamic objects and processes are shown in this paper. The model of a virtual analyzer based on a neural network is proposed. Comparative analysis of mathematical models implemented in MathLab software showed that the most effective from the point of view of the reproducibility of the result is the model with seven neurons in the hidden layer, the training of which was performed using the method of scaled coupled gradients. Comparison of the data from the laboratory analysis and the theoretical model are showed that the root-mean-square error does not exceed 3.5, and the calculated value of the correlation coefficient corresponds to a "strong" connection between the values.

1. Introduction

According to the classical theory of automatic control, the design of automatic control systems is a completely formalized operation sequence. At the initial stage, a mathematical description of the technological object or process and the control device is made using the fundamental laws of the applied field of research. Then, the performance criteria of control (response time, overshoot, phase and amplitude margins, etc.) are refined. At the final stage, the design of the automatic control system is carried out, namely synthesis of the structure and parameters of the control device to ensure the required values of performance criteria of control. The nature of the arising subtasks and the ways of dealing with that kind of problem depends on the results obtained in the previous stages. By the methods of the theory of automatic control, the last two stages of designing the automatic control system are realized, namely the analysis of the object of control and the synthesis of the control mode that maintain an acceptable or optimal operation mode. Analytical calculations are carried out proceeding from the fact that the control system is written in exact terms of the formal model and adequately reflects the actual state and processes of interaction with the external environment. In practice, to hack out the mathematical model, it is required to simplify and linearize the differential equations. As a result, the synthesized ACS is only «workable in theory». Thus, the model of the control system (technological object and control device) is the critical moment, on the basis of which the quality and operability of the real ACS depends.

Models of control systems are created and modified as the quality requirements of the ACS become more stringent and the process and control objects become more complex. Each kind of model generates new control criteria and methods of synthesis of ACS. At the moment, the critical parameters of automatic control systems are high levels of autonomy and adaptability, reliability and quality of functioning in the conditions of uncertainty of external disturbing influences, as well as high requirements for high-speed, precision and other technical characteristics of the control object.

The application of classical methods of the theory of automatic control, developed in the second half of the twentieth century by Ya. Z. Tsypkin [1], A. A. Krasovskii [2], E. P. Popov [3], S. V. Emel'yanov [4], and others ., in the modern formulation of the problem of designing an ACS, it does not always allow to achieve the required functionality and qualitative parameters. The technology of PID-regulation laws allows to provide limited characteristics of accuracy, speed and invariance to external perturbations.

According to the analysis of the modern tendencies in the development of systems for automatic control of complex technological objects established that artificial intelligence methods and technologies should be used to increase the levels adaptability and autonomy of such systems.

2. Formulation of the problem

The modern intellectual information technologies include technologies of expert systems, fuzzy logic, neural network structures, associative memory, etc. [4–7]. Technologies of neural network structures are built on the basis of operating characterictics of biological prototypes. Intellectual system is a homogeneous structure, consisting of a set of interrelated dynamic elements with a given transfer function. The body of knowledge, laid in the learning process, is determined by the adjustment of the weight coefficients of interelement connections. The most important advantage of the technology under consideration is the rapid response time achieved by calculation process paralleling during their hardware implementation.

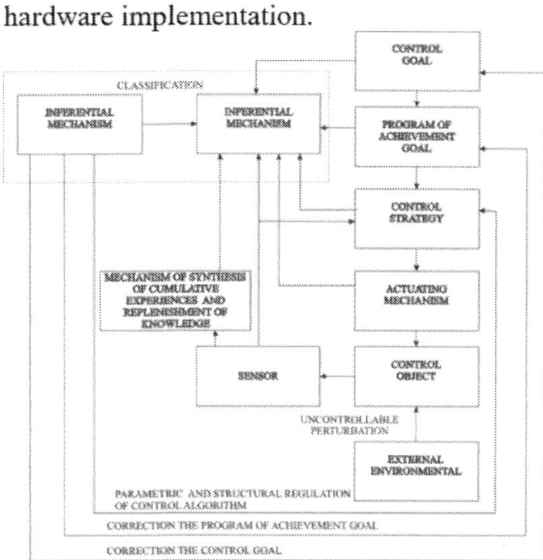

Figure 1. Generalized structure of intellectual management system

The architectural characteristic of intelligent control systems (Figure 1) is connected with the usage of special mechanisms for storage and processing of knowledge to perform the required functions in incomplete (or uncertain) conditions with the random nature of environmental perturbations [5].

The changing control purpose, operating characteristics of the technological object, environmental parameters, etc. relate to uncontrolled (random) perturbations. The structure of the system is included specialized tools for a generalization of the accumulated experience, and as a result, the replenishment of knowledge if necessary. Thus, the control object is a complex construction of a number of functionally-subordinate subsystems.

3. Results of experiments

A promising trend in the development of automatic control systems is the usage of virtual analyzers [8], which can be implemented both by means of classical methods of statistics and intelligent technologies, in particular neural networks. As an example, the control loop for the temperature of the end of boiling of stable alkylation gasoline will be discussed., The parameter should not exceed 205°C to provide the quality of the product (butane-butylene fraction with a high octane number of 96)., The laboratory data are used for verification the adequacy of mathematical models of virtual analyzers. According to the normative documentation the reproducibility of the laboratory analysis is not more than 3.5 °C.

Various models of neural networks differing in the number of neurons in the hidden layer, as well as the learning algorithm are synthesized by the MathLab software (Table 1). A comparative analysis of the results was carried out using statistical indicators, namely the standard deviation (RMS) and the correlation coefficient (CC) and is presented in Table. 1.

Table 1. Neural network simulation results.

№	Learning Algorithm	Number of neurons in the hidden layer	Entire segment		Training segment		Final segment	
			RMS	CC	RMS	CC	RMS	CC
1	Levenberg-Markar method	5	4.11	0.66	4.4	0.59	3.79	0.72
2	The Baess regularization method	3	3.75	0.73	4.14	0.64	3.32	0.79
3	Quasi-Newtonian method	1	4.17	0.66	4.55	0.54	3.74	0.74
4	Method of elastic back propagation	2	3.77	0.72	3.84	0.71	3.68	0.75
5	Method of scaled coupled gradients	7	3.45	0.77	3.61	0.74	3.29	0.79
6	Method of coupled Powell-Bil gradients	7	3.97	0.71	4.27	0.62	3.61	0.79
7	Method of coupled Polak-Ribira gradients	8	3.59	0.75	3.94	0.68	3.2	0.81
8	One-step secant method	4	3.73	0.73	3.7	0.73	3.74	0.75

Since the root-mean-square deviation for model No. 5 satisfies the reproducibility requirements for the process under consideration, the optimal model from the list in Table 1 is recognized the model with seven neurons in the hidden layer and the learning algorithm "The method of scaled coupled gradients" (Fig. 2).

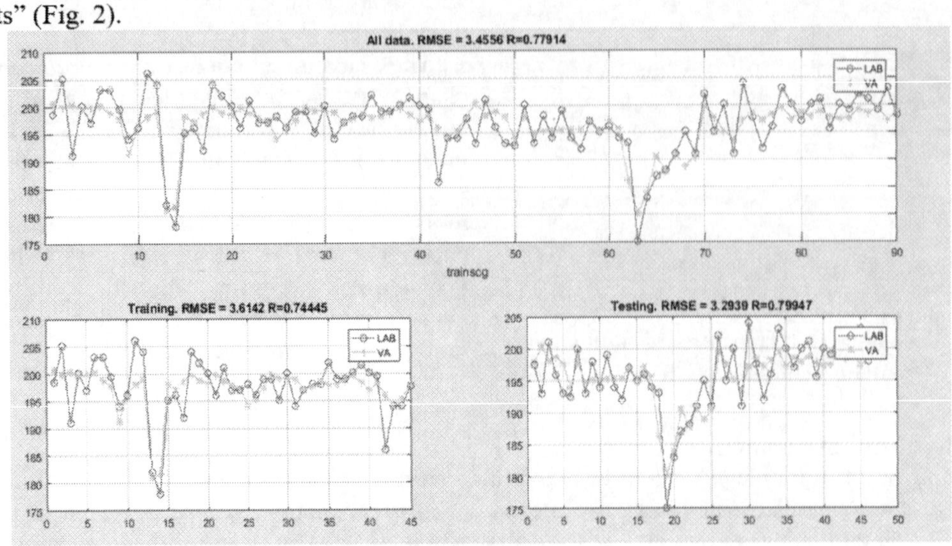

Figure 2. Model No. 5 of the virtual quality analyzer with 7 neurons in the hidden layer.

As a result of the comparison of the laboratory analysis data for the training, final and entire segments with the calculated values of the virtual analyzer, it was established that the theoretical curve has a "good" repeatability, the RMSE does not exceed 3.7. The value of the correlation coefficient R of the model output and the data of the laboratory analysis is declarative of a strong relationship between the factors.

IOP Conf. Series: Journal of Physics: Conf. Series **944** (2017) 012038 doi:10.1088/1742-6596/944/1/012038

Conclusion

Thus, the results of the literature review, experimental and applied research prove the effectiveness of the usage of intelligent technologies for the control of complex technological objects, in particular, the installation of sulfuric acid alkylation. The synthesized mathematical model of a virtual analyzer for a temperature-control loop of the end boiling of stable alkylation gasoline, proved the applicability and adequacy of neural network technologies for control a continuous technological process.

References

[1] Tsypkin Ya 1966 Adaptation, learning and self-learning in automatic systems *Automation and Telemechanics* **1** pp 23–61
[2] Krasovsky A 1987 *Manual on the theory of automatic control* (Publishing House: Science) p 711
[3] Popov E 1989 *Theory of linear systems of automatic regulation and control* (Moscow: Science) p 304
[4] Korovin S, Emelyanov S and Myshlyaev L 2008 *Theory and practice of forecasting in control systems* (Russian Universities)
[5] Makarov I 2006 Principles of constructing intellectual regulators and control systems for weapons and military equipment *Izvestiya TRTU. Thematic issue. Perspective systems and management tasks* **3(58)**
[6] Dehuri S, Misra B, Ghosp A and Cho S.-B 2011 A condensed polynomial neural network for classification using swam *Applied Soft Computing* pp 3107–13
[7] Korolyov L 2010 Evolutionary calculations, neural networks, genetic algorithms - formal statements of problems *Fundamental and applied mathematics* **15 (3)** pp 119–33.
[8] Musayev A 2003 Virtual analyzers: the concept of construction and application in control problems of continuous TP *Automation in industry* **8** pp 28–33.

IOP Conf. Series: Journal of Physics: Conf. Series **944** (2017) 012039 doi:10.1088/1742-6596/944/1/012039

Features of electric drive sucker rod pumps for oil production

F A Gizatullin[1], M I Khakimyanov[2] and F F Khusainov[2]

[1] Ufa State Aviation Technical University, 12, K. Marx ave., Ufa, 450008, Russia
[2] Ufa State Petroleum Technological University, 1, Kosmonavtov ave., Ufa, 450062, Russia

E-mail: hakimyanovmi@gmail.com

Abstract. This article is about modes of operation of electric drives of downhole sucker rod pumps. Downhole oil production processes are very energy intensive. Oil fields contain many oil wells; many of them operate in inefficient modes with significant additional losses. Authors propose technical solutions to improve energy performance of a pump unit drives: counterweight balancing, reducing of electric motor power, replacing induction motors with permanent magnet motors, replacing balancer drives with chain drives, using of variable frequency drives.

1. Introduction

Production of hydrocarbons is one of the most important industries in the structure of Russia's economy. The industry provides a significant portion of foreign exchange earnings in the country's economy. It is also one of the most energy-intensive: up to 5% of all generated electricity consumed by oil-producing enterprises [1].

Downhole mechanized mining is the most energy intensive among all the technological processes of oil production. This explained by the fact that the number of wells is large; it is dispersed over large distances; many of them operate in inefficient modes with additional losses [2].

2. Sucker rod pump electrical drives

Scheme sucker rod pump (SRP) shown in figure 1. The installation consists of a control station 1, a electric motor (EM) 2, a reducing gear 3, a oscillating crank mechanism 4, a walking beam 5, a rod string 6 and a borehole pump 7.

SRP used for driving asynchronous three-phase motors EM voltage of 0.4 kV with a high starting torque. It is made with a squirrel-cage winding or the double squirrel-cage winding. The synchronous speed is 500-1500 rpm. The electric motors power is from 7.5 to 55 kW, but the most widely used EM are 22 and 30 kW [3].

SRP units have low specific power consumption values than other types of submersible pumps, but their electric drives operate in heavy modes [4]. This is due to the following factors:

- EM is selected with a power reserve. It is necessary to provide a large starting torque. After startup, it works in underload mode [5];
- The load of EM is unstable. It has a cyclic character (figure 2) [6].

The cyclic nature of the load is due to principle of operation and the kinematic scheme of SRP unit. The first half of the period is pulling out of the rod string along with the borehole fluid, the second half – down casing of rod string. At the same time, the counterweights are rising upward. When these the counterweights go upward, then the potential energy is stored, which using in the next oscillation cycle [7].

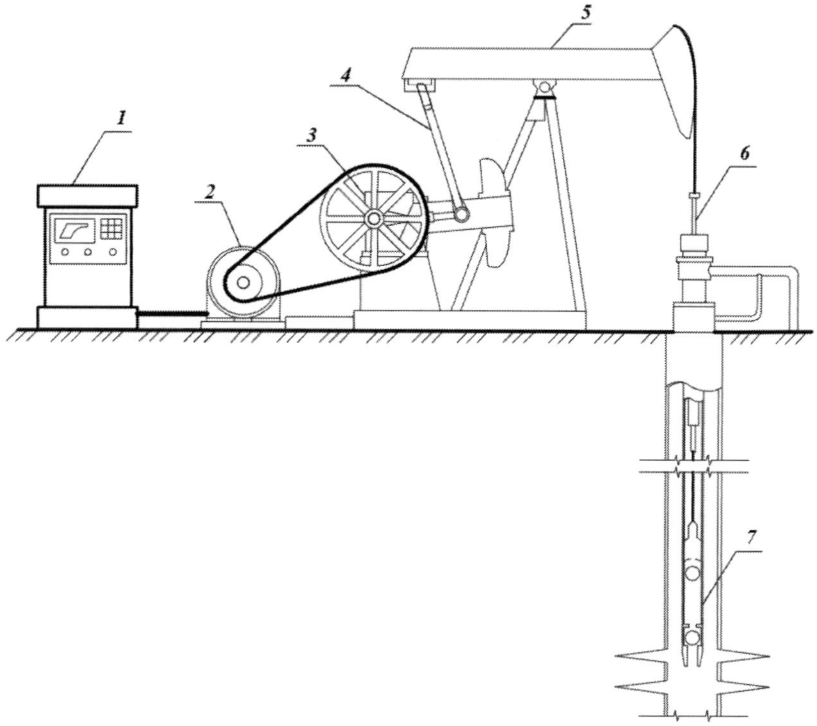

Figure 1. Scheme of sucker rod pump unit: 1 – the control station, 2 – the electric motor, 3 – the reducing gear, 4 – the oscillating crank mechanism, 5 – the walking beam, 6 – the rod string, 7 – the borehole pump.

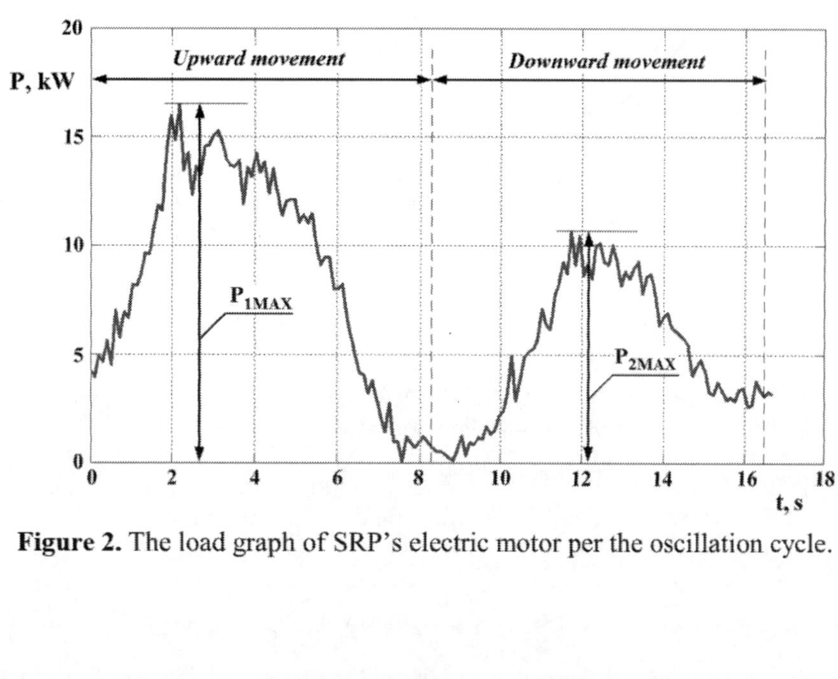

Figure 2. The load graph of SRP's electric motor per the oscillation cycle.

IOP Conf. Series: Journal of Physics: Conf. Series **944** (2017) 012039 doi:10.1088/1742-6596/944/1/012039

The peak values of the power consumed from the network (P_{1MAX}) and (P_{2MAX}) will coincide at an ideal balancing of the installation. The load graph will not be uniform in any case. In practice, up to 90 ... 95% of units are not well balanced [8, 9].

Variability of load leads to changes in all characteristics of EM, including changes in efficiency and power factor cos φ.

3. Investigation of the variation drive parameters within the oscillation cycle

To estimate the effect of the load on the efficiency and the power factor cos φ, the typical dependences of these parameters (table 1) on capacity factor β was used:

$$\beta = \frac{P}{P_{NOM} / \eta_{NOM}}, \quad (1)$$

where P – value of power consumption at the analyzed time; P_{NOM} – nominal capacitance of EM; η_{NOM} – nominal efficiency of EM.

Table 1. Typical depending efficiency and power factor on the capacity factor for asynchronous EM.

β	0.0	0.1	0.2	0.3	0.4	0.5	0.6
Efficiency	0.00	0.70	0.82	0.87	0.89	0.9	0.91
Power Factor	0.07	0.31	0.50	0.62	0.73	0.80	0.84
β	0.7	0.8	0.9	1.0	1.1	1.2	1.5
Efficiency	0.91	0.90	0.89	0.89	0.88	0.87	0.85
Power Factor	0.86	0.87	0.88	0.89	0.89	0.88	0.87

Calculations were performed for digitized wattmeter charts (power variation charts per oscillation cycle) of two SRP units. In both cases of EM have been installed with a rated power P_{NOM} = 30 kW.

The obtained graphs of power change, efficiency and power factor cos φ per oscillation cycle for SRP units № 1 and 2 are shown in figures 3 and 4.

As can be seen from the graphs (figure 3), the electric motor of SRP unit's № 1 is underloaded. The average power is only 6.97 kW, and the maximum - 15.56 kW. Thus the minimum power value is lowered to almost zero. Inequality peak load values at pulling out and down casing of the rod string shows the imbalance unit.

Per one oscillation cycle, the efficiency of EM varies from 0.11 to 0.90. The average value per cycle is 0.75. The power factor cos φ per oscillation cycle varies from 0.08 to 0.81. The average value per cycle is 0.49.

SRP unit № 2 is unbalanced, since the first load peak is completely absent. This is evident from the wattmeter chart of SRP unit № 2 (figure 3). The energy of EM is spent only on the lifting of counterbalancing loads, since the mass of the counterweights is too high. The rise of the borehole fluid is completely due to the potential energy accumulated in them.

The power consumption varies from 1.61 to 6.92 kW. The average power value is 3.52 kW, which indicates a low load of EM.

The electric motor efficiency varies from 0.57 to 0.83. The mean value per the oscillation cycle is 0.70. The power factor cos φ per oscillation cycle varies from 0.21 to 0.53. The mean value is 0.34.

Thus, both SRP unit are not sufficiently balanced, their EMs are underloaded. As a result, the efficiency and power factor cos φ have lower values. A significant change these parameters in each oscillation cycle adversely affects the operation of EM and electricity grid.

IOP Conf. Series: Journal of Physics: Conf. Series **944** (2017) 012039 doi:10.1088/1742-6596/944/1/012039

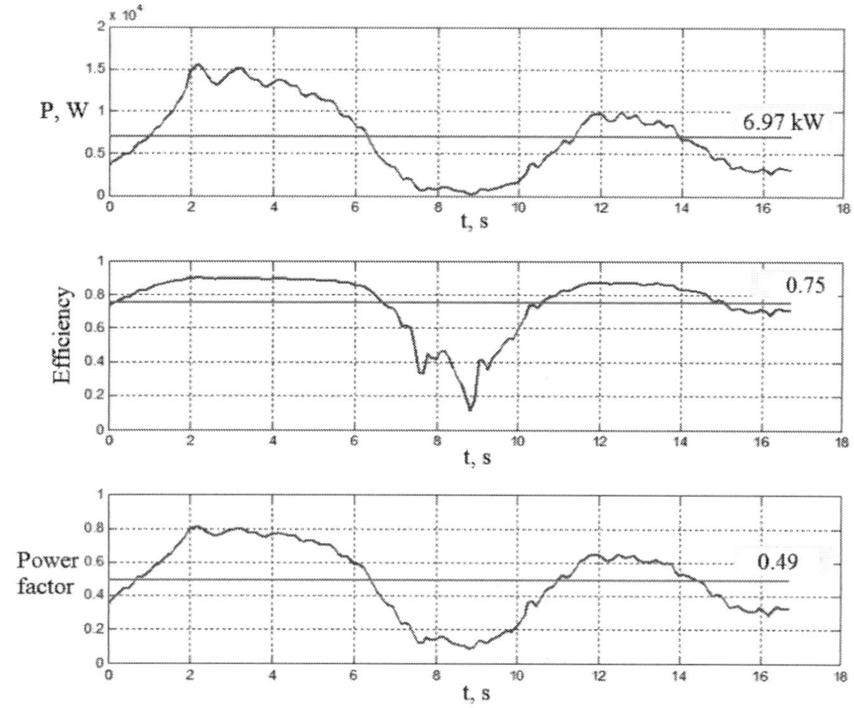

Figure 3. The graphs of power, efficiency and power factor per the oscillation cycle for SRP unit №1.

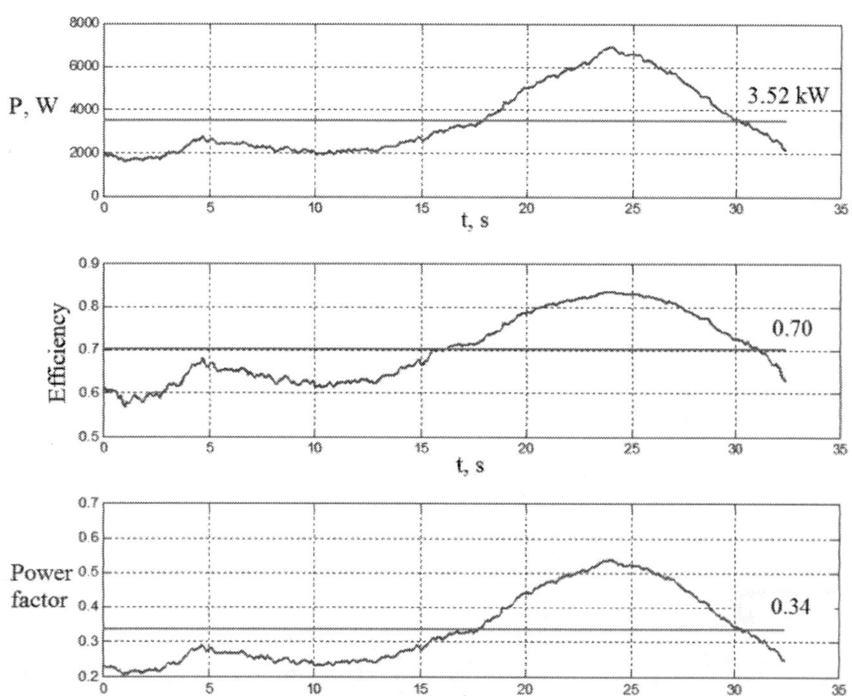

Figure 4. The graphs of power, efficiency and power factor per the oscillation cycle for SRP unit №2.

4. Recommendations for improving the operating modes of electric drives

To improve the energy characteristics of electric drives, it is required to perform their counterbalancing. Also, you need to install an EM of lower power (this requires testing according to the starting conditions, especially in winter conditions). A promising direction is to use Permanent Magnet Motors to SRP installations having high overload capacity. This will increase the loading of EM, and also achieve higher values of efficiency compared to asynchronous EM [10].

The greater stability of the parameters per the oscillation cycle is provided by SRP with chain drives. Therefore, in many fields ones are installed instead of balancing drive to improve the energy efficiency of oil processes [11].

One of the ways of increasing the efficiency of SRP is to use the variable frequency drive. This will allow you to adjust the pump performance according to recovery wellbore and to regulate the rotation speed of EM. This provides more uniform plunger movement and alignment of the load curve EM.

5. Conclusions

The conducted researches allow us to draw the following conclusions:

- The main reason for the low efficiency of SRP electric drive is the low load of motors and imbalance of counterweights.
- The inconstancy of the load per the oscillation cycle leads to changes all the characteristics of EM. This leads to a change in efficiency and power factor cos φ, which may vary within wide limits; the mean values per the cycle are very low. Changing of these parameters per the oscillation cycle negatively affects the operation of EM and the electricity grid.
- To improve the energy characteristics of electric drives SRP it is necessary to accurately balance the counterweights, as well as reduce electric power motors, allowing them to increase their load. In addition, in many cases it is advisable to replace the balancing drives with chain drives, which provide a higher stability of the parameters per the oscillation cycle.
- The promising direction is the use of valve motors for SRP drives. It has a higher efficiency than asynchronous electric motors and high overload capacity. This will increase the load of electric motors at a lower installed power.

Acknowledgments

The work is done in the framework of the project 8.1277.2017/HR "Research, development and introduction of advanced Electromechanical converters for Autonomous objects with a hybrid power plant".

References

[1] Ordonez B, Codas A and Moreno U F 2009 *Control Applications, (CCA) & Intelligent Control (ISIC)* pp 1259-1264

[2] Khakimyanov M I, Shafikov I N and Khusainov F F 2015 *International Siberian Conference on Control and Communications (SIBCON)*

[3] Lindh T, Montonen J H, Grachev M and Niemela M 2015 *IEEE 5th International Conference* pp 416-420

[4] Ordonez B, Codas A, Moreno U F and Teixeira A 2008 *IEEE International Conference* pp 282-289

[5] Khakimyanov M I and Svetlakova S V 2014 *10th International scientific practical conference "Current research and development* Vol. 29, pp 62-65.

[6] Guluyev G, Pashayev A, Pashayev F, Rzayev A and Sabziev E 2012 *Proc. of IV International Conference "Problems of Cybernetics and Informatics"* pp 19-21

[7] Khakimyanov M I, Khusainov F F and Shafikov I N 2015 *Electronic Science magazine "Oil and Gas business"* **1** 533-563

[8] Aliev T, Nusratov O, Guluyev G, Rzayev A, Pashayev A and Pashayev F 2010 *The Third International Conference "Problems of Cybernetics and Informatics"* pp 219-222
[9] Rzayev As H, Rezvan M H and Khakimyanov M I 2013 *Electronic Science magazine "Oil and Gas business"* **5** 19-29
[10] Khakimyanov M I, Shafikov I N, Khusainov F F, Semisynov R A and Bezryadnova E A 2017 Journal of Physics: Conference Series, Vol. 803 **1**
[11] Khakimyanov M I and Khusainov F F 2016 *IX International Conference on Power Drives Systems*

Multicomponent Arcjet plasma Parameters

V Gorbunkov, V V Kositsin, V I Ruban and V V Shalay

Omsk State Technical University, 11, Mira ave., Omsk, 644050, Russia

E-mail:vigorbunkov@list.ru

Abstract. To determine the plasma arc parameters of an arcjet thruster, the kinetic theory of gases is used. We can find a well-known statement about the adiabatic character of the compression process due to the growth of the gas temperature in a change in its spectral composition and in the Doppler effect. The use of tungsten in the nozzle design details explains the appearance of atoms of this element in the plasma volume.The emission spectra of tungsten allow us to indirectly judge the temperature of the arc discharge and its character. Absorption of the long-wavelength wing of the line contour at $\lambda = 465.987$ nm substantiates our conclusion about the consumption of the anode material in the process of operating the arcjet. The Doppler shift of the emission lines of argon allows us to determine the rate of the gas jet escape. The results of the study can be useful in the design of aircraft.

1. Introduction

The small closed volume occupied by the gas plasma is a feature of the correcting propulsion systems of spacecraft (PSSC) [1]. The inter-electrode distance and cavity diameter do not exceed a few millimeters. The walls of the arcjet chamber are heated to a certain temperature. Particles of matter emit and absorb photons, so equilibrium between particles of matter and radiation can be established [2]. The laws of equilibrium radiation play an important role in the gases kinetic theory. They determine the accuracy of propulsion systems plasma parameters.

It was noted [3] that due to the Langmuir probe measurements performed in the jet of the reactive 1 ... 2- kW engine at an argon pressure of $P_{Ar} = 6$ bar, the electron temperature was 1 … 2 eV higher than the temperature of the neutral gas in the flare.

The electron density was in the range 10^{10}–10^{13}cm^{-3}, which is also much higher than the Sakha equilibrium for temperature ($T_{Ar} \approx 1000$K) and gas concentration ($n_{Ar} \approx 10^{17}$cm^{-3}), but much lower than the equilibrium temperature for the electron temperature, which did not allow agreement between the model Thermal equilibrium and experiment. This allowed the authors [3] to come to the conclusion that the temperature $T_e = 1 ... 2$ eV most fully reflects the arc electron temperature.

A simulation of an electro-thermal 240-W arcjet propulsion system in the quasi-neutral plasma approximation was carried out in [4]. A three-component plasma at P = 2.5 bar is considered from argon atoms, their ions and electrons. The closeness of the electron temperature T_e = 16 000K to the maximum axial gas temperatures T_{Ar}, = 11 500 K near the narrowing part of the nozzle, as well as their separate reduction to 5000 K and 3300 K at the nozzle exit is noted. The thermal equilibrium state observed between 30 % of the divergence between the axial electron and gas temperatures is significantly disturbed in the expanding part of the nozzle due to a drop in the electron density in the radial direction.

A high level of electron temperature increases the ionization and electrical conductivity level when cooling the gas layer near the walls of the arcjet chamber. This allows us to regard the plasma as quasineutral, and assume that the surface of the electrodes is adiabatic for T_e. At the same time, it is known [5] that when the gas is heated without heat exchange with surrounding bodies, adiabatic compression is observed due to the growth of the gas temperature and, according to the Doppler effect, its spectral composition should change. The model [6] allows us to see how the emission spectrum of an resistojet plasma changes when passing through a nozzle.

We will use the experimental setup [6] to analyze the parameters of a multicomponent plasma of an resistojet. We will take into account the anode material, since it is not important to establish an equilibrium between matter and radiation, in the collision with which particles the atoms are excited.

2. Experimental setup

The model of an arc electrothermal arcjet was investigated (Fig. 1). The discharge cavity is a quartz tube 1.3 mm thick, 6.4 mm in diameter, geometrically located inside the cylindrical body of the arcjet [6].

Figure 1. Arcjet electrothermal model [6] and its fragment. 1 – nozzle, 2 – anode, 3 – swirler of gas flow, 4 – tube made of quartz glass, 5 – cathode; In the figure on the left, an aperture measuring 5 × 50 mm for optical registrations is clearly visible in the micro-motor wall.

The interelectrode distance between the cathode 5 and the edge of the annular anode 2 is 3 ÷ 5 mm. The material of the electrodes is tungsten. Buffer gas argon is fed into the cavity through a swirler under pressure (1.03–1.08) × 101.3 × 10^3 Pa.

Power was provided by the standard power supply unit «Tetrix 230». The discharge current was measured by a shunt with a resistance of 1 Ohm, and the voltage on the discharge with the help of an ohmic divider 11.9 / 43.8 kΩ with the help of the oscillograph "RIGOL DS 2202". The design of the discharge power supply unit did not provide for a voltage change, therefore the discharge parameters were estimated at the current I_R = 3.8 – 4.2 A and the corresponding voltage drop U_R = 16.8 ÷ 19.0 V.

The working temperature of the quartz tube wall (420 – 450 K) and the outer wall of the nozzle (900-1100 K) was recorded by chromel-alumelium thermocouples.

Spectral measurements were carried out using a five-channel modular calibrated radiometric complex «Ava-Spec-ULS204L-5-RM», whose spectral range was 235 – 815 nm, the optical resolution in the range 296 – 400 nm was 0.07 nm, in the range of 605 – 815 nm not less than 0.15 nm. For simultaneous use of five channels, a 25 mm focusing collimating lens with an SMA adapter was used. The dynamic range of changes in the intensities of the spectral lines reached 60,000. The exposure could vary from 50 ms to 1 min. This ensured the possibility of recording the spectrum in a linear regime of the intensity changes of the spectral lines under study. We investigated the emission spectrum of a discharge through the wall of a quartz tube at an angle of 90 ° to the axis of the micromotor. To observe the glow of the total discharge in the transverse direction relative to the motor housing, the radiation was output to the spectrograph from the aperture. We also analyzed the spectra of the axial radiation of the nozzle of the micromotor model.

The distance between the aperture and the collimation lens was 110 mm, when measuring the longitudinal radiation of the nozzle, the distance to the nozzle was 420 mm. Optical measurements were carried out after the resistojet model was released to the stationary mode of operation.

The laboratory room temperature was 20 ° C.

IOP Conf. Series: Journal of Physics: Conf. Series **944** (2017) 012040 doi:10.1088/1742-6596/944/1/012040

3. Experimental results

With the operating parameters of the power supply unit, measurements were made and the discharge current density was calculated for a known diameter of the positive discharge column d = 0.5…1.0 mm. The current density j = (121-538.5) A / cm^2 found the electron concentration of the arc discharge.

The necessary drift velocity of electrons V_d can be obtained from the consideration that electrons in the gas gap L_e = 0.3 ... 0.5 cm move in an argon atmosphere. The argon concentration is $N_{Ar} = P_{Ar} / kT = 2.45 \times 10^{19}$ cm^{-3}.

It is known that the mobility of electrons depends on the strength of the electric field, and in a number of cases leads to a specific instability of the gas-discharge plasma. Taking into account the dependence of the drift electrons velocity in argon on the reduced electric field strength [7] E / N_{Ar} = 0.18 – 0.3Td in the approximation of low electron density, the sought value is $V_d = (3 ... 6) \times 10^5$ cm s^{-1}. The current density of the arc discharge $j = en_eV_d$ = (121 ... 538.5) A / cm^2 makes it possible to estimate the electron concentration ne = $(1.2 ... 5.6) \times 10^{15}$ cm^{-3}.

The flow of electrons under the action of an external electric field excites the gas discharge plasma. The flow of charged particles heats the electrodes to a very high temperature and, judging by the erosion of the cathode, reaches the melting point of tungsten $T_W \approx 3700$ K, which makes it possible to estimate the concentration of tungsten atoms from the saturated vapor pressure [8].
The concentration of tungsten vapor corresponding to the vapor pressure $P_w = 5 \times 10^{-2}$mm. Hg (6.67 Pa) at T_w = 3655 K is n_w 1.32×10^{14} cm^{-3}.
In Fig. 2 shows the emission spectrum of an arc discharge plasma of the *resistojet model.*

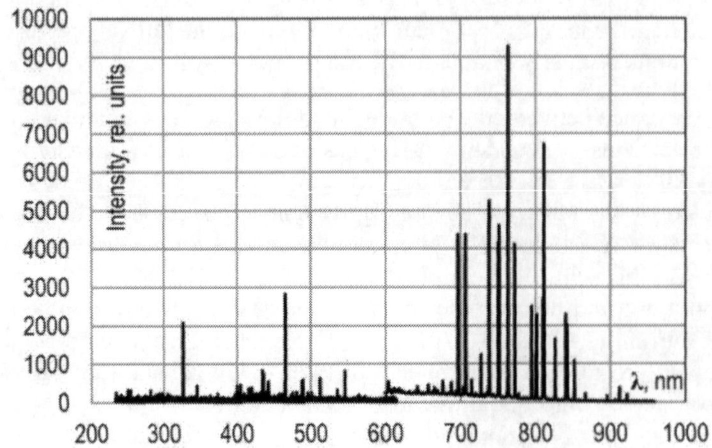

Figure 2.The discharge radiation spectrum observed through the nozzle.

Figure 2 clearly shows a number of spectral lines in the visible range (462 ... 547) nm, which belong to tungsten atoms (Table 1).

All the noticeable spectral lines in Fig. 3 refer to the lines of argon (Table 2).

IOP Conf. Series: Journal of Physics: Conf. Series **944** (2017) 012040 doi:10.1088/1742-6596/944/1/012040

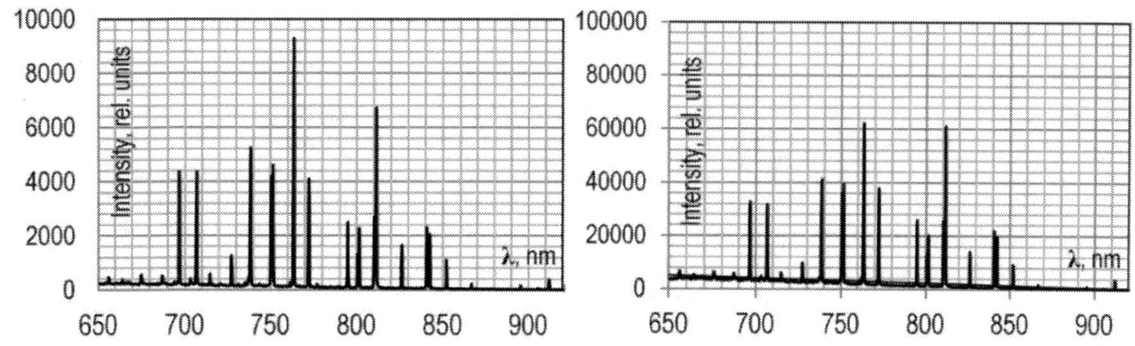

1 2

Figure 3. The spectral region (667 ... 930) nm (1) – taken through the nozzle and (2) – through the aperture.

From the comparison of Fig. 3 (positions 1 – 2) it can be seen that the intensity of emission lines from the nozzle in the 650 ... 850 nm range, compared to radiation through the aperture, decreased by almost an order of magnitude due to the increase in the distance to the registrar.

Table 1. Atomic tungsten spectrum.

№	λ_{tab}, *nm [10]*	λ_1, *nm (aperture)*	λ_2, *nm (nozzle)*
1	462.055	462.0	462.0
2	464.256	464.2	464.2
3	465.987	465.2	465.2
4	487.828	487.8	487.8
5	489.244	489.2	
6	500.616		500.7
7	513.0	512.6	
8	546.9	546.0	

Table 2. Atomic argon spectrum.

Registration of spectrum 1 is performed through an aperture, spectrum 2 through a nozzle

№	λ_{tab}, *nm*[9]	λ_1, *nm (aperture)*	λ_2, *nm (nozzle)*	*Paschen'transition [11]*	L→ S *transition* [11]
1	696.543	696.4	696.4	$^{1}P_1 \rightarrow {}^{3}P_2$ $2p_2 \rightarrow 1s_5$	$4p'^2[1/2]_1 \rightarrow 4s^2[3/2]_2$
2	706.722	706.7	706.7	$^{3}P_2 \rightarrow {}^{3}P_2$ $2p_3 \rightarrow 1s_5$	$4p'^2[3/2]_2 \rightarrow 4s^2[3/2]_2$
3	714.704	714.6		$^{3}P_1 \rightarrow {}^{3}P_2$ $2p_4 \rightarrow 1s_5$	$4p'^2[3/2]_1 \rightarrow 4s^2[3/2]_2$
4	727.294	727.2		$^{1}P_1 \rightarrow {}^{3}P_1$ $2p_2 \rightarrow 1s_4$	$4p'^2[1/2]_1 \rightarrow 4s^2[3/2]_1$
5	738.398	738.4	738.4	$^{3}P_2 \rightarrow {}^{3}P_1$ $2p_3 \rightarrow 1s_4$	$4p'^2[3/2]_2 \rightarrow 4s^2[3/2]_1$
6	750.387	750.3	750.3	$^{1}S_0 \rightarrow {}^{1}P_1$ $2p_1 \rightarrow 1s_2$	$4p'^2[1/2]_0 \rightarrow 4s'^2[1/2]_1$
7	751.465	751.3	751.3	$^{3}P_0 \rightarrow {}^{3}P_1$ $2p_5 \rightarrow 1s_4$	$4p^2[1/2]_0 \rightarrow 4s^2[3/2]_1$
8	763,511	763.5	763.5	$^{3}D_2 \rightarrow {}^{3}P_2$ $2p_6 \rightarrow 1s_5$	$4p^2[3/2]_2 \rightarrow 4s^2[3/2]_2$
9	772,376	772.3	772.3	$^{3}D_1 \rightarrow {}^{3}P_2$ $2p_7 \rightarrow 1s_5$	$4p^2[3/2]_1 \rightarrow 4s^2[3/2]_2$
10	794.815	794.8	794.6	$^{3}P_1 \rightarrow {}^{3}P_0$ $2p_4 \rightarrow 1s_3$	$4p'^2[3/2]_1 \rightarrow 4s'^2[1/2]_0$
11	800,616	800.6	800.4	$^{3}D_2 \rightarrow {}^{3}P_1$ $2p_6 \rightarrow 1s_4$	$4p^2[3/2]_2 \rightarrow 4s^2[3/2]_1$
12	801,479	801.3	801.3	$^{3}D_2 \rightarrow {}^{3}P_2$ $2p_8 \rightarrow 1s_5$	$4p^2[5/2]_2 \rightarrow 4s^2[3/2]_2$
13	810.369	810.3	810.2	$^{3}D_1 \rightarrow {}^{3}P_1$ $2p_7 \rightarrow 1s_4$	$4p^2[3/2]_1 \rightarrow 4s^2[3/2]_1$
14	811.531	811.4	811.4	$^{3}D_3 \rightarrow {}^{3}P_2$ $2p_9 \rightarrow 1s_5$	$4p^2[5/2]_3 \rightarrow 4s^2[3/2]_2$
15	826.452	826.4	826.4	$^{3}P_1 \rightarrow {}^{1}P_1$ $2p_2 \rightarrow 1s_2$	$4p'^2[1/2]_1 \rightarrow 4s'^2[1/2]_1$
16	840.821	840.7	840.7	$^{3}P_2 \rightarrow {}^{1}P_1$ $2p_3 \rightarrow 1s_2$	$4p'^2[3/2]_2 \rightarrow 4s'^2[1/2]_1$
17	842,465	842.4	842.4	$^{3}D_2 \rightarrow {}^{3}P_1$ $2p_8 \rightarrow 1s_4$	$4p^2[5/2]_2 \rightarrow 4s^2[3/2]_1$
18	852.144	851.9	851.9	$^{3}P_1 \rightarrow {}^{1}P_1$ $2p_4 \rightarrow 1s_2$	$4p'^2[3/2]_1 \rightarrow 4s'^2[1/2]_1$

4. Results and discussion

In the arc discharge of the resistojet model, the pressure of saturated tungsten vapors is four orders of magnitude less than the argon pressure. Therefore, argon has a decisive influence on the drift velocity of the electrons V_d and on the thermal conductivity of the plasma, and the role of tungsten with a small ionization potential (E_W = 7.98 eV) reduces to the formation of the ne concentration and the T_e temperature of the electrons.

According to the calculations, the discharge plasma of anresistojet is a multicomponent argon-tungsten plasma having several local temperatures.

The concentration of argon is much greater than the concentration of tungsten vapor and electron concentration, so the width of the radiation contour could be determined, basically, by elastic collisions of argon atoms with each other and with electrons. However, the calculations of the electron temperature based on the shock broadening of the radiation contours of argon atoms in the range 696 ... 852 nm do not allow one to unequivocally state the shock excitation of heavy-particle atoms of the plasma in question.

The emission lines of tungsten provide a white glow discharge, observed through the aperture. The appearance of tungsten atoms in the plasma volume, on the one hand, is not unexpected, since tungsten is part of the structural parts, but, on the other hand, it is a signal of the erosion of structural elements. [12]

At the same time, a strong absorption of the longwave wing of the contour of the intense tungsten line is observed at λ = 465.987 nm. If this effect is determined by self-absorption, then by the parameters of the "tail" of this line (at λ = 465.2 nm) recorded through the nozzle of the engine, it is possible to estimate the consumption of tungsten. Probably, the absence of tungsten lines in the spectrum of the nozzle at λ = 489.2 nm, 513.0 nm and 546.9 nm can also be explained by self-absorption.

It was noted that the spectrum of argon passed through the aperture lost several lines: λ = 714.704 nm, 727.294 nm, 866.794 nm, 912.296 nm, 922.450 nm. This phenomenon was explained in [11] by the high rate of quenching of excited states of argon by nitrogen molecules from the surrounding nozzle of the air medium.

When comparing the wavelengths of argon lines λ = 794.815 nm, 800.616 nm and 810.369 nm, (Table 2), the radiation from the nozzle relative to radiation from the discharge chamber shows that the lines have a Doppler shift of wavelengths.

The confirmation of this shift is found in the fact that the velocities of the argon atoms in the chamber are more chaotic than in the gas jet directed from the nozzle, mainly with an axial velocity.

5. Conclusions

The discharge plasma of the arcjet model is a multicomponent argon-tungsten plasma having several local temperatures.

The discharge glow spectrum observed through the aperture of the discharge chamber is a consequence of the excitation of tungsten and argon atoms. In the spectrum observed through the nozzle of the discharge chamber, a strong absorption of the longwave wing of the contour of the intense tungsten line was observed at λ = 465.987 nm. The registration of partial self-absorption of this tungsten line, observed through the nozzle of the engine, makes it possible to estimate the magnitude of the erosion of the structural elements of the plasma device.

Under the considered excitation regime of the resistojet model in the discharge between the spectral lines of argon atoms at λ = 794.815 nm, 800.616 nm and 810.369 nm, an offset is observed when radiation passes through the discharge chamber nozzle. The marked shift, according to the Doppler effect, can be used to find the velocity of the gas jet outflow.

For more complete representation of the mechanisms of excitation of the DETMD discharge, additional studies are needed.

IOP Conf. Series: Journal of Physics: Conf. Series **944** (2017) 012040 doi:10.1088/1742-6596/944/1/012040

6. References

[1] Mazouffre S 2016 Electric propulsion for satellites and spacecraft: established technologies and novel approaches *Plasma Sources Sci. Technol* **25** pp 1–27
[2] El'yashevich M A 1962 *Atomic and Molecular Spectroscopy* (Moscow) [in Russian].
[3] Brinkman E A Electron densities and temperatures 1997 *J. Appl. Phis* **81** (3) pp 1093–1098
[4] Bijie Yang 2014 Quanhua Sun. Numerical Analysis of the Plasma Flow in an Arcjet Thruster *AIP Conference Proceedings* Vol 1628 pp 1132–1138
[5] Landau L D, Lifshitz E M 1987 *Fluid Mechanics* vol 6 Butterworth–Heinemann. ISBN 978-0-08-033933-7
[6] Blinov V N., Ruban V I. et al. 2016 Design features and experimental researches of an arcjet thruster for small satellite *Dynamics of Systems, Mechanisms and Machines (Dynamics)*
[7] Smirnov B M 2009 Modeling of gas discharge plasma *Physics – Uspekhi* vol 52 pp 559–571
[8] Rosebury Fred 1964 *Handbook of Electron Tube and Vacuum Techniques* (Massachusetts) p 456
[9] Frish S E 1963 *Optical spectra of atoms (*Moscow) p 640
[10] Zaĭdel'A N, Prokof'ev V K, S M Raĭskiĭ et al. 1977 *Tables of Spectral Lines* (Moscow: Nauka)
[11] Cullen P J, Milosavljevic V **2015** Spectroscopic characterization of a radio-frequency argon plasma jet discharge in ambient air *Prog. Theor. Exp. Phys*. 063J01
[12] Smirnov Yu M 2009 Electron impact excitation cross-section of tungsten atom *High temperature* Vol 47 pp 13–21

Acknowledgments

The work was supported by the Russian Foundation for Basic Research, grant No. 15-08-00726.

IOP Conf. Series: Journal of Physics: Conf. Series **944** (2017) 012041 doi:10.1088/1742-6596/944/1/012041

Influence of winding construction on starter-generator thermal processes

P Yu Grachev[1], A A Bazarov and A S Tabachinskiy[2]

Samara State Technical University, Samara, 443100, Russia

e-mail: [1] pgrachov@mail.ru [2] tabachinski.as@ieee.org

Abstract. Dynamic processes in starter-generators features high winding are overcurrent. It can lead to insulation overheating and fault operation mode. For hybrid and electric vehicles, new high efficiency construction of induction machines windings is proposed. Stator thermal processes need be considered in the most difficult operation modes. The article describes construction features of new compact stator windings, electromagnetic and thermal models of processes in stator windings and explains the influence of innovative construction on thermal processes. Models are based on finite element method.

1. Problem Statement

For vehicles application mass and dimension reduction of electric equipment is strongly relevant. In hybrid vehicles, starter-generator, based on induction squirrel-cage machine, is applied [1]. Original stator construction of electric machine for HEVs is designed [2]. It allows to decrease overhang deviation and reduce copper intensity due to rectangular cross-section stator conductors. Greater dimensions reduction can be achieved in case of multipolar induction machine integrated with multiphase semiconductor converter [3, 4]. Electric machine in [3] has squirrel-cage rotor and stator windings. Suggested electric drive based on this electric machine, has real-time control of phase number and wide speed range. However, small number of turns per phase defines extra-low voltages.

Mass and dimensions reduction is also relevant for aviation [5]. Integration of electromechanical converters and engine is prospective. Electric machines operates at high temperatures and rotation speed. Here, as well as in combined power units for HEVs and EVs, attention is paid to cooling system of power converters [6, 7].

Significant dimensions and overhang deviation reduction is able in innovative stator compact winding construction. This winding has rectangular irregular cross-section conductors [8, 9]. Construction feature defines irregular current density in parts with reduced cross-section. Irregular current density influences phase resistance [9]. Improvement of dynamic modes increase starting currents in these electric machines [10]. It needs be considered in electromagnetic and thermal calculation of electromechanical converters.

In fig. 1 time diagrams of HEV internal combustion engine (ICE) start process is illustrated. Starter-generator rated power is 15 kW, and its power supply is 144 V DC source. Induction machine replaces flywheel on the ICE crankshaft.

IOP Conf. Series: Journal of Physics: Conf. Series **944** (2017) 012041 doi:10.1088/1742-6596/944/1/012041

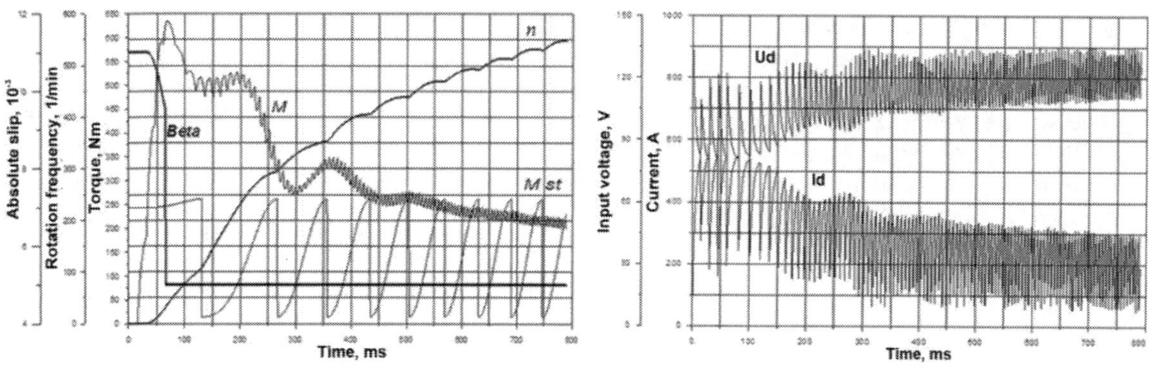

a) b)

Figure 1. Frequency start of ICE time diagrams.

In fig. 1a variations of electromagnetic torque M, crankshaft torque M_{st}, rotation frequency n and absolute slip *Beta* are illustrated. In fig. 1b valve converter input voltage U_d and DC source current I_d are illustrated. For start torque 450 Nm, DC source current I_d reaches 400 A. Stator phase current reaches maximum rating in the first moment of start process, and winding temperature rises rapidly. Insulation aging is significant at high temperatures. Thus, to improve reliability of electric machine in start/stop processes for HEVs and EVs, precise thermal simulation is needed [11].

This article focuses on thermal simulation and calculation for innovative induction machines as parts of autonomous power units. Innovative construction allows to reduce dimensions and improve efficiency, and is protected by patent [8]. Fig. 2 illustrates stator of innovative induction machine.

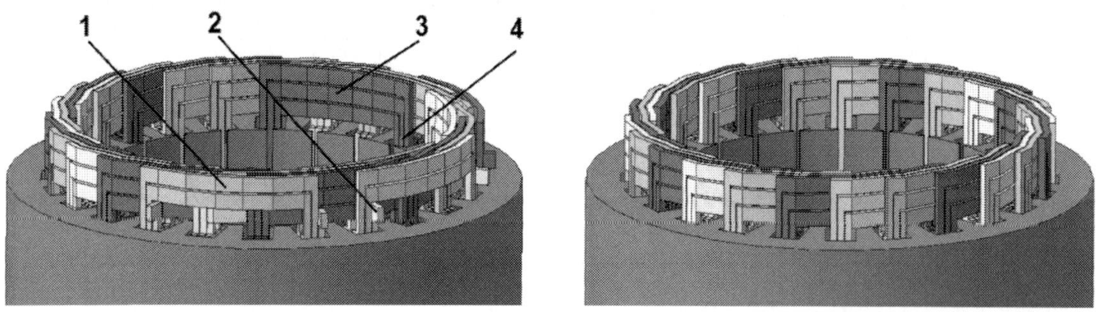

a) b)

Figure 2. Stator compact winding

Let us consider the features of the construction. Stator winding is three-phase and conductors are rectangular cross-section. Each phase is divided on three branches. Overhang parts of compact winding are parallel to stator end-faces. Connection spots of slot and overhang parts has double reduced cross-section. This solution allows to place overhang conductors of different slots in the same layer.

Fig. 2a illustrates compact winding stator end-face with phase outs. The winding in this machine is wave and concentrated, and the branches are connected in series. On the figure, position 1 points conductor, connecting branches of the winding. Pos. 2 points phase outs, connected with slot conductors of the upper layer. Pos. 3 points connection of half-branches. Pos. 4 points connection spots of slot and overhang parts. There is a small air gap between overhang parts and stator end-face. Overhang deviation is significantly reduced.

Fig. 2b illustrates opposite stator end-face. This is the side without phase outs. Overhang conductors connect slot conductors with each other. Overhang construction does not depend on branch connection order.

Due to length and resistance reduction, power losses in stator winding also decrease. Overhang deviation reduction decrease total length of the electric machine, allowing to apply machine in limited space of HEVs and EVs. Nevertheless, irregular cross-section of stator conductors influences thermal processes in stator [12]. This needs be considered especially in dynamic modes, start/stop processes of hybrid and electric vehicles, for example.

In this article, thermal processes for different air gap value between overhang parts and stator end-faces are considered. Research is made for induction squirrel-cage machine with compact stator winding.

2. Theoretical Basement

For electromagnetic and thermal research in electromechanical converters, finite element method is suitable [13, 14].

For 2D flat model, electromagnetic field is described by equation:

$$j\omega\sigma\mathbf{A}+\nabla\times\left(\mu_0^{-1}\mu_r^{-1}\nabla\times\mathbf{A}\right)-\sigma\mathbf{v}\times\left(\nabla\times\mathbf{A}\right)= =\left(\frac{\sigma\Delta V}{L}+J_z^e\right)e_Z \cdot \mathbf{A}=A_z e_z \tag{1}$$

here: J_z^e – current density; **v** – rotor velocity vector; σ – conductivity; A – magnetic vector potential; μ_0 - magnetic constant; μ_r - relative permeability; ΔV – potentials difference on the conductor ends; L – conductor length by the Z axis; ω – cyclical frequency.

Zero magnetic potential on the distant border of computing area is the boundary condition:

$$A_z = 0 \tag{2}$$

Difference between magnetic circuits of rotor and stator is considered with B-H curves.

Definition of limit operation modes, when the conductors and the cores of stator and rotor are not overheated, is one of the design issues [15–22]. For this reason, internal heat sources in stator and rotor conductors are calculated in electromagnetic problem. In squirrel-cage conductors current frequency depends on slip; this leads to skin effect and increased power losses in conductors. Heat sources definition in overhang parts of stator winding is another issue. Broken magnetic circuit gives another current density condition, despite of small conductor dimensions. Even if current penetration depth exceeds conductor cross-section, current density and heat dissipation density conditions vary a lot.

Thermal processes calculation faces problems, concerned with heat transfer coefficient definition between stator and rotor construction and air [23–26], as well as thermal conductivity of multi-layer insulation and air gaps. Heat transfer coefficient is defined in hydrodynamic problem of air transfer inside the housing, or by theoretical calculation within proper criteria. Air gap can be easily taken into account in mathematical model, but in practice air gap depends of many different circumstances, such, as vibration, assembly quality, etc.

The following terms influence induction machine cooling condition:

1. Forced cooling provides air flow and improves heat convection in overhang parts.
2. Reduction of gap between overhang parts and machine housing increases air flow resistance and reduces heat transfer.
3. Overhang parts of stator winding are copper busbars with insulation layer and air gap between each other. Thermal conductivity is low, and overhang conductors in the middle of the overhang layer are in hardest thermal condition.

Thermal calculation in «stator – rotor – stator overhangs – environment» system takes into consideration overhang position. Small air gap prevents air circulation, reducing heat convection between overhangs and stator core. The same situation we can see in air gap between stator and rotor. Without axial air flow, temperature of air between stator and rotor cores is high.

For thermal distribution calculation in stator winding and core, the following 3D mathematical model is given:

$$\frac{\partial T(r,x,t)}{\partial t} = \nabla \cdot (k\nabla T) + \frac{1}{c\gamma} Q(x,y,z,t) \tag{3}$$

Initial and boundary conditions are:

$$\begin{cases} T(x,y,z,0) = T_0 = T_c; \\ n \cdot (k\nabla T) = \alpha(T - T_C), \end{cases} \tag{4}$$

here: n – normal to the outer surface of computing area; c – specific heat capacity of conductors, core steel and insulation; γ – specific density of conductors, core steel and insulation; α – heat transfer coefficient; k – heat conductivity coefficient of conductors, core steel and insulation; $Q(x,y,z,t)$ – internal heat sources distribution function.

Partial differential equation (3) describes conditions of copper conductors, laminated steel, wire and slot insulation. Anisotropy in stator core, defined by laminated stacks of steel, was considered in mathematical model.

3. Computer Simulation Results

Cross-section of conductors is $3{,}4 \times 1{,}74 = 5{,}9$ *mm*², total cross section of conductors in a stator slot – 34,5 *mm*², and rotor slot – 40 *mm*². Rms current is 60 *A*, frequency – 50 *Hz*.

For this current value and stator length 100 *mm*, thermal calculation has given the following results: in each stator conductor average power losses $\Delta P_1^S = 0{,}61 \div 0{,}988$ *W*, total power losses in six slot conductors $\Delta P_6^S = 4{,}963$ *W*, rotor conductor power losses $\Delta P_1^R = 2{,}191$ *W*. Instantaneous values of current and power losses in every slot differ from maximum values, and total power loss in stator conductors $\Delta P_{24}^S = 119{,}1$ *W* and in rotor conductors 43,82 *W*. Bulk power density in stator for this values of power losses $Q = 1{,}4 \cdot 10^6$ *W/m³*.

The same calculations for overhang part show small influence of air gap width on bulk power density (table 1).

Table 1.

Δx, *mm*	0,3	1	10
Q, *W/m³*	998000	997000	968500

Air gap between overhang parts and stator end-face has small impact on stator bulk power. Thus, this variable would have minimum value in case air gap exceeds 10 *mm*. For each 30 *mm* overhang conductor heat power $\Delta P_{1OV}^S = 0{,}177$ *W*, and for array of six conductors heat power is 1,062 *W*.

For stator end-face and the nearest overhang winding parts, air flow is parallel to stator end-face surface. Average air speed equals linear speed of the fan impeller outer parts:

$$v_0 = \omega r, \tag{5}$$

here: ω – rotor cycling speed, r – rotor radius.

For 200 *Hz* source frequency and 4 pole-pairs number, rotation frequency equals 3000 *rpm*. Rotor radius is 0,044 *m*, so air speed reaches 13,86 *m/s*.

Reynolds number is defined by perimeter of average stator circle [21]:

$$\mathrm{Re} = \frac{v_0 l_m}{\nu} \tag{6}$$

Perimeter of stator circle:

$$l_m = 2\pi(r_{\max} + r_{\min})/2 = \pi(0{,}065 + 0{,}044) = 0{,}342m \tag{7}$$

For air kinematic viscosity equals $\nu = 15{,}1 \cdot 10^{-6}$ m^2/s, Reynolds number:

$$\mathrm{Re} = \frac{13{,}86 \cdot 0{,}342}{15{,}1 \cdot 10^{-6}} = 3{,}14 \cdot 10^5 \tag{8}$$

Air flow mode is defined by Reynolds number. For flat surface streamlining, turbulence appears when Reynolds number exceeds $5 \cdot 10^5$. But in case of perturbance by complex form of overhang parts, unstable mode appears. Hence, we take into consideration turbulent air flow, and Nusselt number is [21]:

$$\overline{Nu} = 0.032\,\mathrm{Re}^{0.8} = 0{,}032 \cdot \left(3{,}14 \cdot 10^5\right)^{0{,}8} = 799{,}27 \tag{9}$$

From another hand, Nusselt number is:

$$\overline{Nu} = \frac{\alpha \cdot l_m}{\lambda} \tag{10}$$

From this equation, convective heat transfer coefficient is:

$$\alpha = \frac{Nu \cdot \lambda}{l_m} = \frac{799{,}27 \cdot 0{,}0257}{0{,}342} = 60W / m^2 {}^\circ C \tag{11}$$

here: λ – air heat conductivity, equals 0,0257 *W/m °C* for air temperature 20 °C.

Convective heat transfer coefficient varies from 5 $W/(m^2{}^\circ C)$ without air convection to 60 $W/(m^2{}^\circ C)$ with high intensity air flow. On the surfaces that are closed from the direct air flow transitional coefficient values are set. Every conductor part needs be considered in limit operation modes. Temperature distribution in stator for different overhang positions illustrated on fig. 3, 4.

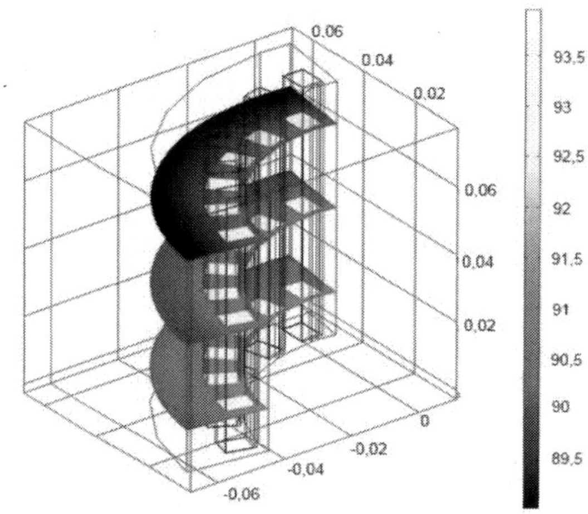

Figure 3. Temperature distribution in stator for 60 *A* stator current and 2 *mm* air gap between overhang parts and stator steel.

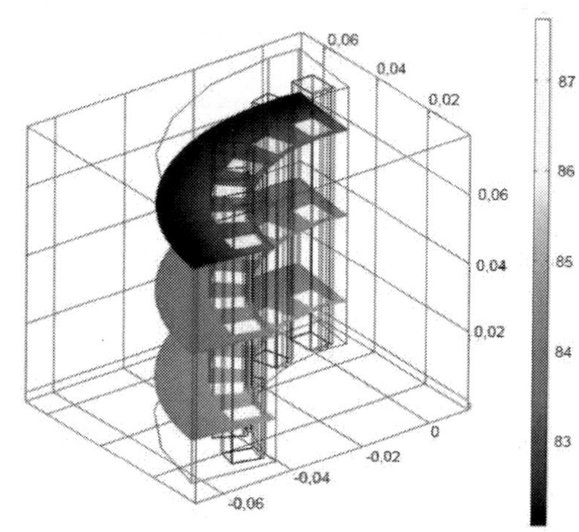

Figure 4. Temperature distribution in stator for 60 *A* stator current and 15 *mm* air gap.

Heat transfer area on stator end-face decreases due to shielding by overhang parts, and stator steel temperature rises. At the same time, heat transfer in enclosed overhang parts decreases due to small air gap between stator and overhang parts.

Overhang conductors could be enclosed with common insulation for greater mechanic solidness and reliability. Thermal conditions for overhang parts are more difficult than for slot parts. This leads to temperature increasing, depicted on fig. 5. Temperature diagram illustrates asymmetrical thermal condition of different overhang conductors.

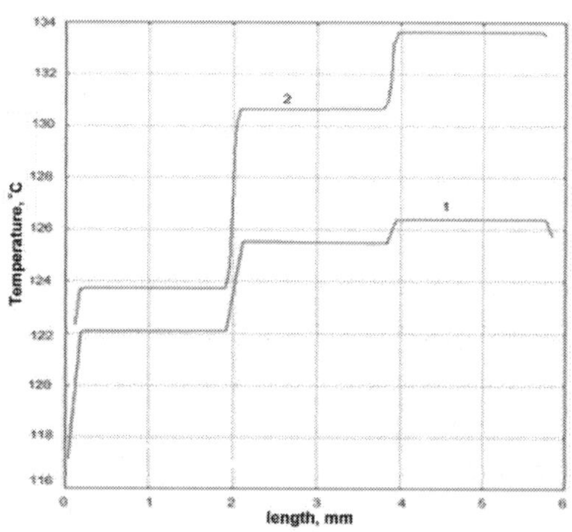

Figure 5. Temperature diagrams in overhang parts for: 1 – with common insulation; 2 – without common insulation.

Hence, to improve heat transfer in overhang parts they need common insulation without air gap between each other.

To increase air flow and heat transfer in stator end-faces, air gap between overhang parts and stator needs be increased. Temperature calculation for wide air gap is given below. Calculations are made for variety of heat transfer coefficients. Average results of temperature distribution in slot and overhang parts are given in table 2.

Table 2. Temperature distribution in stator compact winding.

Heat transfer coefficient, *W/(m²°C)*		Slot conductors temperature, *°C*		Overhang conductors temperature, *°C*	
Stator end-face	**Overhang parts**	**min**	**max**	**min**	**max**
30	29	107.5	109.5	108.3	109.1
30	35	98.8	100.6	99.1	100
30	39	86.8	88.9	86.8	87.7
60	43	73.5	75.8	72.8	73.55
60	29	90.3	92	89.8	90.3

From this results we can see, that temperature gradient in slot and overhang parts becomes lower, as well as maximum temperature in machine. From the other hand, overhang deviation and total length increase.

4. Results Discussion

The given research work explains the possibility of significant reduction of induction machines dimensions. This is able due to original overhang parts construction. Innovative construction needs new methods of thermal calculation and machine design. Design of electric machine in static and dynamic modes should consider reduced heat transfer. Minimum temperature in windings and cores is the best criteria for construction optimization. Improved precision simulation should be held for heat transfer, thermal and electric conductivity determination, with the help of multiphysical models of electromagnetic, thermal fields and aerodynamics.

New construction of compact overhang parts allows to increase air gap between them and stator end-face. Big air gap reduces temperature gradient in stator winding. However, this solution negates positive effect of dimension reduction. It is necessary to conduct further researches of thermal reliability and cooling system optimization.

Conclusions

1. Proposed compact winding construction for induction machines allows to modify overhang deviation and rule thermal condition of machine.
2. To improve thermal conductivity of conductors, overhang parts are placed closely within common insulation.
3. To decrease temperature gradients within induction machine and improve heat transfer, air gap between overhang parts and stator should be increased.
4. Rely thermal condition allows to increase current and power of induction machines and starter-generators in static and dynamic modes.

References

[1] Patzak A and Gerling D 2014 Design of an automotive 48V integrated starter-generator on the basis of an induction machine with concentrated windings *17th International Conference on Electrical Machines and Systems ICEMS* pp 2583–2589

[2] Ishigami T, Tanaka Y and Homma H 2015 Motor stator with thick rectangular wire lap winding for HEVs *IEEE Transactions on Industry Applications* vol 51 issue 4 pp 2917–2923

[3] Patzak A, Bachheibl F, Baumgardt A, Dajaku G, Moroz O and Gerling D 2016 ISCAD – Electric High Performance Drive for Individual Mobility at Extra-Low Voltages *SAE Int. J. Alt. Power* vol 5 pp 148–156

[4] Lysenko O A 2016 The research of axel-box 24-pole generators for communications systems of freight railway wagons *2016 Dynamics of Systems, Mechanisms and Machines (Dynamics)*

[5] Ismagilov F R, Vavilov V E, Karimov R D and Bekuzin V I 2015 High-speed-cal electromechanical energy converters with the possibility of integration of the aviation-engine

UAV *Vestnik UGATU (scientific journal of Ufa State Aviation Technical University)* vol 19 no 3 (69) pp 171–177

[6] Dragomirov M S, Zhuravlyov S A and Zaitsev A M 2016 Analysis of cooling systems efficiency of traction induction motors and motor-generators *Vestnik ISEU* no 3 pp 39–43

[7] Kobelev A S and Makarov L N 2016 Two electric vehicle traction motor design concepts *Proceedings of the 9th International (Russian) Conference on Automated Electric Drive (AED-2016)* pp 414–416

[8] Grachev P Yu and Gorbachev E E RU Patent 2 526 835 Energy-efficient electrical machine 2014

[9] Grachev P Yu, Gorbachev E E and Tabachinsky A S 2016 Power Converters with Compact Winding SCIGs for HEVs, Micro Hydroelectric Power Stations and Wind Turbines *2nd International Conference on Industrial Engineering, Applications and Manufacturing ICIEAM* pp 1–5 DOI: 10.1109/ICIEAM.2016.7911521.

[10] Bubnov A V2006 A comparative analysis of the ways to improve the dynamics of electric drive with phase synchronization *Russian Electro mechanics: Universities` News* no 2 pp 24–29.

[11] Grachev P Yu, Strizhakova E V and Tabachinsky A S 2017 Starter-Generator Design and Analysis of Start/Stop Processes for HEVs *International Conference on Industrial Engineering, (ICIE-2017).*

[12] Grachev P Yu, Bazarov A A and Tabachinsky A S 2017 Compact Windings Design for Traction Induction Motors and SCIGs *International Conference on Industrial Engineering (ICIE-2017).*

[13] Livadaru L, Munteanu A, Simion A and Cantemir C-G 2015 Design and finite element analysis of high-density torque induction motor for traction applications *9th International Symposium on Advanced Topics in Electrical Engineering, ATEE 2015* pp 211–214.

[14] Tatevosyan A A 2017 The calculation of the magnetic field of the synchronous magnetoelectric generator *Dynamics of Systems, Mechanisms and Machines (Dynamics)*, DOI:10.1109/Dynamics.2016.7819095.

[15] Sun J-L, Qiu C-W, Peng Y, Dong Z-K and Chang R-W 2016 Combined electromagnetic and thermal FEM modeling of heavy cylinder during induction heating process *Kang T'ieh/Iron and Steel* vol 51 issue 11 pp 93–100.

[16] Kumar A, Marwaha S, Marwaha A and Kalsi N S 2010 Magnetic field analysis of induction motor for optimal cooling duct design *Simulation Modelling Practice and Theory*, Vol 18 Issue 2 pp 157–164

[17] Iskierka S 1992 Finite element analysis of the performance of a ferromagnetic solid-rotor motor *Finite Elements in Analysis and Design* Vol 11 Issue 1 pp 1–8

[18] Korkmaz S A and Kürüm H 2015 3-dimensional Analysis of Asynchronous Motor with Finite Element Method *Procedia - Social and Behavioral Sciences* Vol 174 pp 4018-4025.

[19] Wu J, Hu J, Chen A, Mei P, Zhou X and Chen Z 2016 Numerical analysis of temperature distribution of motor-refrigerant in a R32 rotary compressor *Applied Thermal Engineering* Vol 95 pp 365–373

[20] Sarkar D and Naskar A K 2013 Computation of thermal condition in an induction motor during 4reactor starting *International Journal of Electrical Power & Energy Systems* vol 44 issue 1 pp 938–948

[21] Mikheev M A and Mikheeva I M 1977 *Osnovy teploperedachi.*

[22] Kim C and Lee K-S 2017 Numerical investigation of the air-gap flow heating phenomena in large-capacity induction motors *International Journal of Heat and Mass Transfer* vol 110 pp 746–752.

[23] Kim C, Lee K-S and Yook S-J 2016 Effect of air-gap fans on cooling of windings in a large-capacity, high-speed induction motor *Applied Thermal Engineering* Vol 100 pp 658–667

[24] Hosseini R, Ramezani M and Mazaheri M R 2009 Experimental study of turbulent forced convection in vertical eccentric annulus *Energy Conversion and Management* Vol 50 Issue 9 pp 2266–2274

IOP Conf. Series: Journal of Physics: Conf. Series **944** (2017) 012041 doi:10.1088/1742-6596/944/1/012041

[25] Poncet S, Haddadi S and Viazz S 2011 Numerical modeling of fluid flow and heat transfer in a narrow Taylor–Couette–Poiseuille system *International Journal of Heat and Fluid Flow* Vol 32 Issue 1 pp 128–144
[26] Boglietti A, Cavagnino A and Staton D 2007 Determination of critical parameters in electrical machine thermal models *42nd Annual Meeting of the IEEE. Industry-Applications-Society* (New Orleans, LA).

Stability of Magnetic Bearings on Lorentz Forces

F R Ismagilov, V Ye Vavilov, and V V Ayguzina

Ufa State Aviation Technical University, 12, K. Markxa St. Ufa, 450008, Russia

e-mail: vtipy@mail.ru

Abstract. In this paper, the rotor stability of the ultra-high-speed electrical rotating machines on the magnetic bearings on Lorentz forces was proved. For this system, the Lyapunov function was obtained. Its study allowed determining the sustainable movement of the rotor on the magnetic bearings on Lorentz forces and making recommendations of the controller coefficient selection. The stability analysis was carried out both for magnetic bearings with a rectangular and cylindrical air gaps. The obtained results were verified on an experimental unit and compared with the results of other authors, which shows the convergence of theoretical and experimental data in both cases.

1. Introduction

The development of high-precision laser systems, high-speed photography, medical equipment and robotics requires an increase in the speed of Electrical Rotating machines (ERMs) with a simultaneous trend towards their miniaturization. For some industries, the slotless 100–500 W ERMs with a rotational speed from 500,000 rpm to 1,000,000 rpm and a rotor diameter of 5-10 mm are perspective.

World corporations are actively working to create and improve the efficiency of ERM. In particular, the IHI Corporation (Japan) develops a slotless 400 W microgenerator with a rotational speed of 400,000 rpm for robotics. The Celeroton Corporation (Switzerland) serially develops slotless 100 W electric motors with a rotational speed of 500,000 rpm. These motors are designed for high-precision optical systems, machine tools and medical equipment. Similar works are being conducted in the USA, Belgium and South Korea in the interests of space agencies and the army of these countries [1–3]. As a long-term goal of all these companies are aiming to exceed 1 000 000 rpm, as this will open up entirely new perspectives in different industries.

One of the main problems of Ultra-High Speed Electrical Machine (UHSEM) design is a problem of bearing unit reliability. Miniature, instrumental mechanical bearings can reliably operate at the rotational speed below 400 000 – 500 000 rpm.

Various types of non-contact bearings, such as active magnetic bearings and gas-dynamic bearings, also do not provide reliable operation of such ERMs.

At high rotational speeds, eddy currents induces in a rotor on the active magnetic bearings, which are the reason of additional rotor heating. Moreover, a rotor cooling is extremely difficult at high rotational speeds. Therefore, the active magnetic bearings are not used at frequencies above 250,000 rpm.

Gas-dynamic bearings ensure a non-contact rotation only at a certain frequency (frequency of ascent). Before reaching a given frequency, they work as a mechanical bearing, which significantly reduces their reliability. In addition, gas-dynamic bearings are not able to operate in a vacuum, which limits their use.

The solution of the bearing reliability problems of ultra-high-speed ERMs can be non-contact bearings on the Lorentz forces [4].

The advantages of these bearings are the lack of the stator core, which significantly improves their performance by reducing the electromagnetic time constant. In addition, it is minimum weight and size parameters, a possibility of using at rotor rotational speeds above 500,000 rpm. The main disadvantage of magnetic bearings on the Lorentz forces are the low bearing capacity. However, since the rotor of ultra-high-speed ERM weighs 5–50 grams, this drawback does not prevent the use of magnetic bearings on Lorentz forces (MBLF) in ultra-high-speed ERMs.

2. Statement of research tasks

With all MBLF advantages for ultra-high-speed ERMs, it began to develop relatively recently [4–7]. The research works of this bearing type are aimed to determination of energetic characteristics of MBLF. In particular, it is studied of their forces [4, 5], new design [6] or the calculation methods [7].
At the same time, in addition to energy characteristics, the control system parameters and the rotor stability of the ultra-high-speed ERM have a significant influence on the efficiency of the MBLF operation.
The rotor stability analysis and the control system parameter selection is presented in [8, 9] for active magnetic bearings, in [10] for hybrid magnetic bearings and in [11] for diamagnetic bearings. However, a significant MBLF differences from another bearing types and their higher performance do not allow a full use of the [8–11] work results to assess the MBLF rotor stability. Moreover, these works are performed for the system with PD controller. And as it is shown in [12], PID controllers are more efficient to use in MBLF.
Thus, the objective of this work is the MBLF rotor stability analysis, the study of the processes in the MBLF control system and the experimental verification of the theoretical results.

3. The research object

The research object if this work is the ultra-high-speed ERM with MBLF (Fig. 1). The structural control scheme of the research object is represented in Fig. 2.

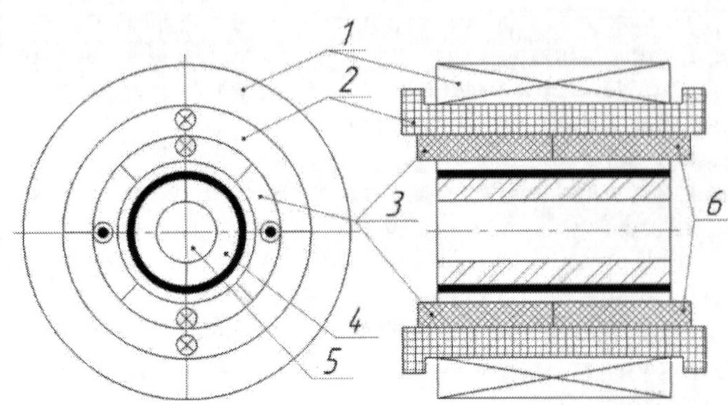

Figure 1. The ultra-high-speed ERM with MBLF: 1 – stator core; 2 – ERM windings; 3 – windings of the left MBLF L; 4 – rotor with permanent magnets; 5– shaft; 6 – windings of the right MBLF.

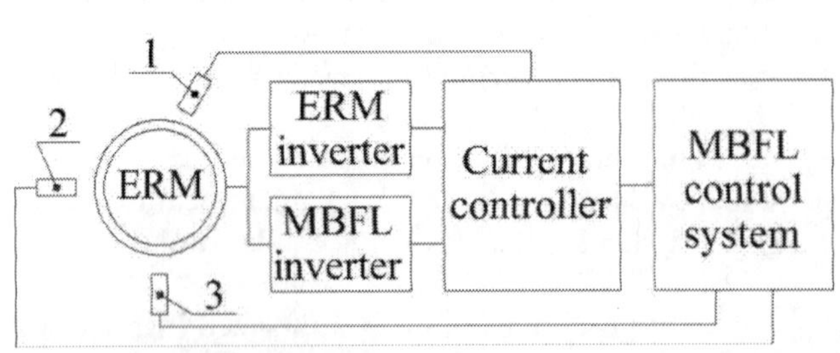

Figure 2. The structural control scheme of the research object: 1 – the rotor position sensor; 2 – the rotor displacement sensor along the x axis; 3 – the rotor displacement sensor along the y axis.

The description of the MBLF operation given on the example of the ERM in the motor mode. The processes occurring in ERM in the generator mode are similar to the processes in the motor mode. The MBLF windings NSL are performed as ordinary polyphase ERM windings. Herewith, the following condition is fulfilled: if the ERM rotor has p pole pairs, the MBLF winding will be carried out for the poles number of $p = p \pm 1$.

In Fig. 1, the ERM rotor has a 1 poles pair, the ERM windings consist of two coils. The MBLF windings consist of 4 coils, i.e. it is made for two poles pairs.

The MBLF contains two windings (right and left), which are installed in the air gap above the rotor with permanent magnets, the rotor position sensors and the control system. To provide a mechanical moment, a current is fed to the motor windings and switched by an ERM inverter. The switching law is determined by the rotor position sensor, the current value is switched in the "current controller". Also current is fed to the MBLF windings and switched by an MBLF inverter. The frequency of switching current is the same with the frequency in the ERM windings.

In each moment of time, between a certain magnet pole and current flowing through the windings, the Lorentz force is arises, which ensures the rotor levitation. And since the MBLF windings are made with the $p = p \pm 1$ poles number, no mechanical moment is created for the ERM. The rotor displacement, as well as in conventional bearings, is estimated by rotor displacement sensors. The signal is fed to the MBLF control system. Based on this signal, a current value is formed and applied to the MBLF.

For generality of this work results, it is considered a two-phase rotor suspension with a plane-parallel air gap (Fig. 3a) and with a cylindrical air gap (Fig. 3b).

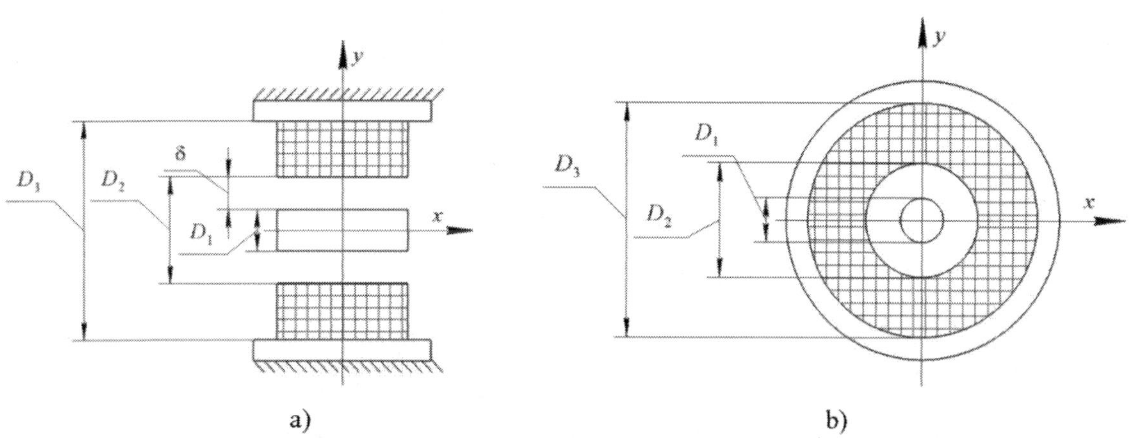

Figure 3. Two-phase rotor suspension of the MBLF with a plane-parallel air gap (a), with a cylindrical air gap (b).

The following assumptions are used to solve the research problems.

– Permanent magnets of the same brand, the same thickness and length are applied. The magnetic flux density on the permanent magnet surface does not change under the external factor influence.

– The environment and winding temperatures are constant throughout the entire working process. Therefore, the MBLF winding resistance is assumed as constant.

– The change in inductance from the current is negligible.

4. Mathematical description of the MBLF

The main characteristic describing the processes in MBLF is the Lorentz force, which act on the rotor with permanent magnets.

For ERM with cylindrical air gap, the resulting Lorentz force can be presented in the matrix form [4]:

$$F_{\mathrm{LF}} = \frac{3}{2} l k_{\mathrm{w}} (I + I_0) B_{\mathrm{M}} \left(\frac{D_1}{2}\right)^2 \ln\left(\frac{D_3}{D_2}\right) \begin{bmatrix} \sin(\varphi - \varphi_{\mathrm{M}}) \\ -\cos(\varphi - \varphi_{\mathrm{M}}) \\ 0 \end{bmatrix}, \tag{1}$$

where l is the MBLF active length; k_{w} is the winding factor; I_0 is the current flowing in the winding; I is the control current; B_{M} is the amplitude of the radial component of the magnetic flux density in the air gap; $D_1 - D_3$ are the geometric dimensions, according to Fig. 3; φ is the phase angle of the MBLF winding; φ_{M} is the rotor rotation angle.

For the MBLF with the plane-parallel air gap (Fig. 3a), the winding phase angle and the rotor rotation angle are equal. And the Lorentz force is determined as following:

$$F_{\mathrm{LF}} = -\frac{3}{2} l k_{\mathrm{w}} (I + I_0) B_{\mathrm{M}} \left(\frac{D_1}{2}\right)^2 \ln\left(\frac{D_3}{D_2}\right), \tag{2}$$

In [13], it is presented equations to determine the magnetic flux density in the air gap, which are functions of the permanent magnet properties and geometry and allow the accurate numerical calculation of the permanent magnet parameters. Since these equations are cumbersome, the more simplified equations for the magnetic flux density calculations were used in this work:

$$B_{\mathrm{M}} = \frac{1}{1 + \frac{(\delta \pm x) B_{\mathrm{r}}}{h \mu_0 H_{\mathrm{cB}} \sigma}} \frac{B_{\mathrm{r}}}{\sigma}, \tag{3}$$

where B_{r} is the residual flux density of the permanent magnet; δ is the air gap of the ERM; H_{cB} is the coercive force of the permanent magnet; h is the height of the permanent magnet; μ_0 is the magnetic permeability of vacuum; σ is the leakage coefficient of the rotor magnetic system; x is the rotor displacement.

The appearance of rotor displacement in the x-axis direction leads to the following forces in the MBLF with cylindrical air gap:

$$F_{\mathrm{LF1}} = -\frac{3}{2} \cdot \frac{l k_{\mathrm{w}} (I_0 - I)}{1 + \frac{(\delta + x) B_{\mathrm{r}}}{h \mu_0 H_{\mathrm{cB}} \sigma}} \cdot \frac{B_{\mathrm{r}}}{\sigma} \cdot \left(\frac{D_1}{2}\right)^2 \cdot \ln\left(\frac{D_3}{D_2}\right) \cdot \begin{bmatrix} \sin(\varphi - \varphi_{\mathrm{M}}) \\ -\cos(\varphi - \varphi_{\mathrm{M}}) \\ 0 \end{bmatrix}, \tag{4}$$

$$F_{\mathrm{LF2}} = \frac{3}{2} \cdot \frac{l k_{\mathrm{w}} (I + I_0)}{1 + \frac{(\delta - x) B_r}{h \mu_0 H_{cB} \sigma}} \cdot \frac{B_r}{\sigma} \cdot \left(\frac{D_1}{2}\right)^2 \cdot \ln\left(\frac{D_3}{D_2}\right) \cdot \begin{bmatrix} -\sin(\varphi - \varphi_{\mathrm{M}}) \\ \cos(\varphi - \varphi_{\mathrm{M}}) \\ 0 \end{bmatrix}, \tag{5}$$

In system with the plane-parallel gap, it is followings:

$$F_{\mathrm{LF1}} = -\frac{3}{2} \cdot \frac{l k_{\mathrm{w}} (I + I_0)}{1 + \frac{(\delta - x) B_r}{h \mu_0 H_{cB} \sigma}} \cdot \frac{B_r}{\sigma} \cdot \left(\frac{D_1}{2}\right)^2 \cdot \ln\left(\frac{D_3}{D_2}\right), \tag{6}$$

$$F_{\mathrm{LF2}} = \frac{3}{2} \cdot \frac{lk_{\mathrm{w}}(I+I_0)}{1+\dfrac{(\delta+x)B_r}{h\mu_0 H_{cB}\sigma}} \cdot \frac{B_r}{\sigma} \cdot \left(\frac{D_1}{2}\right)^2 \cdot \ln\left(\frac{D_3}{D_2}\right), \tag{7}$$

The obtained equations (2.4) – (2.7) show that the forces in the MBLF system linearly dependent on the current unlike the active or hybrid magnetic bearings due to active magnetic bearing and hybrid magnetic bearings have a quadratic dependence. A linear dependence greatly simplifies the control processes.

Next, for simplification of the calculations, a system with plane-parallel gap was considered. With above observations, the resulting solution can be used for systems with cylindrical gap.

To analyze the stability, it is necessary to take into account the MBLF winding inertia:

$$U_{\mathrm{inc}} = R_0 I + \frac{d}{dt}\left(L_0 I + \varphi_{\mathrm{pm}}\right), \tag{8}$$

where U_{inc} is the voltage increment in the MBLF windings; R_0 is the resistance of the MBLF windings; L_0 is the inductance in the MBLF windings; φ_{pm} is the flux linkage of the permanent magnets.

In addition, it is necessary to consider the inertia of the measurement sensors [9]:

$$\gamma \frac{dx_{\mathrm{rp}}}{dt} = x - x_{\mathrm{rp}}, \tag{9}$$

where γ is the time constant; x_{rp} is the rotor position coordinate.

5. The mathematical description of the rotor position control system in the MBLF

As a logical control device for MBLF, the PID controller is used. The regulation current is determined depending on the controller coefficients and the measured error e of the rotor position:

$$I = -(K_1 e + K_2 \frac{de}{dt} + K_3 \int_{t_0}^{t} e d\tau), \tag{10}$$

where $e = q - r$ is the difference between the actual and the desired values of the generalized coordinates; K_1, K_2, K_3 are diagonal matrices of respective amplification factors:

$$K_1 = \mathrm{diag}(K_{11}, K_{11}, K_{12}, K_{12}),$$
$$K_2 = \mathrm{diag}(K_{21}, K_{21}, K_{22}, K_{22}),$$
$$K_3 = \mathrm{diag}(K_{31}, K_{31}, K_{32}, K_{32}).$$

The estimated amplification values are determined by equations:

$$K_{11} = K_{12} = \frac{-2k_{\mathrm{s}1,2}}{k_{\mathrm{i}1,2}},$$

$$K_{21} = K_{22} = \frac{\sqrt{\dfrac{-Mk_{\mathrm{s}1,2}}{2}}}{k_{\mathrm{i}1,2}},$$

$$K_{31} = K_{32} = \frac{Mp}{k_{\mathrm{i}1,2}},$$

where M is the rotor mass, p is a positive number.

The control purpose is to stabilize of the rotor with permanent magnets or the magnet at the equilibrium position, then $r = 0$. Measurement of the magnet position relative to the air gap is carried out using motion sensors such as linear Hall sensors, inductive sensors or eddy-current sensors. Therefore, the control law takes the form:

$$I = K_1 q_s + K_2 \frac{dq_s}{dt} + K_3 \int_{t_0}^{t} q_s dt, \quad (11)$$

where $q_s = (x_{rp}, x_{rp2})^T$ is the matrix of magnet displacement in the sensor measurement area.
A correlation between the generalized coordinates of the levitating magnet mass centre or of the rotor and the matrix q_s is determined by the ratio $q_s = T_s q$, where T_s is the transformation matrix, which is taken in the form:

$$T_s = \begin{vmatrix} 1 & 0 & 0 & -a_{s1} \\ 0 & 1 & a_{s1} & 0 \\ 1 & 0 & 0 & a_{s2} \\ 0 & 1 & -a_{s2} & 0 \end{vmatrix}. \quad (12)$$

6. Stability analysis of the ERM rotor with MBLF

To analyse the stability of the rotor on the MBLF, it is necessary to formulate the equation of the perturbed motion of the rotor on the MBLF, which, taking into account the time constant γ of the position sensors along the x axis, is:

$$m\left(\gamma \frac{d^3 x_{rp}}{dt^3} + \frac{d^2 x_{rp}}{dt^2}\right) = -\frac{3}{2} l k_w (I + I_0) \frac{k_2}{1 + (\delta - x)k_1} \left(\frac{D_1}{2}\right)^2 \ln\left(\frac{D_3}{D_2}\right) + \frac{3}{2} l k_w (I - I_0) \frac{k_2}{1 + (\delta + x)k_1} \left(\frac{D_1}{2}\right)^2 \ln\left(\frac{D_3}{D_2}\right), \quad (13)$$

where $k_1 = \frac{(\delta - x)B_r}{h\mu_0 H_{cB}\sigma}$; $k_2 = \frac{B_r}{\sigma}$; m is the rotor mass.

To account the inertia of the rotor position and winding sensors, expressions (2.7) and (2.8) are introduced in (4.1):

$$m\left(\gamma \frac{d^3 x_{rp}}{dt^3} + \frac{d^2 x_{rp}}{dt^2}\right) = -\frac{3}{2} l k_w \left(\frac{D_1}{2}\right)^2 \ln\left(\frac{D_3}{D_2}\right) k_2 \times \times \left[\frac{\left(\frac{U_{inc}}{R_0} - \frac{dIL_0}{dt} - \frac{d\varphi_{pm}}{dt}\right)}{1 + \left(\delta - \gamma \frac{dx_{rp}}{dt} - x_{rp}\right)k_1} - \frac{\left(I_0 - \frac{U_{inc}}{R_0} + \frac{dIL_0}{dt} + \frac{d\varphi_{pm}}{dt}\right)}{1 + \left(\delta + \gamma \frac{dx_{rp}}{dt} + x_{rp}\right)k_1}\right]. \quad (14)$$

Both parts of expression (4.2) are integrated:

$$m\gamma\frac{d^2x_{rp}}{dt^2}+m\frac{dx_{rp}}{dt}=-\frac{3}{2}lk_w\left(\frac{D_1}{2}\right)^2\ln\left(\frac{D_3}{D_2}\right)k_2\times$$
$$\times\left[\int\frac{\left(\frac{U_{inc}}{R_0}-\frac{dIL_0}{dt}-\frac{d\varphi_{pm}}{dt}\right)}{1+\left(\delta-\gamma\frac{dx_{rp}}{dt}-x_{rp}\right)k_1}-\int\frac{\left(I_0-\frac{U_{inc}}{R_0}+\frac{dIL_0}{dt}+\frac{d\varphi_{pm}}{dt}\right)}{1+\left(\delta+\gamma\frac{dx_{rp}}{dt}+x_{rp}\right)k_1}\right]. \quad (15)$$

In this expression, the control law (3.2) is introduced by multiplying both sides of expression (4.3) to the expression (3.2):

$$m\gamma\frac{d^2x_{rp}}{dt^2}\left(K_1x_{rp}+K_2\frac{dx_{rp}}{dt}+K_3\int x_{rp}dt\right)+m\frac{dx_{rp}}{dt}\left(K_1x_{rp}+K_2\frac{dx_{rp}}{dt}+K_3\int x_{rp}dt\right)=$$
$$=-\frac{3}{2}lk_w\left(\frac{D_1}{2}\right)^2\ln\left(\frac{D_3}{D_2}\right)k_2\left(K_1x_{rp}+K_2\frac{dx_{rp}}{dt}+K_3\int x_{rp}dt\right)\times$$
$$\times\left[\int\frac{\left(\frac{U_{inc}}{R_0}-\frac{dIL_0}{dt}-\frac{d\varphi_{pm}}{dt}\right)}{1+\left(\delta-\gamma\frac{dx_{rp}}{dt}-x_{rp}\right)k_1}-\int\frac{\left(I_0-\frac{U_{inc}}{R_0}+\frac{dIL_0}{dt}+\frac{d\varphi_{pm}}{dt}\right)}{1+\left(\delta+\gamma\frac{dx_{rp}}{dt}+x_{rp}\right)k_1}\right]. \quad (16)$$

A replacement is introduced: $Q_1=\frac{3}{2}lk_w\left(\frac{D_1}{2}\right)^2\ln\left(\frac{D_3}{D_2}\right)k_2$.

Using the known transformations of analogous mathematical expressions for systems with electromagnetic suspension [9], the entire system is reduced to the form:

$$m\gamma K_2\frac{d^2x_{rp}}{dt^2}\frac{dx_{rp}}{dt}+mK_1x_{rp}\frac{dx_{rp}}{dt}+m\gamma K_1\left(x_{rp}\frac{dx_{rp}}{dt}\right)'_t+$$
$$+m\gamma K_3\frac{d^2x_{rp}}{dt^2}\int x_{rp}dt=Q_1\left(K_1x_{rp}+K_2\frac{dx_{rp}}{dt}+K_3\int x_{rp}dt\right)\times$$
$$\times\left[\int\frac{\left(I_0-\frac{U_{inc}}{R_0}+\frac{dIL_0}{dt}+\frac{d\varphi_{pm}}{dt}\right)}{1+\left(\delta+\gamma\frac{dx_{rp}}{dt}+x_{rp}\right)k_1}-\int\frac{\left(I_0+\frac{U_{inc}}{R_0}-\frac{dIL_0}{dt}-\frac{d\varphi_{pm}}{dt}\right)}{1+\left(\delta-\gamma\frac{dx_{rp}}{dt}-x_{rp}\right)k_1}\right]+ \quad (17)$$
$$+m\gamma K_1\left(\frac{dx_{rp}}{dt}\right)^2-mK_2\frac{dx_{rp}}{dt}-mK_3\frac{dx_{rp}}{dt}\int x_{rp}dt.$$

The right part is represented as a time derivative:

$$\frac{d}{dt}\left[\gamma K_2\left(\frac{dx_{rp}}{dt}\right)^2 + K_1 x_{rp}^2 + \gamma K_1\left(x_{rp}\frac{dx_{rp}}{dt}\right) + m\gamma K_3\frac{dx_{rp}}{dt}\int x_{rp}dt\right] =$$

$$\left[\int\frac{\left(I_0 - \frac{U_{inc}}{R_0} + \frac{dIL_0}{dt} + \frac{d\varphi_{pm}}{dt}\right)}{1+\left(\delta+\gamma\frac{dx_{rp}}{dt}+x_{rp}\right)k_1} - \int\frac{\left(I_0 + \frac{U_{inc}}{R_0} - \frac{dIL_0}{dt} - \frac{d\varphi_{pm}}{dt}\right)}{1+\left(\delta-\gamma\frac{dx_{rp}}{dt}-x_{rp}\right)k_1}\right]\times \tag{18}$$

$$\times Q_1\cdot\left(K_1x_{rp} + K_2\frac{dx_{rp}}{dt} + K_3\int x_{rp}dt\right) + m\gamma K_1\left(\frac{dx_{rp}}{dt}\right)^2 - mK_2\frac{dx_{rp}}{dt} - mK_3\frac{dx_{rp}}{dt}\int x_{rp}dt.$$

It is obvious that the right side is Lyapunov's function, and the left side is its derivative. The left part is a sign-definite negative function provided that $\gamma K_1 > K_2 + K_3$. Herewith, the right side is a sign-definite positive function, which is the evidence of the asymptotic system stability according to the Lyapunov theorem. Then, the general rule for the asymptotic rotor stability of the ultra-high-speed ERM is the regulator setting rule: $\gamma K_1 < K_2 + K_3$.

7. Theoretical result verification

To verify the obtained results, an experimental unit with MBFL was developed. It is presented in Fig. 4.

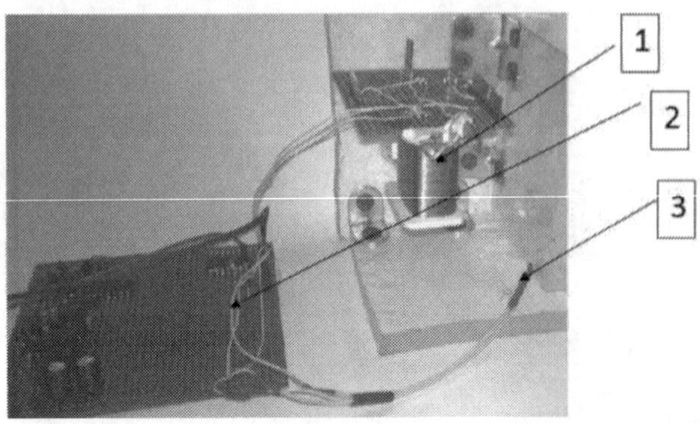

Figure 4. Experimental unit with MBFL: 1 – coil; 2 – control system with PID controller; 3 – Hall sensor

Two Hall sensors with a time constant of 3 ms were used as a rotor displacement sensor. A permanent magnet NdFeB weighing 15 grams was used for suspension. The current control through the coil was carried out by the pulse-width modulation. The PID regulator coefficients were chosen according to the received inequality and accounted for $K_1 = 1$, $K_2 = 5$, $K_3 = 1$. These coefficients allow to achieve a steady permanent magnet suspension during 5-7 seconds. After this time, an error was accumulated in the integrating link and the system emerged from the stability state. To solve this problem, it is necessary to use the cleaning of the integrating link.

When the coefficients were changed by $K_1 = 1000$, $K_2 = 0.5$, $K_3 = 0.1$, the magnet levitation was absent, the magnet adhered to the coil, i.e. the system immediately lost its stability.

It should be noticed that similar experiments were carried out in [11], where the stable rotor suspension in the MBFL was obtained with the coefficients $K_1 = 4000$, $K_2 = 15$, $K_3 = 200$ and the sensor time constant of 0.3 ms. These coefficients also satisfy the obtained inequality $0.00003 \cdot 4000 < 215$.

Thus, the experimental results of [12] confirm theoretical and experimental results if this research.

8. Conclusion

In this paper, the rotor stability of the ultra-high-speed ERM on MBFL was proved. For this system, the Lyapunov function was obtained. Its study allowed determining the sustainable movement of the rotor on the MBFL and making recommendations of the controller coefficient selection. The stability analysis was carried out both for magnetic bearings with a rectangular and cylindrical air gaps. The obtained results were verified on an experimental unit and compared with the results of other authors, which shows the convergence of theoretical and experimental data in both cases.

Acknowledgments

This work was supported by the Russian Foundation for Basic Research (project 16-38-60001).

References

[1] Zwyssig C, Kolar J W and Round S D 2009 Mega-Speed Drive Systems: Pushing Beyond 1 Million RPM *IEEE/ASME Transactions on Mechatronics* **14** (5) pp 564–574

[2] Zwyssig C, Kolar J W and Round S D 2006 Power Electronics Interface for a 100 W, 500000 rpm Gas Turbine Portable Power Unit *USA Applied Power Electronics Conference* pp 283–289

[3] Isomura K, Murayama M and Teramoto S 2006 Experimental Verification of the Feasibility of a 100W Class Micro-scale Gas Turbine at an Impeller Diameter of 10 mm *J. Micromech. Microeng* **16** pp 254–261

[4] Looser A, Baumgartner T, Kolar J W and Zwyssig C 2010 Novel highspeed, Lorentz-type, slotless self-bearing motor *IEEE Energy Conversion Congress and Exposition (ECCE 2010)* pp 3971–3977

[5] Looser A, Baumgartner T, Kolar J W and Zwyssig C 2012 Analysis and Measurement of Three-Dimensional Torque and Forces for Slotless Permanent-Magnet Motors IEEE *Transactions on Industry Applications* **48** (4) pp 1258–1266

[6] Looser A, Baumgartner T, Kolar J W and Zwyssig C 2014 Analysis and Design of a 300-W 500 000-r/min Slotless Self-Bearing Permanent-Magnet Motor *IEEE Trans. Industrial Electronics* **61** pp 4326–4336

[7] Baumgartner T and Kolar J W 2013 Multivariable state feedback control of a 500 000 rpm self-bearing motor *Electric Machines & Drives Conference (IEMDC), 2013 IEEE International* pp 347–353

[8] Balandin D V and Kogan M M 2011 Motion control for a vertical rigid rotor rotating in electromagnetic bearings (in Russian) *Izvestiya Rossiyskoy Akademii Nauk. Teoriya i sistemy upravleniya* **5** pp 3–17

[9] Vostokov V S, Lebedeva S V, 2011 Stability of electromagnetic suspension of a rotor (in Russian) *Izvestiya Rossiyskoy Akademii Nauk. Teoriya i sistemy upravleniya* **2** pp 3–37

[10] Ismagilov F, Gerasin A, Khayrullin I and Vavilov V 2014 Stability analysis of hybrid magnetic bearings *Journal of Computer and Systems Sciences International* **53** (1) pp 130–136

[11] Jie-Yu Chen and Jian-Bin Zhou 2008 Diamagnetic bearings for MEMS: Performance and stability analysis *Mechanics Research Communications* **35** (8) pp 546–552

[12] Ueno S and Kato T 2009 A novel design of a Lorentz-force-type small self-bearing motor *Taipei International Conference on Power Electronics and Drive Systems (PEDS)* pp 926–931

[13] Ismagilov F, Khayrullin I, Vavilov V and Yakupov A 2016 Mathematical Model of High-frequency Electromechanical Energy Transducer with High-coercitive Permanent Magnets *IAENG International Journal of Applied Mathematics* **46** (3) pp 1–9

IOP Conf. Series: Journal of Physics: Conf. Series **944** (2017) 012043 doi:10.1088/1742-6596/944/1/012043

Influence of primary elements parameters on result of voice information protection estimation

A V Ivanov, I L Reva, and D S Kofanov

Novosibirsk State Technical University, 20, Karla Marksa ave., Novosibirsk, 630073, Russia

E-mail: andrej.ivanov@corp.nstu.ru

Abstract. The purpose of the study is to investigate the main characteristics of measurement microphones. An experiment with use of different parameters microphones was conducted. Influence of the parameters on word intelligibility was estimated. All measurements were made in the same fix point by different microphones. Essential influence of inherent noise level of microphones was discovered by protection estimation in a fix point without active protection means (measurements are made on inherent noise level).

1. Introduction

The criteria of voice information protection from leakage through vibro-acoustic channels is word intelligibility, which can be estimated with the Pokrovsky`s approach [1], adjusted for information security problems [2]. The intelligibility is calculated with results of acoustic and vibrational signals. Sound meters, microphones and acceleration meters are used as measuring means. Obviously, the primary elements characteristics can essentially influence on accuracy of physical quantities measurements and therefore on word intelligibility calculation result. The question is how essential the influence.

Conduct an experiment of word intelligibility estimation in the same fix point with use of microphones with different characteristics to estimate the influence quantity of primary elements. Analyze the derived results and make conclusions.

2. The main microphones characteristics

The main microphones characteristics are the following [3]:

- sound pressure level – ratio of output microphone voltage to sound pressure influencing on it;
- dynamic range – difference between limit sound pressure levels and inherent noise;
- frequency range;
- inherent noise level;
- signal-noise ratio.

All the parameters, one way or another, influence on a measurements result. For example, low sound pressure level of a microphone influence on measurement error (it increases) because the signal on preamplifier is low, and amplification of a low signal leads to error increase. Inherent noise level shows what is the minimal useful signal level can be perceived by preamplifier. Signal-noise ratio is connected with dynamic range and inherent noise level (Sometimes the inherent noise level is not mentioned and signal-noise ratio is mentioned instead).

The inherent noise influence is the most essential in fix points without active protection means

IOP Conf. Series: Journal of Physics: Conf. Series **944** (2017) 012043 doi:10.1088/1742-6596/944/1/012043

(noise generators), i.e. the signal is measured on background noise, level of which can be lower than inherent noise level.

3. Experiment

The 3 following microphones were used for the experiment: GRAS 40AE (first accuracy class), ZETLAB BC 501 (third accuracy class) and Behringer ECM8000 used in hardware-software complex for acoustic and vibro-acoustic measurements «Sprout-mini-A».

The main technical characteristics of the compared devices are the same, a big difference is only in inherent noise level. The characteristics are shown in table 1.

Table 1. Technical characteristics of microphones.

Technical characteristics	GRAS 40AE	ZETLAB BC 501	Behringer ECM8000
Frequency range, Hz	3.15 – 20 000	50 – 13 000	20 – 20 000
Sound pressure level, mV/Pa	50	50	10
Dynamic range, dB	19 – 146	35 – 140	-
Inherent noise level, dBA	19	35	-

A fix point (a vestibule with doors), in which word intelligibility (W) with use of Behringer ECM8000 is less than 0.2 (less than 20%). Make the same measurements with use of the other microphones.

4. Experiment results

Word intelligibility calculation was made with use of the approach [2] on 7 octave bands. For word intelligibility estimation the following measurements were made:

- test signal level;
- noise level (background);
- noise+signal mixture (with turned on test signal generator).

The approach [2] consists in next:

- all frequency range of a voice signal separated on 7 octave bands with center frequencies 125, 250, 500, 1000, 2000, 4000, 8000 of Hz respectively;
- for each i-th band the noise level of L_{ni} and level of a compound of signal + noise by $L_{(s+n)i}$ is measured;
- levels of signals of L_{si} for each octave band are calculated;
- feeling levels are calculated;
- on known dependence of P(Q) coefficients of perception of P_i for each octave band are calculated;
- the formant intelligibility is calculated;
- on known dependences transfer from formant intelligibility to speech intelligibility.

The derived results of measurements and calculations are show in tables 2–4.

There is no essential difference in measurement of high-level signals (test signal and signal+noise mixture) for different microphones (the difference is less than 2 dB). But for low-level signals (background noise) there is influence of inherent noise level of each microphone. Make calculations of word intelligibility for each microphone (tables 5–7).

Table 2. Levels of test signal in each octave band.

Average compound octave band frequency F, Hz	Test signal level L_{testi} , dB
125	78.2
250	83.7
500	83.6
1000	89.3
2000	88.4
4000	90.9
8000	88.3

Table 3. Noise levels in each octave band.

Average compound octave band frequency F, Hz	Noise level Lni (microphone ECM8000), dB	Noise level Lni (microphone BC 501), dB	Noise level Lni (microphone 40AE), dB
125	46.3	50.9	45.3
250	38	41.9	36.8
500	33.3	26.5	24.9
1000	30.2	23.1	19.1
2000	29.2	21.8	14.9
4000	28.3	22.0	13.6
8000	32.9	22.1	13.5

Table 4. Signal+noise levels in each octave band.

Average compound octave band frequency F, Hz	Test signal level L_{testi} , dB
125	60.8
250	55.1
500	52
1000	45.7
2000	41
4000	40.2
8000	35

Table 5. Calculated parameters values with use of ECM8000 microphone.

Octave band number	Difference between test signal level and average speech level ΔL_i, dB	Signal level L_{si} , dB	Sensation level Q_i, dB	Perception index p_i	Word intelligibility W
1	25.20	60.64	-35.86	0.0012	
2	17.70	55.01	-18.69	0.0634	
3	17.60	51.94	-12.96	0.1462	
4	28.30	45.58	-21.92	0.0357	0.19
5	32.40	40.70	-26.90	0.0127	
6	37.90	39.91	-31.29	0.0044	
7	39.30	30.84	-45.36	0.0001	

Table 6. Calculated parameters values with use of BC 501 microphone.

Octave band number	Difference between test signal level and average speech level ΔL_i, dB	Signal level L_{si}, dB	Sensation level Q_i, dB	Perception index p_i	Word intelligibility W
1	25.20	60.33	-40.77	0.0003	
2	17.70	54.89	-22.71	0.0307	
3	17.60	51.99	-6.11	0.3092	
4	28.30	45.68	-14.72	0.1156	0,50
5	32.40	40.95	-19.25	0.0577	
6	37.90	40.13	-24.77	0.0203	
7	39.30	34.77	-30.63	0.0052	

Table 7. Calculated parameters values with use of 40 AE microphone.

Octave band number	Difference between test signal level and average speech level ΔL_i, dB	Signal level L_{si}, dB	Sensation level Q_i, dB	Perception index p_i	Word intelligibility W
1	25.20	60.68	-34.82	0.0017	
2	17.70	55.04	-17.46	0.0772	
3	17.60	51.99	-4.51	0.3570	
4	28.30	45.69	-10.71	0.1919	0.79
5	32.40	40.99	-12.31	0.1586	
6	37.90	40.19	-16.31	0.0921	
7	39.30	34.97	-21.83	0.0364	

5. Analysis of the results

Results of the experiment showed that inherent noise of microphones influence on word intelligibility essentially. From value 0.19 (hiding of the private conversations fact in accordance with [4]) to 0.79 (information is not protected, value 0.4 – hiding of the private conversations content [4] – essentially exceeded).

6. Conclusion

Influence of the preamplifier characteristics on word intelligibility was investigated. Influence of microphones inherent noise level was experimentally estimated by the example with three different microphones. The derived results show necessity of specifying requirements for preamplifier parameters, especially – inherent noise level. Also, it is possible to specify requirements for background noise on octave bands. If a chosen measuring mean cannot provide signals measurement on this level, consider the measurement result as signal+noise mixture and consider standardized value instead of noise levels (to make so-called «noise calculation»).

References

[1] Pokrovskiy N B 1962 Raschet i izmerenie razborchivosti rechi Moscow (Svyaz'izdat) p 390
[2] Zheleznyak V K, Makarov Yu K, Khorev A A 2000 Spetsial'naya tekhnika Nekotorye metodicheskie podkhody k otsenke effektivnosti zashchity rechevoy informatsii 4 p 39–45
[3] Petelin R Y, Petelin Y V 1998 Zvukovaya studiya v PC Saint-Petersburg p 256
[4] Khorev A A, Makarov Y K 2001 Zaschita informatsii. Konfident Metody zaschity rechevoy informatsii I otsenki ikh effectivnosty 4 p 22–23

IOP Conf. Series: Journal of Physics: Conf. Series **944** (2017) 012044 doi:10.1088/1742-6596/944/1/012044

Investigation of Influence of Speech Specter Differences on Intelligibility Estimation Results

A V Ivanov, I L Reva, and E E Shemshetdinova

Novosibirsk State Technical University, 20, Karla Marksa ave., Novosibirsk, 630073, Russia

E-mail: andrej.ivanov@corp.nstu.ru

Abstract. The purpose of this study is to investigate influence of difference of speakers` speech specters on word intelligibility. Experiment on estimation of speakers` speech specter difference was conducted with taking into account an effect of a text which a speaker uses. Theoretical calculation of speaker`s speech specter influence on intelligibility was done with use of voice information protection approach. The influence of speaker`s speech specter on intelligibility was confirmed by immediate appreciation tests. Recomendations on possible consideration of the effect are provided.

1. Introduction

Standard approach of estimation of voice information protection [1] is the approach of word intelligibility estimation, based on the Pokrovsky`s calculating-experimental formant approach [2]. The approach was developed for conditions, which are essentially different from information protection. As the result, it led to appearance of essential limitations, many of which are recently discussed by a lot of authors who try to adjust the chosen approach for information security conditions [3-8].

One of the important aspects of the approach is use of average speech specter. However, it is known that different people have different speech specters because every human`s articulation organs are unique, beside essential differences of men`s and women`s speech specters. Therefore, there are questions: is it correct to use the average speech specter in word intelligibility estimation; what is a range of values of word intelligibility with use of different speakers` speech specters?

Thus, the purpose of the study is to investigate how differences of speech specters influence on the result of word intelligibility.

2. Word intelligibility estimation approach

Word intelligibility is the index of voice information protection, which is stated as ratio of correctly received speech elements to the number of transmitted speech elements.

Essence of the approach [1] is estimation of sound pressure levels (vibrational acceleration) of the following signals octave bands:

- test signal, formed with use of white noise generator or tone signals with frequencies corresponding to average compound frequencies of octave bands;
- background noise or noise created by active protection means, measured behind a protective structure with a turned off test signal generator and a turned on active protection mean (if any);

IOP Conf. Series: Journal of Physics: Conf. Series **944** (2017) 012044 doi:10.1088/1742-6596/944/1/012044

- signal-noise mixture, measured behind a protective structure with a turned on test signal generator;

Calculation of word intelligibility is made with results of the measurements. The approach is described in details in [1, 2].

It should be mentioned that test signal is a signal with integrity levels in octave bands corresponding to the average speech level. Distribution of energy in spectrum for integrity level 70 dB is characterized by octave bands` integrity levels, shown in table 1.

Table 1. Octave levels of sound pressure of average speech specter

Frequency, kHz	0.125	0.250	0.5	1	2	4	8
Sound pressure level, dB	53	66	66	61	56	53	49

3. Experiment

1. *Selection of text*

An experiment on estimation of sound pressure levels in octave bands of different speakers was conducted.

The first problem of the experiment is selection of a text for speakers and its wordage. The problem was solved by another experiment. Four speakers (2 men and 2 women) were offered two texts:

- phrasal articulatory tables, presented in Russian national standard «GOST R 50840-95 Speech transmission by communication channels. Estimating methods of quality, intelligibility and awareness (Attachment D)» [9];
- a part of a technical text with the same wordage.

Recording duration was set as 1 minute. Speakers` speech was calm, clear, with average volume – characteristics of speech in real life. Noise level meter ZET 110 and microphone BC501 with 1kHz calibrated frequency using calibrator SVAN SV35 was used for the recording. The recording was made with sampling frequency 50 kHz. Another aim of the experiment was to investigate results repetition, that is why all the texts were recorded twice.

Specters of the recordings were plotted with use of the Adobe Audition functions. Samples of the specters help to calculate octave bands integrity levels. We should mention that audio editors' levels are estimated in dBFS – full scale (dB by voltage, relatively to the maximum A/D Converter value set on a PC). It means that 0 dBFS does not correspond to the limit value of signal level, and there will not be so-called «Clipping». Therefore, the minimum value of signal level fixed by the A/D Converter depends on its number of digits. For example, 16-digit A/D Converter`s minimum signal level is -96 dBFS, 20-digit A/D Converter`s minimum signal level is -120 dBFS, 24-sigit A/D Converter`s minimum signal level is -144 dBFS. Traditional dB, which stand for sound pressure (relatively to 20 μPa) can be derived only if you know the A/D Converter`s number of digits and microphone/ acceleration meter sound pressure level (dB SPL). So, Adobe Audition does not allow to estimate absolute level of signal recording, but allows to compare spectral characteristics of signals.

Results of the experiments are octave bands signal levels for two recordings of one speaker with use of different texts (table 2) of the two recordings

Table 2. Octave levels of sound pressure with use of different texts

Sound pressure levels for different texts	Average compound octave bands` frequencies, Hz						
	125	250	500	1000	2000	4000	8000
Phrasal articulatory tables, dBFS	-36.69	-34.57	-34.72	-41.52	-45.33	-46.16	-57.01
Technical text, dBFS	-37.22	-33.65	-33.95	-40.75	-44.94	-46.43	-56.50

We can see that the difference of octave bands` sound pressure with use of different texts does not exceed 1 dD, so we can conclude that type of the text does not essentially influence on speech specter. Derived divergence (up to 1 dB) appears even when a speaker reads the same text. It means that the divergence appears because it is hard to speak on the same level.

We can try to reduce the effect by averaging levels of 2 texts.

Therefore, text type almost does not influence on speech specter. The technical text was chosen for further experiments because it is more similar with real life speech than special tables from the GOST.

2. Analysis of influence of speech specters differences on intelligibility

After selection of the text we can start statistics collection. 10 speakers (5 men and 5 women) without speech disorders and special characteristics were chosen for the experiment. The result of each speaker is calculated by averaging 2 recordings for increase of accuracy. The results are shown in table 3.

Reduce all levels to 70 dB for easy comparison of the results with average speech specter` level. On this step divergences are not critical, it is another work (estimation of amplitude speech constitution), described in details in [10]. We consider only specter forms with octave bands` levels.

Now we can substitute the results to [1] as test signal levels and analyze their influence on word intelligibility W. Now plot dependencies of W on signal-level ratio q with use of pink noise for all the speakers.

Table 3. Integrity levels of speakers` speech

F, kHz	Sound pressure levels in octave bands for different speakers, dB										
	№1	№2	№3	№4	№5	№6	№7	№8	№9	№10	Average
0.125	64.79	59.13	62.01	60.09	60.69	60.67	64.44	65.62	65.30	63.00	53
0.250	66.06	64.34	65.80	65.21	67.82	66.03	66.06	65.00	63.96	66.15	66
0.500	63.03	66.48	65.95	65.85	63.86	66.25	63.77	63.31	65.00	64.44	66
1	55.23	62.02	55.15	62.49	51.47	54.14	56.80	55.82	55.45	56.21	61
2	50.81	57.65	51.92	58.63	49.54	50.91	51.77	53.28	53.05	55.77	56
4	54.75	53.26	47.84	52.24	48.61	51.97	51.45	55.98	54.15	53.89	53
8	50.22	51.73	44.46	52.46	45.57	48.25	45.91	47.56	51.63	48.24	49

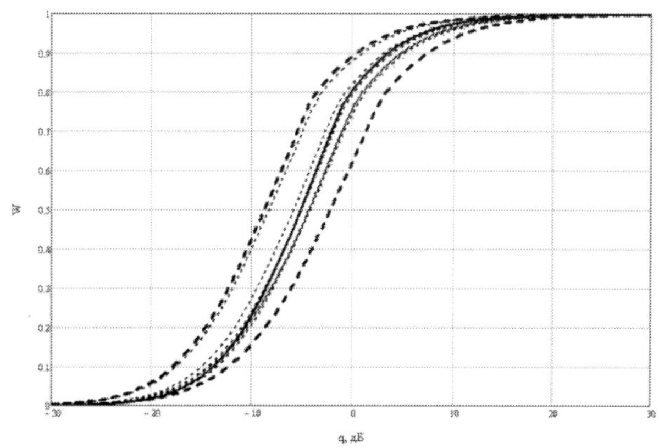

Figure 1. Dependences of word intelligibility *W* on signal-noise ratio *q* for pink noise (different speakers` speech specters). On the diagram: full line – average specter, bold dashed lines – curves with minimum and maximum values.

The diagram shows that, for example, if q = -10 dB for average speech specter, W=0.22. For speech specters of different speakers W falls in the limits from 0.15 to 0.43. Also we can notice that the lines with bigger W belong to men, with smaller W – to women.

3. Check of the results with use of immediate appreciation tests

We have done some appreciation tests to accept the results. Immediate appreciation test is a process of subjective estimation of word intelligibility. Its essence is addition of noises with particular signal-noise ratios to examined recordings, which after that are listened by auditors group.

Choose the speakers` recordings with the best and the worst results. These speakers are the speaker №5 (woman) and the speaker №4 (man), corresponding to table 2. By the way, theoretically (using the approach calculations), the difference between word intelligibility for different speakers is approximately 30%.

Add the pink noise and its generation in audio editor. Signal-noise ratio` value does not have big influence, but it should be the same for both speakers. Choose such a noise signal for the speech of speaker №5 (the minimum word intelligibility) to make only some words from it recognized, i.e. word intelligibility does not exceed 0.1-0.15.

After that we add the same noise with the same level to the speech of speaker №4 (maximal word intelligibility). The examined recordings were adjusted by level (nevertheless divergence between them was 0.3 dB).

As the result, even a small group of auditors admitted that word intelligibility of speaker №4 is essentially bigger (about 0.6-0.7). So, practical influence of speech specter form of speakers` speech have even bigger influence on word intelligibility than theoretical. Also, we should mention that the main differences were between man`s and woman`s voices.

4. Analysis of the results

The derived results show the following:

- speakers` speech specter essentially influence on word intelligibility, even from the approach point of view (difference is about 30%);
- immediate appreciation tests practically confirmed that the effect appears with bigger difference (50-60%);
- the main difference in word intelligibility is between man`s and woman`s speech.

IOP Conf. Series: Journal of Physics: Conf. Series **944** (2017) 012044 doi:10.1088/1742-6596/944/1/012044

The immediate appreciation tests` results can be explained that word intelligibility depends not only on speech specter, but more on distribution of formant in frequency range. The distribution is individual for each speaker and especially for men and women.

5. Conclusion

We can make several conclusions on the results. First of all, speech specter essentially influences on word intelligibility. Also, the work problem about text selection for tests done. The difference between phrasal tables from GOST and technical texts is less than 1 dB.

As the result, we can recommend to choose a speaker based on estimation of all speakers` speech specters.

If it is not possible, it is necessary to estimate protection level with use of the speech specter with the biggest word intelligibility instead of the average. Probably, for derivation of such a specter it is necessary to do additional immediate appreciation tests with bigger auditors` group.

Consideration of the effect in estimation of voice information protection would allow to increase information security level of private conversation rooms.

References

[1] Zheleznyak V K, Makarov Yu K and Khorev A A 2000 *Spetsial'naya tekhnika* 4 39–45
[2] Pokrovskiy N B 1962 *Raschet i izmerenie razborchivosti rechi* (Moscow: Svyaz'izdat) 390
[3] Avdeev V B 2013 *Spetsial'naya tekhnika* 2 1–10
[4] Didkovskiy V S, Didkovskiy M L and Prodeus A N 2008 *Akusticheskaya ekspertiza kanalov rechevoy kommunikatsii* (Kiev: Imeks-LTD) 420
[5] Didkovskiy V S, Prodeus A N, Ladoshko O N and Samoylenko N A 2014 *Radioelektronika* vol 57 2 (620) 55-60
[6] Gerasimenko V G, Lavrukhin Yu N and Tupota V I 2008 *Metody zashchity akusticheskoy rechevoy informatsii ot utechki po tekhnicheskim kanalam* (Moscow: Fakel) 258
[7] Ivanov A V, Reva I L and Trushin V A 2010 (Novosibirsk: Novosibirsk State Technical University Press) 133–136
[8] Ivanov A V, Trushin V A, Beresneva A V and Markelova G V 2014 The experimental research of security of speech in-formation of leakage from channels with technical account effect of forcing speech Actual problems of electronic instrument engineering *2014 12th International Conference on Actual Problems of Electronics Instrument Engineering (APEIE)* 266-272
[9] 1996 GOST R 50840-95 Peredacha rechi po traktam svyazi. Metody otsenki kachestva, razborchivosti i uznavaemosti (Moscow: Izdatel'stvo standartov) 198
[10] Ivanov A V, Trushin V A and Khitsenko V E 2015 On the choice of the model of the test signal in the assessment of speech information security from leak on technical channels *Proc. of SPIIRAS* vol 40 122-133

IOP Conf. Series: Journal of Physics: Conf. Series **944** (2017) 012045 doi:10.1088/1742-6596/944/1/012045

Increasing operational life of brush-contact device in the turbine generator due to using lubricating molybdenum disulphide brushes

A I Izotov, A A Fominykh, S V Nikulin, D K Prokoshev, A B Legoti, N V Timina

Vyatka State University, 11, Moskovskaya street, Kirov, 610000, Russia

e-mail: aa_fominyh@vyatsu.ru

Abstract. A way of reducing irregular current distribution in multi-brush systems of sliding current transfer with its wear reduction due to installing lubricating molybdenum disulphide brushes on slip rings to ensure a greasing nano-sized cover on the slip ring surface is proposed. The authors give the results of industrial tests estimated on the performance effectiveness of lubricating brushes on slip rings of the TBB-320-2UZ-type turbine generator. The results showed that the lubricating brushes reduce a) the wear of 6110 OM^{+M} and EG2AF^{+M} brushes by 1.2 and 2.1 times respectively, b) current distribution irregularity in parallel operating brushes due to stabilizing the contact arc, and c) the temperature of the electrical brush-contact device due to the friction reduction in brushes.

1. Introduction

Multi-brush devices of sliding current transfer are widely used in turbine generators. According to the industrial experience, these devices operate with irregular current distribution in parallel operating brushes followed by overloads of some brushes, then their heating and further destructing. More than the fourth part of all turbine generator faults happen in electrical brush-contact devices [1–12].

The aim of the article is the assessment of capability for a) increasing the operational life of the brush-contact device in the turbine generator by reducing the brush wear, and b) reducing irregular current distribution in parallel operating brushes and temperature in the brush-contact device. The task is solved by installing lubricating molybdenum disulphide non-conducting brushes in extra brush holders of the brush-contact device in the turbine generator. The described way has been developed by the staff at the Department of Electrical machines and Devices of Vyatka State University in partnership with LLC EF-KONTEL (Moscow) [16, 17].

2. Test procedure

The test was carried out on TBB-320-2UZ-type turbine generator with EG2AF^{+M} -type brushes (negative polarity ring) and 6110 OM^{+M} -type brushes (positive polarity ring) being used on slip rings of the brush-contact device. The brush device of the TBB-320-2UZ-type turbine generator has two four-track slip rings and each track has 14 DBU-type brush holders as shown in figure 1.

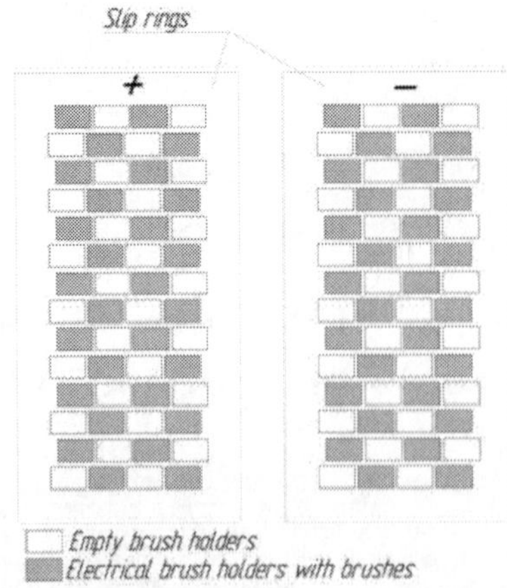

Figure 1. The pattern scheme of brushes on tracks.

The use of +M brushes [14] makes it possible for the contact device to operate with one-half working set (28 items per a ring) due to the unique brass bracing the current distributor and the carbon brush body. Note that the track has seven 20×32×64 mm brushes instead of 14 ones as in the standard device. To carry out the test 4 brush holders were used for lubricating brushes to be installed.

To assess the capability for using lubricating brushes in the brush-contact device in the turbine generator tested the calculation of microtemperatures in the 'brush-slip ring' contact zone was done with the programme 'Contact' modeling flexible electrical contact [18]. To apply the programme complex in-situ testing was carried out to measure the average temperature in the brush-contact device. Then, based on the obtained results the record of microtemperatures in the very 'brush-slip ring' contact zone was done. The simulation results showed that with the temperature in the micro-contact zone not exceeding 300°C, lubricating molybdenum disulphide brushes can be used (at the temperature exceeding 400°C the lubricant loses its properties and reacts to form an oxide - abrasive substance).

Furthermore, two variants of the brush holder device for lubricating brushes to be installed were designed providing desirable lubricant film thickness in the contact zone. Figure 2 depicts variant 1. An extra brush holder 2 made from non-conducting material and with stops on its side surface to protect against whipping is installed into the standard brush holder 1. Into the case of the extra brush holder a composite lubricating brush is installed. Its load-bearing part 4 and contact one 5 are string-applied with two coil springs 5 to control the load in a wide range (50-300 gr). The cover 6 ensures pressing the string 5 which is string-loaded with the standard string of the DBU brush holder. The main advantage of the device is a) a wide range of controlling the pressing on the lubricating brush, and b) insulation of the extra brush against the current-conducting brush holder. The disadvantage of the device is complexity and labour input aspects.

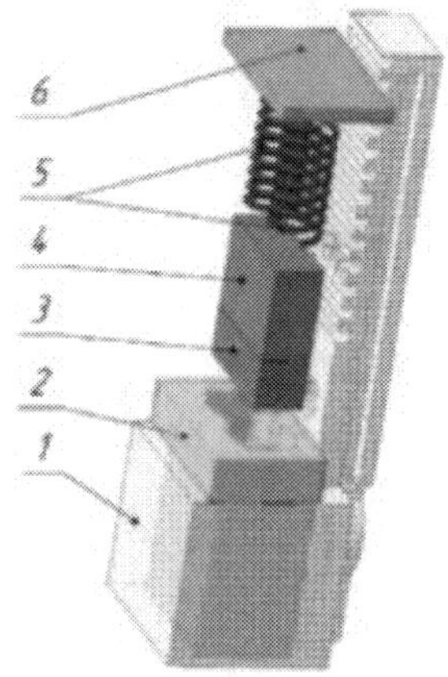

Figure 2. The extra brush holder device for lubricating brushes to be installed in the 320 MW turbine generator: 1 – standard brush holder, 2 – extra brush holder, 3 – contact part of the lubricating brush, 4 – load-bearing part of the lubricating brush, 5 – coil springs, 6 – cover.

Figure 3 depicts variant 2. In this device a channel for the contact part of the lubricating brush to be installed is milled in the standard brush. Then strings for the standard brush holder are selected to provide proper lubricant feed in the contact zone. The device is easy to make, so it is used to carry out the tests.

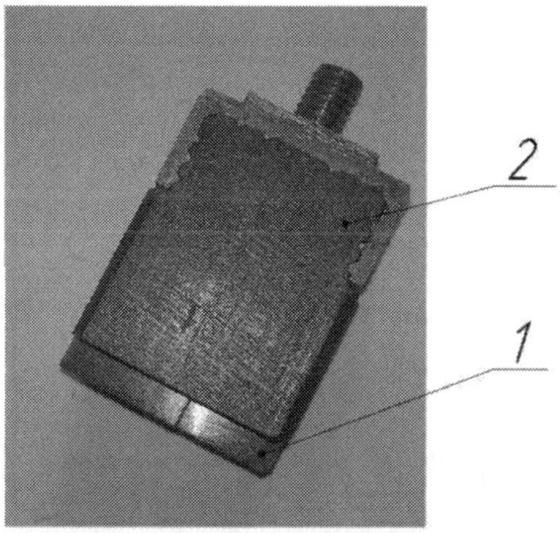

Figure 3. The device for installing lubricating brushes: 1 – lubricating brush; 2 – standard brush body.

The generator fitted with the brush set (28 ones both on the positive and negative rings) was operating for 4 hours at full capacity (exciting current I_e=2,200 A). Then lubricating brushes (Figure 4) were installed on two tracks of the negative ring (one brush was on 2 current tracks). And the tests to assess the effect of lubricating brushes on the stabilization of current transfer in parallel operating brushes at full capacity were being carried out for 5 hours.

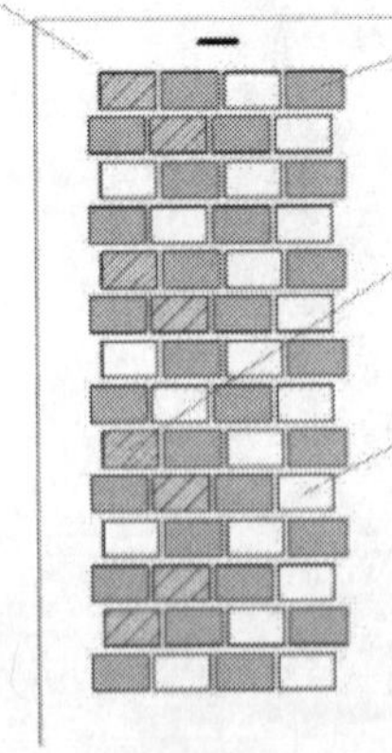

Figure 4. The pattern scheme of lubricating brushes.

As the brushes on slip rings were staggered, the grease slightly covered adjacent tracks.

3. Test Results

The analysis of rotor current distribution in the brushes without lubricating ones showed that current load is significantly irregular (see Table 1).

Tables 2 and 3 give the test results of the brush-contact device in the generator after installing lubricating brushes in 2 and 5 hours. They also show the current distribution results of the rotor, instantaneous current I_{hi} under the brushes on tracks 1 and 2 on the negative ring, net current of the exciting coil $\sum I_B$, average current in brushes I_{AV}, being calculated according to the operations manual of the brush-contact device [15]. Besides, they demonstrate absolute deviations of current in all brushes from the average value δ, calculated as follows (1):

$$\delta = \frac{1}{n}\sum_{i=1}^{n}\left|I_{hi} - I_{AV}\right| ,$$

where n – the number of brushes;

$$I_{AV} = \frac{\sum I_B}{n} .$$

During the tests on the brush-contact device before installing lubricating brushes maximum current load under some brushes was 120 – 140 A (brushes № 2, 6, 9), whereas most brushes were operating at 83 A current load. The three brushes were loaded with no more than 50 A (№ 18, 21, 25, table 1). The average absolute deviation without lubricating brushes was 29,5A.

IOP Conf. Series: Journal of Physics: Conf. Series **944** (2017) 012045 doi:10.1088/1742-6596/944/1/012045

Table 1. Current distribution in the rotor on brushes of the negative ring (100 % full capacity, no lubricating brushes).

№ tracks	$\sum I_B$, A	№ brushes	I_{hi}, A	I_{AV}, A	δ, A
1;2	2.044	1	78	73	29.5
		2	140		
		5	106		
		6	120		
		9	140		
		10	90		
		13	87		
		17	60		
		18	41		
		21	40		
		22	80		
		25	50		
		26	80		
		28	25		

The installation of the lubricating brushes reduced irregular current transfer. In two-hour operating time of the brush-contact device (table 2) maximum current load was 130A (brush № 9), whereas most brushes were operating at 75–80 A current load. The average absolute deviation was 22A.

Table 2. Current distribution in the rotor on brushes of the negative ring (100 % full capacity, operating time of lubricating brushes – 2 hours).

№ tracks	$\sum I_B$, A	№ brushes	I_{hi}, A	I_{AV}, A	δ, A
1;2	2.044	1	77	73	22.0
		2	111		
		5	90		
		6	111		
		9	130		
		10	48		
		13	96		
		17	50		
		18	40		
		21	81		
		22	80		
		25	73		
		26	100		
		28	50		

With increasing the operating time of lubricating brushes further stabilization of current distribution was observed (table 3). At the end of the test in 5-hour operating time maximum current load was 110 – 120 A (№ 6, 9, 22, 26), whereas main current load was 70 – 75 A. The average absolute deviation was 18A.

Table 3.Current distribution in the rotor on brushes of the negative ring (100 % full capacity, operating time of lubricating brushes – 5 hours).

№ tracks	$\sum I_B$, A	№ brushes	I_{hi}, A	I_{AV}, A	δ, A
1;2	2.200	1	86	78.6	17.8
		2	106		
		5	82		

6	100
9	120
10	60
13	90
17	69
18	50
21	85
22	111
25	75
26	100
28	63

Having analyzed the data given above, the installation of lubricating molybdenum disulphide brushes is proved to provide a more regular current distribution between parallel operating brushes. This happens due to the fact that the grease ensures a more stable contact arc providing constant resistance in the transfer layer 'brush-contact ring 5' (figure 5). It considerably exceeds (3–4 times) [13] total resistance comprising a) contact resistance between the terminal and the brush crank 1, b) contact resistance between the brush terminal and the brush holder rocker 2, c) resistance between the crank and the brush body 3, and d) resistance of carbon material (of the brush body) 4. The contact arc is known to considerably change while operating and at certain moments it reduces up to 1/1,000 from the theoretical contact arc [19, 20].

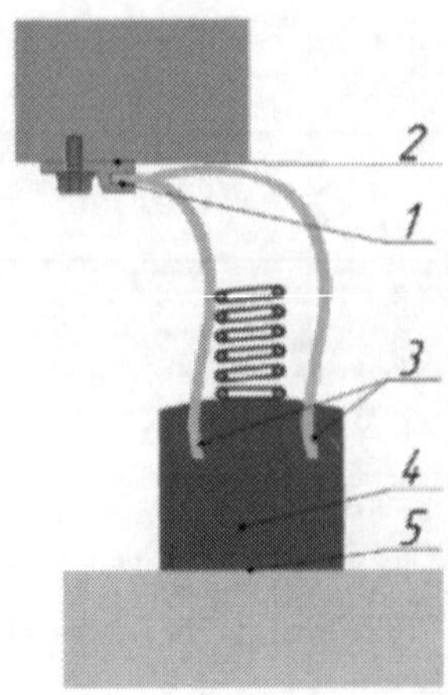

Figure 5. Contact resistance 'brush – slip ring.

Table 4 gives the results of thermal imaging of the brush surface with no lubricating at the 2,044A net current in the rotor. A warmer brush is 140°C, whereas the average temperature (T_{AV}) of all brushes on track №1, 2 is 92°C. The installation of lubricating brushes (table 4) and the temperature measurement in 2-hour exploratory work reduces the average temperature of brushes up to 89.1°C and the maximum temperature - up to 110 °C at the 2.150 A current in the rotor. At the longer operating time of brushes (4 hours) and at the higher net current (2.200 A) in the rotor a favourable temperature reduction occurs: the average temperature of all brushes is 88.8°C and the maximum temperature of the warmest brush is 110°C. Thus, the installation of lubricating molybdenum disulphide brushes has reduced the average temperature of the brush-contact device due to the friction reduction in brushes.

Table 4. The temperature on brushes of the negative ring (100 % full capacity, no lubricating brushes).

№ tracks	№ brushes	I_{hi}, A	$\sum I_B$, A	T_{AV}, °C
1;2	1	78	2.044	92
	2	88		
	5	104		
	6	91		
	9	140		
	10	92		
	13	63		
	17	82		
	18	71		
	21	115		
	22	110		
	25	89		
	26	88		
	28	77		

Similar data on the temperature reduction in the current transfer unit have been obtained in the 2,4 kW-capacity ac collector motor with the PKMS copper layer (copper-silver alloy) on the collector [10]. The installation of lubricating brushes has reduced the temperature in the open circuit collector by 28°C and in the loaded collector by 15°C.

4. Conclusions

The carried out tests on installing lubricating molybdenum disulphide brushes in the brush-contact device of the TBB-320-2UZ-type turbine generator have demonstrated that their use:
– reduces irregular current distribution in parallel operating brushes due to stabilizing the contact arc of current-conducting brushes;
– reduces the temperature in the brush-contact device due to the 'brush-slip ring' friction reduction.

References

[1] Takanezawa M, Hiramatsu D, Kobayashi M, Kakiuchi M, Nagakura K, Morita N, Otaka T, Study of Brush Characteristic of Turbine Generator 2012 *15th International Conference on Electrical Machines and Systems (ICEMS)*

[2] M. Takanezawa, D. Hiramatsu, M. Kobayashi, M. Kakiuchi, K. Nagakura, N. Morita, "Study of carbon brush characteristic of turbine generator (2nd report)," 2012 15th International Conference on Electrical Machines and Systems (ICEMS)

[3] Hall R D and Roberge R P 2010 Carbon brush performance on slip rings *Conference Record of 2010 Annual Pulp & Paper Industry Technical Conference*

[4] SkjØlberg J K, Ohma H F and Runde M 2009 Wear Rates and Current Distribution of Carbon Brushes on Steel Slip Rings *IEEE Transactions on Energy Conversion* Vol 24 pp 835–840

[5] Kim Y J, Han B, Woo C G and Kim H J 2017 Performance of Ultrafine Particle Collection of a Two-Stage ESP Using a Novel Mixing Type Carbon Brush Charger and Parallel Collection Plates *IEEE Transactions on Industry Applications* Vol 53 no 1 pp 466–473

[6] Sawa K, Liu L and Ueno T Influence of Arc Discharge on Carbon Commutator and Brush Wear in DC Motor Driving Fuel Pump *2012 IEEE 58th Holm Conference on Electrical Contacts*

[7] Morita N, Ueno T, Takanezawa M, Otaka T and Hiramatsu D 2010 A Study for Heavily Saturated V-I Characteristics at Carbon Brush / Steel Collector Ring Sliding Contact, on the View Point of Parallel Connected Brush Current Sharing Unbalance *Proceedings of the1 56th IEEE Holm Conference on Electrical Contacts*

[8] Tanaka H, Morita N, Sawa K and Ueno T 2010 Carbon Brush and Flat Commutator Wear of DC Motor Driving Automotive Fuel Pump in Various Fuels *2010 Proceedings of the 56th IEEE Holm Conference on Electrical Contacts* Charleston, SC

[9] Noboru Morita; Takahiro Ueno; Toru Otaka; Masanori Arataб 2007 Comparison of Brush Dynamic Operation Characterisitics for Turbine Generator Steel Collector Ring *Electrical Contacts - Proceedings of the 53rd IEEE Holm Conference on Electrical Contacts* pp 205–210

[10] Dillich S and D. Kuhlmann-Wilsdorf 1980 Effects of Surface Films on the Performance of Silver-Graphite (75 wt% Ag, 25 wt% C) *IEEE Transactions on Components, Hybrids, and Manufacturing Technology* Vol 3 Issue 1 pp 37–41

[11] Xiang Wang, Jun Xu, Li Li, Xuhong Guo, Catrin Junge 2008 Nano-Sized Spherical Polyelectrolyte Brushes: Potential Protein Carriers and Diagnostic Tools *2nd International Conference on Bioinformatics and Biomedical Engineering* pp 1592–1595

[12] Samorodov Y N 2014 Causes and Effects of Turbine Generator Accidents *Unified network energy* 2 (13) pp 70–80

[13] Livshits P S 1974 *Sliding contact of electric machines* (Moscow: Energy) p 321

[14] Company KONTEL *Brush roll cleaner* Russian Federation Patent 2007132129/22 Aug. 24 2007

[15] Manual OPS.460.468 re *Machine contact brush turbogenerators* 2008

[16] Izotov A I The influence of mechanical factors in alternating-current collector machines on level of radio noise 2013 *Russian Electrical Engineering* 84 (8) pp 463–465

[17] *Sliding current transfer unit (variants)*: pat. 112513 Russian Federation MPK H 01 R 39/00 A.I. Izotov № 2011120198/07; applic. 19.05.2011; published. 10.01.2012, Bull. 1 p 2

[18] Plokhov I V 2002 *Technical state diagnostics and prediction of sliding current transfer units in turbine generators: Extended Abstract. Diss. Doctor. Sci. (*St Petersburg: St Petersburg State Technical University) p 36

[19] Deeva V S and Slobodyan S M 2011 Action of isomorphic sliding current-collect destruction *Energetic* Vol 9 pp 36–38

[20] Romanishina S A, Katyuk D Y, Deeva V S and Slobodyan S M 2015 Dynamics layer of the sliding contact collector elements *IEEE 35th International Conference on Electronics and Nanotechnology (ELNANO)* pp 116–118

[21] Timoshenko V N, Izotov A I 2013 Wear reduction of elements in sliding current transfer units due to applying lubricating brushes *International scientific and technical conference «Increase of Operational Effectiveness of Collector Electromechanical Energy Converter* (Omsk: OmSTU) pp 128–13

Obtaining, bulk physical and chemical properties, certification of heterosystem InSb-ZnS solid solutions

I A Kirovskaya[1], E V Mironova[1], D A Polonyankin[1], S N Pogodin[1], L V Novgorodtseva[1], A I Blesman[1], S A Korneev[1] and Yu I Matyash[2]

[1]Omsk State Technical University, 11, Mira ave., Omsk, 644050, Russia
[2]Omsk State Transport University, Omsk, Russia

e-mail: kirovskaya@omgtu.ru

Abstract. Solid solutions $(InSb)_x(ZnS)_{1-x}$ of different compositions have been obtained using a specially developed technology (applied to the InSb-ZnS system under study). Performed X-ray and electron-microscopic studies made it possible to certify the solid solutions obtained as solid substitutional solutions with the sphalerite structure, and also to determine their element composition, which practically coincides with the molar composition.
The influence of the physical and chemical properties of the initial binary compounds on the corresponding physical and chemical properties of the solid solutions has been shown (with the example of two $A^{III}B^{V}$-$A^{II}B^{VI}$ type systems) which manifested in their concentration dependences nature.
The inevitable competition between the relative contributions of Lewis and Bronsted acide sites has been justified.
The predictions of the trend of the surface isoelectric state pH in the series InSb → $(InSb)_x(ZnS)_{1-x}$→ ZnS and their activity with respect to gases of a certain electronic nature have been given.

Keywords: solid solutions, physical and chemical properties, certification, promising materials, semiconductor gas analyzers.

1. Introduction

The research subjects in the present work are solid solutions of the new heterogeneous replacement (type $A^{III}B^{V}$-$A^{II}B^{VI}$) -InSb-ZnS system.
The unique physical and physicochemical properties of the initial binary compounds (InSb, ZnS), the possibility of adjusting and optimizing properties with a change in composition and the discovery of unexpected effects due to the peculiarities of internal processes accompanying the formation of solid solutions allow to expect to obtain promising materials for sophisticated technology and semiconductor instrument engineering, including semiconductor gas analyzers [1, 2].
To accelerate the realization of the expected prospects it is necessary to develop a relatively simple technology for obtaining and knowing their bulk and surface physical and chemical properties.
In terms of the above, at the first stage, the results of obtaining, X-ray, electron microscopic studies and certification of InSb-ZnS system solid solutions are analyzed relying on necessary information on a single-type system (GaSb-ZnTe).

2. Problem statement

IOP Conf. Series: Journal of Physics: Conf. Series **944** (2017) 012046 doi:10.1088/1742-6596/944/1/012046

To develop the technology, to obtain InSb-ZnS system solid solutions certification based on the results of X-ray studies. With the help of electron microscopic studies to determine their elemental composition, surface structure.
Using the example of two systems of the type $A^{III}B^{V}$-$A^{II}B^{VI}$ (InSb-ZnS, GaSb-ZnTe, it is necessary to observe the influence of the physical and chemical properties of the initial binary compounds on the same physical and chemical properties of the formed solid solutions and the nature of the "property-composition" dependences.
To evaluate the relative contributions of Lewis and Bronsted sites and justify the inevitable competition between them.
To predict the nature of the change in the pH_{iso} in the InSb → $(InSb)_x(ZnS)_{1-x}$ → ZnS series and, accordingly, the surfaces activity towards gases of a certain electronic nature.

3. Experimental part

The investigated objects were mainly fine powders InSb, ZnS and their solid solutions $(InSb)_x(ZnS)_{1-x}$ (x = 0.04, 0.08, 0.18, 0.23, 0.90, 0, 95, 0.98 mol). The solid solutions were prepared using a special technology (relative to the InSb-ZnS system), based on the isothermal diffusion of InSb, ZnS (in vacuumed, sealed quartz vessels, at a reasonably selected temperature of 1000 K), using a special program of temperature heating [2]. The completion of the synthesis was concluded from the results of X-ray diffraction studies used for the certification, structure determination and the basic structural characteristics of solid solutions. The results of X-ray studies were supplemented by the results of electron microscopy studies.
X-ray studies were carried out on the D8 Advance diffractometer of Bruker (Germany) in CuKα radiation (λ = 0.15406 nm, T = 293 K) using the large-angle survey technique [3, 4] and the Lynxeye position-sensitive detector. The decoding of the obtained X-ray diffraction patterns was carried out using the powder diffraction database ICDDIPDF-2, refinement of the lattice parameters in the TOPAS 3.0 program (Bruker), using the least squares method.
Electron microscopical study was performed on a scanning electron microscope SCM – 5700, equipped with a nitrogen-free x-ray energy-dispersive spectrometer [5].
The reproducibility and accuracy of the data was verified by the results of parallel measurements using mathematical statistics and processing the results of quantitative analysis. Statistical processing of the obtained numerical values, calculation of measurement errors, construction and processing of graphical dependencies were carried out using computer programs Stat-2, Microsoft Exel and Origin.

4. Results and discussion

Let us consider the results of X-ray studies (fig. 1, 2, table 1). They indicate the formation of solid substitutional solutions with the cubic structure of sphalerite in the InSb-ZnS system. The relative position and distribution of the intensities of the main lines on the x-ray diffraction patterns of binary compounds and solid solutions, the dependence on the composition of the values of the parameter of the crystal lattices (a), the interplanar distances (d_{hkl}), and the density (ρ_r) served as criteria. Deviations from the smooth or linear dependencies a = f (x_{ZnS}), d_{hkl} = f (x_{ZnS}) at the boundary content in the ZnS system (77 mole %), still "allowing" unlimited mutual solubility of the binary components (InSb, ZnS), i.e. formation of solid substitution solutions can be explained by not manifested structure change [2, 6].

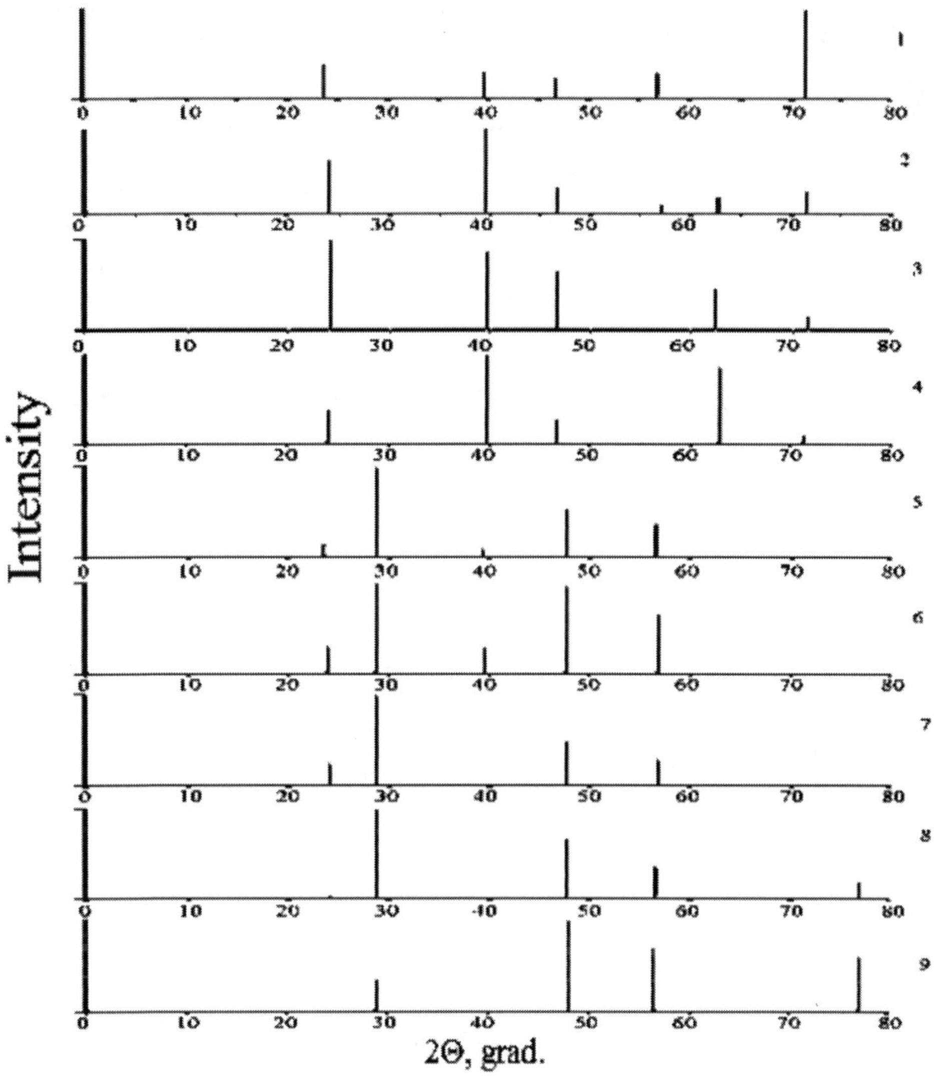

Figure 1. X-ray diffraction patterns diagrams of the InSb-ZnS system components: 1 – InSb, 2 – $InSb_{0.98}ZnS_{0.02}$, 3 – $InSb_{0.95}ZnS_{0.05}$, 4 – $InSb_{0.90}ZnS_{0.10}$, 5 – $InSb_{0.23}ZnS_{0.77}$, 6 – $InSb_{0.18}ZnS_{0.82}$, 7 – $InSb_{0.08}ZnS_{0.92}$, 8 – $InSb_{0.04}ZnS_{0.96}$, 9 – ZnS.

IOP Conf. Series: Journal of Physics: Conf. Series **944** (2017) 012046 doi:10.1088/1742-6596/944/1/012046

Table 1. Crystal lattice parameters value (a,), interplanar distances (d_{hkl}) and theoretical calculated crystal density (ρ_r) of InSb-ZnS system components.

X (Mole fraction ZnS)	*a*, Å	d_{hkl}, Å			ρ_r, г/cm^3
		111	220	311	
0	6.4770±0.001	3.7411	2.291	1.9550	5.7820
0.02	6.4583±0.001	3.7287	2.287	1.9518	5.7571
0.05	6.4507±0.001	3.7243	2.286	1.9514	5.6745
0.10	6.4462±0.001	3.7217	2.283	1.9500	5.5140
0.77	5.3970±0.001	3.1160	-	1.6303	5.4620
0.82	5.4020±0.001	3.1191	-	1.6305	5.1530
0.92	5.4033±0.001	3.1296	-	1.6307	4.5654
0.96	5.4041±0.001	3.1202	-	1.6310	4.3879
1	5.4107±0.001	3.1239	-	1.6314	4.0914

IOP Conf. Series: Journal of Physics: Conf. Series **944** (2017) 012046 doi:10.1088/1742-6596/944/1/012046

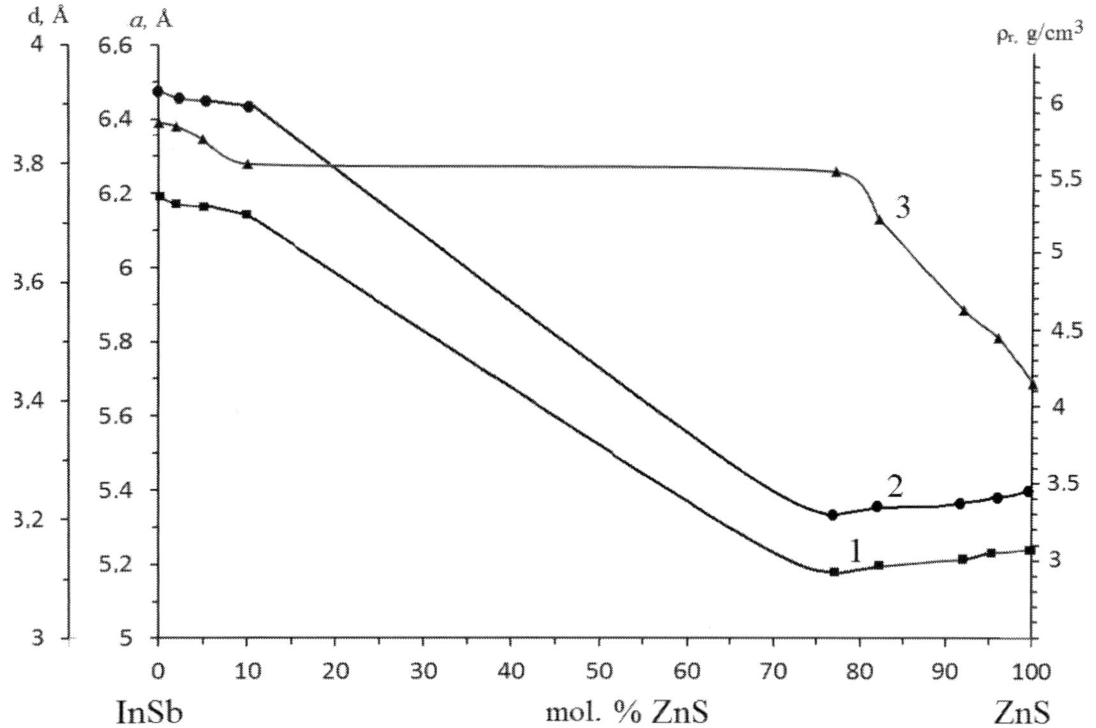

Figure 2. Dependence on the composition of the crystal lattices parameter values *a* (2), interplanar distance d_{111} (1), and theoretical calculated crystal density ρ_r (3) of the InSb-ZnS system components.

Solid solutions formation is also indicated by the absence of additional lines on the X-ray patterns corresponding to unreacted binary components and blurriness of the main lines.

Based on the results of electron microscopic studies (fig. 3–5), the elemental compositions of binary components and solid solutions, the structure of surfaces, the average particle size, the dependence of the number of particles of a certain (average) size on the composition of the system have been determined.

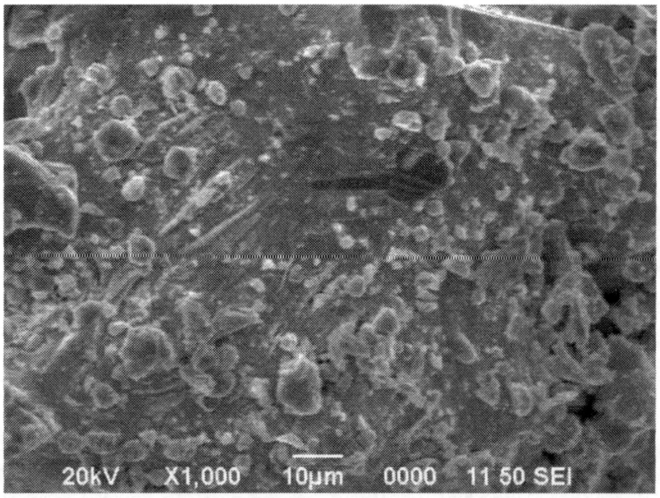

Figure 3. SEM - image of InSb powder in phase contrast regime.

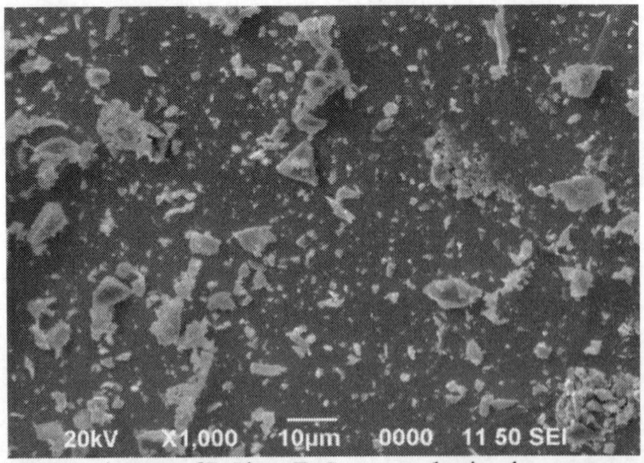

Figure. 4. SEM – image of $InSb_{0,90}ZnS_{0,10}$ powder in phase contrast regime.

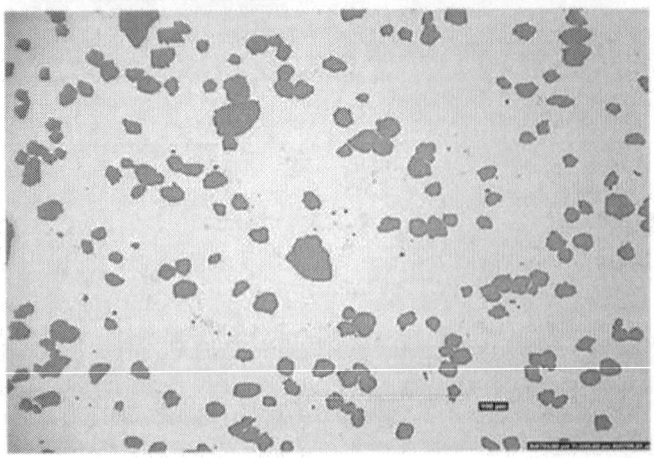

Figure 5. SEM – image ofZnS powder in contrast mode phase.

It is established [7]: the elemental composition of all components has a satisfactory agreement with the mole composition, surfaces have a polycrystalline structure with an inhomogeneous distribution of crystallites that can be associated with agglomerates from grains of various sizes.

In our opinion, *it was interesting to follow* how physical and chemical properties of the initial binary compounds – $A^{III}B^{V}$-$A^{II}B^{VI}$ type system components – affect on the same properties of the formed solid solutions and the dependence of the latter on the systems composition. Let us demonstrate it with the example of two systems: InSb-ZnS and GaSb-ZnTe (tables. 1, 2, fig. 2, 6).

More significant differences in the values of the indicated physical and chemical properties of the binary components of the InSb-ZnS system are noticed rather than the binary components of the GaSb-ZnTe system (table 2). As a result we have an extreme in the first case and the smooth in the second one dependences $a = f(x_{A2B6})$, $d_{hkl} = f(x_{A2B6})$ (fig. 2, 6).

IOP Conf. Series: Journal of Physics: Conf. Series **944** (2017) 012046 doi:10.1088/1742-6596/944/1/012046

Table 2 Basic physico-chemical properties of initial binary components of InSb-ZnS, GaSb-ZnTe systems.

The binary component of systems	Crystal lattice type	Crystal lattice parameter (*a*), Å	Density, (ρ_r), g/cm^3	Melting temperature (T_m), ^{0}C)	Bandgap width (ΔE), eV	Electronegativities difference (ΔX)
InSb	Sphalerite	6.4779	5.7820	525	0.18	0.10
ZnS	Sphalerite	5.4107	4.0914	1830	3.67	0.90
GaSb	Sphalerite	6.096	5.650	710	0.72	0.20
ZnTe	Sphalerite	6.108	5.668	1295	2.12-2.20	0.50

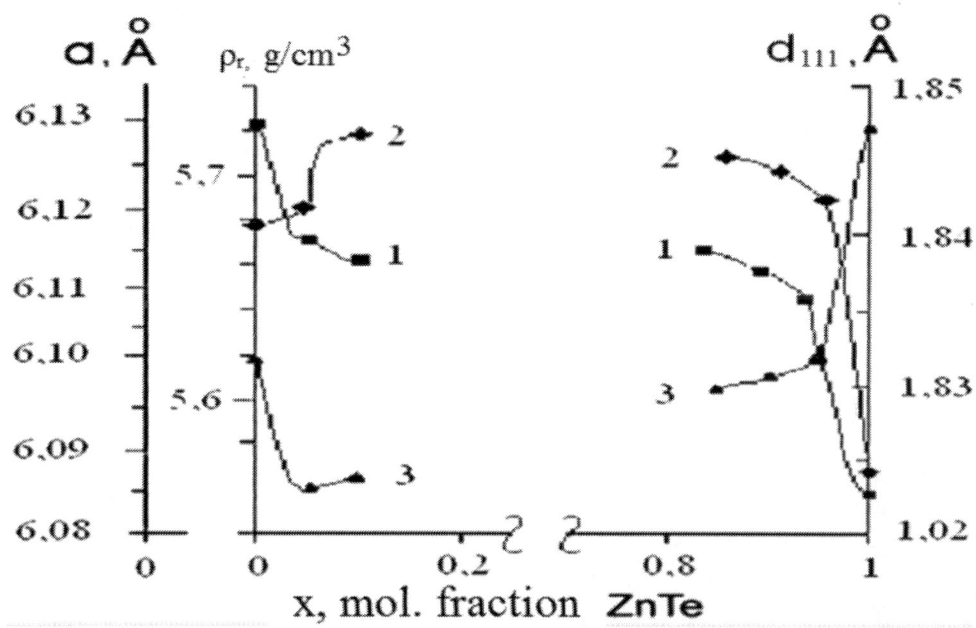

Figure 6. Dependence on the composition of the crystal lattices parameters values *a* (1), interplanar distance d_{111} (2), and theoretical calculated crystal density ρ_r (3) of the system GaSb-ZnTe components.

It should also be noted the density (ρ_r) decreases smoothly with an increase in the content of zinc sulphide in the InSb-ZnS system with parallel smooth increases in the band gap (ΔE) and the difference in electronegativity (Δx) (fig. 2, table 2).

So, we have the case of the inevitable competition between the growth of the coordination unsaturation of surface atoms, contribution of Lewis sites and increase in the bond ionicity, degree of surface hydration and accordingly the contribution of the Bronsted sites. With the predominance of the latter, the hydrogen index of the surfaces isoelectric state (pH_{iso}) in the InSb → $(InSb)_x(ZnS)_{1-x}$ → ZnS series should be expected to increase.

IOP Conf. Series: Journal of Physics: Conf. Series **944** (2017) 012046 doi:10.1088/1742-6596/944/1/012046

5. Conclusions

According to the developed technology solid solutions of the InSb-ZnS system certified on the basis of the results of X-ray studies as solid substitution solutions with a sphalerite structure have been obtained.

As a result of electron microscopic studies the elemental composition of the InSb-ZnS system components have been determined, practically coinciding with the molar composition; polycrystalline structure of surfaces has been established.

The influence of the initial binary components physical and chemical properties on the same physicochemical properties of the formed solid solutions has been shown by the example of two systems of $A^{III}B^{V}$-$A^{II}B^{VI}$ (InSb-ZnS, GaSb-ZnTe) type which manifested in the extreme and smooth character of the dependences $\alpha = f(x_{A2B6})$, $D_{hkl} = f(x_{A2B6})$ respectively for the InSb-ZnS and GaSb-ZnTe systems.

The inevitable competition on the surfaces of the InSb-ZnS system components between the relative contributions of Lewis and Bronsted acid sites has been proved. Predictions have been made about the nature of the pH of the pH in the InSb → $(InSb)_x(ZnS)_{1-x}$ → ZnS series, and hence about the surface activity of gases with certain electronic nature.

References

[1] Kirovskaya I A 2015 Physico-chemical properties of binary and multicomponent diamond-like semiconductors (Novosibirsk, SB RAS Publ) p 367

[2] Kirovskaya I A 2010 Solid solutions of binary and multicomponent semiconductor systems (Omsk: Omsk State Technical University) p 400

[3] Gorelik S S, Rastorguev L N and Yu A Skakov 1970 The radiographic and electron-optical analysis (Moscow, Metallurgy) p 107

[4] Smyslov E F 2006 Rapid X-ray test method in determining lattice constant of nanocrystal materials (*Zavodskaya laboratoriya, Diagnostica materialov*) vol **72** № 7 pp 33–35

[5] Goldstein J, Newbury D E, Joy D C, Lyman C E, Echlin P, Lifshin E, Sawyer L and Michael J R 1984 Scanning electron microscope investigation and X-ray microanalysis Vol **1** (Moscow, Mir) p 303

[6] Cat M V and Tarziu V G 1970 Semiconductor compounds and their solid solutions (Kishinev: Moldova) p 31

[7] Kirovskaya I A, Mironova E V, Grigan A A, Pogodin S N and Zhansakova K S Rapid Surface Sensitivity and Selectivity Determination of New Materials Based on CdS, ZnS *Procedia Engineering*
Vol **152** pp 634–638 (www.elsevier.com/locate/procedia)

Mechanochemical activation and gallium and indiaarsenides surface catalycity

I A Kirovskaya, E V Mironova, I V Umansky, O Yu Brueva, A O Murashova and A V Yureva

Omsk State Technical University, 11, Mira ave., Omsk, 644050, Russia

e-mail: kirovskaya@omgtu.ru

Abstract. The present work has been carried out in terms of determining the possibilities for a clearer identification of the active sites nature, intermediate surface compounds nature, functional groups during adsorption and catalysis, activation of the diamond-like semiconductors surface (in particular, the $A^{III}B^{V}$ type) based on mechanochemical studies of the "reaction medium (H_2O, iso-C_3H_7OH) – dispersible semiconductor (GaAs, InAs)" systems. As a result, according to the read kinetic curves of dispersion in water, both acidification and alkalinization of the medium have been established and explained; increased activity of the newly formed surface has been noted; intermediate surface compounds, functional groups appearing on the real surface and under H_2O adsorption conditions, adsorption and catalytic decomposition of iso-C_3H_7OH have been found (with explanation of the origin). The unconcealed role of coordinatively unsaturated atoms as active sites of these processes has been shown; the relative catalytic activity of the semiconductors studied has been evaluated. Practical recommendations on the preferred use of gallium arsenide in semiconductor gas analysis and semiconductor catalysis have been given.in literature searches, great care should be taken in constructing both.

Keywords: semiconductors, newly formed surface, active sites, relative catalytic activity, semiconductor gas analysis.

1. Introduction

Analyzing the possibilities of using the results of physical and chemical studies of diamond-like semiconductors surfaces in nano- and sensor technology, it is reasonable to consider the preparation and investigation of newly formed surfaces in various reaction media. In a logical way, one expects a clearer identification of the nature of the active sites, nature of the intermediate compounds, mechanochemical activation of the surface [1].

The objects of the present work are semiconductors of $A^{III}B^{V}$ type, isanoionic analogues – GaAs, InAs, water and isopropyl alcohol (propanol-2) act as reaction media.

2. Problem statement

To conduct mechanochemical studies of the newly formed surface of $A^{III}B^{V}$ type (GaAs, InAs) semiconductors in various reaction media (H_2O, iso-C_3H_7OH). As a result to reveal more clearly the nature of active sites, nature of the intermediate compounds, functional groups arising on the real, partially hydrated surface of semiconductors and under the adsorption conditions of water, adsorption and catalytic decomposition of isopropyl alcohol on the newly formed surface; expected surface

IOP Conf. Series: Journal of Physics: Conf. Series **944** (2017) 012047 doi:10.1088/1742-6596/944/1/012047

activation; to estimate the relative catalytic activity of the semiconductors under examination in the dehydration reaction of iso-C_3H_7OH; to assess the possibility of their use in semiconductor gas analysis and semiconductor catalysis.

3. Experimental part

The work was carried out according to a specially developed procedure [2]: polycrystals of semiconductors had been dispersed in the chosen medium (water, isopropyl alcohol) until the constant pH value was established. IR spectroscopic studies of suspensions of medium-dispersible semiconductor systems were conducted simultaneously [2]: IR spectra of multiple frustrated total internal reflection (IRS MFTIR) were recorded on the Fourier spectrometer Infralum FT-02 in wave numbers 800-8000 cm^{-1} area at a room temperature.
Based on the results obtained, the kinetic isotherms of the medium pH changes were made during the dispersal of GaAs and InAs semiconductors, according to which the state of the newly formed surface and the IRS MFTIR were previously evaluated, allowing to describe in details the behavior of the newly formed surface (compared to exposed in the air), reveal its mechanism.
Comparing quantitatively intensities of the IR bands 1383, 3200–3500 and 1650, 3600 cm^{-1} corresponding to the deformation and valence vibrations of OH groups of isopropyl alcohol and water formed by the isopropyl alcohol decomposition, one can judge the catalytic conversion of iso-C_3H_7OH and relative catalytic activity Investigated with the semiconductors. At the same time, the mechanism of catalytic decomposition of isopropyl alcohol on diamond-like semiconductors at room temperature described in [2], including the stages of dehydrogenation and dehydration (with water removal) was the key for that. So the following is expected: in the IR spectra of the «iso-C_3H_7OH-dispersible semiconductor» systems, in comparison with the IR spectrum of isopropyl alcohol, the decrease in the intensity of the 1383, 3200-3500 cm^{-1} bands corresponding to the deformation and valence vibrations of the iso-C_3H_7OH OH groups and the appearance along with the increased (with growth of dispersion time)1650, 3600 cm^{-1} bands intensity corresponding to the deformation and valence vibrations of the OH groups of the water formed and adsorbing on the surface in accordance with the donor-acceptor mechanism.

4. Results and discussion

Dispersing semiconductors GaAs, InA in water, one notes: a decrease in pH (acidification of the medium) in the first case and an increase in pH (alkalinization of the medium) – in the second (fig.1).

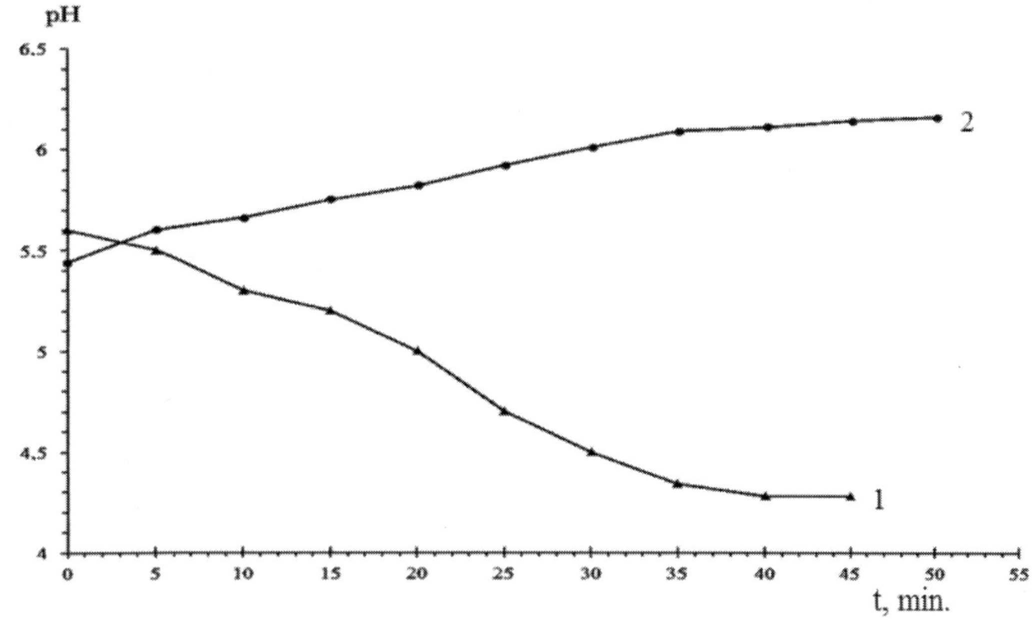

Figure 1. Kinetic isotherms of dispersing semiconductors GaAs (1) and InAs (2) in water.

Taking into account IR spectra containing bands of coordinately bound water (1650, 3600 cm^{-1}) and acid residues of AsO_4^{3-} (825, 850 cm^{-1}) (fig. 2, 3) and earlier made points (for example, in [3, 4]), the acidification of the medium can be explained by the formation of H_3AsO_4 acid – the result of the interaction of the particles H^+ and O^-, appearing during the dissociative desorption of chemically adsorbed H_2O molecules (fig. 2, [2, 3]), with coordinatively unsaturated As atoms, most efficiently seen on a newly formed surface. In the process of mechanochemical action, H_3AsO_4 acid passes into the medium, eliminating AsO_4^{3-}.

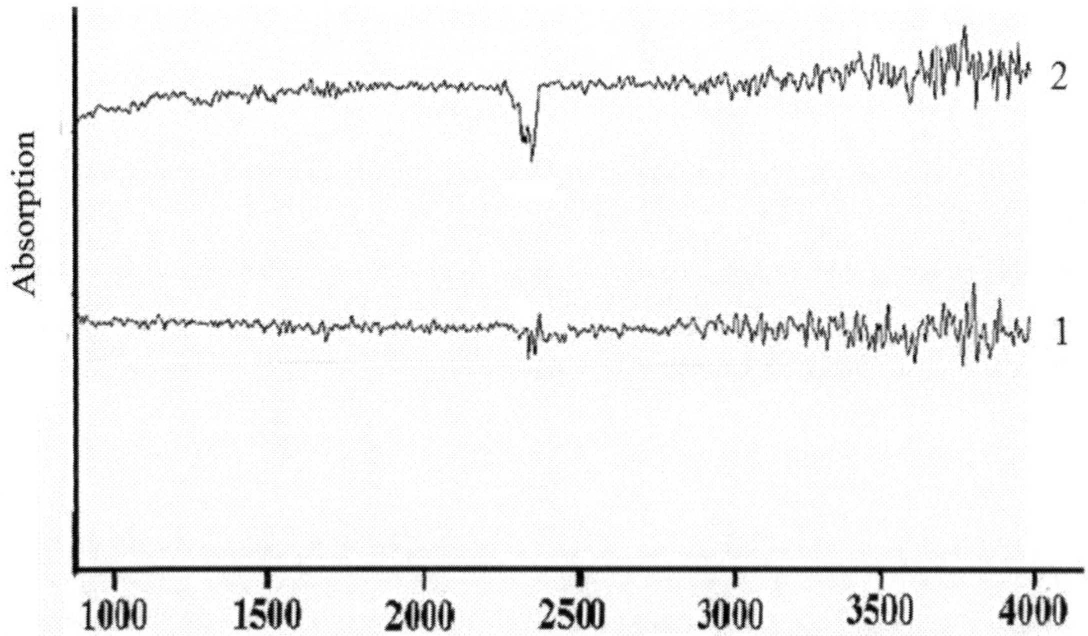

Figure 2. MFTIR IR spectra of GaAs (1) and InAs (2) semiconductors initial surfaces.

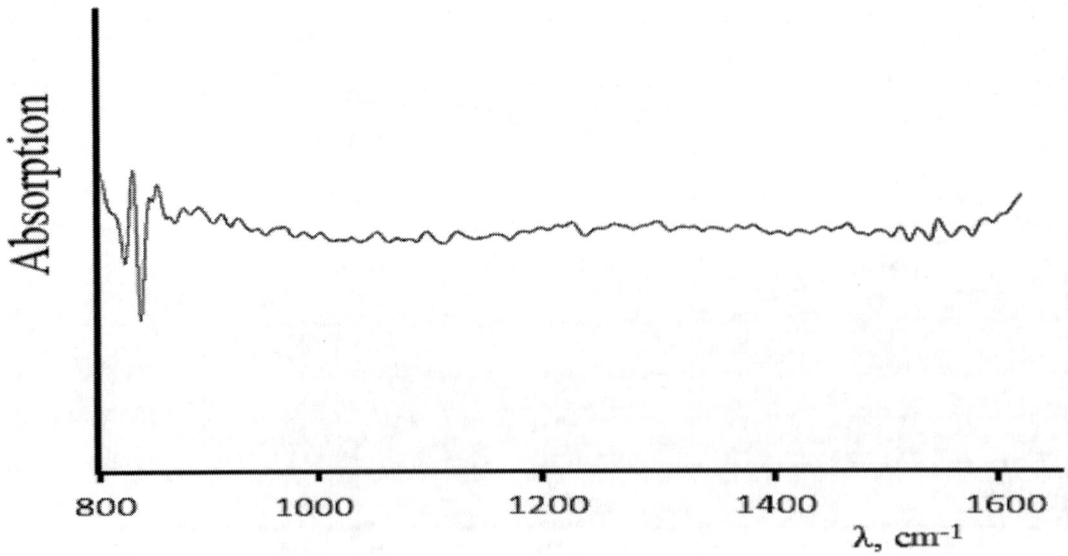

Figure 3. IR spectrum of InAs aqueous suspension after dispersing for 75 min.

An increase in the pH of the medium (when dispersing InAsis) can be related to AsO_4^{3-} ions hydrolysis.
The role of coordinatively unsaturated atoms, which is most effective on a newly formed surface, should also be mentioned when *dispersing GaAs and InAs semiconductors in isopropyl alcohol*: iso-$C_3H_7OH^{+\delta}$ – $As^{-\delta}$ surface compounds are formed, fragments of which (AsO_4^{3-}) are found in IR-spectra (fig. 3).
It is logical to assume that the formation of the marked surface compounds of chemically adsorbed by the donor-acceptor mechanism of iso-C_3H_7OH precedes the multistage reaction of its decomposition. In this case, the activation and subsequent rupture of certain bonds in the parent molecules will cause appropriate selectivity [2].
To confirm such considerations, let us consider the results of a comparative analysis of the pure isopropyl alcohol IR spectra 15 minutes later its contact with polycrystals of semiconductors and after dispersion of these polycrystals in it, taking into account their composition and dispersion time (fig. 4). Here comes under notice the absence of AsO_4^{3-}, MeOH, Ga^{3+}, In^{3+} bands in the second case, which indicates a very low solubility of the semiconductors samples in isopropyl alcohol, and appearance of new absorption bands when they are dispersed in the same medium. Specifically, absorption bands with maxima at 1650 and 3600 cm^{-1} appear in the IR spectra of the «isopropyl alcohol-dispersible semiconductor» systems, corresponding to the deformation and valence vibrations of OH groups resulting from the dehydration of iso-C_3H_7OH water (δ_{OH}) [3] and at the same time the intensities of the absorption bands in areas 1393 and 3200–3500 cm^{-1} corresponding to the deformation and valence vibrations of OH groups of isopropyl alcohol [5].

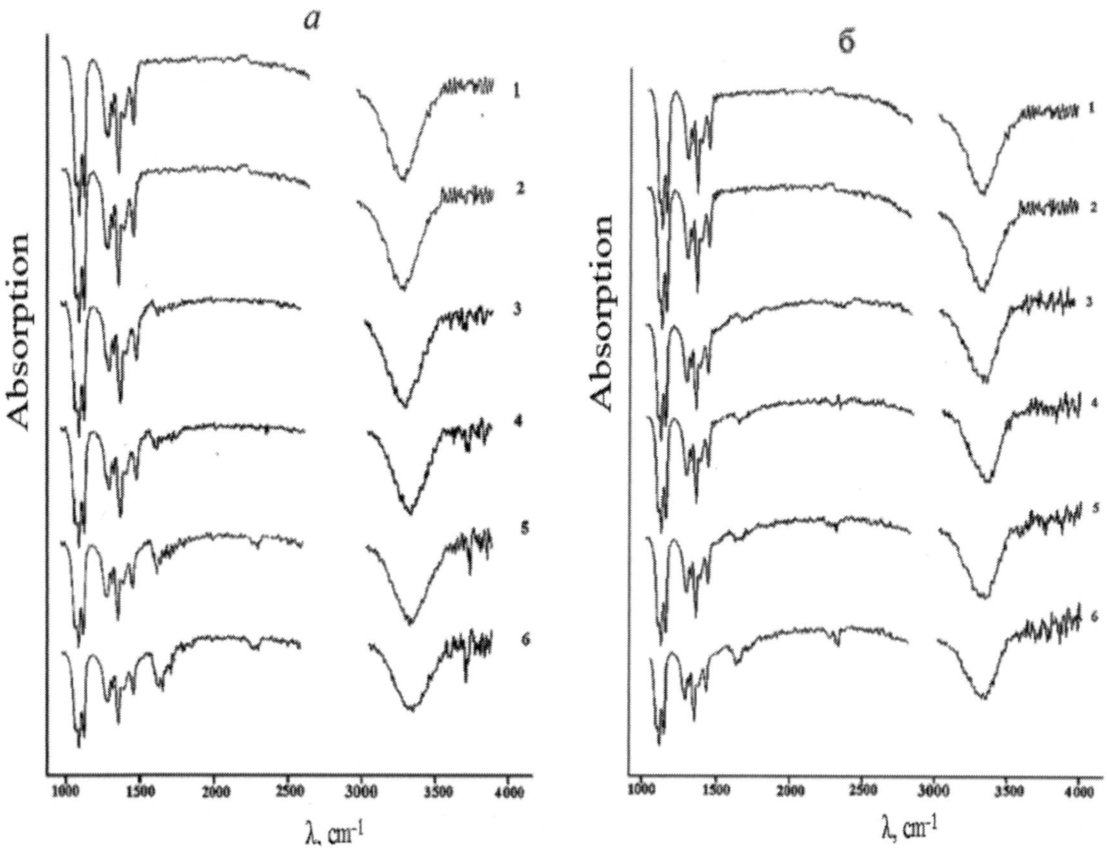

Figure 4. MFTIR IR spectra of pure iso-C_3H_7OH (1) 15 minutes later its contact with polycrystals of GaAs (2a), InAs (2б), and suspensions of «iso-C_3H_7OH-dispersible GaAs» (a) and «iso-C_3H_7OH – Dispersible InAs» (б) systems for different dispersing times: 3 – 15, 4 – 30, 5 – 60, 6 – 75 min.

Different intensity in IR spectra of the examined semiconductors of the deformation and valence vibrations absorption bands of the OH groups of water (1650, 3600 cm^{-1}), formed as a result of the dehydration of iso-C_3H_7OH, (fig. 5, table 1) should also be noted.

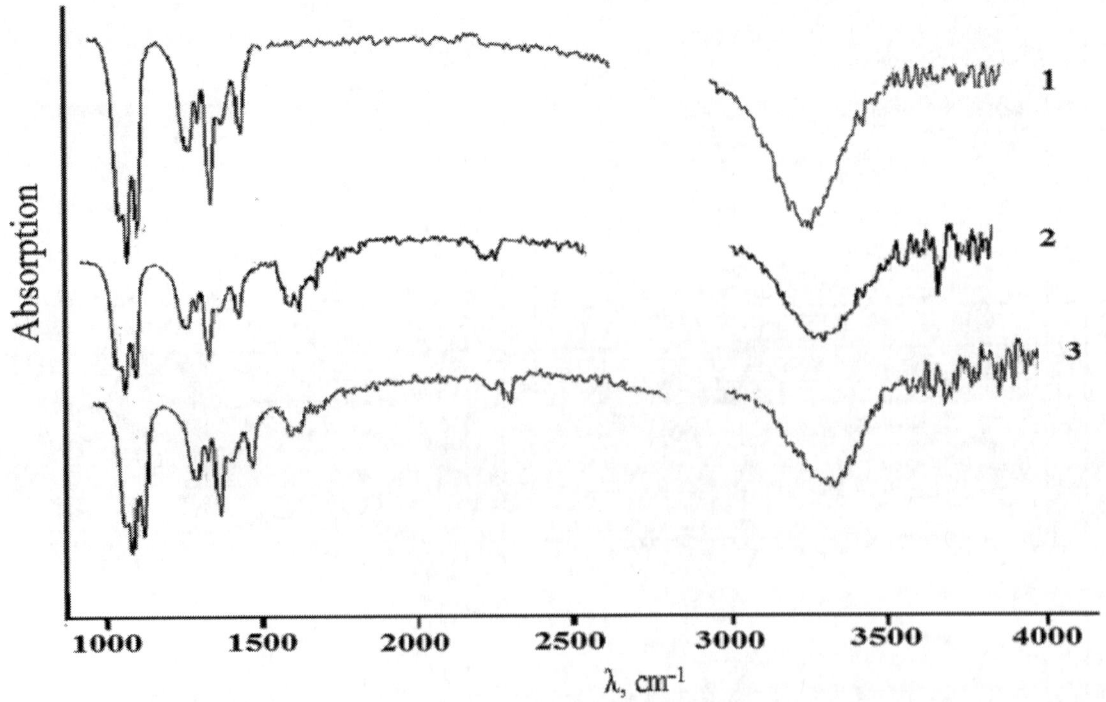

Figure 5. MFTIR IR spectra of pure iso-C_3H_7OH (1), suspensions of the systems «iso-C_3H_7OH – dispersible GaAs» (2) and «iso-C_3H_7OH – dispersible InAs»" (3) at the same dispersion time (75 min)

It allows to estimate the unequal relative catalytic activity and accordingly, the unequal strength and concentration of acid sites [2], responsible for the catalytic reaction proceeding in this case by the acid-base mechanism [1,2]. The intensities of the bands 1650, 3600 cm^{-1} and, accordingly, relative to the catalytic activity, gallium arsenide is noticeably leading (Table 1, Fig. 4). It is proposed as a material for microimpurities of isopropyl alcohol vapor and as an active, selective catalyst for its dehydration.

Table 1. Dependence of ir-absorbing band intensitivity changing on the composition of semiconductors disperratable in iso-C_3H_7OH at the same time of dispersion (75 min).

Dispersible semiconductor	ΔI, %			
	(1650 cm^{-1})	(3600 cm^{-1})	(1393 cm^{-1})	(3400 cm^{-1})
InAs	20	20	25	22
GaAs	40	40	43	45

5. Conclusions

– The kinetic isotherms of gallium and indium arsenides dispersion in water indicating both acidification (decrease in pH) and alkalinization (increase in pH) of the medium. The reasons are:
– intermediate compounds, functional groups have been found appearing on the real, partially hydrated surface of semiconductors and under water adsorption conditions, adsorption and catalytic decomposition of isopropyl alcohol on the newly formed surface;
– increased activity of newly formed surface;

– enrichment of the newly formed surface with coordination-unsaturated atoms, more clearly acting on such surface as active sites during the adsorption of H_2O, iso-C_3H_7OH type molecules and their decomposition has been established;
– the relative catalytic activity of the examined semiconductors (GaAs, InAs) in the dehydration reaction of iso-C_3H_7OH has been estimated;
– practical recommendations on the preferential use of GaAs in semiconductor gas analysis (for measuring cells production) and in semiconductor catalysis (especially for dehydration reactions) have been given.

References

[1] Kirovskaya I A 2015 *Physico-chemical properties of binary and multicomponent diamond-like semiconductors* (Novosibirsk, SB RAS Publ) p 367
[2] Kirovskaya I A 2010 *Solid solutions of binary and multicomponent semiconductor systems* (Omsk: Omsk State Technical University) p 400
[3] Kirovskaya I A 2001 *Surface phenomena* (Omsk: Omsk State Technical University) p 176
[4] Borovikova E Yu, Kurazhkovskaya V S, Boldyrev K N, Sukhanov M V, Petrkov V I and Kokarev S A 2014 Vibrational spectra and factor-group analysis of double arsenates of zirconium and alkali metal $MZr_2(AsO_4)_3$ (M = Li – Cs) *Vibrational Spectroscopy* Vol **73** pp 158–163
[5] Kiselev A V and Lygin V I 1972 *Infrared spectra of surface compounds* (Moscow: Nauka) p 459

Possibilities of new materials surface sensibility express determination based on ZnSe-CdS system by pH isoelectric state measurements of the surface state

I A Kirovskaya[1], E V Mironova[1], O V Ushakov[1], P E Nor[1], A V Yureva[1] and Yu I Matyash[2]

[1]Omsk State Technical University, 11, Mira ave., Omsk, 644050, Russia
[2]Omsk State Transport University, Omsk, Russia

e-mail: kirovskaya@omgtu.ru

Abstract. A method for determining the hydrogen index of the surfaces isoelectric state (pH_{iso}) at various gases pressures - possible components of the surrounding and technological media has been developed. With its use, changes in pH of binary and more complex semiconductors-components of the new system-ZnSe-CdS under the influence of nitrogen dioxide-have been found. The limiting sensitivity of surfaces - minimum P_{NO2}, causing a change in pH has been estimated. The most active components of ZnSe-CdS system, recommended as materials for measuring cells of NO_2, have been revealed.
The relationship between the changing patterns with the composition of surface (acid-base) and bulk (in particular, theoretical calculated crystal density) properties has been established, allowing to find the most effective materials for sensor technology and for semiconductor analysis.
Keywords: binary and multicomponent semiconductors, a new technique, surface sensitivity to gases, the relationship between surface and bulk properties, measuring cells.

1. Introduction

The solid solutions of ZnSe-CdS system (compared to the initial binary compounds, ZnSe, CdS) studied in the present work represent multicomponent diamond-like semiconductors – promising materials for sophisticated technology, in particular, for sensor technology [1].

When finding out possibilities of using new, often obtained for the first time materials for sensor technology, it is necessary to assess their sensitivity and selectivity to microimpurities of the surrounding and technological media. Direct adsorption measurements are perfect in this case, especially with the use of quartz crystal microbalance, which is widely used in works of the authors [1, 2].

In this case, taking into account the labour coefficient of direct adsorption measurements, an attempt has been made to approach to the rapid determination of sensitivity by measuring the surface isoelectric state pH in the respective gaseous media at different contents of the analyzed gas. Subsequent control adsorption measurements showed the reasonability of using such virtually express method. As a result, the limiting sensitivity to the selected gas (NO_2) has been established, the pHiso depends on the pressure and composition of the system components.

2. Problem statement

To determine the reasonability of using new materials based on ZnSe-CdS in sensor technology, it is possible to determine the possibility of evaluating their sensitivity to harmful microimpurities by measuring the surface isoelectric state pH (pH_{iso}) in the respective gaseous media. For this purpose, the pH of the ZnSe-CdS-binary semiconductor system components and solid solutions is determined by the method developed, under the influence, in particular, of nitrogen dioxide at different pressures; to estimate the ultimate sensitivity of surfaces; to establish the dependences of ΔpH_{iso} on the pressure and composition of the system components; to identify the most sensitive components of the system to NO_2 microimpurities, which are recommended for the manufacture of suitable measuring cells.
To facilitate the search for effective materials for sensor technology, to establish the relationship between the changing patterns with the composition of surface (pH) and bulk properties.

3. Experimental part

The investigated objects – solid solutions $(ZnSe)_x(CdS)_{1-x}$ (x = 0.49, 0.55, 0.97 mol) and the initial binary compounds (ZnSe, CdS) are powders (Sd ≤ 1.4 m^2/g) and films (d = 50–70 nm). Powders of solid solutions were obtained by isothermal diffusion of binary compounds (ZnSe, CdS) in vacuum sealed quartz ampoules at a temperature higher than the melting point of easily melted compound (CdS), according to a special program of temperature heating [1, 2], films of solid solutions and binary compounds – by discrete thermal spraying in vacuum [3, 4] (T_{cond} = 298 K , p = 1.53 10^{-4} Pa) mainly on the electrode sites of the piezoquartz resonators of the AT-cut, followed by annealing in the pairs of the objects under study. The thickness of the films was measured by change in the frequency of the piezoquartz resonator, interferometrically and based on spraying conditions [2, 5].
The preparation of solid solutions and their structure was judged by the results of X-ray and indirectly-by the results of IR and UV spectroscopic studies. The relative position and distribution of the intensities of the main lines on the x-ray diffraction patterns of binary compounds and solid solutions, the relative position and distribution of the intensities of the main bands in the IR and UV spectra, the dependence on the composition of the parameters values of the crystal lattices (a, c), Interplanar distances (d_{hkl}), density (ρ_r). According to the data, solid substitution solutions are formed in the ZnSe-CdS system with the specified compositions, which possess both a cubic sphalerite structure (with insignificant CdS content of ≤0.03 mole) and a hexagonal wurtzite structure (at $x_{CdS} \geq 0.45$ mole). In the first case, solid solutions are related in structure to zinc selenide, in the second – to cadmium sulphide [6].
X-ray diffraction studies were performed on diffractometers Dron-3, in $CuK_{\lambda,\ \beta}$-radiation (λ = 0.154178 and 0.139217 nm) and Advance D8 from Bruker (Germany), in CuKλ-radiation (λ = 0.154056 nm), using the large-angle survey technique [7], using the Lynxeye position-sensitive detector, IR *spectroscopic* - on the Fourier spectrophotometer Infralum FT-02, in the range of 830-4000 cm^{-1} [8], UV *spectroscopic* spectrophotometers on the UV-2501 PC spectrophotometer Shimadzu with a diffuse reflection attachment ISR-240 AI and SPECORD-40, in the range 190-900 nm [9].
To assess the surfaces sensitivity, the values of the hydrogen index of the isoelectric state of surfaces, pH_{iso}, were determined using the method of hydrolytic adsorption [1, 8].
The pH values of the surfaces allow to judge the average strength, and the pH changes the degree, nature and mechanism of the gases surface interaction. To realize such possibilities, a technique was developed allowing to determine semiconductor adsorbents surfaces pH_{iso} exposed in the air, in argon and then subjected to the action of adsorbate gas (in particular, NO_2) gas in a wide range of pressures, including the achieved minimum.

4. Results and discussion

As table. 1 shows, according to the values of *initial surfaces isoelectric state pH*, components of the ZnSe-CdS system are arranged in the ZnSe → $(ZnSe)_x(CdS)_{1-x}$→ CdS series, indicating predominantly the main nature of zinc selenide and solid solutions surfaces with practically equimolar and excess contents (pH_{iso} varies from 6.4 to 8.2).

Table 1. pH values of surfaces iseelectric conditions of ZnSe-CdS system components exposed in the air.

System component	pH_{iso}
ZnSe	8.20
$(ZnSe)_{0.97}(CdS)_{0.03}$	7.25
$(ZnSe)_{0.55}(CdS)_{0.45}$	7.85
$(ZnSe)_{0.49}(CdS)_{0.51}$	7.95
CdS	6.55

On the basis of this data, one can speak of a predominant contribution to the acid-base state of the system components surfaces with pH > 7 Bronsted centers and their increased activity to acid gases (NO_2, CO_2 type). The results of direct adsorption studies [1], in particular, $\alpha_{CO2} \setminus \alpha_{NH3} > 1$, can serve as confirmation. In addition, the adsorption of the main ammonia gas has a reversible (chemical) nature and is therefore accompanied by the formation of weak adsorbent-adsorbate bonds [1].

In accordance with facts mentioned above, it is logical to follow the influence of at least one more acid gas on the acid-base state of the surfaces and to assess their ultimate sensitivity to this gas (i.e., sensitivity at minimum pressure). Finally, studies have been made of the pH_{iso} of the system components, pretreated in argon, at various pressures of nitrogen dioxide (Fig. 1, 2).

As fig. 1 shows, the changes in pH_{iso} started at P_{NO2} = 0.05 Pa. The most sensitive to NO_2 were the solid solutions surfaces with practically equimolar and excess content of ZnSe, possessing the base surface (pH_{iso} = 8.2).

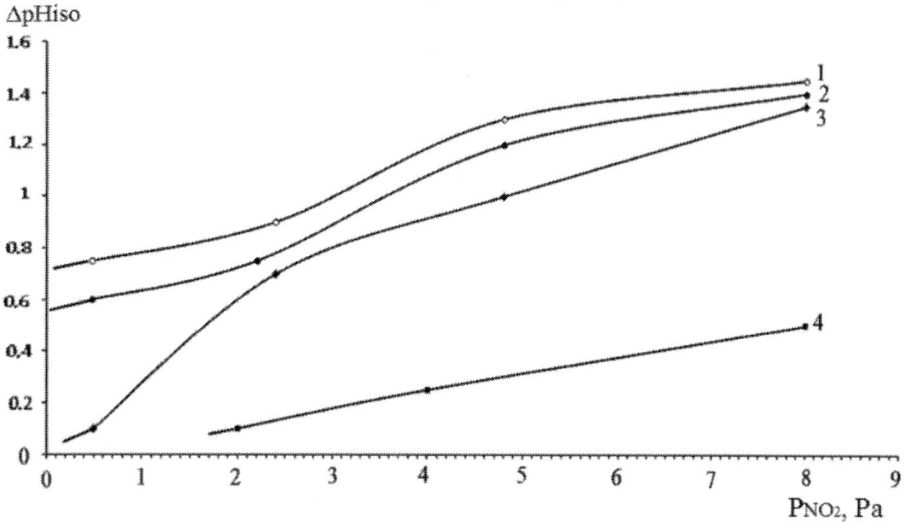

Figure 1. Change dependency of pH isoelectric state of the ZnSe-CdS components system ZnSe-CdS on the partial pressure of NO_2 mixed with argon: 1 – $(ZnSe)_{0.49}(CdS)_{0.51}$, 2 – $(ZnSe)_{0.55}(CdS)_{0.45}$, 3 – ZnSe, 4 – CdS.

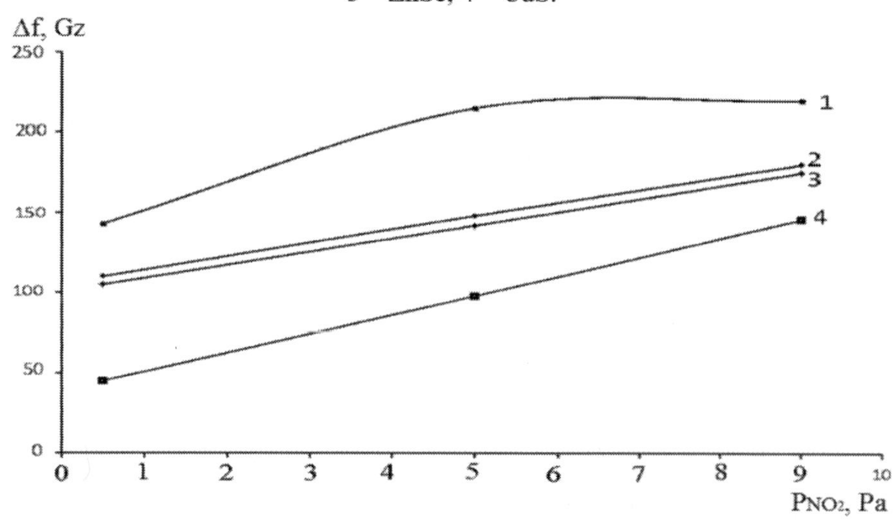

Figure 2. Dependence of piezoquartz resonator frequency variations with the films of the ZnSe-CdS-system components on the partial pressure of NO_2 mixed with argon: 1 – $(ZnSe)_{0.49}(CdS)_{0.51}$, 2 – $(ZnSe)_{0.55}$ $(CdS)_{0.45}$, 3 – ZnSe, 4 – CdS.

Fig. 1 correlates with fig. 2, reflecting the results of the most delicate measurements-piezoquartz resonator frequency variations with the supported nanofilms of the ZnSe-CdS system components, proportional to the adsorption values, from the NO_2 pressure.

Fig. 3 clearly demonstrates *comparative surfaces sensitivity* of the system under study to NO_2 with the dependences of ΔpH_{iso} and Δf on the composition at the same pressure.

On the ground of fig. 1-3 analysis, one can make a conclusion of such possibilities as the evaluation of new materials sensitivity to gases of a certain electronic nature with respect to the values and changes in pH_{iso}, without conducting direct adsorption studies, in this particular case the use of solid solutions of $(ZnSe)_x(CdS)_{1-x}$ with practically equimolar and excessive content of ZnSe for the manufacture of measuring cells for NO_2 microimpurities.

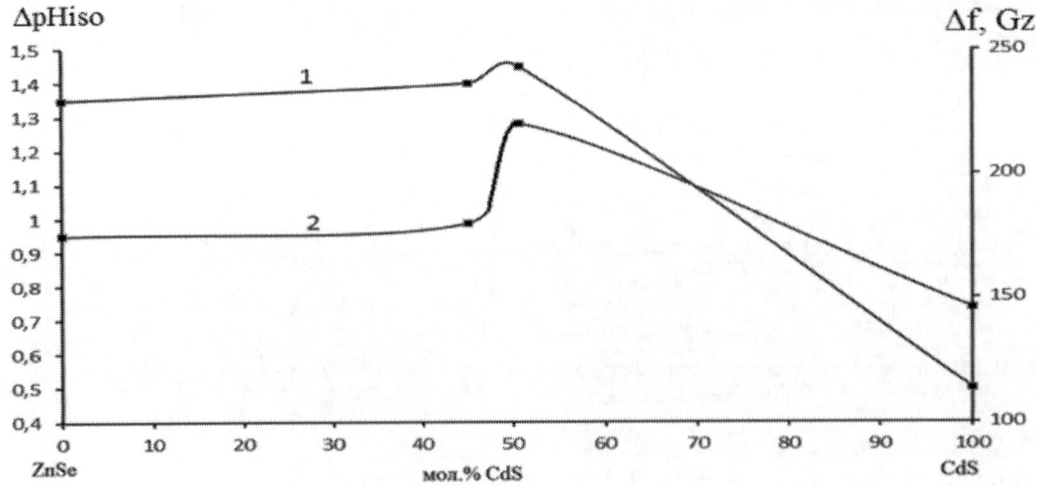

Figure 3. Dependences on the composition of oscillations frequency changes (Δf) (1) and pH index of the surfaces isoelectric state (ΔpH_{iso}) (2) of the ZnSe-CdS system components

A definite relationship between the changing patterns of the surface and bulk properties of the ZnSe-CdS system components with a change in composition is of some interest.

Thus, as the zinc selenide content increases in it, the interplanar distances decrease (d_{hkl}), crystal lattice parameters (a, c), theoretical calculated crystal density (ρ_r), pH_{iso} (fig. 1, 4) change extremely (with an excess of ZnSe). Attention is drawn to the presence of both direct and opposite relationship between ρ_r and pH_{iso}.

Direct relationship is a consequence of decrease incoordination unsaturation of atoms both in the volume and on the surface with increasing ρr, respectively, with an increase in their packing density per unit of the crystal lattice volume. That is bound to cause a decrease in the contribution of the Lewis acid sites, an increase in the contribution of the Brönsted acid sites, and therefore an increase in the pH_{iso} in the series ZnSe → $(ZnSe)_x(CdS)_{1-x}$→ CdS.

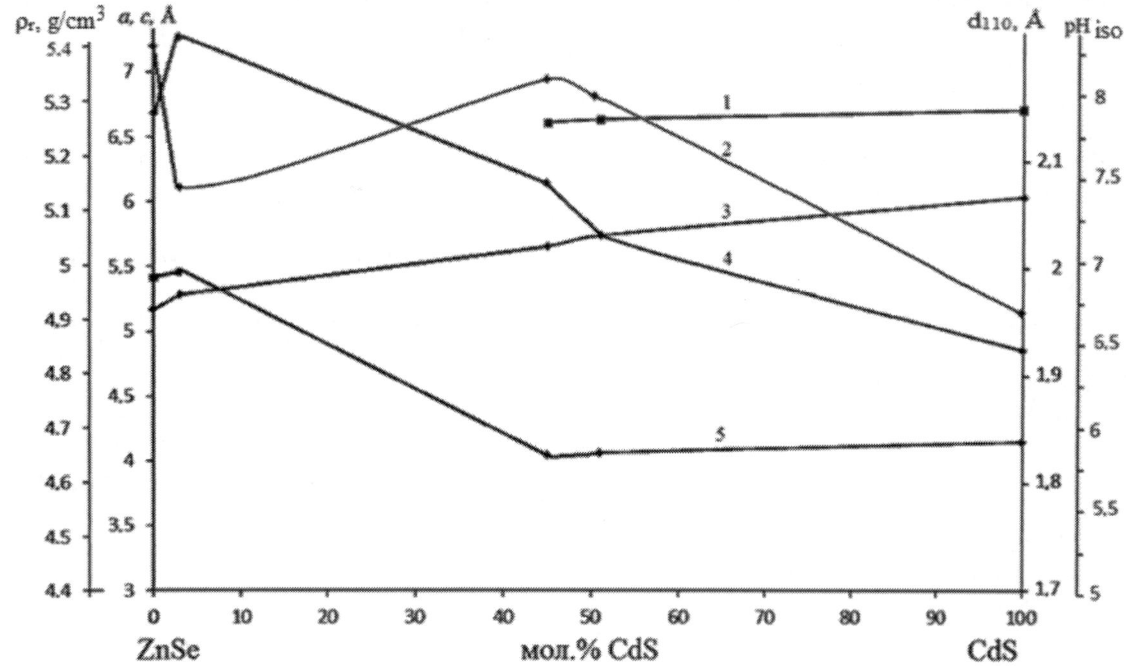

Figure 4. Dependence on the composition of crystal lattices parameters values a (5), c (1), the interplanar distance d_{111} *(3)*, theoretical calculated crystal density ρ_r *(4)*, and the isoelectric state pH of the pH_{iso} (2) surfaces of the ZnSe-CdS system components.

With an excess of zinc selenide (x = 0.97 mol), the packing density of ZnSe atoms per unit volume increases so much and their bond with surface atoms decreases so weakly that a burst in the coordination unsaturation of the latter is revealed, and as a result, the pH_{iso} is reduced (fig.4). Besides, with the other compositions of the ZnSe-CdS system, a predominant effect on the acid-base state of zinc selenide surfaces is observed, which, compared with CdS, has a larger bandgap (ΔE_{ZnSe} = 2.7-2.9 eV, ΔE_{CdS} = 2.42 eV).

Thus, the correlation between the density (ρ_r), the relative contribution of the coordinatively unsaturated atoms (Lewis sites) and the acidity of the surface is clearly traced.

The discovered connections between surface and bulk physicochemical properties (fig. 3, 4), with indisputable scientific significance, offer additional possibilities in the search for effective materials (in this case for measuring cells) on the basis of information on bulk physicochemical properties (in particular , on theoretical calculated crystal density), without carrying out not only labor-intensive adsorption studies, but also determination of acid-base properties.

Specific practical recommendations on the use of obtained materials for the manufacture of measuring cells for the NO_2 microimpurities on the basis of the results of the performed studies of the acid-base surfaces properties is mentioned above.

5. Conclusions

Changes in the surface isoelectric state pH (ΔpH_{iso}) of the system of ZnSe-CdS components – binary semiconductors and their solid solutions under the influence of nitrogen dioxide at various pressures have been determined according to the developed method. The limiting sensitivity of surfaces has been estimated. The dependences of ΔpH_{iso} on the pressure and composition of the system components have been established. The most sensitive components of the system, recommended for the manufacture of appropriate measuring cells have been identified.

Relationships between the changing patterns with the composition of surface (pH_{iso}) and bulk properties making the search for effective materials for the sensor technology easier have been revealed.

References

[1] Kirovskaya I A 2010 *Solid solutions of binary and multicomponent semiconductor systems* (Omsk: Omsk State Technical University) p 400

[2] Kirovskaya I A 1995 *Adsorption processes* (Irkutsk: ISU) p 300

[3] *State diagrams of systems based on semiconductor compounds* $A^{II}B^{VI}$ 1982 (Kiev: Naukova Dumka) p 162

[4] Williamson, W J 1989 Thin films of indium antimonide. Preparation, properties, applications edited by V A Kasian, (Chisinau: Shtiintsa) p 162 https://catalogue.nla.gov.au/Record/2809644

[5] Kirovskaya I A 2006 The development of new materials for analyzers, environmental *Modern problems of science and education* № **3** p 70

[6] Kirovskaya I A, Mironova E V, Ushakov O V, Deeva A A and Yurieva A V 2016 Obtaining Hetero- substituted Semiconductor Materials (ZnSe)x (CdS)1-x and their Crystallochemical and Structural Properties *Procedia Engineering* V **152** pp 681–688

[7] Parkhomenko Y N, Shlenskii A A, Pavlov V F, Shepekina G V and Yugova T G 2010 X-ray diffraction determination of the composition of in x ga1-x sb solid solution *Inorganic Materials* V **46** № 14 pp 1526–1528

[8] Kirovskaya I A 2015 *Physico-chemical properties of binary and multicomponent diamond-like semiconductors* (Novosibirsk: SB RAS p 367

[9] Sobolev V V, Sobolev V Val and Shushkov S V 2011 Optical spectra of six silicon phases *Semiconductors* Vol **45** pp 1247–1250

Surface-active and electrophysical semiconductors properties of the CdTe-CdSe system

I A Kirovskaya[1], T L Bukashkina[1], R V Ekkert[1], O V Ushakov[1], L V Kolesnikov [2] and Yu I Matyash[3]

[1]Omsk State Technical University, 11, Mira ave., Omsk, 644050, Russia
[2]Kemerovo State University, Kemerovo, Russia
[3]Omsk State Transport University, Omsk, Russia

e-mail: kirovskaya@omgtu.ru

Abstract. Surface properties (acid-base, adsorptive, electrophysical) of binary and multicomponent semiconductors of the CdTe-CdSe system have been holistically studied. Changing patterns of the studied surface properties correlated among themselves and with changing patterns of the bulk physical and chemical properties have been revealed. The nature of active centers, acid-base, adsorptive and electronic interactions mechanisms have been determined with due account for local and collective factors The most active discovered adsorbents are recommended as primary transducers of CO trace impurities measuring cells.
Keywords – binary and multicomponent semiconductors, surfaces acidity, adsorption, electrical conductivity, measuring cells.

1. Introduction

The research question here is the surface properties of the binary (CdTe, CdSe) and multicomponent ($(CdTe)_x(CdSe)_{1-x}$) semiconductors of the CdTe-CdSe. Possessing unique properties (electrical, optoelectrical, optical, etc.), binary semiconductors ($A^{II}B^{VI}$ type) they are widely used in engineering: heterojunctions, laser radiating elements, luminescent and electrical luminescence display, diverse photo- and piezoelectric converters, ionizing-radiation detector, etc are based on them.

The subject of special concern is multicomponent semiconductors – substitution solid solutions, implicating even greater possibilities. Preserving many unique properties of the initial binary compounds, they can detect unexpected, specific effects related to the peculiarities of internal processes accompanying solid-solution formation: ordering, structure hardening, combined effect of constituent components as macro- and micro effects and possible high concentrations of impurity centers [1]. Such unexpected, specific effects are useful both for science and sophisticated technology.

At the same time, the potential for use the designated objects in sophisticated technology, in particular in nano- and sensor technology, are considerably determined by awareness of their bulk and surface physico-chemical properties in particular. The determining role of the latter grows with the growing trend to miniaturization of semiconductor devices and facilities.

The issue is acid-base, adsorptive and electrophysical surface properties of the CdTe-CdSe system semiconductors.

2. Problem statement

To study in an integrated manner acid-base, adsorptive (in relation to CO), electrophysical properties of binary and multicomponent semiconductors of the CdTe-CdSe system. To determine the nature of active centers, mechanisms and patterns of surface interactions, their connection with bulk and physico-chemical changing patterns.

To identify the most active adsorbents – possible primary transducers of CO (carbon monoxide) trace impurities measuring cells with subsequent practical guidelines.

3. Experimental part

The research subjects were powders (S_{sp} = 0.405 ... 0.91 m^2/g) and nanosized films (d = 20 ... 100 nm) of CdTe, CdSe, and solid solutions $(CdTe)_x(CdSe)_{1-x}$ (x = 15; 25; 50; 75; 85 mol. %). Solid solutions powders were obtained by isothermal diffusion of binary compounds (CdTe, CdSe) in vacuum-sealed quartz vessel at temperature close to the melting point of more easily melted binary compound (CdTe), followed by homogenizing annealing, according to the created program [1]; binary compounds and solid solutions films – by discrete thermal spraying in vacuum (T_{cond} = 298 K, P = $1.33 \cdot 10^{-3}$ Pa) on electrodes of AT – cut piezoquartz resonators [2, 3]. Simultaneously, on the same film samples the resistance change according to electrical conductivity (σ_s) and the surface charging under adsorption conditions were investigated (using the probe compensation method [3]). The film thickness was determined from frequency deviation of the piezoquartz resonator [3].

To identify solid solutions, the results of X-ray research were mainly used, indirectly supported by the results of IR and UV spectroscopic studies [4].

X-ray studies were carried out on diffractometers DRON-3, in $CuK_{\alpha,\beta}$ – radiation (λ = 0.154178 and и 0.139217 nm), and D8 Advance from «Bruker»(Germany), in CuK_α – radiation (λ = 0.154056 nm), by method of high-angle scattering [5, 6], using the position-sensitive detector Lynxeye, as well as data base ICDIPDF-2 and software ICDIPD 3.0 (Bruker).

Surface properties were judged by the results of acid-base and adsorption properties studies. Acid-base properties were examined by methods of hydrolytic adsorption, mechanochemistry and non-aqueous conductometric titration [1, 7, 8], adsorption properties were examined by methods of piezoquartz crystal microbalance (sensitivity 10^{-11} $g/cm^2 \cdot$ Hz) and by volumetric method [3] at temperature 258 ... 477 K and pressure 6 ... 18 Pa intervals.

Adsorbate gas (CO) *was obtained* according to a well-known method [8].

Reproducibility and accuracy of the experimental data were checked based on the results of parallel measurements using mathematical statistics, quantitative analysis results processing methods and computer programs.

4. Results and discussion

According to X-ray study findings [9], in the system CdTe – CdSe solid substitutional solutions are formed with sphalerite structure (excessive CdTe) and wurtzite (excessive CdSe). This is evidenced by the relative position and distribution by intensity of the basic lines on binary compounds and solid solutions radiographs, dependence on crystalline grids parameter values composition (*a, c*), interplanar spacing (d_{hkl}), density (ρ_r) (for example, see Fig. 1, [9]).

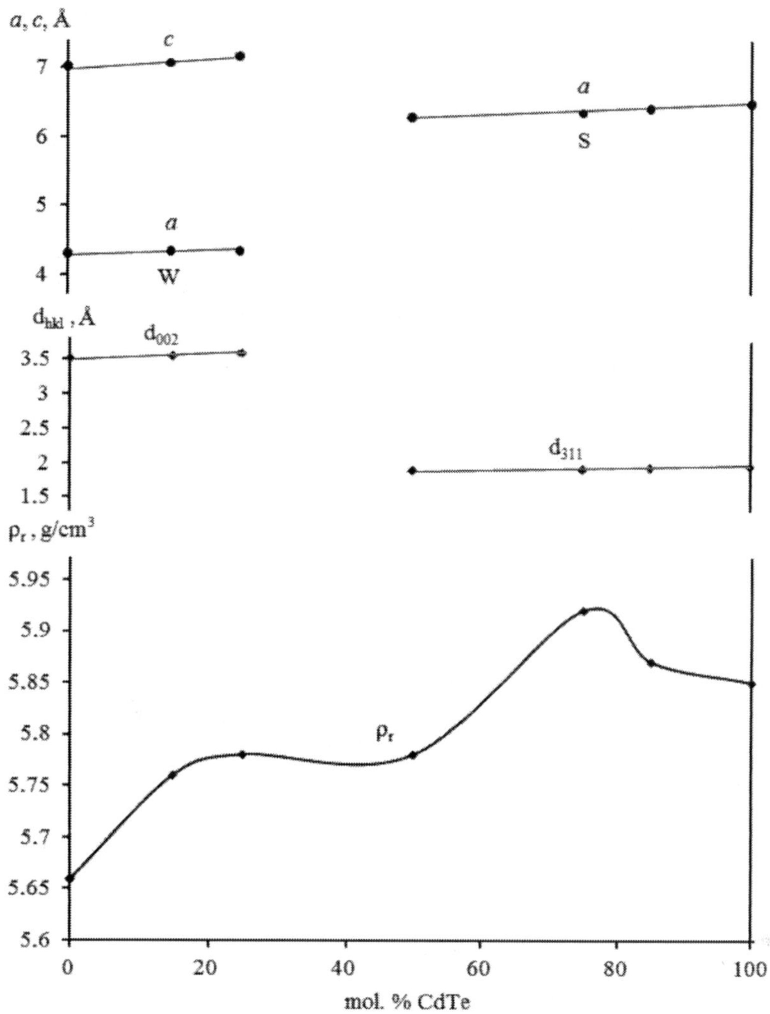

Figure 1. Dependence on crystalline grids parameter values composition (*a, c*), interplanar spacing (d_{hkl}) and theoretical calculated crystal density (ρ_r) of the system CdSe-CdTe components; W – wurtzite, S – sphalerite.

Study of the surfaces acid-base properties made it possible to determine the strength (isoelectric state pH– pH_{iso}), overall concentration ($C_{overall}$) (Table 1), acid sites nature.

Table 1. Acid-basic surface characteristics of the system CdTe – CdSe components.

System component	pH_{iso}	$C_{overall}$
CdTe	6.29	2.0
$CdTe_{0.85}CdSe_{0.15}$	6.13	2.6
$CdTe_{0.75}CdSe_{0.25}$	5.88	2.8
$CdTe_{0.5}CdSe_{0.5}$	5.99	4.3
$CdTe_{0,25}CdSe_{0,75}$	5.84	4.4
$CdTe_{0.15}CdSe_{0.85}$	5.86	3.9
CdSe	6.81	3.4

According to the strength of acid sites (pH_{iso}), CdTe – CdSe system components are arranged on the initial surfaces in the following sequence:

CdTe → $(CdTe)_x(CdSe)_{1-x}$ → CdSe, with changes in pH_{iso} values from 6.29 to 6.81.

The latter indicates the faintly-acid nature of such surfaces and respectively the prevailing contribution of Lewis acid sites charged by coordination unsaturated atoms [1, 7].

The impact of the increased acidity of the CdTe surface (compared to CdSe) on the acidity of the solid solutions surfaces (Table 1) comes under notice. The basis is a regular change in the bulk physical and chemical properties of binary compounds (at transition from CdTe to CdSe). Here it is logical to talk about the melting point (T_m), band gap (ΔE), electronegativity differences (Δx) (Table 2), which determine the A-B bonds strength; possibility of electrons transitions from the valence band to the conduction band and their localization on coordinatively unsaturated atoms (Lewis sires), responsible for the acidity of the surfaces, and ionicity, and the surfaces hydration.

Table 2. Values of bulk (T_m, ΔE, Δx) and surface (pH_{iso}) initial binary compounds properties of the CdTe–CdSe system.

Composition	Melting point (T_m), °C	Band gap (ΔE), eV	Electronegativity differences (Δx)	pH_{iso}
CdTe	1092	1.51	0.4	6.29
CdSe	1258	1.88	0.8	6.81

As the Table 2 shows, growth of the indicated bulk physical and chemical properties provides growth of pH_{iso}. Such parallels in bulk and surface properties changes additionally confirm the defining role of coordinatively unsaturated atoms as the acid Lewis sites in this case. They are also responsible for the adsorption of molecules of CO, CO_2, H_2O [1, 7] type.

The results of the adsorption properties of the CdTe-CdSe system components (relative to CO) studies make it possible to claim of chemical activated adsorption leak at room temperature [10]. The reason for such statement is the appearance of adsorption isobars (see, for example, Fig. 2) the results of the analysis of equilibrium ($\alpha_T = f(p)$) and kinetic ($\alpha_T = f(t)$) adsorption isotherms, performed activation energy calculations (E_α) and adsorption heat (q_α) as well as the results of electrical conductivity measurements (σ_s) under adsorption conditions (Fig. 2, 3).

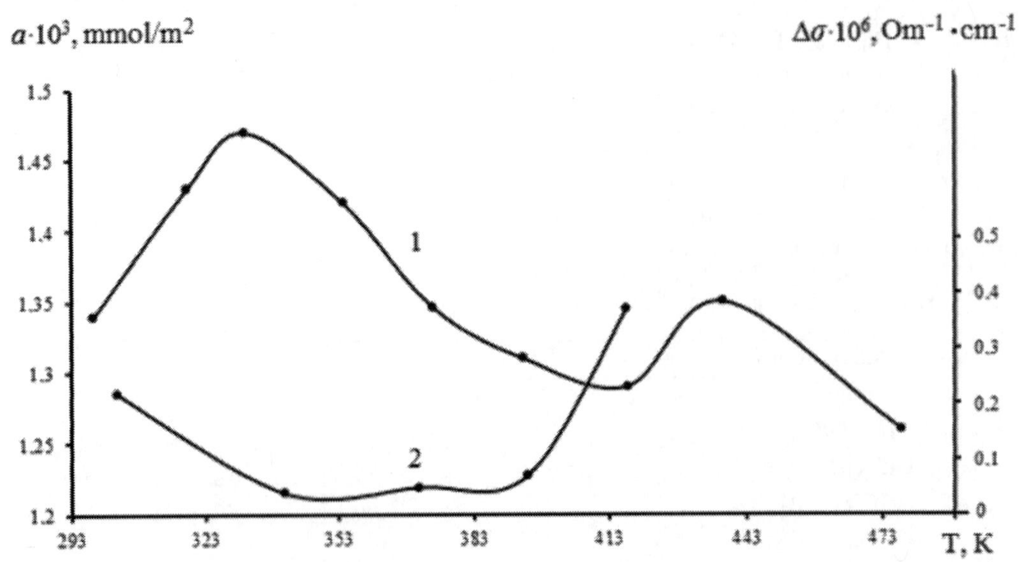

Figure 2. Temperature dependence of CO adsorption amount on the solid solution $(CdTe)_{0.85}(CdSe)_{0.15}$ (1) and the change in its electrical conductivity during CO adsorption at P_b = 16 ... 18 Pa.

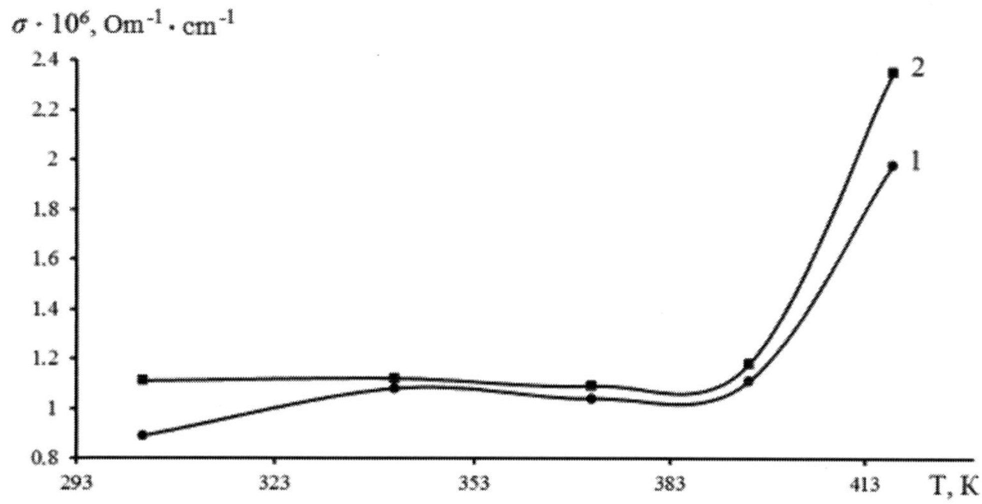

Figure 3. Temperature dependence of the solid solution electrical conductivity $(CdTe)_{0.85}(CdSe)_{0.15}$ in vacuum (1) and during CO adsorption (2) at P_b = 16 ... 18 Pa.

When thoroughly considering the said results, taking into account the acid-base properties of adsorbent surfaces, electronic structure of adsorbate (CO) molecules and their "behavior" on other diamond-like semiconductors [1, 7] it is apparent that CO adsorption proceeds by the donor-acceptor mechanism with predominantly surface A atoms (with free d– and p– orbitals and more evident metallic properties) as acceptors, and adsorbate molecules as donors:

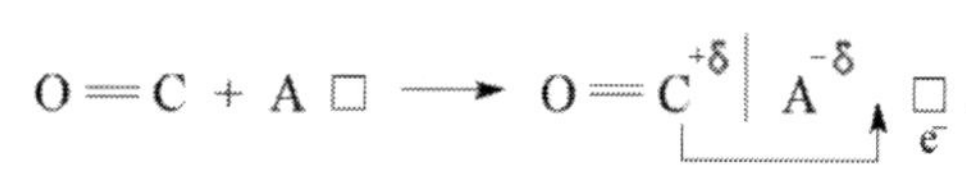

Formation of donor-acceptor bonds is indicated by the presence of the corresponding bands in the IR spectra after CO adsorption [10], and the possible role of CO molecules as electron donors – by the increase in electrical conductivity (σ_s) under CO adsorption conditions (Fig. 2, 3).

Noting the preferential formation of donor-acceptor bonds in the adsorption systems under study one can not exclude due to the peculiarities of the electronic nature of CO molecules the possible definite contribution of back donation and hydrogen bonds [1, 7].

The results of the electrical conductivity measurements under adsorption conditions (Fig. 2, 3), indicating its change and respectively, charging of adsorbent surfaces indicate the dependence of the adsorption capacity of CO molecules not only on the local factor (chemical properties of molecules and active sites) but also on the collective (electronic) factor, which is determined by the position of the Fermi level and the character of the surfaces energy spectrum [1, 7]. That is, with such features of the donor-acceptor bonds $CO^{+\delta}$ – $A^{-\delta}$, the participation of free carriers in the elementary adsorption act is inevitable, although the selectivity of gas adsorption is determined by the local properties of the surface [3, 11].

It is appropriate to recall that the free carriers concentration (electrons - n and holes - p) in crystalline bodies is related to the width of the forbidden band and the position of the Fermi level in it. It is the reason for the statement in the framework of the electron theory of adsorption and catalysis by F.F. Wolkenstein on the role of the Fermi level as a regulator of adsorption and catalytic activity [11].

5. Summary and conclusions

Complex studies of acid-base, adsorption (in relation to CO), electrophysical properties of binary and multicomponent semiconductors of the CdTe-CdSe system have been performed.

Taking into account their physical and chemical properties, the following has been established:

- the nature of active sites and mechanisms of acid-base, adsorption and electronic interactions;
- the role of local and collective factors;
- changing patterns of the studied surface properties correlated among themselves and with changing patterns of the bulk physical and chemical properties.

The most active adsorbents, cadmium telluride and solid solution with its excess have been discovered, proposed as primary transducers of CO trace impurities measuring cells (carbon monoxide). Footnotes should be avoided whenever possible. If required they should be used only for brief notes that do not fit conveniently into the text.

6. References

[1] Kirovskaya I A 2010 *Solid solutions of binary and multicomponent semiconductor systems* (Omsk: Omsk State Technical University) p 400

[2] Kirovskaya I A, Azarova O P, Shubenkova EG and Dubina O N 2002 Synthesis and optical absorption of solid solutions between InSb and II-IV cjmpounds *Inorganic Materials* vol 38 No 2 pp 91–94

[3] Kirovskaya I A and Nor P E 2013 Adsorption properties of CdS-CdTe system semiconductors *Journal of Physical Chemistry* vol 87 No 12 pp 2077–81

[4] Kirovskaya I A Nor P E and Ratushniy A A 2014 Crystal-chemical, spectroscopic and electrical properties of solid solutions and binary components CdS-CdTe system *Dynamics of Systems, Mechanisms and Machines, Dynamics 2014 Proceedings* p 7005666.

[5] Gorelik S S, Rastorguev L N and Skakov Yu A 1970 *Radiography and electronoptical analysis* (Moscow: Metallurgy) p 107

[6] Parkhomenko Y N, Shlenskii A A, Pavlov V F, Shepekina G V and Yugova T G 2010 X-ray diffraction determination of the composition of in x ga1-x sb solid solution *Inorganic Materials* vol 46 No 14 pp 1526–28

[7] Kirovskaya I A 2015 *Physicochemical properties of binary and multicomponent diamond-like semiconductors* (Novosibirsk: SB RAS) p 367

[8] Khudyakova T A and Kreshkov A P 1971 The theory of conductimetric acid-base titrations in non-aqueous solutions *J. of Electroanal. Chem. and Interf. Electrochem* vol 29 No 1 pp 181–194

[9] Kirovskaya I A and Bukashkina T L 2016 Structural properties – precursors of adsorptive activity of cadmium chalcogenide new materials *Dynamics of Systems, Mechanisms and Machines, Dynamics 2016 Proceedings* p 7819022.

[10] Kirovskaya I A, Nor P E, Bukashkina T L and Mironova E V, 2016 Surface properties of semiconductor analogs of CdB^{VI} and their solid substitution solutions *Journal of Physical Chemistry* vol 90 No 3 pp 522–529

[11] Wolkenstein W 1991 *Electronic processes on semiconductor surfaces during chemisorption* (New York: Consultants Bureau) p 444

Basics of identification measurement technology

Yu N Klikushin, V Yu Kobenko and P P Stepanov

Omsk State Technical University, 11 Mira Avenue, Omsk, 644050, Russia

e-mail: Omsk.Petr@gmail.com

Abstract. All available algorithms and suitable for pattern recognition do not give 100% guarantee, therefore there is a field of scientific night activity in this direction, studies are relevant. It is proposed to develop existing technologies for pattern recognition in the form of application of identification measurements. The purpose of the study is to identify the possibility of recognizing images using identification measurement technologies. In solving problems of pattern recognition, neural networks and hidden Markov models are mainly used. A fundamentally new approach to the solution of problems of pattern recognition based on the technology of identification signal measurements (IIS) is proposed. The essence of IIS technology is the quantitative evaluation of the shape of images using special tools and algorithms.

1. Introduction

Recognition, as an information procedure, is inherent in all living natural objects, is their natural, «built-in» function and is aimed at adapting the organism to the conditions of the external and internal environment. However, if in a living nature the recognition function is performed quickly, efficiently and intuitively, then to teach the technical system to perform the same is a complex problem [9]. The concept of recognition can be defined, by analogy with the notion of measurement, as the process of comparing an unknown image with some reference image. If we accept this point of view, then the next step in our reasoning is to extend the whole methodology of the measurement theory to the solution of image recognition problems. In favor of this approach, in addition to general "ideological" considerations, is the fact that the result of comparison in both cases is a quantitative estimate in the form of a number [1, 4]

2. The problem statement

It is known that at the heart of any measurement there are two procedures characterized by concepts: a measuring transducer and a measuring scale. Recall, for example, the measurement of the effective value of an alternating voltage by means of a simple analog device, the arrow (pointer) of which deviates by some angle under the influence of the input measured signalThe first function (conversion of the signal into the angle of the pointer deflection) is realized in the measuring transducer, which can be realized in various versions, for example, by means of an electromagnetic or electrostatic mechanism. The second function – measurement – consists in evaluating the position of the pointer relative to the position of the digital marks of the measuring scale, which is a multi-valued measure of the physical quantity (voltage or current) represented by the input signal. The more digitized scale marks, the more accurate the measurement result can be obtained. If the pointer is between the marks, then the interpolation procedure

IOP Conf. Series: Journal of Physics: Conf. Series **944** (2017) 012050 doi:10.1088/1742-6596/944/1/012050

is applied additionally – rounding the position of the pointer to the nearest mark. Despite the obvious simplicity and convenience of similar, simplest measurements, the results obtained can be very accurate (have small errors).

3. Theory

In the theory and practice of MIS, the function of the measuring transducer is performed by the so-called identification tester (IT). From a programmatic point of view, IT displays an array of numbers into one, a special identification number that has the property of scale invariance-independence of the amplitude and frequency of the input signal-a sample implementation of which is analyzed in the form of an array [1–4]. An example of IT can serve, so-called, S-tester, the program code of which is presented in Fig.1

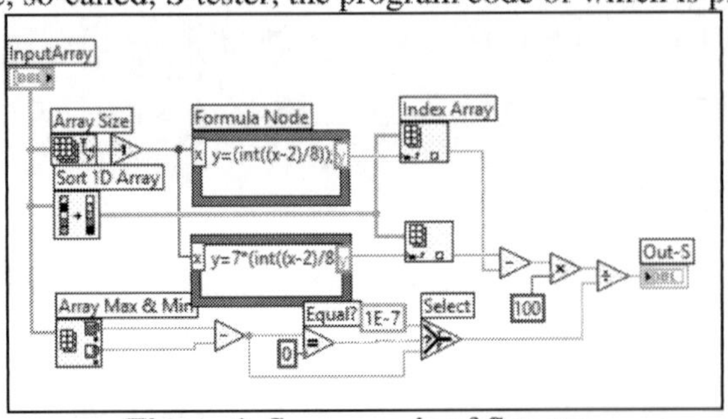

Figure 1. Source code of S-tester.

The principle of operation of the S-tester is based on measuring the steepness of the ranked signal function at its central site relative to the median

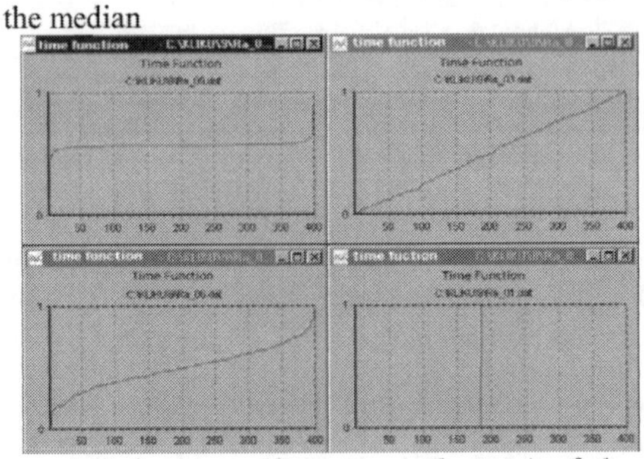

Figure 2. The form of the ranked functions (from top to bottom) of the realizations of signals with (distribution) of CAPS, NORMS, RAVN and 2DMOD distributions (capital).

Table 1. Identification scale of S-type.

IdP=S N=10000, L=100	Type of random signal distribution)								
	cons t	kosh	lapl	gaus	simp	trap	even	asin	2mod
Rank	0	1	2	3	4	5	6	7	8
Mean (S)	0	1	15	32	51	63	75	92	100
Sy = Id[Y(t)]			24.36						

IOP Conf. Series: Journal of Physics: Conf. Series **944** (2017) 012050 doi:10.1088/1742-6596/944/1/012050

4. The results of experiment

Our studies have established that the type of functions presented (pictures to be moved to the results of the experiments) does not depend on the parameters of the shift and scale. The average steepness (S) on the central section gradually increases in the direction from the COSM to the 2MOD distribution, which serves as the basis for "digitizing" the form (s) of the distributions, as illustrated in the examples in Fig. 2 to 4

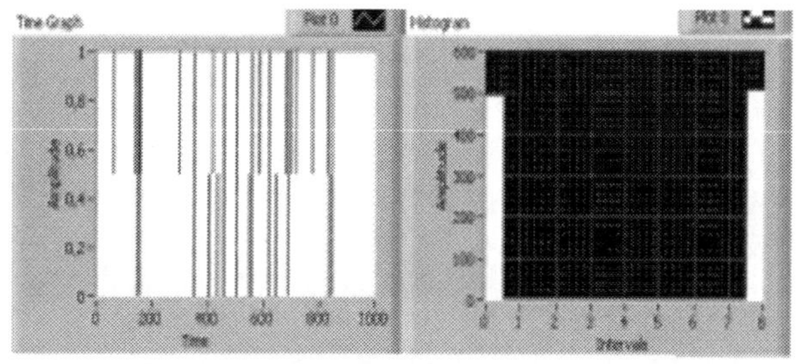

Figure 3. Examples of 2MOD selective realizations of signals and his distributions.

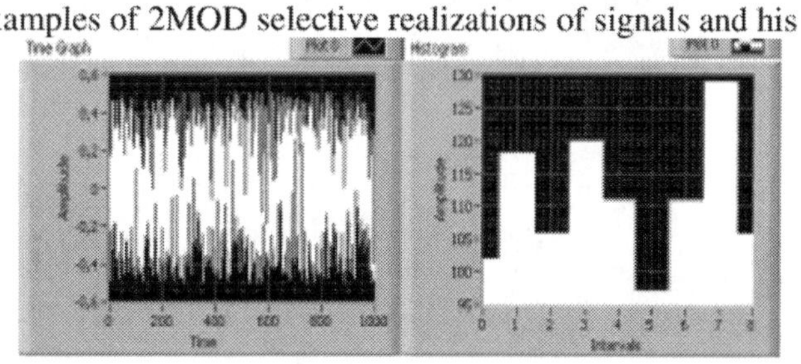

Figure 4. Examples of even distribution selective realizations of signals and his distributions.

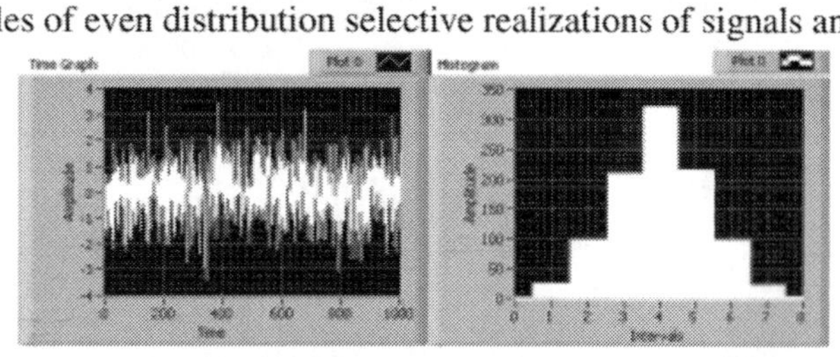

Figure 5. Examples of Normal selective realizations of signals and his distributions.

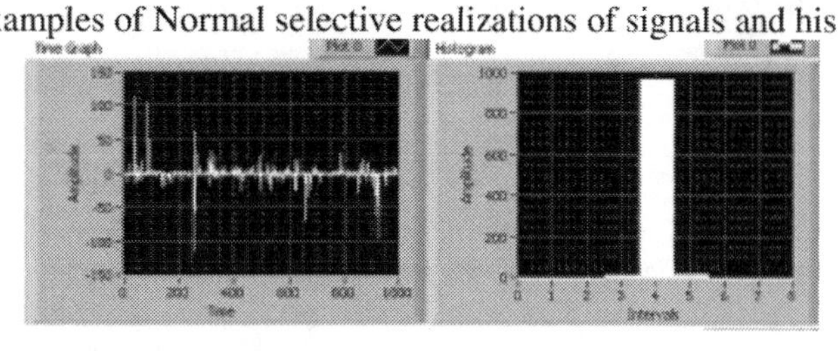

Figure 6. Examples of KOSHI selective realizations of signals and his distributions.

5. The discussion of the results

The analysis of the presented examples allows us to propose a hypothesis in the form of the following generalization: "All possible forms of distributions are enclosed and ordered between 2MOD and Koshi distributions." Hence, in turn, the conclusion fofollows that it is possible to represent distributions in the form of an identification (categorical) scale (IS). For the S-tester, the IS is shown in table 1The reference

IOP Conf. Series: Journal of Physics: Conf. Series **944** (2017) 012050 doi:10.1088/1742-6596/944/1/012050

points of the IS (table 1) have numerical designations in the form of mean values (mean S = 0..100) and ordinal numbers (Rank = 0..8) of marks, as well as qualitative notations - in the form of ordered names (const .. .2mod) distributions. The bottom line shows an example of measuring a signal whose readings for the S-tester were Sy = 24.36. Interpolation of these readings allows us to state that the signal under study has a distribution close to the normal (gaus) in accordance with the obvious rule:

$$\text{Name Sy} \approx \text{Id}^*[\min(\Delta Syg; \Delta Syl)],$$

Where: $\Delta Syg = abs(Sy - Sg) = abs(32 - 24{,}36) = 7{,}64$; $\Delta Syl = abs(Sy - Sl) = abs(24{,}36 - 15) = 9{,}36$.
This meanst: Name Sy ≈ Id * [min (7,64 / gaus; 9,36 / lapl)] = GAUS. Here, the symbol Id * [.] Denotes the operation inverse to the Id [.] Operation of identifying the scale distribution
IS an empirical information object that cannot be obtained formally, based on any theory. However, if a similar IS is created, it can be used to solve problems of recognition (classification) of images of not only signals, but also of any numerical sequences (distributions, arrays).
Categories (classes) of IS can be interpreted in terms of some subject area. For example, in medical diagnosis, different classes can correspond to different stages of a disease.
To reduce the ambiguity of the "array-number" transformation, several identification parameters of the investigated signal should be measured. For this purpose, for example, an S-vector is used (Fig. 5), which measures two identification parameters: for the signal itself (Re-S) and its increments (Im-S). In this case, you can automatically distinguish similar images, for example, triangular signals from saw tooth.
The task of automatic interpolation of the IT indications can be solved not only by rounding the numerical value of the pointer to the nearest ISH mark, but also by using so-called "linguistic" interpolation [5]. A tool, called an interpolator, assigns to the input signal more than one name from Table 1, but all 9. In this case, the order of the distribution names for different signals can be different. Thus, it is possible to compare signals by category lists.

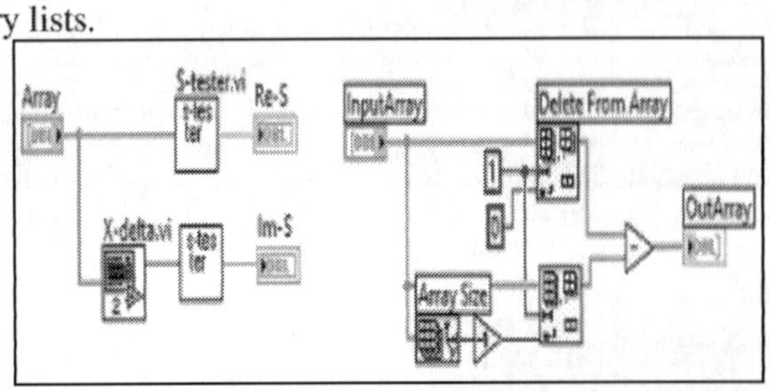

Figure 7. Program codes of the S-vector (left) and its submodule X-delta (right).

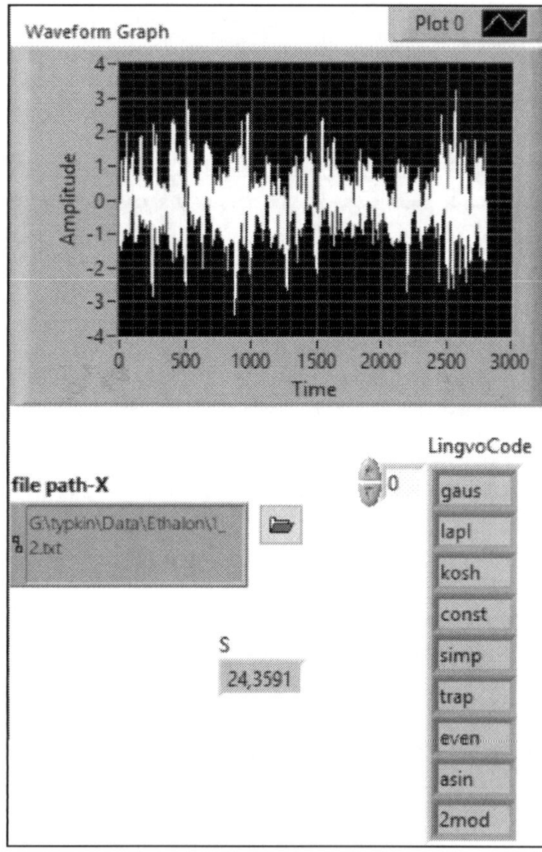

Figure 8. Interpolator control panel view.

The input signal (Waveform Graph) in the form of an array (Fig. 7) is fed to the input of the S-type identification tester, the output of which forms a number (S = 24.36). Using the built-in IS, this number is compared with its reference points (Table 1), resulting in the corresponding differences: Δi = abs (Sx -Si), which are sorted in ascending order and compared with the names of the IS. Thus, the initial sequence (const-kosh-lapl-gaus-simp-trap-even-asin-2mod) of the names in Table 1 is reformatted by the input signal into the sequence: gaus-lapl-kosh-const-simp-trap- even-asin-2mod, shown in Fig. 6 and called the linguistic code (LingvoCode).

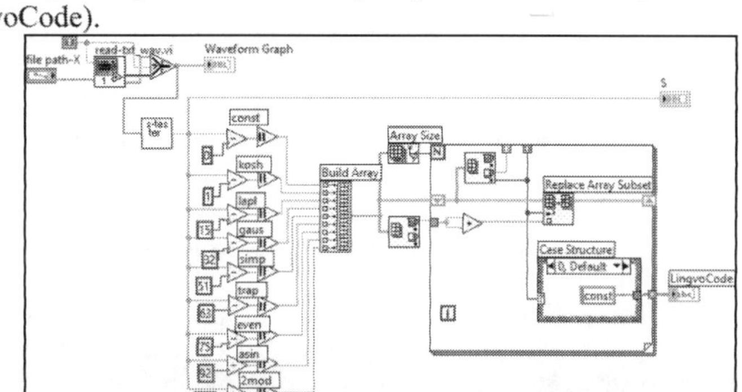

Figure 9. Structure of interpolator code.

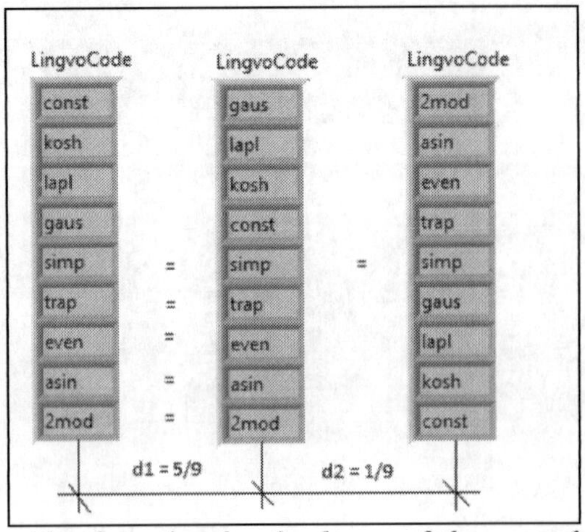

Figure 10. Algorithm for estimating the degree of chaos-regularity of signals.

The idea connected with the possibility of measuring the chaotic-regularity properties is illustrated in Fig.8 Two extreme lists (LingvoCode) are, ordered in ascending and descending order of the identification parameter (S), are the names of the distributions of the IS. Average – unordered – the list is obtained as a result of the analysis of the input signal by an interpolator (Fig. 7).
It's obvious that:

- the extreme, ordered lists specify the full range of existence of all possible signals;
- the list of the signal with the maximum disorder (chaotic) is exactly in the center – at the same distance from the boundary lists, with d1 = d2;
- the offset of the "signal" list to the left or right of the center indicates the presence of some regularity in the analyzed signal, while you can identify the type of regularity, for example, as an increasing or decreasing trend.
-

6. Conclusion
The possibility of quantitative evaluation of the form of signal characteristics, the chaotic-regularity property allows one to recognize and, at the most general level, analyze the state of control objects.

7. References
[1] Gubarev V V, Gorshenkov A A, Klikushin Yu N and Kobenko V Yu 2013 Classification measurements: methods and implementation *Optoelectronics, Instrumentation and Data Processing* pp 171–177

[2] Koshekov K T, Klikushin Yu N, Kobenko V Yu, Evdokimov Yu K and Demyanenko A V 2014 Fuel cell diagnostics using identification measurement *Journal of Fuel Cell Science and Technology* pp 52–54

[3] Klikushin Yu N, Koshekov K T, Kobenko V Yu and Trunin E S 2014 An algorithm for evaluating the state of a generating unit based on the identification measurements of vibrosignals *Russian Journal of Nondestructive Testing* pp 413–418

[4] Koshekov K T, Klikushin Yu N, Kobenko V Yu, Sof'ina Savostin A A and Kashevkin A A 2016 Testing a pump unit by identification measurements of vibration signals *Russian Journal of Nondestructive Testing* pp 280–286

[5] Klikushin Yu N, Kobenko V Yu, Koshekov K T, Belosludtsev O M and Koshekov A *K* 2016 Search of the operational earthquake precursors on the basis of the identification measurements of the seismographic records *2016 Dynamics of Systems, Mechanisms and Machines (Dynamics)* pp 1–6 DOI: 10.1109/Dynamics.2016.7819026

The detection methods of dynamic objects

N L Knyazev and L A Denisova

Omsk State Technical University, 11, Mira ave., Omsk, 644050, Russia

e-mail: sakylax@gmail.com

Abstract. The article deals with the application of cluster analysis methods for solving the task of aircraft detection on the basis of distribution of navigation parameters selection into groups (clusters). The modified method of cluster analysis for search and detection of objects and then iterative combining in clusters with the subsequent count of their quantity for increase in accuracy of the aircraft detection have been suggested. The course of the method operation and the features of implementation have been considered. In the conclusion the noted efficiency of the offered method for exact cluster analysis for finding targets has been shown.

1. Introduction

The main task of the airspace using is the detection of the level of danger of the observed aircraft (A) and the distribution into the appropriate groups: «airplane, in a distress», «unidentified air object», etc. In the conditions of war the detection and the recognition of an air attack weapon is necessary for their ranging in a priority, the subsequent target distribution by means of warning radar.

Detection A in the airspace is the task of radiolocation. The analysis of space and the determination of parameters of objects moving (navigation parameters) are carried out through radar stations (radar station) [1]. For detection of the aircraft and furthermore their identification (the determination of belonging to a certain class or type) information systems working in a radar wave range are created and enhanced. In the case of cluster analysis the tags of source objects initially have not been set, and even the input set can be unknown.

Aircraft is characterized by the following key navigation parameters:

carrier frequency of f_c by which the frequency band of operation of an object is determined;
- power level of signal P by which it's possible to evaluate an object distance;
- aircraft heading v, by which the direction of the airplane in space is evaluated;
- angle of the heel γ и the tangage angle Ψ , necessary for determination of the provision of A concerning its longitudinal and lateral axes.

The complexity of this task is provided by the complexity of the separation of aircraft in a specified frequency range. The difficulties occur then the extensive distribution of networks of the mobile communication 4G working in the same range of the frequency channel has been spread and it complicates the data handling. It compels to improve the algorithms of dynamic objects detection in the air space.

2. The problem statement

Let's assume the sequence of the navigation parameters to be presented in the form of input vector of characteristics of a *p*-object in the airspace $x^p = (x_1^p, x_2^p, .. x_i^p, ..., x_N^p)$, where $p = \overline{1,|X|}$ – index number of the aircraft from the input set of objects X, x_i^p – *i-th* property of the *p-th* aircraft, $i = \overline{1,N}$,

N – input set of the characteristics of the p-object.-The set of the aircraft X, by means of the certain clustering method is dynamically distributed into groups (clusters). The created clusters c_j comprise navigation parameters *p-th* aircraft $c_j = (x^1,...,x^i)$ $x^i \in X$, where $j = \overline{1,M}$, M – quantity of clusters. Each input aircraft (further – object) shall be carried to one of the available clusters c_j. During the mathematical description of the degree of the similarity of objects, the Euclidean metrics is applied [3]. It represents the geometrical distance in a multivariate space.

$$d(x^p, c_j) = \sqrt{\sum (x_i^p - c_m)^2}. \quad (1)$$

Its application is justified if the metric space matches geometrical space. The characterizing parameters of a cluster are: *center* $m(c_j)$, which are defined as follows:

$$m(c_j) = \sum_{i=1}^{N} u_{ij} x^p,$$

where u_{im} – element of the partition matrix $U=(u_{ij})$, such that:

$$u_{ij} = \begin{cases} 1, & for\ d(x^p, c_j) \to \min; \\ 0, & else. \end{cases} \quad (2)$$

and the cluster radius R is the distance between the core of the cluster and the object as far from the core x^p:

$$R = \max(d(x^j - c_j)), x^j \in c_j.$$

The efficiency of the detecting aircraft, owing to a fuzzy clustering on the basis of the concepts described above is planning to increase.

3. Clustering methods of radarspace

In the most modern systems of the navigation parameters processing obtained from the locator, two types of clustering methods are used:

- "learning with a teacher" – differs in the fact that there is an additional information about input objects a priori. The implementation of this type of a clustering takes place in methods of an accurate clustering *k-means*;
- "learning without a teacher" – differs in the fact that the information concerning objects is unknown as well as their exact quantity is not known. Usually the tasks of data analysis of this kind use the methods of fuzzy clustering *c-means*;

Clustering by *k-means* [4] method breaks a set of objects of some (generally vectorial) space into quantity of clusters of M unknown yet. x^p – objects are divided into clusters again and again, according to the new center that is the closest to the selected metrics. The purpose of this method is to minimize the medium square root deviation quantity in the points of each cluster. The cycle has come to an end after the number of iterations set has passed (a priori). However the accurate clustering has the shortcoming connected to the combination of objects on the boundaries of clusters which is corrected in the indistinct methods.

The indistinct clustering by method *c-means* [5] makes partition of an input set of objects in clusters, using the function of accessory $\mu(x^p) \in [0,1]$, which defines the degree of an object belonging to a cluster. The difference of a level of accessory of accurate partition from indistinct is that function values $\mu(x^p)$ from an interval [0,1], but not from a two-element set (0,1) [6]. Taking into account this fact, the method c-means does not require the setting of the specific quantity of clusters with the introduction of a condition of the method stop.

In the case when the operator precisely defines the number of the purposes (dynamic objects in space), by means of the visual analysis the screen of the indicator will not demand the complicated methods of a clustering, such as *c-means*, and the sufficient efficiency will be shown by the *k-means* method. But in a situation when it is visually difficult to define the number of aircraft precisely and the standard methods of recognition become less effective, the use of a method *c-means* will be the best option.

However for the solution of an objective, it is required to introduce some modifications to the existing method, for increase in accuracy of A determination.

4. The modified method FCR_DV

Clustering methods on the basis of a mathematical apparatus of fuzzy logic are effective in search of relatively close metrics of objects [7]. It is explained by the fact that the method *c-means* creates the matrix of distribution of U more precisely owing to the function of accessory $\mu(x^P)$. However, for increase in efficiency of cluster analysis, the required method is to be modified. The modification of a method *FCR_DV* (*Fuzzy Clustering with Recalculation and Deleting Void*) is the following:

- recalculation of centers and radiuses of clusters for distribution of objects in it;
- introduction of criteria of the clusters adjacency and assessment procedure;
- checking of the combining of the similar clusters by special criteria;
- determination and deleting void clusters.

For the determination of an object accessory to a cluster the symmetric Gaussian function of accessory $\mu(x^P)$ is chosen:

$$\mu(x^P) = \exp\left[-\left(\frac{x_i^P - c_j}{2\cdot\delta}\right)^2\right], \tag{3}$$

where δ is a tuning property of the accessory function equal to the estimated quantity of clusters in an initialization step. Vector accessory level x^P to cluster is defined as follows:

$$\mu(x^P) = \max\left[\mu(x_1^P), \mu(x_2^P), \ldots, \mu(x_i^P), \ldots \mu(x_N^P)\right].$$

To consider a level of accessory of objects to two next clusters c_j and c_k, where $k \neq j$, $k \in M$ let's introduce the changes into a formula (1), taking into account a formula (3).
Thereafter function of distance (metrics) will take the following form:

$$d(x^j, c_k) = \sqrt{\sum \mu(x^j)\cdot\left(x_i^j - c_k\right)^2}, x_j \notin c_k.$$

Values increase at a distance decrease between a vector of an object data and a cluster. Therefore the value of the function of accessory is more $\mu(c_j) > \mu(c_k), k \neq j$, where clusters are adjacent located in space c_m и c_k, and as a criterion of two clusters association – cluster radius. The more objects are removed from clusters, the less value for making decisions on association will be.
Let's suppose that it is required to unite two objects with navigation parameters x^1, x^2 in a cluster and at evaluation of the accessory functions (3), their degrees became equal, i.e. $\mu(x^1) = \mu(x^2) = 0{,}5$.
At adjacent values it is possible to consider that a measure of proximity is the expression:

$$\tilde{d} = \left|\mu(x^P) - \mu(x^k)\right|. \tag{4}$$

Thereafter criterion of association of clusters is expression:

$$\alpha = \left(\left(\left|\min(\mu(x^1), \min(\mu(x^2))\right|\right) - 0{,}5\right) \geq \tilde{d}. \tag{5}$$

However, after carrying out a clustering of the next entrance vector some clusters can't have the vectors of data objects adjacent to them. Therefore, they have to be removed from the list, i.e. they are removed from a set of clusters (see Tab. 1). For the identification of empty clusters the variable $V(j)$ satisfying the equality is introduced:

$$\sum_{j=1}^{M} V(j) = L,$$

where L – quantity of objects in initial set. Structurally the scheme of the FCR_DV method is present in fig. 1. Step by step the method looks as follows:

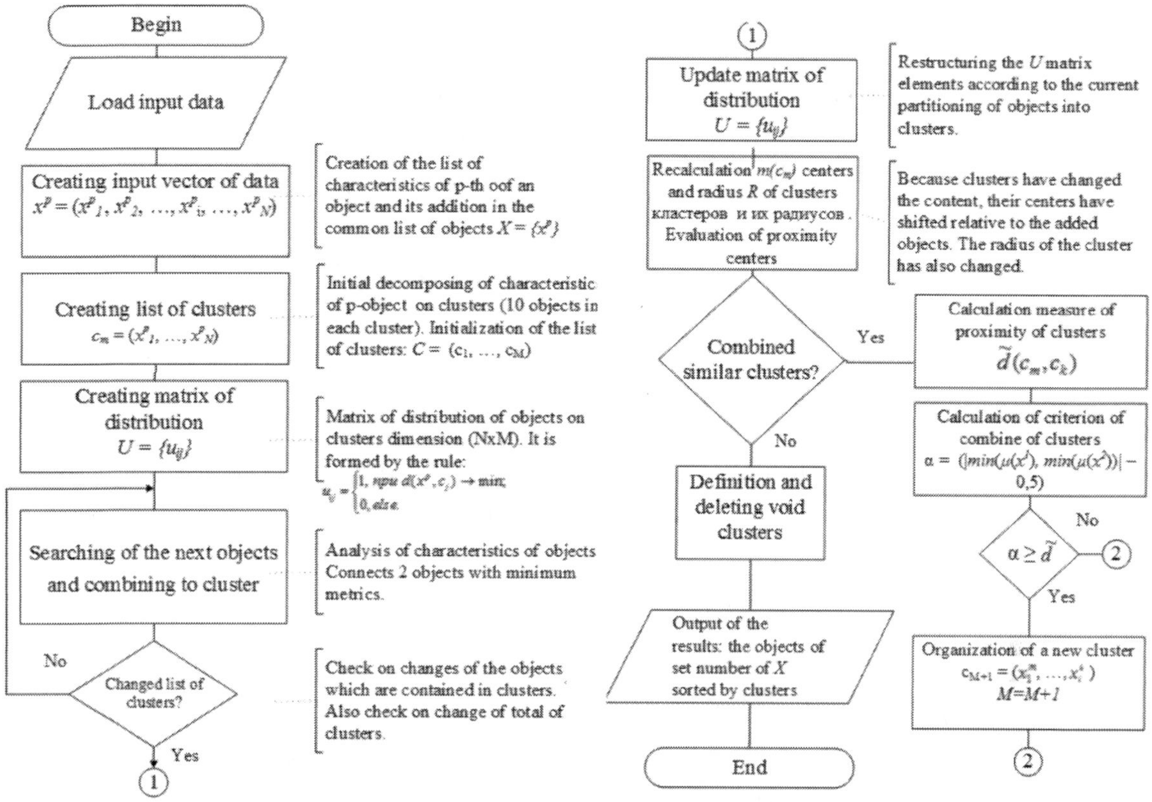

Figure 1. Work scheme of method FCR_DV.

Step 1. To create the list of objects and the list of clusters, initializing the primary splitting according to the scheme: "ten objects in one cluster". On the basis of the accepted splitting to create a matrix of splitting U according to (2).
Step 2. To calculate metrics between objects from a great number of X and the centers of clusters.
Step 3. To update the matrix of splitting U according to the current splitting. To recalculate the centers and radiuses of clusters of R.
Step 4. To check the proximity of clusters on the basis of criteria (4) and (5). In the case of a successful checking to execute association of objects from the close clusters and the subsequent removal of empty clusters.
Step 5. If after passing the previous iteration, the content of the list of clusters and the matrix of splitting U wouldn't change, then go to Step 6, else go to Step 2.
Step 6. The end.

Table 1. Estimated values of criterions of combined clusters

Clusters	Measure of a similar of clusters $\tilde{d}$	Criteria of combining α	Acceptance
1 and 2	0,9	$\lvert(0{,}8-0{,}4)\rvert \le 0{,}9$ or $\lvert(0{,}2-0{,}4)\rvert \le 0{,}8$	Don't combined
2 and 3	0,3	$\lvert(0{,}8-0{,}4)\rvert \ge 0{,}4$ or $\lvert(0{,}2-0{,}4)\rvert \le 0{,}4$	Combined
1 and 3	0,8	$V(3)=0$	Void

5. The results of experiments

To show the effectiveness of the proposed solution the comparative analysis of its work the clustering method c-means is carried out. As input data the navigation parameters are used: carrier frequency f_c, power level of signal P, aircraft heading v, angle of heel υ, tangage angle ψ. Input datas are created by means of a specialized package of the software consisting of two program modules. By means of the first module it's simulated flight trajectories of 40 objects. The second module is intended for formation of a pack of impulses for the modelled trajectory. The realization of the offered *FCR_DV* method is executed in the C# programming language. By virtue of means of program tools STATISTICA 10 (firm StatSoft), method *c-means* is realized.

As a result, the experimental cluster analysis has been conducted (see Tab. 2) by the means of the methods described above (seven experiments).

Table 2. Output of clustering methods.

Navigation parameters	Clustering method	A	Iterations	Count experiments	Relative error of method
Carrier frequency f_c, [MHz]	*c-means*	35	11	5	2.2%
		33	13	2	
	FCR_DV	**43**	**21**	**6**	**1.3%**
		45	23	1	
Power level P, [dBm]	*c-means*	38	12	6	2.7%
		42	13	1	
	FCR_DV	**40**	**21**	**5**	**0.9%**
		39	23	2	
Heading v, [deg.]	*c-means*	42	12	5	4.7%
		37	17	2	
	FCR_DV	**40**	**27**	**5**	**0.9%**
		39	27	2	
Heel υ, [deg.]	*c-means*	39	12	7	0.01%
	FCR_DV	40	20	7	0%
Tangage ψ, [deg.]	*c-means*	39	12	7	0.01%
	FCR_DV	40	20	7	0%

The relative accuracy has been calculated by the means of the *t-test* method [9]. Seven experiments as a result of which statistical data on the recognizable aircraft are obtained have been conducted. The analysis of table 2 has shown that the error of the *FCR_DV* method is less than of *c-means* method.

According to the analysis of a carrier frequency it is visible that both methods do not recognize the precise number of aircraft. The method c-means recognizes in 5 cases 33 air objects, in 2 cases 35 air objects that less than the true quantity of objects |X| = 40.
At the same time, the *FCR_DV* method recognized 43A for six experiments and 45 A for one experiment that exceeds the true quantity. However considering a difference between the number of undetected A and excess A, there is an improvement in results of the *FCR_DV* method since from the point of view of radiolocation it is better to detect three targets than not to find five definite targets.
The cluster analysis of the value of power level of *P* has shown better results, than the previous one (see fig. 2). Owing to the cluster analysis by c-means from 38 to 42 air objects are revealed, and 38 A is found in overwhelming number of experiments. The *FCR_DV* method detects up to 39-40 objects that shows its effectiveness of rather true number of objects.

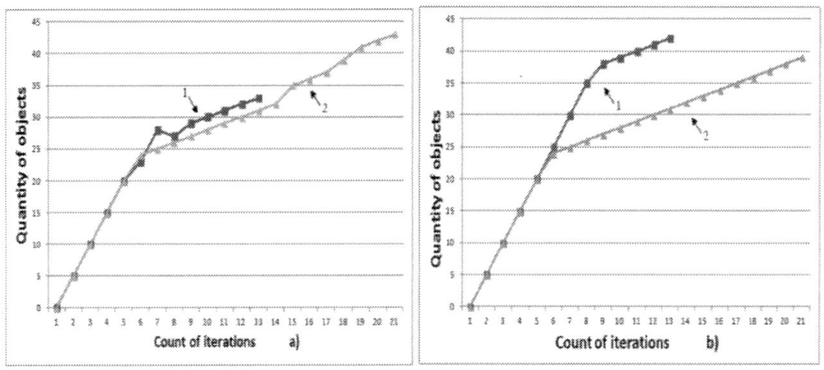

Figure 2. Comparison recognition accuracy of A (*line 1* – method *c-means*; *line 2* – method *FCR_DV*): a) – carrier frequency; b) – power level.

The possible larger count of iterations of the *FCR_DV* method creates the increase in accuracy of discernment. But at the same time, in the presence of other data set, the offered *FCR_DV* method looses in speed of *c-means* method.
The method *FCR_DV* proceeds the deleting and recalculation centers and radiuses of clusters allow to have more precisely objects concerning their «coordinates» in a metric space. In fig. 4 the comparison of two methods which indicates the exceeded quantity of the recognizable objects c-means is connected with the method with the existence of void clusters which were removed in the *FCR_DV* method has been presented.

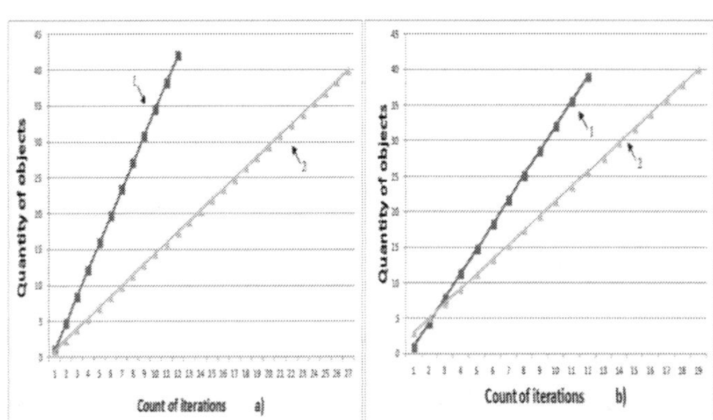

Figure 3. Comparison recognition accuracy of A (*line 1* – method *c-means*; *line 2* – method *FCR_DV*): a) – heading; b) – heel.

The analysis of the aircraft headling, angles heel or tangage (the cluster analysis on angles heel or tangage identical), are united by one common case connecting two schedules (see fig. 3).
The accuracy of an objects detection differs in two undetected A. The cluster analysis of two angles of heel and tangage has shown identical results. Basing on it, it can be said that both methods results on quantity of the found objects are similar, but the distinction in the count of iterations is observed.

6. Conclusion

In the work the results have presented the solution of the problem of dynamic objects detected by means of the modified method FCR_DV. This decision gives an opportunity to define the quantity of clusters in the course of work and to proceed the repeated reorganization of clusters, to define and delete empty clusters for an increase in the accuracy of the distribution.

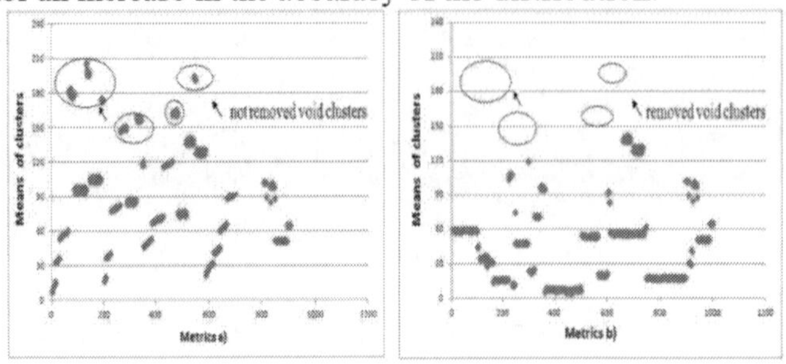

Figure 4. Distribution of objects on clusters: a) – metohd *c-means*; б) – method *FCR_DV*.

By means of the modified *FCR_DV* method it was succeeded to mark out the larger quantity of aircraft, than by means of an algorithm c-means. However the count of iterations of *FCR_DV* is higher, than at c-means method. This fact says that if we have a larger volume of data the cluster analysis will require more time. In the future it is planned to modify clustering methods like "learning without teacher" for applying them at stages of assessment of probabilities of a distribution of objects and making decisions for combining them.

References

[1] Bakulev P 2004 *Radar systems* (Moscow: Radio Engineering) pp 117–136
[2] Ren M, Liu P, Wang Z and Yi J 2016 A Self-Adaptive Fuzzy c-Means Algorithm for Determining the Optimal Number of Clusters *Journal of Computational Intelligence and Neuroscience* pp 5–8
[3] Kotov A and Krasilnikov N *Clustering of data* URL: http://logic.pdmi.ras.ru/~yura/internet/02ia-seminar-note.pdf (reference date: 05.02.2017) pp 13–15
[4] Jiangsheng Y *Clustering Methods Applications of Multivariate Statistical Analysis* (Beijing: Institute of Computational Linguistics) URL: http://www.archive.org/web/20030422001200/icl.pku.edu.cn/yujs/papers/pdf/clster.pdf (reference date: 07.03.2017) pp 15–20
[5] Roubos J and Setnes M 2006 *Learning Fuzzy Classification Rules from Labeled Data* (Delft: Delft University of Technology) URL: http://www.researchgate.net/publication/222657076_Learning_fuzzy_classificaion_rules_from_labeled_data/links/09e4150879582c877f000000 (reference date: 05.05.2017) pp 7–13
[6] Shtovba S 2007 *Designing fuzzy systems using MATLAB* (Moscow: Hot line–Telecom) pp 187–223
[7] Zhang M, Zhang W, Sicotte H and Yang P 2009 A new validity measure for a correlation-based fuzzy c-means clustering algorithm *Annual International Conference of the IEEE Engineering in Medicine and Biology Society* vol 4 pp 3865–3868
[8] Veshkurtsev Yu, Bychkov E and D Titov 2012 *Digital filter based on the theory of fuzzy sets with adaptively varying membership functions* (Moscow: Radioelectrotechnics) 2rd ed vol 3 pp 43–50

[9] Petrov P 2013 *Mathematico-statistical processing and graphic representation of the results of pedagogical research using information technologies* (Izhevsk: Udmurt University) pp 139–170

IOP Conf. Series: Journal of Physics: Conf. Series **944** (2017) 012052 doi:10.1088/1742-6596/944/1/012052

The analysis of the quality of the frequency control of induction motor carried out on the basis of the processes in the rotor circuit

V L Kodkin, A S Anikin, and A A Baldenkov

South Ural State University, 76 Lenin prosp., Chelyabinsk, Russia 454080

Abstract. The results of researches of asynchronous electric drives with the frequency control which are carried out for the purpose of establishment of causes and effect relationships between a control method, the implementable standard frequency converter of the Schneider Electric company (ATV-71, ATV-32) and its efficiency are given in article. Tests with asynchronous motors with wound rotor were for the first time carried out. It allowed registering during the experiments the instantaneous values not only the stator currents, but also rotor currents. Authors for the first time applied spectrum analysis of stator and rotor currents, it showed that «sensorless vector» control leads to origin of high-frequency harmonicas with the considerable amplitude and, as a result of they are non-sinusoidal of the created torque and inefficiency of the electric drive.
The accelerations that are carried out during the researches to 94, 157 and 251 Rad/s confirmed this feature of vector control that appears incapable to linearize the asynchronous electric drive as it was supposed authors of a method. These results do not contradict theoretical provisions if not to neglect assumptions which usually become in case of an output of the equations of vector control. Unfortunately, the modern researchers do not subject these assumptions to doubts.
Continued studies make it possible to create an effective frequency management of asynchronous electric drives required for current technology.
Keywords. squirrel-cage induction motor, slip-ring rotor motor, frequency control, rotor current, spectral analysis, experimental studies, efficiency

1. Introduction

Despite the widespread use of AC electric drives with the frequency control in the industry, theoretical problems remain much. This particularly applies to the implementation of vector control in electric drives with synchronous and asynchronous electric motors. Most articles and papers on vector control [1, 4] are devoted to the description of control algorithms, their advantages and superior characteristics. The complexity of this control method associated with possible errors in generating the control signals are given in these references generally without specific results, methods studies, recommendations for their elimination and any alternative solutions. The operation of the particular industrial drives [1] also rarely touch upon vector control algorithms, because the obtained results may be due to different causes. Furthermore, it should be noted that the control structures of frequency converters of the most well-known manufacturers typically are not disclosed in the documentation on these converters. In this regard, the introduction of asynchronous electric drives in industrial machinery requiring high accuracy of maintaining speed or good dynamics of processes require serious research work.

IOP Conf. Series: Journal of Physics: Conf. Series **944** (2017) 012052 doi:10.1088/1742-6596/944/1/012052

2. Problem definition

Numerous articles and monographs dedicated to these systems describe methods of their implementation, control algorithms, software etc., but usually give just a vague idea of relevant problems – in accuracies of the dynamic equations [1], possible mismatch of the equations of model put in control devices, bulkiness of control algorithms, etc. Practically there are no results of experiments showing mismatch and recommendations about overcoming these problems.

Previously conducted pilot studies showed how operation of sensorless vector control is influenced by external loadings of the torque [2]. But even in these studies it was not possible to establish a clearer cause-and-effect relationship between the control algorithms and the operation of the drive. One of the reasons is a limited information about the processes in a squirrel cage motor; usually these are speed signals from the encoder, the amplitude value of the stator current and the coordinates calculated by the frequency converter - electromagnetic torque M, power P, stator current I, etc. As the structure of model on which these variables are calculated is unknown, it is impossible to consider them authentic (except U, I, f). Instantaneous values of the stator current «filled» with a high frequency are not very informative. The experiments on the stand with two asynchronous motors of different types controlled by frequency converters show that the advantages of vector control over scalar are not always so obvious as in guidelines and advertising articles.

3. Main part

A new information necessary for understanding the processes in frequency-controlled drives may be obtained using a motor with a slip-ring rotor whose current circuits have been installed in the rotor circuits.

Schematic diagram of the drive with the slip-ring induction motor is shown in Figure 1.In addition to the asynchronous motor with a slip-ring rotor (M1), the stand includes the ATV-71 frequency converter from Schneider Electric (UZ1), the speed data were obtained from an encoder (BR) located on the motor shaft. The drive parameters were recorded with a USB oscilloscope (PG) Hantek DSO-2090 and displayed on a personal computer screen.

The stand allowed us to obtain synchronous diagrams of speed and current of the rotor or stator and with the help of software to conduct spectral analysis of these signals.

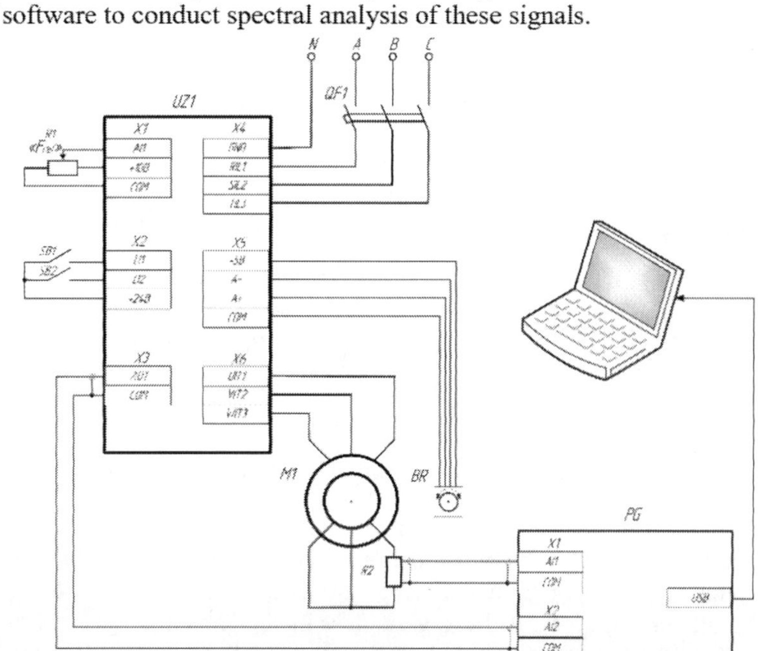

Figure 1. Schematic diagram of the drive with the slip-ring induction motor.

Before proceeding to the description and analysis of the results of the experiment, let us note once again the possible cause of vector control problems. It is sufficient to consider the basic equations of the mathematical description of an induction motor, which are used to compose vector control

IOP Conf. Series: Journal of Physics: Conf. Series **944** (2017) 012052 doi:10.1088/1742-6596/944/1/012052

algorithms. So, to analyze the operation of an asynchronous motor, most specialists [1, 3, 4] use Park's model of a generalized electric machine [5].

Expression of the electromagnetic torque and angular frequency of rotation in case of a choice of vectors of flux linkage of a rotor and current of the stator (the axis d of the synchronous coordinate system is combined with a vector ψ_2) is as follows [3]:

$$m = \frac{3 \cdot z_p \cdot L_m}{2 \cdot L_2} \Psi_{2d} \cdot i_{1q}, \tag{1}$$

where z_p is the number of pole pairs of the motor; ψ_{2d} is the longitudinal component of the rotor's flux-linkage; L_m is the magnetization inductance; i_{1q} is the transverse component of the stator current; L_2 is the inductance of the winding of the rotor.

Angular frequency of current of a rotor in case of a choice of vectors of flux linkage of a rotor and current of the stator is as follows:

$$\omega_2 = i_{1q} \frac{L_m}{\Psi_{2d} \cdot T_2}. \tag{2}$$

The longitudinal component of flux linkage of a rotor is defined as follows [3]:

$$\Psi_{2d} = i_{1d} \frac{L_m}{1 + T_2 \cdot p}, \tag{3}$$

where $T_2 = L_2/r_2$ – electromagnetic time constant; r2 – active resistance of the rotor winding; p – Laplace operator.

Projections of the vector of the stator current, taking into account that the axis d of the synchronous coordinate system is aligned with the vector ψ_2, i.e. ψ_{2q}=0 [3]:

$$i_{1d} = \frac{\Psi_{2d}}{L_m}(1 + T_2 \cdot p);\ i_{1q} = \frac{\Psi_{2d}}{L_m} T_2 \cdot \omega_2. \tag{4}$$

In equation (1), the electromagnetic torque of the motor is formed as the product of the time functions of the longitudinal component of the magnetic flux ψ_{2d} and the transverse stator current component i_{1d}. Further, equations (2, 3 and 4) give expressions for Laplace images of the projections of the magnetic flux, current and frequency of the rotor current ω2. The operator image of the electromagnetic torque is formed as the product of the Laplace functions, whereas in strict accordance with the operator calculus it would be necessary to apply the Parseval equality [6]. The error of this "replacement" is insignificant if the dynamics of one of the factors is substantially lower than the dynamics of the other (in deriving the vector control equations under the assumption: $\frac{d\psi_2}{dt} = 0$).

If a change spectrum ω_2 is compared to a change spectrum of other signals in these equations – U, I, Ψ, etc. the error will be very big since the multiplication of such signals results in considerable signals of higher harmonics that not just brings an error into the electric drive equations, but absolutely «disorganizes» them. I.e. it is impossible to use Laplace transforms, as in the linear differential equations.

It should be noted that in the presence of higher harmonics in current or voltage signals, the formulation of the problem itself is erroneous: to provide a definite position of the current or voltage vectors, since it is impossible to form one channel (slip-ring voltage or current) simultaneously controlling all harmonics. If in the above equations, connecting the longitudinal and transverse components of currents or currents to substitute the sum of the harmonics of the equation lose their meaning, as well as attempts to control these variables. In this way the vector system controls the fundamental harmonic, and the others are formed «on their own», which leads to large errors and pulsations of the torques, and to the final inefficiency of the drive. In this situation, the scalar control reacting to the instantaneous values of the stator currents is fundamentally more effective, since it controls the total current, regardless of the harmonic component. This experiment is very well confirmed.

IOP Conf. Series: Journal of Physics: Conf. Series **944** (2017) 012052 doi:10.1088/1742-6596/944/1/012052

Our assumption that the analysis of processes in the rotor of an asynchronous motor will allow us to approach this problem is based on the fact that the processes in the rotor are essentially low-frequency, with respect to the stator processes, therefore they are more informative.

The engine starts up to various speeds – 94, 157, 251 Rad / s, which corresponds to the frequencies of the supply voltage of 30, 50 and 80 Hz. Figures 2–4 show velocity diagrams (A), stator currents (B) and rotor (C) and their spectra (D, E) with scalar control.

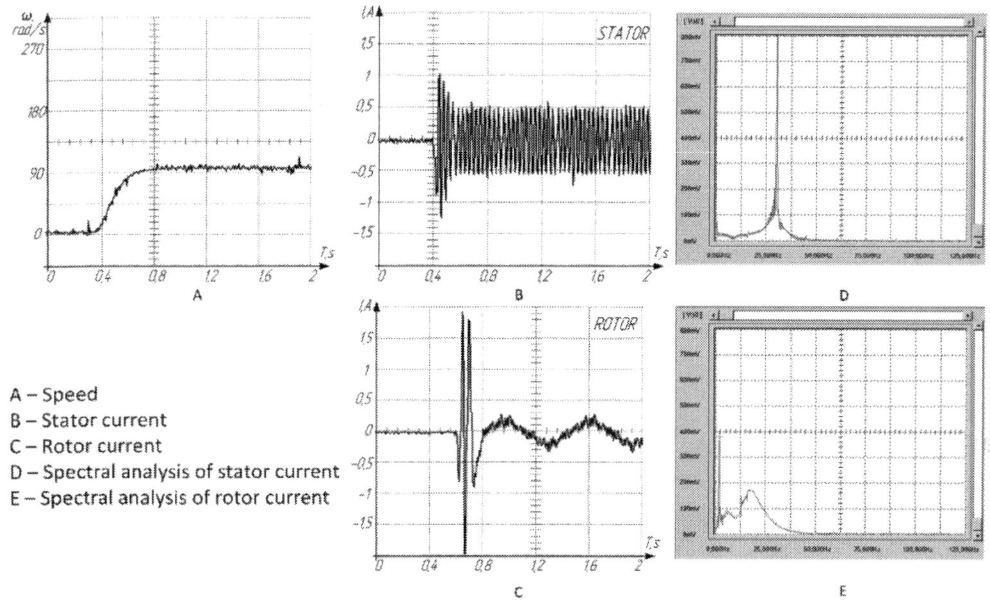

Figure 2. Transient of alternating current drive with scalar control with slip-ring induction motor (ω_{ref}= 94 Rad/s).

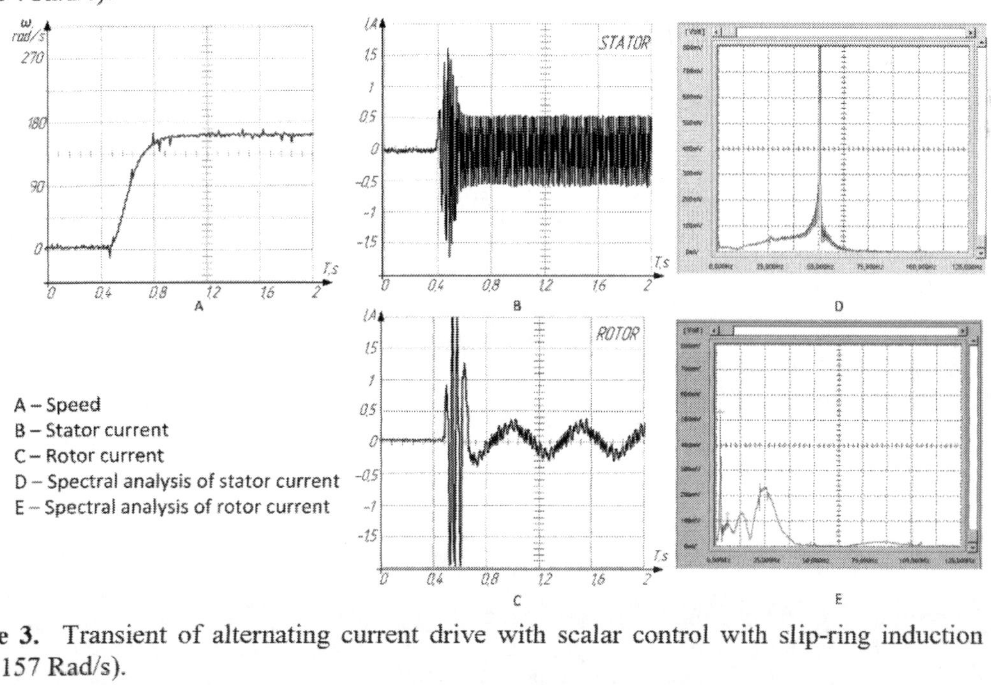

Figure 3. Transient of alternating current drive with scalar control with slip-ring induction motor (ω_{ref}= 157 Rad/s).

IOP Conf. Series: Journal of Physics: Conf. Series **944** (2017) 012052 doi:10.1088/1742-6596/944/1/012052

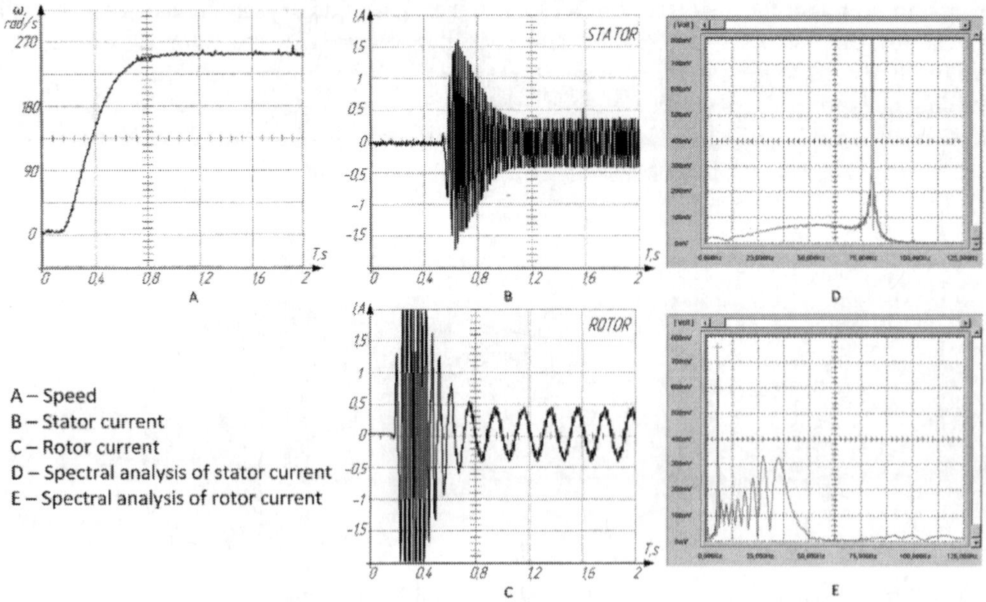

Figure 4. Transient of alternating current drive with scalar control with slip-ring induction motor (ω_{ref}= 251 Rad/s).

Figures 5–7 show similar diagrams, but with vector control.

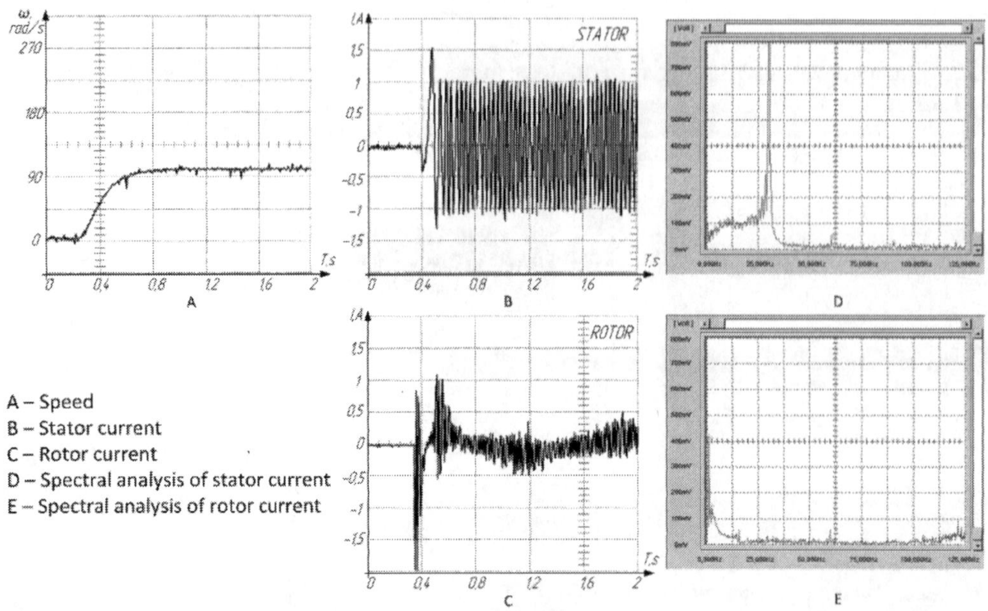

Figure 5. Transient of alternating current drive with vector control with slip-ring induction motor (ω_{ref}= 94 Rad/s).

The spectrum of stator currents contains the fundamental frequency corresponding to the rotational speed of the motor, especially with scalar control. The spectrum of rotor currents under scalar control lies in the low-frequency range, determined by slip and mechanical processes (Fig. 8). In the spectrum of rotor currents, in the vector control, high-frequency components appear that are much higher than the spectrum of mechanical processes (Fig. 9) and form the pulsations of the torque and contribute to the inefficiency of the drive.

IOP Conf. Series: Journal of Physics: Conf. Series **944** (2017) 012052 doi:10.1088/1742-6596/944/1/012052

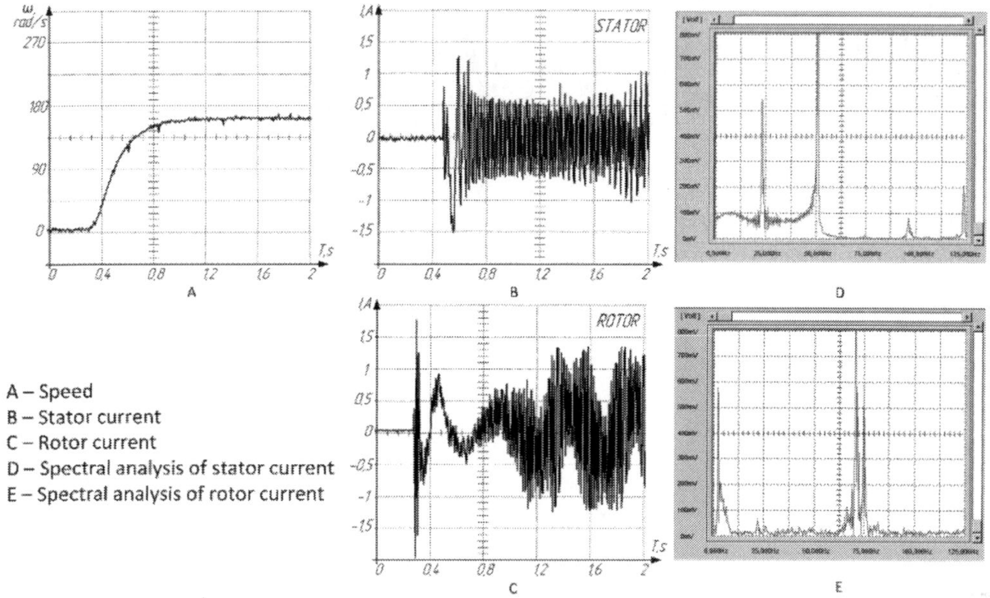

Figure 6. Transient of alternating current drive with vector control with slip-ring induction motor (ω_{ref}= 157 Rad/s).

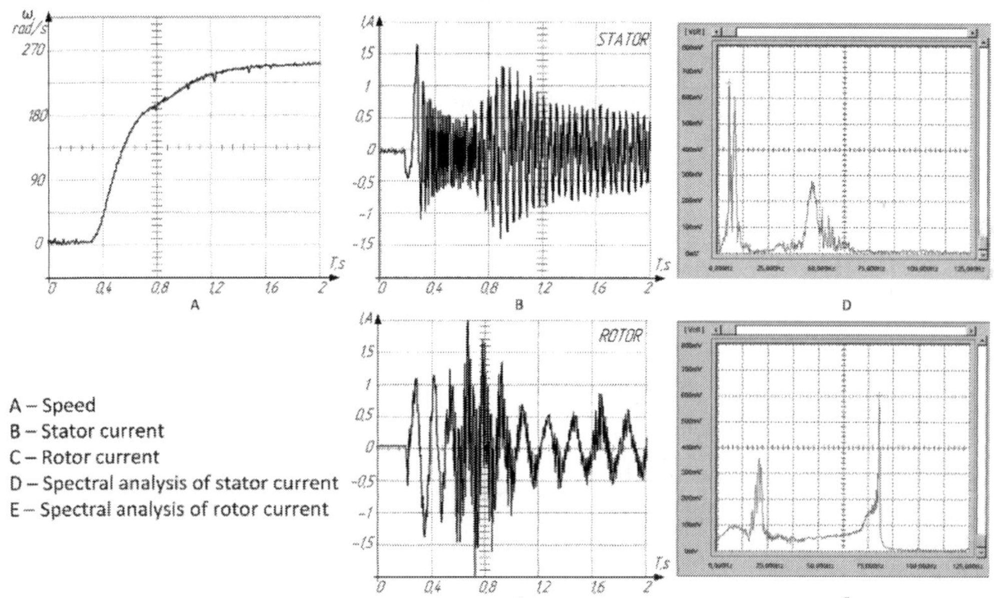

Figure 7. Transient of alternating current drive with vector control with slip-ring induction motor (ω_{ref}= 251 Rad/s).

IOP Conf. Series: Journal of Physics: Conf. Series **944** (2017) 012052 doi:10.1088/1742-6596/944/1/012052

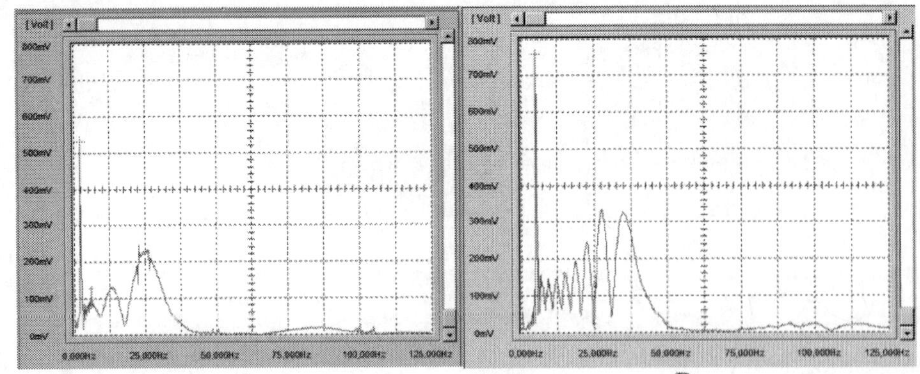

A B

Figure 8. Spectral analysis of rotor current slip-ring induction motor with scalar control (A - ω_{ref}= 157 Rad/s, B - ω_{ref}= 251 Rad/s).

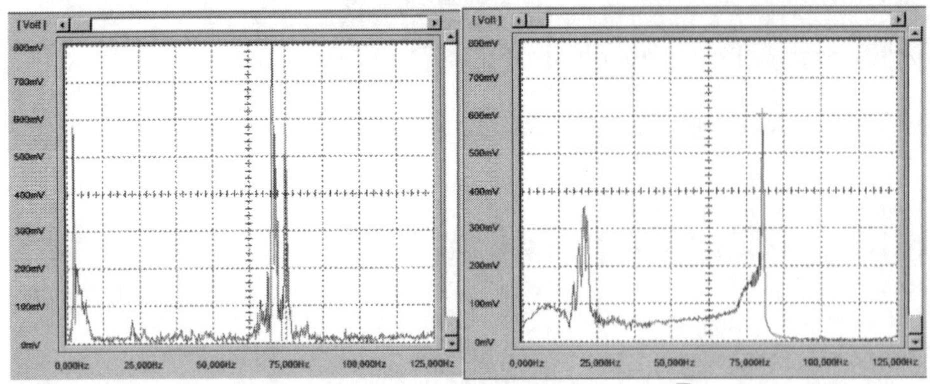

A B

Figure 9. Spectral analysis of rotor current slip-ring induction motor with vector control (A - ω_{ref}= 157 Rad/s, B - ω_{ref}= 251 Rad/s).

In some cases, these pulsations also appear in the spectrum of stator currents. It follows from these figures that when accelerating to velocities higher than 125 Rad/s in the rotor current signals, high-frequency harmonics appear in the vector control, which would not appear if the processes were described by equations similar to (1) – (4), I.e. close to linear. It is obvious that the appearance of these high-frequency harmonics in rotor currents is due to their presence in stator currents, which, despite the high carrier frequencies, are barely noticeable, but in some cases (see Figures 6B, 6D, 7B and 7D) and the nature of this manifestation depends on the settings of the drive.

Figure 10–12 shows the velocity diagrams (A), stator currents (B) and their spectrum (C) for the accelerations up to velocities of 94, 157, 251 Rad/s with the scalar control of an induction motor with a squirrel cage rotor.

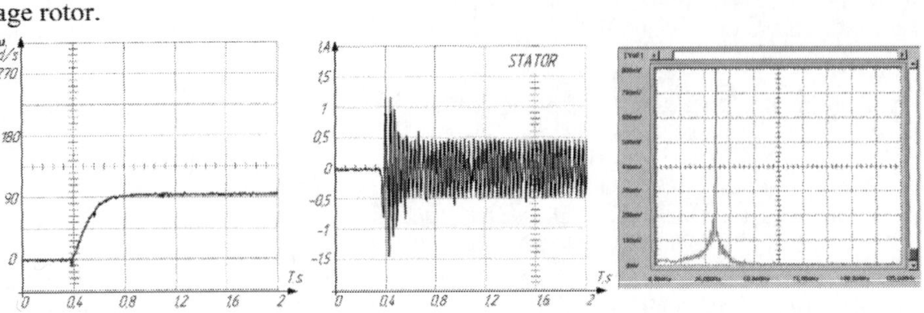

A B C

Figure 10. Transient of alternating current drive with scalar control with squirrel-cage induction motor (ω_{ref}= 94 Rad/s).

IOP Conf. Series: Journal of Physics: Conf. Series **944** (2017) 012052 doi:10.1088/1742-6596/944/1/012052

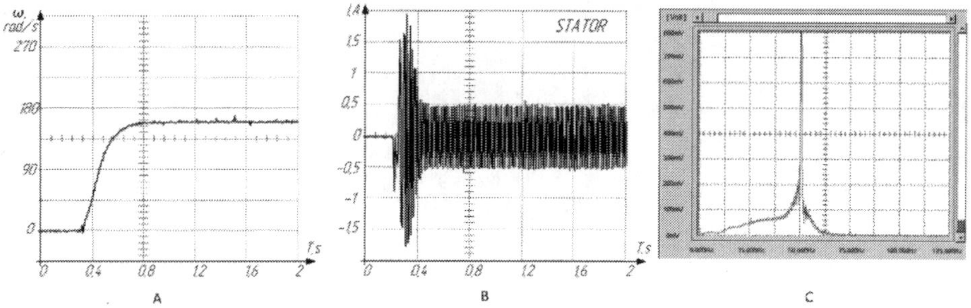

Figure 11. Transient of alternating current drive with scalar control with squirrel-cage induction motor (ω_{ref}= 157 Rad/s).

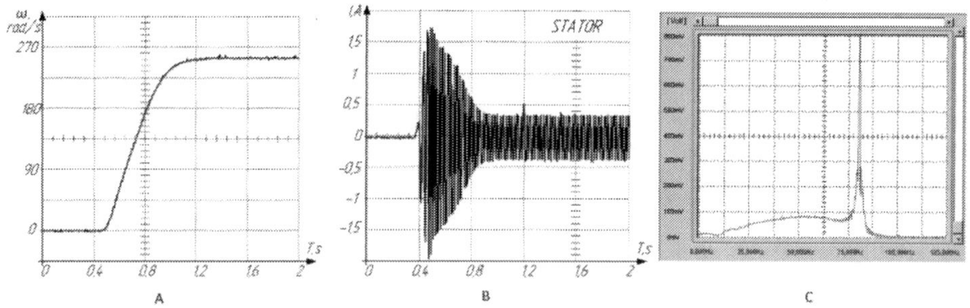

Figure 12. Transient of alternating current drive with scalar control with squirrel-cage induction motor (ω_{ref}= 251 Rad/s)

Figures 13–15 show the speed diagrams (A), stator currents (B) and their spectrum (C) at acceleration to 94, 157, 251 Rad/s with vector control of a squirrel-cage motor.

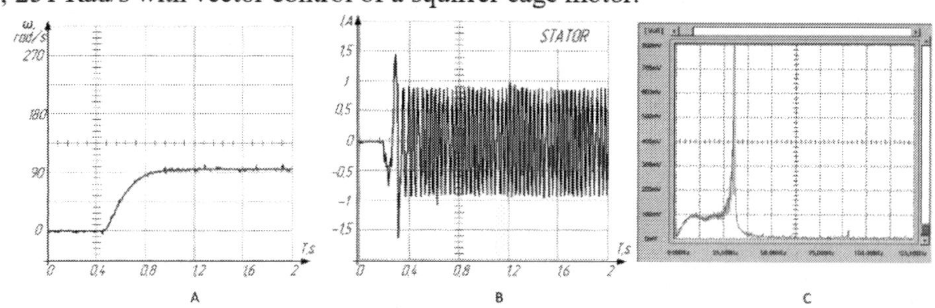

Figure 13. Transient of alternating current drive with vector control with squirrel-cage induction motor (ω_{ref}= 94 Rad/s).

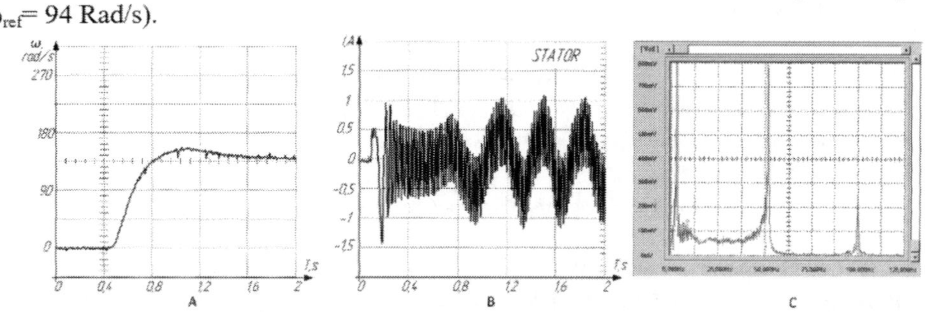

Figure 14. Transient of alternating current drive with vector control with squirrel-cage induction motor (ω_{ref}= 157 Rad/s).

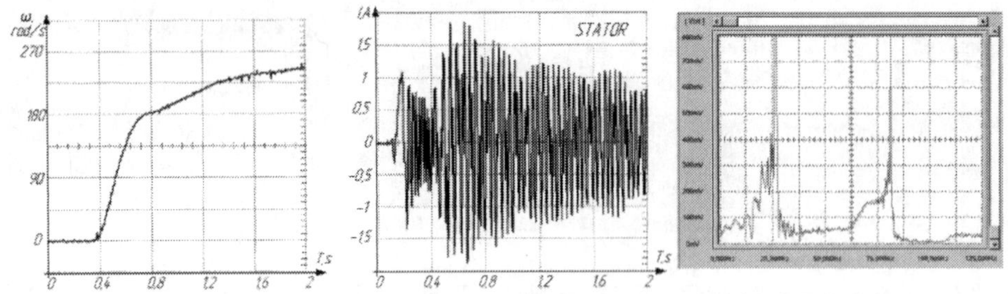

A B C

Figure 15. Transient of alternating current drive with vector control with squirrel-cage induction motor (ω_{ref}= 251 Rad/s).

The comparative analysis of diagrams given in Fig. 9–15 with similar diagrams for slip-ring rotor drives had much the same results. This suggested that the processes (stator current and speed diagrams) occurring in squirrel-cage motors which are used in most industrial equipment were identical to processes in slip-ring rotor drives used in our study in order to obtain the more detailed information about control processes. This allows us to assume that the features of processes indicated in the paper (inefficiency of vector control) are determined not by the electric motor, but by control algorithms. In addition, it will be possible to state that the developed correction methods, which have proved effective in controlling a motor with a slip-ring rotor, will also be effective in electric drives with squirrel cage induction motors.

4. Conclusion

Thus, the analysis of these experiments confirms the assumption that the observation of rotor currents makes it possible to accurately establish the mechanism for the occurrence of vector frequency control problems in asynchronous motors. It should also be noted that the proposed method of spectral analysis of processes in currents turns out to be very promising. Further studies of nonlinear processes in drives based on the proposed methods will allow us to analyze the causes of their origin and methods for their compensation.

5. References

[1] Vasil'ev B Yu and Kozyaruk A E 2014 Structure, composition, and control algorithms of high-efficiency electric drives of gas-compressor units *Russian Electrical Engineering* Vol 84 no 2 pp 94–102 DOI: 10.3103/S1068371213020077

[2] Kodkin V L and Anikin A S 2015 Frequency control of asynchronous electric drives in transport *2015 International Siberian Conference on Control and Communications (SIBCON)* pp 1–4 DOI: 10.1109/SIBCON.2015.7146978

[3] Tomasov V S and Usoltsev A A 2014 Comparative analysis of the energy efficiency of the scalar and space-vector pwm in a three-phase inverter *Russian Electrical Engineering* Vol 85 no 2 pp 111–114 DOI: 10.3103/S1068371214020126

[4] Sokolovskij G G and Stasovskij V M 1995 Speed of response of an electric drive under control of speed of three-mass elastic mechanism *Electricity* **10** pp 48–55

[5] Kozyaruk A E, Khitrov A A and Khitrov A I 2016 An energy-efficient autonomous energy supply system based on an external combustion engine *Russian Electrical Engineering* Vol 87 no 3 pp 119–124 DOI: 10.3103/S1068371216030056

[6] Vinogradov A B 2005 Calculation of losses in the steel, saturation, and the skin effect when simulating dynamic processes in a frequency-controlled asynchronous drive *Russian Electrical Engineering* Vol 76 no 5 pp 63–68

Analysis of parameters for technological equipment of parallel kinematics based on rods of variable length for processing accuracy assurance

A G Koltsov[1], A H Shamutdinov[2], D A Blokhin[1] and E V Krivonos[1]
[1] Omsk State Technical University, Omsk, Russia
[2] Omsk Tank-Automotive Engineering Institute, Omsk, Russia

Abstract. A new classification of parallel kinematics mechanisms on symmetry coefficient, being proportional to mechanism stiffness and accuracy of the processing product using the technological equipment under study, is proposed. A new version of the Stewart platform with a high symmetry coefficient is presented for analysis. The workspace of the mechanism under study is described, this space being a complex solid figure. The workspace end points are reached by the center of the mobile platform which moves in parallel related to the base plate. Parameters affecting the processing accuracy, namely the static and dynamic stiffness, natural vibration frequencies are determined. The capability assessment of the mechanism operation under various loads, taking into account resonance phenomena at different points of the workspace, was conducted. The study proved that stiffness and therefore, processing accuracy with the use of the above mentioned mechanisms are comparable with the stiffness and accuracy of medium-sized series-produced machines.

1. Introduction

Mechanisms of parallel kinematics [1, 3] based on the rods of variable length have been widely used in different technical fields: in operation units of robots, in machine-tools [6, 10], coordinate measuring machines [4], as training devices and simulators [3]. Compared to standard sequential operation mechanisms, parallel mechanisms are of larger stiffness at lower materials intensity and high positioning accuracy [6] as well as larger workspace as opposed to technological equipment with the conventional layout. In modern production the application of technological equipment with multi-drive mechanisms of parallel kinematics is quite restricted. This is due to insufficient previous study of this equipment, namely there is no classification of structures on stiffness parameter, on workspace parameters, static and dynamic characteristics and, therefore, on processing accuracy [5,9]. A specific feature of such system is that particular accuracy values correspond to each point of the workspace due to various stiffness and dynamic characteristics.

2. Problem Statement

Parallel kinematics mechanisms are called multi-movement mechanisms consisting of a mobile platform and base plate, connected at least by two parallel serial kinematic chains [1]. These kinematic chains are called mechanism legs. The investigation tasks are to identify mechanism structures with high symmetry coefficient, to study workspace parameters for the parallel kinematics mechanism in terms of stiffness as the parameter which greatly affects the accuracy of positioning and processing in machine tool industry. Another task is an experimental study of static, dynamic and vibrational characteristics to use them as output data in mathematical modeling of parallel kinematics spatial mechanisms by the parameter of accuracy. The study of dynamic characteristics [11] allows taking into account such structural elements as the mass of actuated parts and work-in-process pieces. Solving dynamic problems, it is necessary to derive equations to describe mechanism behavior with allowance for mass - inertial properties of its elements, in other words in the construction of the dynamic model. These equations allow the operation of equipment under various modes to be modeled. The input data for mathematical modeling are the parameters obtained by the full-scale specimen examination [12].

3. Theoretical Basis

Multi-drive mechanisms of a rod type, mechanisms based on the Stewart platform [1, 2, 3] in particular, are widely used in metal-cutting machines, measuring machines and robotics. There is not much technical literature on theoretical and experimental studies in the field of these mechanisms. Recently, an acute problem has been the classification of these mechanisms in terms of any parameters. In a number of works [5] multi-drive mechanisms were considered, where drivers were symmetrical in space.

It is proposed to synthesize the mechanisms on stiffness parameter [7, 8], these mechanisms being used in machine tool technology. The dynamic properties of the mechanism apart from the external power action and inertial characteristics are determined by kinematic parameters of the mechanism among other factors. Kinematic parameters of such mechanisms are determined by the geometry of the structure, in other words the number, linear dimensions, limit angular span in joints and positional relationship of kinematic pairs.

Optimality of these structures will be estimated by the parameter S_i which is conditionally called the i-th group coefficient of symmetry:

$$S_i = \frac{(K_{n,i})_2}{(K_{n,i})_1}, \tag{1}$$

where $(K_{n,i})_1$ and $(K_{n,i})_2$ are the number of mechanism structures of the i-th group (table 1).

The analysis of the works [7, 8] shows that the most symmetrical group is the group e_{14} which is not inferior in its symmetry (S_{14}=3.2) to the group e_{13} (the Stewart Platform, S_{13} = 4). The groups s_5 and e_3 are slightly inferior. These groups are not six-degree-of-freedom mechanisms. They are unit elements for the six-drive mechanism assembly. And as separate mechanisms they are widely used in the form of three degree-of-freedom manipulators. Kinematic and dynamics properties of the structures e_{13} (the Stewart platform), e_5 and e_3 are described thoroughly in [6, 10, 11]. So, investigation of the poorly studied six-drive group e_{14} with high symmetry and, hence, stiffness is of great interest. The analysis of this system enables the degree of conformity of this structure with the requirements to automated machine-building production to be evaluated.

Table 1. Classification of stiff drive mechanisms.

C

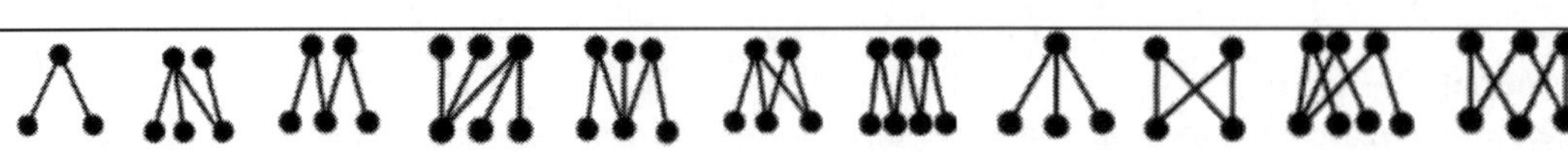

e_i	e_2	e_7	e_6	e_8	e_9	e_{10}	e_{15}	e_3	e_5	e_{14}	e_{13}
S_i	1.33	1.33	1.6	1.6	1.6	1.6	1.78	2	2.29	3.2	4

The stiffness of the parallel kinematics mechanism is directly proportional to the symmetry coefficient [7, 8]. The symmetry and stiffness of the mechanism as well as the Stewart platform are explained by the geometrical arrangement of the structure drives.

Firstly, there are three V-shape links with one node conditionally converging (in fact each drive has its kinematic link with platforms), and this increases the structural stability.

Secondly, spaced through the angle of 120° in terms of stability, they present a stiff structure in the form of a triangle on each platform.

Thirdly, it is sufficient to investigate only one V-shape node in making static and dynamic calculation, and this simplifies the calculations.

The structure shown in Figure 1 is a full-value analogue of the Stewart platform, though it has its specific kinematic and dynamic properties. So, this structure of a mechanism proves to be independent, other than the Stewart mechanism.

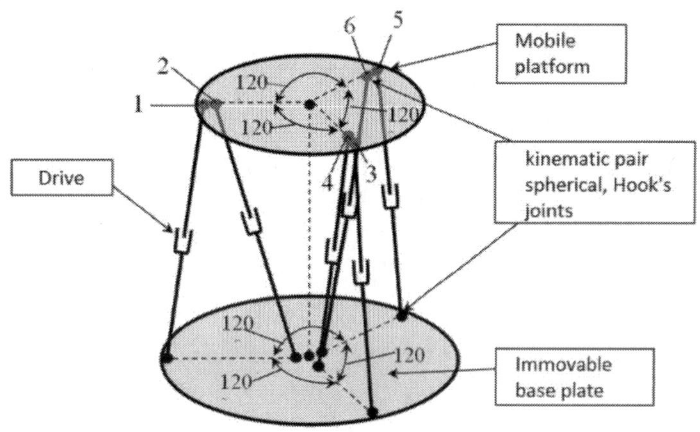

Figure 1. Kinematic diagram of the structure e_{14} (n=6, m=6).

3.1. Work range experimental research

To study the work range, stiffness, free and induced vibrations, the model of parallel kinematics mechanism with the structure e_{14} (n=6, m=6) was constructed [12].

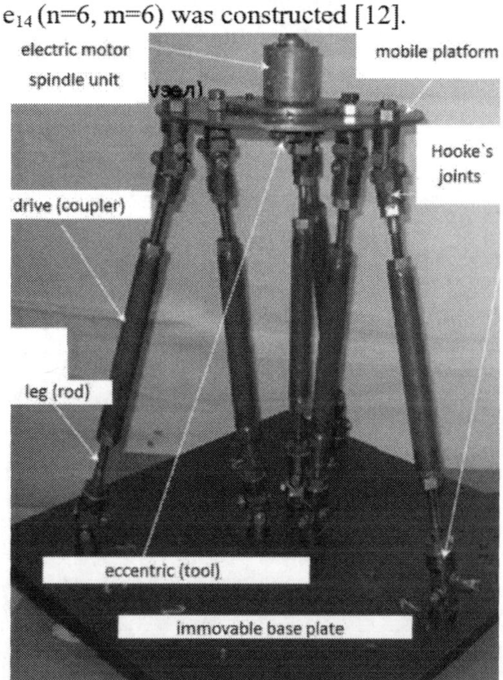

Figure 2. Mechanism under testing.

The mechanism under testing in Figure 2 consists of a mobile platform 1 and fixed base plate 2, six screw leveling legs (rods) 3 linked together by the joints with the possibility to vary length.

Translation and angular movements are possible due to the leg length variation and rotations in two-degree-of-freedom rotating Hooke`s joints 4. In the base central position the mechanism has three vertical planes of symmetry A-A, B-B, C-C, being set at the angle of 120 ^{0} relative to each other and crossed at one point (Figure 3).

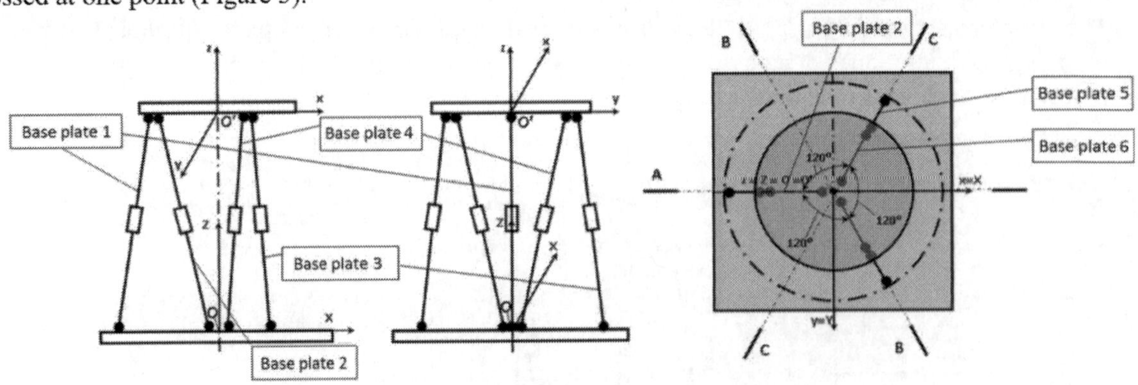

Figure 3. Diagram of the mechanism under testing.

Let us consider the layout of the workspace when the platform moves in the plane A-A. The graphical layout of the work range is stepwise: the platform movement direction is selected, the number of legs is defined, and variation in their length ensures the intended movement. Then the length of the legs is measured (2) and the mobile platform center coordinates are determined (3), the work range curves being plotted in these coordinates. The center coordinates of the mobile platform in absolute coordinate system are determined by the probe (1) being placed in the center hole of the platform. By fixing its length, which corresponds to the coordinate along the axis OX and varying the lengths of legs until it contacts the base plate (4), we determine the center coordinate of the mobile platform along the axis OZ. Based on the measurements of the work range, its curve in one coordinate plane is plotted on the coordinate grid (Fig. 4). The results will be similar for the planes B-B and C-C of the mechanism layout.

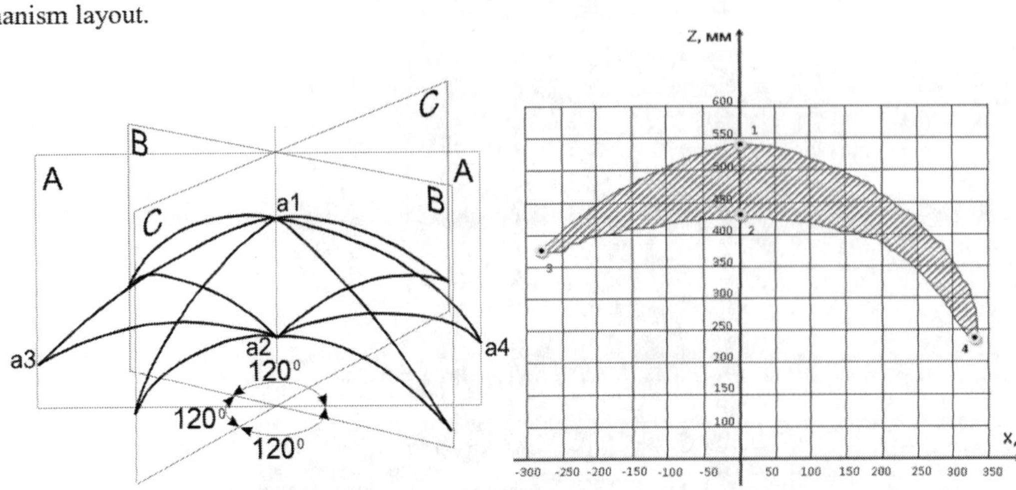

Figure 4. Projection and cross-section of workspace.

3.2. Developing a mathematical model for the dynamic system

The computational model of the dynamic system requires some assumptions [13]. Under these assumptions vibrations along six coordinate directions are examined. The expression of kinetic energy for the system is described by the equation:

$$T = 0{,}5 \cdot (m \cdot (\dot{q}_1 - Y_M \cdot \dot{q}_6 + Z_M \cdot \dot{q}_5)^2 + (\dot{q}_2 + X_M \cdot \dot{q}_6 - Z_M \cdot \dot{q}_4)^2 + (\dot{q}_3 - X_M \cdot \dot{q}_5 + Y_M \cdot \dot{q}_4)^2 + J_X \cdot \dot{q}_4^{\,2} + J_Y \cdot \dot{q}_5^{\,2} + J_Z \cdot \dot{q}_6^{\,2}). \quad (2)$$

The potential energy of the mechanism is defined as the strain energy of six legs and is represented by the expression:

$$\sum_{i=1}^{6} \Pi_i = \sum_{i=1}^{6} 0.5 \cdot c_i \cdot \Delta S_i^{\,2}, \quad (3)$$

where c is the stiffness of the i-th rod; Δ is the strain of the i-th rod, i is the number of rods.

3.3. *Dynamic system vibrations of the mechanism*

To solve the differential equations for system vibrations, let us write them in the form of matrix:

$$\|a\| \cdot \{\ddot{q}_i\} + \|c\| \cdot \{q_i\} = \{P_i(t)\} \quad (4)$$

where $\|a\|$ is the matrix of inertia coefficient; $\|c\|$ is the matrix of stiffness coefficients; $\{P_i(t)\}$ is the matrix of generalized disturbing forces.

The solution of the non-homogeneous differential equations system (4) is sought in the form: $q_{g.\,n.} = q_{g.\,h.} + q_{p.\,n.}$, where: $q_{g.\,h.}$ is the general solution of the homogeneous differential equations system, $q_{p.\,n.}$ is the particular solution of the non-homogeneous differential equations system.

The general solution of the homogeneous differential equations system $q_{g.\,h.}$ describes natural vibration of the system, and it is of interest in studying resonance phenomena of the system and the dependence of the system natural frequency on various parameters of this system.

The solution of the system of particular non-homogeneous differential equations allows the amplitudes of the induced vibrations for the computational model elements to be determined on generalized coordinates. The solution of the given system can be of interest in studying vibrations of the computational model elements due to the action of periodic disturbing force.

3.4. *Studying natural vibrations*

To study the system on the possibility of resonance phenomena under various parameters of the system, let us determine dependencies of natural frequencies on various parameters (mass, leg stiffness, mobile platform projection, leg diameter). The study is based on the solution of the general homogeneous equations using the MATLAB package [13]. The experimental results (data on leg stiffness, geometrical dimensions, and mass) are used as input data for mathematical modeling. The frequency equation of the system is:

$$\lambda^6 - I_1\lambda^5 + I_2\lambda^4 - I_3\lambda^3 + I_4\lambda^2 - I_5\lambda + I_6 = 0, \quad (5)$$

Solving the polynomial by decomposition in the MATLAB program, all values λ_i and natural frequencies $k_i = \lambda_i^{-1}$ can be found.

3.5. *Experimental study of static stiffness*

The definition of stiffening parameters is performed in the end positions of the mobile platform workspace (points 1, 2, 3, 4 in Figure 4), as far as the system has the smallest stiffness in these positions, and, hence, minimum accuracy. Platform loading was performed using a jack with tension and compression force of the value P = 0.2…1.2 kN, controlled by dynamometer ДОСМ-3-02, the movement being recorded by the indicating gage. Table 2 shows the results of stiffness parameters compared to the Stewart platform.

3.6. *Experimental study of dynamic parameters*

Let us define the natural frequencies, forms and decrements of the mechanism vibration. Vibration analyzer "Diana 2M" is used for this analysis.

The process of dynamic testing of the mechanism (Figure 5) with the method of induced vibrations was performed by the step change of the eccentric exciter rotational speed. The readings are recorded by the acceleration gage followed by data processing. To determine disturbing frequency and periodic, it is essential to consider time realization with time signaling.

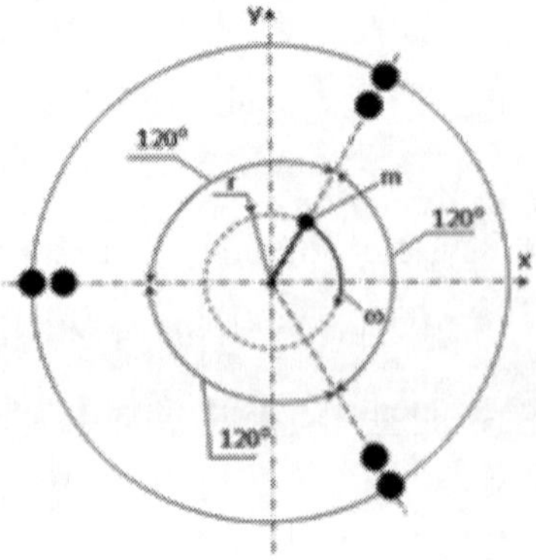

Figure 5. Diagram of the eccentric vibration exciter.

4. Experimental Results

The performed studies allowed to establish that the equipment of parallel kinematics based on rod of variable length with the high symmetry coefficient possesses high stiffness. This coefficient is comparable to the stiffness and accuracy of medium-sized series-produced machines.

Table 2. Comparison of the mechanism under testing with the Stewart platform under the force of loading of 120 kg.

Point of the work range (Fig. 4)	Compression		Tension	
	Stewart platform	Structure e_{14}	Stewart platform	Structure e_{14}
	H_{max}, mm	H_{max}, mm	H_{max}, mm	H_{max}, mm
Point 1	0.3	3.7	0.28	3.43
Point 2	0.28	3.2	0.27	3.06
Point 3	1.01	1.37	1.17	1.85
Point 4	0.77	1.87	0.97	1.69

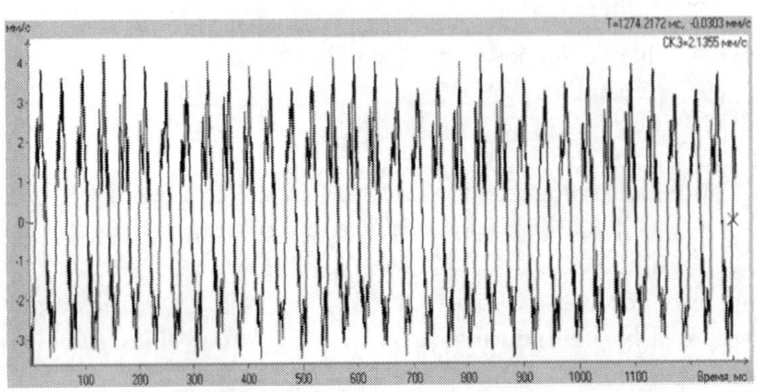

Figure 6. Platform vibration graph.

Having obtained the time realization, it is possible to determine the required frequency and periodic. In Figure 7 the periodic of frequency is $T = 1/n \approx 0.038$sec., $f = 1/T \approx 26$ Hz, where n is the number of waves per 1 second, f *is* the disturbing frequency.

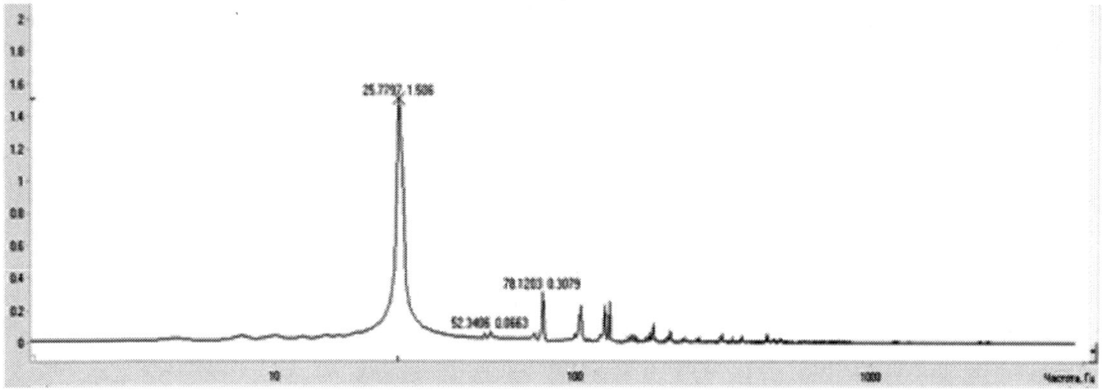

Figure 7. Frequency spectrum.

An important component for mechanical vibrations within the time domain is the dependence of their amplitudes on time. A physical characteristic of mechanical vibrations within the time domain is their frequency spectrum. The time domain having been transformed with the fast Fourier transform (FFT), we obtain the frequency spectrum (Figure 7), where the first harmonic at the frequency of the disturbing force of 25.7 Hz is seen (above we have defined it as ≈26 Hz) as well as all higher harmonics and the value of the disturbing force amplitude. To determine the natural vibration frequencies and logarithmic decrement of the structure under study, it is subjected to a shock load and it results in damping vibrations (Figure 8).

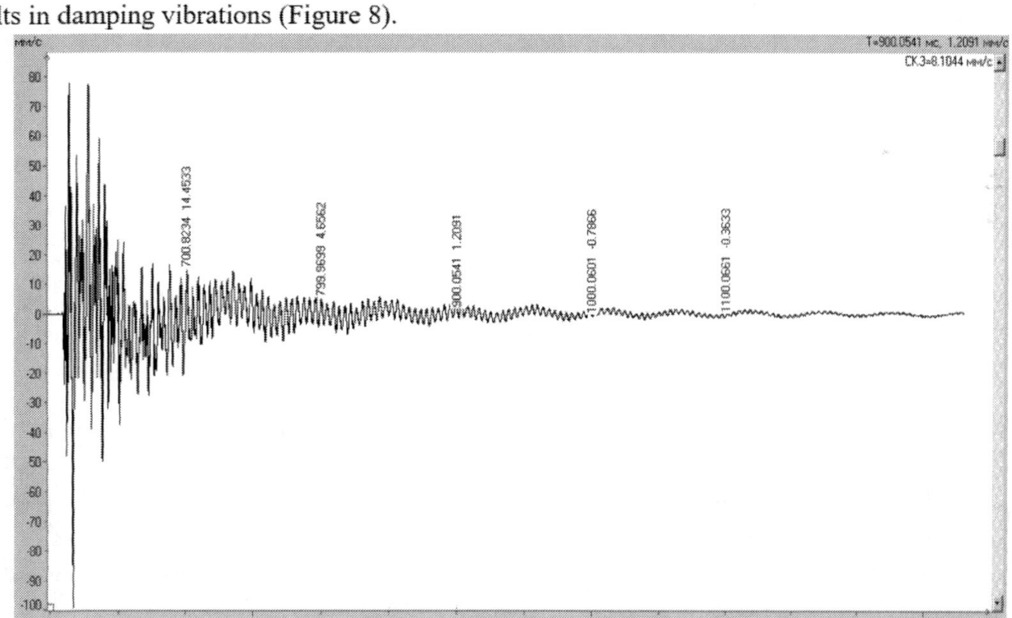

Figure 8. Waveform under the shock load. Platform in the position of maximum stiffness.

The resulting wave having been processed, the frequency and natural period are determined. In Figure 7 the natural period of the mechanism under the maximum stiffness equals $T = 0.1/24 \approx 0.00416$ *sec,* where 0.1 *sec.* is the time interval corresponding to the twenty-four waves, and the natural frequency is $f=1/T=1/0.00416 \approx 240$Hz.

5. Discussion

It is possible to determine the value of the natural frequency and amplitude if the FFT subjects to time realization shown in Figure 7. As a result we obtain the frequency spectrum with clearly defined spike at the resonance frequency of 67.97 Hz (above we have defined it as ≈ 70 Hz) and the value of frequency amplitude.

Resonance frequencies $f_{res.}$ and corresponding logarithmic decrements of vibrations $\delta = \pi \times \Delta f / f_{res.} = 0.1$ are defined on amplitude frequency response, where Δf *is* the width of resonance peak at the level $2^{-0.5}$ from its largest value. Thus, the range of natural frequency variation of 70 – 240 Hz and logarithmic decrement of damping were determined.

6. Conclusion

- The stiffness of mechanism carrying system depends on the geometric parameters, position of the executive element in the work range. The further away from the symmetry axis it is, the smaller is stiffness. The largest stiffness of the system is exhibited vertically.
- The size of the work range is significantly influenced by the constraints imposed by the angular movements in joints and linear length variation of the legs. The smaller the legs are, the smaller are the overall dimensions of the work range.
- The detected natural frequencies enable us to determine the applicability of this mechanism in particular operating conditions and evaluate the possibility of various on-load operations with allowance for resonance.
- At different positions of a mobile platform it possesses the largest vibration stability and the least error from vibration displacements along the axis Z. More careful calculations are required for vibrations along the axes X and Y and torsion vibrations.

7. References

[1] Merlet J P 2000 Parallel robots (second edition). *Solid mechanics and its applications* **128** 394 p.
[2] Gough V E 1956 Contribution to discussion of papers on research in Automobile Stability, Control and Tyre performance *Proc. Auto Div. Inst. Mech. Eng.* 392–394
[3] Stewart D A 1965 platform with six degrees of freedom *Proceedings of the Institution of Mechanical Engineers. London.* **180 \No15.** 371-385
[4] Koltsov A G, Blokhin D A, Khabarov A V and Redorovich D A 2014 The influence of kinematic characteristics of Stewart platform for precision moving measuring mechanism» *Dynamics of Systems, Mechanisms and Machines (Dynamics),* pp. 1-6, DOI: 10.1109/Dynamics.2014.7005669
[5] Gavrilov V A 2005 Classification of mechanisms for technological machines with parallel kinematics *Rus. Eng. Research* **9** 28–31
[6] Astanin V O and Sergienko V M 1993 Research of metal-cutting machine of alternative configuration *Mach. and tools* **3** 5-8
[7] Shamutdinov A H 2011 Studies on multi-drive mechanisms of parallel kinematics classification *Omsk sci. Bulletin* **2 (100)** 85–90
[8] Shamutdinov A H and Koltsov A G 2012 Definition of optimal stiffness parameter for multi-drive mechanism structures of parallel kinematics *Omsk Sci. Bulletin* **3** 184-6
[9] Zhuang H 1997 Self-calibration of parallel mechanisms with a case study on Stewart platform. *IEEE Trans. on Robotics and Automation* **13(3)** 387–397
[10] Vainshein I V 2009 Experimental study of stiffness for five-access machine with parallel kinematics *Rus. Eng. Research* 6-11

[11] Smirnov V A and Petrova L N 2009 Dynamic model of parallel kinematics mechanism *Bulletin SurSTU* **1** 5-8
[12] Koltsov A G 2012 Experimental study of a new structure parallel kinematics mechanism *Omsk Sci. Bulletin* **1** 96-101
[13] Koltsov A G 2012 Experimental study of a new structure parallel kinematics mechanism Sci. J. Bulletin of Ufa St. Aircraft Tech. Uni. 16 **4 (49)** 128 – 137

Mortise terrorism on the main pipelines

V A Komarov[1], N N Nigrey[2], D A Bronnikov[1] and A A Nigrey[1]

[1]Omsk State Transport University, Omsk, Russia
[2]Siberian State Automobile and Highway University, Omsk, Russia

Abstract. The research aim of the work is to analyze the effectiveness of the methods of physical protection of main pipelines proposed in the article from the "mortise terrorism" A mathematical model has been developed that made it possible to predict the dynamics of "mortise terrorism" in the short term. An analysis of the effectiveness of physical protection methods proposed in the article to prevent unauthorized impacts on the objects under investigation is given. A variant of a video analytics system has been developed that allows detecting violators with recognition of the types of work they perform at a distance of 150 meters in conditions of complex natural backgrounds and precipitation. Probability of detection is 0.959.

1. Prediction of terrorist threats in the main pipelines

Accident statistics on pipeline transport facilities indicates presence about problematic issues in this human activity [1]. Along with traditional causes of accidents: corrosion, natural influence, human errors, etc. became an essential role to play external influences on product pipeline (Table 1).

Under external influences refers to third party activities (mechanical action of building and earth-moving equipment, tie-in to theft pumped product, sabotage acts). Noteworthy is a significant violations percentage in Russia. The number of registered unapproved inserts in the country in 2005 exceeded 900, then there was a decrease this level. Incuts became less and less, «but began to steal more». In case we entered the criminal community, who have managed to turn oil theft from pipelines in largest sector of shadow business [3]. In [4] this type activity named «mortise terrorism».

Another reason (at this stage mostly potential) of the growth of mentioned influences – classical terrorism. Along with the terrorist acts aimed at achieving political goals (Budennovsk, New York, Moscow and others.) a technology that would cause economic damage starts to develop (blowing up oil pipelines in Iraq, Nigeria, pipelines in Russia). The question about anti-terrorist protection of the pipeline transportation facilities was raised in «Transneft» [5, 6]. According to predictions in near future intensity of terrorist acts on the objects will increase due to relative safety of their implementation for executors, availability of sales stolen resources market, problems with moral health of society and level of living population, their high efficiency in terms of economical losses and political consequences. The well-known Lancaster's , Volter's and Holling - Tanner's models were used for making predictions. However, these models do not take into account influence of state and economic factors. For a more adequate model should include type of

$$\begin{aligned}\frac{dN_1}{dt} &= \left[f_1(GDP, I/C) - f_2(MH_{90}/MH_t) - \frac{\alpha_1\beta_1}{\alpha_2\beta_2}\gamma N_2 \right] N_1(t),\\ \frac{dN_2}{dt} &= \left[f_3(PC/L) - f_4(DS/L) - \frac{\alpha_2\beta_2}{\alpha_1\beta_1}\gamma N_1 \right] N_2(t),\end{aligned} \tag{1}$$

where N_1 – number of attacks on the protected object, *GDP* – gross domestic product per capita in the past year, *I/C* – ratio of average income of the operation to the cost to carry it out, MH_{90} / MH_t – ratio of population moral health (in the region) in the 90th and current years respectively; α_1 and β_1 level of proficiency and technical equipment of the attacking side, α_2 and β_2 level of proficiency and technical equipment of the defenders, γ – factor of interaction between defenders and attackers (inversely proportional to size / length / area of protected object), N_2 –number of object defenders, *PC* – protection cost, *L* – total losses from attacks on protected object, *DS* – defenders salary.

GDP characterized poverty of society, function $f_1(GDP = \text{const}, I/C)$ – the proportion of people ready to go on offense, depending on expected income and having sufficient knowledge to commit a crime in this activity sphere.

Table 1. Information about accidents reasons on main pipelines in various world regions.

Country (region)	Reasons for accidents on main oil pipelines					
	External influence	Equipment defect	Corrosion	Natural influence	Human error	Other
Russia	63%(17%)	19%(30%)	6%(50%)	–(–)	12% (3%)	–
USA	23%(25%)	20%(19%)	23%(23%)	4%(10%)	7% (2%)	23%(21%)
West Europe	37%(50%)	25%(16%)	28%(15%)	3%(7%)	7% (5%)	– (7%)

Obviously when $(I/C) \leq 1$ value of function $f_1(\ldots) = 1$ (there is no point in carrying out operation without receiving any profit; factor of "revenge" is taken into account). Curve $f_1(\ldots)$ has a saturation point when the resource of society members able to commit an offense specified type is exhausted. For communities, characterized by different ratios of limiting factors, the curve shape is maintained, change its coefficients. $f_1\,(GDP = \text{const}, I/C) = a_1/(1 + b_1\exp(-c_1 I/C))$, where a_1 – coefficient determining part of society, potentially ready to move into camp of offenders from vested interests. Coefficients a_1, b_1, c_1 determined consider the expert technologies obtained in previous year's *GDP*.

Function $f_2(MH_{90}/MH_t) = (MH_{90}/MH_t) f_1\,(I/C)$ describes law-abiding population, its cultural level and legislative base for combating relevant type of crime. MH_t acts as an integral index, evaluated according to statistics (in terms of characterizing social tension in society). Functions f_3, f_4 are special cases of logistic curves.

It is obvious that protection cost PC will be greater, then more N_2. Total losses from attacks L increase with number of attacks, i.e. with increasing N_1. One of possible functions $f_a\,(PC, L, N_1, N_2)$, considering nature of its logistics:

$$f_3(PC, L, N_1, N_2) = a_3 \exp\left[\frac{LN_1}{PC^* N_2} b_3 - c_3\right] \Big/ \left[1 + \exp\left(\frac{LN_1}{PC^* N_2} b_3 - c_3\right)\right],$$

where a_3, b_3, c_3 – coefficients of curve.

By detectable $\frac{dN_1(t)}{dt}$, current expenditure on protection and object losses from attacks on him, the value of I/C, determined by a survey of population adjacent to object and track ratio MH_{90}/MH_t assesses state of protection I_t and its difference from optimum I_{opt} (equality of defense costs and losses caused by attacks to object). By I_{opt} specifies requirements for probability of attacks detection and false alarms, which should provide a security service object.

The results modeling dynamic system, described by equations (1) for different values of function arguments f_1 and α_1, β_1 allowed to make following conclusions.

In the case of low-skill and technical equipment of defenders α_2 and β_2 relatively similar indicators confrontation is periodic. Low discipline and financing of security forces do not allow management of organization to respond quickly to changing situations and entail excessive losses from terrorist activities and maintenance of excess defenders state. Even approach of indicators α_2 and β_2 to maximum possible value does not fully solve the problem of counteraction to terrorist attacks, the system quickly reaches a steady non-zero state. The determining factor in this case is the level of society moral health MH_t. Considering this, as well as exponential growth of content defenders value while increasing α_2 and β_2 we arrive in general case to conclusion about inappropriate choice of maximizing strategy level of protection object against terrorist threats. Optimum protection of economy objects from such threats, the strategy should include the interaction between the state and the object aimed at improving preventive public welfare, moral health of society and fight against crime.

We illustrate the efficiency of use "is not an optimal strategy" in particular example. Given fundamental laws of society in terms of population growth, dwindling natural resources and increasing pollution of environment [7], should be anticipated decline *GDP* and *MH* and increase in growth attacks on sources of resources. This view is justified in [8], in which the major reasons contributing to development of terrorism attributed decline in the living standards of social and professional groups, deepening of social differentiation, provoking political, ethnic and social conflicts. It should be determined by security policy relevant economic facilities from terrorist threats in near future. Solution to this problem for main pipelines emerges when considering the special case arising from the model (1) and reflecting interaction in system "terrorist – main product pipeline protection system – police," which can be represented as a system of four equations

$$\begin{cases} \frac{dN_P}{dt} = r_{PO}\left(1 - \frac{N_P}{N_{PO} * N_{Pm}}\right) N_P - \alpha_P p_v m_P p_M N_P N_M + r_{P2}; \\ \frac{dN_B}{dt} = \alpha_P \left(\frac{N_P}{g_P} - N_B\right) - \frac{N_P}{g_P} p_v \alpha_P; \\ \frac{dN_D}{dt} = r_{SO}\left(\frac{D_B(-t_S)}{v_B D_0}\right)^{k_{SO}} - r_{S1}\left(\frac{D_s}{v_S D_B(-t_S)}\right)^{k_{S1}}; \\ \frac{dN_M}{dt} = \frac{N_P(-t_M)}{g_P} p_B(-t_M)\alpha_P r_{M0} - r_{M1} N_M + r_{M2} N_P, \end{cases} \qquad (2)$$

where $N_p(t)$ –number of terrorist attacks at time t, N_B – number of attacks on the object, N_D – number of pipeline protection elements, N_M – number of employees involved in investigation of criminal activity on main pipelines, r_{po} – coefficient of information distribution (F. Bass [9]) among members of society ("viral" factor), N_{pm}– capacity of environment in which the "idea of product theft" will be distributed [10], α_p – average number of terrorist groups attacks on the object per time, p_v – probability of attack detection, m_p – probability to arrest terrorists, p_M – conviction probability to the detained terrorists, r_{p2} – growth coefficient of the number of product theft ideas adherents, g_p – average size of terrorist group; r_{so}, r_{s1} – coefficients of growth and cost reduction to protect the object; k_{s0}, k_{s1} – protection strategy coefficients that determine FAR and FRR limits of the protection system, D_B – loss from an attack at time t_s, D_s – cost of protection, v_B – loss of income as a result of attacks, v_S – ratio of losses from attacks to cost of object protection, t_M – delay of response to the registered attack, r_{M0}, r_{M1} – respectively, growth and reduction rate of employees number involved in investigation of attacks on the protected object, r_{M2} – coefficient of employees growth at activity changes in sales of stolen product.

Model adequacy (2) is checked by comparing results of process simulation "mortise terrorism" in Samara region in period from 2000 to 2001, with actual data obtained from [11] – [19]. Figures 1, 2 allow us to estimate efficiency of our model to generate predictions in this segment of pipeline transportation safety.

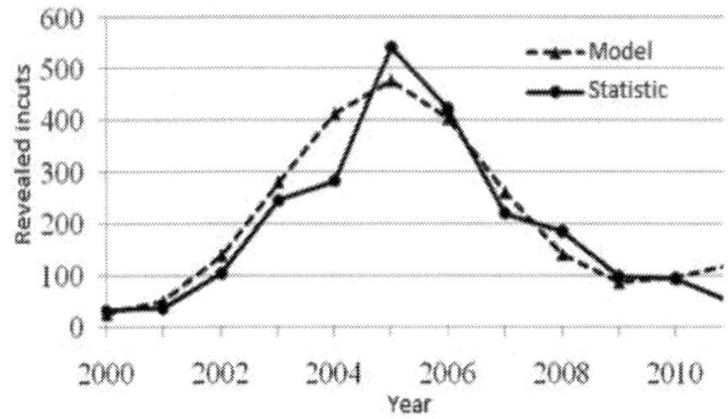

Figure 1. Results of modeling and statistics according to number of detected incuts.

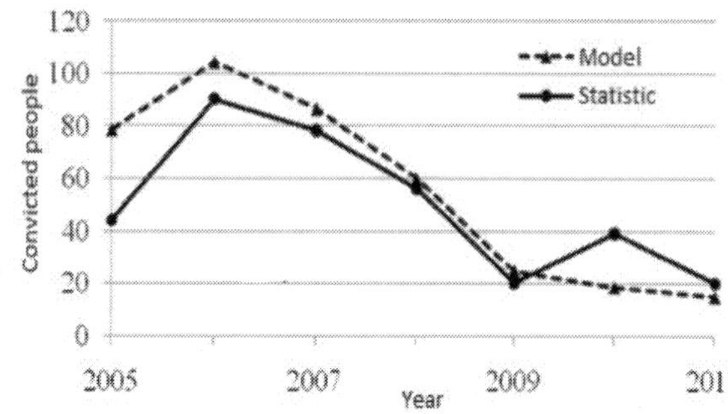

Figure 2. The results of modeling and statistics on number of convicted people.

The model reflects integrated parameters describing interacting parties and does not consider stochastic component of real confrontation process. Therefore, it should only be used for predictions of terrorist attacks in major entities (country, region, major oil company). Long-term predictions based on use of model, predicting a new cycle of significant growth in terrorist attacks on pipelines, followed by recession at the end of third decade (Figure 3).

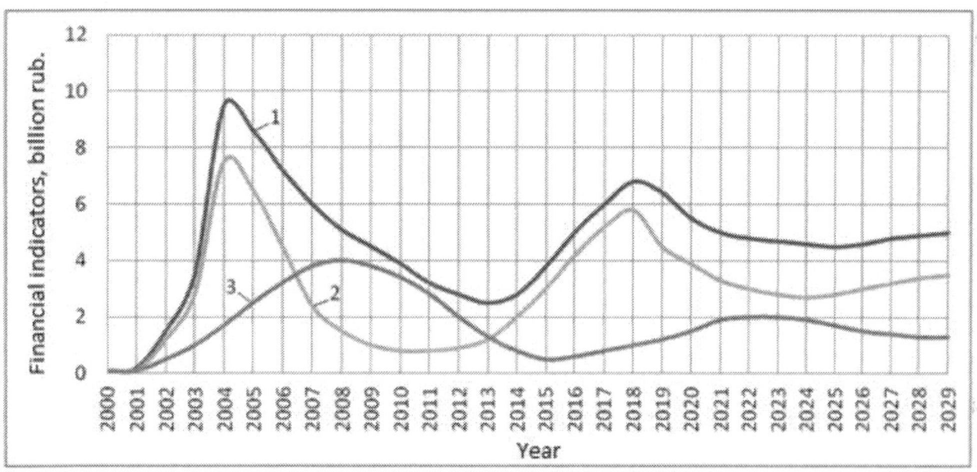

Figure 3. Modelling opposing sides financial indicators until 2029 while maintaining the current level of intrusion detection probability: 1 – the annual total costs, billion rubles, 2 – the annual damage from incuts, billion rubles, 3 – annual defense costs, billion rubles.

Actual protection cost and size of losses from incuts are confidential. Therefore, data in Figure 3 reflect trends in defense costs and damage caused by theft of product in a hypothetically assumed figures at the initial time (2000).

This model provides a single realization of random process at specified rates modeling system. For the degree projected estimate scattering we must move on to development of stochastic models. One embodiment of constructing such models may be based on the use of controlled Markov processes. However, to identify development trends of studied process the information obtained using the model (2) is sufficient.

Computational experiments using model (2) also allow determining the structure of physical protection from terrorist threats. Figure 4 shows financial indicators associated with operation facility, which is subject to terrorist attacks.

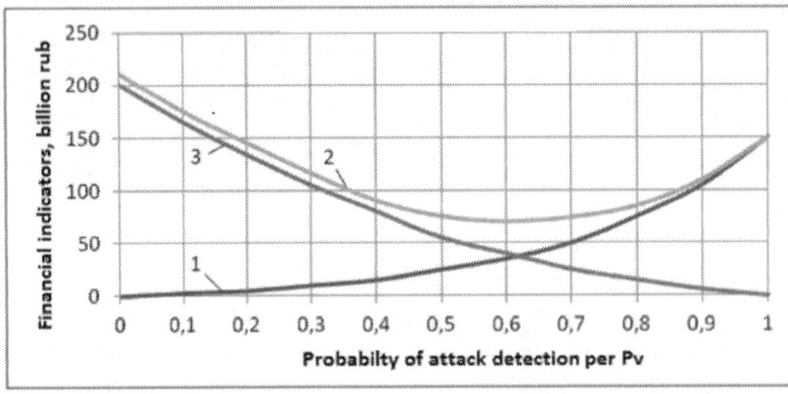

Figure 4. Financial indicators depending from implementation of protection system with a probability of incuts detection p_v: 1 – defense costs, D_S, 2 – damage from attacks D_B, 3 – total loss D_S+D_B.

The argument is taken as probability detected attacks. The curves represent a well-known principle of system analysis: less spending is laid on security created by system, the greater loss to be expected when it is operating. The minimum point «total loss» curve defines a rational protection level.

The curves represent one of possible cost ratio. Reducing cost of object protection or increase losses from attacks reduce minimum point $p_v = 1$.

The probability of attack detection p_v depends on a number of security elements (security number) N_D, their qualification level. In a simplified (illustrative) case value p_v estimated as follows.

We take attitude of «loss from the attack / defense costs» zero. Then $p_v = 0$. The owner of object determines at which excess losses of over-spending q protection is not deemed satisfactory. When a linear dependency p_v (q) for $q = 10$ value p_{vopt} exceeds 0,9; and ratio of cost protection for losses from attacks equals 1. Current costs and losses in organization must be known. All these data allow us to estimate a current level of object protection (p_v), necessary resources for transition to mode p_{vopt}, qualification level and number of defenders, as well as to determine the structure of physical protection system, providing p_{vopt}.

2. The structure of physical protection system from terrorist threats

To resolve the issue of quantitative and qualitative composition of technical protection suggested using a well-known method of hierarchy analysis T. Saaty [20]. Implementation of this method requires knowledge of threat types, probability of their implementation assess effectiveness threat prevention engineering and technical means. For main pipelines question in the open literature is not presented. However, variant indicative promising system for reducing number of attacks, ending positive outcome, emerges from analysis of published materials in recent years. The peculiarity of research problem: occurrence an unauthorized person (group of persons) in a strip ± 25 m from pipeline is considered a precursor to a possible terrorist attack. In addition to human disturbance of physical fields in controlled area caused by animals, vehicles, wind factor (through the vegetation, power support), turbulent motion of pumped oil, etc. It is necessary to solve problems detected in the background noise (interference) and classification of invading object (subject).

If a person is detected and recognized, it solves second part of problem – finding his intentions in contact with product lines. If any such intentions notified security service of the monitored object. Statement of remote subject's intentions determination comes from US Army. This task entrusted to company «Irvine Sensors Corporation» [21]. Automation of detection processes and pattern recognition – a feature of almost all objects for extended protection research. These tasks for detection and pattern recognition offered to solve with a help of video-seismic- vibrotechnology – computer data analysis techniques in order to obtain interesting information about protection without human intervention using optical respectively seismic and vibration signals. During 2008 to 2014 highest number of publications in electronic indexing systems Web of Science, Scopus, RSCI, publications seismoanalysis (115). The diagram in Figure 5a shows distribution of these publications on areas of research, in Figure 5b – t he percentage of different physical channel.

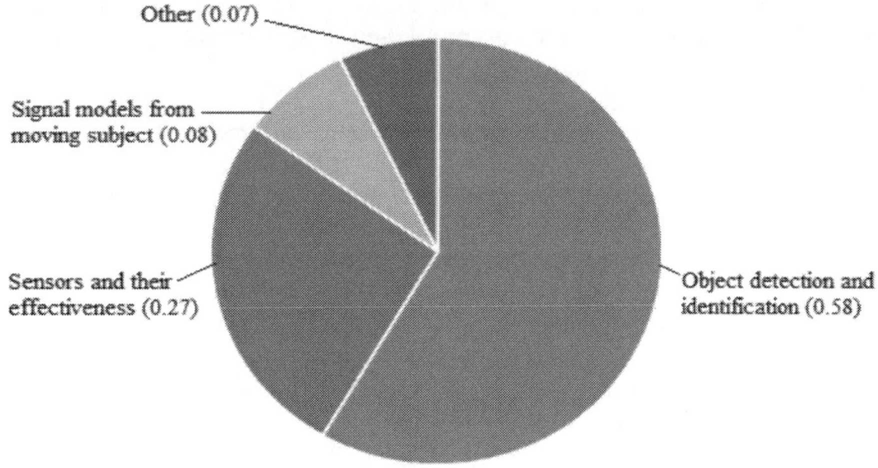

a)

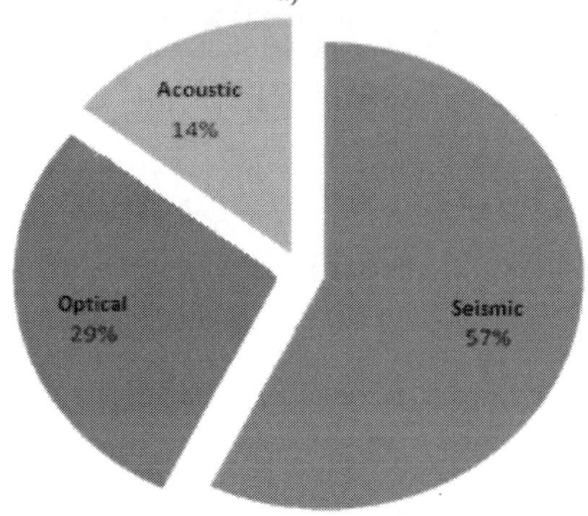

b)

Figure 5. Distribution of publications on seismoanalitics, presented in Web of Science, Scopus, RSCI 2008–2014 g for the period (a), and distribution research by use different channels (b).

Today prevailing view of preference for monitoring use distributed fiber optic sensors. With their help determined spatial distribution of temperature and mechanical strain in pipeline environment.

Local inhomogenities these figures give rise to Raman components (Raman or Brillouin). At time of propagation these components (the Stokes or anti-Stokes) scattering of place (active radar principle) tried on the location of this inhomogeneity.

It is argued that this type of system will help to detect leakage from pipeline an intensity 50 ml / min on highways length 16.85 km. [22] This principle is basis of "leakage detection systems and monitoring activities" Omega. [23].

A detailed analysis of possibility wider use distributed fiber optic systems as problem solution has led to more cautious conclusions. Strong influence of acoustic and seismic noise of natural and artificial origin on decisions taken narrows the scope of their application [24, 25]. Development direction of such systems is focused on complexity data processing algorithms. Australian company Future Fiber Technologies offers a system, which is based on algorithmic support costing over 100 thousand dollars [26].

The diagram in Fig. 5a illustrates the area of interest in development of fiber-optic systems. Is preserved uncertainty about classification problem solving intrusion into protected zone and the species held on her work. Designated in [27] requirements for detection probability of unauthorized crossings protected perimeter (0.95), while ensuring absence of occurrence false alarm for 250-500 hours (average requirements) in the works are not met.

Not worked out issues combating prepared offenders (e.g., installation in ground two seismic generators greatly reduces system parameters in area between them). Classic way overcoming encountered difficulties – use the rule: "If the use of one type detection system is unsatisfactory, need to complexing tools with different physical principles of detection. Applications 2 with mediocre performance is more productive rather than one with higher rates".

On linear part of underground main pipelines main operation is detection formation of pit to access a pipe surface. It can be provided by using vibro-acoustic system based on excitation pulses in pipe shell subsequent analysis of recorded signals at a predetermined distance from place of their formation [28]. Each pulse has a small piece of information about the state of object, but combination of such pulses provides a reasonable probability of errors 1-st and 2-nd kind [29].

Available estimates detection range of pits on control object (including product leakage) separate link «generator – detector» as used by shell method buzzer pipe fluctuate in range of 2–5 km. It will take about 0.5 billion rubles to Russian pipelines to reach this kind of control. This figure is 25 times lower than the estimated economic costs of loss transported product (a significant portion coming into Arctic Ocean from Russia – more than 500 tonnes per year shall be made rivers [30].

The effect of both fiber optic and vibro-acoustic systems can be estimated using a general theorem on repeated experiments. Probability hypothesis if the probability derived from N independent experiments is based on analysis generating function

$$\varphi_N(z) = \prod_{i=1}^{N} (q_{ij} + p_{ij} * z),$$

where $q_{ij} = 1 - p_{ij}$, p_{ij} – probability hypothesis according to j-th channel. Coefficient of Z^0 probability p_{0N} that considered hypothesis is not true. Then, with probability $q_{0N} = 1 - p_{0N}$ it recognized fair. for a combination of p_{ij}= 0,86 and 0,93 total probability detection will be 0.99.

Strong influence of seismic and acoustic noise natural and artificial origin limiting application of fiber optic warning system emergencies in controlled facilities. With reference to pipeline transport as these areas around the pumping station at the intersection of pipeline with a car or a railway line at a close distance from the paved roads or watercourses. For these categories other preferred embodiments of complex systems, «video analytics – vibroacoustic», «video analytics – seismic (at point geophones) – vibroacoustic» and others.

Developed version video analytics systems allow to detect intruders with recognition they perform types of work at a distance in 150 meters in a complex natural backgrounds and precipitation. Installed detection probability – 0.959 [31].The effectiveness of integrated system is illustrated in following figures. If each of subsystems is characterized probabilities 0.5 and 0.9, the integrated system – 0.75 and 0.99 and three subsystems – 0.87 and 0.995.

3. Conclusion

Tendency growth of steal product volumes justified reasons systemic nature. System-wide laws of growth in social structures are exponential. Combating theft can slow down this process, but in short-term trend of growth losses will not change. Aware of need to improve physical protection of pipelines for prevention emergency situations is carried out in a lower pace to meet the existing needs of problem. One reason for this is lack of study basic issues fundamental nature.

A notable step in solving this problem would be creation a comprehensive system of physical protection, based on presentation of concept formation it structure. To do this, assess the potential of fiber identification, vibroacoustic, seismoanalytic systems, estimating their limiting ability to cope with the tasks.

References

[1] Lisanov M V, Savina A V, Degtyarev D V and Samuseva E A 2010 Analysis of Russian and foreign data on accidents on pipeline transportation facilities *Moscow: Safety of work in industry* vol **7** pp 16–22

[2] Mahutov N A, Lisin Yu V, Gadenin M M, Permyakov V N, Fedota V I and Aladinsky V V 2012 Protection of trunk pipelines risk criteria *Moscow: Science & Technologies: Oil and Oil Products Pipeline Transportation* vol **3** pp 10–16

[3] Orlov D 2013 To hit incuts Pipelines and Facilities *Moscow: Pipeline transport of oil* vol **2** pp 12–15

[4] Gurdin K 2012 Oil thieves *Week arguments (Argumenty nedeli)* vol **2** p 5

[5] Dektyarev V A and Rodionov S L 2010 Against the terrorist threat *Moscow: Pipeline transport of oil* vol **9** pp 20–22

[6] Morozov M 2012 Protection of the oil lines is the protection of public interests *Moscow: Pipeline transport of oil* vol **3** pp 42–46

[7] Meadows D H, Meadows D L and Behrens W W 1991 The Limits to growth *Moscow: Moscow State University* p 208

[8] Belonozhkin V I and Ostapenko G A 2011 Information aspects of counter-terrorism *Moscow: Hot line-Telecom* p 112

[9] Bass F M 1969 A new product growth for model consumer durables *Management science*, January vol **15** no 5 (*Printed in U.S.A.*) pp 215–227

[10] Becker G S 1968 Crime and Punishment: An Economic Approach *The Journal of Political Economy* (Apr. 1968, Published by: The University of Chicago) vol **76** no 2 pp 169–217

[11] Media review mountains. Tolyatti and Samara region in June 2004, URL: http://samarskii.narod.ru/mil_krim_hron_0604.html free. The title. from the screen (date accessed: 06.02.2012)

[12] Usoskin M 2010 Considered the average size of bribes *Izvestiya of the Samara regional socially-the political newspaper*, URL:http://samarskieizvestia.EN/document/7820/?day=1&month=3&year=2011, free. The title. from the screen (date accessed: 06.02.2012)

[13] In the Samara region held a solemn ceremony dedicated to the 70th anniversary of the education service bhss–PSU. Ministry of internal Affairs of the Russian Federation, 15.03.2007, URL: http://contrast.mvd.ru/news/show_21379/ free. The title. from the screen (date accessed: 06.02.2012)

[14] On the outcome of the economic crimes of the municipal Department of internal Affairs across the Samara region for 2007 and solve the crime in 2008. Ministry of internal Affairs of the Russian Federation, 14.03.2008, URL: http://contrast.mvd.ru/news/show_41898/ free. The title. from the screen (date accessed: 06.02.2012)

[15] The police Department held a press conference, Deputy police chief for the Samara region the General-the major of justice Yury Suponev. GU MVD of Russia in Samara region, 15.12.2009, URL: http://www.guvd63.ru/news/official/15.12.2009/88/3503/ free. The title. from the screen (date accessed: 06.02.2012)

[16] The results of the Management of BEP police in Samara region in 2009. GU MVD of Russia in Samara region, 21.01.2010, URL: http://www.guvd63.ru/news/official/21.01.2010/88/ 3776/ free. The title. from the screen (date accessed: 06.02.2012)

[17] In the police Department for the Samara region held a press conference the Deputy chief of Central Board on economic security on the theme. Results of work of the economic security service of the police Department in the Samar-tion region in 2010. GU MVD of Russia in Samara region, 27.01.2011, URL: http://www.guvd63.ru/news/official/27.01.2011/88/7565/ free. The title. from the screen (date accessed: 06.02.2012)

[18] Bezverkhov A G 2007 Theft from the pipelines, oil pipelines and gas pipelines: concept, reasons, prevention: scientific and practical complex *Samara: Publishing house GOU VPO "Samara state University* p 304
[19] Information and analytical note to the report of the chief GU MVD of Russia in Samara region before the deputies of the Samara regional Duma. GU MVD of Russia in Samara region, 13.03.2012, URL: http://guvd63.ru/news/official/13.03.2012/88/19524/ free. The title. from the screen (date of access: 17.04.2012)
[20] Borovsky A S and Tarasov AD 2011 A General mathematical model of the physical protection system of objects *Bulletin of computer and information technology* no 10 pp 21–29
[21] Image Analysis for Personnel Intent. Zyn Systems – Sequim Washington, URL: http://zun.com./sbir/sbes/sbir/dod/army/army092–034
[22] Leakage detection using fiber optics distributed temperature monitoring *11th SPIE Annual International Symposium on Smart Structures and Materials of the March 14–18 2004 San Diego, California, USA, Proc. SPIE* vol 5384 pp 18–25
[23] Suvorova E 2012 Out of range *Moscow: Pipeline transport of oil* no 6 pp 34–35
[24] The safety of operation of pipelines in 2010, http://www.torinsk.ru/publication/29-bt. 2010.html
[25] Wang J 2012 FBG Intrusion Recognition Based on SVM Algorithm *Advanced Materials Research*, vol **591–93** pp 1422–1427
[26] Vvedensky B 2006 System perimeter security – new season *Moscow: System security* no 4 pp 110–118
[27] Zverinsky S S 2002 The Problem of choice of perimeter means of detection *Moscow: Special technique* no 4 pp 36–41
[28] Epifansev B N 2014 An Acoustic Method for Diagnostics of the State of Underground Pipelines: New Possibilities *Pleiades Publishing, Ltd: Journal of Nondestructive Testing*, vol **50**, no 5, pp 254–257
[29] Epifantsev B N, Pyatkov A A and Fedotov A A 2015 Evaluation of the sensitivity of vibro-acoustic system for the detection of local perturbations of parameters of the environment in the vicinity of the main pipeline *Ekaterinburg: Defectoscopy* 2015 no 2 pp 17–26
[30] Blocks I P A Brief overview on the breakthroughs of oil and volume of oil spills in Russia URL: http: // www. greenpeace. Org / russia / global / russia/report/Aretic-oil /Oil spills.pdf (date accessed: 11.11.2013)
[31] Epifantsev B N, Pyatkov A A and Fedotov A A 2013 Concept of safe operation of the trunk pipelines in terms of external influences *Moscow: Occupational safety in the industry* no 12 pp 42–49

The procedure for isolation of neoplasms on the retina of the eye

S V Komkova
Murom Institute (branch) of Higher Education "Vladimir State University Alexander G. and Nicholai G. Stoletovs", 23, Orlovskaya st., Murom, 602264, Russia

e-mail: savicheva-svetlana2010@yandex.ru

Abstract. In operation, a computer diagnostic procedure on the human retina neoplasms. The use of this technique in medical institutions in the operation of the ophthalmic practitioner allows early detection of the disease, at periodic inspection of a pictorial pattern of disease progression. The test procedure is performed on a set of real human retinal photographs taken from international STARE database with known diagnoses. Given the numerous experiments which show the possibility of using this technique, developed on the basis of the diagnostic system in a doctor 's office-ophthalmic.

1. Introduction

Currently, diabetic retinopathy is the leading cause of blindness in developed countries of the world population. Statistics show that in the world like disease affects more than 93 million people [1, 2].
This complication is retinopathy of diabetes disease associated with improper blood vessels of the eye fundus. In diabetic retinopathy, retinal vessels become more brittle, permeable, which can lead to hemorrhages.

Firstly, the retinopathy is dangerous in that in the course of the disease may develop acute vascular occlusion of the retina and optic nerve, requiring immediate treatment. The early signs of diabetic retinopathy is characterized by two ways: microaneurysm and hemorrhage. Microaneurysm sometimes leads to narrowing and occlusion of retinal blood vessels, as well as the weakening of the walls of blood vessels. The hemorrhage is the accumulation of blood outflow from blood vessels in the eye cavity or surrounding tissue.

In accordance with the international classification of diabetic retinopathy microaneurysm levels generally present at early stages of the disease, and hemorrhage are manifested in later stages.
Early detection, as well as periodic screening of eyeground potentially assists in reducing disease progression and prevent further vision loss.

In practice, the detection of diabetic retinopathy is a time consuming manual process that requires skilled physician ophthalmic. The physician diagnoses using ophtalmoscopy ophthalmologist 's presence, which is a diagnostic examination of the fundus by the fundus or ophthalmoscope lens. Despite the effectiveness of the manual approach to evaluation of this method is the most labor-intensive, further possible errors due to the influence of the human factor.

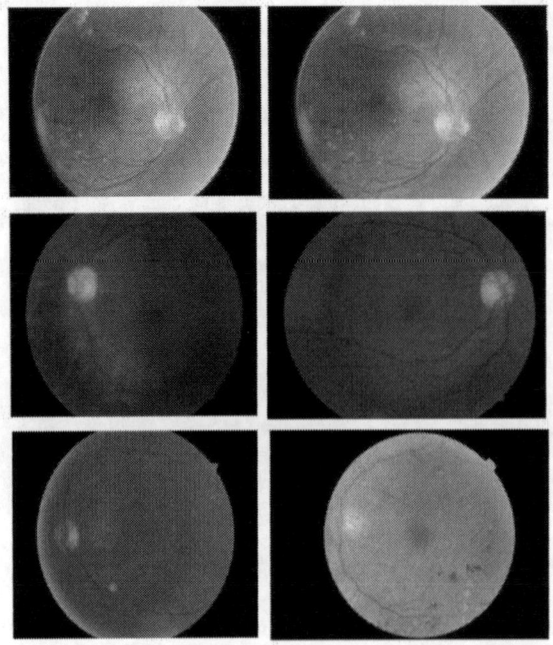

Figure 1. Examples of images of a retina of an eye

The analysis above shows the relevance of neoplastic development computer diagnostic procedures on the retina of the eye on the basis of the images obtained with the simple ophthalmoscope. In addition, it should be noted that the implementation of the methods and systems, constructed based on these measures allow for the driver, is considerably cheaper than the purchase of expensive special digital complexes, and it can be used not only in the large eye centers, but also in all hospitals.

2. The procedure for isolation of neoplasms on the retina of the eye

This approach involves a sequence of steps that convert the original color image of the retina to a form convenient for analysis of informative regions, namely blood vessels and neoplasms.

The procedure consists of the following steps:

1. Mean shift Segmentation method is an iterative process, which consists of the following steps:

Each pixel of the input image is considered in the space environment:

$$(x,y)=\begin{cases}X-sp\le x\le X+sp\\Y-sp\le x\le Y+sp\\(R,G,B)-(r,g,b)\le sr\end{cases},\qquad(1)$$

where sp – a spatial radius, sr – color radius, (X, Y) – surrounding pixels, (R,G,B) and (r,g,b) – pixel color (X, Y) and (x, y) respectively.

For this surrounding mean spatial value and mean value of color of pixel is calculated. These values become surrounding centers on the following iteration.

Iterative process of bypass of the image continues to the required level. As a result of completion of iterative process color of pixels of the input image will accept final value.

2. Conversion of the source image to gray-scale

For convenience of the subsequent processing the image is transferred to a format with 256 gradation of color, gray without indexing [3, 4].

The algorithm of the translation consists in:

- to sequential changeover of brightness of each point of the image of a retina of an eye on a equation:

$$x(i,j) = 0.3r + 0.59g + 0.11b, \tag{2}$$

where r, g, b – values of red, green and blue flowers in each processed point.

- copying of the received value on three channels, that is:

$$r = g = b = x(i,j) \tag{3}$$

3. Filtering image
It is executed by means of the algorithm of median filtering described in [5].
4. Contrast increase
In view of the fact that on the image of a retina of an eye there are many dark sections and details on them become almost indiscernible the algorithm of increase of contrast of the image by alignment of the histogram is applied. This algorithm consists of the following steps:

- creation of the histogram of brightness of H[i] initial eight bit single-channel images f (x, y);
- rating of the histogram in the range [0, 1] by division of each value of the histogram of H[i] into total quantity of pixels N images f (x, y) and creation of the integral histogram:

$$H'[j] = \sum_{0 \le i \le j} H[i] \tag{4}$$

- computation of new value of brightness of pixel according to the received histogram:

$$f'(x,y) = H'(f(x,y) \tag{5}$$

5. Improving of boundaries between a background and blood vessels
This algorithm consists of the following stages:

- the retina mask by means of threshold conversion is calculated:

$$f''(x,y) = \begin{cases} 0, \text{if } f(x,\ y) > 40 \\ 255, \text{otherwise} \end{cases} \tag{6}$$

- the received image is subtracted from rounded $f'(x,y)$:

$$f'''(x,y) = f''(x,y) - f'(x,y) \tag{7}$$

6. Binarization
It is executed by means of Otsu's algorithm described in [6].
7. Separation of circuits
This stage is necessary for separation of walls of blood vessels of a retina of an eye.

As algorithm the morphological gradient was selected. He finds a circuit of objects which extent exceeds the kernel size (the structuring element). By experiments the size of a kernel was selected equal 3x3, the form – a rectangle.

The morphological gradient represents a difference between stretching and an erosion:

$$gradient(f'''(x,y)) = dilate(f'''(x,y)) - erode(f'''(x,y)) \tag{8}$$

8. Marking of spots
Algorithm of marking of spots following:

1) are viewed sequentially all selected on the previous step of a circuit;

2) for each found circuit such parameters as are calculated:

$$perimetr = \sum_{i=1}^{N} \sum_{j=1}^{M} x_{i,j}\ , \tag{9}$$

where *N*, *M* – quantity of points of a circuit of a spot across and down

$$square = \sum_{i=1}^{N} \sum_{j=1}^{M} x_{i,j}, \tag{10}$$

where *N*, *M* – quantity of points in a spot across and down

$$compact = square / perimetr^2 \tag{11}$$

$$area = H * W \tag{12}$$

The decision on circuit belonging to a spot is made on the basis of the following conditions:

$$spot = \begin{cases} square / area < 5 \\ 0/087 \leq compact \geq 0.05 \\ n \geq 10 \end{cases} \tag{13}$$

where *spot* – a spot, *n* – quantity of points in a circuit.

3. Examination of the isolation techniques of neoplasms on the retina of the eye

The offered technique including the considered algorithms was realized and was checked on 100 images of a retina of an eye. Images were taken from basis of STARE. For the analysis the images of sick patients were selected. The part of images of these objects and results of their preliminary processing are given in table 1 and in fig. 2.

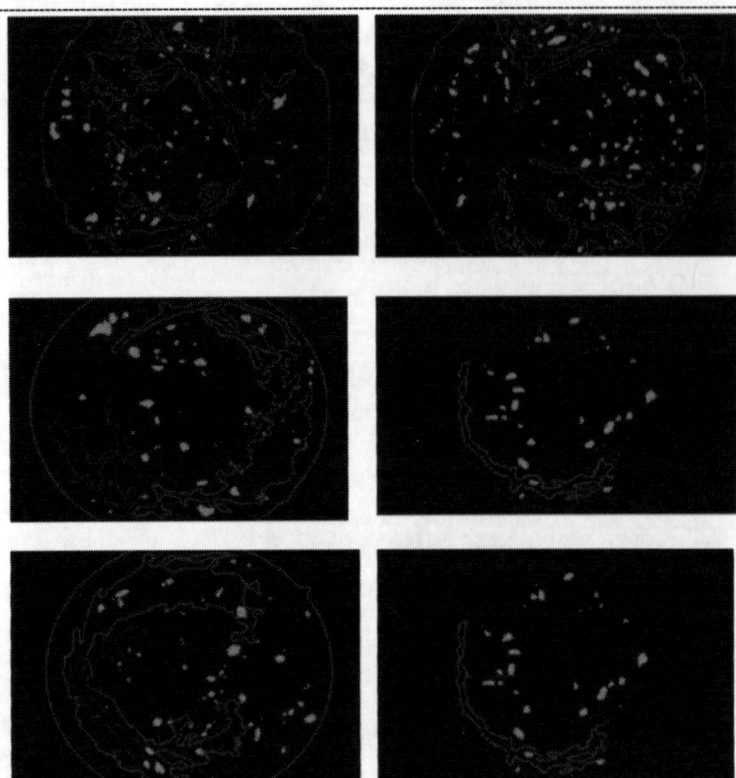

Figure 2. Marking of spots

IOP Conf. Series: Journal of Physics: Conf. Series **944** (2017) 012055 doi:10.1088/1742-6596/944/1/012055

Table 1. Sequential processing retinal images

Segmentation, the conversion into halftone filtering	The contrast enhancement, improvement of the boundaries between the background and blood vessels	Binarization	Contours
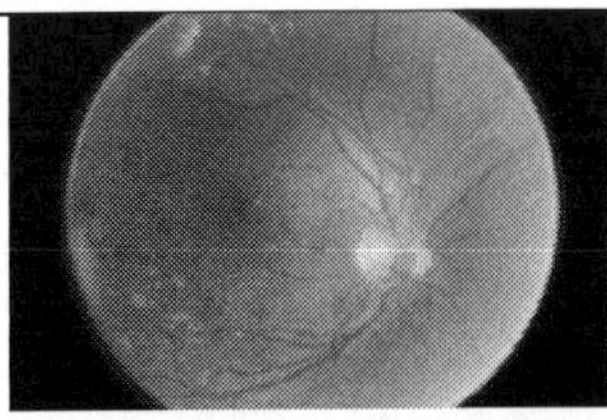	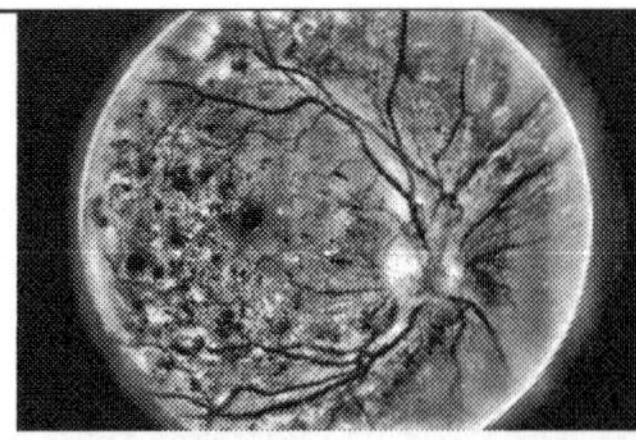	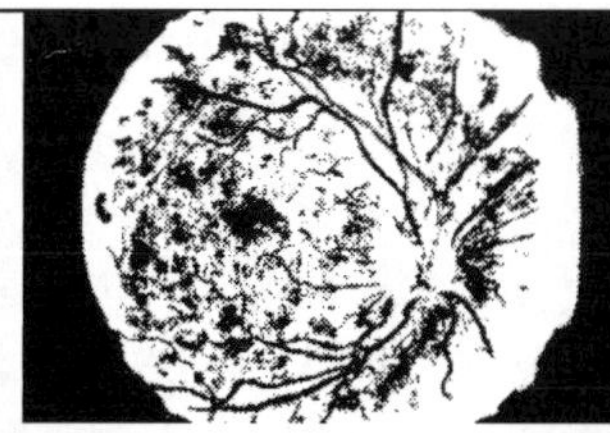	
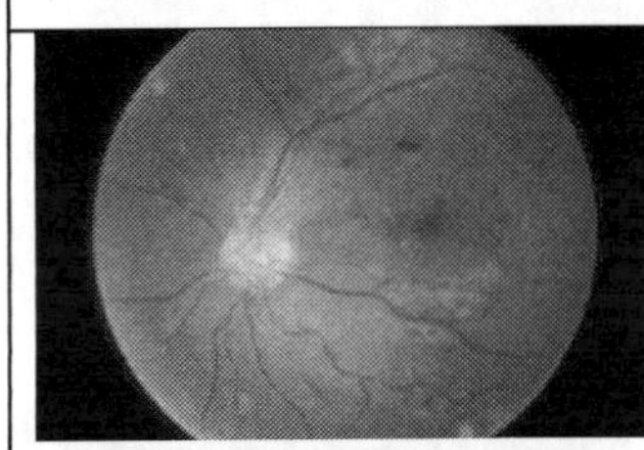	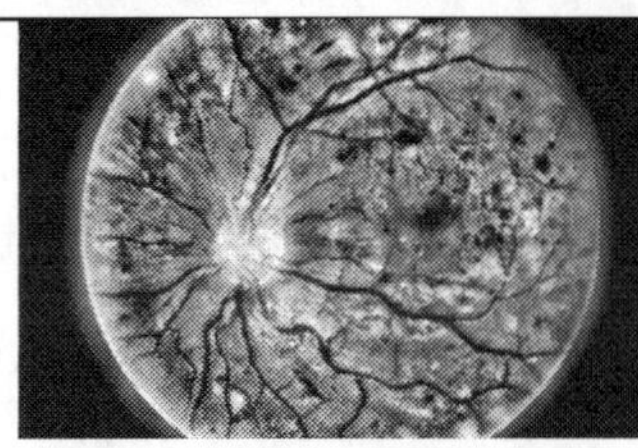	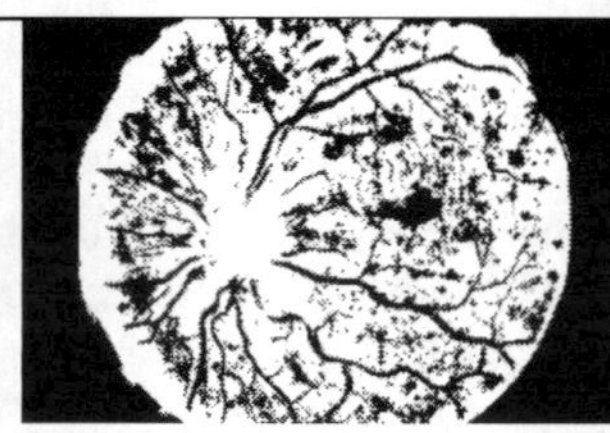	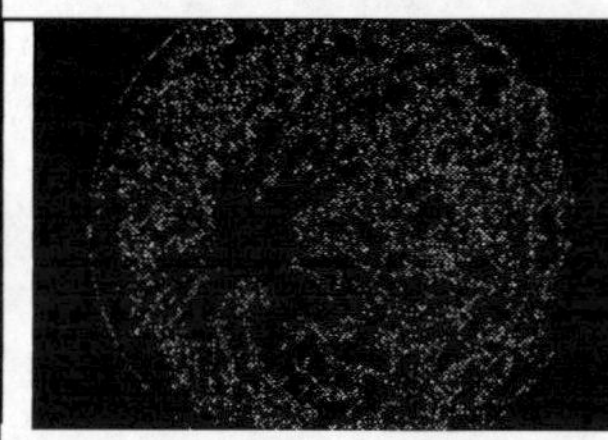
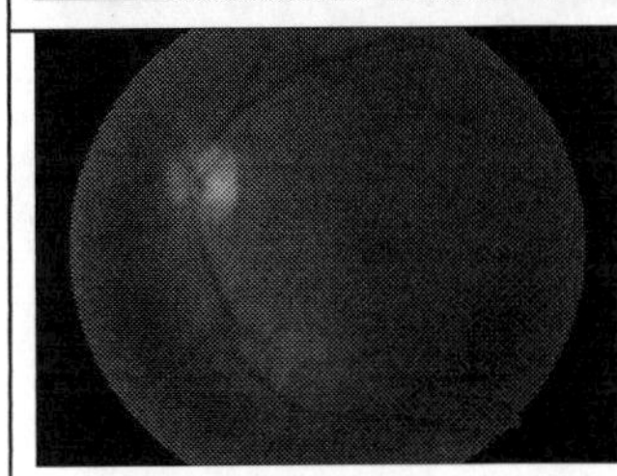	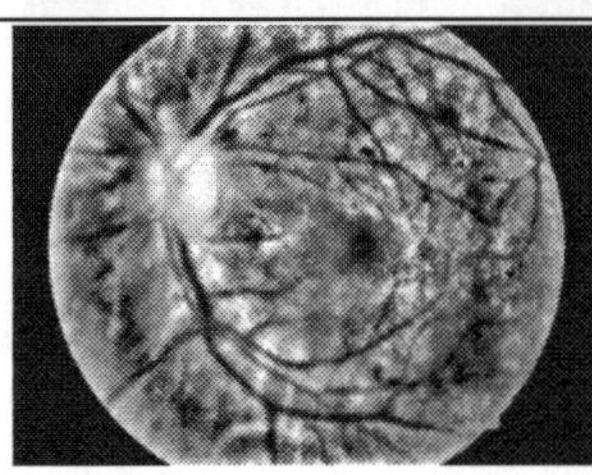	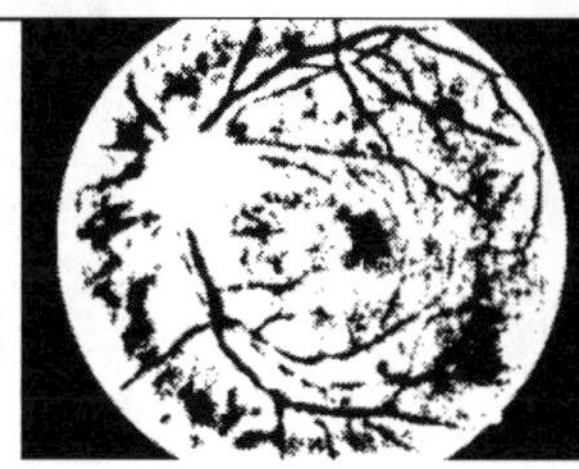	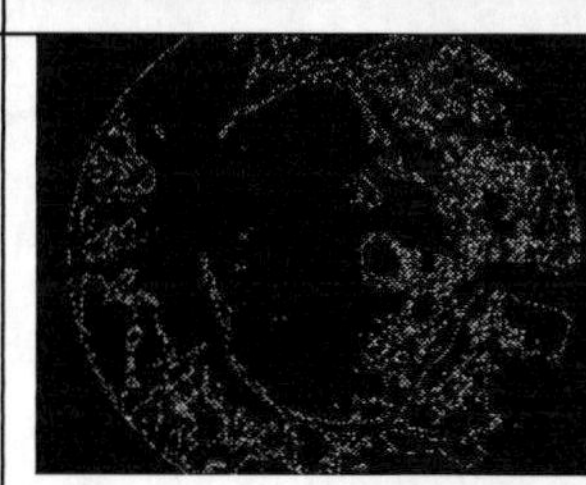
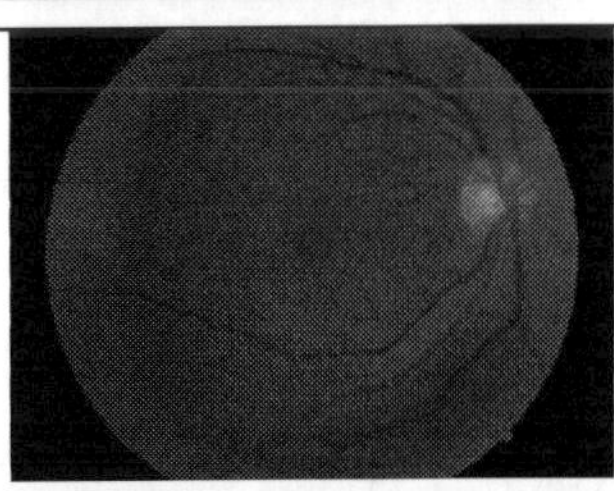	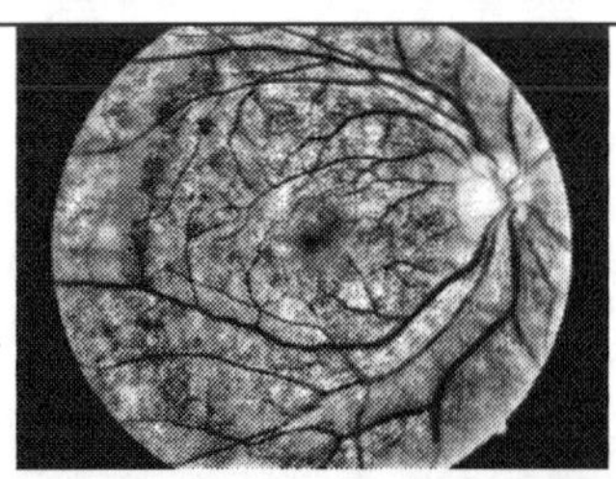	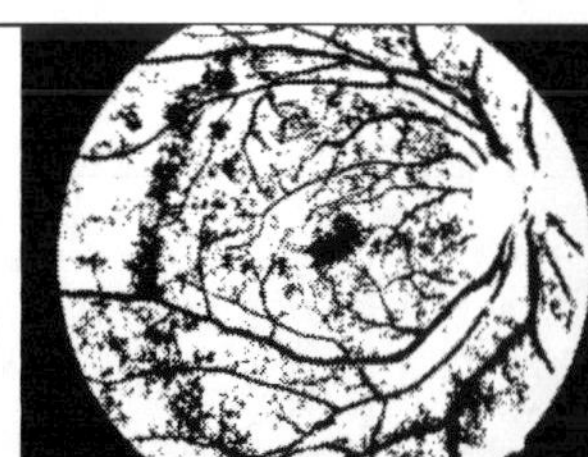	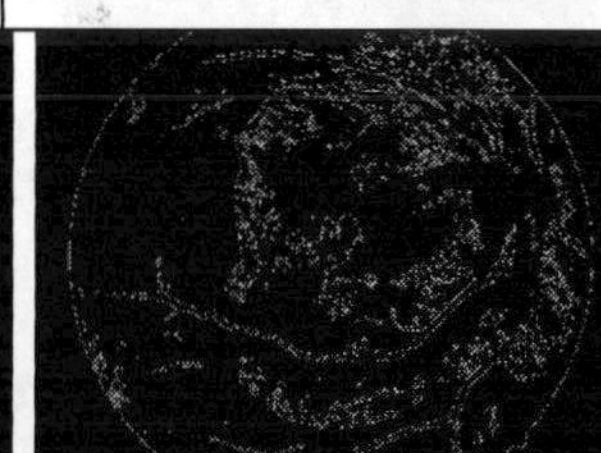
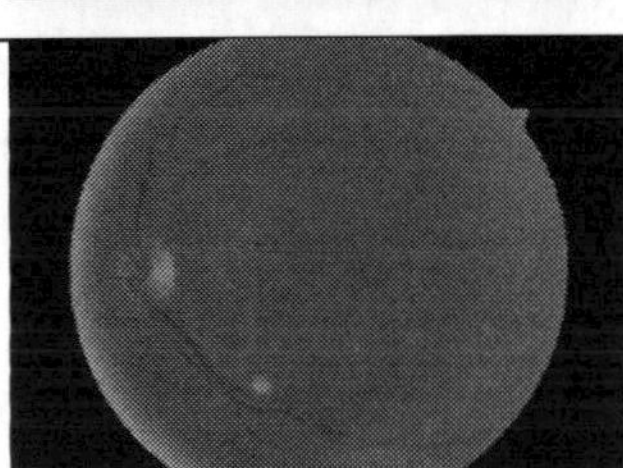	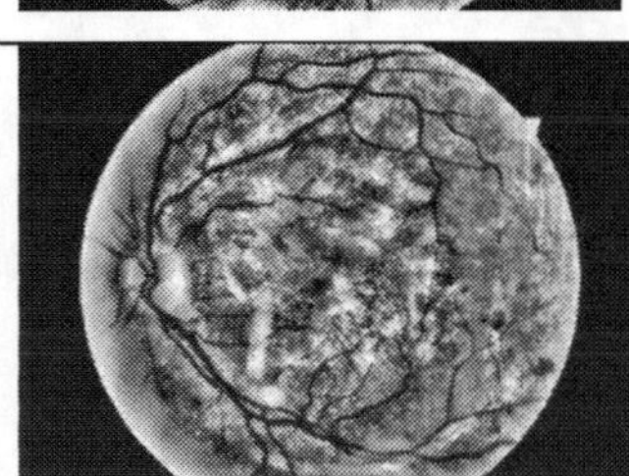	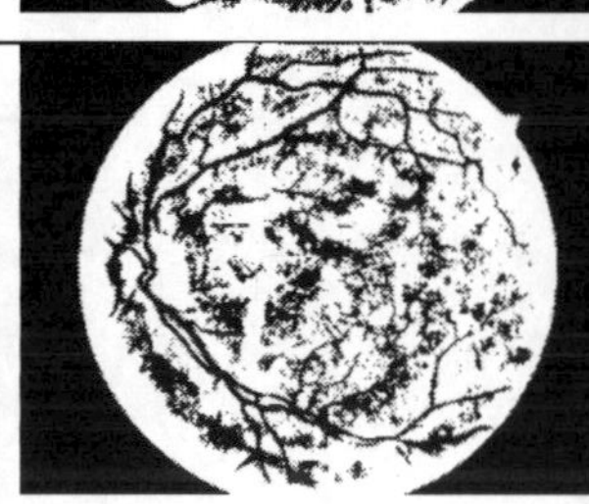	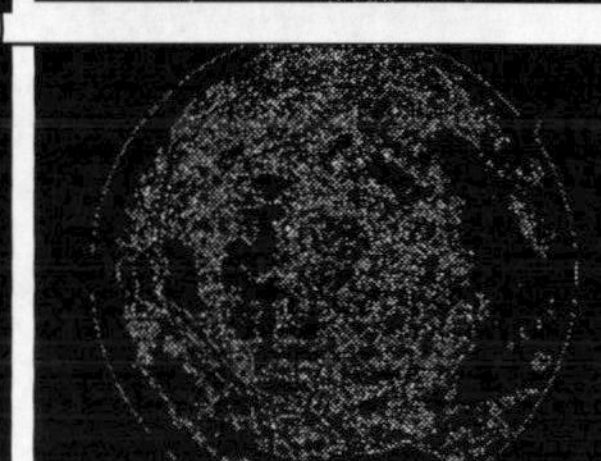

The received results demonstrate that the described approach provides rather high accuracy of marking of tanks and spots. However it is necessary to mark also some shortcomings of this approach. So, faint tanks can be lost as a result of binarization. Besides, in certain cases some spots can be marked as tanks. The specified defects can be eliminated by means of development of the algorithm allowing to select tanks on faint sections of the image.

IOP Conf. Series: Journal of Physics: Conf. Series **944** (2017) 012055 doi:10.1088/1742-6596/944/1/012055

4. Conclusion

The technique of preliminary image processing of a retina of an eye is offered. This technique includes algorithms: segmentations, conversions of the color image in gray-scale, filterings, increases of contrast, improving of boundary between a background and blood vessels, binarizations, and also markings of spots on an eye retina.

The advantage of this technique is accuracy when marking spots. Further development of an analysis algorithm of the selected spots, an assessment of their sizes, quantities is planned. The algorithm will allow to make decisions on existence or not at the patient of a diabetic retinopathy, and also to evaluate its level.

References

[1] Marwan D S 2011 An Automated Blood Vessel Segmentation Algorithm Using Histogram Equalization and Automatic Threshold Selection *Journal of Digital Imaging*, **24(4)** pp 564-57

[2] Siddalingaswamy P C, Gopalakrishna K P 2010 Automatic detection of multiple oriented blood vessels in retinal images *Biomedical Science and Engineering* **3** pp 101-107

[3] Sadykov S S, Savicheva S V 2012 Pretreatment images of flat objects in vision systems *Instrument* **2** pp 19-24

[4] Sadykov S S, Savicheva S V 2012 Evaluation methods recognition capabilitiestion of flat objects *Algorithms, methods, and data processing systems* **4 (22)** pp 110-119

[5] Sadykov S S, Savicheva S V 2011 Identification of the actual plane of objects based on a single feature of their points of external circuits *Information Technology* **8** pp 13-16

[6] Savicheva S V 2015 Recognition of single and overlay of objects on a conveyor belt *International Archives of the Photogrammetry, Remote Sensing and Spatial Information* pp 94-99.

The calculation of rectangular plates on elastic foundation the finite difference method

Komlev A.A., Makeev S.A.
Siberian state automobile and highway University Omsk, Russia

Abstract. The article describes the main advantages and disadvantages existing in the present time of calculation methods for plates on elastic Foundation. Consider automation of the calculation of rectangular plates on elastic basis by finite difference method, on the basis of which received automatic design algorithms. Conducted research of discretization on the accuracy of the calculations. The comparison of the results of strain and effort obtained by the finite element method and the proposed method.

1. Introduction

Despite the fact that currently there are a large number of software complexes for analysis of structures (beams, plates, shells), the bulk calculations are performed manually or by software based on the finite element method (FEM). However, it is not necessary to forget and about other methods of calculation. One such method is the finite differences method (MD).

This method allows to solve a wide range of tasks in the field of construction:

- allows to count the rods [1], rectangular [2 -5], round [6] of the plate as static loads and dynamic loads [7, 8];
- allows to define the stability of the plates [9] and cores;

allows you to set the vibration plates [10, 11];

-allows to make calculations taking into account the nonlinear characteristics of the Foundation [12].

As to the advantages of the method include its high accuracy [13]. However, the MD has its drawbacks, chief among which is the low automation of calculations. In contrast to the FEM software systems implemented in MD just yet. Therefore, all calculations are performed manually. However, it is worth noting that in recent years began to carried out the work on the automation of MD [14]. In these works were developed automated complex for calculation of rectangular plates implemented in a publicly accessible electronic mathematical environment of MS Excel.

This article considers the issue of automation of calculation of rectangular plates on elastic Foundation

2. Theory

The equation of the middle surface of a plate on elastic Foundation, loaded with distributed area load can be described by the differential equation of the following form [15]:

$$\nabla^2\nabla^2 v_{(x;y)} - \frac{C_2}{D}\nabla^2 v_{(x;y)} + \frac{C_1}{D} v_{(x;y)} = \frac{q}{D} \qquad (1)$$

where v is the deflection of an arbitrary point of the middle surface of the plate;

C_1 – the coefficient of compression (kN/m^3);

C_2 – coefficient shear (kN/m);

q – distributed load perpendicular to the middle plane of the plate (kg/m^2);

D – cylindrical stiffness of the plate;

x and y coordinates of the plates in the plan;

$$\nabla^2\nabla^2 v_{(x;y)} = \frac{\partial^4 v}{\partial x^4} + 2\frac{\partial^4 v}{\partial x^2 \partial y^2} + \frac{\partial^4 v}{\partial y^4} \qquad (2)$$

- biharmonic operator;

$$\nabla^2 v_{(x;y)} = \frac{\partial^2 v}{\partial x^2} + \frac{\partial^2 v}{\partial y^2} \qquad (3)$$

IOP Conf. Series: Journal of Physics: Conf. Series **944** (2017) 012056 doi:10.1088/1742-6596/944/1/012056

- Laplace operator.

After substituting (2) and (3) in (1) we get equation

$$\frac{\partial^4 v}{\partial x^4}+2\frac{\partial^4 v}{\partial x^2 \partial y^2}+\frac{\partial^4 v}{\partial y^4}-\frac{C_2}{D}\frac{\partial^2 v}{\partial x^2}-\frac{C_2}{D}\frac{\partial^2 v}{\partial y^2}+\frac{C_1}{D}v=\frac{q}{D} \quad (4)$$

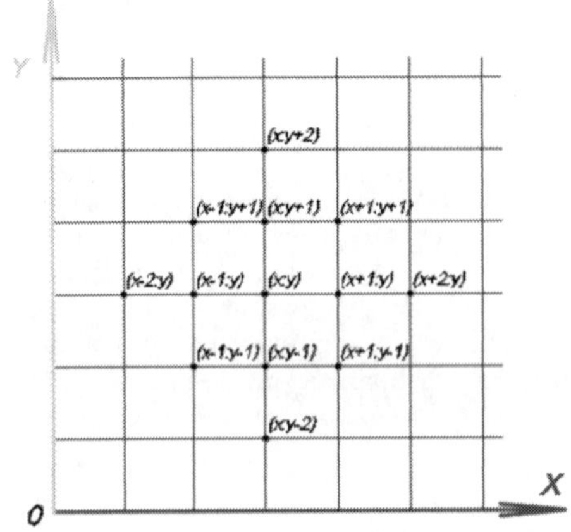

Figure 1. Numbering of the grid points of the plate

For further calculations equation (4) should be submitted in differential form (5), for this we use difference schemes of [16]:

$$\begin{aligned}&\left[6(\alpha+\frac{1}{\alpha})+8\right]v_{(x;y)}-4\left[(1+\alpha)(v_{(x-1;y)}+v_{(x+1;y)})+(1+\frac{1}{\alpha})(v_{(x;y+1)}+v_{(x;y-1)})\right]+\\&+2(v_{(x-1;y+1)}+v_{(x+1;y-1)}+v_{(x-1;y-1)}+v_{(x+1;y+1)})+\alpha(v_{(x-2;y)}+v_{(x+2;y)})+\frac{1}{\alpha}(v_{(x;y+2)}+v_{(x;y-2)})+\\&+\frac{2C_2 v_{(x;y)}(\Delta x^2+\Delta y^2)}{D}-\frac{C_2\Delta y^2(v_{(x-1;y)}+v_{(x+1;y)})}{D}-\frac{C_2\Delta x^2(v_{(x;y-1)}+v_{(x;y+1)})}{D}+\frac{C_1 v_{(x;y)}\Delta x^2\Delta y^2}{D}=\frac{\Delta x^2\Delta y^2}{D}q\end{aligned} \quad (5)$$

where $\alpha=\Delta y^2/\Delta x^2$

Displacements v at the edges of the plate (contour points) and edges (edge points) is determined by the conditions of support according to [17].

The resulting equation is cumbersome, so when manually calculating the designers go on a number of simplifications, most of which is the decision to make payments of a square grid when $\Delta x=\Delta y$. And the second plate is applied to a very large grid (in first approximation) when $\Delta x=\Delta y=L/4$. In the third plate count only two mutually perpendicular cross-sections. Only in this case the system of algebraic equations is easy to calculate. But in this case, neither of which the accuracy of the calculation of the speech cannot go.

The situation began to change with the advent of computers. When using the need for manual calculation anymore and the designers were able to increase the degree of sampling, thereby improving the accuracy of the calculations. However, to date, even when using a computer, on the plates break the mesh not more frequently than $\Delta x=\Delta y=L/8$ [18].

So FDM is currently used only for educational purposes.

Everything changes, however, if the calculations Platin FDM to available mathematical environment of MS Excel. A built-in feature of the iterative calculus allows to solve the system (5) with a substantial degree of sampling.

If the calculated plate spacing grid (Fig.1), expressing a from equation (5) the value of the deflection at point (x:y) and applying this expression to all points of the wafer, it is possible to obtain a system of interrelated equations.

The result is a field where each cell is interconnected with neighboring ones. And thus, formed between them a circular reference. Then, after you activate "Enable iterative calculation" movement in the plate are calculated automatically.
After calculating the movements of effort are calculated by the formulas given in [19].
However, the question of discretization remains open.

3. The results of the calculations

To determine the impact of discretization on the accuracy of the calculations the calculations of the plate lying on the elastic basis for the action of concentrated load FDM and MKE (Lira) with the step of:
- $\Delta x = Lx/10$ and $\Delta y = Ly/10$;
- $\Delta x = Lx/15$ and $\Delta y = Ly/15$;
- $\Delta x = Lx/20$ and $\Delta y = Ly/20$.

Table 1 shows the comparative analysis of the results of calculation of reinforced concrete plate with dimensions in plan 6 × 3 m on an elastic Foundation with a thickness of 200 mm at a point load 1000 kN in the middle of the span in the mathematical environment of MS Excel, and finite element method in Lira. Picture of plate deflections and the bending moments shown in Fig. 2, 3, 4.

Table 1. The results of the calculation

	FDM - Excel			FEM - Lira CAD		
	10x10	15x15	20x20	10x10	15x15	20x20
V_{max}, mm	19.1	18.9	18.6	19.2	18.8	19.0
*M_x, кN*m/m	308.9	225.3	228.2	184.7	267.7	226.9
*M_y, кN*m/m	273.3	182.2	181.5	136.5	205.4	174.7
*M_{xy}, кN*m/m	18.53	18.91	18.56	22.30	24.20	23.50

*calculated values in the table are modulo

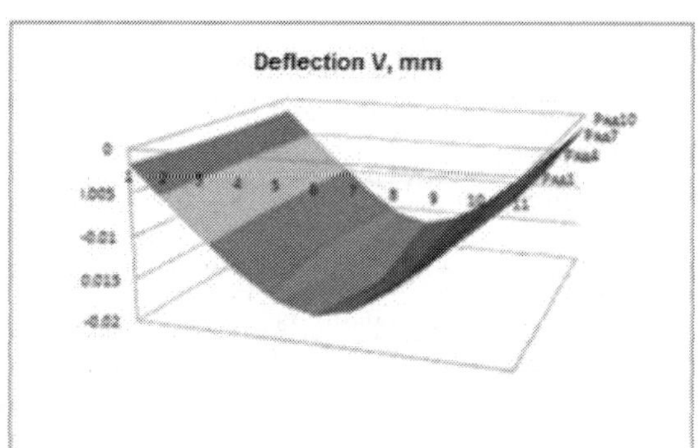

Figure 2. Plot vertical displacement of plate V

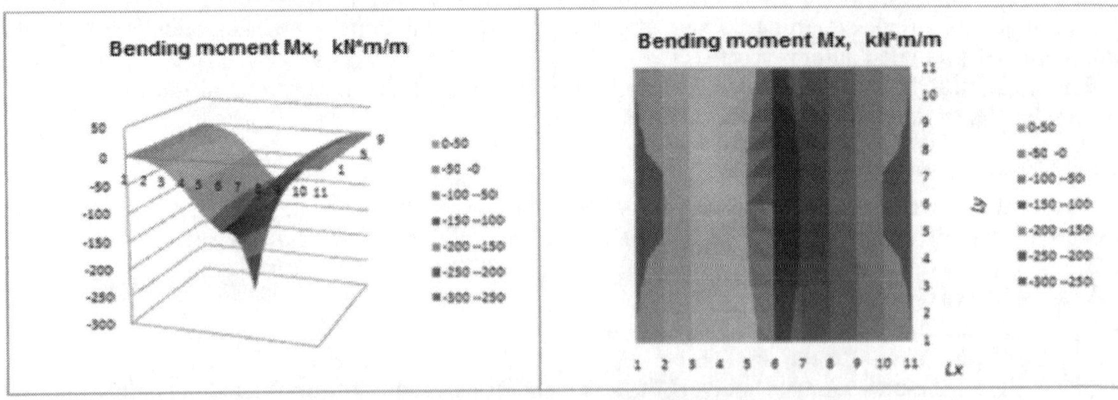

Figure 3. Plot the bending moments *Mx*

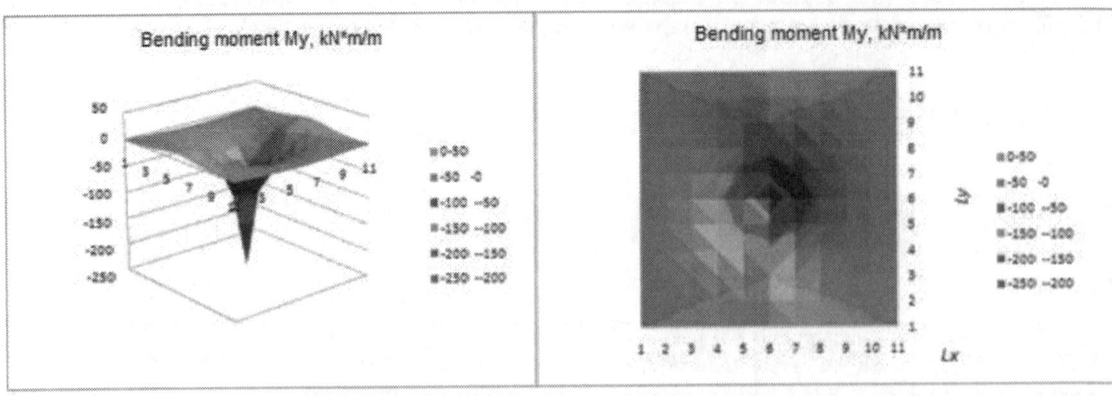

Figure 4. Plot the bending moments *My*

From table 1 it follows that in the calculation of MCR under discretization element more than L/15 the values of the deflections are comparable, and the difference between values of effort does not exceed 2%. In the calculation of the FEM the convergence of the internal forces are not observed. This picture for FEM due to the fact that in him the value of effort is determined by the center of gravity of the elements. And therefore, FEM is more dependent on sampling than FDM.

Next, we conducted a series of comparative calculations in the FDM and FEM plates with various geometry parameters, the reasons and types of loads. Table 2 shows comparative analysis of the results of calculating the maximum deflection for various plates on elastic Foundation. Graphical results of calculated deflections of plates is presented in Fig. 5

Table 2. Comparative analysis the maximum deflections in plates on elastic foundation

№	Deflection	Lira CAD	MS Excel	%	Options
1	V_{max}, mm	10	10	0	Concrete slab 3x3 m with a thickness of 250 mm C1=5000 kN/m3 C2=500 kN/m The uniformly distributed load q=50 kN/m^2
2	V_{max}, mm	12.1	11.94	1.3	Concrete slab 4 x 2 m thickness 100 mm C1=10000 kN/m^3

					C2=1000 kN/m Load strip in the middle of the span along the long side of q=200 kN/m
3	V_{max}, mm	9.24	9.43	2	Concrete slab 5x5 m with a thickness of 150 mm S1=9000 kN/m³ C2=900 kN/m Load in the form of four concentrated forces at the corners of the plate R=150x4 kN
4	V_{max}, mm	3.8	3.91	2.8	Concrete slab 6x2 m and a thickness of 200 mm S1=8000 kN/m³ C2=800 kN/m Load strip in the middle of the span along the short side q=100 kN/m
5	V_{max}, mm	6.76	6.51	1.7	Concrete slab 5x1 m with a thickness of 50 mm C1=7000 kN/m³ C2=700 kN/m Load in the form of three concentrated forces along the length of the plate P=50x3 kN

A comparative analysis of the calculations, the discrepancy of the values of the maximum vertical deflection obtained by FEM and MD are no different. The difference between values of bending moments does not exceed 3%.

4. Conclusion

Calculation of plates on elastic Foundation MD in the electronic environment MS Excel showed his versatility. It allows the calculation of rectangular plates of arbitrary size with the various soil parameters for different types of loads. Also this method is easy to implement the calculation of plates with variable rigidity on elastic basis. MD showed the best convergence in comparison with FEM. However, MD has weaknesses. The studies revealed that MD considers incorrect movements in plates under load located on the edge. This issue requires further study.

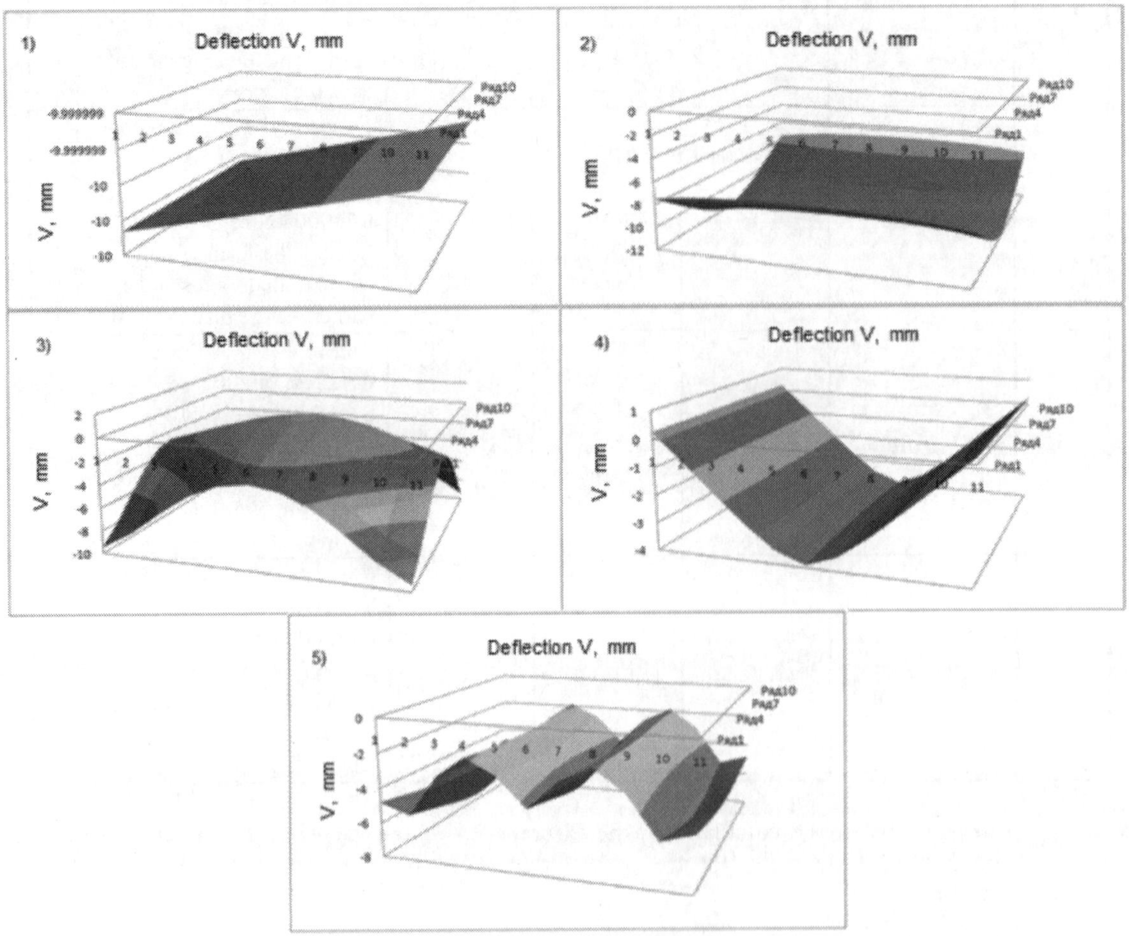

Figure 5. Plot vertical displacement V

References

[1] M. Kamiński, "A generalized version of the perturbation-based stochastic finite difference method for elastic beams" Journal of Theoretical and Applied Mechanics, Warsaw, 2009, No 4, pp. 957-975.

[2] S. P. Timoshengko and S. Woinowsky-Krieger, "Theory of Plates and Shells," 2nd Edition, McGraw-Hill, New York, 1959.

[3] Cheng Xiang-sheng "A free rectangular plate on elastic foundation", Applied Mathematics and Mechanics (English Edition, Vol. 13, No. 10, Oct. 1992)

[4] Karasin A, and Gülkan P, An approximate finite grid solution for plates on elastic foundations, Turkish Journal of Chamber of Civil Engineers, 2008, 19: pp. 4445-4454.

[5] Cheng Xiang-sheng, "A free rectangular plate on elastic foundation" A pplied Mathematics and Mechanics, 1992 No. 10, Vol. 13, pp. 977-982 (English Edition).

[6] A. Karasin, G. Aktas, and P. Gülkan, A finite grid solution for circular plates on elastic foundations, KSCE Journal of Civil Engineering, 2015, 19(4), pp. 1157-1163.

[7] R.F. Gabbasov, T.A. Hoang, M.A. Chikunov, 'Generalized equations of finite difference method in problems of calculation of thin bending plates dynamic load," Vestnik MGSU, 2014, No. 9, pp. 32-38.

[8] Aksu G. & Ali R. Free vibration analysis of stiffened plates using finite-difference method. J. Sound Vibr. 48, 15–25 (1976).
[9] A.V. Lopatin, R.V. Avakumov, 'Stability of orthotropic plates with two free edges, loaded by a bending moment in the plane," Bulletin of Siberian state aerospace University academician M. F. Reshetnev, 2009, No. 4 (25), pp. 28-33.
[10] R.F. Gabbasov, N.B. Uvarov, M.V. Alexander, "Numerical solution of the problem of natural vibrations of bending of orthotropic plates," Industrial and civil construction, 2015, No. 1, pp. 37-39.
[11] R.F. Gabbasov, M. Sali, V.V. Filatov, V. S. Naumov, "To the calculation of beams of variable rigidity on the forced oscillations," Natural and technical Sciences, 2015, No.11, pp. 656-659.
[12] M. Hiroshi, S. Takeshi "Bending analysis of rectangular plate on non-uniform elastic foundations" Japan Society of Civil Engineers Vol. 4, No.1, 1987, pp. 77-85.
[13] R.F. Gabbasov, H.T. Anh, N.X. Anh, "Comparison of results of calculation of thin bending plates using the generalized equations of finite difference method and successive approximations," Industrial and civil construction, 2014, No.1, pp. 62-64.
[14] A.A. Komlev, S.A. Makeev, "Automation of calculation of rectangular plates by the finite difference method," Architectural-construction and road-transport complexes: problems, prospects, innovations, International scientific and practical conference: Electronic resource, 2016, pp. 157-161.
[15] P.L. Pasternak, "Fundamentals of a new calculation method of foundations on elastic foundation by means of two coefficients pastels," Moscov, State publishing house of literature on construction and architecture, 1954 p.56
[16] A. A. Umansky, "Directory of designer industrial, residential and public buildings". Theoretical, In two books, vol.2, 2rd ed., Moscow, Stroiizdat,1973, p.416
[17] S.N. Klepikov, "Calculation of structures on elastic Foundation," / Kiev, Budulinek, p.184
[18] V.V. Kuvatov, "Calculation of plates on the strength and stability of the grid method," Sverdlovsk, UPI Edition, 1972, p.106
[19] P.M. Varvak, L. P. Varvak, The method of grids in problems of design of structures, Moscow, Stroiizdat, 1977, p.160

Hydroelastic Oscillations of a Circular Plate, Resting on Winkler Foundation

D V Kondratov[1], L I Mogilevich[2], V S Popov[2] and A A Popova[2]

[1]Russian Presidential Academy of National Economy and Public Administration, 23/25, Sobornaya St., Saratov, 410031, Russia
[2]Yuri Gagarin State Technical University of Saratov, 77 Politechnicheskaya St., Saratov, 410054, Russia

e-mail: vic_p@bk.ru

Abstract. The forced hydroelastic oscillations of a circular plate resting on elastic foundation are investigated. The oscillations are caused by a stamp vibration under interaction with a plate through a thin layer of viscous incompressible liquid. The axis-symmetric problem for the regime of the steady-state harmonic oscillations is considered. On the basis of hydroelasticity problem solution the laws of plate deflection and pressure in the liquid are found. The functions of the amplitudes deflection distribution and liquid pressure along the plate are constructed. The presented mathematical model provides for investigating viscous liquid layer interaction dynamics with a circular plate resting on an elastic foundation. The above-mentioned model makes it possible to define the plate oscillations resonance frequencies and the corresponding amplitudes of deflection and liquid pressure, as well. Keywords – circular plate; hydroelastic oscillations; vibration stamp; viscous liquid; Winkler foundation

1. Introduction

The investigations of the plates and supporting elastic foundations interaction are of theoretical, as well as of practical interest. For example, the review [1], devoted to this subject matter, considers the items of elastic foundations development models, as well as various approaches to analytical and numerical investigation of beams and plates with elastic foundations interaction. Reference [2] studies free and forced oscillations of a circular plate, the plate resting on Winkler foundation. The investigation is carried out in axis-symmetric setting for the cases of a simply supported edge and a clamped one. Reference [3] studies stability and oscillations of circular polar orthotropic plates of variable thickness resting on Winkler foundation. The cases of the simply supported and clamped edge of the plates are considered. Reference [4] deals with investigation of oscillations of a thin circular plate with elastic fixation along the contour, the plate resting on Winkler elastic foundation. Reference [5] studies free oscillations of thin circular plate on Winkler foundation. The axis-symmetric problem under the condition of foundation rigidity change in radial direction is investigated in the above-mentioned reference. References [6, 7] consider oscillations and stability of multi-layered circular plates on elastic foundation under the impact of local and distributed loadings of various nature. The Winkler and Pasternak models are used for foundation reactions modelling. On the other hand, the investigations of hydroelastic interaction of circular plates with an ideal liquid are of importance, too. For example, reference [8] is one of the first publications on investigation of natural oscillation of a circular plate interacting with an ideal liquid. The study of a circular plate vibration on a free surface of an ideal incompressible liquid is made in [9]. The problem is considered for the case, where liquid volume is limited by a rigid bottom and cylinder surface. Reference [10] studies the oscillations of a circular plate, plunged into an ideal incompressible liquid with a free surface in rigid cylinder. Mathematical modeling and a series of experimental investigations are carried out by the authors of this paper. The damping qualities of the liquid, conditioned by its viscosity are not considered in the references mentioned above. The problem of vibrating discs interaction with viscous incompressible liquid layer between them is solved in [11]. The present paper studies the case, where the discs are considered to be rigid and the case with one of the discs being elastic. The analogous problem in a two-dimensional setting for two plates of finite sizes is solved in [12]. The investigation of a beam hydroelastic oscillations in a viscous liquid flow with an application to piezoelectric elements, aimed at obtaining energy from the flow, is made in [13]. The problem of bending hydroelastic oscillations of the plate forming the narrow channel wall under the impact of viscous liquid pulsating layer is solved in [14]. The forced hydroelastic oscillations of a three-layered circular plate, interacting with viscous incompressible liquid layer under channel foundation vibration, are investigated in [15].

However, evaluating foundation elasticity impact on hydroelastic oscillations is of theoretical and practical interest. Membrane oscillations on Winkler elastic foundation, the membrane situated on the reservoir bottom full of ideal incompressible liquid with a free surface, are investigated in [16]. References [17–19] consider hydroelastic oscillations of rectangular plates, resting on Pasternak foundation and interacting with an ideal incompressible liquid with a free surface. The investigation of rectangular plate, resting on Winkler elastic foundation and interacting with a pulsating layer of viscous incompressible liquid is carried out in [20–23]. Nevertheless, the case of hydroelastic oscillations of a circular plate on an elastic foundation is not considered in the investigations mentioned.

2. Stetting the problem

Let us consider the forced oscillations of a circular plate ***1***, resting on Winkler foundation (see fig. 1). The plate oscillations are caused by its interaction with a vibrating circular stamp ***2*** through viscous liquid ***3***. We will consider the axis-symmetric problem. We bind cylindrical coordinate system $Or\varphi z$ with the medium surface center of the plate in an unperturbed state. Let us consider that the stamp movement takes place according to a set harmonic law in a vertical plane. The plate has thickness h_0, radius R. It is rested on Winkler foundation and clamped at the edge. The viscous liquid fills fully a narrow channel, formed by the plate and stamp, the liquid layer medium thickness in the channel being δ_0. We think that $\delta_0 << R$ and the stamp oscillations amplitude are considerably less than δ_0. The liquid on the plate edge freely leaks into the same liquid with a constant pressure p_0. The presence of viscous liquid between the plate and the stamp leads to quick going out of transitional processes. Therefore, the initial processes impact can be excluded from the very beginning of the investigation, according [24]. Thus, further we will study the forced steady-state harmonic oscillations.

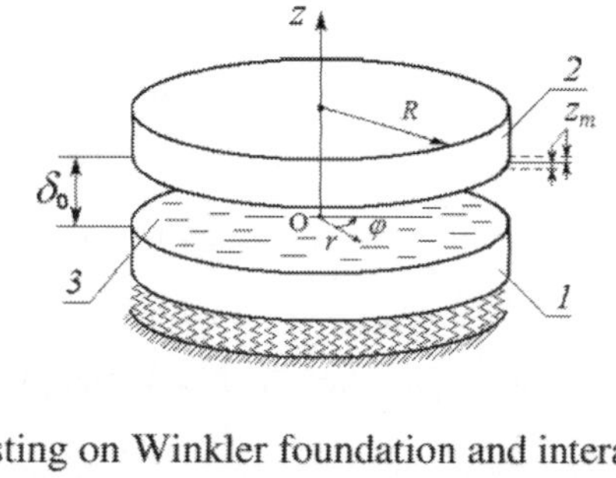

Figure 1. The circular plate **1**, resting on Winkler foundation and interacting with vibrating circular stamp **2** through viscous liquid **3**.

We will present the stamp movement law in the form of:

$$z = z_m f(\omega t),\ \ f(\omega t) = \sin \omega t, \tag{1}$$

where z_m is the stamp oscillations amplitude, ω is the frequency, t is the time.

The liquid movement in a narrow split between the stamp and the plate can be considered to be creeping in [25]. In this case viscous liquid dynamics equations take the form of:

$$\begin{gathered}\frac{1}{\rho}\frac{\partial p}{\partial r} = \nu\left(\frac{\partial^2 u_r}{\partial r^2} + \frac{1}{r}\frac{\partial u_r}{\partial r} + \frac{\partial^2 u_r}{\partial z^2} - \frac{u_r}{r^2}\right),\\ \frac{1}{\rho}\frac{\partial p}{\partial z} = \nu\left(\frac{\partial^2 u_z}{\partial r^2} + \frac{1}{r}\frac{\partial u_z}{\partial r} + \frac{\partial^2 u_z}{\partial z^2}\right),\\ \frac{\partial u_r}{\partial r} + \frac{1}{r}u_r + \frac{\partial u_z}{\partial z} = 0,\end{gathered} \tag{2}$$

where r, z are cylindrical coordinates; u_r, u_z are liquid velocity vector projections on a coordinate axis; p is the pressure; ρ is density of liquid; ν is the kinematic viscosity coefficient of liquid.

The boundary conditions of equation (2) are the ones of liquid movements velocity and a plate velocity shift coincidence:

$$\begin{gathered}u_x = \partial u/\partial t,\ u_z = \partial w/\partial t \text{ at } z = h_0/2 + w,\\ u_x = 0,\ u_z = 0 \text{ at } z = h_0/2 + \delta_0 + z_m f(\omega t),\end{gathered} \tag{3}$$

and also the conditions for the pressure limit on symmetry axis, as well as the ones for the pressure on edge:

$$r\,\partial p/\partial r = 0 \text{ at } r = 0,\ \ p = p_0 \text{ at } r = R. \tag{4}$$

Here u, w are the plate radial shifts an deflection.

The equations of the circular plate oscillations resting on Winkler foundation can be written down, as [1, 6]:

$$D\left(\frac{\partial^4 w}{\partial r^4} + \frac{2}{r}\frac{\partial^3 w}{\partial r^3} - \frac{1}{r^2}\frac{\partial^2 w}{\partial r^2} + \frac{1}{r^3}\frac{\partial w}{\partial r}\right) + \kappa w + \rho_0 h_0 \frac{\partial^2 w}{\partial t^2} = q_{zz}, \tag{5}$$

where D is the flexural rigidity of the plate, h_0 is the plate thickness, ρ_0 is the plate material density, κ is the foundation modulus, $q_{zz} = -p + 2\rho\nu\,\partial u_z/\partial z$ is the normal stress in a liquid layer.

The boundary conditions of (5) take the form of:

$$r\,\partial w/\partial r = 0 \text{ at } r = 0,\ \ w = \partial w/\partial x = 0 \text{ at } r = R. \tag{6}$$

3. The Theory

Let us introduce dimensionless variables and small parameters into the consideration:

$$\begin{gathered}\xi = r/R,\ \zeta = (z - 0{,}5h_1)/\delta_0,\ w = w_m W,\ u = u_m U,\\ u_z = z_m\omega U_\zeta,\ u_x = (z_m\omega/\psi)U_\xi,\ p = p_0 + P\,\nu\rho z_m\omega\psi^{-2}/\delta_0,\\ \tau = \omega t,\ \lambda = z_m/\delta_0 << 1,\ \psi = \delta_0/R << 1.\end{gathered} \tag{7}$$

While taking into account (7), hydromechanics problem in a dimensionless state in a zero approximation in ψ will be written down as:

$$\frac{\partial P}{\partial \xi}=\frac{\partial^2 U_\xi}{\partial \zeta^2},\quad \frac{\partial P}{\partial \zeta}=0,\quad \frac{\partial U_\xi}{\partial \xi}+\frac{1}{\xi}U_\xi+\frac{\partial U_\zeta}{\partial \zeta}=0 \tag{8}$$

With boundary conditions (with accuracy up to ψ and λ):

$$\begin{aligned} &U_\xi=0, U_\zeta=df(\tau)/d\tau \text{ at } \zeta=1,\\ &U_\xi=0, U_\zeta=(w_m/z_m)\partial W/\partial\tau \text{ at } \zeta=0,\\ &P=0 \; at \; \xi=1, \xi\,\partial P/\partial\xi=0 \text{ at } \xi=0. \end{aligned} \tag{9}$$

Normal stress in the liquid with consideration of variables (7) in zero approximation in ψ takes the form of

$$q_{zz}=-p_0-\rho v z_m\omega(\delta_0\psi^2)^{-1}P. \tag{10}$$

Further, by solving (8) with (9), we get:

$$U_\xi=\frac{\zeta^2-\zeta}{2}\frac{\partial P}{\partial \xi},\; U_\zeta=\frac{\partial W}{\partial \tau}-\frac{2\zeta^3-3\zeta^2}{12}\left(\frac{1}{\xi}\frac{\partial P}{\partial \xi}+\frac{\partial^2 P}{\partial \xi^2}\right), \tag{11}$$

$$P=3(\xi^2-1)\frac{df}{d\tau}+12\frac{w_m}{z_m}\int_\xi^1\frac{1}{\xi}\left[\int_0^\xi \xi\frac{\partial W}{\partial \tau}d\xi\right]d\xi, \tag{12}$$

We present the shape of the plate deflection as the series in eigenfunctions of Sturm-Loiville problem:

$$W=\sum_{k=1}^{\infty}(R_k^0+R_k(\tau))\left[J_0(\beta_k\xi)/J_0(\beta_k)-I_0(\beta_k\xi)/I_0(\beta_k)\right], \tag{13}$$

here $R_k(\tau)$ is the time harmonic function, R_k^0 is the constant, J_0 is the Bessel function of the first kind; I_0 is the modified Bessel function of the first kind; β_k is the root of transcendental equation $I_1(\beta_k)/I_0(\beta_k) = -J_1(\beta_k)/J_0(\beta_k)$ (k = 1, 2,…), $J_1(\beta_k)$, $I_1(\beta_k)$ are Bessel function and modified Bessel function of the first kind.

Dimensionless pressure (12) with consideration the plate deflection (13), takes the form:

$$P=3(\xi^2-1)\frac{df}{d\tau}+\frac{w_m}{z_m}\sum_{k=1}^{\infty}\frac{12}{\beta_k^2}\frac{dR_k}{d\tau}\left[\frac{J_0(\beta_k\xi)}{J_0(\beta_k)}+\frac{I_0(\beta_k\xi)}{I_0(\beta_k)}-2\right]. \tag{14}$$

By substituting (14) in (10) and expanding in the series in eigenfunctions of the Sturm-Loiville problem, we get:

$$\begin{aligned} q_{zz}=&-p_0\sum_{k=1}^{\infty}\frac{2}{\beta_k}\frac{J_1(\beta_k)}{J_0(\beta_k)}\left[\frac{J_0(\beta_k\xi)}{J_0(\beta_k)}-\frac{I_0(\beta_k\xi)}{I_0(\beta_k)}\right]-\\ &-\frac{\rho v w_m\omega}{\delta_0\psi^2}\left[12\frac{df}{d\tau}\sum_{k=1}^{\infty}\frac{1}{\beta_k^2}\left[\frac{J_0(\beta_k\xi)}{J_0(\beta_k)}-\frac{I_0(\beta_k\xi)}{I_0(\beta_k)}\right]+\right. \end{aligned} \tag{15}$$

$$+\frac{w_m}{z_m}\sum_{k=1}^{\infty}\left\{\sum_{i=1}^{k-1}\frac{48}{(\beta_i^4-\beta_k^4)\beta_k\beta_i}\left(\beta_k^3\frac{J_1(\beta_i)}{J_0(\beta_i)}-\beta_i^3\frac{J_1(\beta_k)}{J_0(\beta_k)}\right)\left[\frac{J_0(\beta_i\xi)}{J_0(\beta_i)}-\right.\right.$$

$$\left.-\frac{I_0(\beta_i\xi)}{I_0(\beta_i)}\right]\frac{dR_k}{d\tau}+\frac{12}{\beta_k^2}\left(\frac{J_1^2(\beta_k)}{J_0^2(\beta_k)}-\frac{4}{\beta_k}\frac{J_1(\beta_k)}{J_0(\beta_k)}\right)\left[\frac{J_0(\beta_k\xi)}{J_0(\beta_k)}-\right.$$

$$\left.-\frac{I_0(\beta_k\xi)}{I_0(\beta_k)}\right]\frac{dR_k}{d\tau}+\sum_{i=k+1}^{\infty}\frac{48}{(\beta_i^4-\beta_k^4)\beta_k\beta_i}\left(\beta_k^3\frac{J_1(\beta_i)}{J_0(\beta_i)}-\right.$$

$$\left.\left.-\beta_i^3\frac{J_1(\beta_k)}{J_0(\beta_k)}\right)\left[\frac{J_0(\beta_i\xi)}{J_0(\beta_i)}-\frac{I_0(\beta_i\xi)}{I_0(\beta_i)}\right]\frac{dR_k}{d\tau}\right\}\right].$$

Further, we consider that $k = 1,2,\ldots,n$, $i = 1,2,\ldots,n$. By substituting (13), (15) in (5) and equaling the coefficients at identical eigenfunctions of the Sturm-Loiville problem in the obtained equations, we define the expressions for R_k^0:

$$R_k^0=-p_0\frac{2}{w_m\beta_k(\kappa+D\,\beta_k^4/R^4)}\frac{J_1(\beta_k)}{J_0(\beta_k)},\tag{16}$$

and the system of linear ordinary differential equations in time to define $R_k(\tau)$:

$$w_m\left(\frac{D\beta_k^4}{R^4}+\kappa\right)R_1(\tau)+\rho_0h_0w_m\omega^2\frac{d^2R_1(\tau)}{d\tau^2}=$$

$$=-\frac{\rho\nu z_m\omega}{\delta_0\psi^2}\left\{\frac{12}{\beta_1^2}\frac{df}{d\tau}+\frac{w_m}{z_m}\frac{12}{\beta_1^2}\frac{dR_1(\tau)}{d\tau}\left[\frac{J_1^2(\beta_1)}{J_0^2(\beta_1)}-\frac{4}{\beta_1}\frac{J_1(\beta_1)}{J_0(\beta_1)}\right]+\right.$$

$$\left.+\frac{w_m}{z_m}\sum_{i=2}^{n}\frac{48}{(\beta_i^4-\beta_1^4)\beta_1\beta_i}\left[\beta_1^3\frac{J_1(\beta_i)}{J_0(\beta_i)}-\beta_i^3\frac{J_1(\beta_1)}{J_0(\beta_1)}\right]\frac{dR_i(\tau)}{d\tau}\right\},$$

....

$$w_m\left(\frac{D\beta_\ell^4}{R^4}+\kappa\right)R_\ell(\tau)+\rho_0h_0w_m\omega^2\frac{d^2R_\ell(\tau)}{d\tau^2}=\tag{17}$$

$$=-\frac{\rho\nu z_m\omega}{\delta_0\psi^2}\left\{\frac{12}{\beta_\ell^2}\frac{df}{d\tau}+\frac{w_m}{z_m}\sum_{i=1}^{\ell-1}\frac{48}{(\beta_i^4-\beta_\ell^4)\beta_\ell\beta_i}\left[\beta_\ell^3\frac{J_1(\beta_i)}{J_0(\beta_i)}-\right.\right.$$

$$\left.-\beta_i^3\frac{J_1(\beta_\ell)}{J_0(\beta_\ell)}\right]\frac{dR_i(\tau)}{d\tau}+\frac{w_m}{z_m}\frac{12}{\beta_\ell^2}\left[\frac{J_1^2(\beta_\ell)}{J_0^2(\beta_\ell)}-\frac{4}{\beta_\ell}\frac{J_1(\beta_\ell)}{J_0(\beta_\ell)}\right]\frac{dR_\ell(\tau)}{d\tau}+$$

$$\left.+\frac{w_m}{z_m}\sum_{i=\ell+1}^{n}\frac{48}{(\beta_i^4-\beta_\ell^4)\beta_\ell\beta_i}\left[\beta_\ell^3\frac{J_1(\beta_i)}{J_0(\beta_i)}-\beta_i^3\frac{J_1(\beta_\ell)}{J_0(\beta_\ell)}\right]\frac{dR_i(\tau)}{d\tau}\right\},$$

....

$$w_m\left(\frac{D\beta_n^4}{R^4}+\kappa\right)R_n(\tau)+\rho_0h_0w_m\omega^2\frac{d^2R_n(\tau)}{d\tau^2}=-\frac{\rho\nu z_m\omega}{\delta_0\psi^2}\times$$

$$\times\left\{\frac{12}{\beta_n^2}\frac{df}{d\tau}+\frac{w_m}{z_m}\sum_{i=1}^{n-1}\frac{48}{(\beta_i^4-\beta_n^4)\beta_n\beta_i}\left[\beta_n^3\frac{J_1(\beta_i)}{J_0(\beta_i)}-\beta_i^3\frac{J_1(\beta_n)}{J_0(\beta_n)}\right]\frac{dR_i}{d\tau}+\right.$$

$$+\frac{w_m}{z_m}\sum_{i=1}^{n-1}\frac{48}{(\beta_i^4-\beta_n^4)\beta_n\beta_i}\left[\beta_n^3\frac{J_1(\beta_i)}{J_0(\beta_i)}-\beta_i^3\frac{J_1(\beta_n)}{J_0(\beta_n)}\right]\frac{dR_i}{d\tau}+$$

$$+\frac{w_m}{z_m}\frac{12}{\beta_n^2}\left[\frac{J_1^2(\beta_n)}{J_0^2(\beta_n)}-\frac{4}{\beta_n}\frac{J_1(\beta_n)}{J_0(\beta_n)}\right]\frac{dR_n(\tau)}{d\tau}\Bigg\}.$$

Therefore, bearing in mind the quantity of the remaining series members, we can get unknown coefficients R_k^0 and by solving (17), we can find functions $R_k(\tau)$, and define the bending oscillations of a circular plate. Let us consider one-member disintegration as an example. In this case we get the equation for defining $R_1(\tau)$ from (17):

$$\begin{aligned} &w_m\left(\frac{D\beta_1^4}{R^4}+\kappa\right)R_1(\tau)+\rho_0 h_0 w_m\omega^2\frac{d^2R_1(\tau)}{d\tau^2}= \\ &=-\frac{\rho\nu z_m\omega}{\delta_0\psi^2}\frac{12}{\beta_1^2}\left(\frac{df}{d\tau}+\frac{w_m}{z_m}\frac{dR_1(\tau)}{d\tau}\left[\frac{J_1^2(\beta_1)}{J_0^2(\beta_1)}-\frac{4}{\beta_1}\frac{J_1(\beta_1)}{J_0(\beta_1)}\right]\right), \end{aligned} \tag{18}$$

The solution of equation (18) for the regime of the steady-state harmonic oscillations takes the form:

$$\begin{aligned} w_m R_1 &= -z_m\left[\frac{K_1^z K_1^w\omega^2}{(\widetilde{D}_1-\rho_0 h_0\omega^2)^2+(K_1^w\omega)^2}f(\tau)+\right. \\ &\left.+\frac{K_1^z\omega(\widetilde{D}_1-\rho_0 h_0\omega^2)}{(\widetilde{D}_1-\rho_0 h_0\omega^2)^2+(K_1^w\omega)^2}\frac{df(\tau)}{d\tau}\right],\ \widetilde{D}_1=\frac{D\beta_1^4}{R^4}+\kappa, \end{aligned} \tag{19}$$

$$K_k^z=\frac{\rho\nu}{\delta_0\psi^2}\frac{12}{\beta_1^2},\quad K_k^w=\frac{\rho\nu}{\delta_0\psi^2}\frac{12}{\beta_1^2}\left[\frac{J_1^2(\beta_1)}{J_0^2(\beta_1)}-\frac{4}{\beta_1}\frac{J_1(\beta_1)}{J_0(\beta_1)}\right].$$

Finally, with consideration of (16), (19) the law of the plate hydroelastic oscillations caused by the stamp movement, takes the form of:

$$\begin{aligned} w &= -\frac{p_0}{\beta_1(\kappa+D\beta_1^4/R^4)}\frac{2J_1(\beta_1)}{J_0(\beta_1)}\left[\frac{J_0(\beta_1\xi)}{J_0(\beta_1)}-\frac{I_0(\beta_1\xi)}{I_0(\beta_1)}\right]- \\ &-z_m A(\xi,\omega)\sin(\omega t+\varphi(\xi,\omega)), \\ &A(\xi,\omega)=\sqrt{C^2+B^2},\ \varphi(\xi,\omega)=\mathrm{arctg}(B/C), \end{aligned} \tag{20}$$

$$C=\frac{K_1^z K_1^w\omega^2}{(\widetilde{D}_1-\rho_0 h_0\omega^2)^2+(K_1^w\omega)^2}\left[\frac{J_0(\beta_1\xi)}{J_0(\beta_1)}-\frac{I_0(\beta_1\xi)}{I_0(\beta_1)}\right],\ B=\frac{K_1^z\omega(\widetilde{D}_1-\rho_0 h_0\omega^2)}{(\widetilde{D}_1-\rho_0 h_0\omega^2)^2+(K_1^w\omega)^2}\left[\frac{J_0(\beta_1\xi)}{J_0(\beta_1)}-\frac{I_0(\beta_1\xi)}{I_0(\beta_1)}\right].$$

By substituting (19) in (14) with taking into account (7), the pressure in liquid layer between the plates and stamp can be written down as:

$$\begin{aligned} &p=p_0+z_m\Pi(\xi,\omega)\sin(\omega t+\varphi_p(\xi,\omega)), \\ &\Pi(\xi,\omega)=\sqrt{S^2+Q^2},\ \varphi_p(\xi,\omega)=\mathrm{arctg}(Q/S), \end{aligned} \tag{21}$$

$$Q=\frac{(\xi^2-1)}{4}\beta_1^2K_1^z\omega-\frac{\rho\nu}{\delta_0\psi^2}\frac{12}{\beta_1^2}\left[\frac{J_0(\beta_1\xi)}{J_0(\beta_1)}+\frac{I_0(\beta_1\xi)}{I_0(\beta_1)}-2\right]C\omega,\ S=12\frac{\rho\nu}{\delta_0\psi^2}\frac{1}{\beta_1^2}\left[\frac{J_0(\beta_1\xi)}{J_0(\beta_1)}+\frac{I_0(\beta_1\xi)}{I_0(\beta_1)}-2\right]B\omega.$$

4. Summary and Conclusion

The investigation, carried out above, allows making the following conclusions. The mathematical model, allowing to investigate a circular plate hydroelastic oscillations, caused by stamp vibration, is carried out. The model takes into account the impact of Winkler foundation with a plate resting on it and the presence of viscous liquid layer between the plate and stamp. The expressions for the plate deflection and pressure in the liquid in the case of preserving one member of the series in (13) are obtained. However, it is possible to take into account the subsequent members of the series within the framework of the proposed model. To do this, it is necessary to use expression (16) and to define $R_1(\tau)$, $i = 1,2,\ldots,n$ out of (17), and then to write down the corresponding expressions for the plate deflection and the pressure. According to (20) the amplitude of dynamic deflection is defined by the function $A(\xi,\omega)$. This function must be considered as frequency dependent function of deflection amplitudes distribution along the channel. The amplitude of liquid dynamic pressure in the channel is defined by the function $\Pi(\xi,\omega)$. The functions $\varphi(\xi,\omega)$, $\varphi_p(\xi,\omega)$ are frequency dependent functions of phase shift distribution of the plate deflection and the pressure along the channel, accordingly. The investigation of the functions, mentioned above, makes it possible to study dynamic processes in the considered oscillation system. The analysis of the obtained expressions for the plate deflection (20) and pressure (21) allows to make a conclusion, that the foundation rigidity influences the plate static deflection and dynamic deflection amplitude and pressure in the liquid, as well. The obtained solution provides for transition from Winkler foundation to the plate with a clamped edge. To do this, it is necessary to equal the foundation modulus to zero.

Thus, the obtained results can be used for mathematical modeling and analysis of hydroelastic oscillations of circular plates, resting on elastic foundations. The presented model can be used to develop the methods of nondestructive testing of elastic constructions, resting on an elastic foundation and interacting with viscous liquid by their forced oscillations parameters.

References

[1] Wang Y H, Tham L G, Cheung Y K 2005 Beams and plates on elastic foundations: A review *Progress in Structural Engineering and Materials* **7**(4) pp 174–182
[2] Ghosh A K 1997 Axisymmetric dynamic response of a circular plate on an elastic foundation *Journal of Sound and Vibration* **205**(1) pp 112–120
[3] Gupta U S, Ansari A H, Sharma S 2006 Buckling and vibration of polar orthotropic circular plate resting on Winkler foundation *Journal of Sound and Vibration* **297**(3–5) pp 457–476
[4] Bhaskara Rao L, Kameswara Rao C 2013 Vibrations of elastically restrained circular plates resting on Winkler foundation *Arabian Journal for Science and Engineering* **38**(11) pp 3171–3180
[5] Foyouzat M A, Mofid M, Akin J E 2016 Free vibration of thin circular plates resting on an elastic foundation with a variable modulus *Journal of Engineering Mechanics* **142**(4) 04016007
[6] Starovoitov E I, Leonenko D V 2012 Thermal impact on a circular sandwich plate on an elastic foundation *Mechanics of Solids* **47**(1) pp 111–118
[7] Starovoitov E I, Leonenko D V 2016 Vibrations of circular composite plates on an elastic foundation under the action of local loads *Mechanics of Composite Materials* **52**(5) pp 665–672
[8] Lamb H 1921 On the vibrations of an elastic plate in contact with water *Proc. Roy. Soc.* A **98** pp 205–216
[9] Amabili M 2001 Vibrations of Circular Plates Resting on a Sloshing Liquid: Solution of the Fully Coupled Problem *Journal of Sound and Vibration* **245**(2) pp 261–283
[10] Askari E, Jeong K-H, Amabili M 2013 Hydroelastic vibration of circular plates immersed in a liquid-filled container with free surface *Journal of Sound and Vibration* **332**(12) pp 3064–3085
[11] Mogilevich L I, Popov V S 2011 Investigation of the interaction between a viscous incompressible fluid layer and walls of a channel formed by coaxial vibrating discs *Fluid Dynamics* **46**(3) p 375–388

[12] Mogilevich L I, Popov V S, Popova A A 2010 Dynamics of interaction of elastic elements of a vibrating machine with the compressed liquid layer lying between them *Journal of Machinery Manufacture and Reliability* **39**(4) p 322–331
[13] Akcabay D T, Young Y L 2012 Hydroelastic response and energy harvesting potential of flexible piezoelectric beams in viscous flow *Physics of Fluids* **24**(5)
[14] Ageev R V, Kuznetsova E L, Kulikov N I, Mogilevich L I, Popov V S 2014 Mathematical model of movement of a pulsing layer of viscous liquid in the channel with an elastic wall *PNRPU Mechanics Bulletin* **3** pp 17–35
[15] Ageev R V, Mogilevich L I, Popov V S 2014 Vibrations of the walls of a slot channel with a viscous fluid formed by three-layer and solid disks *Journal of Machinery Manufacture and Reliability* **43**(1) pp 1–8
[16] Alekseev V V, Indeitsev D A, Mochalova Yu A 1999 Resonant oscillations of an elastic membrane on the bottom of a tank containing a heavy liquid *Technical Physics* 44(8) pp 903–907
[17] Hosseini-Hashemi S, Karimi M, Hossein Rokni D T 2010 Hydroelastic vibration and buckling of rectangular Mindlin plates on Pasternak foundations under linearly varying in-plane loads *Soil Dynamics and Earthquake Engineering*, **30**(12) pp 1487–1499
[18] Ergin A, Kutlu A, Omurtag M H, Ugurlu B 2012 Dynamic response of Mindlin plates resting on arbitrarily orthotropic Pasternak foundation and partially in contact with fluid *Ocean Engineering* **42** pp 112–125
[19] Ergin A, Kutlu A, Omurtag M H., Ugurlu B 2008 Dynamics of a rectangular plate resting on an elastic foundation and partially in contact with a quiescent fluid *Journal of Sound and Vibration*, **317**(1–2) pp 308–328
[20] Kuznetsova E L, Mogilevich L I, Popov V S, Rabinsky L N 2016 Mathematical model of the plate on elastic foundation interacting with pulsating viscous liquid layer *Applied Mathematical Sciences* **10**(23) pp 1101–1109
[21] Mogilevich L I, Popov V S, Popova A A, Christoforova A V 2016 Mathematical modeling of hydroelastic walls oscillations of the channel on Winkler foundation under vibrations *Vibroengineering PROCEDIA* **8** pp 294–299
[22] Mogilevich L I, Popov V S, Popova A A, Christoforova A V 2016 Mathematical modeling of highly viscous liquid dynamic interaction with walls of channel on elastic foundation *2016 Dynamics of Systems, Mechanisms and Machines (Dynamics)* p 1–4 DOI: 10.1109/Dynamics.2016.7819051
[23] Mogilevich L I, Popov V S, Popova A A 2017 Interaction dynamics of pulsating viscous liquid with the walls of the conduit on an elastic foundation *Journal of Machinery Manufacture and Reliability* **46**(1) pp 12–19
[24] Panovko Y G and Gubanova I I 1965 *Stability and Oscillations of Elastic Systems* (New York: Consultants Bureau Enterprises, Inc.)
[25] Loitsyanskii L G 1966 *Mechanics of Liquids and Gases* (Oxford: Pergamon Press)

Acknowledgments
This study was supported by the Russian Foundation for Basic Research grant 15-01-01604-a and President of Russian Federation Grant MD-6012.2016.8.

IOP Conf. Series: Journal of Physics: Conf. Series **944** (2017) 012058 doi:10.1088/1742-6596/944/1/012058

Electromagnetic Compatibility of Devices on Hybrid Electromagnetic Components

S G Konesev[1], R T Khazieva[1], R V Kirillov[1], I Z Gainutdinov[2], E Y Kondratyev[3]

[1]Ufa State Petroleum Technological University, 1, Kosmonavtov st., Ufa, 450062, Russia
[2]Energodiagnostika, 21, Tolstoy st., Ufa, 450112, Russia
[3]Gazprom dobycha Yamburg, 9, Explorers st., New Urengoy, 629306, Russia

E-mail: khazievart@mail.ru

Abstract. There is a general tendency to reduce the weight and dimensions, the consumption of conductive and electrical insulating materials, increase the reliability and energy efficiency of electrical devices. In recent years, designers have been actively developing devices based on hybrid electromagnetic components (HEMC) such as inductive-capacitive converters (ICC), voltages pulse generators (VPG), secondary power supplies (SPS), capacitive storage devices (CSD), induction heating systems (IHS). Sources of power supplies of similar electrical devices contain, as a rule, links of increased frequency and function in key (pulse) modes, which leads to an increase in electromagnetic interference (EMI). Nonlinear and periodic (impulse) loads, non-sinusoidal (pulsation) of the electromotive force and nonlinearity of the internal parameters of the source and input circuits of consumers distort the shape of the input voltage lead to an increase in thermal losses from the higher harmonic currents, aging of the insulation, increase in the weight of the power supply filter units, resonance at higher harmonics. The most important task is to analyze the operation of electrotechnical devices based on HEMC from the point of view of creating EMIs and assessing their electromagnetic compatibility (EMC) with power supply systems (PSS). The article presents the results of research on the operation of an IHS, the operation principle of a secondary power supply source of which is based on the operation of a half-bridge autonomous inverter, the switching circuit of which is made in the form of a HEMC, called the «multifunctional integrated electromagnetic component»" (MIEC).

1. Introduction

There is a general tendency to reduce the weight and dimensions, the consumption of conductive and electrical insulating materials, increase the reliability and energy efficiency of electrical devices [1, 2]. In recent years, designers have been actively developing devices based on hybrid electromagnetic components (HEMC) such as inductive-capacitive converters (ICC), voltages pulse generators (VPG), secondary power supplies (SPS), capacitive storage devices (CSD), induction heating systems (IHS) [3–5]. Sources of power supplies of similar electrical devices contain, as a rule, links of increased frequency and function in key (pulse) modes, which leads to an increase in electromagnetic interference (EMI) [6–8]. Nonlinear and periodic (impulse) loads, non-sinusoidal (pulsation) of the electromotive force and nonlinearity of the internal parameters of the source and input circuits of consumers distort the shape of the input voltage lead to an increase in thermal losses from the higher harmonic currents, aging of the insulation, increase in the weight of the power supply filter units, resonance at higher harmonics [9–11].

2. Problem statement

The most important task is to analyze the operation of electrotechnical devices based on HEMC from the point of view of creating EMIs and assessing their electromagnetic compatibility (EMC) with power supply systems (PSS) [12, 13]. The article presents the results of research on the operation of an IHS, the operation principle of a secondary power supply source of which is based on the operation of a half-bridge autonomous inverter, the switching circuit of which is made in the form of a HEMC, called the «multifunctional integrated electromagnetic component»" (MIEC) [14, 15]. The implementation of the IHS based on the MIEC allows to reduce the mass, reduce the dimensions of the PSS, and also to ensure the improvement of the quality of electric energy due to a decrease in the EMI level and the non-sinusoidal coefficient [16]. In this article, we present the results of an experimental study confirming this assertion.

With this study, the following tasks we were accomplished:

– the energy quality indicators (non-sinusoidal coefficient, power factor) were analyzed for HEMC operation;

– the parameters of the operating modes of these components are determined;

– an estimation of its EC and influence on the network, including an estimation of the level of the higher harmonics, the ratio of the actual and normative values;

– experimental data on the functioning of a HEMC in an autonomous inverter are obtained.

3. Theory

Disadvantages of electrical devices on discrete elements are a low power factor and the presence of modulation of the active power consumed [17]. The functional integration of electromagnetic elements allows to eliminate of these disadvantages [18, 19].

Consider the circuit of the secondary power supply source (PSS) of IHS, which should also ensure the preservation of the power quality of the supply network (Figure 1).

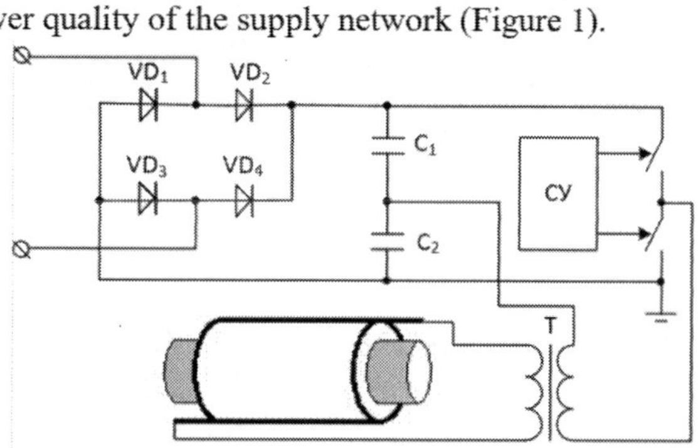

Figure 1. Half-bridge circuit of PSS with ICC on a hybrid component.

The secondary power source consists of the following main units:

– diode bridge providing rectification of alternating voltage;

– transistor half-bridge inverter;

– control system that provides alternate key switching and does not allow simultaneous stay in the open state;

– matching transformer, providing a softer mode of operation of the switches.

The switching circuit is made in the form of a MIEC transformer connected to the secondary winding, which performs the function of local heating.

4. Results of the experiment

Experiments were carried out to study the functioning of the device on a HEMC (IHS) and an EMC estimate of this IHS from SES, the influence of IHS based on the component on the network and the quality of electrical energy. In particular, the level of higher harmonics is determined using the instrument «RESURS-UF2M».

The experiments are performed for four electromagnetic components, the parameters of which are given in Table 1, working separately and in a system consisting of four MIECs connected in series.

Table 1. MIEC and system parameters.

MIEC model	R, mΩ	L, μH	C, μF	ω, turns	fres, kHz
MIEC 10	150	12,8	0,245	23	90
MIEC 11	150	13,7	0,185	22	100
MIEC 12	150	12,8	0,18	22	105
MIEC 24	150	12	0,174	20	110
System	600	19,7	0,076	87	130

The results of the research are data on the indicators of the quality of electrical energy (harmonic amplitude and non-sinusoidal coefficient) and energy parameters (power factor and active power value), presented in Figures 2 – 7.

Figure 2 shows the amplitudes of even harmonics (k = 2, 4, 6, 8, 10, 12, 14, 16) for a system of four MIEKs connected in series at different frequencies; On each of the harmonics, the columns are indicated by colors corresponding to different frequency values.

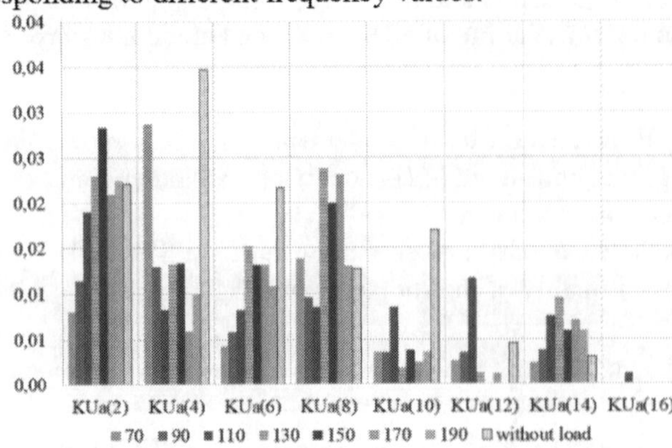

Figure 2. Amplitude of odd harmonics.

Figure 3 shows the amplitudes of odd harmonics (k = 3, 5, 7, 9, 11, 13, 15, 17) for a system of four MIECs connected in series at different frequencies; on each of the harmonics, the columns are indicated by colors corresponding to different frequency values.

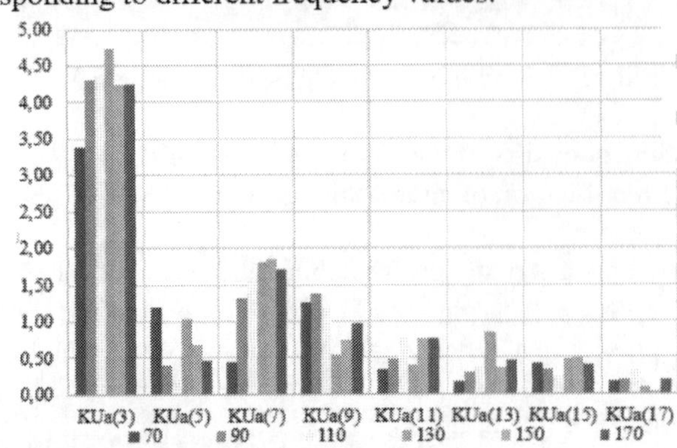

Figure 3. Amplitude of even harmonics at different frequencies.

Figure 4 shows the amplitudes of odd harmonics (k = 3, 5, 7, 9, 11, 13, 15, 17) for a system of four MIECs connected in series, operating in resonance mode; on each of the harmonics, the column to the left shows the normative value, the shaded column in the middle corresponds to the mode under load, the column to the right reflects the amplitude of the harmonics in the system of four MIECs connected in series, operating without load.

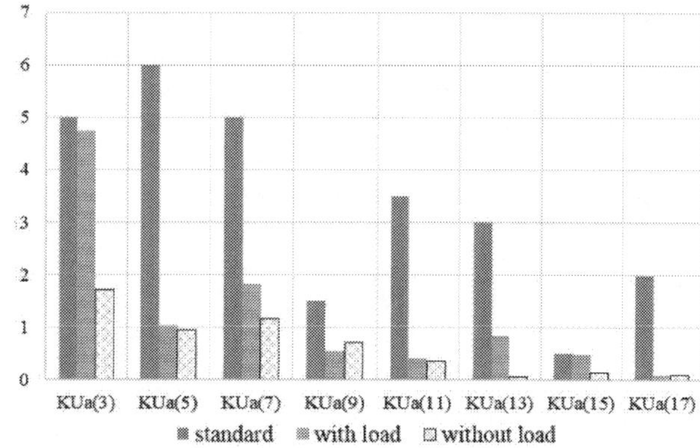

Figure 4. Amplitude of odd harmonics at resonance.

Figure 5 shows a plot of the non-sinusoidal coefficients of phase voltages versus frequency. The column to the left shows the non-sinusoidal coefficient when the MIECs are connected in series, the column in the middle, when «MIEC 10» is working separately, the column to the right reflects the non-sinusoidal coefficient - in the separate work of «MIEC 24».

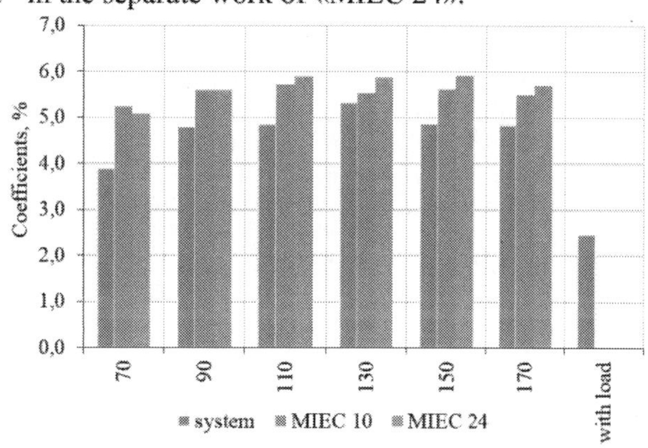

Figure 5. Graph of dependence of non-sinusoidal coefficients on frequency.

Figure 6 shows the dependence of the active power on frequency. The column on the left shows the active capacity of «MIEC 10», the column in the middle corresponds to the work of the MIECs connected in series, the column on the right reflects the active capacity for «MIEC 24».

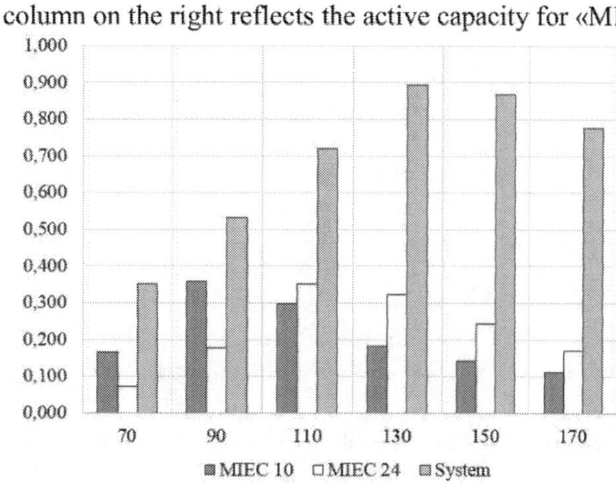

Figure 6. Dependence of active power on frequency.

Figure 7 shows the power factor versus frequency. Figure 6 shows the power factor versus frequency. The column to the left shows the power factor of «MIEC 10», the column in the middle corresponds to the work of the MIECs connected in series, the column to the right reflects the power factor for «MIEK 24».

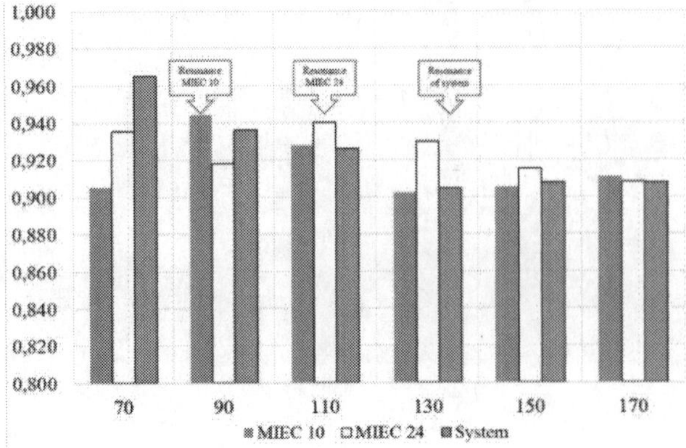

Figure 7. Power factor versus frequency.

Figure 8 shows the dependence of the active and reactive powers on frequency. The columns on top show the active capacity of MIEK 10 and MIEK 24, respectively, the columns from below show the reactive power of MIEK 10 and MIEK 24, respectively.

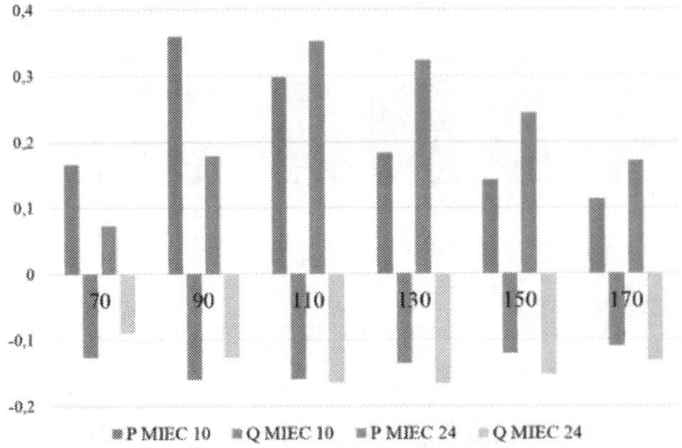

Figure 8. Dependence of active and reactive power on frequency.

5. Discussion

Even harmonic components are practically absent (Figure 2).

At the individual MIEC, the fifth harmonic is maximal, the other odd harmonics gradually decrease. In the system of four connected MIECs working together, the third harmonic is maximum, while the other odd harmonics are much lower (Figures 3 and 4).

The non-sinusoidal coefficient is below the normative value equal to 8 %, while in the case of the system from the MIEC connected in series, it is lower than for the individual MIECs (Figure 5).

The maximum power of «MIEC 10» is achieved at a resonance frequency of 90 kHz, the maximum power of «MIEC 24» is achieved at a resonant frequency of 110 kHz, the maximum system power from the series-connected four MIECs is achieved at a resonant frequency of 130 kHz (Figure 6).

The power factor as a whole is not lower than 0.9, for some MIEC at the maximum power the coefficient is closer to 1, while MIEC's work in the system is on the contrary further (Figure 7).

Reactive power is negative, which indicates the capacitive nature of the load (Figure 8). Multifunctional integrated electromagnetic component acts as a reactive power compensator, provides compensation of reactive power.

6. Conclusions

According to the results of the study:

– indicators of power quality (PQ), which are the most important for devices on hybrid components;

– we estimated of IHS EMC based on a HEMC from PSS, including the level of higher harmonics, the ratio of actual and normative values.

The operation of the HEMC in resonance mode provides a power factor above 0.9.

The maximum power is achieved at the frequency of free oscillations, both for the system and for individual MIECs.

Integral performance of electromagnetic components operating in resonant mode, provides acceptable levels of EMI (higher harmonics).

References

[1] Morrow K, Karner D and Francfort J 2008 *Vehicle Technologies Program*
[2] Grassi F, Manfredi P, Liu X, Sun J, Wu X, Ginste D V and Pignari S A 2017 IEEE *Transactions on Electromagnetic Compatibility* **99** pp 1–12
[3] Konesev S G, Khazieva R T, Strigev D A Compact device for testing insulation by increased voltage *Patent 108856*
[4] Sadeghipour K, Dopkin J A 1996 *Computers in Industry* **3** pp 195–205
[5] Musavi F, Eberle W, Dunford W G 2011 *Twenty-Sixth Annual IEEE* pp 821–828
[6] Li B, Månsson D 2017 *IEEE Transactions on Electromagnetic Compatibility* **99** pp 1–9
[7] Rahmani M, Mazzola M 2016 *IEEE Electrical Design of Advanced Packaging and Systems (EDAPS)* pp 179–181
[8] Kit Sum K 1998 *PCIM Magazine*
[9] Zhang Y 2016 *IEEE Electrical Design of Advanced Packaging and Systems (EDAPS)* 15
[10] Leferink F, Keyer C and Melentjev A 2016 *IEEE Electromagnetic Compatibility Magazine* **4** pp 49–55
[11] Ding J, Li F, Zhao J and Zhao P 2016 *7th IEEE International Conference on Software Engineering and Service Science (ICSESS)* pp 551–554
[12] Pan J, Qi F, Cai H and Xu L 2016 *IEEE Energy Conversion Congress and Exposition (ECCE)* pp 1–5
[13] Margaris I D, Papathanassiou S A, Hatziargyriou N D, Hansen A D and Sorensen P 2012 *IEEE Transactions on Sustainable Energy* **2** pp 189–199
[14] Konesev S G 2016 Multifunctional integrated electromagnetic component *Patent 2585248* № 2012114845/08 13.04.2012
[15] Konesev S G, Khazieva R T and Kirillov R V 2016 The research of stabilization properties of inductive-capacitive converters based on the two-sections hybrid electromagnetic elements *Dynamics of Systems, Mechanisms and Machines (Dynamics)* pp 1–7 DOI: 10.1109/Dynamics.2016.7819030
[16] Konesev S G, Khazieva R T, Kirillov R V, Konev A A 2017 Название *Journal of Physics: Conference Series (JPCS)* **1**
[17] Konesev S G, Khazieva R T 2015 Modern problems of science and education *Electronic scientific journal (electronic scientific journal)* **1**
[18] Konesev S G, Khazieva R T 2014 Improving the reliability and efficiency of electrical systems and complexes *J. Young Scientist* pp 135–138
[19] Konesev S G, Khazieva R T and Kirillov R V 2015 *J. of USATU 4(70)* **19** pp 66–71

Improvement of human operator vibroprotection system in the utility machine

P A Korchagin, I A Teterina and L F Rahuba
Siberian State Automobile and Highway University, 5 Mira pr., Omsk, 644080, Russia

e-mail: iateterina@mail.ru

Abstract. The article is devoted to an urgent problem of improving efficiency of road-building utility machines in terms of improving human operator vibroprotection system by determining acceptable values of the rigidity coefficients and resistance coefficients of operator's cab suspension system elements and those of operator's seat. Negative effects of vibration result in labour productivity decrease and occupational diseases. Besides, structure vibrations have a damaging impact on the machine units and mechanisms, which leads to reducing an overall service life of the machine. Results of experimental and theoretical research of operator vibroprotection system in the road-building utility machine are presented. An algorithm for the program to calculate dynamic impacts on the operator in terms of different structural and performance parameters of the machine and considering combination of external pertrubation influences was proposed. Keywords – vibration, vibroprotection, utility machine, performance parameters of the machine.

1. Introduction

In spite of current achievements in the field of development of modern antivibration systems in the last decades, which provide vibration decrease up to the safe level, issues concerning comfortable working conditions for operators of road-building machines are still relevant. Current international regulations [1, 2] do not prescribe acceptable values, but offer recommendations to obtain minimum possible values of vibration acceleration at the operator's workplace while designing antivibration systems. During operation process operators of utility machines are subjected to vibrating impacts from the power plant and chassis elements contacting uneven microrelief as well as impacts from the brush, which is in contact with the surface under treatment.

2. Materials

It is possible to improve operator vibroprotection system by determining efficient parameters of coefficients of rigidity and coefficients of resistance of operator's cab suspension system elements and those of operator's seat. To accomplish this task a theoretical research was undertaken aimed at investigating effect of machine structural parameters on the magnitude of dynamic impact at the operator's workplace. MATLAB software product was used as a research tool which enabled development of DUPM application. Geometric dimensions, mass of units, coefficients of rigidity and resistance of elastoviscous elements of the system were taken as source data for the program. The mathematical model was based on the algorithm of calculating dynamic impact magnitude at the operator's workplace (Fig. 1) [3].

The algorithm proposed is universal. It enables calculation of parameters of vibroprotection system elements for a number of road-building machines if the input data are known. Except for the research of machine structural parameters effects on the mean-square values of vibration acceleration at the operator's workplace, this algorithm allows to study the impact of performance parameters, such as operating conditions, machine movement speed, peculiarities of the uneven microrelief, etc.

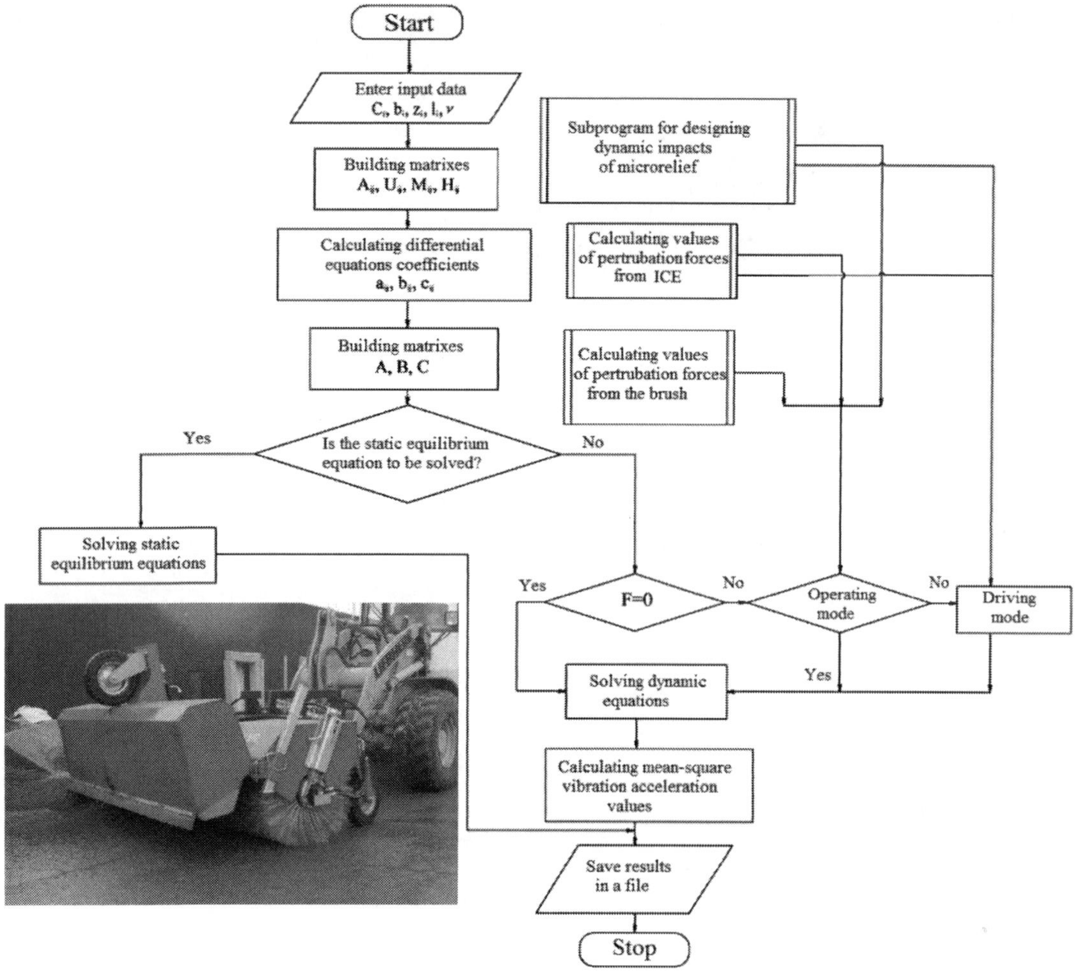

Figure 1. Algorithm of DUPM application for determining the magnitude of dynamic impacts at the operator's workplace.

3. Theory

During mathematical modeling of a complex dynamic system «Pertrubation Influence – Municipal Machine – Operator» effect of machine structural parameters, such as coefficients of rigidity and coefficients of resistance of operator's cab suspension system elements and those of operator's seat, on the dynamic impact magnitude was investigated [3]. The relations obtained demonstrate vividly the influence of these parameters on the magnitude of the dynamic impacts at the operator's workplace (Fig. 2–4). They enable calculation of appropriate intervals between operator's cab suspension system elements and operator's seat to obtain the minimum possible magnitude of dynamic impacts at the operator's workplace [6].

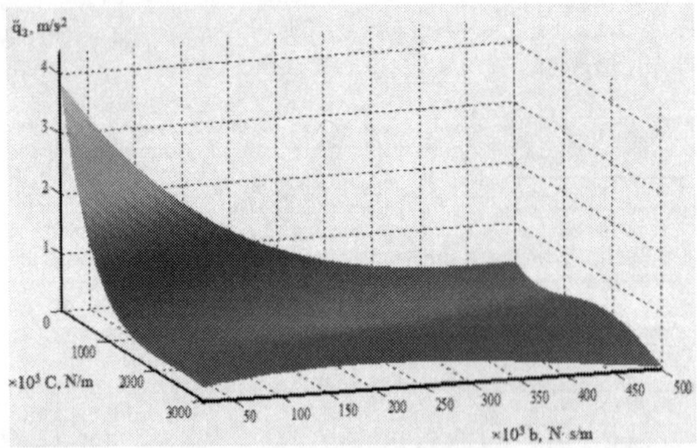

Figure 2. Relations between vibration acceleration values at the cab floor and coefficients of rigidity and those of resistance of operator's cab suspension system elements.

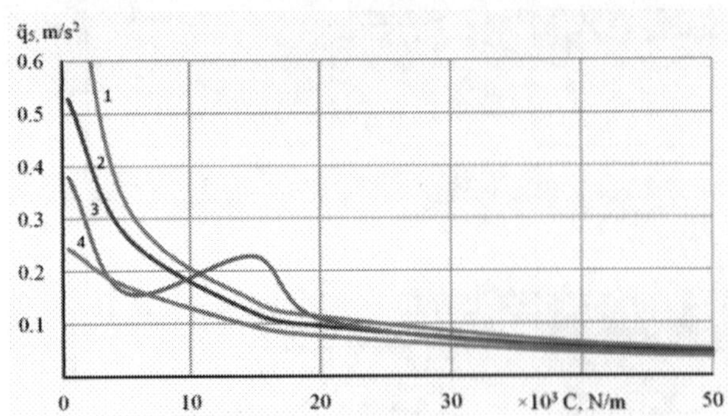

Figure 3. Relations between vibration acceleration values at the operator's seat and coefficients of rigidity of seat suspension with the following coefficients of resistance: 1) b = 4·103 N·s/m; 2) b = 8·103 N·s/m; 3) b = 12·103 N·s/m; 4) b = 20·103 N·s/m.

On analyzing the results of mathematical modeling it was shown that in terms of vibroprotection the acceptable values for the coefficients of rigidity of operator's cab suspension system elements should be within 1.2·106 – 1.8·106 N/m. Values for the coefficients of resistance are to be determined within 2.0·105 – 5.0·105 N·s/m [4].

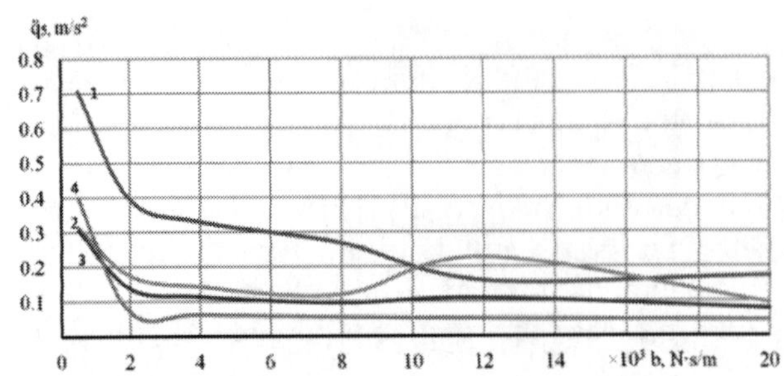

Figure 4. Relations between vibration acceleration values at the operator's seat and coefficients of resistance of seat suspension with the following coefficients of rigidity: 1) C = 5·103 N/m; 2) C = 20·103 N/m; 3) C = 40·103 N/m; 4) C = 80·103 N/m.

Besides, theoretical research made it possible to draw a conclusion that coefficients of rigidity and coefficients of resistance of the seat suspension affect the magnitude of dynamic impacts at the

IOP Conf. Series: Journal of Physics: Conf. Series **944** (2017) 012059 doi:10.1088/1742-6596/944/1/012059

operator's seat (Fig. 5). Due to this the values of coefficients of rigidity and coefficients of resistance of the cab suspension system elements were defined [4].

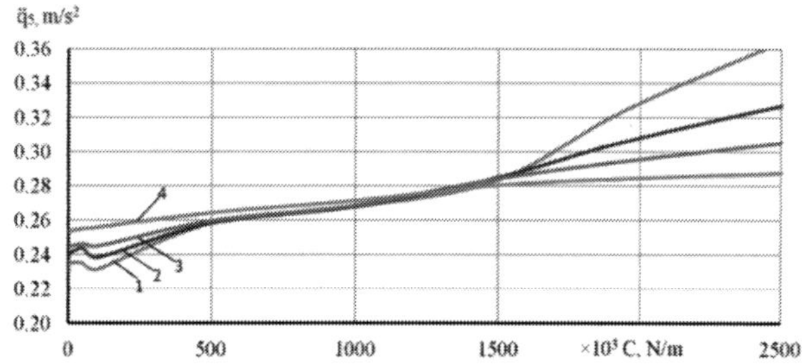

Figure 5. Relations between vibration acceleration values at the operator's seat and coefficients of rigidity of the cab suspension system elements with the following coefficients of resistance: 1) b = 50·103 N·s/m; 2) b = 100·103 N·s/m; 3) b = 250·103 N·s/m; 4) b = 500·103 N·s/m.

On analyzing mathematical modeling results with different combinations of external pertrubation influences and performance parameters a range of acceptable values of suspension system elements was defined: coefficients of rigidity of the cab suspension system elements should vary within 1.3·106 – 1.5·106 N/m, coefficients of resistance of the cab suspension system elements should be within 2.5·105 – 5.0·105 N·s/m [7]. At such values the magnitude of the mean-square vibration acceleration at the operator's workplace is minimal.

4. Experimental results

Based on the results obtained by theoretical research an engineering solution for the problem of decreasing dynamic impacts was designed – a pneumatic suspension for the operator's seat with a rubber-cord cover (Fig. 6) [5].

a b c

Figure 6. Operator's cab suspension system elements: a) metal base with a shoulder; b) rubber-cord cover in the seat suspension design; c) general view of the seat.

The mechanism operates as follows: vibrations from a unit to be vibro-isolated are transmitted to the shock absorber with the rubber-cord cover being subjected to vertical dynamic impacts, which reduces dynamic impacts at the operator's workplace [5]. When the machine moves taking up vibrations is provided by the rubber-cord cover deformation. The suspension rigidity is changed depending on the internal pressure in the rubber-cord cover [5, 6].

Quantitative indicators in assessing performance of the proposed vibroprotection system is the magnitude of vibration at the operator's workplace before and after the implementation of the proposed vibroprotection system (Fig. 7), particularly, decrease in the magnitude of dynamic impacts. Based on the comparative analysis of vibration magnitude before and after the implementation of the proposed vibroprotection system a conclusion was made that the decrease in the magnitude of dynamic impacts was at an average 13-16% depending on the frequency considered.

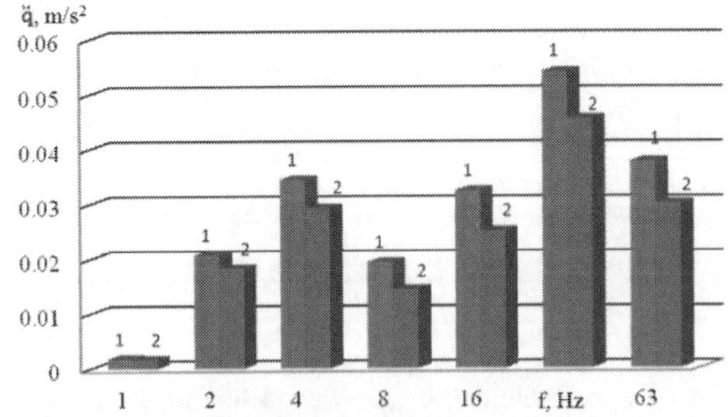

Figure 7. Magnitude of vibration acceleration values at the seat place in operating conditions with the speed of 10 km/hr: 1 – for the machine with the original seat suspension; 2 – for the machine with the modernized seat suspension.

5. Summary and conclusions

Research provided by the use of the mathematical model of a complex dynamic system "Pertrubation Influence – Municipal Machine – Operator" allowed to reveal the effect of coefficients of rigidity and resistance on the magnitude of vibration acceleration at the operator's workplace and to define the range of acceptable values [5].

The proposed algorithm of calculating the magnitude of vibration acceleration values at the operator's workplace made it possible to define the range of acceptable values for the coefficients of rigidity and resistance of operator's cab suspension system elements and those of operator's seat in the utility machine.

The results obtained in the theoretical study provide an opportunity to design new prospective vibroprotection systems not just for the utility machine, but also for a number of other road-building machinery. The algorithm was certified as an electronic resource (RF certificate). The operator's seat suspension proposed was patented as a useful model (RF patent).

References

[1] Burdzik R 2014 Identification of structure and directional distribution of vibration transferred to car-body from road roughness *Journal of Vibroengineering* 1 pp 324–33

[2] Burdzik R, Konieczny L 2013 Research on structure, structure, propagation and exposure to general vibration in passenger car for different damping parameters *Journal of Vibroengineering* 4 pp 1680–1688

[3] Gerts M E and Gerts M M 2015 Drive effect on vibratory maintenance of rotation *J. Mach. Manufact. Reliab.* 3 pp 20–216

[4] Sergeev V I 2016 Dynamics in the small for nonlinear oscillation systems in precision theory problems *J. Mach. Manufact. Reliab.* 3 pp 26–222

[5] Teterina I A, Korchagin P A and Letopolsky A B Patent RF 171043, MPK B 60 N 2/52 *Pneumatic suspension of operator's seat with rubber-cord cover* No 2016136885 filed 14.09.2016 published 17.05.2017 bulletin No 14

[6] Maciejewski I, Meyer L and Krzyzynski T 2009 Modeling and multi-criteria optimization of passive seat suspension vibrasolating properties *Journal of Sound and Vibration* 324 3pp 520–538

[7] Yong C, Virech W. and David G Z 2011 Investigation of helicopter seat structural dynamics for aircrew vibration mitigation *Journal of the American Helicopter Society* 56 1 pp 12006

IOP Conf. Series: Journal of Physics: Conf. Series **944** (2017) 012060 doi:10.1088/1742-6596/944/1/012060

Rationale choosing interval of a piecewise-constant approximation of input rate of non-stationary queue system

Ivan A Korelin[1] **and Sergey V Porshnev**[1]
[1] The Institute of Radioelectronics and Information Technologies, Ural Federal University, Mira st. 32, Yekaterinburg, Russia 620002

Abstract. The paper demonstrates the possibility of calculating the characteristics of the flow of visitors to objects carrying out mass events passing through checkpoints. The mathematical model is based on the non-stationary queuing system (NQS) where dependence of requests input rate from time is described by the function. This function was chosen in such way that its properties were similar to the real dependencies of speed of visitors arrival on football matches to the stadium. A piecewise-constant approximation of the function is used when statistical modeling of NQS performing. Authors calculated the dependencies of the queue length and waiting time for visitors to service (time in queue) on time for different laws. Time required to service the entire queue and the number of visitors entering the stadium at the beginning of the match were calculated too. We found the dependence for macroscopic quantitative characteristics of NQS from the number of averaging sections of the input rate.

1. Introduction

Conclusions and regularities obtained as a result of the development of queuing theory (QT) have found application in a variety of fields and sciences: astronomy, agronomy, geology, hydrology, teletraffic theory, chemistry, economics, ecology, physics and statistics [1]. However, most of the analytical results in QT are obtained for stationary streams of requests (first of all, a Poisson flow) [1, 2]. Such streams of requests are extremely rare in practice, because in reality, the assumption of their stationarity is not met, in most cases. For example this is the case of access control systems that provide access to mass events [4], passenger control devices at airports and railway stations, etc. The study of the features of their operation is of undoubted interest from a practical point of view, since the results obtained can be used as a scientific justification for the design decisions made at the modernization or design stage of non-stationary queuing systems (NQS).

In the conducted research, authors used an approach of the quantitative characteristics of the NQS study, which was previously demonstrated by the authors in [3]. In this approach a piecewise constant approximation of the function $\lambda = \lambda(t)$ is used. This is mean that at each interval of the piecewise constant approximation of the function $\lambda = \lambda(t)$, a stationary stream of requests with rate λ_k arrives at the input of the QS. The lengths of the queue at the entrance to the QS are calculated by accounting the applications received at the input of the QS at the previous intervals of approximation and not served till the time τ_k. It is clear that one of the main questions arising in the practical use of this approach is to chose of the length of the interval of the piecewise constant approximation of the function $\lambda = \lambda(t)$.

This article presents the results of statistical modeling of the NQS in accordance with the approach described above and provides recommendations on the choice of the duration of the interval of piecewise-constant approximation, depending on the parameters of the function $\lambda = \lambda(t)$.

2. Mathematical model of the non-stationary queue system

The block diagram of the model of a non-stationary single-channel SMO with an unlimited queue is presented in Figure 1.

Figure 1. The scheme model of non-stationary QS

As you can see from figure 1, in this model QS processes a flow of requests with the rate $\lambda = \lambda(t)$ varying with time. The main characteristic of the input stream is the instantaneous flux density $\lambda(t)$.This characteristic means the limit of the ratio of the average number of events for a time (t, t+Δt), to the length of this section, provided that the length of the segment tends to zero:

$$\lambda(t) = \lim_{\Delta t \to 0} \frac{M(t+\Delta t)+M(t)}{\Delta t} = \frac{dM(t)}{dt} \quad (1)$$

The M(t) is expected value of the number of events on the site (0, t).

The FIFO policy (the first one in is the first one out) used to serve the queue of visitors. The service speed of incoming requests is determined by the service rate $\bar{\mu}$, which is a random variable with a probability density

$$p\{\xi\} = \begin{cases} 0, \text{ then } \xi < 1, \\ \dfrac{2}{9(M[\xi]-1)}(\xi-1), \text{ then } 1 \le \xi < M[\xi], \\ \dfrac{2}{9(M[\xi]-10)}(\xi-10), \text{ then } M[\xi] < \xi \le 10, \\ 0, \text{ then } \xi > 10, \end{cases} \quad (2)$$

then $\xi \in [1,10]$. In this study, a distribution law is chosen such that $M[\xi]=4$ and $\bar{\mu}=15$.

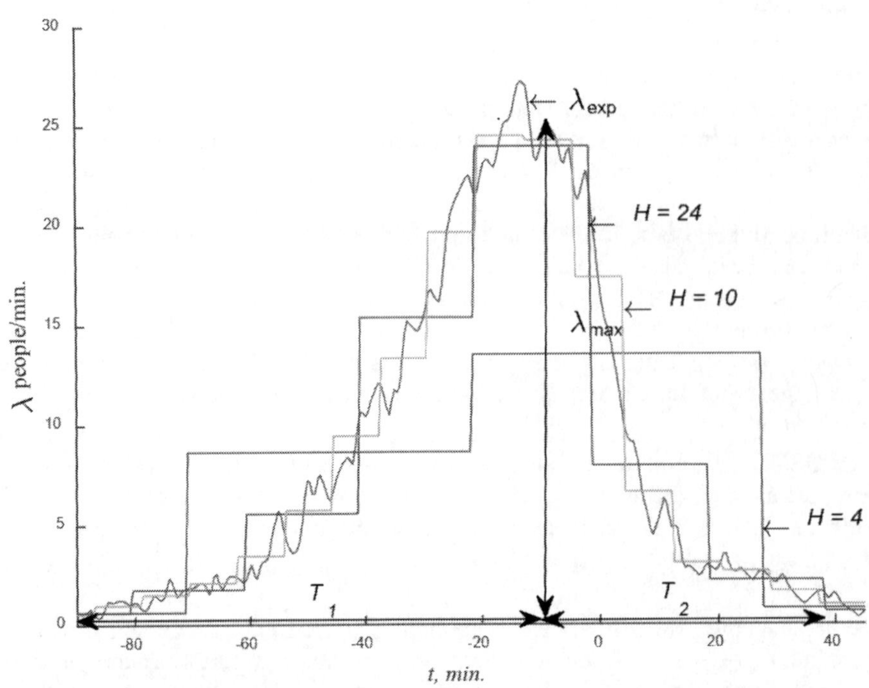

Figure 2. The dependencies $\lambda(t)$ for non-stationary QS (beginning of match $t = 0$).

On figure 2 there is a typical dependence $\lambda_{exp}(t)$, wich was received during the football match between the football clubs "Krylia Sovetov" and "Dynamo" at the stadium "Metallurg" in Samara 05.05.2013 [4].

Figure 2 shows that access to the stadium "Metallurg" was opened 1.5 hours before the start of the football match. After the opening of the turnstiles during the time $T_1\approx 70$ min. input rate of requests increased from 0 people/min. to λ_{max}=28 people/min. Approximately 15 minutes before the match, the intensity decreased from λ_{max}=28 people/min. to 0 people/min. during the time interval $T_2\approx 50$ min. Thus mean that total number N of visitors entered through one access control device (turnstile) at the stadium "Metallurg"

$$N = \int_{-80}^{40} \lambda(t)dt \tag{3}$$

is 1400 people.

The characteristics of the NQS for the family of functions $\lambda = \lambda(t)$ were studied. These functions on time interval T_1 [−80; −10] minutes are Increased by a quadratic law from 0 to λ_{max}, afterwards on time interval T_2 [−10;40] minutes quadratically decreasing from λ_{max} to 0. Parameter λ_{max} varied from 18 people/min. from 32,4 people/min. with step 0,8, T_1 and T_2 remained unchanged, and the parameters of parabolas were chosen in such a way that the total number of entered N for each λ_{max} would be 1400 people. (Figure 3).

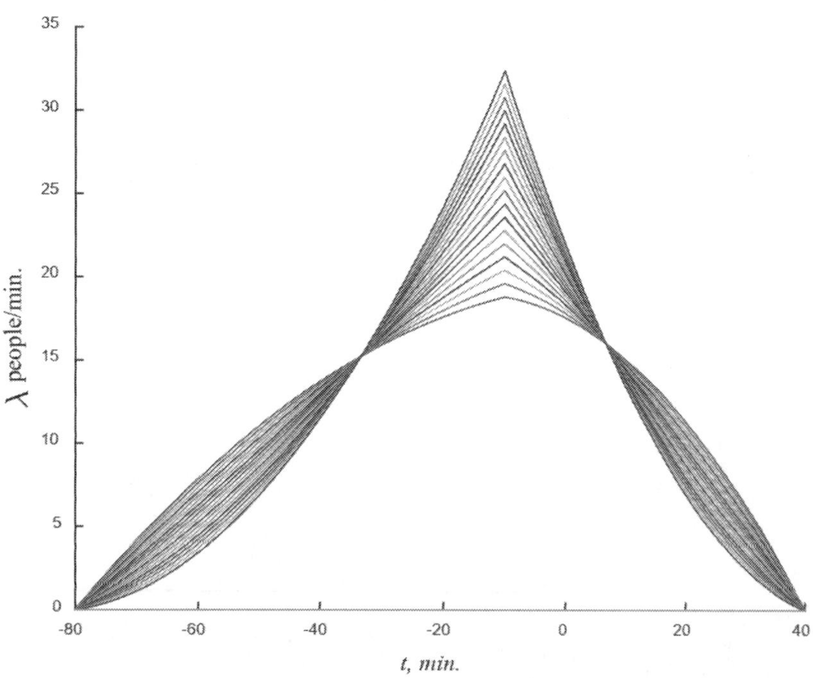

Figure 3. The dependencies of $\lambda(t)$ on the non-stationary QS (the beginning of the match $t = 0$)

In accordance with the chosen approach in carried out computational experiments a piecewise constant approximation of the functions $\lambda = \lambda(t)$ on interval $[T_1, T_2]$ was used. Number of intervals was varied in set $H \in \{2,4,8,16,32,64,100,154,205,308,616,1232\}$.

3. Method of computer experiments

The flowchart of the algorithm used in performing the statistical simulation described in detail in [3]. Requests are generating at each interval of the piecewise constant approximation during time interval. The request arrival times t_A have the exponential distribution law with the rate equal to the average value of the input rate of requests on this interval. The service time interval was generated according to formula (2). Further the time for setting up the service was calculated by alternately viewing all the requests that stood on this interval of the service queue:

$$t_{E_i} = \begin{cases} t_{A_i}, \text{then } t_{A_i} > t_{E_{i-1}} + \tau_{S_{i-1}}, \\ t_{A_i} + (t_{E_{i-1}} + \tau_{S_{i-1}} - t_{A_i}), \text{then } t_{A_i} < t_{E_{i-1}} + \tau_{S_{i-1}}. \end{cases} \tag{4}$$

Note that for the first request $t_{E_1} = t_{A_1}$ it is obvious.

The analysis of the characteristics used to describe the features of the operation of NQS [3] showed that for their quantitative description it is possible, for example, to use the dependence of the length of the queue of visitors (in terms of QS - queue length of requests) from time:

$$L = L(\lambda_{max}, t_k) = |Q|, \text{ then } Q = \{q_n : (t_{A_n} < t_k) \cap (t_{E_n} > t_k)\}, \tag{5}$$

Average waiting time in queue for the visitor (in terms of QS - request) in the queue from time is:

$$\tau_w = \tau_w(\lambda_{max}, t_k) = \frac{\sum_{i=1}^{|Q|} (t_{E_i} - t_{A_i})}{|Q|}, \text{ then } Q = \{q_n : (t_{E_n} \le t_k) \cap (t_{E_n} > t_{k-1})\}, \tag{6}$$

then $t_k = T_1 + \frac{T_2 - T_1}{K}(k-1)$, $k = \overline{1, K}$, K is the number of intervals of piecewise-constant approximation $\lambda(t)$. Also the number of visitors entering at the beginning of the match:

$$N_0 = N_0(\lambda_{max}) = |Q|, \text{ then } Q = \{q_n : t_{E_n} < t = 0\}, \tag{7}$$

And the time required to service all incoming visitors:

$$T_{All} = T_{All}(\lambda_{max}) = \{t : N(t) \ge 0.97 \cdot N_{max}\}, \text{ then } N_{max} = \max(N(t)). \tag{8}$$

Since the Monte Carlo method was used in the simulation, the values of the characteristic function of the non-stationary SMO were taken to be their mean values in the ensemble of independent realizations:

$$\overline{\Phi}(\lambda_{max}, t_k) = \frac{1}{m} \sum_{j=1}^{m} \Phi_j(\lambda_{max}, t_k), \tag{9}$$

then m is the number of independent tests in the Monte Carlo method, Φ is element of set $\{L, \tau_w\}$,

$$\overline{\Psi}(\lambda_{max}) = \frac{1}{m} \sum_{j=1}^{m} \Psi_j(\lambda_{max}), \tag{10}$$

then Ψ is element of set $\{N_0, T_{All}\}$.

4. Analysis of experimental results

Let us consider the dependences $\overline{L}(\lambda_{max}, t_k)$ for different values λ_{max}, $H = 1232$ presented in Fig. 4. It is seen from figure that these dependencies can be characterized by two parameters: the maximum value of the function $\overline{L}_{max} = \max\left(\overline{L}(\lambda_{max}, t_k)\right)$ and the abscissa value of the function $\overline{L}(\lambda_{max}, t_k)$ in which it reaches its maximum value $\overline{\tau}_{max} = \arg\max_{t_k}\left(\overline{L}(\lambda_{max}, t_k)\right)$. These parameters are some functions that depend on the λ_{max}.

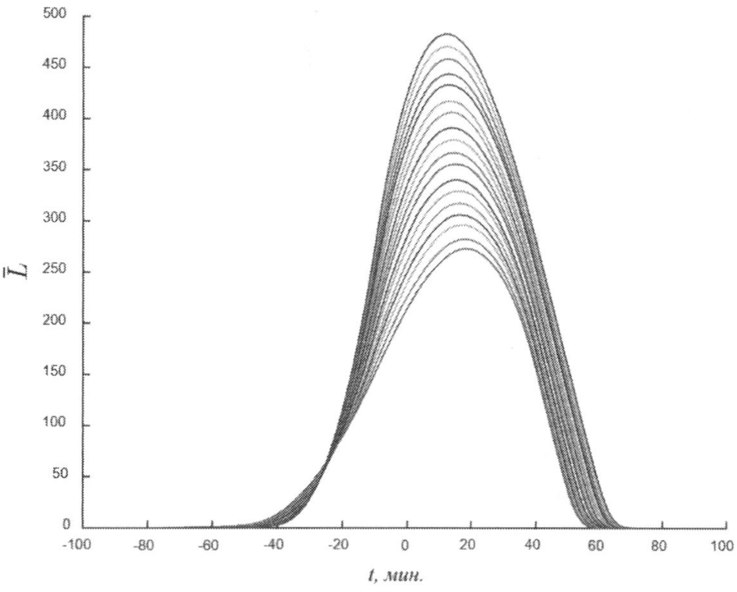

Figure 4. The dependencies $\bar{L}(\lambda_{max},t)$ (the beginning of the match t = 0) for different values λ_{max}

In this connection, the dependencies of the selected parameters on the number of intervals of the piecewise constant approximation H was calculated. Figure 5 and figure 6, respectively shows dependencies $\bar{L}_{\max} = f(\lambda_{\max},H)$ and $\bar{\tau}_{\max} = f(\lambda_{\max},H)$. In addition, there are dependencies $N_0 = f(\lambda_{\max},H)$ and $T_{All} = f(\lambda_{\max},H)$ on figure 7 and figure 8.

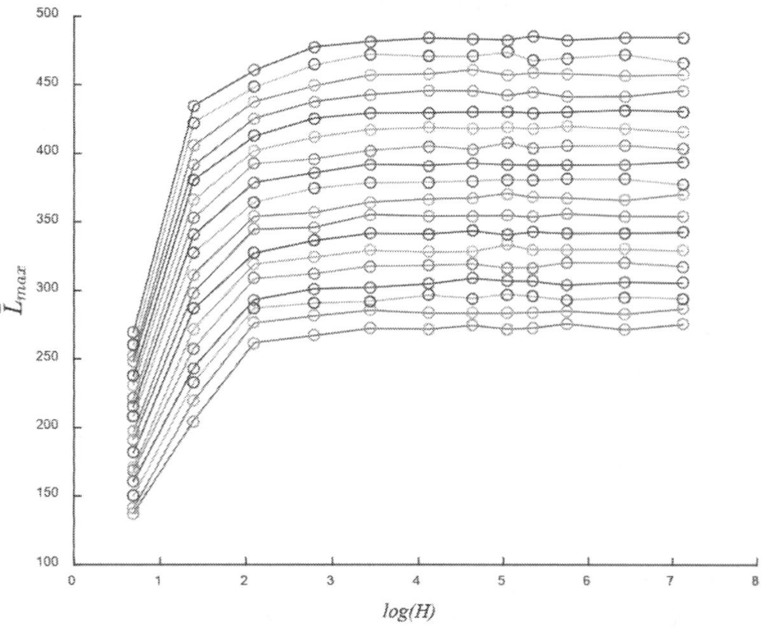

Figure 5. The dependencies $\bar{L}_{\max} = f(\lambda_{\max},H)$ for the above parameter values λ_{max}

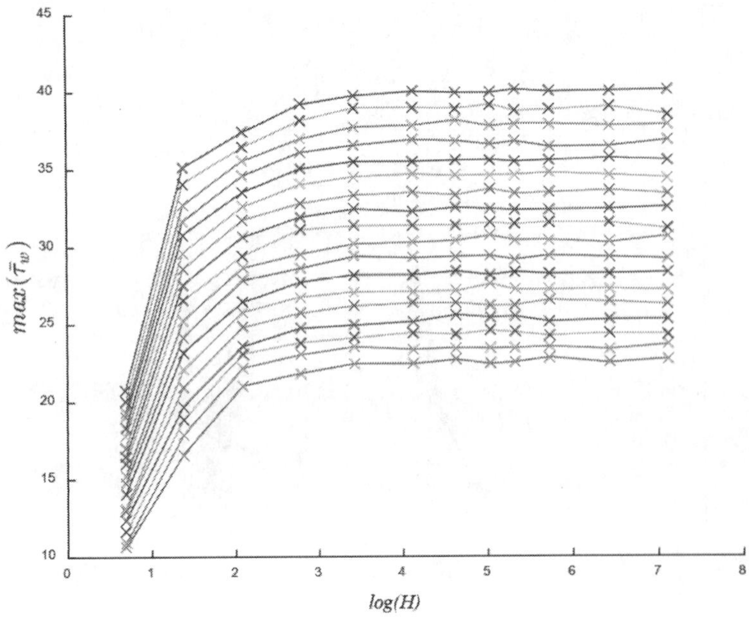

Figure 6. The dependencies $\max(\bar{\tau}_w) = f(\lambda_{\max}, H)$ for the above parameter values λ_{max}

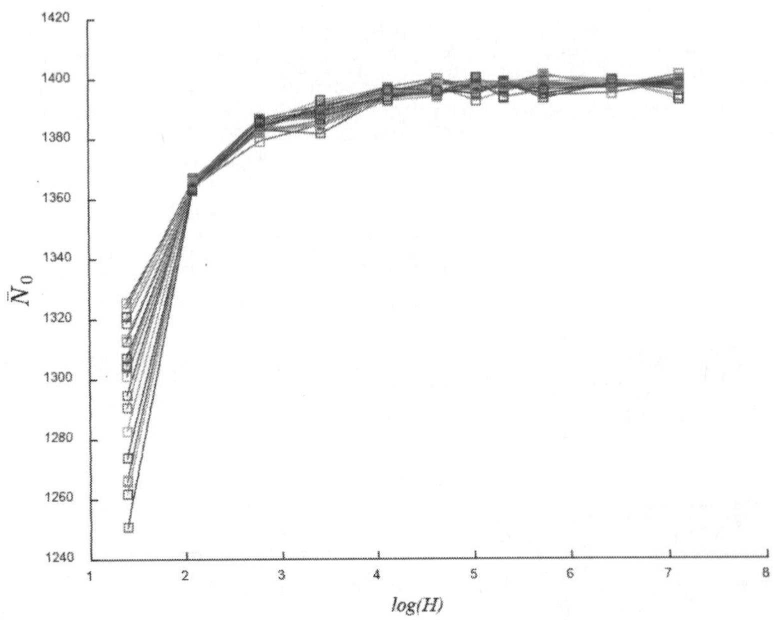

Figure 7. The dependencies $N_0 = f(\lambda_{\max}, H)$ for the above parameter values λ_{max}

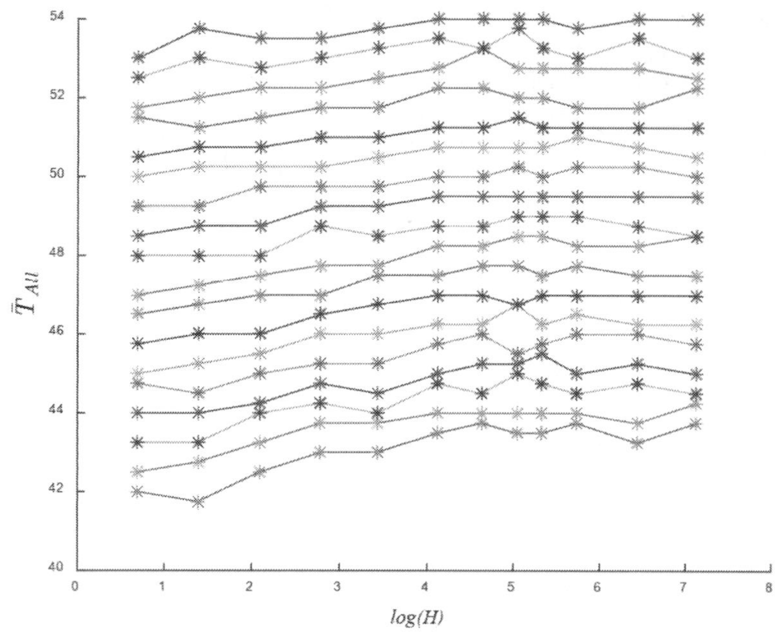

Figure 8. The dependencies $T_{All} = f(\lambda_{\max}, H)$ for the above parameter values λ_{max}

It can be seen from Figures 5-8 that the dependencies $\bar{L}_{\max} = f(\lambda_{\max}, H)$, $\max(\bar{\tau}_w) = f(\lambda_{\max}, H)$, $N_0 = f(\lambda_{\max}, H)$, $T_{All} = f(\lambda_{\max}, H)$ for each of the above values λ_{max} turn out to be similar to each other for values of H greater than 100 (the corresponding length of the interval of the piecewise linear interpolation of the function $\lambda = \lambda(t)$ - 7.5 minutes). The value of these functions practically does not change with even more increasing *H*. This is mean, when simulating the studied NQS, it is sufficient to use intervals of a piecewise-constant approximation H of less than 7.5 minutes. This ensures that estimates $\bar{L}_{\max}, \max(\bar{\tau}_w), N_0, T_{all}$ independent of *H*.

At the same time, it is clear that each of these characteristics, being calculated as the mean value over an ensemble of independent realizations, is some random variable. In this connection, the further direction of research will be the study of the influence of the size of the window of piecewise linear approximation on the laws of distribution of random variables.

5. Conclusion

In this paper estimated the number of intervals and, correspondingly, the duration of intervals of piecewise-linear approximation of the function $\lambda = \lambda(t)$, describing the input rate of requests at input of NQS from time. Also counted value number of intervals providing stable estimates of the maximum length of the queue, the maximum waiting time in the queue, the number of visitors entered by the time the function $\lambda = \lambda(t)$, reaches the maximum value, the time to serve all visitors.

In further publications it is need to be valuate effect the size of window piecewise-constant approximation to the laws of the distribution of random variables $\bar{L}_{\max}, \bar{\tau}_{\max}, \bar{N}_0, \bar{T}_{all}$.

6. References

[1] Gnedenko B V and Kovalenko I N 1966 *Vvedenie v teoriju massovogo obsluzhivanija* [Introduction to queuing theory] Moscow: Nauka Publ. p 432

[2] Kendall D G 1953 Stochastic process occurring in the theory of queues and their analysis by the method of the imbedded markov chain *The Annals of Mathematical Statistics.* **24** (3): p 338 doi:10.1214/aoms/1177728975. JSTOR 2236285

[3] Porshnev S and Korelin I, 2017 Non-Stationary Single-Channel Queuing System Features Research in Context of Number of Served Queries *ITM Web Conf.* **10** 03006 p 4 doi: 10.1051/itmconf/20171003006

[4] Porshnev S V, Jakob D A 2014 *Issledovanie osobennostej funkcionirovanija informacionnyh kontrol'no-propusknyh sistem ob'ektov provedenija massovyh meroprijatij* [The study of the functioning of information systems checkpoints objects of mass events], Ekaterinburg: Institut jekonomiki UrO RAN Publ. p 216.

IOP Conf. Series: Journal of Physics: Conf. Series **944** (2017) 012061 doi:10.1088/1742-6596/944/1/012061

Optimization of Control Points Number at Coordinate Measurements based on the Monte-Carlo Method

A A Korolev[1], A V Kochetkov[2], and O V Zakharov[1,3,4]
[1]Yuri Gagarin State Technical University, Saratov, Russia
[2]Perm National Research Polytechnic University, Perm, Russia
[3]Samara National Research University, Samara, Russia
[4]Institute of Precision Mechanics and Control of RAS, Saratov, Russia

E-mail: zov20@mail.ru

Abstract. Improving the quality of products causes an increase in the requirements for the accuracy of the dimensions and shape of the surfaces of the workpieces. This, in turn, raises the requirements for accuracy and productivity of measuring of the workpieces. The use of coordinate measuring machines is currently the most effective measuring tool for solving similar problems. The article proposes a method for optimizing the number of control points using Monte Carlo simulation. Based on the measurement of a small sample from batches of workpieces, statistical modeling is performed, which allows one to obtain interval estimates of the measurement error. This approach is demonstrated by examples of applications for flatness, cylindricity and sphericity. Four options of uniform and uneven arrangement of control points are considered and their comparison is given. It is revealed that when the number of control points decreases, the arithmetic mean decreases, the standard deviation of the measurement error increases and the probability of the measurement α-error increases. In general, it has been established that it is possible to repeatedly reduce the number of control points while maintaining the required measurement accuracy.

1. Introduction

Coordinate measurement traditionally involves measuring individual points on the workpiece surfaces, replacing the cloud of control points with the nominal geometry of the workpiece, and comparing with the tolerance for the corresponding dimensions. As a result, a decision is made to accept or reject the defective workpiece. Evaluation of the error in the form of the surface consists of several components: the form error after machining, the systematic and random measurement errors. The systematic measurement error is related to the error of basing and the sampling error (the number of control points). The random measurement error primarily depends on the errors of the touch sensor of the coordinate measuring machine.

Obviously, increasing the accuracy requires the use of a large number of control points on the surfaces. However, the measurement productivity is significantly reduced. Therefore, the question arises of the optimal choice of the number of control points. The idea of the need to reduce sample sizes in the measurement and the possibility of a priori estimating the possible error or uncertainty of measurement was repeatedly mentioned in a number of papers [1–12]. The information given in the studies is of unquestionable scientific and practical value, but only concern certain surfaces or aspects of the problem. Therefore, in practice this issue is solved directly by the CMM operator and the result largely depends on its qualification. In this paper, we propose a new optimization method for determining the number and location of control points on surfaces, which is based on statistical simulations Monte Carlo.

2. The formulation of the problem

The initial problem is formulated as follows – to measure the given surface with a certain error and maximum productivity. The maximum productivity is achieved, firstly, by minimizing the number of control points, and secondly, the location of control points, which ensures their best sequencing. As an estimate of measurement error can be used absolute or relative error values, the rejection value for a given tolerance, the probabilities of measuring I and II errors.

The new strategy is based on computer modeling of measurement errors for batches of workpieces using the Monte Carlo method. The fundamentals of the application of the Monte Carlo method for various applied problems are presented in [13]. The advantage of using the Monte Carlo method is to obtain non-point, but interval error estimates for certain batches of workpieces.

Simulation is carried out in the following sequence. At the first stage, sequences of pseudo-random numbers with given correlation and probability distribution laws modeling the random values of error components at each test are generated. With a stable technological process, the error distribution law does not change, and its parameters change insignificantly. Therefore, it becomes possible to determine the law and the error distribution parameters of various surfaces on the basis of a small sample from a batch of workpieces. At the second stage, multiple calculations are performed for surfaces with simulated deviations of the form, determining the values of the size, form or location errors. The number of iterations in modeling can be related to the number of workpieces in a batch or taken large for statistical stability. At the third stage, statistical processing of the simulation results is performed - establish the law and determine the distribution parameters or calculate the statistical moments. Thus, an interval estimation of the measurement error is obtained to compare the variants with different numbers and locations of control points on the surfaces.

The statistical modeling technique was previously described in detail in our paper [14].

3. Mathematical Modeling

As objects of research three types of surfaces are chosen: flatness, cylinder, sphere. It is these surfaces that make up the overwhelming majority on various workpieces of machines and mechanisms. The surfaces are described in a parametric form, where the simulated errors δ_1, δ_2, δ_3 are additionally introduced into the equations.

The equation of the flatness with errors is:

$$\left.\begin{aligned} X_1 &= x_i; \\ Y_1 &= y_i; \\ Z_1 &= c+\delta_1; \end{aligned}\right\} \tag{1}$$

where $x_i = [x_1; x_2]$, $y_i = [y_1; y_2]$ is the linear parameters that determine the dimensions of the workpiece; c is the position of the plane.

The equation of the cylinder with errors is:

$$\left.\begin{aligned} X_2 &= (r_2+\delta_2)\cos v; \\ Y_2 &= (r_2+\delta_2)\sin v; \\ Z_2 &= z_i; \end{aligned}\right\} \tag{2}$$

where $z_i = [z_1; z_2]$ is length of the cylinder; $v = [0; 2\pi]$ is angular parameter; r_2 is radius of the cylinder.

The equation of the sphere with errors is:

$$\left.\begin{aligned} X_3 &= (r_3+\delta_3)\cos\vartheta\cos\varphi; \\ Y_3 &= (r_3+\delta_3)\cos\vartheta\sin\varphi; \\ Z_3 &= (r_3+\delta_3)\sin\vartheta; \end{aligned}\right\} \tag{3}$$

where $\vartheta = [0; 2\pi]$, $\varphi = [0; \pi]$ are angular parameters; r_3 is radius of the sphere.

The initial data for the simulation was the results of measuring the workpieces on the mobile CMM Faro Arm Edge (FARO Swiss Holding GmbH, Switzerland). The measurements were carried out with the following parameters: flatness of size 100×100 mm, a cylinder 50 mm in diameter and 100 mm in length, a hemisphere of radius 60 mm. Was used uniform grid for linear and angular coordinates for control points: 121 for the plane, 110 for the cylinder, and 101 for the sphere. For this, three-dimensional models of workpieces with measurable surfaces with marking of control points were created in the program Power Inspect.

The measurement was made at given points 30 times. The re-basing procedure was carried out 3 times. For the flatness, the array of measured points of all the experiments is shown in Fig. 1. The deviation of the shape at the point of the flatness is the distance between the nominal coordinates of the points and the measured coordinates taken with the sign taken into account. The deviation of the shape for the flatness as a whole is the difference between the maximum and minimum deviation of the shape at the points of the flatness. In the case of ideal basing, the deviations of the shape are the coordinates along the *Z*-axis (this axis is the normal to the flatness). In Fig. 2 shows the calculated deviations in 110 points of 30 experiments after elimination of the error of basing.

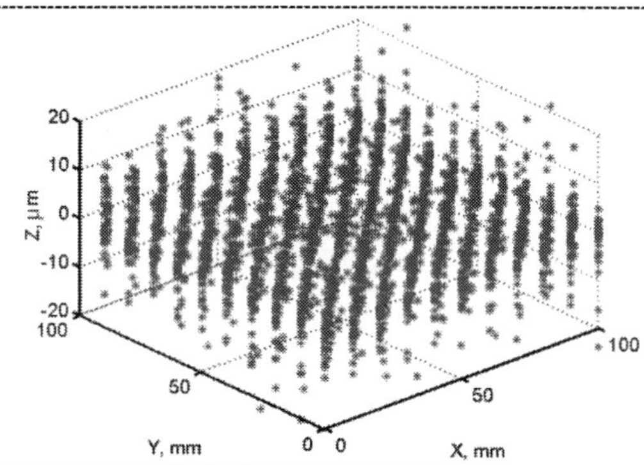

Figure 1. An array of measured points for the flatness.

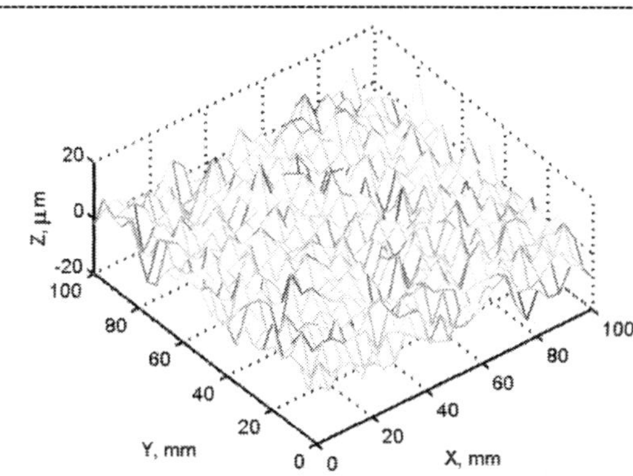

Figure 2. The errors of the flatness after elimination of the error of basing.

On the basis of the data obtained, Monte Carlo simulations were carried out for batches of 100 workpieces with uniform and normal distribution of error components. Then, according to the standard method, the error values are calculated: flatness, cylindricity, sphericity. Modeling and processing of results was carried out according to the developed program in the Matlab environment.

In the simulation, four options of the location of the control points are considered:

- option 1 – uniform grid of linear and angular parameters of the surface, the number of control points 121 for the plane, 110 for the cylinder and 101 for the sphere;
- option 2 – uneven mesh, sparse for the plane along the *X* axis, for the cylinder along the length, for the sphere along the zenith angle, the number of control points 66 for the plane, 60 for the cylinder, 61 for the sphere;

- option 3 – non-uniform grid, sparse for the plane along the *Y* axis, for the cylinder in the angle of the circle, for the sphere in the azimuth angle, the number of control points 66 for the plane, 55 for the cylinder, 51 for the sphere;
- option 4 – uniform grid, sparse in angular and linear coordinates, the number of control points 36 for the plane, 30 for the cylinder, 31 for the sphere.

Option 1 was considered to be a basic one for modeling. All other variants were obtained from the basic by eliminating a number of control points according to the above algorithm. The results of one of the simulation implementations are shown in Fig. 3–5. Here, for a cylinder and a hemisphere, an error scale of 500 times is used.

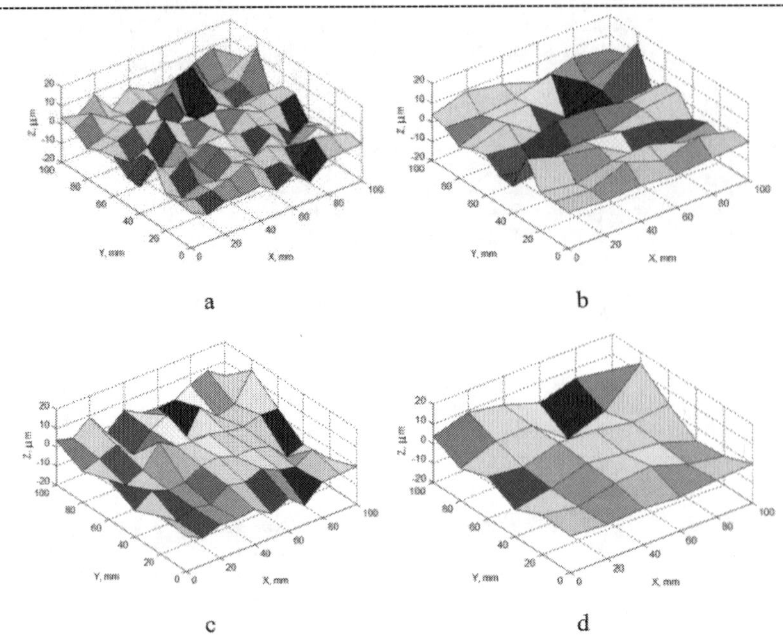

Figure 3. Simulation of the flatness error: a – option 1, b – option 2, c – option 3, d – option 4.

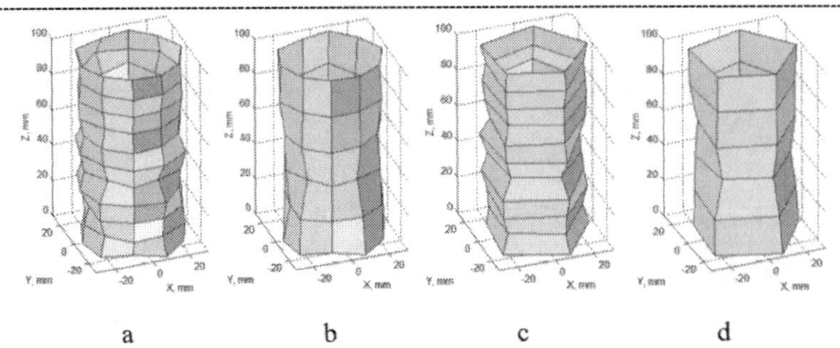

Figure 4. Simulation of the cylinder error: a – option 1, b – option 2, c – option 3, d – option 4.

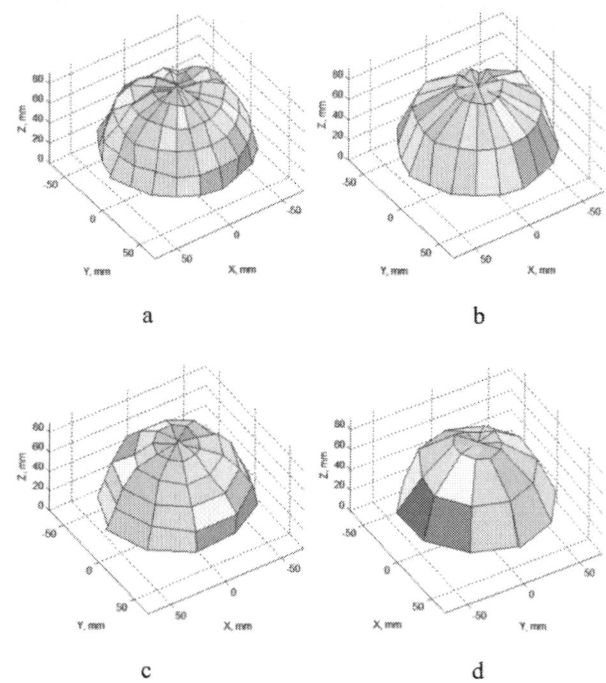

Figure 5. Simulation of the sphere error: a – option 1, b – option 2, c – option 3, d – option 4.

4. Analysis of measurement simulation results

Statistical processing of the data resulting from the simulation showed that the measurement error is well described by the normal distribution law. The evaluation was carried out by Pearson's agreement criterion with a confidence probability of 95%. As an example, Fig. 6 gives empirical distributions of the measurement error for flatness (a) and cylindricity (b).

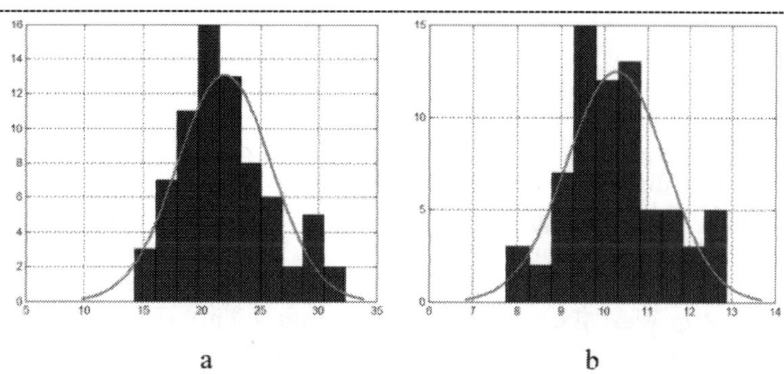

Figure 6. Histogram of the measurement error: a – flatness, b – cylindricity.

Therefore, the comparison of options was carried out on the arithmetic mean value and the standard deviation of the measurement error. Also, the tolerances for the corresponding deviations of the form were conventionally given and calculated the values of the defective workpieces in percent and measuring I and II errors. The results of the calculations are given in Fig. 7 for the flatness, Fig. 8 – for the cylinder, Fig. 9 for the sphere. Shown: I – the arithmetic mean, μm, II – the standard deviation, μm.

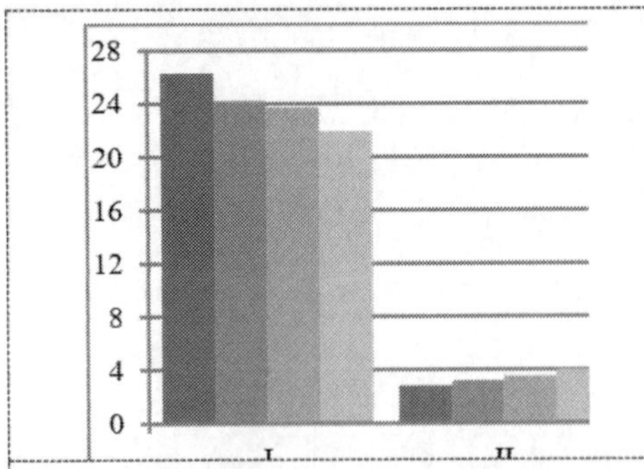

Figure 7. Measuring error for the flatness for option 1–4.

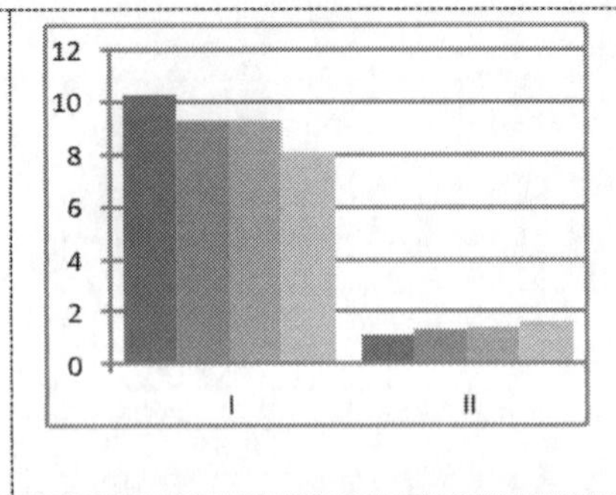

Figure 8. Measuring error for the cylindricity for option 1–4.

It is established that for the considered variants only β-error is characteristic. The results of calculating the β-error in percent are shown in Fig. 10: I – flatness, II – cylinder, III – sphere.

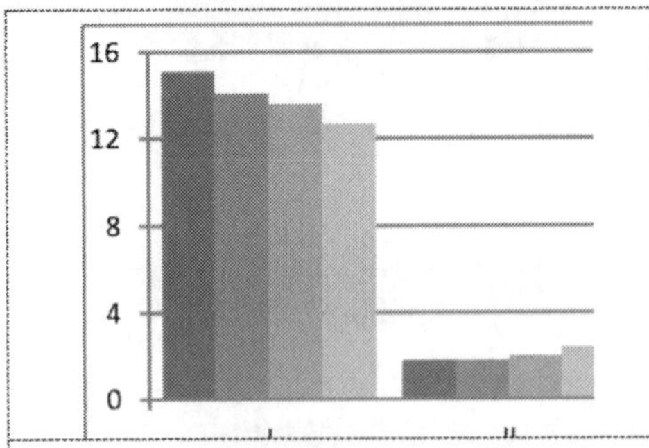

Figure 9. Measuring error for the sphericity for option 1–4.

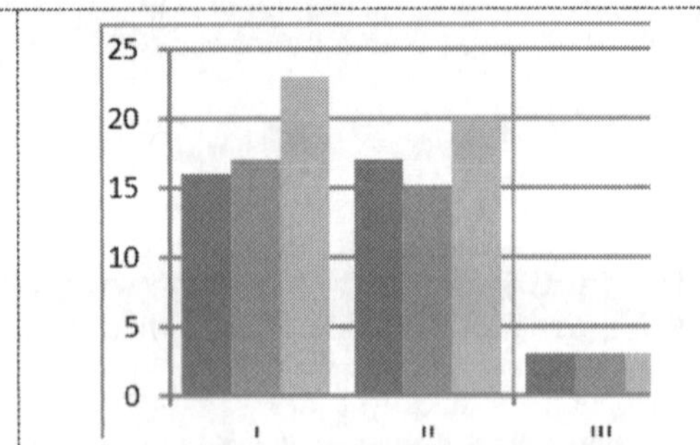

Figure 10. Measuring β-error error for option 2–4: I – flatness, II – cylindricity, III – sphericity.

Analysis of the obtained modeling data allows us to draw the following conclusions:

1) when using a smaller number of control points, the measured value of the surface error always decreases, the largest variation with the base option 1 is given by option 4;

2) for all options only the measurement β-error is characteristic, that is, the probability increases that the defective workpieces is fit for measurement;

3) when measuring the plane, a satisfactory result gives a decrease in control points from 121 to 66, while their location is not of fundamental importance (options 2, 3);

4) when measuring the cylinder, a satisfactory result corresponds to a non-uniform arrangement of control points along the angular step, with a decrease in their number from 110 to 55 (option 3);

5) when measuring the sphere, a satisfactory result was obtained for all variants; therefore, a preference should be given to increase the productivity of variant 4 with the number of points 31, and to ensure accuracy – to the variant with a decrease in the number of control points along the zenith angle to 61 (option 2).

5. Conclusion

The method of optimization of the number of control points with the use of Monte Carlo simulation is considered in the article. Based on the measurement of a small sample from batches of workpieces, statistical modeling is performed, which allows one to obtain interval estimates of the measurement error. The conducted researches established that in the measurement and analysis of flatness, cylindricity and sphericity, the number and location of the control points significantly affects the magnitude of the error. It is revealed that when the number of control points decreases, the arithmetic mean decreases, the standard deviation of the measurement error increases and the probability of the measurement β-error increases. In general, it has been established that it is possible to repeatedly reduce the number of control points while maintaining the required measurement accuracy.

Acknowledgments
The study was performed by a grant from the Russian Science Foundation (project № 16-19-10204).

References

[1] Badar M A , Raman S and Pulat P S 2005 Experimental verification of manufacturing error pattern and its utilization in form tolerance sampling *International Journal of Machine Tools and Manufacture* **45** pp 63–73
[2] Obeidat Suleiman M, Abu Jadayil Wisam M and Mandahawi Nabeel F 2013 Intelligent Sampling for Inspecting Milled Flat Plates *International Review of Mechanical Engineering* **7** pp 767–773
[3] Capello E and Semeraro Q 2001 The harmonic fitting method for the assessment of the substitute geometry estimate error. Part II: statistical approach, machining process analysis and inspection plan optimisation *International Journal of Machine Tools and Manufacture* **41** pp 1103–1129
[4] Jones S D and Ulsoy A G 1995 Optimization strategy for maximizing coordinate measuring machine productivity, part 2: Problem formulation, solution, and experimental results *Journal of engineering for industry* **117** pp 610–618
[5] Pechenin V A and Bolotov M A 2016 Basing error in coordinate measurements of cylindrical gears *Russian Engineering Research* **36** pp 630–634
[6] K. Carr and P. Ferreira, Verification of Form Tolerances Part I: Basic Issues, Flatness, and Straightness *Precision Engineering* vol 17 1995 pp 131–143
[7] Poniatowska Małgorzata and Werner Andrzej 2012 Simulation tests of the method for determining a cad model of free-form surface deterministic deviations *Metrol. Meas. Syst.* **XIX** pp 151–158
[8] Lee G, Mou J and Shen Y 1997 Sampling strategy design for dimensional measurement of geometric features using coordinate measuring machine *International Journal of Machine Tools and Manufacture* **37** pp 917–934
[9] Raghunandan R and Rao P V 2007 Selection of an optimum sample size for flatness error estimation while using coordinate measuring machine *International Journal of Machine Tools and Manufacture* **47** pp 477–482
[10] Jalid A, Hariri S and Laghzale N E 2015 Influence of sample size on flatness estimation and uncertainty in three-dimensional measurement *International Journal of Metrology and Quality Engineering* **6** pp 102
[11] Ascione R, Moroni G, Petrò S and Romano D 2013 Adaptive inspection in coordinate metrology based on kriging models *Precision Engineering* **37** pp 44–60
[12] Colosimo B M, Moroni G and Petrò S 2010 A tolerance interval based criterion for optimizing discrete point sampling strategies *Precision Engineering* **34** pp 745–754
[13] Liu J S 2004 *Monte Carlo Strategies in Scientific Computing* (New York: Springer)
[14] Zakharov O V and Kochetkov A V 2016 Minimization of the systematic error in centerless measurement of the roundness of parts *Measurement Techniques* **58** pp 1317–1321

IOP Conf. Series: Journal of Physics: Conf. Series **944** (2017) 012062 doi:10.1088/1742-6596/944/1/012062

Analytical solution of the problem of acceleration of cargo by a bridge crane with constant acceleration at elimination of swings of a cargo rope

M S Korytov[1], V S Shcherbakov[2] and V V Titenko[3]

[1]Department of Automobiles, Construction Materials and Technologies, Siberian automobile and highway university, Omsk, Russia
[2]Department of Automation of Production Processes and Electrical Engineering, Siberian automobile and highway university, Omsk, Russia
[3]Omsk State Technical University, 11, Mira ave., Omsk, 644050, Russia

Abstract. Limitation of the swing of the bridge crane cargo rope is a matter of urgency, as it can significantly improve the efficiency and safety of the work performed. In order to completely dampen the pendulum swing after the break-up of a bridge or a bridge-crane freight cart to maximum speed, it is necessary, in the normal repulsion control of the electric motor, to split the process of dispersion into a minimum of three gaps. For a dynamic system of swinging of a bridge crane on a flexible cable hanger in a separate vertical plane, an analytical solution was obtained to determine the temporal dependence of the cargo rope angle relative to the gravitational vertical when the cargo suspension point moves with constant acceleration. The resulting analytical dependence of the cargo rope angle and its first derivative can break the process of dispersing the cargo suspension point into three stages of dispersal and braking with various accelerations and enter maximum speed of movement of the cargo suspension point. In doing so, the condition of eliminating the swings of the cargo rope relative to the gravitational vertical is fulfilled. Provides examples of the maximum speed output constraints-to-time when removing the rope swing.

1. Introduction

The movement of the cargo, suspended by a cargo rope, is carried out by means of a bridge crane (BC), working in warehouses and workshops. The movement of the cargo is possible in the complete elimination of the unmanaged components of the pendulum swings. That is, with the vertical position of the cargo rope in the steady-state driving mode. This will not only reduce the time of the BC cycle, but also increase the efficiency and safety of the work performed [1, 2, 3]. Known methods for synthesizing the path of a suspension point [3, 5, 6, 7, 8, 9, 10, 11, 12] have a general disadvantage in the form of a relatively large margin of error in implementing both the vertical and the linear coordinates of the moving cargo. In general, the angle of deflection of the BC freight cable from the vertical is not monitored or controlled. The time of movement of the cargo is increasing.

In order to completely dampen the pendulum swings after the dispersion of the bridge or the BC freight cart to the maximum speed, it is necessary to move the point of the cargo suspension (the bridge or the freight cart of the BC) with a variable acceleration. This requires the use of a frequency-controlled electric drive. Or, if you are using conventional relay management, you need to break the acceleration process by a minimum of three gaps.

2. Formulation of the problem

Accept as an assumption that the dispersion of the bridge (or the BC freight cart) to a certain maximum speed when the appropriate electric motor is turned on, is performed with a constant acceleration of a_1. Break the process of reaching the bridge maximum speed by three gaps. At the end of the third gap, the bridge should be dispersed to the maximum speed v_{max}, and the suspension point acceleration changes to zero. The angle of deflection of the freight cable q and its first derivative $\dot{q}$ at that time is zero.

In the first and third periods, the racing acceleration of a_1 will be carried out. It is created by the bridge drive. The second period between the first and the third can be described as moving the bridge in inertia with slow braking. Braking is done with a constant acceleration of negative sign a_2, under the influence of friction forces in the drive. There is no positive acceleration in the dispersion of the bridge drive. A relatively small deceleration acceleration of a_2 is not created by the bridge braking mechanisms. It is created by friction forces arising in the transmissions of the kinematic steam of the drive and the transmission engine.

3. Theoretical provisions

For a mathematical description of the swings of the BC cargo on the cable hanger, a known mathematical model of the flat pendulum with fading vibrations is used. For the small corners of the cargo rope, it is described by the linear differential equation (DE) of the second order [13, 14, 15, 16]:

$$\ddot{q} + \ddot{x} / L + b\dot{q} + g \cdot q / L = 0, \qquad (1)$$

where L is the length of the BC's cargo rope from the movable point of the freight trolley (chain centre of the roller block) to the centre of the mass of the cargo, m; $b=2b_0/m$; b_0 – the coefficient of viscous friction, given to the angular coordinate, gives the measure of the energy dissipation, N·m·s/Rad; m – mass of cargo, kg; q, $\dot{q}$, $\ddot{q}$ – the angle of deflection of the BC's cargo rope from the gravitational vertical and its first two derivatives in time, respectively, Rad, Rad/s, Rad/s^2; g = 9.81 – acceleration of free fall, m/s^2; $\ddot{x}$ – linear acceleration of the cargo suspension point in the horizontal direction of the freight trolley, m/s^2.

Assumptions were made about: - the minor angle of deflection of the cargo rope from the gravitational vertical (not more than 1-3°); - the constancy length of the cargo rope L in the process of moving the cargo; - the constancy racing acceleration a_1 and constancy acceleration the coast down under the forces of inertia and friction a_2; - the negligible of the small influence of the mass of the transportable cargo and the moving links of the BC to the controlled acceleration parameter of the suspension point that takes values from the discrete series $\ddot{x} = [a_1; a_2]$.

Analytic decision DE (1) has been obtained for constant acceleration:

$$q(t) = C_1 \cdot e^{\left(-t\left(Lb-\sqrt{-L\left(4g-Lb^2\right)}\right)/2L\right)} + C_2 \cdot e^{\left(-t\left(Lb+\sqrt{-L\left(4g-Lb^2\right)}\right)/2L\right)} -$$

$$-\frac{a_1 \cdot e^{\left(-t\left(Lb-\sqrt{-L\left(4g-Lb^2\right)}\right)/2L\right)} \cdot e^{\left(-t\sqrt{L^2b^2-4Lg}/2L\right)} \cdot e^{(bt/2)} \cdot \left(L \cdot b + \sqrt{L^2 \cdot b^2 - 4 \cdot L \cdot g}\right)}{2 \cdot g\sqrt{L^2 \cdot b^2 - 4 \cdot L \cdot g}} + \qquad (2)$$

$$+\frac{a_1 \cdot e^{\left(t\sqrt{L^2b^2-4Lg}/2L\right)} \cdot e^{(bt/2)} \cdot e^{\left(-t\left(Lb+\sqrt{-L\left(4g-Lb^2\right)}\right)/2L\right)} \cdot \left(L \cdot b - \sqrt{L^2 \cdot b^2 - 4 \cdot L \cdot g}\right)}{2 \cdot g\sqrt{L^2 \cdot b^2 - 4 \cdot L \cdot g}}.$$

The expression (2) resulted in an expression of the speed of change in the inclination of the BC's cargo cable also in analytical form. The expression $\dot{q}(t)$ is not given due to its cumbersomeness.

To define permanent integrations C_1 and C_2, a system of two equations of function angle $q(t)$ and its first derivative $\dot{q}(t)$ at time $t = 0$ (initial conditions) was drawn:

$$\begin{cases} q_0 = C_1 + C_2 - \dfrac{a_1\left(Lb + \sqrt{L^2b^2 - 4Lg}\right)}{2g\sqrt{L^2b^2 - 4Lg}} + \dfrac{a_1\left(Lb - \sqrt{L^2b^2 - 4Lg}\right)}{2g\sqrt{L^2b^2 - 4Lg}}; \\ \dot{q}_0 = \dfrac{a_1\left(Lb + \sqrt{L^2b^2 - 4Lg}\right)}{4Lg} - \dfrac{C_1\left(Lb - \sqrt{-L\left(4g - Lb^2\right)}\right)}{2L} - \dfrac{C_2\left(Lb + \sqrt{-L\left(4g - Lb^2\right)}\right)}{2L} + \\ + \dfrac{a_1\left(Lb - \sqrt{L^2b^2 - 4Lg}\right)}{4Lg} + \dfrac{ba_1\left(Lb - \sqrt{L^2b^2 - 4Lg}\right)}{4g\sqrt{L^2b^2 - 4Lg}} - \dfrac{ba_1\left(Lb + \sqrt{L^2b^2 - 4Lg}\right)}{4g\sqrt{L^2b^2 - 4Lg}} - \\ - \dfrac{a_1\left(Lb + \sqrt{-L\left(4g - Lb^2\right)}\right)\left(Lb - \sqrt{L^2b^2 - 4Lg}\right)}{4Lg\sqrt{L^2b^2 - 4Lg}} + \dfrac{a_1\left(Lb + \sqrt{L^2b^2 - 4Lg}\right)\left(Lb - \sqrt{-L\left(4g - Lb^2\right)}\right)}{4Lg\sqrt{L^2b^2 - 4Lg}}, \end{cases} \tag{3}$$

Where q_0 is the value of the angle of the cargo rope at time $t = 0$; $\dot{q}_0$ – the value of the first derivative of the angle of the cargo rope at the time $t = 0$.
The solution of the system (3) relatively unknown C_1 and C_2 allowed their analytical expressions to be obtained:

$$C_1 = \frac{a_1\sqrt{-L\left(4g - Lb^2\right)} + 2Lg\dot{q}_0 + q_0 g\sqrt{-L\left(4g - Lb^2\right)} + Lbq_0 g + \dfrac{Lba_1\sqrt{-L\left(4g - Lb^2\right)}}{\sqrt{L^2b^2 - 4Lg}}}{2g\sqrt{-L\left(4g - Lb^2\right)}};$$

$$C_2 = -\frac{\left(2Lg\dot{q}_0 - a_1\sqrt{-L\left(4g - Lb^2\right)} - q_0 g\sqrt{-L\left(4g - Lb^2\right)} + Lbq_0 g + \dfrac{Lba_1\sqrt{-L\left(4g - Lb^2\right)}}{\sqrt{L^2b^2 - 4Lg}}\right)}{2g\sqrt{-L\left(4g - Lb^2\right)}}. \tag{4}$$

The use of the resulting analytical constraints (2), (4) opens the possibility of optimizing the process of dispersing the BC bridge without using simulation modelling and numerical solution DE (1). The acceleration is carried out in three periods, with the condition of the total elimination of the swings at the end of the third period of dispersal (on the output to a specified constant speed).

4. Experimental results

As an example of the practical use of the analytic expressions received, the values of the three time intervals (racing acceleration – moving the bridge in inertia with slow braking – racing acceleration) T_1, T_2 and T_3, respectively, are optimized to the extent that the target function is minimized

$$f = |q_{end}| + |\dot{q}_{end}|, \tag{5}$$

where q_{end} is the value of the deflection angle of the cargo rope at the end of the time of the dispersion $t{=}T_1{+}T_2{+}T_3$; $\dot{q}_{end}$ – the value of the first derivative of the angle of deflection of the cargo rope at the final time of the dispersion of $t{=}T_1{+}T_2{+}T_3$.

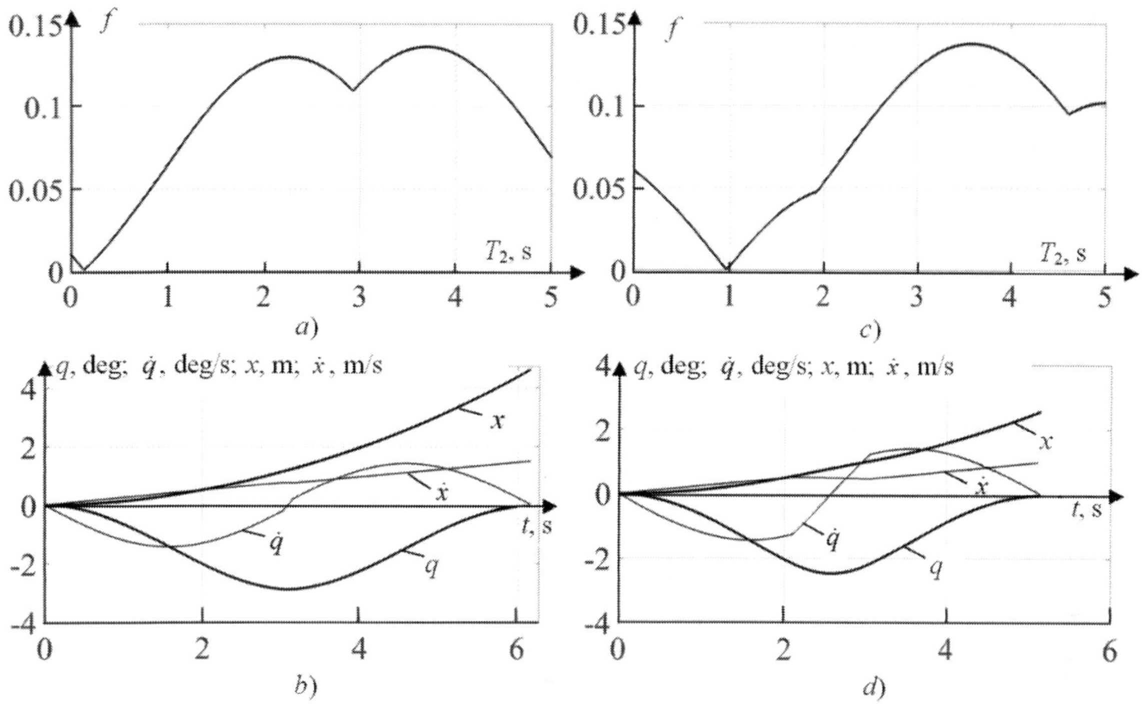

Figure 1. The results of a computational experiment on bridge racing in the mode of eliminating cargo swings upon reaching the maximum speed: the dependence of the target function (*a*) on the time T_2 at v_{max}=1.5 m/s; the optimum process of intermittent racing (*b*) at v_{max}=1.5 m/s; the dependency of the target function (*c*) from time T_2 at v_{max}=1 m/s; the optimum process of intermittent racing (*d*) at v_{max}=1 m/s

The acceleration of the cargo suspension point during the racing and slow braking of the bridge assumed values a_1=0.25 m/s; a_2=–0.05 m/s. In order to reduce the size of the task and the number of independent factors in the computational experiment, an additional condition was adopted for the equality of the first and third periods (racing). In this case, the values of the maximum velocity v_{max}, accelerations and time intervals are related to each other by the dependence

$$2a_1 T_1 + a_2 T_2 = v_{max}. \quad (6)$$

When taking the value of the time of slow braking T_2 between two equal racing periods as the only independent experiment factor, the value of the latter will be calculated from the dependence

$$T_1 = T_3 = (v_{max} - a_2 T_2)/(2 a_1). \quad (7)$$

The values of the angle of the cargo rope *q*, as well as its first derivative, were calculated by (2) for a finite time of each of the three periods and served as initial conditions in (4) for the following period. Some results of the computational experiment are shown in Figure 1 for the values of the maximum velocity v_{max}=1.5 m/s (*a*, *b*) and v_{max}=1 m/s (*c*, *d*). The optimal process of intermittent racing at v_{max}=1.5 m/s is achieved at T_2=0.142 s, T_1=T_3=3.014 s. At v_{max}=1 m/s, the analogous optimal values of the time periods were T_2=0.959 s, T_1=T_3=2.096 s. The total discontinuous acceleration time was 6.17 and 5.151 s for velocities v_{max}=1.5 m/s and v_{max}=1 m/s, respectively.

5. Conclusions

Based on the known DE of swings of a plane pendulum with a movable suspension point and attenuation, the analytical dependences of the slope angle of the cargo rope and its first derivative on time are obtained with constant acceleration of the suspension point. They allow for the separation of the process of acceleration of the bridge or the BC cargo trolley at least three stages (racing with constant acceleration – forward motion with slow braking – further racing with constant acceleration)

to enter the mode of the specified maximum speed of the cargo suspension point when the condition of complete elimination of the swings of the cargo rope around the gravitational vertical.
To advantages of the developed method it is necessary to attribute shortness and limited character of deviations of corners of a cargo rope from a vertical. The use of the method for optimizing the acceleration process of a bridge or a BC freight trolley does not require the implementation of simulation and numerical integration of DE. After the acceleration, the cargo rope and the cargo are in an upright position. At the same time, there are no deviations and no cargo rope swings.
The method can be used in automatic control systems of BC with relay control of moving links for the acceleration of cargo in the mode of eliminating the residual swings of the cargo rope.

References

[1] Shchedrinov A, Serikov S and Kolmykov V 2007 Automatic system of calming the oscillations of cargo for a bridge crane *Devices and Systems. Management, Control, Diagnostics* **8** 13–17
[2] Tolochko O and Bazhutin D 2010 Comparative analysis of methods for damping the oscillations of a load suspended to the mechanism of translational movement of a bridge crane *Electric Machine Building and Electrical Equipment* **75** 22–28
[3] Shvedova O 2014 Algorithms for suppressing oscillations of goods of hoisting and transport mechanisms using fuzzy logic of functioning *Reports of BSUIR* **79** 65–71
[4] Chernousko F, Akulenko L and Sokolov B 1980 *Control of Oscillations* (Moscow: Nauka) p 383
[5] Ridout A 1989 Anti-swing control of the overhead crane using linear feedback *J. of Electrical and Electronics Engineering*. **9** 17–26
[6] Omar H 2003 *Control of Gantry and Tower Cranes : PhD Dissertation* (Blacksburg, Virginia: Virginia Polytechnic Institute and State University) p 100
[7] Korytov M, Shcherbakov V and Volf E 2015 Impact sigmoidal cargo movement paths on the efficiency of bridge cranes *International Journal of Mechanics and Control* vol 16 **2** 3–8
[8] Shcherbakov V, Korytov M, Volf E and Breus I 2015 The reduction of errors of bridge crane loads movements by means of optimization of the spatial trajectory size *Applied Mechanics and Materials* **811** 99–103
[9] Shcherbakov V, Korytov M, Sukharev R and Volf E 2015 Mathematical modeling of process moving cargo by overhead crane *Applied Mechanics and Materials* **701-702** 715–20
[10] Kim Y et al 2003 A new vision-sensorless anti-sway control system for container cranes *Industry Applications Conference* vol 1 pp 262–269
[11] Blackburn D, Singhose W, Kitchen J, Patrangenaru V and Lawrence J 2010 Command shaping for nonlinear crane dynamics *Journal of Vibration and Control* **16** 477–501
[12] Singer N, Singhose W and Seering W 1999Comparison of filtering methods for reducing residual vibration *European Journal of Control* **5** 208–218
[13] Blekhman I 1994 *Vibration mechanics* (Moscow: Fizmatlit) p 400
[14] Shcherbakov V, Korytov M and Wolf E 2015 Algorithm for compensating uncontrolled spatial oscillations of the cargo and increasing the accuracy of its trajectory with a crane *Bulletin of Machine Building* **3** 16–18
[15] Butikov E 2008 Unusual behavior of the pendulum with sinusoidal external action *Computer Tools in Education* **2** 24–36
[16] Kilchevsky N 1972 *Course of theoretical mechanics T 1 (kinematics, statics, point dynamics)* (Moscow: Nauka) p 456

Assessment of Characteristic Function Modulus of Vibroacoustic Signal Given a Limit State Parameter of Diagnosed Equipment

V N Kostyukov[1], A P Naumenko[2], and I S Kudryavtseva[2]

[1]Dynamics SPC, Omsk, Russia

[2]Omsk State Technical University, 11, Mira ave., Omsk, 64450, Russia

E-mail: post@dynamics.ru

Abstract. Improvement of distinguishing criteria, determining defects of machinery and mechanisms, by vibroacoustic signals is a recent problem for technical diagnostics. The work objective is assessment of instantaneous values by methods of statistical decision making theory and risk of regulatory values of characteristic function modulus. The modulus of the characteristic function is determined given a fixed parameter of the characteristic function. It is possible to determine the limits of the modulus, which correspond to different machine's condition. The data of the modulus values are used as diagnostic features in the vibration diagnostics and monitoring systems. Using such static decision-making methods as: minimum number of wrong decisions, maximum likelihood, minimax, Neumann-Pearson characteristic function modulus limits are determined, separating conditions of a diagnosed object.

1. Introduction

The development of the system for defects characteristic criterion is based on a group of diagnostic signals features, which allow accurately determine machinery condition and its change cause [1, 11, 3] has been a recent problem for technical diagnostics. One of the essential parts of the diagnostics is distinguishing of possible technical conditions (diagnoses). In the particular case, one of the two diagnoses are to be distinguished (differential diagnosis or dichotomy), for instance: operating condition (ACCEPTABLE or ACTION REQUIRED conditions) and failure condition (UNACCEPTABLE condition) [4, 5, 6]. The solution of the problem is increasingly based on the static decision-making theory. The work objective is to asses' regulatory values of characteristic function features, used as diagnostic features in systems for vibration diagnostics and monitoring, by methods of static decision-making and risk theory [7, 8, 9, 10, 11, 12, 19, 20, 21].

2. Statement of problem

One of the strategies for diagnostics is static decision-making method [4, 15, 16, 18]. Therewith, a decision procedure is selected on the basis of optimum conditions.

The objective is to select $\boldsymbol{x_0}$ value of x parameter, which is a failure diagnostic feature and distinguishes the characteristic function, particularly, it is the characteristic function modulus given the feature, so if $\boldsymbol{x} < \boldsymbol{x_0}$ it is required to stop the operation and if $\boldsymbol{x} > \boldsymbol{x_0}$ to continue. The features are distinguished on two classes: $\boldsymbol{D_1}$ is operating condition, $\boldsymbol{D_2}$ is failure condition. Then the decision procedure means the following:

$$x \in D_1, \text{if } x > x_0$$
$$x \in D_2, \text{if } x < x_0 \quad (1)$$

Density of probability distribution of the x parameter for different objects condition is shown in figure 1.

The areas of operating (D_1) and failure (D_2) conditions are crossed, so the x_0 value cannot be determined, at which there would be no wrong decisions. The objective is to select the most suitable x_0, so the number of wrong decisions amounts to minimum.

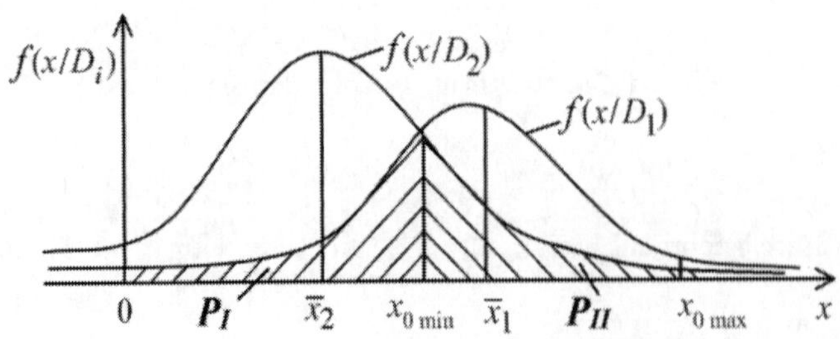

Figure 1. Probability density distribution of diagnostic feature for operating D_1 and failure D_2 conditions; $x_{0\,min}$, $x_{0\,max}$ are the extrimum points of average risk of wrong decisions; P_{II} and P_I are the probabilities of not detecting the defect and false alarm respectively.

3. Theory

There are the possible faults that could appear if decisions made: false alarm (error of first kind P_I): an operative object considered defected (D_1 considered as D_2); not detecting or missing of defect or failure (error of second kind P_{II}): defective object considered operative (D_2 considered D_1). Possible solutions according to the rule (hypothesis) are denoted as H_{ij} (ij = 1, 2), where index i corresponds to the accepted diagnosis; j corresponds to the actual diagnosis.

Thereafter:

H_{21} is false alarm (a type I error or error of first kind);
H_{12} is missing (a type II error or error of second kind);
H_{11} is correct diagnosis (operative condition);
H_{22} is correct diagnosis (faulty condition).

The possibility of the false alarm equals to the possibility of two events: availability of operative condition and $x<x_0$ values for the operative condition is determined as follows:

$$P(H_{21}) = P(D_1)\cdot P(x < x_0 / D_1) = P_1 \cdot \int_{-\infty}^{x_0} f(x/D_1)\,dx = P_1 \cdot [1 - F(x_0/D_1)] \quad (2)$$

The possibility of failure missing

$$P(H_{12}) = P(D_2)\cdot P(x > x_0 / D_2) = P_2 \cdot \int_{x_0}^{\infty} f(x/D_2)\,dx = P_2 \cdot F(x_0/D_2), \quad (3)$$

where $f(x/D_1)$ is probability density of feature for operative condition; $f(x/D_2)$ is probability density of feature for faulty condition; $P_1 = P(D_1)$ и $P_2 = P(D_2)$ is prior probability of the D_1 (operative condition) and D_2 (faulty condition) relatively, which are considered to be known on the basis of preliminary static data: in this case, it is the condition probabilities given the x feature of the specified x_0 value [1];

$F(x_0/D_1) = \int_{-\infty}^{x_0} f(x/D_1)\,dx$ and $F(x_0/D_2) = \int_{x_0}^{\infty} f(x/D_2)\,dx$ is possibilities of operative and faulty conditions relatively at the corresponding intervals from $-\infty$ to x_0 and from x_0 to ∞[4, 15, 16, 18].

The probability of making an erroneous decision is specified by the possibilities of false alarm and of not detecting the failure.

If failures are to be attributed to the costs, and the costs of the correct decisions are to be defined as C_{11} and C_{22}, the formula for average risk (expected amount of loss) is deduced as follows [4, 15, 16, 18]:

$$R = C_{11}P_1\int_{x_0}^{\infty} f(x/D_1)dx + C_{21}P_1\int_{-\infty}^{x_0} f(x/D_1)dx + C_{12}P_2\int_{x_0}^{\infty} f(x/D_2)dx + C_{22}P_2\int_{-\infty}^{x_0} f(x/D_2)dx \quad (4)$$

where C_{21} is cost of false alarm; C_{12} is cost of not detecting the failure (the first index is accepted condition, the second index is current condition); the average formula is $C_{12} >> C_{21}$.

Due to the costs of correct decisions $C_{11}=C_{22}=0$, the formula for average risk is deduced as follows:

$$R = C_{12}P(H_{12}) + C_{21}P(H_{21}) = C_{12}P_2\left[F(x_0/D_2)\right] + C_{21}P_1\left[1 - F(x_0/D_1)\right] \quad (5)$$

The x value is a current (measured) value of a diagnostic feature, and is a random one, so the specified formulae amount to the average value (expectation value) of risk.

4. Outcome of experiment

In order to describe statistic prosperities of diagnostic features, it is required to obtain empiric and theoretical distribution functions and their possibilities for different conditions of diagnosed machinery [1, 11, 17]. The research is the first step to assess limits of diagnostic features, distinguishing machinery condition to the classes.

The Weibull-Gnedenko distribution is used for description of the statistic prosperities of diagnostic features. The distribution is quite cross-functional as if the features are varied it describes a broad range of probabilistic characteristic of different processes [1, 17].

Considering the fact, that some parameters of diagnostic features has Weibull-Gnedenko distribution rule [1], as the rule is transformed to different distribution rules given the different parameters, it is requires to aim at approximation of empiric distribution function (EDF) by two-parameter distribution of Weibull-Gnedenko distribution:

$$F(x) = \begin{cases} 0 & ; x \le 0, \\ 1 - \exp\left\{-\left(\dfrac{x}{c}\right)^b\right\} & ; x > 0 \end{cases} \quad (6)$$

In such case the probability density function is deduced as follows:

$$f(x) = \begin{cases} 0 & ; x \le 0, \\ \dfrac{b}{c}\left(\dfrac{x}{c}\right)^{b-1} \exp\left\{-\left(\dfrac{x}{c}\right)^b\right\} & ; x > 0 \end{cases} \quad (7)$$

Where c is scale parameter, b *is* shape parameter и x is random variable.

Using the method [17] of function approximation distribution by experimental sampled values, the data was processed and theoretical distribution function (TDF) and probability density of characteristic functions parameters of instantaneous values of vibroacoustic signals was determined [22, 23, 24, 25], which was received from such assets of piston compressors as: intake and pressure valve, axial and radial direction of cylinder, slider-crank mechanism, base bearings – for such assets and details conditions (table 1) as UNACCEPTABLE (UAC), ACTIONS REQUIRED (ARQ) and ACCEPTABLE.

Table 1. Parameters of distribution functions.

TDF parameter	CONDITIONS		
	ACCEPTABLE	ACTIONS REQUIRED	UNACCEPTABLE
b	5.050	5.99	5.04
c	0.2849	0.1119	0.0415

Limit value of x_0 is determined by the formula:

$$\frac{f(x_0/D_1)}{f(x_0/D_2)} = \frac{P_2 \cdot (C_{12} - C_{22})}{P_1 \cdot (C_{21} - C_{11})} = \lambda \quad (8)$$

where λ is likelihood ratio: decision threshold.
According to the formula (7) we get:

$$f(x_0/D_1) = \frac{b_1}{c_1}\left(\frac{x_0}{c_1}\right)^{b_1-1} \exp\left\{-\left(\frac{x_0}{c_1}\right)^{b_1}\right\};\ f(x_0/D_2) = \frac{b_2}{c_2}\left(\frac{x_0}{c_2}\right)^{b_2-1} \exp\left\{-\left(\frac{x_0}{c_2}\right)^{b_2}\right\}. \quad (9)$$

The x_0 is understood as a value of a diagnostic feature.
Substituting the values of probability density function in (7) and take up $\mathbf{C_{11}=C_{22}=0}$ the following formula is deduced:

$$\frac{\frac{b_1}{c_1}\left(\frac{x_0}{c_1}\right)^{b_1-1} \exp\left\{-\left(\frac{x_0}{c_1}\right)^{b_1}\right\}}{\frac{b_2}{c_2}\left(\frac{x_0}{c_2}\right)^{b_2-1} \exp\left\{-\left(\frac{x_0}{c_2}\right)^{b_2}\right\}} = \frac{P_2 \cdot C_{12}}{P_1 \cdot C_{21}}. \quad (10)$$

There are the method factors for decision making when determining the threshold value of $\mathbf{x_0}$:

– minimum risk method targeted on minimum achievement of average risk for likelihood ratio $\boldsymbol{\lambda=[(C_{12}-C_{22})P_2]/[(C_{21}-C_{11})P_1]}$;

– minimum number of wrong decision method targeted on the cost of false alarm and of not detecting the failure is equal when $\boldsymbol{\lambda=[P_2]/[P_1]}$;

– maximum likelihood method means the cost and possibility of not detecting the failure approximately equals to the cost and possibility of false alarm given $\boldsymbol{\lambda=1}$;

– minimax method means the risk volume becomes minimum among the maximum values caused by “negative” $\boldsymbol{P_i}$ figure, in such case $\boldsymbol{\lambda=[(C_{12}-C_{22})(1-P_1)]/[(C_{21}-C_{11})P_1]}$;

– Neumann-Pearson method minimizes the possibility of not detecting the failure given the permissible $\boldsymbol{A}$ level of false alarm:

$$P_1 \int_{x_0}^{\infty} f(x/D_1)dx \le A \quad (11)$$

The calculation results of given the prior values $P_1=0.97$, $P_2=0.03$, $C_{12}=1$, $C_{21}=0.01$ for distinguishing the ACTIONS REQUIRED and UNACCEPTABLE conditions are denoted in table 2; the results for ACCEPTABLE and UNACCEPTABLE conditions are denoted in table 3.

Table 2. Calculation results of feature threshold value (ACTIONS REQUIRED – ACCEPTABLE).

Methods	Probabilities and risks			
	x_0	$P(H_{21})$	$P(H_{12})$	R
Minimum risk	0.088	0.013	0.000682	0.000814
Minimum number of wrong decisions	0.077	0.000115	0.000126	0.000126
Maximum likelihood	0.084	0.000409	0.000833	0.000874
Minimax	0.119	0.271	0.002259	0.004973
Neumann-Pearson	0.093	0.049	0.000284	0.000773

Table 3. Calculation results of feature threshold value (ACTIONS REQUIRED – ACCEPTABLE).

Methods	Probabilities and risks			
	x_0	$P(H_{21})$	$P(H_{12})$	R
Minimum risk	0.216	0.017	0.000682	0.000748
Minimum number of wrong decisions	0.186	0.0001875	0.000126	0.000139
Maximum likelihood	0.206	0.005637	0.000833	0.000823
Minimax	0.285	0.2	0.001686	0.003688
Neumann-Pearson	0.227	0.043	0.000284	0.000636

5. Discussion

The calculation results (table 2, figure 2, 3) show that the minimum number of erroneous decision method provide the lowest values of decision-making risk $\boldsymbol{R}$, the possibility of not detecting the failure $\boldsymbol{P(H_{12})}$ and false alarm $\boldsymbol{P(H_{21})}$ for characteristic function modulus $\boldsymbol{\Theta}(0.8)$ of vibroacoustic signal, in case of selecting between UNACCEPTABLE and ACTIONS REQUIRED conditions. The second place takes the Neumann-Pearson method, which provides minimum values of decision-making risk $\boldsymbol{R}$, the possibility of not detecting the failure $\boldsymbol{P(H_{12})}$. The minimax method provides the highest values of decision-making risk $\boldsymbol{R}$, the possibility of not detecting the failure $\boldsymbol{P(H_{12})}$ and false alarm $\boldsymbol{P(H_{21})}$. The third place takes the minimum risk method.

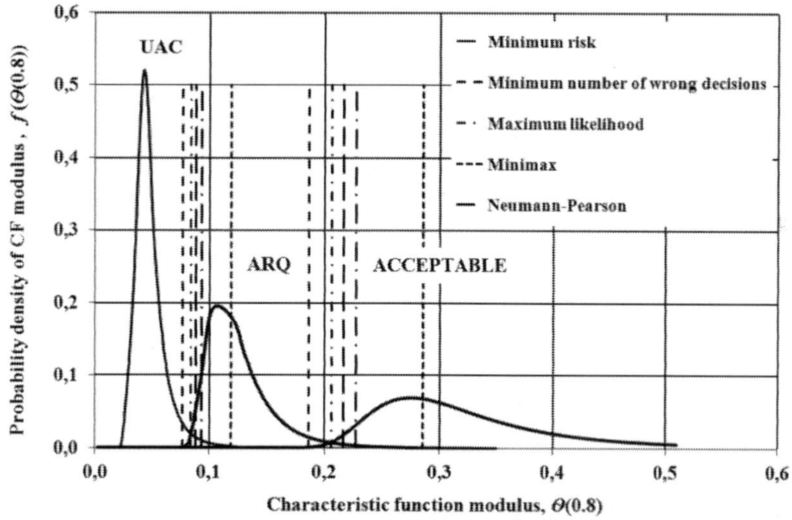

Figure 2. Probability density function of characteristic function modulus value given parameter of characteristic function for different condition of diagnosed machine and modulus threshold values (x_0 from table 2 and 3), determined by different methods of decision-making.

The minimum number of erroneous decision method provides the lowest values of the decision-making risk ($\boldsymbol{R}$) and the possibility of not detecting the failure ($\boldsymbol{P(H_{12})}$), when determining the threshold values for ACTIONS REQUIRED and ACCEPTABLE conditions (table 3, figure 2). The calculation by the Neumann-Pearson methods are reliable as well, and provide lower values of the decision-making risk and the possibility of not detecting the failure. The worst results are given by the minimax method.

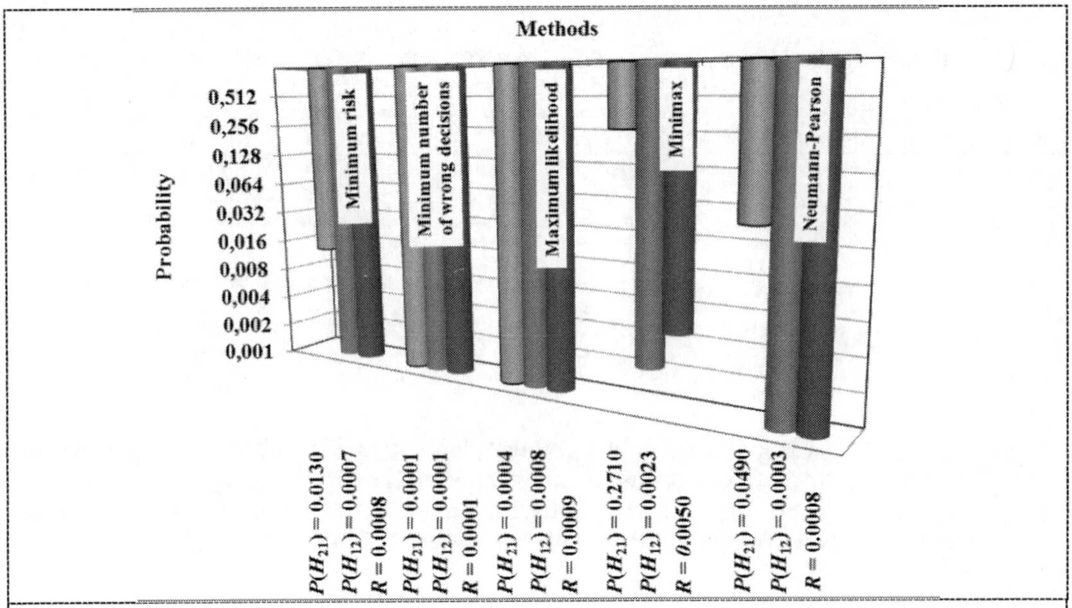

Figure 3. Possibilities of failure skipping $P(H_{12})$, false alarm $P(H_{21})$ decision-making risk R obtained by different decision-making methods when distinguishing ACTIONS REQUIRED and UNACCEPTABLE conditions.

6. Summary and conclusion

The results of the research show that in purpose to determine the limits of the characteristic function modulus of instantaneous value of vibroacoustic signals, corresponding to different conditions of diagnosed machinery; it is required to apply statistical methods for decision-making. That allows determining threshold values of diagnostic features on the basis of decision-making risk assessment. The suggested calculation method is targeted on the diagnostic features and based on the parameters of the characteristic function. Using the statistical method of decision-making the threshold values of the characteristic function modulus was obtained given the characteristic function parameter for distinguishing the ACCPERABLE and ACTION REQUIRED conditions, ACTIONS REQUIRED and UNACCEPTABLE conditions. The vibroacoustic signal and the characteristic function of its instantaneous value were used as the diagnostic signal.

References

[1] Kostyukov V N and Naumenko A P 2016 About the experience in operation of reciprocating compressors under control of the vibration monitoring system *Procedia Engineering* **152 (201)** pp 497–504

[2] Naumenko A P 2014 Basic principles of technology for diagnosing and monitoring of real-time piston machines *XX All-Russian Conf. on Nondestructive Testing and Technical Diagnostics* (*Moscow, 3–6 March, 2014*) (Moscow: "Spectrum" Publishing House) pp 358–362

[3] Kostyukov V N and Naumenko A P 2002 Vibration diagnostics of piston compressors *Compressor Engineering and Pneumatics* **3** pp 30–31

[4] Birger I A 1978 *Technical Diagnostics* (Moscow: Mashinostroenie)

[5] Kostyukov V N and Naumenko A P [et al.] 2012 Standards in the field of technical condition of hazardous equipment *Operational Safety in Industry* **7** pp 30–36

[6] Kostyukov V N and Naumenko A P 2013 Normative and methodical support of diagnostics and monitoring of piston compressors *Operational Safety in Industry* **5** pp 66–70

[7] Kostyukov V N Naumenko A P and Sidorenko (Kudryavtseva) I S 2009 Use of the characteristic function for diagnosis of piston machines *Dynamics of Systems, Mechanisms and Machines* (Omsk: Omsk State Technical University) **2** pp 32–35
[8] Sidorenko (Kudryavtseva) I S and Naumenko A P 2010 Analysis of characteristic functions of vibroacoustic signals of piston compressor valves *Science, Education, Business* (Omsk: Publishing House "Polygraphic Center KAN") pp 259–264
[9] Kudryavtseva I S and Naumenko A P 2011 Evaluation of possibility of using the characteristic function for the diagnostics of water hammers *Science, Education, Business* (Omsk: Publishing House "Polygraphic Center KAN") pp 215–217
[10] Kostyukov V N, Naumenko A P and Kudryavtseva I S 2014 Diagnosis of rolling bearings according to the parameters of the characteristic function *Dynamics of Systems, Mechanisms and Machines* (*Dynamics*) **4** pp 142–145
[11] Kostyukov V N, Naumenko A P and Kudryavtseva I S 2016 Characteristic function as an informative diagnostic feature *Technique and Technology of Petrochemical and Oil and Gas Production* (Omsk: Omsk State Technical University Publishing) pp 113–114
[12] Kostyukov V N, Naumenko A P, Kudryavtseva I S [et al.] 2016 Formation of the vector of diagnostic features on the basis of the characteristic function of the vibroacoustic signal *Testing. Diagnostics* **8** pp 22–29
[13] Kostyukov V N and Naumenko A P 2008 Problems and solutions for the safe operation of reciprocating compressors *Compressor Engineering and Pneumatics* **3** pp 21–28
[14] Kostyukov V N and Naumenko A P 2007 Monitoring system for technical condition of reciprocating machines *Testing. Diagnostics* **3** pp 50–58
[15] Harkevich A A 2009 *Fighting Interference* (Moscow: URSS)
[16] Bogdan N V, ZHilevich M I and Krasnevskij L G 2000 *Technical Diagnostics of Hydraulic Systems* (Minsk: Belavtotraktorostroenie)
[17] Naumenko A P 2011 The technique of statistical analysis of diagnostic features *Science, Education, Business* (Omsk: Publishing House «Polygraphic Center KAN») pp 188–195
[18] Kostyukov V N and Naumenko A P 2014 Assessment of the risk of choosing the normative values of diagnostic features *Dynamics of Systems, Mechanisms and Machines* **4** pp 150–154
[19] Kostyukov V N, Naumenko A P, Kudryavtseva I S and Boychenko S N 2015 Diagnostics of rolling bearings by the parameters of the characteristic function *12th Int. Conf. on Condition Monitoring and Machinery Failure Prevention Technologies (CM 2015 /MFPT 2015)* (The British Institute of Non-Destructive Testing. Printed by Curran Associates, Inc.) pp 246–249
[20] Kostyukov V N, Naumenko A P, Kudryavtseva I S and Boychenko S N 2016 Formation of diagnostic features vector based on characteristic function of vibroacoustic signal *13th Int. Conf. on Condition Monitoring and Machinery Failure Prevention Technologies (CM2016 / MFPT2016)* (British Institute of Non-Destructive Testing. Printed by Curran Associates, Inc.) pp 171–181
[21] Kostyukov V N, Naumenko A P and Kudryavtseva I S 2016 Usage of characteristic function as informative diagnostic feature *Procedia Engineering* **152** (**201**) pp 521–526
[22] Yu J 2004 Empirical Characteristic function estimation and its applications *Econometric Reviews* **23** (2) pp 93–123
[23] Feuerverger A and Mureika R A 1977 The empirical characteristic function and its applications *Ann. Statist.* **5** pp 88–97
[24] Knight J L, Yu J 2002 The empirical characteristic function in time series estimation *Econometric Theory* **18** pp 691–721
[25] Jiang G and Knight J L 2002 Estimation of continuous time processes via the empirical characteristic function *J. Business Economic Statist.* **20** pp 198–212

IOP Conf. Series: Journal of Physics: Conf. Series **944** (2017) 012064 doi:10.1088/1742-6596/944/1/012064

Electrical engineering unit for the reactive power control of the load bus at the voltage instability

A V Kotenev, V I Kotenev, V V Kochetkov and D A Elkin[1]

Electrical Engineering Department, Samara State Technical University, 244, Molodogvardeyskaya St, Samara, 443100, Russia

[1] Corresponding author e-mail: elkin.dmirty.94@gmail.com

Abstract. For the purpose of reactive power control error reduction and decrease of the voltage sags in the electric power system caused by the asynchronous motors started the mathematical model of the load bus was developed. The model was built up of the sub-models of the following elements: a transformer, a transmission line, a synchronous and an asynchronous loads and a capacitor bank load, and represents the automatic reactive power control system taking into account electromagnetic processes of the asynchronous motors started and reactive power changing of the electric power system elements caused by the voltage fluctuation. The active power/time and reactive power/time characteristics based on the recommended procedure of the equivalent electric circuit parameters calculation were obtained. The derived automatic reactive power control system was shown to eliminate the voltage sags in the electric power system caused by the asynchronous motors started.

1. Introduction

The electric power system is intended to supply consumers timely with the power of the appropriate quality and in sufficient quantity provided that the transmission losses are minimized. Actually, these losses accounts for the substantial amount of the electric power as both bulk power systems and distribution systems of Russia operate at the low power factor.

The main reactive power consumers are transformers and asynchronous motors operating at the power factor that is equal to 0.75-0.9. The total reactive power losses account for over 50% of the total reactive power produced. One of the ways to reduce these losses is the reactive power compensation in the load buses.

In the presence of the synchronous motors in the load buses the reactive power compensation can be carried out by means of the automatic excitation control continuously monitoring such parameters as reactive power of the asynchronous load, voltage of the load bus, etc. [1, 2] and applying them as functions to build the control system.

Sufficient number of papers [3-10] covers mathematical models and excitation control algorithms of the synchronous machines operating as generators. Much fewer papers [2, 11-14] address the issue about the mathematical models of the synchronous motors. Moreover, the voltage fluctuation of the electric power system and transients during asynchronous motor breaking away, accelerating and re-accelerating are not taken into consideration in those models.

2. Models and methods

2.1. The mathematical model of the load bus

The single-line diagram of the load bus is shown in figure 1.

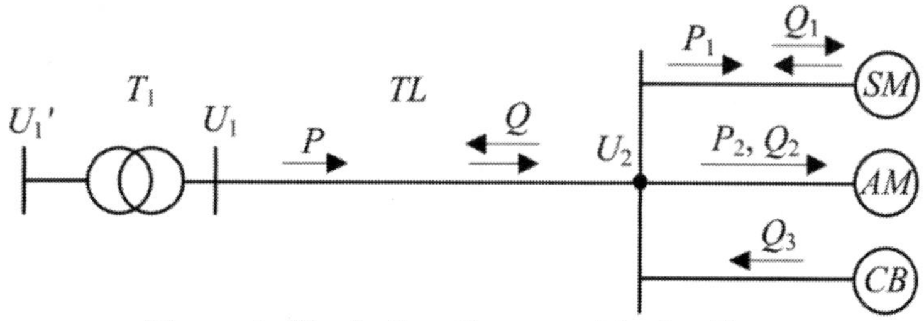

Figure 1. Single-line diagram of the load bus.

The elements of the single-line diagram are as follows: the transformer (T_1); the transmission line (TL); the asynchronous motors (AM) with the total rated power P_r=3000 kW; the synchronous motor (SM); the capacitor bank (CB).

The letter symbols of the diagram are following: P, P_1, P_2 – active powers of the transmission line, the synchronous and the asynchronous motors, respectively; Q, Q_1, Q_2 – reactive power of the transmission line, the synchronous and the asynchronous motors, respectively; U_1', U_1, U_2 – primary and secondary voltages of the transformer and voltage at the input terminals of the motors and the capacitor bank, respectively.

The mathematical model of the load bus is represented by the operator equations (1) [15, 16]:

$$\begin{gathered}
\Delta E_q(p) = \left(\frac{1}{R_f} \Delta U_f(p) + T_3 p \Delta\theta(p) - T_4 p \Delta U_2(p) \right) W_2(p) \\
\Delta\theta(p) = \left(\frac{k_1}{E_{q0}} \Delta E_q(p) + k_2 \Delta m(p) + \frac{k_1}{U_{02}} \Delta U_2(p) \right) W_1(p) \\
\Delta Q_1(p) = k_3 \Delta U_2(p) + k_4 \Delta\theta(p) + k_5 \Delta E_q(p) \\
\Delta Q(p) = \Delta Q_1(p) - \Delta Q_2(p) + \Delta Q_3(p) \\
\Delta U_2(p) = \left[k_{10} \Delta Q(p) + \Delta U_1(p) + (k_6 \Delta m(p) - \Delta P_2(p)) k_9 \right] k_{11}
\end{gathered} \tag{1}$$

The parameters of the synchronous motor:

$$\begin{gathered}
T_3 = \frac{3U_{02} X_{ad} \sin\theta_0}{2\omega_0 R_f x_d}; X_{ad} = \sqrt{3} x_{ad}; T_4 = \frac{x_{ad} \cos\theta_0}{\omega_0 R_f x_d}; \omega_0 = 2\pi f \\
W_2(p) = \frac{1}{T_2 p + 1}; \; T_2 = \frac{x'_d}{x_d} T_f; x'_d = x_d - \frac{3X_{ad}^2}{2x_f}; T_f = \frac{x_f}{\omega_0 R_f} \\
k_1 = -tg\theta_0; k_2 = \frac{1}{\cos\theta_0}; W_1(p) = \frac{1}{T_1^2 p^2 + 2\varepsilon_1 T_1 p + 1}; \; T_1 = \sqrt{\frac{J}{p_p m_{em} M_r \cos\theta_0}} \\
\varepsilon_1 = \frac{m_{bd} M_r}{2 s_{bd} \omega_r \sqrt{p_p m_{em} M_r J \cos\theta_0}}; m_{bd} = \frac{M_{bd}}{M_{em}}; \; m_{em} = \frac{M_{em}}{M_r}; k_3 = \frac{E_{q0} \cos\theta_0}{x_d} \\
k_4 = -\frac{U_{02} E_{q0}}{x_d} \sin\theta_0; \; k_5 = \frac{U_{02} \cos\theta_0}{x_d}; k_6 = P_{em}
\end{gathered} \tag{2}$$

where R_f – excitation winding resistance; E_{q0} – quadrature-axis electromotive force of the synchronous motor; U_{02} – steady-state load bus voltage; x_d – synchronous reactance; θ_0 – steady-state load angle; x_{ad} – mutual reactance; J – moment of inertia of the synchronous drive; p_p – number of pairs of poles of the synchronous motor; M_r, M_{bd}, M_{em} – rated, breakdown and maximum electromagnetic torques, respectively; s_{bd} – breakdown slip; x'_d – direct-axis transient reactance; ω_r – rated angular frequency of the synchronous motor; x_f – excitation winding reactance; P_{em} – ceiling electromagnetic power.
Transfer coefficients of the following load bus elements:
Asynchronous motors:

$$k_7 = \frac{2.3}{U_{02}} \tag{3}$$

Capacitor bank:

$$k_8 = \frac{2Q_{03}}{U_{02}} \tag{4}$$

Transmission line:

$$k_9 = -\frac{R_{TL} + R_T}{2U_{02} - U_{01}};\ k_{10} = -\frac{x_{TL} + x_T}{2U_{02} - U_{01}};\ k_{11} = -\frac{U_{02}}{2U_{02} - U_{01}} \tag{5}$$

where R_{TL}, x_{TL} – resistance and reactance of the transmission line; R_T, x_T – resistance and reactance of the transformer.
The transfer coefficient k_{12} of the element representing the negative feedback by means of k_7 and positive feedback by means of k_8 is equal to:

$$k_{12} = \frac{1}{1 + (k_7 - k_8)k_{10}} \tag{6}$$

The set of equations (1)-(5) is represented in the form of the block-diagram in figure 2 where the voltage feedback ΔU_2 of the synchronous motor is expressed by the elements with the coefficients k_3, k_1/U_{02} and the element with the transfer function T_4p.

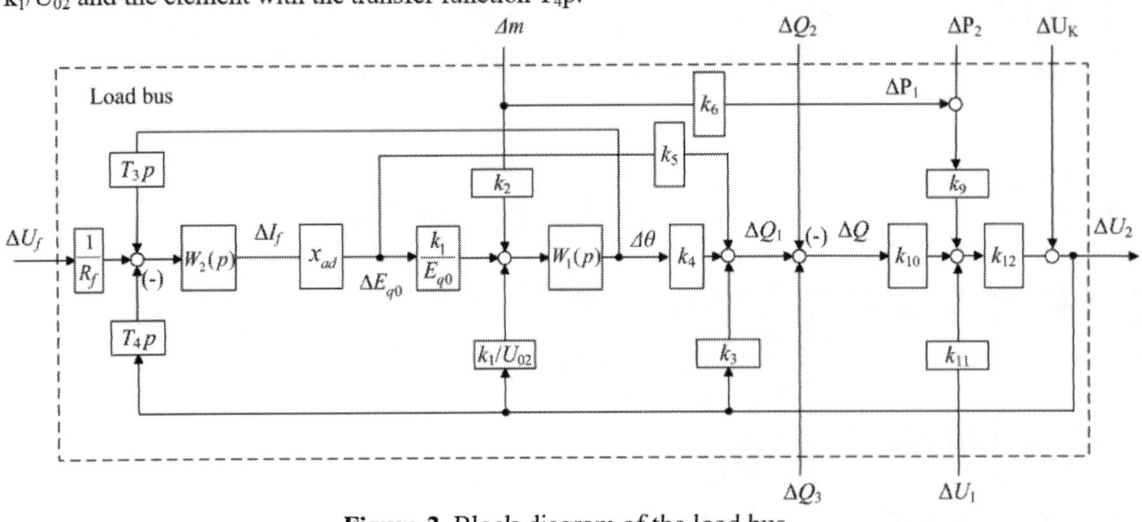

Figure 2. Block-diagram of the load bus.

The parameters of the above-mentioned block-diagram equal: T_0=0.31 s, ε_1=0.1, R_f=0.18 Ohms, X_{ad}=52 Ohms, T_2=0.367 s, U_{02}=6000 V, E_{q0}=11,800 V, T_3=162 A·s, T4=0.034 s/Ohms, k_1= –0.75, k_2=1.25, k_3= –0.078 kA, k_4= –1.29 kVA, k_5= 0.154 kA, k_6=2,500 kW, k_7= –0.075 kA^{-1}, k_9= –1.1, k_{10}=1.19.

2.2. Automatic reactive power control system model

Functional diagram of the automatic reactive power control system is shown in figure 3.

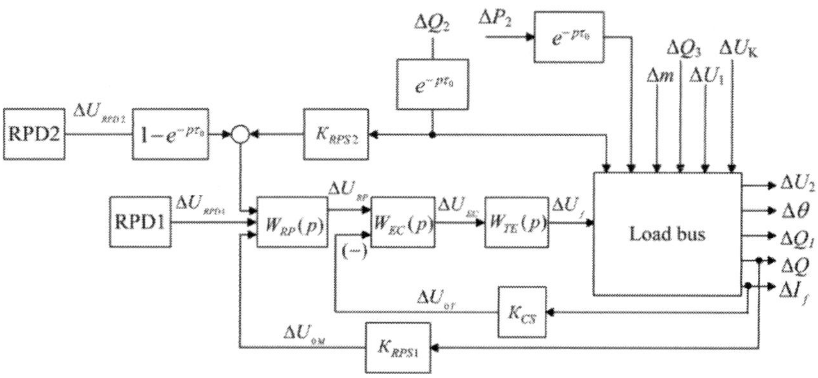

Figure 3. Functional diagram of the automatic reactive power control system.

The letter symbols of the diagram are following: RPD1, RPD2 – reactive power definers; W_{RP}(p), W_{EC}(p), W_{TE}(p) – transfer functions of the reactive power, excitation current and thyristor excitation, respectively; K_{CS}, K_{RPS1}, K_{RPS2}, – transfer functions of the current sensor, reactive power sensor of the load bus and reactive power sensor of the asynchronous motors, respectively.

The set of equations includes both the negative feedbacks of the excitation current ΔIf and reactive power ΔQ and the positive feedback of the reactive power ΔQ_2 that is the most contributing disturbance.

The asynchronous motors are started with delay of $t=\tau_0$. During the same period of time $t=\tau_0$ the RPD2 in collaboration with RPD1 produces the maximum power over this time. Thus, the feedforward control is provided by means of the elements with the transfer functions $e^{-p\tau_0}$ and $1-e^{-p\tau_0}$.

2.3. The procedure of the reactive power/time and active power/time characteristics obtaining during the squirrel cage asynchronous motor starting

Active and reactive powers of the asynchronous motor started are calculated by equations:

$$P_2(s) = \frac{U_2^2 R(s)}{R^2(s) + x^2(s)}$$
$$Q_2(s) = \frac{U_2^2 x(s)}{R^2(s) + x^2(s)} \qquad (7)$$

Slip 's' can be found applying the next formula:

$$\frac{\mathrm{d}s(t)}{\mathrm{d}t} + \frac{2M_{bd}s_{bd}s(t)}{J(s^2(t) - s_{bd}^2)} = \frac{M_L}{J} \qquad (8)$$

where M_L – load torque.

Slip-resistance and slip-reactance relationships can be derived from the equivalent electric circuit [17]:

$$R(s) = R_1 + \frac{x_\mu^2 R_2(s)}{sA(s)};\ x(s) = \frac{x_2(s)x_\mu^2 + x_\mu\left(\frac{R^2(s)}{s} + x_2^2(s)\right)}{A(s)}$$
$$A(s) = (x_\mu + x_2(s))^2 + \frac{R^2(s)}{s^2} \tag{9}$$

where x_1, R_1 – reactance and resistance of the stator winding; x_2, R_2 – reactance and resistance of the rotor winding referred to the stator; x_μ – magnetizing reactance (that represents the reactance of the air gap).

The reactances $x_1(s)$, $x_2(s)$ and resistance $R_2(s)$ are not constant and depend on the slip 's' as when the rotor accelerates the skin effect and rotor teeth saturation effect appear to a greater extent and result in decrease of $x_1(s)$, $x_2(s)$ and increase of $R_2(s)$.

The relationships between $R_2(s)$, $x_K(s)$ and s are better to be represented by the approximated nonlinear function of the form [17]:

$$R_2(s) = \begin{cases} R_2(s_r),\ \forall 0 \le s \le s_{bd}; \\ R_2(0) + (R_2(1) - R_2(0))\sqrt{s},\ \forall s_{bd} < s \le 1 \end{cases}$$
$$R_2(0) = \frac{R_2(s_r) - R_2(1)\sqrt{s_{bd}}}{1 - \sqrt{s_{bd}}} \tag{10}$$
$$x_K(s) = \begin{cases} x_K(s_r),\ \forall 0 \le s \le s_{bd}; \\ x_K(s_r) - \frac{(x_K(s_r) - x_K(1))}{0.25 - s_{bd}}(s - s_{bd}),\ \forall s_{bd} < s \le 0.25; \\ x_K(1),\ \forall 0.25 < s \le 1 \end{cases}$$

where $x_K(s)$ – total reactance of the stator and the rotor.

The stator reactance and the rotor reactance are set to as in [17] and calculated by the formulas:

$$x_1(s) = 0.42x_K(s)$$
$$x_2(s) = 0.58x_K(s) \tag{11}$$

In the existing papers the unknown values R_1, $R_2(s_r)$, $x_K(s_r)$ mentioned above are typically obtained from the solution of two equations derived from the electromagnetic power equation at $s=s_r$ and $s=s_{bd}$. But as long as there are three unknown values some authors [18, 19] predetermine unjustifiably one of them to solve the equation set, and the others [20, 21] apply different iterative algorithms to find a solution. In the first case the result is unambiguous and has unacceptable error if the unknown value is predetermined incorrectly, in the second one, the calculation is extremely complicated and hardly applicable to the practice application.

In this paper the third equation [22]:

$$R_2 = 2a_0 s_r b_{bd} x_K \tag{12}$$

is added to the existing two ones:

$$R_1^2 + \frac{R_2^2}{s_r^2} - \frac{a_1}{s_r}R_2 + \frac{2}{s_r}R_1R_2 + x_k^2 = 0 \tag{13}$$
$$R_1 = a_2 - a_3x_K^2 \tag{14}$$

where $a_1 = \frac{U_r^2}{a_0 P_r}$, $a_0 = \frac{1,02}{1-s_r}$, $a_2 = \frac{a_0 a_1}{4b_{bd}}$, $a_3 = \frac{b_{bd}}{a_0 a_1}$, $R_2 = R_2(s_r)$, $x_K = x_K(s_r)$, b_{bd} – torque ratio at the breakdown slip.

These two equations are obtained from the electromagnetic power equation at $s=s_r$ and $s=s_{bd}$:

$$a_0 P_r = \frac{U_r^2 R_2 s_r^{-1}}{(R_1 + R_2 s_r^{-1})^2 + x_K^2}$$
$$b_{bd} P_r = \frac{U_r^2}{R_1 + \sqrt{R_1^2 + x_K^2}} \quad (15)$$

This third equation is derived from the reactive power rotor dissipation equation:

$$\frac{P_r}{2b_{bd}} = \frac{U_r x_K}{(R_1 + R_2 s_r^{-1})^2 + x_K^2} \quad (16)$$

and the equation (15).

The equations (12) and (14) plugged in the equation (13) give the following one-variable equation:

$$\sum_{j=0}^{4} c_j x_K^j = 0 \quad (17)$$

where $c_0 = \left(\frac{a_0 a_1}{2b_{bd}}\right)^2$, $c_1 = \frac{2a_2 - a_1}{s_r a_3^2}$, $c_2 = \frac{1 - 2a_2 a_3 + a_4^2 s_r^{-2}}{a_3^2}$, $c_3 = -\frac{2a_4}{a_3 s_r}$, $c_4 = 1$.

Magnetizing reactance [22]:

$$x_\mu = \frac{U_r^2}{P_r(\eta_r^{-1} tg\varphi_r - 0,5b_r^2)} - x_1 \quad (18)$$

where η_r – rated efficiency of the motor; φ_r – phase difference between the stator current and the voltage under rated operating conditions.

The locked-rotor resistance and reactance (at s=1):

$$R_2(1) = \frac{K_t a_0 P_r}{3K_i^2 I_r^2} \quad (19)$$

$$x_K(1) = \sqrt{\frac{U_r^2}{K_t a_0 P_r} R_2(1) - (R_1 + R_2(1))^2} \quad (20)$$

where K_i, K_t – locked-rotor current ratio and locked-rotor torque ratio; I_r – rated current of the motor.

3. Experiment results

The reactive power $q_1(t)=Q_2(t)/P_{2r}$ against time and the active power $p_1(t)=p_2(t)/P_{2r}$ against time characteristics are obtained under the conditions of the 125 kW asynchronous motor starting at the load torque of $M_L=1.1M_r$ applying the formulas (6)-(16) and shown in figure 4. The derived characteristics have closely corresponded to the analogous ones q_2, p_2 calculated by the procedure [23] taking into consideration electromagnetic processes.

The error of the resistances and reactances calculation does not exceed 4% (the calculation algorithm can be found in [22]).

The reactive power $\Delta Q(t)$ against time and the voltage $\Delta U_2(t)$ against time characteristics are shown in figure 5. First characteristic is obtained during the 125 kW asynchronous motor starting, but without any control; second one – with control system (non-feedforward); third one – with feedforward control system at τ_0=0.22 s; fourth one – with feedforward control system at τ_0=0.4 s.

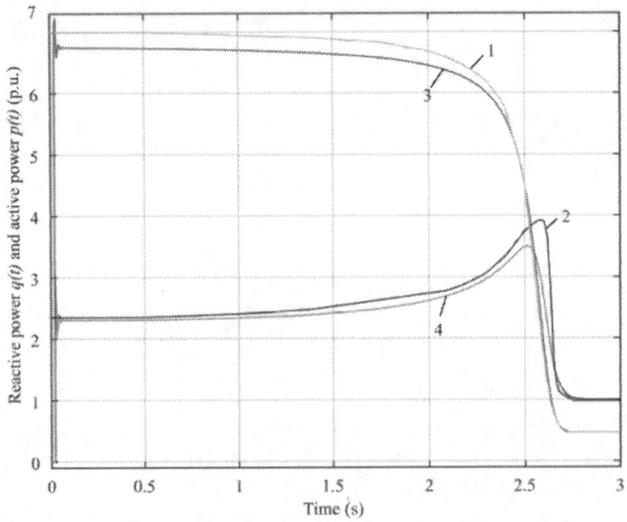

Figure 4. The reactive power and the active power against time characteristics: 1 – $q_1(t)$; 2 – $p_1(t)$, 3 – $q_2(t)$; 4 – $p_2(t)$.

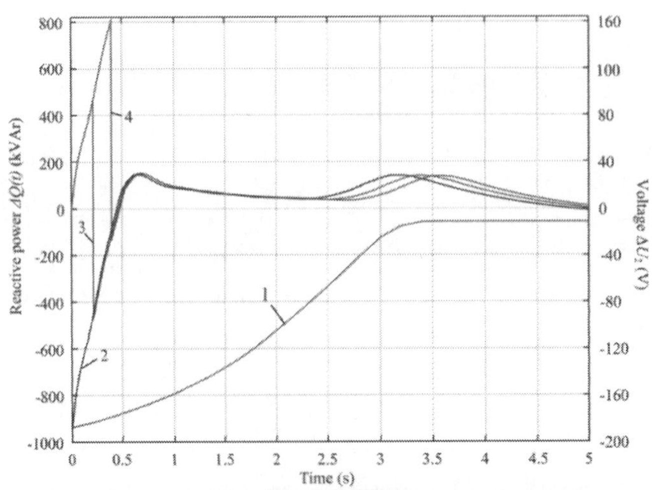

Figure 5. The characteristics $\Delta Q(t)$, $\Delta U_2(t)$ during the 125 kW asynchronous motor starting: 1 – without any control system; 2 – with control system (non-feedforward); 3 – with feedforward control system at τ_0=0.22 s; 3 – with feedforward control system at τ_0=0.4 s.

4. Conclusion

The automatic reactive power control system of the load bus was developed. The main parameters of the load bus elements were taken into consideration in the mathematical model of this system, in contrast to the counterparts.

The reactive and active power calculation of the squirrel cage asynchronous motor started was carried out in accordance with the proposed procedure that has smaller error than other ones.

The proposed automatic reactive power control system with the feedforward control during the asynchronous motors starting improved the control quality substantially.

The reactive power recovery time at the load bus during the motor starting was reduced from 2.8 s to 0.3 s under non-feedforward control. The voltage sag duration was reduced in half under feedforward control at τ_0=0.22 s, but at τ_0=0.4 s the voltage sag was eliminated almost completely. However, the amplitude of the controlled values increased, but still remained within the acceptable limits (5-10% of the rated value).

The calculated results were checked and proved on the laboratory test bench with the 5 kW synchronous motor and asynchronous motor with the total power 6 kW.

The obtained control system can be applied in the electric power systems of the oil fields, oil refining, chemical, machining and other plants, where the synchronous and asynchronous motors are used.

References

[1] Venikov V A 1985 *Electromechanical transients in the electric power systems* (Moscow, Vysshaya Shkola Publication) pp 380-400
[2] Petelin D P 1988 *Automatic control of the synchronous electric drives* (Moscow, Energiya Publication) pp 114-129
[3] Glebov I A 1979 *Excitation control systems of the heavy synchronous machines* (Nauka Publication), pp 190-198
[4] Gorev A N 1985 *Transients of the synchronous machine* (Leningrad, Nauka Publication) pp 211-220
[5] Zhdanov P S 1979 *Stability issues of the electric power systems* (Moscow, Energiya Publication) pp 355-359
[6] Kryuchkov I P 2009 *Transients in the electric power systems* (Moscow, ID MPEI Publication) pp 338-341
[7] Matyukhin V M 1957 *J. USSR Izvestia* **7** pp 3-7
[8] Mikhnevich G V 1978 *Structure synthesis of the synchronous machine automatic excitation control system* (Moscow, Vysshaya Shkola Publication) pp 173-176
[9] Jones H 1996 *J. Gas World International* **4858** pp 30-32
[10] Lyle C and Marshall R 1995 *J. Gas Engineering and Management* **2** pp 34-37
[11] Shvarts G R Abakumov A M Migacheva L A Rasskazov F N and Kuznetsov P K 2008 *Application of the controlled electric drive in the oil and gas transportation* (Moscow, Mashinostroyeniye Pabliction) pp 110-114
[12] Shvarts I G 1992 *J. Industry Power Engineering* **2** pp 38-39
[13] Sazykin V G 1989 *J. Industry Power Engineering* **11**, p 49
[14] Sazykin V G 1993 *J. Industry Power Engineering* **1** p 15
[15] Osipov V S Kotenev V I and Kochetkov V I 2013 *J. Vestnik of Samara State Technical University* **39** pp 175-184
[16] Kotenev V I Osipov V S and Kochetkov V V 2014 *J. Electromechanics* **5** pp 103-106
[17] Kotenev V I Kotenev A V and Kochetkov V V 2014 *J. Vestnik of Samara State Technical University* **4** pp 122-130
[18] Moshchinskiy Yu A Bespalov V Ya and Kiryakin A A 1998 *J. Electrisity* **4** pp 38-42
[19] Gridin V M 2012 *J. Electrisity* **5** pp 40-44
[20] Kachin S I Chernyshov A Yu and Kachin O S 2010 *Automatically-controlled drive* (Tomsk, Tomsk Polytechnical University Publication) pp 99-102
[21] Syromyatnikov I A 1984 *Duties of the asynchronous and synchronous motors* (Energoatomizdat Publication) pp 188-199
[22] Kotenev V I Kotenev A V and Kochetkov V V 2016 *J. Electromechanics*, 2016 **6** pp 13-16
[23] Semenov A S 2014 *J. Vestik of North-Eastern Federal University* **1** pp 51-59

Influence of the Microchannel Height on the Impedance of a Flow Electrochemical Cell with Planar Interdigitated Electrodes

A G Kozlov, E A Fadina
Omsk State Technical University, 11, Mira ave., Omsk, 64450, Russia

kozlov1407@gmail.com

Abstract: The article considers an analytical approach for determining the impedance of an electrochemical cell with planar interdigitated microelectrodes located in a microchannel and for estimation of the effect on the impedance characteristics of the microchannel height. The proposed approach is based on the use of an electric equivalent circuit of the electrochemical cell for determining its impedance and a mathematical simulation of the distribution of the electrical potential in the structure of interdigital microelectrodes located in a microchannel to determine the constant electrochemical cell. Using this approach, the analysis of impedance dependencies is performed and the frequency dependences of the real and imaginary parts, the modulus and the argument of the impedance of the electrochemical cell with the specific geometry of the interdigital microelectrodes for different microchannel heights are determined.

Keywords – electrochemical cell, interdigitated microelectrodes, impedance, cell constant, impedance dependence

1. Introduction

Presently, non-destructive methods for investigating substances and materials become more and more urgent. One of these methods is the method of impedance spectroscopy. Sensors that implement impedance measurements are used in a wide variety of fields, such as medical and microbiological researches, measurement and control of substance concentrations, study of corrosion inhibitors of metals and nanostructures, monitoring of energy carriers (batteries, solar cells, fuel cells) [1–4] .

The method of impedance spectroscopy is based on the study of the response of the system to the external electric effect and the measurement of its impedance at various frequencies. For measuring the impedance, special electrochemical cells with two or three electrodes are used, whose design, depending on the object of control and the purposes of the investigation, can be different. A large number of electrodes for the impedance measurement are known, such as coaxial electrodes, disk and ring-shaped, planar and volume, symmetric and asymmetric ones [5]. Among the microelectrodes the planar interdigitated ones were most widely used [6–10].

Presently, impedance methods are widely used in microsystem technology to develop microanalytical systems. In these systems, flow electrochemical cells are used for the analysis of liquid substances, in which the microelectrode system is located in the microchannel and contacts with the flow of the liquid being analyzed. For this structure of the electrochemical cell, the results of the analysis are affected by a number of additional factors that require detailed consideration and must be taken into account in the development of microsystems for measuring impedance and interpretation and using analysis results.

2. Formulation of the Problem

Measurement of the impedance in a flow electrochemical cell has a number of features, among which the main ones are the following:

1) the test substance (liquid) is constantly shifted relative to the electrode system, and this can affect the magnitude of the impedance;

2) the channel of the electrochemical cell in the electrodes region has limited dimensions, which affect the magnitude of the impedance.

At a low flow velocity, the first feature can be ignored, due to the small effect on the magnitude of the impedance. The second feature relates with small channel sizes and leads to an increase in the measured value of the impedance.

Currently, flow electrochemical cells with interdigital electrodes obtained by the methods of microsystem technology are beginning to be widely used in measuring impedance. In these cells, the microelectrode system is located in a microchannel, the height of which is limited and slightly different from the size of the microelectrode dimensions. For these systems, the estimation of the influence of the microchannel dimensions on the measured impedance is an actual problem.

In this regard, the aim of this paper is to develop the analytical approach that allows one to estimate the influence of the microchannel height on the impedance of a flow electrochemical cell with planar interdigitated electrodes.

3. Theory

Figure 1(a) shows the design of the flow electrochemical cell with planar interdigitated microelectrodes (IME). To analyze the impedance of the electrochemical cell, one should consider its equivalent circuit, which is shown in figure 1(b). The elements of this circuit characterize the various physical processes occurring in the cell. The resistances R_e represent the resistances of the interdigitated microelectrodes themselves. The active resistance of the electrolyte solution between the microelectrodes is characterized by the resistance R_{se}. The capacitances of the double electrical layers at the boundary of the microelectrodes with the electrolyte and the capacity of the electrochemical cell are represented by capacitances C_{dl} and C_c, respectively.

Based on the equivalent electrical circuit of the electrochemical cell, its impedance can be represented as follows

$$Z_c = \frac{C_{dl}(2C_c + C_{dl})R_{se} - j\omega C_c C_{dl}^2 R_{se}^2}{\omega^2 C_c^2 C_{dl}^2 R_{se}^2 + (2C_c + C_{dl})^2} + 2R_e, \quad (1)$$

where Z_c is the impedance of the electrochemical cell; ω is the frequency.

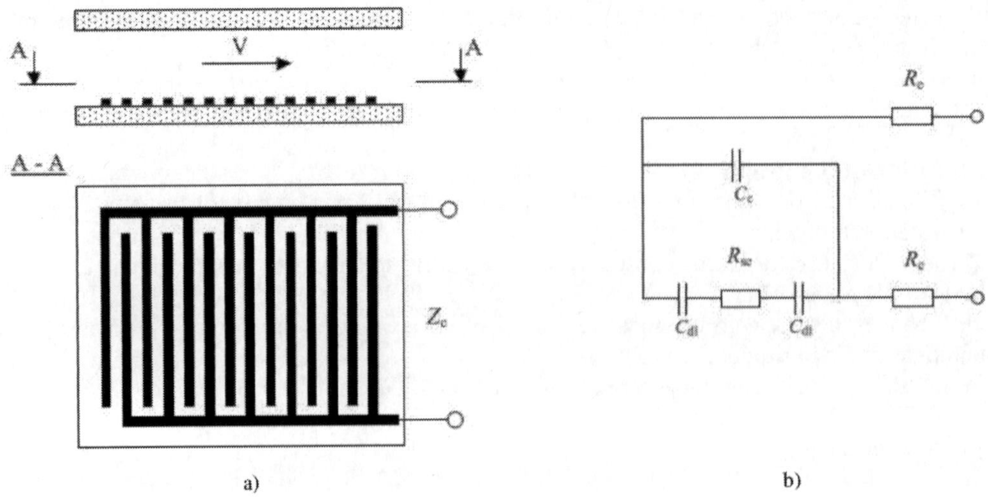

Figure 1. Structure (a) and equivalent electrical device (b) of the electrochemical cell.

The parameters included in the expression for the impedance of the electrochemical cell, depending on the influence of the microchannel height on them, can be divided into two groups:

1) parameters that do not depend on the microchannel height;

2) parameters that depend on the microchannel height.

The parameters that are independent of the microchannel height include the resistances of the interdigitated microelectrodes and the capacitances of the double electrical layers at the boundary of the microelectrodes with the electrolyte solution. The resistance of the interdigitated microelectrode can approximately be determined as the sum of the resistance of the connecting film conductor and the resistance of the series of parallel-connected finger electrodes

$$R_e \approx R_{be} + \frac{R_f}{N}, \tag{2}$$

where R_{be} and R_f are the resistance of the connecting film conductor and the resistance of the finger electrode, respectively; N is the number of fingers in the microelectrode.

The resistances R_{be} and R_{be} can be determined using the characteristics of the film layer from which the microelectrodes are made and the geometric dimensions of these microelectrodes, as follows

$$R_{be} = \rho_s \frac{L_{be}}{b_{be}}, \tag{3}$$

$$R_f = \rho_s \frac{L_f}{b_f}, \tag{4}$$

where ρ_s is the specific surface resistance of the film layer; L_{be} and L_f is the length of the connecting film conductor and the finger electrode, respectively; b_{be} and b_f is the width of the connecting film conductor and the finger electrode, respectively.

The capacity of the double electrical layer at the interface between the electrolyte solution and the electrode is equal to

$$C_{dl} = (L_{be}b_{be} + NL_f b_f)C_0, \quad (5)$$

where C_0 is the specific capacitance of the double electrical layer at the interface between the electrolyte solution and the electrode.

The parameters whose values depend on the microchannel height are the resistance of the electrolyte solution between the microelectrodes and the capacity of the electrochemical cell. For electrochemical cells, these parameters depend on the properties of the electrolyte, the shape, the geometric dimensions and location of the microelectrodes and the shape and dimensions of the cell itself. The dependencies of above-mentioned parameters on these factors can be represented as follows

$$R_e = K_c \rho_e, \quad (6)$$

$$C_c = \frac{\varepsilon_0 \varepsilon_r}{K_c}, \quad (7)$$

where K_c is the cell constant with IME; ρ_e is the specific resistance of electrolyte; ε_0 is the electrical constant; ε_r is the relative permittivity of the electrolyte solution.

In expressions (6) and (7), the parameter ρ_e and the product $\varepsilon_0\varepsilon_r$ characterize the properties of the electrolyte solution. The cell constant K_c entering into these expressions is determined by the geometrical characteristics of the cell and the electrodes, that is, their shape, dimensions and location. The microchannel height is a geometric characteristic of the flow electrochemical cell and affects the value of the cell constant and, consequently, the resistance of the electrolyte solution between the electrodes.

The values of the resistance of the electrolyte solution between the electrodes and the capacitance of the electrochemical cell can be determined analytically using expressions (6) and (7). The following algorithm is proposed for this:

1) the potential distribution in the cell with the IME at a constant electric field is determined;

2) based on the potential distribution in the cell with IME, the resistance of the electrolyte solution between the electrodes at a constant current is determined;

3) using the expression (6), the value of the cell constant is found;

4) using the expression (7), the value of the capacitance of the electrochemical cell is determined.

Consider the steps of the proposed algorithm in detail.

3.1. Determination of the potential distribution in the cell with the IME at a constant electric field

To determine the potential distribution in the electrochemical cell at a constant electric field, the analytical method proposed in [11] is used. According to this method, an elementary cell is allocated in the system of the IME, the translational transfer of which can form the entire system of the IME. The allocated unit cell is the modeling domain and is shown in figure 2(a). For this cell, the distribution of the electric potential is determined by replacing the modeling domain with an equivalent structure and dividing this structure into three regions (figure 2(b)).

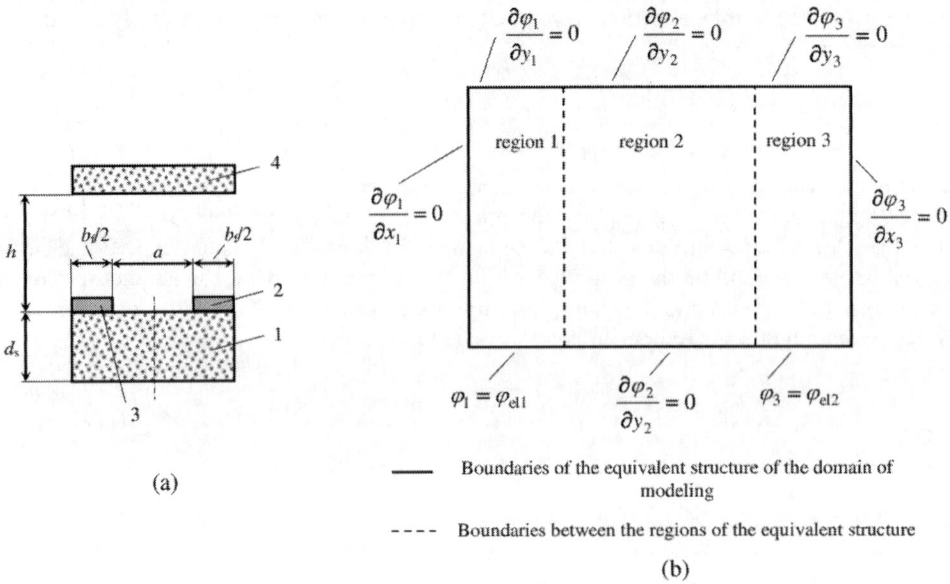

Figure 2. Elementary cell of the interdigitated microelectrode system (a) and the equivalent structure of the domain of modeling (b): 1 – substrate; 2, 3 – sections of the neighbor fingers of the microelectrodes; 4 – upper wall of the microchannel; h – microchannel height; d_s – thickness of the substrate.

The potential distribution in each region according to [11] can be represented as follows

$$\varphi_j=\frac{1}{l_j b_j\sigma_e}\left[-\delta_0^{(j,s)}-\delta_0^{(j,t)}+\delta_0^{(j,u)}+\delta_0^{(j,v)}\right]+\frac{2}{l_j b_j\sigma_e}\sum_{k=1}^{\infty}\left[-(-1)^k\delta_0^{(j,s)}-\delta_k^{(j,t)}+\delta_0^{(j,u)}+\delta_k^{(j,v)}\right]$$

$$\cdot\frac{1}{(k\pi/l_j)^2}\cos\left(\frac{k\pi x_j}{l_j}\right)+\frac{2}{l_j b_j\sigma_e}\cdot\sum_{m=1}^{\infty}\left[-\delta_m^{(j,s)}-(-1)^m\delta_0^{(j,t)}+\delta_m^{(j,u)}+\delta_0^{(j,v)}\right]$$

$$\cdot\frac{1}{(m\pi/b_j)^2}\cos\left(\frac{m\pi y_j}{b_j}\right)+\frac{4}{l_j b_j\sigma_e}\cdot\sum_{k=1}^{\infty}\sum_{m=1}^{\infty}\left[-(-1)^k\delta_m^{(j,s)}-(-1)^m\delta_k^{(j,t)}+\delta_m^{(j,u)}+\delta_k^{(j,v)}\right]$$

$$\cdot\frac{1}{(k\pi/l_j)^2+(m\pi/b_j)^2}\cos\left(\frac{k\pi x_j}{l_j}\right)\cos\left(\frac{m\pi y_j}{b_j}\right),\qquad(8)$$

where φ_j is the potential in the region j; l_j and b_j are the length and width of the region j, respectively; σ_e is the specific electrical conductivity of the electrolyte; x_j, y_j, - coordinates of the region j; k and m are summation indices with respect to the coordinates x_j and y_j, respectively; s, t, u, v are the indices of regions adjacent to the region j; δ are the weight coefficients that characterize the components of the current density on the boundaries between the region.

IOP Conf. Series: Journal of Physics: Conf. Series **944** (2017) 012065 doi:10.1088/1742-6596/944/1/012065

In order to determine the unknown weight coefficients in the expressions for the potential distribution in the regions of the unit cell, the adjoint boundary conditions are used on the boundaries between adjacent regions. A detailed description of this technique is presented in [11].

3.2. Determination of the resistance of the electrolyte solution between the electrodes at a constant current.

Using the electric potential distribution in the unit cell, the linear resistance between the electrodes located in the cell is determined

$$\overline{R}_{a} = \frac{\varphi_{el2} - \varphi_{el1}}{\int_{0}^{b_f/2} j_1 dx_1}, \tag{9}$$

where $\overline{R}_{a}$ is the linear resistance of the electrolyte between the electrodes in the unit cell at a constant current; φ_{el1}, φ_{el2} are the potentials of the electrodes 1 and 2, respectively; j_1 is the current density on the electrode 1, which is determined from the potential distribution in region 1.

Using the found value of the linear resistance between the electrodes in the unit cell, the resistance of the electrolyte solution at a constant current is calculated for the entire electrochemical cell with IME according to the following formula

$$\overline{R}_{e} = \frac{\overline{R}_{a}}{L_f(2N-1)}, \tag{10}$$

where $\overline{R}_{e}$ is the resistance of the electrolyte between the electrodes in the electrochemical cell at a constant current.

3.3. Determining the value of the cell constant

The value of the cell constant is determined on the basis of the relation (6), using the following expression

$$K_c = \overline{R}_e \sigma_e \tag{11}$$

3.4. Determination of the capacitance of the electrochemical cell

The capacitance of the electrochemical cell is determined using the relationship (7) and the found value of the cell constant (11).

4. Results of Modeling

The obtained model was used to investigate the effect of microchannel height on the impedance of an electrochemical cell with interdigitated microelectrodes. As the material of microelectrodes, platinum with a specific resistivity equal to $1.07 \cdot 10^{-6}$ Ω·m was chosen. The geometric dimensions of the microelectrodes fingers had the following values: thickness – 0.5 μm, length – 1000 μm, width – 150 μm. The distance between the microelectrodes fingers was chosen to be 150 μm. The number of the fingers in one microelectrode was assumed to be 14. The specific capacity of the double electrical layer at the interface of the electrolyte solution-electrode was chosen to be equal to 10 μF/cm^2.

Figure 3 shows the impedance dependencies of the electrochemical cell with IME for the five microchannel height values: 50 µm, 100 µm, 150 µm, 200 µm and 500 µm for the fixed values of the conductivity (σe=5·10-5 S/m) and the relative dielectric permeability (εr=27) of the electrolyte solution. Figures 4 and 5 present the frequency dependences of the real and imaginary parts of the impedance of the electrochemical cell with IME for different values of the microchannel height and for the values of the electrolyte solution parameters indicated above. Figures 6 and 7 show the frequency dependences of the modulus and the argument of the impedance of the electrochemical cell with IME for the different values of the microchannel height and for the electrolyte solution parameters indicated above.

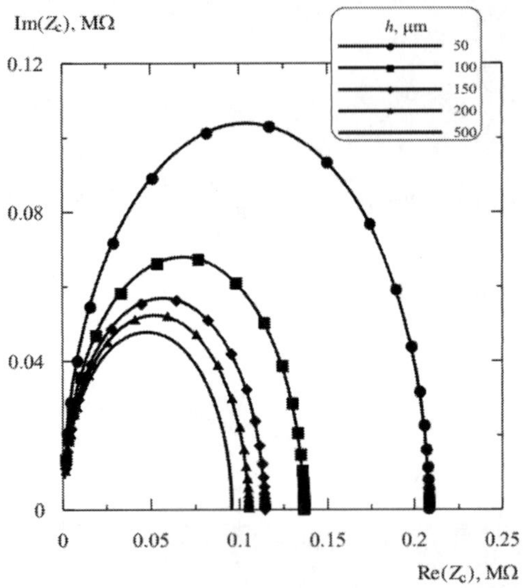

Figure 3. Impedance dependences of the electrochemical cell for the different values of microchannel height *h*: σ_e=5·10^{-5} S/m; ε_r = 27.

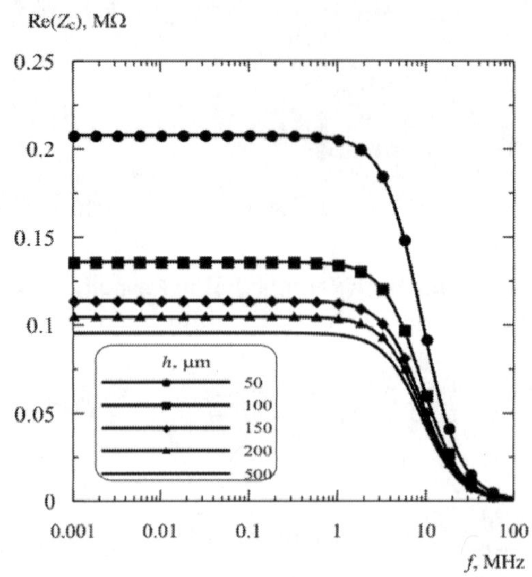

Figure 4. Frequency dependences of the real part of the electrochemical cell impedance for the different values of microchannel height *h*: σ_e=5·10^{-5} S/m; ε_r = 27.

5. The results

The analysis of the impedance dependences of the electrochemical cell with IME, presented in figure 3, shows that, for small values of the microchannel height, the impedance of the cell strongly depends on the microchannel height . In the range of the microchannel height equal to 50...150 µm, a significant change of the impedance of the electrochemical cell from the microchannel height is observed. Beginning with the microchannel height of 200 µm, the impedance depends little on the height.

The frequency dependences of the real part of the impedance of the electrochemical cell with IME (figure 4) have the following features. With increasing the microchannel height, the real part of the impedance of the electrochemical cell with IME decreases. In the frequency range from 1 kHz to 1 MHz, the real part of the impedance does not depend on the frequency for the all considered microchannel height values. It is determined by the resistance of the electrolyte solution in the cell with IME. At frequencies above 1 MHz, the real part of the impedance decreases with increasing the frequency.

For the frequency dependences of the imaginary part of the impedance of the electrochemical cell with IME (figure 5), the following features are characteristic. At low frequencies (up to 15 kHz), the imaginary part of the impedance decreases with increasing the frequency. This is due to the effect of the capacitance of the double electrical layers on the boundaries between the electrolyte solution and the electrodes. Further, the imaginary part of the impedance of the electrochemical cell increases with increasing the frequency and reaches a maximum at 8...10 MHz. After reaching the maximum, the imaginary part of the impedance of the electrochemical cell decreases. The main influence on the change of the imaginary part of the impedance in the high-frequency region is provided by the capacitance of the electrochemical cell. With increasing the channel height, the imaginary part of the cell impedance decreases throughout the frequency range.

The analysis of the frequency dependences of the electrochemical cell impedance module, presented in figure 6, shows that the impedance module at frequencies from 1 kHz to 1 MHz is determined by the real part of the impedance. This follows from a comparison of the frequency dependences of the real part of the impedance (figure 4) and the impedance module (figure 6). In the frequency range from 1 to 10 MHz, the impedance module is determined by the real and imaginary parts of the impedance. Above 10 MHz, the value of the impedance module is mainly due to its imaginary part.

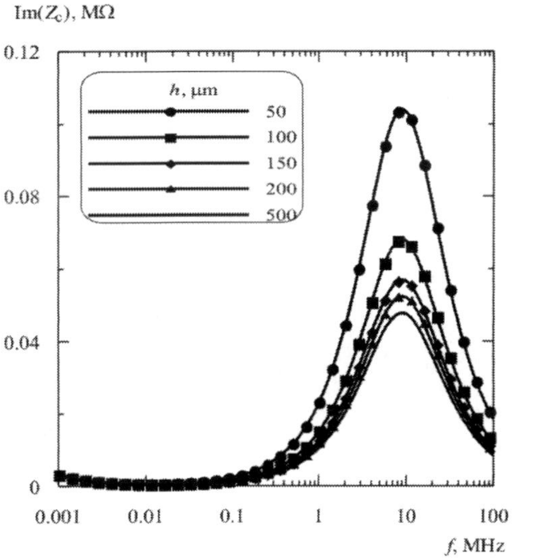

Figure 5. Frequency dependences of the imaginary part of the electrochemical cell impedance for the different values of microchannel height h: σ_e=5·10^{-5} S/m; ε_r = 27.

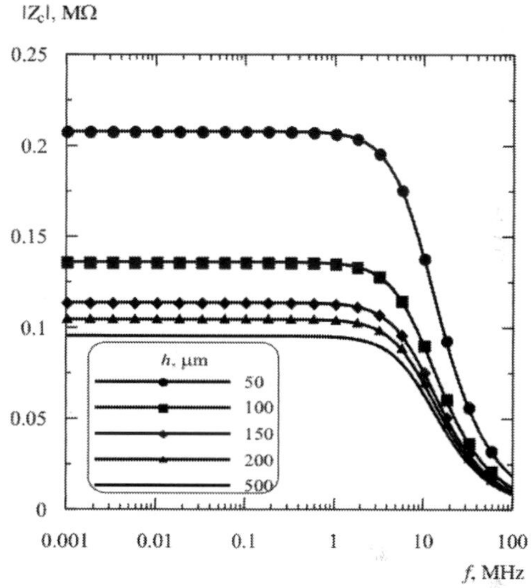

Figure 6. Frequency dependences of the module of the electrochemical cell impedance for the different values of microchannel height h: σ_e=5·10^{-5} S/m; ε_r = 27.

The frequency dependences of the argument of the impedance of the electrochemical cell with IME (figure 7) have the following features. Firstly, the argument of the electrochemical cell impedance at the frequencies above 10 kHz is practically independent of the microchannel height. For the all considered microchannel heights, the frequency dependences of the argument at these frequencies have the same view. The second feature is that the argument of the impedance has in the certain frequency range (approximately 5 to 60 kHz) the maximum value. Below the lower cutoff frequency (5 kHz) the argument decreases due to the influence of the capacitance of the electric double layers on the boundaries of the electrolyte solution - electrodes. In this frequency range, the dependence of the argument on the microchannel height of the electrochemical cell is observed. Above the upper limit of the indicated frequency range, the decrease in the impedance argument is due to the influence of the capacitance of the electrochemical cell.

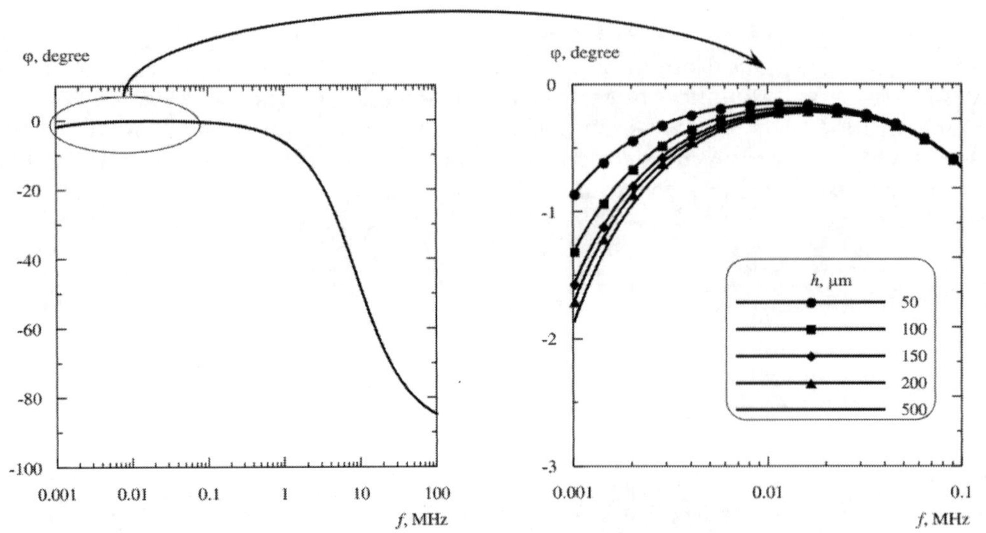

Figure 7. Frequency dependences of the argument of the electrochemical cell impedance for the different values of microchannel height h: σ_e=5·10^{-5} S/m; ε_r = 27.

Using the obtained results, it is possible to indicate the frequency ranges in which the parameters of the electrolyte solutions under investigation should be determined from the results of the impedance measurements. The conductivity of the electrolyte solution should be determined from the frequency dependence of the real part of the impedance in the frequency range where the real part of the impedance is constant. For the electrochemical cell under consideration, this frequency range lies below 1 MHz. In this case it is necessary to take into account that the real part of the impedance depends on the microchannel height. The relative permittivity of the electrolyte solution should be determined from the frequency dependence of the impedance argument with the known conductivity of this solution found from the frequency dependence of the real part of the impedance. The frequency range at which the relative permittivity of the electrolyte solution should be determined must be chosen above 10 MHz.

6. Conclusion

The approach is proposed for determining the impedance dependence of the electrochemical cell with interdigitated microelectrodes and the frequency dependences of the impedance characteristics of the cell. The account of the electrochemical processes taking place in the electrochemical cell is carried out on the basis of its equivalent electric circuit. To find the cell constant, an analytical model is used allowing one to determine the potential distribution in the interdigitated microelectrode system and the current density at the boundary between the electrode and the electrolyte solution.

Using the proposed approach, the impedance dependences of the specific electrochemical cell with interdigitated microelectrodes, as well as the frequency dependences of the real and imaginary parts of the impedance, the module and the argument of the impedance have been determined for the different microchannel heights. The results of modeling these dependences allow one to specify the frequency ranges on which the impedance characteristics of the electrochemical cell can be used to determine the parameters of the electrolyte solution under study.

IOP Conf. Series: Journal of Physics: Conf. Series **944** (2017) 012065 doi:10.1088/1742-6596/944/1/012065

References

[1] Drokin N A, Fedotova A V, Glushchenko G A and Churilov G N 2010 Impedance spectroscopy of high-molecular poly(ethylene) with carbon nanotubes *Physics of the Solid State* vol **52** no 3 pp 657–661

[2] Lvovich V F 2012 *Impedance spectroscopy: applications to electrochemical and dielectric phenomena* (New Jersey: John Wiley & Sons) p 353

[3] Vigdorovich V I, Tsygankova L E, Siniutina S E and Kichigin V I 2008 Estimation of inhibitor efficiency against hydrogen sulfide and carbonic acid corrosion of carbon steel by impedance spectroscopy method *ECS Transactions* vol **6 (24)** pp 67–72

[4] Zoski C G 2006 *Handbook of Electrochemistry* (Ed. Amsterdam: Elsevier) p 935

[5] MacKay S, Hermansen P, Wishart D and Chen J 2015 Simulations of interdigitated electrode interactions with gold nanoparticles for impedance-based biosensing applications *Sensors* vol **15** pp 22192-22208

[6] Ferrigno R Josserand J, Brevet P F and Girault H H 1998 Coplanar interdigitated band electrodes for electrosynthesis. Part 5: finite element simulation of paired reactions *Electrochimica Acta* vol **44** pp 587–595

[7] Stulik K, Amatore C, Holub K, Marecek V and Kutner W 2000 Microelectrodes. Definitions, characterization, and applications *Pure Appl. Chem.* vol **72** pp 1365-3075

[8] Olthuis W, Streekstra W and Bergveld P 1995 Theoretical and experimental determination of cell constants of planar-interdigitated electrolyte conductivity sensors *Sens. Actuators B* vol **25** pp 252–256

[9] Yang L, Li Y, Griffis C L and Johnson M G 2004 Interdigitated microelectrode (IME) impedance sensor for the detection of viable Salmonella typhimurium *Biosensors and Bioelectronics* vol **19** pp 1139–1147

[10] Ibrahim M, Claudel J, Kourtiche D and Nadi M 2013 Geometric parameters optimization of planar interdigitated electrodes for bioimpedance spectroscopy *J. Electr. Bioimp.* vol **4** pp 13–22

[11] Kozlov A G and Fadina E A 2016 Analysis of electrophysical processes in system of interdigitated microelectrodes used in microchannels *Dynamics of Systems, Mechanisms and Machines (Dynamics)* pp 1–5

Dynamic characteristics of numerical systems for induction surface heating

T Kozulina, K Blinov and S Galunin

St. Petersburg Electrotechnical University, 5, Prof. Popova St., St Petersburg, 197373, Russia

E-mail: kozulina.tatiana@mail.ru

Abstract. This paper describes the development, investigation and optimization of induction surface heating process of a workpice taking into account its rotation by numerical simulation in a commercial package ANSYS® [1]. During the investigation it is necessary to select the optimal dynamic characteristics of the heating process in order to obtain the required temperature level and uniform temperature distribution on the surface of the workpiece.

1. Introduction

Induction heating is one of the most popular and effective methods of metal heat treatment. Preheating of metal workpieces is carried out before the next actions:

- quenching (cooling);
- rolling;
- forging;
- stamping;
- other plastic deformations.

Induction heating is superior to heat treatment in furnace or by open flame by many parameters.

By the way of contactless energy transfer, it is possible to heat a selected area of the workpiece to the required depth during a short period of time. This method of heat treatment makes it possible to control the uniformity of the temperature distribution in the workpiece and to avoid oxidation and decarbonization of its surface.

The automated process of induction heating reduces time expenditure and energy cost for heat treatment of workpieces, increases the number of heated workpieces per unit of time and reduces their cost price.

The application of numerical simulation in the investigation and optimization of the high-frequency heating process allows to obtain enough amount of information for the development of induction installation and to reduce the time and material cost for these investigations.

2. Numerical simulation in a commercial package ANSYS®

Any material has electro- and thermophysical properties. These properties of the workpiece material positioned in an alternating magnetic field depend on intensity of magnetic field, the temperature of the workpiece and its heat exchange with the environment. It is the reason for the nonlinearity of the electromagnetic and thermal processes of induction heating, which can influence the temperature distribution in the workpiece.

Up to date numerical simulation in modern commercial packages is used to solve electromagnetic and thermal tasks of induction heating, which takes into account the nonlinear properties of the workpiece material. These packages are based on three numerical methods for solving differential equations: the finite element method (FEM), the boundary element method (BEM), and the finite difference method (FDM).

One of such package is the universal commercial package ANSYS®, based on the finite element method (FEM). It allows to find solutions of electromagnetic and electrothermal problems and to investigate various problems of mechanics and acoustics. Figure 1 shows an example of an algorithm for numerical simulation of the induction hardening process in a commercial package ANSYS®.

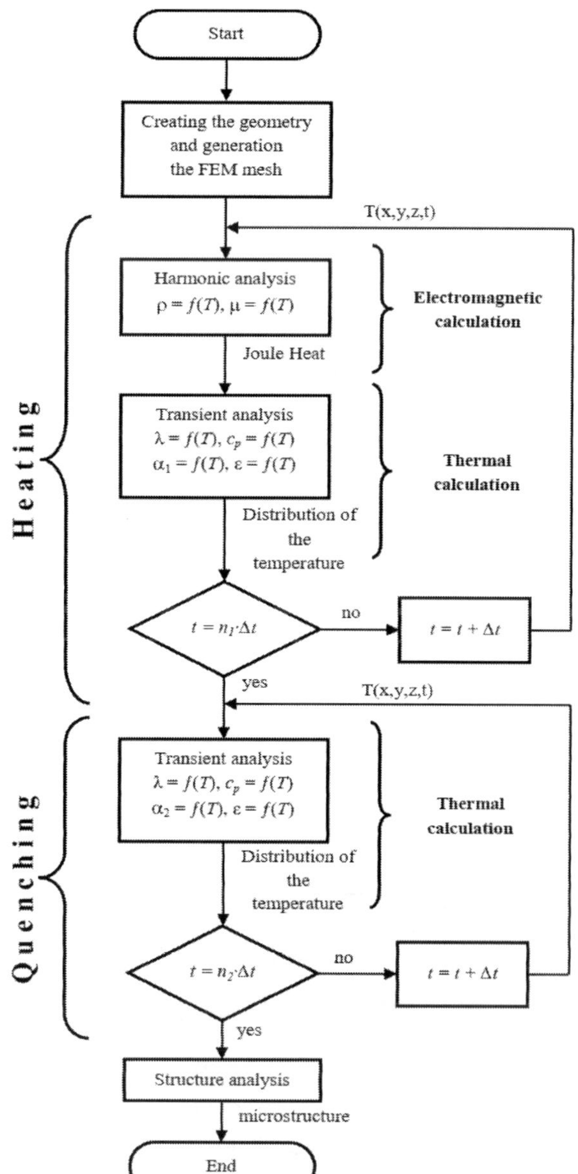

Figure 1. The algorithm of the numerical model for induction hardening process simulation based on ANSYS® package.

To begin with, the geometry of the induction system is developed: the geometry of the workpiece, the inductor and, if it is necessary, the magnetic core. An air space is simulated around it. Then the FEM mesh is generated in this model.

The next stage is the numerical simulation of the heating process, after which the process of the workpiece quenching is simulated [2].

The heating process includes electromagnetic and thermal tasks. Electromagnetic calculation allows determining the parameters of the magnetic field with the known specific electrical resistivity and magnetic permeability of each element of the workpiece, and also with the known initial temperature distribution in it. The result of the electromagnetic calculation is the distribution of Joule heat in the workpiece. On the basis of these results, the following thermal task will be solved.

At the end of the heat calculation, the final temperature distribution in the workpiece is formed at the next time step. If the result of the calculation does not satisfy the requirements, then the electromagnetic task is restarted, the initial conditions of which is the final temperature distribution in the workpiece after last thermal calculation. If the result of the calculation satisfies the conditions of the task, the resulting temperature distribution in the workpiece is the result of numerical simulation of the heating process.

After heating for next hardening, the workpiece is quenched in water, oil or air environment. To simulate the quenching process, only thermal calculation is used, where the initial temperature distribution in the workpiece is the result of the previous calculation of the heating process. The cooling conditions, which are indicated for the thermal calculation of the quenching process, take into account the temperature dependence of the heat-transfer coefficients [3].

The final result of the numerical simulation of the induction hardening process is the temperature distribution in the workpiece after the thermal calculation of the quenching process.

After the quenching process, a structural analysis of the quenched metal can be carried out. The structure of the hardened metal in the workpiece is estimated under the known main conditions of structural transformation, the austenitization temperature, the cooling rate, and the heating and cooling curves.

3. Numerical simulation of the induction heating process with change of dynamic characteristic

Figure 2 shows the induction system, which will demonstrate the numerical simulation of the induction heating process for the hot forming of the workpiece.

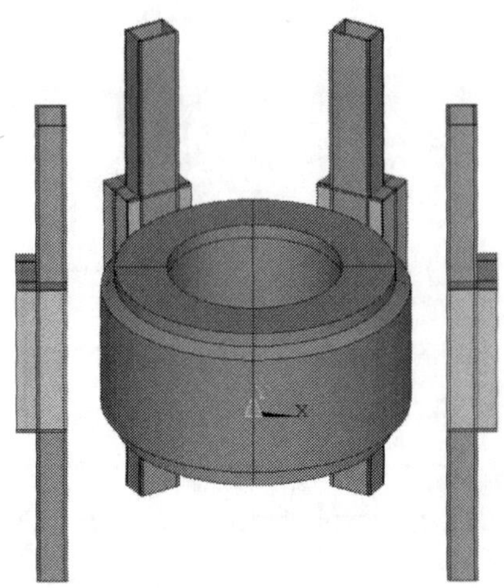

Figure 2. The geometry of the investigated induction system.

During heating of the metal some structural transformations and changes of its properties take place inside it. Most metals and alloys are well deformable after heating to high temperatures, because in the hot state the ductility of metals much increases. Before the hot forming process, the metal workpiece is heated to certain temperatures to impart the necessary properties to the metal, which allow deforming the workpiece without deteriorating the quality of the finished product.

The system under investigation consists of a workpiece and an inductor with a magnetic core and copper plates.

The workpiece is a steel hollow cylinder with a height of about 70 mm with a different wall thickness along its entire height. The maximum diameter of the workpiece is about 106 mm.

The inductor consists of four copper hollow line conductors with a cross-section of 10x15 and a wall thickness of 1 mm. The inductor is cooled from the inside by water. Each of the conductors is equipped with a U-shaped magnetic core, the height of which is 3 mm above and below the height of the zone subjected to heating. The top and bottom of each magnetic core are placed copper plates with an area that repeats the cross-section of the magnetic core. The inductor and the magnetic core are located along the workpiece at an angle of 60 degrees relative to each other.

This induction system should provide local heating of the steel workpiece (heating of the area having a bigger thickness) to the required depth up to a temperature of 840–850 °C for further hot forming of the workpiece. Also this system has to promote achievement of uniform temperature distribution on the set area due to rotation of workpiece in the inductor.

Numerical simulation facilitates the designing of induction systems for heating of rotating workpiece. Therefore, in the commercial package ANSYS®, numerical 3D models of electromagnetic and thermal processes in the induction heating system for the hot forming of workpiece were developed and investigated.

Due to the fact that the induction system is symmetrical about the YZ axis, all the investigations were carried out in the numerical 3D electrothermal model of the ½ part of the system depicted in figure 3.

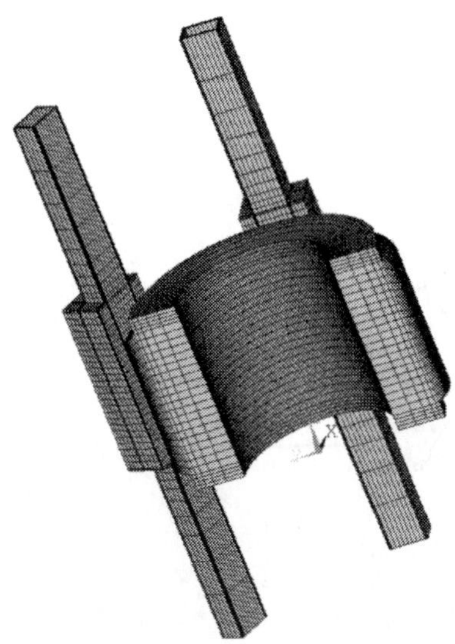

Figure 3. 3D numerical model of ½ part of the system.

During the investigations for achievement of necessary value and uniform distribution of temperature in the workpiece before further mechannical treatment, values of a dynamic characteristic of inductor installation, such as current in the inductor were measured. One of the heat conditions is the difference between the maximum and minimum temperatures in the workpiece after heating, it has to be no more than 50°C. The big difference of temperatures in the workpiece can lead to emergence of cracks during process of hot forming.

During the first 38 seconds of heating the workpiece, the current in the inductor was 7 kA, during which time the surface layer of the workpiece was heated up to 944°C, the minimum temperature was 566°C. Then, with a smooth transition of 5.25 kA from 38th to 40th seconds, the current in the inductor was reduced to 3.5 kA. In this mode, the workpiece was heated from 40th to 82th seconds, and at the end of the heating process the maximum temperature in the workpiece reached 887°C and the minimum temperature in the workpiece being 781°C.

At the 84th second the current was reduced to 1.75 kA, and then the next 5 seconds the current wasn't supplied in the inductor. Thus, the situation when the workpiece was moved from the induction heating installation to a hot forming installation was simulated. During this time, the temperature in the workpiece was leveled due to thermal conductivity, convection and radiation, and the difference between the maximum and minimum temperature values is 51°C, which satisfies one of the heating conditions.

Figures 4 and 5 show the final temperature distribution in the workpiece: the maximum temperature is 854°C and the minimum temperature is 803°C. The temperature distribution is practically uniformly along the outer diameter of the workpiece and its value reaches the required value.

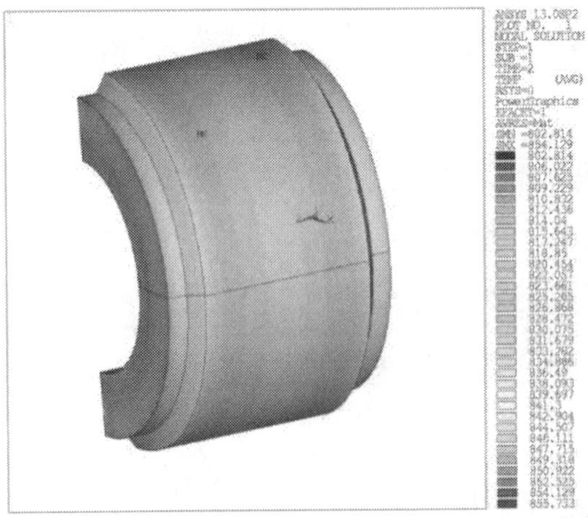

Figure 4. The finally temperature distribution along the outer diameter of the workpiece.

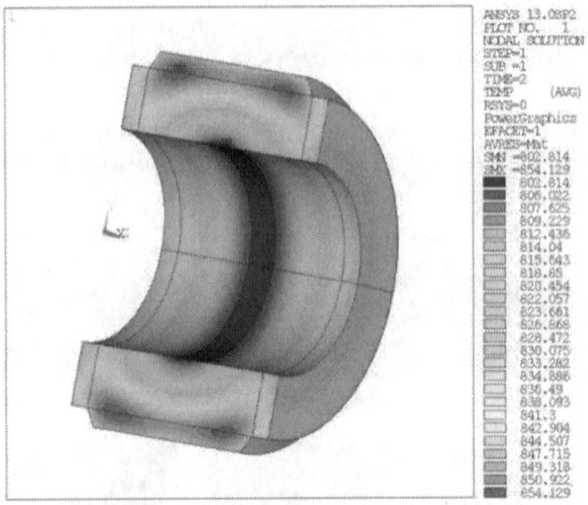

Figure 5. The finally temperature distribution along the inside diameter of the workpiece.

The entire process of heating the workpiece with changes in the values of the power released in the workpiece, the current in the inductor, and the maximum and minimum temperatures in the workpiece can be seen from the diagram shown in figure 6.

The diagram shows that the power in the workpiece increases during 16 seconds and reaches a maximum value of 67 kW, and then it begins to fall. This is because steel is heated up to a Curie temperature of 835°C and loses its magnetic properties.

The temperature in the workpiece also increases very rapidly, until it reaches the Curie point. Then the temperature slowly increases. When the current of the inductor decreases, the temperature also falls slowly. The minimum temperature in the workpiece increases during the heating time, independently of the reduction in current and power. This is due to the heat transmission in the workpiece by thermal conductivity from the area with the maximum temperature to area with minimum temperature.

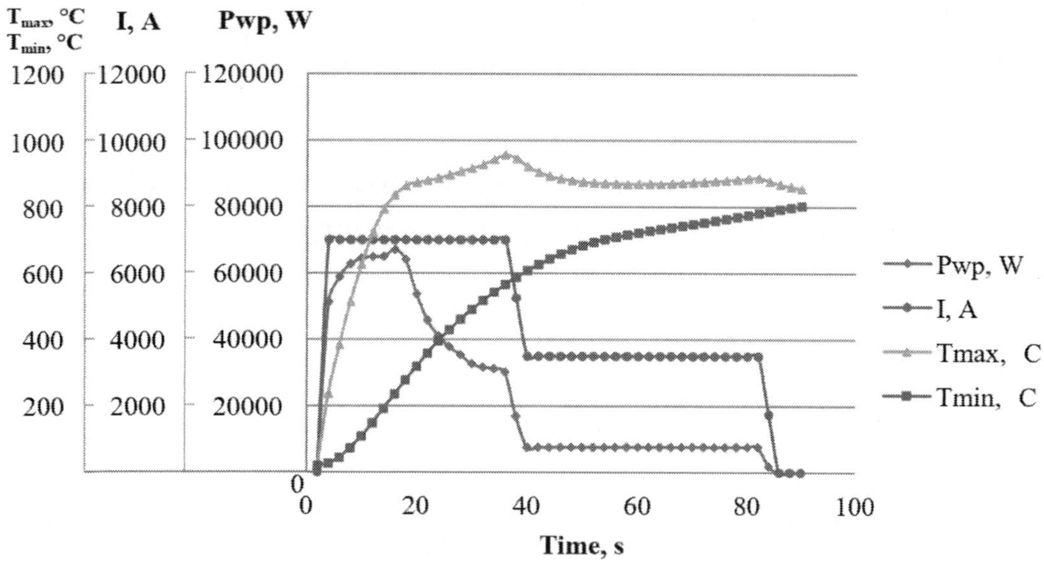

Figure 6. The diagram of the power released in the workpiece, the current in the inductor, and the maximum and minimum temperatures in the workpiece.

4. Conclusions

In the paper the use of numerical simulation for the development of the induction heating system of the workpiece with taking into account its rotation was demonstrated using the example of induction heating for hot forming. For this purpose, in the commercial ANSYS® package, numerical 3D electromagnetic and thermal models of this induction system were developed and investigation was performed with a change in the dynamic characteristics of the induction installation.

References

[1] http://www.ansys.com/

[2] Galunin S, Kudryash M, Kozulina T and Blinov Yu 2016 Numerical simulation of induction heater for hardening of tapered leading gear wheel *Proceedings of the International Symposium HES-16 "Heating by Electromagnetic Sources"* (Padua) pp 203-206

[3] Schlesselmann D, Nacke B, Nikanorov A and Galunin S 2015 Coupled numerical multiphysics simulation methods in induction surface hardening *Proceedings of the 6th International Conference on Coupled Problems in Science and Engineering* 6 pp 392-403

IOP Conf. Series: Journal of Physics: Conf. Series **944** (2017) 012067 doi:10.1088/1742-6596/944/1/012067

On providing the fault-tolerant operation of information systems based on open content management systems

Sergey Kratov

Foundation of Algorithms and Programs SB RAS, Institute of Computational Mathematics and Mathematical Geophysics SB RAS, Novosibirsk, Russia

Abstract. Modern information systems designed to service a wide range of users, regardless of their subject area, are increasingly based on Web technologies and are available to users via Internet. The article discusses the issues of providing the fault-tolerant operation of such information systems, based on free and open source content management systems. The toolkit available to administrators of similar systems is shown; the scenarios for using these tools are described. Options for organizing backups and restoring the operability of systems after failures are suggested. Application of the proposed methods and approaches allows providing continuous monitoring of the state of systems, timely response to the emergence of possible problems and their prompt solution.

1. Introduction

A growing number of information systems designed to serve a wide range of users are based on Web technologies nowadays. Users of such systems get access to them through Internet. One of the most optimal ways of developing such Web-oriented information systems in terms of time and money is to use content management systems or, in more complex cases, frameworks for designing CMS as their basis [1]. Using CMS or frameworks allows significantly reducing the time to develop basic (common for all) subsystems, such as subsystems for authorizing users, granting rights, structuring content, connecting external libraries, etc. Today there are a large number of free and open CMS/frameworks, additional modules and libraries, significantly expanding their functionality, which in turn allows significantly reducing the financial costs of development [2, 3].

At the same time, the use of open CMS is associated with the risk, since the source code of such systems is publicly available and, in case of errors detection in it, can be exploited by intruders [4-7]. That in turn can lead to their compromising and hacking [8-10] in the case of the general availability information systems based on such CMS.

In this article, the author would like to describe the process of supporting the fault-tolerant operation of an information system based on an open and free CMS, using the example of the Foundation of Algorithms and Programs of the SB RAS (http://fap.sbras.ru) [11]. The author would like to show the set of tools provided to the system administrator by the developers of its kernel and modules, through which it will be able to monitor its state and provide trouble-free and uninterrupted operation.

The Foundation's information system (figure 1) [12, 13] is based on the free CMS Drupal, which means it has all the advantages as well as potentially being exposed to all the problems described above.

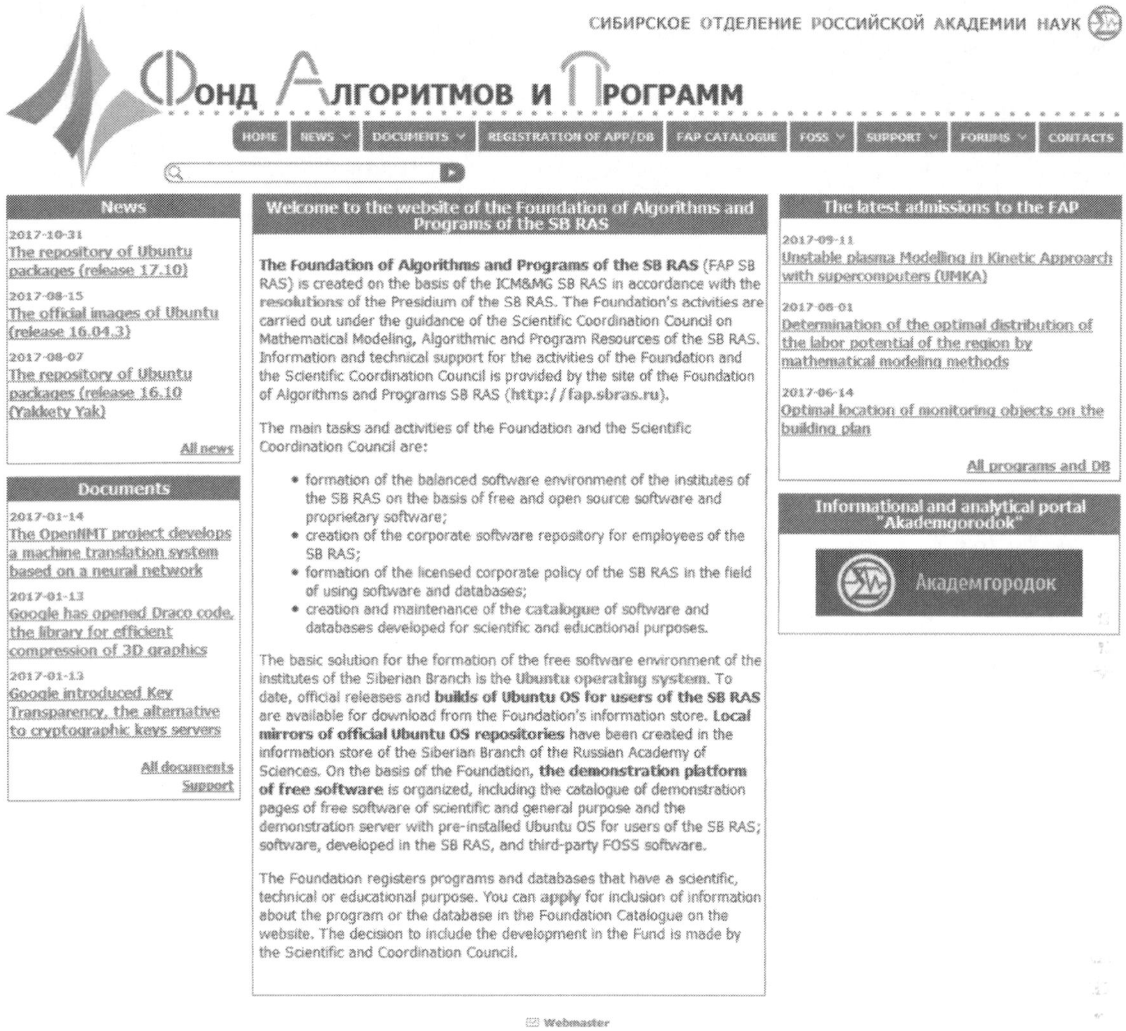

Figure 1. FAP SB RAS.

2. The general information panel

The basic administrative tool is the status report. It is the brief display of the main parameters of the information system state, critical for its overall functioning, and detected problems. For example, in figure 2 are displayed:

- the kernel version of the system and its update status. It can be seen that a kernel update of the system is available, non-critical in terms of security,
- the status of the file system storage, the protection of the main CMS configuration files,
- the time of the last run of scheduled tasks (cron),
- type and version of DBMS, its support for text encodings, system database state.

Figure 3 lists:

- PHP interpreter version, PHP modules, required to run the information system kernel, RAM limits for executing PHP scripts, etc.,
- availability of updates for modules, themes and translations of interfaces,
- availability and versions of system graphical (GD) and javascript libraries (jQuery),
- Web server version.

IOP Conf. Series: Journal of Physics: Conf. Series **944** (2017) 012067 doi:10.1088/1742-6596/944/1/012067

Drupal	[illegible]
Access to update.php	Protected
CAPTCHA	Already 17031 blocked form submissions
CKEditor	[illegible]
Colorbox plugin	[illegible]
Configuration file	Protected
Cron maintenance tasks You can run cron manually. To run cron from outside the site, go to [illegible]	Last run 17 min 11 sec ago
CTools CSS Cache	Exists
Database 4 byte UTF-8 support 4 byte UTF-8 for mysql is enabled.	Enabled
Database system	MySQL, MariaDB, or equivalent
Database system version	[illegible]
Database updates	Up to date
Date API The timezone has been set to Asia/Novosibirsk. The first day of the week has been set to Monday. The medium date format type has been set to Thu, 08/06/2017 - 16:04. You may find it helpful to add new format types like Date, Time, Month, or Year, with appropriate formats, at Date and time settings.	System date settings
⚠ Drupal core update status There are updates available for your version of Drupal. To ensure the proper functioning of your site, you should update as soon as possible. See the available updates page for more information and to install your missing updates.	Out of date (version [illegible] available)
File system	Writable (*public* download method)
GD library PNG support	[illegible]
GD library rotate and desaturate effects	[illegible]

Figure 2. Part 1 of the status report.

Using this information panel, the system administrator can evaluate the overall state of the system at any given time. In case of any problems, he can go down to the level below, going to more specialized reports.

jQuery Update	jQuery [redacted] (configure) and jQuery UI [redacted]
Module and theme update status	Up to date
Node Access Permissions	Disabled
If the site is experiencing problems with permissions to content, you may have to rebuild the permissions cache. Rebuilding will remove all privileges to content and replace them with permissions based on the current modules and settings. Rebuilding may take some time if there is a lot of content or complex permission settings. After rebuilding has completed, content will automatically use the new permissions. Rebuild permissions	
PHP	[redacted] (more information)
PHP extensions	Enabled
PHP memory limit	128M
PHP register globals	Disabled
Translation update status	There are available updates
There are new or updated translations available for currently installed modules and themes. To check for updates, you can visit the translation update page.	
Unicode library	PHP Mbstring Extension
Update notifications	Enabled
Upload progress	Enabled (PECL uploadprogress)
Web server	Apache/[redacted] (Ubuntu)
XML sitemap	Last attempted generation on Thu, 08/06/2017 - 14:45 (1 hour 18 min ago).
XML sitemap cache directory	Writable

Figure 3. Part 2 of the status report.

3. The «available updates» report

CMS Drupal consists of the kernel that provides its basic functions, and has an API for connecting additional modules that extend its functionality. Modules are developed by third-party developers and are publicly available to all web developers (https://www.drupal.org/project/project_module). The authors of the modules release regular updates, that expanding their functionality, adding support for new versions of browsers, the implementation of new standards, and correcting errors found in the modules. Due to the fact that after correcting the errors, information about them becomes available to the public after a while, it is recommended to immediately update the modules used in the information system after the release of their new versions. There is the tool to track updates in the system, which monitors the release of new versions of modules and notifies the system administrator about it (both through the interface of the system itself and via e-mail). Tracking occurs at the start of cron, which allows system administrators to set any frequency of it. Administrator can also check for updates at

IOP Conf. Series: Journal of Physics: Conf. Series **944** (2017) 012067 doi:10.1088/1742-6596/944/1/012067

any time manually. The list of modules with their update status and the time of the last update check is shown in figure 4.

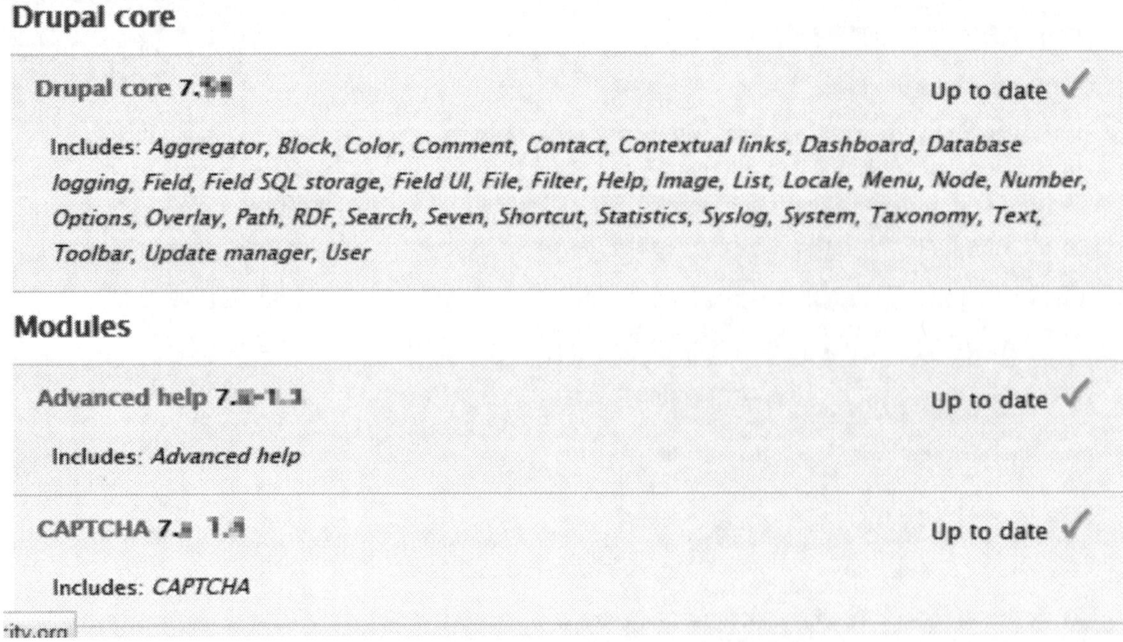

Figure 4. Available updates.

4. The recent log messages

The information system keeps a living log of events. Messages that fall into the log are saved in it, followed by rotation for the period selected by the system administrator. A user with the appropriate rights can view the system event log at any time. The log is the list containing information messages about usage data, performance data, errors, warnings and operational information. It is possible to filter messages in the log according to their source (kernel or module) and the significance level (emergency, alert, critical, error, warning, notice, info, debug). In figure 5 the part of the log is shown in which there are displayed as the ordinary message about the cron start and the critical error of one of the system components functioning.

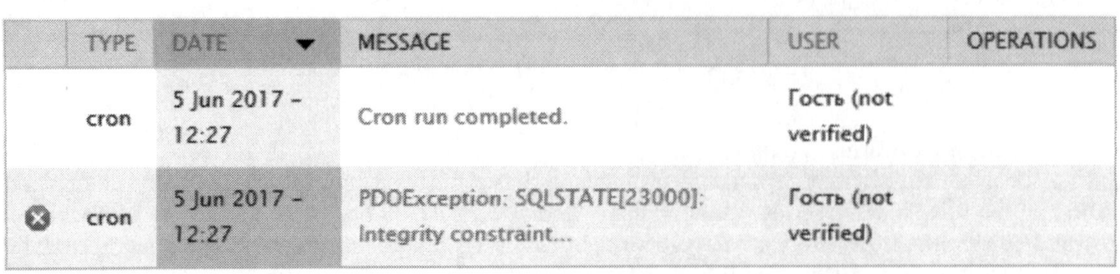

	TYPE	DATE	MESSAGE	USER	OPERATIONS
	cron	5 Jun 2017 - 12:27	Cron run completed.	Гость (not verified)	
	cron	5 Jun 2017 - 12:27	PDOException: SQLSTATE[23000]: Integrity constraint...	Гость (not verified)	

Figure 5. The recent log messages.

Figure 6 shows detailed information about this critical error, including the maximum possible description, the address of the page on which it occurred, the user, the time moment, etc.

Regular viewing of the system log is critically important, as it is often the only way to early notice and fix a system failure.

5. The external libraries

To extend the functions of the information system in addition to installed modules, web developers can connect third-party libraries that will be shared by different modules. Figure 7 lists the libraries used in the system with their versions. Versions of libraries are also recommended to keep up to date - first, old versions can contain security-critical errors; secondly, support for new versions of browsers, standards, etc. are implemented in new versions of libraries.

TYPE	cron
DATE	Monday, 5 June, 2017 - 15:28
USER	Гость (not verified)
LOCATION	[redacted]
REFERRER	
MESSAGE	*PDOException*: SQLSTATE[23000]: Integrity constraint violation: 1062 Duplicate entry 'ynews-ru' for key 'PRIMARY': INSERT INTO {l10n_update_file} (project, language, filename, version, status, last_checked) VALUES (:db_insert_placeholder_0, :db_insert_placeholder_1, :db_insert_placeholder_2, :db_insert_placeholder_3, :db_insert_placeholder_4, :db_insert_placeholder_5); Array ([:db_insert_placeholder_0] => ynews [:db_insert_placeholder_1] => ru [:db_insert_placeholder_2] => ynews-7.x-1.x-dev.ru.po [:db_insert_placeholder_3] => 7.x-1.x-dev [:db_insert_placeholder_4] => 1 [:db_insert_placeholder_5] => 1496651288) in *drupal_write_record()* (line *7383* of [redacted]).
SEVERITY	error
HOSTNAME	141.8.142.16
OPERATIONS	

Figure 6. Details about an error.

NAME	STATUS	INSTALLED VERSION	PROVIDER	LINKS
Colorbox plugin	OK	[redacted]	*Colorbox* module	Homepage | Download
MailChimp API	OK	[redacted]	*MailChimp* module	Homepage | Download

Figure 7. The libraries list.

Figure 8 provides detailed information about one of the libraries connected to the system, indicating its current version, the installation path in the system, and the download page, where it is possible to check the version's relevance.

6. Top 'access denied' and 'page not found' errors

Pages with 'access denied' (figure 9) and 'page not found' (figure 10) errors also carry important information about the health of the system. They allow administrators to track erroneous links on the system pages, as well as attempts to hack it using automated scanners for common vulnerabilities. That in turn allows administrators to quickly track them and take actions to protect of the system.

IOP Conf. Series: Journal of Physics: Conf. Series **944** (2017) 012067 doi:10.1088/1742-6596/944/1/012067

GENERAL INFORMATION	
Name	Colorbox plugin
Machine name	colorbox
Vendor URL	http://www.jacklmoore.com/colorbox
Download URL	https://github.com/jackmoore/colorbox/archive/1.x.zip
Provider	*Colorbox* module
Library path	sites/all/libraries/colorbox
Version	[illegible]
Variants	minified, source

Figure 8. Status report for the library.

COUNT ▼	MESSAGE
2	admin/reports/event/406724
2	admin/config.php
1	user/1645

Figure 9. Top 'access denied' errors.

COUNT ▼	MESSAGE
6	vremenno-aktualnye-rubriki/80-let-sibniirs
3	wp-login.php
2	content/v-novosibirske-do-sih-por-doedayut-sovetskuyu-energetiku
1	sites/default/files/js/js_FvM1alErE930ZiQ6dh2VnUkPAFS739apuHmc11fKTYw.js
1	administrator/index.php

Figure 10. Top 'page not found' errors.

7. Top pages

The main purpose of this report (shown in figure 11) for the administrator can be to track the average time of generation of individual pages, as well as the total time spent by the server on the generation of the particular page. Pages with an abnormally long generation time potentially contain some errors in the page code, database queries, connected external scripts, etc. By correcting such errors or by optimizing queries it is possible to significantly reduce the load of the information system server(s).

IOP Conf. Series: Journal of Physics: Conf. Series **944** (2017) 012067 doi:10.1088/1742-6596/944/1/012067

Pages with an abnormally large number of hits can also indicate the presence of vulnerabilities used by bots.

HITS ▼	PAGE	AVERAGE PAGE GENERATION TIME	TOTAL PAGE GENERATION TIME
15592	News home	*99* ms	25 min 46 sec
2860	The list of employees content/spisok-sotrudnikov	*153* ms	7 min 18 sec
1865	404 node/4385	*76* ms	2 min 21 sec
939	Web-mail node/2720	*37* ms	34 sec
769	The news about conferences conf_archive	*71* ms	54 sec
571	All taxonomy/term/all/all	*141* ms	1 min 20 sec

Figure 11. Top pages in the past 2 weeks.

8. Top visitors

List of registered visitors to the information system, or (for anonymous visitors) IP-addresses that viewed the largest number of pages during the reporting period is shown in figure 12. It can indicate to both a search bot indexing the system's content, and the presence of an open vulnerability that using by bots. In the case of making heavy load on the system server, it is possible to block an individual IP address or an entire subnet, which can be useful in organizing the reflection of attacks aimed at denial of system service.

HITS ▼	VISITOR	TOTAL PAGE GENERATION TIME	OPERATIONS
21681	141.8.132.81	27 min 4 sec	block IP address
3706	88.198.158.233	5 min 35 sec	block IP address
3391	34.253.95.150	5 min 9 sec	block IP address

Figure 12. Top visitors in the past 4 weeks.

9. Backup

Despite all the measures taken to organize a fault-tolerant functioning of the information system, the system administrator also needs to back up both the program code of the system and the user information contained therein. One of the simplest methods is writing shell scripts that dump the database and create an archive of the file system storage. Under the condition that such scripts are launched by cron, this will ensure the creation of regular snapshots of the system state with the possibility of their rapid deployment in case of failures. Examples of such scripts for creating a database dump and an archive of file system storage can be seen in figure 13 and figure 14.

```
#!/bin/sh

USER=username
PASSWD=password
DB_PROD=databasename

if [ -z "$1" ]; then
  echo "USAGE: $0 <file_name>";
  exit 0;
fi

echo "Dumping $DB_PROD to $1"
mysqldump -u $USER -p$PASSWD $DB_PROD > $1
```

Figure 13. Shell script for dumping system database.

```
tar -czf `date +%Y%m%d`.tar.gz /path/to/files/directory

gzip DB-dumpname.sql `date +%Y%m%d`.sql.gz
```

Figure 14. Shell scripts for archiving DB-dump and file directory.

10. Conclusion

The article raised the problem of supporting the functioning of information systems developed on the basis of publicly available free and open CMS. A number of tools provided to the administrator of the system kernel were described, variants and scenarios for their use are shown. Also, options were offered for organizing a regular backup of the database and file storage of the system. The system administrator can use the proposed tools to keep the system up-to-date, to detect and correct the error in time, and to ensure its quickest recovery in case of failure.

References

[1] Drucker J and Svensson P B O 2016 The why and how of Middleware *Digital Humanities Quarterly* iss. 2 **10**

[2] Atayero A A, Chijioke-Keme and Ogunjobi B 2014 FOSS implementation of an educational virtual office suite *EDULEARN14: 6th Int. Conf. on Education and New Learning Technologies* 5018–27

[3] Tramullas J 2013 Content management with Drupal: a review of modules specific to libraries, archives and museums *Profesional De La Informacion* iss. 5 **22** 425–31

[4] Mainka C, Mladenov V and Schwenk J 2016 Do not trust me: using malicious IdPs for analyzing and attacking single sign-on *1st IEEE European Symp. on Security and Privacy* 321–6

[5] Zhang Y, Lo D, Xia X et al. 2015 Combining software metrics and text features for vulnerable file prediction *2015 20th Int. Conf. on Engineering of Complex Computer Systems (ICECCS)* 40–9

[6] Eshkevari L, Dos Santos F, Cordy J R et al. Are PHP applications ready for hack? *2015 22nd Int. Conf. on Software Analysis, Evolution, and Reengineering (SANER)* 63–72

[7] Javed A and Schwenk J 2014 Towards elimination of cross-site scripting on mobile versions of Web applications *Information Security Applications, WISA 2013* **8267** 103–23

[8] Sanchez A B, Segura S, Parejo J A et al. 2017 Variability testing in the wild: the Drupal case study *Software And Systems Modeling* iss. 1 **16** 173–94

[9] Chimos K and Loumpardas P 2014 Identifying security issues in a higher education's institute CMS lab site *EDULEARN14: 6th Int. Conf. on Education and New Learning Technologies Proc.* 3349–54

[10] Patel S K, Rathod V R and Prajapati J B 2013 Comparative analysis of Web security in open source content management system *2013 Int. Conf. on Intelligent Systems and Signal Processing (ISSP)* 344–9

[11] Sokolova O D and Kratov S V 2016 Information systems for popularization of scientific and knowledge-based software *2016 13th Int. Scientific-Technical Conf. on Actual Problems of Electronic Instrument Engineering (APEIE) – 39281 Proc.* **1** 519–22

[12] Kratov S and Bukhtiarov I 2013 The technological platform for software development in the SB RAS *8th Int. Forum on Strategic Technology 2013, IFOST 2013 – Proc.* 353–5

[13] Zybarev Y and Kratov S 2012 The information support system of corporate software resources development *Proc. - 2012 7th Int. Forum on Strategic Technology, IFOST 2012* 683–5

IOP Conf. Series: Journal of Physics: Conf. Series **944** (2017) 012068 doi:10.1088/1742-6596/944/1/012068

Methodological aspects of fuel performance system analysis at raw hydrocarbon processing plants

A V Kulbjakina[1] and I V Dolotovskij[1]

[1] Yuri Gagarin State Technical University of Saratov (SSTU), 77 Politechnicheskaya street, Saratov 410054, Russia

Abstract. The article discusses the methodological aspects of fuel performance system analysis at raw hydrocarbon (RH) processing plants. Modern RH processing facilities are the major consumers of energy resources (ER) for their own needs. To reduce ER, including fuel consumption, and to develop rational fuel system structure are complex and relevant scientific tasks that can only be done using system analysis and complex system synthesis. In accordance with the principles of system analysis, the hierarchical structure of the fuel system, the block scheme for the synthesis of the most efficient alternative of the fuel system using mathematical models and the set of performance criteria have been developed on the main stages of the study. The results from the introduction of specific engineering solutions to develop their own energy supply sources for RH processing facilities have been provided.

1. Introduction

Modern raw hydrocarbon processing plants (RHPP) - gas, gas condensate, oil - are the major consumers of all energy resources (ER) for their own needs. The energy component in the final product cost reaches 11-15% with a negative trend of increase in recent years. Primary ERs for RHPP are fuel (liquid and gaseous), electric and thermal energy. On average, fuel consumption accounts for up to 60%, and electrical energy up to 40%. This distribution of primary ER consumption is related to the large amount of thermal energy produced in its own power engineering plants, thereby replacing consumption from a third-party source. The consumption of ER occurs in the technological system (TS) and the energy complex (EC) of the facility. The technological system is the basis for the formation of the EC and its subsystems, the bases of which are the heat technological, electrotechnological and fuel. The fuel system is characterized by the largest number of linkages with TS and external power supply systems.

2. Statement of problem

The solution of the problems of improving the efficiency of the RHPP fuel system and the synthesis of its optimal structure, taking into account all in-plant and external factors: *technological, operational, economic, environmental, climatic, organizational,* etc. can be only based on the system approach to study complex systems. The fuel system analysis is carried out in stages, including the development of a mathematical link between the original state of the system and its subsystems, as well as EC, TS, external energy systems at all levels of the hierarchy. The analysis also includes the efficiency assessment of the RHPP fuel structure, its comparing to alternatives.

3. Theory

3.1. The characteristics of the RHPP fuel systems

The main fuel users at RHPP are: fire engineering equipment, thermal waste treatment plants, flaring systems, boilers. To provide liquid or gaseous fuels, at the factories processing RH there is a fuel-handling facility, including fuel preparation shops, fuel loopback pipes with continuously circulating liquid fuel, with laid lines for separate technical facilities. There is also heat exchange equipment for the preheating of fuel before it is fed to the furnace burner. The principle scheme for generating and consuming gas and liquid flows in the most general form for gas treatment plants (GTP) is shown in figure 1, for oil refinery plants (ORP) in figure 2.

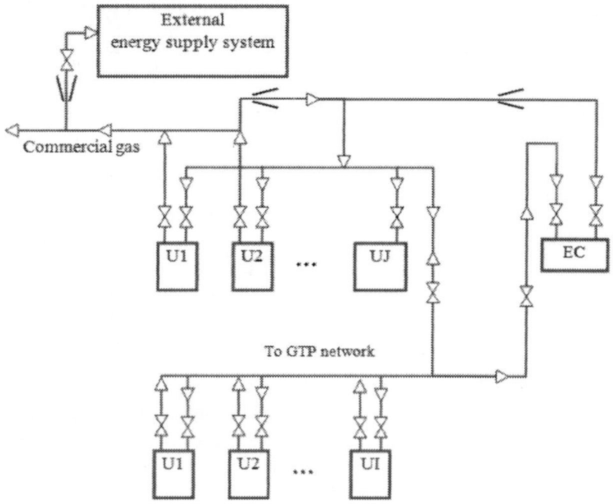

Figure 1. The principle scheme for generating and consuming gas flows at GTP:
У1–УJ - gas plants; У1–УJ - gas condensate plants.

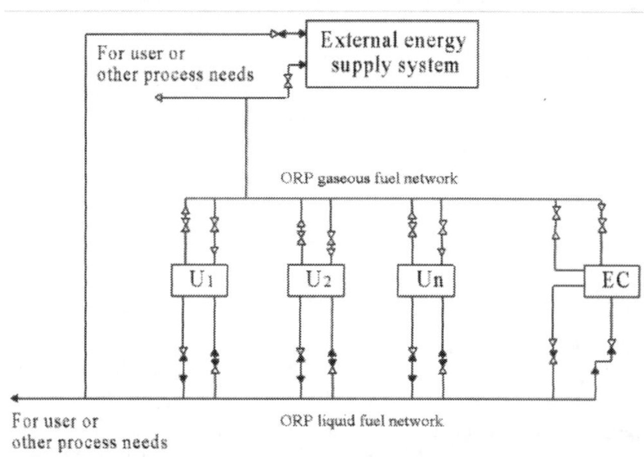

Figure 2. The principle scheme for generating and consuming gas and liquid flows at ORP:
У1–Уn - oil refining units.

Despite the significant similarities in the process of GTP and ORP, their fuel supply systems have substantial differences. Thus, gas from the commercial gas network, as well as hydrocarbon gases obtained during the processing of raw materials, are the main fuel for GTP. The ORP has a large amount of own generation of liquid fuel, refinery gases, and the commercial gas is usually a third-party ER.

For the current domestic ORP, the percentage distribution of the fuel used in the main processes is shown in table 1.

Table 1. Percentage of fuel used in the ORP processes.

Technological process	Own gas,%	Natural gas,%	Fuel oil,%	Coke (conversed fuel),%
Primary distillation (crude vacuum unit)	75	15	10	-
Catalytic cracking	43	6	25	26
Hydrotreating	99.5	0.4	0.1	-
Thermal cracking	80	5	15	-

As it can be seen from the table, its own fuel (refinery gas, fuel oil and conversed fuel) accounts for a large share of the consumption, while the share of natural gas ranges from 0.5 to 20% for all technological processes. In this light, the scientific developments in maximizing the potential of combustible secondary energy resources (SER) that are in large quantities at RHPP are promising. They allow reducing the consumption of commercial or purchased fuel gas for the technological needs of the facility.

The analysis of ER consumption over the past few years reveals a trend of increasing specific consumption. This is due to objective factors such as: increasing RH processing yield, improving the quality of products produced; composition changing and increasing the volume of recycled RH; the introduction of secondary energy-intensive processes. On the other hand, there is a number of negative in-plant and external factors influencing on the absolute indexes and multiples of ER consumption. In-plant factors are: the outdated plant and the resulting low efficiency of energy-using and power generation equipment, the increased intensity of the processes; the low level of heat recovery and regeneration of the technological processes; ER losses. The most significant external factors can include the increased intensity in regional fuel and energy balances and the instability of electricity supply.

The improvement of RHPP fuel system efficiency and the synthesis of their optimum structures is possible with the integrated approach, based on the principles of system analysis. It makes internal and external connections between the facility, the systems and subsystems of RHPP, the external power systems in time variable dynamics, the most significant influencing factors.

3.2. The system analysis of RHPP fuel systems

The system analysis of the fuel system and its components involves the following stages: defining the structure of the object and its operating parameters: developing and calculating characteristics and efficiency factors; the identification of influential factors and their impact on modes of operation. The synthesis of the fuel system is primarily to define the principles underlying its operation, to develop a rational structure and the parameters of the fuel system; to model mathematically structure alternatives; to develop a set of engineering solutions to improve the efficiency of the fuel system.

The first point of RHPP fuel system analysis is to design its structure on the basis of block-hierarchical approach with complex deterministic relationships with ER and TS (figure 3). The analysis of the developed structure indicates that the fuel system is characterized by a large number of relationships with the TS and the heat engineering subsystem in a dynamic continuous ER consumption and generating process.

A formal description of the structure of the RHPP power engineering balance (figure 3), to solve the problems of improving the efficiency of the fuel system, represents the aggregation of multiple structure models:

N TS processing

$$P_{TS} = \left\{ P_{TS}^{1} ... P_{TS}^{N} \right\}; \qquad (1)$$

M is EC subsystems

$$P_{EC} = \left\{P_{EC}^{1} ... P_{EC}^{M}\right\}; \tag{2}$$

RHPP as a whole

$$RHPP = P_{TS} \cup P_{EC} \tag{3}$$

where $P_{TS}^{1} ... P_{TS}^{N}$ is the mathematical description of the structure of RH preparation and processing, waste recycling, conditioning and storage of intermediate and final technological flows, transportation of products;

$P_{EC}^{1} ... P_{EC}^{M}$ is the mathematical description of the structure of the EC subsystems that generate, transform, transport fuel, electricity, thermal energy, water, in-plant energy supplies.

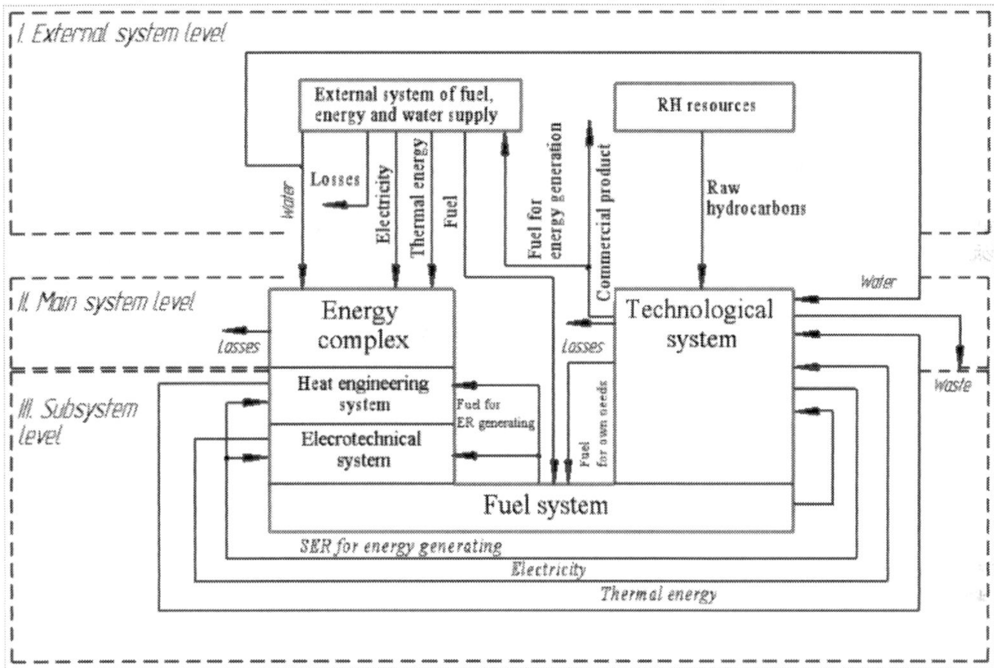

Figure 3. The scheme of RHPP power engineering balance at the hierarchical level of the external systems, the main plant systems and the subsystems.

The synthesis of a rational structure and definition of the operating modes of the RHPP fuel system on the basis of the system performance vector test:

$$\overline{U} = \sum_{i=1}^{n} \gamma_i \cdot \overline{u_i} \tag{4}$$

where $\overline{u_i}$ is the value of i efficiency criterion u_i assigned to its extreme value u_{extr} in the technical solution alternatives; $\overline{u_i} = u_i / u_{extr}$;

n is the number of partial performance criteria of solutions being developed to increase fuel energy efficiency;

γ_i are the vector elements of partial criteria relative importance (rank) factors.

The priority vector $W=(\gamma_1,\gamma_2,...,\gamma_n)$ is determined by calculating the geometric mean of each matrix row.

$$\gamma_i^/ = \sqrt[n]{\prod_{j=1}^{n} g_{ij}} \tag{5}$$

where g_{ij} is the element of i row in j column in the matrix of pairwise comparisons of private criteria. The elements of the priority vector are obtained by dividing by the sum of the geometric means

$$\gamma_i = \gamma_i^/ \Big/ \sum_{i=1}^{n} \gamma_i^/ \tag{6}$$

The criteria for the rationalization of balances as well as technical and economical for efficiency have been taken as partial one.

Mathematical models for calculating and optimizing the balances of the fuel system have been established on the basis of the developed fuel system structure, the set of performance indicators.

3.3. *Mathematical modelling of RHPP fuel systems*

In general terms, the mathematical model of the fuel system, depending on the influencing factors can be written:

$$\overline{Y}=\overline{F}(\overline{X},T,\overline{D},\overline{P},\overline{O},\tau), \tag{7}$$

where $\overline{Y},\overline{X}$ is the vector of output and input variables, respectively;

$\overline{F}$ - the vector function of vector arguments $\overline{X},\overline{D},\overline{P},\overline{O}$ and scalar argument τ (time); T - process topology; $\overline{D}$ - the vector of structural parameters; $\overline{P}$ - the vector of the technological parameters; $\overline{O}$ - the parameter vector of the element process mode.

The mathematical model of the fuel system efficiency, which is a quality indicator of its operation, is written in general terms as follows:

$$k=k(T,\overline{D},\overline{P},\overline{O},\overline{S},\tau), \tag{8}$$

where k is the efficiency coefficient, $\overline{S}$ is the vector of system properties;

The use of mathematical modelling, the strategy for applying the principles and methods of system analysis, synthesis and optimization to solve the problems of synthesizing the optimum structure of the fuel system are based on the mathematical models of the fuel system (7, 8) (Figure 4). The use of mathematical models with specified vector of input parameters X and chosen efficiency criteria with required accuracy allows defining output variables Y, evaluating the system properties and values of the fuel system efficiency.

Thus, the task of synthesizing the optimum fuel system is a boundary value problem that will determine the process topology of the system, the parameters of its elements, the optimum parameters for its functioning in accordance with selected efficiency factors.

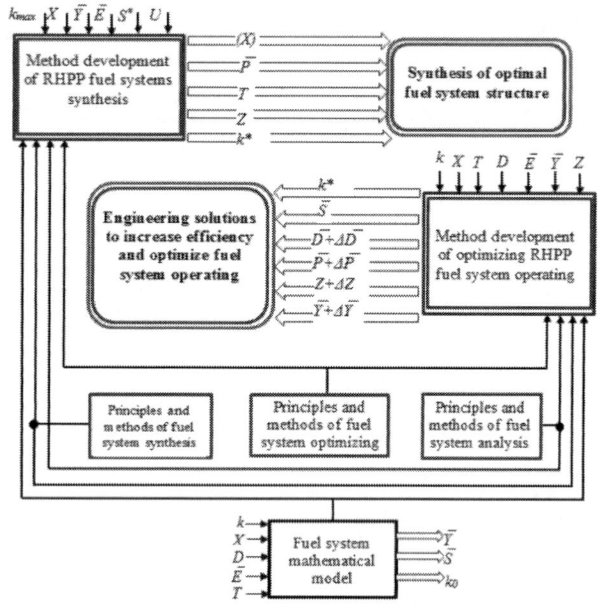

Figure 4. Block diagram of the synthesis of the fuel structure using mathematical modelling.

$\overline{E}$ - environmental vector; S^* - the extreme (favorable) values of system properties; U -modern-level hardware; k_0 - some value of the efficiency test; k^* - optimum value of the efficiency coefficient; $k_{\max}{}^*$ - the extreme optimum value of the efficiency coefficient of the existing fuel systems; Δ-variation of vector change.

4. Results of numerical experiments

The considered theoretical aspects of system analysis allowed developing a number of technical solutions for optimizing the fuel system structure as a part of RHPP EC. For RH production, processing and storage plants, the system of fuel, energy and water supply with maximum recycling of hot waste and effluents was proposed [1].

Also, options for the energy supply systems of individual units of the large RHPP with house set on the basis of combined cycle power plants have been developed [2, 3]. For example, for integrated oil refining unit GK-3 with productivity 3000000 t/y, gas turbine GTE-6u, KU-42 TKZ, *R-2.7-4.5/0.6 KTZ* was considered for energy supply:

Capacity, MW:	
• electric	9.2
• -thermal	14.2
Fuel consumption, ton of reference fuel/h	2.844
Profitability index rub/rub	1.963
Integral effect (for 10 years), mln. rub.	990
Payback period, years	no more than 5

Options for a refinery tankage with the units RBS -5000 located in the medium climatic zone have been proposed [4]. The feed capacity is 9.1 million t/y by *gas turbine power-plant* GTES-2.5.*Exhaust heat boiler* Г-250 *Drive condensing turbine* K-2,5-3,4П:

Capacity, MW:	
• electric	4.2
• thermal	8.0
Fuel consumption, ton of reference fuel/h.	1.393
Integral effect (for 10 years), mln. rub.	372.9
Payback period, years	~4,0

5. Result discussion

House set options were evaluated in accordance with developed set of efficiency criteria. The data analysis allows concluding about their use as part of RHPP EC, a variety of process topology. For individual processes in RHPP, the application of the proposed energy supply options with the optimization of the fuel system results in a significant economic impact and reduces the ER consumption, including fuel for its own needs. The synthesis of the fuel system structure as part of RHPP was carried out in accordance with the developed principles of the system approach and based on a block diagram, using mathematical modelling techniques.

6. Summary and conclusion

1. Methodological aspects for system analysis of raw hydrocarbon processing plants fuel systems have been defined including the development of the hierarchical structure for the fuel system, the set of performance criteria and the mathematical model of its elements.
2. The developed hierarchical structure of the fuel system has resulted in the identification of quantitative and qualitative relationships between the systems and subsystems of the raw hydrocarbon processing plants and the external energy supply systems.
3. The block scheme to synthesize the optimum structure of the fuel system has been developed on the basis of the set of efficiency criteria, the fuel system structure, the mathematical models of the elements using the principles of system analysis, synthesis and optimization.
4. The technical and economic indicators for the implementation of solutions to develop efficient fuel systems as part of energy complex at raw hydrocarbon processing plants have been given.

Acknowledgment

The work was supported by the grant of the Russian Science Foundation (project №17-79-10134).

Reference

[1] Dolotovskij I V, Larin E.A., Dolotovskaja N V and Kulbjakina (Lenkova) A V Pat. No. 134993 Russian Federation, IGC f01k 17/02. Energy-fuel-water supply apparatus B33
[2] Kulbjakina (Lenkova) A V, Larin E A and Dolotovskij I V 2016 Efficient energy supply system for combined oil units *The problems of improvement of fuel and energy complex* **8** 233-38
[3] Kulbjakina (Lenkova) A V and Dolotovskij I V 2013 A plant for absorbent regeneration utilizing fuel wastes *Chemical and Petroleum Engineering* **49** 517-521
[4] Dolotovskij I V, Larin E.A., Dolotovskaja N V and Kulbjakina (Lenkova) A V 2017 Increasing the Energy Efficiency of Heating and Cooling Systems for the Tank Farms of Oil and Gas Enterprises. *Chemical and Petroleum Engineering* **52** (11) 803-809

Manufacture of conical springs with elastic medium technology improvement

S A Kurguzov, U V Mikhailova and O B Kalugina

Nosov Magnitogorsk State Technical University, 38 Lenin Avenu, Magnitogorsk, 455000, Russia

Abstract. This article considers the manufacturing technology improvement by using an elastic medium in the stamping tool forming space to improve the conical springs performance characteristics and reduce the costs of their production. Estimation technique of disk spring operational properties is developed by mathematical modeling of the compression process during the operation of a spring. A technique for optimizing the design parameters of a conical spring is developed, which ensures a minimum voltage value when operated in the edge of the spring opening.

Keywords – conical springs, elastic medium, stamping, elastic stress-strain state, elastic deformation, solid mechanics.

1. Introduction

Conical springs are widely used in engineering for damping dynamic and shock loads arising during the operation of machines (shock absorbers, buffer devices, etc.).

However, according to numerous studies and practical tests, a large percentage of springs in the course of operation under cyclic loading does not withstand the number of cycles specified in the industry standard ($2 \cdot 10^6$). Destruction occurs due to the formation of fatigue cracks in the edge of the hole and wear on their contact surfaces. Defects of the spring edges surface, appearing during the typical stamping, accelerate the fatigue cracks creation. In the case of standard springs, significant voltages arise in the operation therefore the resource decreases. The contact surface of the spring wears out quickly due to its small area. To increase the life of a conical spring, it is necessary to reduce stresses during operational deformation and to increase the contact surface area and to prevent the surface defects during its manufacturing .

Reduction of stresses is possible by optimizing the design parameters of the spring, based on the results of the analysis of stress strain behavior during compression in operation. Flanging of the spring edge will increase the area of its contact surface. To produce an improved spring with optimal design parameters, it is necessary to change the design of the die tool. Most FEM systems use the methods of calculating the stress-strain state on the basis of variational methods created by the works of many scientists, such as R. Hill [1], D.Banabic [2], F. Barlat [3–7], J.L. Bassani [8], et al. However, the use of FEM codes for complex shape parts requires considerable computer resources. One possible solution to this problem is a preliminary study in the 2D model, and then on the problem areas more accurate 3D modeling [14].

A large number of works have been devoted to the research of the conical springs sheet stamping process [7-8, 10]. The usual methods of sheet-metal stamping , widely used in large-scale and mass production, can be insufficient in the conditions of small-scale and fast-retooled production, since it takes a long time to produce structurally complex and expensive stamps. In some cases, the investment is not return. For the rapid development of new types of products, it is necessary to use new technological processes in the specified conditions using cheap universal or partially universal equipment. Improving the technology of manufacturing disc springs, eliminating the operation of the mechanical processing of the edges, will reduce production costs and improve the performance characteristics of the spring.

2. Optimization of the conical springs design parameters

Optimization of the conical springs design parameters consists of three stages. The first stage: determination and investigation of the spring surface areas, on which stresses state arises, causing fatigue failure, based on the finite element method. The second stage: improvement of the spring shape to reduce the tensile stresses in the area determined in the first stage. The third stage: the calculation of the optimal design parameters of the disk spring, based on the methods of linear programming.

The problem of calculating the elastic stress strain stage of a conical spring is solved on the basis of the potential energy minimum principle. The solution considered an axisymmetric stress strain stage in cylindrical coordinates, since the conical spring has axisymmetric shape.

Due to the symmetry of the body and the external forces symmetry, the directions of all radii r are equal. The stress components do not depend on the angle φ. The tangential stresses on the planes φ-τ_rφ and τ_zφ, as well as the τ_rφ and τ_zφ equal to them, will be zero. Neglecting infinitesimals of higher orders, taking into account the equilibrium equations, the potential energy functional takes the form:

$$\chi = \frac{1}{2}\int_0^{2\pi} d\varphi \int_{R_2}^{R_1}\int_{z_2}^{z_1} \varepsilon^T \cdot D \cdot \varepsilon \cdot r dr dz - k \cdot p_z \delta \int_0^{2\pi} R_2 \cdot u d\varphi \cdot \quad (1)$$

where $z_1 = H_1 - \frac{H_1}{R_1} r$;

$z_2 = H_2 - \frac{H_1}{R_1} r$;

$H_1 = R_1 \cdot tg\alpha$;

$H_2 = H_1 + \frac{S}{\cos\alpha}$;

$p_z = \frac{F}{2\pi R_2 \cdot \delta}$;

$$D = \frac{E}{1-\mu^2}\begin{bmatrix} 1 & \mu & 0 \\ \mu & 1 & 0 \\ 0 & 0 & (1-\mu)/2 \end{bmatrix}$$

— k is the constant of friction;
— μ is the Poisson ratio;
— ε is the deformation matrix,
— u is the displacement vector,
— δ is the width of the contact surface

As a result of the displacement approximation on each of the elements by linear functions and using the rules of matrix differentiation, we calculated the partial derivatives of elemental energies in terms of the nodal displacements vectors and obtained an elementary matrix equation from the equation of the functional (1):

$$\frac{\partial \chi^e}{\partial U^e} = 2\pi\left(B^T \cdot D \cdot B \cdot U^e \cdot I_{D^e} - k \cdot p_z \cdot I_{L^e} \right), \tag{2}$$

where D^e is the area of the triangular element;
L^e - length of the contact side, loaded boundary elements;
U^e - approximation of the each finite elements displacements by linear functions

$$I_{D^e} = \int_{D^e} r dD^e;$$

$$I_{L^e} = \int_{L^e} \begin{bmatrix} 0 \\ N_i^e \\ 0 \\ N_j^e \\ 0 \\ N_k^e \end{bmatrix} r dL^e;$$

$$B = \begin{bmatrix} \frac{\partial}{\partial r} & 0 \\ 0 & \frac{\partial}{\partial z} \\ \frac{\partial}{\partial z} & \frac{\partial}{\partial r} \end{bmatrix} N^e;$$

N^e are the basis functions.

Further, after using the extremum condition for the function (1) and combining the elemental matrix equations (2), we obtain a system of linear equations[15]:

$$\sum_{e=1}^{48} K^e \cdot U^e - \frac{F \cdot S \cdot (R_1 - R_2)}{288 \cdot \pi \cdot R_2 \cdot \cos\alpha} = 0, \tag{3}$$

where $K^e = B^T \cdot D \cdot B \cdot I_{D^e}$ - is conical springs stiffness matrix.

To numerical implementation of this model, a software product in the C# was developed[16]. To check the obtained results, simulation in the CAE "ANSYS" system was conducted.

3. Experiment results

As a result of the study of the stress strain stage of standard springs, it is determined that the region with the greatest tensile stresses is the edge of the 2 holes (Fig. 1). These stresses create favorable conditions for the appearance and development of fatigue cracks.

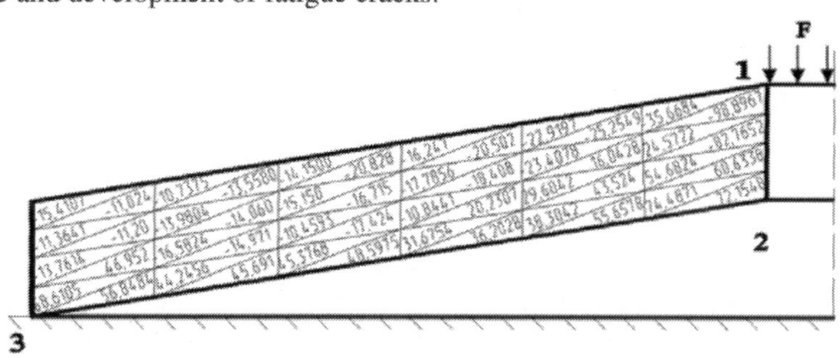

Figure1. Stresses σφ in a loaded conical spring

IOP Conf. Series: Journal of Physics: Conf. Series **944** (2017) 012069 doi:10.1088/1742-6596/944/1/012069

With the purpose of reducing the value of stress in the edge 2, a technique for calculating the optimal design parameters of a disk spring was developed[17]. The technique based on the calculation formulas given in national standard 3057-90 "Conical Springs".

As a target criterion, the minimum stresses at the edge 2 for compression are chosen. As a limitation, the geometric dimensions of the conical spring are adopted, which must correspond to the location of its installation and its rigidity. To implement the proposed methodology for calculating the optimal design parameters of the disk spring, application software was developed.

The disk spring was also improved by a change in shape, forming a generatric curvilinear of the cone in the longitudinal section (Fig. 2). This led to a reduction of the stress value in the edge 2.

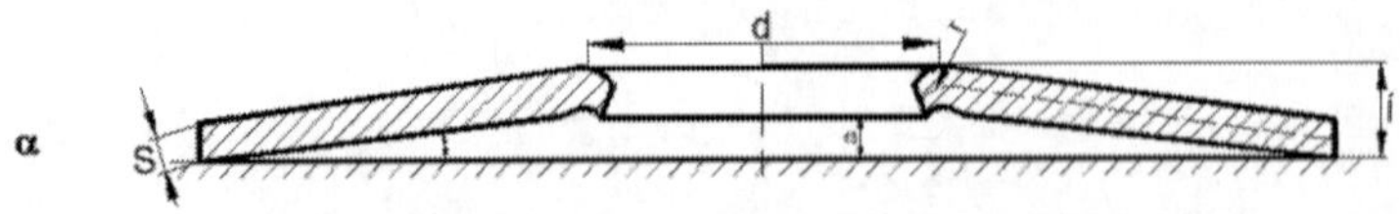

Figure 2. Improved conical spring cross section

The design of the conical springs using this technique of calculating and improving the shape leads to a decrease of the stress value in the edge of hole 2 to 30%.

4. Discussion

Using the developed software product, a series of test calculations of design parameters of standard and improved conical springs were done. In table 1 the results of calculations of the stress in edge 2 at 0.6 f3 spring deformation shown (Fig. 2) for standard springs taken from standard tables and optimized springs.

Table 1. Results of test calculations

Spring type	*D*, mm	*d*, mm	*S*, mm	f_3, mm	*C*, H/ mm	σ_2, MPa	r_1, mm	r_2, mm
standard spring	31.5	12.5	1.05	1	1094	532.07	-	-
optimized spring	31.5	15.75	0.97	0.83	1200	459.20	1.22	0.5

The analysis the of test calculations results showed that for a given diameter of the spring D and the stiffness range, a decrease in stress σ2 (stress in edge 2 (Fig. 1)) was recorded by 14%. For other design parameters of the conical springs, a reduction in the stress σ2 to 30% was recorded. The results were discussed and approved at conferences [18, 19].

5. Conclusions

A technique for calculating the optimal design parameters and the shape of the plate springs is developed, consisting of three stages. The first stage: with the help of the developed mathematical model of the elastic compression process during the operation of the spring, a surface area of the spring is defined, where the greatest stresses occur that promote fatigue failure.

The second stage: improvement of the shape of the Belleville spring in order to reduce the stresses on the found area. The most loaded of them is the lower edge of the hole. The third stage: optimizing the design parameters of the spring, based on linear programming methods, providing a minimum value of the voltage at the lower edge of the hole during operation.

The use of the proposed calculation procedure led to decrease in the value of the stresses in the lower edge of the hole to 30% in comparison with the standard spring.

References

[1] Hill R 1990 Constitutive modeling of orthotropic plasticity in sheet metals *Journal of the Mechanics and Physics of Solids* **38**(3) pp. 405–417
[2] Banabic D, Aretz H, Comsa D S [et al.] 2005 An improved analytical description of orthotropy in metallic sheets *International Journal of Plasticity* **21** pp 493–512
[3] Barlat F, Lege D J and Brem J C 1991 A six-component yield function for anisotropic materials *International Journal of Plasticity* **7** pp 693–712
[4] Barlat F, Maeda Y, Chung K [et al.] 1997 Yield function development for aluminum alloy sheet *Journal of the Mechanics and Physics of Solids* **45** pp 1727–1763
[5] Barlat F, Brem J C, Yoon J W [et al.] 2003 Plane stress yield function for aluminum alloy sheets *Part I: theory. Int. J. Plast.* **19** pp 1297–1319
[6] Barlat F, Aretz H, Yoon J W [et al.] 2005 Linear transformation based anisotropic yield functions. *International Journal of Plasticity* **21** pp 1009–1039
[7] Barlat F, Yoon J W and O Cazacu 2007 On linear transformation of stress tensors for the description of plastic anisotropy *International Journal of Plasticity* **23** pp 876–896
[8] Bassani J L 1977 Yield characterisation of metals with transversally isotropic plastic properties *Int. J. Mech. Sci.* **19** pp 651–660
[9] Wang J and Sun J 2012 Plane strain transversely anisotropic analysis in sheet metal forming simulation using 6-component Barlat yield function *International Journal of Mechanics and Materials in Design* **8**(4) pp 327–333
[10] Mikhailova U V 2011 Improvement technology manufacturing special conical spring with elastic medium *Dissertation for the degree of technical sciences candidate* (Magnitogorsk: Nosov Magnitogorsk state technical university)
[11] Zhelezkov O S, Kochukov S V and Mikhailova U V 2006 Modeling compress the conical springs by using finite element method *Processes and equipment of metallurgical production* (Magnitogorsk: NMSTU) pp 239–243
[12] Mikhailova U V 2006 Application of finite element method for the metal stress-strain state computation Mathematics *Mathematics applications in the economic, technical and pedagogical research* (Magnitogorsk: NMSTU) pp 258–260
[13] Mikhailova U V 2007 *The elastic stress-strain state conical spring modeling using the finite element method The certificate of registration of the branch development* № 8128 Russian Scientific and Technical Center № 50200700_808.
[14] Norenkov I P 2002 Fundamentals of computer-aided design (Moscow: Bauman MSTU)
[15] Korinchenko G M, Mikhailova U V and Kislitsin K A 2007 Sparse matrix application for thedesign solution analysisdesign elementsin mechanical engineering *The application of mathematics to economic and Technical Studies* **1** (1) pp 142–145
[16] Mikhailova U V 2007 Elastic stress strain state of conical spring modeling using finite element method Certificate of industrial design registration № 8128 Russian Scientific and Technical Center № 50200700_808
[17] Mikhailova U V 2007 Optimization of calculation of conical spring design parameters Certificate of industrial design registration № 8129 Russian Scientific and Technical Center № 50200700_753
[18] I I Barankova I I, Mikhailova U V and Demidovich V B 2016 Application the sparse matrix method to calculate the metal elastic stress-strain state, using the finite element method *ECCOMAS Congress 2016 – Proceedings of the 7th European Congress on Computational Methods in Applied Sciences and Engineering* pp. 7581–7588
[19] V Demidovich V, Mikhlyuk A, Barankova I and Mikhailova U 2016 Advanced induction coils for heating steel products and their computer models *Proceedings of IFOST-2016 11th International Forum on Strategic Technology* pp 73–77

IOP Conf. Series: Journal of Physics: Conf. Series **944** (2017) 012070 doi:10.1088/1742-6596/944/1/012070

Reduction of Dynamic Loads in Mine Lifting Installations

N K Kuznetsov[1], S V Eliseev[2] and A Yu Perelygina[1]

[1] State Budget Educational Institution of Higher Education «Irkutsk National Research Technical University», Irkutsk, Russian Federation
[2] State Budget Educational Institution of Higher Education «Irkutsk State Transport University», Irkutsk, Russian Federation

Abstract. Article is devoted to a problem of decrease in the dynamic loadings arising in transitional operating modes of the mine lifting installations leading to heavy oscillating motions of lifting vessels and decrease in efficiency and reliability of work. The known methods and means of decrease in dynamic loadings and oscillating motions of the similar equipment are analysed. It is shown that an approach based on the concept of the inverse problems of dynamics can be effective method of the solution of this problem. The article describes the design model of a one-ended lifting installation in the form of a two-mass oscillation system, in which the inertial elements are the mass of the lifting vessel and the reduced mass of the engine, reducer, drum and pulley. The simplified mathematical model of this system and results of an efficiency research of an active way of reduction of dynamic loadings of lifting installation on the basis of the concept of the inverse problems of dynamics are given.

1. Introduction

The high speeds of movement of mine lifting installations lead to the emergence of large dynamic loads, elastic deformations of structural elements and oscillating movements, causing excessive stress, fatigue, increased wear and, as a result, deterioration of durability and safety in operation. The tighten of characteristics of safety and reliability of modern lifting installations places high demands on the level of their dynamic calculations and necessitates taking into account the elastic properties of the design and development of methods and tools for reducing dynamic loads and vibrations [1–3].

To solve this problem, in [4–7] proposed a special mathematical model that takes into account all important features of the mechanics of mine lifting installation under braking, and in [8] considered the impact of different parameters on the dynamic behavior of mine lifts during extreme braking. The various circuit designs that can be configured to implement dynamic braking on a mine hoisting system are offered and selection of the load resistance for the desired dynamic braking system performance is addressed [9]. The present state of knowledge of dynamics of hoisting systems and available modern computer hardware enable to develop and analyze simulation models used for optimizing of dynamic characteristics of the system of mine lifting installations [10–12].

In reference [13], computer monitoring systems and registers of the parameters of the lifting installations are proposed, which allow to determine the occurrence causes of additional loads and to develop measures for their elimination. The system of automatically controlled safety braking is proposed in [14] for minimize dynamic overloads in the mechanical part and increase the safety of the mine lifting installations. This system is designed for use on all types of inclined drum and vertical multi-channel lifts with radial and disc brakes. In reference [15–17] the use the smooth laws of motion of the lifting vessels are encouraged for reduce the dynamic loads in mine hoisting installations. The possibilities of application of the retarding force on the vessel during the working of mine lifts in mode of lifting are investigated in [18]. The main disadvantages of these methods reduce the dynamic loads are either low efficiency under varying operating conditions or the performance of the mining installations.

In this regard, particular interest is the use of the method of reduction of dynamic loads and elastic vibrations of the links of managed machines based on the use of the concept of inverse problems of dynamics, proposed in [19–23]. The results of research, related to the implementation of this method to reduce of dynamic loads and damping of elastic vibrations on the example of one-ended lifting installation, are given in this article.

2. Problem statement

Currently the developed methods of dynamic synthesis, which would allow to realize a controlled movement with the desired dynamic properties, are absent in the dynamics of controlled machines with elastic links. Most of the known solutions in this field based on engineering experience and intuition than on scientific – methodological basis. Analysis of known work in this area has shown that an effective method of problems solving of control synthesis, providing the mechanical movement with the desired properties, is the use of methods of solution of inverse problems of dynamics for a given type of vibrational motion. A principal scheme of one-ended lifting installation is shown in figure 1, for which considered the implementation of this approach for problem solving of reducing of dynamic loads.

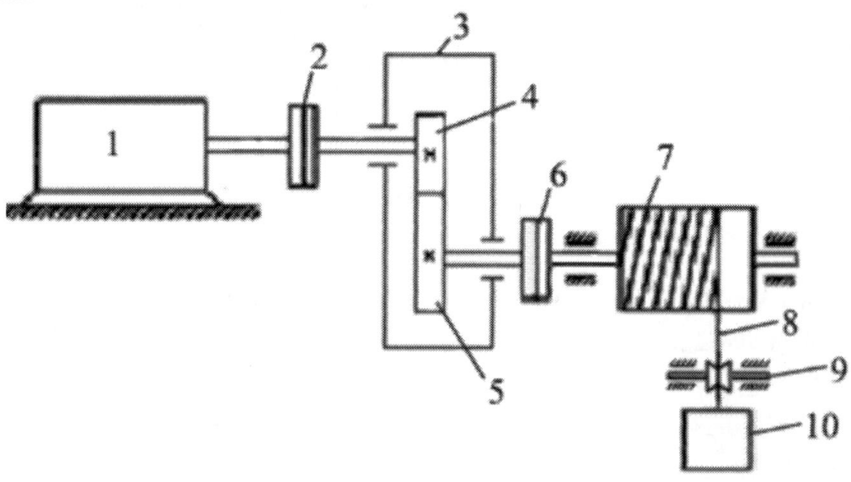

Figure 1. A principal scheme of lifting installation: *1* – electric engine; *2, 6* – clutch; *3* – reducer; 4 – gear; *5* – gear wheel; *7* – drum; *8* – rope; *9* – pulley; *10* – lifting vessel.

The following assumptions are adopted to obtain the dynamic model of the mine lifting installation:

- the rigidity of the gear reducer 3, the couplings 2, 6 and shafts is much higher than the stiffness of the rope 8;
- the mass of the rope 8 is concentrated at the drum 7 and lifting vessel 10, the rope is a weightless and non-viscous elastic system with a constant stiffness coefficient;
- the aerodynamic air resistance and friction force in the conductors of the shaft are absent;
- the movement is considered from the position of static equilibrium of the lifting vessel;

- the influence of dynamic processes in the drive motor or braking device is not taken into account.

A design scheme of the mine lifting installation can be represented as a two-mass mechanical system, shown in figure 2.

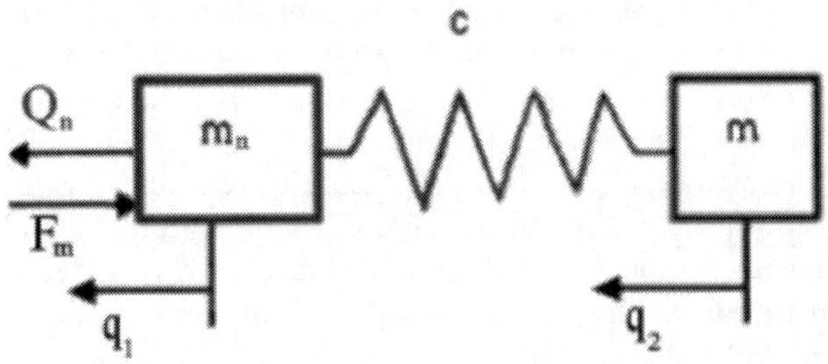

Figure 2. Two-mass design scheme of the lifting installation.

An inertial elements of the lift installation are the mass of the lifting vessel with part of the rope mass and the resulting mass of the engine, reducer, drum, parts of a rope and pulley, connected by a weightless elastic rope. The differential equations of motion of two-mass system, obtained on the basis of the D'alembert principle, are as

$$m_n\ddot{q}_1 + c(q_1 - q_2) = Q_n - F_m; \quad (1)$$

$$m\ddot{q}_2 + c(q_2 - q_1) = 0, \quad (2)$$

where m_n – the reduced mass of the engine, reducer, drum, parts of a rope and pulley; m – the reduced mass of the lifting vessel with part of the rope mass; c – the reduced coefficient of rope stiffness; Q_n, F_m – the reduced values of the driving force of the actuator and the braking force; q_1 – the generalized coordinate of the mass movement m_n; q_2 – the generalized coordinate of the lifting vessel movement.

Summing to equations (1) and (2) and denoting the elastic deformation of the rope $\Delta q = q_2 - q_1$, obtain the equation of motion of the first mass

$$(m + m_n)\ddot{q}_1 + m\Delta\ddot{q} = Q_n - F_m. \quad (3)$$

The expression (2) takes the form

$$m\Delta\ddot{q} + m\ddot{q}_1 + c\Delta q = 0. \quad (4)$$

It is required to determine the law of change of motive (Q_n) or braking (F_m) forces in expression (1), providing a minimum dynamic load and compensation of elastic vibrations of the lifting vessel, when it goes up, by solving the inverse dynamic problem for a given law of change of elastic deformation of the rope and limiting the movement acceleration of the first mass value $|\ddot{q}_1| \le a$.

3. Theory

Find the control action Q_n, by which the elastic vibrations of the lifting vessel is fully extinguished for any valid values of acceleration time (deceleration) and the motion acceleration of the first mass does not exceed a predetermined value $|\ddot{q}_1| \le a$. Consider the mode of the vessel lifting, assuming that the elastic vibrations are absent in the initial time, i.e. $\Delta q(0) = \Delta\dot{q}(0) = 0$, in this case the elastic deformation of the rope is changed under the harmonic law

$$\Delta q = \gamma_p \cdot \left(1 - \cos\frac{2\pi}{T_p}t\right), \quad (5)$$

where γ_p – a constant value; T_p – acceleration time (deceleration). Then the conditions $\Delta q = \Delta\dot{q} = 0$ will be implemented when $t = 0$ and $t = T_p$ and on the basis expression (5). After double differentiating in time (5), the law of acceleration change of the elastic deformation of the rope is obtained

$$\Delta\ddot{q} = \frac{4\pi^2\gamma_p}{T_p}\cos\frac{2\pi}{T_p}t. \tag{6}$$

Express from expression (1) when $F_m = 0$ the acceleration

$$\ddot{q}_1 = \frac{Q_n + c\Delta q}{m_n}. \tag{7}$$

Combining expressions (7) with (4), is obtained

$$\Delta\ddot{q} + \omega^2\Delta q = -Q_n/m_n, \tag{8}$$

where $\omega = \sqrt{\frac{c(m+m_n)}{m m_n}}$ – the frequency of free oscillations of the system.

Substituting expressions (5) and (6) into equation (8), the required law of change of driving (braking) force is determined

$$Q_n = -m_n\gamma_p\left[\omega^2\left(1-\cos\frac{2\pi}{T_p}t\right) + \frac{4\pi^2}{T_p^2}\cos\frac{2\pi}{T_p}t\right]. \tag{9}$$

The definition of constant γ_p. Substituting expressions (5) and (9) into equation (7) is obtained

$$\ddot{q}_1 = -\gamma_p\left[\frac{4\pi^2}{T_p^2}\cos\frac{2\pi}{T_p}t + \omega_0^2\left(1-\cos\frac{2\pi}{T_p}t\right)\right], \tag{10}$$

where $\omega_0 = \sqrt{c/m}$ – the partial oscillation frequency of the mass of the lifting vessel.

After double integrating (10) the speed and coordinate the movements of the first mass are

$$\dot{q}_1 = -\gamma_p\left[\frac{2\pi}{T_p}\sin\frac{2\pi}{T_p}t + \omega_0^2\left(t - \frac{T_p}{2\pi}\sin\frac{2\pi}{T_p}t\right)\right] + C_1; \tag{11}$$

$$q_1 = \gamma_p\cos\frac{2\pi}{T_p}t - \omega_0^2\gamma_p\left(\frac{t^2}{2} + \frac{T_p^2}{4\pi^2}\cos\frac{2\pi}{T_p}t\right) + C_1 t + C_2. \tag{12}$$

The constants of integration C_1 and C_2 are found using the initial conditions $q_1(0) = \dot{q}_1(0) = 0$

$$C_1 = 0;\ C_2 = -\gamma_p\left(1 - \frac{\omega_0^2 T_p^2}{4\pi^2}\right). \tag{13}$$

Constant γ_p is defined using the boundary conditions $\dot{q}_1(T_p) = V_{ycm}$, where V_{ycm} is the steady-state value of the movement speed. Substituting the last condition in expression (11) is obtained

$$\gamma_p = -\frac{V_{ycm}}{\omega_0 T_p}. \tag{14}$$

Given expression (14) the laws of change of the driving (braking) forces and acceleration of the first mass are of the form

$$Q_n = \frac{(m+m_n)V_{ycm}}{T_p}\left(1-\cos\frac{2\pi}{T_p}t\right) + \frac{4\pi^2 m_n V_{ycm}}{\omega_0^2 T_p^3}\cos\frac{2\pi}{T_p}t; \tag{15}$$

$$\ddot{q}_1 = \frac{V_{ycm}}{T_p}\left(1-\cos\frac{2\pi}{T_p}t\right) + \frac{4\pi^2 V_{ycm}}{\omega_0^2 T_p^3}\cos\frac{2\pi}{T_p}t. \tag{16}$$

Substituting in expression (16) instead $\ddot{q}_1$ of its limit value $\ddot{q}_1^{np} = a$, the expression for determining of the permissible value of acceleration time (deceleration) is

$$a = \frac{V_{ycm}}{T_p}\left[1-\left(1-\frac{4\pi^2}{\omega_0^2 T_p^2}\right)\cos\frac{2\pi}{T_p}t\right]. \tag{17}$$

It is obvious that the maximum acceleration will occur in start and end moment of acceleration time ($t=0$ and $t=T_p$). In this case expression (17) takes the form

$$a = \frac{4\pi^2 V_{ycm}}{\omega_0^2 T_p^3},$$

where permissible acceleration time

$$T_p \geq \frac{2\pi}{\omega_0}. \tag{18}$$

4. Discussion of results

A comparison of the effectiveness of the obtained compensating exposure with the optimal movement of two-mass system at a predetermined distance is held. We define the control providing a system movement from resting state

$$q_*(0) = \dot{q}_*(0) = \Delta q(0) = \Delta\dot{q}(0) \tag{19}$$

at a predetermined distance q_0, with vibration damping

$$q_*(T_p) = q_0;\ \dot{q}_*(T_p) = \Delta q(T_p) = \Delta\dot{q}(T_p) = 0. \tag{20}$$

When $F_m = 0$ from expression (3) and (4) the acceleration is

$$\Delta\ddot{q} = -\omega^2\Delta q + u(t), \tag{21}$$

where $u(t) = -Q_n/m_n$ – the driving force.

Assume that the driving force is limited in absolute value

$$|u(t)| \leq u_0. \tag{22}$$

Neglecting the influence of elastic oscillations on the actuator movement, the driving force according to expression (1) for the minimum time T_p taking into account (22) is

$$u \approx \ddot{q}_*. \tag{23}$$

Present expressions (21) and (23) in the form

$$\dot{z}_1 = z_2;\quad \dot{z}_2 = u;\quad \dot{z}_3 = z_4;\quad \dot{z}_4 = -\omega^2 z_3 + u, \tag{24}$$

where $z_1 = q_*$; $z_2 = \Delta q$.

The Hamilton function for the system (24) can be written

$$H = \psi_1 z_2 + \psi_2 u + \psi_3 z_4 + \psi_4(u - \omega^2 z_3), \tag{25}$$

where $\psi_1, \psi_2, \psi_3, \psi_4$ are auxiliary variables that can be determined from the system of equations

$$\dot{\psi}_1 = 0;\ \dot{\psi}_2 = -\psi_1;\ \dot{\psi}_3 = \psi_4;\ \dot{\psi}_4 = -\psi_3. \tag{26}$$

The optimum control, providing the maximum of the function (26), has the form

$$u = u_0\,\mathrm{sgn}(\psi_2 - \psi_4). \tag{27}$$

According to expression (27) the optimal control $u(t)$ is a relay function taking values $\pm u_0$. The number of switching points and their positions are unknown. As an example we consider the control with one switching point $t = T_p/2$ of function $u(t)$

$$\begin{cases} u(t)=u_0 \quad npu \quad 0 \le t < T_p/2; \\ u(t)=-u_0 \quad npu \quad T_p/2 < t \le T_p. \end{cases} \tag{28}$$

Integrating the motion equation (24) with $u = u_0$ and initial conditions (19), we obtain

$$\begin{gathered} q_{1*} = z_1 = \frac{u_0 t^2}{2}; \qquad \dot{q}_{1*} = z_2 = u_0 t; \\ \Delta q = \frac{u_0}{\omega^2(1-\cos\omega t)}; \quad \Delta\dot{q} = \frac{u_0}{\omega \sin \omega t}. \end{gathered} \tag{29}$$

Similarly we find the solution for $u = -u_0$ on the interval $\lfloor T_p/2, T_p \rfloor$ and the end conditions (20)

$$\begin{gathered} q_{1*} = q_0 - \left(\frac{u_0(T_p - t)^2}{2}\right); \quad \dot{q}_{1*} = u_0(T_p - t); \\ \Delta q = \frac{u_0}{\omega^2\left[\cos\omega(T_p - t) - 1\right]}; \quad \Delta\dot{q} = -\frac{u_0}{\omega \sin \omega(T_p - t)}. \end{gathered} \tag{30}$$

The found solutions expressions (29) and (30) must be continuous at $t = T_p/2$. The continuity of the values of $\dot{q}_*$ and $\Delta\dot{q}$ takes place at any values of the coordinate q_0, the continuity condition for $\dot{q}_*$ and $\Delta\dot{q}$ is secured if

$$q_0 = \frac{u_0 T_p^2}{4}; \quad \cos\left(\frac{\omega T_p}{2}\right) = 1. \tag{31}$$

On the basis of expression (31) are

$$T_p = \frac{4\pi}{\omega k}; \quad q_0 = \frac{4\pi^2 u_0 k}{\omega^2}, \tag{32}$$

where $k = 1,2,3...$

Thus, the control mode with one switching point satisfies the maximum principle only for magnitude values of the displacement q_0, defined by expression (31). For prove the optimality of the obtained control, it is sufficient to verify the existence of a nonzero vector of auxiliary variables, that providing $H(T_p) \ge 0$ and satisfy system (26). Solving this system, we define

$$\psi_1 = -C; \quad \psi_2 = C\left(t - \frac{T_p}{2}\right); \quad \psi_3 = 0; \quad \psi_4 = 0, \tag{33}$$

where $C = const < 0$.

Substituting expression (33) into (27), we have

$$u = u_0 \operatorname{sgn}\left[C\left(t - \frac{T_p}{2}\right)\right]. \tag{34}$$

The calculation of the Hamilton function according to expression (25) leads to $H(T_p) > 0$.

We compare the acceleration time, expended for control with one switching point, with time expression (18), received for action, which provides the harmonic nature of the changes in the elastic coordinates. On the basis of expression (32) the minimum acceleration time can't be less than the value $T_{p_{\min}} = 2\pi/\omega$. Given $\omega = \omega_0\sqrt{\frac{m+m_n}{m_n}} < \omega_0$, we have $T_{p_{\min}} < T_p$, defined from the expression (18). In this case, the time difference grows with decreasing mass m_n. Thus, management (34) provides higher performance compared to the found control (15). However, management (34) requires instantaneous change in force, which makes difficult its practical feasibility using real drives.

The found the control law (15) is a software, independent on the regulator structure of the engine or the brakes. As a result, the possibility of problem solving, not only parametric but also structural synthesis of motion control system of the lifting installation appears, by using the expression of obtained time dependence on generalized coordinates. Using expressions (5) and (14), trigonometric function through elastic coordinate is expressed

$$\cos\frac{2\pi}{T} = 1 + \frac{\omega_0^2 T}{V_{ycm}}\Delta q. \tag{35}$$

Substituting expression (35) into (15), we obtain

$$Q_n(t) = Q_0 - c^*\Delta q, \tag{36}$$

where $c^* = \left[\left(\frac{4\pi^2}{T_p^2} - \omega_0^2\right)m_n - c\right]$; $Q_0 = \frac{4\pi^2 m_n V_{ycm}}{\omega_0^2 T_p^3}$ – the constant component of the force.

We find the control as a function of acceleration of elastic vibrations. After double differentiating expression (5) in time subject to (14), is obtained

$$\Delta\ddot{q} = -\frac{4\pi^2 V_{ycm}}{\omega_0^2 T_p^3}\cos\frac{2\pi}{T_p}t,$$

from which

$$\cos\frac{2\pi}{T_p}t = -\frac{\omega_0^2 T_p^3}{4\pi^2 V_{ycm}}\Delta\ddot{q}. \tag{37}$$

Substituting expression (37) into (15), we have

$$Q_n = Q_0' - m^*\Delta\ddot{q}, \tag{38}$$

where $Q_0' = \frac{(m + m_n)V_{ycm}}{T_p}$ – the constant component of the force; $m^* = \frac{[m\omega_0^2 T_P^2 + m_n(\omega_0^2 T_P^2 - 4\pi^2)]}{4\pi^2}$.

The coefficients c^*, m^* in expressions (36) and (38) can be interpreted as an amplification coefficients of additional feedback on the elastic coordinate Δq and its derivatives. Structural scheme of two-mass system with an additional feedback of the form (36) is shown in figure 3.

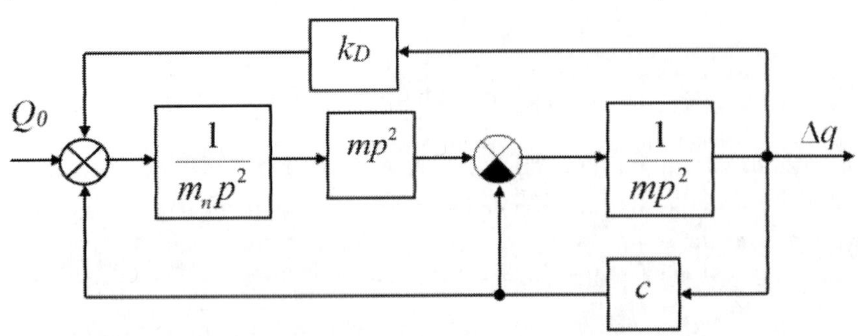

Figure 3. Structural scheme of the system with additional feedback on the elastic coordinate

In this diagram, the amplification coefficient of the additional feedback on the elastic coordinate are equal $k_D = -c^*$. Thus, the found control can be implemented using a closed-loop by the elastic deformation of the rope of control system providing for active damping of oscillations and the required level of dynamic load of mine lifting installation.

References

[1] Stepanov A G 1994 *Dynamics of Mine Lifting Installations* (Moscow: Science) p 203
[2] Stepanov A G 1999 *Dynamics of Machines* (Ekaterinburg: publishing house of Russian Academy of Sciences/ Ural branch) p 392
[3] Kuznetsov N K 2015 The Evaluation of the Safety Braking Modes of Mine Hoist Installations *Proc. of ISTU* (Irkutsk) vol 11 pp 31–34
[4] Tejszerska D and Wojnarowski J 1989 Mathematical Model of Vibrations of Mine Hoists, Coupled with Longitudinal *Zeszity naukowe Politechniki slaskiej* chapter 1 pp 107–118
[5] Kaczmarczyk S and Ostachowicz W 2003 Transient Vibration Phenomena in Deep Mine Hoisting Cables (Part 1: Mathematical model) *Journal of Sound and Vibration* vol 262 pp 219–244
[6] Ilin S R, Ilina S S and Samusia V I 2014 *Mechanics of Mine Hoist* (Dnepropetrovsk: National mining University) p 247
[7] Srebnaia E G and Kondrakhin V P 2015 Mathematical Modeling of Dynamic Processes in one-ended Lifting Machine Ropes of Great Length *Materials handling equipment and logistics: V regional student conf.* (Donetsk: DonNTU) 20–21 may 2015
[8] Wolny S 2017 Emergency Braking of a Mine Hoist in the Context of the Braking System Selection *Archives of Mining Sciences* **1** pp 45–54
[9] Barkand T D and Helfrich W J 1988 Application of Dynamic Braking to Mine Hoisting Systems *IEEE Transactions on Industry Applications* **24** n 5 pp 884–896
[10] Tejszerska D and Switonski E 2000 Modelling of Vibrations of Hoisting Systems 2000 *European Congress on Computational Methods in Applied Sciences and Engineering ECCOMAS* (Barcelona 11-14 September)
[11] Tejszerska D 2001 Optimisation of Dynamic Features of Mine Hoisting Systems *10th Jubilee Int. Scientific Conf. Achievements in Mechanical & Materials Engineering.*
[12] Osypova T N and Nesterov A P 2014 On the Dynamics and Optimization of Mine Hoists *Machinery Manufacture: Collected Works* **13** pp 74-81
[13] Zverev V Yu, Trifanov G D and Strelkov M A 2015 Analysing of Dynamic Loads Acting on the Mine Hoisting Ropes *Actual Problems of Increase of Efficiency and Safety of Mining and Neftaraka-cost Equipment: Materials II Int. Scientific-Practical Conf. Mining and Petroleum Electrical Engineering* (Perm) pp 26–32
[14] Vasiliev V I 2010 The ways of the Dynamic Loads Reducing in Ropes of Mine Lifting Installations by Systems Automatically Controlled Safety Brake *Steel ropes: collection of proceedings* (Odessa: Astroprint) **8** pp 18–29
[15] Korniakov M V 2007 *Protection of Mine Lifting Installations from Dynamic Loads during Movement of the Lifting Vessel in the Shaft* (Irkutsk: Publishing house IrGTU) p 164
[16] Stepanov A G and Korniakov M V 2014 *Dynamics of Machines* (Irkutsk: Publishing house IrGTU) p 412
[17] Stepanov A G 2013 The Theoretical Basis of the Dynamics of Shaft Hoisting *Mining Equipment and Electrical Engineering* **7** pp 31–40
[18] Kuskildin R B 2014 The Vessel Slowdown with Cargo during Safety Braking in Mode of Lifting the Load on the Mine Hoist Installations *Proceedings of the of II Int. Scientific-Practical Conf.* (St. Petersburg: National mineral resources University "Mining") **1** pp 156–160.
[19] Krutko P D 1987 *Inverse Dynamics Problems of Controlled System* (Moscow: Science) p 304

[20] Krutko P D 1991 Motion Control of Elastic Multimass Systems *Journal of Machinery Manufacture and Reliability* (Moscow) **4** pp 90–96
[21] Krutko P D 2004 *Inverse Problems of Dynamics in the Theory of Automatic Control* (Moscow: Machinery Manufacture) p 576
[22] Dovgan S M and Samoilenko A A 2003 Formation of the Control Laws of the Mine Lifting Installations on the Basis of Solutions of Inverse Problems of Mechanics *Mining Information-Analytical Bulletin (Scientific and Technical Journal)* **11** pp 188–191
[23] Kuznetsov N K 2009 *Dynamics of Controlled Machines with Additional Links* (Irkutsk: Publishing house IrGTU) p 288

Three dimensional movement control system applying CBR-technology

O N Kuzyakov and M A Andreeva

Tyumen Industrial University, 625000, Tyumen, Volodarskogo Street, 38, Russia

e-mail: mayandr72@yandex.ru

Abstract. In this paper the basic aspects of elaborating the three dimensional movement control device are presented. This device is used if the movement control of the suspended object is required. A multipoint light emitter is firmly attached to the controlled object. A set of light receivers registering the light emitters' signals determines the object location. The aspects of elaborating the three dimensional movement control system structure are presented. The system is elaborated upon the movement control device. Case-Based Reasoning technology is used to determine the location of the controlled subject.

Key words – device, three dimensional movement, light emitter, light receiver, system, Case-Based Reasoning.

1. Introduction

Currently, movement parameters control of unstable objects represents a task. Exploiting automatic movement control systems provides with a solution for this task. Such systems are based on MEMS-accelerometers, MEMS-gyroscopes, inclinometers, etc. [1-3]. For example, a two dimensional tilt control device is elaborated [4]. However, the disability of such devices to measure three dimensional movements and their mechanical drawbacks constrain their functionality. Moreover, the quality of sensor values decreases because of vibrations and influence of external environment [5-9].

Thus, a problem of developing three dimensional movement control system marked by high sensibility and reliability and quick response is currently important. The main part of the system is to be a device measuring three dimensional movement of the object.

2. Methods

The device adequate to requirements mentioned was elaborated. Its structure is presented at figure 1. Elements of the device are multipoint light emitter (MLE) and signal receiver block (SRB). The MLE 4 is firmly attached to researched object 1 by bar-bell 2. Connecting point of bar-bell 2 and MLE 4 is marked as 3. MLE which is put into space confined by cylindrical hull 6 consists of four laser ray emitters 5. The hull includes SRB which is represented by light receiver matrix on its side. Sensitive side of each light receiver is directed inside the hull. Light receiver matrix on the side of the hull has a structure presented at figure 2. The side of the hull has a rectangular form, the vertical and horizontal distances between two next light emitter rows equal respectively l and h so that the side of the hull has a grid structure. Accuracy of the measurement is inversely proportional to the grid step (l and h values).

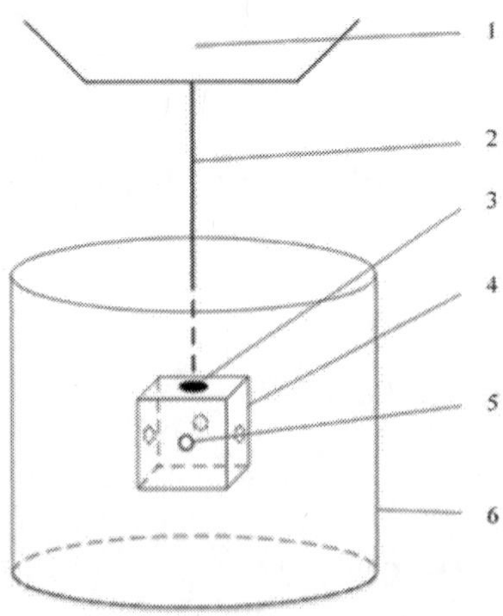

Figure 1. Structure diagram of three dimensional movement control device elaborated

The common light emitters number in the device depends on measurement accuracy demanded and is calculated upon the formula [10]:

$$N = \frac{2\pi R}{l} \cdot \frac{H - 2h}{h},$$

where l is a horizontal distance value between light emitter rows, h is a vertical distance value between light emitter rows, R is a cylindrical hull radius value, H is a cylindrical hull height value.

Target function derived has a formula:

$$F = \frac{MA \cdot NA}{P \cdot Z \cdot S} \to max.$$

Calculations undertaken of target function value show that optical principle of action of device elaborated is an advantage because it provides with lack of extra data filters. Meanwhile, using different measuring devices results in necessary application of such filters.

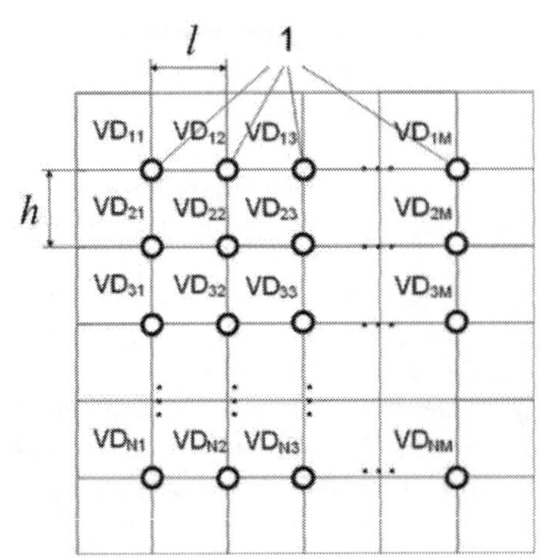

Figure 2. Light receiver matrix structure

A target function was created to justify selection of the device elaborated from a range of measurement devices. In common such function has a formula presented below:

$$F = f(u_1, u_2, \dots, u_n),$$

where u_i is a parameter analyzed.

The comparative analysis was undertaken for following cases: combined usage of accelerometer and gyroscope, usage of inclinometer and usage of device elaborated. The parameters analyzed were assumed to be market availability *MA* (on a scale of 1 to 5), price *P*, zero drift *Z* and sensitivity drift *S* (temperature characteristics), number of measurement axis *NA*.

Thus, device elaborated is an optical movement control device. Its principle of action is based on the usage of MLE (four laser ray emitters) attached firmly to object analyzed and SRB (light receiver matrix on the internal side of cylindrical hull). The device can be used while studying following parameters of object analyzed: tilt, three dimensional displacement, velocity, etc. Movement tendency research of the object can be carried out if its position dependence on time is identified.

The device can be a part of three dimensional movement control system, which consists of the following elements:

- three dimensional movement control device;
- two signal multiplexers;
- microcontroller.

3. Results and discussion

Functional diagram of the system is presented at figure 3. System executes as follows: after system run initialization is implemented, which considers setting reference values, downloading constants to microcontroller SMC, resetting light receivers' values, switching on light emitters, setting reference position of object controlled.

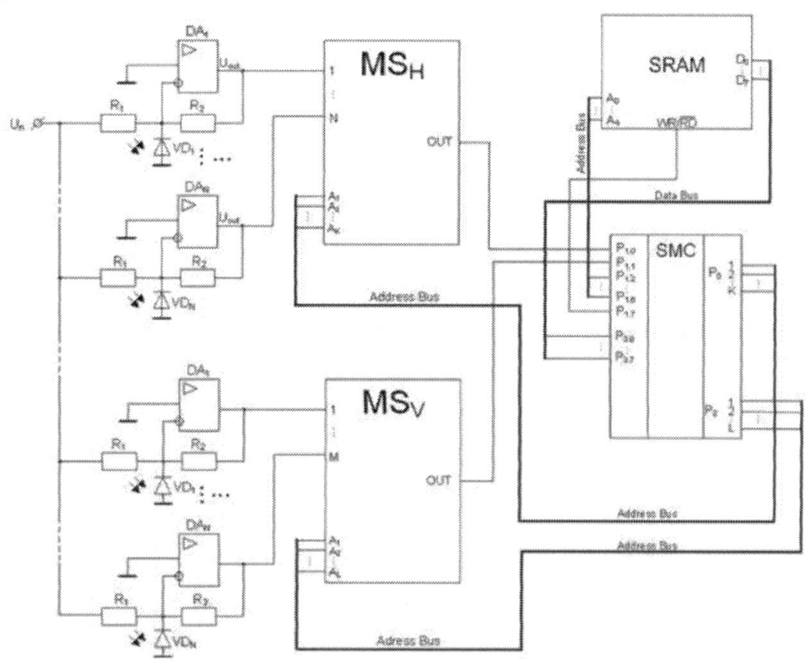

Figure 3. System functional diagram

A continuous questioning of SRB takes place thereafter. Signals flow from light receivers $VD_1 \dots VD_N$ to current converting devices, each of them converts photocurrent from one light receiver to output voltage U_{out} and includes photodiode VD, resistors R_1 and R_2 and operational amplifier DA. After that signals flow from current converting devices through channels $1..N$ (from horizontal matrix rows) and $1..M$ (from vertical matrix rows) to corresponding multiplexers MS_H and MS_V. Signals from horizontal and vertical matrix rows are processed separately there and then flow to SMC. For instance, AVR by Atmel (8-bit device with RISC-architecture and speed of 16 MIPS) may serve as SMC.

Addresses of channels chosen for data transmission flow from SMC outputs 0 and 2 through address lines to the inputs $A_1 - A_K$ and $A_1 - A_L$ of MS_H and MS_V respectively. Signals from light receivers flow from MS_H and MS_V outputs marked as OUT to SMC inputs P1.0 and P1.1 through data lines and are stored in SMC RAM. Addresses of units chosen for rewriting data on current condition of the controlled object flow from SMC outputs P1.2-P1.6 through address lines to inputs $A_0 - A_4$, whereas data exchange with SRAM is performed through SMC outputs P3.0-P3.7. Write-read managing output P1.7 of SMC is connected to corresponding input of WR/RD SRAM. Program stored in SMC Program Flash processes signals from light receiver matrix and establishes a set of light receivers, which are registering signals of light emitters at the moment.

It is suggested to use Case-Based Reasoning (CBR) method [11-13] to solve a task on adapting decision to experience related to solving similar tasks. The process of searching for information required is a subsequent implementation of following actions: obtaining (retrieving) a precedent, searching for the closest decision in Case Base (CB) considering similarity degree on criteria chosen, adjustment and saving, decision making. System structure considering application of CBR is given at figure 4. CBR algorithm respective to this is given at figure 5.

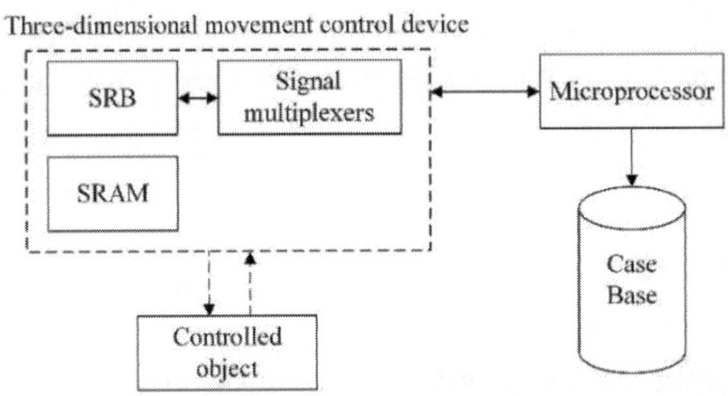

Figure 4. System structure diagram with Case-Based Reasoning applied

A precedent may be presented as follows:

$$CASE = (VD_i, VD_j, VD_k, VD_l, R)$$

where VD_i, VD_j, VD_k, VD_l is a set of light receivers, which register signals from the first, second, third, fourth light emitters, which define current precedent, R corresponds to decisions (managing recommendations).

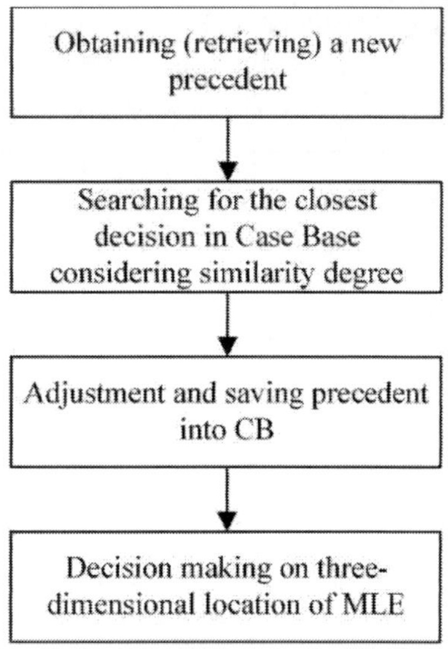

Figure 5. Case-Based Reasoning algorithm

Meanwhile, *S(P, T)* is a similarity degree, which is defined as follows:

$$S(P,T) = \frac{N_{PT}}{N},$$

where NPT is a number of matching features (signals) for precedent P and situation T, N is a common number of features.

CB is formed before system commissioning. Usage of exact system parameters (object controlled, length of bar-bell and location of connecting point of bar-bell and MLE (figure 1), MLE size and object controlled size) as will be used while exploiting system while forming CB is vitally important.

During system execution a similarity degree of current situation (a set of light receivers registering light emitters signals) to situations stored in CB is estimated. This is implemented with the use of Nearest method and failed decisions, related to incorrect system elements functioning, make up a special base of problematic conditions and are put into SMC SRAM.

In our view, for determining similarity degree of current situation it is more desirable to use a modified method. In this case, decision is to be chosen on the basis of a few closest neighbors (precedents), and it is required to take into account a possibility of changing h parameter (distance between light receivers) in a set metric, which characterizes similarity degree.

It is obvious that when h → min, the number of light receivers in light receiver matrix N → max, and common number of combinations RΣ of active light receivers, which register light signals from light emitters, increases.

Meanwhile, the number of these light receivers depends on the directional pattern of light emitters and may be minimized if laser ray is used. The location of nearby light receivers VD relatively to basis point i is given at figure 6. Horizontal and vertical distance values between light receivers nearby to i-th light receiver correspond to h.

The light receiver response condition is defined by boolean function:

$$B^n \to B,$$

where B = {0, 1} is boolean set, n is non-negative number (arity).

Boolean function for light receivers, which register light emitters signals, is defined as truth table given in table 1.

Table 1. Truth table for boolean function

VD_1	VD_2	VD_3	VD_4	$f(VD_1, \ldots, VD_4)$
1	1	1	0	f(1, 1, 1, 0)
1	1	0	1	f(1, 1, 0, 1)
1	0	1	1	f(1, 0, 1, 1)
0	1	1	1	f(0, 1, 1, 1)
1	1	1	1	f(1, 1, 1, 1)

Thus, a satisfactory condition of decision-making on choosing the nearest precedent is $M \geq 3$ considering weight coefficient λ. This value corresponds to reliable response of at least three light receivers in system with four light emitters.

Conclusion

It should be noted that two main components constitute a precedent structure: identifying (allows reuse) and tutoring (characterizes experience, includes proves for decisions and alternative and failed decisions). Object coordinates, its tilt, velocity and displacement value are determined if a decision is found with the use of CBR (considering parameters matching current situation).

As a result, system elaborated allows to determine the following parameters of object controlled unequivocally with accuracy demanded: object displacement value, tilt, velocity, acceleration, etc.

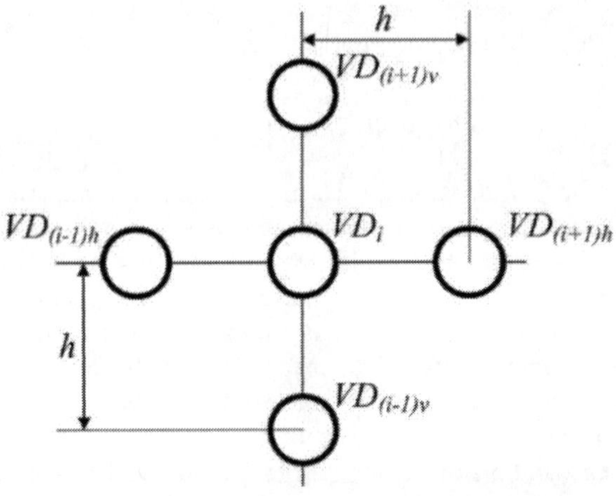

Figure 6. Scheme of nearby photodetectors

References

[1] Varadan V K and Varadan V V 1999 Microsensors, microelectromechanical systems (MEMS) and electronics for smart structures and systems *Smart Mater. Struct.* **9** 953-72
[2] Gupta A and Ahmad A 2007 Microsensors based on MEMS technology *DSJ* **3** 225-32
[3] Gardner Julian W 1994 *Microsensors: principles and applications* Wiley
[4] Makarov A V and Kuzyakov O N 2000 *Movement control device* Patent RF 2150086
[5] Agarwall M, Park K K, Hopcroft M, Chandorkar S, Candler R N, Kim B *et al* 2006 Effects of mechanical vibrations and bias noise on phase of MEMS resonator based oscillators *Proc. IEEE International Conference on Micro Electro Mechanical Systems* 154-7
[6] Annovazzi-Lodi V and Merlo S 1999 Mechanical-thermal noise in micromachined gyros *Microelectr. J.* **30** 1227-30
[7] Mohd-Yasin F, Korman C E and Nagel D J 2003 Measurement of noise characteristics of MEMS accelerometers *Solid State Electron.* **47** 357-60

[8] Mohd-Yasin F, Nagel D J, Ong D S, Korman C E and Chuan H T 2007 Low frequency noise measurement and analysis of capacitive micro-accelerometers *Microelectron. Eng.* **84** 1788-91
[9] Mohd-Yasin F, Zaiyadi N N, Nagel D J, Ong D S, Korman C E and Faidz A R 2009 Noise and reliability measurement of a three axis micro-accelerometer *Microelectron. Eng.* **86** 991-5
[10] Kuzyakov O N and Andreeva M A 2016 Elaboration of three dimensional movement control device *Proc. VI International Sci.-Prac. Conf. Modern Issues of Development Fundamental and Applied Sciences* (Praha, Czech Republic: Printing house "Maestro") 108-11 (in Russian)
[11] Kolodner J L 1983 Reconstructive memory: a computer model *Cognitive Science* **7** 281-328
[12] Aamodt A and Plaza E 1994 Case-based reasoning: foundational issues, methodological variations, and system approaches *AI Communications* **7** 39-59
[13] Watson I D and Marir F 1994 Case-based reasoning: a review *Knowl. Eng. Rev.* 4 355-81

IOP Conf. Series: Journal of Physics: Conf. Series **944** (2017) 012072 doi:10.1088/1742-6596/944/1/012072

Electronic-projecting Moire method applying CBR-technology

O N Kuzyakov, U V Lapteva and M A Andreeva

Tyumen Industrial University, 625000, Tyumen, Volodarskogo Street, 38, Russia
mayandr@yandex.ru

Abstract. Electronic-projecting method based on Moire effect for examining surface topology is suggested. Conditions of forming Moire fringes and their parameters' dependence on reference parameters of object and virtual grids are analyzed. Control system structure and decision-making subsystem are elaborated. Subsystem execution includes CBR-technology, based on applying case base. The approach related to analysing and forming decision for each separate local area with consequent formation of common topology map is applied.
Keywords – surface topology, Moire method, system, control, virtual grid, case, algorithm, decision-making.

1. Introduction

Statistical analysis of emergency causes on oil transport and storage facilities showed that the average percentage of accidents caused by a range of defects related to manufacturing and constructing is 32%; percentage of accidents caused by corrosion is 28%; percentage of accidents caused by external factors is 31%. External factors include influence of anthropogenic processes, occasional damage by external organizations and influence of ongoing physical-geological processes of various kinds.

Meanwhile, visual and measurement ways of examining are implemented at almost all stages of life cycle of definite components, part elements and items.

Specificities of particular objects diagnosis are reflected in specialized documents: "The rules of structuring and safe maintenance" and construction codes and regulations.

According to requirements listed in guidance documents, visual and measurement examine is executed at both internal and external sides of the items, which gets significantly complicated because of lack of access to inside cavities of the items.

Technical endoscopic systems (endoscopes, video-endoscope, etc.) or self-moving robotized systems for diagnosing extended objects of large diameter of 1.5 meters and more (for instance, oil intra-field and transport pipelines) are used as examining means during technical diagnosis in oil and gas companies.

Further development and improvement are obtained nowadays by methods of examining technical state of oil and gas industrial facilities. Such methods allow to increase remaining resource of these facilities' elements and constructions. Ultrasonic, electromagnetic, acoustic, optical and other types of methods are highly spread. One of the perspective and not fully studied method is based on the usage of Moire effect [1-4]. It means formation of Moire pattern of dark and light fringes by overlay of two grids. The pattern may serve as a qualitative and quantitative feature of object surface deformational process.

The important tasks ahead in diagnosing oil and gas industrial facilities are remote visual examining of technical state of items, evaluation of condition of surfaces and welded joints, build quality. Visual monitoring necessity is regulated by guidance documents of Russian Federal Service for Ecological, Technical and Atomic Supervision (Rostehnadzor). One of these documents is

"Instruction on visual and measurement control" (Guidance Document 03-606-03), which establishes the order of implementing visual and measurement control of main material and welded joints during manufacturing, building, assembling, repair, reconstruction, exploitation, technical diagnosing of technical device and buildings, which are used and exploited at hazardous production facilities controlled by Rostehnadzor.

2. Methods

Differential electronic-projecting method of measuring the form of object surface is available. Its essence is projecting computer-synthesized sample grid containing alternating dark and light fringes, scanning grid by digital camera and comparing picture obtained to sample grid, which results in defining surface parameters in each point [1].

Theoretical prerequisites for elaborating method mentioned are listed below. Forming Moire fringes (phenomenon of mechanical interference) is possible in the following cases:

- overlay of two grids of parallel fringes with equal step values α, angle between the grids is φ;
- overlay of two grids with unequal step values (α_1 and α_2), zero angle between the grids;
- overlay of two grids with unequal step values (α_1 and α_2), angle between the grids is φ (presented at figure 1).

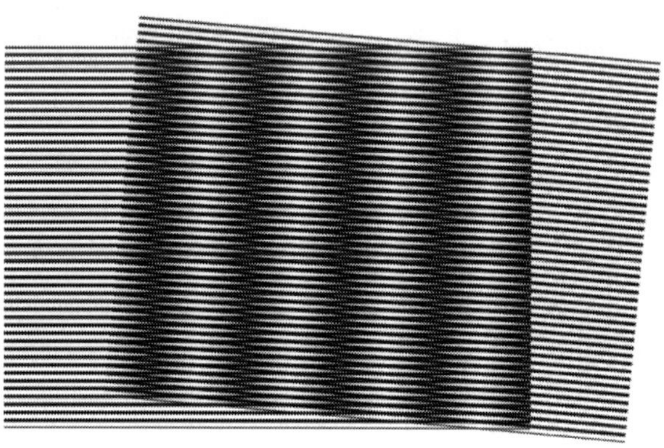

Figure 1. Forming Moire fringes

In case of overlay of two grids with unequal step values (α_1 and α_2 respectively) and angles φ_1 and φ_2 between the grids and the x-axis, Moire fringes are formed with step S and angle β between the fringes and the x-axis.

Angle β is defined as follows:

$$\tan\beta = \frac{\alpha_1 * \sin\varphi_2 - \alpha_2 * \cos\varphi_1}{\alpha_1 * \cos\varphi_2 - \alpha_2 * \sin\varphi_1}, \tag{1}$$

where α_1, α_2 are step values of the grids overlaid.

It's obvious that $\tan\beta$ does not depend on distances c_1 and c_2 between the origin coordinate and the first fringe of respective grid. Thus, after a parallel displacement of one or both systems of lines the angle β does not change. The turn of Moire fringes may be obtained only by a relative turn of overlaid systems of lines.

X-intercept of Moire fringe is defined as:

$$b = \frac{p * \alpha_1 * \alpha_2 + c_2 * \alpha_1 - c_1 * \alpha_2}{\alpha_1 * \cos\varphi_2 - \alpha_2 * \sin\varphi_1}. \tag{2}$$

Distance between the Moire fringe and the origin of coordinates equals

$$l = b * \cos\beta = \frac{b}{\sqrt{1+\tan^2\beta}}. \quad (3)$$

Fringes step S is defined upon the formula:

$$S = \frac{\alpha_1 * \alpha_2}{\sqrt{\alpha_1^2+\alpha_2^2-2*\alpha_1*\alpha_2*\cos(\varphi_1-\varphi_2)}}, \quad (4)$$

i.e. distances between all nearby Moire fringes are equal.
And if $\Delta\varphi = \varphi_1 - \varphi_2 = 0$, the Moire fringes step equals

$$S = \frac{\alpha_1 * \alpha_2}{\alpha_1 - \alpha_2}, \quad (5)$$

i.e. the less $\alpha_1 - \alpha_2$ is, the bigger Moire fringes step is. That is why it is desirable to decrease step value S for increasing examining accuracy. In case there is an angle between the grids $(\Delta\varphi \neq 0)$, increasing this angle will result in decreasing step value S. Thus, by programmatically varying parameters α_1, α_2, $\Delta\varphi$ it is possible to increase the resolution ability of examining method.

3. Results and discussion

Analytical research showed that measurement accuracy is under influence of the following parameters:

- step values α_1, α_2 of grids projected;
- tilt angles φ_1, φ_2 of grids projected.

Variation of Moire fringes step value is presented at diagrams (figure 2) in case of period (step) value of the first grid hold fixed α_1=0.1 cm and step value of the second grid increasing α_2 = (0,1; 0,2; 0,3; 0,4; 0,5; 0,6; 0,7; 0,8; 0,9; 1,0) cm; φ=(φ_1 - φ_2) is varying from –90° to 90°.

At a more detailed examination of inclined line, it presents a gradual line, which consequently means that increasing grid step value will influence inaccuracy. That is why, for decreasing inaccuracy it is crucial to determine the most optimal angle $\varphi = (\varphi_1 - \varphi_2)$ and therefore select φ_1 and φ_2 values.

Diagrams at figure 2 show that at angle φ hold fixed, Moire fringe step value S may be alternated by alternating the second grid step value. Angles φ = 15° and 20° are the more efficient, because they allow step value S to alternate significantly while alternating step value α_2.

In case φ = 15° it is applicable to have the second grid step value α_2 = 0.2 cm and α_2 = 0.7 cm, so that S equals respectively 0.187 cm and 0.116 cm.

In case φ = 20° and α_2 = 0.2 cm and α_2 = 0.7 cm, S equals respectively 0.18 cm and 0.115 cm. Thus, at a small angle φ=15°..20° Moire fringes step value for similar grids alternates insignificantly.

Computational experiment on revealing dependency of Moire fringes step value at an angle hold fixed allows to make a conclusion on Moire fringes step value S decreasing with the increasing of the second grid step value α_2.

Analyze undertaken showed that considering α_1 = 0.1 cm, the most desirable parameters are α_2 = (0.2 cm; 0.7 cm), φ = (15°; 20°), accordingly φ_1 and φ_2 values are to be found. In this case, φ_2 is defined as $\varphi_2 = \varphi_1 - \varphi$, meanwhile it is recommended that φ = (15°; 20°). Besides, Moire fringes step value S dependency on natural brightness level Φ is proved experimentally. If brightness alternates within the range $16\times10^4 \div 175\times10^4$ Lux, the number of pixels in the image decreases from 9926 to 5256.

Therefore, image quality decreases from 100% (16×10^4 Lux) to 52% (175×10^4 Lux). For minimizing this inaccuracy, it is recommended to operate at minimal brightness or to use projecting facilities with greater level of luminous flux. For implementing method described it is offered it improve system previously elaborated in order to minimize complexity of processing collected data at low resulting accuracy.

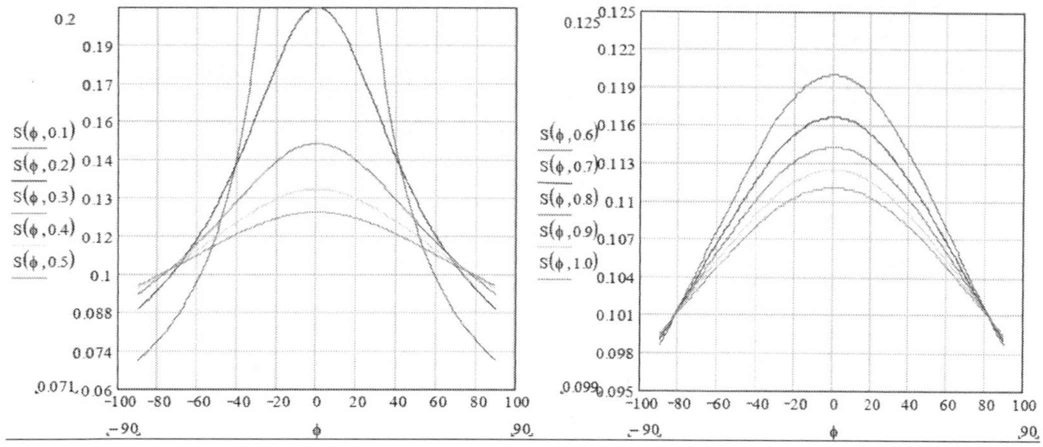

Figure 2. Variation of Moire fringes step value at α_l=0.1 cm

For achieving this goal, it is necessary to elaborate system for defining surface topology, which uses up-to-date, high-accurate contactless methods and ways of intellectual data processing. Structure of such system is presented at figure 3.

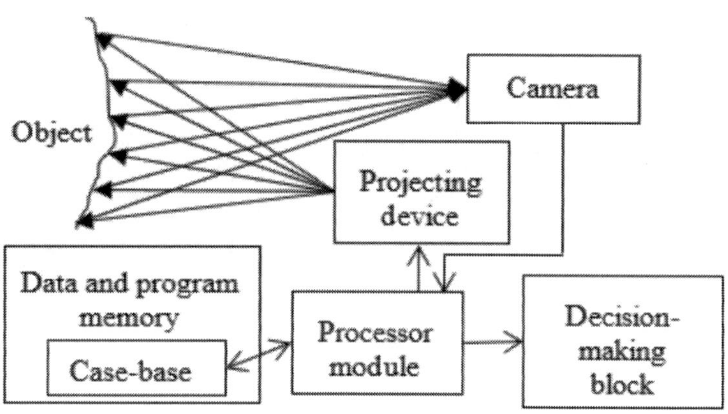

Figure 3. System structure diagram

The system consists of processor module with downloaded program implementing electronic-projecting Moire approach for defining surface topology; read-only memory storing programs and data, including case-base; digital projector and digital video-camera for implementing projecting sample grid and taking picture of grid projected.

System execution algorithm is presented at figure 4.

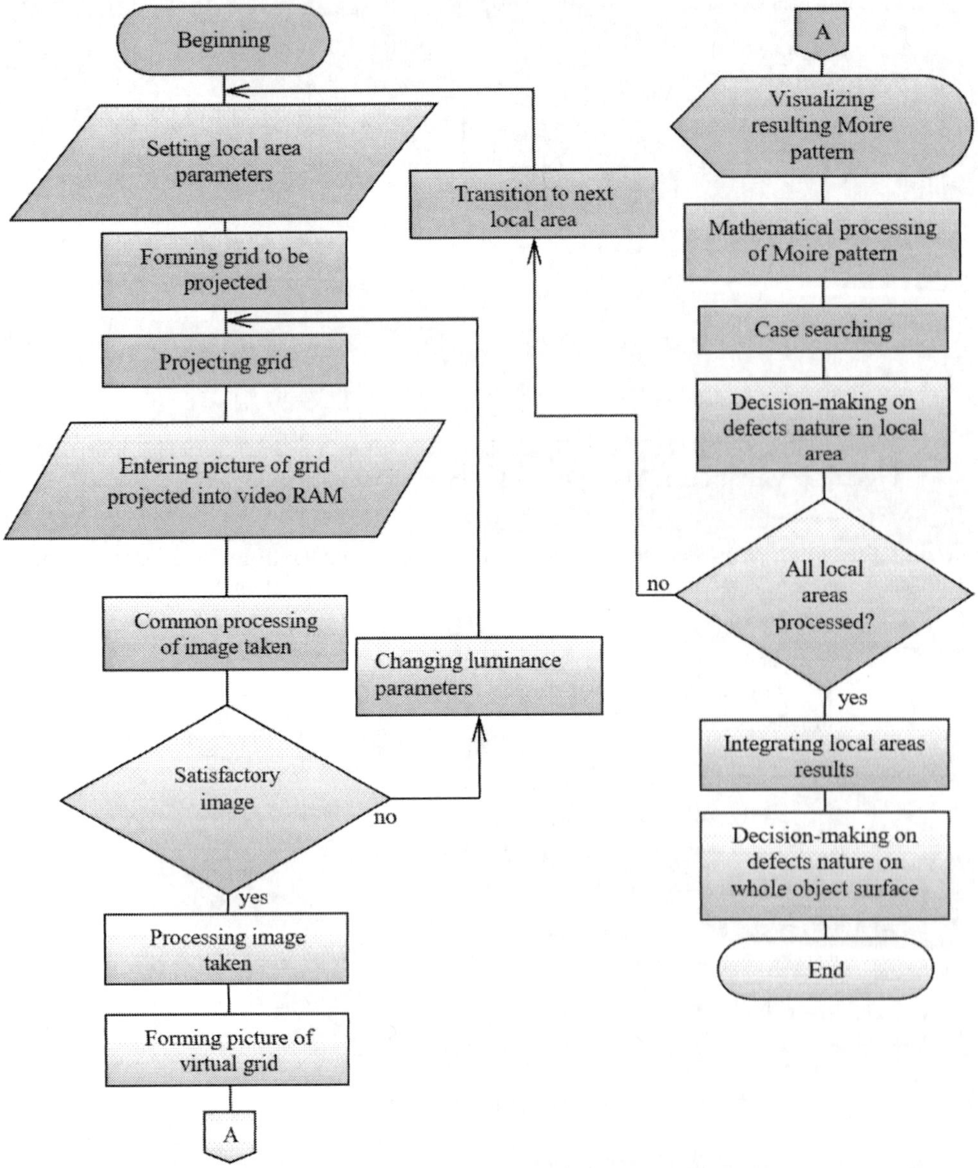

Figure 4. System execution algorithm

Setting parameters and forming grid to be projected on object surface are implemented at the first stage. A picture of grid projected taken is entered to video RAM of processor module thereafter by using digital video camera. Preliminary processing and analyzing image taken allows either to continue processing or redo projecting grid. If image taken is sufficient, processing of this image is performed according to a precise program, which disposes of shades and increases contrast. Virtual grid generation in processor module, formation and visualization of Moire pattern on the display of processor module are implemented at the next stage. Further mathematical processing of image taken is to define coordinates of Moire fringes centers, develop matrix of heights for each local area of object examined. The next stage is precedent (case) search on the basis of Cased-based reasoning (CBR) technology [5-13], i.e. search of case from case base according to similarity criteria set. CBR allows solving task by using solutions already known, i.e. adapting solution of already known task considering collected experience to match current state.

If case base adjustment or extra examining hypothesis is required, algorithmic loop is performed all over again. Decision-making block execution is presented at figure 5 in more detail.

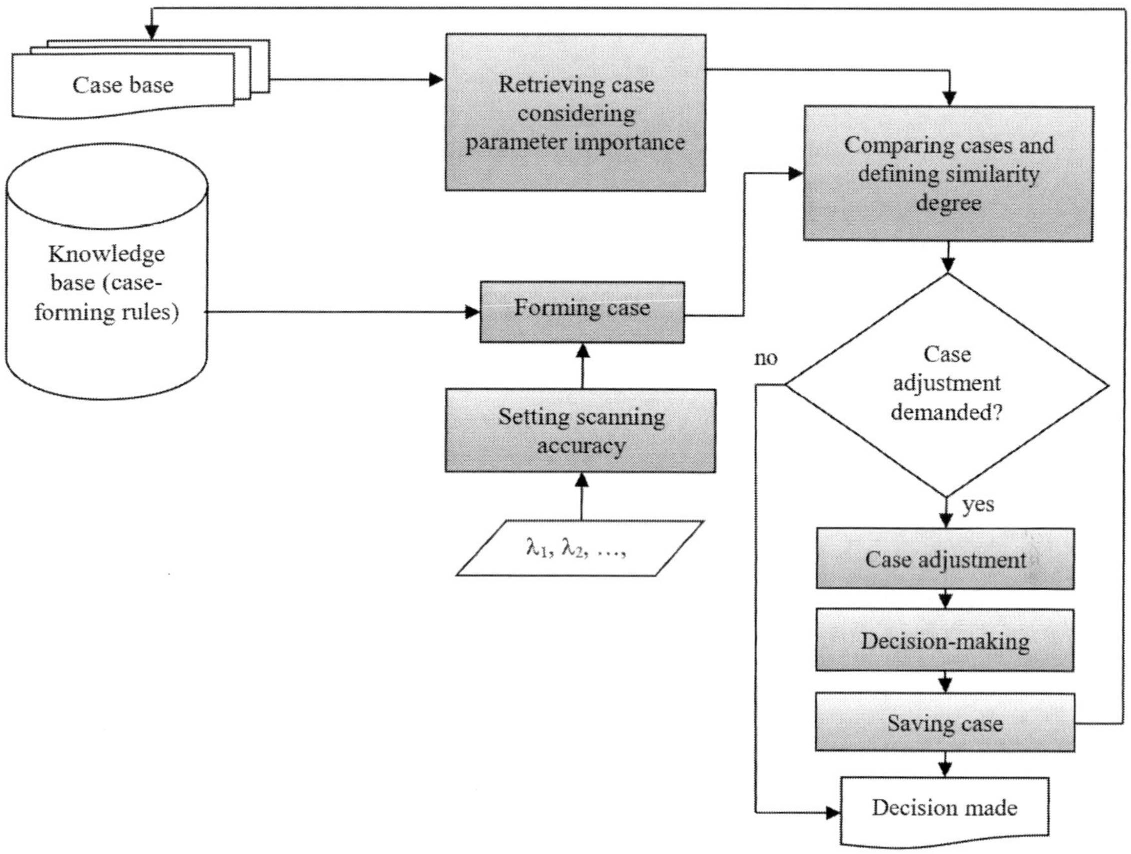

Figure 5. Functioning algorithm of intellectual decision-suppuort subsystem

At the first stage scanning accuracy is set, and case P is formed by using case building model stored in knowledge base as a set of rules. Case P defines a problematic situation in a formalized language. Case formed in this way enters block, which performs comparison and definition of its similarity degree to case C retrieved from base storing ready cases. Search of case is performed on the basis of nearest neighbor method. Metric d_{CP} serves as alignment criteria; d_{CP} is defined by formula of Euclidean distance (distance between two points):

$$d_{CP} = \sqrt{\sum_{i=1}^{n}(x_i^C - x_i^P)^2}, \tag{6}$$

where n is a number of features compared, x_i^C is value of i-th feature of case C, x_i^P is value of i-th feature of case P. If d_{CP} value meets set alignment criteria, then solution is formed. Otherwise, adjustment of case retrieved is implemented in accordance with set scanning parameters. Case adjusted is wrote thereafter into case base and solution is formed.

It is noteworthy that processing of video taken requires consideration of zoom coefficient K_z, which may be defined by a program on the basis of sample (reference) and projected (object) grids comparison. Automatic adjustment of fringes step value and width, which considers accuracy value, angle between optical axes of projecting and taking image subsystem components, allows to increase level of confident identification of data collected.

In the long run, it is possible to increase complexity of cases structure and to change algorithm of their retrievement.

Final stage is program integrity of separated local areas and forming common topology map, which makes up a base of defect nature over the entire surface area.

4. Conclusion

To sum up, using electronic-projecting Moire method including CBR technology allows to obtain high-quality surface topology map considering experience (data) previously collected, and widen base of cases, which describe separate local areas of examined surface as well.

References

[1] Kucheryuk V I, Kuzyakov O N and Dubatovka U V 2005 *Surface topology control device applying Moire method* Patent RU 2267087 (in Russian)
[2] De Nicola S *et al* 2000 Fourier transform method of fringe analysis for moiré interferometry *J. Opt. A: Pure Appl. Opt.* **2** 228
[3] Hubing Du *et al* 2011 *J. Opt.* **13** 035405
[4] D'Acquisto L *et al* 2002 *Meas. Sci. Technol.* **13** 613
[5] Varshavskiy P R and Yeremeev A P 2009 Modeling of case-based reasoning in intelligent decision support systems *Artif. Intel. And Decision Support* 2 45-57 (in Russian)
[6] Watson I D and Marir F 1994 Case-based reasoning: A review *Knowl. Eng. Rev.* 4 **9** 355-81
[7] Kolodner J L 1992 An Introduction to Case-Based Reasoning *Artif. Intell. Rev.* **6** 3-34
[8] Naderpajouh N and Afshar A 2008 A case-based reasoning approach to application of value engineering methodology in the construction industry *Construction Management and Economics* **26** 363-72
[9] Althof K-D, Auriol E, Barlette R and Manago M 1995 *A review of industrial case-based reasoning tools* (A1 Intelligence)
[10] Aamodt A and Plaza E 1994 Case-based reasoning: foundational issues, methodological variations, and system approaches *AI Communications* (IOS Press) **7** 39-59
[11] Alterman R 1989 Panel discussion on case representation *Proc. Second Workshop on Case-Based Reasoning* (US)
[12] David B S 1991 Principles for case representation in a case based aiding system for lesson planning *Proc. Workshop on Case-Based Reasoning* (Washington)
[13] Kuzyakov O N and Lapteva U V 2013 Surface topology control system based on Moire effect *Proc. System Analysis and IT RAS etc.* 362-7

IOP Conf. Series: Journal of Physics: Conf. Series **944** (2017) 012073 doi:10.1088/1742-6596/944/1/012073

The active control devices of the size of products based on sapphire measuring tips with three degrees of freedom

E V Leun[1], V I Leun[2], V K Sysoev[1], K A Zanin[1], A V Shulepov[3] and P A Vyatlev[1]

[1]Lavochkin Association, 24 uliza Leningradskaya, Khimki, 141402, Russia
[2]Omsk State Technical University, 11 Pr. Mira, Omsk, 644050, Russia
[3]Moscow State Technological University «STANKIN», 11 Vadkovsky pereulok, Moscow, 127055, Russia

Abstract. The article presents the results of the calculation of the load capacity of the active control devices (ACD) sapphire tip, which showed nearly 30-fold margin of safety to shock loads and experimental researches in mechanical contact with 5 cogs cutter 15 mm in diameter rotating with a frequency of 1000 rpm, which confirmed the calculations, determined the surface roughness R_z of the contact area of no more than 0.15 μm. Conditions have been created for recording without distortion of the image through a sapphire tip in contact with the processed article.

A ACD design with new functionality is proposed: with one, two and three degrees of freedom of the sapphire tip and allows measuring the taper of the article and measurements on the chord. It is shown that with the implementation of their fixed head like the frame of the gyroscope with the rotations around the axes OY and OZ. It is shown that the rotation of the tip around the axis OX can be replaced more convenient for the implementation of the angular offset of the transferred image due to rotation of the output end of the flexible optical waveguide relative to the input. This makes it possible to reduce the "blurring of the image" during registration of the fast moving product profile when the slope of the recorder lines coincides with the slope of the edges of the image elements of the selected moving elements of the article.

1. Introduction

It is known that the use of active control devices (ACD) of products sizes allows to increase productivity up to 100% or more in the manufacture of products on a variety of machine tool applications in the aerospace industry, instrument, machine and machine tool.

ACD has been actively developed in the 60-ies of XX century [1,2] and until today actively developing in Omsk State Technical University [2,3]. Now one of the leading manufacturers of ACD for quality machine tool is an Italian firm, Marposs [4]. However, these ACD, with high resolution, up to ≈0.4 μm for smooth cylindrical products, significantly reducing their metrological characteristics when measuring the dimensions of products with intermittent surfaces. And use metal measuring tips (the tips) do not allow the use of optical methods of measuring and registration of images of the surface. Russia is probably the only manufacturer ACD is AS "Niiizmereniya". So, they produced the instrument BV-4304 for control of parts of a continuous surface (with hinged bracket) readability of digital reading is 1 μm [5].

Since 2013, a direction has appeared in Russia [6] associated with laser ACDs using sapphire (ruby) contact tips (Fig. 1), and later also contactless [7] and hydrojet [8] ACDs were developed. However, despite this rapid development, the ACD still has large reserves of improvement. Such promising directions associated with the measurement of taper products or dimensions along the chord, causing the possibility of angular rotations of the tips with one, two and/or three degrees of freedom. Discussion one of the options ACD with these features and this article.

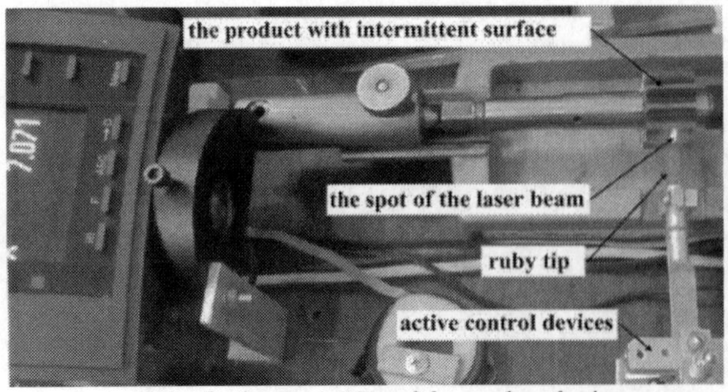

Figure 1. A model of a laser ACD with a ruby tip in contact with a product having a discontinuous surface.

2. Formulation of the problem

The tasks of this research are to analyze the results of theoretical and experimental studies of their load capacity when mechanically contacting sapphire tips with a product having a discontinuous surface. Analysis of transmission conditions without distortion through a sapphire tip in contact with the workpiece, images of its surface. Also, the research task is to consider the developed ACD with the realization of one two and three degrees of freedom of the tip due to their rotations around the axes OX, OY and OZ.

3. Theory

To confirm the possibility of using sapphire as a high-strength and optically transparent material for arrowheads in the ACD was the calculations of the load capacity to blow loads when contact measurements [6] and experimental studies [9], the results of which are presented next.

3.1. Calculation of load capacity of sapphire tips

The loading capacity of the sapphire tip were determined through calculation of margin of safety k_{ms}, determined by the ratio of impact loading F''_{bl} sapphire rod does not lead to its mechanical failure (cracking, chipping or other defects), to the specific pressure generated in the process of leaving the tip F'_{bl} with depression on a ledge in the control of dimensions of products with intermittent surface:

$$k_{ms}=F''_{bl}/F'_{bl} \quad (1)$$

The calculation carried out in [6] showed that when using a tip of sapphire will be provided with almost 30-fold safety margin and confirmed the high blow strength of this material to mechanical contact with the goods having a discontinuous surface. measurements [6] and experimental studies [9], the results of which are presented next.

3.2. Experimental study of mechanical characteristics of the contact sapphire tip with a workpiece having a discontinuous surface

In addition, the experiments verified the correctness of the calculations and confirmed by the actual highly shock-resistant sapphire crystal rod of rectangular cross-section 5x6 mm and a length of ~20 mm at its mechanical contacting flat end with 5 cogs cutter 15 mm in diameter rotating with a frequency of -1000 rpm exceeds the empirical value up to 50 times the standard pressure of ~3 H and no return movements of the tip, mandatory for high accuracy contact measurement.

After conducting this experiment for two sections of flat end face in contact with the mill were removed profilogram (Fig.2), the results of the analysis are defined roughness R_z of no more than 0.15 μm. with these features and this article.

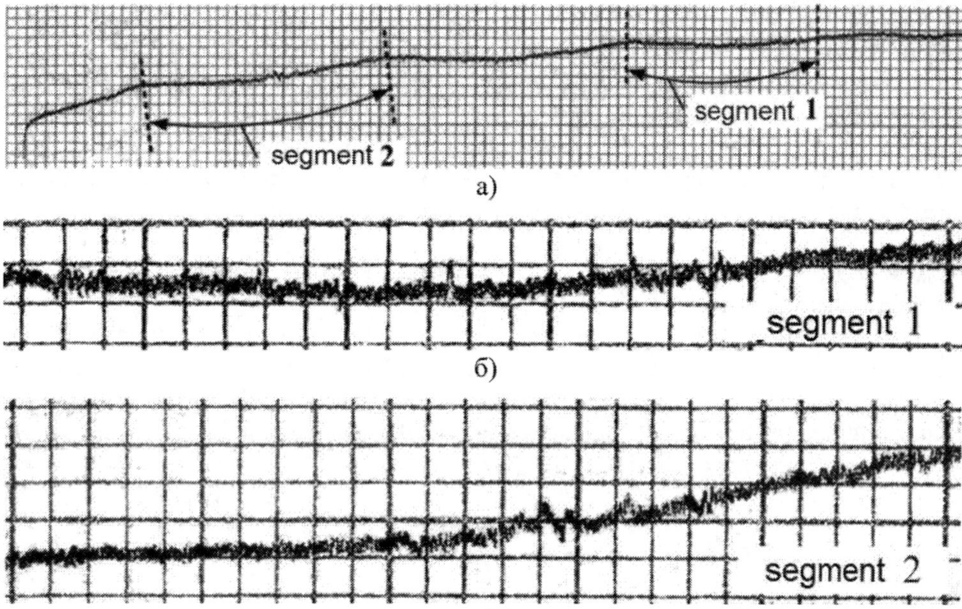

Figure 2. The profilograms of the two segments (a) end of sapphire rod after mechanical contact (b,c – with zoom, the value 1 cell: height ≈0.38 μm, length ≈33 μm).

3.3. Conditions of registration of the image without distortion of the product surface through contact with her sapphire tip

It is known that the influence of surface roughness R_z on the changes of passing (or reflected) optical flows or violation of the coherence of laser radiation from spurious diffuse scattering can be neglected under the condition of $\lambda/6 \geq R_z$, And taking into account the obtained experimental results can be generated for this condition:

$$\lambda \geq 0.9\ \mu m, \tag{2}$$

where λ is the wavelength.

The resulting condition means the transition from a visible range to an infrared radiation (IR), invisible to human eyes.

Enhancing the functionality of the ACD with the sapphire tips can be carried out by measuring the temperature of the product pyrometric method. For different modes of operation, cooling conditions and material of the product surface temperature usually ranges up 20° to 450° C, forming IR radiation in the range $\Delta\lambda_t$~1.0 to 1.6 μm. This range included in the wavelength range defined by condition (2), and the possibility of spurious cross-effects of "dimensional" and pyrometric measurements on each other. This disadvantage can be eliminated in the following ways:

3.3.1 The use two different recorders for separate pyrometric and "dimensional" measurements. Use for pyrometric and "dimensional" measurements on individual recorders with different spectra of sensitivity spaced relative to each other is not less than 1 μm, described by the system of inequalities:

$$\begin{cases} 2{,}0\ \text{мкм} \leq \Delta\lambda_l \leq 2{,}6\ \text{мкм}; \\ 1{,}0\ \text{мкм} \leq \Delta\lambda_t \leq 1{,}6\ \text{мкм}. \end{cases} \tag{3}$$

3.3.2 The use general recorder for pyrometric and "dimensional" measurements. Use one common recorder for the single wavelength range of infrared radiation $1{,}0\ \mu m \leq \Delta\lambda_l \approx \Delta\lambda_t \leq 1{,}6\ \mu m$, but with amplitude modulation of optical flow for "dimension" and pyrometric measurements at different

frequencies. Subsequent frequency separation of the optical signals is possible through the use of digital signal processing.
The rotation of the products with discontinuous surface, such as a drill or milling cutter, leads to a natural amplitude modulation of the optical flow is defined by a discontinuous surface and the number of teeth n and the rotational speed N to write the expression:

$$f_t = n/N, \tag{4}$$

while, as a rule, in the low frequency range: 50 to 200 Hz. For example, for a rotation frequency of 1000 rpm 5-cogs cutter, the frequency of passage of the teeth past the tip amounted to f~85 Hz [9]. So the frequency of the backlight f_{lig} used when registering image should be significantly higher for subsequent efficient frequency separation of the signals of low-frequency modulation f_t, the generated heat flux, in writing it in the form of empirical equations: $f_{lig} = kf_t$, where $k > 5$.

3.4. ACD based sapphire tip with three degrees of freedom.

In 2016 we developed a family ACD with sapphire tips with one, two and three degrees of freedom. In all these technical solutions for fixing the video image surface with the possibility of subsequent processing and the so-called telemetry is used recorder (CCD-camera). It can be installed from the back side of the movable tip or to be fixed optically coupled with the tip through a light guide. Scheme ACD with three degrees of freedom and a general view of the tip shown in Fig.3 and 4, respectively.

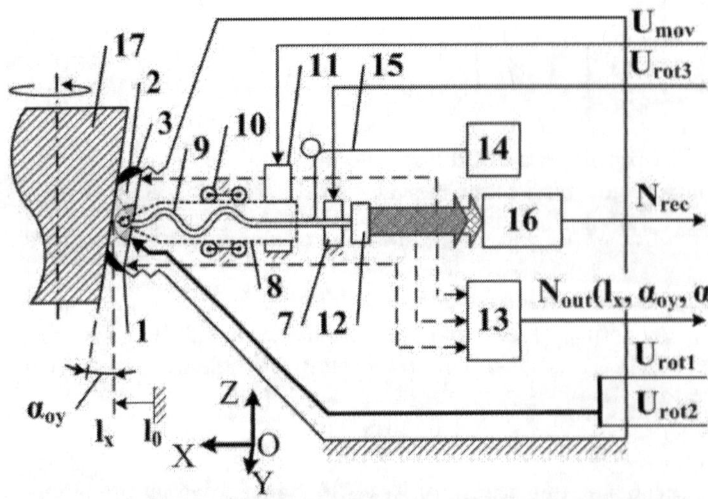

Figure 3. Scheme of the developed ACD with three degrees of freedom.

The developed ACD consists of a transparent tip 1 with an optical guide 2 and a protective coating 3, a frame 4 with two degrees of freedom, a first 5, a second 6 and a third 7 axial drives for rotations around the axes OY, OZ and OX, respectively, a measuring rod 8, a flexible fiber 9 , the guides 10, the linear actuator 11, the collimator 12, the two-axis meter 13, the radiator 14, the fiber set 15, the recorder 16. It is assumed in the article that the drives can also perform the brake functions.
The main components developed by ACD dimensions 17 have the following features. The protective coating 3 is opaque and installed on the front and (fully or partially) the side surfaces of the transparent light guide pads 2. It forms on the surface of the hole and the size of the zone of contact with the workpiece 17 tip 1 window for input/output of optical radiation and protects the light guide pads 2 mechanical shock impacts.
Frame 4 is a so-called gyroscope frame, inside which due to the axial actuators 5 and 6 managed signals and U_{rot1} U_{rot2} can be rotated around the axes OY and OZ tip 1 together with an associated mechanical and optical front end of a flexible optical waveguide 9.

The rod 5, being a hollow tube with a cross section mainly of a polygon with a through inner hole is inserted the front part of the guides 10, and the rear linear actuator 11 that is controlled by the signal U_{mov}. Front of the measuring rod 8 is designed forked end with a movable connection of the handpiece 1.
Flexible light guide 9 is a fiber-optic cable, the minimum diameter of which can reach only about 0.37 mm [10], consisting of a set of fibers with a regular stacking and preservation of the structure of the transmitted image, is bendable, at least for the connection phase with transparent cap 1 with the stock length, providing the possibility of angular displacements up to 180° rear end. Its front end is attached via the optical elements to the rear side of the light guide pads 2, with possibility of turning with him, lighting the surface of the product 17 and transceiver reflected optical and thermal flows to the recorder 16. The rear end of the flexible light guide 9 accommodated within the third axial actuator 7 for rotations around the x-axis (equivalent to the third degree of freedom) of the images transmitted through the collimator 12 to the recorder 16.

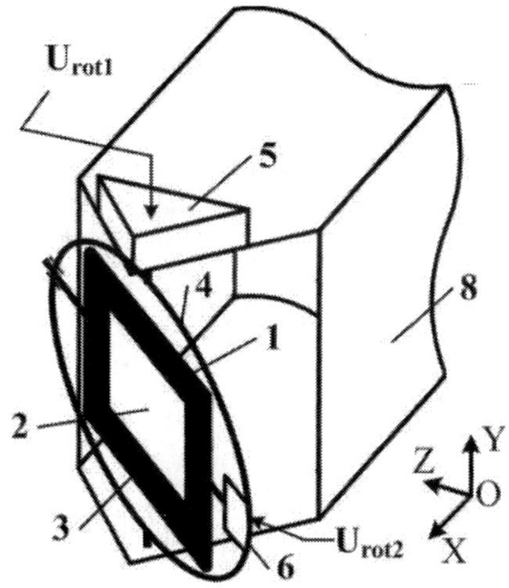

Figure 4. Tip with two degrees of freedom, fixed in the measuring rod (flexible fiber and guides are not shown).

Along a flexible optical fiber 9 is laid a set of fibers 15 used for switching from the emitter 14 illumination modulated with a frequency $f_{п}$ illuminating optical radiation angle of the surface 17 so that the emerging from it, the cut-off was in the field of view of the flexible optical waveguide 9 and, accordingly, the recorder 16.
Dual-axis meter 13 is used to measure the position lx and angles α_{oy} and α_{oz} tip 1 with the formation of the output signal $N_{out}(l_x, \alpha_{oy}, \alpha_{oz})$. The most convenient two-axis measuring 13 implemented on the basis of three laser triangulation linear encoders that measure the spatial parameters of the tip 1 by the coordinates of its three points. For each such point, the sensor illuminates a laser beam tip 1 with the formation on it of the light spot in the form of so-called "Bunny" and records the position of this Bunny coordinate-sensitive receiver.
Recorder 16 is a CCD-matrix [11], the registered image of the surface of the product 17 and its thermal radiation in a single IR range $\Delta\lambda_l \approx \Delta\lambda_t \approx 1{,}0 \ldots 1{,}6$ μm with an output digital signal N_{out}. For the recorder 16 introduces the concept of axes of the recorder, the appropriate one of the lines of the rows of its cells. It is used to achieve matching it with the tangent to the element 16 (the cutting edge of the tool, thread, etc.) to enhance the resolution of the image by eliminating smudging of the recorded image according to the technology of the time delay accumulation [12].
The principle of the developed ACD is the implementation of contact and contactless measurements.
Contact measurements are realized when approaching the tip 1 to the surface of the product 17 to a mechanical contact with it. Current position l_x of the tip 1 forced turns α_{oy} and α_{oz} due to the size and

taper of the 17 items measured in our two axis filament meter 13, for example, through the use of the three triangulation sensors with the formation of the output signal $N_{out}(l_x, \alpha_{oy}, \alpha_{oz})$.
In the process of contacting one of the methods of optical inspection of surface profile is also implemented – shadow method. To this end, the emitter 14 is formed of a modulated optical signal of the backlight, the illuminating angle of the surface 17 so that the emerging at the same cut-off is field of view optical input of the flexible light guide 9. Any surface irregularities, for example, the thread pitch for screw surface (Fig.5), forming near the cut-off individual pattern of shadows, passed down a flexible optical fiber 9 to the input of the recorder 16.
With them is also transmitted thermal radiation generated by the product 17, is proportional to its temperature. The subsequent digital signal processing N_{rec} with a different frequency separation of the optical signal is information about the surface roughness of the product and the temperature of the product of T_{det}.

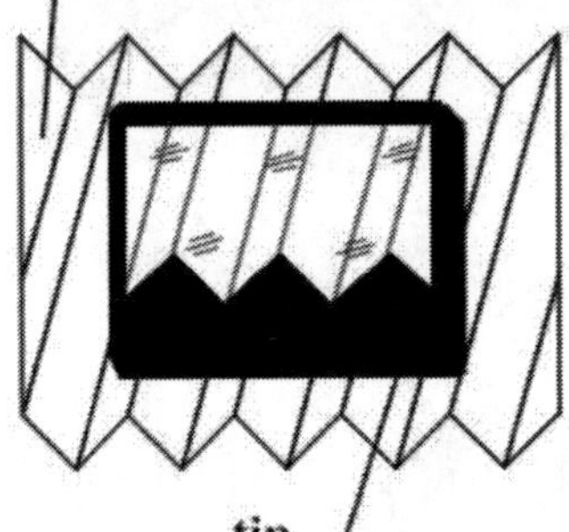

Figure 5. An example of the implementation of the shadow method developed by the ACD when controlling the thread parameters (the image of the shade of the "cut-off" border transmitted to the input of the recorder).

4. The discussion of the results

4.1. The calculation of the load capacity of the sapphire tip to the ACD
The results of the calculation of the load capacity of the sapphire tip to the ACD, ensuring a nearly 30-fold safety factor for shock loads and experimental researches in mechanical contact with 5 cogs cutter 15 mm in diameter rotating with a frequency of 1000 rpm, which confirmed the calculations, determined the surface roughness R_z of the contact area of no more than 0.15 μm. This value determines the transmission of optical flow in the infrared range with wavelengths of more than 0.9 μm to eliminate the effect of spurious diffuse scattering.

4.2. The calculation of the load capacity of the sapphire tip to the ACD
The design of the ACD allows to implement new functionalities: one, two and three degrees of freedom of the sapphire tip in the form of its angular movements. In one embodiment, this tip is similar to the frame of the gyroscope, in which the tip can be rotated around axis OY and OZ with the possibilities of the measurements of the taper of the product and measurement "chord". To reduce "motion blur" during the registration profile's fast movement products technology time delay of accumulation of rotation of the tip around the OX axis in the present device it is necessary to substitute the angular displacements of the transmitted image using the by turning the end of the flexible optical waveguide.

4.3. The visualization and registration image the surface of the product by using shadow method

IOP Conf. Series: Journal of Physics: Conf. Series **944** (2017) 012073 doi:10.1088/1742-6596/944/1/012073

Designed ACD allows also to visualize the surface of the product with the registration image, shadow method, by measuring the temperature of the product pyrometric method.

5. Conclusion

5.1. The use of sapphire for high-strength and optically transparent ACD tips

Studies have confirmed the possibility of using sapphire as a material for high strength and optically transparent ferrules ACD for contact measurement of dimensions of products with intermittent surfaces.

5.2. Sapphire tips with one, two and three degrees of freedom for ACD

The developed design ACD demonstrates the possibility of creating sapphire handpieces with one, two and three degrees of freedom in the form of angular offsets. One degree of freedom associated with the rotation of the tip around the axis OY, is primarily applicable to measurements of size and taper of the workpiece. The second degree of freedom defined by angular displacements of the tip around the axis OZ and can be used for measuring the "chord". The need for the third degree of freedom of the tip due to the desire to improve the image resolution fast-moving processes to reduce "motion blur" according to the technology of the time delay of the accumulation by selection of the angle of inclination of the recorder, namely, rows of photocells of the CCD coincident with the angle of the picture elements of the end product.

5.3 The sapphire tips for wide functionality possibility ACD

The conducted research provide the basis for the establishment of the contact and contactless ACD with sapphire tips with greater functionality.

References

[1] Kondashevskiy V and Lotze V 1976 *Active control of part dimensions on machine tools* (Omsk: West Siberian book publishing house) p 431

[2] Sobolev M 2005 *Automatic dimensional inspection on machine tools* (Smolensk:Oikumena) p 300

[3] Leun V and Tignibidin A 2010 New principles of construction of active control devices for products of tool production and engineering *The Journal Omsk Scientific Bulletin* 2(90) pp 165-169

[4] The website of the company Marposs 2017 URL:http://www.marposs.com product.php/eng/grinders_control (date accessed: 31.05.2017).

[5] The website of JSC Niizmereniya 2017 http://www.micron.ru/production/active_instruments/bv-4304/ (accessed: 31.05.2017)

[6] Leun E 2016 Development of active control devices dimensional parameters of products by using sapphire measuring tips *The Journal Omsk Scientific Bulletin* 4(148), pp 123-127

[7] Leun E 2016 *Method of measuring the linear dimensions of products made with depressions and protrusions on the surface*“ (Moscow:Patent of the RF 2603516, IPC B24B 49/04, 49/12 B, publ. 27.11., bul. 33)

[8] Leun E 2016 *Hydro-interferometric method control dimensions of details Dynamics of Systems, Mechanisms and Machines* **2** pp 101–109

[9] Leun E and Shulepov A 2017 Research and development of sapphire crystal measuring tips for active control devices dimensional parameters of products *The Journal Omsk Scientific Bulletin* № 3(153) pp 91-95

[10] Ultrathin microborescope (Made in USA) 2017 http://www.fiberscope.net/ru/ultra-fine - micro-borescope.html (date accessed: 31.05.2017)

[11] Burlakov I, Grinchenko L, Dirochka A and Zaletaev N 2014 Detectors short-wave infrared band based on the In GaAss *Uc. of applied physics* vol 22 pp 131-162

[12] Petroshenko M and Szolomicki D 2016 *Technology of company e2v* http://www.npk - photonica.ru/images/tehnologii_e2v.pdf (date accessed: 31.05.2017)

IOP Conf. Series: Journal of Physics: Conf. Series **944** (2017) 012074 doi:10.1088/1742-6596/944/1/012074

The Basic Principles and Methods of the System Approach to Compression of Telemetry Data

A V Levenets

Pacific National University, 136, Tihookeanskaya St., Khabarovsk, 680035, Russia

levalvi@bk.ru

Abstract. The task of data compressing of measurement data is still urgent for information-measurement systems. In paper the basic principles necessary for designing of highly effective systems of compression of telemetric information are offered. A basis of the offered principles is representation of a telemetric frame as whole information space where we can find of existing correlation. The methods of data transformation and compressing algorithms realizing the offered principles are described. The compression ratio for offered compression algorithm is about 1.8 times higher, than for a classic algorithm. Thus, results of a research of methods and algorithms showing their good perspectives.

1. Modern Data Compression Techniques

The problem of efficient data coding with minimum redundancy is still urgent due to prevalence of information-measurement systems in different areas of activity. The problem is enhanced due to the wide use of data in audio and video systems, particularly, high quality ones, which results in increasing volumes of transmitted data [1–4]. So, the development of data compressing methods and devices has not lost its urgency.

An amount of work has been done in this area. The majority of the proposed compression methods are applicable to media data compression [5, 6], which can be explained by their prevalence over other types of data in the telecommunications traffic. The peculiarity of such data compression consists in reducing their information content by reference to specific features of human perceptive organs. It should be noted that media data for information-measuring systems are not always considered high priority.

For telemetry systems the most important methods for measuring data compression are those ones which are characterized by reciprocity. Nowadays there is a little information in literature available on the subject, and the decisions offered are oriented to specific fields, e.g. aerospace engineering, sensor networks, etc. At the basis of the majority of such methods there lies the modification of already existing methods for data compression. For example, discrete cosine transform for integer data compression, different realizations of the Burrows-Wheeler transform and the peculiarities of the LZW algorithm is offered. The paper [7] is of special interest since it offers to use the procedure of lossless data compression incorporating two stages: decorrelation and entropy coding. Decorrelation of data was carried out with the use of a few variants of linear prediction and bitwise operation XOR. Entropy coding comprised the well-known Huffman method, Rice coding, arithmetic coding and the Deflate algorithm. The data received via telemetry compression in IRIG-106 format showed, in particular, that the frame with a complex structure built on the ground of the two-stage commutation undergoes more substantial compression if decorrelated in each communication channel.

A some authors for data compression offer to use a neural nets [8] and fuzzy logic [9], although there is no significant advance in this line of research, and the methods offered can be applied to a restricted tasks. It should be noted that the suggested methods for data compression are usually oriented to text-based information.

Alongside with compression methods proper a number of scholars offer data preprocessing methods which lead to a better structured mode, thus increasing the probability of their efficient compression. For example, in [10] the authors offer to consider the frames of telemechanical data as one-dimensional bit sequences which, after correlation dependence assessment, form structures of higher dimensions having a better ordering as compared to initial data. The disadvantage of the suggested method is its focus on quasi-steady zones of the telemetry system. It is necessary to mention that in some cases preprocessing is closely connected with the compressing procedure, which makes it impossible to set them apart.

Alongside with the development of new methods and modification of already existing ones there was developed hardware enabling to increase high-speed performance of data compression and decompression subsystems. Also, some of scholars tried to formalize data compression and preprocessing tasks and view them together with general tasks of the information-measuring system. So, article [11] views the questions of task formalization of adaptive compression and provides a classification of compression algorithm selection criteria. Paper [12] is devoted to the methodology for building distributed information-measuring subsystems, collecting and processing information of complex technical systems whose characteristics can vary over time. In paper [13] there was made an attempt to develop an adaptive compression system whose peculiarity consists in the effective automatic choice of data compression algorithms which results in enhancing data-exchange network characteristics. The offered program system is aimed at improving Internet efficiency by use of three compression algorithms (LZO, Bzip, and Zlib). It uses a number of methods for assessing the transmission time of compressed and uncompressed data and takes into consideration free resources. A decision about the use of a compression algorithm based upon the assessment of received data is made.

However, the overwhelming majority of the developed procedures for data compression are oriented toward the processing of a restricted data class. It should also be noted that there has not been developed a general approach to the building of data compression systems which solves the problems of data preprocessing and compression together with the problems of data adequacy, the efficient use of the communication channel, etc. This integrated approach would make it possible to enhance efficiency of data compression systems as well as the telemetry system on the whole.

A significant point is that the majority of the suggested methods and techniques for data preprocessing and compression are aimed at processing the data of each source (sensor) as a separate object, and do not consider the telemetry frame as a whole. It should be noted that parameter values transmitted through the telemetry frame, as a rule, refer to one complex technical system so the data inside the telemetry frame are correlated even if belonging to one measurement object. Thus, viewing the telemetry frame as a whole and using both existing and new methods, we can achieve a higher level of data compression efficiency. We can assume that the application of this approach can be efficient under certain conditions, which makes the case for conducting research in this area.

2. Basic Principles of System Approach to Compression of Telemetry Data

As it follows from the current status of research, to date, the overwhelming majority of methods for data preprocessing and compression are being developed within the framework of the traditional approach to the tasks, when each data source is processed independently from other sources, without searching for cross-correlation between them. An exception to this are visual data processing methods characterized by lossy compression, which makes direct application of such methods to compressing measuring data of little promise.

On the other hand, telemetry information most commonly comes from one complex technical object whose measurement parameters can be cross-correlated, at times implicitly. Thus, we can argue that when the compression method considers the frame as a whole, its efficiency is higher than if otherwise. We can also assume that the compressed frame can be viewed as a single object (sample) whose substructure can be changed inconvertibly.

The most demonstrable transformation of this type is the presentation of all signal frames as a bit sequence which can later be transformed by given parameters into more complex geometrical patterns. Such patterns, provided they are easily described, can undergo compression more efficiently than the source data frame. It should be noted that we can expect better compression of these patterns using traditional compression methods. However, the maximum effect can be achieved when developing application-specific compression methods and algorithms. In general, we can assume that the dimension of the data structure depends significantly both on the number of data sources, stored in the frame, and their correlation degree.

It should also be noted that telemetry systems are characterized by the status change of the communication channel which links all constituent parts of the system. In this case adaptive control of the communication channel is required, particularly the regulation of the power of the error correcting code. In this instance in order to prevent data loss the communication channel bandwidth is chosen wide enough to provide normal operation under worst conditions.

To lower requirements to the communication channel affected by a short-term impairment, I can offer word length reduction of the transmitted data with a following transmission of the cut-off bits into periods of “normal” channel status. Thus, both the compression subsystem and the data transmission system should be considered as a unified control system, which is to be reflected in the principles of constructing such subsystems. Moreover, the widespread occurrence of distributed information-measuring systems, e.g. sensor networks, makes it necessary to develop such methods for data processing that would require lower computational costs and, consequently, low energy costs, which has a substantial impact on weight-size parameters and self-contained operation time of the elements of such systems.

Thus, considering the aforesaid, I offer the following principles of designing of systems of compression of telemetry information:

1. The telemetry frame is proposed to be considered as a single information space which has its internal cross-correlation relationships including the ones on the bit level.

2. The source data frame is offered to be considered as a bit sequence with a possibility of constructing structures of a better ordering and higher dimensions. To achieve this, we offer to use autocorrelation computation modified with allowance for the binarity of the processed data. The application of this approach will make it possible to develop data preprocessing methods to enhance compression efficiency, including that of the existing methods.

3. The codeword bound violation principle is understood as the absence of the rigid support of initial word length. This approach is supposed to build n-dimensional structures having a better ordering as compared to the structures built with the consideration of the length of the source word.

4. Data preprocessing methods are supposed to be divided into two classes: those for the stationary (quasi-stationary) operation condition and those for the non-stationary operation condition of the system. The first case is based on singling out trends including harmonic ones, dynamic range compression (differential method) and the construction of homogeneous structures (compression method). In the second case may be used the “reduction to stationary” principle, the basis of which is the description of the source data with the use of quasi-random sequences.

5. The development of the methods and algorithms on the basis of the above mentioned principles implies compressing the whole frame or the whole series of sequential frames. By doing so, we exclude the correlation of traditional compression methods with a single data source, which supposedly gives advantages in compression efficiency, at least, for stationary modes.

6. The structures for data preprocessing and compression subsystems should provide minimum equipment, computational and energy costs aiming at ensuring the possibility of their application in different areas of modern technology, including sensor networks and information-measuring systems of different purposes.

For full realization of possibilities of offered approaches, the compression subsystem should be organized in the form of the multilevel system presented on Figure 1. The first level is formed by the well-known compression algorithms providing the guaranteed compression ratio. They work when you turn on the system, and also in case of an inefficiency of algorithms of higher levels. In process of accumulation of the statistical data, the compression subsystem starts to work with algorithms of higher level which consider an implicit correlation both in a frame of the data, and in the data flow from one source.

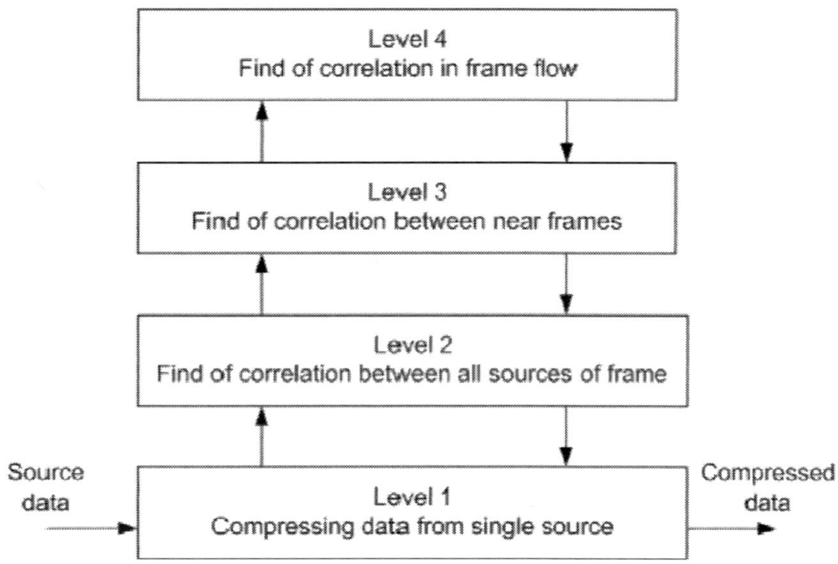

Figure 1. Multilevel organisation of data compressing system.

In case of compression of telemetry it is necessary to consider that they are transferred by frames of the fixed size with invariable (as a rule) structure. Besides, wide using of the distributed information measuring systems, in particular, of sensor networks, leads to the need to developing of methods of data handling with low computing expenses and small energetic expenses.

3. Methods and Algorithms

In this paper a stream of frame, for which the mean of an arithmetical difference between the same samples in two next frames will make no more than four units (without a sign) we will designate as stationary. It is obvious that for such a flow as a method of compression of the first level will be effective differencing algorithm, the essence of which is the element by element calculation of the arithmetic difference between the same counts in the current and the previous (next) frames.

An important task is the decomposition of the source data frame to two-dimensional structure and highlight in it the regions that are maximally correlated in the time. The application of difference algorithm to these blocks will provide a high compression ratio. Thus if we have the completely identical blocks in neighboring frames, to describe them we need only a few bits (a flag of the immutability of the data block and a relative address in the frame). An example of such algorithm and results of a study of its effectiveness for the problem of compression of telemetry data is describing in [14]. This algorithm provides an average compression ratio of telemetry data about 4.0 and can works in real-time mode. Some methods for higher levels of compression are described below.

3.1. Word Length Convertion

This method belongs to methods of preprocessing of measuring data before holding a procedure of compression and is necessary for transformation of statistical characteristics of the compressed data for compressing them with maximum efficiency. It should be noted that the measuring data from the object operating in normal mode have a stationary parameters. These data are well compressed by delta-encoding algorithm. However, when we need a compression of non-stationary data, the compression efficiency may reduce to unacceptable values.

As a method of data preprocessing, which will allow avoiding (reducing) the loss of efficiency of the compression algorithm, it is proposed to change the word length of the compressed data. It is based on a representation of the source data by one-dimensional bits sequence that is further breaking onto words of a predetermined bit width. An example of a procedure for a test sequence of 8-bit unsigned samples is shown on Figure 2.

Here the most optimal division of source sequence of the data was by 5-bit words. However, the offered method will not be so effective for all possible data sets.

It is obvious that high tetrads of words of the source sequence are changed less than low tetrads. Therefore in case of such break it is possible to expect the maximal quantity of repetitions of one value. Therefore, it is possible to expect increase in efficiency of compression.

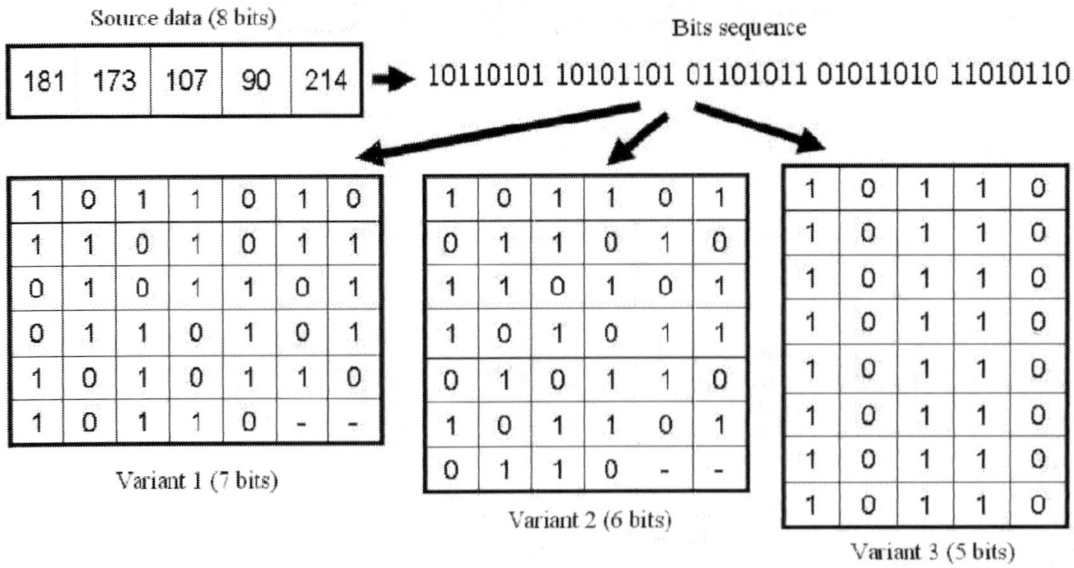

Figure 2. Example of word-length convertion

For verification of such assumption the data from system of telemechanics of energetic object were studied. Word length of data is 8, and the volume of data is 10 000 – 20 000 samples. The typical histograms received before and after preprocessing of 19 285 samples are shown on Figure 3.

The estimation of transform efficiency was carried out using a compression algorithm with codes of variable length. It was assumed that the most frequent value is encoded by bit "0", a three less frequent values - bit sequence "1xx", and other values – a bit sequence "111xx".

The histogram value "1" corresponds to a maximum repetition rate of 14 046, three less frequently occurring values "0", "2" and "3" correspond to the frequency 7 594, 621 1 559 and 1, respectively. Thus, the total volume of the transformed data stream is 14 046 (7 594 1 621 1 559)·3 13 750·5 = 115 118 bit, the volume of the original flow is 154 280 bits, so the value of compression ratio is approximately 1.33. Compressing of the source sequence (not transformed) in the same way provide of value of compression ratio is approximately 0.2.

In the process of the study were reviewed of 150 sets of data, and for 87% of the sets were obtained similar results. The average value of the compression ratio is approximately 1.4. Thus we can say what the proposed method is useable for the tasks of compression of the measuring data. In my opinion, a further increase of the compression ratio we can obtain only with a specialized compression algorithm.

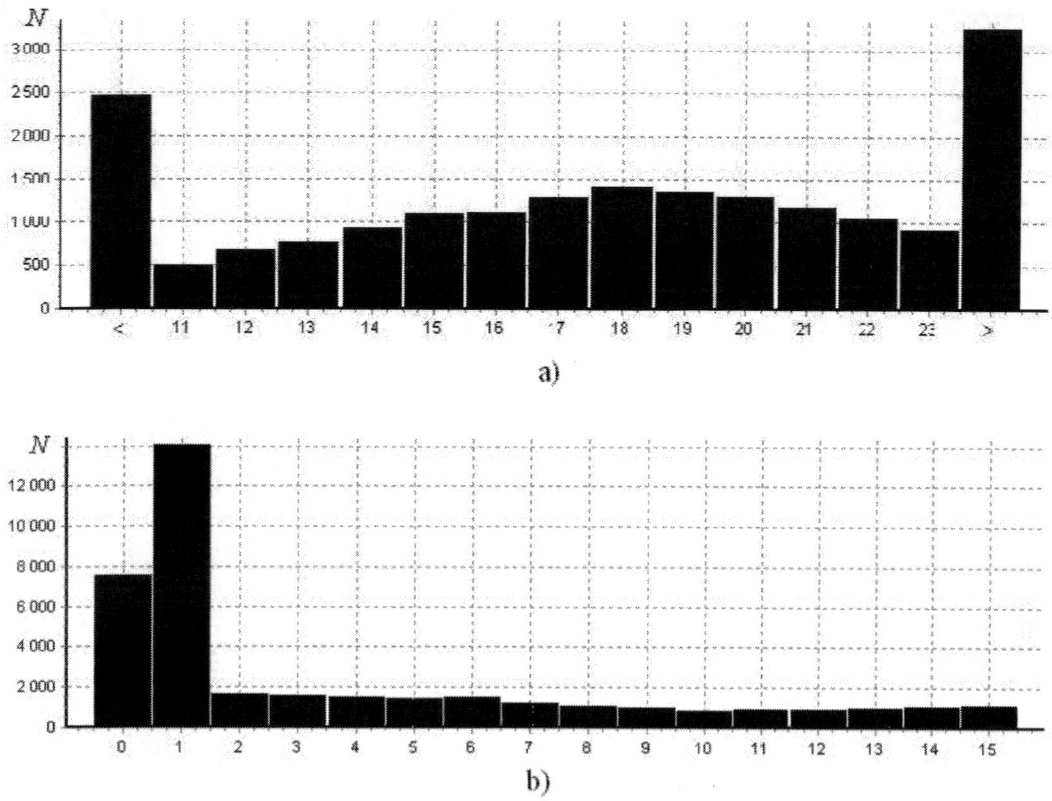

Figure 3. Example of histogram before (a) and after (b) data preprocessing

3.2. Representation of Input Data

An important task is to find the optimal representation (for compression) of the source sequence of digital data. As the source data is used a random data with uniform distribution. The optimal transformation can be estimated by the number of trees. In this case the term "tree" refers to the set of closely spaced identical elements.

Search of optimum representation happens as follows:

- The initial sequence of measuring data will be transformed to the bit sequence;
- The bit's autocorrelation function of a data is calculated. Function $R(n)$ is defined as number of coincidence of values of the current bit of the signal and bit with given distance n. For bit sequence S with length N this function defines as:

$$R(n) = \frac{1}{N}\left(\sum_{i=0}^{N-1} S[i] \leftrightarrow S[i+n] - \sum_{i=0}^{N-1} \overline{S[i] \leftrightarrow S[i+n]}\right),$$

  where $\leftrightarrow$ is a logic operation of equivalence.
- The autocorrelation function breaks into blocks on m bits. In each block the searching of the maximum value is run. Maxima collect in a special array. Width of two-dimensional representation is calculated as the ratio of an index of the current maximum to an autocorrelation function of number of its block;
- Matrixes of values of width of the unit are built and for each matrix the number of trees is counted. To optimum width of the block there corresponds the smallest number of trees.

The analysis of obtained data shows that mathematical expectation of optimum width of the block for different amount of samples of a measuring signal (from 250 to 4000) lies in the range 7–13 while boundary values of optimum length of block lies in the range 5-30.

The average length of a tree can be used for estimating of compression ratio of the processed data. The conducted researches showed that this value is about 6.0 and practically doesn't changing. It allows speaking on the one hand, about good potentiality of compression, and with another – about need of search of more effective methods of determination of optimum parameters of data representation.

It is necessary to note that the offered data transformation isn't restricted by two-dimensional representation, i.e. its representation by objects of higher dimensionality is possible that can lead to even more essential abbreviation of information redundancy.

3.3. Compressing Algorithm

Considering features of the telemetry information, it is possible to offer the following compressing algorithm of the second level. The source data frame is presented as the two-dimensional structure which breaks onto the rectangular blocks which are slightly changing in time. Then a block-by-block compression by the delta-encoding algorithm is executing.

For the offered algorithm it is necessary to note the two main operations: a choice of the amount of two-dimensional representation of a frame and its break onto rectangular blocks with the maximum correlation. Estimation of correlation of blocks happens by computation of a difference between adjacent blocks with a choice of the minimum value. Estimation was made for some given period.

The algorithm was tested by means of telemetry data from energetic objects. Mean value of a compression ratio lies in the range 7.5–8.0 that is about 1.8 times higher, than for a test algorithm. In case of stationary data mean value of a compression ratio makes about 9.5.

Thus, the offered algorithm in simple variant, without of optimization of parameters is more effective, than the algorithms with classical technology of compression. Shortcoming of this algorithm is that it doesn't consider dependence of more than for two adjacent frames. This is especially noticeable for data with the monotonic increase and decrease of values. It is also necessary to consider that the algorithm potentially allows working in real time, possessing rather low computing expenses.

3.4. Conclusion

Methods for preprocessing before the compression procedure based on the geometrical approach to data structuring and resulting in improving the existing compression algorithms. The result corresponds to the international research level.

Methods for telemetry data compression considering cross-correlation in the telemetry frame among data coming from different sources (sensors). The result corresponds to the international research level.

Thus, the offered approaches potentially can be used by developers of efficient information-measuring multi-purpose systems, including telemetry systems, where combine the requirements of operation in real time and necessity of transmissions of large volumes of information by communication channels with limited throughput.

References

[1] Wootton C 2005 *A practical guide to video and audio compression* New York Springer 787.

[2] Salomon D 2006 *Data compression: the complete reference* New York Springer 900

[3] Candes E J and Wakin M B 2008 *IEEE Signal Processing Magazine* **25** 21-30

[4] Krintz C and Sucu S 2006 *IEEE Trans. Parall. and Distrib. Syst.* **17** 15-24

[5] Miano J 2002 *Compressed Image File Formats* New York ACM Press 336

[6] Pujar J H and Kadlaskar L M 2010 *J. of Theoretical and Applied Information Technology* **15** 18–23

[7] El`shafei M A, Sidyakin I M, Haritonov S V and Vornychev D S 2014 *Vestnik MGTU im. N.E. Baumana. Ser. "Priborostroenie"* **3** 92-104

[8] Talalaev A A, Tishchenko I P, Fralenko V P and Hachumov V M 2008 *Iskusstvennyy intellekt i prinyatie resheniy* **2** 24-33

[9] Ibrahim Wael R Anis and Morcos Medhat M 2005 *IEEE Trans. Power. Deliv.* **20** 2136-43
[10] Ch'e En Un, Levenets A V and Nilga V V 2011 *Informatsionno-upravlyayushchie sistemy (Information and Control Systems)* **6** 7-10.
[11] Filippov A K and Rufitskiy V A 2006 *Proektirovanie i tekhnologiya elektronnykh sredstv* **3** 14-19
[12] Vasil`ev A M 2014 *Vestnik Moskovskogo gosudarstvennogo universiteta priborostroeniya i informatiki. Seriya: Priborostroenie i informatsionnye tekhnologii* **50** 5-16.
[13] Krintz C and Sucu S 2006 *IEEE Trans. Parall. and Distrib. Syst.* **17** 15-24
[14] Levenets A V and Chye En Un 2012 *Proc. Int. Conf. European Science and Technology* (Munich) 181-4

Research of paste transition to substrate in LTCC-technology

S N Litunov and V Y Yurkov

Omsk State Technical University, 11, Mira ave., Omsk, 644050, Russia

E-mail: viktor_yurkov@mail.ru

Abstract. The electronics development demands for accuracy of printed technologies, in particular, to screen printing. Under a flat blade operation the print form is deformed and the image is distorted relative to the original. A squeegee in a form of a smooth cylinder reduces distortion, but it allows obtaining satisfactory print quality only when using high density grids. The paper shows findings of using roller squeegee with dosed ink supply. The roller squeegee is provided with an elastic layer. Dosage is carried out due to the cells on the elastic layer surface. There were used meshes 100-31 and 120-34 for the stencil. The experiments were carried out with layers of photopolymers and rubber. The carried out calculations made possible to choose the optimum printing pressure. Under the selected conditions, the printed image had minimal distortion. The findings allow drawing a conclusion about the possibility of roller squeegee using in chips manufacture according to LTCC-technology.

Keywords – screen printing, electronics, roller squeegee, mesh surface.

1. Introduction

When printing a squeegee is usually used in an elastic plate form.

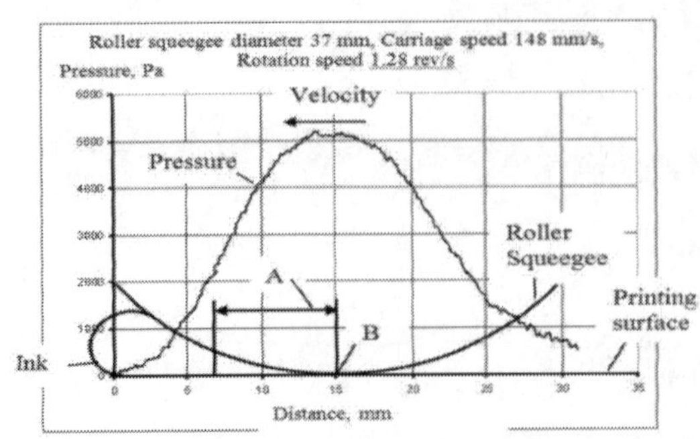

Figure 1. Schematic representation of screen printing with flat squeegee.

Screen printing allows obtaining a layer of paste with thickness of up to 100 μm. This feature made screen printing irreplaceable in manufacturing of microchips for thick-film technology, sensors and meters. The transition of paste to the substrate mainly depends on pressure in paste under the squeegee operation. Studies of pressure distribution in paste and solder paste under a flat blade operation were carried out by Dietrich E. Riemer (1988) [1, 2], Ekere N. N. and Lo E. K. (1991) [3], Mannan, S. H. et al (1993) [4], Glinski G. P. et al (2001) [5].

Disadvantages of using a flat blade:

1. mesh deformation due to friction forces between screen and squeegee. The standard stencil elongation is approximately 100 μm for a 450×450 mm screen frame with an area of 150×150 mm [6];
2. of-contact, necessary to separate the mesh from the substrate.

I. J. Fox, T. C. Claypole and D. T. Gethin (2003) [7] examined the possibility of screen printing with a roller squeegee. The work revealed a lack of roller squeegee - a satisfactory print quality is obtained on a grid of 150-34 or more. If lower grid density is used, paste spreads on the substrate. Its reason is especially the pressure distribution in the printing paste when the roller squeegee moves. Figure 1 shows the pressure distribution obtained experimentally by Litunov S.N. (2007) [8]. Distance A is marked on Figure 2 where paste is forced through the mesh with the roller squeegee action. The distance is measured from the contact point B between the roller squeegee and the printing surface. Sagging paste is accumulated on the back side of the screen when roller squeegee moving and it is crushed along the substrate.

2. The object of study

A squeegee roller with paste dosing has been offered (Litunov S.N. *et al.*, 2006) [9]. It is a metal cylinder with an elastic layer. There are cells on the surface of the elastic layer and they are filled with paste during the printing process. Excess paste is removed with a metal blade (Figure 2). Printing is carried out through the flow of paste through the stencil when the cells are deformed under the action of pressure. The object of the study is an experimental study of printing with this device and the transition of paste from the elastic layer cells to the substrate.

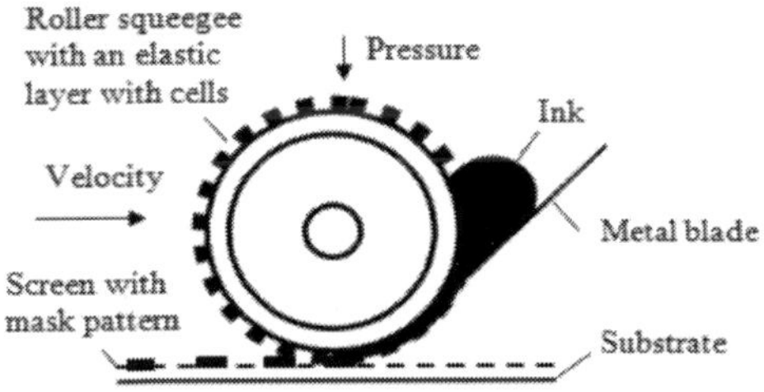

Figure 2. Schematic image of screen printing with a roller squeegee.

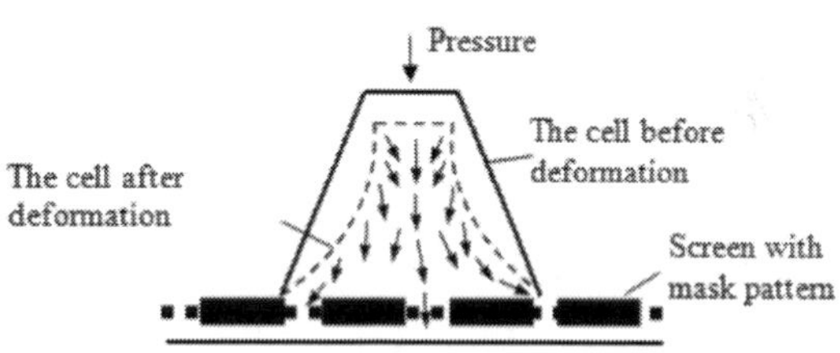

Figure 3. Scheme of the cell deformation and the paste flow through the stencil

3. Methods

3.1. Roller squeegee material parameters

When printing the elastic layer deforms in the contact strip. The cell volume decreases and paste is forced through the screen (Figure 3). Thus, the quantity of paste passing through the screen depends on the cell deformation.

37 mm diameter steel roller was used in our study. Photopolymer materials ACE of BASF firm (Germany) and NSF of TOYOBO firm (Japan) are used in flexographic printing. The cells on the photopolymer material were obtained using technology adopted in flexographic printing.
Also oil-and-petrol resistant (OPR) rubber was used. Cells were obtained with laser engraving on the rubber. Cells on photopolymer material (A, C) and rubber (B, D) are shown on figure 4. Mechanical

characteristics were measured on a tensile breaker IP 5158. The cell size obtained with photographs was used to calculate the volume of the cells.

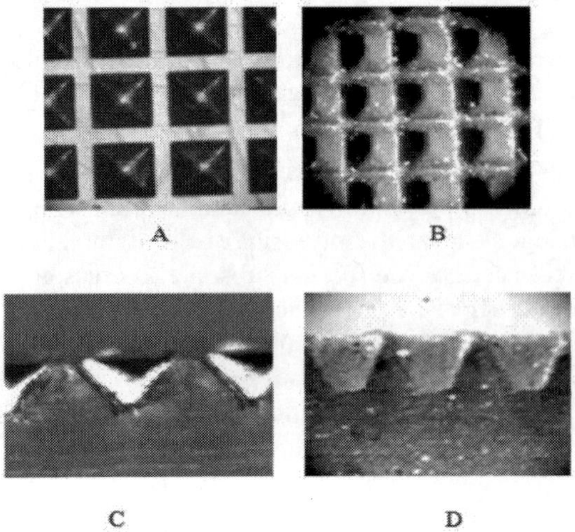

Figure 4. The image of the cells from the top (top) and the section (bottom): A and C are polymer material; B, D are rubber.

The characteristics of the selected materials are given in Table 1.

Table 1. Characteristics of Materials.

Name of Material	Thickness, mm	Depth of the cell, mm	Elastic moduls, MPa	Poisson`s ratio	Cell volume, mm^3
ACE, BASF (Germany)	1.52	0.16	28	0.485	0.0098
NSF, TOYOBO (Japan)	2.66	0.18	37	0.435	0.01103
Oil-and-petrol resistant rubber	1.75	0.22	7	0.5	0.0229

3.2. Mechanical parameters of the elastic layer

To select the printing pressure experiments were conducted to determine the deformation of the elastic layer material depending on the load. The measurement scheme is shown in figure 5.

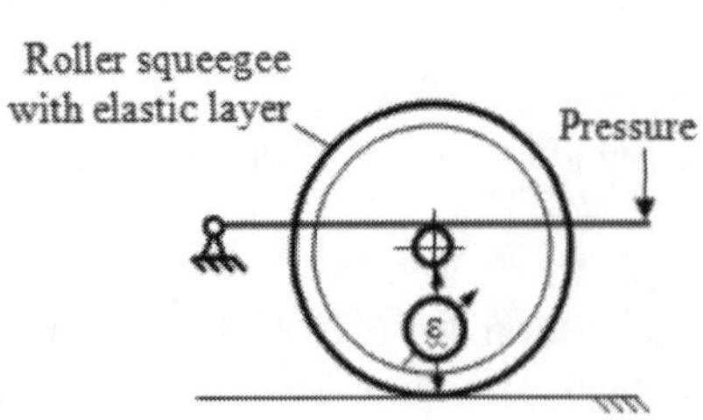

Figure 5. Scheme of measuring the elastic layer material deformation.

The deformation dependence on the load for the materials under study is shown in figure 6.

Each chart has an inflection point. The deformation of the material develops linearly before and after this point. Mainly the cells on the material surface are deformed before the inflection point. After the inflection point, both the cell and the material are deformed as a unit. It is not clear in what range the optimal print pressure is. To select the printing pressure during the experiments, the deformation of the material with the cells on its surface was calculated.

The deformation was calculated for the element (Figure 7) considered an elastic rod on an elastic base. When calculating, it was assumed that the pressure was evenly distributed over the surface of the element; the material is isotropic and incompressible; each vertical end of the element can move only in a plane that coincides with it; the rigidity of the element and the elastic base are equal.

The deformation of the rod was determined with the equation:

$$dW/dz = \varepsilon_z, \qquad (1)$$

where W(z) is the longitudinal displacement of the cross section with the coordinate z; εz is z cross sectional points elongation.

The equation of equilibrium of the rod element separated with two sections has the form N(z) = - P, where N(z) is the normal force in the cross section z; (z)=σ(z)·F(z); σ(z) is the normal tension at points of the cross section z; F(z) is is the cross-sectional area of the rod z.
The accepted boundary conditions are:

$$W(0) = 0;\ \ N(z) = -W(c) / \delta, \qquad (2)$$

where δ is the compliance of the base where the rod is located.

Rod relative deformation ε_z was determined using the system of equations (1) and (2). Using ratio $\varepsilon_z = \nu\varepsilon_x = \nu\varepsilon_y$, where ν is Poisson's ratio, the rod relative deformation ε_x, ε_y was defined in x and y directions respectively.

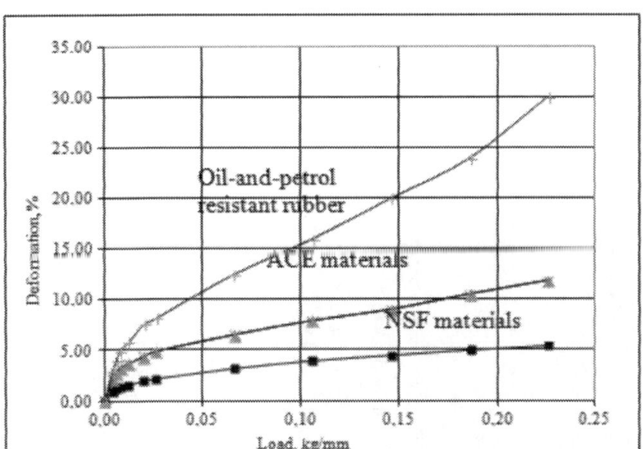

Figure 6. Deformation dependence on the load of elastic layer materials.

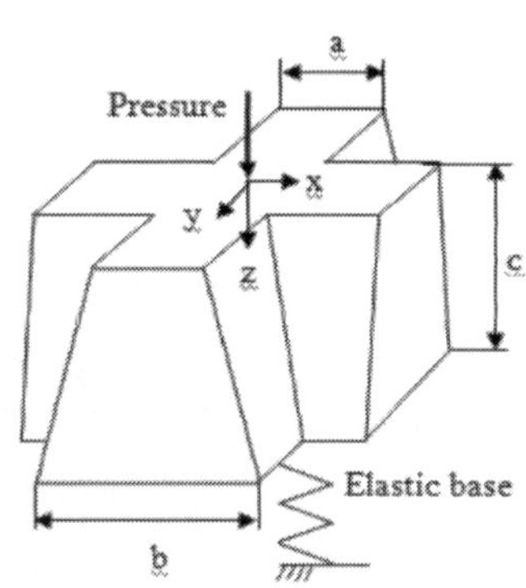

Figure 7. An element for calculating the cell deformation.

Equation (1) in a form of finite differences is

$$W[z(i+1)] = W[z(i)] + N[z(i)]\cdot\Delta c / (E\cdot F[z(i)]),$$

where W[z(i)] and W[z(i+1)] are displacements in the longitudinal direction of the cross-section at the i-th and (i+1) step; N[z(i)] is the force normal to the transverse i-th cross-section; F[z(i)] is the cross-sectional area at the i-th step. Newton's method was used for the calculation. The calculation was carried out in the EXEL software. A trend line was obtained for each cell profile in a form of an exponential dependence. Having integrated from zero to the depth of the cell, considering the Poisson's ratio, the volume of the deformed cell was obtained.

According to the results of calculations, knowing the amount of paste to be passed through the substrate during the cell deformation, the optimum load on the roller squeegee was determined. It was 0.11 kg/mm for NFS material, it was 0.07 kg/mm for ACE material, it was 0.02 kg/mm for oil-and-petrol resistant rubber. The cell profile on the ACE material deformed with a load 0.07 kg/mm is shown in figure 8.

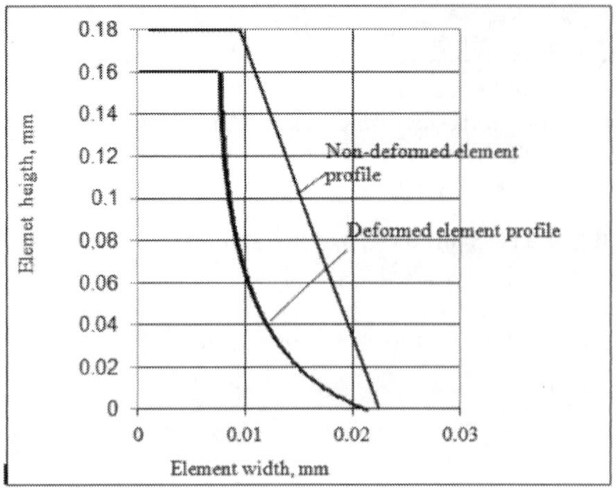

Figure 8. Cell profiles on ACE material.

3.3. Printing devices

A printing device is shown in figure 9 it has a carriage 1, a roller squeegee 2 with an elastic casing 3, a metal blade 4, adjusting screws 5. The stencil is mounted in clamps and is not shown in figure 9.

Figure 9. Printing device with a roller squeegee.

The carriage speed was 148 mm/s, 193 mm/s; 232 mm/s. The diameter of the roller squeegee is 40.04 mm for ACE material, 42.32 mm for NSF material, 40.5 mm for oil-and-petrol resistant (OPR) rubber. The polyester meshes SAATI 100-31 нит/sm and 120-34 нит/sm were used. The net tension is 21 н/sm.

The image on the stencil test was a scale with lines 05-0.5 mm thick, that are along and across the roller squeegee movement and plots with raster fields.

During the experiments varied parameters were the elastic material layer, stencil mesh and carriage speed with a roller squeegee. As a paint, plastisol paint with a metallic pigment of viscosity 700 Pa·s was used. Such paint by its properties is closest to the pastes used for the production of LTCC-microcircuits. On the obtained prints there were monitored changing line lengths along and line thickness disposed transversely roller squeegee motion. The measurements were carried out using a universally measuring microscope UIM-21. The substrate was organic glass sheets 4 mm thick.

Results and experiments
Increase in speed in the selected speed range slightly affects the change in the size (0.5-1.0%). The hardest of the subjects is the NFS material, it gives the least change in size, and the least hard is OPR rubber and it gives the biggest change. This is because the less rigid material is deformed more, and more paste passes onto the substrate. However, no dependence of the size change on the cell volume was detected. Hence, it can be concluded that the optimal print pressure is selected correctly. The deviation of the length of the lines on the print from the length of the lines on the original has been increased by 0.01-0.015 mm with a line length of 150 mm and a frame size 450 mm. This is less than the standard extension by 3-4 times. The deviations along the line width had the same values. It allows to conclude that the increase in the lines size does not occur from the deformation of the stencil during printing, but from the flowing of the paste under the whitewashed elements, and it is typical for screen printing as a whole. This defect is eliminated at the prepress stage. The findings allow to draw a conclusion about the possibility of using roller squeegee in the manufacture of microcircuits using LTCC-technology where the incompatibility between the layers of the chip should not exceed 20 % from the minimum diameter of the transition hole [10].

Among other features of the printing device operation with a roller blade type, the phe-nomenon of self-sticking, which allows to reduce of-contact to minimum can be noted. During the experiments, the gap was zero. At the same time, the substrate was on the printing table freely, without fixation.

References
[1] Dietrich E Riemer August 1988 *Analytical Engineering Model of the Screen Printing Process Part 1 Solid State Technol.* pp 107-111
[2] Dietrich E Riemer September 1988 *Analytical Engineering Model of the Screen Printing Process Part 2 Solid State Technol.* pp 85-90
[3] Ekere N N and Lo E K 1991 New challenges in solder paste printing *J Electronics Mfg* **1** pp 29-40
[4] Mannan S H, Ekere N N Lo E K and Ismail I 1993 Application of ink screening models to solder paste printing in SMT assembly *J. Electronics Mfg* **3** pp 113-120
[5] Glinski G P, Bailey C and Pericleous K A June 2000 A non-Newtonian computational fluid dynamics study of the printing process *Proc Instn Mech Engrs.* **215** Part C pp 437-446
[6] Shurygina V Printed Electronics. What is it, how is it created, what to expect from it? Part I. *Electronics NTB* [Electronic resource] Issue № 3/2010
[7] Fox I J, Claypole T C and Gethin D T April 2003 An experimental investigation into ink transfer using a roller squeegee in high-speed screen printing *Proc Inst Mech. Engrs.* **217** Part E:J Process Mechanical Engineering pp 307-321

[8] Litunov S N 2007 *Methods of calculating printing machines for screen printing machines: monograph* (Omsk: OmSTU) pp 132-137
[9] Pat. 70198 Russian Federation IPC Colorful screen printing machine S N Litunov no 2006107089/22; appl. 06.03.2006; publ. 20.01.2008 Bul. 2
[10] Litunov S N, Batischeva M V, Skitchenko V V and Serdyk O E 2015 Features of filling paste holes in the workpiece LTCC-IC *The Journal Omsk Scientific Bulletin* 140 pp 126-129

IOP Conf. Series: Journal of Physics: Conf. Series **944** (2017) 012076 doi:10.1088/1742-6596/944/1/012076

Utilization of Variable Consumption Biofuel in Diesel Engine

V A Markov[1], V G Kamaltdinov[2] and A A Savastenko[3]

[1]Bauman Moscow State Technical University, 5, 2-ya Baumanskaya str., Moscow, 105005, Russia
[2]South Ural State University, 76, Lenin ave., Chelyabinsk, 454080, Russia
[3]Russian University of Peoples' Friendship, 6, Miklukho-Maklaya str., Moscow, 117198, Russia

e-mail: vladimir.markov58@yandex.ru

Abstract. The depletion of oil fields and the deteriorating environmental situation leads to the need for the search of new alternative sources of energy. Actuality of the article due to the need for greater use of the alternative fuels in internal combustion engines is necessary. The advantages of vegetables origin fuels using as engine fuels are shown. Diesel engine operation on mixtures of petroleum diesel and rapeseed oil is researched. A fuel delivery system of mixture biofuel with a control system of the fuel compound is considered. The results of the system experimental researches of fuel delivery of mixture biofuel are led. Keywords — internal combustion engine, diesel engine, diesel fuel, rapeseed oil, mixture biofuel.

1. Features and advantages of biofuel using in internal combustion engines

The researches of internal combustion engines on alternative fuels become more and more actual. It is because these is a necessity of petroleum motor fuels substitution by fuels produced from alternative raw materials. Fuels produced from vegetable raw materials (biofuel) are the most attractive among the alternative fuels. These fuels are obtained from renewable raw material base – vegetable raw materials. The applying these fuels - oxygen circulation in the atmosphere can be provided and the main greenhouse gas (carbon dioxide) emission into the atmosphere can be reduced.

Bioethanol became widely used in internal combustion engines with forced ignition of operating mixture. In the context of diesel engines liquid fuels based on vegetable oils have great prospects. The rapeseed oil is considered as the most perspective source of energy [1, 2, 3]. Mixtures of petroleum diesel fuel with rapeseed oil have the greatest benefits [4, 5, 6]. Regulation of these mixtures allows improving diesel engines indexes significantly during speed and loading modes. In this way the fuel delivery system of petroleum diesel fuel and rapeseed oil mixtures should expediently be equipped with biofuel mixture compound control system.

2. Fuel delivery system of mixture biofuels in diesels

The feeding valve of high pressure line is considered to be applied as such a control system (initial pressure regulator valve). The rapeseed oil delivery into high-pressure fuel pipe is occurred through this valve [7, 8, 9]. The valve mounting scheme into the high-pressure line is shown in figure 1. The fuel system consists of two lines: diesel fuel and rapeseed oil. Every line has its own fuel-feed pump (1) and fine filter (2). Diesel fuel line operates in a conventional manner and fuel enters into fuel injection pump assembly (3) after the fine filter passing. After that fuel goes through the high-pressure line (4) to the nozzle (5). The nozzle (5) is equipped with initial pressure regulator (6) into the high-pressure line. The high-pressure line is fed with rapeseed oil which is submitted from fuel-feed pump through the initial pressure regulator.

The regulator design is shown in figure 1, b. The system operates in the following way. Pressure control valve into fuel injection pump assembly closes after fuel cut-off, relieving the high-pressure line to avoid reinjection. Discharge value depends on uploading run of the pressure control valve (uploading volume) and operation mode of the fuel equipment. This operation mode is characterized by a rotating speed of engine crankshaft (or camshaft of the fuel injection pump assembly) and injection rate. Residual volumes form in some modes during the uploading in the high-pressure line. These volumes can be filled with an additional fuel (e.g. rapeseed oil) between feeding phases through the replenishment valve. The amount of submitted rapeseed oil through the replenishment valve depends on residual volume quantity in the high-pressure pipe. This value formed after cut-off process of the feeding.

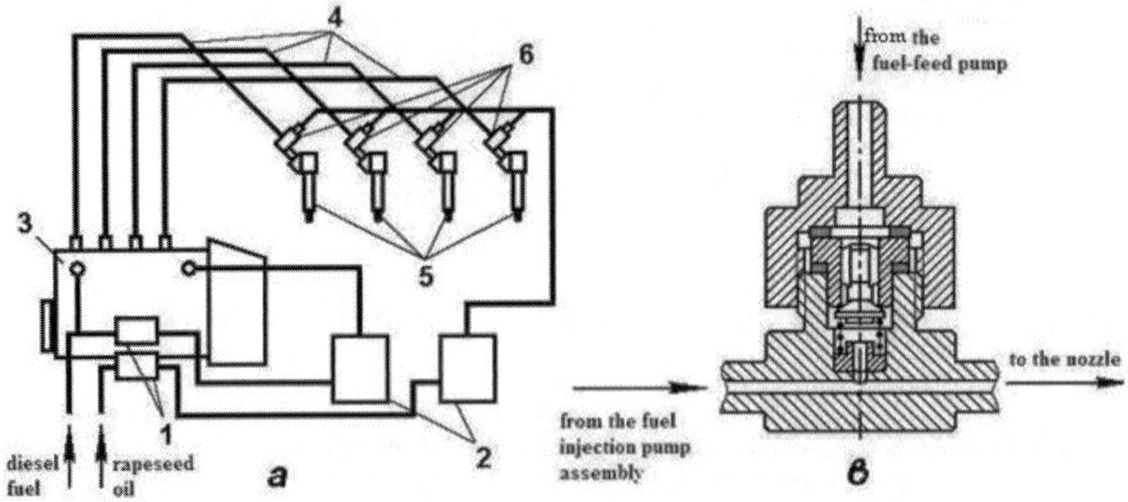

Figure 1. Feed system of high-pressure line through initial pressure regulator valve: a – system scheme; b – Initial pressure regulator; 1 – fuel-feed pump; 2 – fine filter; 3 – fuel injection pump assembly; 4 – high-pressure line; 5 – nozzle; 6 – initial pressure regulator.

3. Fuel delivery system experiments of mixture biofuels

The series of motor-less tests of fuel pump have been carried out. Fuel pump type – 4UTNM with FD-22 nozzle (sprayer type 145.1112110), 62 cm – light fuel pipe and 2 mm inner diameter, pressure control valves with a 3mm uploading stroke. During the experiment diesel fuel was submitted through the fuel injection pump assembly under 0.1-0.2 MPa over-pressure. This pressure was created with the fuel-feed pump. Speed characteristics of the fuel pump are shown in figure 2. They are obtained at different rail positions of the fuel injection pump assembly during the high-pressure line replenishment of diesel fuel. The experimental results show that speed characteristics of fuel feed with growing branches are formed during the line replenishment and the rotating speed increasing of fuel injection pump assembly camshaft. Fuel-feed correction and corresponding adjuster setting are necessary for diesel speed characteristic ensuring with a torque reserve. Correction level is determined with a diesel performance and it depends on a composition of mixture fuel (rapeseed oil proportion having a lower calorific ability).

Mixture fuel composition depends on replenishment value of high-pressure line through the initial pressure regulator. The peculiar experiments with fuel equipment of D-245-type diesel engine have been carried out. It has been made with the aim of replenishment value verification entering through the initial pressure regulator. Fuel amount measurement has been carried through a burette and a stopwatch. This fuel was entering the fuel-feed pump and out it one to the high-pressure line. The dependences of fuel amount entering the high-pressure line through the replenishment valve in a cycle are shown in figure 3. These dependences are rearranged according to the cycle fuel feed at different camshaft rotation speed of the fuel injection pump assembly. According to the presented dependences it follows that the fuel amount submitted through the initial pressure regulator depends on both rail position of the fuel injection pump assembly (diesel load) and camshaft rotating speed of this one.

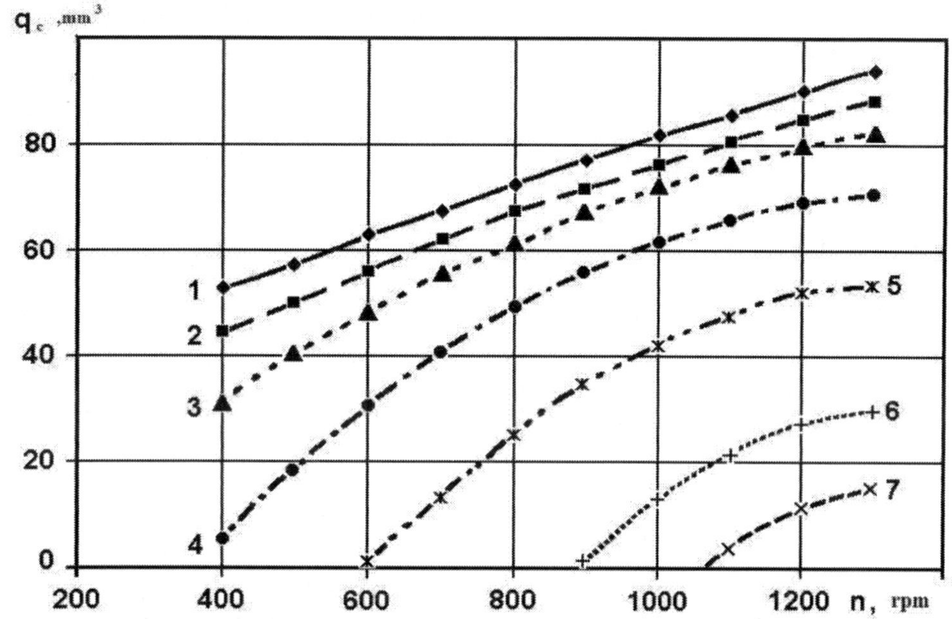

Figure 2. Fuel injection pump assembly speed characteristics with the high-pressure line replenishment through the initial pressure regulator at different positions of apportion controller hr (rail of fuel injection pump assembly) [mm]: 1 – hr=6; 2 – 5; 3 – 4; 4 – 3; 5 – 2; 6 – 1; 7 – 0.

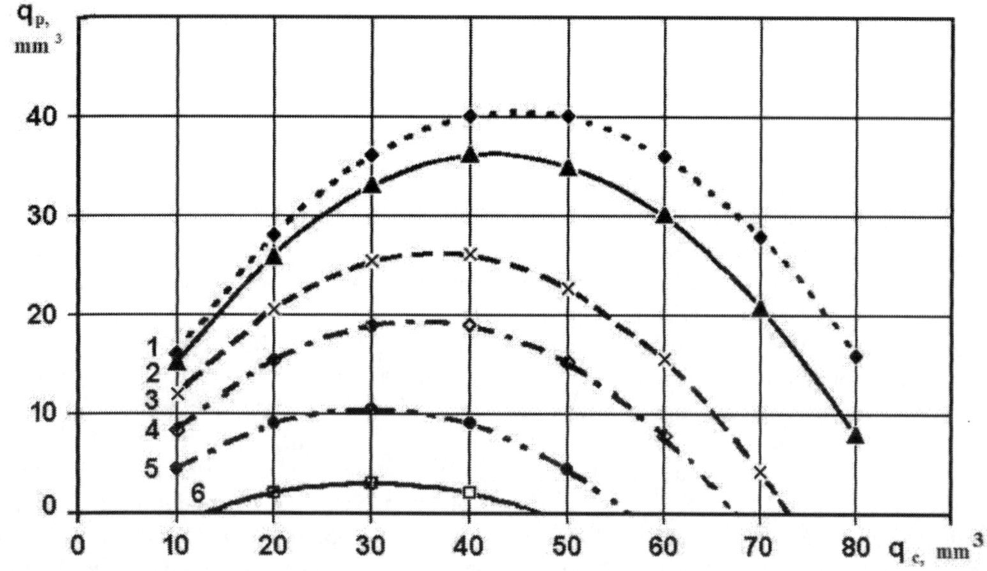

Figure 3. The replenishment volume change of the high-pressure line (qp) depending on the cycle fuel feed (qc) and camshaft rotating speed of the fuel injection pump assembly nTN [rev-1]: 1 – nTN=1300; 2 – 1200; 3 – 1000; 4 – 800; 5 – 600; 6 – 400.

Mixture fuel with a different concentration of rapeseed oil will be formed in the fuel pipe depending on an engine operation mode, during the rapeseed oil feed through the initial pressure regulator. The obtained dependences allow to measure rapeseed oil concentrations change at the mixture in the area of diesel operating modes. Such characteristic for diesel engine D-245-type is shown in figure 4. Assessment of obtained mixture fuel concentrations (that is submitted into the engine cylinder in different operation modes) can be realized with the characteristic. According to the characteristic, maximum concentration of rapeseed oil in the mixture is obtained in modes with a low loads and a high rotating speeds of diesel crankshaft. Rapeseed oil concentration in a high-loads zone reduces till 25% at nominal conditions and to the zero in a maximum torque zone. Rapeseed oil concentration does not exceed 15% on idle. The replenishment into the high-pressure line is absent in the launching mode and clean diesel fuel comes to the nozzle.

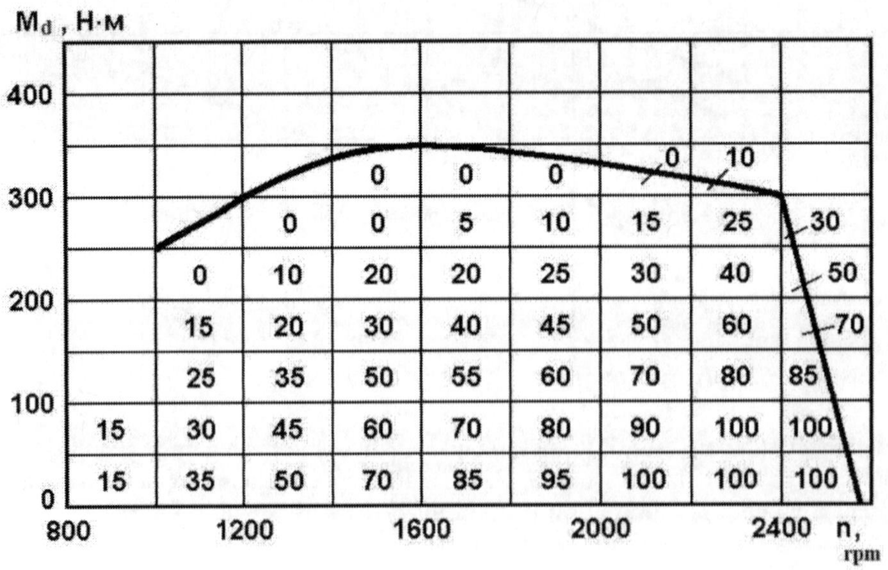

Figure 4. Rapeseed oil concentration change in mixture fuel for D-245-type diesel engine with the replenishment system of the high-pressure line through the initial pressure regulator valve.

4. Conclusion

The carried out experiments of mixture fuel preparation system in the high-pressure line through the replenishment valve showed an opportunity of its application at D-242 and D-245-type diesel engines. The system allows providing the rapeseed oil concentration change in mixture fuel from 0 to 100% depending on the operation mode. Oil concentrations in fuel below are provided during the operation mode due to the proposed system (shown in figure 4):

- rapeseed oil concentration in mixture changes from 0 to 25% in full-load modes. What is more, high values of the concentration are obtained in maximum rotating speeds of the engine crankshaft. Such value concentrations will not impair power and economical diesel data in maximum loads;
- rapeseed oil concentration in mixture on idle modes changes from 15% at low rotating speeds of engine crankshaft to 100% at maximum rotating speed. Such concentrations will not affect to diesel operation in crankshaft low rotating speed modes. But the rapeseed oil concentration increase till 100% in the mixture fuel should not significantly impair diesel operation. It is because a quality of mixture formation in these modes is joined with an intensive air charge movement and high values of air-fuel ratio;
- fuel with 15% rapeseed oil concentration is in the system after minimum rotating speed modes of the engine crankshaft on idle (before stopping). And engine stop in this mode will save the same oil concentration in a start mode (i.e. 15%). It will ensure good starting properties of the engine;

- an abrupt rapeseed oil concentration change was not observed in a mixture during transition from one operation mode to another. So a smooth diesel power change should be provided during its operation.

Thereby the fuel based on rapeseed oil can be used in diesels without special changes. This fuel is one of the most perspective through alternative fuels for diesel engines. The applying of high-pressure line replenishment system allows obtaining mixture fuel directly in the diesel engine fuel system through a special replenishment valve. The obtaining rapeseed oil concentrations in a mixture fuel provide saving of power and economical diesel indexes by using this system and its starting qualities saving after engine stop.

References

[1] Hemmerlein N, Korte V, Richter H, Schröder G 1991 *SAE Technical Paper Series* Performance, Exhaust Emissions and Durability of Modern Diesel Engines Running on Rapeseed Oil, **910848** pp 1-16, DOI: https://doi.org/10.4271/910848.

[2] Hashimoto M, Dan T, Asano I, Arakawa T 2002 *SAE Technical Paper Series* Combustion of the Rape-Seed Oil in a Diesel Engine, **2002-01-0867** pp 1-12, DOI: https://doi.org/10.4271/2002-01-0867.

[3] Spessert B M, Schleicher A 2007 *MTZ Worldwide* Influence of biofuels on exhaust gas and noise emissions of small industrial diesel engines, **68(3)** pp 17-20, DOI: https://doi.org/10.1007/BF03226815.

[4] Markov V A, Devyanin S N, Semenov V G, Shahov A V, Bagrov V V 2011 Using Vegetable Oils and Vegetable Oils Based Fuels in Diesel Engines (Moscow: LLC Scientific Research Centre Engineer, LLC Oniko-M Publ.) p 527, URL: http://www.rfbr.ru/rffi/ru/books/o_1783214.

[5] Aleksandrov A A, Arkharov I A, Bagrov V V, Gayvoronskiy A I, Grekhov L V, Devyanin S N, Ivaschenko N A, Markov V A 2012 Alternative Fuels for Internal Combustion Engines (Moscow: LLC Scientific Research Centre Engineer, LLC Oniko-M Publ.) p 790.

[6] Markov V A, Bashirov R M, Gabitov I I 2002 Toxicity of Diesel Exhaust Gases (Moscow: Publ. of Bauman Moscow State University) p 376.

[7] Grekhov L V, Ivaschenko N A, Markov V A 2005 Fuel Supply Systems and Diesel Engine Control Systems (Moscow: Legion-Avtodata Publ.) p 344.

[8] Gayvoronskiy A I, Markov V A, Ilatovskiy Yu V 2007 Use of Natural Gas and other Alternative Fuels in Diesel Engines (Moscow: LLC IRTS Gazprom Publ.) p 478.

[9] Patrakhal'tsev N N 2008 Increasing of Economic and Environmental Qualities of Internal Combustion Engines by Using Alternative Fuels, (Moscow: Publ. of Russian University of Peoples' Friendship) p 248.

Petroleum Diesel Fuel and Linseed Oil Mixtures as Engine Fuels

V A Markov[1], V G Kamaltdinov[2] and A A Savastenko[3]

[1]Bauman Moscow State Technical University, 5, 2-ya Baumanskaya str., Moscow, 105005, Russia
[2]South Ural State University, 76, Lenin ave., Chelyabinsk, 454080, Russia
[3]Russian University of Peoples' Friendship, 6, Miklukho-Maklaya str., Moscow, 117198, Russia

e-mail: vladimir.markov58@yandex.ru

Abstract. The actual problem is the use of alternative biofuels in automotive diesel engines. Insufficiently studied are the indicators of toxicity of exhaust gases of these engines operating on biofuel. The aim of the study is to identify indicators of the toxicity of exhaust gases when using of petroleum diesel fuel and linseed oil mixtures as a fuel for automotive diesel engines. Physical and chemical properties of linseed oil and its mixtures with petroleum diesel fuel are considered. Experimental researches of D-245.12C diesel are carried out on mixtures of diesel fuel and corn oil with a different composition. An opportunity of exhaust toxicity indexes improvement using these mixtures as a fuel for automobiles engine is shown.

1. Physical and chemical properties of linseed oil and its mixtures with petroleum diesel fuel

Modern development stage of engine construction is characterized by a wider alternative engine fuels using [1]. The choice of one or another alternative fuel is firstly determined with a significant source of raw materials for its production. Byproducts are often formed during an agricultural manufacture. They can be used for alternative engine fuels yield. For example these are vegetable oils (sunflower, rapeseed, soybean) that are remained after protein food obtaining for cattle – oilcake, meal and etc. Another example of complex agricultural products using is flax growing. Its fiber goes to different cloth manufacture and linseed oil is produced from seeds. This oil is used as edible oil. Linseed oil is one of the best drying oils and has a big technical appliance at the same time. Attractiveness of raw materials source above for engine oils production is determined with com-parably low price of vegetable oils and this charge source renewability [1, 2].

The opportunity of vegetable oils using as motor oils determines with molecules composition and structure. These oils are mainly consist of triacyglycerides (95-97%) – organic compounds, esters of glycerin and different fatty acids, and also mono- and diacyglycerides. Acyglycerides in its turn contain molecules of various fatty acids which are joined with $C_3H_5(OH)_3$ glycerin molecule [1, 2]. Vegetable oils contain mainly fatty acids with even number of carbon atoms (for example C_{14}, C_{16}, C_{18} b and etc.). Vegetable oils consist both of unsaturated fatty acids (oleic-, linoleic-, linolenic- and etc.) and saturated fatty acids (tetradecanoic, palmitic-, stearic- and etc.). Fatty-acid compound of linseed oil differs from similar composition sunflower oil which is the most common in Russia. So if sunflower oil is rich in linoleic acid, linseed oil contains linolenic acid having three unsaturated bonds. Thereby linseed oil is less stable in oxidizing processes compared with sunflower oil.

The carried out analysis of linseed oil physical and chemical properties confirmed an opportunity of its using as motor oil for diesels. However such appliance is not always possible because of linseed oil properties distinctions from similar petroleum diesel fuel characteristics. Thereby linseed oil is advantageous to use as engine oil only after its being refining or in mixture with diesel- or alternative fuels. The analysis of specified linseed oil appliance opportunities shows that its using in mixtures with petroleum diesel fuel is the most advisable. Un-refined linseed oil made by OOO "Aromavita" (Kurilovo village, Podolsky district, Moscow region) is researched for opportunity assessment of such linseed oil appliance. And its mixture with petroleum diesel fuel (L-mark under state standard № 305-82) of 5 and 9% (by volume) linseed oil is also explored (table 1).

Linseed oil differ noticeably with high density and viscosity compared with petroleum diesel fuel. Therefore diesel operation is advisable on diesel fuel and linseed oil mixtures. These components are well mixed between themselves forming stable mixtures. And besides, acceptable physical properties of mixture can be achieved by the way of mixture fuel composition selection. The considered mixture biofuels, having 5 and 9% of linseed oil, have heightened density and viscosity. But these properties are closer to the similar diesel fuel characteristics (mixture biofuel densities, having 9% of linseed oil, and diesel fuel ones are equal 837 and 830 kg/m^3 respectively, and their viscosities – 6.0 and 3.8 mm^2/s, table 1). Such physical properties differences of linseed oil (and also its mixtures with diesel fuel) from petroleum diesel fuel properties affect process parameters of diesel fuel supply, fuel spraying and mixing characteristics. Specified mixture biofuels differ from petroleum diesel fuel by a row of other physical and chemical properties: auto-ignition temperature, cetane number, calorific efficiency (low heat value) and etc. All of the factors affect indexes of fuel economy and diesel exhaust toxicity operating on indicated fuels above.

Table 1. Physical and chemical properties of researched fuels.

Physical and chemical properties	Fuel			
	Diesel fuel	Linseed oil	95% diesel fuel mixture and 5% linseed oil	91% diesel fuel mixture and 9% linseed oil
Density at 20 °C, kg/м^3	830	912	834	837
Kinematic viscosity, mm^2/s				
at 20 °C	3.8	59.6	4.5	6.0
at 40 °C	2.3	23.9	-	-
Surface tension coefficient s at 20 °C, mH/m	27.1	30.0	-	-
Low heat value, кJ/kg	42500	37600	42200	42000
Cetane number	45	38	-	-
Auto-ignition temperature, °C	250	300	-	-
Cloud point temperature, °C	-25	-12	-	-
Chilling temperature, °C	-35	-20	-	-
Air amount which is necessary for 1 kg substance combustion, kg	14.3	12.62	14.23	14.16
%-content by mass				
C	87.0	77.8	86.54	86.17
H	12.6	12.0	12.57	12.55
O	0.4	10.2	0.89	1.28
Common sulfur content (%) by mass	0.20	0.002	0.190	0.182

Note: "-" – properties were not determined; there is specified volumetric percentage content of components for mixtures

The row of published works is known. These works are dedicated to linseed oil using as motor fuels for diesel engines both in pure form [3, 4, 5] and after methyl ether output of linseed oil from it [6, 7, 8, 9]. Moreover diesel operation on diesel fuel mixtures and linseed oil is advisable in connection with indicated distinctions of linseed oil physical and chemical properties from similar characteristics of petroleum diesel fuel. But the problem of definition of indicators of toxicity of exhaust gases when using of petroleum diesel fuel and linseed oil mixture as a fuel for automotive diesel engines is poorly understood.

2. Researches of the diesel operating on mixtures of petroleum diesel fuel and linseed oil

Experimental researches of D-245.12C-type diesel (4CN 11/12.5) were carried out. It was made for the opportunity confirmation of petroleum diesel fuel and linseed oil using as a motor fuel. This diesel was made by Minsk Motor Factory and it was mounted on light commercial vehicles "Bychok", on buses of Pavlovsky Bus Factory also and on "Belarus" tractors.

The diesel is researched at the motor stand on full-load curve mode and 13-mode test cycle ECE R49. Diesel settings by adjusting fuel injection advance angle (13° of crankshaft rotation to the top dead center) and by dosaging rail stop position (maximum fuel feed stop) were stayed constant at all considered fuel types using. Experiences of D-245.12C-type diesel have been carried out at the first researches stage. It has been made on clean diesel fuel and on mixture of 91%-diesel fuel and 9%-linseed oil in full-load curve modes. The results are shown in figure 1 and table 2.

It should be noted according to these data that availability of oxygen atoms in linseed oil molecules led to perceptible reduction of diesel exhaust smokiness operating on mixture biofuel. So transition from diesel fuel to mixture of 91%-diesel fuel and 9%-linseed oil accompanied exhaust smokiness reduction (K_X) from 16 to 11% on the Hartrige scale in maximum capacity mode at n=2400rpm. And opacity was decreased from 43 to 36% on the Hartrige scale in maximum torque mode at n=1500rpm.

The experimental researches results of D-245.12.C in 13-mode testing cycle regimes of ECE R49 are shown in figure 2. According to the data it should be noticed that there is a trend of content reduction in exhaust of nitric oxides C_{NOx} (figure 2, b), carbon monoxide C_{CO} (figure 2, c) and unburned hydrocarbons C_{CHx} (figure 2, d).

The indexes of fuel economy and integrate specific mass exhausts of toxic components in 13-mode cycle (e_{NOx}, e_{CO} and e_{CHx} respectively) are calculated. It is realized in agreement with given characteristics (figure 2) of fuel consumption per hour G_T and content of normed toxicity components in exhaust (nitric oxides C_{NOx}, carbon monoxide C_{CO}, unburned hydrocarbons C_{CHx}) with appliance of practical standards. Toxic components concentrations in exhaust (C_{NOx}, C_{CO}, C_{CHx}) are determined and their hourly mass emissions (E_{NOx}, E_{CO}, E_{CHx}) are calculated at the integral toxicity estimation of engine exhaust in such cycle regimes in every mode. The obtained values of emissions are summarized for all cycle by every component with an allowance for weight coefficients K_i reflecting time fraction of every regime. And after that specific mass emissions of harmful substances are determined according to the formulas with division by conditional average capacity of the diesel for experimental cycle $\sum(N_{ei} \cdot K_i)$ [2]:

$$e_{NOx} = \frac{\sum_{i=1}^{13} E_{NOx\,i} \cdot K_i}{\sum_{i=1}^{13} N_{e\,i} \cdot K_i}, \quad e_{CO} = \frac{\sum_{i=1}^{13} E_{CO\,i} \cdot K_i}{\sum_{i=1}^{13} N_{e\,i} \cdot K_i}, \quad e_{CHx} = \frac{\sum_{i=1}^{13} E_{CHx\,i} \cdot K_i}{\sum_{i=1}^{13} N_{e\,i} \cdot K_i}. \qquad (1)$$

Assessment of operational fuel consumption in 13-mode cycle regimes is appraised by average (conditional) specific effective fuel consumption and conditional effective performance. They were determined using dependencies [2]

$$g_{e\,cond.} = \frac{\sum_{i=1}^{13} G_{Ti} \cdot K_i}{\sum_{i=1}^{13} N_{ei} \cdot K_i}, \quad \eta_{e\,cond.} = \frac{3600}{H_U \cdot g_{e\,cond.}}, \tag{2}$$

where G_{Ti} is fuel consumption per hour in i-mode; H_U is inferior calorific value, MJ/kg. The calculations results of showed parameters are represented in table 2.

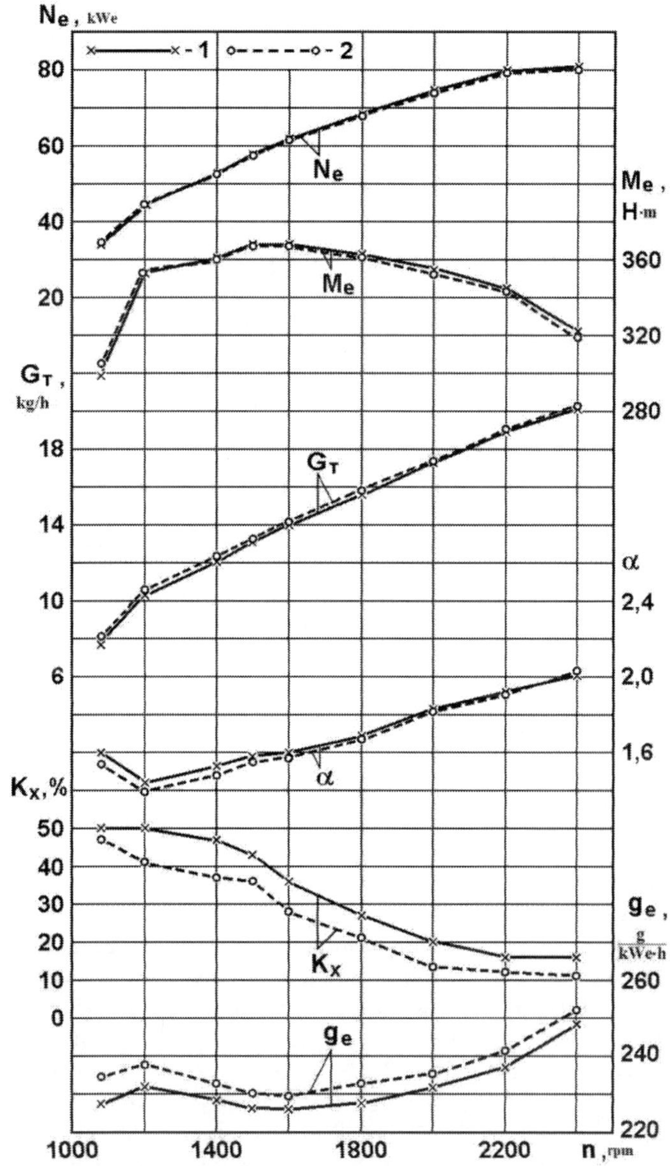

Figure 1. Dependency of effective power (N_e), torque (M_e), fuel consumption per hour (G_T), air-fuel ratio (α), exhaust smokiness (K_X) and specific effective fuel consumption (g_e) of D-245.12C-type diesel from crankshaft rotating speed (n). It happens in full-load curve mode using different fuels: 1-diesel fuel; 2-mixture of 91%-diesel fuel and 9%-linseed oil.

Table 2. D-245.12C-type diesel indexes operating in different fuels.

Diesel indexes	Sort of fuel		
	Diesel fuel	95% diesel fuel mixture and 5% linseed oil	91% diesel fuel mixture and 9% linseed oil
Specific effective fuel consumption g_e, g/(kWe·h):			
- in a maximum output mode	248.4	250.8	252.1
- in a maximum torque mode	226.2	228.9	230.1
Effective diesel performance η_e:			
- in a maximum output mode	0.341	0.340	0.340
- in a maximum torque mode	0.374	0.373	0.373
Exhaust smokiness K_X, % on the Hartrige scale:			
- in a maximum output mode	16.0	12.0	11.0
- in a maximum torque mode	43.0	37.5	36.0
Integrates in 13-mode cycle regimes:			
- effective fuel consumption $g_{e\ cond}$, g/(kWe·h)	247.97	248.72	252.26
- effective performance $\eta_{e\ cond}$	0.341	0.343	0.340
Integrates in 13-mode cycle regimes, effective mass exhaust, g/(kWe·h):			
- nitric oxides e_{NOx}	7.018	6.230	6.441
- carbon monoxide e_{CO}	1.723	1.631	1.511
- unburned hydrocarbons e_{CHx}	0.788	0.695	0.664

The presented data in table 2 confirm an improvement opportunity of D-245.12C diesel ecological indexes which is transferred from diesel fuel to mixture of 91%-diesel fuel and 9%-linseed oil. So exhaust smokiness reduced at 16-31% compared with petroleum diesel fuel appliance. It occurred during submission into diesel combustion chamber of experimental mixture biofuel in maximum capacity and maximum torque modes. Specific mass emission of unburned hydrocarbons e_{CHx} reduced from 0.788 to 0.664 g/(kWe·h), i.e. at 15.7%, in 13-mode cycle regimes. Specific mass emission of carbon monoxide e_{CO} decreased from 1.723 to 1.511 g/(kWe·h), i.e. at 12.3%. Wherein specific mass emission of nitric oxides e_{NOx} reduced from 7.018 to 6.441 g/(kWe·h), i.e. at 8.2%, and conditional diesel performance $\eta_{e\ cond}$ remained practically constant (it decreased from 0.341 to 0.340 that was within accuracy limits of determination of this index).

The experimental results of D-245.12.C-type diesel are presented in figure 1 and 2. They are obtained during this diesel operation on mixture biofuel of 91%-diesel fuel and 9%-linseed oil. But the question about mixture biofuel composition influence on diesel characteristics is the cause of particular interest. Thereby the analysis have been carried out of D-245.12C-type diesel indexes operating on clean petroleum diesel, mixture of 95%-diesel fuel and 5%-linseed oil, mixture of 91%-diesel fuel and 9%-linseed oil.

The carried out experimental researches results of D-245.12C-type diesel on indicated mixture fuels are presented in table 2 and figure 3, a. Linseed oil concentration increasing in mixture biofuel C_{LO} from 0 to 5 and 9% leads to specific effective fuel consumption growth (g_e increases from 248.4 to 250.8 and 252.1 g/(kWe·h)) and exhaust smokiness reduction K_X from 16.0 to 12.0 and 11.0% in a maximum capacity mode at n=2400rpm. Similar increasing of C_{LO} was accompanied with growth of g_e from 226.2 to 228.9 and 230.1 g/(kWe·h) and reduction of K_X from 43.0 to 37.5 and 36.0% in a maximum torque mode at n=1500rpm. The presented data proof that considered mixture biofuel appliance allowed reducing exhaust smokiness noticeably. Indicated increase of specific mixture biofuel consumption is explained with their lower inferior calorific value H_U (table 2). Wherein the efficiency of combustion process was changing relatively weakly, during the linseed oil content increasing in mixture biofuel. So conditional effective diesel performance $\eta_{e\,\text{cond}}$ was 0.341, 0.340 and 0.340 in a maximum capacity mode at n=2400 rpm, having such concentration of linseed oil in mixture biofuel: C_{LO}=0; 5 and 9%. And $\eta_{e\,\text{cond}}$ was equal to 0.374, 0.373 and 0.373 in a maximum torque mode at n=1500rpm (table 2).

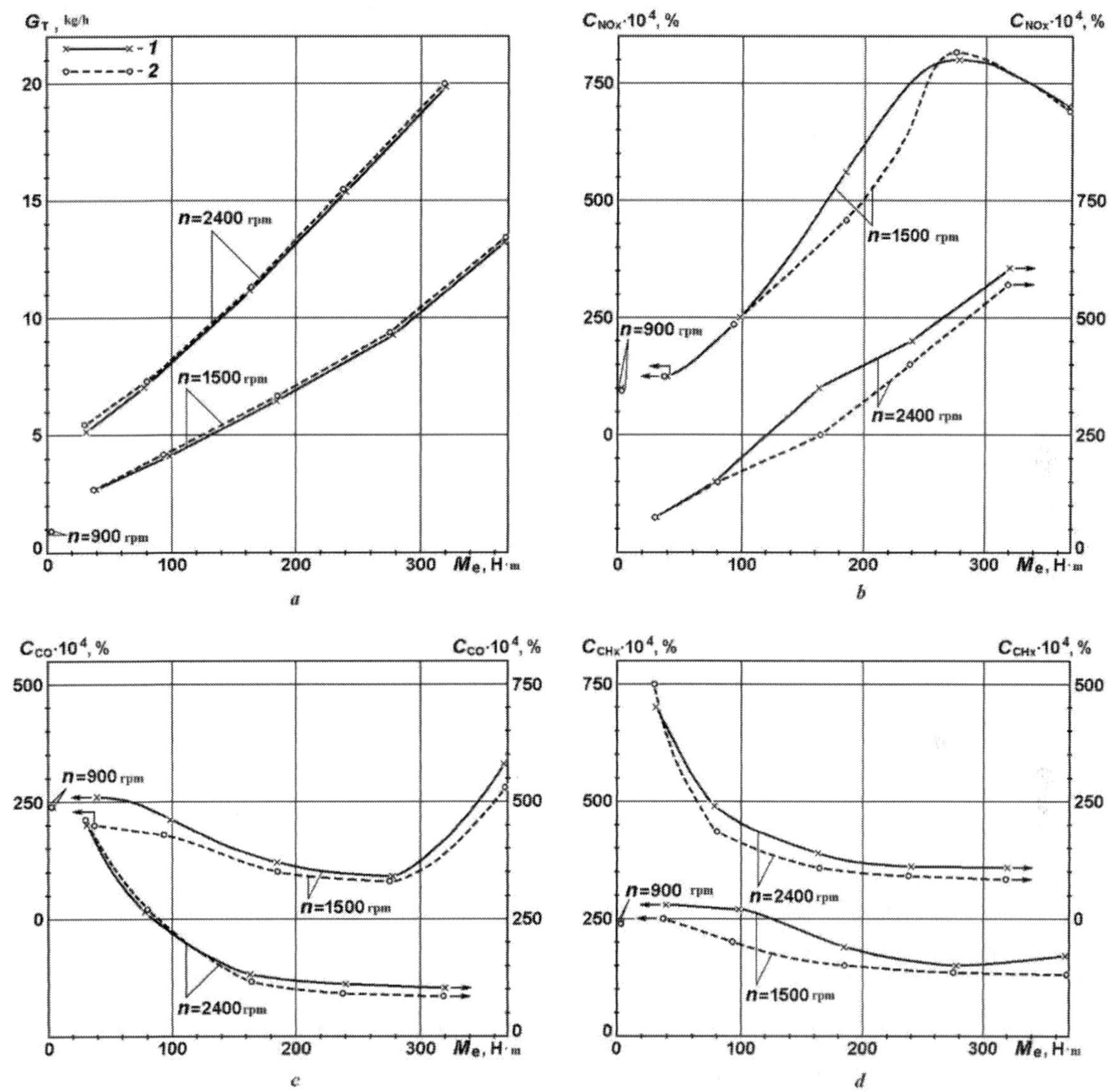

Figure 2. Dependence of fuel consumption per hour G_T (a), volumetric concentration of nitric oxides C_{NOx} (b), carbon monoxide C_{CO} (c) and unburned hydrocarbons C_{CHx} (d) in exhaust gases of D-245.12C-type diesel from rotating speed n and torque M_e using different fuels: 1 – diesel fuel; 2 – mixture of 91%- diesel fuel and 9%-linseed oil.

Integral values in 13-mode experimental cycle regimes of specific mass emissions of toxic components are illustrated in table 2 and figure 3, b. They confirmed an opportunity of perceptible ecological indexes improvement of researched diesel using mixture biofuels. Application of them allowed reducing emissions of all toxic diesel exhaust components: nitric oxides NO_x, carbon monoxide CO, unburned hydrocarbons CH_x. Emissions of nitric oxides e_{NOx} were respectively 7.018,

6.230 and 6.441 g/(kWe·h); carbon monoxide emissions e_{CO} – 1.723, 1.631 and 1.511 g/(kWe·h); unburned hydrocarbon emissions e_{CHx} – 0.788, 0.695 and 0.664 g/(kWe·h) at linseed oil concentration in mixture biofuel C_{LM}=0, 5 and 9%. Figure 3 data prove optimization conducting advisability of explored mixture biofuels composition with the aim of toxic components emissions minimization of diesel exhaust.

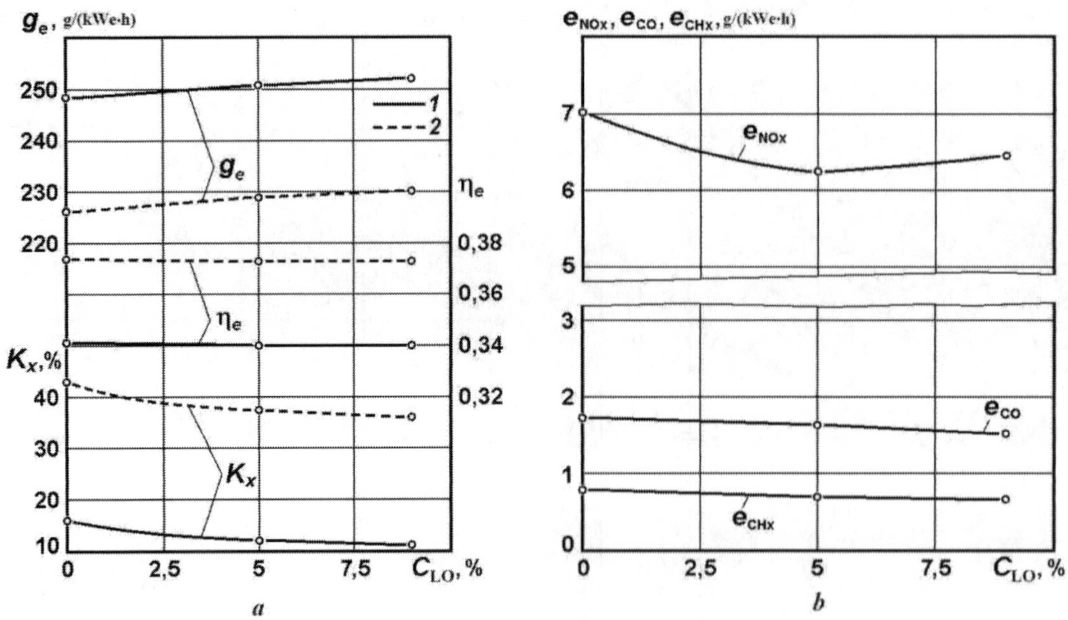

Figure 3. Dependency of specific effective fuel consumption g_e, effective performance η_e, exhaust smokiness K_X in full-load curve modes (*a*), specific mass emissions of nitric oxides e_{NOx}, carbon monoxide e_{CO} and hydrocarbons e_{CHx} in 13-mode regimes of diesel experimental cycle (*b*). The relation of D-245.12C-type diesel is shown from lin-seed oil content C_{LO} in mixture biofuel: 1 - in maximum capacity mode at n=2400rpm; 2 - in maximum torque mode at n=1500rpm.

3. Conclusion

- Physical and chemical properties of diesel fuel and linseed oil mixtures with low content of the last (10%) are close to petroleum diesel fuels characteristics.
- Carried out experimental researches of D-245.12C diesel on mixtures of petroleum diesel fuel with linseed oil confirmed an opportunity of considerable ecological indexes improvement of researched diesel.
- Appliance of these mixtures as motor fuel allowed reducing all normed toxic components emissions of researched diesel exhaust – nitric oxides NO_x, carbon monoxide CO, unburned hydrocarbons CH_x.
- During the transfer of researched diesel from diesel fuel to mixture of 91%-diesel fuel and 9%-linseed oil nitric oxides emissions e_{NOx} reduced from 7.018 to 6.441 g/(kWe·h), carbon monoxide emissions e_{CO} reduced from 1.723 to 1.511 g/(kWe·h), unburned hydrocarbons emissions e_{CHx} reduced from 0.788 to 0.664 g/(kWe·h).
- Increase of linseed oil concentration in mixture biofuel C_{LM} from 0 to 9% in maximum capacity mode at n=2400rpm leads to exhaust smokiness reduction K_X from 16.0 to 11.0% on the Hartrige scale, and from 43.0 to 36.0% in maximum torque mode at n=1500rpm on the Hartrige scale.
- Content optimization conducting of researched mixture biofuels is advisable with a view to emissions minimization of exhaust toxic components.

References

[1] Aleksandrov A A, Arkharov I A, Bagrov V V, Gayvoronskiy A I, Grekhov L V, Devyanin S N, Ivaschenko N A, Markov V A 2012 *Alternative Fuels for Internal Combustion Engines* (Moscow: LLC Scientific Research Centre Engineer, LLC Oniko-M Publ.) p 790.

[2] Markov V A, Devyanin S N, Semenov V G, Shahov A V, Bagrov V V 2011 *Using Vegetable Oils and Vegetable Oils Based Fuels in Diesel Engines* (Moscow: LLC Scientific Research Centre Engineer, LLC Oniko-M Publ.) p 527, URL: http://www.rfbr.ru/rffi/ru/books/o_1783214.

[3] Goering C E, Daugherty M J 1982 *Transactions of the ASAE* Energy Accounting for Eleven Vegetable Oil Fuels **25(5)** p 1209-1215. DOI: https://doi.org/10.13031/2013.33698.

[4] Goering C E, Schwab A W, Daugherty M J, Pryde E H, Heakin A J 1982 *Transactions of the ASAE* Fuel Properties of Eleven Vegetable Oils **25(6)** p 1472-1477, DOI: https://doi.org/10.13031/2013.33748.

[5] Knorr W, Daute P, Grutzmacher R, Hofer R 1995 *Fat Science Technology* Development of New Fields of Application for Linseed Oil **97(5)** p 165-169, DOI: https://doi.org/10.1002/lipi.19950970502.

[6] Agarwal A K, Das L M 2001 *Transactions of the ASME. Journal of Engineering for Gas Turbines and Power* Biodiesel Development and Characterization for Use as a Fuel in Compression Igni-tion Engines **123(2)** p 440-447, DOI: https://doi.org/10.1115/1.1364522.

[7] Agarwal A K, Bijwe J, Das L M 2003 *Transactions of the ASME. Journal of Engineering for Gas Turbines and Power* Effect of Biodiesel Utilization of Wear of Vital Parts in Compression Igni-tion Engine **125(2)** p 604-611, DOI: https://doi.org/10.1115/1.1454114.

[8] Agarwal A K, Bijwe J, Das L M 2003 *Transactions of the ASME. Journal of Engineering for Gas Turbines and Power* Wear Assessment in a Biodiesel Fueled Compression Ignition Engine **125(3)** p 820-826, DOI: https://doi.org/10.1115/1.1501079.

[9] Sebedio J L, Prevost J, Grandgirard A 1987 *Journal of the American Oil Chemists' Society* Heat Treatment of Vegetable Oils. I. Isolation of the Cyclic Fatty Acid Monomers From Heated Sunflower and Linseed Oils **64(7)** p 1026-1032, DOI: https://doi.org/10.1007/BF02542443.

PTFE-nanocomposites structure and wear-resistance changing in various methods of structural modification

Yu K Mashkov[1], A S Ruban[2], E A Rogachev[1] and O V Chemisenko[1]

[1]Omsk State Technical University, 11, Mira ave., Omsk, 644050, Russia
[2]Siberian State Automobile and Highway University,5, Mira ave., Omsk, 644080, Russia

e-mail: rubananna@mail.ru

Abstract. Conditions of polymer materials usage containing nanoelements as modifiers significantly affect the requirements for their physic-mechanical and tribological properties. However, the mechanisms of nanoparticles effect to the polymers tribotechnical properties have not been studied enough. The article aim is to analyze the results of studying polytetrafluoroethylene modified with cryptocrystalline graphite and silicon dioxide and to determine the effectiveness of the modification methods used and methods for further improving filled PTFE mechanical and tribotechnical properties. The effect of modifiers to PCM supramolecular structure was analyzed with SEM methods. The results of modifying the PCM samples surface by depositing a copper film with ion-vacuum deposition methods and changing the structural-phase composition and tribological characteristics are considered. The findings make possible to characterize the physicochemical processes under frictional interaction in metal polymer tribosystems.

1. Introduction

Theoretical and experimental results of the researches connected to the effect of metal-polymer tribosystems modifying polymer materials to the tribotechnical properties of polymer composites have shown that the problem of predicting tribotechnical properties of multicomponent polymer composite materials is still actual [1,2]. Modifications are subject to both as separate components of the mixture as finished composite materials by changing the surface layer structure and properties [3-8]. The comparative analysis of seven PTFE-based composites presented in [4] shows that it is possible to improve PTFE properties, even if the most attractive characteristic of low friction is lost due to the presence of solid micro- and nanoparticles in the polymer matrix. The introduction of both as soft as solid phases into the polymer matrix improves its self-lubricating and supporting properties increasing PTFE tribological properties. In particular, the PTFE nanocomposite film wear resistance was increased with nanodiamonds addition [2]. All this shows that even small additions of nanofillers can change polymers mechanical and tribotechnical properties radically. The same time mechanism peculiarities of nanoparticles effect to polymers and polymer composites frictional properties had been studied insufficiently. Therefore, the task of selecting fillers type and number is being solved mostly experimentally now [2,4,6].

To optimize the structure and characteristics of material operational properties the modes of PCM sample sintering (heat treatment) technological process are changed. PCM properties are established to be significantly improved, for example, by imposing compression pressure to the sample during sintering process. One method of the surface modification is a method of thin films deposition. Work [8] studied, for example, the molecular organization of polytetrafluoroethylene (PTFE) thin coatings with included Ag, Cu and Mo nanoparticles, they are deposited from an active gas component. Due to its unique physical and mechanical properties PTFE remains one of the leading non-metallic materials used in mechanical engineering and instrument making. As a result of PTFE modification with different types of fillers it is possible to change its properties in a wide range. Thus, the analysis of metal-polymer tribosystems materials properties shows the relevance of studies according to the proved choice of modifiers and a PTFE modifying method depending on the planned operating conditions for the metal-polymer tribosystem.

2. Statement of the problem

The basic method of PTFE structural modification and PCM synthesis is introduction fillers - modifiers of various kinds into a polymer matrix. The choice of the investigated samples component composition in this work was made on the basis of the tests presented in [6].

To conduct research the samples containing polytetrafluoroethylene (PTFE)-89 % wt, cryptocrystalline graphite (SCG)-8 % wt, dioxide silicon (white carbon black WCB-120)-3 % wt were made. The samples were manufactured in two different technologies. The first technology is a free sintering, the second technology is sintering of clips with limitation of thermal expansion in the heat treatment process.

Manufactured with various methods the samples were tested for friction and wear, when contact pressure and sliding velocity were variation factors. Samples that passed the test on the friction machine are further modified by applying a copper film, by an ion-vacuum deposition method.

The following tasks were set in this work:

1. Using methods of electron microscopy to determine the regularities of modifier or filler influence for different types and concentration and interaction modes on the structural-phase transformations in the sample.

2. To investigate the changes of PCM structural-phase state and tribological properties parameters in the case of modifying the samples surface using ion-vacuum deposition

3. Experimental results

Wear rate dependences of the nanocomposite studied (PTFE-89% wt, SCG-8% wt, WCB-3% wt) on load parameters were obtained on a universal friction machine "UMT-2168" according to the "finger-disk" friction scheme. Contact pressure and sliding velocity were adopted as variable loading parameters. Wear rate dependences graphs of the samples on the contact pressure and sliding velocity are shown in Figure 1 and Figure 2. The contact pressure varied from 1.0 MPa to 3.0 MPa with a step 0.5 MPa, the sliding velocity is 1.2 m/sec to 2.4 m/sec with an interval 0.3 m/sec.

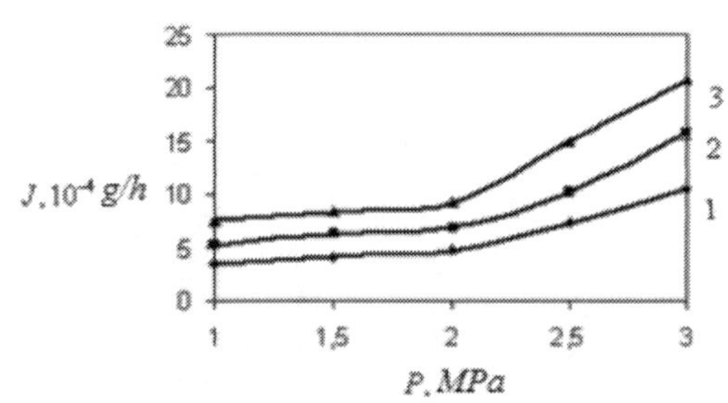

Figure 1. The sample wear rate dependence on contact pressure at sliding velocity: 1 - 1.2 m/sec, 2-1.8 m/sec, 3-2.4 m/sec.

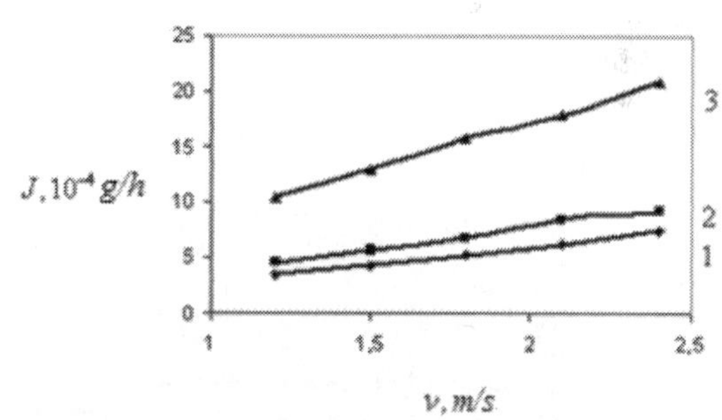

Figure 2. The sample wear rate dependence on the sliding velocity at contact pressure values: 1 - 1MPa, 2-2MPa, 3-3MPa.

To study physicochemical processes developing in the PCM surface layer under conditions of frictional interaction and their effect on the PCM wear resistance, the friction surface elemental composition was studied before and after friction interaction with an energy-dispersive microanalysis using JEOL JSM-5700 scanning electron microscope attachment. The results of the study are given in Table 1.

Table 1. PCM containing components, % wt

PTFE-89WT%, SCG-8WT%, WCB-3WT%	F	C	O	Si	Ni	Al
before interaction	60,60	24,35	10,02	3,84	-	1,19
after interaction	20,29	51,33	15,44	-	7,42	-

A friction surface topography micrograph and simulation of the surface profile for the analysis of PCM height field samples (Figure 3) were obtained using atomic force microscopy (NTEGRA microscope).

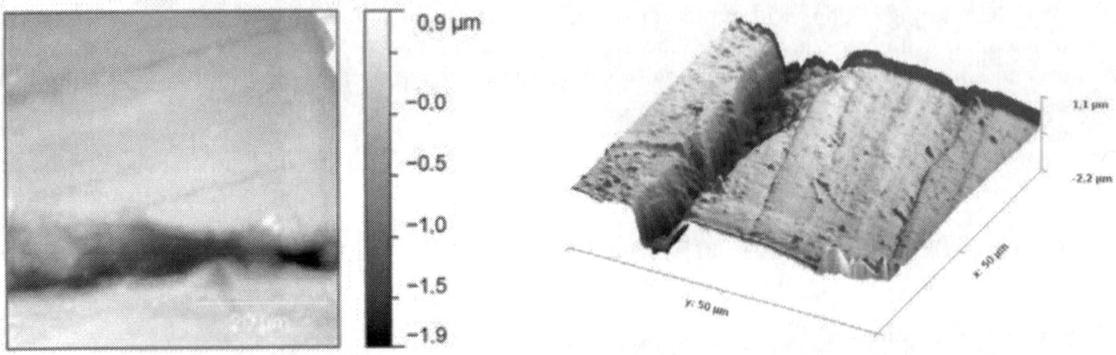

Figure 3. Micrograph of the friction surface topography and simulation of the surface profile containing: PTFE-89wt%, SCG-8wt%, WCB-3wt%.

To modify the PCM surface and to continue the obtained samples structural-phase studies the polymer samples manufactured with technology of sintering in the terminals and containing PTFE-89% wt, SCG-8% wt, WCB-3% wt were selected. A part of the samples was previously subjected to frictional interaction of the friction machine "UMT -2168" according to the friction "finger-disk" scheme with contact pressure values 2.66 MPa and the sliding velocity 1.2 m/sec., during 90 minutes.

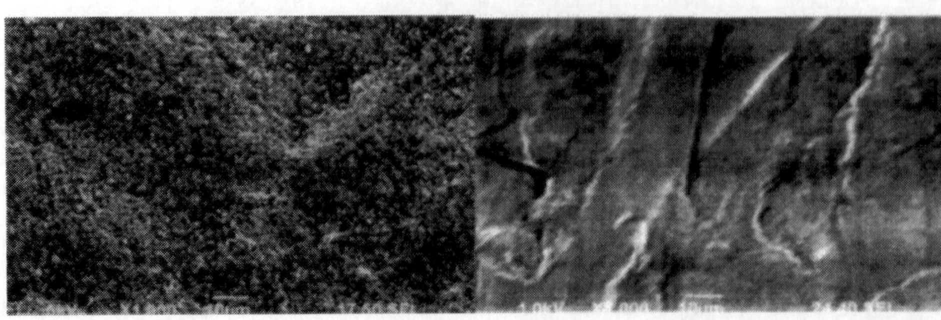

a) b)

Figure 4. Microphotographs of the PCM surface: a) - the surface of the cold chip; b)- surface modified with copper ion-vacuum deposition.

Deposition of copper films with thickness 1 µm was produced with ADVAVAC unit on the PCM samples subjected to frictional interaction. The substrate heating was carried out at temperature 50°C. The structure of the modified ion-vacuum coating on the PCM surface and cold chip inner surface, i.e. in the sample volume (Figure 4) was investigated with electron microscopy

To study the effect of the compressive statistical load on changes in the surface layer of a polymer sample and strength of a sputtered copper film, a different force effect was exerted on a part of the samples during 10 sec in the form of a load 98N and 294N, it was developed with a tip in the form of a pyramid used in the Vickers method. When 98N load was applied to the surface of the sample, a pyramid print was characteristic for this method, while under 294 N load, a gap in the sputtered copper film was observed. To analyze the structural-phase changes and the elemental composition with this kind of external loading, 10 sections were analyzed on three samples subjected to various force levels. The average values of elements mass fractions determined in the course of JEOL JSM-5700 scanning electron microscope are shown in Table 2.

Table 2. Content of elements in PCM with copper film surface, % wt

Forse P , H	F	C	O	Si	Fe	Cu
98	57,25	32,89	-	1,66	-	7,20
294	43,75	23,23	15,61	1,61	9,27	6,70

Tribotechnical properties research of a nanocomposite modified with copper ions was carried out on a universal friction machine "UMT -2168". The obtained results in comparison with unmodified samples are given in Table 3.

Table 3. Characteristics of nanocomposites samples tribotechnical properties

Sample composition, %	Coefficient of friction, μ	Wear rate, $J, 10^{-4} g/h$
PTFE-89% wt, SCG-8% wt, WCB-3% wt	0,092	3,80
PTFE-89% wt, SCG-8% wt, WCB-3% wt with copper ion-vacuum deposition.	0,12	1,46

4. Discussion of the results

The results showed that when PTFE is introduced with complex and polydisperse nanosized modifiers including alongside JCG nano powder SiO_2 brand WCB-120, there is a significant change in the morphology of the polymer matrix similar to the one shown in [2,8]. In the process of friction interaction, there is a significant change in the concentration of elements decreases the fluorine content in 3 times, the carbon content is increased more than two times and oxygen is increased in half.

Analysis of the elemental composition and friction surfaces profile allows concluding that chemical and diffusion processes caused with temperature gradient in the friction zone due to frictional interaction develop on the friction surface and in thin surface layers. Under the temperature gradient effect, structural-phase changes are in the transfer film formed during the frictional interaction and in the polymer sample surface layer (i.e., where the temperature gradient is maximum) affecting the surface topology.

It follows from Figure 1 that with an increase in the contact pressure from 1MPa to 2MPa, an increase in the wear rate at all sliding velocities on average by 30% is observed, with a further increase in the contact pressure to 3MPa, a considerable increase in the wear rate is observed, 2-2.2 times compared with the wear rate at the initial value of the contact pressure.

The maximum increase in the wear rate (see Figure 2) is typical for the contact pressure 3MPa, the change in the sliding velocity is characterized with a uniform change in the wear rate for all the considered contact pressure values.

The increase in the load and, as a consequence, substantial deformation of the PCM surface layer and the copper coating applied to the samples, makes possible to determine the presence of iron and oxygen in the contact zone of the PCM sample surface and the copper film, characteristic of the transfer film appeared during the frictional interaction and thus forms an intermediate layer between the test sample and the sprayed coating. With an increase in the load acting on the samples under study, a substantial change in the elemental composition is observed: a decrease in the fluorine content by 20% and carbon by 30%. At the same time, the content of silicon and copper will change insignificantly in the samples.

In this case the cold chip surface is characteristic for PTFE modified with dispersed nano-modifier WCB-120 (Figure 4). It is a loosely packed lamellar structure of a polymer containing micropores. The PCM surface modified with ion-vacuum deposition has a denser structure, and it indicates that copper ions have filled the defective parts of the surface. The difference in structure is reflected in the PCM tribotechnical characteristics values. A small increase in the friction coefficient and a significant decrease in the wear rate (Table 3) are established in comparison with a similar sample without ion-vacuum deposition. Change in the friction coefficient is within the experimental error, the wear rate decreases in 2.6 times.

5. Conclusions

The PCM synthesis theory, as it considers variety and complexity of the processes occurring with materials both during the synthesis of a polymer composite and a process of frictional interaction, is not fully developed now. The fillers choice and modification of polymers with various methods makes possible to change radically properties of synthesized materials and expand significantly their application field.

As a result of the conducted studies, the following is established:

1. Analysis of changes in the chemical composition and PCM samples surface micro profile indicates the development of complex physic-chemical processes in the PCM frictional interaction conditions with a metal counter body.
2. PCM wear rate dependence on the contact pressure is non-linear, regardless of the sliding velocity and fillers concentration.
3. PCM surface modification with applying of a thin copper film by an ion-vacuum deposition method is accompanied with changes in the structural-phase state of the surface layer, it provides an increase in PTFE nanocomposites wear resistance by 2.5 times

References

[1] Fuzhi Song, Qihua Wang and Tingmei Wang 2016 The effects of crystallinity on the mechanical properties and the limiting PV (pressure×velocity) value of PTFE *Tribology International* **93** A pp 1-10

[2] Jeong-Yeop Lee, Dong-Phill Lim and Dae-Soon Lim 2007 Tribological behavior of PTFE nanocomposite films reinforced with carbon nanoparticles *Composites* B Engineering **38** pp 810-16

[3] Yijun Shi Xin, Feng Huaiyuan Wang and Xiaohua Lu 2008 The effect of surface modification on the friction and wear behavior of carbon nanofiber-filled PTFE composites *Wear* **264** pp 934-39

[4] Conte M and Igartua A.2012 Study of PTFE composites tribological behavior *Wear* **296** pp 568-74

[5] Nguyen X T, Panin S V, Kornienko L A, Ivanova L P, Korchagin M A and Shilko S V 2014 Mechanical and tribotechnical properties of polymer-polymeric composites based on "UHMWPE-PTFE" blends *2014 International Conference on Mechanical Engineering, Automation and Control Systems (MEACS)* pp.1-7.

[6] Mashkov Y K, Kropotin O V, Chemisenko O V and Shilko SV 2015 Wear-resistant silica-containing PTFE nanocomposites for metal-polymeric frictional units *J. of Friction and wear* **36** pp 476-80.

[7] Deborah D.L. Chung 2017 *Polymer-Matrix Composites: Mechanical* Properties *and Thermal Performance.Carbon Composites (Second Edition)* pp218-55.

[8] Rahachou A V, Rogachev A A, Yarmolenko M A, Jiang Xiao-Hong and Liu Zhu Bo 2012 Molecular structure and optical properties of PTFE-based nanocomposite polymer–metal coatings *Applied Surface Science* **258** pp1976-80

Modelling of deformation process for the layer of elastoviscoplastic media under surface action of periodic force of arbitrary type

V V Mikheyev[1] **S V Saveliev**[2]
[1]Omsk State Technical University, 11, Mira ave., Omsk, 644050, Russia
[2]Siberian state road and highway university, 5/2, Mira ave., Omsk, 644080, Russia

e-mail:vvm125@mail.ru

Abstract. Description of deflected mode for different types of materials under action of external force plays special role for wide variety of applications – from construction mechanics to circuits engineering. This article con-siders the problem of plastic deformation of the layer of elastoviscolastic soil under surface periodic force. The problem was solved with use of the modified lumped parameters approach which takes into account close to real distribution of normal stress in the depth of the layer along with changes in local mechanical properties of the material taking place during plastic deformation. Special numeric algorithm was worked out for computer modeling of the process. As an example of application suggested algorithm was realized for the deformation of the layer of elasoviscoplastic material by the source of external lateral force with the parameters of real technological process of soil compaction.

1. Introduction

Modelling of deformational processes in materials has important applications in theoretical researches as well as for the solution of practical problems in numerous fields of science and engineering. Analysis of evolution of deflected mode of material under dynamic action is necessary for design of complex mechanical systems and vibration protection systems, is in use for building construction and optimization of working regimes of different machines and mechanisms.

In case of dynamic deformation of elastovicsoplastic materials numeric modelling can describe accumulation of plastic strain which can cause the change of mechanical properties of the material and in consequence can lead to strengthening of the material. New properties may sufficiently widen the field of technical applications for modified material..

Classic results for the problems in the given field in case of isotropic elastic medium were obtained by Boussinesq and later generalized by Westergaard [1]. Approaches and methods developed by by Boussinesq and Westergaard are quite useful even now for computations of bases and foundations of constructions as well as in other problems of construction mechanics.

Dynamic deformation with accumulation of plastic strain and following modification of elastoviscoplastic media is a case of special interest of the investigators. The cause of the constant interest is wide spectrum of properties of materials and sources of surface loading which occur in technical applications [2], [3], [4], [5].

2. Formulation of the problem

The object of investigation chosen in this paper is the process of deformation of the layer of elastoviscoplastic material under the action of external surface force of given characteristics (amplitude, time of loading, dependence on time) acting on the contact spot of given shape and dependence on deformation. Patterns of accumulation of plastic strains will sufficiently differ for the material of the layer at different depths. Since that we suggest to considerate different zones of layer as massive elements with different elastic and viscous characteristics which determine their replies on stress.

Choice of the shape of zones is to be done according to the shape of equal normal stress surfaces in the depth of the soil. Surfaces we choose to restrict elements correspond to the values of yield point, plastic limit and elastic limit of the material. Their exact values may be obtained from the stress-strain diagram. Calculation of lumped mechanical characteristics are to be done taking into account the dependences of characteristics of the material on the density depending on plastic strain accumulated by the layer.

Numeric solution of equation of motion for the active zone of the material under action of the surface loading will give the values of plastic strain and density of the material achieved by the time when dynamic loading finishes

3. Theory

The main point of suggested approach to theoretical description of deformation processes in layers of elastoviscoplastic media is that different parts of active zone of the layer are in different states in respect to their response on deforming stress. Boundaries of the parts may be chosen as equal stress surfaces corresponding to limit values from the stress-strain diagram. Boussinesq theory for point vertical loading Q gives following result for stress distribution

$$\sigma_z = \frac{3Q}{2\pi z^2}\frac{1}{(1+(r/z)^2)^{5/2}}, \tag{1}$$

Westergaard's theory for the same case of point vertical load Q states

$$\sigma_z = \frac{Q}{2\pi z^2}\frac{\gamma}{(\gamma^2+(r/z)^2)^{3/2}}, \tag{2}$$

Where coefficient γ accounts redistribution of stress in the depth of the soil

$$\gamma = \left(\frac{1-2\mu}{2-\mu}\right)^{\frac{1}{2}}, \tag{3}$$

μ - Poisson coefficient.

In all of the expressions above r - stands for the distance from the point of observation to the vertical line going deep from the point of loading, variable z - stands for depth of observation point in respect to the surface of the layer.

Expression for stress caused by surface point load allow to find the stress from distributed surface loads. One of the most interesting cases here is infinite stripe $2b$ wide vertically loaded with surface stress σ_0. Distribution of stress in the depth of the material looks like (Boussinesq) looks as follows

$$\sigma_z = \frac{\sigma_0}{\pi}\left(\arctan(\frac{x+b}{z}) - \arctan(\frac{x-b}{z}) - \frac{2bz(x^2-b^2-z^2)}{(x^2-b^2+z^2)^2+4b^2z^2}\right), \tag{4}$$

Where after transition to the dimensionless variables $\zeta = z/b, \xi = x/b$, one gets

$$\sigma_z = \frac{\sigma_0}{\pi}\left(\arctan(\frac{\xi+1}{\zeta}) - \arctan(\frac{\xi-1}{\zeta}) - \frac{2\zeta(\xi^2-1-\zeta^2)}{(\xi^2-1+\zeta^2)^2+4\zeta^2}\right) \tag{5}$$

Graphs which represent sections of equal stress surfaces $f(\xi,\zeta) = \frac{\sigma_z}{\sigma_0} = C$, for different values of C are given on Figure1. Function $f(\xi,\zeta) = C$ connects variables ξ,ζ implicitly and it's impossible to find analytic dependence $\xi = \xi(\zeta)$ because of transcendental equations to be solves for that purpose.

The result for Westergaard case can be written down as

$$\sigma_z = \int_{-b}^{+b} d\sigma_z(t) = \frac{\sigma_0\gamma}{\pi}\int_{-b}^{+b}\frac{zdt}{(\gamma^2z^2+(x-t)^2)} = \frac{\sigma_0}{\pi}\left(\arctan(\frac{x+b}{\gamma z}) - \arctan(\frac{x-b}{\gamma z})\right), \tag{6}$$

IOP Conf. Series: Journal of Physics: Conf. Series **944** (2017) 012079 doi:10.1088/1742-6596/944/1/012079

which in dimensionless variables $\zeta = z/b, \xi = x/b$ looks like

$$\sigma_z(\xi,\zeta) = \frac{\sigma_0}{\pi}\left(\arctan(\frac{\xi+1}{\gamma\zeta}) - \arctan(\frac{\xi-1}{\gamma\zeta}) \right). \quad (7)$$

Graphs which represent sections of equal stress surfaces for Westergaard distribution is given on Figure 2 in coordinates $\zeta = z/b, \xi = x/b$.

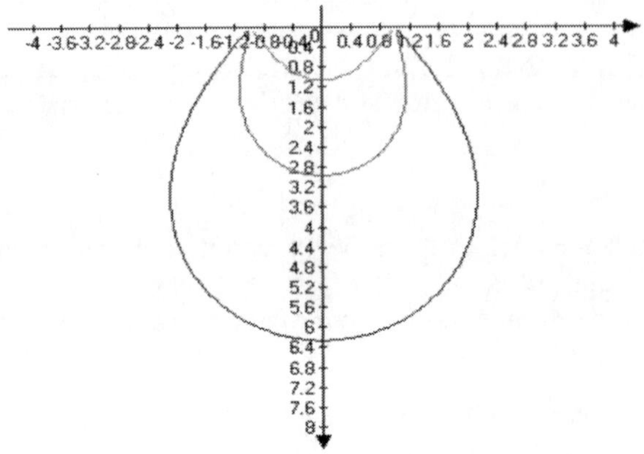

Figure 1. Equal stress surfaces distribution according to Boussinesque (uniformly distributed load on infinitely long stripe) for stress equal 0.1 , 0.2 ,0.4 of surface value (in variables ζ, ξ)

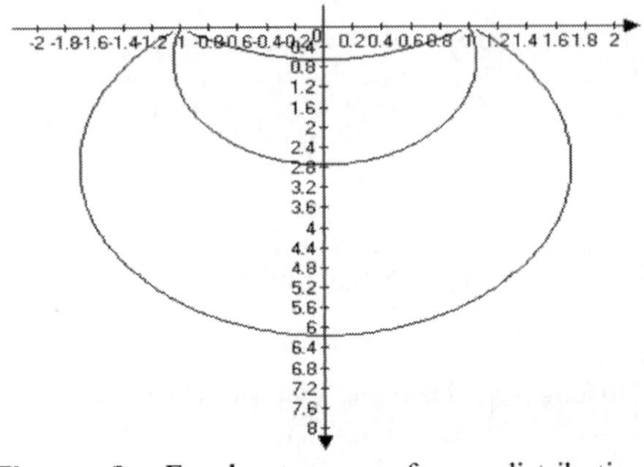

Figure 2. Equal stress surfaces distribution according to Westergaard (uniformly distributed load on infinitely long stripe) for stress equal 0.1,0.2,0.4 of surface value (in variables ζ, ξ)

The sections of Westergaad distribution may have analytic representation and appear to be ellipses going through points (-1,0) and (1,0). Since that surfaces of equal stress represent family of intersecting elliptic cylinders. Let's write down equation of sections of equal stress surfaces as

$$\frac{\pi\sigma_z}{\sigma_0} = \arctan(\frac{\xi+1}{\gamma\zeta}) - \arctan(\frac{\xi-1}{\gamma\zeta})$$

Let's transform it to canonical view, assuming $C = \frac{\pi\sigma_z}{\sigma_0} = const$

$$\frac{(\varsigma - \frac{1}{C})^2}{\frac{1}{\gamma^2}(1+\frac{\gamma^2}{C^2})} + \frac{\xi^2}{(1+\frac{\gamma^2}{C^2})} = 1 \, .$$

Distribution of equal stress surfaces as a family of intersecting cylinders is good approximation for different parts of active zone in interesting case of loading when force is applied to the rectangular contact spot with length much greater then width.
Let's apply to the mechanical system of active zone of material modified Lagrange formalism which makes it possible to take into account dissipative forces of viscous friction and dry friction as described in [5].

$$\frac{d}{dt}\left(\frac{\partial T}{\partial \dot{x}}\right) - \frac{\partial T}{\partial x} = F(x,\dot{x},t) + F_{mp}, \tag{8}$$

where $T(x,\dot{x}) = \frac{m(x)\dot{x}^2}{2}$ - kinetic energy of active zone,
$F(x,\dot{x},t)$ - internal forces of elasticity and viscous friction and external force of surface loading,
F_{mp} - separately taken force of dry friction to model plastic element when contact stress exceeds yield point.
At the start of loading process material of the layer while stress does not exceed value 1 on Figure 3 undergo only elastic deformation (it's known that if the material is not dense enough then point 1 coincides with origin of coordinates); between points 1 and 2 deformations recover partially after deloading. If stress takes values greater then at 2 deformations are becoming plastic but stress must grow in order to obtain greater strain. Point 3 corresponds to the yield limit and after it material is gaining plastic deformations without change in external stress.

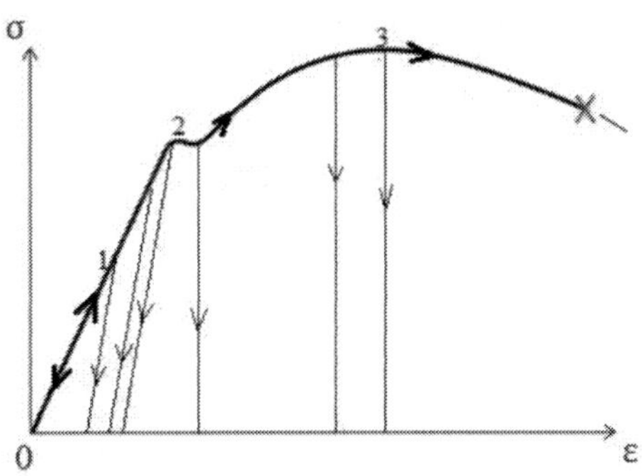

Figure 3. "Strain-stress" diagram for elastoviscoplastic material for normal axial loading with consequent reload (thin lines stand for reload)

The difference of suggested method from usual is consideration of dynamics of the mass of active zone under loading. In this case right side of (8) looks like

$$\frac{d}{dt}\left(\frac{\partial T}{\partial \dot{x}}\right) - \frac{\partial T}{\partial x} = \frac{d}{dt}(m(x)\dot{x}) - \frac{\partial m(x)}{\partial x}\frac{\dot{x}^2}{2} = m(x)\ddot{x} + \frac{\partial m(x)}{\partial x}\frac{\dot{x}^2}{2}, \tag{9}$$

and modifies equation of motion of active zone, obtained in [6].
Elastic or quasi-elastic force for the layer of elastoviscoplastic media is also assumed to be nonlinear in respect to the value of deformation (strain). Nonlinear behaviour is to be modelled by introduction of non-constant stiffness of active zone

$$F_{ynp} = c(x)x \, . \tag{10}$$

IOP Conf. Series: Journal of Physics: Conf. Series **944** (2017) 012079 doi:10.1088/1742-6596/944/1/012079

Unequal deformational properties for loading and deloading are to be taken into account by following model description:

$$\sigma_{def} = E\varepsilon,\ E = (E_{pl}\theta(\dot{\sigma}) + E_{el})\theta(\sigma_{pl}), \tag{11}$$

where E_{pl} - plastic deformation modulus of the material, E_{el} - Young modulus, $\dot{\sigma}$ - velocity of change in contact stress, $\theta(z)$ - θ - Heaviside function, σ_{pl} - limit of plasticity of the material. Transition to the lumped elastic element gives:

$$F_{def} = c(x)x,\ c(x) = (c_{pl}(x)\theta(\dot{\sigma}) + c_{el}(x))\theta(\sigma_{pl}) \tag{12}$$

It must be mentioned that force and elasticity must depend on coordinate of the centre of mass of active zone not on surface coordinate and for transition to coordinate of the surface must be recalculated.
Viscous friction force of the active zone is modelled by lumped viscosity element

$$F_{visc} = b(x)\dot{x},$$

And it must be pointed that dependence of viscous force on the coordinate is quite slow and mostly is influenced by the change in viscosity with accumulation of plastic strain.
In the framework of given approach behaviour of the material after yield point is exceeded must be modelled by the element of dry friction which "turns on" when the following condition is satisfied

$$\sigma(t) \geq \sigma_{pl}, \tag{13}$$

where $\sigma(t)$ - value of contact stress. Corresponding constant resistance force does not depend on deformation

$$F_{fr}(t) = S_{cont}(t)\sigma_{pl}$$

where $S_{cont}(t)$ - contact spot area between the source of loading and material surface in the moment t. At the same moment of time elastic force (12) "turns off".
Since that equation of motion for active zone (9) may be rewritten under model assumption made above

$$m(x)\ddot{x} + \frac{\partial m(x)}{\partial x}\frac{\dot{x}^2}{2} = -(c_{pl}(x)\theta(\dot{\sigma}) + c_{el}(x))(1-\theta(\sigma_{pl}))x - b(x)\dot{x} - \frac{\dot{x}}{|x|}\theta(\sigma_{pl})S_{cont}(t)\sigma_{pl} + F_{ext}(t). \tag{14}$$

Here $F_{ext}(t)$ - is total external surface force generated by the compactor. Time dependence in it is determined by the technical properties of the machine and may be chosen arbitrary. Lumped parameter model of corresponding mechanical system is given on Figure 4.
More detailed consideration of the process is possible in the framework of multy-body approach which can show the evolution of the stressed state in the depth of the layer. The general form of the equation of motion (14) for interacting parts of the active zone must be corrected according to their properties of the reply on deforming stress. Lumped parameter model of multy-body system is given on Figure 5Suggested multy-body model generates system of differential equations which requires simultaneous solution in order to calculate plastic strain accumulated by the layer. Index 0 stands for the source of surface loading, indices 1,2,3 – for different parts of the active zone.

$$\begin{cases} m_0\ddot{x}_0 = -c_0(x_0)(x_{PO} - x_2) - b_0(x_0)\dot{x}_0 + F_{ext}(t) \\ m_1(x_1)\ddot{x}_1 + \frac{\partial m_1(x_1)}{\partial x_1}\frac{\dot{x}_1^2}{2} = c_0(x_0)(x_{PO} - x_2) - b_1(x_1 - x_2)(\dot{x}_1 - \dot{x}_2) - \frac{\dot{x}_1 - \dot{x}_2}{|\dot{x}_1 - \dot{x}_2|}\theta(\sigma_{pl})S_{cont}(t)\sigma_{pl} \\ m_2(x_2)\ddot{x}_2 + \frac{\partial m_2(x_2)}{\partial x_2}\frac{\dot{x}_2^2}{2} = \frac{\dot{x}_1 - \dot{x}_2}{|\dot{x}_1 - \dot{x}_2|}\theta(\sigma_{pl})S_{cont}(t)\sigma_{pl} - (c_{2pl}(x_2)\theta(\dot{\sigma}) + c_{2el}(x_2))(1-\theta(\sigma_{pl}))(x_2 - x_3) - b_2(x_2 - x_3)(\dot{x}_2 - \dot{x}_3) \\ m_3(x_3)\ddot{x}_3 + \frac{\partial m_3(x_4)}{\partial x_3}\frac{\dot{x}_3^2}{2} = (c_{2pl}(x_2)\theta(\dot{\sigma}) + c_{2el}(x_2))(1-\theta(\sigma_{pl}))(x_2 - x_3) - (c_{3pl}(x)\theta(\dot{\sigma}) + c_{3el}(x))(1-\theta(\sigma_{pl}))x_3 - b_3(x_3)\dot{x}_3 \end{cases} \tag{15}$$

In the case under consideration index 0 corresponds to the work tool of the compactor, indices 1,2,3 consequently to the part of soil layer without elastic response to the stress, then to the part of soil with different properties for loading and deloading and, at last for usual viscoelastic part.

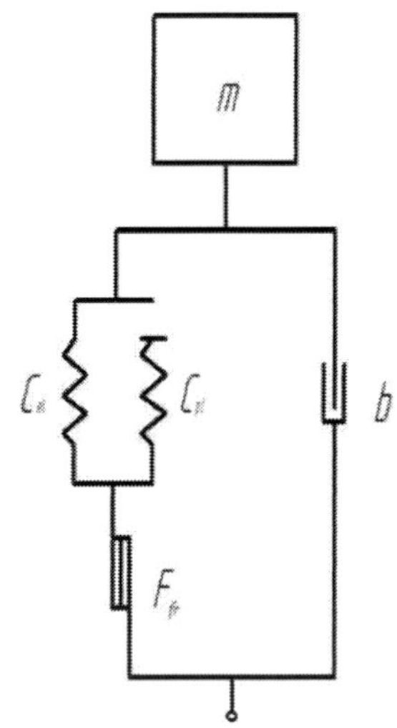

Figure 4. One-body model for active zone of material layer under surface loading

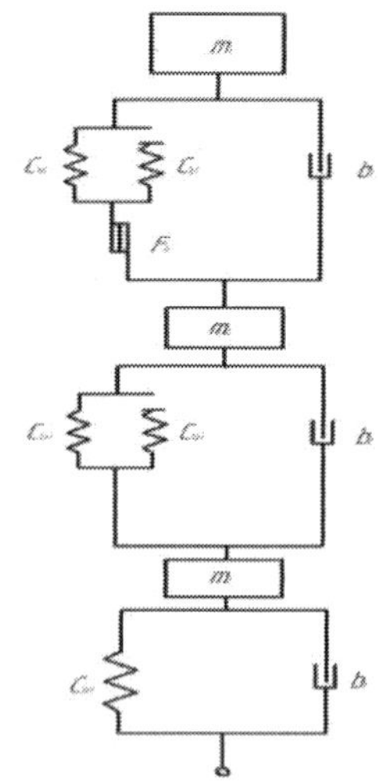

Figure 5. Three-body model for active zone of material layer under surface loading

System (15) is definitely nonlinear and requires numeric methods which must allow taking into account motion of the surface under deforming loading and redistribution of masses and mechanical characteristics between different parts of active zone with accumulation of plastic strain.

Very important issue for program realization of given numeric algorithm for the solution of system (15) is selection of partition for time interval. Transition to the finite elements is to be performer the usual way by substitution for first and second derivatives of strain in respect to time following expressions

$$\dot{x}_i^{(k+1)} = \frac{x_i^{(k+1)} - x_i^{(k)}}{\Delta t},$$

$$\ddot{x}_i^{(k+2)} = \frac{x_i^{(k+2)} - 2x_i^{(k+1)} + x_i^{(k)}}{\Delta t^2}$$

Where upper index numerates step of partition of time interval and lower index numerates part of active zone and therefore corresponding model. Values of strain for each part allow to calculate full amount of work that was spent on the deformation and work that was spent to provide accumulated plastic strain and since that to determine net power of extern source al dynamic action.

Numeric approach suggested in the article makes the motion equations linear because coefficients which may cause significant non-linear effects and lead to the problems are constant and are to be recalculated on each step of scheme in order to take into account modification of elastoviscoplastic material when density grows.

4. Results of numeric experiments

The algorithm developed above for numeric integration of non-linear equation of motion for multy-body model of active zone of the material was realized in the framework of computer algebra package Maple 2016.

It must be pointed out that numeric methods as well as experimental investigations are the most common and effective research tools for interaction of different types of media with source of surface force. Numeric methods can be used for revealing of the properties of interaction process [4], [7], as well as for solution of applied problems [8], [9], [10]. Case of periodic vibratory surface loading is one of the most interesting from both points. During numeric modelling media were represented by widely used in construction clay loam and sandy loam and work tool (drum) of the vibratory road roller represented source of surface loading. The case of soil media quite common for investigation of the vibratory loading [11], [12], [13], [14]. Numeric modelling was performed with use of parameters of external force normal for midsize vibratory road roller manufactured by Rybinsk road machine plant DM-13-VD acting on layer of sandy loam (required density 1700 kg/m^3) and clay loam (required density 1900 кg/m^3) 0.3 thick. Static loading on vibratory drum is 65 kN; vibratory force for sandy loam was assumed to be of amplitude 62 kN and frequency 40 Hz and for clay loam of amplitude 62 kN and frequency 40 Hz. Results of modelling in graphic form are presented on figures 6,7,8, 9. Initial parameters of the modelling along with properties of the roller included roller speed which varied from 0.2 to 1.6 m/c and initial density of the soil defined by the value of compaction degree – rate between initial density and required one.

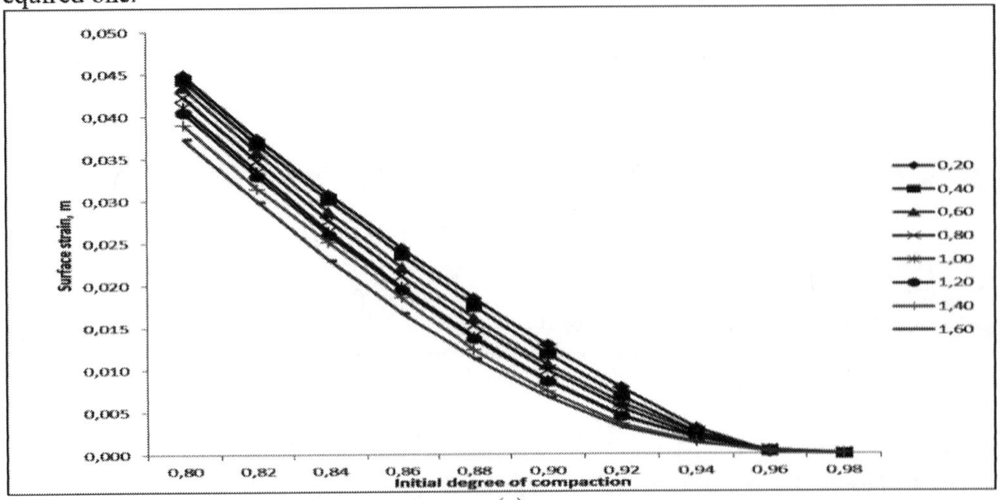

(a)

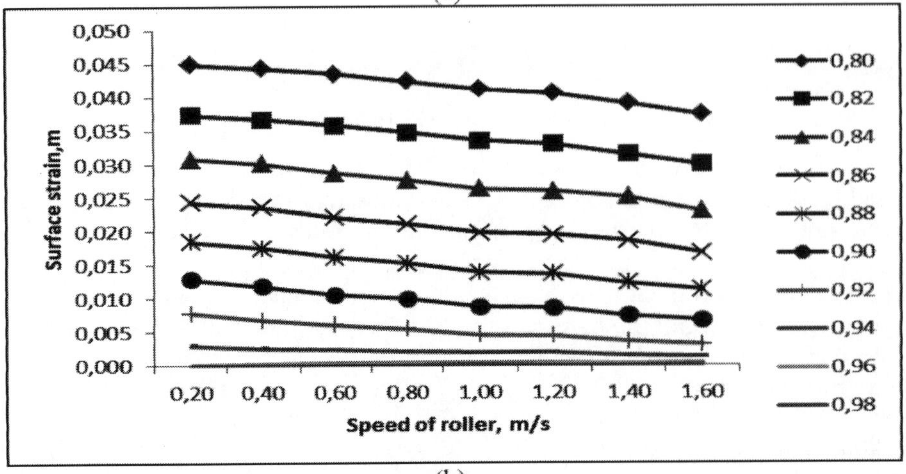

(b)

Figure 6. Accumulated plastic strain for one roller pass depending on initial compaction degree of a soil (a) and speed of the roller (b) (sandy loam)

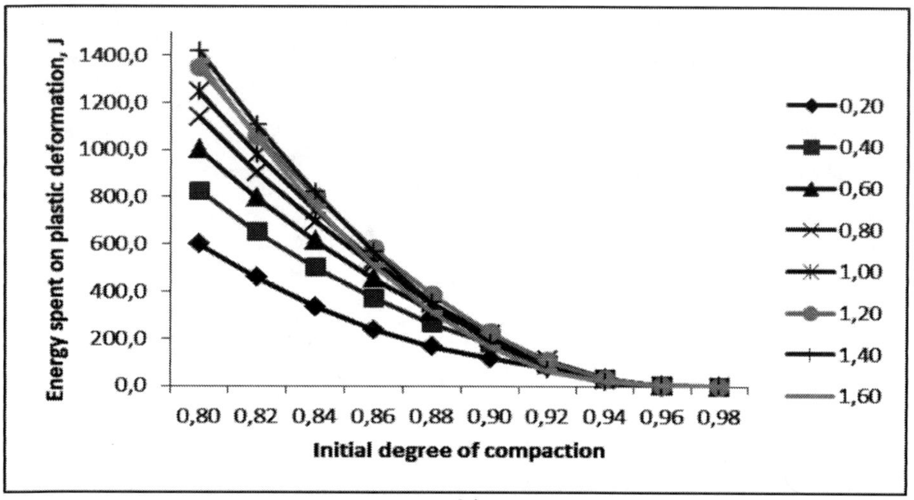

(a)

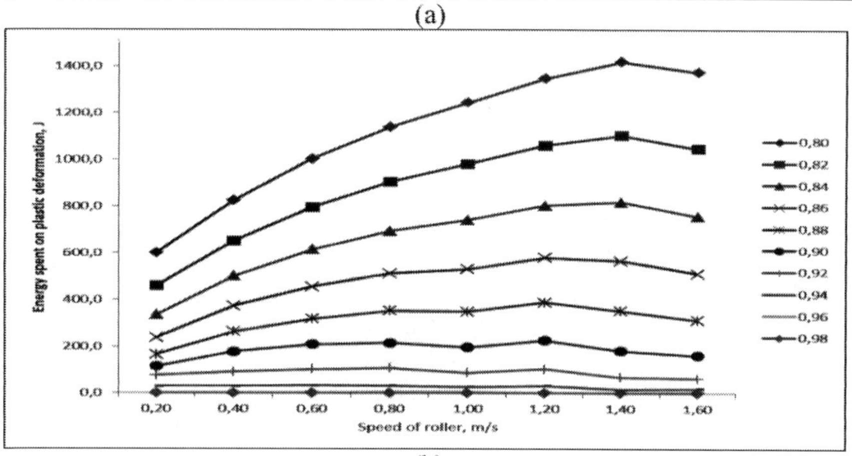

(b)

Figure 7. Energy spent on plastic deformation of the soil layer depending on initial compaction degree of a soil a) and speed of the roller b) (sandy loam)

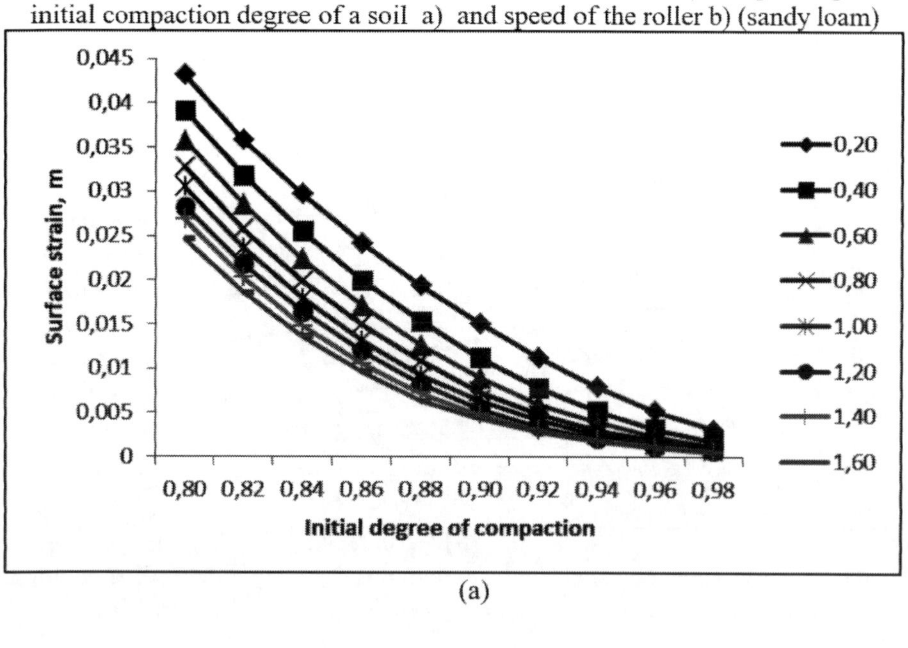

(a)

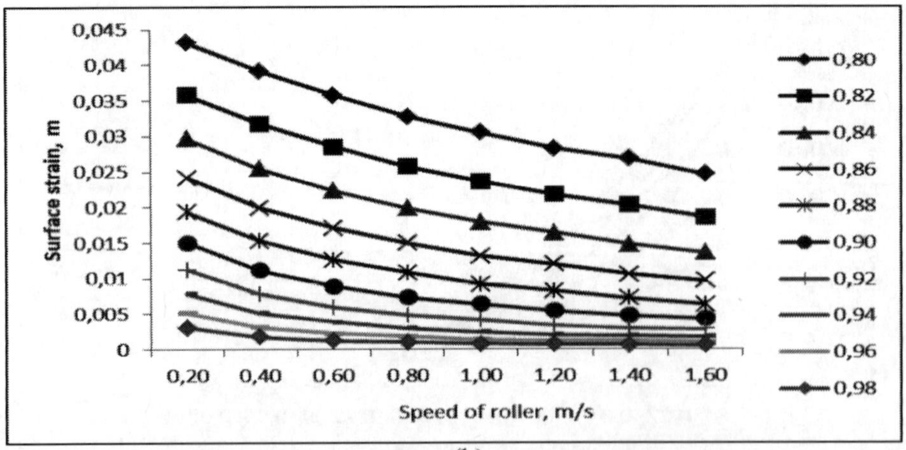

(b)

Figure 8. Accumulated plastic strain for one roller pass depending on initial compaction degree of a soil a) and speed of the roller b) (clay loam)

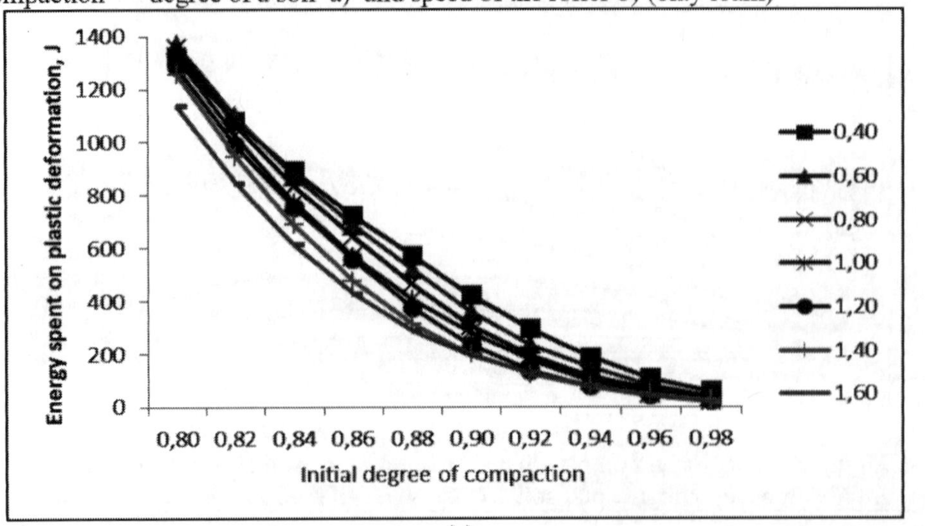

(a)

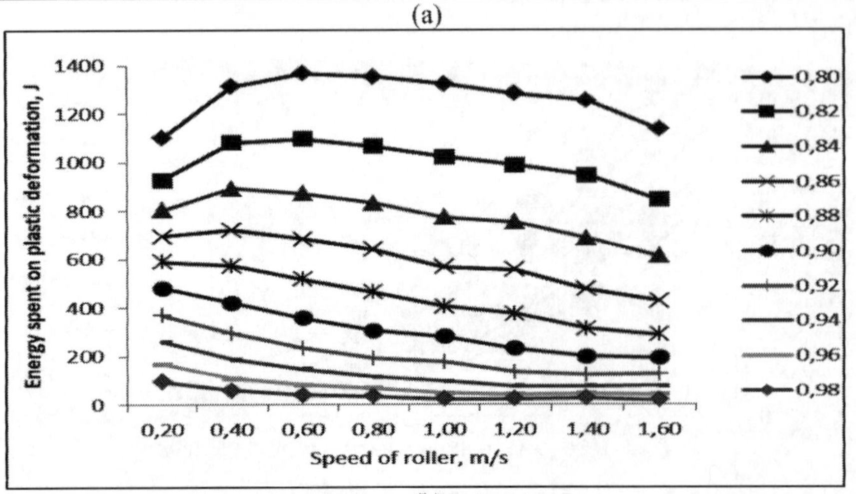

(b)

Figure 9. Energy spent on plastic deformation of the soil layer depending on initial compaction degree of a soil a) and speed of the roller b) (clay loam)

Accumulated plastic strain of the layer and amount of mechanical work for that on one roller pass may be considered parameters for the choice of energy and time efficient regime of soil compaction by the given type of the roller for given initial state of the soil layer .

5. Discussion

Results of computer modelling and numeric experiments well correspond with results of contemporary researches and show that new modified approach of lumped parameters for description of accumulation of plastic strains in elastoviscoplastic media. Mathematical model developed in the work for evolution of deformation processes in depth of material layer under surface force as well as corresponding numeric algorithm allow effective application for evaluation of strain accumulation dynamics in case of arbitrary force and evaluation of efficiency of power transmission to the material layer under surface loading.

6. Conclusion

Accumulation of plastic strain by elastoviscoplasic media is still a problem of constant attention for specialists in the field of construction mechanics, road and highway engineering because of its interest from the point of compaction processes in soils and other media. Effective theoretical solution of the problem can allow to perform so called Intelligent Soil Compaction technology in order to achieve required densities of soil media with least possible spent time and resources [15], [16], [17], [3]. Approach applied in the article for the first time for description of interaction between elastoviscoplastic material layer and work tool of compactor makes it possible to predict necessary parameters of surface force by given physical properties of the material in order to provide the most efficient accumulation of plastic strain. Functional dependences revealed by the numeric experiment presented in the article in graphic form may be used for design of energy efficient work regimes of road rollers and construction of new work tools.

References

[1] Murthy V N S *Geotechnical engineering*. London, CRC Press, 2002.

[2] Hong Z, Yu S B, Fei W P, Yao S The technique of paralleling motivation with impact and vibration for impact roller, 2006 International Technology and Innovation Conference (ITIC 2006), Pages: 676 - 680, DOI: 10.1049/cp:20060845

[3] Townsend F C, Anderson B 2004, *A Compendium of Ground Modification Techniques*, Research Report BC-354, pp. 16-60. Florida Department of Transportation (FDOT).

[4] Hanbing Liu; Jing Wang; Weichao He; Jingke He Study on shear strength of compacted subgrade soils by test and numerical simulation, World Automation Congress 2012, pp.: 1-4

[5] Kozlov V V Lagrange mechanics and dry friction, *Nonlinear Dynamics, 2010,* **vol. 6, issue 4**, pp. 855-868 DOI:10.20537/nd1004009\

[6] Mikheyev V V, Saveliev S V Investigation of influence of deformation of road roller adaptive work tools on deformation process of the soil under compacting action. *Stroitel'nye I dorozhnye mashiny*. **№7**, 2013. -pp. 45-51

[7] Rouveure R, Bacconnet C, Chanet M, Monod M O Simulation of realistic soils for 3-D computational modelsIGARSS 2003. 2003 IEEE International Geoscience and Remote Sensing Symposium. Proceedings (IEEE Cat. No.03CH37477) Pages: 2900 - 2902 vol.4, DOI: 10.1109/IGARSS.2003.1294625

[8] Elaoud A, Chehaibi S, Abrougui K Simulating the effects of the passage of tractors on agricultural land. 2013 5th International Conference on Modeling, Simulation and Applied Optimization (ICMSAO). Pages: 1 -6, DOI: 10.1109/ICMSAO.2013.6552661

[9] Hong Z, Yu S B, Fei W P, Yao S The technique of paralleling motivation with impact and vibration for impact roller, 2006 International Technology and Innovation Conference (ITIC 2006), Pages: 676 - 680, DOI: 10.1049/cp:20060845

[10] Jun Li, Yan Li, Zhili Zhou, Fengkui Cui. Study on continuous detection of vibro-roller compactness, 2009 International Conference on Mechatronics and Automation, Pages: 2637 - 2641, DOI: 10.1109/ICMA.2009.5246709

[11] Forssblad L 1977. *Vibratory Compaction in the Construction of Roads, Airfields, Dams, and Other Projects*, Research Report No. 8222, Dynapac, S-171, No. 22, Solna, Sweden.

[12] Kézdi A 1962. *Erddrucktheorien*. Springer Verlag, Berlin/ Göttingen/ Heidelberg.

[13] Massarsch K R; Fellenius B H, 2002. Vibratory compaction of coarse-grained soils. *Canadian Geotechnical Journal,* **39(3)** 695-709.

[14] Yoo T S 1975, *A Theory for Vibratory Compaction of Soil*, The Dissertation for Degree of Doctor of Philosophy to University of New York at Buffalo

[15] Kloubert H J 2001, *New intelligent compaction system for vibratory rollers*. Report for IRF in Paris.

[16] Shen Peihui Mathematic modeling and characteristic analysis for the intelligent compactor. 2011 IEEE International Conference on Mechatronics and Automation, Pages: 2249 - 2253, DOI: 10.1109/ICMA.2011.5986289

[17] Mooney, M.A., Rinehart, R.V., Facas, N.W., Musimbi, O.M., White, D.J., Vennapusa P.K.R, 2010, *Intelligent Soil Compaction Systems*. NCHRP REPORT 67 6. Washington, D.C.

Modeling and prototyping of biometric systems using dataflow programming

N Minakova, I Petrov

Altai State University, 61 Lenin av., Barnaul, 656049, Russia

E-mail: PetrovIV90@gmail.com

Abstract. The development of biometric systems is one of the labor-intensive processes. Therefore, the creation and analysis of approaches and techniques is an urgent task at present. This article presents a technique of modeling and prototyping biometric systems based on dataflow programming. The technique includes three main stages: the development of functional blocks, the creation of a dataflow graph and the generation of a prototype. A specially developed software modeling environment that implements this technique is described. As an example of the use of this technique, an example of the implementation of the iris localization subsystem is demonstrated. A variant of modification of dataflow programming is suggested to solve the problem related to the undefined order of block activation. The main advantage of the presented technique is the ability to visually display and design the model of the biometric system, the rapid creation of a working prototype and the reuse of the previously developed functional blocks.

1. Introduction

We can see rapid growth and development of biometrics systems in recent years. Biometrics technologies got width spreading in various areas of human being. The last version of smartphones from the several producers has fingerprint readers and even iris scanners. Despite on rapid growth and width spreading there are many problems related to biometrics such as [1-3]:

- Possibility to fake biometrics features. Biometrics features can be spoofing (for example, rubber pads on the fingers). Many producers and developers of biometric devices react to this by adding additional technologies for detecting spoof. However, intruders find the new ways to counterfeits.
- Compromising. Unlike the password, biometric characteristic can't change in case of compromising. Compromised biometric characteristic cannot used for identification person anymore.
- Bandwidth of biometric systems. The developers constantly strive to accelerate recognition using more optimal algorithms and more advanced scanners.
- Confidence. Biometric systems can prove a personality only with some probability. The basic characteristics of any biometrics systems are false acceptance and false rejection rates (FAR and FRR respectively).
- Convenience. Some biometrics systems demand high level of cooperation, which reduces the convenience of using this system.

There are no specific methodologies of biometrics systems development. Assessments of performance and reliability based on empirical data. The basic characteristics of biometrics systems such as FAR and FRR are calculated from results of testing by biometric samples. A public databases of biometrics samples are used for reproducibility purposes during the technological tests. Developers can preliminary reconfigure his recognition algorithms under the specific sampling during the tests on such databases [3]. So, early making of a working prototype helps to reveal many different problems and fix them.
The technique of modeling and prototyping of biometric systems based on dataflow programming is proposed in this work.

2. Dataflow programming

A dataflow program is a directed graph with the functions in the nodes which transform input data from the input arcs to output data which is transmitting then over the outputs arcs to other nodes [4-6]. Arcs connect output and input ports of blocks. Data performing occur as result of block activation after their input ports receipt the data. Important components of a biometric system are a feature extraction subsystem and a feature comparison subsystem. These subsystems represent a pattern recognition system. The internal structure of such systems is in good agreement with the structure of the data flow program. The data in this case are raw, intermediate and processed biometric samples, and nodes are processing functions, filters, classifiers, etc. This approach has many advantages in developing of systems oriented on data processing:

- Visualization. Dataflow graphs can easily visualize as dataflow diagrams.
- Modularity. Concept of nodes is in good agreement with modularity which is allowing reuse of the earlier developed components.
- Extension. The modules have simple and uniform interface which facilitates the expansion of the system.
- Parallelization. Ease to parallelize a program.

The commercial systems have much more complex structure. Beside the primary functionality, it is necessary to solve many other tasks including integrity protection of system, data encryption, storage the big data, integration with external systems, etc. Implementation of all of this with the dataflow programming is not expedient. However, dataflow programming allows significantly reduce developing time and reveal previously hidden problems and features when creating prototypes of biometric systems, or individual components of biometric systems that perform basic functions. Also, dataflow architecture leads to natural parallelization, which can has positively affect to system performance in multi-core architectures.
The modeling environment based on dataflow programming has been developed in frames of this work. Functional logic of individual blocks is placed in dynamic plugged modules. Special open interfaces for integration of various third-party algorithms in one system are provided.
Proposed technique of developing includes three stages:

1. Functional blocks developing;
2. Biometric system's model creating;
3. Prototype generating.

3. Functional blocks developing

Header files are provided with the abstract Block class, which contains virtual functions that provide the basic functionality:

1. InternalPerform is a function which processing data from the inputs ports and transmitting results directly to the outputs ports.
2. Init is the function which is necessary to create ports which will be used in performing function.

To extend the functionality of the system it is necessary to inherit from the base class Block.
An example code with the implementation of the block's functions for this system shown in figure 1.

IOP Conf. Series: Journal of Physics: Conf. Series **944** (2017) 012080 doi:10.1088/1742-6596/944/1/012080

```
using namespace eMaSE::core;

class SummBlock: public Block
{
    Q_OBJECT
public:
    bool Init() override;
protected:
    bool InternalPerform() override;
};
```

Figure 1. Header file of the function block.

4. Biometric system's model creation

The second stage is the process of modeling. Using the functional blocks created in the first stage, a data flow graph is created with adding these blocks to the special scene and creating links between the input and output ports of the added blocks. A distinctive feature of this approach is the possibility of using composite blocks. Such blocks allow to store other functional blocks inside it, including composite ones. This makes it possible to simplify the system by grouping elementary functional blocks into containers that serve as separate subsystems.

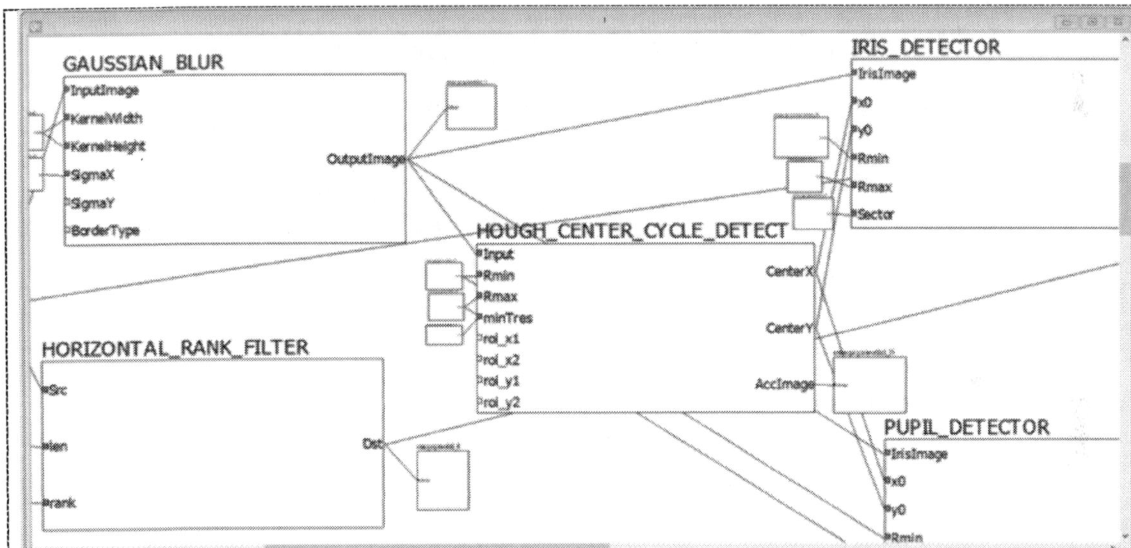

Figure 2. Screenshot of the program, the stage of modeling the biometric system, using the example of the subsystem of the localization of the iris image.

Figure 2 shows the screenshots of the program at the modeling stage. This screenshot shows an example of a localization subsystem of iris image loaded from a file. Small blocks on the left represent source blocks and allow users input data into the program through the graphical interface. Small blocks on the right are receivers and allow users output data to the graphical interface. Blocks can be placed in special container (or composite) blocks, which allows to combine a group of functional blocks into a separate subsystem. Large blocks perform specials algorithms. This iris localization subsystem unites several different algorithms such as low-pass filtering [7], rank filtration [8], Hough transforms (to search for the center of the pupil) [9], an algorithm for pre-localization of the iris [10], Daugman's integro-differential operators for searching for internal and the outer boundaries of the iris [11] and the eyelid detector. The result of this subsystem is the attributes of the localized boundary of the iris. The following steps in the operation of the iris recognition system should include the normalization of the iris and the obtaining of the code (as an option using the methods of texture analysis used to study heterogeneous structures [12]). An example of a graphical user interface is shown in Figure 3.

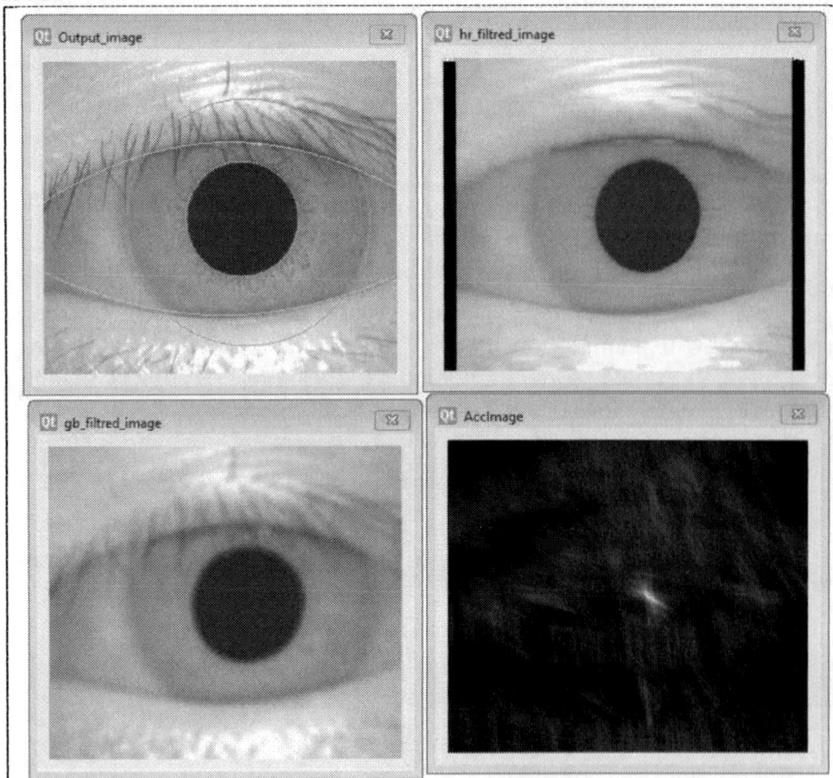

Figure 3. Screenshot of the program. The graphical user interface of modeling environment.

In the screenshot placed the display of various intermediate data (visualization of the results of the localization of the iris, the result of the work of the rank filter for eliminating the eyelashes, the result of the transformation of Hough for the search for the center of the pupil).

5. Communication with biometric sensors

Biometric sensors are an external data source for a data flow program. To communicate with them, special blocks of data sources are implemented. Inside these blocks there is an interaction with the device with using drivers.

6. Prototype generation

At the third stage, the prototype is automatically generated from the model. The prototype is a file containing the structure of a dataflow graph and dynamically plugged binary modules with the logic of function blocks. The prototype is executed in a special interpreter, which reads the structure of the graph from the file, loads the necessary modules, and starts the execution of the program.

7. Modification of the approach. Use priorities for block activation

Block activation occurs after data is set on all input ports in dataflow programs. Two (or more) blocks which is ready for activation can be executed at the same time in different threads [7]. In some cases, this approach may be unacceptable. For example, when it is necessary to turn on the biometric sensor, read the data from it and turn it off. If the data is read in the program at once, it is possible to include the reading of biometric data and switching on/off the sensor into one functional unit. However, in the case when sensor reading occurs repeatedly during the execution of the program of data streams, the permanent switching-off of the sensor is inadmissible. In such a case, it is necessary to create the three blocks for each operation of the sensor: the block for starting, stopping and reading. These blocks in the program will be executed at a time as they are sources, but it is necessary that the switching unit works first, then the read block (repeatedly) and at the end the block for closing. In this paper, the introduction of such a block parameter as the execution priority is proposed. In this case, blocks ready for activation at one time will be ranked by this priority and activated according to it.

8. Comparison with analogues

A literature review of known solutions based on data flow programming was conducted in this work. Most of the considered solutions were specialized for a specific application, for example, vvvv [13] was developed for prototyping interfaces, KNIME [14] was developed for working with large data and analytics. Such solutions as Lucid [15] are a separate non-visual programming language that requires the mastering of specific language constructs. The RiDE method [16] is also considered. The main disadvantage of RiDE can be attributed to the fact that individual computing nodes are supposed to be implemented as separate programs. Such an approach can affect performance, as it requires the use of interprocess communication tools when data is being transferring between computing nodes.

9. Conclusion

Thus, in this paper, a method for developing biometric systems based on modeling and prototyping was presented. The developed model can be used to generate prototypes, or integrated into the biometric system as a separate component for the parametrization and comparison of biometric samples.

The main stages of prototype development were described. A test system was developed to evaluate this approach. The advantages include:

- Rapid prototype development, testing in conditions close to real. Rapid development is achieved using a graphical development method.
- Allows to reveal new aspects of the developed biometric system. Due to the graphical way of representing the system, one can consider in general and analyse the developed model.
- Reusing the designed units. Blocks are stored in separate dynamically connected modules.

References

[1] Jain A K, Flynn P and Ross A A 2008 *Handbook of Biometrics* (Springer-Verlag London)
[2] Boulgouris N V and Konstantinos N P 2009 *Biometrics:Theory, Methods, and Applications* (Wiley-IEEE Press)
[3] Petrovska-Delacrétaz D and Chollet G 2009 *Guide to Biometric Reference Systems and Performance Evaluation* (Springer-Verlag London)
[4] Sousa T B 2012 Dataflow Programming Concept, Languages and Applications *Dr. Symp. Informatics Eng* **7** p 13
[5] Carkci M 2014 *Dataflow and Reactive Programming Systems* (Leanpub book)
[6] Hils D D 1992 Visual languages and computing survey: Data flow visual programming languages *J. Vis. Lang. Comput* **3(1)** pp 69–101
[7] Nixon M S and Alberto S A 2008 *Feature Extraction and Image Processing* (Academic Press) p 88
[8] He Z Tan T Sun Z and Qiu X 2008 Robust eyelid, eyelash and shadow localization for iris recognition *Proc. - Int. Conf. Image Process. ICIP* pp 265–268
[9] Hilal A Daya B and Beauseroy P 2012 Hough Transform and Active Contour for Enhanced Iris Segmentation *IJCSI Int. J. Comput. Sci. Issues* **9(6)** pp 1–10
[10] Minacova N N and Petrov I V 2015 Method of preliminary localization of the iris in biometric access control systems *IOP Conf. Ser. Mater. Sci. Eng.* **93(1)** p 12056
[11] Daugman J 2004 How iris recognition works *Circuits Syst. Video Technol. IEEE Trans.* **14(1)** pp 21–30

[12] Minakova N N Karpov S A and Ushakov V Ya 2002 Texture analysis of the disperse structure of composite elastomers with a modified carbon filler *Russian Physics Journal* **45(2)** pp 185-189
[13] vvvv - a multipurpose toolkit. https://vvvv.org/
[14] KNIME, https://www.knime.org/
[15] Ashcroft E and Wadge E 1985 *Lucid, the dataflow programming language* (APIC Stud. data Process)
[16] Bakhterev M Kazantzev A Vasev P and Albrekht I 2011 Dataflow-Based Distributed *Computing System Proceedings of the Euromicro PDP* pp 6-7

Mathematical Modeling of Hydroelastic Oscillations of the Stamp and the Plate, Resting on Pasternak Foundation

L I Mogilevich[1], V S Popov[1], A A Popova[1] and A V Christoforova[2]

[1]Yuri Gagarin State Technical University of Saratov, 77 Politechnicheskaya Street, Saratov, 410054, Russia
[2] Saratov State University, 83 Astrakhanskaya Street, Saratov, 410012, Russia

e-mail: vic_p@bk.ru

Abstract. The forced oscillations of the elastic fixed stamp and the plate, resting on Pasternak foundation are studied. The oscillations are caused by pressure pulsation in liquid layer between the stamp and the plate. Pasternak model is chosen as an elastic foundation. The laws of the stamp movement, the plate deflection and pressure in the liquid are discovered on the basis of hydroelasticity problem analytical solution. The functions of amplitude deflection distribution and liquid pressure along the plate are constructed, as well as the stamp amplitude-frequency characteristic. The obtained mathematical model allows to investigate the dynamics of hydroelastic interaction of the stamp with the plate, resting on elastic foundation, to define resonance frequencies of the plate and the stamp and corresponding deflections amplitudes, as well as liquid presser amplitudes. Keywords – elastic plate; hydroelastic oscillations; vibration stamp; viscous liquid, Pasternak foundation

1. Introduction

The investigation of the plate interaction with liquid is one of the important problem of contemporary mechanics for considering dynamic processes in complicated mechanical systems. For example, the problem of bending beam oscillations under interaction with an ideal liquid is solved in [1]. This solution made it possible to construct the mathematical model for defining the cylinder cavitations resource of the combustion engine with water cooling. Reference [2] deals with modeling of oscillations of a rectangular plate plunged into an ideal incompressible liquid with a free surface. Reference [3] studied the interaction of an ideal incompressible liquid with a plate, making forced oscillations due to its contact with a vibrating machine. The plane problem of acoustic wave's radiation by vibrating plate is considered. The free oscillations of the cantilevered plates partly plunged into an ideal incompressible liquid with free surface are investigated in [4]. This investigation made it possible to evaluate the liquid movement inertia impact on the natural oscillations of the plate. Analogous study is carried out in [5] for cantilevered composite plate. The numerical investigation of free oscillations rectangular plates, fully plunged in a motionless ideal liquid or floating on its free surface, is made in [6]. Analogous investigation for rectangular plates, interacting with flow of ideal fluid, is considered in [7]. The paper's authors carried out the analysis of the plate eigenfrequencies and found critical velocities corresponding to the loss of stability. Reference [8] studied chaotic oscillations of the plate, interacting with ideal incompressible fluid flow.

However, the liquid damping characteristics caused by its viscosity are excluded from the above mentioned investigations. Reference [9] considered bounding vibrations of infinitely long beam on the viscous liquid layer. Hydroelastic oscillations of cantilevered beam plunged into a viscous incompressible fluid are studied in [11]. Reference [10] is devoted to the investigation of transverse oscillations of elastically fixed rigid wall of the flat channel, the channel having finite sizes. The study of vibrating discs, one of them being elastic, with viscous incompressible liquid between them is made in [12]. The analogous problem for two vibrating plates is considered in [13]. Reference [14] studies hydroelastic beam oscillations in viscous liquid flow for piezoelectric elements. Reference [15] solves the problem of bending hydroelastic oscillations of the plate, forming the narrow channel wall under pulsating viscous liquid layer impact. The forced hydroelastic oscillations of three-layered circular plate, interacting with a viscous incompressible liquid layer are investigated in [16]. On the other hand, the investigation of the constructions elastic elements oscillations with consideration of foundation elastic pliability is of great importance for contemporary engineering [17]. For instance, references [18-21] consider oscillations and stability of multi-layered beams and plates resting on an elastic foundation under the impact of local and distributed loadings of various natures. Winkler and Pasternak models are used for foundation reactions. That is why, the evaluation of foundation elasticity impact on hydroelastic oscillations of the plate is important. There are few papers devoted to this subject matter: e.g. the oscillations of the membrane resting on Winkler foundation and situated on the bottom of a tank filled with an ideal incompressible liquid with a free surface are investigated in [22]. Hydroelastic oscillations of rectangular plates, resting on Pasternak foundation and interacting with an ideal incompressible liquid with a free surface, are studied in [23-25]. The oscillations of the plate, resting on Winkler foundation and interacting with viscous incompressible liquid layer, are studied in [26-29]. Reference [30] considers the oscillations of the shell interacting with an annular layer of strongly viscous liquid inside it and surrounded by Winkler elastic medium.

2. Stetting the problem

This paper deals with the interaction dynamics of the stamp ***1*** with the elastic plate ***2*** through viscous liquid layer ***3*** (Fig. 1). The stamp oscillations and bending oscillations of the plate are caused by pressure pulsation in a thin layer of viscous incompressible liquid between them, the plate resting on Pasternak foundation. Let us connect Cartesian coordinate system Oxz with a medium surface of the plate in an unperturbed state.

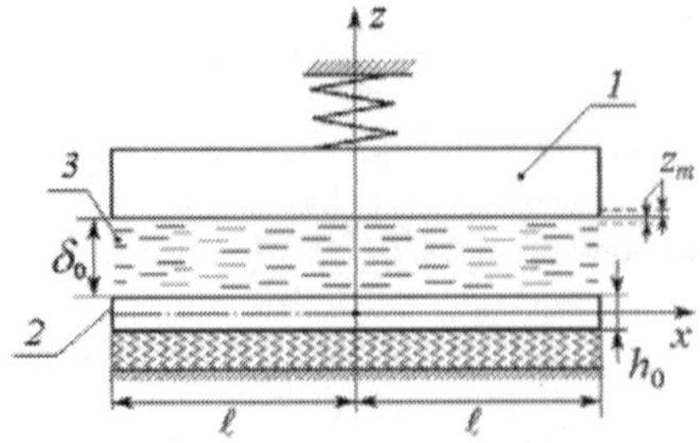

Figure 1. The plate, resting on Pasternak foundation and interacting with the stamp through layer of pulsating viscous liquid

We consider the stamp to possess a spring suspension and to be able to move only in vertical direction. The plate's length is 2ℓ, the thickness is h_0. The plate is simply supported on the edges. We assume the pressure pulsation at the edges acting in terms of the harmonic process. The viscous liquid layer possesses middle thickness $\delta_0 \ll \ell$, while the stamp oscillations amplitude z_m and the plate deflection w_m are considerably less than δ_0. The liquid at the edges freely leaks into the same liquid where the assumed law of pressure pulsation is maintained. The layer of viscous liquid between the stamp and the plate possesses strong damping characteristics. Therefore transitional processes in the considered oscillating system quickly go out and the steady-state harmonic oscillations appear. Consequently, the initial conditions impact may be excluded from consideration, according to [31]. We will continue investigating the steady-state harmonic oscillation regime, taking into account this remark.

The assumed law of pressure pulsation at the edges takes the form of:

$$p = p_0 + p^*(\omega t),\ \ p^*(\omega t) = p_m f_p(\omega t),\ \ f_p(\omega t) = \sin \omega t \tag{1}$$

where p_0 is the constant pressure, p_m is the amplitude of pressure pulsation, ω is the frequency, t is the time.

The equation of the stamp oscillations takes the form of:

$$m d^2 z_1 / dt^2 + n z_1 = N \tag{2}$$

where $z_1 = z_m f(\omega t)$ is the law of stamp motion, z_m is the stamp oscillations amplitude, m is the stamp mass, n is the spring constant, N is the force influencing the stamp due to viscous liquid.

The expression for force N takes the form of:

$$N = -b\int_{-\ell}^{\ell} q_{zz} dx,\ q_{zz} = -p + 2\rho\nu\, \partial u_z / \partial z \text{ at } z = \delta_0 + z_m f(\omega t) \tag{3}$$

where q_{zz} is the normal stress, u_z is the liquid velocity vector projection on a coordinate axis, p is the pressure; ρ is density of the liquid; ν is the kinematic viscosity coefficient of liquid.

The equation of bending oscillations of the plate resting on Pasternak foundation can be written down as [29, 32]:

$$D\partial^4 w/\partial x^4 + \kappa w - \eta\, \partial^2 w/\partial x^2 + \rho_0 h_0\, \partial^2 w/\partial t^2 = q_{zz}\big|_{z=h_0/2+w}. \tag{4}$$

where w is the plate deflection, D is the flexural rigidity of the plate, ρ_0 is the density of the plate material, κ is the foundation normal stiffness, η is the foundation shear coefficient.

The plate dynamic equation is added with the simply supported boundary conditions, i.e.:

$$w = \partial^2 w/\partial x^2 = 0, \text{ at } x = \pm\ell. \tag{5}$$

The movement of viscous liquid layer between the stamp and the plate can be considered as a creeping one [33]. Thus, the liquid dynamic equations represent the Navier-Stokes equation, wherein the inertia members being excluded from the consideration. As a result we have:

$$\begin{gathered}\frac{1}{\rho}\frac{\partial p}{\partial x} = \nu\left(\frac{\partial^2 u_x}{\partial x^2} + \frac{\partial^2 u_x}{\partial z^2}\right),\ \frac{1}{\rho}\frac{\partial p}{\partial z} = \nu\left(\frac{\partial^2 u_z}{\partial x^2} + \frac{\partial^2 u_z}{\partial z^2}\right),\\ \frac{\partial u_x}{\partial x} + \frac{\partial u_z}{\partial z} = 0,\end{gathered} \tag{6}$$

where u_x, u_z are liquid velocity projections on the coordinate axis.

The boundary conditions of (6) include: no-slip conditions

$$\begin{gathered}u_x = \partial u/\partial t,\ u_z = \partial w/\partial t \text{ at } z = h_0/2 + w,\\ u_x = 0,\ u_z = z_m\, df(\omega t)/dt \text{ at } z = h_0/2 + \delta_0 + z_m f(\omega t),\end{gathered} \tag{7}$$

and pressure at the edges conditions

$$p = p_0 + p_m f_p(\omega t) \text{ at } x = \pm\ell . \tag{8}$$

Here u, w are longitudinal movement and deflection plate laws.

3. The Theory

Let us introduce dimensionless variable and small parameters into consideration:

$$\begin{gathered}\xi = x/\ell, \zeta = (z - 0{,}5h_1)/\delta_0, \tau = \omega t, u_z = z_m\omega U_\zeta, \\ u_x = (z_m\omega/\psi)U_\xi, w = w_m W, u = u_m U, \lambda = z_m/\delta_0 << 1, \\ p = p_0 + p^*(\tau) + P\nu\rho z_m\omega\psi^{-2}/\delta_0, \ \psi = \delta_0/\ell << 1.\end{gathered} \tag{9}$$

The problem of liquid dynamics in dimensionless variable (9) in zero approximation in ψ and λ written down as:

$$\partial P/\partial\xi = \partial^2 U_\xi/\partial\zeta^2, \partial P/\partial\zeta = 0, \ \partial U_\xi/\partial\xi + \partial U_\zeta/\partial\zeta = 0, \tag{10}$$

with boundary conditions:

$$\begin{gathered}U_\xi = 0, U_\zeta = df(\tau)/d\tau \text{ at } \zeta = 1, \\ U_\xi = 0, U_\zeta = (w_m/z_m)\partial W/\partial\tau \text{ at } \zeta = 0, P = 0 \text{ at } \xi = \pm 1.\end{gathered} \tag{11}$$

Normal stress in the liquid in zero approximation in ψ takes the form:

$$q_{zz} = -p_0 - \rho\nu z_m\omega(\delta_0\psi^2)^{-1}P, \tag{12}$$

and the expression for force (3) takes the form:

$$N = 2\ell b(p_0 + p^*(\tau)) + b\ell\rho\nu z_m\omega(\delta_0\psi^2)^{-1}\int_{-1}^{1} Pd\xi . \tag{13}$$

By solving (10) with boundary conditions (11) we get:

$$U_\xi = \frac{\zeta^2 - \zeta}{2}\frac{\partial P}{\partial\xi}, U_\zeta = \frac{w_m}{z_m}\frac{\partial W}{\partial\tau} - \frac{2\zeta^3 - 3\zeta^2}{12}\frac{\partial^2 P}{\partial\xi^2} \tag{14}$$

and the equation for pressure:

$$\partial^2 P/\partial\xi^2 = 12(df/d\tau - (w_m/z_m)\partial W/\partial\tau). \tag{15}$$

Further, by solving (15) and satisfying the boundary conditions for the pressure at the edges (11) we get:

$$P = 6(\xi^2 - 1)\frac{df}{d\tau} + 12\frac{w_m}{z_m}\int_\xi^1\int \partial W/\partial\tau \, d\xi d\xi + 6(\xi - 1)\frac{w_m}{z_m}\int_{-1}^{1}\int \partial W/\partial\tau \, d\xi d\xi ., \tag{16}$$

According to the boundary conditions (5) we present the plate deflection dependence in the series of trigonometric functions:

$$w = w_m W = w_m\sum_{k=1}^{\infty}\left(R_k^0 + R_k(\tau)\right)\cos((2k-1)\pi\xi/2), \tag{17}$$

here $R_k(\tau)$ is the time harmonic function, R_k^0 is the constant.

By substituting (16), (17) in (12) and making disintegration of the remaining members in the series of the chosen trigonometric functions of longitudinal coordinate ξ, we get:

$$q_{zz} = (p_0 + p^*(\tau))\sum_{k=1}^{\infty}\frac{4(-1)^k}{(2k-1)\pi}\cos\frac{2k-1}{2}\pi\xi -$$
$$-\frac{\rho v z_m \omega}{\delta_0 \psi^2}\left[12\frac{w_m}{z_m}\sum_{k=1}^{\infty}\left(\frac{2}{(2k-1)\pi}\right)^2\frac{dR_k}{d\tau}\cos\frac{2k-1}{2}\pi\xi + \right. \tag{18}$$
$$\left. +12\frac{df}{d\tau}\sum_{k=1}^{\infty}\frac{4(-1)^k}{(2k-1)\pi}\left(\frac{2}{(2k-1)\pi}\right)^2\cos\frac{2k-1}{2}\pi\xi\right].$$

Equation (4) with consideration of (17), (18) can be written down as:

$$\left(\frac{D}{\ell^4}\left(\frac{(2k-1)\pi}{2}\right)^4 + \kappa + \frac{\eta}{\ell^2}\left(\frac{(2k-1)\pi}{2}\right)^2\right)w_m(R_k^0 + R_k) +$$
$$+\rho_0 h_0 \omega^2 w_m \frac{d^2R_k}{d\tau^2} = -12\frac{\rho v \omega}{\delta_0\psi^2}\left(\frac{2}{(2k-1)\pi}\right)^2 w_m\frac{dR_k}{d\tau} + \tag{19}$$
$$+12\frac{\rho v \omega}{\delta_0\psi^2}\frac{4(-1)^k}{(2k-1)\pi}\left(\frac{2}{(2k-1)\pi}\right)^2 z_m\frac{df}{d\tau} + (p_0 + p^*(\tau))\frac{4(-1)^k}{(2k-1)\pi}.$$

Due to linearity of (19) for the constant pressure component we get the following expression:

$$w_m R_k^0 = \frac{4(-1)^k p_0}{(2k-1)\pi}\left(\frac{D}{\ell^4}\left(\frac{(2k-1)\pi}{2}\right)^4 + \frac{\eta}{\ell^2}\left(\frac{(2k-1)\pi}{2}\right)^2 + \kappa\right)^{-1}. \tag{20}$$

The solution of (19) for the regime of steady-state harmonic oscillations is written down as:

$$w_m R_k = \frac{4(-1)^k}{(2k-1)\pi}\left[p_m\left(C_{pk}f_p(\tau) + B_{pk}\frac{df_p(\tau)}{d\tau}\right) + z_m\left(C_{zk}\frac{df(\tau)}{d\tau} + B_{zk}\frac{d^2f(\tau)}{d\tau^2}\right)\right], \tag{21}$$

$$C_{zk} = \frac{a_{2k}^2}{a_{1k}^2 + a_{2k}^2},\ C_{pk} = -\frac{a_{2k}}{a_{1k}^2 + a_{2k}^2},\ B_{zk} = -\frac{a_{2k}a_{1k}}{a_{1k}^2 + a_{2k}^2},\ B_{pk} = \frac{a_{1k}}{a_{1k}^2 + a_{2k}^2},$$

$$a_{1k} = \widetilde{D}_k - \rho_0 h_0 \omega^2,\ a_{2k} = 12\frac{\rho v \omega}{\delta_0 \psi^2}\left(\frac{2}{(2k-1)\pi}\right)^2,\ \widetilde{D}_k = \frac{D}{\ell^4}\left(\frac{(2k-1)\pi}{2}\right)^4 + \frac{\eta}{\ell^2}\left(\frac{(2k-1)\pi}{2}\right)^2 + \kappa.$$

By substituting (21), (16) in the expression (13), and then taking into account the obtained expression in (2), we get:

$$(m + M_z)\frac{d^2z_1}{dt^2} + nz_1 + K_z\frac{dz_1}{dt} = 2\ell b p_0 + (2\ell b + M_p)p^* + K_p\frac{dp^*}{dt}, \tag{22}$$

here we introduce the symbols:

$$K_z = 8\ell b\frac{\rho v}{\delta_0\psi^2}\left[1 - 6\sum_{k=1}^{\infty}\left(\frac{2}{(2k-1)\pi}\right)^4 B_{zk}\right],\ K_p = -48\ell b\frac{\rho v}{\delta_0\psi^2}\sum_{k=1}^{\infty}\left(\frac{2}{(2k-1)\pi}\right)^4 B_{pk},$$

$$M_z = 48\ell b \frac{\rho v}{\omega \delta_0 \psi^2} \sum_{k=1}^{\infty} \left(\frac{2}{(2k-1)\pi} \right)^4 C_{zk}, \; M_p = 48\ell b \frac{\rho v \omega}{\delta_0 \psi^2} \sum_{k=1}^{\infty} \left(\frac{2}{(2k-1)\pi} \right)^4 C_{pk}.$$

The solution of (22) for the regime of steady-state harmonic oscillations takes the form of:

$$z_1 = \frac{2\ell b p_0}{n} + p_m A_z(\omega) \sin(\omega t + \varphi_z(\omega)),$$
$$A_z(\omega) = \sqrt{\tilde{C}^2 + \tilde{B}^2}, \; \varphi_z(\omega) = \mathrm{arctg}(\tilde{B}/\tilde{C}), \quad (23)$$

$$\tilde{C} = \frac{e_1 d_1 - e_2 d_2}{d_1^2 + d_1^2}, \; \tilde{B} = \frac{e_1 d_2 + e_2 d_1}{d_1^2 + d_1^2}, \; d_1 = n - (m + M_z)\omega^2, \; d_2 = K_z \omega, \; e_1 = 2\ell b + M_p, \; e_2 = K_p \omega.$$

Finally, with consideration of (20), (21) and (23) the law of the plate hydroelastic deflection (17) takes the form of:

$$w = \frac{p_0 \ell^4}{D} \sum_{k=1}^{\infty} \frac{4(-1)^k}{(2k-1)\pi} \left(\left(\frac{(2k-1)\pi}{2} \right)^4 + \frac{\eta \ell^2}{D} \left(\frac{(2k-1)\pi}{2} \right)^2 + \frac{\kappa \ell^4}{D} \right)^{-1} \cos\left(\frac{2k-1}{2\ell} \pi x \right) -$$
$$- p_m A_w(x,\omega) \sin(\omega t + \varphi_w(x,\omega)), \; A_w(x,\omega) = \sqrt{T^2 + E^2}, \; \varphi_w(x,\omega) = \mathrm{arctg}(E/T), \quad (24)$$

$$T = \sum_{k=1}^{\infty} \frac{4(-1)^k}{(2k-1)\pi} \left(B_{pk} - (B_{zk}\tilde{C} + C_{zk}\tilde{B}) \right) \cos\frac{2k-1}{2\ell} \pi x, \; E = \sum_{k=1}^{\infty} \frac{4(-1)^k}{(2k-1)\pi} \left(C_{pk} - (B_{zk}\tilde{B} - C_{zk}\tilde{C}) \right) \cos\frac{2k-1}{2\ell} \pi x.$$

By substituting (21) in (16) and taking into account (7), the pressure in the liquid layer between the stamp and the plate is written down as:

$$p = p_0 + p_m \sin \omega t + p_m \Pi(x,\omega) \sin(\omega t + \varphi_p(x,\omega)),$$
$$\Pi(x,\omega) = \sqrt{S^2 + Q^2}, \; \varphi_p(x,\omega) = \mathrm{arctg}(Q/S), \quad (25)$$

$$Q = 6\left(\left(\frac{x}{\ell} \right)^2 - 1 \right) \tilde{C} + 24 \sum_{k=1}^{\infty} \left(\frac{2}{(2k-1)\pi} \right)^3 (-1)^k (B_{pk} - C_{zk}\tilde{B} - B_{zk}\tilde{C}) \cos\frac{2k-1}{2\ell} \pi x,$$

$$S = 6\left(1 - \left(\frac{x}{\ell} \right)^2 \right) \tilde{B} + 24 \sum_{k=1}^{\infty} \left(\frac{2}{(2k-1)\pi} \right)^3 (-1)^k (B_{zk}\tilde{B} - C_{zk}\tilde{C} - C_{pk}) \cos\frac{2k-1}{2\ell} \pi x.$$

4. Summary and Conclusion

The obtained analytical solution allows making the following conclusions. The function $A_z(\omega)$ of expression (23) is the stamp amplitude-frequency characteristic and $\varphi_z(\omega)$ is the stamp phase response. The first element of expression (24) for the deflection presents the plate static deflection, conditioned by static pressure in liquid p_0, the second element is a dynamic deflection, conditioned by the plate and stamp interaction through a viscous liquid layer. The values of the given deflection component are defined by function $A_w(x,\omega)$, which can be considered as frequency dependent function of deflection amplitude distribution along the plate.

The analogous remarks can be made in relation to the law of pressure change in the channel (25). The first and the second components in the pressure law present the assumed static pressure and the law of pressure pulsation at the edges. The third component is dynamic pressure in the channel, conditioned by the stamp motion and oscillation of the plate resting on Pasternak foundation. The value of the third component is defined by function $\Pi(x,\omega)$, which is the frequency dependent function of dynamic pressure amplitudes distribution along the channel. Function $\varphi_w(x,\omega)$ is the frequency dependent function of the phase response distribution of plate deflection, and function $\varphi_p(x,\omega)$ is the frequency dependent function of the phase response distribution of pressure along the channel. The investigation of the mentioned above functions behavior makes it possible to study dynamic processes in the oscillation system under consideration. The analysis of the obtained expression for the plate deflection (24) allows to confirm, that the coefficients depending on Pasternak foundation reaction will affect as the static plate deflection, as the dynamic plate deflection. Also, the obtained solution provides transition from Pasternak foundation to Winkler one, i.e. in the case then the foundation shear coefficient η equaling zero. Thus, the results of the considered study can be used for mathematical modeling and analyzing hydroelastic oscillations of the constructions elastic elements interacting with viscous liquid and elastic foundation.

Acknowledgments
This study was supported by the Russian Foundation for Basic Research grants 15-01-01604-a and 16-01-00175-a.

References

[1] Indeitsev D A, Polypanov I S, Sokolov S K 1994 Calculation of Cavitation Life-Time of Ship Engine Liner *Problemy Mashinostraeniya i Nadezhnos'ti Mashin* **4** pp 59-64.
[2] Haddara M R, Cao S A 1996 Study of the Dynamic Response of Submerged Rectangular Flat Plates *Marine Structures* **9**(10) pp 913-933
[3] Chapman C J, Sorokin S V 2005 The forced vibration of an elastic plate under significant fluid loading *Journal of Sound and Vibration* **281**(3) pp 719-741.
[4] Ergin A, Ugurlu B 2003 Linear vibration analysis of cantilever plates partially submerged in fluid *Journal of Fluids and Structures* **17** pp 927-939.
[5] Kramer M R, Liu Z, Young Y L 2013 Free vibration of cantilevered composite plates in air and in water *Composite Structures* **95** pp 254-263.
[6] Kerboua Y, Lakis A A, Thomas M, Marcouiller L 2008 Vibration analysis of rectangular plates coupled with fluid *Applied Mathematical Modelling* **32**(12) pp 2570-2586.
[7] Bochkarev S A, Lekomtsev S V, Matveenko V P 2016 Hydroelastic stability of a rectangular plate interacting with a layer of ideal flowing fluid *Fluid Dynamics* **51**(6) pp 821-833.
[8] Avramov K V, Strel'nikova E A 2014 Chaotic oscillations of plates interacting on both sides with a fluid flow *International Applied Mechanics* **50**(3) pp 303-309.
[9] Önsay T 1993 Effects of layer thickness on the vibration response of a plate-fluid layer system *Journal of Sound and Vibration* **163**(2) pp 231-259.
[10] Ageev R V, Mogilevich L I, Popov V S, Popova A A, Kondratov D V 2014 Mathematical Model of Pulsating Viscous Liquid Layer Movement in a Flat Channel with Elastically Fixed Wall *Applied Mathematical Sciences* **8**(159) pp 7899-7908.
[11] Faria C T, Inman D J 2014 Modeling energy transport in a cantilevered Euler-Bernoulli beam actively vibrating in Newtonian fluid *Mechanical Systems and Signal Processing* **45**(2) pp 317-329.
[12] Mogilevich L I, Popov V S 2011 Investigation of the interaction between a viscous incompressible fluid layer and walls of a channel formed by coaxial vibrating discs *Fluid Dynamics* **46**(3) p 375-388.
[13] Mogilevich L I, Popov V S, Popova A A 2010 Dynamics of interaction of elastic elements of a vibrating machine with the compressed liquid layer lying between them *Journal of Machinery Manufacture and Reliability* **39**(4) p 322-331.
[14] Akcabay D T, Young Y L 2012 Hydroelastic response and energy harvesting potential of flexible piezoelectric beams in viscous flow *Physics of Fluids* **24**(5).

[15] Ageev R V, Kuznetsova E L, Kulikov N I, Mogilevich L I, Popov V S 2014 Mathematical model of movement of a pulsing layer of viscous liquid in the channel with an elastic wall *PNRPU Mechanics Bulletin* **3** pp 17-35.

[16] Ageev R V, Mogilevich L I, Popov V S 2014 Vibrations of the walls of a slot channel with a viscous fluid formed by three-layer and solid disks *Journal of Machinery Manufacture and Reliability* **43**(1) pp 1-8.

[17] Wang Y H, Tham L G, Cheung Y K 2005 Beams and plates on elastic foundations: A review *Progress in Structural Engineering and Materials* **7**(4) pp 174-182.

[18] Kubenko V D, Pleskachevskii Yu M, Starovoitov E I, Leonenko D V 2006 Natural vibration of a sandwich beam on an elastic foundation *International Applied Mechanics* **42**(5) pp 541-547.

[19] Starovoitov E I, Leonenko D V 2012 Thermal impact on a circular sandwich plate on an elastic foundation *Mechanics of Solids* **47**(1) pp 111-118.

[20] Starovoitov E I, Leonenko D V 2016 Vibrations of circular composite plates on an elastic foundation under the action of local loads *Mechanics of Composite Materials* **52**(5) pp 665-672.

[21] Dash P R, Pradhan M and Pradhan P K 2016 Static and dynamic stability analysis of an asymmetric sandwich beam resting on a variable pasternak foundation subjected to thermal gradient *Meccanica*, **51**(3) pp 725-739.

[22] Alekseev V V, Indeitsev D A, Mochalova Yu A 1999 Resonant oscillations of an elastic membrane on the bottom of a tank containing a heavy liquid *Technical Physics* **44**(8) pp 903-907.

[23] Hosseini-Hashemi S, Karimi M, Hossein Rokni D T 2010 Hydroelastic vibration and buckling of rectangular Mindlin plates on Pasternak foundations under linearly varying in-plane loads *Soil Dynamics and Earthquake Engineering*, **30**(12) pp 1487-1499.

[24] Ergin A, Kutlu A, Omurtag M H, Ugurlu B 2012 Dynamic response of Mindlin plates resting on arbitrarily orthotropic Pasternak foundation and partially in contact with fluid *Ocean Engineering* **42** pp 112-125.

[25] Ergin A, Kutlu A, Omurtag M H., Ugurlu B 2008 Dynamics of a rectangular plate resting on an elastic foundation and partially in contact with a quiescent fluid *Journal of Sound and Vibration*, **317**(1-2) pp 308-328.

[26] Kuznetsova E L, Mogilevich L I, Popov V S, Rabinsky L N 2016 Mathematical model of the plate on elastic foundation interacting with pulsating viscous liquid layer *Applied Mathematical Sciences* **10**(23) pp 1101-1109.

[27] Mogilevich L I, Popov V S and Popova A A 2017 Interaction dynamics of pulsating viscous liquid with the walls of the conduit on an elastic foundation *Journal of Machinery Manufacture and Reliability*, **46**(1) pp 12-19.

[28] Mogilevich L I, Popov V S, Popova A A, Christoforova A V 2016 Mathematical modeling of highly viscous liquid dynamic interaction with walls of channel on elastic foundation *IEEE Conference 2016 Dynamics of Systems, Mechanisms and Machines (Omsk, 2016)*.

[29] Mogilevich L I, Popov V S, Popova A A, Christoforova A V 2016 Mathematical modeling of hydroelastic walls oscillations of the channel on Winkler foundation under vibrations *Vibroengineering PROCEDIA* **8** pp 294-299.

[30] Mogilevich L I and Popov V S 2016 Mathematical modeling of incompressible viscous liquid layer interaction dynamics in an annular slit with its wall, surrounded by elastic medium *IEEE Conference 2016 Dynamics of Systems, Mechanisms and Machines (Omsk, 2016)*.

[31] Panovko Y G and Gubanova I I 1965 *Stability and Oscillations of Elastic Systems* (New York: Consultants Bureau Enterprises, Inc.).

[32] Pasternak P L 1954 *On a New Method of Analysis of an Elastic Foundation by Means of Two Foundation Constants* (Moscow: Gosudarstvennoe Izdatelstvo Literaturi po Stroitelstvu i Architekture)

[33] Loitsyanskii L G 1966 *Mechanics of Liquids and Gases* (Oxford: Pergamon Press).

IOP Conf. Series: Journal of Physics: Conf. Series **944** (2017) 012082 doi:10.1088/1742-6596/944/1/012082

The chemical energy unit partial oxidation reactor operation simulation modeling

A N Mrakin[1], A A Selivanov[1], P A Batrakov[2] and D G Sotnikov[1]

[1]Yuri Gagarin State Technical University of Saratov, 77, Politechnicheskaya Str., Saratov 410054, Russia
[2]Omsk State Technical University, 11, Mira Ave., Omsk 644050, Russia

e-mail: peter_1cool@mail.ru

Abstract. The chemical energy unit scheme for synthesis gas, electric and heat energy production which is possible to be used both for the chemical industry on-site facilities and under field conditions is represented in the paper. The partial oxidation reactor gasification process mathematical model is described and reaction products composition and temperature determining algorithm flow diagram is shown. The developed software product verification showed good convergence of the experimental values and calculations according to the other programmes: the temperature determining relative discrepancy amounted from 4 to 5 %, while the absolute composition discrepancy ranged from 1 to 3%. The synthesis gas composition was found out practically not to depend on the supplied into the partial oxidation reactor (POR) water vapour enthalpy and compressor air pressure increase ratio. Moreover, air consumption coefficient α increase from 0.7 to 0.9 was found out to decrease synthesis gas target components (carbon and hydrogen oxides) specific yield by nearly 2 times and synthesis gas target components required ratio was revealed to be seen in the water vapour specific consumption area (from 5 to 6 kg/kg of fuel).

1. Introduction

The energy resources and petrochemical feedstock providing is the indispensable prerequisite of every country economy successful development. The natural organic raw materials main sources are oil, natural gas, coal, oil shale and biomass. Nowadays, the most demanded chemical and energy resources are oil and natural gas [1]. Their production is reaching the highest level possible and in the near future is to reduce. Consequently, the importance of the organic raw materials alternative sources such as oil sludge, gas condensate, etc., increases. To overcome the petroleum feedstock shortage, the synthesis gas (syngas) large-scale production according to the energy and technology principle with the pyrolysis and gasification technologies application has started [2-4]. Practically all the most important petrochemical products are assumed to be obtained by the synthesis gas chemical processing.

2. The problem statement

Taking into account the foregoing, the consideration of energy and technology units for energy carriers (heat and electric power) and synthesis gas combined generation by gas condensate gasification seems to be quite relevant. The chemical energy units with fuel partial oxidation (gasification) establishment near hydrocarbon deposits and synthesis gas major consumers will make possible not only to provide them with energy carriers, but also with the process gas for various industries. The unit heat diagram is shown in figure 1. In case of economic expediency, the chemical energy unit operation including the third party consumers supply with carbon dioxide and gaseous nitrogen is possible to be organized.

The unit operates in the following way. The partial oxidation reactor 3 consumes air, process oxygen (98% O_2) or their mixture (oxygen enriched air) from the compressor 1 and/or air separation unit 2 and water (blow) vapour from the vapour waste heat boiler 6 for liquid or gaseous fuel gasification process performing. The generated reaction products enter the gas turbine 4, which drives the electric generator 5.

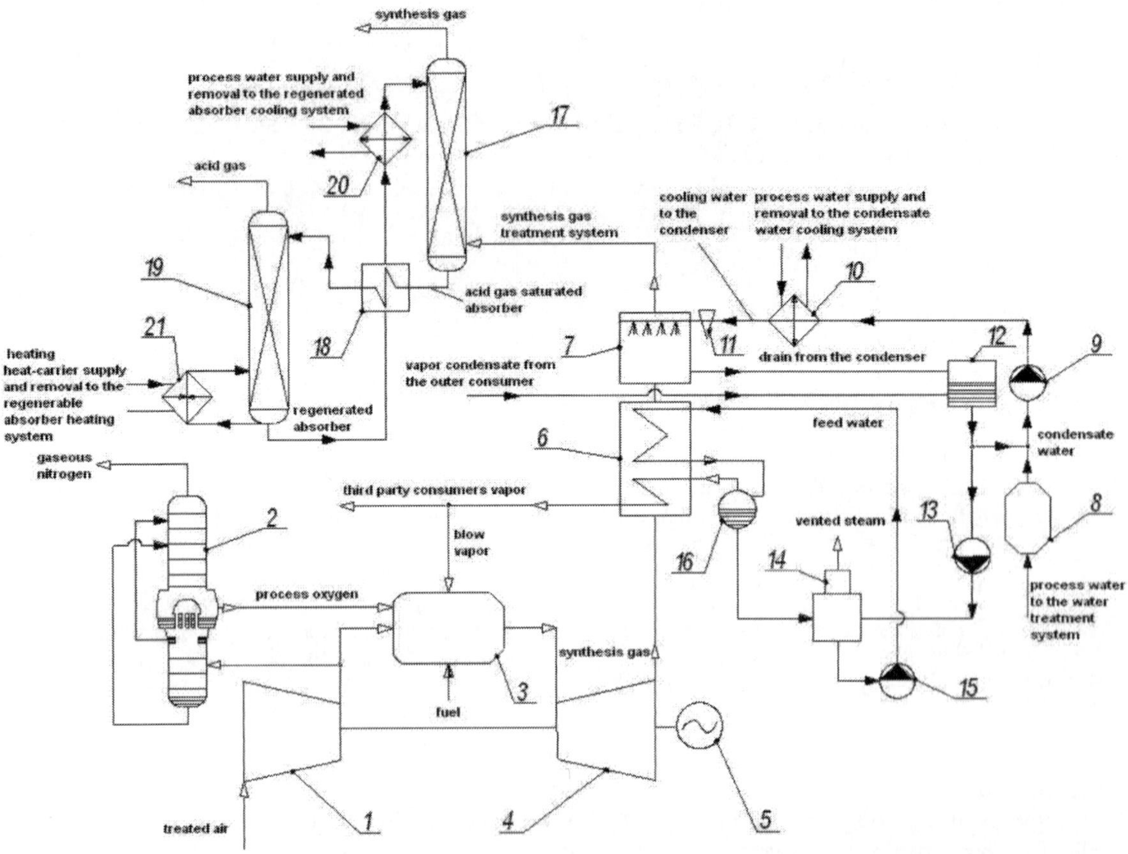

Figure 1. The basic heat diagram of the chemical energy unit including process gas, electric and heat energy producing: 1 – compressor; 2 – air separation unit; 3 – partial oxidation reactor; 4 – gas turbine; 5 – electric generator; 6 – vapour waste heat boiler; 7 – contact condenser; 8 – water treatment system; 9 – condensate water pump; 10 – condensate water cooler; 11 – condensate water filter; 12 – water storage tank; 13 – condensate pump; 14 – deaerator; 15 – feed pump; 16 – separator; 17 – scrubber for the acid gas absorbing by the liquid absorber; 18 – regenerative heat exchanger; 19 – liquid absorber regenerator; 20 – refrigerator; 21 – liquid absorber heater for regeneration.

The exhaust synthesis gas with the pressure required for processing equipment aerodynamic resistance overcoming and synthesis processes normal conducting is fed to the waste heat boiler 6 where it is cooled and then directed to the contact condenser 7 where water vapour which enters the water storage tank 12 condenses. Then synthesis gas is supplied into the scrubber for the acid gas absorbing by the liquid absorber 17 for the further treatment and equivalence ratio bringing to the required values according to the synthesis gas further use technology regulations. The synthesis gas prepared in this way, consisting of nitrogen, hydrogen and carbon monoxide, is transferred to the technological consumer or can be used to generate energy carriers (heat and electric energy). Treated as above synthesis gas consisting of nitrogen, hydrogen and carbon oxide is supplied to the process consumer or

is possible to be used for energy carriers (heat and electric power) generation. The part of the vapour produced in the waste heat boiler 6 is supplied into the partial oxidation reactor 3, while the remained vapour is sold to the consumers. Gas turbine unit electric generator 5 generates electric power transferred to the end-users. From the water storage tank 12 through the condensate water cooler 10, where condensate water is cooled by the process water from the recycling water supply system and condensate water filter 11 by applying the condensate water pump 9 and water treatment system 8 connected with the process water supply line to the water treatment system, cooled water is supplied to the contact condenser 7. The condensate from the consumer is returned into the water storage tank 12, where from through the deaerator 14 by using the condensate pump 13 and feed pump 15, the feed water passes through the waste heat boiler 6 economizer and evaporative section and enters the separator 16 where it is decomposed into vapour and water. Water vapour passes through the waste heat boiler 6 superheater zone and enters the partial oxidation reactor 3 as well as it is supplied to the consumer, while water returns to the deaerator 14. The absorber saturated with acid gas from the scrubber for acid gas absorbing by the liquid absorber 17 is supplied to the regenerative heat exchanger 18 where it receives heat from the regenerated absorber and to the liquid absorber regenerator 19 where the regeneration occurs. It should be noted that the part of the regenerable absorber is bled from the liquid absorber regenerator bottom 19 and passes through the liquid absorber heater for regeneration 21 where it is heated by the heating neat-carrier from the regenerable absorber heating system until it returns to the regenerator. The regenerated absorber is fed to the heat exchanger 18 where the heat is transferred to the absorber saturated with acid gas, then to the refrigerator 20 where it is cooled by the process water from the liquid absorber cooling system before returns to the scrubber for acid gas absorbing by the liquid absorber 17.
The most interesting from the simulation modeling point of view is the partial oxidation reactor (POR) represented in figure 2 which can be considered as the plug-flow reactor. Fuel, air and/or oxygen as well as water vapour are supplied to this unit for H_2 content in the synthesis gas bringing to the required by the consumer values. Oxidizing agent is fed less than the stoichiometric value. The POR operating process is similar to gas turbine combustion chambers one, however, it has its own peculiarities and is defined by the specific conditions of aerodynamics, chemically-reacting system, heat exchange, temperature level, structural and geometric dimensions.

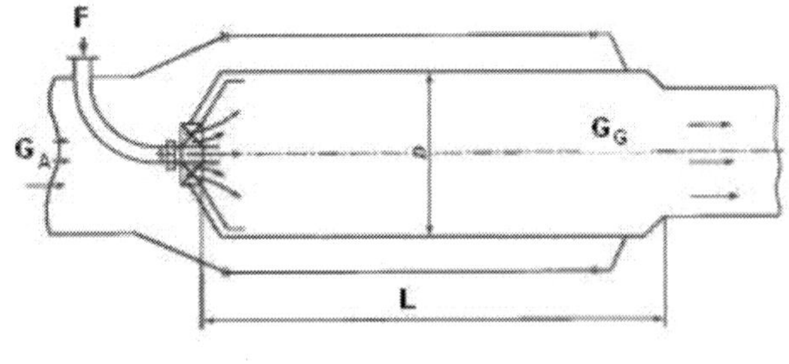

Figure 2. The chemical energy unit partial oxidation reactor sketch: G_A – air and water vapour mixture entering the POR chamber; G_G – reaction products entering the gas turbine; F – fuel entering the atomizer; D – POR chamber diameter; L – POR chamber length.

Burning due to lack of oxidizer and reaction products composition calculation as well as the theoretical reaction temperature determination are more difficult compared to the complete oxidation since gases should contain such components as CO and H_2 along with the complete combustion products when equilibrium state is reached.

3. Theory

To evaluate the reaction products quantitative and qualitative characteristics and POR mass-dimensional parameters, it is necessary to develop the mathematical model. The gas-generating process calculation algorithm basic flow diagram is presented in figure 3.

IOP Conf. Series: Journal of Physics: Conf. Series **944** (2017) 012082 doi:10.1088/1742-6596/944/1/012082

When developing the mathematical model, the following assumptions not impacting the research results accuracy were admitted: the reaction products are evenly (perfectly) mixed and in the equilibrium state, in this case at any point of the reaction mixture the reaction products composition will be the pressure and temperature function. The assumption is characteristic of the other authors works, for example [4, 5, 6].

The reaction products equilibrium composition defining reduced to solving the nonlinear algebraic equations system including the material balance equations in chemical elements constituting the initial fuel, the equilibrium constant equation for which the defining reaction is CO conversion reaction by water vapour (water gas reaction [5]) as well as Dalton equation defining the total pressure connection with the synthesis gas individual components partial pressures. The obtained synthesis gas composition was calculated by the following main components: carbon oxide, carbon dioxide, methane, hydrogen, water vapour and nitrogen.

IOP Conf. Series: Journal of Physics: Conf. Series **944** (2017) 012082 doi:10.1088/1742-6596/944/1/012082

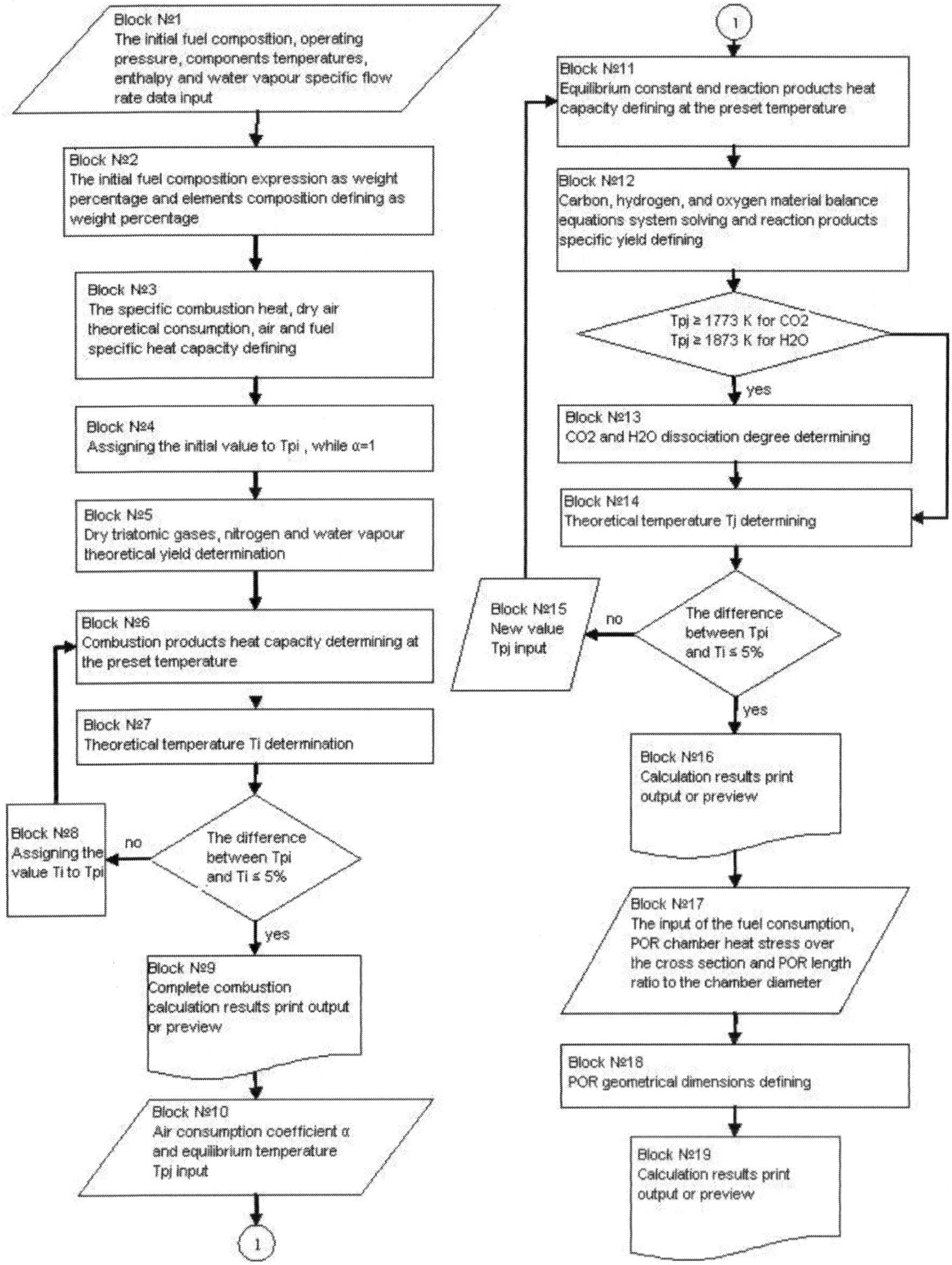

Figure 3. Gas-generating process calculation algorithm flow diagram.

Taking into account the presented methodological postulates and materials [7], the reaction products composition, their temperature and flow calculating program [8] considering the initial fuel process composition impact, fuel flows and blowing vapour ratios, oxidizing agent as well as POR pressure and temperature was developed in C++ environment and registered.
The programme operating principle is based on the iterative calculations by using the least squares method and relies on theoretical temperature determining followed by its comparison and necessary basic data preliminary determination in the corresponding programme blocks.
The program makes possible to take into account CO_2 and H_2O dissociation (more than 1773 K and 1873 K correspondingly [9]) at the high temperatures. Carbon dioxide α_D and water vapour β_D dissociation degrees depending on their partial pressures and temperature Tp_j are identified according to the reference data presented in [9] before heat losses from the carbon dioxide and water vapour dissociation are determined, kJ/kg:

$$q_D = 12640 \cdot \alpha_D \cdot v_{CO_2} + 10800 \cdot \beta_D \cdot v_{H_2O} \quad (1)$$

Reaction products theoretical temperature is determined taking into account the formula (1) by the expression, K:

$$T_j = \frac{Q_I^o + \alpha \cdot L_0 \cdot t_{air} + c_{cond.} \cdot t_{cond.} + G \cdot h - q_D - 10800 \cdot v_{H_2} - 12640 \cdot v_{CO}}{v_{N_2} \cdot c_{N_2} + v_{CO_2} \cdot c_{CO_2} + v_{H_2O} \cdot c_{H_2O} + v_{CO} \cdot c_{CO} + v_{H_2} \cdot c_{H_2} + v_{SO_2} \cdot c_{SO_2}} + 273 \quad (2)$$

where Q_I^o is the gas condensate combustion specific heat, kJ/kg; α is the oxidizer flow coefficient (air); L_0 is the dry air specific consumption for the fuel unit complete combustion, m^3/kg; t_{air} is the oxidizer temperature (air), °C; $c_{cond.}$ is the fuel heat capacity (gas condensate), kJ/(kg·K); $t_{cond.}$ is the fuel temperature (gas condensate), °C; G is the specific consumption of vapour supplied to POR, kg/kg of gas condensate; h is the enthalpy of vapour supplied to POR, kJ/kg; *10800* is the hydrogen low heating value, kJ/m^3 [9]; *12640* is the carbon oxide low heating value, kJ/m^3 [9]; v_i is the synthesis gas individual components theoretical yield, m^3/kg; c_i is the synthesis gas individual components heat capacity, kJ/(kg·K).

4. Experimental results

To verify the proposed mathematical model, the calculations by using [8] and the Aspen HYSYS oil and gas processing manufacturing procedures accurate modeling system representing the software package for the chemical and engineering productions modeling, designing, optimization and business planning in the field of hydrocarbons production and processing and petrochemicals were performed. In this case the software product Aspen HYSYS performs the plug-flow, perfect-mixing, equilibrium, stoichiometric reactors calculation. The results obtained by using mathematical models and experimental data are presented in tables 1 and 2.
Table 1 shows the results of methane conversion to air blowing at the air consumption coefficient of α = 0.34, the initial methane-air mixture heating to 400 °C and the pressure of 101.325 kPa; table 2 presents the experience and calculations results of the initial methane-air mixture heating combining to 315 °C and air enrichment to 26% of oxygen at α = 0.42 and the pressure of 101.325 kPa.

Table 1. POR calculation results comparison with the factual data.

Parameter, UM	**Value**		
	Experimental data [10]	The calculation by applying [8]	The calculation by applying the software product Aspen HYSYS
Synthesis gas composition, volume %			
N_2	46.5	46.2	45.9
H_2O	5.7	7.2	4.9
CO_2	2.6	0.7	1.4

CO	13.8	16.8	16.6
H_2	30.0	28.4	31.1
CH_4	1.2	0.6	0.0
Synthesis gas temperature, °C	1000	1049	1039

Table 2. POR calculation results comparison with the factual data.

Parameter, UM	**Value**		
	Experimental data [11]	The calculation by applying [8]	The calculation by applying the software product Aspen HYSYS
Synthesis gas composition, volume			
N_2	43.0	49.2	49.6
H_2O	9.9	16.0	15.3
CO_2	3.8	3.0	2.7
CO	14.8	13.7	14.1
H_2	26.0	18.1	18.3
CH_4	1.4	0.0	0.0
C_nH_m	1.1	0.0	0.0
Synthesis gas temperature, °C	1187	1569	1440

The results presented in tables 1 and 2 demonstrate the obtained POR research tool adequacy as a software product which calculation results showed good convergence of the experimental values and calculations according to the other programmes: the temperature determining relative discrepancy amounted from 5 to 25 %, while the absolute composition discrepancy ranged from 3 to 6%. Such a high synthesis gas temperature discrepancy is likely explained by not taking into consideration the heat losses into the environment in the mathematical models which in fact constitutes a high value and ranges from 250 to 380 °C.

The composition and parameters calculation results of the produced gas from the condensate of the following composition: CH_4=16.1%; C_2H_6=3.1%; C_3H_8=1.6%; i-C_4H_{10}=1.1%; n-C_4H_{10}=1.0%; i-C_5H_{12}=0.5%; n-C_5H_{12}=0.9%; C_{6+}=73.8%; CO_2=1.8%; CH_3OH=0.1% at h=2681 kJ/kg; $t_{condensate}$=–27 °C are represented in figures 4 and 5.

Moreover, during the preliminary calculations performing [7] the synthesis gas composition was found out practically not to depend on the supplied to POR water vapour enthalpy and compressor pressure increase ratio.

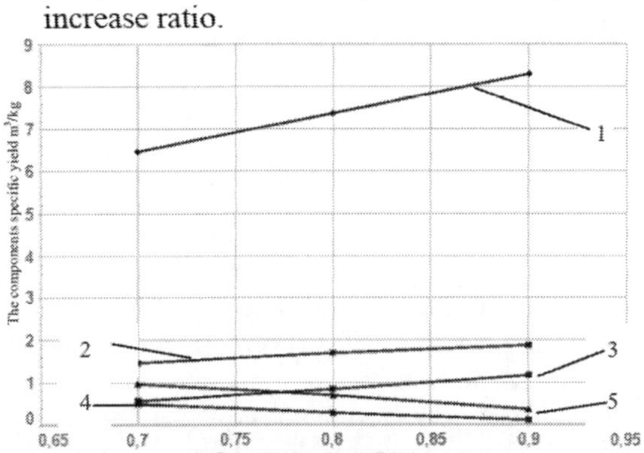

Figure 4. The synthesis gas components specific yield

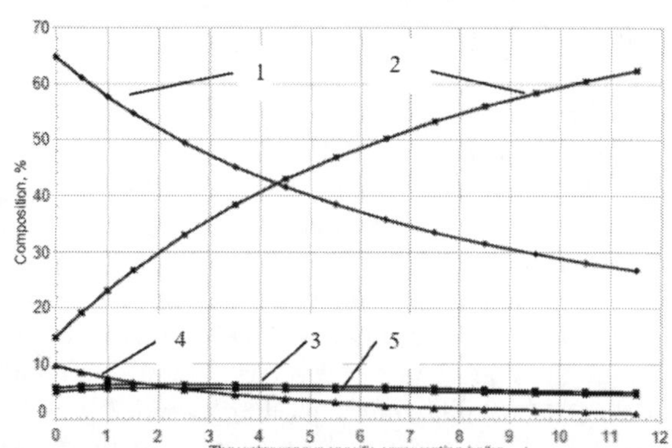

Figure 5. The synthesis gas composition dependence on

dependence on the air consumption coefficient at π_c=5 and G=0 kg/kg: 1 is nitrogen; 2 is water vapour; 3 is carbon dioxide; 4 is carbon oxide; 5 is hydrogen.

the water vapour specific consumption G, kg/kg at α=0.7 and π_c=5: 1 is nitrogen; 2 is water vapour; 3 is carbon dioxide; 4 is carbon oxide; 5 is hydrogen.

Figure 4 analysis shows that the air consumption coefficient α increase from 0.7 to 0.9 reduces the synthesis gas target components specific yield (carbon oxide and hydrogen) by almost 2 times, while as α reaches the stoichiometric value (value 1.0), the reaction products become the combustion products. Figure 5 shows that the synthesis gas target components required ratio is observed in the field of the water vapour specific consumption from 5 to 6 kg/kg, however, there will be the increased value of the unreacted water vapour content in the synthesis gas from 45 to 49 % that will impose certain peculiarities on the gas turbine, vapour waste heat boiler, and contact condenser operation.
The reactor geometry calculation results depending on the pressure created by the chemical energy unit compressor at the gas condensate oxidation of 1400 kg/h with the specific combustion heat of Q_l^o=44847 kJ/kg for the various POR chamber heat stresses corresponding to the forced (lines 4 and 5) and unforced (lines 1, 2 and 3) operation are represented in figure 6.

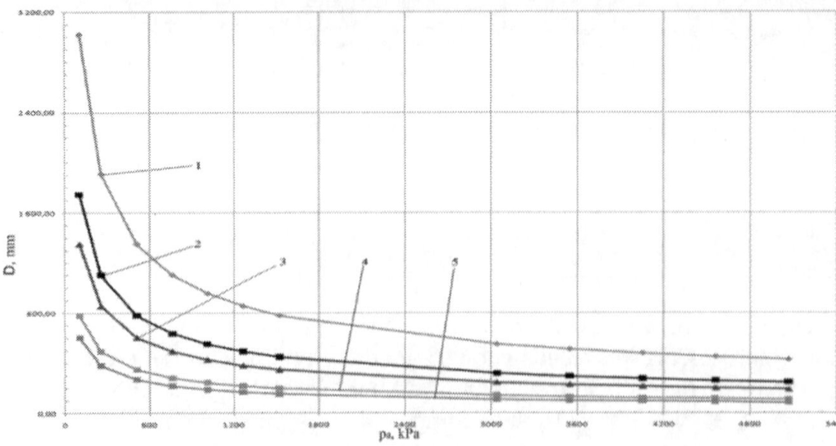

Figure 6. POR diameter dependence on pressure: 1 is 24 W/(m^2·Pa); 2 is 72 W/(m^2·Pa); 3 is 120 W/(m^2·Pa); 4 is 360 W/(m^2·Pa); 5 is600 W/(m^2·Pa).

As can be seen in figure 6 the transfer from the atmospheric pressure to the increased one (π_c = 5) makes it possible to reduce the POR diameter by almost 2.2 times from 3050 mm to 1350 for the unforced operation (line 1, figure 6) and by 2.4 times for the forced one (line 5, figure 6). Thus, it can be concluded that under the same POR pressure range, the most preferred operation mode is the forced one reducing the unit specific amount of metal (capital investments). Furthermore, figure 6 shows that the considerable pressure influence on POR diameter practically ceases in the range of π_c = 30-36 for the unforced operation mode (line 1, figure 6) and π_c = 10-15 for the forced one (line 5, figure 6).

5. Results and discussion

Thus, according to the results of the carried out preliminary studies, the POR calculating programme which is possible to be integrated into the complete chemical energy unit complex mathematical description was developed. On the basis of this programme the synthesis gas composition was found out practically not to depend on the enthalpy of the supplied to POR water vapour and the compressor air pressure ratio, and consequently these parameters can be optimized on the assumption of the system thermodynamic analysis and economic efficiency defining principles. Based on the tend to the synthesis gas target components content maximum, it is necessary to approach the air consumption coefficient minimum value by the soot formation condition, i.e. to α = 0.7 for which the water vapour specific consumption appropriate value at the range from 5 to 6 kg/kg of gas condensate was chosen. Furthermore, relying on the works performed at the given stage it is quite obvious that the POR forced operation mode will provide its geometric dimensions minimization, besides the dimensions need to be specified by using the chemical kinetics, heat and mass transfer and aerodynamics methods.

6. Conclusions

It can be concluded that the developed fuel partial oxidation products composition calculation procedure as part of the chemical energy units is applicable at the stage of the main and auxiliary equipment implementation and operating parameters optimization variants preliminary scientific investigations and feasibility elaborations.
The air consumption coefficient increase α from 0.7 to 0.9 was found out to reduce the synthesis gas target components specific yield by 2 times from 1 m^3/kg of gas condensate on CO and from 0.5 m^3/kg of gas condensate on H_2 by means of calculation. For the considered gas condensate composition, the synthesis gas target components required ratio is observed in the water vapour specific consumption range from 5 to 6 kg/kg of gas condensate that can be explained by the low hydrogen content of the initial fuel.
The POR diameter variants calculations results showed that it is possible to recommend the most preferred operation mode, namely, the forced one decreasing the unit specific amount of metal, in this case the considerable pressure influence on POR diameter practically ceases in the range of π_c = 10-15 in the above mentioned mode.

Acknowledgment

The work was supported by the grant of the Russian Science Foundation (project №17-79-10036).

References

[1] Lapidus A L, Golubeva I A, Krylov I F and Zhagfarov F G 2009 Production of alternative motor fuels based on natural gas *Chem Tech Fuels Oils* **45** 305-12
[2] Lapidus A L, Zhagfarov F G, Grigoreva N A, Khudyakov D S and Kozlov A M 2010 Chemical processing of natural gas and gas condensate *Chem Tech Fuels Oils* **46** 141-47
[3] Kaddour O, Sherbi A, Busenna A and Kolesnikov I M 2009 Production of olefin and aromatic hydrocarbons by pyrolysis of gas condensate *Chem Tech Fuels* Oils **45** 440-42
[4] Malkov Yu P, Molchanov O N, Britov B K. and Fedorov I A *2016* A converter for producing a hydrogen-containing synthesis gas *J Eng Phys Thermophys* **89** 1545-53
[5] Ryzhkov A F, Gordeev S I and Bogatova T F 2015 Selecting the process arrangement for preparing the gas turbine working fluid for an integrated gasification combined-cycle power plant *Thermal engineering* **62** 796-801
[6] Baskakov A P, Volkova Yu V and Plotnikov N S 2014 Optimum chemical regeneration of the gases burnt in solid oxide fuel cells *J Eng Phys Thermophys* **87** 763-72
[7] Mrakin A N, Sotnikov D G and Selivanov A A 2016 Calculation Model Construction for the Fuel Partial Oxidation Reactor Geometric Dimensions Defining Fuel And Energy Complex Improving Problems *Proc. Int. Conf* (Russia: Saratov) pp 318–22
[8] Sotnikov D G and Mrakin A N *Fuel Partial Oxidation Reactor Calculating Programme* Computer programme no 2017615007
[9] Grigorieva V A and Zorina V M 1988 *Heat Engineering Theoretical Foundations Thermotechnical Experiment* (Moscow: Energoatomizdat)
[10] Lunkin V N, Udalov V P and Zhebrak Yu A. 1986 *Air And Oxygen Conversion of Natural Gas* (Saratov: Publishing House Sarat. University)
[11] Lunkin V N 1974 *Investigation Of natural Gas Partial Oxidation High Temperature Process In The Cyclonic Reactor* PhD dissertation (Saratov)

IOP Conf. Series: Journal of Physics: Conf. Series **944** (2017) 012083 doi:10.1088/1742-6596/944/1/012083

The procedure for determining the residual life of high-temperature aggregates

A S Nikiforov, E V Prihodko, A K Kinzhibekova, A E Karmanov
Pavlodar State University, 64, Lomov str., Pavlodar, 140008 Kazakhstan

e-mail: john1380@mail.ru

Abstract. One of the main reasons for the withdrawal of high-temperature aggregates for repairs is the destruction of enclosing structures due to the occurrence of temperature stresses. A wide range of refractory materials used, a large number of product names, a difference in the operation of even the same aggregates makes it impossible to apply general principles for determining the residual resource of high-temperature aggregates, which is based, as a rule, on the determination of temperature stresses.
In the article there is suggested a technique based on the method of simulation modeling, allowing to estimate the remaining resource and reliability of the operating equipment. There are given data on the calculation of these indicators for a 25-ton steel-casting ladle. The values obtained make it possible to evaluate the rationality of the further operation of the high-temperature unit by the condition of reliability of the enclosing structures.

1. Introduction

The modern concept of "industrial security" is closely interconnected with the concepts: risk analysis, operability and reliability of equipment, as well as determining its residual resource. Determination of the reliability of the system as a whole is done by analyzing the reliability of its individual components. The analysis uses actual data on equipment operation in recent years. So, the authors [1] determine the reliability of the work of a complex of conveyor systems and robots based on the operation of robots. An analysis of the failure rate and the average time between failures leads to a calculation of reliability with the assumption of a constant frequency of failure. This allows you to consider options for replacing robots with the least reliability.

An important problem of reliability theory is the estimation of the parameters of the expected failure models. Evaluation can be based on data collected during operation. At the same time, the collected data array can incorrectly display the state of the aggregate for further data processing. In work [2] the application of the Bayes theorem for evaluative analyzes for competing risk systems is considered. For this distribution, a simulation is performed to check models and estimates and to study the effectiveness of estimates for a different sample size. Conclusions are given about the possibility of Bayes' theorem application for risk assessment.

The main problem in determining residual wear in such cases is the determination of the actual state of the constituent elements of the unit during its operation. Naturally, for a number of units, measuring the physical parameters of the elements in their operation is an impossible task at this stage in the development of technology. In other cases, the results will be characterized by the accuracy of the data obtained.

Let's consider definition of a residual resource of protecting constructions of high-temperature units. One of the main types of control of protecting constructions of high-temperature aggregates for determining their actual state is the thermal type of control. All methods of thermal control are based on the interaction of heated sections of protecting structures with sensors for measuring temperature or on the measurement of these temperatures by a non-contact method [3]. According to the temperature, the condition of the protecting structure is judged.

The presence of defective areas is determined by increasing the temperature at control points, which allows to judge the condition of the protecting structure and carry out repair work before the onset of emergency situations (for example, metal leaks for metallurgical machines).

These thermal control methods have significant drawbacks. First, there is no possibility of estimating the remaining resource. The use of these methods provides an assessment of the current wear of the enclosing structures without an analysis of previous processes and the subsequent development of the situation as a whole. It should be noted that when a critical value is reached, an assessment of the subsequent development of the situation is made - forecasting the emergency situation, but this is only at the final stages.

Secondly, the installation of temperature sensors in enclosing structures is a very undesirable process, as this breaks the integrity of part of its layer, and when thermal monitoring in metallurgical equipment there are limitations on the depth of installation of sensors.

2. Formulation of the problem

Thus, an important task is to obtain accurate data on the current state of the enclosing structure with minimal disruption of its integrity and use of this data to predict the residual resource.

3. Theory

The residual life of linings of a number of high-temperature aggregates, during which the thickness of the liner does not change (or there is a topography of wear of the enclosing structure during the working campaign of the unit), depends on a number of factors, and primarily on the value of the temperature stresses exceeding the permissible, their duration, and also on the lengths of compression and expansion zones in which the temperature stresses exceed the allowable ones.

The effect of other factors - the aggressive action of the environment, the quality of the materials used and the work performed, the level of vibration, etc., can be assumed constant for this unit under constant operating conditions. Let's consider high-temperature aggregates of periodic action, in which the work campaign consists of several cycles of heating enclosing structures.

The authors offer the following method of estimating the residual life of the enclosing structures of a high-temperature unit. The first step is to place the temperature sensors in the refractory layer, taking into account the limitation on the depth of installation of the sensors. The number of sensors and the distance from the inner surface are chosen on the basis of operational limitations related to the possibility of emergency situations (leakage, etc.). Also their number and location will affect the accuracy of calculations.

According to the technique stated in the article [4], the temperature fields in the enclosing structure can be determined.

The second stage is the arising stress σ by the formula:

$$\sigma = -\frac{\alpha \cdot \mathrm{E}}{1-\nu} \cdot (T_{av} - T_i) \tag{1}$$

where α is the coefficient of thermal expansion, (1/°C);

T_{av} - is the average temperature of the refractory layer, °C;

T_i - temperature of the point at which the temperature stress is calculated, °C;

E - modulus of elasticity of material, MPa;

ν - Poisson's ratio.

We introduce two concepts: total compression rate ($N_{compr.}$) and total expansion rate ($N_{exp.}$). If the obtained value of the temperature stress σexp. or σcompr. exceeds the value of the tensile strength for a given type of refractory material, its value is taken into account in determining the corresponding total value, which is calculated as follows. Next, the step in time and coordinate for these conditions is determined.

To take into account the lengths of compression and expansion zones in which the temperature stresses exceed the allowable ones, we introduce a coefficient that considers these lengths. By the length of the compression zone we mean the distance between two neighboring points in which the stresses are determined, provided that the compression stresses in these two points exceed the allowable ones. That is to say, if the compressive stresses at a given step exceed the permissible values at three points, then z = 2. The formula for calculating the total compression criterion ($N_{compr.}$) at a given time step (for example, for a time of 1:40) would look like this:

$$N_{compr}^{1:40} = \sigma_{compr} \cdot (z+1), \qquad (2)$$

where z is the number of compression zones in which the temperature stresses exceed the allowable ones.

A similar formula is used for tensile stresses. Next, we summarize all the obtained values of the summary indicators (separately for compression and expansion) for all time steps. This amount, when calculated for the whole work campaign of the unit, is the maximum total compression index ($\Sigma N_{compr.}$) and the maximum total expansion index ($\Sigma N_{exp.}$). These totals take into account only the stresses exceeding the ultimate strength and leading to the destruction of the enclosing structures.

To introduce the residual resource criteria, we introduce the concept of total admissible indicators that will consider the stresses arising at the time and at points where, in reality, the stresses exceed the allowable values. Only when calculating the total allowable values, instead of real values of stresses, we substitute the strength values of the material. For example, at the time 0:40 a temperature stress of 113 MPa (with the strength value of 40 MPa) occurs at the point, which we consider for calculating the total index. To calculate the total allowable values, it is necessary to take into account the value of the stress at the same point at the same time, but at the value of 40 MPa.

If we divide the total values obtained by the total allowable values, we obtain a criterion of reliability for strength (for example, for compression):

$$k_{compr} = \frac{\Sigma N_{compr}}{\Sigma N_{compr}^{add}}. \qquad (3)$$

In this case, the step in time and coordinate in the determination of stresses will determine the accuracy of the calculation of reliability criteria. To estimate the remaining resource, it is also necessary to calculate two arithmetic average values of the total indices for one cycle (for compression - $\Sigma N_{compr.av}$, for expansion – $\Sigma N_{exp.av}$).

The resulting residual resource criteria will determine the residual resource for subsequent work campaigns of this unit under the unchanged operating conditions of the unit and the use of the same materials, layer thicknesses, etc. That is, after the next major repair, temperature measurement, calculation of temperature stresses ($\sigma_{exp.}$ and $\sigma_{compr.}$), total indicators and reliability criteria for strength are performed. The obligatory condition is the using in these calculations of the same step in time that was chosen to calculate the residual resource criteria.

After the first cycle of the unit operation, we have two values: $N_{compr.1}$ and $N_{exp.1}$. After the second and subsequent cycles, we add the existing total indicators with the newly obtained ones for the cycle n and obtain the total values after the cycle n ($\Sigma N_{compr.\ n}$ and $\Sigma N_{exp.\ n}$). At the same time, for each next cycle n we estimate the possibility of its accident-free operation by adding to the available sum $\Sigma N_{compr.n}$ and $\Sigma N_{exp.n}$ the values $\Sigma N_{compr.av}$ and $\Sigma N_{exp.av}$, respectively. If the obtained values are greater than the maximum total compression indices ($\Sigma N_{compr.}$) or the model total tensile indices ($\Sigma N_{exp.}$), then it is not rational to use the unit in the next cycle.

To determine the remaining life of the enclosing structures of the high-temperature unit, steel-pouring ladles with a capacity of 25 tons were chosen. The operation of these ladles is carried out at the enterprise PF LLP "Casting" (Pavlodar, Kazakhstan).

The cycle of the ladle operation is as follows: heating of its lining at the stand with a gas burner to the temperature of 950 - 1000 °C; discharge of the metal from the arc steel casting furnace into it; further processing on the installation ladle-furnace to the required chemical composition; transportation of the metal to the machine of continuous casting of blanks and discharge of the metal into the installation.

The ladle lining consists of four layers. The first layer (working), directly in contact with the molten metal, consists of refractory bricks 135 mm thick. The second layer - heat-insulating, is made of mullite-corundum ramming mass, 30 mm thick. The third layer is made of chamotte bricks of grade ShB-5 with the thickness of 65 mm and the last layer 4, in contact with the ladle cover, represents sheets of asbestos board 10 mm thick (Figure 1).

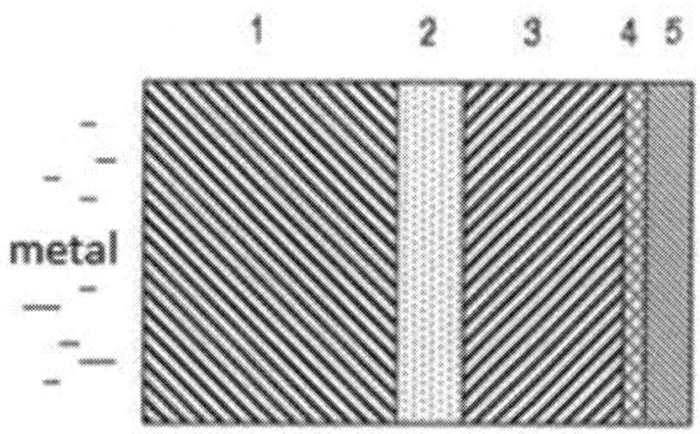

Figure 1. Lining of walls of a 25-ton steel-pouring ladle

The arising stresses will be considered only in the refractory layer, for which the compressive strength is 40 MPa; tensile strength: 25 MPa [5].

Analyzing the thermal processes in the lining of the steel-pouring ladle, we can say that the process that determines the resistance of the lining of the ladle will be heating it with gas on the stand. It is this process, controlled manually and based only on the experience of the personnel (there are no technical methods for controlling the surface temperatures of the lining), is individual for each cycle. Thus, the achievement of the final result in this process - the surface temperature of the lining 950 - 1000 °C - each time is carried out by similar, but different curves. The remaining processes: discharge of metal into a ladle, transporting it, etc. - practically identical in different cycles.

4. Results

To study the arising temperature stresses, measurements were made of the temperatures of the inner surface of the lining at two characteristic points: 2/3 of the height of the ladle – the level of the core of the torch (curve 1) and on the bottom (curve 2) (Fig. 2) [6].

It can be seen from these graphs that there are two points in time at which the surface temperature of the lining rises sharply. This corresponds to an increase in the fuel supply to the burner at the stand for heating the ladles. The withdrawal of the ladles for repair is carried out with a residual brick thickness of the refractory layer of at least 70 mm, while the ladle lifetime before the overhaul is 40 melts (or cycles).

Further, according to the equation (1), it is possible, by specifying the step in time and coordinate, to calculate the arising temperature stresses. After calculating the temperature stresses in the working layer of the lining, we obtain the following picture (Figure 3). There is a clear relationship between the temperature jumps (Figure 2) and the values of stresses that exceed the permissible values.

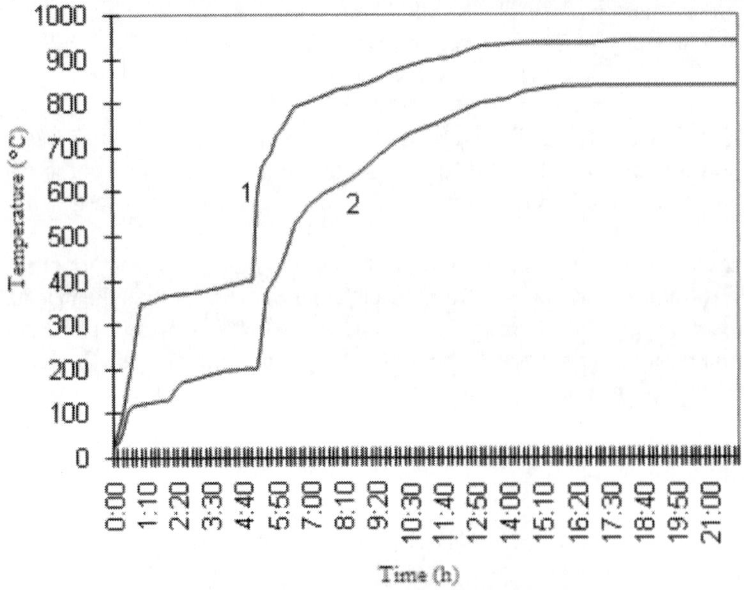

Figure 2. Internal Ladle Temperature

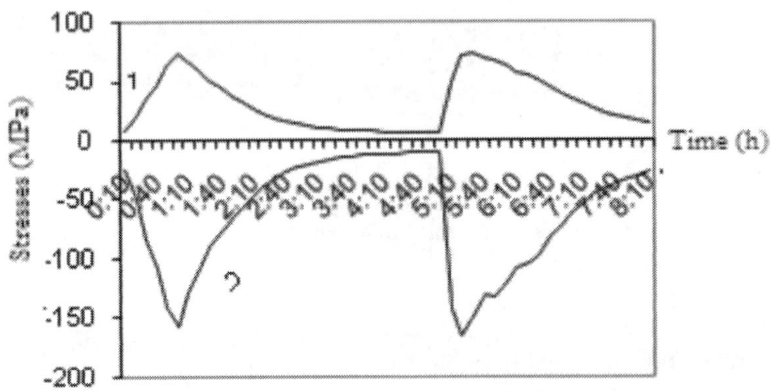

Figure 3. Stresses at the point of the level of the core of the torch on the outer (1) and inner (2) surfaces of the working layer of the lining

In the process of heating, two periods of time are allocated in graph 1 (Figure 3), the temperature stresses tension in which exceed the permissible: (0h 20m - 2h 10m, 5h 10m - 7h 30m); on the graph 2 (compressive stresses): (0h 30m - 0h 50m, 5h 20m - 8h 10m).

Now, having the values of the real temperature stresses that arise during the heating process of the lining of the steel-pouring ladle, taking into account the step in coordinate and time, it is possible to calculate the residual resource criteria. We give the following data: when determining the temperature stresses with a time step of 10 minutes and by breaking the thickness of the refractory layer of the liner into six sections, with the given heating curve, we obtain the values of the maximum total values for one cycle for compression – $\Sigma N_{compr.\ av} = 6951.215$; for the expansion $\Sigma N_{exp.\ av} = 8516.46$.

By taking the temperature readings for the working campaign of the steel-casting ladle in general (before the overhaul) for 40 cycles and calculating the total values, we obtain the following data: for compression $\Sigma N_{compr.} = 279883.314$; for the expansion $\Sigma N_{exp.} = 342826.321$. The calculated reliability criteria for strength are: for compression $k_{compr.} = 2.17$; for the expansion $k_{exp.} = 1.91$.

The criteria show the following: if during the operation of the unit the criterion for compressing $k_{comr.}$ and the criterion for expanding the $k_{exp.}$ exceed the values obtained, then the ladle must be taken out for repair (the thickness of the refractory layer is about 70 mm). If during the working campaign of the ladle $k > 1$, the operation of the ladle is performed with stresses exceeding the permissible limit, and if $k < 1$, then the resulting stresses do not exceed the strength limit of the materials used.

The residual resource n_{rem} (in quantities of smelting) can be estimated by the following formula:

$$n_{rem} = \frac{\Sigma N_{comp}}{\Sigma N_{comp.ad}} - n.$$

5. Discussion of results

The obtained results of the values of the total indices for compression and expansion allow calculating the residual resource at any time (after any cycle n) by comparing the values ($\Sigma N_{compr.\ n}$ and $\Sigma N_{exp.\ n}$) with the values of the total indices for compressive stresses ($\Sigma N_{compr.}$) or tensile stresses ($\Sigma N_{exp.}$), respectively.

The values obtained for the reliability criteria for strength show the limiting state of the criterion, beyond which the destruction of the liner will be critical for these operating conditions.

The main problem is to obtain data on the temperature fields of the refractory lining layer during the heating process. These data can be obtained in any accessible and technologically safe way and used to calculate the residual resource criteria. The methodology can be translated into a computer programming language and, when connected to a computer and temperature sensors can be used at workplaces of high-temperature unit operators.

References

[1] Fudzin A F and M A A Majid Reliability and availability analysis for robot subsystem in automotive assembly plant: a case study (2015) IOP Conf. Series: Materials Science and Engineering 100

[2] Ismed Iskandar and Yudi Satria Gondokaryono Competing risk models in reliability systems, a weibull distribution model with bayesian analysis approach (2016) IOP Conf. Series: Materials Science and Engineering 114.

[3] GOST R (Russian National Standard) 56511-2015. Non-destructive testing.Methods of thermal form.General requirements.

[4] Nikiforov A.S., PrikhodkoE.V., Kinzhibekova A.K., Kucher E.O., Nikonov G.N. Method for determining the thermal state of the lining of a thermal unit. Innovative patent № 26932, Republic of Kazahstan, MKI G 01 K 13/00, published 15.05.2013, bul. № 5.

[5] Refractories for industrial units and furnaces. Directory. Ed. Kashcheeva I.D. Moscow: Intermet Engineering, 2000.

[6] PrikhodkoE.V. Increase of the efficiency of the heat-consuming installations on account of new technologies of drying and heating of the lining: Thesis for the scientific degree of candidate of technical sciences. Omsk: OmGTU, 2006.

Analysis of the stress-deformed condition of the disassembly parabolic antenna

M N Odinets, N V Kaygorodtseva and I V Krysova

Omsk State Technical University, 11, Mira ave., Omsk, 644050, Russia

e-mail: aomnl@mail.ru

Abstract. Active development of satellite communications and computer-aided design systems raises the problem of designing parabolic antennas on a new round of development. The aim of the work was to investigate the influence of the design of the mirror of a parabolic antenna on its endurance under wind load. The research task was an automated analysis of the stress-deformed condition of various designs of computer models of a paraboloid mirror (segmented or holistic) at modeling the exploitation conditions.
The peculiarity of the research was that the assembly model of the antenna's mirror was subjected to rigid connections on the contacting surfaces of the segments and only then the finite element grid was generated. The analysis showed the advantage of the design of the demountable antenna, which consists of cyclic segments, in front of the construction of the holistic antenna. Calculation of the stress-deformed condition of the antennas allows us to conclude that dividing the design of the antenna's mirror on parabolic and cyclic segments increases it strength and rigidity. In the future, this can be used to minimize the mass of antenna and the dimensions of the disassembled antenna. The presented way of modeling a mirror of a parabolic antenna using to the method of the finite-element analysis can be used in the production of antennas.

1. Introduction

Today the question of antenna design is important: Sectional? Holistic? Or folding? At the same time, the construction should preserve its characteristics to the maximum under various conditions of its operation: wind, vibration, icing, etc. Presently, there are many special programs for different industries. At the same time, the search and implementation of new methods and techniques for the development of complex devices and structures, for example, such as radio and communications equipment is still being carried out.

Parabolic antennas are characterized by complex geometry, the presence of absorbing and nonlinear elements. This is of interest for the application of various methods of mathematical modeling. Existing specialized programs, along with the universal software capabilities of design automation, use basically the same tools to create a computer model and analyze it. In particular, this applies to the method of finite-elements, which underlies most programs for analyzing models of objects under load. With certain settings and a correctly created scheme of loading, the finite element method is also successfully used for the analysis of mirrors of parabolic antennas. Despite the large number of existing special programs for designing antennas, many authors [1–6] note that at present, the actual task remains to improve and optimize the mechanical characteristics of satellite and terrestrial navigation systems. In particular, the actual problems remain the preservation of efficiency in various unfavorable external exploitation conditions [3, 5, 7, etc.].

In accordance with the results presented in [8] and with the above-mentioned problems, it is proposed to use the method of mathematical modeling when the mirror of a parabolic antenna is represented as a set of segmental components for further computer modeling and finite-element analysis. The computer model is built on the basis of a mathematical model, which is based on the method of separating the paraboloid describing the antenna mirror, as the surfaces of parallel transport, along the forming lines. This separating of the surface of the paraboloid along the forming line - circle and along the forming line -parabola, gives more variants for the subsequent computer simulation and analysis.

2. Formilation of the problem

The functional purpose of the antenna determines sufficiently stringent requirements for the mechanical characteristics of the structure. This largely determines the operability and radar characteristics of the antenna.

Because the quality of the received signal by a parabolic antenna is affected by any deviation of the mirror and its distortion, accordingly, an important place in the design is the task of analyzing the construct under load. The main stage of the design is research, as a result of which it becomes possible to adjust the construct of the antenna's mirror taking into account the deviation under wind and vibration effects, snow load, icing, flight overloads, own weight of the structure, etc. [9, 10]. Therefore, a large number of scientific studies are aimed at studying such characteristics as the change in emissivity due to deformation of the surface of the antenna mirror during exploitation [3, 5, 10–12]. At the same time, in most cases, the surface of the antenna is modeled as a holistic surface, although in the production of parabolic antennas the mirror is created segmented. In the article [8], the authors carried out a geometric modeling of the surface of a parabolic antenna. The study allowed choosing a way of dividing the surface of the antenna into segments. The article describes the separation of the surface of a paraboloid along a forming line - a circle (this will allow to divide the antenna into cyclic segments) and along the forming line - a parabola (this will allow dividing the antenna into parabolic segments). On the basis of the obtained mathematical model, the work of the authors has a development. The development consists in investigating the analysis of the stress-deformed condition of three variants of the parabolic antenna design, which is under the influence of wind load: antennas without segmentation, with division into cyclic segments and with division into parabolic segments (figure 1). The finite-element analysis of the presented antenna models was carried out using the system of strength analysis APM FEM integrated in COMPASS-3D.

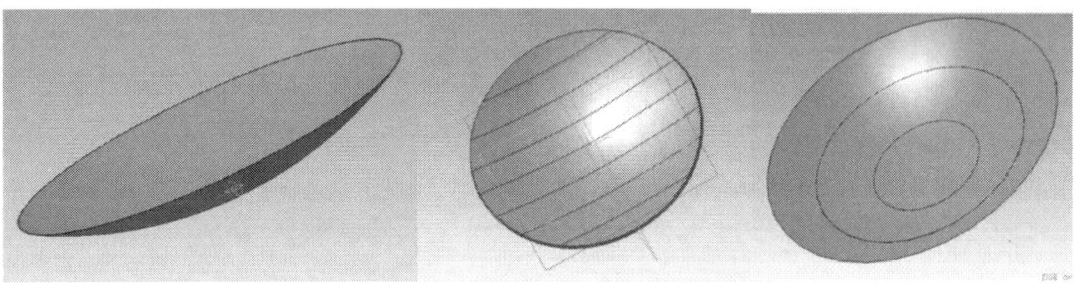

(a) (b) (c)

Figure 1. Models of a smooth mirror of a parabolic antenna:
(a) without division into segments; (b) with division into parabolic; (c) with division into cyclic segments.

3. Theory

3. 1 Disadvantages of existing methods of analysis

There are no fundamental differences in the use of finite-element analysis programs, except for the difference in their functionality. At the same time, the question of the degree of elaboration used in the procedure for analyzing models is often not considered. The antenna model, created as a holistic solid, is subjected to analysis, although in reality, to facilitate production, transportation, the antenna mirror is collapsible. The use of analysis programs based on the finite element method allows you to change the frequency of the partition and the size of the finite elements themselves. At the same time, the question of designing a reflector for radar antennas with regard to the segmentation of the paraboloid mirror remains poorly studied [3, 4]. The mathematical model, which was obtained earlier in [8], allows perform computer modeling of the surface of a parabolic antenna, taking into account it's partitioning into segments. And since each mathematical model must have a certain degree of accuracy, adequacy and profitability, the issues of increasing accuracy without loss of profitability of the model of the projected construct of parabolic antennas remain highly relevant.

3.2 The main provisions of the study

Three models of a smooth mirror of a parabolic antenna were created using the program «COMPASS-3D» (figure 1). The diameter of the mirror is 1330 mm, the depth is 190 mm. The mirror of the antenna was modeled as Solid. In cases where the antenna is divided into segments, the model was assembled into an assembly unit using linking over the coupling surfaces (figure 2).

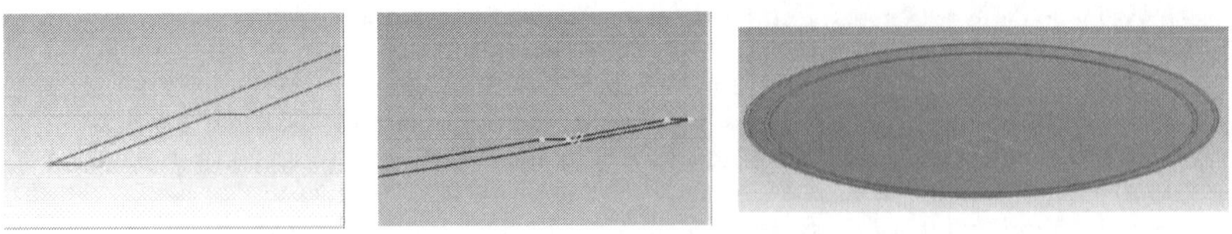

Figure 2. Outline of attachment and 3D-model of one parts of the antenna mirror.

The peculiarity of the study was that rigid constraints were imposed on the model on the contacting surfaces of the antenna mirror segments. And only then a finite element grid was created (figure 3).

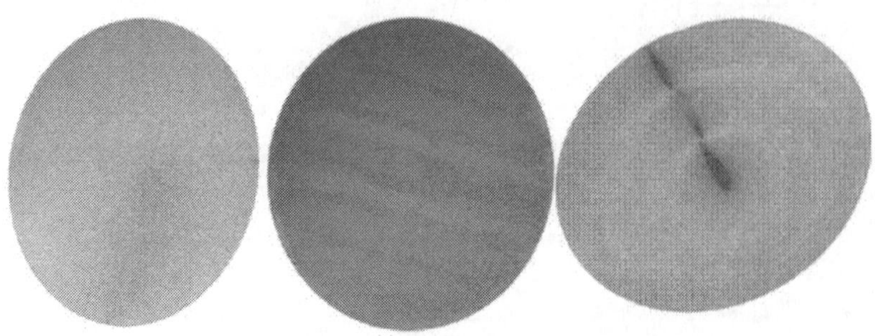

Figure 3. General view of the finite-element model of the antenna mirror surface: a) without division into segments; b) with division into parabolic segments; c) with division into cyclic segments.

The wind load was set as the specific force by the area, which was calculated according to the formula 1 [13]:

$$W_0 \cdot k \cdot C_x, \tag{1}$$

where W_0 – the normative value of wind pressure, which is adopted depending on the wind district,
k – the coefficient that takes into account the change in wind pressure as a function of altitude,
C_x – the aerodynamic coefficient of frontal drag.

The wind load was set taking into account the operation of the antenna in the fifth wind district, in the terrain of type C [14]. The value of the aerodynamic coefficient was taken equal to 1.4 in accordance with the guidance [15]. The pulsating component of the wind load was also taken into account.

For the analysis and comparison of the received models of parabolic antennas (without segmentation, with division into cyclic segments and division into parabolic segments under load), they were subjected to strength and wind resistance tests. To determine the strength characteristics of separable parabolic antenna, in comparison with the non-separable one, model calculations were performed under equal conditions of fixation and loading.

4. Experimental results

The results of computer analysis showed that the use of a mathematical model of the surface of a parabolic antenna and on the basis of its separation along a forming line - a circle, allows not only technologically to divide the antenna into cyclic segments, but also to create a more accurate and adequate computer model. For the model with segmentation, the stresses in the construct are reduced, the coefficients of reserve by fluidity and strength increase. The best strength characteristics were shown by an antenna with cyclic segments. In addition, the maximum stresses with this variant of partitioning are concentrated on one ring, which allows, if necessary, to improve the strength of only one ring of the antenna. In the future, this can be used to minimize the mass-dimensional characteristics of the antenna.

5. Discussion of the results

Analysis of load maps revealed deviations of the deformed surface from the approximating surface, as well as the voltage difference of the holistic and segmental models of the surface of the antenna mirror. In the case where the reflector is made as one solid model, the maximum voltage is 42 MPa. If the antenna is divided into parabolic segments, the maximum voltage is 24 MPa, which is almost two times less than in the model without partitioning. Calculation of the antenna with cyclic segments showed that the maximum value of the voltage is equal to 17 MPa. This is almost 2.5 times less than in the model without partitioning. Herewith the maximum stress values are concentrated at the points of attachment of the antenna segments, but closer to the antenna edge, the strains are evenly reduced. In the case of cyclic separation of the antenna, the maximum voltages are fixed at the points of attachment of segment and are concentrated along the ring. This indicates that the cyclic segment with the points of attachment of the antenna is most loaded, and the central and outer rings have minimal stresses. Therefore, in the case of increased loads (antenna output to orbit or other tasks) with unsatisfactory strength of the antenna it will be sufficient to «strengthen» only the middle ring.

Visualization of the values of equivalent stresses at each point of the construct is shown on the maps (figure 4).

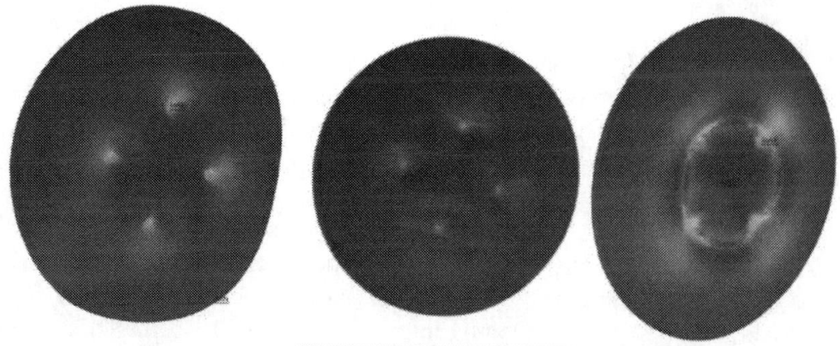

(a) (b) (c)

Figure 4. Maps of equivalent stresses in the construct of a parabolic antenna: (a) without division into segments; (b) with division into parabolic segments; (c) with division into cyclic segments.

Strength characteristics of the antenna can be estimated from the coefficients of reserve by strength and fluidity. Visualization of the values of the coefficients of reserve at each point of the construct is shown on the maps (figure 5, 6).

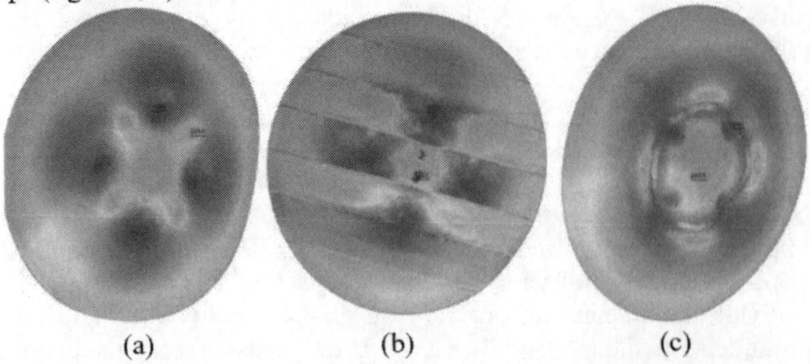

(a) (b) (c)

Figure 5. Maps of the distribution of values of the coefficient of reserve by fluidity in the model of a parabolic antenna: (a) without division into segments; (b) with division into parabolic segments; (c) with division into cyclic segments.

The minimum value of the coefficients of reserve by strength in the model of a parabolic antenna without segmentation and with division into parabolic and cyclic segments is 9, 19 and 24, respectively.
The minimum value of the coefficients of reserve by fluidity in the model of a parabolic antenna without segmentation and with division into parabolic and cyclic segments is 5; 11.3 and 13.7 respectively.

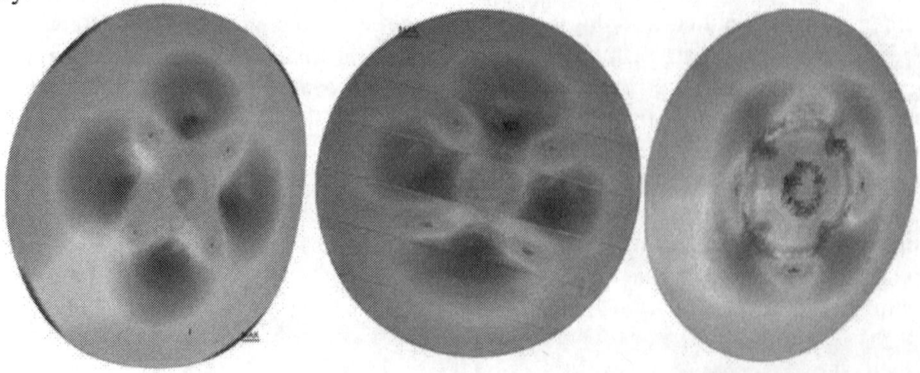

(a) (b) (c)

Figure 6. Maps of the distribution of values of the coefficient of reserve by strength in the model of a parabolic antenna: (a) without division into segments; (b) with division into parabolic segments; (c) with division into cyclic segments.

In a comparative analysis of the maps of the model of a parabolic antenna without separation into segments (figure 6, a) and with the division into parabolic segments (figure 6, b), it is evident that the dimensions of the region with the smallest value of the coefficients of reserve by fluidity are almost identical. And in the case of dividing the antenna into cyclic segments (figure 6, c), the dimensions of the region become much smaller. The same trend is also observed for coefficients of reserve by strength.
To eliminate the damage associated with resonance phenomena, the first frequency of oscillations of three versions of the mirrors of a parabolic antenna was calculated. The results of the calculation are shown in table 1.

Table 1. Value of first own frequency.

Frequency	Models of the mirror of the parabolic antenna		
	Without separati on	***With separation into parabolic segments***	***With separation into cyclic segments***
1st natural frequency, Hz	25	102	191

For antenna structures, if the main external load is wind, it is necessary that the natural frequencies of the structure go beyond 5.9 Hz, since resonance phenomena from wind loads arise up to the specified value [14]. The considered variants of models satisfy this condition. The results of the calculation showed that sectional antennas in comparison with non-separable antennas have greater rigidity. Maximum rigidity of the construct is achieved when separation the antenna into cyclic segments. The obtained results (stress values at each point of the model) allow determining the elements and methods of connecting the antenna segments to each other, as well as fixing the antenna design at the site of exploitation.

6. Conlusions

Along with modeling the radio-technical characteristics of antennas, it is important to model its stress-deformed condition, which directly affects the technical parameters. This article shows one of the stages of computer-aided design of a parabolic antenna – computer analysis of the surface of an antenna subjected to wind load. Conducted automated engineering analysis allows us to evaluate the strength characteristics of the designed antenna, optimize its design, improve quality and shorten the development time. Carried out computer simulation and comparative analysis of antenna stress condition allows to conclude that the separation of the design of antenna on parabolic and cyclic segments increases its strength and stiffness in comparison with the non-separable model. The separation of a complex curved surface into segments is carried out without replacing it with a grid of finite planar elements, so the proposed methods for separating a parabolic antenna allow preserving its technical characteristics. In general, a comparative analysis of the stress-deformed condition showed the expediency of separation the model on segments.

Thus, the method presented in the article for modeling a construct of mirror of a parabolic antenna using the finite-element analysis method allows using the results obtained in the design of parabolic antennas.

References

[1] Sarsikeev E J, Lyarunov D Yu, Bobihov R S and Petryusyov A S 2014 Visual modeling of the wind load of the reflector of the parabolic communication antenna in the program COMSOL MULTIPHYSICS *Modern problems of science and education* **3** p 137
[2] Qiu D, Sun M, Wang Z, Wang Y and Chen Z 2014 Practical wind-disturbance rejection for large deep space observatory antenna *IEEE Transactions on Control Systems Technology* **22** pp 1983-90 DOI: 10.1109/TCST.2013.2296935.
[3] Evdokimov A S, Ponomaryov S V and Buyanov Yu I 2011 A joint calculation of the stress-deformed condition and the directivity pattern of cosmic reflectors *Bulletin of the Tomsk State University «Mathematics and Mechanics»* **1** pp 74-82

[4] Shinohara K and Takaki R 2013 Contact finite element analysis of hinge joints for large deployable antenna in space satellite *2013 IEEE Aerospace Conference* **20** pp 1-20 DOI: 10.1109/AERO.2013.6496896.

[5] Khrushchelev P A and Pel A N 2016 Numerical solution of the problem of deformation of a parabolic antenna *Science. Industry. Defense: Proceedings of the 17th All-Russian scientifical-technical conference* (Novosibirsk: Publishing house of NSTU) **1** pp 209-214

[6] Babkina L A and Sorokin D V 2015 Geometric modeling and modal analysis of a parabolic antenna of a spacecraft with a multivariate attachment scheme *Theoretical and applied questions of modern information technologies: Collection of scientific papers* (East Siberian State University of Technology and Management) pp 41-45

[7] Gimpilevich B Yu and Shirokov I B 2002 Control of microwave aerospace antenna service condition *Geoscience and Remote Sensing Sympozium IGARSS'02 2002 IEEE International* **6** pp 3644-45 DOI: 10.1109/IGARSS.2002.1027277

[8] Odinets M N, Kaygorodtseva N V, Odinets A I and Kaygorodtseva T N 2016 Computer modeling of a collapsible surface of the mirror of the parabolic antenna, while maintaining the technical characteristics *Dynamics of Systems, Mechanisms and Machines* **6** pp 1-5 DOI: 10.1109/ Dynamics.2016.7819054.

[9] Lopatin A V, Pasechnik K A, Vlasov A Yu and Shatov A V 2013 Development of precision antenna reflectors from polymer composite materials: finite element simulation of the structure *Bulletin of the Siberian State Aerospace University named after academician M. F. Reshetnev* **3 (49)** pp 73-78

[10] Astafev I V 2010 Analysis of the possibilities of design packages for modeling thermal effects on parabolic antennas *Proceedings of the International Symposium «Reliability and Quality»* (Penza: Penza State University) **1** pp 269-271

[11] Ludick D J, Venter M, Davidson D B and Venter G 2016 A multiphysics analysis of dish reflector antennas for radio astronomy applications *10th European Conference on Antennas and Propagation (EuCAP)* **4** pp 1-4 DOI: 10.1109/EuCAP.2016.7481745.

[12] Sukharevsky O I, Nechitaylo S V, Khlopov G I and Voitovych O A 2013 Influence of the snow cover on radiation characteristics of reflector antennas *IX Internatioal Conference on Antenna Theory and Techniques* **3** pp 447-449 DOI: 10.1109/ICATT.2013.6650807.

[13] Savitckiy G A 1962 *Antennas-mast constructions* (Moscow: State Publishing House of Literature on Communications and Radio) p 231

[14] SP 20.13330.2011 *Loads and impacts* Updated version of sanitary norms and regulations 2.01.07-85 (Introduced 2011-05-20 Ministry of Regional Development of Russia)

[15] *Guide to the calculation of buildings and structures to the effect of wind* 1978 (Moscow: Stroyizdat) p 216

IOP Conf. Series: Journal of Physics: Conf. Series **944** (2017) 012085 doi:10.1088/1742-6596/944/1/012085

Time Factor in the Theory of Anthropogenic Risk Prediction in Complex Dynamic Systems

V A Ostreikovsky[1], Ye N Shevchenko[1], N K Yurkov[2], I I Kochegarov[2] and A K Grishko[2]

[1]Surgut State University, Lenin Avenue, 1, Surgut, 628400, Russia
[2]Penza State University, Krasnaya Str. 40 , Penza, 440026, Russia

e-mail: kipra@mail.ru

Abstract. The article overviews the anthropogenic risk models that take into consideration the development of different factors in time that influence the complex system. Three classes of mathematical models have been analyzed for the use in assessing the anthropogenic risk of complex dynamic systems. These models take into consideration time factor in determining the prospect of safety change of critical systems. The originality of the study is in the analysis of five time postulates in the theory of anthropogenic risk and the safety of highly important objects. It has to be stressed that the given postulates are still rarely used in practical assessment of equipment service life of critically important systems. That is why, the results of study presented in the article can be used in safety engineering and analysis of critically important complex technical systems.

1. Introduction

According to studies [1,2] the anthropogenic risk will be used in this article in the sense of damage (negative consequences) as a result of a possible initial event (failure, disaster, catastrophe) in a stochastic situation of applying a complex dynamic system. Anthropogenic risk, being a random value, is assigned on a probability space (Ω, F, P) which includes:
Ω – non-empty *set of possible outcomes;*
F – σ-*algebra* of events, its elements are sub-sets of set Ω and are called *events*, $\Omega \in F$;
Q – σ-additive set function Q on F and restricted by the condition $Q(\Omega) = 1$ is a probability measure or probability.
For the event $A \in F$ the value of $Q(A)$ is called the probability of possible initial event A.
Triple (Ω, F, Q) is called *probability space* or *probabilistic model.*
As the risk is connected with actual indeterminate situations, then to describe it in mathematical terms of probability theory, it is necessary to satisfy certain parameters and conditions:

- "unpredictability: the outcome of a stochastic situation is impossible to predict with absolute precision. This quality is apparent, because if the outcome of the situation can be predicted exactly, then there is no need to resort to the methods of the probability theory;
- reproducibility: this parameter is a key one in the assuredness of the success of applying the methods of the probability theory to the description of stochastic processes. Probability theory and mathematical statistics are aimed at studying mass phenomena. Considering weak reproducibility of rare events, many attempts at applying the probability theory to the analysis of unique complex systems can yield controversial results, distant from actual events;

IOP Conf. Series: Journal of Physics: Conf. Series **944** (2017) 012085 doi:10.1088/1742-6596/944/1/012085

- stability of event frequency: the frequency of the event, linked with the situation under study, with multiple repetition of the situation is fluctuating near some number, being closer to it only with more frequent occurrences of the situations" [1].

Risk is always connected with the occurrence of an event, a random event A, which is called a risk event (initial event in case of disaster, emergency) from the multiple of possible F events, that describe the given risk situation. These events are also usually distributed in time in some way and are accompanied with certain material or other damages, also random ones. In this way, risk is characterized by two variables - time T of the occurrence of risk even and value C of damages caused by it. From this point of view, risk is understood as a probabilistic model (probabilistic space)

$$R = \{\Omega, F, Q\} \quad (1)$$

where the two-component random value (T, C) is determined. The first component T is the time of risk event A occurrence, calculated from a fixed moment, and the second component C shows damage brought about by this risk event:

$$R(p, c, t) = R\{P < p_i, C < c_i, T < t\} \quad (2)$$

where $i = \overline{1, N}$ is the number of possible risk events.

Figure 1 shows the generalized dynamic model of anthropogenic risk of system operation Let us analyze the sets: $Q = \{q_1, q_2, \ldots, q_n\}$, $q_i \in Q$, $i = \overline{1, n}$ is the set of possible initial events (failures, emergencies, disasters), $C = \{c_1, c_2, \ldots, c_n\}$, $c_i \in C$, $i = \overline{1, n}$ – is the set of consequences (damage) of i-th initial events happening, $t_i \in T$ is the set of times, $r_i \in R$ – is the set of possible risks $R = \sum_{i=1}^{n} R_i$.

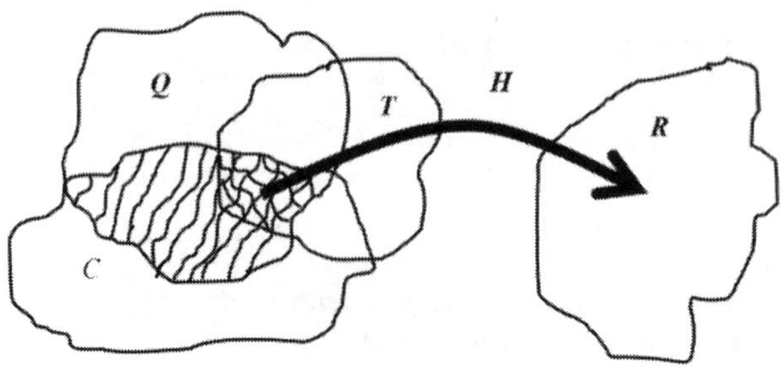

Figure 1. Graphic visualization of the relations between the set of risks R, probabilities of initial events Q, damages C during operating time T of a complex dynamic system.

It is evident, that

$$R = H\{Q \times C \times T\} \quad (3)$$

or in scalar form

$$R(q, c, t) = \sum_{i=1}^{n} q_i(t) c_i(t), \quad (4)$$

where H is the operator that implements the relations

$$Q \times C \times T \rightarrow R. \quad (5)$$

Or

$$R(q, c, t) = H\{t, t_0, R_0(q_0, c_0, t_0), R(q, c)^{t}_{t_0}\}, \quad (6)$$

where t is the time during which the risk is determined;

t_0 – the starting moment of observing the state of the system, $t \geq t_0$;

q_0, c_0, R_0 –probability of initial events of the dynamic system, damage and risk at the starting moment of observing the state of the system, correspondingly.

2. Problem statement

In studies [2,3,7–9] the authors developed the classification of models of anthropogenic risks in complex dynamic systems, giving detailed theoretical basis to the analysis of characteristics of sets R, Q and C when performing the mathematical modelling of the anthropogenic risk.

From a statistical point of view, the components $R(t)$, $Q(t)$ are considered as random time functions, vector ones in the general case. If we regard risk as a possible damage, then the time model $R = f(T)$ should have the following properties:

1) the process of change in time $R = f(T)$ is considered in general as an unsteady stochastic process;
2) the time of observing the system during T is finite;
3) energy spectrum of the random process $R(T)$ is continuous and other than zero on the whole frequency axis $-\infty < \varepsilon < +\infty$;
4) the correlation interval τ_0 of the random function $R = f(T)$ is limited. The correlation interval is understood as a time interval during which the correlation links between separate components $R(t)$, $Q(t)$ disappear completely. With non-stationary character of random function change $R = f(T)$, the correlation interval in general case depends on the examined time period., i.e. $\tau_0 = f(t)$.

Non-stationary stochastic processes of change $R = f(T)$ should be presented as a sum of several processes $R = f(T)$:

$$R(t) = A(t) + B(t) + \varepsilon(t) \tag{7}$$

where $A(t)$ is a non-stationary stochastic process that characterizes irreversible changes in the system due to aging, deterioration and adjustment; $B(t)$ is a stationary stochastic process that characterizes reversible changes $R(t)$ caused by fluctuations of external operating conditions of the system; $\varepsilon(t)$ is a stationary stochastic process of measurement errors $Q(t)$.

Earlier work [4] describes the methods of evaluating the time factor when planning multi-parameter reliability tests of complex systems (parameter Q):

1. Method employing time factor when modelling risk in the conditions of orthogonal shift.
2. Method assessing time factor by transforming the regression parameter from time function to numerical coefficients.
3. Method of assessing time factor in the experiment when one of the controlled variables is time.
4. Method of assessing time factor using spectrum-correlation theory of random processes.

This can be partially explained by the fact that the risks theory for complex dynamic systems took their roots from the classic formula introduced by F.R. Farmer $R=QC$, where both the probability of initial events (emergencies and disasters) Q and the damage caused by them C were understood as sets which are not dependent analytically from time. At present this would seem outdated, but in the 60s of the previous century time parameter was not being given proper attention. Although at that time already it was specially noted that reliability and failure-free operation of complex systems are inherently time related. Considering the above said, the present article is primarily oriented at considering time factor when modelling risk R as a pivotal point in the theory of anthropogenic risk.

3. Mathematical models that allow to consider the time factor in predicting the anthropogenic risk

Class 1. Probabilities of initial events and damage are random time functions (random processes), in general case both dependent and independent.
Then for independent random processes with distributions $f_Q(q, t)$ and $f_C(c, t)$
$R = H_3 \{q, c, t\}$;

$$F_R(r) = \iint\limits_{W_3} f_Q(q, t) f_C(c, t)\, dq(t)\, dc(t) \tag{8}$$

Class 2. Probability density of initial events are described by partial differential equations.
It is worthwhile to use the methods where the output (complex, generalized) parameter of the object is regarded as a random function. To determine the probability $q_i(t)$ during time t, the random process is $Y = Y(X, t)$ (X – vector of external and internal influences on the object), then $Y(t)$ is described by n-dimensional density, where n depends on the value of $t \in T$ the rate of change of the random process $Y(t)$ and demands precise calculations. The drawbacks of this class of models are very complex input data in the form of multi-dimensional distribution laws, the process of obtaining this data is very difficult. The use of one-dimensional distribution densities instead of multi-dimensional ones can significantly alter the calculation results. That is why, it was suggested in studies [7,10] to use, when determining probability $q_i(t)$, the additional limitations on the random process $Y(t)$.
Suppose $Y(t)$ is a continuous one-dimensional homogeneous Markov process with finite Euclid phase space. This process is described by the function $Q(\theta, y, t, Y)$ - the probability that the object at the time θ ($\theta > 0$) is in state y, then at the time $t (t > \theta)$ it will be in one of the states $Y \subset \Omega$, where Ω – θ is the algebra of subsets of phase space. Function $Q(\theta, y, t, Y)$ satisfies the Chapman–Kolmogorov equation, and the probability density of the transition $f(\theta, y, t, Y)$ satisfies the partial differential equations (both forward and backward Kolmogorov equations).

$$\left.\begin{aligned} &\frac{\partial f(\theta,y,t,Y)}{\partial \theta} + \sum_{k=1}^{n} \alpha_k(\theta,y) \frac{\partial f(\theta,y,t,Y)}{\partial y_k} + \frac{1}{2}\sum_{i,k=1}^{n} \beta_{i,k}(\theta,y) \frac{\partial^2 f(\theta,y,t,Y)}{\partial y_k^2} = 0 \\ &\frac{\partial f(\theta,y,t,Y)}{\partial t} + \sum_{k=1}^{n} \frac{\partial}{\partial y_k}\left[\alpha_k(\theta,y) f(\theta,y,t,Y)\right] - \frac{1}{2}\sum_{i,k=1}^{n} \frac{\partial^2}{\partial y_i \partial y_k}\left[\beta_{i,k}(\theta,y) f(\theta,y,t,Y)\right] = 0 \end{aligned}\right\} \tag{9}$$

where coefficients $\alpha(\theta,y)$ and $\beta^2(\theta,y)$ are drift and diffusion coefficients, correspondingly, that equal

$$\left.\begin{aligned} \alpha(\theta,y) &= \lim_{t \to 0} \frac{M[Y(t+\theta) - Y(\theta)\big|_{Y(\theta)=y}]}{t}; \\ \beta^2(\theta,y) &= \lim_{t \to 0} \frac{D[Y(t+\theta) - Y(\theta)\big|_{Y(\theta)=y}]}{t}. \end{aligned}\right\} \tag{10}$$

To find the coefficients $\alpha(\theta,y)$, $\beta^2(\theta,y)$ of equation (10) the following system of equations must be solved

$$\frac{\partial m_1(\theta,y)}{\partial\theta}=\alpha(\theta,y)\frac{\partial m_1(\theta,y)}{\partial y}+\frac{\beta^2(\theta,y)}{2}\cdot\frac{\partial^2 m_1(\theta,y)}{\partial y^2};$$
$$\frac{\partial m_2(\theta,y)}{\partial\theta}=\alpha(\theta,y)\frac{\partial m_2(\theta,y)}{\partial y}+\frac{\beta^2(\theta,y)}{2}\cdot\frac{\partial^2 m_2(\theta,y)}{\partial y^2}, \tag{11}$$

where m_1, m_2 are, correspondingly, conditional expected value and dispersion of the process:

$$m_1(\theta,y)=M\,[Y(t)]\Big|_{Y(\theta)=y}=\alpha(t)Y=y\int_{-\infty}^{\infty}ydF_y(\theta,y); \tag{12}$$

$$m_2(\theta,y)=M[Y^2(t)]\Big|_{Y(\theta)=y}=\beta(t)Y-\alpha^2(t)Y^2=$$
$$=y\int_{-\infty}^{\infty}\varphi(\theta,y)y^2dy-y^2\left(\int_{-\infty}^{\infty}ydF_y(\theta,y)\right)^2. \tag{13}$$

It is could be seen from equations (12) and (13) that mathematical expectation and dispersion of the process are functions of time and depend on type and values of the laws of dispersion in the slices of the random process *Y(t)*.
Therefore, infinitesimal characteristics of the random process *Y(t)* allow to find probability of the system transition from one state to another during time interval $[t_0, t_k]$, and calculate distribution of the other functionals of the process. In particular time for process to reach the certain area and distribution of the process *Y(t)* value in the continual area until moment t_k.

Risk models with partial differential equations with leaps of state change of dynamic systems

Class 3. It should be noted that equations (9)–(11) are well-set models of known fundamental preservation laws [10]:

$$\frac{\partial f^{\omega}(x,z,y,t)}{\partial t}+\sum_{j=1}^{n}\frac{\partial A_j^{(\omega)}(f,x,z,y,t)}{\partial x_j}=B^{*(\omega)}(f,x,z,y,t),$$
$$x,y,z\in R^n,\ t>0,\ \omega\in\Omega, \tag{14}$$

where $f=\{f^{(\omega)}\}$ is an unknown vector-function (in our case it is $q_i(t)$);
$x, y, z \in R^n$ spatial coordinates;
A_j и B^* – operators that are considered as specified by the character of modelled physical processes in objects;
Ω – set of parameters ω, numbering equations (14).
Preservation principles (14) reflect the objects functioning, described in the general case by systems of non-linear differential and integral-differential equations.
For the cases when the right part of equation (14) is discontinued, it is necessary to refer to the models of Liouville-Vlasov equations [10]

$$\left.\begin{aligned}&\frac{\partial f(z,t)}{\partial t}+\frac{\partial\,[f(z,t)p(z)]}{\partial z}=0;\\&f(z,t)\big|_{t=0}=f^0(z),\end{aligned}\right\} \tag{15}$$

where $f(z,t)$ is probability density of distribution of system state in phase space in time t;
$p(z)$ – field of velocities of state change of the system in phase space;
$f^0(z)$ – initial probability density of distribution of system state in phase space.

Although Liouville's equation is an equation of continuity and the major conservation principle determining the statistical solutions of equations for dynamic systems, but there are possible applications of this equation with leaps of state change of dynamic systems. In other words, Liouville-Vlasov equation, with leaps of state change, is a solution of a differential equation system with discontinuous right part. The use of the equation (15) with discontinuous coefficient leads to functional solutions. It has been found in [10] that the numerical modelling of equations of Liouville's type with discontinuous coefficients is a significant advantage, in comparison with difference schemes, as the latter in our case are approximating.

The experience of calculating the anthropogenic risk shows that determining the damage $C_i(t)$ does not present any significant difficulties, apart from organizing ones, connected with subjective factors. The majority of problems arise with assigning the values of probabilities of initial events of failures, emergencies, and disasters. Currently, safety theory employs and applies various logical-and-probabilistic models for the assessment of $q_i(t)$. These models are based on the methods of Fault Tree Analysis - Event Tree Analysis, diagrams of functional integrity, general logical-and-probabilistic methodology, with the use of topological methods, logical models and others. Detailed theory of many of these models have been given by scientists both in Russia and abroad. The majority of the models have been automated and implemented in a computerized form [3,5]. They have been recommended by many national and international organizations to be used for practical assessments and calculations in probabilistic safety analysis of complex critically important dynamic systems. However, the overwhelming number of logical-and-probabilistic methods when assessing the risk and safety have to use the equipment reliability parameters in the form of probability or failure rate. This is connected with solving such difficult problems as ensuring high reliability of equipment, low number of failures, heterogeneous and reduced sampling, variability of elements base, differing technological plans, etc.

Equipment elements with lumped parameters are less problematic in obtaining the safety characteristics than the systems with distributed parameters.

There are other possible forms of interrelations between sets Q, C and T. For example, one might consider the previous history of values of anthropogenic risk and others.

4. Postulates of time phenomenon in the theory of anthropogenic risk in in in complex dynamic systems

Postulate evolving from Markov processes theory. Markov process is a stochastic process, the evolution of which after each prescribed value of time parameter t does not depend on the evolution preceding t, given that the process is fixed in time (i.e. 'future' and 'past' of the process do not depend from each other with the given 'present') This postulate is conventionally called Markov postulate [11]. It was first formulated by A.A. Markov in "Proceedings of Physics and Mathematics Society of Kazan University' in 1906.

Markov processes of Brownian motion type are closely linked with differential equations of parabolic type. The transitional density of $f(\theta,y,t,Y)$ diffusion process $Y(t)$ satisfies, with some additional assumptions, Kolmogorov backward equation [9].

In other words, Markov process is a process during which the future evolving of object state depends only from the present current state. For objects with continuous time this means that the process is local in time, that is memory effects are not working.

Shannon's postulate The difficulty of predicting the equipment lifespan of complex dynamic systems lies not only in building the model of changing in time deterioration processes of equipment materials, that will correspond the predicted processes on the interval of previous history, but in preserving this correspondence at the interval of prevention. It is evident, that any mathematical prediction tools render of little use if the physical properties of predicted processes are not taken into account. Determining the regularities of changes in physical processes of equipment deterioration of complex dynamic systems in prediction problems must be based on one of the fundamental laws of physics on preserving the essence of physical processes when inferring the object behavior, that was formulated by C. Shannon that states that major regularities observed in the past will be preserved in the future.

The functioning of complex dynamic systems should be considered, in this situation, as a consecutive change of its states under the influence of external and internal factors (Figure 2).

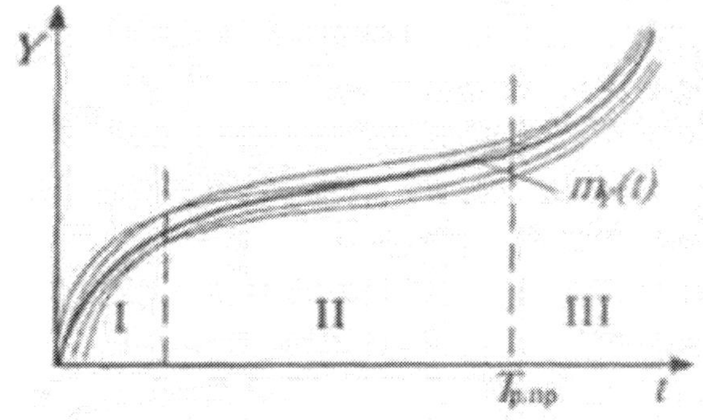

Figure 2. Change of deterioration processes of materials and equipment of complex dynamic systems during operation: my(t) — expected value of the predicted lifespan; I — running in zone; II — area of normal operation period; III — zone of intensive wear and deterioration.

Sedyakin's postulate It is known [4,12], that notwithstanding the law of distribution of random value of object operating time the probability of failure-free operation equals

$$p(t) = e^{-\int_0^t \lambda(\tau)d\tau} \tag{16}$$

where $\lambda(t)$ us the object failure rate.
The right part of equation (16) is the function of the integral, specified by N.M. Sekyakin [12] through

$$r(t) = \int_0^t \lambda(\tau)d\tau \,. \tag{17}$$

"Let us consider the value of this integral as a measure of service lifespan used by the element in the time interval (0, t). Evidently, the concept of service lifespan used by the element in the period (0, t) agrees with out intuitive understanding. Indeed, if $t = 0$, then $r(0)= 0$. That corresponds the case when the lifespan equals zero.
In another extreme case, when the element works failure-free for indefinitely long period ($t \to \infty$), then, if the function $\lambda(z)$ is nondecreasing, then $r(t) \to \infty$. This means that the elements lifespan is not limited. In actual life, if the definite element is capable of working during finite time $\tau^*=\tau$, and its full available lifespan is

$$r(\tau) = \int_0^\tau \lambda(z)dz \,, \tag{18}$$

where τ is the object failure-free operation time.
As the object's failure-free operation time can not be predicted unambiguously in advance, its full lifespan is a random value.
The element's operation mode is understood as the amount of load.

$$\varepsilon = \frac{H}{H_0}, \tag{19}$$

where H and H_0 are acting and nominal loads of the element.
Depending on the value ε function $w(\tau)$, can change, as a matter of fact, within a wide range. However, if $r(\tau)$ is determined with a definite value ε, then the object failure rate will, according to (16)-(17), present a well-defined function

$$\lambda(t) = \lambda(t, \varepsilon)\,, \tag{20}$$

IOP Conf. Series: Journal of Physics: Conf. Series **944** (2017) 012085 doi:10.1088/1742-6596/944/1/012085

where ε is determined from equation (19).
Approximate dependency $\lambda(t, \varepsilon)$ from t for different values of ε is shown in Figure 3.

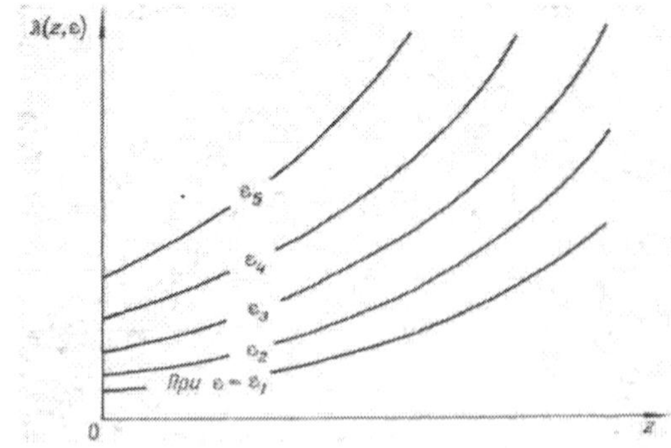

Figure 3. Approximate dependencies of possible failures from time, with operation conditions εk , k = 1,2,3, ...

Using the introduced concept of available lifespan $r(t)$, it becomes possible to formulate one of the laws of the theory of reliability, Sedyakin's postulate, the meaning of which is that the reliability of an object depends on the value of used lifespan in the past and does not depend on how this available lifespan has been used [12], i.e.

$$P\left(\frac{t}{r}\right)=p^{(1)}\left(\frac{t}{t_1}\right)=p^{(2)}\left(\frac{t}{t_2}\right), \tag{21}$$

where t_1 and t_2 satisfy the integral relation

$$r=\int_0^{t_1}\lambda(z,\varepsilon_1)dz=\int_0^{t_2}\lambda(z,\varepsilon_2)dz. \tag{22}$$

It should be noted, that if this law is true, then the sphere of its application has to be limited by those physical processes which do not lead to the qualitative change of the structure of object materials. In following this condition it is natural to expect, that the object reliability in the future must depend on the amount of the lifespan used in the past. Here, the left part of the equation (22) corresponds to the lifespan, used by the object for the period x_1 in conditions $\varepsilon = \varepsilon_1$, and the right part of the equation is the lifespan of the object used for the period x_2 in the conditions ε_2 . It should be born in mind, that equations (21) and (22), being of statistical nature, show the laws of object behavior in mass sense" [12].

Lyapunov's postulate In chaotic dynamic systems the concept of trajectory loses its sense after a certain characteristic time (Lyapunov's time) [13]. Chaos will be understood as the system behavior when the initially close trajectories diverge exponentially in a certain time. The mode is called chaotic if the distance between any two points, however small originally, grows exponentially with time. A.M. Lyapunov's postulate is formulated as follows: the divergence of trajectories is described by the function $exp(t/\tau)$, where $1/\tau$ is, for chaotic systems, a positive value. Value $1/\tau$ is called Lyapunov's index, and τ Lyapunov's time [13, 14].

Prigogine postulates. There are three types of the laws of nature:
1. the first type operates with the trajectories in classic Newtonian mechanics and wave functions in quantum mechanics;
2. the second type deals with statistical framework of laws of nature (J.Gibbs, A.Einstein), it is 'deduced' or 'reduced';
3. the third type of laws of nature states that the laws of chaos are probabilistic, yet unreliable.

The laws that govern the behavior of stable systems are deterministic and reversible in time. And, vice versa, the laws describing the chaotic systems, relate to probabilities and include irreversibility.

The evolution of chaotic systems in time demands irreducible probabilistic description, in terms of ensembles and probability distributions. The evolution of probability distribution has to be described in space that is time-dependent.
The solution of time paradox is only possible that the space becomes 'temporally-related' as the past and the future do not play the same role.
Prigogine postulates can be formulated as follows:
1. Chaos leads to including the arrow of time in fundamental dynamic description of system behavior. The evolution of systems with $t \rightarrow +\infty$ and $t \rightarrow -\infty$ is different.
2. All systems that allow irreducible probabilistic description are considered chaotic.
3. Laws describing chaotic systems relate to probabilities and are characterized by irreversibility.

5. Conclusion

1. Complex dynamic systems, as a rule, have a long life cycle. That is why, it is of great importance, in the theory of anthropogenic risk, to work out and develop the methods and models of predicting the changes in time of risk quantitative values.
2. The article describes three classes of models for predicting the anthropogenic risk, which, according to the authors' view, are most promising in terms of the solution of the problems described in the first part of the article.
3. Emergencies and disasters risks are characteristic of such critically important complex dynamic objects as nuclear power systems, aircrafts, systems of hydrocarbons transportations, chemical industry plants, and others. That is why, the article focuses on the models of anthropogenic risk, described in terms of non-linear differential and integral differential equations, with jumps and discontinuous coefficients.
4. This is the first time that the postulates have been defined that will allow to work further on time phenomenon application in the theory of anthropogenic risk, which is one of the basic fields of modern mathematical theory of safety of critically important objects.

References

[1] Ostreikovsky V A, Genyush A O, Shevchenko Ye N 2010 *Mathematical Modelling of Anthropogenic Risk* Surgut p 96
[2] Ostreikovsky V A 2013 Theory of Anthropogenic Risk: Mathematical Methods and Models (Surgut) p 320
[3] Galkin V A 2009 *Analysis of Mathematical Models: Systems of Conservation Laws, Boltzman and Smoluchowsky Equations* (Moscow: Akademia) p 408
[4] Sedyakin N M *On a Physical Principal of Reliability Theory and Some of Its Applications* (Leningrad) 1965 p 41
[5] Lyapunov A M 1956 Collected Works (Leningrad: Gostekhizdat) p 263.
[6] Jensen U. 2002 Probabilistic risk analysis: foundations and methods
[7] Grishko A, Kochegarov I, Yurkov N. 2017 Structural and Parameter Optimization of the System of Interconnected Processes of Building Complex Radio-Electronic Devices *14th International Conference The Experience of Designing and Application of CAD Systems in Microelectronics (CADSM)* (Polyana, Svalyava, (Zakarpattya), Ukraine, February 21–25, 2017) pp 192–194 DOI: 10.1109/CADSM.2017.7916112
[8] Ostreikovsky V A Fedorov A S 2000 Mathematical Models for Quantitative Estimation of Object Reliability with Catastrophe Theory Bifurcational Set Methods Applied in *Second International Conference on Mathematical Methods in Reliability (MMR'2000) : Methodology, Practice and Inference* (Bordeaux, France, Jule 4–7, 2000) vol 2 p 841-842
[9] Grishko A, Goryachev N, Yurkov N. 2015Adaptive Control of Functional Elements of Complex Radio Electronic Systems". *International Journal of Applied Engineering Research* Vol 10 pp 43842–43845
[10] Grishko A, Goryachev N, Kochegarov I, Brostilov S, Yurkov N. 2016 Management of Structural Components Complex Electronic Systems on the Basis of Adaptive Model *13th International Conference on Modern Problems of Radio Engineering, Telecommunications, and*

Computer Science (TCSET) (Lviv-Slavsko, Ukraine, February 23-26, 2016) pp 214–218 DOI: 10.1109/TCSET.2016.7452017
[11] Prigogine I, Stengers I, *Time Chaos and Quantum Theory* (Moscow) p 240
[12] Grishko A., Goryachev N., Kochegarov I., Yurkov N. Dynamic Analysis and Optimization of Parameter Control in Radio Systems in Conditions of Interference *2016 International Siberian Conference on Control and Communications (SIBCON)* (Moscow, Russia, May 12-14, 2016) DOI: 10.1109/SIBCON.2016.7491674.
[13] Grishko A, Yurkov N, Goryachev N. 2017 Reliability Analysis of Complex Systems Based on the Probability Dynamics of Subsystem Failures and Deviation of Parameters *2017 14th International Conference The Experience of Designing and Application of CAD Systems in Microelectronics (CADSM)* pp 179–182 DOI: 10.1109/CADSM.2017.7916109
[14] Yurkov N K, Trusov V A and Lysenko A V 2016 Methods of providing reliability of avionics and aerospace equipment at the design stage *13th International Scientific-Technical Conference on Actual Problems of Electronics Instrument Engineering (APEIE)* vol 03 DOI: 10.1109/APEIE.2016.7806899.

Surface triads with optical properties

K L Panchuk, E V Lyubchinov, I V Krysova

Omsk State Technical University, 11, Mira ave., Omsk, 644050, Russia

e-mail: panchuk_kl@mail.ru

Abstract. A geometric model of formation of surfaces comprising an interconnected triple of emitter, reflector and receiver is presented in the paper. The model is based on cyclographic mapping of a spatial curve to the plane. In such map any given point *(x, y, z)* of the curve corresponds to a cycle with center *(x, y)* and radius equal to z applicate. The entire curve corresponds to a directed envelope of cycles consisting, in the general case, of two branches.
It is shown that the triad of curves consisting of two branches of the envelope and the orthogonal projection of the original curve within the plane *(xy)* corresponds to a triad of developable surfaces. The triad of curves in the plane *(xy)* and the original curve together form a triad of ruled surfaces. Both triads have an optical property. Any ray of light emerging from the point of the emitter surface along the normal to it and falling on the surface of the reflector afterwards is directed along the normal vector to the surface of the receiver.
The direct and inverse problems of formation of the triad of surfaces are solved. In the first case, a one-parameter set of triads of surfaces is defined from a given spatial curve. In the second case, a single triad of surfaces is defined from a pair of curves "emitter-receiver" defined on the plane *(xy)*. Numerical examples of solutions of the direct and inverse problems are considered and the corresponding visualizations are given.
The results of the work can be used in the design of reflector antennas in radar systems and systems for converting solar energy into electric and thermal energy. Keywords — geometric model, cyclographic mapping, triad of surfaces, optical property.

1. Introduction

Reflection and focusing of electromagnetic waves are studied and applied in geometric and computer optics [1, 2, 3]. The use of reflective surfaces in modern antennas in radar-location allows solving problems of telemetry, navigation, cartography, detection of minerals, determining the coordinates and shape of the target and its support. Traditionally, parabolic, cylindrical, spherical surfaces and their combinations are used as reflective surfaces of reflector antennas [4]. It is known to use wide-angle helicoidal type antennas with parabolic generators, which have a number of advantages in comparison with antennas with a simple reflector profile [5, 6].

An important and promising direction of reflective surfaces application is development of solar radiation concentrators with such surfaces [7, 8, 9] for solving a number of energy-consuming space problems, such as destruction of space debris, energy generation in space with its subsequent transmission to the Earth, lighting the Earth from the orbit [7]. The concentrator receives solar radiation, reflects it, multiplies the density of reflected solar flux by many times, and directs it to the heat receiver in which solar energy conversion system functions. The efficiency of a space solar power plant largely depends on the geometry of concentrator reflective surface and heat receiver surface [7]. Thus, there is a need to solve the problem of research and development of reflective surfaces with new geometric shapes.

2. Formilation of the problem

The task is to develop a geometric model for forming reflective surfaces of the reflectors with the simplest, after plane, form and the simplest geometry, namely, linear developable surfaces and at the same time to obtain the same simplest form of the heat receiver surface. At the same time, the heat receiver surface should be correlated with the reflector surface, while variation of shape and position of the heat receiver, allowed by the model, should allow achieving effective operating characteristics of heat receiver. Thus, for example, reflector antenna emitter is often moved beyond the opening of the reflector and placed on the focal arc of the reflective surface to avoid shading effect of emitter and reflector [5], and heat receiver of space solar power plant is often placed in the concentrator focal plane to achieve necessary heat transfer parameters on heat receiver surface [7].

3. Theory

3.1 Mathematical model of a triad of surfaces including the surface of reflection and radiation.

A generalized cyclographic projection of a spatial curve

$$\bar{P}(t)=\{x(t),y(t),z(t)\},\bar{P}'(t)\neq 0, t\in R: T_0 \leq t \leq T\text{ , (1)}$$

can be formed on a plane $\Pi_1(xy)$. Mathematical model of such cyclographic projection is a system of parametric equations:

$$x_\beta(t)=x(t)+z(t)\cdot e(t)\cdot\frac{-x'(t)\cdot\mu(t)\mp y'(t)\sqrt{\lambda(t)-\mu^2(t)}}{\lambda(t)},$$
$$y_\beta(t)=y(t)+z(t)\cdot e(t)\cdot\frac{-y'(t)\cdot\mu(t)\pm x'(t)\sqrt{\lambda(t)-\mu^2(t)}}{\lambda(t)},\quad (2)$$

where $\mu(t)=e(t)\cdot z'(t)+z(t)\cdot e'(t)$,

$\lambda(t)=x'(t)^2+y'(t)^2, e(t)=tg\beta(t)$,

$\bar{P}'=\frac{d\bar{P}}{dt}, x'(t)=\frac{dx}{dt}, y'(t)=\frac{dy}{dt}$,

$z'(t)=\frac{dz}{dt}, e'(t)=\frac{de}{dt}$.

In case $e(t)=const$, the equations of the envelope of directed circles can be obtained from the equations (2) for the case of $\beta=const$, and in case $e(t)=1$ – the known envelope equations for cyclographic projection for the case of $\beta=45°$ [2,10].

A geometric scheme of spatial curve $\bar{P}(t)$ cyclographic projection formation is shown in figure 1.

A cyclographic projection of any point of the spatial curve $\bar{P}(t)$ is a cycle in the plane Π_1 (xy). The center of the cycle has coordinates (x,y) and the radius of the cycle $R=\pm|z|$, where $R=|z|$ corresponds to location of the point above the plane Π_1 and $R=-|z|$ – to location below the plane Π_1. In the first case the cycle has a positive direction and in the second case - a negative one.

In general case the envelope (2) of cycles in the plane Π_1 consists of two branches $\bar{P}_{\beta(1)}$ and $\bar{P}_{\beta(2)}$. These lines together with the line $\bar{P}_1$ of cycle centers form a triad of cylindrical surfaces Φ_1, $\Phi_{(1)}$ and $\Phi_{(2)}$, which have projecting location relative to the plane Π_1 (figure 2).

Such triad of surfaces has the following optical property: a ray of light emerging from one of the surfaces, $\Phi_{(1)}$ or $\Phi_{(2)}$, along the normal to it, after reflection from the medial surface Φ_1 is directed along the normal vector to the other. This optical property of the surface triad under consideration follows from known optical property of lines $\bar{P}_1$, $\bar{P}_{\beta(1)}$ and $\bar{P}_{\beta(2)}$ [2], which are directing lines of surfaces Φ_1, $\Phi_{(1)}$ and $\Phi_{(2)}$ respectively.

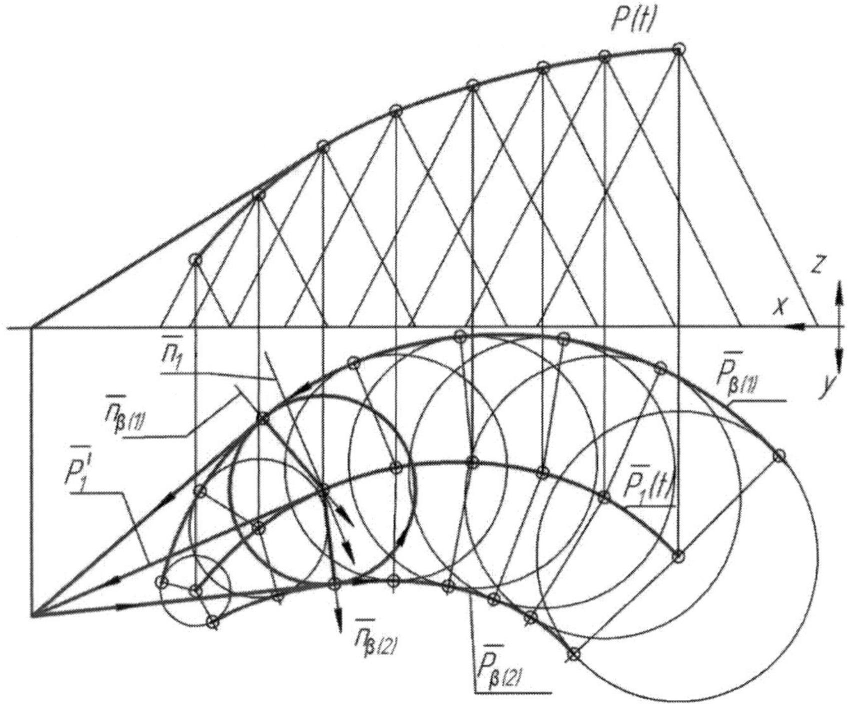

Figure 1. A geometric scheme of spatial curve $\overline{P}(t)$ cyclographic projection formation.

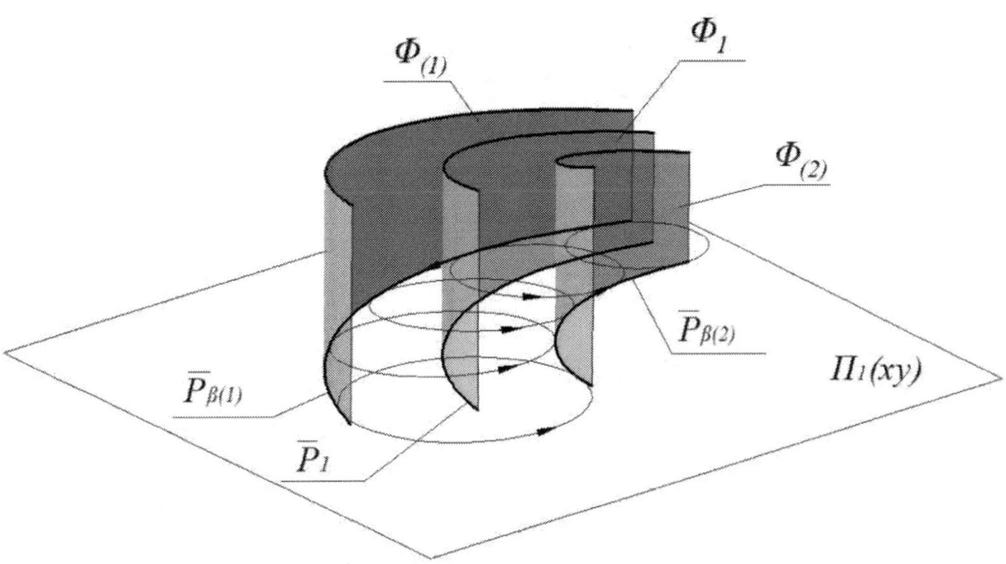

Figure 2. The triad of cylindrical surfaces.

Consider pairs of lines $\overline{P}, \overline{P}_{\beta(1)}$ and $\overline{P}, \overline{P}_{\beta(2)}$ as directing lines of expanding surfaces $\Psi_{(1)}$ and $\Psi_{(2)}$ respectively. The mathematical model of these surfaces in the form of a system of parametric equations is obtained as follows:

$$\begin{aligned} &X(t,l) = x(t) + l \cdot \left[x_\beta(t) - x(t)\right], \\ &Y(t,l) = y(t) + l \cdot \left[y_\beta(t) - y(t)\right], \\ &Z(t,l) = z(t) \cdot (1 - l), \\ &t, l \in R : T_0 \le t \le T, L_0 \le l \le L. \end{aligned} \tag{3}$$

The ruled surfaces $\Psi_{(1)}$ and $\Psi_{(2)}$ represent two parts of the envelope surface of a one-parameter set of cyclographic mapping cones, vertices of which belong to the spatial line $\bar{P}(t)$, and bases of which belong to the plane of the projections Π_1 and represent cycles. The half-angles β at the vertexes of cones can be constant $\beta = const$ for all points of the spatial line $\bar{P}(t)$, or can be expressed by a certain function $\beta = f(t) : f(t) \subset C^1, t \in R : T_0 \le t \le T$.

The triad of ruled surfaces Φ_1, $\Psi_{(1)}$ and $\Psi_{(2)}$ also has the optical property: a ray of light emerging from one of them along the normal to it after a reflection from the medial surface $\Phi_1$is directed along the normal vector to the other.

For half-angle $\beta \in R : B_0 \le \beta \le B$ a one-parameter set of triads of ruled surfaces with constant reflection surface Φ_1 is obtained (figure 3). A similar situation occurs for triads composed of triples of cylindrical surfaces with constant reflection surface Φ_1 (see figure 2).

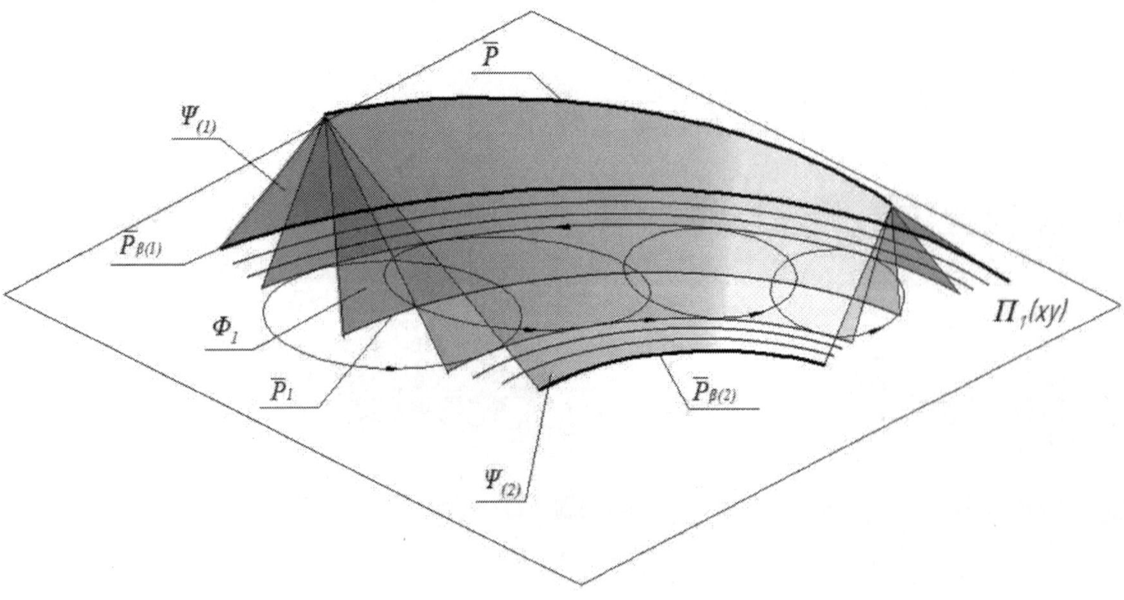

Figure 3. A system of triads of ruled surfaces of a spatial line.

The presence of a set of surface triads with constant medial reflection surface is convenient for constructing the surfaces $\Phi_{(1)}$ and $\Phi_{(2)}$ or $\Psi_{(1)}$ and $\Psi_{(2)}$ as emitter and heat receiver surfaces. This requires additional conditions that determine the geometric shape and position of these surfaces relative to the reflective surface Φ_1.

3.2 Algorithm for solving the inverse surface triads formation problem.

The problem is formulated as follows: given the guiding curves $\bar{P}_{(1)}(\lambda)$ and $\bar{P}_{(2)}(\mu)$ of emitter and receiver surfaces within the plane, where $\lambda, \mu \in R : \Lambda_0 \le \lambda \le \Lambda,\ M_0 \le \mu \le M$, it is required to determine the reflective surface of the triad.

In order to simplify the solution of the problem, it is assumed that the half-angles β at the vertices of the mapping cones have constant values. Then the algorithm for solving the inverse problem comes to the following:

1. Evolutes $\overline{Q}_{(1)}(\lambda)$ and $\overline{G}_{(2)}(\mu)$ of the curves $\overline{P}_{(1)}(\lambda)$ and $\overline{P}_{(2)}(\mu)$ within the plane $\Pi_1(xy)$ are plotted respectively (figure 4).

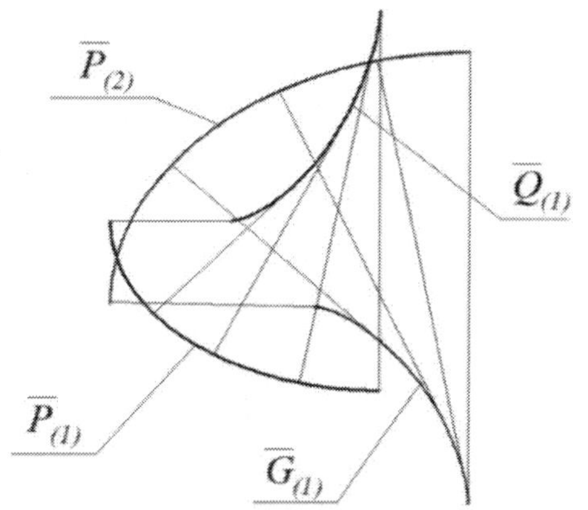

Figure 4. Initial curves and their evolutes.

2. At each point of both evolutes the perpendiculars to the plane Π_1 are plotted with length equal to curvature radius of corresponding initial curve. All perpendiculars plotted from one evolute have to be aligned in the same direction along the z axis.

On figure 5 all perpendiculars to the plane Π_1 are plotted from both evolutes in the positive direction of the z axis. Thus for each evolute a corresponding curve in space is obtained: $\overline{Q}_{(1)}(\lambda) \to \overline{Q}(\lambda), \overline{G}_{(1)}(\mu) \to \overline{G}(\mu)$.

3. Consider pairs of curves $\overline{P}_{(1)}(\lambda)$, $\overline{Q}(\lambda)$ and $\overline{P}_{(2)}(\mu)$, $\overline{G}(\mu)$. As follows from the formation scheme of the curves $\overline{Q}(\lambda)$ and $\overline{G}(\mu)$, their geometric form is determined by the form parameter λ and μ of the corresponding curve $\overline{P}_{(1)}(\lambda)$ or $\overline{P}_{(2)}(\mu)$. The pairs of curves $\overline{P}_{(1)}(\lambda) \to \overline{Q}(\lambda)$ and $\overline{P}_{(2)}(\mu) \to \overline{G}(\mu)$ form the ruled surfaces $\Psi_{(1)}$ and $\Psi_{(2)}$ of emitter (or receiver).

4. The curve of intersection of the surfaces $\Psi_{(1)} \cap \Psi_{(2)} = \overline{P}$ and its horizontal projection $\overline{P}_1$ are defined. The curves $\overline{P}$ and $\overline{P}_1$ are the guiding curves of the desired reflective surface Φ_1 with projecting generatrix lines with respect to $\Pi_1(xy)$.

4. Experimental results

To perform a numerical experiment in order to construct a set of triads of ruled surfaces along a given spatial curve $\overline{P}(t)$, the arc of the helical curve was used:

$$\overline{P}(t) = \{10\cos(t), 10\sin(t), 2t\}, t \in R : \frac{\pi}{2} \le t \le \pi .$$

The calculations were performed by means of computer algebra on the basis of equations (2) and (3) and their visualization was obtained (figure 3).

To solve the inverse problem the curves (arcs of ellipses) were taken as initial data:

$$\overline{P}_{(1)}(t_1) = \{5 + 3\cos(t_1), 3 + 2\sin(t_1)\}, t_1 \in R : -\pi \le t_1 \le -\frac{\pi}{2},$$

$$\overline{P}_{(2)}(t_2) = \{6 + 4\cos(t_2), 2 + 3\sin(t_2)\}, t_2 \in R : \frac{\pi}{2} \le t_2 \le \pi.$$

On the basis of the algorithm for solving the inverse problem the results of a numerical experiment were obtained and their visualization was performed (figure 5).

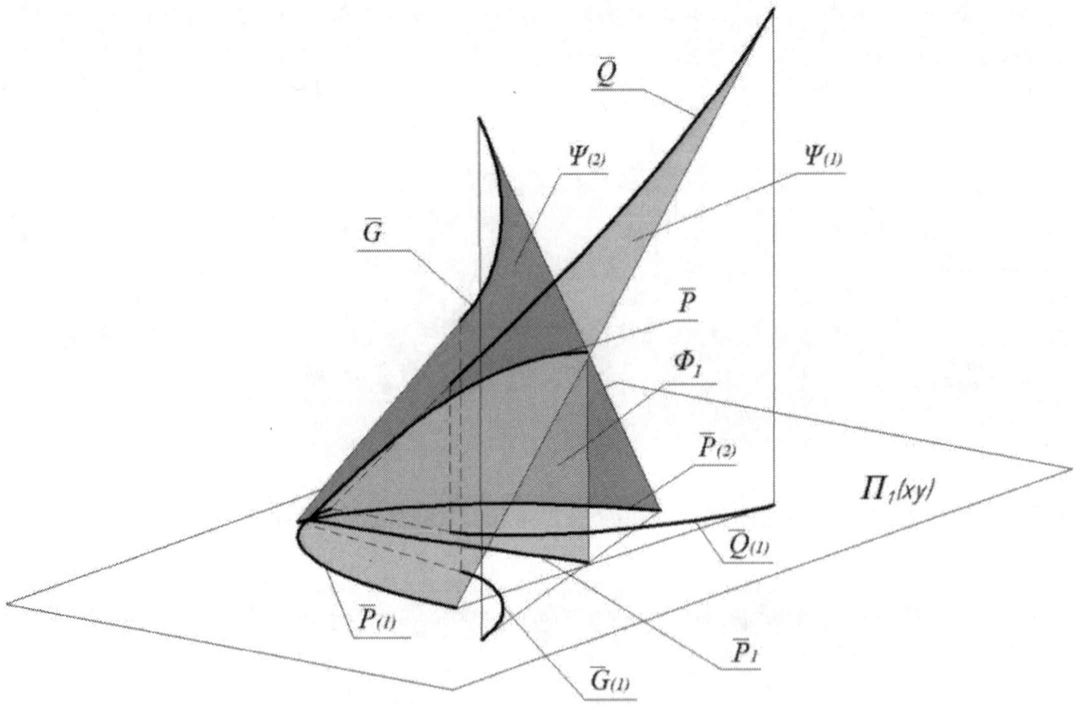

Figure 5. Formation of the reflective surface Φ_1.

5. Discussion of the results

The analysis of the results of numerical experiments and computational algorithms capabilities that obtain them shows that the definition of triads of surfaces along a given curve $\overline{P}(t)$ is a simple task from computational point of view. This facilitates making the optimal choice of a triad from a one-parameter set of triads designed for a certain technical application.

The solution of the inverse problem, i.e. the definition of a triad by a set of two curves $\overline{P}_{(1)}(\lambda)$ and $\overline{P}_{(2)}(\mu)$ in a plane, is much more complicated from the computational point of view. As a rule, numerical methods have to be applied in the mathematical model of solving this problem.

6. Conlusions

The paper proposes a geometric model and presents its mathematical description for the formation of triad surfaces possessing an optical property. The surfaces of the triad are geometrically interconnected, mathematically formalized and can be used, for example, in technical systems of radar as reflector, receiver and emitter surfaces.

References

[1] Farouki R T and Chastang J-C A 1992 Curves and surface in geometrical optics *Mathem. Meth. in Comp. Aided Geom. Des.* **2** pp 239-260

[2] Pottmann H and Wallner J 2001 *Computational Line Geometry* (Berlin Heidelberg: Springer Verlag)

[3] Borisova K V, Moiseev M A, Doskolovich L L and Andreev E S 2014 Design of reflective surface focusing light flux into an arbitrary curve *Computer optics* **3** pp 449-455

[4] Skolnik Merrill I 2008 *Radar Handbook* ed Chief (McGraw-Hill Companies)

[5] Pelevin V 2012 Wide-angle scanning by reflecting antennas *Bull. of the Polotsk State Un.* **S** pp 88-98
[6] Barabashov B G and Pelevin O Yu 1989 Estimate of the effectiveness of interference-error averaging algorithms of an amplitude-type direction finder *Telecom. and Radio Eng.* **44(5)** pp 45-47
[7] Knysh L I 2011 Influence of concentrator geometry on the energy parameters of the reception system of the space solar power plant *Space Science and Tech* (Dnepropetrovsk National Un.) **17(5)** pp 19-23
[8] Knysh L I and Gabrinets V A 2013 Assessment of PVT-technology efficiency in combined solar power plants *Scien. Bul. of the Nat. Mining Un.* **2** pp 74-78
[9] Simon A 1968 Calculation of the solar energy concentration in the focal spot of parabolic reflector *J. Solar Energy Sci. and Engineering* **2** pp 25-28
[10] Choi H I, Han C Y, Moon H P, Roh K H and Wee N-S 1999 Medial axis transformand offset curves by Minkowski Pythagorean hodograph curves *Comput. Aided Design* **31** pp 59-72

Contact interaction of flexible Timoshenko beams with small deflections

I V Papkova[1], A V Krysko[1,2], O A Saltykova[1,2], A A Zakharova[2], V A Krysko[1]

[1]Department of Mathematic and modelling, Saratov State Technical University, Politehnicheskaya str, 77, Saratov, Russia
[2]Department of engineering graphics and industrial design, Tomsk Polytechnic University, Sovetskaya str, 84/3, Tomsk, Russia

Abstract. In this work chaotic dynamics contact interaction of two flexible Tymoshenko beams, under the action of a transversal alternating load is investigated. The contact interaction of the beams is taken into account by the Kantor model. The geometric nonlinearity is taken into account by the model of T. von Karman.The system of partial differential equations of the twelfth order reduces to the system of ordinary differential equations by the method of finite differences of the second order. The resulting system by methods of Runge-Kutta type of the second, fourth and eighth orders was solved.Our theoretical/numerical analysis is supported by methods of nonlinear dynamics and the qualitative theory of differential equations. Chaotic vibrations of two flexible beams of Timoshenko were investigated and the optimal step values over the spatial coordinate and the time steps for the numerical experiment were found. Convergence for all applicable numerical methods have been achieved and shown that chaotic signals are true.

Keywords: Contact interaction, Timoshenko's beam, chaos, finite difference method, Runge-Kutta type methods, geometric nonlinearity

1. Introduction

Although nonlinear dynamics and contact interaction of beams based on hypothesis of different approximations have been successively used for many years, there still are present open problems originated from both engineering and science. Several hypotheses are known that describe the dynamics of a beam element: first, second, third approximation theories, and others. The hypothesis of the first approximation is the Euler-Bernoulli hypothesis [1], the second-approximation hypothesis or the Tymoshenko hypothesis [2], allows us to take into account the normal rotation to the midline after deformation. This theory allows more accurate description of the dynamics of the beam element. The nonlinear dynamics of beams, based on various hypotheses, is investigated in articles [3-5]. This article allows to prove the truth of the chaotic vibrations of two beams of Timoshenko with a small gap, under action of the transverse alternating load, taking into account the contact interaction. In the known literature, there are no solutions to such problems. This is a fundamentally important issue. When solving such complex non-linear systems of equations by numerical methods, there is a probability of obtaining an incorrect solution due to errors in numerical methods. In this paper we shall use the definition of chaos given by Gulik [6]. Gulik believes that chaos exists when either there is a significant dependence on the initial conditions or the function has a positive Lyapunov exponent at each point of the region. In this case, it is not periodic. As initial conditions, in addition to the conditions imposed on the functions entering into the system of differential equations, we mean the number of partitions with respect to the spatial coordinate, the order of the Runge-Kutta method, and the kinematic hypothesis. To prove the reliability and truth of solutions in chaos methods of analysis of nonlinear dynamics were used. These methods include the construction of phase portraits, Fourier power spectra, Poincaré pseudo-mappings, wavelet spectra, signals, calculation of the value of the highest Lyapunov exponent in three different algorithms Kantz [7], Wolff [8] and Rosenstein [9]. If the results obtained by all these methods gave the same result, then we can assume that the results obtained are true. In addition, the question of the convergence of the results, depending on the number of partitions along the spatial coordinate and on the time step for the finite-difference method, has been investigated. In contrast to previous studies [10], in this paper it was required to achieve convergence of results in chaos not only in Fourier power spectra, but also in signals. A separate point of the study was the choice of the Runge-Kutta method for solving the contact problem.

2. Mathematical model

We study a two-layer beam, where the layers can contact each other as it is shown in Figure 1. We consider the case when sticking between the layers of the beam is not possible because contact pressure is small. The Cartesian Coordinate system introduced on the Fig. 1. The equations of motion of the beams, as well as the boundary and initial conditions, are obtained from energy principle of the Hamilton-Ostrogradskiy $\int_{t_0}^{t_1}(\delta K - \delta\Pi + \delta' W)dt = 0$, where K – is the kinetic energy, Π – is the potential energy, $\delta' W$ - is the sum of elementary work of external forces. In this coordinate system, a structure of two beams, like a two-dimensional domain Ω, is defined as follows $\Omega = \{x \in [0, a\} - h \le z \le h_k + 3h\}$, $0 \le t \le \infty$.

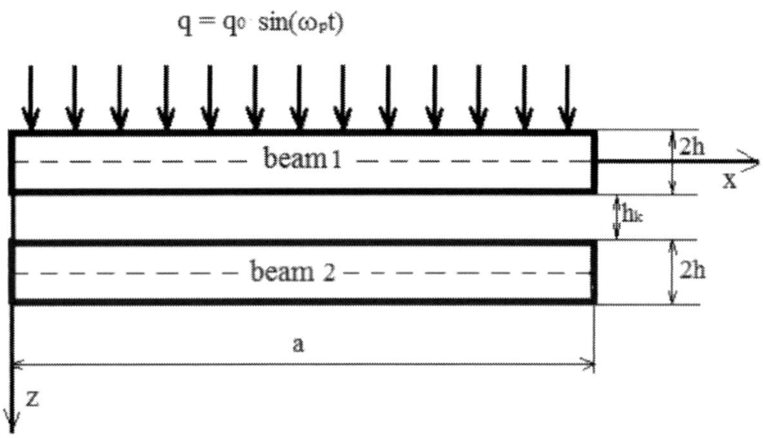

Figure 1. Scheme of the analyzed beam structure.

To simulate the contact interaction of beams according to the model of Kantor B.Ya. in the equations of the deflection beams, it is necessary to introduce the term $(-1)^i K(w_1 - w_2 - h_k)\Psi$, $i = 1,2$ - is a beam number, function Ψ is defined by $\Psi = \frac{1}{2}[1 + sign(w_1 - h_k - w_2)]$, where $\Psi = 1$, if there is contact between the beams - $w_1 > w_2 + h_k$, otherwise no contact [11], K – is a coefficient of rigidity of transverse compression of the structure in zone of the contact , h_k – is a gap between the beams. Tangential displacement u^z, w^z are distributed on thickness $\{-h \le z \le h\}$ by linear law $u_z = u + z\gamma_x$; $w_z = w$, where $\gamma_x = \gamma_x(x,t)$ — angle of rotation of the normal to the line $z = 0$. Then $e_{xx}^z = \frac{\partial u}{\partial x} + \frac{1}{2}\left(\frac{\partial w}{\partial x}\right)^2 + z\frac{\partial \gamma_x}{\partial x}$; $e_{xz}^z = \gamma_x + \frac{\partial w}{\partial x}$ and if denoted by ε_{11} - tangential deformations of the midline, bending deformation through $H_{11} = \frac{\partial \gamma_x}{\partial x}$, deformation of shear $\varepsilon_{13} = \gamma_x + \frac{\partial w}{\partial x}$, then the expressions for the deformations will be represented in form of a linear expansion by degrees z: $e_{xx}^z = \varepsilon_{11} + zH_{11}$; $e_{xz}^z = \varepsilon_{13}$; $z \in (\delta_i - \Delta, \delta_{i+1} - \Delta)$. We introduce a shear deformation by the formula: $\varepsilon_{13}^z = \frac{1}{2hG_{13}} Q_1 f(z)$, where Q_1 - transverse force; G_{13} - shear modulus, $f(z)$ - function characterizing the law of distribution of tangential stresses along the thickness. Here $Q_1 = 2hG_{13}k^2\varepsilon_{13}$, where $\frac{1}{k^2} = \frac{1}{2h}\int_{-h}^{h} f^2(z)dz$ .Value k^2 for such a function will be equal to $\frac{5}{6}$.

Taking into account Hooke's law, the expression for the stresses can be written in the form $\sigma_{xx}^i = E^i\varepsilon_{11}^i + E^i zH_{11}^i$, $\sigma_{xz}^i = G_{13}^i\varepsilon_{13}^i$. Then $T_{11}^i = \int_{-h}^{h}\sigma_{xx}^i dz$ — is an internal efforts; $Q_{11}^i = \int_{-h}^{h}\sigma_{xz}^i dz$ — is a cutting forces; $M_{11}^i = \int_{-h}^{h}\sigma_{xx}^i z dz$ — is a bending moments. We write down the Hamilton-Ostrogradskiy principle $\delta S^i = \delta\int_{t_0}^{t}(K^i - \Pi^i)dt = 0$, where

$\Pi^i = \Pi^i_c + \Pi^i_u$. The energy of the middle surface is $\Pi^i_c = \frac{1}{2}\int_e T^i_{11}\varepsilon^i_{11}dx = \frac{1}{2}\int_e T^i_{11}\left(\frac{\partial u_i}{\partial t} + \frac{1}{2}\left(\frac{\partial w_i}{\partial t}\right)^2\right)dx$. The bending energy is expressed as follows $\Pi^i_u = \frac{1}{2}\int_e \left[M^i_{11}H^i_{11} + Q^i_{11}\left(\gamma_{xi} + \frac{\partial w_i}{\partial t}\right)\right]dx$, but the kinetic energy $K^i = \frac{1}{2}\frac{(2h)\gamma_i}{g_i}\int_e \left[\left(\frac{\partial u_i}{\partial t}\right)^2 + \left(\frac{\partial w_i}{\partial t}\right)^2 + \left(\frac{\partial \gamma_{xi}}{\partial t}\right)^2\right]dx$.

From the Hamilton-Ostrogradskiy principle, we write the equations of motion of a structure of two Timoshenko beams in displacements taking energy dissipation into account in the dimensionless form as follows:

$$\begin{cases} \frac{1}{3}\left(\frac{\partial^2 w_i}{\partial x^2} + \frac{\partial \gamma_{xi}}{\partial x}\right) + \frac{1}{\lambda^2}\left(L_1(w_i,u_i) + \frac{3}{2}L_2(w_i,w_i) + L_3(w_i,u_i)\right) + \\ +(-1)^i K(w_1 - w_2 - h_k)\Psi + q(t) - \frac{\partial^2 w_i}{\partial t^2} - \varepsilon_1\frac{\partial w_i}{\partial t} = 0; \\ \frac{\partial^2 u_i}{\partial x^2} + L_4(w_i,w_i) - \frac{\partial^2 u_i}{\partial t^2} = 0; \\ \frac{\partial^2 \gamma_{xi}}{\partial x^2} - 8\lambda^2\left(\frac{\partial w_i}{\partial x} + \gamma_{xi}\right) - \frac{\partial^2 \gamma_{xi}}{\partial t^2} = 0;\ i = 1, 2, \end{cases} \tag{1}$$

Here $i = 1,2$ - is a serial number beams, $L_1(w_i,u_i) = \frac{\partial^2 w_i}{\partial x^2}\frac{\partial u_i}{\partial x}$, $L_2(w_i,w_i) = \frac{\partial^2 w_i}{\partial x^2}\left(\frac{\partial w_i}{\partial x}\right)^2$, $L_3(w_i,u_i) = \frac{\partial w_i}{\partial x}\frac{\partial^2 u_i}{\partial x^2}$, $L_4(w_i,w_i) = \frac{\partial w_i}{\partial x}\frac{\partial^2 w_i}{\partial x^2}$ - are the non-linear operators, γ_{xi} - is a function transverse shear, w_i, u_i – are the functions of deflections and displacements of beams, respectively. To the system of differential equations (1), we must add boundary conditions and initial conditions.

The system of equations (1), the boundary and initial conditions are reduced to the dimensionless form by means of variables:

$$\bar{w} = \frac{w}{2h},\ \bar{u} = \frac{ua}{(2h)^2},\ \bar{x} = \frac{x}{a},\ \lambda = \frac{a}{(2h)},\ \bar{q} = q\frac{a^4(1-\nu^2)}{(2h)^4 E},$$
$$\bar{t} = \frac{t}{\tau},\ \tau = \frac{a}{c},\ c = \sqrt{\frac{Eg}{(1-\nu^2)\rho}},\ \bar{\varepsilon}_1 = \varepsilon_1\frac{a}{c},\ \bar{\gamma}_x = \frac{\gamma_x a}{(2h)}.$$

The resulting system of nonlinear partial differential equations (1) together with the boundary and initial conditions reduces to the system of ordinary differential equations by the finite differences method with approximation $O(c^2)$, where c – step in the spatial coordinate. At each point of the grid, we obtain a system of ordinary differential equations. The Cauchy problem obtained in time is solved by methods of the Runge-Kutta type. In the work, different Runge-Kutta methods are compared: the Runge-Kutta of the 4th (rk4), 2nd (rk2) orders, the Runge-Kutta-Felberg 4th order (rkf45), 4th order Kesh-Karp (rkck), the Runge-Kutta Prince-Dormand of the eighth order (rk8pd), the implicit Runge-Kutta method of the second order (rk2imp) and the 4th order (rk4imp). On basis of this algorithm was created programs, which allow to solve the problem depending on the control parameters $\{q_0, \omega_p\}$. Great attention in work was given to the question of not penetrating the elements of the structure into each other. As noted above, the problems under study are highly nonlinear, so the question arises about the reliability of the results obtained.

3. Numerical results

The boundary conditions for sealing both ends of the beams:

$$w_i(0,t) = w_i(1,t) = u_i(0,t) = u_i(1,t) = \frac{\partial w_i(0,t)}{\partial x} = \frac{\partial w_i(1,t)}{\partial x} = 0, i = 1, 2$$

Initial conditions:

$$w_i(x,t)|_{t=0} = u_i(x,t)|_{t=0} = \gamma_{xi}(x,t)|_{t=0} = 0,\ \frac{\partial w_i(x,t)}{\partial t}\Big|_{t=0} = \frac{\partial u_i(x,t)}{\partial t}\Big|_{t=0} = \frac{\partial \gamma_{xi}(x,t)}{\partial t}\Big|_{t=0} = 0.$$

On the first beam is acted transverse distributed over the surface sign-variable load type: $q = q_0\sin(\omega_p t)$, where q_0 - amplitude, ω_p - frequency of driving vibrations.

For the numerical experiment we consider: $\omega_p = 5.1$, $q_0 = 5000$, $h_k = 0.1$, $\lambda = a/2h = 50$, $\varepsilon_1 = 1$. The frequency of the driving vibrations is close to the natural frequency of the beam. A preliminary question was investigated about convergence of the finite differences method. In Fig. 2 (a, б) signals are presented, calculated for a different number of points division of a segment $n = 40; 80; 120; 240; 360; 400; 440$. For the second beam, the convergence by the number of divisions of the segment is much worse and does not completely come. The error between signals calculated at n = 360 и n = 400 isc 3%, However, the signals coincide in shape over the entire time interval. The results were obtained using the Runge-Kutta method of the 8th order in the Prince-Dormand modification (rk8pd).

a) b)

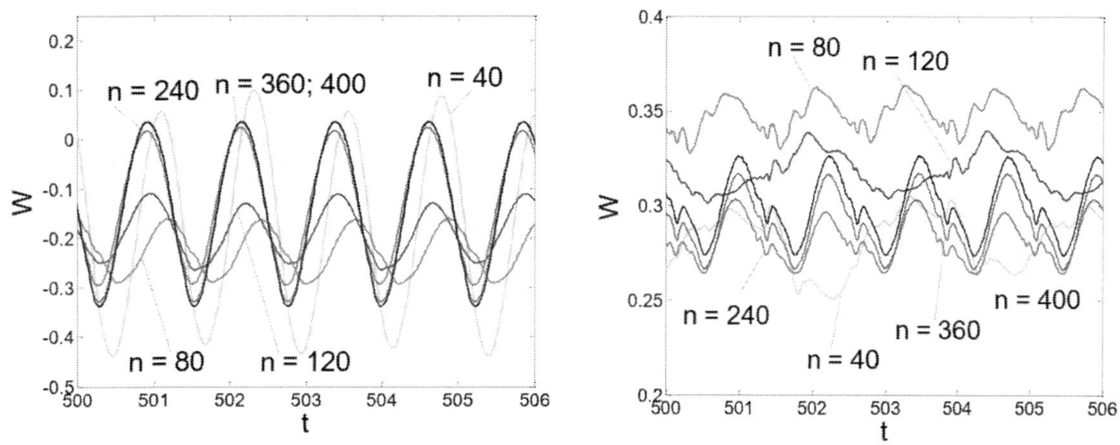

Figure 2. Beam deflection of signals $w_t(0.5,t)$ at $t \in [500..506]$ for $n = 40; 360; 400$ a) First beam, b) Second beam.

In [10], it was considered sufficient to show convergence with respect to the Fourier power spectra for chaotic vibrations. By signal, it was impossible to achieve convergence. At $n = 400$ the convergence of results for chaotic results, even with a signal, will be achieved.

Further, the convergence of the signals was investigated depending on the type of the Runge-Kutta method. For both beams, the results of the Runge-Kutta methods of the second, fourth and eighth orders coincided completely, however, it was decided to use the Prince-Dormand 8-th order method (rk8pd) in further calculations, since this method allows automatic step-by time.

Let us investigate the dynamic characteristics of beams for different number of partitions with respect to the spatial coordinate. In Table 1 we give graphs of Fourier power spectra, 3D phase portraits, and Poincaré pseudo-mappings for both beams.

Table 1. Dynamic characteristics of beams

n	№	Power spectrum	Phase portrait 3D $w(w'_t, w''_{tt})$	Pseudo Poincare map
40	1	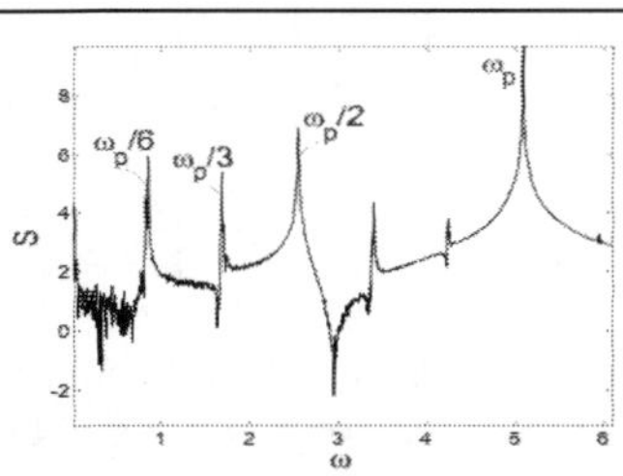	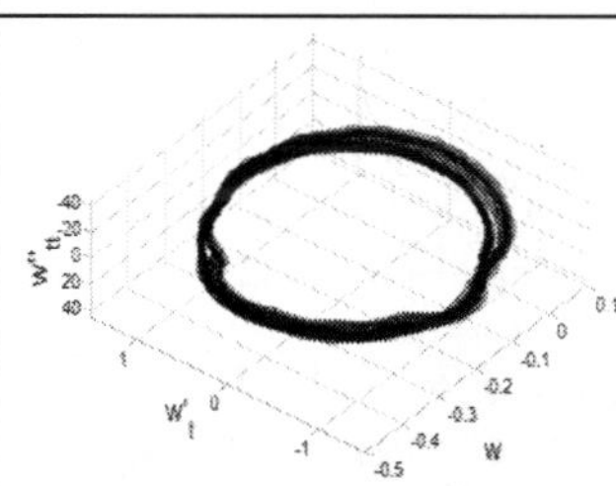	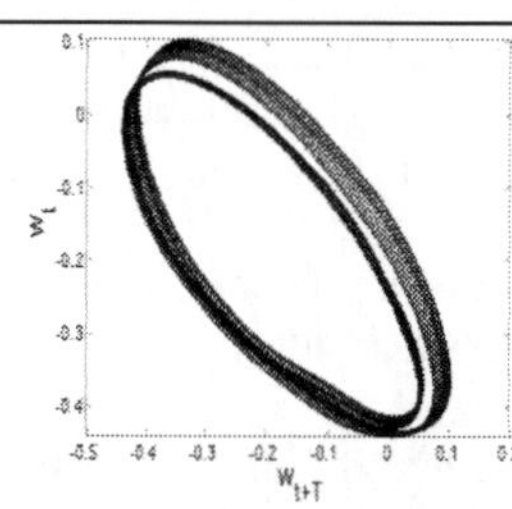

	2	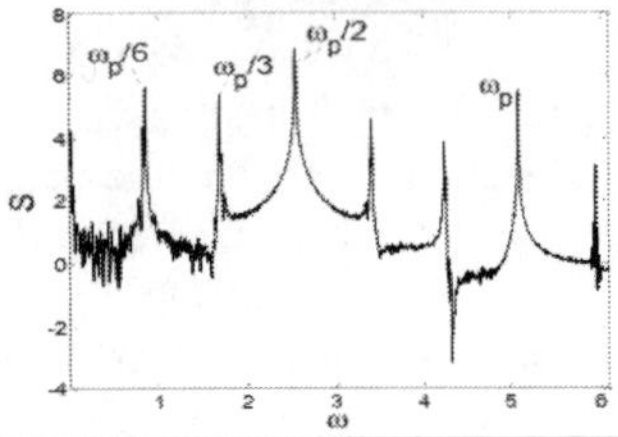	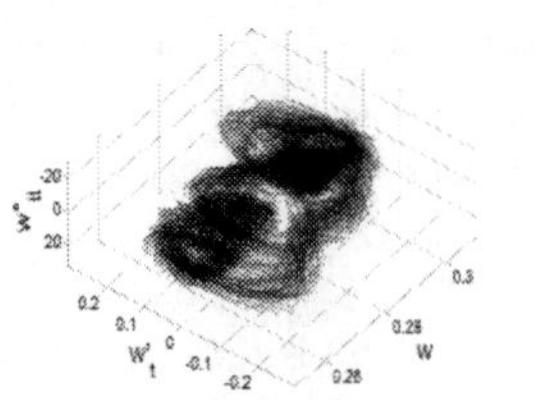	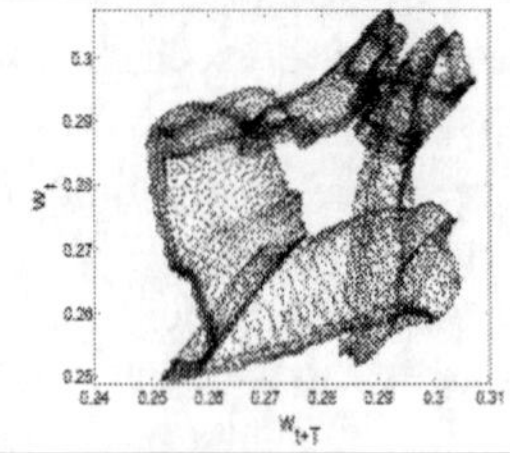
360,400	1	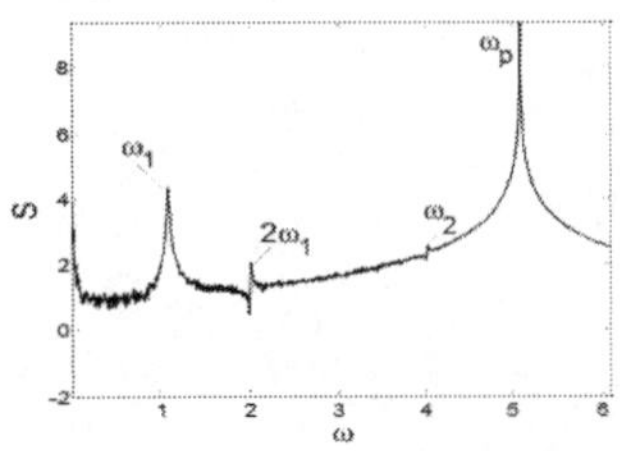	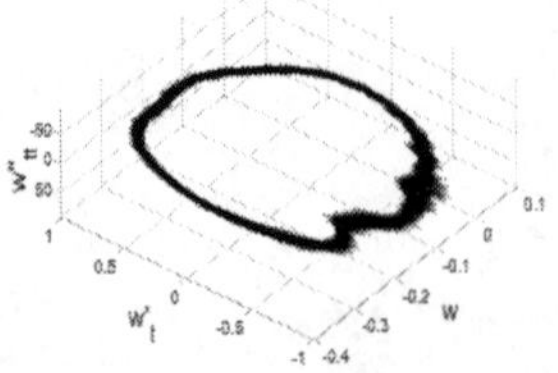	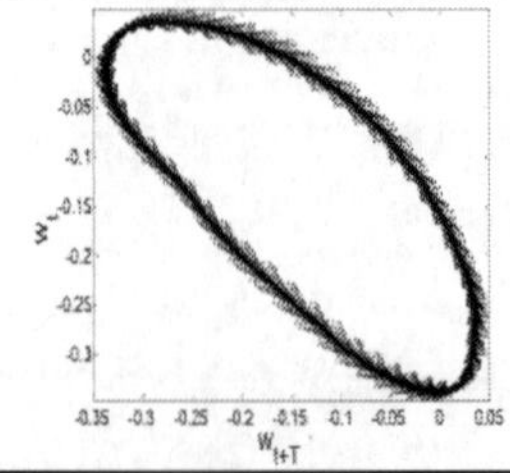
	2	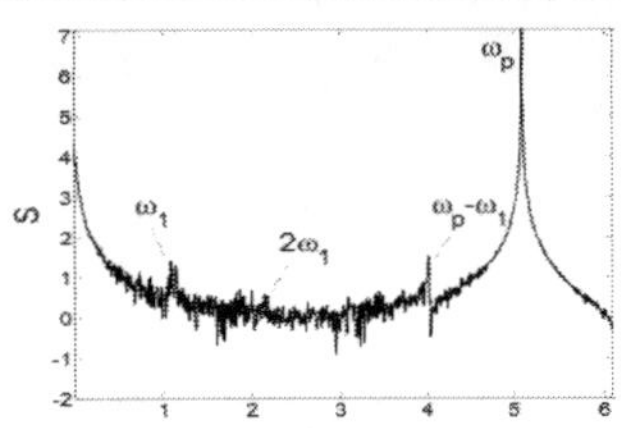	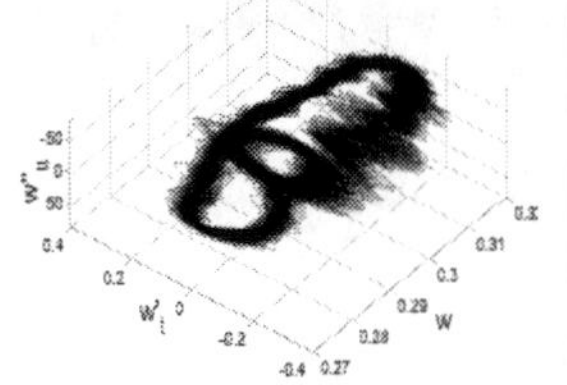	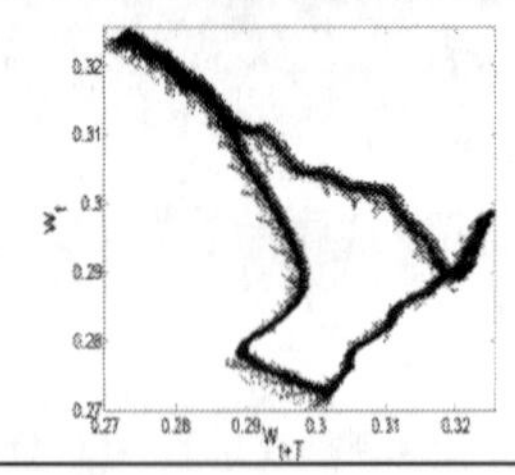

4. The discussion of the results.

Analyzing the Fourier power spectra for different numbers of partitions n with respect to the spatial coordinate, for beam 1, we can first note an increase, and then a reduction in the number of frequencies and a reduction in the noise component with n. At $n=40$ the power spectra of both beams demonstrate vibrations at linearly dependent frequencies $\omega_p/2$, $\omega_p/3$, $\omega_p/6$ and the presence of a noise component at low frequencies. Both beams vibrate at the same frequencies, i.e. frequency synchronization of vibrations occurs. An increase in n doubles the overall chaotic pedestal. At $n=120$ in the power spectrum of both beams, the frequency again appears $\omega_p/2$, also a decrease in the chaotic component as compared to $n=80$. At $n=240$ in the signal of the first and second beams there are frequencies в $\omega_p/2$ и $\omega_p/4$. При $n=360$ и $n=400$ power spectra are cleaned from the noise component, in the signal of both beams there are frequencies $\omega_1=\omega_p/5=1.02$, $2\omega_1$ и $\omega_2=\omega_p-\omega_1$, then we can talk about synchronization of oscillations at these frequencies.

As a rule, 2D phase portraits are considered, but in this paper it is proposed to consider 3D phase portraits $w(w_t', w''_{tt})$, in this case we will have information about all the characteristics of the dynamics. Consider 3D phase portraits $w(w_t', w''_{tt})$. At the minimum number of nodes n the phase portrait for beam 1 gives a ring, but in space it is clear that this ring has a thickness and is non-uniform. For the second beam, the phase portrait is a solid spot, which corresponds to chaotic vibrations. With an increase in the number of equations, the phase portrait and the appearance of rings are compressed.

The pseudo-Poincare map for a beam 1 for all n has the shape of an oval, but with small changes in its thickness. Beginning from $n=360$ their convergence is noted, as well as for beam 2, where for $n=40$ the pseudo-Poincaré map is scattered.

Complex analysis of frequency characteristics, it allows us to make the conclusion about the sufficiency n = 400, for the study of nonlinear dynamics and the contact interaction of two beams with a gap, when both beams are described by Timoshenko's model. In this paper, the values of the highest Lyapunov exponent for n = 400 are calculated, using the methods Wolff, Rosenstein and Kantz. The values are obtained on the basis of solutions of the Cauchy problem by the Runge-Kutta method of the 8th order (rk8pd). Different methods of calculating Lyapunov's indices must be used to determine the true chaos. When the number of points of division of the beam

on n = 40; 80, 120; 240, 360, 400 segments in the method of finite differences, the Lyapunov exponents converge at any counting method to the second or third decimal place.
In Table 2, we give the values of the highest Lyapunov exponent for both beams at n = 400, calculated using the methods of Kantz, Wolf, and Rosenstein. In order to avoid obtaining erroneous conclusions in the study of chaotic oscillations, the Lyapunov exponents are obtained by several methods. At present there is no reliable method for determining such.

Table 2. Lyapunov exponent

n		Beam 1			Beam 2		
	method	Wolf	Rosenstein	Kantz	Wolf	Rosenstein	Kantz
$n = 400$	Rk8pd	0.01658	0.05646	0.02191	0.02835	0.04617	0.02363

All the values of the highest Lyapunov exponent, regardless of the method of solving the Cauchy problem, from the number of intervals of the beam partition, from the calculation algorithm are positive. That is, we are dealing with the true chaotic oscillations of the investigated beam structure.

5. Conclusions

In this article the reliability of the numerical results of the solution for the problem of the two beams contact interaction, described by the kinematic hypothesis of Tymoshenko, with a small gap between them, was given and defended. A comprehensive study of the nonlinear dynamics of the contact interaction of Timoshenko beams under action of a transversal alternating load. Selection number of points of partitions on the spatial coordinate (n = 400) and the choice of the method for solving the Cauchy problem (the Runge-Kutta method of the 8th order of Prince-Dormand) was justified. On basis of carried out research of frequency characteristics, we can talk about the phenomenon of chaotic frequency synchronization of beams vibration.

6. Acknowledgments

This work was supported by the Ministry of Education and Science of the Russian Federation 2.1642.2017 / 4.6

References

[1] Euler L 1757 Sur la force des colones *Memories de L'Academie de Berlin* Vol 13 p 252–282.
[2] Timoshenko S P 1921On the correction for shear of differential equation for transverse vibration of prismatic bar *Philosophical Magazin* 41 pp 744-746
[3] Zhigalov M V, Pavlov S P, Krysko V 2016 A Nonlinear Dynamics of Multilayer Beam Structures with Gaps between them taking into Account the Geometrical and Physical Nonlinearities *International Journal of Applied Engineering Research* Vol 11 pp 10427-10432
[4] A.V. Krysko, J. Awrejcewicz, M.V. Zhigalov, S.P. Pavlov, V.A. Krysko 2017 Nonlinear behaviour of different flexible size-dependent beams models based on the modified couple stress theory. Part 1. Governing equations and static analysis of flexible beams *International Journal of Non-Linear Mechanics* 93 96-105
[5] Krysko A V, Awrejcewicz J, Zhigalov M V, Pavlov S P, Krysko V A Nonlinear behaviour of different flexible size-dependent beams models based on the modified couple stress theory. Part 2. *Chaotic dynamics of flexible beams, International Journal of Non-Linear Mechanics* 93 pp 106-121
[6] Denny Gulick *Encounters with Chaos* (New York: McGraw-Hill) 1992
[7] Kantz H. 1994A robust method to estimate the maximum Lyapunov exponent of a time series *Phys. Lett. A* 185. P. 77-87
[8] Wolf A., Swift J.B., Swinney H.L., Vastano J.A. 1985 Determining Lyapunov Exponents from a time series *Physica 16 D* pp 285-317
[9] Rosenstein M T, Collins J J, Carlo J. De Luca A practical method for calculating largest Lyapunov exponents from small data sets *Neuro Muscular Research Center and Department of Biomedical Engineering (*Boston University) 1992

IOP Conf. Series: Journal of Physics: Conf. Series **944** (2017) 012087 doi:10.1088/1742-6596/944/1/012087

[10] Awrejcewicz J., Krysko V.A., Papkova I.V., Krysko A.V. 2016 *Deterministic Chaos in One-Dimentional Continuous Systems (*Singapur: World Scientific series on Nonlinear Science Series) p 561
[11] Kantor B Ya, Bogatyrenko T L 1986 A method for solving contact problems in the nonlinear theory of shells *Dokl. Ukrain Academy of Sciences. Ser A* 1 pp 18-21 (In Russian)

Method of transmission of dynamic multibit digital images from micro-unmanned aerial vehicles

E P Petrov, N L Kharina

Vyatka State University, str. Moscow, 36, Kirov, 610000, Russia

e-mail: eppetrov@mail.ru[1], natal_res@mail.ru[2]

Abstract. In connection with successful usage of nanotechnologies in remote sensing great attention is paid to the systems in micro-unmanned aerial vehicles (MUAVs) capable to provide high spatial resolution of dynamic multibit digital images (MDI). Limited energy resources on board the MUAV do not allow transferring a large amount of video information in the shortest possible time. It keeps back the broad development of MUAV. The search for methods to shorten the transmission time of dynamic MDIs from MUAV over the radio channel leads to the methods of MDI compression without computational operations onboard the MUAV. The known compression codecs of video information can not be applied because of the limited energy resources. In this paper we propose a method for reducing the transmission time of dynamic MDIs without computational operations and distortions onboard the MUAV. To develop the method a mathematical apparatus of the theory of conditional Markov processes with discrete arguments was used. On its basis a mathematical model for the transformation of the MDI represented by binary images (BI) in the MDI, consisting of groups of neighboring BIs (GBI) transmitted by multiphase (MP) signals, is constructed. The algorithm for multidimensional nonlinear filtering of MP signals is synthesized, realizing the statistical redundancy of the MDI to compensate for the noise stability losses caused by the use of MP signals.

1. Introduction

Due to the progress in the development of unmanned aerial vehicles on small platforms and the development of means for obtaining multibit digital images (MDI) of high spatial resolution, the creation of remote sensing on micro unmanned aerial vehicles (MUAV) is of great interest. According to experts, 50 percent of useful video information in near real-time mode is obtained from the MUAV at the operational level [1]. A significant disadvantage of MUAV is the "tight" restrictions on energy resources, therefore one of the actual tasks to be solved when creating an MUAV is the minimization of energy resources onboard of the MUAV for the transfer of dynamic MDI in the shortest possible time.

The search for methods to reduce the transmission time of dynamic high-resolution MDI has led to the development of methods for compressing dynamic MDIs without computational operations that are not critical to the resolution of dynamic MDIs, which exceed the known video compression methods, such as MPEG2, MPEG4 [2,3]. These codecs require the implementation of large time and energy resources, so they can not be implemented onboard the MUAV. Therefore, the development of new methods and improvement of the known ones for reducing the time of transfer of originals of dynamic MDIs without distortion and computational operations onboard the MUAV is urgent.

2. Description of Method

The paper suggests a method for solving the problem to reduce the transmission time of dynamic MDI from the MUAV board. The method consists of two parts. The first part consists in the transformation of the originals of dynamic MDI, consisting of binary images (BI) (Figure 1) in the MDI, consisting of groups of neighboring BIs (GBIs) comprising two or more BIs (Figure 2) obtained onboard the MUAV.

Figure 1. An example of dividing a dynamic MDI frame into a BI

The number of GBIs is smaller than the BI, the transition from BI to GBI is carried out in every MDI without distortion and computational operations on board of the MUAV, which allows to reduce the time of transfer of dynamic MDI without the cost of energy resources onboard of the MUAV. For the transmission of GBI via the radio channel, multiphase manipulated (MP) signals are used. It is possible to use other types of signals, for example, multifrequency (MF) sygnals as well.

A significant drawback of the use of MP signals for the transmission of dynamic MDI, consisting of BI, is the noise immunity of receiving MP signals by 3 dB in power for each division of the phase of the MP signal by 2. For example, in the transition from MP signals with two phase states to MP signals with four phase states, the transmission time and interference immunity of reception of MDI, consisting of BI, is reduced by half.

Figure 2. The combination of two BIs in GBIs

The second part, more complicated, consists in developing a method for realizing the statistical redundancy of originals of dynamic MDIs to improve the interference stability of MP signals receiving sufficient to compensate for the loss of noise stability caused by the use of MP signals for the transmission of dynamic MDIs.

Based on the representation of the transformed dynamic MDIs by three-dimensional Markov processes with discrete arguments with several (more than two) states and the theory of multidimensional nonlinear filtration, there have been developed a mathematical model of dynamic MDI consisting of BI with two or more BIs and an algorithm for three-dimensional nonlinear filtering of the transformed dynamic MDI transmitted MP signals, effectively realizing the statistical redundancy of the originals of dynamic MDIs for improving the noise stability of receiving MP signals. It allowed to reduce the transmission time of dynamic MDI by several times without the loss of noise stability and without the cost of energy resources onboard the MUAV.

We will assume that in each frame of the BI g BIs is combined and the dynamic MDI is approximated by three-dimensional Markov processes with discrete arguments with several states $N = 2^g$ with the probability vector of the initial states $P = \|p_1, p_2, ..., p_N\|^T$ and the transition probability matrix (MTP) ${}^1\Pi$, ${}^2\Pi$ and ${}^4\Pi$ from the state M_i to the neighboring state M_j $(i, j \in N)$ along the horizontal, vertical and between frames, respectively [4]:

$$
{}^1\Pi = \begin{Vmatrix} {}^1\pi_{11} & {}^1\pi_{12} & \cdots & {}^1\pi_{1N} \\ {}^1\pi_{21} & {}^1\pi_{22} & \cdots & {}^1\pi_{2N} \\ \cdots & \cdots & \cdots & \cdots \\ {}^1\pi_{N1} & {}^1\pi_{N2} & \cdots & {}^1\pi_{NN} \end{Vmatrix}, \quad {}^2\Pi = \begin{Vmatrix} {}^2\pi_{11} & {}^2\pi_{12} & \cdots & {}^2\pi_{1N} \\ {}^2\pi_{21} & {}^2\pi_{22} & \cdots & {}^2\pi_{2N} \\ \cdots & \cdots & \cdots & \cdots \\ {}^2\pi_{N1} & {}^2\pi_{N2} & \cdots & {}^2\pi_{NN} \end{Vmatrix},
$$

$$
{}^4\Pi = \begin{Vmatrix} {}^4\pi_{11} & {}^4\pi_{12} & \cdots & {}^4\pi_{1N} \\ {}^4\pi_{21} & {}^4\pi_{22} & \cdots & {}^4\pi_{2N} \\ \cdots & \cdots & \cdots & \cdots \\ {}^4\pi_{N1} & {}^4\pi_{N2} & \cdots & {}^4\pi_{NN} \end{Vmatrix}. \quad (1)
$$

The mathematical model of dynamic BI is presented in figure 3.

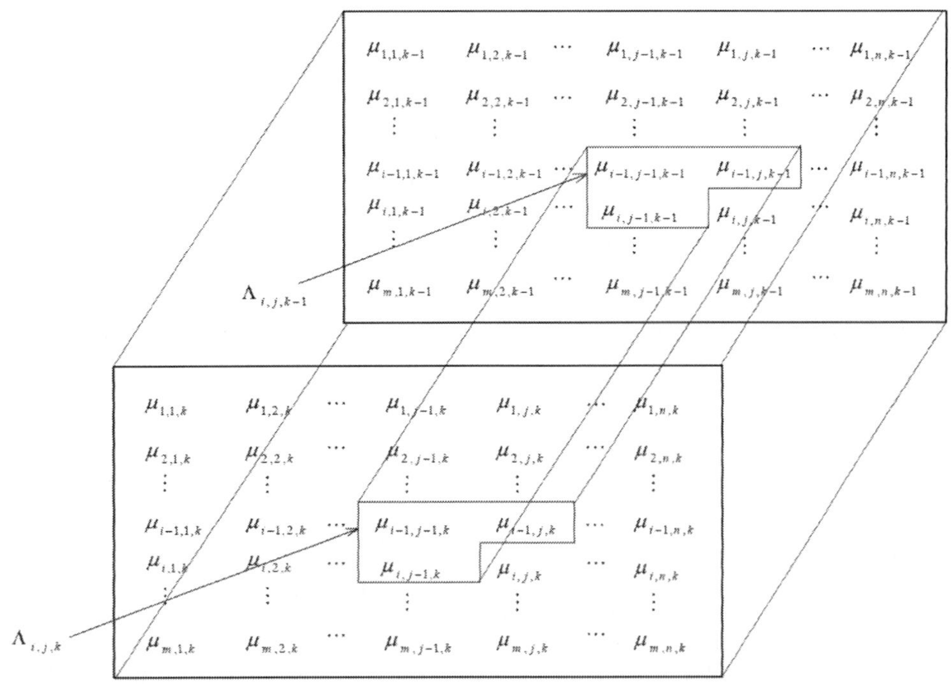

Figure 3. MM of dynamic BI

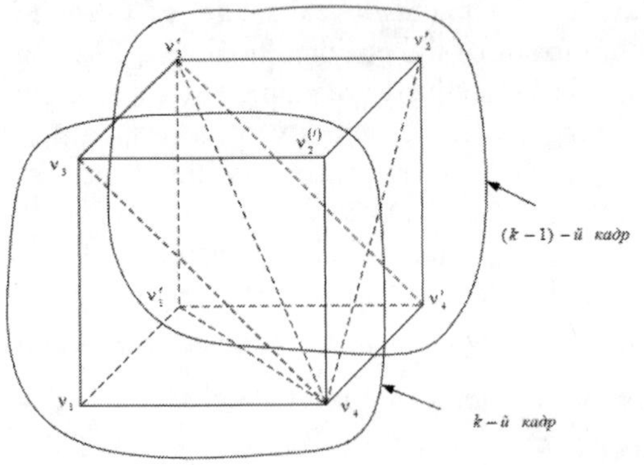

$$v_1 = \mu_{i,j-1,k}, \quad v_1' = \mu_{i,j-1,k-1},$$
$$v_2 = v_{i-1,j,k}, \quad v_2' = \mu_{i-1,j,k-1},$$
$$v_3 = \mu_{i-1,j-1,k}, \quad v_3' = \mu_{i-1,j-1,k-1},$$
$$v_4 = \mu_{i,j,k}, \quad v_4' = \mu_{i,j,k-1}.$$

Figure 4. The neighborhood of the dynamic MDI pixel, consisting of BI and g BI

The state of the element v_4 depends only on the known elements of a certain subset $\Lambda_{i,j,k}$, called the neighborhood of the element. The best way to satisfy the condition of causality is the configuration of the neighborhood (Figure 4)

$$\Lambda_{i,j,k} = \{v_1, v_2, v_3, v_1', v_2', v_3', v_4\} \tag{2}$$

The transition probabilities from the states of the elements $\Lambda_{i,j,k}$ to the state v_4 form a MTP of the form [4]:

$$\Pi = \begin{bmatrix} \pi_{iiii} & \pi_{iiij} & \cdots & \pi_{iiil} & \cdots & \pi_{iiiN} \\ \pi_{iiji} & \pi_{iijj} & \cdots & \pi_{iijl} & \cdots & \pi_{iijN} \\ \vdots & \vdots & \cdots & \vdots & \cdots & \vdots \\ \pi_{ijki} & \pi_{ijkj} & \cdots & \pi_{ijkl} & \cdots & \pi_{ijkN} \\ \vdots & \vdots & \cdots & \vdots & \cdots & \vdots \\ \pi_{NNNi} & \pi_{NNNj} & \cdots & \pi_{NNNl} & \cdots & \pi_{NNNN} \end{bmatrix}_{N^3 \times N} \quad i,j,k,l = \overline{1,N} \tag{3}$$

With known matrices ${}^1\Pi$, ${}^2\Pi$, ${}^4\Pi$ to calculate the elements of the matrix Π (3), it is necessary to first calculate the matrices

$$\begin{gathered} {}^3\Pi = {}^1\Pi \cdot {}^2\Pi; \quad {}^5\Pi = {}^1\Pi \cdot {}^4\Pi; \quad {}^6\Pi = {}^2\Pi \cdot {}^4\Pi; \\ {}^7\Pi = {}^3\Pi \cdot {}^4\Pi = {}^1\Pi \cdot {}^2\Pi \cdot {}^4\Pi. \end{gathered} \tag{4}$$

which define statistical relationships between the elements of the neighborhood $\Lambda_{i,j,k}$ with element v_4.

The values of the elements of the matrix Π (4) can be calculated as follows:

$$\begin{gathered} \pi_{iiii} = \frac{{}^1\pi_{ii} \cdot {}^2\pi_{ii} \cdot {}^4\pi_{ii} \cdot {}^7\pi_{ii}}{{}^3\pi_{ii} \cdot {}^5\pi_{ii} \cdot {}^6\pi_{ii}}, \quad \pi_{iiij} = \frac{{}^1\pi_{ij} \cdot {}^2\pi_{ij} \cdot {}^4\pi_{ij} \cdot {}^7\pi_{ij}}{{}^3\pi_{ii} \cdot {}^5\pi_{ii} \cdot {}^6\pi_{ii}}, \quad \ldots, \\ \pi_{iiil} = \frac{{}^1\pi_{il} \cdot {}^2\pi_{il} \cdot {}^4\pi_{il} \cdot {}^7\pi_{il}}{{}^3\pi_{ii} \cdot {}^5\pi_{ii} \cdot {}^6\pi_{ii}}, \quad \ldots, \pi_{iiiN} = \frac{{}^1\pi_{iN} \cdot {}^2\pi_{iN} \cdot {}^4\pi_{iN} \cdot {}^7\pi_{iN}}{{}^3\pi_{ii} \cdot {}^5\pi_{ii} \cdot {}^6\pi_{ii}}, \end{gathered} \tag{5}$$

The expressions for calculating the elements of the first row of the matrix Π are given. The calculation of the remaining matrix elements for various combinations of pixel values of the neighborhood $\Lambda_{i,j,k}$ is carried out in a similar way.

The algorithm for multidimensional nonlinear filtering of BI from g BI

The system of equations for multidimensional nonlinear filtering of dynamic MDI, consisting of BI with g BI, transmitted by MP signals, can be represented in the form [5-7]:

$$u_{1(k+1)}(\nu_4)=\left[f(M_1(\nu_4))-f(M_N(\nu_4))\right]+u_{1(k)}(\nu_1)+z_1(u(\nu_1),{}^1\pi_{ij})+$$
$$+u_{1(k)}(\nu_2)+z_1(u(\nu_2),{}^2\pi_{ij})+u_{1(k)}(\nu_4')+z_1(u(\nu_4'),{}^4\pi_{ij})+u_{1(k)}(\nu_3')+z_1(u(\nu_3'),{}^7\pi_{ij})-$$
$$-u_{1(k)}(\nu_3)-z_1(u(\nu_3),{}^3\pi_{ij})-u_{1(k)}(\nu_1')-z_1(u(\nu_1'),{}^5\pi_{ij})-u_{1(k)}(\nu_2')-z_1(u(\nu_2'),{}^6\pi_{ij})$$

$$u_{2(k+1)}(\nu_4)=\left[f(M_2(\nu_4))-f(M_N(\nu_4))\right]+u_{2(k)}(\nu_1)+z_2(u(\nu_1),{}^1\pi_{ij})+$$
$$+u_{2(k)}(\nu_2)+z_2(u(\nu_2),{}^2\pi_{ij})+u_{2(k)}(\nu_4')+z_2(u(\nu_4'),{}^4\pi_{ij})+u_{2(k)}(\nu_3')+z_2(u(\nu_3'),{}^7\pi_{ij})-$$
$$-u_{2(k)}(\nu_3)-z_2(u(\nu_3),{}^3\pi_{ij})-u_{2(k)}(\nu_1')-z_2(u(\nu_1'),{}^5\pi_{ij})-u_{2(k)}(\nu_2')-z_2(u(\nu_2'),{}^6\pi_{ij})$$
$$\vdots$$
$$u_{(N-1)(k+1)}(\nu_4)=\left[f(M_{(N-1)}(\nu_4))-f(M_N(\nu_4))\right]+u_{(N-1)(k)}(\nu_1)+z_{(N-1)}(u(\nu_1),{}^1\pi_{ij})+$$
$$+u_{(N-1)(k)}(\nu_2)+z_{(N-1)}(u(\nu_2),{}^2\pi_{ij})+u_{(N-1)(k)}(\nu_4')+z_{(N-1)}(u(\nu_4'),{}^4\pi_{ij})+u_{(N-1)(k)}(\nu_3')+z_{(N-1)}(u(\nu_3'),{}^7\pi_{ij})-$$
$$-u_{(N-1)(k)}(\nu_3)-z_{(N-1)}(u(\nu_3),{}^3\pi_{ij})-u_{(N-1)(k)}(\nu_1')-z_{(N-1)}(u(\nu_1'),{}^5\pi_{ij})-u_{(N-1)(k)}(\nu_2')-z_{(N-1)}(u(\nu_2'),{}^6\pi_{ij}),$$

where $u_j(\nu_4)=\ln\left[\frac{p_j(\nu_4)}{p_N(\nu_4)}\right]\left(j=\overline{1,N-1}\right)$; $p_j(\nu_4)\left(j=\overline{1,N}\right)$ – A posteriori probability of the state of the discrete parameter of the IMF signal in the pixel ν_4 MM GBI (Fig. 4), $f(M_j(\nu_4))-f(M_N(\nu_4))\left(j=\overline{1,N-1}\right)$ is the difference between the logarithms of the likelihood function of the states of the discrete parameter of the MP signal in the GBI pixel; $z_j(\cdot)$ is a nonlinear function of the form:

$$z_q\left(u(\nu_l),{}^l\pi_{ij}\right)=$$
$$=\ln\left[\frac{\sum_{i=1,i\neq j}^{N-1}\left\{\exp\left(u_i(\nu_l)-u_j(\nu_l)\right){}^l\pi_{ij}\right\}+\exp\left(-u_j(\nu_l)\right){}^l\pi_{Nj}+\pi_{jj}}{\sum_{i=1}^{N-1}\left\{\exp\left(u_i(\nu_l)\right){}^l\pi_{iN}\right\}+\pi_{NN}}\right],\ \left(q=\overline{1,N-1},\ l=\overline{1,7}\right)$$

As a criterion for distinguishing the states of the elements of the CBI, we take the criterion of the maximum of the logarithm of the ratio of a posteriori probabilities, $u_j(\nu_4)$, in accordance with which, if $u_j(\nu_4)>u_i(\nu_4)$, $i,j=\overline{1,N-1}$; $i\neq j$, then a decision is taken on the state of the image element $\nu_4=M_j$; if all values $u_j(\nu_4)\leq 0\left(j=\overline{1,N-1}\right)$, then a decision is taken on the state of the image element $\nu_4=M_N$.

3. Experimental results

To demonstrate the proposed method we simulated the process of transmission and reconstruction of the dynamic MDI, consisting of BI with g = 4 BI at a signal-to-noise ratio of the signal power at the input of the non-linear filter of 0 dB. The pictures show a frame of a dynamic MDI represented by a set of two GBIs, each of which includes four neighboring BIs (Figure 5), distorted (Figure 6) and a reconstructed NDI frame (Figure 7). Fig. 8 and 9 show the distorted and reconstructed frame of the dynamic MDI with a signal-to-noise ratio at the input of the non-linear filter of 3 dB. To assess the quality of nonlinear filtering of dynamic MDI, the standard error MSE was taken.

IOP Conf. Series: Journal of Physics: Conf. Series **944** (2017) 012088 doi:10.1088/1742-6596/944/1/012088

Figure 5. Dynamic MDI frame onboard of the MUAV

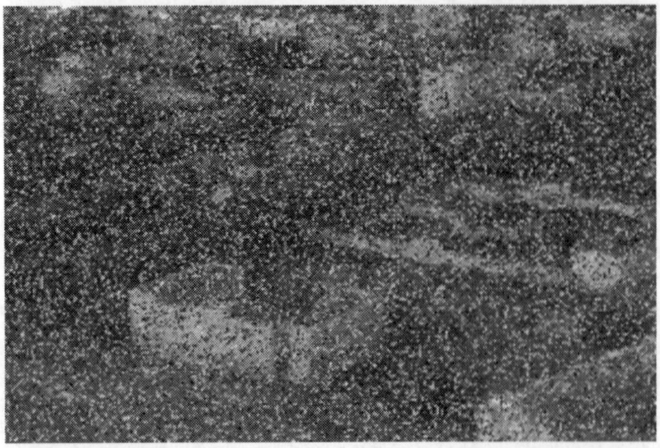

Figure 6. Dynamic MDI frame at the input of the receiver, MSE = 37×10^2

Figure 7. Recovered frame of dynamic MDI, MSE = $2,75\times10^2$

Figure 8. Dynamic MDI frame at the input of the receiver, MSE = $15{,}9\times10^2$

Figure 9. Reconstructed frame of dynamic ISC, MSE = $0{,}73\times10^2$

4. Conclusions

The analysis of the results shows that, despite the decrease in noise immunity arising due to the use of MP signals, the losses in noise immunity are completely compensated by the statistical redundancy of the MDI. The transmission time was reduced by a factor of four compared to the direct transmission of MDI by binary signals. Losses in noise stability of reception of MDI, consisting of GBIs, are completely compensated by the implementation of statistical redundancy of MDI.

References

[1] Slyusar V 2010 Data transfer from UAV: NATO standards. *ELECTRONICS: Science, Technology, Business* **3** 80-86

[2] Dvorkovich V P and Dvorkovich A P 2012 *Digital video information systems (theory and practice).* Moscow: Technosphere

[3] Dvorkovich V P and Dvorkovich A P 2015 *Metrological Support of Video Information Systems.* Moscow: Technosphere

[4] Petrov E P, Smolsky S M and Kharina N L 2007 Synthesis of models of multidimensional multivalued Markov processes. *Bulletin of the Moscow energetic institute* **1** 147-152

[5] Petrov E P, Kharina N L and Rzhanikova E D 2015 Non-linear processing of multi-digit digital halftone images in remote sensing systems of the Earth. *Nonlinear World* **6** 18-23

[6] Petrov E P, Kharina N L and Rzhanikova E D 2017 Method of reducing transmission time of multi-digit digital images from unmanned aerial vehicles. *Current problems in remote sensing of the earth from space* **14(1)** 71–79.
[7] Petrov E P, Kharina N L and Rzhanikova E D Method of simultaneous transmission of video frames from mini unmanned aerial vehicles. *Current problems in remote sensing of the earth from space* **14(2)** 49–57

IOP Conf. Series: Journal of Physics: Conf. Series **944** (2017) 012089 doi:10.1088/1742-6596/944/1/012089

Analysis of pre-heated fuel combustion and heat-emission dynamics in a diesel engine

S A Plotnikov, A N Kartashevich and S V Buzikov

Vyatka State University, 36, Moscow Street, Kirov, 610000, Russia

e-mail: Shamilvb@mail.ru

Abstract. The article explores the feasibility of diesel fuel pre-heating. The research goal was to obtain and analyze the performance diagrams of a diesel engine fed with pre-heated fuel. The engine was tested in two modes: at rated RPMs and at maximum torque. To process the diagrams the authors used technique developed by the Central Diesel Research Institute (CDRI). The diesel engine's heat emission curves were obtained. The authors concluded that fuel pre-heating shortened the initial phase of the combustion process and moderated the loads, thus making it possible to boost a diesel engine's mean effective pressure.

1. Introduction

Nowadays, the diesel engine is the primary power source used in tractor-manufacturing industry. The diesel engine's characteristics are determined by the law of heat input and the nature of combustion process [1-4, 5, 23]. With the fuel injection pump (FIP) diesel engine heat input as well as heat emission are determined by the injection characteristics. This explains the limitations for boosting the diesel engine's mean effective pressure, as the increase in the cycle's fuel supply shifts the combustion process beyond the top dead center (TDC). [11-13]. Currently, one of the most promising ways to boost the diesel engine's mean effective pressure is to shorten the ignition delay period (IDP). By now various ways to shorten the IDP have been researched. Among such ways there is the use of ignition plugs, glow plugs, use of heat-resistant inserts, ceramic coating of the combustion chamber walls, fuel charge turbulization, use of heaters, various fuel additives, alternative fuels, and the combination of the above-mentioned measures [6, 14, 16, 17]. The most promising of them the authors consider fuel preheating (heating diesel fuel prior to its injection into the engine cylinders) [2-4, 14, 19]. It is a well-established fact that for optimal performance of a diesel engine the charge's temperature at the end of the compression cycle should be about 623 K, as well as high degree of fuel dispersion. At the same time the diameter of a fuel drop should not exceed 40 mkm, and the conditions should minimize heat loss of the air charge while the injected fuel fumes are heated, vaporized, and over-heated in the CC [1-4, 18, 21-23].When injected the fuel drops are satisfactorily spread throughout the complete CC volume. However, due to low fuel concentration in the air-fuel mix, the transition from pre-burning oxidation processes to self-accelerating reactions takes too long. While the heat input from outermost regions of the air charge to fuel jets is somewhat complicated [2, 20]. Hence the IDP value. To eliminate all the above-mentioned negative effects it is necessary to minimize the volume of accumulated fuel fumes in the diesel engine's CC within the period from fuel injection to its self-ignition. In modern fuel-feed systems this issue is solved by dividing the injection process into two stages: the preliminary and main ones.

2. Research objectives

The objective of this research was to study the processes that make it possible to shorten the IDP through the increase of fuel jets' energy by pre-heating the fuel. Another task was to improve start properties of a diesel engine. Thus, the following tasks were set: to study the influence of fuel preheating on IDP, the explosiveness of the combustion process, speed of active heat emission, average cycle temperature, and start properties of the diesel engine as a whole.

3. Materials and methods

Researchers of the Vyatka State University (Russia) in cooperation with State Agrarian and Engineering University in Podilya (Ukraine) studied the effects of fuel preheating on the operation of 2Ch 10.5/12.0 (D-120) diesel engine. Generally, the study was performed within the framework of GOST 18509-88 requirements. The authors used grade L diesel fuel as per GOST 305-82. The diesel engine mounted on the testing bench is shown in Figure 1.

Figure 1. The 2Ch 10.5/12.0 diesel engine mounted on the testing bench.

The authors resorted to the indicator diagram technique to study the combustion process in a pre-heated fuel diesel engine in two operational modes: the rated one at RPM of $n=2000\ min^{-1}$ and the maximum torque mode at RPM of $n=1400\ min^{-1}$. The indicator diagrams were drawn at the fuel injection advance angle of $\theta inj=30^0$ and constant mean effective pressure values in every operational mode. The indicator diagram data were processed using the CDRI developed technique [14]. The start of the serial produced diesel engine (at $\varepsilon=16.5$) was performed with CT-212 starter motor at $n=230\ldots240\ min^{-1}$ RPM. The time required to achieve stable RPMs and the RPM values themselves were recorded.

4. Results and discussion

The analysis of the indicator diagrams demonstrated that the increase in fuel temperature resulted in the decrease of φ_i angle, corresponding to the ignition delay period (Figure 2) [7, 10]. Thus, at RPM of the diesel engine $n=2000\ min^{-1}$ and no fuel pre-heating the angle was $\varphi_{60}=25.7$ degrees, however, when the diesel fuel was heated to 150°C and 300°C the angle was $\varphi_{150}=23.2$ degrees and $\varphi_{300}=20.5$ degrees respectively (Figure 2, a). When the diesel engine operated in the maximum torque mode, then the IDP angle value also went down: $\varphi_{60}=22.7$ degrees, $\varphi_{150}=20.5$ degrees, and $\varphi_{300}=18.1$ degrees respectively (Figure 2, b). It was possible to suggest that the decrease of the φi angle, corresponding to the IDP, explained shortening of the thermo-chemical combustion reaction's induction time and lessening of the fuel amount injected during the IDP. As a result the explosiveness of combustion was moderated [1-4, 22, 23]. When the fuel was pre-heated, some drop in a cycle maximum pressure was

IOP Conf. Series: Journal of Physics: Conf. Series **944** (2017) 012089 doi:10.1088/1742-6596/944/1/012089

observed Pz (Figure 2). Thus, at RPM equal to n = 2000 min^{-1} the cycle maximum pressure was Pz_{60}=6.955 MPa, Pz_{150}=6.683 MPa and Pz_{300}=6.391 MPa respectively (Figure 2, a).

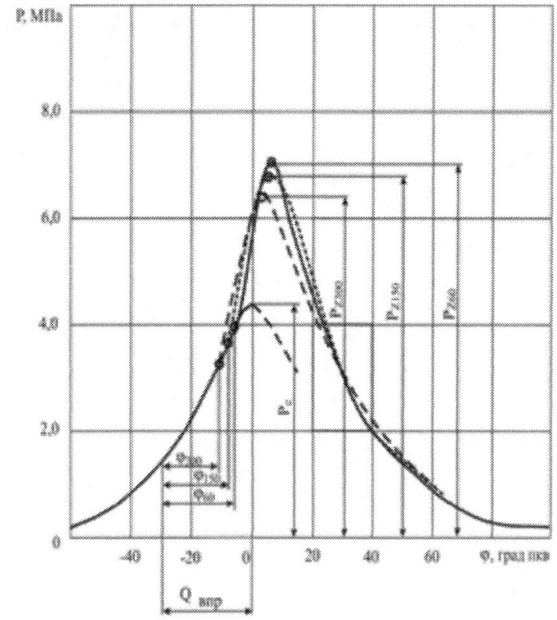

a

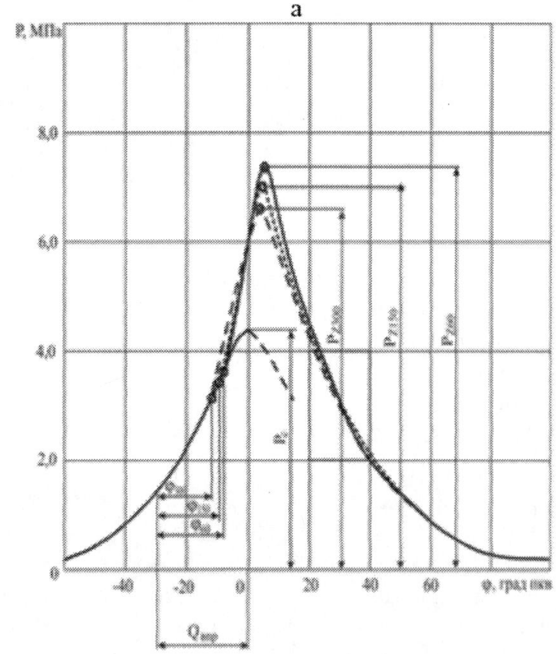

b

Figure 2. Indicator diagrams of 2Ch 10.5/12.0 diesel engine at: a – n=2000 min-1, b – n=1400 min-1

diesel fuel pre-heated to 60°C;
diesel fuel pre-heated to 150°C;
diesel fuel pre-heated to 300°C;

At RPM equal to n=1400 min^{-1} those values were Pz_{60}=7.369 MPa, Pz_{150}=6.964 MPa and Pz_{300}=6.614 MPa respectively (Figure 2, b). The drop in the cycle maximum pressure Pz was explained by the smaller amount of fuel injected into the cylinder of the diesel engine and eventually evaporated during the IDP. The analysis of the indicator diagrams using the CDRI technique proved that the diesel engine, when fed pre-heated fuel, demonstrated certain changes in its heat emission characteristics (Figure 3) [9].

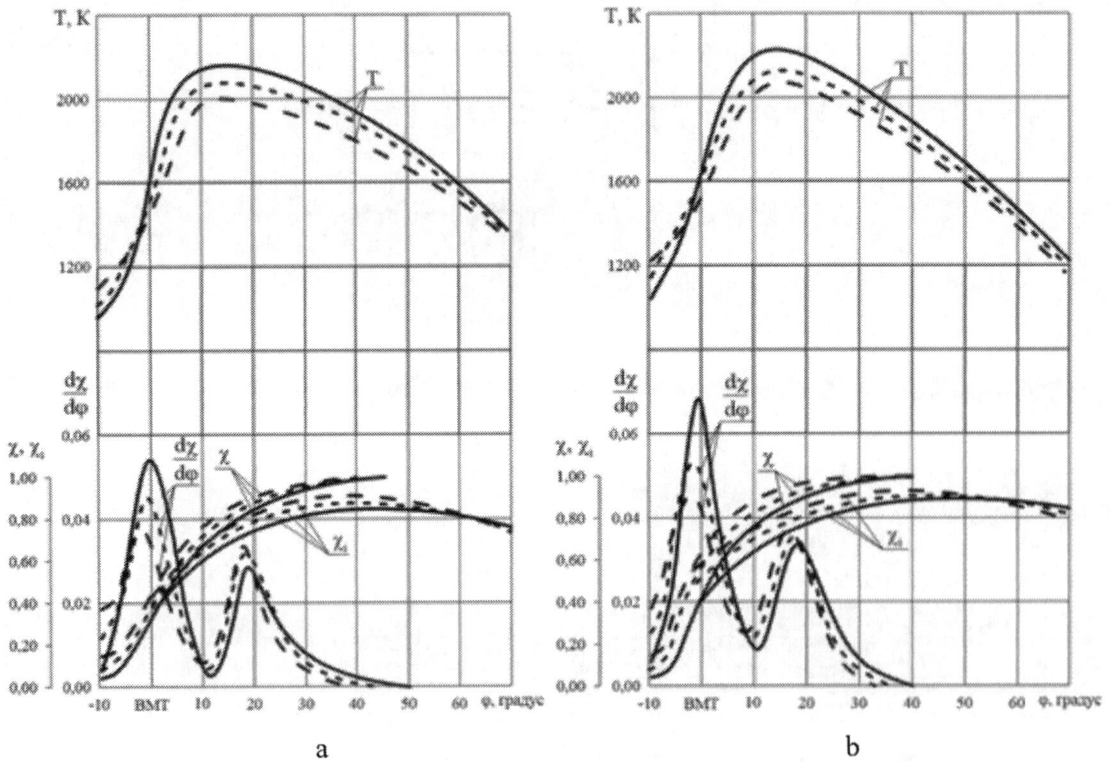

Figure 3. Graphs of heat emission dynamics dχ/dφ, of average gases temperature in a cylinder T, of heat emission χ and active heat emission χ_i in relation to the diesel engine's crankshaft rotation angle φ at: a – n=2000 min^{-1}; b – n=1400 min^{-1}
——— diesel fuel pre-heated to 60°C;
- - - - diesel fuel pre-heated to 150°C;
— — · diesel fuel pre-heated to 300°C;

Thus at RPM n=2000 min^{-1} and no fuel pre-heating the maximum average gases temperature in the cylinder was recorded at $T_{max,60}$=2184 K, when the fuel was heated to 150°C and 300°C, the temperatures recorded were $T_{max,150}$=2092 K and $T_{max,300}$=2003 K respectively. At the same time, the maximum temperature values corresponded to $\varphi_{60}=14^0$ of the crankshaft rotation (CSR), $\varphi_{150}=12.5^0$ CSR and $\varphi_{300}=11^0$ CSR, however, by the end of the combustion process at φ=60…70^0 CSR, the values were practically equal (Figure 3, a). The increase of the average cycle temperature could be explained by the following. One could suggest that when the fuel was pre-heated, the fuel-air temperature gradient along the jet axis diminished. Thus, fuel pre-heating could simultaneously increase total heat content of air/fuel mix at the end of the compression stroke [2]. The drop of the average maximum cycle temperature with its peak shifting towards TDC was determined by lower heat emission speed in the kinetic phase and its peak's shift in the same direction [9]. Comparison of the heat emission curve χ and active heat emission curve χ_i in the engine's rated operational mode at RPM of n=2000 min$_{-1}$ revealed the fact that, when the fuel was pre-heated, the combustion started earlier. So, at the TDC the recorded values were χ_{i60}=0.32, χ_{i150}=0.39 and χ_{i300}=0.44. The heat emission process ended practically at once at φ=42…46^0 CSR angle (Figure 3, a). Growth of heat emission values as well as those of active heat generation, when the diesel fuel was preheated, could be explained by the shortening of thermo-chemical combustion reaction's induction time due to faster achievement of the fuel's self-ignition threshold during the IDP [15]. Once the diesel fuel was pre-heated, the maximum heat generation rate $d\chi/d\varphi$ in the kinetic phase of the combustion process was recorded at $(d\chi/d\varphi)_{max60}$=0.053, $(d\chi/d\varphi)_{max150}$=0.046, $(d\chi/d\varphi)_{max300}$=0.037, while in the diffuse phase it was $(d\chi/d\varphi)_{max60}$=0.029, $(d\chi/d\varphi)_{max150}$=0.031, $(d\chi/d\varphi)_{max300}$=0.033 respectively. At the same time, the maximum heat generation rate in the kinetic phase corresponded to φ_{60}=1 , φ_{150}=2 and φ_{300}=4 prior the TDC, while in the diffuse phase it measured φ_{60}=18 , φ_{150}=17 and φ_{300}=16.5, once the TDC was passed. The heat emission rate droped to zero practically at once at φ=42…46^0 CSR angle (Figure 3, a). With the diesel fuel pre-heated, the reduction of the heat emission rate in the kinetic phase along with shift of the rate's maximum value regarding the CSR was determined by the increase of fuel drops' initial temperature and the subsequent shortening of the ignition delay period. Moreover, the authors observed a drop in the pressure increase rate depending on the engine's crankshaft rotation angle. Another recorded fact was thermal destruction of fuel molecules with the formation of unsaturated hydrocarbons resulting in the rise of pre-combustion reactions' speeds, that subsequently gave start to yellow combustion flame [2, 3, 13, 20]. With the diesel fuel pre-heated, the increase of the heat emission rate in the diffuse phase along with the shift of the rate's maximum value regarding the CSR was determined by the increase of fuel drops' initial temperature and the subsequent shortening of the ignition delay period. Once the diesel engine was run in the maximum torque mode at n=1400 min^{-1}, no pre-heating, the average maximum gases temperature in the cylinder was recorded $T_{max,60}$=2242 K, $T_{max,150}$=2132 K and $T_{max,300}$=2074 K respectively. At the same time the maximum temperature values corresponded to φ=14^0 CSR and by the end of the combustion φ=60…70^0 CSR, the values were practically equal (Figure 3, b). Analysis of the heat emission curve χ and active heat generation curve χ_i demonstrated, that, when the diesel fuel was pre-heated, the combustion started earlier.. So, at the TDC the recorded values were χ_{i60}=0.40, χ_{i150}=0.51 and χ_{i300}=0.62. The heat emission process ended practically at once at φ=32…40^0 CSR angle (Figure 3, b). Once the diesel fuel was pre-heated, the maximum heat generation rate $d\chi/d\varphi$ in the kinetic phase of the combustion process went down. Thus, the authors recorded the following data at fuel temperature 60^0 C $(d\chi/d\varphi)_{max60}$=0.068, and, when the fuel was pre-heated $(d\chi/d\varphi)_{max150}$=0.052, $(d\chi/d\varphi)_{max300}$=0.044. In the diffuse phase it was $(d\chi/d\varphi)_{max60}$=0.034, $(d\chi/d\varphi)_{max150}$=0.035 and $(d\chi/d\varphi)_{max300}$=0.036. At the same time, the maximum heat generation rate in the kinetic phase corresponded to φ_{60}=1.5, φ_{150}=2.5 and φ_{300}=3 prior the TDC, while in the diffuse phase it measured φ_{60}=18 , φ_{150}=17 and φ_{300}=15.5 once the TDC was passed. The heat emission rate dropped to zero practically at once at φ=32…40 once the TDC was passed (Figure 3, b). Therefore, the authors concluded that such changes in the combustion characteristics were to be expected for the maximum torque mode as well. Certain changes in the values of heat emission and in the timing of their reaching the maximum values were determined by a drop in the engine's RPM. The tests unambiguously established direct connection and interdependence between the combustion and heat emission characteristics of the processes occurring in the cylinders of a diesel engine and the temperature of the pre-heated diesel fuel [7-10]. The start characteristics were monitored and recorded to check on the operability of the diesel engine when it was fed pre-heated fuel (Figure 4).

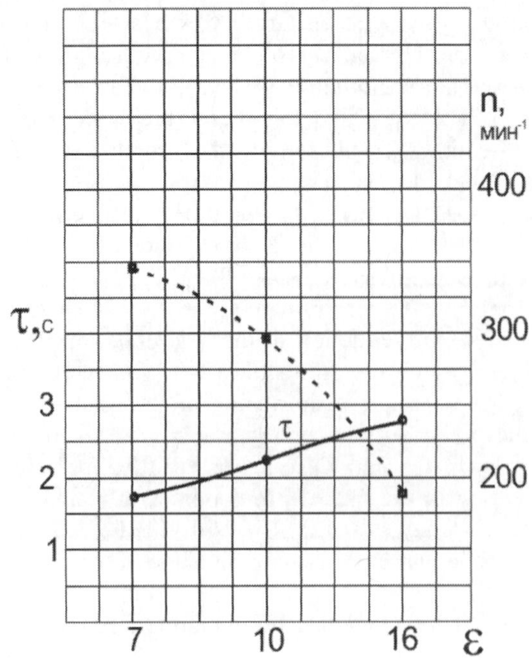

Figure 4. Start characteristics of 2Ch 10.5/12.0 diesel engine

As the data in Figure 4 demonstrate, the diesel engine fed pre-heated fuel started up easier than a serial production one. The authors believe that the most obvious cause was the increase of start RPM due to much lower resistance momentum.

5. Conclusion

1. Fuel pre-heating (thermal boost) makes the ignition delay period shorter and moderates the explosiveness of the combustion process.
2. Fuel pre-heating (thermal boost) lowers the speed of active heat generation.
3. Lowering the cycle average temperature, shortening the IDP, and moderating the explosiveness make it possible to boost a diese engine's average effective pressure.
4. Fuel pre-heating improves a diesel engine's start characteristics and facilitates the operation of starter motors.

References

[1] Alekseyev V P and Vyrubov D N 1977 *Physical Basis of Processes in Combustion Chambers of Reciprocating Internal Combustion Engines* (Moskva: "Bauman. N. Uh. Bauman")
[2] Voinov A N, Skorodelov D I and Sokolov F P 1964 *Kinetics and catalysis* **5** 388-398
[3] Gavrilov B G and Bulin Ye N 1963 *Chemical basis of diesel engine thermo-forcing Vol 36 Is 2*, (Moskva: Publ. House of the USSR Acad. of Sci.) p 2498
[4] Kamfer G M and Plotnikov S A 2005 *Mathematical Modeling of the Combustion Process of Alcohol-Containing Fuel in a Diesel Engine* (Kirov: "Avangard")
[5] Buzikov Sh V 2014 Effect of the initial fuel temperature on the characteristics of fuel supply and the period of ignition delay in a diesel engine *Scientific-methodical electronic journal «Concept»* 10(2014)6–10 Available: http://e-koncept.ru/2014/14617.htm
[6] Plotnikov S A 2014 Creation of new alternative fuels *Scientific-methodical electronic journal «Concept»* 10(2014)26-30 Available: http://e-koncept.ru/2014/14621.htm
[7] Plotnikov S A, Buzikov Sh V, and Atamanyuk V F 2012 Forecasting the processes of ignition and combustion of heated fuel in a diesel engine *Problems of intensification of animal*

husbandry taking into account spatial infrastructure of agriculture and environmental protection (Falenty-Warsaw), p 2040
[8] Plotnikov S A, Buzikov Sh V, and Atamanyuk V F 2016 A way to reduce the combustion explosiveness in a diesel engine *Inventions Utility models* **31** 1-5
[9] Plotnikov S A, Buzikov Sh V and Atamanyuk V F 2014 *Tractors and farm machinery* **7** p 25-27
[10] Plotnikov S A, Buzikov Sh V and Atamanyuk V F 2015 *Vestnik of Mari state University* **2** 39-43
[11] Sviridov Yu B 1972 *Mixing and Combustion in Diesel Engines* (Leningrad: "Mechanical engineering")
[12] Semenov V I 1970 *Study of the Indicator Period of Ignition Delay for a High-Speed Multi-Fuel Diesel Engine With an in-Piston Chamber* (Moskva: "Mechanical engineering")
[13] Plotnikov S A and Cheremisinov P N 2015 Disadvantages of the use of rapeseed-oil fuels in diesel engines *Actual directions of scientific researches of the XXI century: theory and practice, Sb. nauch. tr. po mat. zaoch. nauch.-prakt. konf.* (Voronezh: Nauchnaya kniga) p 1001
[14] Lazurko V P and Kudryavtsev V A 1975 Indicator charts processing software in basic Fortran algorithmic language *Sbornik tr CSRDI* (Leningrad) p 182
[15] Khachiyan A S and Labetskas G S 1982 *Dvigatelestroyeniye* **6** 7-11
[16] Khachiyan A S and Bagdasarov I G 1986 *Dvigatelestroyeniye* **7** 23-26
[17] Shirlin I I 2004 *Vestnik SibADI* **1** 85-87
[18] Chiu H H, Kirn H Y, and Croke E J 1982 Internal group combustion of liquid droplets *19th Symp. (Intl.) Comb.* (Pittsburgh: The Combustion Institute) p 971
[19] Cho S Y, Yetter R A, and Dryer F L 1992 *J. Comp. Phys* **102** pp 160-179
[20] Fenimore C P and Jones G W 1967 *J.Phys.Chem* **71** pp 593–597
[21] Hottel H C and Hawthorne W R 1979 Diffusion in laminar flame jets *3rd Symp. (Intl.) Comb.* (Baltimore: Williams and Wilkins) p 254
[22] Stapf P, Maas U and Warnatz J 1991 Detaillierte mathematische Modellierung der Tropfcen-verbrennung *7. TECFLAM-Seminar «Partikel in Verbrennungsvergangen»* (Stuttgart: Karls-ruhe. DLR) p 125
[23] Williams F A 1984 *Combustion Theory* (Benjamin: Cummings Menlo Park) p 2010

Synthesis and surface physicochemical properties of $(CdTe)_{0.03}(ZnSe)_{0.97}$ solid solution

S O Podgornyi[1], O T Podgornaya[1], E D Skutin[1], I P Demesko[1], O V Lukoyanova[1] and I V Muromtsev[2]

[1]Omsk State Technical University, 11 Mira Pr., Omsk 644050, Russia
[2]Laboratory of analytical and physicochemical research methods, Institute of Hydrocarbons Processing of SB RAS, 54 Neftezavodskaya Str., Omsk 644040, Russia

e-mail: skutin@omgtu.ru

Abstract. The research is centered on $(CdTe)_{0.03}(ZnSe)_{0.97}$ solid solutions. The article is aimed at developing innovative primary transducer material for semi-conductor sensors, investigating their surface physicochemical properties and evaluation their applicability in carbon monoxide diagnostics. Powders and nanofilms of $(CdTe)_{0.03}(ZnSe)_{0.97}$ solid solutions were obtained by isothermal diffusion and discrete thermal evaporation in vacuum. $(CdTe)_{0.03}(ZnSe)_{0.97}$ applicability in gas analysis was investigated. IR spectroscopy of multiple disturbed complete internal reflections and hydrolytic adsorption were used to study chemical composition and acid-base properties of $(CdTe)_{0.03}(ZnSe)_{0.97}$. Adsorption properties of the given material for carbon oxide (II) and oxygen were studied by the piezoquartz microweighing and volumetrically. The principles of adsorption, depending on the process conditions, were established. Based on the obtained experimental data, CO micro-impurities sensors were developed, the laboratory tests passed successfully

1. Introduction

Sensor systems application is one of the most effective ways for express control of process and surrounding media composition. Special attention is to be given to the sensors with sensitive element represented by A^2B^6 type diamond-like semiconductors polycrystalline films. This allows for toxic gases micro-impurities to be detected at low, even room, temperatures [1].

Substantial improvement of such sensor characteristics can be achieved by using solid solutions of the type A^2B^6–A^2B^6 as well as nanomaterials as gas-sensing materials.

Thus, publications [1, 2] demonstrate that the surface of solid solutions $(ZnSe)_x(CdTe)_{1-x}$ (x=0.05-0.15) exhibit increased adsorptive sensitivity towards CO, O_2 and their mixture as compared to CdTe. Similar principles are registered in case of solid solutions based on ZnSe [3] – the sensitivity of $(CdTe)_{0.05}(ZnSe)_{0.95}$ surface is higher towards CO than that of ZnSe. Whereas the highest CO detection efficiency is characteristic of the sensors with the least thickness of the sorbing nanofilm CdTe [4].

It should be noted, that currently, the binary components of the ZnSe-CdTe system are widely used in optoelectronics and microelectronics, in laser [5], solar cells developments [6], X-ray detectors [7] and γ-ray detectors [8]. The unique surface physicochemical properties of ZnSe and CdTe determine the perspectives of the given materials application in gas analysis [1].

The present article follows the cycle of the investigations devoted to obtaining and studying the surface physicochemical properties for the ZnSe–CdTe system components.

The article is aimed at developing innovative primary transducer materials for semi-conductor sensors, investigating their surface physicochemical properties and evaluation their applicability in carbon monoxide diagnostics.

2. Experimental technique

Powders of solid solutions were obtained by isothermal diffusion method by using a specially developed heating program (Snol 6.7/1300 laboratory oven). To obtain $(CdTe)_{0.03}(ZnSe)_{0.97}$ nanofilms (d=30…100 nm), the method of discrete thermal evaporation in vacuum was used (vacuum universal post VUP-5).

Certification of the solid solutions was performed from results of X-ray analysis (Bruker D8 Advance diffractometer with Cu-Kα radiation, λ=1.5418 Å, using the LynxEye position sensitive detector). Phase identification was carried out using Bruker EVA software paired with the ICDD PDF-2 2006 database.

The chemical composition of the surface of $(CdTe)_{0.03}(ZnSe)_{0.97}$ was determined by IR spectroscopy of multiple disturbed complete internal reflections (Infra LUM FT-02 Fourier spectrometer, PIKE Technologies HATR, crystal type - Ge, spectral range - 830…4000 cm^{-1}). The acid-base properties of surface were studied by hydrolytic adsorption (determination of the isoelectric state pH).

The evaluation of $(CdTe)_{0.03}(ZnSe)_{0.97}$ applicability in carbon monoxide diagnostics was carried out on the basis of adsorption research results. The research having been performed with the piezoquartz microweighing and volumetrically in the temperature range of 243 to 453 K and pressures of 3 to 15 Pa. Adsorbates (CO and O_2) were obtained according to the familiar procedures [2].

3. Results discussion

3.1. Synthesis and identification of the solid solution

As shown in Figure 1 the X-ray diffraction results indicate that the substitutional solid solution with sphalerite cubic structure is formed. The X-ray diffraction lines of the $(CdTe)_{0.03}(ZnSe)_{0.97}$ solid solution are characteristically shifted relative to the lines of the ZnSe, while their number remains unchanged.

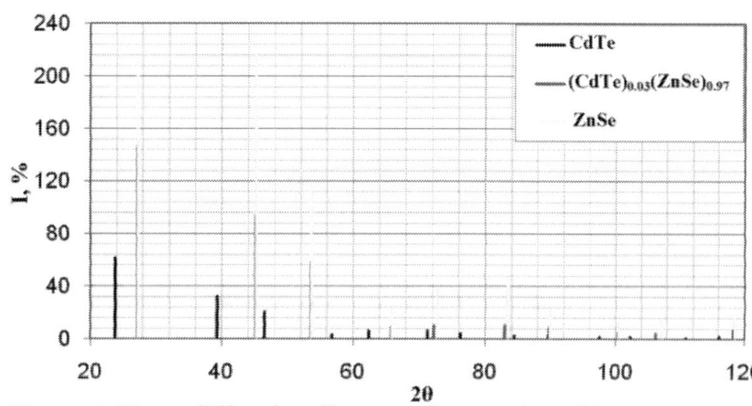

Figure 1. X-ray diffraction diagrams for samples of the ZnSe–CdTe system

As shown in Table 1 the value of unit cell parameter of the $(CdTe)_{0.03}(ZnSe)_{0.97}$ solid solution occupy an intermediate position between the values of the binary components of the ZnSe-CdTe system.

The values of the X-ray coherent scattering regions (CSR) and X-ray density are minimum in case of solid solution. The absence of extra X-ray diffraction lines from unreacted binary components proves that the synthesis of the $(CdTe)_{0.03}(ZnSe)_{0.97}$ solid solution is complete.

Table 1. The values of unit cell parameters, a_0, X-ray density, ρ_r, and CSR of the components of the ZnSe-CdTe system

№	Composition	a_0, Å	ρ_r, g/cm^3	$CSR_{(111)}$, Å
1	CdTe	6.4796	5.860	836.4
2	$(CdTe)_{0.03}(ZnSe)_{0.97}$	5.6763	5.242	399.4
3	ZnSe	5.6688	5.263	1328.8

3.2. Chemical composition and acid-base properties of the surface

According to the data obtained with IR spectroscopy of multiple disturbed complete internal reflections (Figure 2), the surface of the solid solution $(CdTe)_{0.03}(ZnSe)_{0.97}$ exposed to air contains mostly adsorbed molecules of water, OH groups, as well as carbon dioxide, this being characteristic of diamond-like semiconductors [9, 10].

After the thermal vacuum treatment till the minimal gas release, the surface of $(CdTe)_{0.03}(ZnSe)_{0.97}$ becomes almost free from the adsorbed impurities and this is reflected in a prominent decrease of all bands intensity.

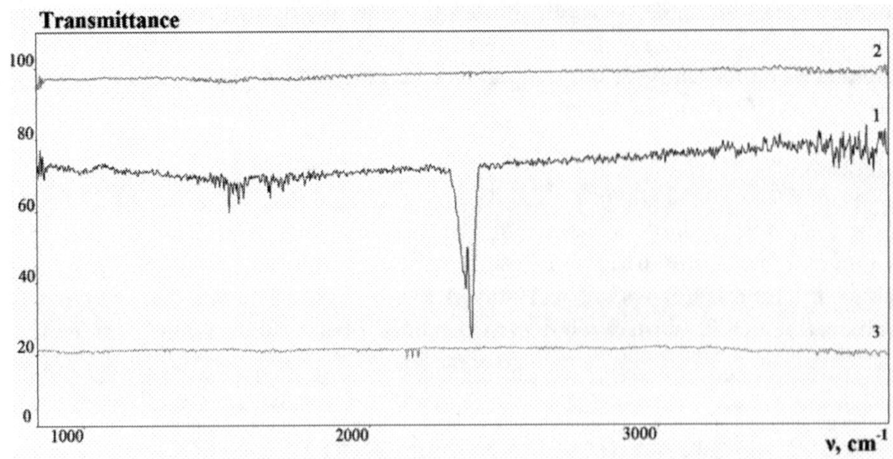

Figure 2. Fourier IR spectra (MDCIR) of the surface of the $(CdTe)_{0.03}(ZnSe)_{0.97}$ solid solution exposed in air (1), vacuum (2) and CO (3)

It is noteworthy that after the CO adsorption $(CdTe)_{0.03}(ZnSe)_{0.97}$ IR spectra contain bands in the range of 2140…2174 cm^{-1} which can be attributed to linear carbonyls.

The values of pH_{iso} for $(CdTe)_{0.03}(ZnSe)_{0.97}$, exposed to air for a long time, are less than 7, signifying the prevalence of acid sites on the surface.

The value of pH_{iso} $(CdTe)_{0.03}(ZnSe)_{0.97}$ (6.51) takes an intermediate position between those of pH_{iso} ZnSe (6.67) and CdTe (5.99), and this is another evidence of the solid substitution solutions formation in ZnSe–CdTe system.

3.3. Adsorption investigation

The research of the $(CdTe)_{0.03}(ZnSe)_{0.97}$ solid solution adsorptive properties has demonstrated high selective sensitivity of the material to the chosen adsorbates (Figure 3, 4). The observed order of CO and O_2 adsorption values (10^{-7} mol./m2) is typical for adsorption processes on powders of diamond-like semiconductors. The adsorption value temperature dependence curves allow us to claim that physical adsorption takes place in the range of 257…298 K, while chemical activated adsorption began at 298 K. It should be noted that activated adsorption seems to be realized in one form, as indicated by the presence of one ascending sector on isobars.

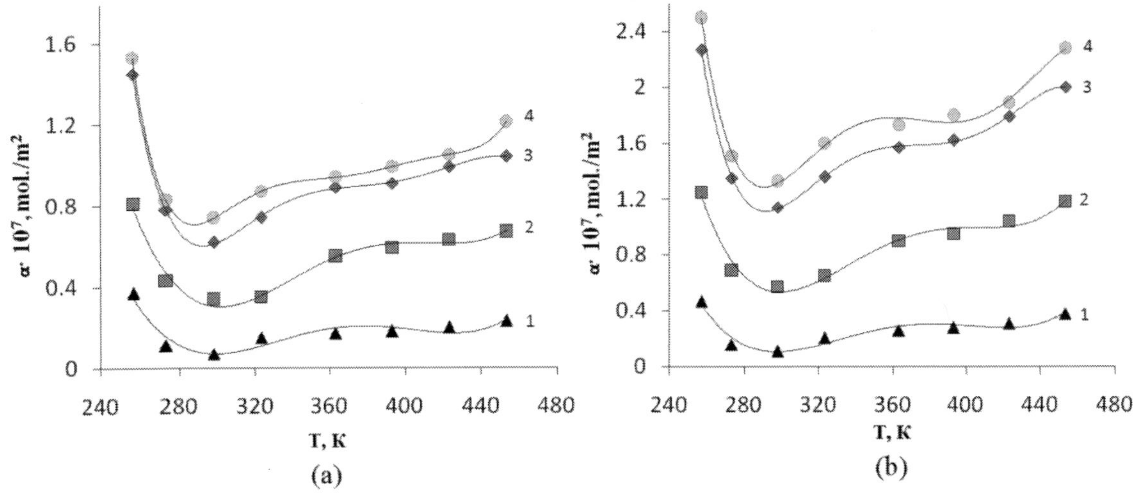

Figure 3. Temperature dependencies of CO (a) and O_2 (b) adsorption values on $(CdTe)_{0.03}(ZnSe)_{0.97}$ solid solution at P_n: 3 (1), 6,5 (2), 12 (3) and 15 (4) Pa

The calculated values of differential heat (q) of CO and O_2 adsorption, amounted to 4.5 and 5.1 kJ /mol. respectively, and changes in the entropy ($-\Delta S$) amounted to 63,0…80,1 J/(mol.·K), are typical for the chemical nature of the adsorption processes on diamond-like semiconductors [1].

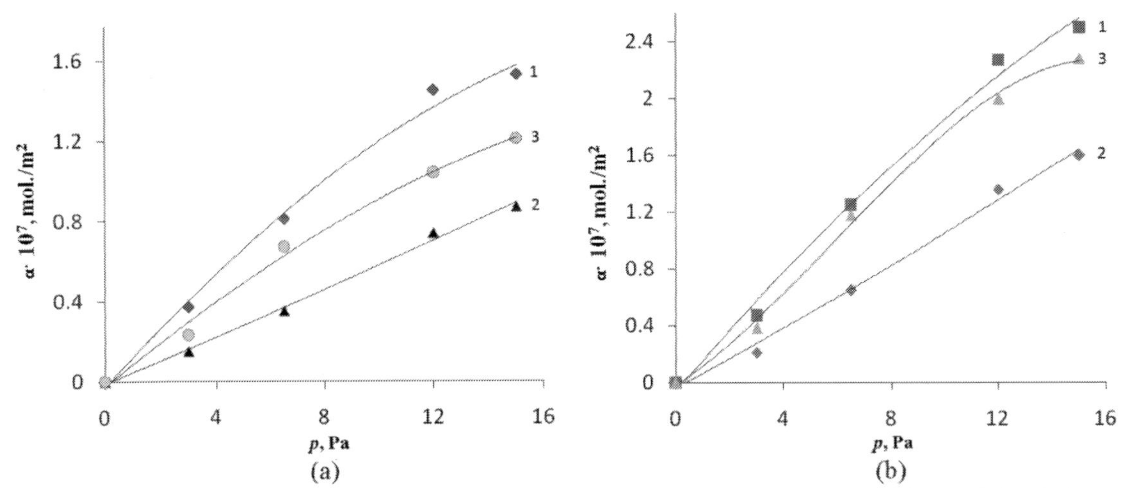

Figure 4. CO (a) and O_2 (b) equilibrium adsorption isotherms on $(CdTe)_{0.03}(ZnSe)_{0.97}$ solid solution at T: 257 (1), 323 (2), 453 (3) K

The analysis of $(CdTe)_{0.03}(ZnSe)_{0.97}$ solid solution adsorptive properties leads to the conclusion that this adsorbent surface is highly sensitive to O_2 in contrast to CO in all investigated temperature and pressure ranges (Figure 5). It should be noted that similar ratios of CO and O_2 adsorption values are also typical for ZnSe [1].

The heat of adsorption declines upon an increase in the degree of occupation (Figure 6), indicating the heterogeneous character of the surface, the presence active sites that differ in strength, and the energetic state.

IOP Conf. Series: Journal of Physics: Conf. Series **944** (2017) 012090 doi:10.1088/1742-6596/944/1/012090

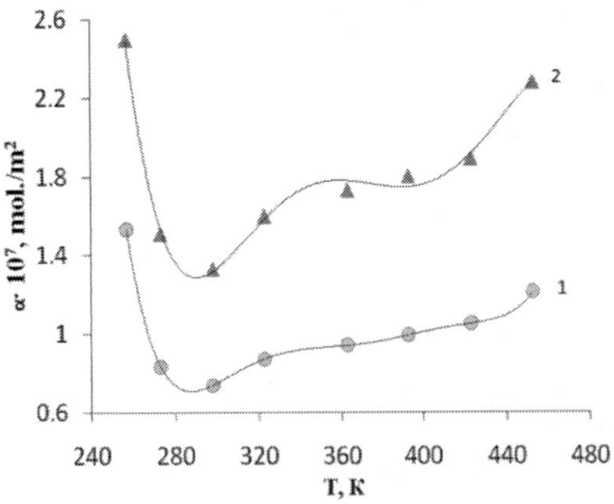

Figure 5. Temperature dependencies of CO (1) and O_2 (2) adsorption values on $(CdTe)_{0.03}(ZnSe)_{0.97}$ solid solution at Pn=15 Pa

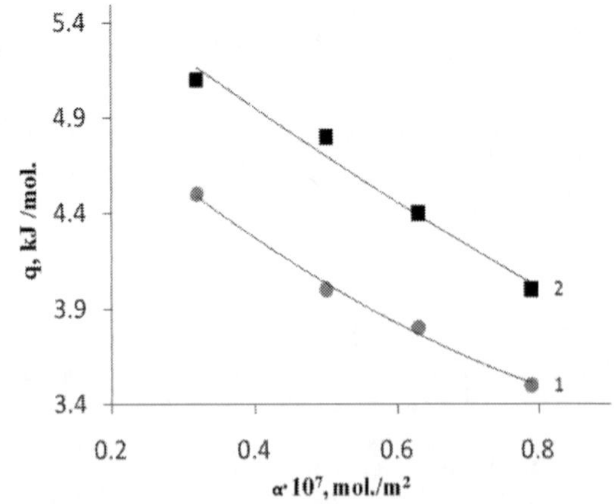

Figure 6. Dependencies of the heat of CO (1) and O_2 (2) adsorption on degree of occupation of $(CdTe)_{0.03}(ZnSe)_{0.97}$ solid solution

Coordination-unsaturated atoms of Zn and Cd are largely responsible for the CO actives sites of the investigated $(CdTe)_{0.03}(ZnSe)_{0.97}$ surface, at the same time F centers are largely responsible for the O_2 actives sites, as well as other components of the ZnSe-CdTe system [1,2].
Comparing the characteristics of $(CdTe)_{0.03}(ZnSe)_{0.97}$ and the previously studied components of ZnSe–CdTe system [1-3] makes us conclude that the surface of the solid solutions $(CdTe)_x(ZnSe)_{1-x}$ possesses an increased sensitivity towards CO as compared to ZnSe. Thus, we notice: in the row ZnSe→$(CdTe)_{0.03}(ZnSe)_{0.97}$→$(CdTe)_{0.05}(ZnSe)_{0.95}$ the adsorption values increase and, consequently, CO adsorption heat does as well. The given principle correlated with the data of X-ray investigations as well as with those of acid-base properties studies (Figure 7).

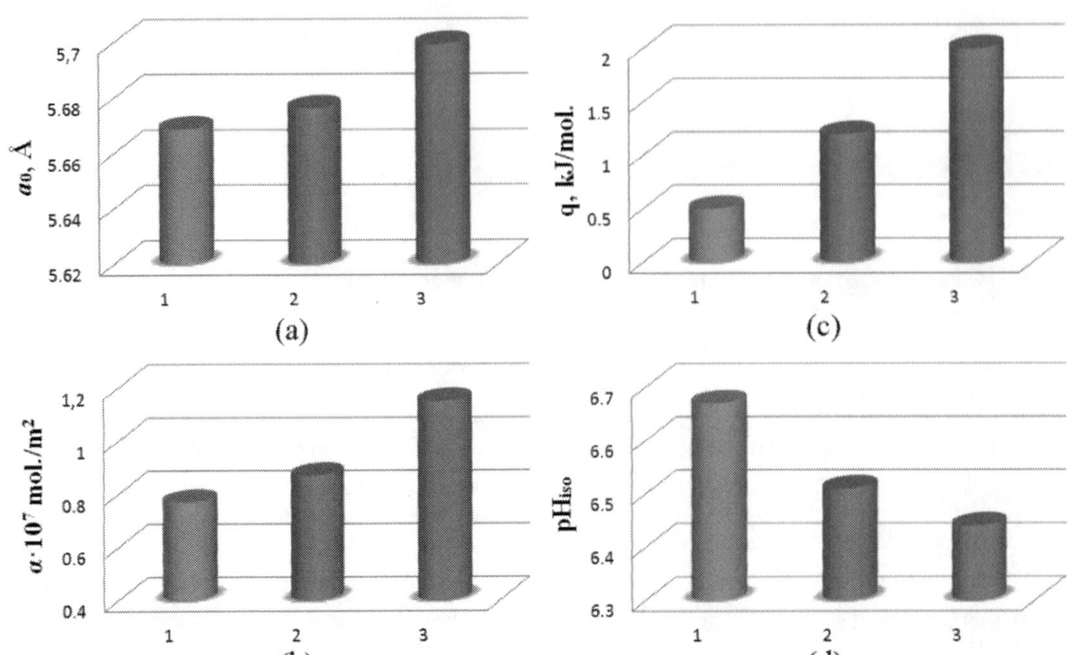

Figure 7. The values of unit cell parameters (a), CO adsorption value (b), CO adsorption heat (c) and pH_{iso} (d) for ZnSe (1), $(CdTe)_{0.03}(ZnSe)_{0.97}$ (2) and $(CdTe)_{0.05}(ZnSe)_{0.95}$ (3)

3.4. Carbon monoxide micro-impurities sensor

The detected high selective sensitivity of $(CdTe)_{0.03}(ZnSe)_{0.97}$ solid solution surface towards CO, particularly at room temperatures, demonstrates the viability of the studied material application in semiconductor gas analysis.

CO micro-impurities sensor was developed based on the obtained data. The sensor primary transducer represents an AT-cut piezoquartz resonator with applied nanoscale (d=30 nm) sorbing $(CdTe)_{0.03}(ZnSe)_{0.97}$ solid solution film.

The sensor response is the changes in piezoquartz resonator oscillation frequency due to the analyte adsorption (desorption) on $(CdTe)_{0.03}(ZnSe)_{0.97}$ nanofilm. The designed sensor passed the laboratory tests successfully, enhanced detection efficiency compared with sensors based on ZnSe [11] is noted.

4. Conclusions

Powders and nanofilms of $(CdTe)_{0.03}(ZnSe)_{0.97}$ solid solutions were obtained. The chemical composition and acid-base properties of $(CdTe)_{0.03}(ZnSe)_{0.97}$ were studied by IR spectroscopy MDCIR and hydrolytic adsorption.

Adsorption properties of the given material for CO and O_2 were investigated by the piezoquartz microweighing and volumetrically. Carbon monoxide micro-impurities sensor was developed. The laboratory tests passed successfully.

5. References

[1] Kirovskaya I A and Podgornyi S O 2011 *Rus. J. Phys. Chem. A* **85** (11) 1971–1976
[2] Kirovskaya I A and Podgornyi S O 2012 *Rus. J. Phys. Chem. A* **86** (1) 14–18
[3] Podgornyi S O, Podgornaya O T, Demeshko I P, Lukoyanova O V and Muromtsev I V 2017 *AIP Conference Proceedings* **1876** (1) 020085
[4] Podgornyi S O, Podgornaya O T, Skutin E D, Demeshko I P, Lukoyanova O V, Fedotova K I 2016 *2016 IEEE Dynamics of Systems, Mechanisms and Machines (Dynamics)* 7819060
[5] Firsov K N, Gavrishchuk E M, Ikonnikov V B, Kazantsev S Yu, Kononov I G, Kotereva T V, Savi D V and Timofeeva N A *Laser Physics Letters* 2016 **13** (5) 1–5
[6] Kephart J M, McCamy J V, Ma Z, Ganjoo A, Alamgir F M and Sampath W S *Solar Energy Materials & Solar Cells* 2016 **157** 266–275
[7] Ishikawa S, Katsuragawa M, Watanabe S, Uchida Y, Takeda S, Takahashi T, Saito S, Glesener L, Buitrago-Casas J, Krucker S and Christe S *Journal of Geophysical Research: Space Physics* 2016 **121** (7) 6009–6016
[8] Niraula M, Yasuda K, Takai N, Matsumoto M, Suzuki Y, Tsukamoto Y, Ito Y, Sugimoto S, Kouno S, Yamazaki D and Agata Y *IEEE Electron device letters* 2015 **36** (8) 856–858
[9] Kirovskaya I A and Filatova T N *Rus. J. Phys. Chem. A* 2012 **86** (3) 503– 507
[10] Kirovskaya I A, Timoshenko O T and Shubenkova E G *Rus. J. Phys. Chem. A* 2010 **84** (4) 661–667
[11] S.O. Podgornyi, O.T. Podgornaya, E.D. Skutin, I.P. Demeshko, O.V. Lukoyanova and K.I. Fedotova *Procedia Engineering* 2016 **152** 474–477

IOP Conf. Series: Journal of Physics: Conf. Series **944** (2017) 012091 doi:10.1088/1742-6596/944/1/012091

Research of thread rolling on difficult-to-cut material workpieces

A Yu Popov[1], I A Bugay[1], P V Nazarov[1], O P Evdokimova[1], P E Popov[1] and E V Vasilyev[1]

[1] Omsk State Technical University, Omsk, Russia

e-mail: popov_a_u @list.ru

Abstract. In medicine production Ti-6Al-4V Grade 5 alloys are used. One of the most important tasks is to increase the strength of the products and decrease in value. The possibility to roll special thread on Ti-6Al-4V Grade 5 alloy workpiece on 2-roller thread rolling machine has been studied. This is wrought alloy, treatment of which in cold condition causes difficulties due to low plasticity. To obtain Ti-6Al-4V Grade 5 alloy product with thread by rolling is rather difficult. This is due to large axial workpiece displacements resulting from large alloy resistance to cold plastic deformation. The provision of adequate kinematics requires experimental researches and the selection of modes - speed of rolling and pressure on the movable roller. The purpose of the work is to determine the optimal modes for rolling thread on titanium alloy workpiece. It has been stated that, after rolling, the product strength has increased up to 30%. As a result of the work, the unit has been made and recommendations to choose the optimal rolling process modes have been offered. Keywords — plastic deformation, rolling thread, knurls, titanium alloy, thread-rolling machine.

1. Introduction

A very productive method to obtain a thread is rolling in cold condition in 2-roller thread rolling machines, instead of cutting. Medical implants for osteosynthesis periosteal equipped with special screws, conditions which require high accuracy and quality of the thread. The formation of threads on such pieces by cutting is difficult, time-consuming, and does not provide the necessary physical and mechanical properties. Therefore, processing by rolling is more efficient.

The rolling process, depending on the piece design, can be performed on different equipment: both special thread-rolling machines and versatile machine tools using a variety of technological equipment and thread-rolling tool designs. The most common is rolling on double-roller machines, where thread rolling occurs when knurls approach gradually and roll the workpieces.

As a result, the redistribution of the workpiece material from the surface into the roots of the roller threads happens.

The process continues till full thread turns filling (on closed profile tool) or obtaining full form thread on the part (on open-loop roller operating). Thus, upon rolling threads with knurls the plastic deformation of outer surface of the workpiece occurs. This causes the bend of outer metal fibers equidistantly to thread profile and the metal fibers are not cut and rolling is performed without chips. The rolled thread in compare with cut one has the increased surface hardness (up to 30%).

The papers of T.A. Sultanov, Yu.L. Frumin, I.Ya. Shnajderman and M.Z. Hostikoev et al. [1-2] take the important place in the study and practical implementation of rolling techniques.

IOP Conf. Series: Journal of Physics: Conf. Series **944** (2017) 012091 doi:10.1088/1742-6596/944/1/012091

2. Statement of problem

A number of papers are devoted to the study of titanium properties and structure during the cold plastic deformation process and its structure. In most cases, however, the papers of this kind deal with the microstructural analysis of titanium after rolling or extruding. In another case, titanium properties are given taking into account thermal effects. In this paper the possibility to roll special thread on BT 6 titanium alloy workpiece on 2-roller thread rolling machine has been studied.

Rolling is a process of plastic forming on the given workpiece surface, as opposed to the whole workpiece impact, or stamping when much smaller parts of the workpiece are acted.

The titanium alloy workpiece of 3 mm diameter was used to determine the possibility of obtaining threads. Rolling was produced on 2-roller machine. This method makes it possible to form threads on a freely rotating workpiece and is independent of the length of the rod. The objective of the research is to define optimal parameters for rolling which which will allow the formation of high quality thread.

3. Theory

Figure 1 shows the dependency of the basic mechanical titanium properties, such as: yield stress, ultimate strength on the deformation speed during workpiece tensile. This dependence demonstrates that the values σ0.2 and σE are gradually increasing upon speed magnification. Therefore, upon deformation rate magnification, the mechanical properties increase, making machining process more difficult.

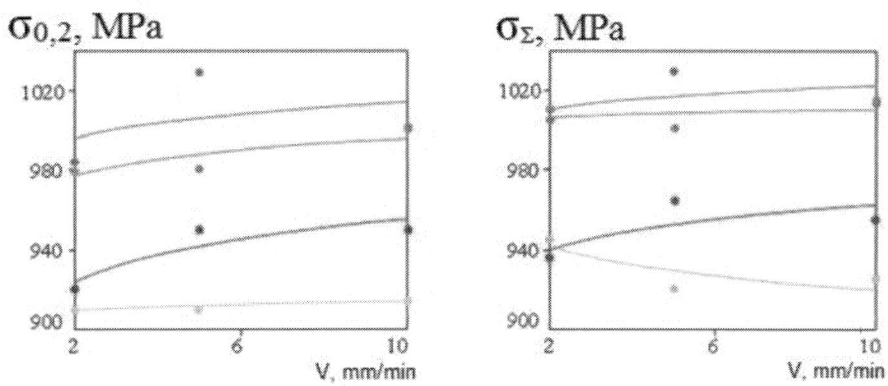

a) b)

Figure 1. Dependence of yield stress (a), ultimate strength (b) on the tensile speed of Ti-6Al-4V Grade 5 workpiece heated to different temperatures.

In the case of rolling thread on the rod, the following can be noted:

1. The stresses in the rod depend on rolling speed. With the die rotation speed magnification the deformation resistance increases. At higher turns, the deformation of the rod slows down due to hardening. The workpiece begins to move along the surface of the knurl cutter because the rollers interrupt introduction in the workpiece and thread helix angles on the rollers and workpiece do not match. This is a organic disadvantage of this rolling technology.

2. In the process of through-feed rolling (the rollers interrupt introduction in the workpiece due to thread surface hardening), the ply separation of the workpiece surface occurs and the husks appear, making the surface roughness of the part unacceptable and results in the destruction and folding of the thread sections.

In light of the above, it can be concluded that, with the increase of rolling speed, the indicators of thread rolling process are deteriorating. The workpiece begins to move more actively along the axis and often slides upward from the rolling zone.

4. Experiment results

To carry out the work it was necessary to modify 2-roller thread rolling machine. Frequency controller was installed on the machine to reduce head rotation speed by four times, up to 5 -7 m/min. To determine the feasibility of heating the workpiece, an analysis was made of the dependence of the mechanical properties of the three materials on temperature.

We have obtained hardening graphs depending on temperature:

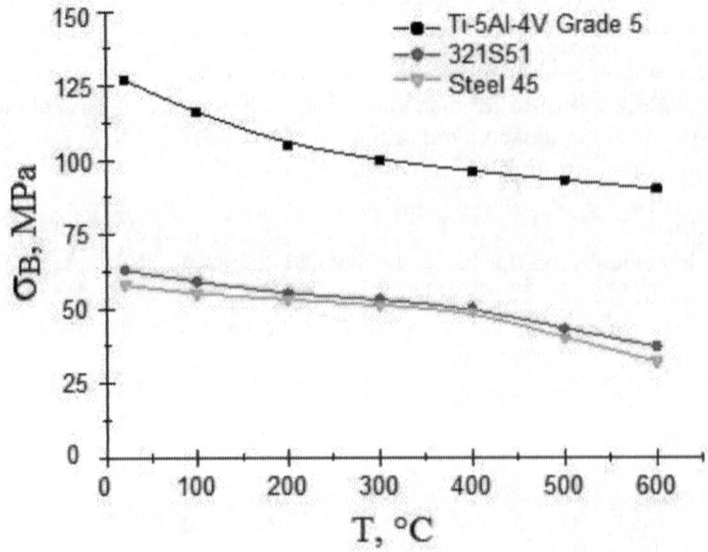

Figure 2. Hardening of given materials by temperature.

In operating temperature range (200-600°C), steel temporary resistance reduces and remains high for titanium alloys, despite non-significant decrease (Figure 2).

For the Ti-5Al-4V Grade 5 and steel 321S51, the temperature rise affects the hardening in the same way. To a temperature of 300 - 400°C, the drop in strength is not significant, and the use of rolling liquid is not possible. But, since at high rolling speed the tool material is additionally heated in the contact zone, an irreversible drop in hardness of the tool from Cr12-M steel is unavoidable, from which it follows that, with all other similar parameters and temperature changes, it is expedient to process titanium alloys at lower speeds using rolling liquid and under higher pressure in the roller hydraulic cylinder.

5. Results and discussion

During the work process, a round thread was rolled on 3 mm diameter titanium workpiece at low speeds (Figure 2). In detailed examination, on the tread crests one can see the groove specific to rolling.

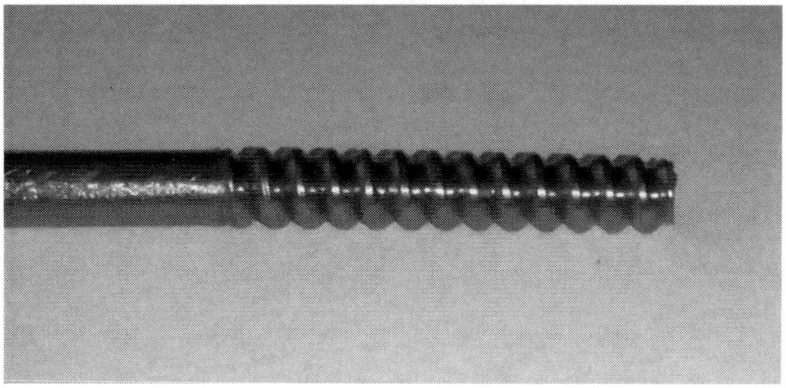

Figure 3. Rolled thread.

During carrying out the work it has been experimentally proved that to roll it is necessary to reduce the speed lower than the limits of the machine to 5-7 m/min. It is also worth noting that the power of roller reduction was more than rolling thread on similar steel and alloy workpieces.

Taking into account these recommendations, production of titanium screws can be implemented in this way.

6. Summary and conclusion

The features of selection processing modes of rolling thread on Ti-5Al-4V Grade 5alloy have been defined.

To roll thread on titanium alloys parts, it is useful to reduce the velocity of rolling 5-7 m/min at maximum pressure in the hydraulic-circuit system of roller feeding.

7. References

[1] Lyashkov A A, Vasil`ev E V and Popov A Y 2017 Development of 3D modeling technology for manufacturing finned ribbons from heat-resistant steels *IOP Conf. Series: Earth and Enviromental Science. IPDME* **Vol. 23**. pp. 1-12. doi:10.1088/1742-6596/858/1/012017.

[2] Korovin G I, Filippov A V, Proskokov A V and Gorbatenko V V 2015 Cutting Edge Geometry Effect on Plastic Deformation of Titanium Alloy *IOP Conf. Series: Materials Science and Engineering* **Vol. 125.** pp. 58-64. doi:10.1088/1757-899X/125/1/012012.

Automatic system for estimating the volume of the left ventricle based on two-dimensional MRI images of the heart along the long axis

S.V. Porshnev, A.O. Bobkova, V.V. Zyuzin, A.A. Mukhtarov, M.A. Chernyshev, D. M. Akhmetov

Department of Information Technology and Automation UrFU, IRIT-RTF
Yekaterinburg, Russia

Abstract. The article is devoted to the development of a system that allows automatic estimation of systole, diastole and left ventricular ejection fraction of the heart based on the sequence of MRI images from the apical two-chamber and four-chamber positions. The implemented system was tested on the images of the heart of 200 patients, and its accuracy and operability was assessed. Keywords – left ventricle; final systolic volume; final diastolic volume; ejection fraction; magnetic resonance imaging; two-chamber projection; four-chamber projection; neural networks; recovery of volume

1. Introduction

In the practice of cardiac research, magnetic resonance imaging (MRI) is used, which allows the cardiologist to assess the functional state of the human heart muscle. For this purpose, on the basis of the analysis of MRI images, a specialist evaluates, usually manually, the change in the volume of the left ventricle of the heart (LV) and the ejection fraction (EF) of the LV (the ratio of the difference between the maximum (diastole) and minimum (systole) LV volumes to diastolic LV volume). As a result, the process of assessing the functional state of the heart muscle is very laborious. In this regard, the development of an automatic system for estimating LV volume based on MRI images is relevant.

The article describes an automatic system for recovering the volume of the LV based on two-dimensional images of the heart along the long axis.

2. Problem Definition

In the study of the heart muscle, MRI images of the heart are traditionally obtained in three projections: an apical two-chamber projection, an apical four-chamber projection, and a parasternal position of the short axis. In this connection, it is clear that the task of automatically estimating the volume of the LV based on MRI images can be

IOP Conf. Series: Journal of Physics: Conf. Series **944** (2017) 012092 doi:10.1088/1742-6596/944/1/012092

decomposed into two independent tasks in accordance with the types of projections used: 1) projections along the long axis of the LV [1]; 2) projections along the short axis of the LV. In the studies carried out, projections on the long axis were used. In the course of solving the task, the following problems were identified and solved:

- Formation of a set of MRI images of the heart in selected projections.
- Transformation of MRI images into data suitable for training neural networks (NN).
- Choice of architecture of neural networks and their training.
- Selection of postprocessing methods for the results of processing MRI frames obtained at the output of the neural network.
- Evaluation of volume and EF of LV.
- Analysis of the results of the automatic system.

3. Main part

3.1. The structure of the source data

In the study, data was used on the competitive platform Kaggle [9]. They are a collection of more than 100.000 MRI images of more than 700 patients in the DICOM format. For the training of neural networks, MRI images of 500 patients were used, the remaining images were used to assess the quality of the work of the automated evaluation system. Examples of the two-chambered projection of the heart along the long axis and the four-chamber projection in the successive phases of one cardiac cycle are shown in Figure 1, 2, respectively.

3.2. Data preparation

3.2.1. Marking of MRI images For the training of neural networks, a pre-prepared set of data representing MRI frames was used, on which the areas corresponding to the LV were manually marked (Figure 3)

About 11.000 MRI frames were hand-marked in equal parts for a two-chamber and four-chamber projection.

3.2.2. Artificail increase in the data set Due to the fact that the set of 5500 images was insufficient for training the neural network and obtaining satisfactory estimates of LV volume, the original volume of MRI frames was artificially increased by the distorted versions of the original set [4]. This technique will increase the network's resistance to various shifts and rotations of the left ventricular contour on MRI images. To obtain distorted source images, we used affine transformations (mirror reflection with respect to oX and oY axes, shifts, rotations by a random angle in and counterclockwise, and also compression of the image along coordinate axes with a random compression ratio). The described affine transformations were applied simultaneously for both MRI frames and corresponding images obtained as a result of manual marking (Figure 4).

Figure 1. Sequence of frames of one cardiac cycle of a two-chamber projection of the heart along the long axis.

As a result, the original set of images was enlarged from 11.000 to 10^5 images. Further increase in the number of images was found to be inexpedient, since with a large number of MRI frames, the training time on the graphic card GTX 1060 was unacceptably large.

3.3. Architectures and learning outcomes of neural networks

The accumulated experience of using neural networks in visual recognition problems shows that with the task of isolating the contour of the left ventricle on biomedical images, deep convolutional NN are best able to cope [6], since they are able to assign a class label to each pixel separately [8]. In this connection, the FCN-8s [7] and UNET [10] architectures were used.

3.3.1. Modification of NN architectures Due to the fact that the FCN-8s architecture is designed to segment the image into 21 classes, which is redundant when the left ventricle area for which is redundant when the left ventricle area identification needs

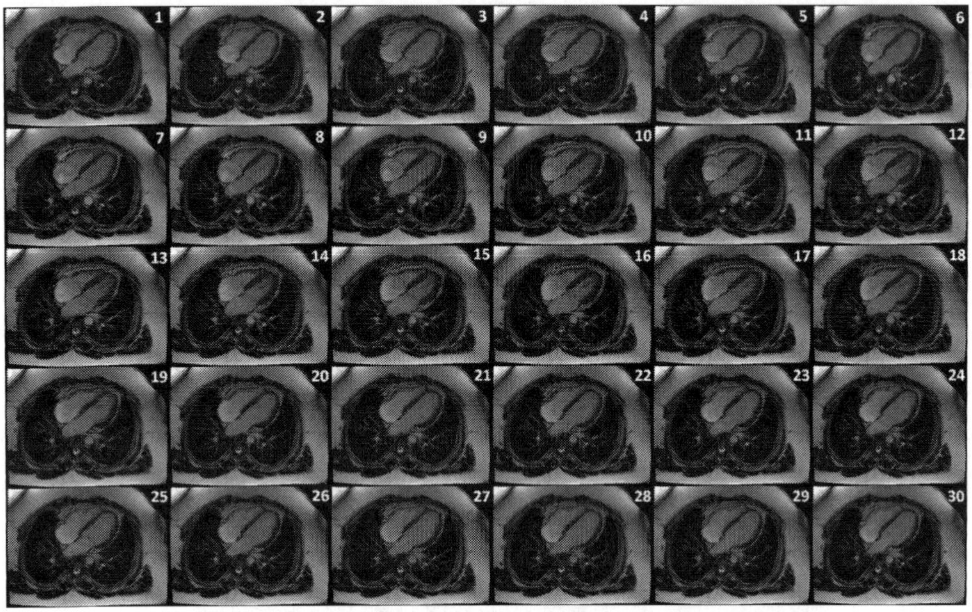

Figure 2. Sequence of frames of one cardiac cycle of a four-chambered projection of the heart along the long axis.

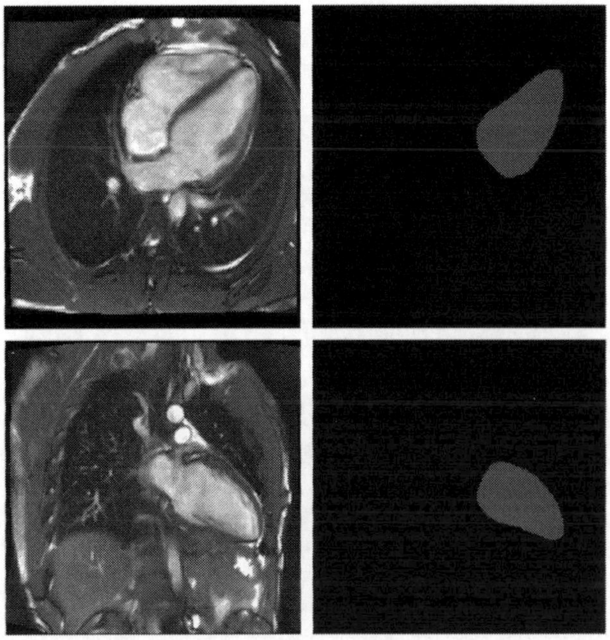

Figure 3. Results of marking MRI images: from above – four-chamber projection, from below – two-chamber projection.

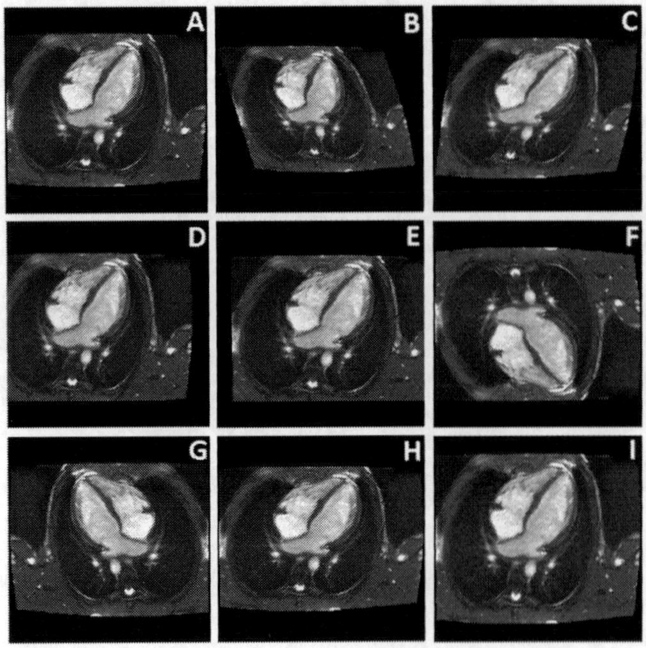

Figure 4. Affine transformations: A – original image, B – slope to the left, C – slope to the right, D – random shift, E – random shift, F – reflection along the x axis, G – reflection along the y axis, H – compression along the x, I – compression along the y.

only 2 classes (background and LV area). A new layer has been added to the network architecture, allowing the neural network to make a prediction for two classes. The introduction of a new layer allowed the used NN to find the region of the LV using information about the background and area of the LV.

Due to the fact that at the output of the NN of UNET architecture the image is smaller than the input one, therefore the UNET architecture was modified by introducing an additional layer, which provided an increase in the size of the input image by 92 pixels from each side by means of mirror reflection.

To prevent the overfitting of the NN with the FCN-8s architecture, the regularization method – dropout, which was located in the two lower full-connected layers was used (the coefficients used in this method were chosen equal to 0.5). Implement an analog approach for a NN with the UNET architecture failed due to insufficient video memory.

Also additionally, in the architecture of both NN, a layer is added that provides the calculation of the metric used to estimate the accuracy:

$$dice = \frac{2 \times |P \times T|}{|P|^2 + |T|^2}, \tag{1}$$

where P – network prediction, T – true image.

2017 IEEE Dynamics of Systems, Mechanisms and Mahines (Dynamics) (Omsk, Russia) 14 Nov–16 Nov 2017

3.3.2. Selecting the values of the hyperparameters of the NN The following settings of the hyperparameters of the NN were used in the studies:

- base_lr – 0.0001.
- lr_policy – step.
- training_epochs – 10.
- weight_decay – base_lr / 100.
- type – Adam.
- momentum – 0.9.

whose values were selected experimentally, based on an analysis of the values of the coefficient dice, calculated in accordance with equation 1.

3.3.3. Refinement of the estimination of the probability of the pixel belonging to the region of the LV To correct the values of the probability of belonging pixels to the LV areas, the weighted average probabilities of the LV pixel area belonging to the FCN-8s (0.55 weight coefficient) and UNET (0.45 weight coefficient) neural networks were calculated [5]. The values of the weight coefficients were chosen experimentally.

3.3.4. Learning outcomes of NNs Training of neural networks was carried out in Caffe [2] framework with graphical shell DIGITS.

As a result of training, the following models of deep convolutional NNs were obtained:

- The model, trained on a data set with an apical two-chamber projection and based on the FCN-8s architecture.
- A model trained on a set of data with an apical two-chamber projection and based on the UNET architecture.
- The model, trained on a data set with an apical four-chamber projection and based on the FCN-8s architecture.
- The model, trained on a data set with an apical four-chamber projection and based on the UNET architecture.

3.4. Postprocessing

3.4.1. Removing faulty areas identified using NNs The analysis of LV areas manually processed by MRI images showed that the minimum size of the LV image in both the two-chamber and the four-chambered projection always exceeded 400 pixels. In this connection, objects allocated with a NN with an area of less than 300 pixels were considered erroneous and were removed from the binary image [3] (Figure 5).

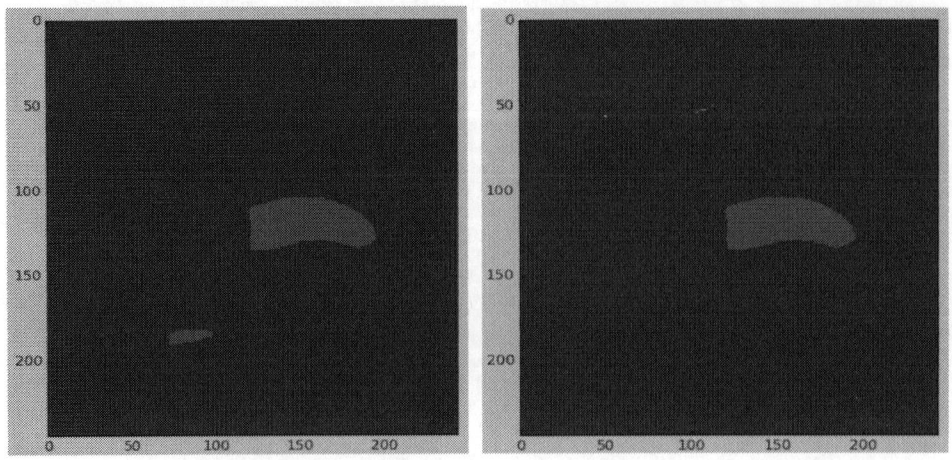

Figure 5. Example of deleting mistakenly identified areas on a binary image.

3.5. Evaluation of the volume of the LV

3.5.1. Calculation of finite volumes using an ensemble of neural networks To obtain LV volume estimation, the modified Simpson method [11] was used, the illustration of which is shown in Figure 6.

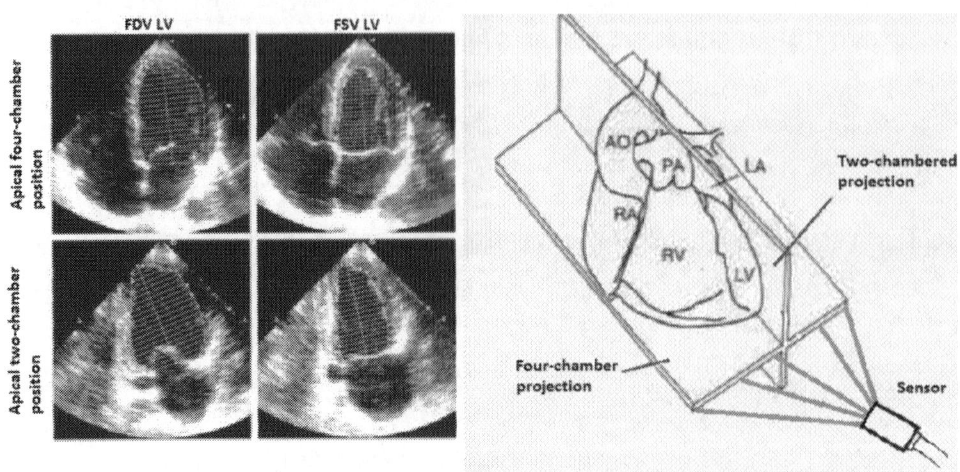

Figure 6. Modified Simpson method.

This method was applied to each of the frames MRI-film, the duration of which corresponded to the duration of one cardiac cycle. Further, the diastolic (maximal) and systolic (minimal) volumes of the LV were determined from the obtained time series containing the "instantaneous values" of the volume of the left ventricle.

3.6. Elimination of a systematic error in estimates of LV volume and EF

The analysis of the volume values on a sample of data used for training neural networks revealed the difference between these estimates and similar expert estimates. From a mathematical point of view, the differences between the automatically determined values of the volumes of the LV and the corresponding expert values, as well as between the EF calculated by these volumes, represent some samples of the random variables presented in Figure 7.

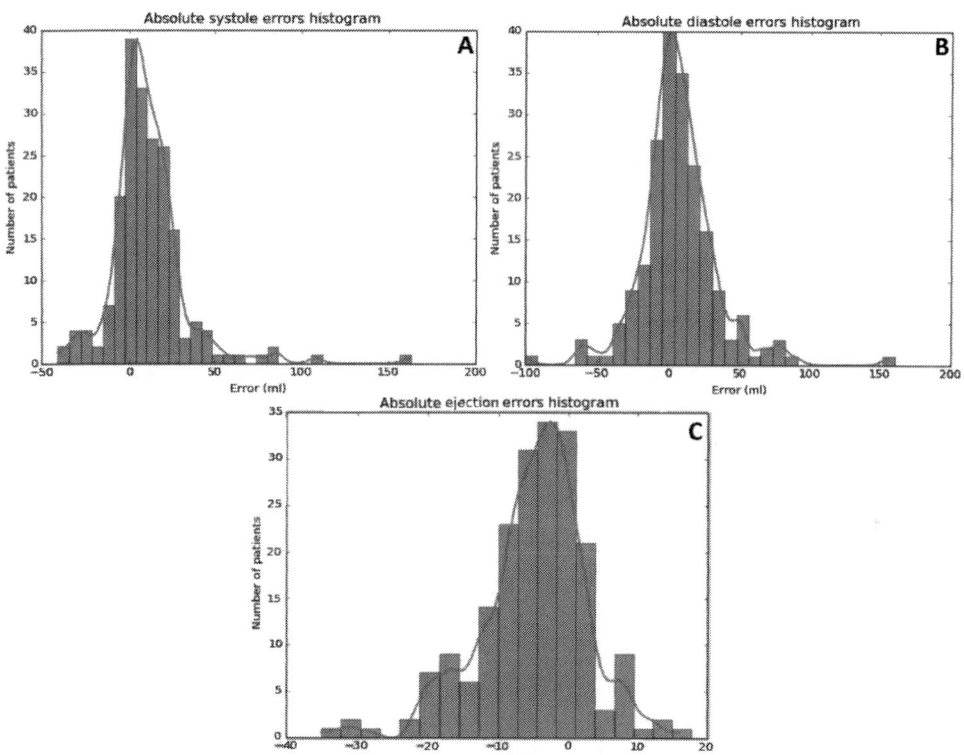

Figure 7. System operation results before calibration: A – histogram of error distribution of estimates of systolic volume, B – histogram of error distribution of estimates of diastolic volume, C – histogram of error distribution of emissions fraction estimates.

It can be seen from Figure 7 that the maximum of the distribution density, the discussed random sample samples are reached at the following points: 0.8 ml (diastolic volume), 4.4 ml (systolic volume), -2.5% (EF), which are estimates of the systematic error of diastolic, systolic values and EF. In the future, the calculation of the ejection fraction was preceded by the removal of the obtained estimates of the corresponding values of the systematic errors from the values of the systolic and diastolic volumes.

2017 IEEE Dynamics of Systems, Mechanisms and Mahines (Dynamics) (Omsk, Russia) 14 Nov–16 Nov 2017

Table 1. Comparison of received results

Id	Systole(ml)		Diastole(ml)		Ejection fraction(%)	
	Received	Expert	Received	Expert	Received	Expert
501	84.58	104.30	186.08	212.50	54.55	50.92
521	78.24	58.30	189.15	165.80	58.64	64.84
541	47.79	45.40	122.04	116.20	60.84	60.93
561	75.78	54.50	186.24	158.00	59.31	65.51
581	14.91	8.00	19.55	19.00	23.73	57.89
601	66.67	73.00	175.00	176.60	61.90	58.66
621	34.14	65.80	96.40	139.90	64.59	52.97
641	97.18	104.30	209.59	234.20	53.63	55.47
661	87.26	94.20	161.31	173.80	45.91	45.80
681	35.29	32.60	71.14	84.00	50.39	61.19
700	77.25	57.50	135.39	114.7	42.94	49.87

4. Analysis of the results

Consider the results obtained with the use of an automatic system for estimating the volume of the LV. Table 1 presents the results of systolic, diastolic volumes and EF for 11 patients.

From table 1 it can be seen that the expert and automatic values are close enough to each other. The discrepancy between the values of the ejection fraction does not exceed 12 percent. Analysis of similar data obtained for all patients showed that only one patient had an EF that differed from the expert value by more than 30%. An analysis of this patient's MRI film that the frames of this film are of inadequate quality.

Histograms of differences in expert estimates of systolic and diastolic volumes and EF and similar estimates obtained automatically are presented in Figures 8, 9, 10.

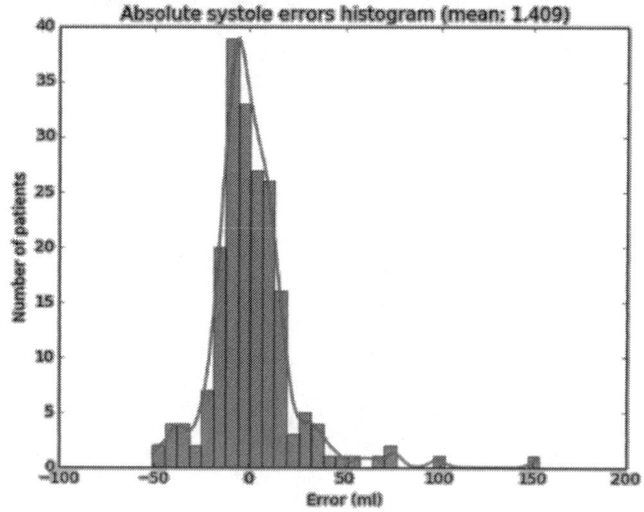

Figure 8. Histogram of error distribution of estimates of systolic volume.

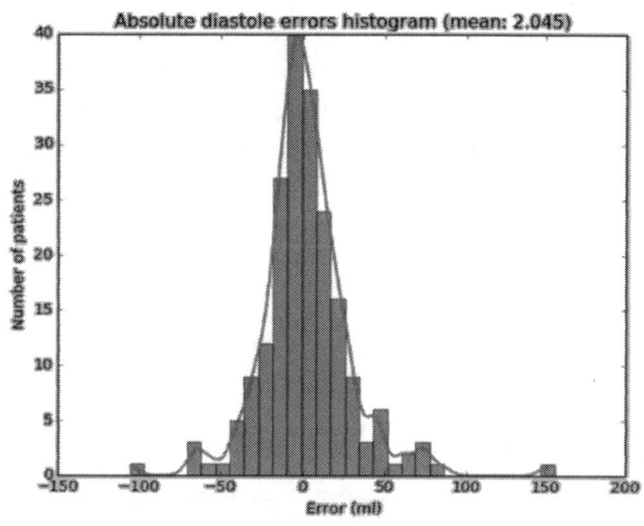

Figure 9. Histogram of error distribution of estimates of diastolic volume.

It can be seen from Figures 8, 9, 10 that the maxima of the distribution densities of the differences under consideration are reached at points with abscissa equal to 0. This indicates that there is no systematic bias of automatically obtained estimates of diastolic, systolic volumes and EF.

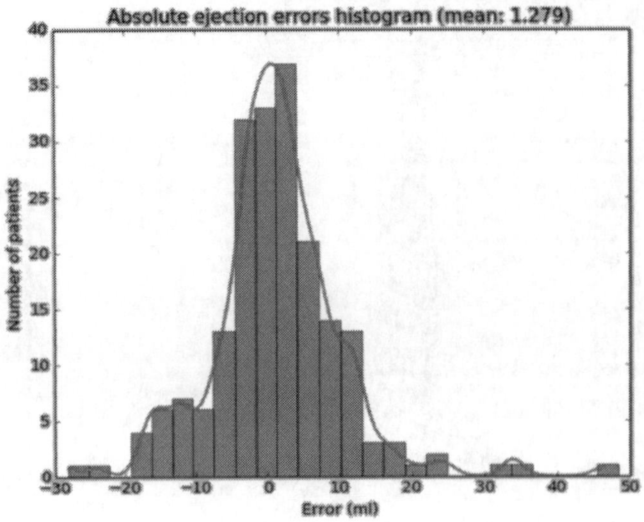

Figure 10. Histogram of error distribution of emissions fraction estimates.

Conclusions

The developed automatic system of volume estimation of the LV of the heart on the basis of the analysis of two-dimensional MRI images of two-chamber and four-chamber projections along the long axis with the use of NNs demonstrated its efficiency. A more detailed study of the statistical properties of the errors in the systolic and diastolic volumes of the ejection fraction is the subject of subsequent publications.

References

[1] Automatic Left Ventricle Detection in MRI Images Using Marginal Space Learning and Component-Based Voting [Electronic resource]. – Access mode: `https://www.umiacs.umd.edu/ zhengyf/LV_Detection_MRI_SPIE09.pdf`
[2] Caffe — Layer Catalogue [Electronic resource]. – Access mode: `http://caffe.berkeleyvision.org/tutorial/layers.html`
[3] Deep Learning for Medical Image Segmentation [Electronic resource]. – Access mode: `https://arxiv.org/pdf/1505.02000.pdf`
[4] Diagnosing Heart Diseases in the Data Science Bowl [Electronic resource]. – Access mode: `http://blog.kaggle.com/2016/04/13/diagnosing-heart-diseases-with-deep-neural-networks-2nd-place-ira-korshunova`
[5] Ensemble learning [Electronic resource]. – Access mode: `http://www.scholarpedia.org/article/Ensemble_learning`
[6] Estimating the volume of the left ventricle from MRI images using deep neural networks [Electronic resource]. – Access mode: `https://arxiv.org/pdf/1702.03833.pdf`
[7] Fully Convolutional Networks for Semantic Segmentation [Electronic resource]. – Access mode: `https://arxiv.org/pdf/1605.06211.pdf`
[8] How do Convolution Neural Networks work? [Electronic resource]. – Access mode: `https://brohrer.github.io/how_convolutional_neural_networks_work.html`

IOP Conf. Series: Journal of Physics: Conf. Series **944** (2017) 012092 doi:10.1088/1742-6596/944/1/012092

[9] Second Annual Data Science Bowl [Electronic resource]. – Access mode: `https://www.kaggle.com/c/second-annual-data-science-bowl`

[10] U-Net: Convolutional Networks for Biomedical Image Segmentation [Electronic resource]. – Access mode: `https://arxiv.org/pdf/1505.04597.pdf`

[11] Recommendations on the quantification, structure and function of the heart chambers [Electronic resource]. – Access mode: `http://webmed.irkutsk.ru/doc/pdf/echo.pdf`

Content and structure of knowledge base used for virtual control of android arm motion in specified environment

F N Pritykin and V I Nebritov

Omsk State Technical University, 11, Mira ave., Omsk, 644050, Russia

E-mail: pritykin@mail.ru

Abstract. The paper presents the configuration of knowledge base necessary for intelligent control of android arm mechanism motion with different positions of certain forbidden regions taken into account. The present structure of the knowledge base characterizes the past experience of arm motion synthesis in the vector of velocities with due regard for the known obstacles. This structure also specifies its intrinsic properties. Knowledge base generation is based on the study of the arm mechanism instantaneous states implementations. Computational experiments connected with the virtual control of android arm motion with known forbidden regions using the developed knowledge base are introduced. Using the developed knowledge base to control virtually the arm motion reduces the time of test assignments calculation. The results of the research can be used in developing control systems of autonomous android robots in the known in advance environment.

1. Introduction

Building algorithms for the control process of autonomous android robot operating in a planned environment requires a number of factors to be considered. The given control algorithm is implemented as a set of rules and a relevant logic decision mechanism. Also within the assigned time intervals throughout the control process proper estimation of certain parameters should occur. These parameters define the conditions that the specified algorithm should understand. Typically, the algorithm of android control faces unplanned events and unknown situations, and it is desirable that it behaves intelligently in these situations. The behaviour of an autonomous android robot can be improved if the control system takes into account and uses the knowledge base or the past experience of motion synthesis with due account for positions of the known in advance forbidden regions.

2. Problem statement

Let us consider the development technique and the structure of the past experience knowledge base used for the virtual modelling of robot motions in the vector of velocities. Let us remark here that the virtual control of the motion allows the arm mechanism capability to be evaluated before implementing motions in drives with the prescribed position of forbidden regions and the start and end points of the synthesized motion path of the output link centre (OL) [3-7]. The knowledge base of the past experience is the parameter part arranged in arrays defining intrinsic properties of an android arm mechanism as well as properties with due account for the position of the known forbidden regions. These arrays can be calculated in advance to determine the past experience related to the motion synthesis in the vector of velocities. The motion synthesis can be studied in advance at distinct points of the configuration space (defining the service angle along the whole set of possible values of generalized coordinates by instantaneous states implementation [8, 9]). The motion synthesis along the specified paths of OL (of the arm) in the working envelop with consideration for object obstacles position leading to deadlocks can also be studied [10].

The most frequent movement task for an android is putting on a working object or taking it off the tool rack. Therefore, let us examine the structure of the knowledge base in virtual control of the android arm motion in this example. Suppose we need to model the motion of the arm mechanism to transfer the manipulation object from the point A^S (the start point of the synthesized path) to the point A^E (the end point) considering the location of tool racks P_1 and P_2 (figure 1).

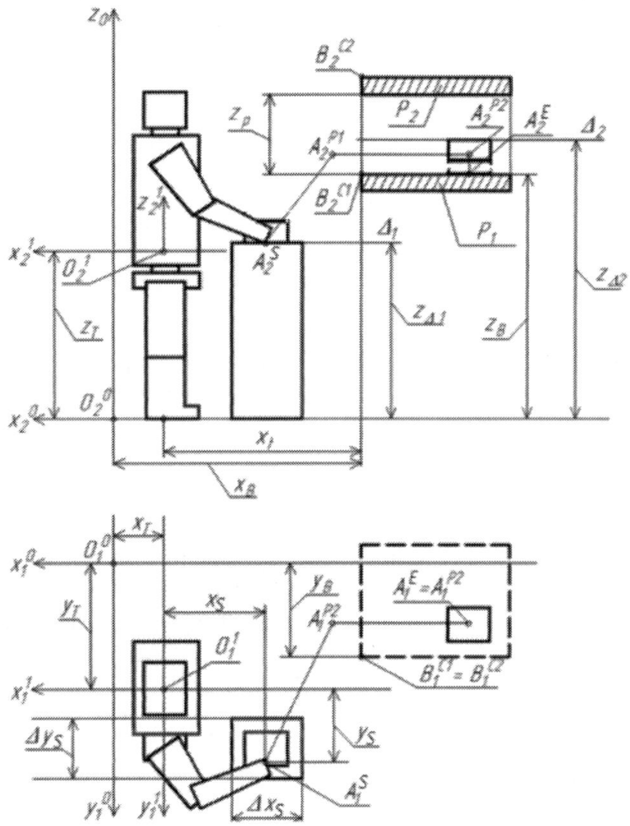

Figure 1. Start and end position of the manipulation object being mounted on the tool racks by android

The position of the points A^S and A^E in figure 1 is assigned by their projections $A^S{}_1$, $A^S{}_2$, $A^E{}_1$ and $A^E{}_2$. In this figure subscript one or two respectively means that the points belong to the horizontal plane of projection (or the top view) or the frontal plane of projection (or the front view). The general view and the kinematic diagram of an android robot arm mechanism are represented in the paper [8]. Here, it is necessary to clarify whether the arm can reach the target point of the synthesized path using virtual modelling or not. In figure 1 the system of coordinates $O^1x^1y^1z^1$ is connected with the body of the android, the parameters x_T, y_T and z_T defining the origin position of this system relative to the inertial coordinate system $O^0x^0y^0z^0$. The parameter x_t determines the minimal assigned safe distance from the android body to the tool racks. The position of the reference point B^{C1} of the bottom tool rack is defined by the coordinates x_B, y_B and z_B. The virtual modelling problem is solved using the knowledge base of the past experience.

3. Theoretical basis

At the first stage of solving the motion problem related to the transfer of the manipulation object from the point A^S to the point A^{P1}, and then to the points A^{P2} and A^E, it is necessary to evaluate the working envelope with proper account of the location of tool racks. The point O^1 is considered to be fixed. The construction of the working envelope with regard to the location of tool racks is studied in paper [11]. The parameters that define the working envelope projections with proper allowance made for the position of tool racks in this paper can be used by one of the component parts of the knowledge base about the past experience.

Let us agree to use the first component of the knowledge base to calculate the optimal rest position (configuration) of the android arm relative to the manipulation object (with the account of the start point A^H of the OL path). The optimal position in motion synthesis in the vector of velocities characterized by the parameters x_s, y_s (figure 1) and generalized coordinates q_i specifies the position of the android arm and body relative to the point A^S at which the solid angle U_s obtained by the instantaneous states implementation takes the maximum value [8]. The specified position of the android robot is calculated for each certain plane of level Δ_i, assigned with a definite step of the coordinate $z_{\Delta 1}$, $z_{\Delta 2}$ and so on (figure 1), the curve of the function being plotted for this coordinate:

$$U_s = f(x_s, y_s). \tag{1}$$

Let us consider the procedure to determine the graph of the function (1) for the plane Δ_1. To determine the graph of the function (1) and array of configurations where the OL centre coincides with some points A^S of the plane Δ_1, specified with a certain grid space, the starting point $A_S^{\Delta 1}$ being prescribed. The parameters x_S and y_S determine the rest position of the start point $A_S^{\Delta 1} \in \Delta_1$ by the paths of the OL in the system $O^1x^1y^1z^1$, with the centre of the OL moving in the plane Δ_1 (symbol $\in$ defines that the point belongs to the geometric object Δ_1). Figure 1 does not show the position of the point $A_S^{\Delta 1}$. This position coincides with one of the rectangle vertices of the plane Δ_1, its sides being determined by the parameters Δx_S и Δy_S. The set of point positions $A^S \in \Delta_1$, constructed within the specified interval Δx_S and Δy_S (figure 1) on the plane Δ_1, is obtained by the arm motion synthesis with the range of motion minimization criterion [2, 10]. The specified grid step for definition of points A^S is determined by the modulus of the vector of linear velocity for the OL centre motion in the plane Δ_1. The initial agreed value of the vector $\boldsymbol{q}_{\Delta i}(q_1, \ldots, q_5)$ corresponding to the OL centre position at the point $A_S^{\Delta 1}$ provides the maximum angle U_s value. This value is derived from the analysis of a number of arm configurations that provide the assigned values of parameters $z_{\Delta 1}$ and x_t. The maximum angle U_s value configuration, obtained by the instantaneous states implementation, is determined from the set of specified configurations [8]. The components of the vector $\boldsymbol{q}_{\Delta i}$ or the values of generalized coordinates $q_1, \ldots, q_5$ specify the initial configuration relative to the position of the OL centre at the point $A_S^{\Delta 1}$. The values of the vector $\boldsymbol{q}_{\Delta i}$ and the values x_S and y_S are read after the motion synthesis for each point A^S of the plane Δ_i. The solid angle U_s obtained by the instantaneous states implementation for the distinct points A^S and values of the vector $\boldsymbol{q}_\Delta$ is approximately determined from the formula:

$$U_S = (U_S^{\text{hor}} + U_S^{\text{fr}} + U_S^{\text{prof}}) \tfrac{1}{3}, \tag{2}$$

where angles U_s^{hor}, U_s^{fr} and U_s^{prof} specify angle projections U_s on the horizontal, front and profile planes of projection respectively [8]. The first component of the knowledge base assigned by the array Ψ_1 specifies the compliance of the values $z_{\Delta i}$ and parameters x_s, y_s and $\boldsymbol{q}_{\Delta i}$, the angle value U_s being the maximum. This knowledge base is used to ensure the optimal rest position of the arm relative to the start point of the transfer path for the manipulation object. This component of the knowledge base characterizes the intrinsic properties of the android arm mechanism.

Data on the start and end positions of points specifying the synthesized path of the OL motion and the forbidden region positions (of the tool racks P_1 and P_2) arise as the second component of the knowledge base Ψ_2, this knowledge base being the cause of deadlocks [10]. The example of the forbidden regions position P_1, P_2 and the path of the android robot arm motion, being the cause of the deadlock, is shown in figure 2a. The motion synthesis along the path $A^SA^{P1}A^{P2}$ with the point-to-point accuracy being $\delta < 10$ mm is presented in figure 2a. In that case there is a deadlock. Figure 2b shows the motion synthesis along the same path with $\delta < 60$ mm.

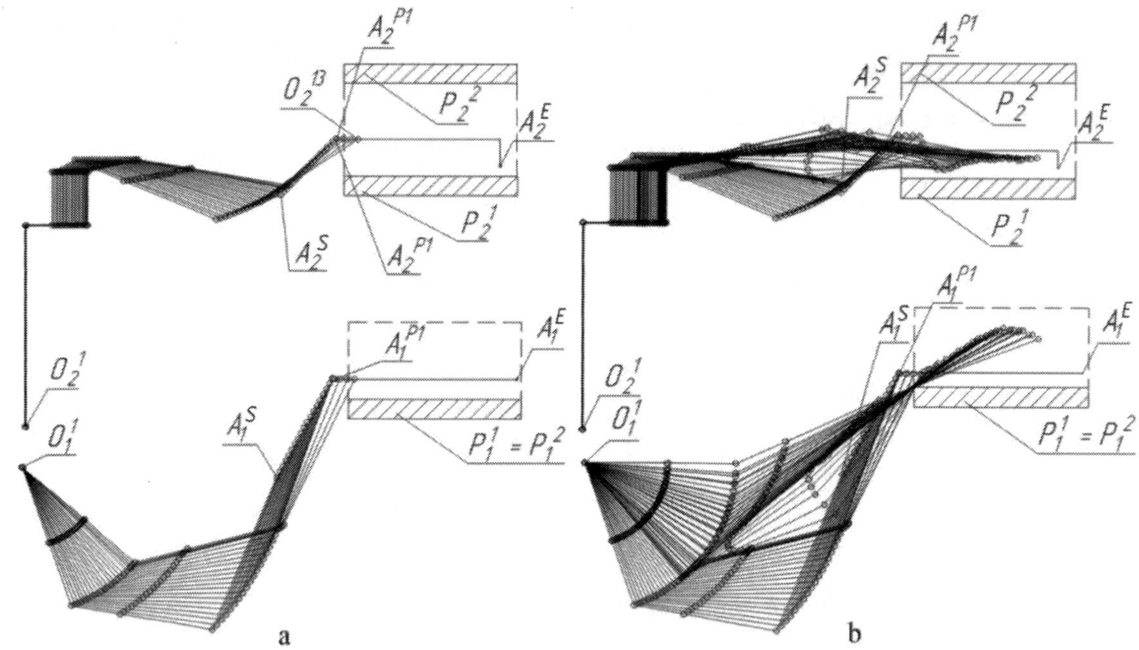

Figure 2. Motion synthesis along the path prescribed by the segments A^sA^{P1} and $A^{P1}A^{P2}$: a — with the deadlock occurrence at δ < 10 mm, b — motion synthesis at δ < 60 mm

Figure 2b shows that in the second case a significant deviation from the path occurs. Generally this can lead to the collision of the manipulation object and tool racks. In this case under the prescribed type of configuration, the motion synthesis in the vector of velocities for δ < 60 mm at the path segment $A^{P1}A^{P2}$ is impossible. Hence, providing the configuration at the point A^{P1}, first it is necessary to move the arm from the point A^S to the point A^{P1} with the type of configuration to be changed using the values of weighting factors for generalized velocity changes not being equal to unity.

It is also possible to change the type of configuration by fine motions synthesis where the centre of the output link is in the assigned region defined by the parameter δ < 10 mm [9]. The use of the third component of the knowledge base Ψ_3 is necessary in reducing the calculation time for the vector of generalized velocities where calculated interim configurations intersect the forbidden regions P_1 and P_2. The vector of generalized velocities Q^N for the case is calculated by the equation:

$$Q_N = Q_M + \sum_{i=1}^{p} k_i m Q_i \ , \qquad (3)$$

where $\boldsymbol{Q_M}$ is the vector specifying the point $M^Q \in \Gamma^Q$ conforming with the range of motion minimization criterion [2, 10]. The superscript ^{Q} designates the membership of the geometric object to the five-dimensional space of generalized velocities. The point M^Q prescribes the check point centre connected with the p-plane Γ^Q; k_i is the coordinates of the point N^Q in the p-plane Γ^Q (particular instant state of the manipulator mechanism corresponds to each point N^Q); m is the length of a unit check point segment of the p-plane Γ^Q, $\boldsymbol{Q_i}$ is unit director vectors of the check point axes, p is the dimension of the p-plane Γ^Q. For the considered example $p = 2$.

The knowledge base Ψ_3 is used to calculate parameter limits k_i^{max} in the relationship (3) for different arm configurations. The array of parameters Ψ_3 determines the value of coefficients a_j^{ki} for hyper surface equations which reflect functions $k_i^{max} = f(q_3, q_4, q_5)$ within prescribed ultimate value intervals of generalized coordinates q_1 and q_2 [12]. The parameters k_i^{max} allow the regions in space of generalized velocities to be specified, these regions assigning the permissible values of vector $\boldsymbol{Q_N}$. These values ensure the prescribed positional accuracy δ < 10 mm of the OL centre.

The algorithm scheme of the android arm motion path synthesis based on the developed knowledge base is shown in figure 3.

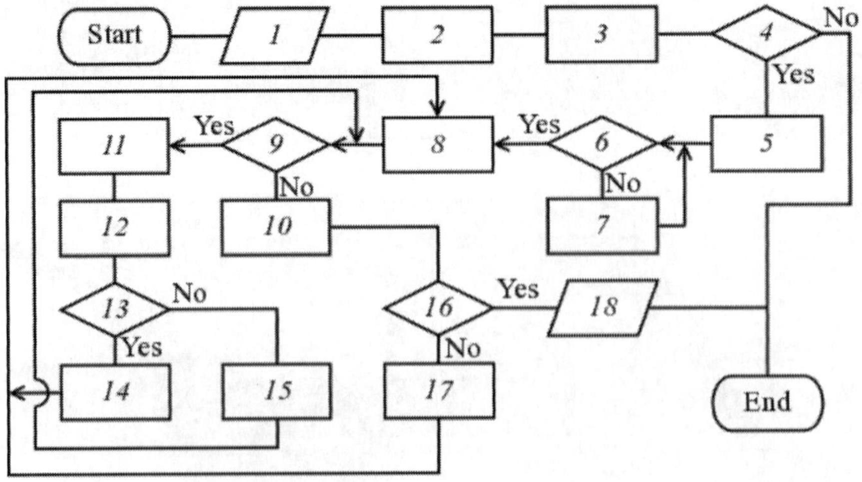

Figure 3. Virtual modelling algorithm scheme of android arm motion based on knowledge base

The following notations are taken in figure 3: 1 is the input of arrays data l_i, *smi*, *Kod* and variables $n_k = 0$, $A^S(x^S, y^S)$, $A^E(x^E, y^E)$, $B^{C1}(x_{BC1}, y_{BC1})$, B^{C2} (x_{BC2}, y_{BC2}) and x_t, where arrays determine: l_i are the lengths of mechanism links, *smi* are displacements along a coordinate axes, *Kod* are codes of the coordinate system transformations [8]. Input of array data Ψ_1, Ψ_2, and Ψ_3 specifying the knowledge base of the past experience; 2 is the calculation of via points A^{P1} and A^{P2} specifying the position of a synthesized path for the OL centre point movement, these via points providing the absence of the path and forbidden regions P_1 and P_2 intersection (provision of minimum offset to forbidden regions); 3 is the determination of parameters specifying icons for projection of arm working envelope $\lambda^0{}_i$ based on the known coordinate points B^{C1}, B^{C2} [11]; 4 is the recognition of points A^S, A^{P1}, A^{P2} and A^E belonging to the working envelop [11]; 5 is the calculation of vector $\mathbf{q}_{A1}$ and values x_S, y_S specifying an optimal rest position of the android arm and body (at U_s = max), relative to the manipulation object based on the parameter z_{Ai} and array Ψ_1; 6 is the analysis of value compliance $\mathbf{q}_{A1}$, point coordinates A^S, A^{P1}, A^{P2}, A^E, B^{C1}, B^{C2} with array values Ψ_2, which prescribes deadlock; 7 is the motion synthesis to change the configuration type and determine a new value **q** [9]; 8 is the calculation of the generalized velocity vector $Q_M(\dot{q}_{1M}, \dot{q}_{2M}, \ldots, \dot{q}_{iM})$based on the range of motion minimization criterion on the path sections prescribed by the segments A^SA^{P1}, $A^{P1}A^{P2}$ and $A^{P2}A^E$ [2,10]; 9 is the determination of the arm configuration condition assigned by the values and forbidden regions (here the assumption is made that $\Delta q_i \approx \dot{q}_{iM}$), $\dot{q}_{iM}$ are the components of the vector Q_M; 10 is the construction of the next configuration or $q_i = q_i + \dot{q}_{iN}$; 11 is the determination of maximum values k_i^{max} based on q_i and the array Ψ_3 [12]; 12 is the change of values $k_i = k_i + 1$ used in the vector equation (3); 13 are the values k_i satisfying the maximum values k_i^{max}; 14 is the synthesis of motions with invariant OL centre position [9]. The motion synthesis using weighting factors of generalized velocity values. The arm motion to change the type of configuration; 15 is the calculation of the generalized velocity vector $Q_N(\dot{q}_{1N}, \dot{q}_{2N}, \ldots, \dot{q}_{iN})$ (3) [8]; 16 is the target point reached in the prescribed segment; 17 is the change in the value of the next configuration $n_k = n_k + 1$; 18 is the result output of the motion synthesis along the complete path section determined by the segments A^SA^{P1}, $A^{P1}A^{P2}$ and $A^{P2}A^E$.

4. Experimental results

In figure 4 the results of the motion synthesis using the knowledge bases to transfer the manipulation object from the point A^S to the point A^E are shown. In figure 4 the motion synthesis with the type of configuration being changed and using weighting factors of generalized velocities values is shown on the path section prescribed by the points A^SA^{P1}.

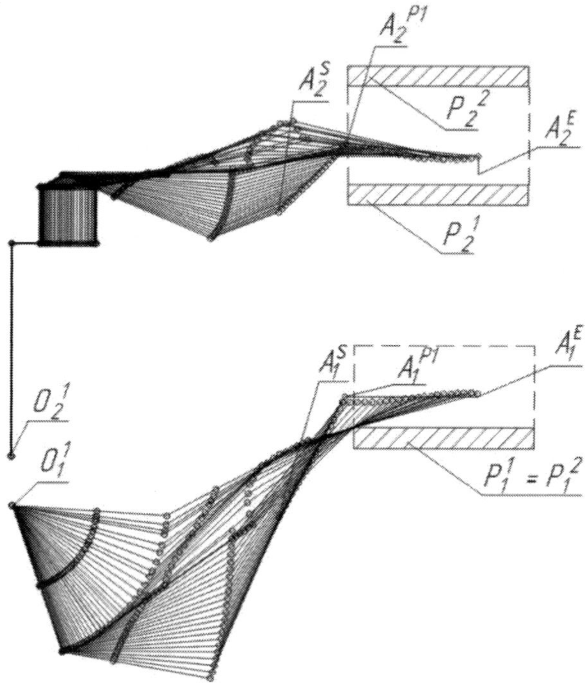

Figure 4. Result of motion synthesis to the end point of the synthesized path.

5. Conclusion

The developed algorithm for virtual control of android motion using the developed knowledge bases allows the complex evaluation of current situations to be performed and from this the optimal logic choice to be implemented. Minimal total change of generalized coordinates occurs when making this choice.

The results of computational experiments demonstrate computation time decimation of test assignments related to the putting the manipulation objects on the tool racks and taking them off using the developed knowledge base. The number of cases for motion synthesis resulting in deadlocks is considerably reduced.

The results of the research can be used in developing intelligent control systems of autonomous android robots in the known in advance environment.

References

[1] Wihtney D E, 1972 The mathematics of coordinated control of prosthetic arms and manipulators *Tpons. ASME, Ser G, J Dymamic, Just, Meas and Control* **94** PP 19–27.

[2] Kobrinsky A A and Kobrinsky A A 1976 Construction of optimal motions for manipulation systems *Mechanical engineering* **1** pp. 12–18.

[3] Pratt J, Dilworth P and Pratt G 1997 Virtual model control of a bipedal walking robot *Proceedings of Int. Conf. on Robotics and Automation* **1** pp 193–198

[4] You B, Zou Y, Xiao W and Wang J 2010 Telerobot control system based on dual-virtual model and virtual force *Int. Forum on Strategic Technology* PP 246–250

[5] Hrr J, Pratt J, Chew C-M, Herr H and Pratt G 1998 Adaptive virtual model control of a bipedal walking robot *Proceedings of Int. Joint Symp. on Intelligence and Systems* pp 245–251

[6] Tsukamoto H K, Takubo T, Ohara K, Mae Y and Arai T 2010 Virtual impedance model for obstacle avoidance in a limb mechanism robot *Int. Conf. on Information and Automation* pp 729-734

[7] Hasegawa T, Suehiro T and Takase K 1992 A model-based manipulation system with skill-based execution *IEEE Transactions on Robotics and Automation* **8** pp 535–544

[8] Pritykin F N and Nebritov V I 2016 Studying tolerance range of generalized velocities vector under android motion synthesis *Dynamics of Systems, Mechanisms and Machines (Dynamics)* DOI:10.1109/Dynamics.2016.7819064
[9] Pritykin F and Gordeev O 2016 Defining a service angle for planar mechanisms of manipulators based on the instantaneous states analysis *IOP Conf. Series: Materials Science and Engineering* DOI:10.1088/1757-899X/124/1/012025
[10] Pritykin F N and Tevlin A M 1987 Procedure for construction of manipulator motions from a given local grip path in the presence of obstacles *Soviet machine science* **4** pp 30–33
[11] Pritykin F N and Nebritov V I 2017 Construction of android arm mechanism working envelope taking into account forbidden zones position *Omsk Science Bulletin* **1** pp. 5–9
[12] Pritykin F N 2016 Research in the region size and form in multidimensional space of generalized velocities specifying permissible instantaneous states of android mechanism *Omsk Science Bulletin* **5** pp. 29–34

IOP Conf. Series: Journal of Physics: Conf. Series **944** (2017) 012094 doi:10.1088/1742-6596/944/1/012094

Using fuzzy fractal features of digital images for the material surface analisys

Privezentsev D.G., Zhiznyakov A.L., Astafiev A.V., Pugin E.V.
Vladimir State University named after Alexander and Nikolay Stoletovs Vladimir, Russia
E-mail: dgprivezencev@mail.ru

Abstract. Edge detection is an important task in image processing. There are a lot of approaches in this area: Sobel, Canny operators and others. One of the perspective techniques in image processing is the use of fuzzy logic and fuzzy sets theory. They allow us to increase processing quality by representing information in its fuzzy form. Most of the existing fuzzy image processing methods switch to fuzzy sets on very late stages, so this leads to some useful information loss. In this paper, a novel method of edge detection based on fuzzy image representation and fuzzy pixels is proposed. With this approach, we convert the image to fuzzy form on the first step. Different approaches to this conversion are described. Several membership functions for fuzzy pixel description and requirements for their form and view are given. A novel approach to edge detection based on Sobel operator and fuzzy image representation is proposed. Experimental testing of developed method was performed on remote sensing images.

1. Introduction

To extract information from remote sensing image different methods of image processing are used. Edge detection is one of these methods. It can be used to further extraction of interesting objects. There are a lot of algorithms developed in this area. The most popular are Canny, Sobel, Prewitt, Roberts operators and some others [1]. One of the perspective techniques in image processing is the use of fuzzy logic and fuzzy sets theory [2]. Different algorithms and methods use different approaches to fuzzy processing. Linguistic variables are useful when the results are well distinguishable [3], but this is not very common situation. On the other hand it is better to process the results using analytical methods. Most papers suggest to switch to fuzzy sets on very late stages of processing when the source image is enhanced, converted or somehow preprocessed. Papers that extract fuzzy properties based on existing crisp features can be taken to this category [4]. Early defuzzification of the results has also negative influence on fuzzy processing [5]. All these things lower flexibility of fuzzy approach.

Other drawback is the use of fuzzy sets of first type (Type-1 Fuzzy Sets or T1FS), that have very little amount of uncertainty. To solve this issue type-2 fuzzy sets (T2FS) and other more high-type fuzzy sets were introduced [6]. Also other types of sets based on fuzzy sets are evolving: rough sets, soft sets, soft rough sets, blurry sets and others [7].

In this paper we propose a novel approach to image processing based on fuzzy sets, that suggest a transition to fuzzy image representation with fuzzy pixels on the earliest stages of processing. With this approach all extracted features are fuzzy by definition and defuzzification process at best should be done only when retrieving information from the computer system.

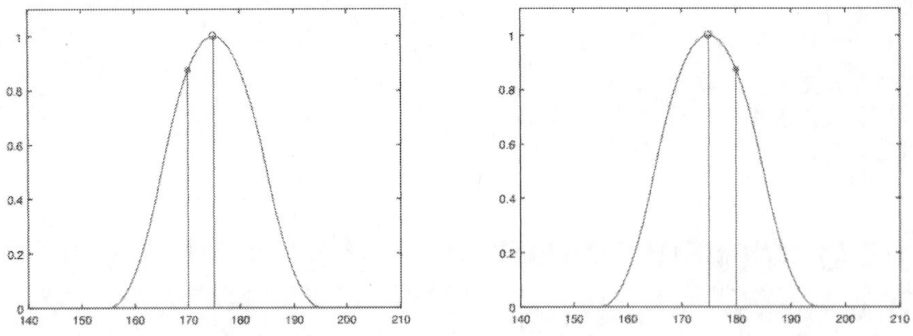

(a) (b)

Figure 1: Membership function of image fuzzy pixel. Real intensity level at $\mu(F(x, y)) = 1$: a) greater, b) less than current.

2. Fuzzy image representation

Continuous image can be described as two-dimensional signal $f(x, y)$, where x and y – coordinates. During formation of digital image transition to discrete coordinates and values of intensities are performed:

$$F(x, y) = D[f(x, y)],$$

where $D[\cdot]$ – transformation operator from continuous signal to discrete, that is implemented on hardware, $F(x, y)$ – the discrete image. Obviously that with insufficient level of quantization, signal levels are rounded to integers. Then real intensity level of point $R(x, y)$ could be computed as

$$R(x, y) = F(x, y) + d(x, y),$$

where $d(x, y)$ – round error. More noticible distortion is brought by different noises $\nu(x, y)$, that could be greated than 1. We will use the simplest noise model – additive noise. More complex models of noise could be investigated accordingly. Then the real intensity level could be calculated as

$$R(x, y) = F(x, y) + d(x, y) + \nu(x, y).$$

Computations that does not take into account these details, soon could accumulate big error related to source continuous image. Methods of fuzzy sets theory allow us to save the uncertainty till the latest stages of image processing and analysis. To do it we should switch to fuzzy image representation. $U(F, x, y)$

$$U(F, x, y) = \mu(F(x, y)),$$

where $\mu(F(x, y))$ – membership function of a pixel with coordinates (x, y) to intensity level $F(x, y)$. Graphically this can be represented as shown on Fig. 1.

There are a lot of membership functions known. One must select those which satisfy the following conditions

$$\lim_{l \to \infty} \mu(F(x, y)) = 0,$$

$$\int_0^{L-1} \mu(F(x, y))dl > 0, \quad l \in [0; L-1].$$

3. Edge detection

Most of the edge detection operators uses gradient operator that has modulo $|\nabla G|$ and direction θ

$$|\nabla G| = \sqrt{G_x^2 + G_y^2},$$

$$\theta = \arctan \frac{G_y}{G_x},$$

where $G_x = M_x * G$, $G_y = M_y * G$ – the result of convolution operator with horizontal and vertical matrices.

Let us consider possibility of the use of fuzzy pixels in Sobel and Prewitt edge detection operator. To do this we must process the membership function values $\pi(F(x, y))$ in addition to normal intensity levels. Calculations of $G_{x\mu}$ and $G_{y\mu}$ can be done analogously. The difference here is in gradient computations of π-function.

$$|\nabla \pi| = 1 - \sqrt{G_{x\mu}^2 + G_{y\mu}^2}.$$

We take complementary value because after squaring, sum and square root operations from values on interval $[0; 1]$, the result is near to 0. Final value of gradient will be

$$|\nabla G_\pi| = |\nabla G||\nabla \pi|.$$

Threshold value of gradient can be selected manually or with different binarization techniques (e.g. with Otsu method [8]).

4. Testing

One of the most important application domains of nondestructive testing methods is the evaluation of defects. This problem is generally solved by recognizing visual patterns of defects on defectoscopy images and by determining their geometric characteristics (length, width, and area). To detect flat steel defects of arbitrary forms and sizes (for example, dents and blisters), we use the algorithm for detecting uncharacteristic image segments, with the maximum error being the average error of the rank block approximation. [9, 10].

We found that the developed algorithms for uncharacteristic segment detection identify most defects, including dents, blisters, and shears. Further analysis of the found defects shows that the number and geometric characteristics of defects are almost identical to the corresponding values that were obtained by a nondestructive testing inspector. This supports the usability of the proposed algorithms in practical problems of nondestructive testing. The detection of groups of defects that consist of a set of single defects is an important defectoscopy problem because, according to certain standards, such defects may be unacceptable.

5. Conclusion

Proposed algorithm showed their applicability in edge detection task during image processing. Main feature is the use of fuzzy image representation based on fuzzy pixels. This approach is very perspective because it saves the uncertainty much better rather other existing algorithms. In the future this model could be extended to type-2 and higher fuzzy sets and also to other kinds of fuzzy sets. The results of comparing the developed algorithm for finding flat steel defects on the image with well known algorithms allow us to make the following conclusions. The proposed algorithm enables the automatic detection of objects without preprocessing the source image; given a preprocessed image, this algo rithm is inferior to the wellknown ones. Further analysis of the defects that were found showed that the numbers and geometric characteristics of defects are almost identical to the corresponding values that were obtained by a nondestructive

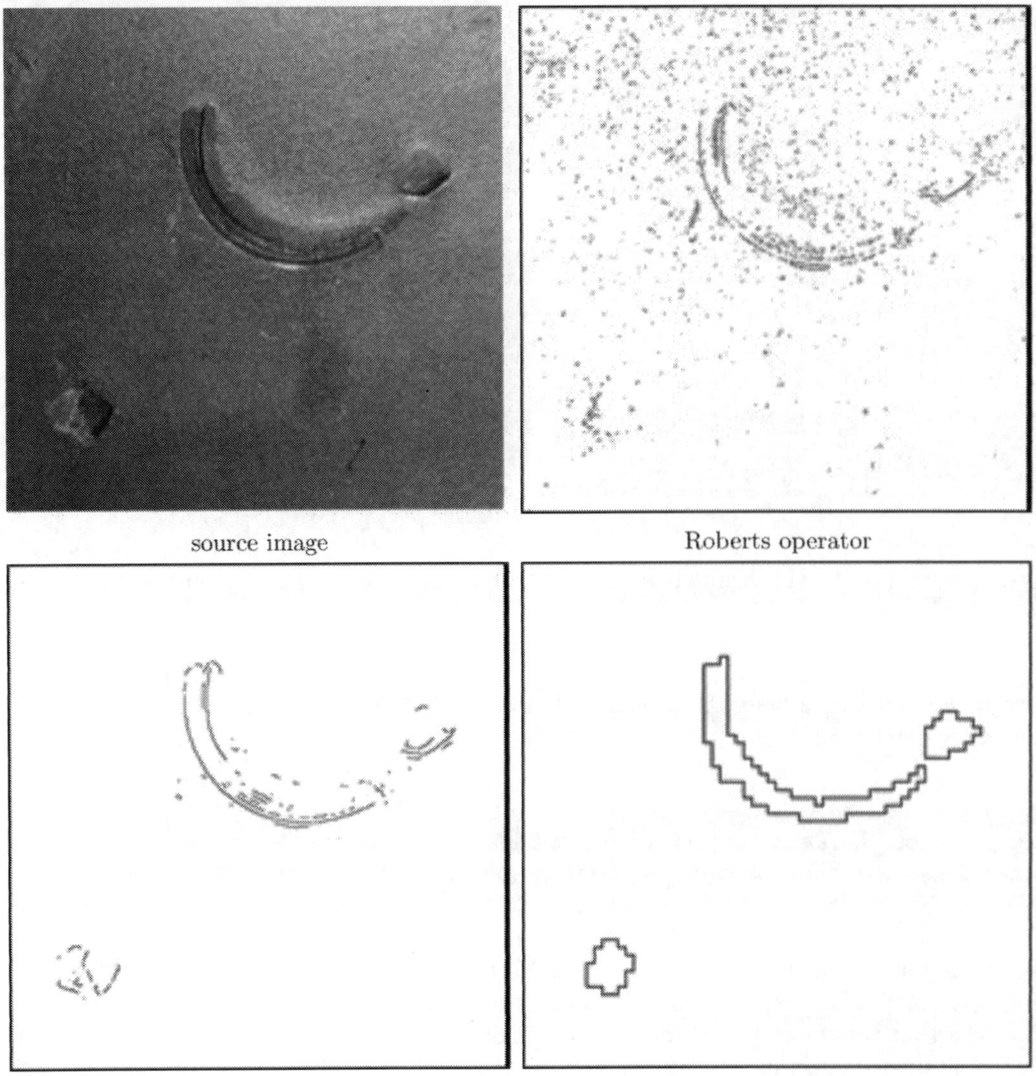

source image Roberts operator

Prewitt operator our method

Figure 2: The results of the detection of a sheartype defect.

testing inspector. This supports the usability of the proposed algorithms in practical problems of nondestructive testing.

This work was supported by the project number 2.1950.2017/ in the framework of the basic tasks of the state of the Russian Ministry of Education.

The reported study was funded by RFBR according to the research project 17-47-330073

[1] Canny, J. A computational approach to edge detection / John Canny // Pattern Analysis and Machine Intelligence, IEEE Transactions on. 1986. 11. Vol. PAMI-8, no. 6. P. 679698.

[2] Kuo, Y.-H. A new fuzzy edge detection method for image enhancement / Yau-Hwang Kuo, Chang-Shing Lee, Chao-Chin Liu // Proceedings of 6th International Fuzzy Systems Conference. [S. l.] : Institute of Electrical and Electronics Engineers (IEEE), 1997.

[3] Becerikli, Y. A new fuzzy approach for edge detection / Yasar Becerikli, Tayfun M. Karan // Computational Intelligence and Bioinspired Systems. [S. l.] : Springer Nature, 2005. P. 943951.

[4] Interval type-2 fuzzy sets constructed from several membership functions: Application to the fuzzy thresholding algorithm / Miguel Pagola, Carlos Lopez-Molina, Javier Fernandez [et al.] // IEEE Transactions

on Fuzzy Systems. 2013. apr. Vol. 21, no. 2. P. 230244.

[5] Fuzzy Techniques in Image Processing / Ed. by Etienne E. Kerre, Mike Nachtegael. [S. l.] : Physica-Verlag HD, 2000.

[6] Edge-detection method for image processing based on generalized type-2 fuzzy logic / Patricia Melin, Claudia I. Gonzalez, Juan R. Castro [et al.] // IEEE Transactions on Fuzzy Systems. 2014. dec. Vol. 22, no. 6. P. 15151525.

[7] Molodtsov, D. Soft set theoryfirst results / D. Molodtsov // Computers & Mathematics with Applications. 1999. Vol. 37, no. 4. P. 19 31. URL: http://www.sciencedirect.com/science/article/pii/S0898122199000565.

[8] Otsu, N. A threshold selection method from gray-level histograms / N Otsu // IEEE Trans. Sys., Man., Cyber. 1979. Vol. 9. P. 6266.

[9] Privezentsev, D. G. Use of characteristic image segments in tasks of digital image processing / Denis G. Privezentsev, Arkady L. Zhiznyakov // 2015 International Conference Stability and Control Processes in Memory of V.I. Zubov (SCP). [S. l.] : Institute of Electrical and Electronics Engineers (IEEE), 2015. oct.

[10] Zhiznyakov, A. L. Using fractal features of digital images for the detection of surface defects / A. L. Zhiznyakov, D. G. Privezentsev, A. A. Zakharov // Pattern Recognition and Image Analysis. 2015. jan. Vol. 25, no. 1. P. 122131.

Prevention of unorganized emissions of ammonia in installations of dewaxing of oils

E O Rehovskaya[1], I Yu Nagibina and A Yu Ivanov

Omsk State Technical University, Mira ave.,11, Omsk, 644050, Russia

e-mail: helena10_04@mail.ru

Abstract. The problem of lack of automation devices in oil dewaxing units is considered in this work. As a result, fugitive ammonia emissions that exceed the maximum permissible concentration, which adversely affect the health of personnel and the environment, can occur in the atmospheric air. The device and the operating principle of the automatic air separator are shown.

1. Introduction

At present, the interest of modern society to environmental problems, compliance with the requirements of environmental legislation is growing. Industrial organizations of various scales seek to reduce the negative impact on the environment, by achieving favorable characteristics of the equipment. All measures indicate effective control of the environmental impact. The main principle in the work of any industrial company should be the conservation of natural resources and the environment. This cannot be done without proper control over the activities of the enterprise at all stages of production.

Now in industrial enterprises widely used in refrigeration plants operating on ammonia (refrigerant R717), which has good thermodynamic properties. Ammonia – a colorless gas having a pungent suffocating odor belonging to the IV class of hazard. Maximum permissible concentration in the air of working zone is 20 mg/m^2.If the ammonia vapor content in the air is from 20%, an explosion is likely. From the chemical point of view, ammonia is inert to ferrous metals, therefore, in the manufacture of refrigeration equipment, pipelines designed for working with ammonia, various grades of steel and cast iron are used [1]. Ammonia interacts with copper, zinc, dissolves the rubber, so it is possible to use only aluminum-silicon, high-tin bronzes, babbits from alloys of non-ferrous metals.

Due to the fact that ammonia plants that use dangerous refrigerant, are working at high pressure during operation must be complied with safety requirements.

2. Formulation of the problem

The lack of automation equipment of air-purge drum has a number of disadvantages:

– the operation of the air separator is not constant, because ammonia losses occur;

– untimely inclusion of the air separator leads to the accumulation of air in the system, resulting in increased pressure, discharge temperature, deterioration of condensation, there is an increase in electricity costs and as a result, increased wear and tear of equipment;

– untimely shutdown leads to an unorganized release of ammonia into the atmospheric air;

– the quality and timeliness of air removal from the refrigeration system depends on the human factor;

The main task is to replace existing air separators with automatic ones. As a result of this work using modern means of automation, the air separator was selected, the technical and economic parameters of the refrigeration department were improved, the risk and likelihood of exposure to ammonia and emerging harmful factors of production for personnel were reduced.

3. Main part

At the dewaxing unit of oils type 39 / 8№5, the pour point of the residual and distillate oil fractions is reduced due to the removal of solid paraffins from them. In this case, crystallization is used, based on the cooling of the raw material in a mixture with the solvent. Crystallization of paraffinic hydrocarbons from the solution takes place, followed by separation by filtration from the liquid. The solvent is a mixture of methyl ethyl ketone and toluene. The components of the oils obtained on the unit are subsequently used for the production of lubricating oils. As a result, after dewaxing turns slack wax is used as a raw material, component and petrolatum, required for obtaining no aging material.

The oil dewaxing unit includes the following parts: separation of cooling of the raw material and solvent mixture, separation of filtration, separation of solvent regeneration from solutions and its dehydration, refrigeration unit and commodity park with heating oil heating system. The refrigerating compartment is designed to cool a mixture of raw materials with a solvent and is an ammonia refrigeration unit of two-stage compression based on ammonia compressors (ADK-73/40, AGK-73, condenser condensation unit for ammonia KTV, plate heat exchangers KB-1,2,3, AV3, AVG, crystallizers for cooling of a mix of raw material and solvent and other auxiliary devices). The purpose of cooling is the crystallization of solid paraffinic hydrocarbons from the low-viscosity, medium-viscous and viscous epoxy fraction [2]. For a single filling of all ammonia units of the installation, about 30 tons of ammonia is required, it is intended for cooling the mixture of raw materials with a solvent up to -35 °C.

The cycle follows the pattern: 1) a pair of ammonia from the molds come in liquid separators, where the deposition of mechanically entrained droplets of ammonia; 2) while increasing the level of ammonia up to 60% alarm system compressor, if the level is about 90% automatically locks all compressors; 3) from the liquid separators of the pair are sucked into a two-stage compressor; 4) the precipitated liquid ammonia flows down into the drainage receiver; 5) pairs of ammonia, from the first stage of ammonia compressors, enter the intermediate vessels for cooling; 6) in containers, thanks to direct injection, a constant level of liquid ammonia is maintained; 7) further, saturated vapors are sucked by the second stage of the compressor, compressed and sent through the oil separators for condensation to air-cooling units, water condensers, plate heat exchangers; 8) condensed ammonia flows into the receivers, then enters the coils and enters the evaporator system; 9) fresh ammonia is taken from the reagent farm on a special line to the working receivers, while a vacuum is created in the receiver receiving ammonia, and in the tank from which ammonia is taken, pressure is created by means of ammonia compressors; 10) along the same line, ammonia can be delivered from working receivers to any facility in the workshop or to a reagent farm.

Air from the system is removed by means of air separators, representing a small vertical heat exchanger of a rigid type (Figure1). Its work consists in condensing ammonia from the air-ammonia mixture under pressure, due to the evaporation of the liquid refrigerant. Ammonia compressors create a vacuum in the ammonia evaporation system. Air entering the system is possible only through insufficiently tight connections, for example, through gland seals or at the opening of individual elements of the installation at the time of repair work, as well as when filling with a refrigeration reagent. Air, creating additional pressure in the system, prevents normal condensation of ammonia vapors.

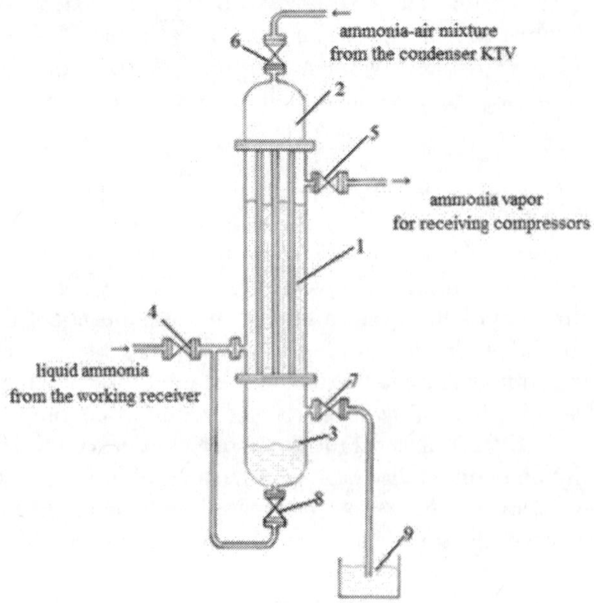

Figure 1. Ammonia air separator:
1– annular space; 2 – tube space; 3 – the bottom cap; 4 – valve supplying liquid ammonia from the working receiver; 5 – the valve for the extraction of ammonia vapors from the shell space; 6 – valve supplying the air-ammonia mixture into the pipe space; 7 – air venting valve; 8 – valve for supplying condensed ammonia to the shell side of the air separator body; 9 – a tank of water.

The refrigeration unit includes ammonia compressors ADK-73/40 and AGK-73, the technical characteristics of which are given in table 1.

Table 1. Technical specifications of compressors.

Compressor brand	AGK-73		ADK-73/40	
	Low grade	High degree	Low grade	High degree
Diameter of the cylinder, mm	730	450	730	400
Cylinder piston stroke, mm	550	550	550	550
Stem diameter, mm	100	100	100	100
Number of revolutions per minute	167	167	167	167
Theoretical volume of the compressor, m^3 / h	4570	1710	4570	1340
Cooling capacity, kcal / h	about 900000		about 400000	
Indicator power consumption Ni, hp	669	669	669	669
Power consumption on shaft Ne, hp	786	786	786	786
Diameter of pipelines, mm	325/8	325/8	325/8	-
Diameter of stop valves, mm	300	300	300	-
Motor type	SDK 760-167		SDK 760-167	
Motor power, kW	623		623	
Voltage, V	6000/3000		6000/3000	
Number of revolutions per minute	167		167	
Weight of compressor without motor, kg	22600		23700	
Weight of motor, kg	12700		12700	

In case of emergencies, the following measures should be taken:

1. When the pressure of the second stage of the compressors is increased because of the large accumulation of air in the ammonia system, it is necessary to turn on the air separator and eliminate the leaks through which air is sucked.

2. If the air separator is not working properly:

2.1. If an insufficient amount of liquid ammonia is supplied, then open between the tube space;

2.2. If ammonia is high, then stop the flow by opening the suction valve to freeze the air separator;

2.3. At low pressure, fully open the valve at the inlet of the ammonia-air mixture into the air separator.

3. When the air-to-air discharge tube from the air separator freezes because of the mercury tank with liquid ammonia, it is necessary to restart the excess ammonia into the annular space of the air separator.

Ammonia horizontal compressor brand AGK-73 is a twin-stage compressor of double-stage compression with direct drive from a synchronous motor. Cooling capacity of the compressor is about 900,000 kcal / h at evaporation temperature totк= +35 °C, condensation temperature tк = +35 °C and subcooling temperature before the control valve ti= +30 °C.

Ammonia horizontal compressor brand ADK-73/40 is also a twin compressor of two-stage compression of a double action with direct drive from a synchronous motor. The cooling capacity of the compressor is about 400,000 kcal / h at evaporation temperature to= -43°C, condensation temperature tк= +38°C and subcooling temperature before the control valve ti= +35 °C.

4. Results

In total, the cooling capacity of the working compressors is 1980 kW. For this value, an automatic air separator of the brand AV-2 (Figure 2), developed by VNIHI (All-Russian Scientific Research Institute of the Refrigeration Industry), is suitable. They are exploited for many years at various enterprises.

The cooler, appliances, fittings and pipelines are compactly mounted on a common frame. Thus, the automatic air separator brand AV-2 is a unit ready for inclusion in the system of an ammonia refrigeration unit [3, 4].

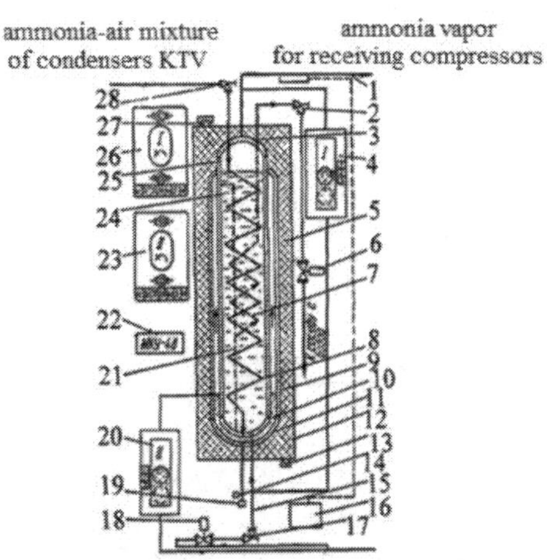

Figure 2. Automatic air separator brand AV-2:
1 – the suction pipe; 2, 17,28 – shut-off valves; 3, 10, 11 – bottoms; 4,20 – level sensors; 5 – insulation; 6, 18 – solenoid valves; 7, 8 – coils; 9 – an external pipe; 12 – a casing; 13.27– plugs; 14.15 – branch pipes; 16 – temperature relay; 19 – stopper; 21 – secondary cooling coil tube; 22 – intermediate MKU – 48; 23, 26 – amplifiers; 24 – inner tube; 25 – housing.

The principle of operation is based on cooling the air-ammonia mixture under the condensing pressure, by boiling ammonia in the inner tube. The supply of ammonia to it is automatically controlled by the level 4 float switch of the PRU-2 mark in combination with the solenoid valve 18. Liquid ammonia produced by the air separator is permanently removed from the annular space through the float level sensor chamber into the linear receiver [5-7].

5. Conclusion

Replacing an existing air separator with an automatic one will result in a reduction in ammonia emissions to the atmosphere, thus minimizing the likelihood of negative impact of ammonia on workers, and in the case of large emissions on the population living near the territory of the enterprise.

On the given technological block the following types of damaging factors are possible: shock wave, thermal radiation and toxic damage.

The most dangerous scenario for the development of an accident at a dewaxing plant is the complete depressurization of one of the crystallizers with the expiration of a dangerous substance - ammonia, the damaging factor is toxic damage. The probability of occurrence of the most dangerous scenario of an emergency at the deep dewaxing unit of oils is 4.6×10^{-05} 1 / year. In creating the damaging factors, up to 2180 kg of ammonia can participate. In addition to ammonia, the main harmful substances released into the atmosphere from sources are hydrocarbons and toluene.

Reduction to or below a regulated level (maximum permissible concentration) or complete elimination of air pollution by emissions along with other technological indicators is one of the main criteria for the quality of the plant's operation. To reduce emissions of harmful substances into the environment, the following measures are envisaged:

– draining of oil products from heat exchangers and steaming of heat exchangers for repair is carried out in a drainage tank with subsequent processing;

– the inert gas thrown into the atmosphere is first cleaned in the absorber by an oil component, which is then pumped out for regeneration;

– air discharged to the atmosphere from the ammonia system is preliminarily cooled by ammonia in an air separator with condensation of ammonia and its return to the system.

Continuous operation of the automatic air separator will ensure a constant removal of air from the system, which will lead to an improvement in the following indicators:

1. Practically complete absence of unorganized ammonia emissions into the atmosphere, respectively, reduction of environmental pollution, and a reduction in the likelihood of exceeding the MPC standards in the work area of personnel;

2. Timely removal of air from the system, respectively, reducing the load on ammonia compressors, which in turn will lead to a reduction in energy consumption, a reduction in the amount of water consumed in water-cooled condensers and less equipment wear [6].

Thus, it becomes clear that there is an interest in minimizing the presence of air in the system, which will certainly lead to an increase in the technical and economic performance of the installation. The feasibility of introducing an automatic air separator is confirmed not only by technical advantages, but also by economic efficiency.

The obtained results can be used as a rationalization proposal for the introduction of an automatic air separator brand AV-2 on the dewaxing unit of oil type 39/8№5, the production of oils, additives and lubricants.

The work was carried out within the framework of research work No. 17079B at the Omsk State Technical University.

References

[1] 2016 *Rules for the protection of labor in the operation of refrigeration* (Moscow: Publishing house Alpha-Press) p 44
[2] Briganti A 2004 *Maintenance Manual for Refrigeration and Air Conditioning Systems* (Moscow: Euroclimate) p 311

[3] Polevoy A A 2010 *Automation of refrigeration and air conditioning systems* (St. Petersburg: Publishing house Profession) p 244

[4] Chandrappa R and Kulshrestha U C 2015 Sustainable Industrial Air Pollution Management *Sustainable Air Pollution Management* (New York: Springer New York consultants bureau) pp 207-290

[5] Yakovlev S P , Kerm L Y and Os'mushnikov V A 2014 Results of addition of pulsating crystallizer to dewaxing unit at "Omsk lubricant plant" - a subsidiary of OOO "Gazprom-sm" *Chemistry and Technology of Fuels and Oils (*vol 50, issue 4*)* (New York: Springer New York consultants bureau) pp 273-280

[6] 2013 *Federal rules and regulations in the field of industrial safety "Safety rules in the oil and gas industry"*(Moscow: Scientific and Technical Center for Research on Industrial Safety Problems) p 288

[7] Aneja V P, Bunton B, Walker J T and Malik B P 2001 Measurement and analysis of atmospheric ammonia emissions from anaerobic lagoons *Atmospheric Environment (*vol 35, issue 11*)* (Amsterdam: Elsevier Science Publishing Company, Inc.) pp 1949-1958

IOP Conf. Series: Journal of Physics: Conf. Series **944** (2017) 012096 doi:10.1088/1742-6596/944/1/012096

Dynamics analysis in the design of turbomachinery using sensitivity coefficients

O Repetckii[1], I Ryzhikov[2,*], Tien Quyet Nguyen[2]

[1] Irkutsk State Agrarian University n.a. A.A. Ezhevsky, 664038 Irkutsk, Russia

[2] Irkutsk National Research Technical University, 664074 Irkutsk, Russia

[*] E-mail: rin111@list.ru

Abstract. Solving the optimization problems based on numerical models is an important step in the design of machinery and equipment. Special attention is given to the calculation gradient of the target and limitation functions behavior with the variation of design variables (known as sensitivity). Using the sensitivity function and varying the design variables the designer gets the opportunity to analyze various basic design variants quickly and efficiently without time-consuming calculations for each of them. This allows speeding up greatly the design process and reducing its time input. In this regard the numerical models development using algorithms of sensitivity calculation for the strength and vibration analysis of structures is an urgent task.

1. Introduction

Strength or vibration characteristics (displacement, stresses, frequencies) dependences on the geometric or mass variables parameters may act as the target function for aircraft structures. The rotor blade of a gas turbine engine can be specified as an example. Due to cantilever design the blade vibrates and the characteristics of this vibration are very sensitive to mass or geometric parameters changes. Using the sensitivity coefficients in blade natural frequencies analysis provide quality improvements of resonance detuning.

Many authors describe numerical methods of analysis of turbomachines rotors dynamics and strength [1-7]. Development of analysis methods using sensitivity algorithms is described in following publications [8-12].

2. Fundamentals

The statics equation in finite element method for constant speed of rotation and temperature is:
linear formulation:

$$K\delta = f\,, \tag{1}$$

nonlinear formulation:

$$(K + K_G + K_R)\delta = f\,, \tag{2}$$

where K is stiffness matrix, K_G is matrix of a geometrical stiffness, K_R is supplementary stiffness matrix arising from rotation, δ is node displacements, f is vector of loadings.

The dynamic equation in finite element method in case of the free vibration without damping is [5]:
linear formulation:

$$M\ddot{\delta}+K\delta=0, \tag{3}$$

nonlinear formulation:

$$M\ddot{\delta}+M_C\dot{\delta}+(K+K_G+K_R)\delta=0, \tag{4}$$

where M_C is Coriolis matrix.

The solution of the equation (3) takes the form of harmonic oscillation:

$$\delta=y\cos(\omega-\beta), \tag{5}$$

substituting this expression in the equation (3), we obtain the following result:

$$(\omega^2M+K)y\cos(\omega t-\beta)=0 \text{ or } (K-\lambda M)y=0, \tag{6}$$

where ω is angular frequency. λ is eigenvalue.

This equation makes sense when the determinant is equals to zero:
linear formulation:

$$\det(K-\lambda M)=0, \tag{7}$$

nonlinear formulation:

$$\det(K+K_G+K_R)-\lambda M=0. \tag{8}$$

Solving the equation (8) in the linear formulation, we can define eigenvalues:

$$\lambda=\{\lambda_1,\lambda_1,...\lambda_N\}^T, \tag{9}$$

or modal displacements (mode shapes):

$$y^{(i)}=\delta^{(i)^T}=\{\delta_{i1},\delta_{i2},...,\delta_{iN}\}^T, i=1,2,...,N, \tag{10}$$

and vibration frequencies:

$$f=\frac{1}{2\pi}\sqrt{\lambda}=\{f_1,f_2,...,f_N\}^T. \tag{11}$$

Differentiating equations (1, 2) with respect to thickness, we obtain the equation of static displacements sensitivity:
linear formulation:

$$K\frac{d\delta}{db_i}=\frac{df}{db_i}-\frac{dK}{db_i}\delta, \tag{12}$$

nonlinear formulation:

$$(K_E+K_G(\delta)+K_R)\frac{d\delta^k}{db_i}=\frac{df}{db_i}-(\frac{dK_E}{db_i}+\frac{\partial K_G(\delta)}{\partial b_i}+\frac{dK_R}{db_i})\delta-\tilde{K}_G(\frac{\partial\delta^{k-1}}{\partial\delta})\delta. \tag{13}$$

To solve the equation (4) an iterative algorithm is used.

The sensitivity of the eigenvalues for free vibration without and with taking into account the static stress-strained state is respectively:

$$\frac{d\lambda}{db_i}=y^T\left[\frac{dK}{db_i}-\lambda\frac{dM}{db_i}\right]y, \tag{14}$$

$$\frac{d\lambda}{db_i}=\frac{y^T\left[\dfrac{dK_E}{db_i}+\dfrac{dK_G(\delta)}{db_i}+\dfrac{dK_R}{db_i}-\lambda\dfrac{dM}{db_i}\right]y}{y^T My}. \tag{15}$$

Taking into account that $\lambda=(2\pi f)^2$ the sensitivity of the free vibration frequency is:

$$\frac{df}{db_i}=\frac{df}{d\lambda}\frac{d\lambda}{db_i}=\frac{1}{4\pi\sqrt{\lambda}}\frac{d\lambda}{db_i}. \tag{16}$$

The sensitivity of frequency to changes of structures thickness in the nodes is:

$$\frac{dF}{db}=\sum\sum K_f(i,j). \tag{17}$$

A large absolute value of frequencies sensitivity means a greater intensity of frequency change while changing thickness. Positive or negative sensitivity means that the frequency decreases or increases (the trend) when changing the structure thickness. Two methods can be used to predict the trend and intensity of each frequency plate oscillation change. In the first method, the thickness change ratio for each individual node with the corresponding expected frequency changes is selected. In the second method given coefficient is chosen for each separate section.

Frequency change F_0 according to the frequencies sensitivity method is calculated by the formula:

$$\Delta F=\sum\sum\Delta b(i,j)K_f(i,j),\ i=1,NCN,\ j=1,NE, \tag{18}$$

where $\Delta b(i,j)$ is thickness change value of node i in element j.

Prediction of frequencies according to the method of sensitivity:

$$F(b-\Delta b)=F_0+\Delta F, \tag{19}$$

where F is natural frequencies of original model(F_1, F_2 , F_3 , F_4).

Analysis of frequencies resonant detuning possibilities means the calculation of frequency change range while changing design parameters. This is the basis for eliminating any possibility of blade resonance.

Sensitivity is positive when the function reaches a maximum value in the upper range of design parameters variations and, on the contrary, sensitivity is negative when the function reaches a maximum value in the lower range of design parameters variation. A large absolute sensitivity value means a greater degree of changes in the experimental parameters (e.g., frequency) when changing the design parameters (e.g., thickness).

The use of sensitivity coefficients allows achieving a more effective result and at the same time assessing the impact of selected option of thickness changes overall calculated frequencies range. The error estimation is sufficient for engineering calculations.

3. Numerical results

The analysis of compressor blade free vibration with the use of sensitivity coefficients was carried out. The parameters of the blade: length L=0,186 m; chord X_k = 0,167 m; Young's modulus E = 2,16*10^5 MPa; Poisson's ratio μ = 0,3; density ρ = 7,85*10^3 kg/m^3; blade root radius R_0 = 0,236 m. The finite element blade model is shown in Figure 1. If the speed of blades rotation, gas stream pressure and temperature are constant, then it is possible to say that the aviation engine compressor is in a stationary mode (Figure 2). In our example, the rotational speed of the rotor is equal to 100 s^{-1}.

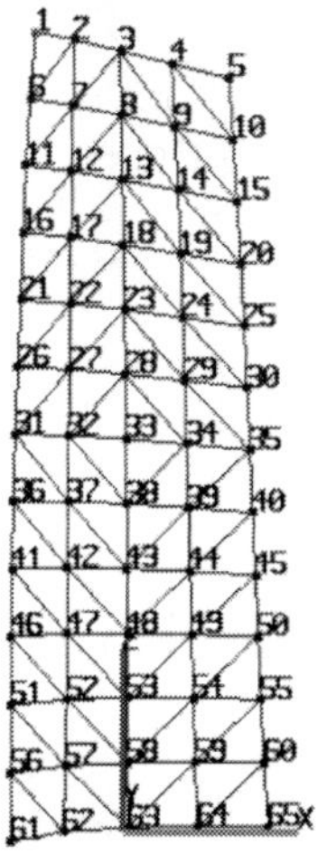

Figure 1. Finite element model of the rotor blade.

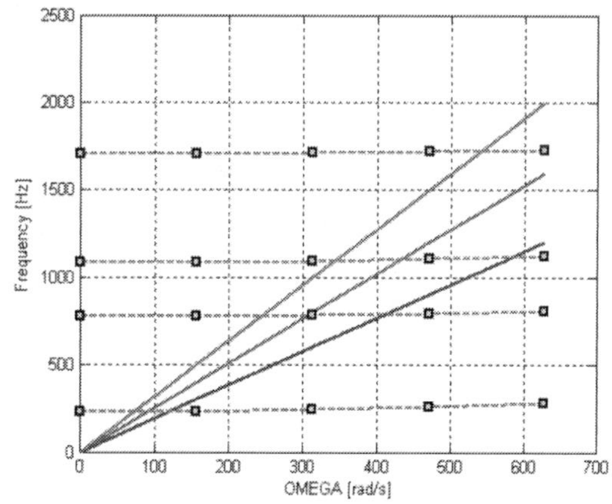

Figure 2. Resonance diagram of the rotor blade.

Frequencies sensitivity distribution for the nodes at coordinates (x_0, y_0, z_0, b_0) is shown in Figure 3. Frequencies sensitivity distribution for the blade cross sections is shown in Figure 4.

Table 1. The effect of rotation on the natural frequency.

Mode	$n = 0\ s^{-1}$	$n = 25\ s^{-1}$		$n = 50\ s^{-1}$		$n = 755\ s^{-1}$		$n = 100\ s^{-1}$	
	F_0, Hz	F, Hz	ΔF, %	F, Hz	ΔF, %	F, Hz	ΔF, %	F, Hz	ΔF, %
1	229.36	232.89	1.54	243.76	6.28	260.73	13.68	282.53	23.18
2	775.73	777.08	0.17	783.98	1.06	794.84	2.46	808.81	4.26
3	1082.62	1091.34	0.81	1091.34	0.81	1104.28	2.00	1122.70	3.70
4	1705.22	1703.34	-0.11	1708.49	0.19	1716.69	0.67	1727.45	1.30

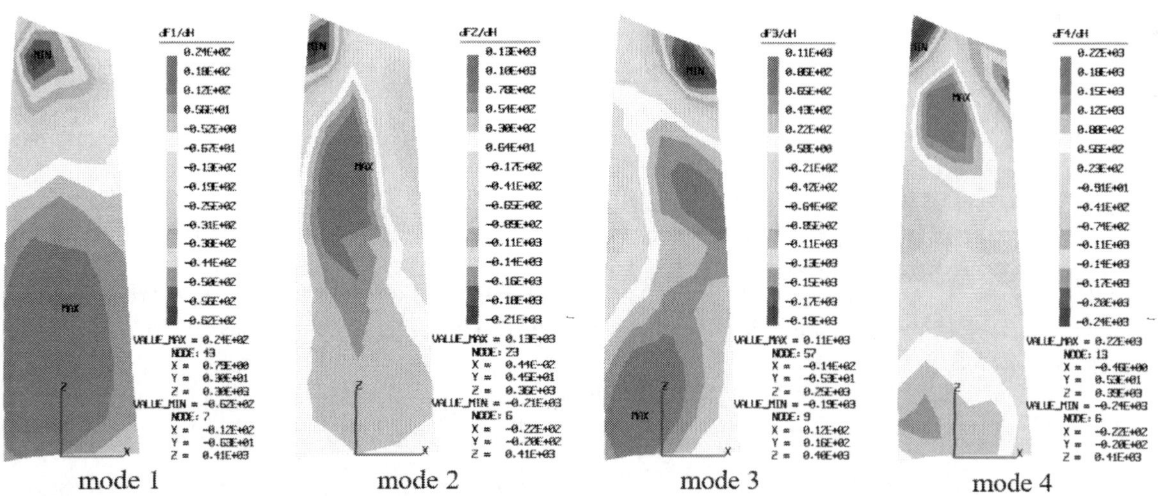

mode 1 mode 2 mode 3 mode 4

Figure 3. Sensitivity of the blade natural frequencies.

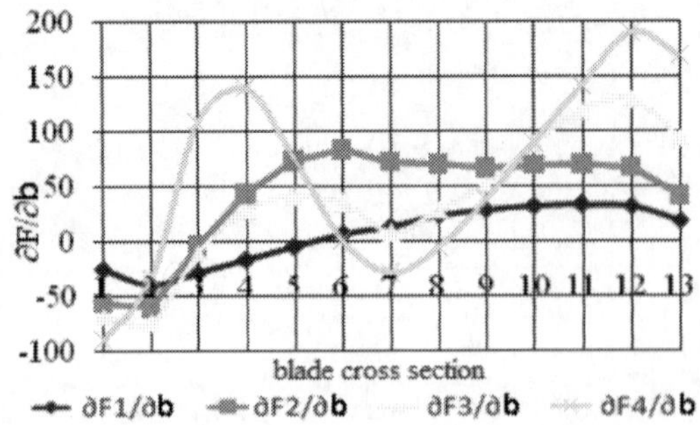

Figure 4. Sensitivity of the frequencies according to the cross sections of the blade.

In addition, forecast accuracy of blade natural frequencies changes at a thickening of 10% along the entire surface of the blade or in certain zones was analyzed. The error of frequencies forecast by sensitivity coefficients is small in comparison with calculations results of the modified models, and it enables using sensitivity coefficients to assess the possible projects of the blade.

4. Conclusions

The mentioned results of the calculations demonstrate a sufficiently small error, that allows of applying the developed method for the analysis of natural frequencies changes using sensitivity coefficients for optimal design of the gas turbine engines rotor blades.

Acknowledgments
The authors grateful to Dr. Zainchkovsky K S and Dr. Nguyen Dinh Duong for their valuable comments and suggestions.

References

[1] Repetckii O V 1990 Automation of strength calculations of turbomachines (Irkutsk: publishing house) p103
[2] Repetckii O V 1990 Computer analysis of the dynamics and strength of machines (Irkutsk: Irkutsk State Technical University Publishing House) p 301
[3] Irretier H Repetckii O V 1998 Vibration and Life Estimation of Rotor Structure *IFToMM Conf. on Rotor Dynamics* (Darmstadt)
[4] Heiman B Gerdt V Popp K Repetckii O V 2010 Mechatronics: components, methods, examples (Novosibirsk: Publishing House of the SB RAS) p 602
[5] Repetckii O V and Buy Manh Cuong 2010 To a question of the choice of a numerical method of the analysis of stresses at an assessment of a multi-cycle fatigue of blades of transport turbomachines *ISEA News 6* pp 153–158
[6] Elovenko D A Repetckii O V 2011 Analysis of thermophysical properties of heat insulation materials for new constructions of cylindrical walls in high-pressure autoclaves *BSU News 6* pp 201–206
[7] Elovenko D A Repetckii O V 2011 Analysis of stress state of elastic half-plane loaded with constant pressure at limited intermediate sections with specified period by finite element method based on MSC Mars software *BSU News 5* pp 171–175
[8] Repetckii O V Phan Van Tuan 2012 Construction of mathematical model for analysis of friction dampers influences on vibration of gas turbine engines blades *BSU News 1* pp 200–205
[9] Repetckii O V Do Manh Tung 2012 Mathematical model operation and numerical analysis of vibrations of ideal cyclic and symmetric systems finite element method *ISEA News 3* pp 149–153
[10] Repetckii O V Do Manh Tung 2014 Investigation of the characteristics of mistuned turbomachinery bladed discs vibration based on the reduced-order modeling by finite element method *Bulletin SibSAU 1 (53)* pp 60–66

IOP Conf. Series: Journal of Physics: Conf. Series **944** (2017) 012096 doi:10.1088/1742-6596/944/1/012096

[11] Repetckii O V Ryzhikov I N Nguyen Tien Quyet 2015 An approach to turbomachinery bladed disc durability estimation *Bulletin ISTU 5 (100)* pp 22-28

[12] Repetckii O V Ryzhikov I N Nguyen Tien Quyet 2016 Dynamics of gas turbine engines rotors taking into account non-linear effects *Vibroengineering PROCEDIA. 8* pp 361–365

Investigation of mistuning impact on vibration of rotor bladed disks

O Repetckii[1], **I Ryzhikov**[2,*], **Tien Quyet Nguyen**[2]

[1] Irkutsk State Agrarian University n.a. A.A. Ezhevsky, 664038 Irkutsk, Russia

[2] Irkutsk National Research Technical University, 664074 Irkutsk, Russia

[*] E-mail: rin111@list.ru

Abstract. Mistuning often reduces the fatigue life of bladed disks. The objective of this study is to determine the degree of influence of various types of mistuning on bladed disk vibration. It is also important to determine how the position of the detuned blades in the bladed disk affects the vibrations. The results of experimental and numerical analysis of mistuned bladed disk vibration are presented. The authors investigated the effect of various types of mistuning (geometry, mass, etc.) on the free vibrations of the bladed disk. The worst cases with minimum mistuning and maximum localization were identified. The developed algorithms for calculating of mistuned bladed disks vibration and obtained results can be used, when designing turbomachines rotors.

1. Introduction

The real bladed disks are not absolutely symmetrical constructions. There are always small differences in the blades in size, weight, their fixation in the disk. These differences are called mistuning. Mistuning arise due to inaccuracies in manufacturing, assembly, and due to operational defects. The mistuning leads to a change in the vibration spectrum and modes, to their localization, to stress increase and as a result to their fatigue life decreasing. Therefore, investigation of vibration behaviour of mistuned bladed disks is a very actual task.

Mistuning has been an active area of research for past more than 40 years. Analysis of mistuned bladed disks described in papers [1-17].

2. Fe analysis for cyclic symmetry systems with mistuning

Most vibration analyses of tuned bladed discs assume a model of only a single blade and it's associated segment of the disc. Often attempts are made to minimise the computation effort required to solve the mistuned assembly case and have been found to be effective tools, capable of representing the dynamics of real assemblies once their tuned system properties are known. Such approach is used in papers by Griffin [12]. Analyzing the eigenvalue problem of the system with cyclic symmetry without mistuning may be proceeded by solving the next equation:

$$K - \omega^2 M\delta = 0, \tag{1}$$

where K is stiffness matrix, δ is node displacements, ω is eigenfrequency of vibration system.

For a cyclic symmetry system (CSS) the difference of deflection between the two boundaries of finite element [18-20] is given by the equation:

$$\delta_{S+1} = e^{\mu}, \ \mu = i2\pi m / N, \tag{2}$$

where $i=\sqrt{-1}$, $S = 1,\ldots, N$, $m = 0, 1, 2\ldots, N/2$ for evens N and $m = 0, 1, 2\ldots, (N-1)/2$ for odds N.

To analyze the CSS with a mistuning effect the modified perturbation method was suggested in the work [11]. In the free vibration case for this method, we have equation:

$$\ddot{q} + \omega_B^2 A q = 0, \tag{3}$$

where $A = \begin{bmatrix} 1+2R^2+\Delta f_1 & -R^2 & 0 & \ldots & -R^2 \\ -R^2 & 1+2R^2+\Delta f_2 & -R^2 & 0 & \ldots \\ \ldots & \ldots & \ldots & \ldots & \ldots \\ \ldots & \ldots & \ldots & \ldots & \ldots \\ -R^2 & 0 & 0 & -R^2 & 1+2R^2+\Delta f_N \end{bmatrix}$,

where ω_B is nominal blade natural frequency without mistuning, ω_{B_i} is blade natural frequency of i-th blade, ω_C is coupling frequency, $R^2 = \omega_C^2 / \omega_B^2$ is dimensionless coupling, $\Delta f_i = (\omega_{Bi}{}^2 - \omega_B{}^2) / \omega_B{}^2$ is mistuning of i-th blade.

$$\omega_i^* = \omega_i + \Delta\omega_i + \Delta^2\omega_i. \tag{4}$$

Then eigenvalue of a mistuned system may be obtained as:

$$\omega_i^* = \frac{\omega_i^2}{\omega_{b_i}^2}, \tag{5}$$

where $i = 1,\ldots,N$, ω_i - vector of blade frequency without mistuning (for an ideal structure), $\Delta\omega_i$ - increasing vector of blade frequency with the first order mistuning, $\Delta^2\omega_i$ - increasing vector of blade frequency with the second order mistuning.

3. Numerical results

For case of geometrical mistuning we use blisk, presents on Figure 1 [16]. The length of one blade was changed.

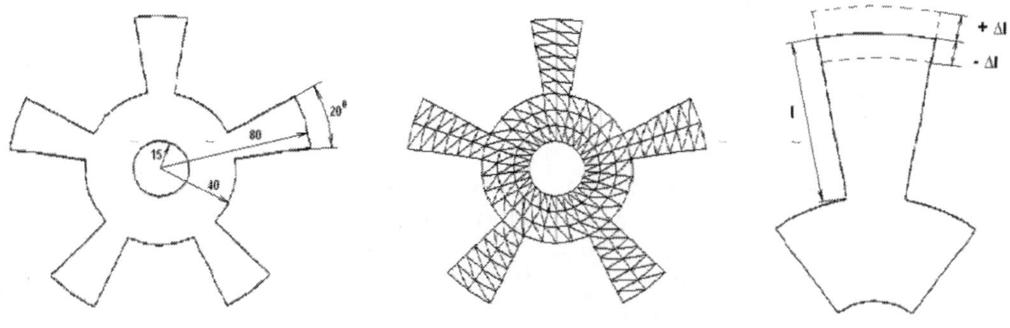

Figure 1. Geometrical mistuning of blisk.

For tuned blisk each group of modes has double modes with equal eigenfrequencies. For mistuned blisk these eigenfrequencies are splitted. One of these modes does not change. For other we can see changes of frequencies (Fig.2). We also obtain vibration localization even on 0,5% mistuning. With increase of mistuning strength the localization become more and more. And in case with 5% mistuning we have "worst case", when one blade vibrate, but other blades stay immovable (Figure 2).

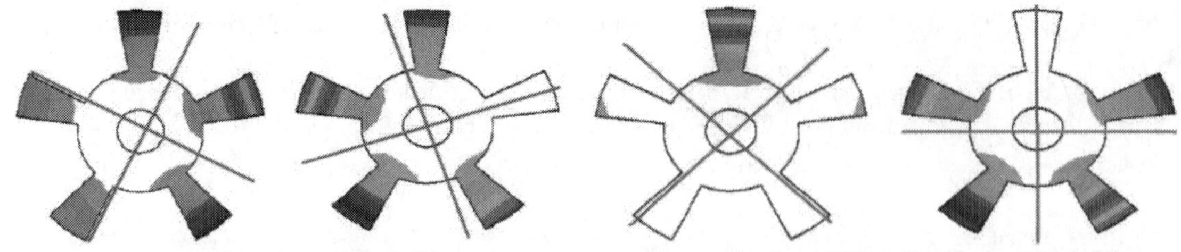

tuned ($f_1 = f_{2'} = 702{,}7\ Hz$) mistuned ($f_1 = 668{,}5\ Hz, f_{2'} = 702{,}7\ Hz$)

Figure 2. Geometrical mistuning of blisk.

Usually in real conditions not only one blade has mistuning, but all blades have differences from nominal eigenfrequency. On Fig. 3 there are results of calculations of blisk modes and eigen-frequencies of various variants of blades location on the disk. The variant with "saw-tooth" distribution has minimal differences in comparison with tuned blisk.

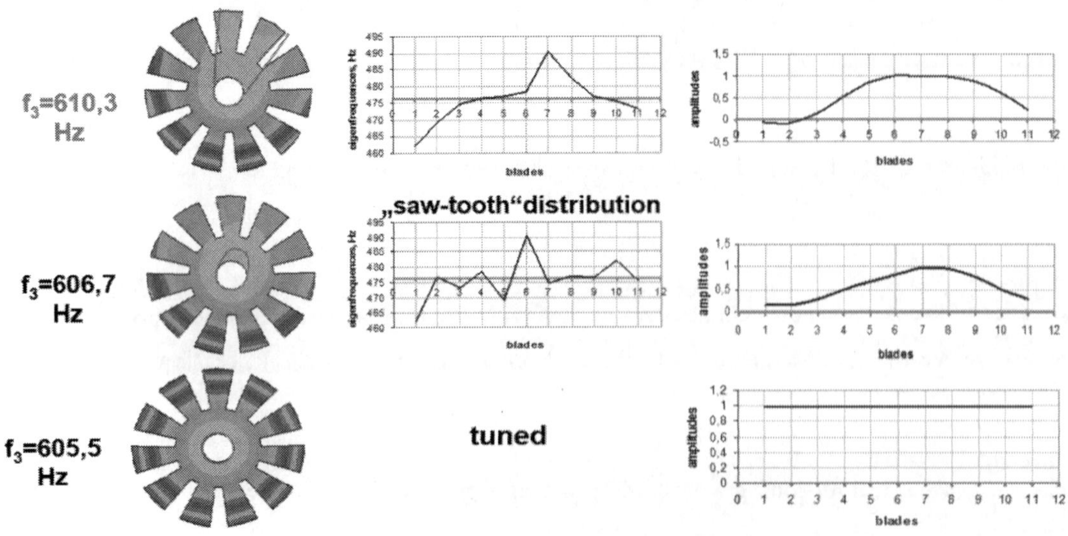

Figure 3. Sensitivity of the blade natural frequencies.

4. Experimental measurements of free vibration of mistuned blisk

An experiment on the vibration analysis of the mistuned blisk model was conducted. The experiment was conducted in the laboratory of the Institute of solid mechanics of TU Dresden. To measure vibration characteristics the laser scanning vibrometer PSV-400 was used. A flat impeller having a central hole was cut on high-precision laser equipment from sheet steel with a thickness of 2 mm. The model had 7 blades. Vibration measurements of the model without mistuning showed that the accuracy of its manufacture and fixation did not affect vibrations and model is almost symmetrical (Figure 4). The experimental results are in good agreement with the results of calculations in the programs ANSYS and BLADIS+ (Table 1).

Mistuning was introduced by joining the peripheral part of the blades of the segments of magnetic tape. The mass of each segment was measured on electronic scales to the nearest 0.001 g.

At the beginning, the additional mass was attached to only one of the blisk blade. The measurement results showed that even a small additional mass (0.1 g) introduces significant changes in the frequency spectrum and mode shapes of the model. For example, on the 2nd form there is a significant localization of the oscillations when only one blade vibrates, and the remaining almost are stationary (Figure 5).

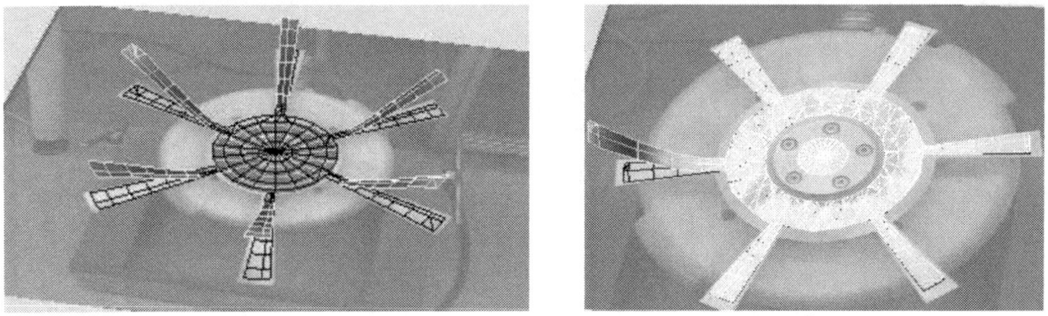

a b

Figure 4. (a) - first mode of tuned blisk vibration; (b) - localization on second mode of mistuned blisk vibration.

Was also investigated 2variants, where the additional mass was added to all blades of the blisk. In the first case, the magnitude of the additional mass uniformly increased from 0.2 to 1.2 g in the circumferential direction from blade to blade (0.2-0.4-0.6-0.8-1.0-1.2). In the second case, the mass in the circumferential direction was varied "sawtooth" (0.2-1.0-0.6-1.2-0.4-0.8). Comparison of results showed that in the first case, there is a significant localization of vibrations in contrast to the second case. Also for the first case the difference in frequency for the pair of mode shapes is more significant than for the second (Table 2).

Table 1. Natural frequencies of tuned model.

Mode	Experiment, *Hz*	ANSYS, *Hz*	BLADIS+, *Hz*
f_1	85.219	88.220	86.710
f_2	86.975	88.274	86.726
$f_{2'}$	87.250	88.274	86.726
f_3	92.375	90.268	89.007
$f_{3'}$	92.094	90.268	89.007

Table 2. Natural frequencies of mistuned model.

Mode	Variant 1, *Hz*	Variant 2, *Hz*
f_1	79.844	80.031
f_2	81.750	82.032
$f_{2'}$	83.063	82.344
f_3	88.719	89.219
$f_{3'}$	90.313	90.094

The actual continuation of this work is the sensitivity analysis and the optimal arrangement of the detuned blades in the axial bladed disk (Rolls-Royce plc. [17]), shown in Figure 5.

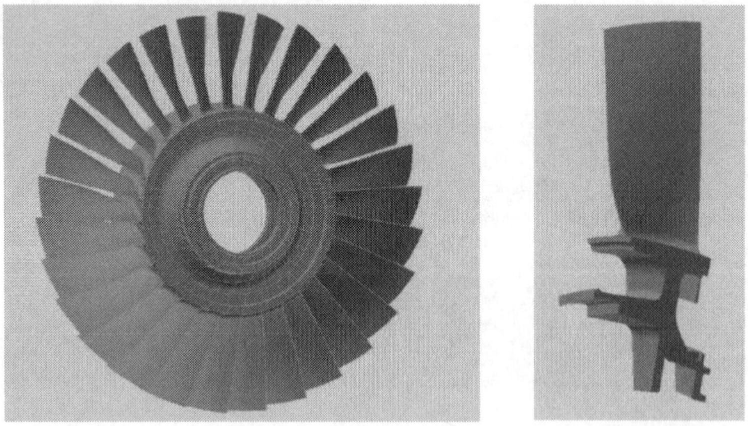

Figure 5. Rolls-Royce bladed disk and its single sector.

5. Conclusions

The mentioned results of the calculations and measurements demonstrate, that even very small mistuning (0,5%) lead to splitted modes and localization, when distortion of harmonic distribution law for circumferential amplitudes of blades is observed. Nodal diameters take broken line form. In case of the set of mistuned blades, minimal mistuning impact on blisk vibration is achieved by "saw-tooth" placement of blades in the disk.

Acknowledgments
The authors grateful to Prof. R. Schmidt (TU Dresden) for his valuable comments and suggestions.

References

[1] Yang M T and Griffin J H 2001 A Reduced-Order Model of Mistuning Using a Subset of Nominal System Modes *Journal of Engineering for Gas Turbines and Power* 123 pp 893-900
[2] Repetckii O V and Buy Manh Cuong 2010 To a question of the choice of a numerical method of the analysis of stresses at an assessment of a multi-cycle fatigue of blades of transport turbomachines *ISEA News 6* pp 153-158
[3] Repetckii O V Do Manh Tung 2014 Investigation of the characteristics of mistuned turbomachinery bladed discs vibration based on the reduced-order modeling by finite element method *Bulletin SibSAU 1 (53)* pp 60-66
[4] Repetckii O V Phan Van Tuan 2012 Construction of mathematical model for analysis of friction dampers influences on vibration of gas turbine engines blades *BSU News 1* pp 200-205
[5] Elovenko D A Repetckii O V 2011 Analysis of thermophysical properties of heat insulation materials for new constructions of cylindrical walls in high-pressure autoclaves *BSU News 6* pp 201-206
[6] Elovenko D A Repetckii O V 2011 Analysis of stress state of elastic half-plane loaded with constant pressure at limited intermediate sections with specified period by finite element method based on MSC Mars software *BSU News 5* pp 171-175
[7] Repetckii O V Ryzhikov I N Nguyen Tien Quyet 2015 An approach to turbomachinery bladed disc durability estimation *Bulletin ISTU 5 (100)* pp 22-28
[8] Heiman B Gerdt V Popp K Repetckii O V 2010 Mechatronics: components, methods, examples (Novosibirsk: Publishing House of the SB RAS) p 602
[9] Irretier H Repetckii O V 1998 Vibration and Life Estimation of Rotor Structure *IFToMM Conf. on Rotor Dynamics* (Darmstadt)
[10] Repetckii O V Ryzhikov I N Nguyen Tien Quyet 2016 Dynamics of gas turbine engines rotors taking into account non-linear effects *Vibroengineering PROCEDIA 8* pp 361–365

[11] Bladh R Castanier M P Pierre C 1999 Reduced Order Modelling and Vibration Analysis of Mistuned Bladed Disk Assemblies With Shrouds *Journal of Engineering for Gas Turbines and Power 121 July* pp 515-522
[12] Griffin J H 1988 On Predicting the Resonant Response of Bladed Disk Assemblies *Journal of Engineering for Gas Turbines and Power 110* pp 45-50
[13] Irretier H and Schmidt K J 1982 Mistuned Bladed Disks – Dynamical Behaviour and Computation *Proc. IFToMM Conference of Rotordynamics Problems in Power Plants* (Rome, Italy) pp 215-226
[14] Rao J S 1991 Turbomachine Blade Vibration (New Delhi: Wiley Eastern Limited)
[15] Sinha A 1986 Calculating the Statistics of Forced Response of a Mistuned Bladed Disk Assembly *AIAA Journal 24 11* pp 1797-1801
[16] Repetckii O V Ryzhikov I N Schmidt R 2010 Investigations of the influence of various types of mistuning on the vibrations and fatigue life of turbomachine bladed disks *Letter of the Baikal DAAD Fellowship Union 1* pp 18–23
[17] Repetckii O V and Beirow B 2016 Application of the analysis of sensitivity for the research of systems with mistuning *Baikal Letter DAAD 1* pp 88-97
[18] Repetckii O V 1990 Automation of strength calculations of turbomachines (Irkutsk: publishing house) p103
[19] Repetckii O V 1990 Computer analysis of the dynamics and strength of machines (Irkutsk: Irkutsk State Technical University Publishing House) p 301
[20] Repetckii O V Do Manh Tung 2012 Mathematical model operation and numerical analysis of vibrations of ideal cyclic and symmetric systems finite element method *ISEA News 3* pp 149-153

IOP Conf. Series: Journal of Physics: Conf. Series **944** (2017) 012098 doi:10.1088/1742-6596/944/1/012098

Solution of the problem of superposing image and digital map for detection of new objects

I S Rizaev, D I Miftakhutdinov and E G Takhavova

Kazan National Research Technical University named after A.N.Tupolev-KAI, Kazan, 420111, Russia

e-mail: isr4110@mail.ru, mdi_55@mail.ru, elzzy@yandex.ru

Abstract. The problem of superposing the map of the terrain with the image of the terrain is considered. The image of the terrain may be represented in different frequency bands. Further analysis of the results of collation the digital map with the image of the appropriate terrain is described. Also the approach to detection of differences between information represented on the digital map and information of the image of the appropriate area is offered. The algorithm for calculating the values of brightness of the converted image area on the original picture is offered. The calculation is based on using information about the navigation parameters and information according to arranged bench marks. For solving the posed problem the experiments were performed. The results of the experiments are shown in this paper. The presented algorithms are applicable to the ground complex of remote sensing data to assess differences between resulting images and accurate geopositional data. They are also suitable for detecting new objects in the image, based on the analysis of the matching the digital map and the image of corresponding locality.

1. Introduction

Currently, obtaining images of a terrain by different systems is not a problem: there are available the digital maps built sufficiently detail. However, if we have a fairly accurate map of a particular area and the image of this area, the task can be formed which is linked with discovering new objects in the region described by given map. These might be objects that were previously unknown. Let we have the map of certain area on the one hand and the images of this area made in different frequency bands. An important task may be formulated as follows: to do analysis by superposing them and to find out the differences. To improve the reliability of the results, the image of the area which has been obtained by various sensors may also be superposed with each other. The sensors may be operating on different physical principles or images can be received at different times. Also the following task is solved: to estimate various ways to superpose the obtained images [1, 2]. They may be the fragments of the images which are represented in different spectral ranges regarding the digital local map.

2. Mathematical description of the problem

In practice, the images of the same object or area obtained at different times or by different sensors can differ considerably from one another and from the digital local maps (DLM). Hence several important problems appear. They are the problem of superposing, problem of pegging and problem of precise mutual geometric correction and brightness correction. The solving of these problems should be the base for the subsequent joint analysis. In any case, this requires the establishment of the correspondence between the elements of the original images. It results in need to allocate support points in the image. In other words, they are so called bench mark or conjugate points. Pegging pictures on coordinates with accompanied simultaneous geometric correction [3,4,5] can be accomplished on using these conjugate points. Pegging may be done on the navigation parameters and using search algorithms that establish a correspondence between elements of images. Let us consider ways of superposing images from the DLM. Since the earth surface image can be obtained at a range of hundreds of kilometers it is necessary to consider the influence of the earth's sphericity [6] as it is shown and demonstrated on Fig.1. As was shown in [7], the angle of the center of the frame - γ, slant range to the center of the frame DLM and flight height of carrier (H) are linked each to other with expression (1), where D_{CF} denotes the center of the frame and R_E is the Earth's radius of the refraction, it is equal to 8,400 km.(1):

$$\sin(\gamma) = \frac{H}{D_{CF}} + \frac{D_{CF}}{2R_E} \tag{1}$$

Thus, the height of the carrier with respect to the center of the frame is given, according to [8], by expression (2):

$$H_C = D_{CF}\sin(\gamma) = D_{CF}\left(\frac{H}{D_{CF}} + \frac{D_{CF}}{2R_E}\right) = H + \frac{D_{CF}^2}{2R_E} \tag{2}$$

The value of the belittling of the point corresponding to the center of the frame [8] may be calculated by formula (3).

$$\Delta H = \frac{D_{CF}^2}{2R_E} \qquad (2)$$

Influence of the sphericity of the Earth must be considered when calculating the center of the view area position specified on the MSC, as well as when recalculating area image from one coordinate system to another to display on the monitor.

3. The algorithm for calculating the brightness values of the converted image on the base of navigation parameters

Let we have a rectangular image, which was received in the coordinates "slant range - azimuth". The offered algorithm is reduced to the transformation of this rectangular picture into the polar coordinate system, taking into account the sphericity of the Earth. In this coordinate system, the starting point of the coordinate system corresponds to the position of the carrier and the angular position of the element of image is counted relatively the north [9,10]. As the result of transformation, the image conversion from the Cartesian coordinate system to the polar the image on the screen is formed as the sector represented in Fig.2. To make the conversion for each screen pixel (r, c) the point on the earth's surface with the coordinates (x, y) is calculated and is associated with this pixel. For this point on the earth's surface the corresponding slant range and azimuth of targets are calculated. Based on the calculated value of the azimuth and slant range corresponding values of the number of the strobe and number of the azimuth reference number are calculated.

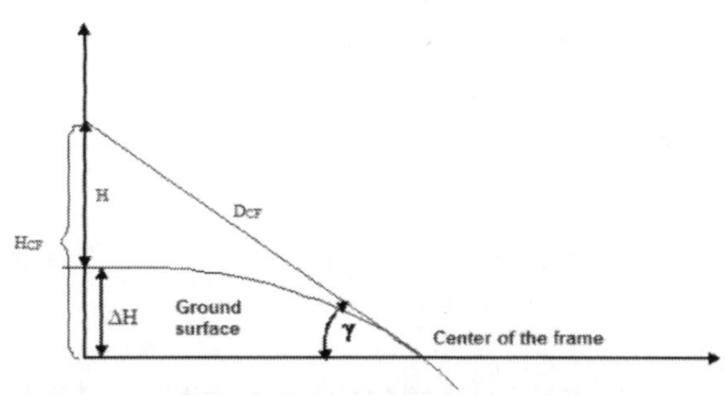

Figure 1. Influence of the sphericity of the Earth.

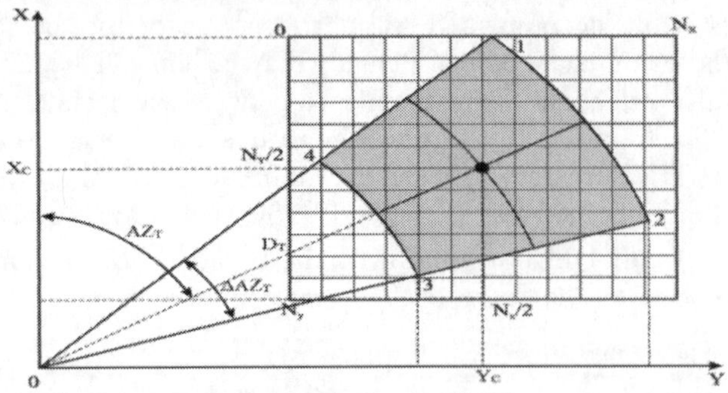

Figure 2. Conversion image of Cartesian coordinate system to the polar.

A strobe is time interval in the monitor's reamer to monitor, control, or post-processing. The calculation [11,12,13] is made as the following sequence of steps.

Step 1. The pixel's number is counted from the upper right corner. The position, which corresponds to it on the ground, may be calculated according to formulas (4) and (5)

$$X_{сзк} = X_I + \frac{zM}{100}\left(r - \frac{N_X}{2}\right), r \in [1; N_X] \tag{4}$$

$$Y_{cur} = Y_I + \frac{zM}{100}\left(c - \frac{N_Y}{2}\right), c \in [1; N_Y] \tag{5}$$

Step 2. On the basis of current coordinates of the position the current horizontal distance to the point is calculated [17] by using formulas (6) and (7):

$$D_{cur}^2 = (X_{cur} - X_C)^2 + (Y_{cur} - Y_C)^2 \tag{6}$$

$$D_{cur} = \sqrt{\left(D_{horcur}^2 + H^2 + \frac{D_{horcur}^4}{4R_E^2}\right)\frac{R_E}{R_E - H}} \tag{7}$$

The angular position of a point in the imaging sector counting from the left corner of the image is equal to the result of (8).

$$AZ_{cur} = arctg2(x, y) - AZ_T + \frac{\Delta AZ_T}{2} \tag{8}$$

Step 3. The value of image reference corresponding to the current point is calculated according to the formulas (9) and (10) [14].

$$N_D = \frac{D_{cur} - D_I}{\delta D} + \frac{N_{CN}}{2} \tag{9}$$

$$N_\varphi = \frac{AZ_{curN}}{\Delta AZ_T} N_{AZ} \tag{10}$$

In formulas (9) and (10) X_I and Y_I are values of the coordinates of the center of the selected area to form an image in the cartographic coordinates. In the above formulas, the following notations are used:

- X_C, Y_C – coordinates of a carrier in a plane of rectangular coordinates in meters;
- Z_T - the azimuth of targeting relatively the direction of the north;
- D_T - slant range of targeting until the image center of the frame;
- H – the height of the flight the carrier relative to the surface of the Earth;
- N_X, N_Y – the dimensions of the image area of cartographical information displayed in pixels;
- Z – resolution of the indicator;

IOP Conf. Series: Journal of Physics: Conf. Series **944** (2017) 012098 doi:10.1088/1742-6596/944/1/012098

- ∂D – image resolution of the range;
- N_{CN} – the number of the channels of the range;
- N_{AZ} – counts the number of images in azimuth;
- ΔAZ_T – the size of viewing area in azimuth;
- M_{DLM} – the scale of display the map.

If the number of reference on the range or azimuth is outside the existing image values there no image is at the given local point. In this case this pixel is set to 0 [15]. Since the calculated values of the respective sample images are real numbers, the resulted transformed coordinates do not fall mostly in the discrete lattice sites. Therefore, to determine the brightness of the image at the given point of the screen it is necessary to make the appropriate interpolation values in the array "slant range - azimuth" by one of the existing methods [16].

4. The algorithm for calculating the brightness values of the converted image on the base of bench marks

The placement of reference points is carried out manually by the operator. Example of arrangement bench marks is shown in Fig. 3. Consistently placing the appropriate bench marks in the DLM are (X_{RIi}, Y_{RIi}). On the image of the area they are represented as (X_{Ii}, Y_{Ii}), so we get arrays of coordinates of bench marks [7]. Let we have k conjugate points which are placed on the image. On the basis of k-spaced conjugate points vectors of coordinates of reference points for local area are formed represented by expressions (11) and (12) [9, 17].

$$X_{RI} = \begin{pmatrix} X_{RI1} & X_{RI2} & \dots & X_{RIk} \end{pmatrix}^T \tag{11}$$

$$Y_{RI} = \begin{pmatrix} Y_{RI1} & Y_{RI2} & \dots & Y_{RIk} \end{pmatrix}^T \tag{12}$$

Also a special matrix of bench marks of the DLM is formed as it is shown in expression (13).

$$H = \begin{pmatrix} 1 & x_1 & y_1 & x_1^2 & x_1y_1 & y_1^2 \\ 1 & x_2 & y_2 & x_2^2 & x_2y_2 & y_2^2 \\ 1 & x_3 & y_3 & x_3^2 & x_3y_3 & y_3^2 \\ \vdots & \vdots & \vdots & \vdots & \vdots & \vdots \\ 1 & x_k & y_k & x_A^2 & x_ky_k & y_k^2 \end{pmatrix} \tag{13}$$

Thus the placement of reference points and formation of the vector coordinates of the image points of the terrain and formation of the matrix of the points of DLM are finished. Thereafter a calculation of the corresponding point of the image for each screen pixel by the formulas (14), (15) is performed [20,21].

$$N_D = a_0 + a_1r + a_2c + a_3r^2 + a_4rc + a_5c^2 \tag{14}$$

$$N_\varphi = b_0 + b_1r + b_2c + b_3r^2 + b_4rc + b_5c^2 \tag{15}$$

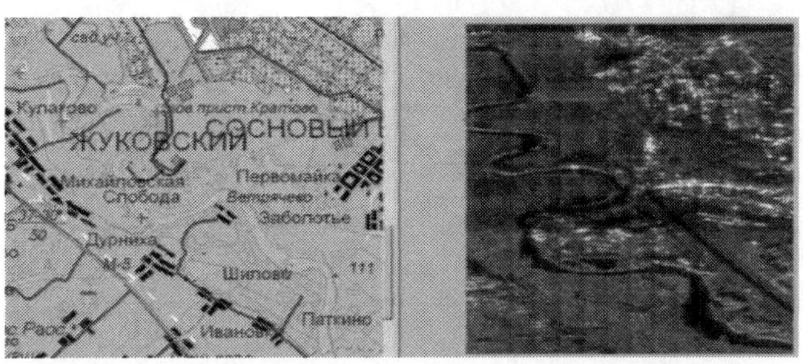

Figure 3. Example of arrangement bench marks.

The transformation parameters are the polynomial coefficients as represented in expressions (16):

$$A = (a_0 \quad a_1 \quad \dots \quad a_5)^T \qquad B = (b_0 \quad b_1 \quad \dots \quad b_5)^T \tag{16}$$

They are calculated according to formulas (17), where the superscript T denotes transposition of matrix.

$$A = (H^T H)^{-1} H^T X_{RI}, \qquad B = (H^T H)^{-1} H^T Y_{RI} \tag{17}$$

Similarly, the image conversion algorithm for navigation settings to obtain the integer values of samples, an interpolation of image samples is necessary to do [18,19].

5. Experimental results

The case of superposing the converted image area in a given coordinate system with a choice of transparency [20,21] is considered. It was performed by using two described algorithms. When superposing the converted image of the area and the DLM, the operator may choose the level of transparency of image layers, it is equal to α. The level of transparency of image layers may be ranged from 0 to 1. The value 0 means fully transparent, when only DLM is available. The value 1 means fully opaque, when only the image area is available. Fig. 4 shows the examples of the superposition of DLM and the local image of the image of the same area with different transparency on the base of navigation parameters. Selection of the transparency coefficient is often complicated by the fact that when superposing each of the layers prevents to the image of other layer [22,23,24]. When transparency of the image is greater, DLM significantly interferes with perception. Low transparency is hampered analysis of DLM. This mode is appropriate particularly in the review of terrains when a detailed, thorough analysis of objects is not required. In this case, the entire object structure and is displayed and there is a possibility to analyze the entire image. Consider overlay vector layers of DLM onto converted bitmap of areas by the algorithm calculating the brightness values of image on the base of placed bench marks. In this case, the image appears opaque, and above it layers of the electronic map are presented.

Fig.5 shows cases overlay maps in the image of the area. On the right frame on Fig.5 some of the layers, preventing analysis of the image, are not displayed. Thus, it can be argued that for superposing the image of a local area and DLM using of the algorithm for calculating brightness values of the converted image on the original image on the navigation parameters is permissible for cases when a detailed clarification of the location of objects is not required. For specification position of the image of area on the DLM it is better to apply the algorithm for calculating the brightness values of the converted image on the original image on the base of placed bench marks.

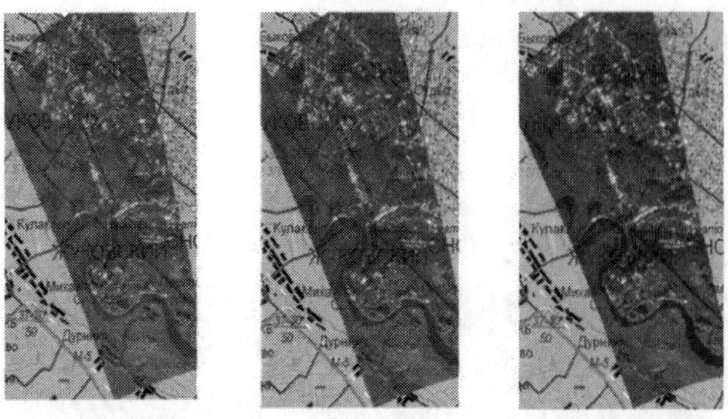

a) b) c)

Figure 4. Examples of superposing DLM and the image with different coefficients of transparency: a) α= 0,7); b) α= 0,8; c)α= 0,9.

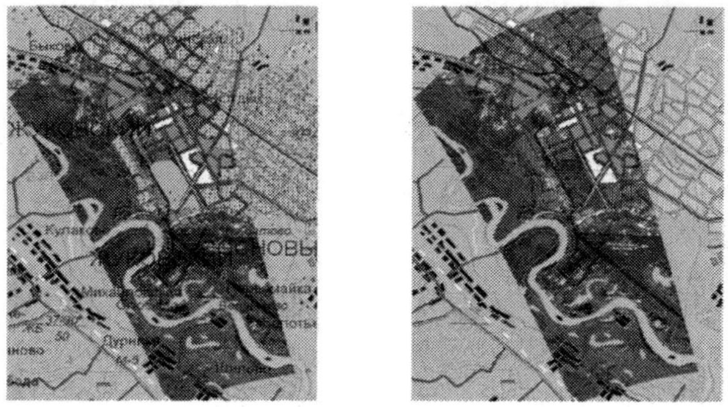

Figure 5. The options imposition of map on the image of the area.

6. Conclusion

For high-quality superposing images obtained by various means of observation with map and among themselves it is necessary to eliminate linear and non-linear distortions of the images associated with the features of the systems functioning. To eliminate the distortions it is advisable to use a polynomial transformation, the parameters of which are formed on the basis of current navigation parameters and characteristics of observing systems.

In the calculation of the image transformation parameters obtained at considerable distances (tens hundreds of kilometers), it is necessary to take into account the sphericity of the Earth. Influence of the sphericity of the Earth must be considered when calculating the position of the center of the view area specified on the DLM, as well as in recalculating image from one coordinate system to another when displaying on the monitor. The transformation parameters by placing the conjugate points by an operator and run on the basis of their additional polynomial transformations may be updated. This gives opportunity to do analysis of superposing digital maps and information given by observing systems.

7. References

[1] Ablameyko S V 2000 *Image processing: technology, methods, applications* ed S V Ablameyko, D M Lagunovsky (Minsk: Amalfeia) p 304
[2] Gashnikov M V, Glumov N I, Ilyasova N Yu and Myasnikov V V 2003 *Methods of computer image processing* ed. V A Soifer (Moscow: FIZMATLIT) p 784
[3] Kirpichnikov A P, Miftahutdinov D I and Rizaev I S Solution of the problem geopositioning by correlation matching method 2015 *Bulletin of Kazan Technological University* **18** (Kazan) pp 226-228
[4] Kirpichnikov A P and Miftahutdinov D I 2015 The solution of the correlation problem snapping images and digital maps *Bulletin of Kazan Technological University* **18** (Kazan) pp 186-189
[5] Miftahutdinov D I and Rizaev I S 2015 Features of the image matching algorithms with digital maps *The II International scientific-practical conference* (Stavropol) pp 44-48
[6] Rizaev I S 2013 *Geographic information systems* (Kazan: Publishing house Kazan state tehn.un Press) p 139
[7] Baklitskiy V K 2009 *Correlation-extreme methods of navigation and aiming* (Tver: TO "Book Club") p 360
[8] Gonzalez R and Woods K 2012 *Digital Image Processing* (Moscow: Technosphere) p 1104
[9] Pyt'ev Y P and Chulichkov A I 2010 *Methods of morphological image analysis* (Moscow: FIZMATLIT) p 336
[10] Umnyashkin S V *Theoretical Foundations of digital signal processing and presentation* 2009 (Moscow: Technosphere) p 368

[11] Uelstid S 2003 *Fractals and wavelets for image compression in action* (Moscow: Triumph) p 320
[12] Fissenko V T and Fissenko T Y 2008 *Computer processing and image recognition* (St. Petersburg: ITMO) p 192
[13] Bykov R E, Fryer R, Ivanov K V and Mantsvetov A A 2003 *Digital image conversion* ed R E Bykov (Moscow: Hotline-Telecom) p 228
[14] Jane B 2007 *Digital image processing* (Moscow: Technosphere) p 584
[15] Stewenius H, Engels C and Nister D 2006 Recent developments on direct relative orientation. *ISPRS Journ. of Photogrammetry and Remote Sensing* **60** pp 284–294
[16] Longuet-Higgins H C 1981 A computer algorithm for reconstructing a scene from two projections *Nature* **293** pp 133–135
[17] Miftahutdinov D I and Rizaev I S 2016 Refinement of parameters superposing images by placing control points *Proc. of the International Scientific and Practical Conference January 20, 2016, Kurgan* (Ufa: AETERNA) pp 51-53
[18] Berthold Horn K P 1990 Recovering Baseline and Orientation from 'Essential' Matrix *Journ. of the Optical Society of America* (Citeseer, January) pp 1–10
[19] Konstantinov K A, Talalaev A A, Tishchenko I P, and Khachumov V M 2009 The concept of architecture of neuronet space appointment system of control, diagnostics and image processing. *Aerospace instrumentation* **5** pp 39–47
[20] Tishchenko I P, Emelyanova Yu G, Timoshina N E, Gvozdyakov Yu A, Poltavets Yu I and Krylov I A 2011 Design of intelligent graphic interface to control and diagnose space subsystems *Proc. of the 3rd All-Russian conf. "Relevant problems of missile and space instrumentation and information technology"* (Moscow) pp. 272–281
[21] Tishchenko I P, Stepanov D N and Fralenko V P 2012 The development of modeling system of unmanned aerial vehicle autonomous flight *Program Systems: Theory and Applications* **3** 3 pp 3–21
[22] http://cgm.computergraphics.ru/issues/issue17/invariant_features
[23] Bradski G and Kaehler A 2008 *Learning OpenCV* (Sebastopol: O'Reilly Media, Inc.) p 576.
[24] Bay H, Ess A, Tuytelaars T and Van Gool L 2008 Speeded Up Robust Features (SURF) *Computer Vision and Image Understanding* **110** 3 pp 346–359

The dynamic improvement methods of energy efficiency and reliability of oil production submersible electric motors

V S Romanov[1,2] and V G Goldstein[1]

[1]Samara State Technical University, 141, Galaktionovskaja ave., Samara, 443000, Russia
[2]Romanov_v.s@mail.ru

Abstract. In the organization of production and operation of submersible electric motors (ESP), as the most essential element of electric submersible plants (ESP) in the oil industry, it is necessary to consider specific operating conditions. The submersible electric motors (SEM) as most essential element of electrosubmersible installations (EI) in oil branch accounting of operation specific conditions is necessary in the process production and operation. They are determined by the conditions under which the EPU is operated. They are defined by the EPU operation conditions. For a complete picture the current state of the SED fleet in oil production, the results of its statistical analysis are given. For a comprehensive idea of the SEM park current state the results of statistical analysis are given in oil production. Currently, assessed the performance of submersible equipment produced by major manufacturers. Currently the operational characteristics assessment of the submersible equipment released by the main producers is given. It is stated that standard equipment cannot fully ensure efficient operation with the help of serial EIs, therefore new technologies and corresponding equipment are required to be developed. It is noted that the standard equipment could not provide fully effective operation by means of serial ESP therefore new technologies development and the corresponding equipment are required.

1. Introduction

The total volume of oil produced in many regions of Russia by means of electric submersible units (EI), driven by a submersible electric motor (SEM), is more than 80%. The application of this method is economically justified both for new deposits and for those that are under intensive exploitation for a long time. When choosing the method of operation of wells, many factors, such as viscosity, the content of mechanical impurities, the gas factor, the economic feasibility of the equipment used, must be taken into account. Based on the classification and comparative analysis of statistics on the park of the main SEM in domestic oil production, it can be argued that more than 20% of the wells from the entire fund in Russia are idle, since standard equipment has no opportunity to operateeffectively using serial EPUs of both domestic and expensive foreign Production. Therefore, the development of new technologies and corresponding equipment is required. The main goal of the work is to study the modern SED fleet in the oil industry of Russia with the development of specific proposals for increasing the energy efficiency and reliability of the entire technological process of oil production, the key element of which is the energy industry. The study is based on the current scientific approaches to the use of EPU in oil production. The theoretical basis for studying the issues of the work presented was the work of domestic and foreign specialists. The study of questions of theory and methodology in the EPU of oil extraction was devoted to the studies of such authors as A.N. Kitabov, V.Ya. Charonov, T.A. Atakishev, RV Babayev, A.A. Baryudin, V.P. Tokarev, S.S. Shubin. From the point of view of the analysis of the condition of the submerged electrical equipment park, it is worth

IOP Conf. Series: Journal of Physics: Conf. Series **944** (2017) 012099 doi:10.1088/1742-6596/944/1/012099

noting the large and painstaking work to collect data on the operation of the SEM, which was carried out at JSC "Samaraneftegaz"by V.P. Sotnikov.

2. Formulation of the problem

Proceeding from the foregoing, there is a need to create and study EPUs for operation in complicated conditions with high efficiency in the basic modes, with a wide range of feed rates and the possibility of adjusting the head of the facility, which ensures the synchronization of the parameters of the system "formation-well-facility". Reduction of the mass-dimensions parameters of the underground part of the installation will significantly increase the possibility of unobstructed passage of curved sections of wells with complex geometry and small casing dimensions. Most of the regions of oil production in Russia are remote areas with sharply continental climatic conditions. Remoteness of oil fields from equipment manufacturers entails an increase in the costs of delivering and storing a large range of bulky EPUs, which once again [1, 2], emphasizes the need and relevance of reducing weight and size characteristics. It is also necessary to minimize losses in submersible motor line supplying from the surface, to ensure automatic adaptation to changing operating conditions and optimization of an operating mode of a pumping unit. For this, an effective system for monitoring the parameters of the EI is required, which allows reducing the intervals between repair intervals. It should be noted that an important aspect in the selection and introduction into production of the equipment is a minimization of the cost price of SEM.

3. Theory. Technical analysis of ESP structures

For a more complete and effective presentation of the current state of SED used in domestic oil production at present, it is necessary to analyze the operating data of their fleet. The diagram in figure 1 clearly shows the distribution of the main producers of the SEM market, used in Russia.

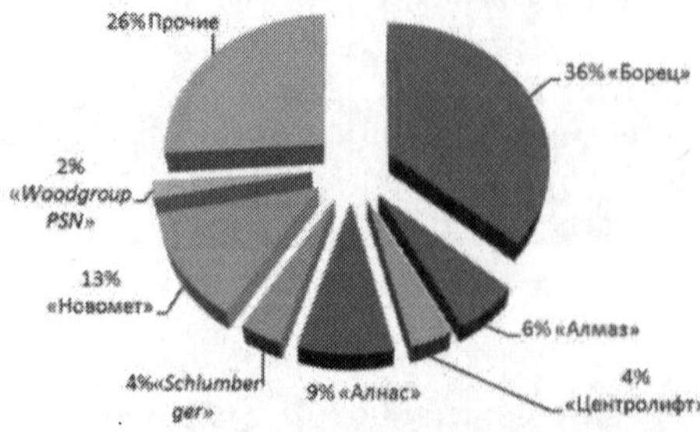

Figure1. The main producers of the SEM market in Russia.

The FED of Borets is a three-phase short-circuited oil-filled bipolar manual winding motor. The rotation speed of the asynchronous bipolar motor at full load is 3000 rpm at a power frequency of 50 Hz and 3600 rpm at a power frequency of 60 Hz. The engines are filled with a high level of cleaning dielectric oil, which provides lubrication of bearings and thermal conductivity. The engine is able to withstand extreme temperatures both on the surface and in the well, which, as a rule, are a reason of damage to the EI and can even cause premature failure of a pump.

"Almaz" company produces oil-filled SEM for EI of SEM-Y series with a power from 22 to 360 kW and parametric submersible three-phase oil-filled series RPPED-YA (Russian parametric SED designed by Professor NV Yalovega). The main differences between the RPPED-Y and the conventional ones are smaller dimensions, stable operation over a wide range of supply voltages, regulation of a rotor speed in a wide range by changing a supply voltage, increased power, large starting torque, stable operation with significant changes in the supply voltage. The engines are available in sizes 117 and 103 mm, in heat-resistant and conventional designs. Sectional engines are present in the nomenclature series. Power of the engines is from 28 to 340 kW.

Novomet CJSC produces three-phase, short-circuited, oil-filled SEM versions of various modifications (for operation in complicated operating conditions, with installed immersion control units of the plant parameters, etc.). SEMs are available in one-, two- and three-sectional layouts and are in the form of a design version according to the method of mounting 1M3631, designed for continuous operation of S1 GOST R 52776-2007 from AC mains with a frequency of 50 Hz. The engine consists of a stator, a rotor, a head, and a base. The stator is made of a tube into which the magnet core is pressed. The magnet core is made of sheets of electrical steel with heat-resistant coating. The stator winding is single-layered, stretched, reeled, made of heat-resistant winding wire with insulation from polyimide-fluoroplastic films. The phases of the winding are connectedas a star. The rotor is squirrel-cage, sectional. Advantages of SEM: reliability of construction, application of high-temperature materials, long operating time, provision of technical requirements of customers, continuous improvement of production processes.

OJSC Alnas produces certified complete submersible pumping units for oil production, reservoir pressure maintenance, various ground equipment, including control systems. The certificate for this equipment is issued by BureauVeritasQuality. OJSC "Alnas" confirms its compliance with the requirements of ISO 9001: 2000. The manufacturing technology determines a high quality and reliability of the SEM. The stator is executed with the closed groove increasing the purity of the internal cavity of the engine, it allows successfully use grooving insulation in the form of a tube. Winding wire of the stator insulation company "DuPont". In the SED rotor, the original bearings are used, which have a mechanical fixation from the rotation and thus retain the possibility of easy movement along the axis of the shaft. Using special electrotechnical materials makes it possible to operate SED at a formation fluid temperature of up to 120 ° C, and in superthermally resistant version - up to 160 ° C.

Woodgroup PSN produces SED and multistage centrifugal pumps. The main advantages: high resistance to abrasive wear, shaft material is made of nitronic 50, which significantly increases the corrosion resistance of the unit, a number of innovative measures aimed at reducing the failure of the EPO. Weatherford Company, whose representative offices operate in all major oil and gas regions of the world, produces highly efficient electric motors and multi-stage centrifugal pumps, which use modern speed controllers for a drive. The main advantage is the minimum requirements for an installation on the surface, especially in low-temperature conditions. The percentage of EPL failures is significantly reduced, as a result of the use of innovative materials and structures.

The company Reda (Schlumberger) produces an innovative quick-connect design of SED, hydroprotection and sensors of the REDA Maximus system. Main advantages: there is no need for responsible assembly operations, that can be affected by weather conditions; Execution of installation operations, such as adjustment of shafting, oil filling of electric motors and hydraulic protection in controlled conditions of the company's production and service centers; There is no need to install a current lead isolation on an outfall; components of the Maximus system are ready for installation to a well; Reliable installation in an extremely low ambient temperature; Reduction of the work time for a facility installation; Increased period of operation of the facility; Optimum performance.

BakerHughes (a division of Centrilift) produces EPU for operation in complicated conditions. Advantages of these designs are: modularity; Possibility of replacing individual units of the installation without disturbing the alignment of the shafts; A significant reduction in repair work; Ecological compatibility, in sum of absence of leaks in oil-filled EI units, low noise level and absence of system vibration; Work in high-temperature conditions; Effective monitoring systems; The possibility of exploitation in conditions of high sand content, etc.

4. The discussion of the results

Typical principles of innovative changes in modern EIs and, above all, submersible electric motors (SEM) are: a use of a valve drive, an increase in the rotational speed of 3000-6000 rpm and nominal voltage, schematic and parametric changes in structures, etc. They are strategically aimed at increasing their Energy efficiency and energy saving, increase in between-repair intervals. Each of the selected strategies of innovation changes in the EI and in particular, SEM has its advantages and disadvantages. However, when choosing a method for operating a well, it is necessary to individually consider each of the proposed methods based on local conditions for obtaining the maximum economic effect. One of the most promising areas of energy saving and energy efficiency improvement for oil and gas producing enterprises is the increase in between-repair periods (MCI) of the EPU [1] and, first of all, for SEM is a minimization a number of emergency failures and violations. Reliability and efficiency

IOP Conf. Series: Journal of Physics: Conf. Series **944** (2017) 012099 doi:10.1088/1742-6596/944/1/012099

of operation of SEM [2, 3] depend to a large extent on the organization of maintenance and repair (MRO), which are implemented in the oil industry by two main approaches:
• by the operation time of the EPU (or its individual components) with reference to a known standard group (for example, SEM 63-117) with terms and composition of work established for it [2];
• according to the current technical condition, which is determined by the comparison of the continuous monitoring data (which is currently being introduced in the industry) and technical diagnostics with an established boundary data of the diagnostic parameters [3 ± 5].

In the first case, MRO includes routine maintenance and preventive maintenance of the equipment in accordance with regulatory deadlines, which are established on a basis of guidance documents and operational experience for a particular type of SED and its life cycle phase. At the same time, a condition of a SEM is checking according to the preventive principle when performing maintenance, current and major repairs.

This approach has a number of significant drawbacks, primarily related to technically unjustified costs. Thus, for the formation of a volume and composition of preventive maintenance and repair works [2] for individual SEMs and their groups, works for MRO and SED, and their components with different time schedules, are often combined for organizational and technical reasons. A number of objects are included in the MRO total works and deadlines are assigned. This leads to underutilization of electrical installation resources and, in addition, to a downtime due to unnecessary long running and lifting operations [2].

At the same time, here are quite typical a shortage of oil, a significant reduction in MCI and, as a result, an increasing losses. It can be stated that MRO on the operating time as a whole is determined by the type of SEM and does not take into account the individual features of the operation of SEM in the conditions of specific wells. In addition, even with high-quality maintenance, with careful disassembly, detailed analysis and replacement of worn out components and parts, if necessary, it is not possible to avoid violations of their compatibility (which increases the possibility of new defects and violations) and ensure the absence of sudden failures in MCI. Efficiency of EPU operation can be improved by generally organizing MRO with the definition of time, composition and schedule for the current state of SEM. In other words, taking into account a rational use of internal reserves of durability of its individual elements and in general of the whole EPU, as a technical resource. It should be noted that in this organization of maintenance and repair, as in the previous case, underutilization of electrical installations takes place, since when performing repair of a specific element of EPU, it is expedient to expand the composition of work for efficient components that have significant wear. At the same time, it is necessary to provide a full-scale determination of an actual technical condition of the SEM by an available composition of diagnostic parameters (DP), sufficient in content, quality and reliability of information [4-6]. Further, based on comparison of the DP with their boundary permissible values, takes a decision on the possibility of further operation of the installation and the required volume of maintenance.

The exception of this approach is an elimination of the consequences of sudden failures. In these situations, the analysis of a pre-emergency condition of the SEM is especially important. Since it allows obtaining necessary information on dynamics of defect development with the purpose of their operative diagnosis, localization, prediction and, as a result, reduction of emergency damage [3-6].

5. Conclusions

Based on the analysis of operation data for the SAM fleet of OJSC Samaraneftegaz, it has been established that the planned repairs and preventive maintenance of the equipment in accordance with the scheduled terms have technically unjustified costs. They are caused by the fact that a number of objects are included in the overall composition of MROs earlier than previously assigned terms. This leads to underutilization of electrical installation resources and shortage of oil. Organizing maintenance for the current state check of SEM, taking into account the rational use of their internal reserves of durability, in a number of cases, the resources of the submersible equipment are also insufficiently use. To increase efficiency, a full-scale determination of an actual technical condition of the SEM is necessary on the basis of detailed diagnostics.

The foregoing can be generalized by the following conclusions.

1. Analysis of serial EPU produced by domestic and foreign manufacturers shows that in complex conditions of modern oil production they do not fully meet the high demands of modern efficient operation. It is necessary to develop new technologies and related equipment.

2. It is necessary to widely introduce EPU for operation in complicated conditions with high efficiency in modes with a wide range of feed rates and possibility of regulating the head of the installation with the provision of parameters synchronization of the "formation-well-installation" system.

3. Reduction of mass-dimensions parameters of an underground part of EPO will significantly increase a possibility of unimpeded passage of curved sections of wells with complex geometry and small casing dimensions.

4. In the organization of operation and management of SEM life cycle parameters of the petroleum industry, the integrated organization of MRO is appropriate, using jointly both the operating data and the current status as the main provisions.

The high energy intensity of the oil production sector, the need to reduce energy costs and increase the efficiency of the oil industry, as well as the short service life of the SEM require system accounting and processing of operational data for specific submersible electrical equipment based on information technologies.

References

[1] Brinner T R and Bulmer J 1985 *Kellg Lighting protection for submergible oilwell pumps*vol 34 (Houston tex.: 32-nd Annu. Petrol. and Chem. Ind. Conf.) p 91

[2] Colombo S, Costa G andPiccarreta L1988 Results of an investigation on the overvoltages due to a vacuum circuit breaker when switching an H.V. motor*IEEE Trans. On Power Delivery***3** p 205

[3] Ford D V 1972 The British Electricity Boards national fault and interruption reporting scheme – objectives and operating experience *IEEE Trans. Power Appar. Syst.* **5** pp 34-52

[4] Gibbs J D, Koch D, Malkin D P and Cornick K L 1989 Comparison of performance of switching technologies on E CIGRE motor simulation circuit *IEEE Transaction on Power Delivery***4** pp 89-100

[5] Hurstell V L and West M G 1959 Shielding 13,8 kV. Distribution Circuits *Power Appar. Syst.***45** pp 134-162

[6] IEEE Working group progress report 1981 Impulse voltage strength of ac rotating machines*IEEE Trans. on PAS***8** pp 89–100

[7] Ivanovo V N 2011 Oil production: the main directions of optimization of energy consumption *Engineering practice***6** 196

[8] Kitabov A N and Tokarev V P 2011 *Information-measuring system for diagnostics of submersible electric motor* (Ufa: Bulletin of the UGATU. Electronics, measuring equipment, radio engineering and communications) pp 163-164

[9] Matsui T, Yokoyama E and Umeya E 1988 Resignation current interruption characteristics of the vacuum interrupters*IEEE Transactionon Power Delivery***3** pp 1672-1677

[10] Rano R, Haramada H and Kurosawa Y 1975 Switching surge phenomena in induction motor windings and their endurance *Hitachi Review***24** pp 225-232

[11] Short T A and Ammon R H 1999 Monitoring Results of the Effectiveness of Surge ArreserSpacins on Distribution Line Protection*IEEE Trans. On Power Delivery***14**pp 1142-1150

[12] Telander S H, Wilhelm M R andStump K B 1987 Surge limiters for vacuum circuit breaker switchgear *IEEE Transaction on Power Delivery***2** pp 458-462

[13] Veno K, Toya H and Murai Y 1984 Monte–Carlo simulation of overvoltage generation in the inductive current interruption by vacuum interrupters *IEEE Trans.***103** pp 498-504

[14] Zamihovsky L M and Kalyavin V P 1999 *Technical diagnostics of submersible electrical installations for oil production*(Ivano-Frankivsk: Hospitality University of Oil and Gas) p 234

[15] Zotos P A 1988 Motor failures due to steep fronted switching surges: The need for surge protection – user's experience*IEEE Transaction on Industry Appl.***30** pp 886-895

IOP Conf. Series: Journal of Physics: Conf. Series **944** (2017) 012100 doi:10.1088/1742-6596/944/1/012100

The techniques of quality operations computational and experimental researches of the launch vehicles in the drawing-board stage

K Rozhaeva

Omsk State Technical University, 11, Mira ave., Omsk, 644050, Russia

e-mail:rozhaevakseniya@gmail.com

Abstract. The aim of the researchis the quality operations of the design process at the stage of research works on the development of active on-Board system of the launch vehicles spent stages descent with liquid propellant rocket engines by simulating the gasification process of undeveloped residues of fuel in the tanks. The design techniques of the gasification process of liquid rocket propellant components residues in the tank to the expense of finding and fixing errors in the algorithm calculation to increase the accuracy of calculation results is proposed. Experimental modelling of the model liquid evaporation in a limited reservoir of the experimental stand, allowing due to the false measurements rejection based on given criteria and detected faults to enhance the results reliability of the experimental studies; to reduce the experiments cost..

1. Introduction

In this study, at the stage of scientific research as the main characteristics of the object, which is included in its quality, the reliability criterion of scientific resultsis applied.

To assess the reliability of the research results is to establish the degree of closeness of numerical and experimental studies results. Taking into account such factors affecting the final result as: different conversion, assumptions and simplifications in the mathematical and physical models, using the wrong dimensions, human factors, external conditions the carrying out of the experiment, which lead to inaccuracy and various sorts of errors that require minimization.

The methods of ensuring the theoretical results reliability were used at the stage of research works by using known solutions for the process under study and the licensed software, that confirmed with experiments and a satisfactory convergence of the calculated and experimental data.

The analysis of the current level of theoretical and practical achievements in the field of quality operations in the early design stages of complex technical systems by Russian scientists: V. Vasil'ev, A. Akhtulov, A. Al'breht, D. Golovin, and foreign scientists: B. Anderson , V. Roshan Joseph , Harriet Black Nembhard, showed that there is a need for improved approaches, concepts and methodologies of quality operations theoretical and experimental studies, that is, the creation of the generalized characteristics of the research results, to evaluate their objectivity, depth, novelty and usefulness [1-5].

When creating the elements of space-rocket technologies, not having prototypes, the basic requirements to the system are in the formative stage, and appropriate research methods require additional refinement. Accordingly, the base for comparisons, as a rule, is absent, and respectively the probability of incorrect design decisions is large enough [6].

Thus, there is a need for experimental studies to substantiate the objectivity and reliability of the developed techniques of research and development based on that the technical specification for the next design stage will form, in particular, on the preliminary design.

The practical implementation involves considerable financial and time costs, so the minimization of error (methodological, mechanical and others) is extremely important.

2. Theory

When conducting theoretical and experimental studies of the functioning process of the active on-Board descent system is proposed to the systematize and adapt existing methods of error search for the task are proposed; use the system of reliability criteria of the obtained results (theoretical and experimental) at each stage of the research based on fundamental laws that identify methodological, technical dysfunctions in the preparation of mathematical and physical models; during theoretical and experimental research to provide parallelism for their comparability (e.g., in parallel with the experiment on the stand real to perform calculations in the computer using results from the physical measurements from the stand); in the preparation of mathematical and physical models with the introduction of assumptions into account to make this error; mandatory verification of the mathematical model.

Figure 1 shows the procedure algorithm for ensuring the quality of theoretical and experimental studies. The proposed techniques covers the mathematical and physical modelling, each of which presents a set of mechanisms to enhance the reliability of research results, which ultimately ensures the quality of design techniques of the real system.

During the equations integration of the mathematical model for the control of the current results, the criteria ensure the reliability of the results obtained based on respect for fundamental laws is used: the first law of thermodynamics, the second law of thermodynamics [7].

1. The temperature of the heat transfer is subject to the second law of thermodynamics showing the direction of the processes:

$$T_{HC} \geq T_{gas} > T_{liq} > T_{plate} > T_{glass} > T_{steel}, \quad (1)$$

where T is the temperature of the heat carrier, gas, plate, glass and steel walls of the experimental model stand.

Since the heat carrier has the highest temperature and is the only source of heat transfer.

2. The heat carrier thermal energy is equally distributed between the participants of heat transfer, according to the law of internal energy change of each participant:

$$U_{HC} > U_{gas} > U_{liq} > U_{plate} > U_{glass} > U_{steel}, \quad (2)$$

where $U=cm\Delta T$ – is the internal energy of the heat transfer participants: heat carrier, gas, glass and steel walls.

3. According to the energy conservation law, all deposited energy of the heat carrier in an experimental model stand is consumed for the internal energy change of the heat transfer participants and between the participants:

$$Q_{HC} = Q_{gas} + Q_{liq} + Q_{glass} + Q_{steel} + Q_{plate} - Q_{dr}, \quad (3)$$

4. The gas pressure inside the experimental model stand should be more atmospheric:

$$p_o < p(t) < p_o + \Delta p \quad (4)$$

where p_o = 101325 Pa,, $\Delta p = \frac{G^2 * p}{2\mu^2 * F^2}$

5. There will be no the mass growth of the liquid inside the stand:

$$\dot{m}_{liq_{max}} > \dot{m}_{HC} - \dot{m}_{dr} - \dot{m}_{liq} > 0 \quad (5)$$

where $\dot{m}_{HC}$ – the mass flow rate of the heat carrier. $\dot{m}_{dr}$– is the mass flow rate of the gas at the outlet of the stand, $\dot{m}_{liq}$ – istherateofliquidevaporation.

The proposed criteria are the basic conditions of reliability of calculated results and allow us to judge the correctness of the mathematical model and the basis for the termination of simulation and search, elimination of mistakes in the calculation algorithm.

The term "gasification" is used in the development of active on-Board descent system of the launch vehicle spent stage, and the term "evaporation" – when the process modelling at the experimental stand.

The proposed techniques includes the following steps aimed at bringing into line the given conditions of experimental studies and evaluation criteria process [8-9].

1. We provide additional measurement of the heat carrier flow velocity in the characteristic points of the experimental model stand, gas humidity at the outlet of the experimental model stand and calculated based on the measurements the values of the total heat received by the plant size during the experiment according to the formula:

$$Q_{\Sigma}^{0} = c_{HC} \dot{m}_{HC} T_{HC} \tau, \tag{6}$$

where c_{HC} – is the heat carrier specific heat (a table value); $\dot{m}_{HC}$ – is the measured heat carrier flow rate; T_{HC} – the measured temperature of the heat carrier,τ – is the measured time of the experiment.

2. Calculated total value of heat spent for heating each object model of the experimental stand involved in the heat transfer during the experiment according to the formula:

$$Q_{\Sigma}^{*}=Q_{\Sigma}^{*}=Q_{gas} + Q_{liq} + Q_{plate} + Q_{glass}\,, \tag{7}$$

and compare the calculated values of Q_{Σ}^{0} with $Q_{\Sigma}^{0} c Q_{\Sigma}^{*}$.

In the case of the condition implementation:

$$\left|Q_{\Sigma}^{0} - Q_{\Sigma}^{*}\right| \geq E_{Q}, \tag{8}$$

where E_Q includes instrumental and methodical errors, the results of the experiment admit the invalid, stop the experiment and identify faults in the measurement system and data logging.

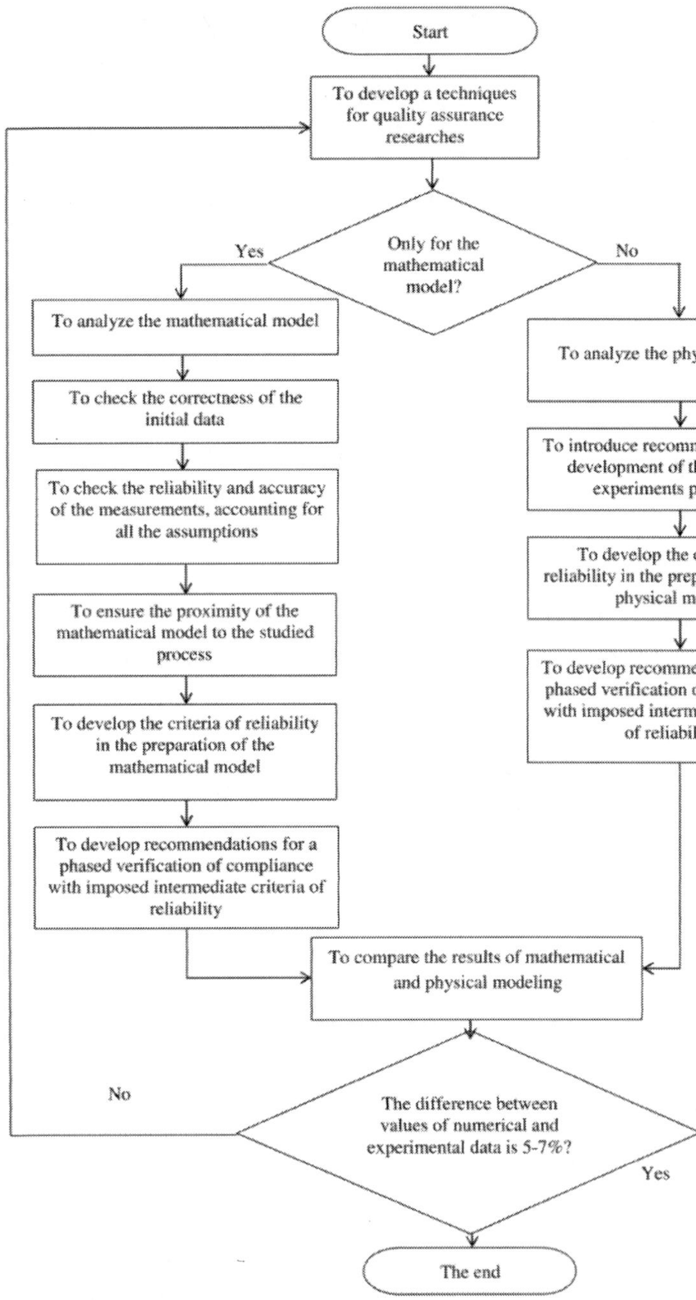

Figure 1. The algorithm of the techniques to ensure quality of theoretical and experimental studies.

3. In the experiment at the current time must be adhered to the second law of thermodynamics, to check used the temperature sensor, compares the temperature of the heat carrier T_{HC} and T_i elements of the experimental model stand:

$$T_{HC} \geq T,\ i=1,\ldots,N, \qquad (9)$$

where N – is the number of installed temperature sensors inside the experimental model stand.

In case of conditions dysfunction (9) the experiment is terminated and the detected fault in the measurement system and data logging.

4. For steady state convective gasification of the liquid in which the gas temperature inside the

experimental model stand is permanent $\frac{dT_{gas}}{d\tau} = 0$, sensors determined the gas temperature T_{gas} and the gas pressure p_{gas} inside the stand.

In case of adiabaticity conditions dysfunction p_{gas}/T_{gas} the experiment is terminated and the detected fault in the measurement system and data logging.

5. Based on the measurements of relative gas humidity ϕ in the volume of the experimental model stand the experimental partial pressure at the gas temperature T_{gas}in the experimental model stand is calculated:

$$p_p^* = \varphi p_{gv} \tag{10}$$

where p_p^* – is the partial gas pressure in the experimental model stand;p_{gv}– is the gas vapour pressure in the experimental model stand, which is calculated in accordance with the actions.

Received the partial gas pressure in the experimental model stand is compared to the tablep_p^0, defined in accordance with the actions:

$$|p_p^0 - p_p^*| \geq E_p \tag{11}$$

where E_p includes instrumental and methodical imprecisions.

In case of conditions dysfunction (11) the experiment is terminated and the detected fault in the measurement system and data logging.

3. Discussion

Proceeding from the above, achieved the following results:

- Due to rejection of invalid measurements and the time-detected faults increases the accuracy of research results;
- Timely detected fault;
- Reduces the cost of experiments.

Energy costs are decreased by 20-25%. This reduced work time consumption.

Reduced labour cost due to:

- The termination of the "gas-liquid" system experiment when the temperature becomes stationary;
- Reducing the time of preparation of the heat carrier (the heating time to a predetermined temperature) due to the insulation of the electric heater, and connecting fittings.

Thus, the proposed quality ensure techniques of research can significantly improve the accuracy of the results.

4. Conclusion

The use of the proposed techniques of quality ensure the experimental studies on early stages of design to avoid a number of methodological (mathematical model), algorithmic (the compilation of algorithm and program of calculation) and mechanical errors (use of incorrect values of physical constants, dimensions, and other mechanical errors in the preparation of the initial data).

Acknowledgements

The research was supported by the grant of the Ministry of education and science of the Russian Federation "Improvement of environmental safety and economic efficiency of launch vehicles with main liquid rocket engines", No. 9.1023.2017 PCH

Scientific supervisor – doctor of technical Sciences, Professor, Professor of the 'Aircraft and rocket building' Department of Omsk state technical University V. Trushlyakov.

The author thanks doctor of technical Sciences, Professor of the 'Life Safety' Department of Omsk state technical University A. Akhtulov

References

[1] Anderson B and Haller H 2011Combining computational fluid dynamics analysis data withexperimentally measured data *Quality Engineering*vol **23**no 1 pp 46–58

[2] Black-Nembhard H andValverde-Ventura R 2003Integrating experimental design and statistical control for quality improvement*Journal Of Quality Technology*vol **35**no 4 pp 406–426
[3] Akhtulov A, Leonova A and Akhtulova L 2013 Methods of assessing the quality of design processes of complex technical devicesvol **3***The Journal Omsk Scientific Bulletin*pp 87–91
[4] Golovin D and Zyong K 2007 *Quality assurance in the technological design of complex aircraft structures* vol **11** (Omsk: Polet) pp 24–29
[5] Vasil'ev V, Kobzar A and Sholom A 2012 Management of quality parameters of promising rocket and space technologyvol **8***Industrial and manufacturing engineering* pp 61–65
[6] Trushlyakov V, Kudentsov V and Lesniak V2012 *A method for modelling of gasification residues of liquid rocket fuel in the tanks of the separated part of the tread of the launch vehicle and device for its implementation*Patent Russian Federation no 2461890 p 9
[7] Rozhaeva K2015 Method of improving the quality of research. Mathematical modelling*Standards and quality*vol **1** p 75
[8] Rozhaeva K 2017Evaporation of a Model Liquid vol **37***Russian Engineering Research* pp 1–4
[9] Rozhaeva K and Trushlyakov V2016 Developing a Test Stand to Verify Accuracy of Conducted Research at the Early Design Stages of Active Descent Systems of Spent Launcher Stages vol **9***Indian Journal of Science and Technology*pp 1–7

IOP Conf. Series: Journal of Physics: Conf. Series **944** (2017) 012101 doi:10.1088/1742-6596/944/1/012101

Kinetic synergistic transitions in the Ostwald ripening processes

I N Sachkov, V F Turygina and A N Dolganov

Ural Federal University named after the first President of Russia B.N.Yeltsin, Yekaterinburg, Russia

e-mail: v.f.volodina@urfu.ru

Abstract. There is proposed approach to mathematical description of the kinetic transitions in Ostwald ripening processes of volatile substance in nonuniformly heated porous materials. It is based upon the finite element method. There are implemented computer software. The main feature of the software is to calculate evaporation and condensation fluxes on the walls of a nonuniformly heated cylindrical capillary. Kinetic transitions are detected for three modes of volatile substances migration which are different by condensation zones location. There are controlling dimensionless parameters of the kinetic transition which are revealed during research. There is phase diagram of the Ostwald ripening process modes realization.

1. Introduction

Thermal Ostwald ripening process is one of the essential mechanisms of substance migration. The process consists of the fact that material located in the hot zone of temperature heterogeneous systems evaporates, and its vapors migrate and then condense on cooler surfaces. The effect of the Ostwald ripening on layer formation was found in technologies such as microelectronics [7], laser radiation effect [8], protoplanetary cloud material formation [9], nuclear power objects [10], etc.

However, several features of process associated with diffusion interaction of evaporation and precipitation zones remained outside the field of view of researchers. Heretofore, there are no approaches that allow to predict the kinetic synergistic transitions of Ostwald ripening process in non-uniformly heated capillary-porous medium.

2. Task setting

A non-uniformly heated cylindrical cavity is considered as an element of the capillary-porous medium. A volatile matter is placed on the hot base of the cavity. This matter vaporizes and diffuses into the volume of the cavity and in background gas medium. Then the vapor condenses on the cooler areas of cavity surface.

The cavity radius is equal to R_0 and its length – L_0. Using a cylindrical coordinate system (R,Y) the origin of coordinates is aligned with the center of more heated cylinder base. The Y axis is directed along the cavity axis. In this case we proceed to dimensionless variables in equations (1), (2), (3) shown in figure 1. Limitations in the present section are given by studying the features of the processes which have changing gas and cavity surface temperature, T, monotonically changes along Y from T_e (evaporating end) to T_{con} (condensing end) and does not depend on the radius.

$$r = \frac{R}{L_0} \quad (1)$$

$$y = \frac{Y}{L_0} \quad (2)$$

$$r_0 = \frac{R_0}{L_0} \tag{3}$$

3. Theory

The features of Ostwald ripening are being researched which realized in a cylindrical cavity with a quadratic temperature changing law along the Y axis as in equation (4).

$$\theta(y) = \theta_{ey} - \Delta\theta(1 - A_2)y + A_2 y^2 \tag{4}$$

where A_2 – quadratic rate parameter ($-1 \leq A_2 \leq 1$).

Three ways of vapor diffusion are expected and shown in figure 1. γ is a condensation only on the cold end of the cavity, α – precipitation on the base and on the whole side surface of the cylinder, β – at the end and part of the side surface. Let us determine the conditions under which the discussed modes α, β and γ are realized. We establish distribution of precipitation and condensation fluxes on the cavity surface.

For the purpose it is necessary to calculate concentration field of the volatile component (its partial pressure) P (r, y). The solution of the problem is possible by using the finite element method. Previously, the method shows great flexibility in describing processes of transport and phase transformation realized in spatially heterogeneous systems [1-5]. Algorithms is the basis of developed computer software presented in excellent book referenced by [6].

Taking the variation approach, according to which the diffusion equation is equivalent to extreme functionality condition as in equation (5), [1-6].

$$X = \int_{V_c} D\,(grad\ P)^2\, d\,V \tag{5}$$

where V_c – cylindrical cavity volume, D – diffusion rate of volatile component vapors in background gas medium as in equation (6)

$$\mathrm{D} = \mathrm{D_0} \times \mathrm{T}^{3/2} \tag{6}$$

D_0 – practically independent of temperature constant.

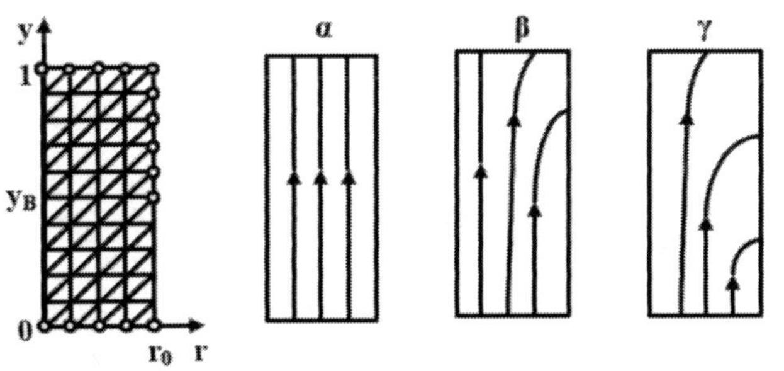

Figure 1. Dividing workspace on simplex-elements schemes and diffusion lines of Ostwald ripening modes patterns α, β and γ. White circles – boundary conditions implementing of the first type; y_B – condensation and clean surface areas boundary.

The finite element method [1-6] is used to find the field function P(r, y) that minimizes the function in equation (5). The area of two-dimensional space is sampled by triangular simplex elements. The uniform grid divides volume cross-section into triangular elements in figure 1. The pressure is approximated by a linear function within each triangular element as in equation (7).

$$P^e = \alpha^e\, r + \beta^e\, y + \gamma^e \tag{7}$$

The volume differential stated in equation (8) helps to take into account the problem spherical symmetry.

$$\mathrm{dV} = 2\pi r^e\, \mathrm{dr\, dy} \tag{8}$$

where r^e – radius of gravity center of triangle "e". Thus, the contribution of each element out of "e" volume to the functional X is defined by equation (9).

$$X^e = 2\pi\, r^e\, D^e \int_{S^e} (\mathrm{grad}\, P^e)^2\, dr\, dy \tag{9}$$

Here S^e – square of the triangular element, $D^e = D\,(r^e, y^e)$ – diffusion coefficient calculated at the center area of "e"-element temperature.
There are three types of inner surfaces: evaporating, condensing and free from migrating matter. The boundary conditions are established for them. The boundary conditions of the first type are established for nodes which are located on evaporating and condensing end (i.e. at $y = 0$ and $y = 1$ respectively) and also on the side surface at $r = r_0$ and $0 < y < y_B$ (where condensing process takes place – white circles in figure 1). It means that nodal values P_j are equal to saturated vapor pressure P_{ev}. The polyterma P_{ev} (T) are described by Clapeyron's equation (10).

$$P_{ev} = B_o \exp(-1/\theta) \tag{10}$$

$$\theta = \frac{R_g T}{Q_{ev}} \tag{11}$$

where θ – dimensionless temperature, Q_{ev} – evaporation heat, R_g – gas constant, T – local surface temperature.
Functional extremum conditions of equation (5) lead to a system of linear algebraic equations for the set of nodal pressure values $\{P^i\}$. The procedure of its solution described pretty well by the reference [6].
As a result, continuous distribution P(r,y) is replaced by the piecewise linear relationship. The values of fluxes density, especially precipitation and evaporation, determined by value $D\, grad\, P^e$.
There is a problem in using of the approach. The problem is to find condensing area boundary and "clean" surface on the side surface of cylinder under condition of adiabatic process. An iterative process is used to find the value of y_B. It is important that the amount of density of condensing flux is a continuous function of coordinates and goes to zero at $y = y_B$. The start value of boundary must be defined as well. Next step is to calculate the pressure P (R, Y) and the flux density D (P/R) at the point of surface with coordinate y_B. If the flux value is not equal to zero, the value y_B will be changed by a certain amount (for example, by 5%) and estimated at new boundary point. Further, new approximate value y_B is calculated using two coordinates values, their respective fluxes and pruner's method. The iterative process continues until the flux density is equal to zero on the boundary (or becomes less than defined value). It is necessary to transform partition grid on each step of process to increase accuracy of calculations by the following way: one of grid lines passed through the boundary node at y_B coordinate.
The above statements become the basis of computer software which allows to investigate profiles of condensing and evaporation fluxes on the ends and the side surface of the cavity, to gain the information about irregularity of thickness of formed layers and about mass flow ratio condensing on the side surface and the cold end. Source code is written in the Fortran language.

4. Results of computer experiments

Calculations of concentration fields and flux density of condensation and evaporation on the inner surfaces of the cavity are performed at the varying values θ_{ev}, θ_{con}, A_2 and r_0. The calculation results show that they greatly depend on the relative cavity radius r_0 and temperature parameters $\Delta\theta$ and θ_{ev} as in equation (12). If the control dimensionless parameter is the same the results are slightly different.

$$\Delta\theta = \theta_{ev} - \theta_{con} \tag{12}$$

where $\Delta\theta$ – relative maximal temperature differential.
The calculation results are summarized in the form of a chart in figure 2. It represents stability conditions of the three modes of the Ostwald ripening.

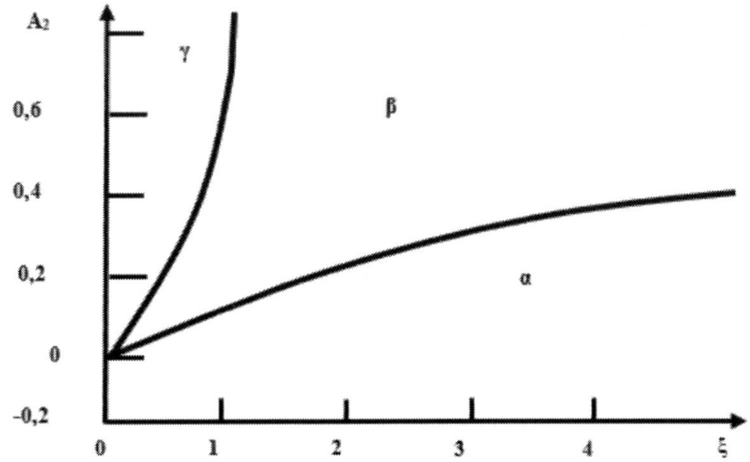

Figure 2. Charts of stability of the Ostwald ripening modes; A_2 – temperature field quadratic parameter, ξ – dimensionless temperature differential as in equation (13), $r_0 = 0,1$.

$$\varepsilon = \frac{\Delta\theta}{\theta_{ev}^2} \quad (13)$$

5. Discussion of results

Analysis of result shows that the mode α is only possible if $A_2 = 0$ is strictly linear law of temperature variation along Y-axis. The condensing process takes place along the side surface of the capillary. If $A_2 > 0$ then the movement of vapors takes place along the capillary axis without condensing on its side surface at low relative temperature gradients ξ. The increase of ξ results to the formation of condensation zone on the side surface adjacent to the cooler end of the capillary. Further increase of temperature gradient makes mode α more stable. In the similar way kinetic transitions may occur as a result of variations of the parameter square of the temperature field A_2.

Thus, changes in the control parameters ξ and A_2 lead to kinetic transitions of the Ostwald ripening modes α, β and γ.

The features of their implementation are in figure 3. The representative profiles of relative deposition flux densities on the side surface of capillary are in figure 3 and in equation (14).

$$q_B = -D\frac{\partial P/\partial r}{q_0} \quad (14)$$

where D – diffusion coefficient, P – pressure of diffusing vapors, q_0 – density of evaporation flux at the center of evaporating end.

In figure 3 curves 1-4 correspond to values of dimensionless temperature differences ξ which is equal to 0.5, 1.0, 2 and 4 for curves 1-4 respectively. Square parameter is $A_2 = 0.2$.

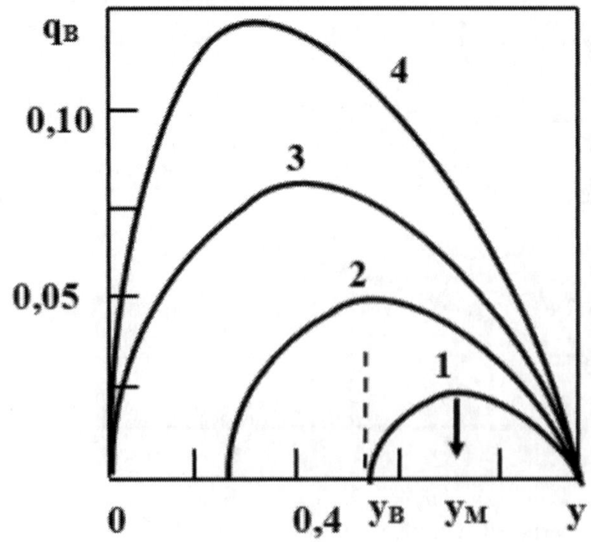

Figure 3. Transformation of profile forms of relative flux density of vapor deposition from the relative distance from the hot end depended on relative temperature difference value.

Thus, dependencies 1-2 correspond to β mode and 3-4 – to α mode. In case of small values of dimensionless temperature difference ξ there is no deposition on the side surface of the capillary – γ mode. Specific points are highlighted in the plots. Dashed line indicates position of the boundary of the condensation zone of the side surface y_B. The coordinate defines the formation of process modes. The arrow indicates axial coordinate corresponding to the maximum vapor deposition rate. Its value sets the position of the capillary section on which its occlusion is possible.
The results are illustrated by analyzing the conditions of condensing process on the side surface using simplified one-dimensional model. It is realized when the relative radius of the capillary is $r_0 \rightarrow 0$. In this case the diffusion process is grown to one-dimensional one, the vapor pressure of migrating matter linearly depends on axial coordinate as in equation (15) and the value of saturated vapor pressure is described by equation (10).

$$\frac{\partial P_{dif}}{\partial y} = -D \quad (15)$$

The ratio $P_{ev} > P_{dif}$ is the condition of condensation process. It means that the pressure of saturated vapors should be higher than the pressure provided by the diffusion.
Relationship schemes of discussed vapor pressures from the distance to the evaporating end corresponding to different process modes are stated in figure 4. In situation α the pressure of saturated vapors is higher than the pressure of diffusion and the deposition over the whole length of the capillary takes place on the side surfaces. In situation β the diffusion occurs at the beginning of the capillary and the deposition takes place on the cooler side of the capillary. In situation γ the mode of free escape of vapors from the capillary is formed.

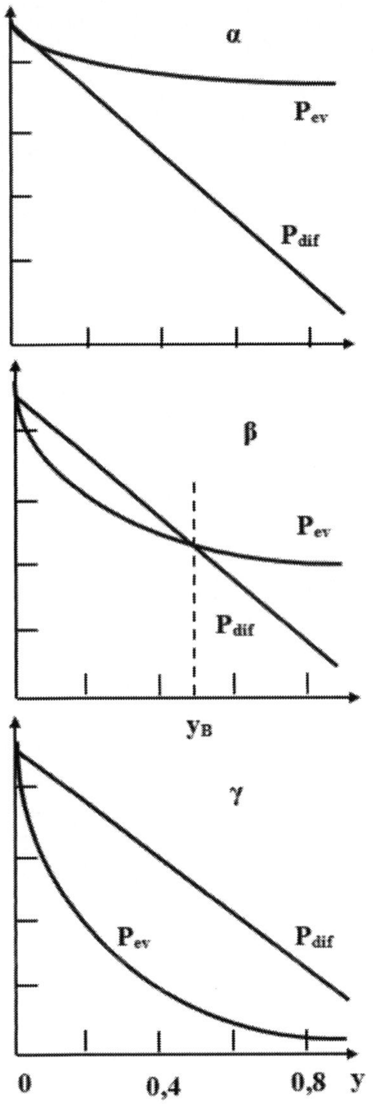

Figure 4. Dependence of vapor concentration on distance from evaporating end under conditions of free diffusion P_{dif} and saturation P_{ev}.

Thus, non-uniformly heated capillary-porous body from which evaporation of volatile matter takes place represents nonlinear dynamic system. During the growth of the temperature differential playing the role of the driving force of the process, critical phenomena occur. As the result of the flow of kinetic transitions, the picture and the intensity of vapor motion are significantly change.
Even more significant process transformations may be expected if the channel plugging capability is taken into account. The discussed phenomena are possible in α and β modes. In this case, the capillary is divided into two volumes by lengths $y_1 = y_M$ and $y_2 = (y_0 - y_M)$. It should be noted that new values of dimensionless temperature differentials ξ_1 and ξ_2 are characteristic of them. In new conditions, the processes of evaporation and condensation are started. The discussion of their laws goes beyond the scope of the present work.

6. Conclusions
The proposed approach to the mathematical description of synergistic kinetic transitions can be used to describe the processes of the migration of volatile substances in non-uniformly heated porous systems. The possibility of existence is observed in the following laws.

- It should be noted that in real porous medium there may be inclusions of different thermal conductivity. This can lead to substantial non-uniformity of the temperature field, i.e. to variations in the values ξ and A_2, and to local changes of migration mode.
- Realization of α and β modes should lead to blockage of capillaries as a result of growing of their walls. Therefore, it is possible to form extended condensation zones tied to certain points of capillary body in non-uniform porous medium.
- During the developing of α mode, it is possible for "plugs" to appear which divide the original capillary into two or more shorter ones.

Further development of research in the discussed directions is planned.

References

[1] Sachkov I N 1995 *J. High temperature* **33**:5, 759
[2] Sachkov I N 1996 *Russian J. of Phys. Chemistry* **70**:1, 111
[3] Sachkov I N and Chemerinskaya L S 2008 *Russian Phys. J.* 11(2) 165
[4] Sachkov I N, Marinova O, Turygina V F and Turygin E E 2015 The effect of the geometry of the micro pores on the effective permeability of soil (Rodos, Greece) p 4
[5] Sachkov I N, Marinova O, Turygina V F, Turygin E E and Dolganov A N 2016 Regularities of non-uniform heating of the two-phase environment during electric current *16th International Multidisciplinary Scientific GeoConference SGEM* vol 3(1) pp 599–606 DOI: 10.5593/SGEM2016/B13/S05.076
[6] Segerlind L 1984 *Applied finite element analysis* (New York: Wiley)
[7] Pavlyukevich N V, Gorelik G E and Levdansky V V 1980 *Physical kinetics and transport processes during phase transformations* (Minsk) p 208
[8] Anisimov S I, Imas Ya A, Romanov G S and Khodyko Yu V 1970 *The effect of high-power radiation on metals* (Moscow: Nauka) p 272
[9] Dorofeeva V A, Makalkin A V and Mironenko M V 1993 *Recondensation of matter in a protoplanetary disk* (Moscow: Nauka) pp 45–73
[10] Degaltsev Yu G, Ponomaryov-Stepnoy N N and Kuznetsov V F 1977 *The behavior of high-temperature fuel upon irradiation* (Moscow) p 145

IOP Conf. Series: Journal of Physics: Conf. Series **944** (2017) 012102 doi:10.1088/1742-6596/944/1/012102

Chaotic dynamics of two beams described by the kinematic hypothesis of the third approximation in the case of small clearance

O A Saltykova[1,2], I V Papkova[1] and V.A.Krysko[1]

[1]Saratov State Technical University, Politehnicheskaya str, 77, Saratov, Russia
[2]Cybernetic Institute, National Research Tomsk Polytechnic University, 634050 Tomsk, Lenin Avenue, 30, Russia

Abstract. The chaotic dynamics of the contact interaction of two beams described by the third-approximation hypothesis (the Pelekh--Sheremetyev model) are studied in the paper. There is a small clearance between the beams. One of the beams is subjected to the action of a transversal harmonic load. Contact pressure is determined by Kantor's method. The mathematical model of the beam structure taking into account the geometric nonlinearity and contact interaction. The system of partial differential equations reduces to the ordinary differential equations by the finite differences method with the approximation of the second order. The resulting system is solved by the Runge–Kutta methods of various orders. The chaotic vibrations of two beams were investigated by the methods of nonlinear dynamics. The reliability of the obtained results was grounded. The Lyapunov exponents are calculated by three different algorithms – Kantz, Wolf, Rosenstein. The 2D and 3D phase portraits and the Fourier power spectra, the Poincaré pseudo section were constructed. The phenomenon of frequency synchronization has been detected.

Keywords – Contact interaction, 3d approximation beam, nonlinear dynamics, finite difference method, Runge–Kutta methods, geometric nonlinearity.

1. Introduction

Beams and beam structures are widely used in modern industry, machine building, rocket engineering, etc. Often such structures are subjected to various external dynamic influences. The study of nonlinear dynamics and contact interaction of beam structures is a very important issue at the present stage development of science. In the Russian and foreign scientific literature we can find a large number of works devoted to the study of Euler–Bernoulli [1], Timoshenko [2], Peleh–Sheremetev [3, 4] beams models, but there are no papers on the contact interaction of beams of third-order. A third-order mechanics model is used in this paper. The hypothesis of the third approximation is commonly called the Reddy hypothesis [5] in the foreign literature. But this theory was first described in [6]. It was shown in [4]. The Pelekh–Sheremetev hypothesis makes it possible to maximize the model to a three-dimensional and to obtain the most accurate results. In 1964 two Ukrainian scientists M.P. Sheremetev and B.L. Pelekh [6] proposed a third-order theory. This theory was applied to the calculation of beams of plates and shells in publications [7]. Twenty-seven years later, in the works of Levinson [8] and Reddy [5, 9], this model was reopened. Thus, the following situation has developed. In the Russian scientific press and in publications of scientists from several countries of Eastern Europe, the approximation of the deflection function by a polynomial of the third degree is associated with the names of M.P. Sheremetev and B.L. Pelekh. In the other works this theory is called the Reddy-Levinson theory. We will adhere to the historical name – the Pelekh–Sheremetev model.

The equations describing the dynamic processes are very difficult, so it is not possible to find an analytical exact solution of the nonlinear dynamics problems.

The solution by numerical methods of such problems is the correct solution this problem. But the question about the reliability of the obtained solutions is appears. The numerical errors that accumulate when solving problems by numerical methods are often mistaken for chaotic vibrations.

We will try to determine the truth of chaos for problems of nonlinear dynamics and contact interaction of beams, depending on applied numerical methods and initial conditions.
We will follow the chaos definition according to the Gulik work [10] in our studies. Chaos exists when there is a significant dependence on the initial conditions, or the function has a positive Lyapunov exponent at each point of the domain of its definition, and therefore is not ultimately periodic
As the initial conditions, we assume: kinematic hypotheses, boundary and initial conditions, the number of intervals of integration in the finite differences method, the methods for solving the Cauchy problem in the form of Runge–Kutta methods, the time step in solving dynamic problems.
To reduce the infinite-dimensional problem to the Cauchy problem, we used the finite–difference method with second-order approximation. The parameters of this method and the method itself remain constant throughout the work.
In the present paper, we propose a comprehensive study to reveal the truth of chaotic vibrations in the numerical solution of the problem of nonlinear vibrations of two beams structure described by the third approximation kinematic hypothesis with a gap under the action of a transverse alternating load:
1. Ensure complete convergence of the results for n and n * 2 obtained by the finite difference method.
2. Check the convergence of the results depending on the used method for solve the Cauchy problem.
3. The time histories, phase portraits, Fast Fourier transform (FFT), displacement diagram, and Poincare's section are constructed for each step in the spatial coordinate and each method for solving the Cauchy problem.
4. For each step in the spatial coordinate, the highest Lyapunov exponent is calculated using three methods: Kants [11], Wolf [12], and Rosenstein [13].
The final decision on the validity of chaotic vibrations and on the convergence of the results is taken in the case when all the points of complex study described were checked.
The gap between the beams is a tenth of the beam thickness. The beams displacement are small and can be described within the framework of the linear theory of elasticity. However, geometric nonlinearity according to T. von Karman model is taking into account.

2. Formulation of the problem

We will study the nonlinear dynamics and contact interaction of two beams with a gap. The beam 1 is under the action of a transverse alternating load, and the beam 2, comes into motion due to contact with the beam 1. Both beams are described by the kinematic hypothesis of the third approximation. The contact interaction of the beams is taken into account by the B.Ya. Cantor model [14].
We are focused on the investigation of two beams structure being as 2D object in space R^2 with rectangular co-ordinates introduced in the following way. In the body of beam 1 a certain arbitrary reference curve $z=0$ is fixed; the OX goes along main curvature of the reference curve, whereas the axis OZ is directed to the reference curvature center. In the given coordinates the beam structure as a 2D object Ω is defined as $\Omega=\{x\in[0,a]-h\leq z\leq h_k+3h\}$, $0\leq t\leq\infty$, where $[0,a]$ defines a straight beam line.

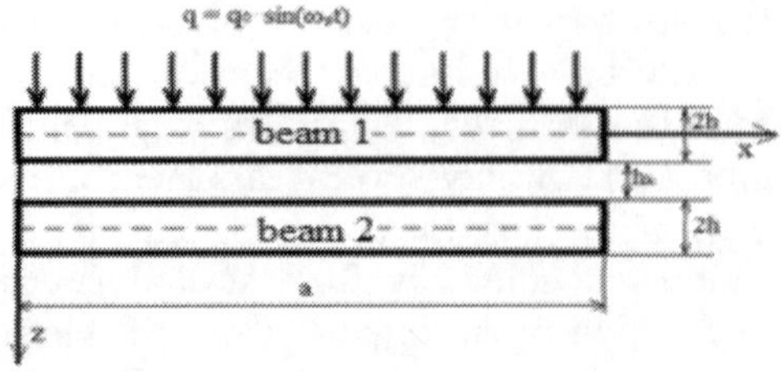

Figure 1. The settlement scheme

The equations of beams motion, as well as the boundary and initial conditions, are obtained from the Hamilton-Ostrogradskiy principle.

For construction of a Pelekh–Sheremetev beam mathematical problem, we will formulate the hypothesis [4, 6] (the third approximation hypothesis):

– cross sections don't remain flat and perpendicular to the deformed axis of a beam;
– a turn and a normal curvature for a beam we will set in a form: $u^z = u + z\gamma + z^2 u^T + z^3 \gamma^T$, $w^z = w$;
– the inertial components associated with the rotation of the sections are taken into account.
It is necessary to introduce the term $(-1)^i K(w_1 - w_2 - h_k)\Psi$ in the equations, where $i = 1,2$ is a beam's number for simulate the contact interaction according to the Cantor model. The function $\Psi = \frac{1}{2}[1 + sign(w_1 - h_k - w_2)]$ describes the contact pressure between beams. We begin with detection of contact between beams, i.e., $\Psi = 1$ if $w_1 > w_2 + h_k$ and there is a contact between beams; otherwise $\Psi = 0$ [14]. K is stiffness coefficient of a contact zone, h_k – denotes the clearance between the beams. The contact problem is solved at each time step of the computational procedure.

3. Theory

After application of a variation principle of virtual work we receive system of nonlinear partial differential equations for two Pelekhe-Sheremetev beams in the displacements taking into account energy dissipation:

$$\begin{cases} \dfrac{1}{\lambda^2 63}\left[\dfrac{4}{5}\dfrac{\partial^3 \gamma_{xi}}{\partial x^3} - \dfrac{1}{4}\dfrac{\partial^4 w_i}{\partial x^4}\right] + k^2 \dfrac{G_{13}}{E_1}\left[\dfrac{\partial \gamma_{xi}}{\partial x} + \dfrac{\partial^2 w_i}{\partial x^2}\right] + \\ \dfrac{1}{\lambda^2}\left[L_3(w_i,u_i) + L_1(w_i,u_i) + \dfrac{3}{2}L_2(w_i,w_i)\right] - (-1)^i K(w_1 - w_2 - h_k)\Psi - \dfrac{\partial^2 w_i}{\partial t^2} - \varepsilon_1 \dfrac{\partial w_i}{\partial t} = 0; \\ \dfrac{\partial^2 u_i}{\partial x^2} + L_4(w_i,w_i) - \dfrac{\partial^2 u_i}{\partial t^2} = 0; \\ \dfrac{204}{315}\dfrac{\partial^2 \gamma_{xi}}{\partial x^2} - \dfrac{48}{315}\dfrac{\partial^3 w_i}{\partial x^3} - 12\lambda^2 k^2 \dfrac{G_{13}}{E_1}\left[\gamma_{xi} + \dfrac{\partial w_i}{\partial x}\right] - \dfrac{\partial^2 \gamma_{xi}}{\partial t^2} = 0. \end{cases} \tag{1}$$

$i = 1,2$ – beam's number,

$L_1(w_i,u_i) = \dfrac{\partial^2 w_i}{\partial x^2}\dfrac{\partial u_i}{\partial x}$, $L_2(w_i,w_i) = \dfrac{\partial^2 w_i}{\partial x^2}\left(\dfrac{\partial w_i}{\partial x}\right)^2$, $L_3(w_i,u_i) = \dfrac{\partial w_i}{\partial x}\dfrac{\partial^2 u_i}{\partial x^2}$, $L_4(w_i,w_i) = \dfrac{\partial w_i}{\partial x}\dfrac{\partial^2 w_i}{\partial x^2}$ are the nonlinear operators, γ_{xi} – the transverse shear function, w_i, u_i, – the deflection and displacement functions of the beams, respectively. We must add boundary and initial conditions to the system of differential equations (1). In equations (1) bars are omitted.

Equations (1) contain the fourth derivative for w_i. This fact improves the convergence of the applied numerical methods to solve the problem and facilitates the proof of the existence theorem for the solution, in comparison with the second-approximation hypothesis.

The system of governing equations (1), together with the boundary and initial conditions are reduced to the non-dimensional form using the following relations:

$$\bar{w} = \frac{w}{2h},\ \bar{u} = \frac{ua}{(2h)^2},\ \bar{x} = \frac{x}{a},\ \lambda = \frac{a}{(2h)},\ \bar{q} = q\frac{a^4(1-\nu^2)}{(2h)^4 E},$$
$$\bar{t} = \frac{t}{\tau},\ \tau = \frac{a}{c},\ c = \sqrt{\frac{Eg}{(1-\nu^2)\rho}},\ \bar{\varepsilon}_1 = \varepsilon_1\frac{a}{c},\ \bar{\gamma}_x = \frac{\gamma_x a}{(2h)}. \tag{2}$$

The resulting system of nonlinear partial differential equations (1), together with the boundary and initial conditions, reduces to a system of ordinary differential equations by the finite differences method with approximation $O(c^2)$. Where c is a step along the spatial coordinate. In each grid node we obtain the following system of ordinary differential equations. The Cauchy problem is solved by Runge–Kutta type methods. In the work was used the Runge–Kutta methods of the second, fourth and eighth orders are compared in explicit and implicit form (the second and fourth orders Runge–Kutta method (rk2 and rk4), the fourth-order Runge–Kutta–Felberg method (Rkf45), the 4th order Kesh–

Karp Runge-Kutta method (rkck), the 8th Prince–Dormand Runge–Kutta method (rk8pd), the implicit 2nd and 4th order Runge–Kutta method (rk2imp and rk4imp)). The explicit method is characterized by the fact that the matrix of coefficients has a lower triangular form (including the zero main diagonal) – in contrast to the implicit method, where the matrix has an arbitrary form. The methods rkf45, rkck, rk8pd provide for automatic step change, as well as the possibility of controlling the integration error. A software package has been created for solve the task posed depending on the control parameters $\{q_0, \omega_p\}$.Much attention was paid to the problem of non-penetration of structural elements into each other. As noted above, we study the problems nonlinear, so there is a question about the reliability of the results.

4.Experimental results

To carry out a numerical experiment, we put : $\omega_p = 5.1$, $q_0 = 5000$, $h_k = 0.1$, $\lambda = a/2h = 50$, $\varepsilon_1 = 1$.

The ω_p is a natural beam frequency. Beams are rigidly fixed from both ends. The boundary conditions for clamping both ends of the beams:

$$\begin{gathered} w_i(0,t) = w_i(1,t) = 0;\ u_i(0,t) = u_i(1,t) = 0;\ \gamma_{xi}(0,t) = \gamma_{xi}(1,t) = 0; \\ \frac{\partial w_i(0,t)}{\partial x} = \frac{\partial w_i(1,t)}{\partial x} = 0;\ \frac{\partial u_i(0,t)}{\partial x} = \frac{\partial u_i(1,t)}{\partial x} = 0; \\ \frac{16}{5}\frac{\partial^2 \gamma_{xi}(0,t)}{\partial x^2} - \frac{\partial^3 w_i(0,t)}{\partial x^3} = 0;\ \frac{16}{5}\frac{\partial^2 \gamma_{xi}(1,t)}{\partial x^2} - \frac{\partial^3 w_i(1,t)}{\partial x^3} = 0; \\ \frac{136}{315}\frac{\partial \gamma_{xi}(0,t)}{\partial x} - 0.038\frac{\partial^2 w_i(0,t)}{\partial x^2} = 0;\ \frac{136}{315}\frac{\partial \gamma_{xi}(1,t)}{\partial x} - 0.038\frac{\partial^2 w_i(1,t)}{\partial x^2} = 0. \end{gathered} \tag{3}$$

Initial conditions:

$$w_i(x,t)|_{t=0} = u_i(x,t)|_{t=0} = \gamma_{xi}(x,t)|_{t=0} = 0,\ \left.\frac{\partial w_i(x,t)}{\partial t}\right|_{t=0} = \left.\frac{\partial u_i(x,t)}{\partial t}\right|_{t=0} = \left.\frac{\partial \gamma_{xi}(x,t)}{\partial t}\right|_{t=0} = 0. \tag{4}$$

The following harmonic load is applied on the beam 1:

$$q = q_0 \sin(\omega_p t) \tag{5}$$

Where q_0 is the amplitude, ω_p is the excitation frequency.

We study the convergence of results at the different $n = 40;80;120;240;360;400;440$ in this paper. n is a number of points of division of an interval. We used the Prince–Dormand Runge–Kutta method of the 8th order for solving the Cauchy problem. The time histories for $n = 40;80;120;240$ are very different from the time histories for $n = 360;400$. Since $n = 360;400$ the time histories for the first and second beams are coincide. In Figure 2 we give the time histories $w_i(0.5,t)$, $t \in [500..506]$ in case $n = 40;360;400$ for beams 1 and 2. The short dotted line corresponds to $n = 40$, the long dotted line corresponds to $n = 360$, and the solid line corresponds to $n = 400$.

a) b)

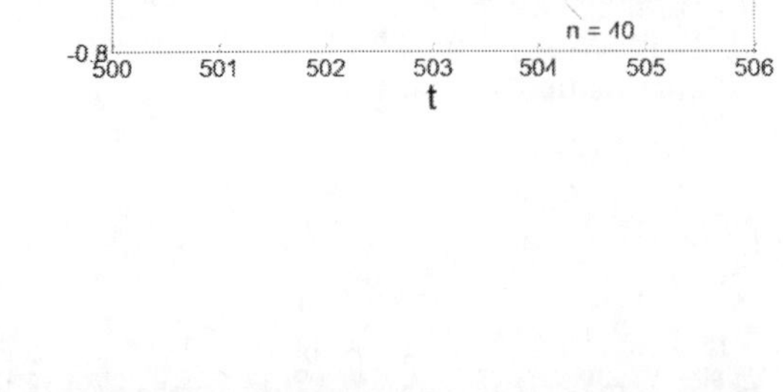

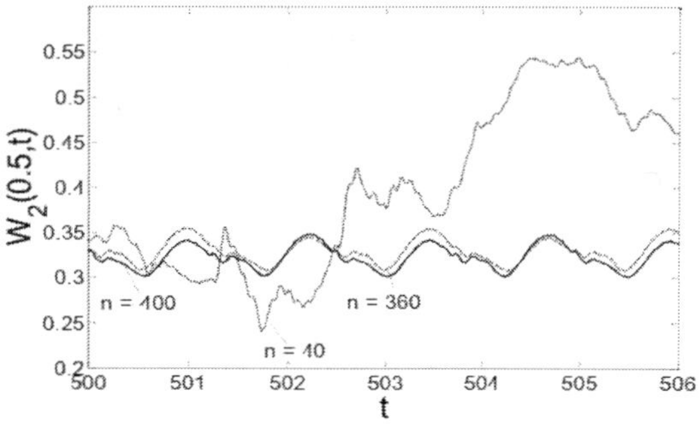

Figure 2. Time histories $w_i(0.5,t)$, $t \in [500..506]$ for $n = 40;360;400$ a) beam 1, b) beam 2.

We note that it is not possible to ensure complete convergence with respect to the time histories for chaotic vibrations for beam 2, in contrast to beam 1. As a result, the convergence of chaotic signals is achieved. It was considered sufficient to show convergence with respect to the FFT for chaotic vibrations[15]. By signals, it was impossible to achieve convergence.
Further, the convergence of the signals was investigated depending on the type of the Runge–Kutta method. In Figure 3 we give the time histories $w_i(0.5,t)$, $t \in [500..506]$ obtained by the methods rk2 (dashed line) and rk4 (solid line).

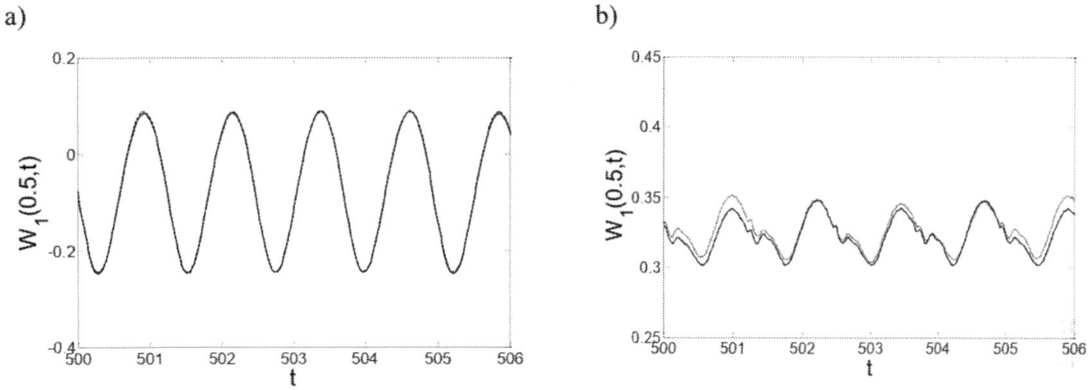

Figure 3. Time histories $w_i(0.5,t)$, $t \in [500..506]$ obtained by the rk2 and rk4 methods a) beam 1, b) beam 2.

For the beam 1, the results of the Runge-Kutta methods of the second, fourth and eighth orders coincided completely. For beam 2 the convergence is worse, but the difference in the values of the deflection does not exceed 0.5%. The 8th order Prince–Dormand Runge–Kutta method chosen as the reference.
According to the above algorithm, we further investigate the dynamic characteristics of beams for different number of partitions along the spatial coordinate. In Table 1, we give graphs of FFT, 2D phase portraits, and Poincare's section for both beams.

Table 1. Dynamic characteristics of beams.

n	№	FFT	2D Phase portrait $w(w'_t)$	Poincare's section
40	1	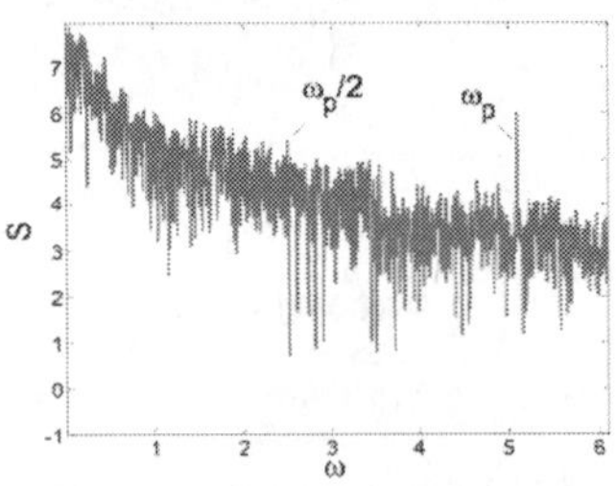	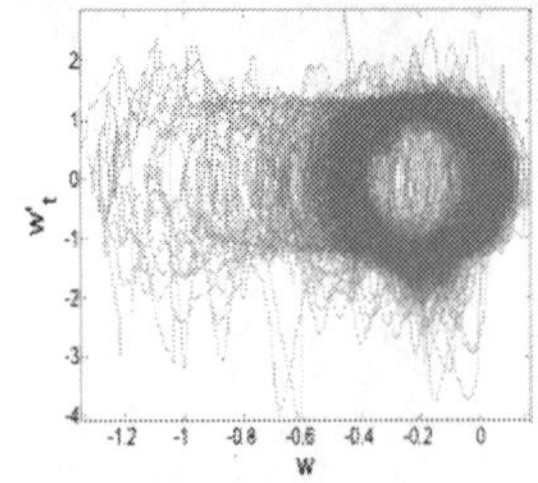	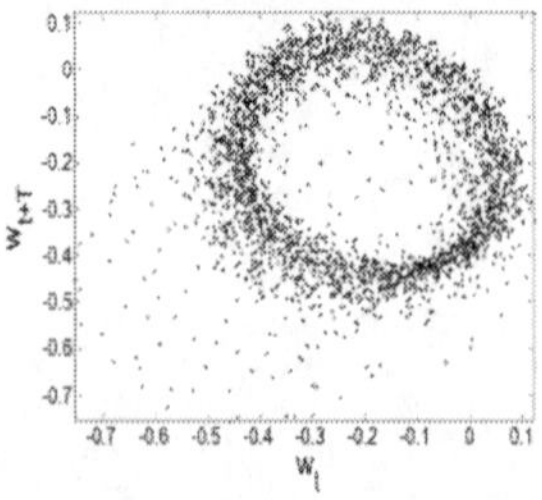
	2	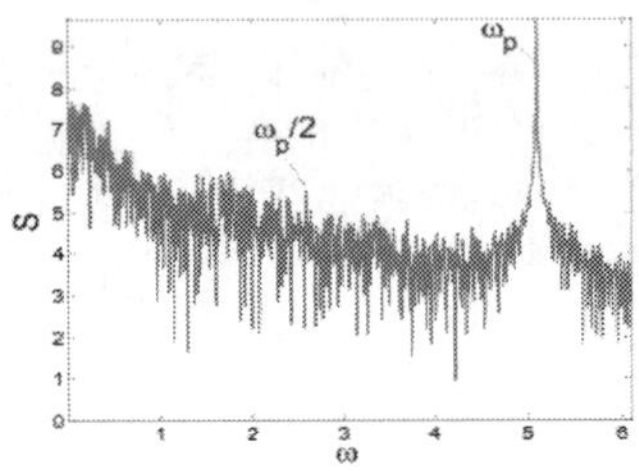	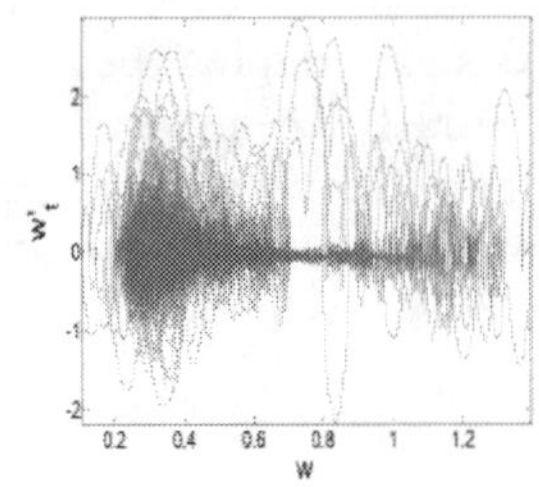	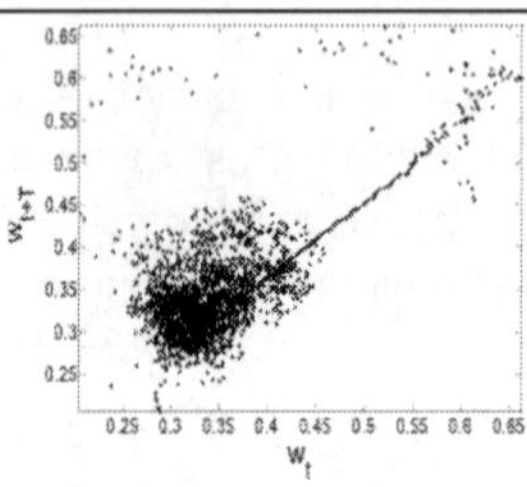
360, 400	1	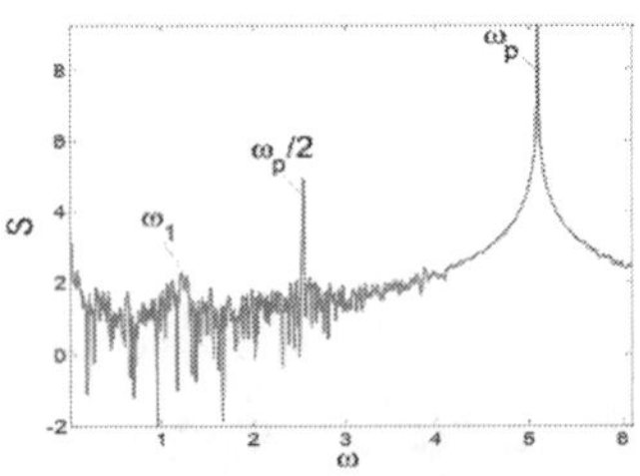	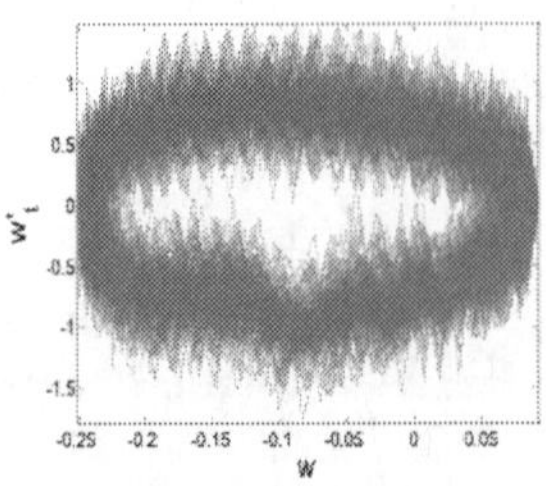	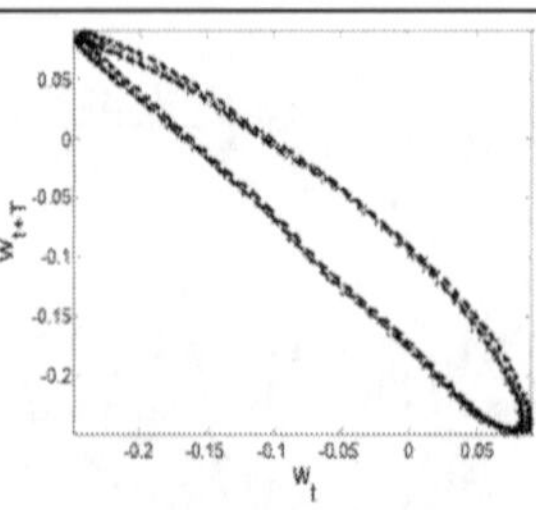
	2	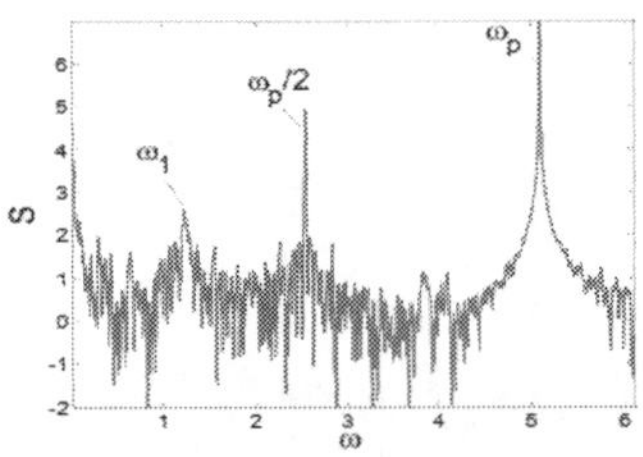	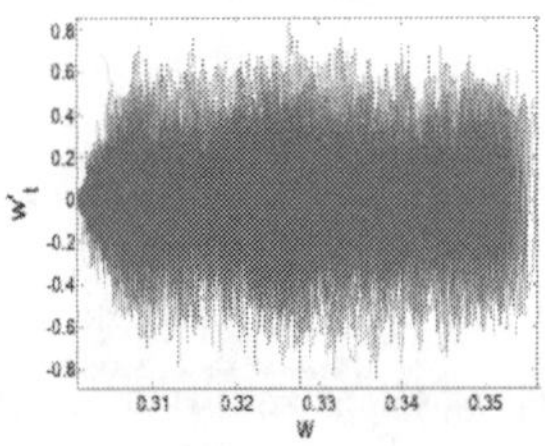	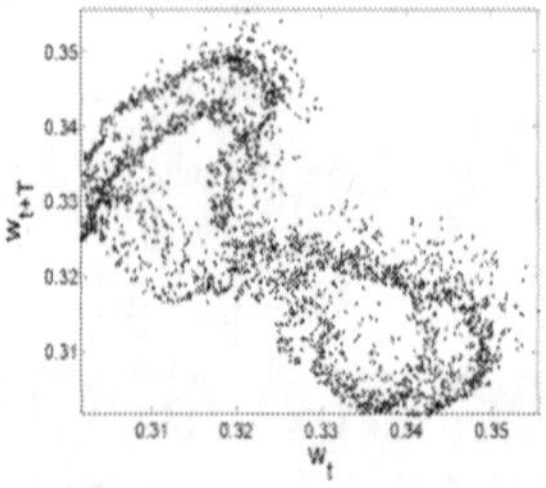

5. Discussion of the results

The FFT for any n have a chaotic component. With an increase in n, the noise harmonics become smaller in the signal. For $n = 240$ appears the bifurcation frequency $\omega_p/2$. A further increase in n leads to the appearance of a linearly dependent frequency $\omega_1 = \omega_p/4$ for both beams.

2D phase portraits for beam 1 represent, for a small n = 40, a continuous spot, for beam 2 there are two attracting sets, and for large values n = 360; 400 is a thickened oval for the beam 1 and a spot for the beam 2. The Poincare's section qualitatively changes and becomes more pronounced with increasing number of segments of the partitions along the spatial coordinate. This indicates a decrease in the chaotic component in the signal.

For $n = 360$ and $n = 400$ all dynamic characteristics are completely coincide, both for the first and for the second beam. On FFT are present bifurcation frequencies $\omega_p, \omega_p/2, \omega_p/4$. There is a synchronization of beam vibrations at the frequencies of the first and second bifurcations. With increasing n, the noise component in the signal decreases. This can be caused by the error of numerical

methods, which can be mistaken for true chaos. However, the signal remains chaotic, which is confirmed by Lyapunov exponents and other dynamics characteristics.
The values of the highest Lyapunov exponent for all n are calculated by the Wolff, Rosenstein, and Kants methods. The values are obtained on the basis of solutions of the Cauchy problem by the 8th-order Renge–Kutta method. Different methods of calculating Lyapunov's indices must be used for determine the true chaos. When the number of divisions of the beam is n = 40; 80, 120; 240, 360, 400 segments in the finite differences method, the Lyapunov exponents converge at any decimal point to the second decimal point. The values for the highest Lyapunov exponent are mostly positive. That is, we are dealing with the true chaotic vibrations of the investigated beam structure.

6. Conclutions

The reliability of the solution of chaotic dynamics problem and the contact interaction of two beams described by the Pelekh–Sheremetev kinematic hypothesis is proved and justified in the paper.
A complex study of the nonlinear dynamics of the contact interaction of beams is carried out. The choice of the number of partitions with respect to the spatial coordinate (n = 400) is justified. The choice of the method for solving the Cauchy problem (the Runge-Kutta method of the 8th order of Prince–Dormand) is justified.
The values of the highest Lyapunov exponent are calculated for the three methods (Kantz, Wolf, and Rosenstein). The sign of the highest exponent coincides for each method, which allows us to speak about the truth of the chaotic vibrations.
The phenomenon of frequency synchronization has been detected.
It was noted that the presence of the fourth derivative of the deflection function with respect to the spatial coordinate x in the beam motion equations promotes rapid convergence in solving the finite difference method, in contrast to the equations derived with S.P.Timoshenko hypothesis, in which the highest derivative has a second order.

7. Acknowledgment

The work is executed at financial support of RSF 16-11-10138.

References

[1] Desyatova A S, Zhigalov M V, Krys'ko V A and Saltykova O A 2008 Dissipative dynamics of geometrically nonlinear Bernoulli-Euler beams *Mechanics of Solids* 43 (**6**) pp 56–948
[2] Saltykova O A and Krys'ko V A 2017 The contact interaction of two Timoshenko beams *Nelineinaya Dinamika* **1** pp 41–55
[3] Krysko V A, Zhigalov M V, Saltykova O A and Soldatov V V 2010 Investigations of complex vibrations of beams within the framework of the Sheremet'ev-Pelekh kinematic model using the wavelet transform *Journal of Machinery Manufacture and Reliability* 39 (**4**) pp 17–313
[4] Krysko A V, Awrejcewicz J, Pavlov S P, Zhigalov M V and Krysko V A 2017 Chaotic dynamics of the size-dependent non-linear micro-beam model *Communications in Nonlinear Science and Numerical Simulation* **50** pp 16–28
[5] Reddy J N 1984 A simple higher-order theory for laminated composite plates *J Appl Mech* **51** pp 745–752
[6] Sheremet'ev M P, Pelekh B L and Dyachina O P 1972 A study of the effect of shear strains on the bending of a square slab by a concentrated force *Source of the Document Soviet Applied Mechanics*
[7] Hanuška A 1967 Contribution to the Reissnerian algorithm in the theory of bending of elastic plates *Aplikace matematiky* 12 (**6**) pp 67–449
[8] Levinson M 1981 A new rectangular beam theory *J Sound Vib* **74** pp 81–8787
[9] Ma H M, Gao X L and Reddy J N 2010 A nonclassical Reddy–Levinson beam model based on a modified couple stress theory *Int J Multiscale Comput Eng* **8** pp 167–180
[10] Denny Gulick 1992 Encounters with Chaos *McGraw-Hill* (New York)
[11] Kantz H 1994 A robust method to estimate the maximum Lyapunov exponent of a time series *Phys. Lett. A* **185** pp 77–87

[12] Wolf A, Swift J B, Swinney H L and Vastano J A 1985 Determining Lyapunov Exponents from a time series *Physica* **16** pp 285–317
[13] Rosenstein M T, Collins J J and Carlo J De Luca 1992 A practical method for calculating largest Lyapunov exponents from small data sets *Neuro Muscular Research Center and Department of Biomedical Engineering (Boston University)* November 20
[14] Bogatyrenko T L and Cantor B Ya T L 1986 A method for solving contact problems of nonlinear shell theory *Dokl. Akad. Nauk Ukr. SSR Ser A* 5 pp 18–21
[15] Awrejcewicz J, Krysko V A, Papkova I V, Krysko A V 2016 Deterministic Chaos in One-Dimentional Continuous Systems *World Scientific series on Nonlinear Science: Series A* p 561

Approach to Operational Experimental Estimation of Static Stresses of Elements of Mechanical Structures

A V Sedov[1,2], V V Kalinchuk[1] and O V Bocharova[1]

[1] Department of Mechanics, Physics and Nanotechnology, Southern Scientific Center of Russian Academy of Sciences, Rostov-on-don, Russian Federation
[2] Department of information technology and control, Platov South-Russian State Polytechnic University (NPI), Novocherkassk, Russian Federation

e-mail: Sedov07@list.ru

Abstract. The evaluation of static stresses and strength of units and components is a crucial task for increasing reliability in the operation of vehicles and equipment, to prevent emergencies, especially in structures made of metal and composite materials. At the stage of creation and commissioning of structures to control the quality of manufacturing of individual elements and components, diagnostic control methods are widely used. They are acoustic, ultrasonic, X-ray, radiation methods and others. The using of these methods to control the residual life and the degree of static stresses of units and parts during operation is fraught with great difficulties both in methodology and in instrumentation. In this paper, the authors propose an effective approach of operative control of the degree of static stresses of units and parts of mechanical structures which are in working condition, based on recording the changing in the surface wave properties of a system consisting of a sensor and a controlled environment (unit, part). The proposed approach of low-frequency diagnostics of static stresses presupposes a new adaptive-spectral analysis of a surface wave created by external action (impact). It is possible to estimate implicit stresses of structures in the experiment due to this approach.

1. Introduction

Various types of diagnostic control methods are widely used at the stages of the creation, commissioning and using of mechanical structures to evaluate the workmanship and the functioning of individual elements and components. A rather complete review of these methods, their merits and demerits is given in [1-5].

There are known works in which attempts were made to determine the stress states of structures from the analysis of responses to serial shock effects [6,7].

The features of the determination of the state (without a defect or with a defect) of a vertically suspended physical model of a periodic welded rod structure were established in [6]. This problem is solved by constructing the Fourier transforms of linear and nonlinear deformation responses to serial shock effects and comparing the phase diagrams of deformation behavior of an intact and damaged model.

The qualitative and quantitative features of the structure staying in the stressed state are experimentally substantiated based on the analysis of the nonlinear and linear regions of the constructed Fourier transforms in [7]. The using of Fourier transforms limited the possibilities for accuracy of these methods, as the possibility of adaptation, self-learning and tuning to a specific type of response functions.

Successful application of these approaches, ensuring the required accuracy, are directly related to the used mathematical methods of processing of the measured response mechanical wave - reactions. Thus,

the most effective approaches of mathematical processing of reactions are traditional spectral analysis [6-10]; the correlation analysis of signals [9-11]; the artificial neural networks [12,13]; statistical approaches and models [14,15,16]; wavelet transform signals [9,15,17]; optimal orthogonal decompositions of signals on the adaptively adjusted basis based on training sample [11].
Exactly the latter approach is a detail considered in the present work as a promising one with a number of advantages. The main advantage is the adaptive adjustment of the orthogonal basis, in accordance with the criteria for the best recognition of the stressed state of the structure.
In the work, the evaluation of the diagnostic observability [18,19] of static stresses of a mechanical structure with the using of the method of low-frequency diagnostics of static stress and recording of the changing in the surface wave properties of a system consisting of a sensor and a controlled environment (unit, part) was made. In this case, a set of diagnostic state variables in the form of the coordinates of the orthogonal diagnostic space was defined.

2. The scheme of the research of the approach to estimation of static stresses

A multifunctional measuring complex and a package of mathematical modeling of mechanical structures, which makes it possible to conduct research, to compare signals and to build spectral characteristics with sensors of various types, were used for the research and evaluation of the proposed approach.
The features of the dynamic process on the surface of the parallelepiped made of aluminum material were researched (see figure 1). Influences forming static stresses in the sample were applied along three perpendicular axes x_1, x_2, x_3.
The effect was determined by the quantity of shortening (compression) $-\Delta_{x_i}$ or extension (stretching) $+\Delta_{x_i}$ of the sample along the axis. The compression and stretching of the sample along the axes were varied. Each of the quantities set its static stress in the sample. The probing surface oscillations in the sample were excited by a shock force action $\boldsymbol{F}$ of constant type at one of the ends. The occurred oscillations $f(t), t \in [0, T_H]$ were fixed by the sensor $\boldsymbol{S}$ during a time interval T_H sufficient for the arrival of reflected waves from the opposite end of the structure.
A general view of the fixed response functions $f(t)$ for samples with different stress along the axes x_1, x_2, x_3 of the structure is shown in figure 2. As follows from figure 2, their shape depends on the presence and magnitude of the stress in the structure and its distribution along the axes x_1, x_2, x_3.

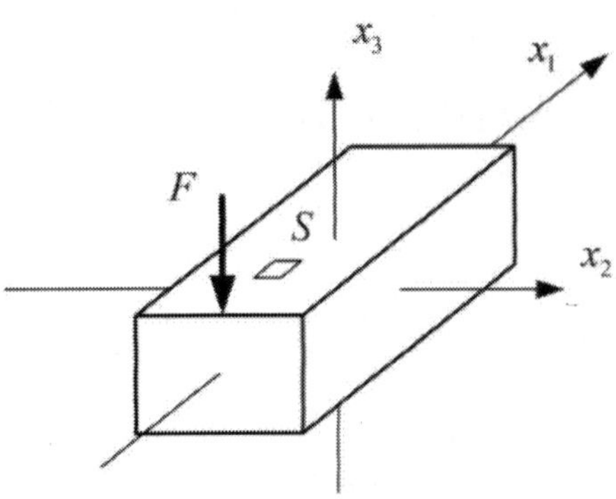

Figure 1. Three-dimensional representation of the parallelepipedal experimental structure.

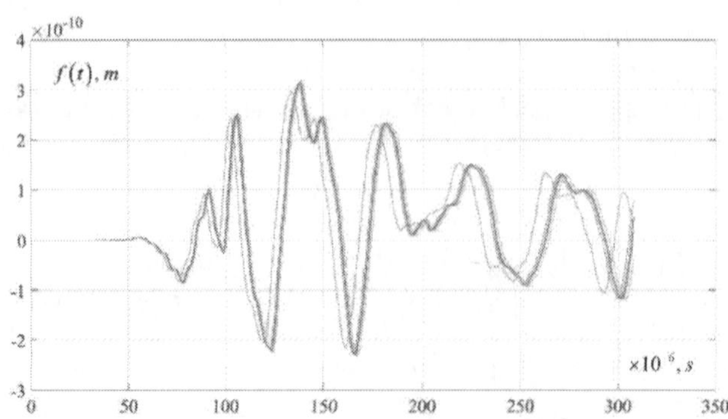

Figure 2. Graphs of response functions $f(t)$.

3. Adaptive-spectral analysis of the surface wave in the estimation of static stresses

In the case of a discrete implementation of the recognition of the magnitude of the stress from the response functions $f(t)$, the construction of the adaptive feature diagnostic space ξ is carried out as follows. A similar mathematical apparatus was used for modeling in power engineering [10] and low-frequency nondestructive testing of mechanical structures [17].

Each graph $f_i\,(t)$ of the functions $f(t)$ of the low-frequency response to external impact $\boldsymbol{F}$ in the diagnostics of structures can be considered as a vector of real values $\vec{f}_i=[f_{i1},f_{i2},f_{i3},\ldots f_{iN}]^T\in\mathbf{R}^N$ of the change of response in time. The sampling step Δt in time or the number of samples N in the observation interval is chosen in accordance with the sampling theorem [8-10].

Conditionally, the number of vectors $\vec{f}_i$ in the sample $\{\vec{f}_1,\vec{f}_2,\vec{f}_3,\ldots,\vec{f}_n\}\in\mathbf{R}^{N\times n}$ is determined by the required accuracy of stress identification in the structure and is dependent on the stress measurement range. And it is determined, as a rule, by the number $n=5\div15$ of uniformly distributed stresses throughout the measured range in the structure.

Initially, the set of vectors $\{\vec{f}_1,\vec{f}_2,\vec{f}_3,\ldots,\vec{f}_n\}$ is linearly dependent. This is due to the excessive dimensionality N of the vectors $\vec{f}_i$, the proximity of the shapes of the graphs and is confirmed by the value of the rank of the matrix $\mathbf{f}=\left[\vec{f}_1,\vec{f}_2,\vec{f}_3,\ldots,\vec{f}_n\right]^{\mathrm{T}}\in\mathbf{R}^{n\times N}$ for which $\operatorname{rank}(\mathbf{f})<<N$.

The basis of stress identification in a structure is the finding of such orthogonal transformation $\Xi\in\mathbf{R}^{N\times m}$ of a matrix $\mathbf{f}\in\mathbf{R}^{n\times N}$ into a matrix $\mathbf{A}\in\mathbf{R}^{n\times m}$, $\mathbf{A}=\mathbf{f}\,\Xi$, that excludes an excessive dimensionality $\mathbf{f}$ associated with poorly informative, often random variations in the graphs. The matrix $\Xi=[\xi_1,\xi_2,\xi_3,\ldots,\xi_m]$ in this case defines a linear subspace $\mathbf{R}^N$.

Moreover, the vectors $\xi_1,\xi_2,\xi_3,\ldots,\xi_m$ form an orthonormal basis in $\mathbf{R}^N$, and for the matrix Ξ, $\Xi^T\Xi=\boldsymbol{I}_m$ is true. Conditionally, a matrix Ξ can be considered as a matrix of orthogonal compressing of a linear space $\mathbf{R}^N$ into a space $\mathbf{R}^m$, $m<N$. Vectors $\vec{f}_i\in\mathbf{R}^N$ of responses are transformed by means of a matrix Ξ into images $\boldsymbol{A}_i\in\mathbf{R}^m$ and, thus, the entire matrix of responses $\mathbf{f}$ is transformed into an image matrix $\mathbf{A}=[\boldsymbol{A}_1,\boldsymbol{A}_2,\boldsymbol{A}_3,\ldots\boldsymbol{A}_n]^{\mathrm{T}}\in\mathbf{R}^{n\times m}$.

In the general form, the orthogonal decomposition of the initial vectors $\vec{f}_i\in\mathbf{R}^N$ in terms of the basis Ξ can be represented in the form $\vec{f}_i=\Xi\boldsymbol{A}_i+\boldsymbol{A}_0$ where $\boldsymbol{A}_0$ is the constant component of the transformation.

An orthonormal basis $\xi_1,\xi_2,\xi_3,\ldots,\xi_m$ is adaptive, trained and dependent on a recognizable sample $\{\vec{f}_1,\vec{f}_2,\vec{f}_3,\ldots,\vec{f}_n\}\in\mathbf{R}^{N\times n}$. In determination of the basis, a complex of optimization problems is solved, in particular:

- providing the best reproducibility: $\left\|\mathbf{f}-\mathbf{A}\Xi^T-\mathbf{A}_0\right\|_2\to\min$, where $\mathbf{A}_0\in\mathbf{R}^{n\times N}$ - the matrix of constant components of the transformation, consisting of vectors $\boldsymbol{A}_0$.
- ensuring the orthonormality of the basis:

$$\left\|\Xi^T\Xi-\boldsymbol{I}_m\right\|_2\to\min$$

- realizing the best discernibility:

$$d^2(\mathbf{A})=\frac{1}{m^2-m}\sum_{\substack{i,j=1\\ i\neq j}}^{m}\left\|\boldsymbol{A}_i-\boldsymbol{A}_j\right\|_2\to\max.$$

The discrete realization of the method of recognizing the magnitude of the stress from the response functions makes it possible to simplify the algorithmization of the processing of graphs.

For the simplicity of the graphic and physical interpretation, the recognizing of the stresses of the samples will be carried out in the two-dimensional feature space by images $\boldsymbol{A}_i=(a_{i1},a_{i2})\in\mathbf{R}^2$. As the

IOP Conf. Series: Journal of Physics: Conf. Series **944** (2017) 012103 doi:10.1088/1742-6596/944/1/012103

two coordinates of the image, let us choose those that have the maximum varieties of projections of the images.

4. Experimental research of approach of estimation of the static stress

The research showed that a two-dimensional feature space is sufficient for qualitative estimation of stressed states of structures. Different in magnitude and sign stresses causing displacement such as compression $-\Delta_{x_i}$ or stretching $+\Delta_{x_i}$ of the sample along three mutually perpendicular axes x_1, x_2, x_3 were considered (see figure 1). The values of the displacements were specified from the series $\Delta_{x_i} = 0.5\cdot10^{-3}; 1.0\cdot10^{-3}; 1.5\cdot10^{-3}\ m$. The shock impact was formed and the emerging surface waves were recorded as a response function. The set of 19 response functions was received:

$$\{f(t)\} = \left\{ \begin{matrix} f_{bd}(t), f_{1,5}^{s}(t), f_{1,0}^{s}(t), f_{0,5}^{s}(t), f_{1,5}^{r}(t), \\ f_{1,0}^{r}(t), f_{0,5}^{r}(t), f_{1,5}^{ss}(t), f_{1,0}^{ss}(t), f_{0,5}^{ss}(t), \\ f_{1,5}^{rr}(t), f_{1,0}^{rr}(t), f_{0,5}^{rr}(t), f_{1,5}^{sss}(t), f_{1,0}^{sss}(t), \\ f_{0,5}^{sss}(t), f_{1,5}^{rrr}(t), f_{1,0}^{rrr}(t), f_{0,5}^{rrr}(t) \end{matrix} \right\},$$

where the designations *bd* - without stress; *s, ss, sss* – the compression relative to the axes x_1, x_2, x_3 accordingly; *r, rr, rrr* - the stretching relative to the axes x_1, x_2, x_3 accordingly; 1.5; 1.0 and 0.5 - the values of the displacements along these axis.

The result of the construction of a two-dimensional feature diagnostic space and the location of the images for the specified samples of the response functions is shown in figure 3. An important fact is that all given images corresponding to different stresses along each of the axes x_1, x_2, x_3 are located on their ideal parabolas:

$$a_{2,x_1} = \alpha_1 a_1^2 + \beta_1 a_1 + \gamma_1;\quad a_{2,x_2} = \alpha_2 a_1^2 + \beta_2 a_1 + \gamma_2;\quad a_{2,x_3} = \alpha_3 a_1^2 + \beta_3 a_1 + \gamma_3.$$

This localization of images greatly simplifies the identification of the current value of the stress in the structure by the location of the image. The parabolic shape of the image localization curve is determined by the shape of the experimental structure and the features of the construction of the diagnostic space. The images of the compression $f_{1,5}^{s}(t)$ and stretching $f_{1,5}^{r}(t)$ graphs along the axis x_1 are removed from figure 3 to increase the scale and to improve detailing. These images $s1.5$ and $r1.5$ are also located higher on the axis parabola x_1.

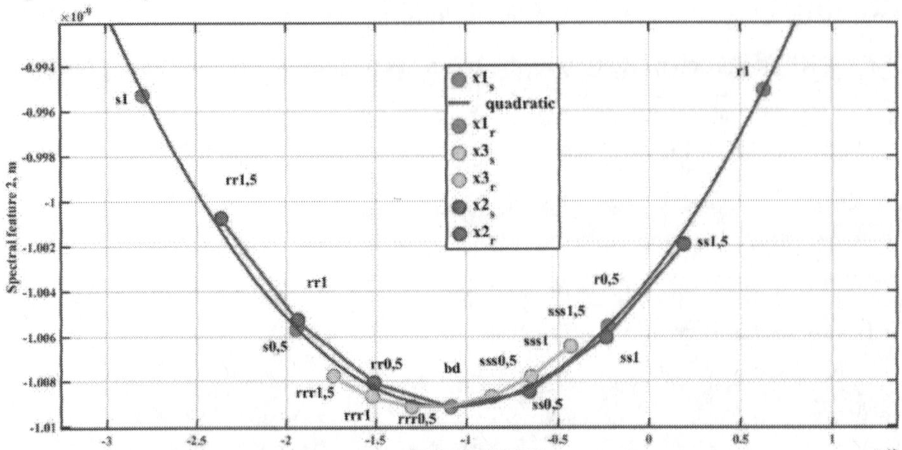

Figure 3. Two-dimensional feature diagnostic space and images of response functions corresponding to compression and stretching of the structure along the axes x_1, x_2, x_3.

Each parabola can be viewed as a separate scalable stress transformation along each of the three axes onto the main axis x_1. The axis x_1 is the main one because it is the shortest path between the probing impact point ***F*** and the location of the sensor ***S***. This axis covers the longest part of the structure and the signal propagated along this axis is the main measuring one containing the greatest information about the stresses; therefore, the axis will be called the *main projection axis*.
The fact of the inversion of the arrangement of the images on the parabola takes place for the axes x_2 and x_3 in relation to the images obtained for the axis x_1. By inversion the authors mean the change in the arrangement along the branches of the parabola of the stretching and compression patterns. This can be explained by the scheme of the changing of the structure at the deformation for the axes x_1 and x_2, considered in figure 4. This inversion can also be interpreted by the projection of stresses along the axis x_2 (or x_3) onto the main projection axis x_1.

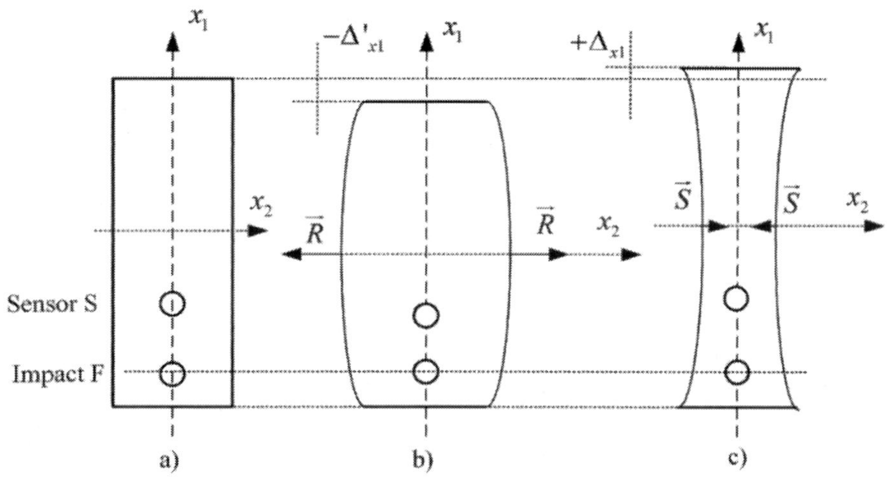

Figure 4. The relationship between compression and stretching of the structure along perpendicular axes x_1, x_2.

The following notation are used in figure 4: a) - the initial state of the structure; b) - an axial x_2 stretching $\vec{R}$, leading to an axial x_1 compression $-\Delta'_{x1}$; c) - the compression $\vec{S}$ along the axis x_2 leading to axial x_1 stretching $+\Delta_{x1}$. The presence of two parabolas in figure 3 for the axes x_1 and x_2 is explained by the indicated nonlinear character of the interrelation between compressions and stretching along perpendicular axes. The relationship between compression and stretching between the axes x_1 and x_3 can be explained similarly.

5. Conclusions

A new adaptive-spectral approach to the operative experimental estimation of static stresses in mechanical structures is proposed. It is based on the analysis of the response functions to the impact low-frequency nondestructive action, which allows estimating both the magnitude of the stresses and the direction of the action.
The principle of constructing spaces for the analysis of response signals in the operative experimental estimation of static stress based on an adaptive orthogonal basis is proposed in accordance with the criteria for the best recognition of the stressed state of a structure.
The property of the parabolic distribution of images of the response functions for structures is revealed at various stresses caused by the stretching or compressing structure along one or another axis. The property of fluctuations of the parabolas of the image distribution is revealed depending on the direction of the applied action causing the stress.

The proposed method and the principle of constructing diagnostic spaces can be the basis for building trained systems of operational control of stress for various types of structures.

6. References

[1] Sharpa R 1972 Non-destructive monitoring methods (Moscow: Mir)
[2] Kershenbaum V Y 1992 *Non-destructive testing methods* (Moscow: Nauka i tekhnika)
[3] McMaster R 1965 *Non-Destructive Testing* (Moscow: Energiya)
[4] Glagovskii B L and Moskovenko I B 1977 *Low-frequency acoustic methods of control in mechanical engineering* (Leningrad: Mashinostroyeniye)
[5] Guz A N, Mahort F G, Guscha O I and Lebedev V K 1974 Fundamentals of ultrasonic non-destructive method for determining stresses in solids (Kiev: Naukova Dumka)
[6] Esipov Y V, Mukhortov V M and Andzhikovich I E 2010 *Ecological bulletin of BSEC scientific centers* **4** 29-35
[7] Esipov Y V, Sazanov S N and Uzunyan A M 2010 *Measuring technique* **12** 32-5
[8] Mallat S A 1999 *Wavelet Tour of Signal Processing* (New York: Academic Press)
[9] Gonzales R C and Woods E W 2002 *Digital Image Processing* (New Jersey: Prentice Hall Publishers)
[10] Tung R, Garg A, Kovacs A, Peroulis D and Raman A 2013 *J. Micromech. Microeng.* **23** 045009
[11] Sedov A V 2010 Modelling Objects with Discrete-Distributed Parameters: Decomposition Approach (Moscow: Nauka)
[12] Dervilis N, Choi M and Taylor S G 2014 *J. Sound Vib.* **333-6** 1833–50
[13] Salazar A, Vergara L and Llinares R 2010 *Mech. Syst. Signal Process.* **24-6** 1870–86
[14] Tondreau G and Deraemaeker A 2013 *Mech. Syst. Signal Process* **39 1–2** 162–80
[15] Cogranne R and Retraint F 2014 *Signal Process.* **96** 173–89
[16] Martinez M, Rocha B, Li M, Shi G, Beltempo A, Rutledge R and Yanishevsky M 2012 *Smart Mater. Struct.* **21** 115001
[17] Tsui P C and Basir O A 2006 *Ultrasonics* **45 1–4** 1–14.
[18] Sedov A V 2017 The concept and the principle of the diagnostic observability of the object in problems of monitoring and non-destructive testing *Materials Science and Engineering* **177** 012034
[19] Bocharova O V, Sedov A V, Andzhikovich I E and Kalinchuk V V 2016 *Russian Journal of Non-destructive Testing* **52-7** 377–82.

Acknowledgments
The work was supported by the Russian Foundation for basic research (grant 15-08-08352) and the Russian Scientific Foundation (project 14-19-01676).

Research Results Of Stress-Strain State Of Cutting Tool When Aviation Materials Turning

A G Serebrennikova[1], E P Nikolaeva[2], A V Savilov[2], S A Timofeev[2] and A S Pyatykh[2]

[1]Komsomolsk-na-Amure State Technical University, 27 Lenin Prospect, Komsomolsk-na-Amure 681013, Russia
[2] Irkutsk National Research Technical University, 83 Lermontov street, Irkutsk 664074, Russia

Abstract. Titanium alloys and stainless steels are hard-to-machine of all the machining types. Cutting edge state of turning tool after machining titanium and high-strength aluminium alloys and corrosion-resistant high-alloy steel has been studied. Cutting forces and chip contact arears with the rake surface of cutter has been measured. The relationship of cutting forces and residual stresses are shown. Cutting forces and residual stresses vs value of cutting tool rake angle relation were obtained. Measurements of residual stresses were performed by x-ray diffraction.
Keywords – rake angle, x-ray diffractometer, hard alloy, turning tool, aviation materials cutting

1. Introduction

Most of published scientific works are devoted to study of stress-strain state (SSS) of processed material [1–7].

Residual stresses influence on the durability, corrosion resistance, fatigue strength, working life and other properties of the workpiece material. If material has high rigidity, and value of residual stresses is more than tensile strength, surface cracks appear. Deformation occurs with low rigidity and large residual stresses. Numerous studies of different authors showed that residual stresses could be controlled through adjustment of technical processes. A value, sign and distribution of residual stresses along the surface layer depth depend on types and modes of machining of workpiece. Changing the parameters of machining and the sequence of processing techniques, it is possible to eventually obtain the required value and distribution of residual stresses along the depth of a surface layer.

In aeronautical materials machining (titanium alloys, stainless steels) large cutting forces are applied, which lead to intense heat generation. Temperatures in the cutting zone reach high values [8]. Tool and workpiece material heating leads to negative changes in the microstructure and properties of them both. Impact of high cutting forces and temperatures cause the accelerated wear of the cutting edges of the tool itself. Hereupon, the permissible cutting speeds and process performance are reduced. Optimization of cutting tool configuration [9, 10] and other measures are taken to enhance machinability. Hardness increase of a cutting tool and whole technological system leads to a decrease of vibrations amplitude [11], improves the accuracy and surface quality of the workpiece [12].

In turning or milling operations residual stresses form in the surface layers of a cutting tool as well as in the material machined. Modern instrumental methods are rarely used for measuring residual stresses in cutting tools. Residual stresses in cutting tools are often evaluated on the basis of mathematical calculations and/or simulations [13, 14]. In the works [15, 16] x-ray method was used to study the SSS of tool. It is confirmed that residual stresses are associated with manufacturing techniques of tool materials and cutting tools. Many researchers suppose the residual compressive stresses contribute to the improvement of cutting tool wear resistance. However, a defining influence of the sign and value of residual stresses on the efficiency of tool is discussed.

2. Statement of the problem

Authors studied the stress-strain state of brazed inserts cutting edge of turning tools with different rake angle "gamma". The objective of the study was to determine the relationship between the cutter geometrics, cutting forces and a value and sign of residual stresses in the cutting edge. Taking to account the hypothesis about influence of a value and sign of residual stresses on tool life the indicated relationship may be the basis for further research on optimization of cutting edge geometrics and optimal cutting conditions for different materials. Turning was performed on lathe machining center DMG NEF 400. Straight-turning tool was used as a metal cutter. The cutter was brazed a cemented-carbide tip to (8% WC-Co). This hard alloy is a composite tool material. Its mechanical properties are hardness 88 HRA; bending strength 1670 MPa; compressive strength 4410 MPa.

Rake angle of turning tools («gamma») was ranged from -2 to -8 and from +2 to +8 degrees. Piped workpieces with a diameter of 46 mm and a thickness of 3 mm were machined. The workpiece was mounted in the three-jaw chuck. Workpiece weight was measured before and after turning to determine the shrinkage factor and chip thickness. Cutting data are shown in Table 1.

Table 1. Cutting data.

Machined material	Depth of cut, a_p, mm	Cutting speed, V_c, m/min	Feed, f_n, mm/rev	Feedrate, V_f, mm/min	Spindle RPM, n, rpm
Titanum alloys	3	60	0.16	76.48	478
Steel	3	100	0.16	127.36	796
Aluminium alloys	3	250	0.20	397.80	1989

Materials: aluminum alloys, titanium alloys and high-alloy stainless steel was machined sequentially by each cutter. Sections of insert cutting edge where there has been a turning of alloys are shown in figure 1.

Figure 1. Contact area of the chip with the rake surface of insert.

First, high-strength aluminum alloy Al-6%Zn-2.3% Mg-1.7%Cu-0.4%Mn and Al-6% Zn-2% Mg-1% Cu and duralumin Al-4% Cu and 1.4% Mg-0,6% Mn were machined on the area 1. Then, titanium alloy Ti-5%Al-5%Mo-5%V was machined by the same area of the cutter.

On the area 2 corrosion-resistant high-strength steel Fe-14%Cr-6%Ni-1.5%Mo-1%Mn-0.06%C were machined.

On the area 3 titanium alloy Ti-6%Al-2%Zr were machined.

Contact area of a chip with the rake surface was measured by the universal measuring machine for cutting tools Zoller Genius Pilot 3.0 using a 20-power magnification camera.

3. Theory

3.1. Determination of residual stresses by x-ray diffraction

The effect of residual stresses in the surface layer of insert causes displacement of the atomic planes in the material. Stresses corresponding to the deformation of the crystal lattice are calculated from the elasticity equations of isotropic, continuous and homogeneous mediums. X-ray method of residual stresses determination is based on the Wulf-Bragg law [17]:

$$2d \cdot \sin\theta = n\lambda \quad (1)$$

where λ is the wavelength, d is the interplanar distance, θ is the angle of rays diffraction, n is an integer number, diffraction order, n = 1, 2,..

Relation (2) is given while differentiation of equation (1), it shows the relationship of material deformation with changes in the interplanar distances of material crystal lattice:

$$\varepsilon_{\varphi,\psi} = \frac{d_{\varphi,\psi} - d_{\circ}}{d_{\circ}} = -(\theta_{\varphi,\psi} - \theta_{\circ}) \cdot \mathrm{ctg}\theta_{\circ} \quad (2)$$

where do, dφ, ψ are the interplanar distances of the crystal lattice for the undeformed and deformed material, θo, θφ, ψ are the Wolf-Bragg angles for the undeformed and deformed material, φ is the azimuthal angle, ψ is the angle between the normals to sample surface and to the plane of the crystal lattice (hkl).

The relationship between stress and strain in any direction φ takes into account the $sin^2\psi$ −method.

Residual stresses were measured by x-ray diffractometer XStress 3000 G3/G3R. Measurements of residual stresses were carried out in "Modified χ" mode. Standard Cr-K-α radiation was used. The angle of diffraction is 139 degree. Young's modulus (598 000 MPa) and Poisson's ratio (0.22) were inserted as material parameters for automatic calculation of stresses. We use cross correlation method for residual stresses calculation. Residual stresses were measured on the main and the side surface of insert. Analysis of insert SSS was performed using the sin^2-method.

3.2. Estimation of cutting forces

Measurement of cutting forces was carried out on a three-component dynamometer Kistler 9129AA, which has a measured range Fx, Fy, Fz ±10 kN. As the scheme of free rectangular cutting was considered, the cutter had been exposed to two forces Ft and Ff – tangential and axial [18]:

$$F_t = \frac{f_n a_p \tau\cos(\beta-\gamma)}{\sin\varphi\cos(\varphi+\beta-\gamma)} \quad (3)$$

$$F_f = \frac{f_n a_p \sin(\beta-\gamma)}{\sin\varphi\cos(\varphi+\beta-\gamma)} \quad (4)$$

where τ is the tangential stresses in shear plane, φ is the shear angle, γ is the rake angle of the cutter, a_p is the cutting depth, f_n is the feed, β is the tool-chip interface friction angle.

In formulas (3) and (4) when calculating F_t, and F_f it has been taken into account the effect of the rake angle of insert, which allows to assess rake angle factor to the pressure on the front surface.

Resultant cutting force was calculated according to the formula (5) for further interconnection (interrelation) analysis:

$$F = \sqrt{F_t^2 + F_f^2} \quad (5)$$

4. Experimental results

Figures 2 and 3 demonstrate function of residual stress changes and insert rake angle γ (gamma) for different materials machined.

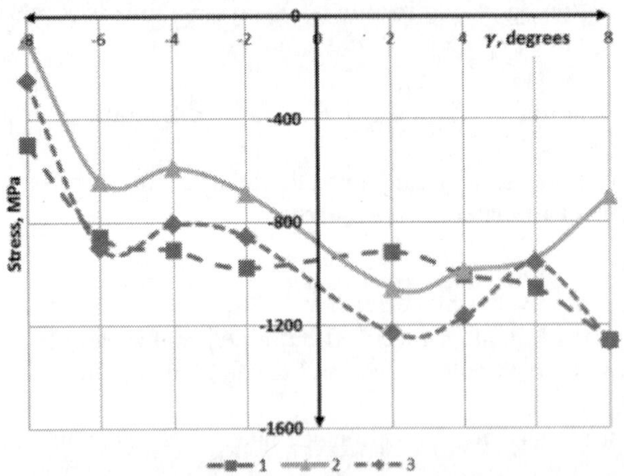

Figure. 2. «Gamma» angular dependence of residual stresses on the tool rake surface.

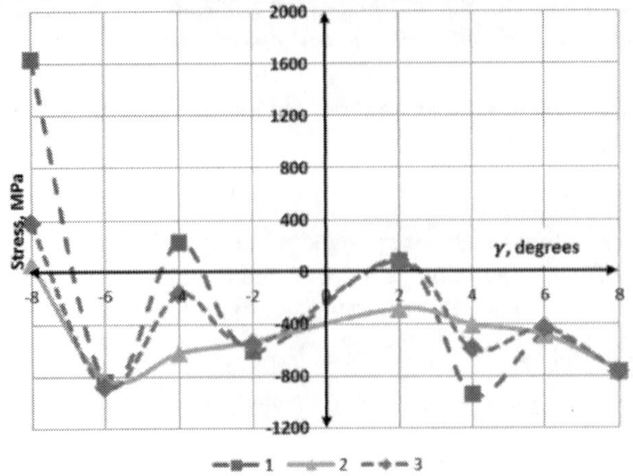

Figure. 3. «Gamma» angular dependence of residual stresses on the tool flank surface.

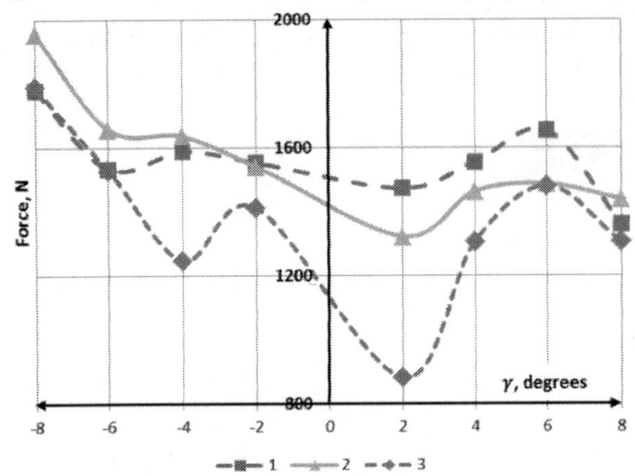

Figure. 4. «Gamma» angular dependence of the resultant cutting forces.

Graphs numeration (1, 2 and 3) corresponds to site number in figure 1. Cutting force vs plate angle γ relationship is shown in figure 4.

In figures 5–8 the chip contact areas with the rake surface of insert after machining of aviation materials are shown. After aluminum alloys' turning a small contact area is revealed on the insert (figure 5), where the build-up edge has been formed.

After steel and titanium alloys machining, defects are found on all inserts: edge chipping (figure 7, a), chips and cracks (figure 6, d).

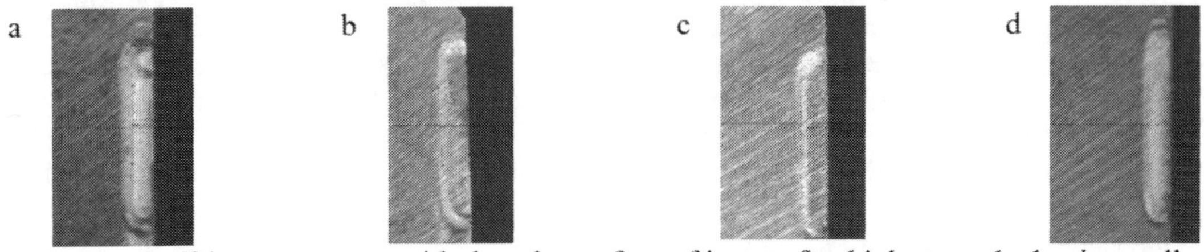

Figure 5. Chip contact area with the rake surface of insert after high strength aluminum alloys machining. Rake angle, gamma degrees: a) -8; b) -4; c) +2; d) +6.

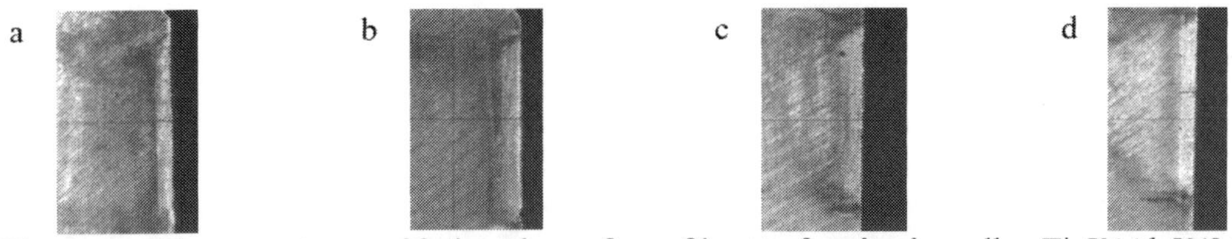

Figure 6. Chip contact area with the rake surface of insert after titanium alloy Ti-5%Al-5%Mo-5%V machining. Rake angle, gamma degrees: a) -8; b) -4; c) +2; d) +6.

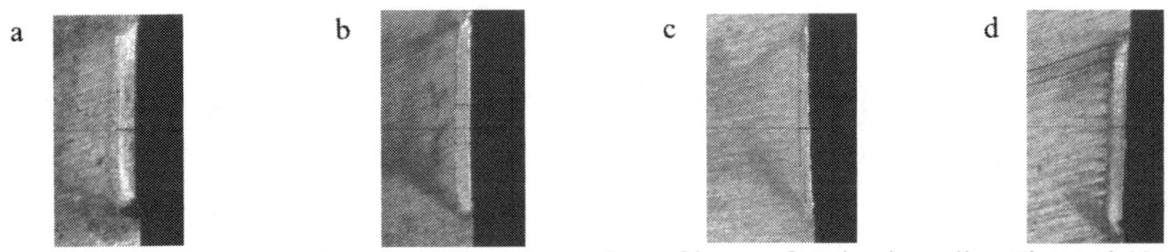

Figure 7. Chip contact area with the rake surface of insert after titanium alloy Ti-6%Al-2%Zr machining. Rake angle, gamma degrees: a) -8; b) -4; c) +2; d) +6.

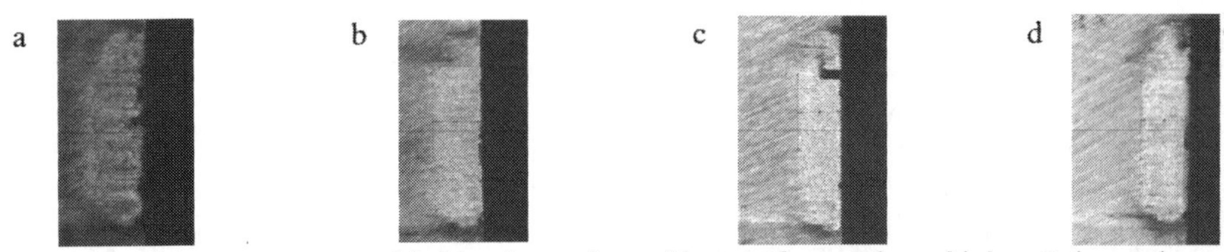

Figure 8. Chip contact area with the rake surface of insert after steel machining. Rake angle, gamma degrees: a) -8; b) -4; c) +2; d) +6.

5. Discussion of results

Measurements on the grinded, not being turned insert surface have shown that surface residual stresses are compressive at about -901 MPa. Sign and value of residual stress is very close to the stresses obtained by the same method in the work [15].

The rake surface of insert lied in direct contact with the material being treated. Residual stresses pattern on the rake surface of the plane (Fig. 2) correlates with cutting forces changes (Fig. 4). Stresses in the insert with the rake angle "-8" are close to tensile ones. Analysis of resultant cutting forces shows (Fig. 4) when the cutter rake angle is «-8» the cutter is exposed to maximum force, and if the rake angle is «+2», there is noticeably decrease in cutting forces during machining the all materials involved in experiment.

Residual stresses on the flank surface of insert change ambiguous. In cutters with rake angle «-8», «-4», «+2» residual stresses are tensile (Fig. 2). Contact area increases after alloy Ti-5 % Al-5 %

Mo-5 %V turning. Insert heating occurs in the contact with material machined and forms a heat-affected zone (HAZ) which extends from the cutting edge deep into the insert. Also very long HAZ was discovered in inserts after of titanium alloys (Fig. 6, 7) and high-strength steel (Fig. 8) machining.

Residual stresses were measured in the area from the edge up to 2 mm. Therefore, we have analyzed the whole HAZ, and residual stress, which were defined on the flank surface of insert; characterize a destructive state of insert material. Complex experimental and theoretical studies allow us to formulate the following conclusions.

The impact of high cutting forces and temperatures cause the accelerated wear of cutting edges. Structural-phase changes in insert material yet to be explored. Minimum cutting forces were registered at rake angle value $\gamma = +2$ when all of materials machining.

That means that given feed angle provides the best parameters of power inputs per volume unit removed when material cutting. The findings can be further used for optimization of cutting modes in turning of aviation materials.

References

[1] Zamashchikov Y I 2006 Duality in Metal Cutting: Impact to the Surface Layer Residual Stress *Mat. and Manuf. Proc.* **21** pp 66–551
[2] Zamashchikov Y I 2007 Surface Residual Stress Measurements By Layer Removal Method *Int. Jour. of Mach. and Mach. of Mat.* **16** pp 187–211
[3] Zamashchikov Y I 2007 Machining Residual Stresses And Part Distortions *Int. Jour. of Mach. and Mach. of Mat.* **2** p 378
[4] Zamachtchikov Y, Breaban F, Vantomme P and Deffontaine A 2002 Method To Evaluate Residual Stresses In Laser Cutting Process *Lasers in Engineering* **12** p 27
[5] Vaskin K Ya 2012 Calculation of residual pressure at turning surfaces with wearproof coverings *Vekt. Nauki of TSU* p 5–60
[6] Kabaldin Yu G, Kuzmishina A M and Sery S V 2013 Atomicdislocation approach to the process of metal cutting. *Trans of NNSTU* p 98–189
[7] Yablokova N A and Trofimov V V 2013 XRA Study of the Stress-Strain State of the Compressor Blades *Ind. Lab.* **1** p 36–44
[8] Oteniy Y N and Martynenko O V 2015 The treatment process features research by cutting the diametrical surfaces components *Int. Jour. of App. and Fun. Res* p 6–452
[9] Savilov A V, Nikulin D S, Nikolaeva E P and Rodygina A E 2013 Current state of manufacturing high performance cutting tools from powdered metal high-speed steels and hard alloys *Proc. of ISTU* p 26–33
[10] Bogdanov K V, Nikulin D S, Savilov A V, Nikolaeva E P and Rodyigina A E 2013 Proizvodstvo vyisokoproizvoditelnogo rezhuschego instrumenta v usloviyah IAZ. *Sci., and Tech.* p 91–5
[11] Savilov A V and Ahmadi K 2015 Modeling the mechanics and dynamics of arbitrary edge drills *Int. Jour. of Mach. T. and Man.* **89** p 20–208
[12] Nikolaev A Yu 2017 Simulation of the plain milling process *IOP Conf. Series: Mat. Sci. and Eng.* [Online]. **177** Available: http://iopscience.iop.org/article/10.1088/1757-899X/177/1/012080/pdf
[13] Kabaldin Y G and Vlasov E E 2015 Nanostructured Processes at the Destruction and Wear of Carbide Cutting Tools *Proc. of HEI. Mach. Build.* p 59–66
[14] Oteniy Y N, Muravoiv O P, Epov A A, Tymenov T N, Zhunusova A S, Tkacheva Y O, Pleshakova E A and Erahtina A V 2012 Forces and stresses on the surface cutting blade in cutting metals *Fund. Res.* p 7–155
[15] Collins A and Uzunyan M D 2014 Issledovanie i analiz ostatochnyih napryazheniy pri almazno-iskrovom shlifovanii tverdyih splavov s primeneniem tehnologii minimalnoy smazki. *Cut. and Tool. in Tech. Syst. NTU KhPI.* [Online]. **84** p 7–11 Available: http://repository.kpi.kharkov.ua/bitstream/KhPI-Press/16729/1/RITS_2014_84_Kollinz_Issledovanie.pdf

[16] Nikolaeva E P and Nikulin D S 2016 The application of innovative means for quality control of the high-speed steel tools. *Mod. Tech. Syst. An. Mod.* [Online] p 73–80 Available: https://www.irgups.ru/sites/default/files/irgups/journal/no_2_50_2016.pdf
[17] Hauk V 1997 Structural and residual stress analysis by nondestructive methods: evaluation, application, assessment (Amsterdam, Netherlands: Elsevier) p 640
[18] Armarego E J A and Brown R H 1977 (Moscow: *Mechanical engineering)* p 325

Acknowledgments
The research was carried out with the support of the Engineering center «Innovation materials and technolo-gies» of the FSBEO HE of «K-na-ASTU» Komsomolsk-on-Amur, Vice-rector of scientific research and innovation Belyh S. V. All studies were conducted in the research laboratories of «Technology of high-duty machining, forming and hardening of machinery» and «Study of technological residual stresses and deformations» INRTU. The authors express their gratitude to the staff of the laboratories.

IOP Conf. Series: Journal of Physics: Conf. Series **944** (2017) 012105 doi:10.1088/1742-6596/944/1/012105

Development of unauthorized airborne emission source identification procedure

L O Shtripling[1], V V Bazhenov[1] and N S Varakina[1], N P Kupriyanova[2]

[1]Omsk State Technical University, 11, Mira ave., Omsk, 644050, Russia
[2]Federal Service for Surveillance on Consumer Rights Protection and Human Wellbeing in Omsk Region, Omsk, Russia

e-mail: losht59@mail.ru

Abstract. The paper presents the procedure for searching sources of unauthorized airborne emissions. To make reasonable regulation decisions on airborne pollutant emissions and to ensure the environmental safety of population, the procedure provides for the determination of a pollutant mass emission value from the source being the cause of high pollution level and the search of a previously unrecognized contamination source in a specified area. To determine the true value of mass emission from the source, the minimum of the mean-root-square mismatch criterion between the computed and measured pollutant concentration in the given location is used.

1. Introduction

High pollution level is one of the most important environmental problems in cities with developed industries. To estimate the atmospheric contamination level, monitoring systems comprising of priority pollutants stationary sites are arranged in the cities. However, the capability of such stations is limited. All we can know is the pollutants concentration reading at the site of the observational station. To make the atmospheric contamination level estimates and forecasts more accurate, the systems of monitoring include mathematical models of pollutants transfer to determine pollutant concentration in any location under specified meteorological conditions [1, 2].

At the same time, an important task in decision making in the field of ecological safety is to determine the sources of unauthorized emissions. This task can be solved in two ways: improvement of instrumental base for the network of monitoring and development of mathematical models for measuring different source contribution to the overall level of atmospheric contamination. Using lidar systems for control and detection of contamination source [3], wireless sensors [4, 5] and portable laboratories in monitoring networks possesses one significant drawback, which is a high cost of implementation. In addition, the capability for lidar systems of control in case of ground level concentration determination is restricted by dense industrial development.

Numerical simulation of atmospheric contamination processes is widely used in the search for contamination sources. For instance, the Gauss and the Lagrange model algorithms are applied in works [2, 7] to determine contamination sources. However, the emission source identification in this case is limited to determination of its contribution to the overall level of contamination. But the options for producing extreme high levels of air pollution from a previously unknown emission source or from an unauthorized mode operation source are not considered. A more reliable method of positive matrix factorization (PMF) is used in papers [8, 9] for pollution source detection. This method is used with success to solve local problems such as defining the cause-effect relationships of urban pollution and to analyze global problems, for instance, reasons for black carbon pollution in the northern hemisphere [10]. A significant drawback of the PMF method is that it can be used solely for the analysis of pollution by solid substances such as carbon, PM 2.5, PM10 and other solid particles, while most emissions of industrial enterprises include gaseous impurities.

Thus, the problem of source detection of unauthorized atmospheric pollution by gaseous and solid substances is vital in case of insufficient information about its location and operation parameters in particular.

2. Task description

The purpose of the study is to develop the pollution source identification procedure for unauthorized airborne pollutant emission detection and localization of the source. The following steps are essential in achieving this objective. The first step is to develop the detection procedure for pollutant mass emission from the source resulted in unauthorized atmospheric contamination. The next step is to develop an algorithm for search and location of a previously unknown pollution source. The final step is to develop software for decision-making in the field of ecological safety.

3. Theoretical Basis

This work assumes that most elevated air pollution episodes are primarily related to familiar pollution sources operation under unauthorized modes. In this case, it is necessary to prove that a particular source or group of sources have exceeded the enforceable emission standards to take informed decisions on compliance measures or source operation constraints.

Pollutant concentration measured at the observational station (C_{ss}) can be represented as follows:

$$C_{ss} = \sum_{i=1}^{n} C_i \left(x, y, z = 2\right), \tag{1}$$

where C_i is the substance concentration from the i-th pollution source at the observation point above ground level of 2 meters, µg/m^3.

To calculate the concentration C_i, any mathematical model can be used. For instance, the Gauss model determines the pollutant concentration of a single point source at the assigned area from the formula:

$$C_i = \frac{Q \cdot K \cdot V}{2 \cdot \pi \cdot u_s \cdot \sigma_y \cdot \sigma_z} \cdot \exp\left(-0.5 \cdot \frac{y^2}{\sigma_y^2}\right), \tag{2}$$

where Q is the substance emission, g/s; K is the conversion factor being equal to 10^6; V is vertical dispersion conditions; σ_y, σ_z are standard deviations of horizontal and vertical dispersion, m; u_s is the wind velocity of the effective source height, m/s.

The use of the Gauss model and its derivatives to determine pollutant concentration is broadly examined, for example in [1], and it is outside the scope of this paper.

The parameters of unauthorized emission source are determined as follows:

1. Pollutant concentrations (C_i) caused by emission of each i-th pollution source at the observation point are determined from the formula (2). The calculation takes into account meteorological conditions (wind direction and velocity, ambient temperature, etc.). It is assumed that Q is the quantity of the authorized substance emission from the source.
2. The criterion for expediency of pollution source accounting is determined:

$$C_{bc} = 0.4 \cdot \max(C_i). \tag{3}$$

The determination of the true quantity of mass emission is performed among multiple pollution sources satisfying the condition:

$$C_i \geq C_{bc}. \tag{4}$$

This allows reducing further calculations and taking into account emission sources that contribute to pollution at the specified point, the contribution being not less than 80 %. The parameter G is determined for the remaining sources:

$$G = \sum_{j=1}^{n} C_i. \tag{5}$$

IOP Conf. Series: Journal of Physics: Conf. Series **944** (2017) 012105 doi:10.1088/1742-6596/944/1/012105

For the pollution sources satisfying the condition (4), the functional J is calculated from the formula:

$$J = \sum_{k=1}^{n} \left(\left(C_{ss}^{k} - G^{k} \right) - C^{k} \right)^{2} , \tag{6}$$

where C^k is the total computed pollutant concentration from all pollution sources at the k-th observation point, μg/m^3; C_{ss}^k is the measured substance concentration at the k-th point.

Subsequently, varying the value Q in the formula (2), we minimize the functional J, for instance, by applying the search method of global extremum. As the true values Q for pollution sources we take the values with minimum functional J being attained.

If the following condition is fulfilled for the i-th pollution source:

$$Q_i > Q_i^{norm} , \tag{7}$$

where Q_i^{norm} is the mass emission of the i-th source under its normal operation condition, this source is admitted to produce an unauthorized emission, and it is necessary to take additional regulatory measures towards this source.

It is obvious that calculation of mass emission by the formulas above brings to a misleading result if the unauthorized emission is conducted by previously unrecognized contamination source. In such a case the value Q exceeds manifold Q_i^{norm}. In this instance it is essential to identify the possible location of the pollution source to take decisions on elimination of contamination.

To solve this problem we use the data on measured concentrations (C_{ss}^k) at a minimum three ground points as well as the value of pollutant concentration C^k defined by the formula (2). The search area is divided into equal-area rectangles with the sides of 500 m. For each rectangle the possible parameters of virtual atmospheric contamination source such as height, mass emission and other parameters are determined. With these parameters the particular source may become a cause of the unauthorized pollution and under specified meteorological conditions produce the concentration:

$$C_n^k + C^k \approx C_{ss}^k , \tag{8}$$

where C_n^k is the concentration produced by the virtual contamination source.

Provided that such source parameters are detected and the concentration measured at three points differs from the computed one within 25 per cent, such rectangle defines the most probable site for the unauthorized pollution source. If the deviation of computational concentration from the measured one is within 25-35 per cent, the rectangle determines the possible site for the pollution source.

Subsequent to the computational results, a map of the possible site for the pollution source is generated to continue ground location search and take measures on pollution elimination.

4. Experimental Results

The procedure considered above is software-implemented. This software was used to determine the previously unspecified hydrocarbon emission source in April, 2017 in the city of Omsk, Russia. An excess of pollution standard was recorded on April, 22 by the stationary site in the industrial area of the city. The search data for the unauthorized emission are shown in figure 1.

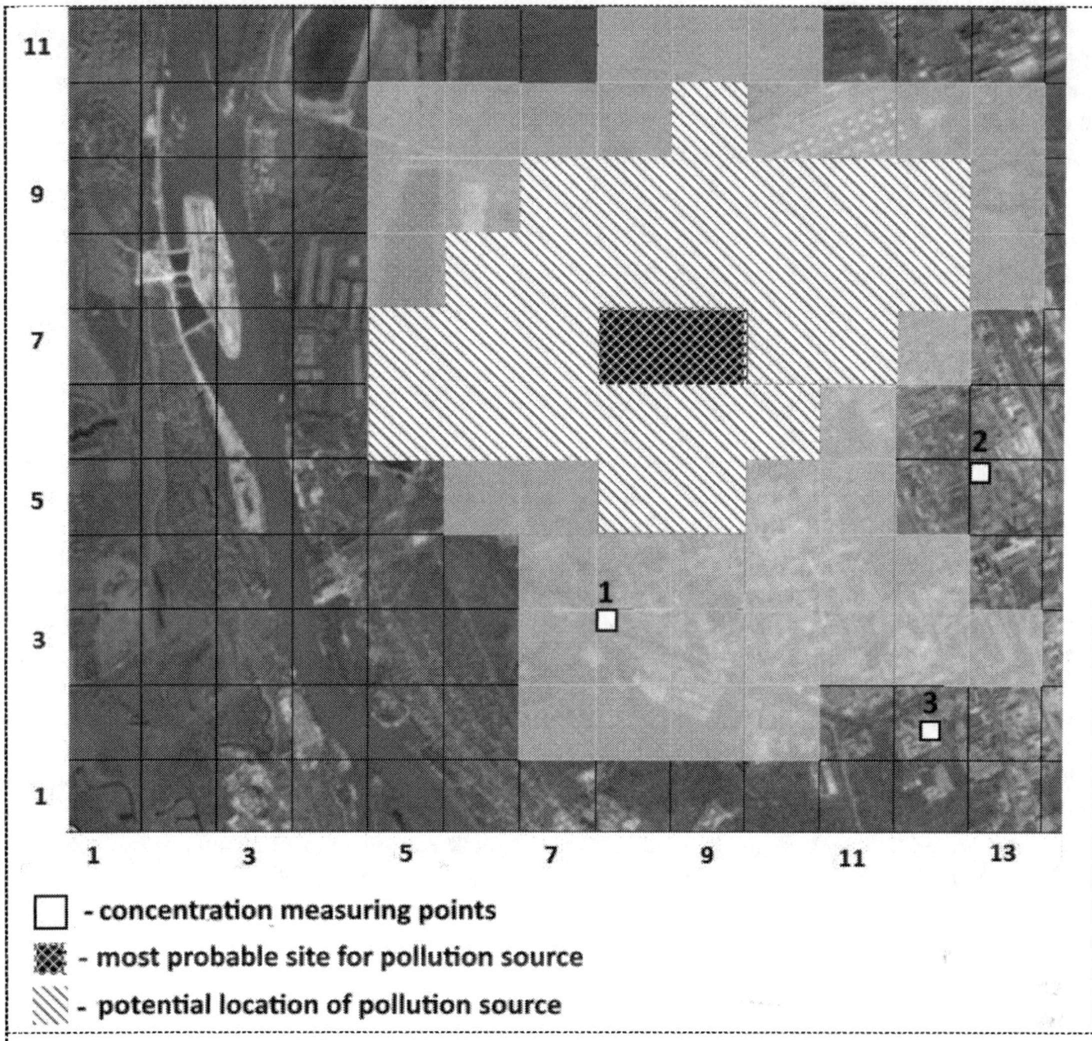

Figure 1. Location map for unauthorized emission source

Table 1 shows the calculation data for one of the rectangles in the area of the most probable location of the pollution source.

Table 1. Concentration of ethyl benzene at computational points

Rectangle position			Concentrations, μg/m^3			
Vertical	Horizontal	Measuring point	C_n^k, μg/m^3	C_{ss}^k	$\lvert\Delta\rvert$, %	Q, g/s
		1	0.013	0.016	18.75	3.9
7	8	2	0.01	0.013	23.07	3.9
		3	0.009	0.008	11.11	3.9

Table 1 and figure 1 represent data on the source of atmospheric contamination, with ethyl benzene not being previously accounted there.

5. Conclusions

The developed procedure to determine an unauthorized emission source allows one to define accurately the source location and its mass emission. The technique was approved in the city of Omsk, Russia, when ethyl benzene atmospheric contamination source was detected. The developed software

IOP Conf. Series: Journal of Physics: Conf. Series **944** (2017) 012105 doi:10.1088/1742-6596/944/1/012105

allows detecting the source of atmospheric contamination with the previously unaccounted emission of the pollutant (ethyl benzene).

However, for further improvement of the presented procedure, it is necessary to conduct additional analyses aimed at defining optimum location for pollutant concentration points and optimizing the algorithm for the search of the pollution source location.

References

[1] Zhu J Y, Zheng Y and Yi X, Li V 2016 Gaussian Bayesian model to identify spatio-temporal causalities for air pollution based on urban big data *Computer Communications Workshops (INFOCOM WKSHPS) IEEE Conf.* (San Francisco, CA, USA: IEEE) pp 112–7

[2] Li L, Liu Y and Wang Y 2017 Monitoring an air pollution episode in Shenzhen by combining MODIS satellite images and the HYSPLIT model *IOP Conf. Ser.: Earth Environ. Sci. 74 012010* pp 1–4

[3] Gaudio P, Gelfusa M, Malizia A, Parracino S, Richetta M, De Leo L, Perrimezzi C and Bellecci C 2015 Detection and monitoring of pollutant sources with Lidar/Dial techniques *J. Phys.: Conf. Ser. 658 012004* pp 1–9

[4] He T and Zhang S 2011 Application of localization algorithm in monitor system of pollution emergency *Remote Sensing, Environment and Transportation Engineering (RSETE)* (Nanjing, China: IEEE) pp 7402–05

[5] Bathiya B, Srivastava S and Mishra B 2016 Air Pollution Monitoring using Wireless Sensor Network *Electrical and Computer Engineering (WIECON-ECE)* (Pune, India: IEEE) pp 112–7

[6] Bayat B, Crasta N and Li H, Ijspeert A 2016 Optimal search strategies for pollutant source localization *Intelligent Robots and Systems (IROS)* (New York: IEEE) pp 1801–07

[7] Ensora K, Ray B and Charltona S 2014 Point source influence on observed extreme pollution levels in a monitoring network *Atmospheric Environment* **92** pp 8–191

[8] Vargas F, Rojas N, Pachon J and Russell A PM_{10} characterization and source apportionment at two residential areas in Bogota *Atmospheric Pollution Research* **3** pp 72–80

[9] Pabroa P, Santos F, Morco R, Racho J, Bautista VII A, Bucal C 2011 Receptor modeling studies for the characterization of air particulate lead pollution sources in Valenzuela sampling site (Philippines) *Atmospheric Pollution Research* **2** pp 8–213

[10] Ma P, Gattiker J, Liu X and Rasch P A novel approach for determining source–receptor relationships in model simulations: a case study of black carbon transport in northern hemisphere winter 2013 *Environ. Res. Lett.* **8** 024042 pp 1–8

Geometric parameters determination of the installation for oil-contaminated soils decontamination in Russia, the Siberian region and the Arctic zones climatic conditions with reagent encapsulating

L O Shtripling, E G Kholkin

Omsk State Technical University, 11, Mira ave., Omsk, 644050, Russia

e-mail: holkin555@mail.ru

Abstract. The article presents the procedure for determining the basic geometrical setting parameters for the oil-contaminated soils decontamination with reagent encapsulation method. An installation is considered for the operational elimination of the emergency consequences accompanied with oil spills, and the installation is adapted to winter conditions. In the installations exothermic process thermal energy of chemical neutralization of oil-contaminated soils released during the decontamination is used to thaw frozen subsequent portions of oil-contaminated soil. Installation for oil-contaminated soil decontamination as compared with other units has an important advantage, and it is, if necessary (e.g., in winter) in using the heat energy released at each decontamination process stage of oil-contaminated soil, in normal conditions the heat is dispersed into the environment. In addition, the short-term forced carbon dioxide delivery at the decontamination process final stage to a high concentration directly into the installation allows replacing the long process of microcapsule shells formation and hardening that occur in natural conditions in the open air.

Key words – Equipment for decontamination, oil spills, oil-contaminated soil, oil sludge, decontamination, reagent encapsulation.

1. Introduction

The problem of overcoming the consequences of accidental soils contamination with oil products does not lose its relevance. Potential sources of emergencies are operated machines, equipment for the extraction, storage and transportation of hydrocarbons [1, 2].

Various technologies for reducing pollution effects have been developed and are being applied, but practically all of them are effective only at positive temperatures [3, 4, 5].

Siberia and the Arctic zone of the Russian Federation occupy a significant territory of the country and differ from other regions with special natural and climatic conditions, in particular a long period of negative temperatures and relatively undeveloped infrastructure. The main problem of neutralizing soils contaminated with oil products in conditions of negative ambient temperature is that the contaminated soil is in a frozen state, and it prevents the normal course of the neutralization process, so additional energy is required for preparing the soil, but the problem is in its lack.

Considering the short vegetation period and the extreme natural and climatic conditions, it is very important to develop and to use equipment and technologies adapted to arctic conditions.

For regions with winter time more than six months, the most promising technology for the operative elimination of emergency consequences involving oil spills is based on the pollutant encapsulation (reagent encapsulation technology) using an alkaline reagent based on calcium [6]. In this case, the end product of oil-contaminated soil decontamination is encapsulated material which looks like fine mixture similar to ordinary sand.
The authors conducted studies to clarify the formulation of soil decontamination, contaminated with petroleum products, using for neutralization the reagent encapsulation technology, including adapted to "winter conditions" [7]. Also a patent for utility model for decontamination of oil-contaminated soil , and oil sludge was obtained [8], which allows, if necessary (e.g., in winter conditions) to use the heat energy dissipated at each stage of decontamination process of oil-contaminated soil, which in normal conditions is dispersed into environment. In addition, a short-term forced carbon dioxide delivery at the final stage of the decontamination process to a high concentration directly into the installation allows replacing the long process of microcapsule shells formation and hardening that occur in natural conditions in the open air. An assessment of energy potential of an exothermic process chemical reaction of chemical decontamination of soils contaminated with petroleum products, indicates that the thermal energy dissipated with an exothermic process chemical reaction of chemical decontamination of soils contaminated with petroleum products, is guaranteed enough for successful implementation of the oil-contaminated soils decontamination process, which allowed to implement the oil-contaminated soil decontamination process in winter conditions [6].

2. The problem setting

The study object is to determine the main geometric parameters of the installation for decontamination of oil-contaminated soils with reactant encapsulation method when the thermal energy dissipated at each stage of decontamination process, is guaranteed enough for thawing frozen soil portions, contaminated with petroleum products.

3. Theory

All the energy released at each stage of the oil-contaminated soil decontamination process is used for thawing the frozen pieces of oil-contaminated soil, heating the coolant located in the inner cavity of the unit casing, and part of the energy is dissipated into the environment and is a loss. Then, in a first approximation, the heat balance equation looks as follows:

$$\Sigma Q_{CH.R.} = Q_{HEATING.} + Q_{COOLANT} + Q_{LOSSES}, \qquad (1)$$

where $Q_{CH.R.}$ is energy released during a chemical reaction;
$Q_{HEATING}$ is energy required to thaw the frozen pieces of oil-contaminated soil;
$Q_{COOLANT}$ is energy transferred to the coolant located in the inner cavity of the unit casing;
Q_{LOSSES} is energy dissipated into the environment (approximation $Q_{LOSSES} = 0.5 \cdot Q_{CH.R.}$).
The main constructive elements of the installation for oil-contaminated soils decontamination are shown on the diagram scheme (Fig.1). The inner cavity made in the wall of the unit housing is filled with coolant Dixis-65, as it has a sufficiently high specific heat capacity (c=2.974 kJ/kg·°C) and a freezing point 65°C below zero. The main geometrical parameters of the installation, that affect its operability and performance are internal d and the outer mixer diameter D, the hopper wall height H for thawing frozen oil-contaminated soils, the installation length L and cavity width t in the housing wall filled with coolant Dixis-65.

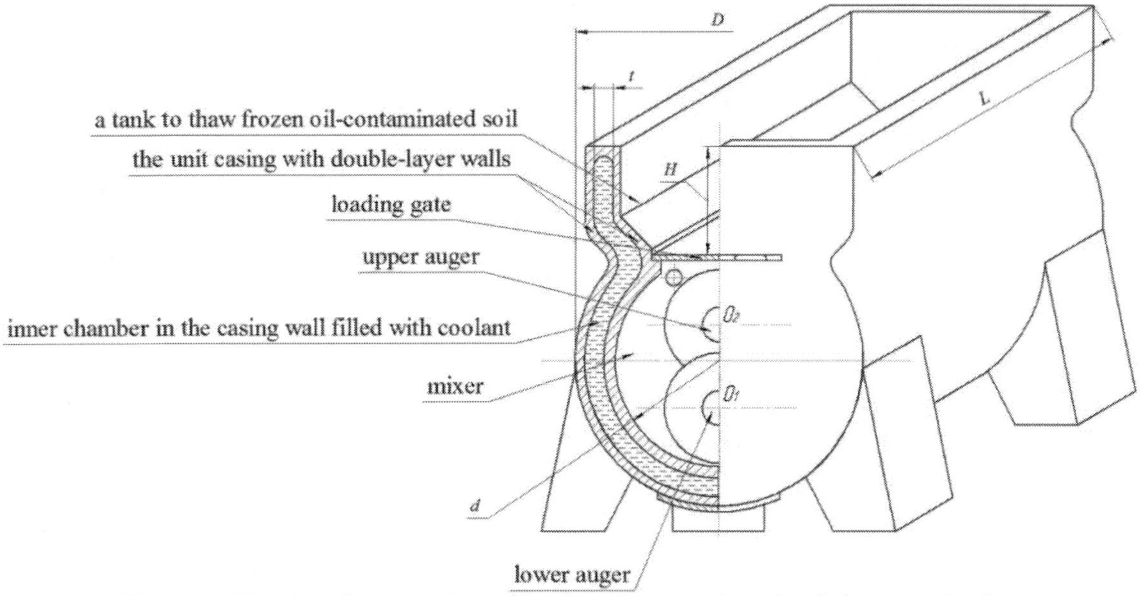

Figure 1. The plant lay-out scheme for the oil-contaminated soil decontamination.

The width of the cavity *t* in the housing wall is defined starting from the necessary amount of the coolant that is required for the accumulating and transferring heat to thaw the frozen subsequent portions of oil-contaminated material into the tank. The thickness of the unit walls is assumed equal to 5 mm. To determine the geometric parameters of the remaining parameters one introduces the following ratios: $D=d+2\cdot t+10$mm; $H=0.6\cdot d$; $L=2.5\cdot d$.

In determining the geometric parameters, conditionally there were distinguished three types of installations: small, medium and large. Small installations allow loading the mixer up to 100 kg of oil-contaminated soil, medium – up to 1,000 kg, large– up to 20,000 kg. Experimental studies [7] show that for the guaranteed oil-contaminated soil decontamination process on average is necessary to use lime as a reagent, in an amount not exceeding 80% by weight of the oil-contaminated soil. Calculations [6] showed that due to exothermic process chemical reaction of chemical decontamination of soil contaminated with petroleum products, using 1 kg reagent is dissipated $Q_{CH.R.}$ = 2332.25 kJ, and for thawing 1 kg frozen pieces of oil-contaminated soil is necessary $Q_{COOLANT.}$ = 424.4 kJ. Using equation (1), the quantity of energy is determined ($Q_{COOLANT}$), transferable to the coolant in the inner cavity of the installation housing. The mass of coolant is determined, using the formula for determining the heat amount during the unit heating:

$$Q_{COOLANT.} = c \cdot m \cdot \left(t_f - t_i\right), \qquad (2)$$

where *c* is specific heat capacity of the coolant (for coolant Dixis-65 c=2.974 kJ/kg·°C);

m is the coolant mass, kg;

t_f and t_i are final and initial temperatures of the unit, respectively, °C.

Then, knowing the coolant density (ρ=1091 kg/m³), the volume of coolant and a width of the inner cavity *t* in the housing wall have been determined. Parameter setting calculations for oil-contaminated soil decontamination were conducted using software Microsoft Excel, and results of calculations are presented in Table 1.

Table 1. Geometric parameters results calculations of the installation for oil-contaminated soils decontamination.

Oil-contaminated soil mass, kg	d	D	H	L	t
	mm				
100	450	640	270	1125	90
1000	1000	1350	600	2500	170
20000	2500	3700	1500	6250	580

4. The calculation results

The main geometric parameters for three types of the installation for oil-contaminated soil decontamination with reagent encapsulation method (Table I), when thermal energy released at each stage neutralization process is guaranteed enough for thawing of frozen portions of soils contaminated with petroleum.

5. Results and discussions

Installation for oil-contaminated soil decontamination as compared with the other units has an important advantage, which is to use, if necessary (e.g., in winter) the heat energy released at each stage of oil-contaminated soil decontamination process, in normal conditions it is dissipated into the environment. In addition, a short-term forced carbon dioxide delivery at the final stage of the decontamination process to a high concentration directly into the plant allows replacing the long process of microcapsule shells formation and hardening that occur in natural conditions in the open air.

6. Conclusion

In the installation model the exothermic process heat energy of oil-contaminated soil decontamination is accumulated in a coolant which is in the inner cavity wall of the installation housing, and then, if necessary (e.g., in winter conditions) it is used for the subsequent thawing of frozen portions of oil-contaminated soil placed in a tank, snow to obtain water needed for encapsulation reaction.

The installation allows decontaminating not only the oil-contaminated soils, but saturated with oil ice or snow.

The installation can be used both in stationary conditions and in the field at low (negative) ambient temperatures.

It is possible to design the considered installation model for almost any performance.

References

[1] Adebiyia F M, Adeyemia A F 2010 Characterisation of the petroleum hydrocarbons-contaminated soils around a petroleum products depot *Chemistry and Ecology* Vol 26 Issue 2 pp 137–146 (http://doi:10.1080/02757541003627712)

[2] Adebiyia F M and Afediaa M O 2011 The Ecological Impact of Used Petrochemical Oils on Soil Properties with Special Reference to Physicochemical and Total Petroleum Hydrocarbon Contents of Soils around Automobile Repair Workshops Energy Sources *Part A: Recovery, Utilization and Environmental Effects* Vol 33 Iss 16 pp 1556–1565 (http://doi:10.1080/15567030903397883)

[3] Tomei M C and Daugulis A J 2013 Ex Situ Bioremediation of Contaminated Soils: An Overview of Conventional and Innovative Technologies, Critical Rev *Environmental Science and Technology* Vol 43 Is 20 pp 2107–2139 (http://doi:10.1080/10643389.2012.672056)

[4] Manzetti S 2014 Remediation technologies for oil-drilling activities in the Arctic: oil-spill containment and remediation in open water *Environmental Technology Rev*. Vol 3 Is 1 pp 49–60 (http://doi:10.1080/21622515.2014.966156)

[5] Welander U 2005 Microbial Degradation of Organic Pollutants in Soil in a Cold Climate *Soil and Sediment Contamination: An Intern. J.* Vol 14 Is 3 pp 281–291 (http://doi:10.1080/15320380590928339)

[6] Shtripling L O and Kholkin E G 2017 Development of provisions for oil contaminated soil neutralizing in the conditions of Siberia and the Arctic *AIP Conference Proceedings* Is 1 August 23 p 4 (http://doi.org/10.1063/1.4998848)

[7] Kholkin E G, Shtripling L O and Larionov K S 2017 Liquidation of oil spills consequences in the Arctic zone of Russia using the reagent encapsulation technology *Arktika: ecology and economics* № 1 (25) pp 120–129

[8] Pat. 157884 Russian Federation IPC B 09C 1/08 Installation for the decontamination of oil-contaminated soil, and sludge Larionov K S, Kholkin E G, Shtripling L O № 2015130193/13 appl. 21.07.2015 publ. 20.12.2015 Bull. № 35

Improvement of heat and mass transfer process, a mixture of methane and blast furnace gas

R A Shtykov
Muromskiy institute Vladimirskogo state university, 11, K Marksa, Murom, 602263, Russia

ipmrroman@yandex.ru

Abstract. This article deals with the aspects that lead to more intensive mixing and combustion of turbulent jet mixture of methane and blast furnace gas used in combustion and other combustion processes. Keywords — heat and mass transfer, combustion, turbulent jet gas

1. Introduction

As practice shows, flammable gases used in combustion and other processes have a complex composition, including several combustible, passive (in terms of chemical reaction) components and oxygen. There are combustible elements in the composition of the air, albeit in small quantities. In addition, a small preliminary addition of fuel to air and air (oxygen) to the combustible gas can be a good help in controlling heat and mass transfer processes in turbulent flows, which will lead to more intensive mixing and combustion.

2. Materials

It is believed that the fuel in its composition has two combustible components (A_2 и A_3), gross-combustion reactions, which have the form:

$$\begin{cases} A_2 + \nu_{1'} O_2 \rightarrow \nu_{4'} A_4 + \nu_{5''} A_5 + h_2^*, \\ A_3 + \nu_{1''} O_2 \rightarrow \nu_{4''} A_3 + \nu_{5''} A_5 + h_3^*; \end{cases}$$

Where: ν_i и ν_k - stoichiometric coefficients of the reactants and reaction products; h_i^* - calorific value of the i-th component .

Two products of combustion A_4 и A_5 and, as well as chemically passive gas A_6. Oxidizing agent has in its composition O_2 (A_1)and A_6 is chemically passive gas. Therefore, the boundary conditions in the input sections have the form $\overline{X} = \overline{X}_{+0}$:

$$C_i = \begin{cases} < C_i >_2 & i = 2,3,4,5 \text{ и } 6, \\ 0 & i = 1; \end{cases}$$

Where: C_i - mass concentration of the gas component of the gas in the mixture (kg kg-1);

And when $\overline{X} = \overline{X}_{+0}$:

$$C_i = \begin{cases} 0 & i = 2,3,4,5, \\ < C_i >_1 & i = 1 \text{ и } i = 6. \end{cases}$$

The equations of transport and conservation of the masses of the components have the form $L(C_i) = \omega_i$,

Moreover $\omega_6 = 0$.

Where: - ω_i mass rate of formation / disappearance of the th component of the gas.

3. Theory

We determine the concentration of the introduced combustion products and inert gases as a function of the conservative function

$$\begin{cases} C_4' = <C_4>_1 + \overset{\approx}{C}(<C_4>_2 - <C_4>_1), \\ C_5' = <C_5>_1 + \overset{\approx}{C}(<C_5>_2 - <C_5>_1), \\ C_6 = <C_6>_1 + \overset{\approx}{C}(<C_6>_2 - <C_6>_1). \end{cases}$$

It remains to understand the active gases and products of combustion, where it is possible to use, in particular, the provision that part of the oxygen $C_{1O} = <C_1>_1 - \sum_-$ is distributed between the combustible $C_{1O} = \sum_{i=2}^{N_\alpha+1} C_{1O,i}$.

Then the normalized Schwab-Zel'dovich functions are introduced for each individual fuel component, as well as the corresponding parts of oxygen and formed combustion products, the latter of which are equivalent both to each other and to functions introduced for inert gases. The place of the flame front, determined according to the model of Ya.B. Zeldovich [1,2,3], for each of the combustible components will be

$$\overset{\approx}{C}{}^*_{iO} = \frac{\nu_{i.1} m_i C_{1O,i}}{\nu_{i.1} m_i C_{1O,i} + \nu_{1,i} m_1 (1-k_+) <C_i>_2}$$

But in one zone of displacement there can not be several flame fronts; After the first front of the flame from the side of the fuel, the access of oxygen to other combustible components ceases. Such a judgment leads to dependencies:

$$. \overset{\approx}{C}{}^*_{2O} = \ldots \overset{\approx}{C}{}^*_{(N_\alpha+1)O}$$

These three systems of linear equations uniquely determine the unknown to us $\overset{\approx}{C}_{1O,i}$.

We believe that the initial oxygen in the oxidizer is distributed between the combustible A_2 and A_3

$$<C_1>_1 = <C_{12}>_1 + <C_{13}>_1,$$

And the concentrations of some components consist of three parts

$$C_i = C_i^O + C_i^' + C_i^"$$

Where C_i^O - introduced parts, $C_i^'$ and $C_i^"$ - corresponding to the first and second combustible components of the concentration part.

The Schwab-Zel'dovich functions for the first fuel are introduced

$$\widetilde{C}_2 = C_2 + \Omega_{24'} C_4^', \quad \widetilde{C}_1^' = C_1^' + \Omega_{1'4'} C_4^', \quad \widetilde{C}_5^' = C_2 + \Omega_{25'} C_5^'. \ (1)$$

Where: $\Omega_{ik} = \nu_k m_k / (\nu_i m_i)$ is the ratio of the moduli of the chemical reaction rates of the first and second components.

Similarly for the second fuel we have:

$$\widetilde{C}_3 = C_3 + \Omega_{34"} C_4^", \quad \widetilde{C}_1^" = C_1^" + \Omega_{1"4"} C_4^", \quad \widetilde{C}_5^" = C_3 + \Omega_{35"} C_5^".$$

We introduce the Schwab-Zeldovich normalized functions

$$\tilde{\tilde{C}}' = \frac{\widetilde{z}_i^' - <\widetilde{z}_i^'>_1}{<\widetilde{z}_i^'>_2 - <\widetilde{z}_i^'>_1} \qquad z_i^' = C_2,\ C_1^',\ C_5^'$$

$$\tilde{\tilde{C}}" = \frac{\widetilde{z}_i^" - <\widetilde{z}_i^">_1}{<\widetilde{z}_i^">_2 - <\widetilde{z}_i^">_1} \qquad z_i^" = C_3,\ C_1^",\ C_5^"$$

Excluding intermediate functions, the concentration of components in the fuel zone is determined.

For the first fuel:

$$\begin{cases} C_2 = \left(<C_2>_2 + \Omega_{24'} <C_1^'>_1\right)\tilde{\tilde{C}} - \Omega_{24'} <C_1^'>_1, \\ C_4^' = <C_1^'>_1 \Omega_{4'1'}\left(1-\tilde{\tilde{C}}\right), \quad C_5^' = <C_1^'>_1 \Omega_{5'4'}\left(1-\tilde{\tilde{C}}\right); \end{cases}$$

For the second fuel:

$$\begin{cases} C_3 = \left(<C_3>_2 + \Omega_{34"} <C_1^">_1\right)\tilde{\tilde{C}} - \Omega_{24'} <C_1^">_1 \\ C_4^" = <C_1^">_1 \Omega_{4"1"}\left(1-\tilde{\tilde{C}}\right), \quad C_5^" = <C_1^">_1 \Omega_{5"4"}\left(1-\tilde{\tilde{C}}\right). \end{cases}$$

Concentrations in the oxidant zone for the first fuel:

$$\begin{cases} C_1^' = <C_1^'>_1 - \left(<C_1^'>_1 + <C_2>_2 \Omega_{1'2}\right)\tilde{\tilde{C}}; \\ C_4^' = <C_2>_2 \Omega_{4'2} \tilde{\tilde{C}}, \quad C_5^' = <C_2>_2 \Omega_{5'2} \tilde{\tilde{C}}; \end{cases}$$

For the second fuel

$$\begin{cases} C_1'' = <C_1''>_1 - \left(<C_1''>_1 + <C_3>_2 \Omega_{1''3}\right)\tilde{\tilde{C}}; \\ C_4'' = <C_3>_2 \Omega_{4''3}\tilde{\tilde{C}}; \quad C_5'' = <C_3>_2 \Omega_{5''2}\tilde{\tilde{C}}. \end{cases}$$

The flame front for these combustible materials is defined as

$$\tilde{\tilde{C}}'* = \frac{1}{1 + \Omega_{21'} <C_2>_2 / <C_1'>_1}, \quad (2)$$

$$\tilde{\tilde{C}}''* = \frac{1}{1 + \Omega_{31''} <C_3>_2 / <C_1''>_1}. \quad (3)$$

Since the flame front for combustible components is single, the values of (2) and (3) must coincide: $\tilde{\tilde{C}}'* = \tilde{\tilde{C}}''*$.

4. Results

Hence, the proportions of the combustible components corresponding to the initial oxygen concentration

$$<C_1'>_1 = \frac{\Omega_{1''3} <C_1>_1 <C_2>_2}{\Omega_{1'2} <C_3>_2 + \Omega_{1''3} <C_2>_2},$$

$$<C_1''>_1 = \frac{\Omega_{1'2} <C_1>_1 <C_3>_2}{\Omega_{1'2} <C_3>_2 + \Omega_{1''3} <C_2>_2}.$$

Therefore

$$\tilde{\tilde{C}}* = 1/\left(1 + <C_3>_2 \Omega_{31''} + <C_2>_2 \Omega_{21'}\right)$$

And the concentrations of the components are determined in the fuel zone as

$$\begin{cases} C_1 = 0, \\ C_2 = \left(<C_2>_2 + \Omega_{24'} <C_1'>_1\right)\tilde{\tilde{C}} - <C_1'>_1 \Omega_{24'}, \\ C_3 = \left(<C_3>_2 + <C_1''>_1 \Omega_{34''}\right)\tilde{\tilde{C}} - <C_1''>_1 \Omega_{34''}, \\ C_4 = \left(<C_1'>_1 \Omega_{4'1} + <C_1''>_1 \Omega_{4''1''}\right)\left(1 - \tilde{\tilde{C}}\right) + <C_4>_1 + \left(<C_4>_2 - <C_4>_1\right)\tilde{\tilde{C}}, \\ C_5 = \left(<C_1'>_1 \Omega_{5'4} + <C_1''>_1 \Omega_{5''4''}\right)\left(1 - \tilde{\tilde{C}}\right) + <C_5>_1 + \left(<C_5>_2 - <C_5>_1\right)\tilde{\tilde{C}}; \end{cases}$$

nd in the oxidant zone -

$$\begin{cases} C_1 = <C_1>_1 - (<C_1>_1 + <C_1>_1 \Omega_{1'2} + <C_3>_2 \Omega_{1''3})\tilde{\tilde{C}} \\ C_2 = 0, \qquad C_3 = 0, \\ C_4 = (<C_2^{'}>_2 \Omega_{4'2} + <C_3>_2 \Omega_{4''3})(1-\tilde{\tilde{C}}) + <C_4>_1 + (<C_4>_2 - <C_4>_1)\tilde{\tilde{C}}, \\ C_5 = (<C_2>_2 \Omega_{5'2} + <C_3>_2 \Omega_{5''3})(1-\tilde{\tilde{C}}) + <C_5>_1 + (<C_5>_2 - <C_5>_1)\tilde{\tilde{C}}. \end{cases}$$

The mass concentration of chemically passive gas is determined by a single formula throughout the heat and mass transfer region:

$$C_6 = <C_6>_1 + (<C_6>_2 - <C_6>_1)\tilde{\tilde{C}}.$$

In favor of the reliability of the expressions obtained for the concentrations of components, the following fact can be cited: the sum of the mass concentrations of the components in each of the zones (fuel and oxidizer) is equal to unity.

5. Discussion of results

As an example, combustion of a homogeneous mixture of natural and blast furnace gases flowing into an infinite space of resting and confused airflow is considered. For simplicity, it is assumed that natural gas consists only of methane CH_4; Blast furnace gas has a mass composition: 30% - carbon dioxide CO, 10% - carbon dioxide CO_2, 60% - nitrogen N_2; Air has in the composition one oxidizer – oxygen O_2 (23.2%) and inert gas - nitrogen (76.8%).

$C_П$, $C_Д$ - the mass fractions of natural and blast-furnace gases in the fuel composition are denoted by and denoted by $C_П + C_Д = 1$

those. The fuel consists only of natural and blast furnace gases.

Each fuel has its own stoichiometric equation:

$$CH_4 + 2O_2 \rightarrow CO_2 + 2H_2O + h_1^*, \quad (4)$$

$$CO + 0.5O_2 \rightarrow CO_2 + h_2^*. \quad (5)$$

In calculations for calorific values, the values are: for methane is equal (h_1^*) to 11985 kcal / kg, and for carbon dioxide (h_2^*) - 2450 kcal / kg.

The initial concentration of oxygen $<C_O>_1$ is distributed between methane $(<C'_O>_1)$ and carbon dioxide $(<C''_O>_1)$. Therefore

$$<C_O>_1 = <C'_O>_1 + <C''_O>_1,$$

Where, on the basis of the generalization of the model described above,

$$<C'_O>_1 = \frac{<C_1>_1 <C_O>_1}{<C_1>_2 + <C_2>_2 /7}.$$

Multiplying equation (4) by $< C'_O >_1 / (\nu_{O1} < C_O >_1)$, (5) by $< C''_O >_1 / (\nu_{O2} < C_O >_1)$ and adding the obtained equations, we arrive at

$$\nu_0 O_2 + \nu_1 CH_4 + \nu_2 CO \rightarrow \nu_3 CO_2 + \nu_4 H_2O + \nu_1 h_1^* + \nu_2 h_2^*$$

Where

$$\nu_0 = 1, \quad \nu_1 = < C'_O >_1 / (2 < C_O >_1), \quad \nu_2 = 2\,(1 - < C'_O >_1 / < C_O >_1),$$
$$\nu_3 = 2\,(1 - 0.75 < C'_O >_1 / < C_O >_1), \quad \nu_4 = < C'_O >_1 / < C_O >_1$$

According to the provisions of the generalized model, instead of the N transport and conservation equations for the i-th component (for reagents), we obtain a single equation $L(\tilde{\tilde{C}}) = 0$ with boundary conditions in the input section $< \tilde{\tilde{C}} >_1 = 0, \ < \tilde{\tilde{C}} >_2 = 1$.

6. Summary and conclusions

In the case of diffusion combustion, the maximum temperature takes place at the flame front. At moderate flow velocities, the temperature at the flame front remains the same.

Figure 1 shows the values of the temperature at the flame front for different values T_1 and T_2 in the form of a function of $C_П$. At a mass ratio of methane to blast furnace gas of 1: 1, the temperature at the flame front is reduced by no more than 50 degrees, compared to the combustion of pure methane, and more than 7 percent of natural gas is saved. If the share of natural gas in the fuel mixture is 20% of the total mass, the temperature of the torch is reduced by 100 degrees and the fuel economy is 20%. Along with this, a social effect is achieved, which is expressed in reducing the amount of harmful emissions to the atmosphere.

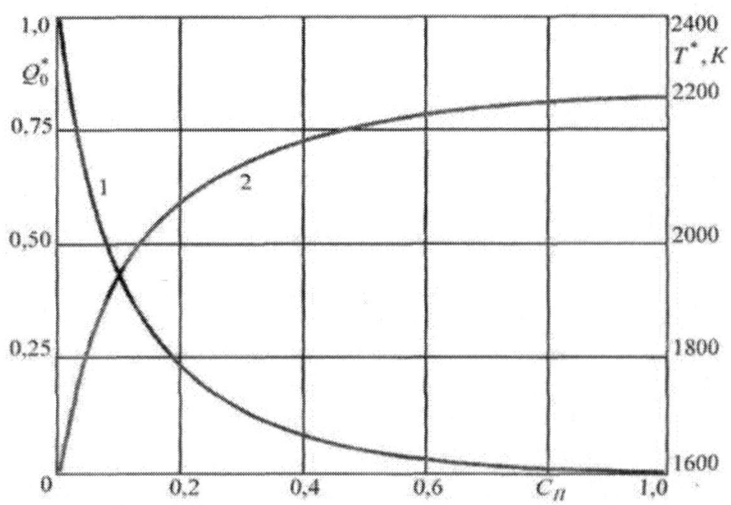

Figure 1. Changes in the fraction of blast furnace gas as a function of the mass fraction of methane in the mixture methane + blast furnace gas

IOP Conf. Series: Journal of Physics: Conf. Series **944** (2017) 012107 doi:10.1088/1742-6596/944/1/012107

References

[1] Johnson C.R. Lectures on adaptive parameter estimation. Englewood Cliffs. - N.J. Prentice Hall, 1998. - 160 p.

[2] Shtykov R.A. Improvement of a method of estimating the heat and mass transfer of the linear part of the pipeline network based on its static functioning mode. Measurement Techniques. 2013. T. 56. № 3. C. 297-300

[3] Walter E., Pronzato L. Identification of parametric models from experimental data. - London: Spriger, 1997.- p 413

Improvement of automatic control system for high-speed current collectors

O A Sidorov[1], V N Goryunov[2] and A S Golubkov[1]

[1] Omsk State Transport University, 35, Marksa ave., Omsk, 644046, Russia
[2] Omsk State Technical University, 11, Mira ave., Omsk, 644050, Russia

e-mail: anton.golubkov@omgups.com

Abstract. The article considers the ways of regulation of pantographs to provide quality and reliability of current collection at high speeds. To assess impact of regulation was proposed integral criterion of the quality of current collection, taking into account efficiency and reliability of operation of the pantograph. The study was carried out using mathematical model of interaction of pantograph and catenary system, allowing to assess contact force and intensity of arcing at the contact zone at different movement speeds. The simulation results allowed us to estimate the efficiency of different methods of regulation of pantographs and determine the best option.

1. Introduction
One of the major problems at higher speeds on the electrified Railways is provision of reliable current collection in the context of significant electrical current, mechanical and aerodynamic loads. To improve the quality of the current collection at high speeds we can use automatic control systems that allows to regulate contact force between pantograph and contact wire with respect to conditions of the interaction process.

2. Statement of the problem
To achieve high quality current collection it is necessary to provide continuous contact of pantograph strips with contact wires, the pressing force of strips on contact wire should be in the narrow range of acceptable values.

Reducing the contact force leads to sparking and arcing in the contact that causes excessive wear of the current collecting strips and overheating of the pantograph. The increase of force leads to increased mechanical wear due to friction forces and heating of the pantograph.

With the increase of speed of electric rolling stock dispersion of contact force is increased due to raise of aerodynamic effects and vibration of vehicle. [1].

The study was conducted to assess the effectiveness of the various automatic control systems in the design of modern high-speed pantographs that operates on railroads of Russia.

3. Theory
When designing a system of automatic control of the pantograph we need to define quality criterion of regulation, which will allow us to select the optimal implementation of control system. The main criterion are the efficiency of the current collection associated with the service life of the contact materials and reliability, that depends on probability of failure of pantograph or catenary.

The current collection efficiency can be estimated by the wear rate of the contact strips of the pantograph, which at nominal current depends on the contact force. With increasing contact force increases the intensity of wear, reduction of force leads to arcing and electrical erosion of materials. Dependence of wear rate from contact force $\gamma(P_{cf})$ is a U-shaped curve, the shape of the curve is determined by the contact materials, current density in the contact and other factors. For each pair of materials in contact, the U-shaped curve is determined by the results of experimental studies [2].

The reliability of current collection is influenced by a big number of random factors, such as the technical condition of overhead lines and pantograph, the impact of weather events and many others. It is not possible to take into account impact of all factors in assessing the quality of regulation, therefore, as the main criterion of reliability of current collection we used requirements for safe operation of power equipment, among which was selected main criterion – maximum allowable value of the contact force $P_{cf\ max}$. Exceeding this value can lead to failures of elements of the catenary network. To account for the excess of the contact force over the allowable value in quality criterion of current collection used penalty coefficient $k_{c\ max}$. The final quality criterion of current collection can be expressed by the following relationship:

$$k_c(P_{cf}) = \begin{cases} \gamma(P_{cf}), P_{cf} \le P_{cf\ max} \\ k_{cmax}, P_{cf} > P_{cf\ max} \end{cases}. \tag{1}$$

Evaluation of quality of automatic regulation should be performed by measuring the integral quality criterion I_{kc} obtained by integrating criterion $k_c(P_{cf})$ by the time of experiment:

$$I_{kc} = \int_0^T k_c(P_{cf}(t))\mathrm{dt} \tag{2}$$

For optimal regulation of contact force it is necessary to minimize the value of the integral quality criterion of current collection.

One of design feature of most modern pantographs is rubber-cord pneumatic lifting mechanism. This mechanism allows to change the contact force of the pantograph by adjusting the pressure in the pneumatic system. To implement the automatic regulation system of pantograph pneumatic mechanism can be equipped with additional electro-pneumatic valves, that will allow to change pressure in the rubber-cord of the pantograph. Depending on the operating conditions, the controller may switch one of the valves, raising or lowering the lifting force of a pantograph. Figure 1 shows a diagram of the pneumatic control with two pressure steps [3].

The disadvantage of such a system is a significant duration of the transition process when valve is switched caused by inertia of the pantograph lifting mechanism.

The most effective input signal for automatic control system is contact force, since it directly affects the quality criterion of current collection. Difficulty of measurement of contact force caused by a number of factors such as high cost of equipment, the necessity to modify the design of a pantograph for mounting the sensors and the requirements for electrical safety. In this regard other input signals was considered in development of automatic control system. As such signals is considered the speed of the electric rolling stock and the ratio of arcing in sliding contact.

The ratio of arcing is defined as the ratio of arcing duration to the total duration of the movement, it indicates loss of contact and low quality of current collection. Measurement of arcing can be executed with optical or electromagnetic sensors. Sensors can be placed on the roof of electric rolling stock and don`t require modification of the pantograph [4].

To evaluate process of automatic control of contact force we developed a mathematical model of the pantograph in mathematical environment MATLAB/Simulink. The scheme of the pantograph is an uniaxial two-mass model with two degrees of freedom. The model consists of a point-mass system of pantograph parts, the frequency and amplitude of pantograph base vibrations have a dependency on the speed of the electric rolling stock.

The static uplift force of the pantograph is set by the model of rubber-cord lifting mechanism, its input connected to the signal of pressure in the pneumatic system. Lifting and lowering mechanism is modeled with a transfer function which is determined experimentally for the chosen design of the pantograph [5].

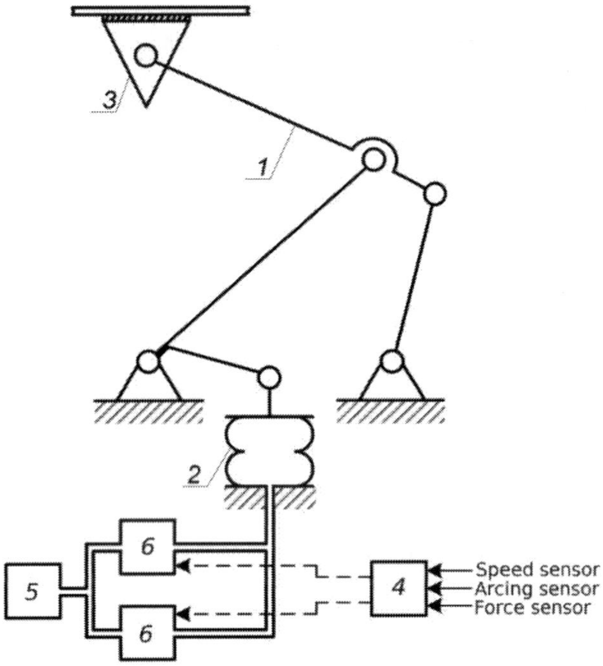

Figure 1. System pneumatic control of the pressing of current collector: 1 – system of movable frames of the pantograph; 2 – rubber-cord; 3 – contact strips; 4 – control unit; 5 – source of compressed air; 6 – electro-pneumatic dispenser.

Arcing at the contact is modeled with a uniform random function. The value of the function is set to zero when contact force exceed minimum threshold values $P_{\mathrm{cf\,min}}$.

Figure 2 shows scheme of a mathematical model of automatic control system of a pantograph constructed in Simulink.

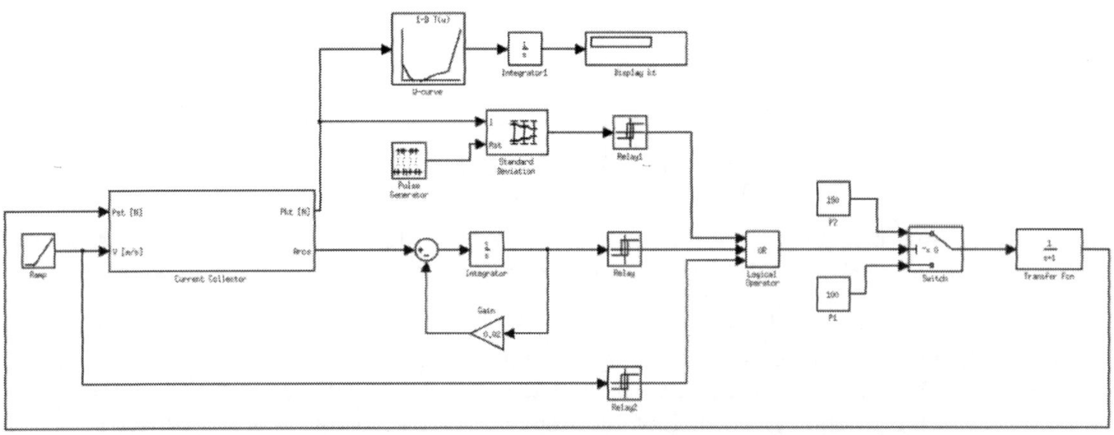

Figure 2. Scheme of mathematical model of automatic control of a pantograph in the Simulink environment.

The movement speed control subsystem uses Relay element, which produces the control signal for switching to the high pressure when the threshold speed value v_t is reached.

Since the lifting and lowering mechanism is characterized by long transient process, we can`t use rapidly changing contact force signal directly in control system, so the contrct force automatic control subsystem performs calculation of the standard deviation of contact force σP_{cf} at a fixed time window. System switches pressure when standard deviation of contact force reaches threshold limit $\sigma P_{cf\,t}$.

Arcing control subsystem performs integration of the arcing intensity $i(t)$. When integrated signal reaches threshold value $I_{\Sigma t}$, the vale is switched to high pressure. After the cessation of sparking, the system switches pressure to previous level, providing a return of depression to the pre-set level.

The control signals of all the controllers are combined by using Boolean function "OR", which allows parallel work of each subsystem.

4. The results of the experiments

The aim of the study was to determine the optimal control scheme of pantograph and suitable threshold values for each control subsystem. This was accomplished through a series of experiments in which we simulated the process of current collection during acceleration of electric rolling stock up to the maximum speed. Figure 3 shows results of the simulation of current collection with automatic control system. Changes of static contact force P_{st} corresponds to the switching instants of the pressure valves caused by signal the automatic control system. Graph of intensity of arcing $i(t)$ shows the relative intensity of arcing at the points of the low contact force. Quality criterion of current collection k_c have U-shaped form that resembles the shape of the intensity of wear of the contact material; the minimum value of the criterion $k_{c\,min}$ corresponds to the optimal value of the contact force $P_{cf\,opt}$, the maximum value of the criterion corresponds to the zero value of contact force (loss of contact) or exceeding maximal allowable value for safe operation $P_{cf\,max}$.

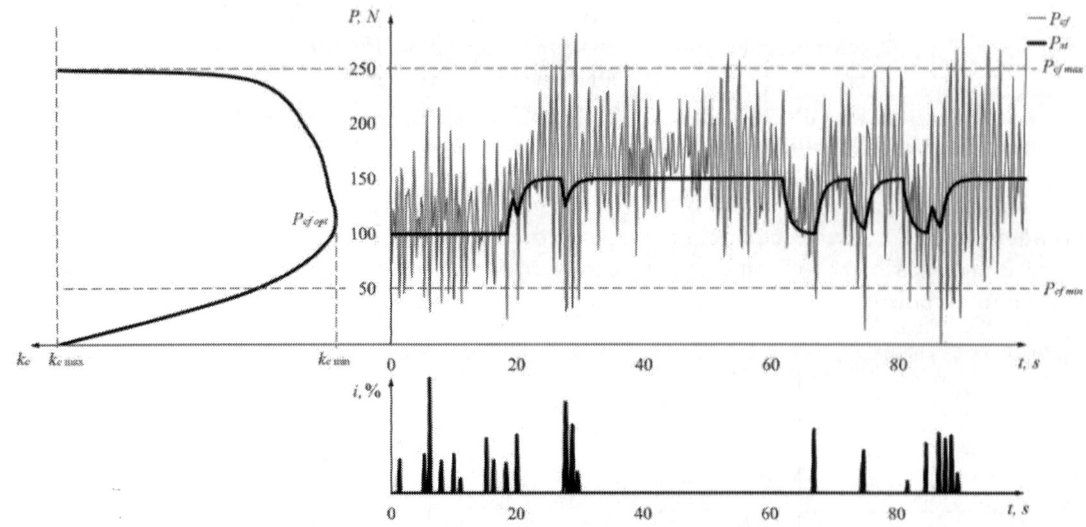

Figure 3. The results of simulation of automatic control system of pantograph: k_c – quality criterion of current collection, P_{cf} – contact force, P_{st} – static contact force, $i(t)$ – intensity of the arcing.

Table 1 shows the resulting values of integral quality criterion of current collection while using each control subsystem individually. For easy comparison of the effectiveness of each controller subsystem table shows relative values of a gain of integral quality criterion $\Delta I_{kc} = I_{kc\,b} - I_{ct}$ relative to the base value of the criterion $I_{kc\,b}$, obtained by simulation of the process without automatic control system.

The simulation results show that the most effective regulation of pantograph can be achieved by using input signal of contact force and arcing. Controlling by movement speed signal have lower efficiency due to the lack of feedback loop. Figure 4 shows relative increment of integral quality criterion of current collection with different threshold parameters of the regulators.

For evaluation of simultaneous work of control subsystems we conducted a series of experiments with different combinations of threshold values of contact force deviation and arcing rate. The results of the experiments are shown in table 2.

The results of the experiments showed a slight increase in the quality of current collection when combined control scheme is used. The maximum quality increase above separated control subsystem amounted to 0.3 %.

Table 1. Quality criterion of current collection with different regulation schemes.

Control by contact force			Control by arcing			Control by movement speed		
$\sigma P_{cf\,t}$, N	I_{kc}	$\Delta I_{kc}/I_{kc\,b}$	$I_{\Sigma t}$	I_{kc}	$\Delta I_{kc}/I_{kc\,b}$	v_t, м/с	I_{kc}	$\Delta I_{kc}/I_{kc\,b}$
10	31.5	0.6 %	5	31.0	2.2 %	25	31.5	0.6 %
20	31.2	1.6 %	10	30.2	4.7 %	30	31.3	1.3 %
30	31.1	1.9 %	15	30.1	5.0 %	35	30.7	3.2 %
40	30.2	4.7 %	20	29.6	6.6 %	40	30.2	4.7 %
50	29.2	**7.9 %**	25	29.5	6.9 %	45	30.1	**5.0 %**
60	29.7	6.3 %	30	29.1	**8.2 %**	50	30.4	4.1 %
70	30.9	2.5 %	35	29.4	7.3 %	55	30.5	3.8 %
80	31.6	0.3 %	40	29.6	6.6 %	60	31.6	0.3 %

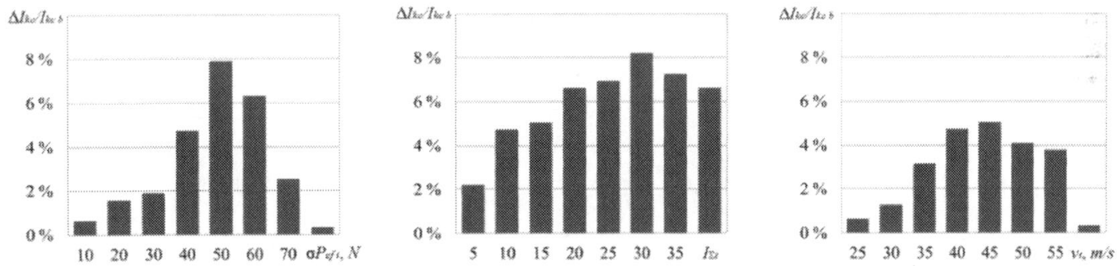

a) b) c)

Figure 4. The relative increment of the integral quality criterion of current collection, when using control system with following input signals: contact force (a), arcing rate (b) and movement speed (c).

Table 2. Increase of integral quality criterion of current collection δI_{kc} / $I_{kc\ b}$ (%) with combined control subsystems.

$\sigma P_{cf\,t}$, N	$I_{\Sigma t}$							
	5	**10**	**15**	**20**	**25**	**30**	**35**	**40**
10	0.6	0.6	0.9	1.6	1.3	1.3	0.9	0.9
20	1.6	1.6	1.6	1.6	1.6	1.6	1.6	1.6
30	1.9	1.9	1.9	1.9	1.9	1.9	1.9	1.9
40	1.9	1.9	1.9	4.4	4.7	4.7	4.7	4.4
50	2.2	4.7	6.9	7.9	7.6	7.6	7.3	6.6
60	2.2	3.8	4.4	6.6	7.6	**8.5**	7.9	7.3
70	2.2	3.5	4.7	5.4	6.3	8.2	7.3	6.3
80	2.2	4.7	5.0	6.3	6.6	8.2	7.3	6.6

5. Discussion of results

The experimental results show that the use automatic control system in design of pantographs can improve the quality and reliability of the current collection at high speeds. Using movement speed as input signal for control system is less effective compared to other options, since the control signal is not linked with contact force directly and lacks feedback loop. Use of arcing rate as input signal for automatic control system can be almost as effective as using direct contact force measurement.

6. Conclusion

Considered automatic control system with stepping change of pressure in pneumatic lifting mechanism of pantograph showed the possibility of effective application for improve the quality of current

IOP Conf. Series: Journal of Physics: Conf. Series **944** (2017) 012108 doi:10.1088/1742-6596/944/1/012108

collection at high speeds. Use arcing rate as input signal for control system appears to be most promising judging by high effectiveness and low implementation costs

To further improve automatic control systems of pantographs it is possible to consider increasing the number of steps of pressure regulation and optimization of control algorithms. To achieve a substantial improvement of the current collection it is necessary to solve the problem of long duration of transients with a step-like pressure control, this can be achieved by means of special control mechanisms embedded in the design of the pantograph.

References

[1] Sidorov O A, Smerdin A N and Zhdanov V A 2012 Evaluation procedure of current collection system readiness at railway mainlines' high-speed sections *Vniizht Bulletin (Railway Research Institute Bulletin)* **2** pp 31–35

[2] Sidorov O, Filippov V and Stupakov S 2015 Research of Electromechanical Wear of Contact Pairs of Electric Transport Current Collection Devices *Friction And Wear* **36** pp 511–517

[3] Sidorov O A, Arkashev A E, Larkin I V and Sosnovskiy S Y 2011 Patent of Russian Federation No. 105862, B 60 L 5/00. Current collector of electric rolling stock Claim date 02/11/2011; Publication date 06/27/2011

[4] Sidorov O A, Smerdin A N, Chertkov I E and Golubkov A S 2008 Patent of Russian Federation No. 82445, B 60 L 5/00. Pantograph arcing registration device / Claim date 11/19/2008; Publication date 04/27/2009

[5] Sidorov O A, Smerdin A N and Golubkov A S 2015 Experimental studies of pantographs *Railways Transport* **11** pp 69–70

IOP Conf. Series: Journal of Physics: Conf. Series **944** (2017) 012109 doi:10.1088/1742-6596/944/1/012109

Backup Mechanical Brake System of the Wind Turbine

E A Sirotkin[1], E V Solomin[1] and S A Gandzha[1], I M Kirpichnikova[1]

[1]International innovative centre «Alternative Energy», South Ural State University (National Research University), 76a, Lenina prospect, Chelyabinsk, 454080, Russia

E-mail: easirotkin@susu.ru

Abstract. Paper clarifies the necessity of the emergency mechanical brake systems usage for wind turbines. We made a deep analysis of the wind turbine braking methods available on the market, identifying their strengths and weaknesses. The electromechanical braking appeared the most technically reasonable and economically attractive. We described the developed combined electromechanical brake system for vertical axis wind turbine driven from electric drive with variable torque enough to brake over the turbine even on the storm wind speed up to 45 m/s. The progress was made due to the development of specific kinematic brake system diagram and intelligent control system managed by special operation algorithm.

1. Introduction

During the last decade several countries have been actively engaged in the development of the Arctic territories. The northern territories are full of undiscovered mineral deposits, and that was the main reason of this activity. According to the UN estimates, the cost of the mineral reserves on the RF Arctic territories comprises over 21 trillion dollars [1, 2]. About 90 % of the Russian gas are produced on the Yamal [3, 4]. Besides, the Arctic territories are treated as strategic locations. Military facilities (for example, air defense complexes) or navigation facilities (GLONASS stations) can be placed there [5]. The Yamal peninsula is also unique because it is equidistant from Paris and Beijing. Thus, in the existing geopolitical environment the region can become a central link in the energy reserves supply chain throughout the entire continent.

Currently, two thirds of the Russian territory are still not electrified by the centralized network. [6, 7]. It can be easily explained economically – the cost of installing power electric lines and their operation and maintenance expenses will be very high, and it may take too long for them to pay back. However, it should be noted that even now the territories connected to the common electric grid are to a certain extent covered by decentralized and self-contained power supply. Yet, the cost of 1 kW-h at some facilities can reach 100 rubles and more [8–10]. Such a high price is predetermined by a low development level of the applied energy technologies. According to different estimates, from 10 to 25 mln. people live on the energetically decentralized territory of the country [11–14]. Over 50 regions are considered energy-deficient, which jeopardizes their own energy security [15].

Figure 1 shows a night satellite image of the Eurasian continent [16]. It is clear from the image that the European (western) and the southern parts of Russia are most electrified, while «the darkest» are the northern and the far-eastern territories. For a stable development of these territories it is necessary to have a relevant infrastructure with a sufficient power supply. Currently, diesel fuel is the main energy carrier in these regions [17]. However, this type is environmentally harmful, and the fuel transportation cost increases the cost of 1 kW-h making it swollen [18].

Figure 1. Eurasian continent screenshot during nighttime.

The noted problem can be solved by using wind turbines (WT). Moreover, it will be expedient to use small wind turbines with the power of up to 100 kW, which can operate under self-sustainment (in a non-networked mode) [19]. Figure 2 shows the distribution of the average annual wind speeds on the RF territory and the adjacent territories [20]. If we compare figure 1 and 2, we can see that the maximum average annual wind speeds are registered on the «unilluminated» northern and far-eastern territories. The northern coastal regions of the continent are within the green area of the wind load, which corresponds to the wind speed range of 5…10 m/s. Some far-eastern regions, in particular, the Kamchatka peninsula, are within the area of stably high wind speeds of 10…15 m/s (marked yellow). Thus, the use of wind turbines on the said territories will be maximally profitable for the economic and technical reasons.

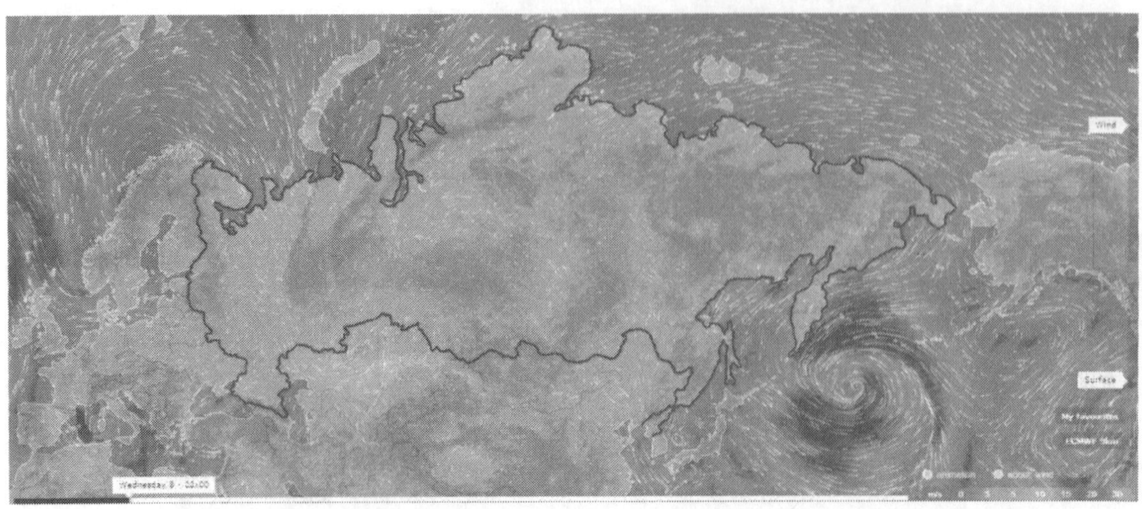

Figure 2. Annual wind speeds spherical map covering Eurasian continent (RF border is marked red).

2. Methods of the WT Power Control

An overwhelming majority of wind turbine manufacturers design their products in such a way, that the nominal power is generated at the wind speed of 11 m/s [21]. At the same time, they do not consider climatic and geographical factors of the future wind turbine operation. Based on such design features of wind turbines, at the wind speed of over 11 m/s it is necessary to limit power at the wind wheel [22].

To date, there are several traditional ways to limit power at the wind turbine rotor:

1. Increase of the electric load on the generator. When the wind speed grows, the WT controller increases the electric load on the generator, thus, decreasing the rotor speed [23]. This power control method is most energy-efficient from the perspective of power takeoff, however, it is not meant for operation at continuous high wind speeds [24]. It is because the generator windings will be heated with an increasing electric load and at some time reach the critical temperature, after which they have to be cooled down. Periods of a high wind load generally last from one day to several weeks.

2. Closure of the generator windings using the pulse-width modulation. Such braking method enables to brake the wind turbine rotor within the necessary rotation frequency ranges. However, a weak point of this method is a quick overheating of the electrical generator, whereas the kinetic energy of the wind wheel is generated into the heat energy on the generator windings [25]. As a result, this method can be used only for a short period.

3. Airbraking. There are many configurations of aerodynamic regulators, including: blades rotating around the flap axis, special brake blades, which turn around depending on the rotation frequency, devices moving the wind wheel away from the approach flow. However, in practice such regulators were efficient only at large megawatt-class WTs.

4. Mechanical brake systems. This type of systems is most primitive out of all the above-listed. However, at this technical and technological development stage it is most reliable and most economically sound. Firstly, mechanical braking of the wind wheel implies turning of the kinetic energy into the heat one generated on the brake shoes. Modern friction materials preserve their operational characteristics at the temperatures of up to 800 oC, while it is not recommended to heat the generator windings above 100 oC [26]. Secondly, such mechanisms have maximum maintainability, whereas worked-out mechanical parts can be replaced without opening the wind turbine body. Thirdly, mechanical brake systems play the part of a compulsory backup brake system. According to the standards, all the WT, which power exceeds 5 kW, must be fitted with at least one backup brake system [27].

3. WT Electromechanical Brake System

This paper considers an electromechanical brake system with an actuating mechanism in the form of a three-pawl unit for a 3 kW vertical-axis WT. The strong points of such wind turbine include a lack of necessity for wind orientation, low vibration and acoustic irradiation indices, relatively high wind energy efficiency (up to 0,4) [28].

The electromechanical brake system is integrated into the WT structure in the mating place of the mast and the hub. Figure 3 shows a 3-d model of the actuating three-pawl mechanism directly used for braking. Each pawl is fitted with a replaceable friction lining. A limit switch for timely system disabling in the reverse mode is located inside the pawl body.

Due to the multiplicity of advantages of this wind turbine can be installed near residential properties. And also, it may be mounted on the roof of the building. In this case the structure will not threaten the vibrational oscillations of the wind turbine. Given the close location of wind turbines to the residential infrastructure necessary to provide adequate operational safety equipment. First and foremost, we need to prevent the destruction of the windmill blades, since the rotor speed 180 rpm variation of the blades can be up to 50 m in radius. Therefore, the braking system should prevent such situations and to ensure safe work of the personnel.

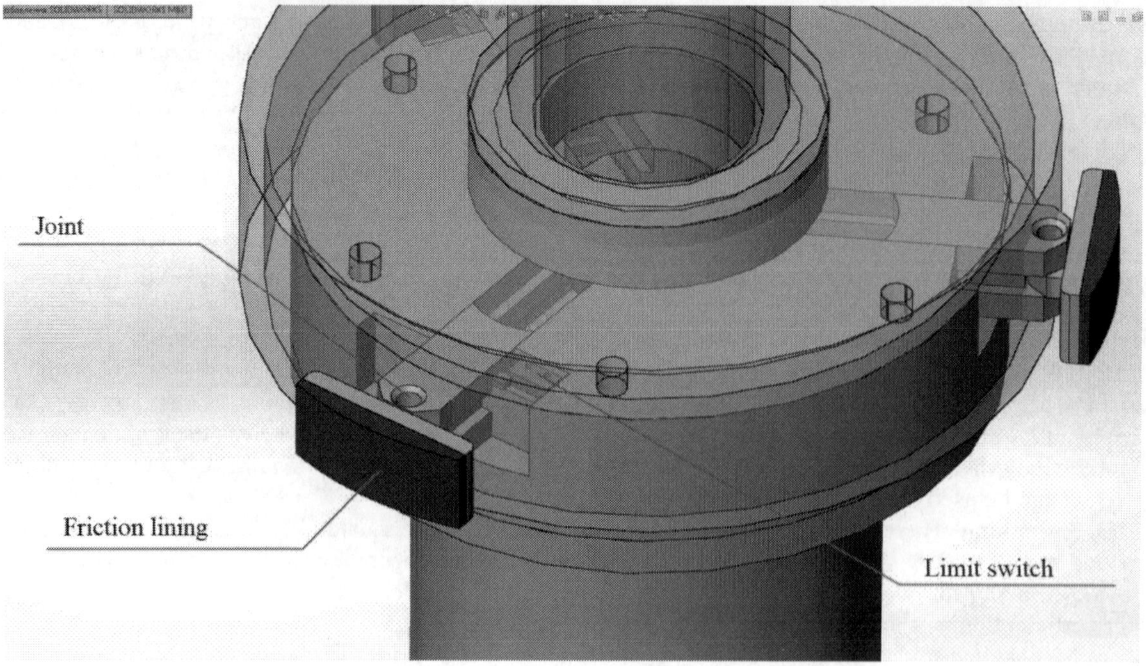

Figure 3. Computer model of the actuating three-pawl assembly of braking system.

Figure 4 shows a kinematic diagram of the entire brake system. When the permissible rotation frequency of the WT rotor or the generator's maximum operating temperature is exceeded, or in case of vibrations, the system's electric drive is activated. Then, power is supplied from the drive through the planetary reduction gear to the scroll gear (with an Archimedian spiral). The scroll gear is fitted with friction pawls with special combs cut in the bottom part. When the scroll gear rotates, the pawl combs begin to move along the spiral, and the pawls move radially. The pawls moving from the center to the periphery bump into the contact wall, which is fixed on the hub and rotates together with it. Thus, when the pawls touch the contact wall, the sliding friction process begins, and the wind wheel brakes.

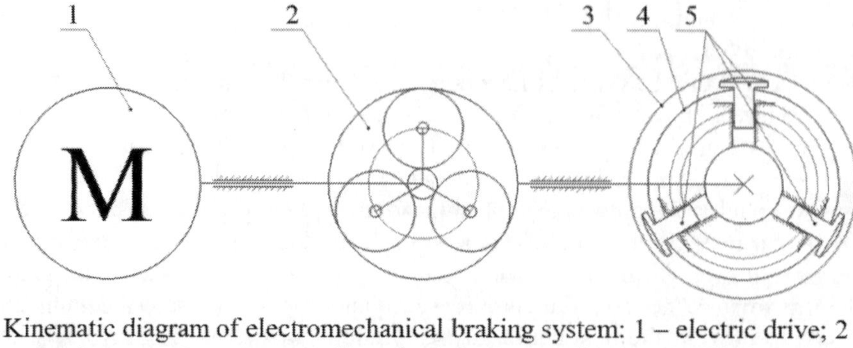

Figure 4. Kinematic diagram of electromechanical braking system: 1 – electric drive; 2 – planetary reduction gear; 3 – contact wall on the hub; 4 – scroll gear; 5 – friction pawls.

Such electromechanical brake system is to be used when the regular control system is unable to carry out the control process (for example, in case of a hurricane force wind) or in case of emergency [29]. These can be technical failures of the regular control system and external effects (lightning stroke or influence of electrical appliances) [30]. Therefore, the designated microcontroller, being a part of the top-level controller, is meant to control the system, however, it can also operate under self-sustainment. Besides, this controller and the system's electric drive are also fed by a separate autonomous accumulator. The charge of this accumulator passes parallel to the charge of the main accumulator bank, being a part of the entire energy complex.

The control system tracks the status of the WT's three main parameters: generator temperature, wind wheel rotation speed and metal structure's vibration oscillations. If necessary, the system activates the brake's electric drive as per the algorithm and scheme shown in fig. 5 and 6.

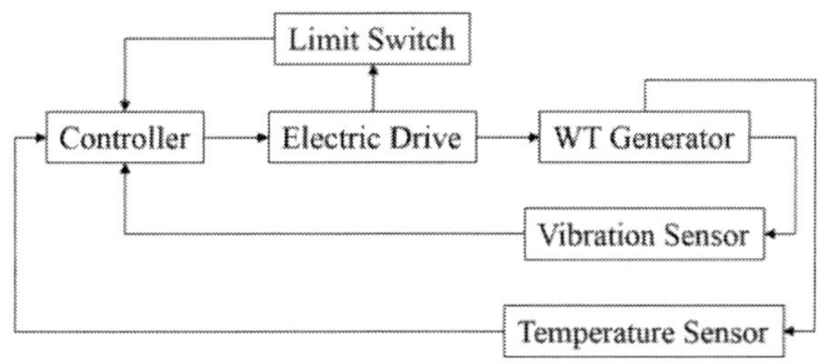

Figure 5. The flow chart of interconnection between braking system electric components.

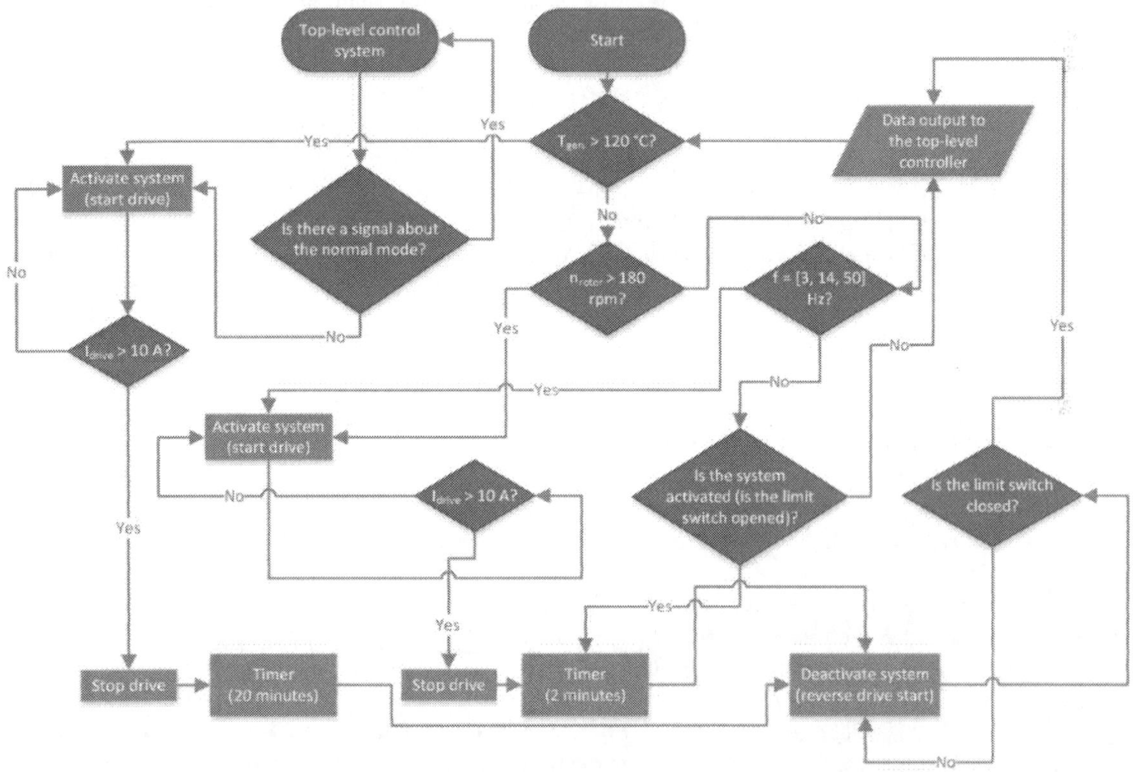

Figure 6. Electromechanical braking system operation algorithm.

Actuation and operation of the electric drive continue, until the current intensity in the drive reaches 10 A. It proves that the pawls press the contact wall with the force of 30,000 N. Such pressure must be applied to brake the vertical-axis wind turbine. Then, the rotor stops, and the system waits for 2 minutes, after which the electric drive is re-activated but in the reverse mode. The drive keeps operating, until the pawls return to the initial position and press the limit switch. A brake signal can be given to the electromechanical brake control system from the top-level control system (for example, it can be user-activated remote braking of the WT).

4. Conclusion

The considered electromechanical brake system has the outstanding competitive advantages both on economic and technical sides. Reliable mechanical links withstand high temperature loads absorbing more excessive power at the WT rotor area, comparing with analogs. The mechanical links are operable at a high load long term operation, which enables braking of the wind turbine even at storm wind speeds. Thus, at this technological development stage, the electromechanical method (as a duplicating one) is most reasonable.

5. Acknowledgments

The work was supported by Act 211 Government of the Russian Federation, contract №02.A03.21.0011.

References

[1] Gogoberidze G, Abramov V M, Karlin L N, Lednova J and Malakhova J 2014 Marine economic potential assessment for environmental management in the Russian arctic and subarctic coastal regions *International Multidisciplinary Scientific GeoConference Surveying Geology and Mining Ecology Management* SGEM 3 (5) pp 253–260

[2] Tatarkin A I and Loginov V G 2015 Estimation of potential for natural resources and production in northern and Arctic areas: Conditions and prospects for use *Studies on Russian Economic Development* 26 (1) pp 22–31

[3] Dudin M N, Sekerin V D, Gorohova A E, Bank SV and Danko T P 2016 Arctic zone: Global strategic priorities for integrated development and infrastructure policy (Man in India) 96 (7) pp 2297–2313

[4] Yoon J R and Kim Y 2001 Reviews on Natural Resources in the Arctic: Petroleum, Gas, Gas Hydrates and Minerals *Ocean and Polar Research* 23 (1) pp 51–62

[5] Dudin M N, Ivashchenko N P, Frolova E E, Abashidze A H and Smbatyan A S 2016 Innovative approach to the development of the logistics system of supply of the Arctic region space *International Journal of Economics and Financial Issues* 6 (4) pp 1965–1972

[6] Boute A 2013 Renewable energy federalism in Russia: Regions as new actors for the promotion of clean energy *Journal of Environmental Law* 25 (2) pp 261–291

[7] Milov V, Coburn L L and Danchenko I 2006 Russia's energy policy 1992–2005 *Eurasian Geography and Economics* 47 (3) pp 285–313

[8] Gore O, Viljainen S, Makkonen M and Kuleshov D 2012 Russian electricity market reform: Deregulation or re-regulation? *Energy Policy* 41 pp 676–685

[9] Bashmakov I 2009 Resource of energy efficiency in Russia: Scale, costs, and benefits *Energy Efficiency* 2 (4) pp 369–386

[10] Lombardi P, Sokolnikova T, Suslov K, Voropai N and Styczynski Z A 2016 Isolated power system in Russia: A chance for renewable energies? *Renewable Energy* 90 pp 532–541

[11] Larionov A and Nezhnikova E 2016 Role of energy efficiency in improving the quality of housing projects *International Journal of Applied Engineering Research* 11 (6) pp 4433–4439

[12] Larionov A and Nezhnikova E 2016 Energy efficiency and the quality of housing projects *ARPN Journal of Engineering and Applied Sciences* 11 (3) pp 2023–2029

[13] Boute A 2016 Off-grid renewable energy in remote Arctic areas: An analysis of the Russian Far East *Renewable and Sustainable Energy Reviews* 59 pp 1029–1037

[14] Efendiev A M, Nikolaev Y E, Evstaf'ev D P 2016 Opportunities of energy supply of farm holdings on the basis of small-scale renewable energy sources *Thermal Engineering* (English translation of Teploenergetika) 63 (2) pp 114–121

[15] Runkle B R K, Wille C, Gažovič M, Wilmking M and Kutzbach L 2014 The surface energy balance and its drivers in a boreal peatland fen of northwestern Russia *Journal of Hydrology* 511 pp 359–373

[16] https://www.nasa.gov/mission_pages/NPP/news/earth-at-night.html

[17] Overland I 2010 2017 Russia's Arctic energy policy *International Journal* 65 (4) pp 865–878

[18] Sevastyanov S and Kravchuk A 2017 The Russian approach to national security in the arctic *Korean Journal of Defense Analysis* 29 (1) pp 131–150

[19] Morais H, Kádár P, Faria P, Vale Z A and Khodr H M 2010 Optimal scheduling of a renewable micro-grid in an isolated load area using mixed-integer linear programming *Renewable Energy* 35 (1) pp 151–156

[20] http://explorer.worldwind.earth/

[21] Khan M J and Iqbal M T 2005 Dynamic modeling and simulation of a small wind-fuel cell hybrid energy system *Renewable Energy* 30 (3) pp 421–439

[22] Datta R and Ranganathan V T 2003 A method of tracking the peak power points for a variable speed wind energy conversion system *IEEE Transactions on Energy Conversion* 18 (1) pp 163–168

[23] Hansen A D, Sørensen P, Iov F and Blaabjerg F 2006 Centralised power control of wind farm with doubly fed induction generators *Renewable Energy* 31 (7) pp 935–951

[24] Bossanyi E A 2003 Wind turbine control for load reduction *Wind Energy* 6 (3) pp 229–244

[25] Teodorescu R and Blaabjerg F 2004 Flexible control of small wind turbines with grid failure detection operating in stand-alone and grid-connected mode *IEEE Transactions on Power Electronics* 19 (5) pp 1323–1332

[26] Abrahamsen A B, Mijatovic N, Seiler E, Sørensen M P, Koch M, Nørgard P B, Pedersen N F, Træholt C, Andersen N H and Østergard J 2009 Design study of 10 kW superconducting generator for wind turbine applications *IEEE Transactions on Applied Superconductivity* 19 (3),art. no. 5067256, pp 1678–1682

[27] Solomin E V, Sirotkin E A and Martyanov A S , 2015 Adaptive control over the permanent characteristics of a wind turbine *Procedia Engineering* 129 pp 640–646

[28] Korobatov D V, Sirotkin E A, Troickiy A O and Solomin E V 2017 Wind turbine power plant control *2016 Dynamics of Systems, Mechanisms and Machines, Dynamics 2016* art. no. 7819031

[29] Wang K, Hansen M O L and Moan T 2014 Dynamic analysis of a floating vertical axis wind turbine under emergency shutdown using hydrodynamic brake *Energy Procedia* 53 (C) pp 56–69

[30] Yankov P V, Van Den Bossche A and Valchev V C 2010 Successive resistive braking circuit for permanent magnet wind turbine generators *Proceedings of EPE-PEMC 2010 - 14th International Power Electronics and Motion Control Conference* art. no. 5606915 pp T1127-T1131

Gas sensitivity of polymeric composites with carbon nanotubes

E D Skutin, S O Podgorniy, O T Podgornaya, N S Kolyanichev and A A Katkov

Omsk State Technical University, 11 Mira Pr., Omsk 644050, Russia

e-mail: skutin@omgtu.ru

Abstract. The possibility of application for carbon nanotubes DEALTOM as a conductive filler for composite sensors based on polysiloxanes is shown. Sensors show predicted results on sensitivity and selectivity. Nanotubes according to chemical, physical and mechanical characteristics are suitable for modifying of specified composites and are applicable for work with rather aggressive reagents.

1. Introduction

Combination of nano- and biotechnologies, computer, information and cognitive technologies in a single chain enables the emergence of new scientific trends on creation of inanimate matter intellectual systems, which can be used in a wide number of areas from industrial production to ecology and medicine. One of a recent trends is a chemical analysis with the use of multi-touch gas-analysis systems called «electronic nose» [1, 2]. The basis of these systems is an array of sensors characterized by the sensitivity to a broad range or the whole class of analyzed substances (analytes).

Nowadays a wide range of various sensors can be used in sensor arrays, but chemoresistors based on composites with the application of nanotubes are of special interest. Carbon nanotubes have a number of unique properties making their usage in chemical sensors very prospective. A large surface area of nanotubes provides efficient adsorption of molecules in many substances, and modification of carbon nanotubes surface by polymers can increase both selectivity, and sensitivity of sensors [3].

Nanocomposite materials can also provide important advantages for sensors: the possibility of sensitivity control by alteration of a quantitative relationship and the nature of filler and polymer interaction, little response time owing to a fast diffusion of analyte molecules into the volume of a film and the expansion in a variety of sensitive layers [4].

The purpose of this work is to examine gas sensitivity in a hybrid sensor array comprising polymeric composite sensors based on three variously modified polysiloxanes with electroconductive filler in the form of carbon black and DEALTOM nanotubes.

2. Problem statement

DEALTOM multi-wall carbon nanotubes (MWCNT) are polycrystalline graphite carbon filamentary structures of a cylindrical shape with the internal channel. MWCNTs without cleaning (immediately after reactor) are the powder consisting of nanotubes in two sizes with an external diameter D_{out1} = 49.3 nm or D_{out2} = 72.0 nm, respectively according to Transmission Electron Microscope (TEM) data. The inner diameter of nanotubes according to TEM is D_{in} = 13.3 nm. The rough estimate of upper bound for nanotubes' length based on Scanning Electron Microscope (SEM) data gives the value of L = 5 μm. Multi-wall carbon nanotubes DEALTOM are synthesized by the original method that is a low-temperature thermocatalytic pyrolysis of hydrocarbons, which allows one to create low cost carbon nanotubes.

The low cost, portability, mobility and low power consumption are necessary requirements to multi-touch gas analyzers in which therefore the usage of simple sensors, like chemoresistors based on the polymeric composites is preferable. Chemoresistive composite-based sensors using stationary phases for a gas chromatography (GC) as a polymeric matrix are objects of special interest. Distinctive features of these polymeric phases are a higher polymer purity with eliminated traces of catalysts, monomers, solvents and other undesirable impurities, and a narrower molecular-weight distribution compared to general purpose industrial polymers. Liquid GC phases in the form of three variously modified polysiloxanes such as Polydimethyl-siloxane (PDMS), Polycyanopropyl methyl-siloxane (PCPMS) and Polyphenylmethyl-siloxane (PPhMS) have been used as an isolating polymeric matrix of composites in this paper.

The experience in using systems with uniform sensor arrays shows, however, that they are capable to detect rather limited range of analytes. This results in the need to create hybrid arrays with the sensors of different functioning principles, or to apply specific operating modes in systems [5]. In particular, the application of a pulse mode in analytes' supply makes possible to use sensory response kinetic features as an additional information factor about the analyzed object composition [6].

3. Theory

The text of your paper should be formatted as follows: The intensity of composite sensors' response is first of all determined by a thermodynamic activity of analyte in a polymer matrix and by its ability to bring the matrix to swelling. To estimate the swelling extent in polymeric matrix of composites in solvent vapors it is possible to use such characteristic as solubility parameter δ, introduced by Hildebrand to describe nonelectrolyte solutions [6].We can propose the polymer solubility and its swelling ability if values of these parameters are close for a polymer and a solvent $\delta_p \approx \delta_s$.

The solubility parameter is calculated from the relationship

$$\delta = (\Delta E / V)^{1/2}, \qquad (1)$$

where ΔE is a cohesive energy defined as $\Delta E = \Delta H - RT$; ΔH is a latent heat of a substance evaporation; R is a universal gas constant; T is an absolute temperature; V is a molar substance volume.

The square of solubility parameter represents a cohesive energy density, i.e. its value in a molar volume unit ($\delta^2 = \Delta E/V$). The cohesive energy value ΔE is additive and is presented as the contribution sum $\sum_i \Delta E_i^*$ from each atom and kind of intermolecular interaction. The calculation of cohesive energy density by the increment method can be accomplished by the equation:

$$\delta^2 = \frac{\sum_i \Delta E_i^*}{N_A \sum_i \Delta V_i}, \qquad (2)$$

where $\sum_i \Delta E_i^*$ is the increments sum of cohesive energy for all atoms and bonds in a solvent molecule or in a repeating polymer macromolecule link; $N_A \sum_i \Delta V_i$ is the increments sum of van der Waals molecule volume or the repeating macromolecule link. The values of increments ΔE_i^* and ΔV_i are given in [6].

Table 1 provides the values of solubility parameters δ calculated by the method of increments for the solvents used in this paper as analytes, and for the polymers used in this work as the basis of polymeric composites.

Table 1. The values solubility parameters δ for analytes and polymers.

№	Name	$\sum_i \Delta E_i^*$ (J/mol)	$N_A \sum_i \Delta V_i$ (cm³/mol)	δ (MPa$^{1/2}$)
		Solvents		
1	Petroleum ether	-	-	14.3
2	n-Hexane	16642	114.8	15.5
3	Chloroform	13310	71.4	17.5
4	Toluene	20740	96.0	18.3
5	Acetone	15515	59.2	20.9
6	Ethanol	29669	51.0	26.1
7	2-Propanol	32382	68.2	28.1
8	Butanol	28219	88.1	23.1
		Polymers		
1	PDMS	23580.9	125.3	17.7
2	PCPMS	15483.6	59.6	20.7
3	PPhMS	21081.3	96.7	19.0

4. Experimental technique

The experimental multi-touch system represents a structure of three functional units: a sensor array, a sampling subsystem and a signal processing unit of sensor array.

4.1. Sensor array

The sensor array basis comprised 8 polymercomposite (PC) sensors. Composite chemoresistors can be considered as narrow selective among resistive sensors. The liquid phases for gas chromatography, presented in Table 2, were used as the isolating polymeric matrix of composites, and structural formulas of monomeric links of these polymers are given in fig. 1. The carbon black (CB) N472 with particle size of 20…25 nm, and DEALTOM MWCNTs were used as electroconductive fillers of composites.

PC sensors were made in the form of film chemoresistors with applying a toluene solution of a proper composite on a mica substrate, with a preliminary vacuum deposition of contact pads and Ag electrodes on the substrate. The production ended with drying of finished sensors in the open air until a complete solvent elimination that was controlled according to the stabilization of electric resistance R_0.

Table 2. Materials of polymeric matrices for electroconductive composites.

№	Notation	Polymer	Brand of GC phase
1	PDMS	Poly(dimethyl-siloxane)	SE-30
2	PCPMS	Poly(cyanopropylmethyl-siloxane)	OV-105
3	PPhMS	Poly(phenylmethyl-siloxane)	OV-17

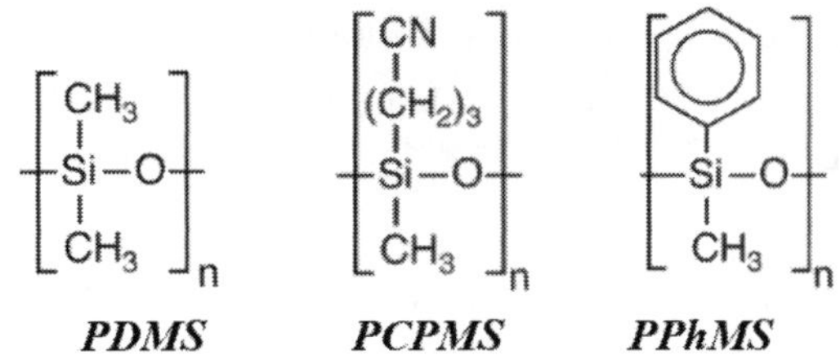

Figure 1. Structural formulas of polymers' monomeric links.

Wide sensitivity sensors, i.e. one thermo-catalytic (TC) sensor and two metal-oxide (MO) sensors with the various operation temperature were additionally implemented into the structure of system sensor array. The characteristics of both additional TC and MO sensors, and fabricated PC sensors with the filler in the form of CB and MWCNTs are given in Table 3.

TC sensors operation is based on a measurement of thermal effect in a reaction of catalytic oxidation of analytes by air oxygen with a heated thermoresistor covered by a catalytic layer of platinum or palladium deposited on aluminum oxide. The peculiarity of these sensors is a wide sensitivity to almost any combustible substances capable to oxidation on active centers of the catalyst [8].

Table 3. Characteristics of the used sensors.

№	Type	Basic material	Filler (*% wt*)	
			CB	MWCNT
1	TC	Al_2O_3+Pd	-	-
2	MO	SnO_2	-	-
3	MO	SnO_2	-	-
4	PC	PCPMS (OV – 105)	10	-
5	PC	PPhMS (OV – 17)	10	-
6	PC	PPhMS (OV – 17)	-	25
7	PC	PCPMS (OV – 105)	-	25
8	PC	PCPMS (OV – 105)	-	20
9	PC	PDMS (SE – 30)	10	-
10	PC	PDMS (SE – 30)	-	25
11	PC	PPhMS (OV – 17)	-	10

Two thin-film (~1 μm) metal-oxide SnO_2 sensors have been made on mica substrates by the method of thermal deposition of metal tin with its further oxidation at the temperature of 773 K in the air. MO sensors among resistive devices have an average selectivity [9].

4.2. Sampling subsystem

The sensor array was placed in a 21 ml measuring cell, which was blown by the air pump with 50 ml/min speed. Pulse samples' dosing was performed with 3…5 % accuracy, by air-vapor mixture injecting of analytes' saturated vapors in the air line by means of a syringe. The results of chromatographic calibration of these samples, which was carried out on the gas chromatograph «Chromos GH-1000», are presented in [10].

4.3. Signal processing unit of sensor array

Collection and processing of analog data from the sensor array were conducted by E14-140 measuring module produced by *«L-Card»* under control of *LGraph2* program. The sample identification problem was solved by processing of obtained signal sampling in the whole sensor array by means of artificial neuron network made in the form of a multilayer perceptron. The necessary network inputs n_i

corresponded to the product $n_i = n_s \times n_t$, where n_s is a number of sensors in array and n_t is a number of points in a discrete kinetic sampling of their signals. Every analyte in an output layer of neuron network was assigned with a proper neuron, its signal level showed the analyte content in the sample.

5. Results discussion

Characteristic property of polymeric nanocomposites is a percolation nature of their electroconductivity. The nanocomposite conductivity sharply increases achieving a threshold concentration of a conductive filler (a percolation threshold). In this paper, the percolation threshold of nanocomposites based on polysiloxanes was achieved by a stepwise 5 % increase in mass concentration of MWCNTs in a polymeric matrix up to the demonstration of sensitivity to analytes by the chemoresistors. Noticeable gas sensitivity appeared as a rule under decreasing of electric resistance to 1…2 MΩ in chemoresistors. At the same time, a further increase of MWCNT concentration led to a noticeable decrease in resistance and removal from the percolation threshold with the sensitivity loss.

The values of solubility parameters calculated by the method of increments for the solvents used in this paper as analytes, and for the polymers used in this paper as a polymeric composites basis are given in Table 1. The obtained values of indicators allow us to predict a good sensitivity to nonpolar analytes (petroleum ether, n-hexane and acetone) and a weak sensitivity to polar alcohols for PC sensors based on polysiloxanes (SE-30, OV-17 and OV-105).

The example of experimental data with responses of hybrid sensor array to n-hexane sample obtained by the measuring module E14-140 under *LGraph 2* program control is presented in fig. 2. Numbers of curves in the figures correspond to numbers of sensors in Table 3.

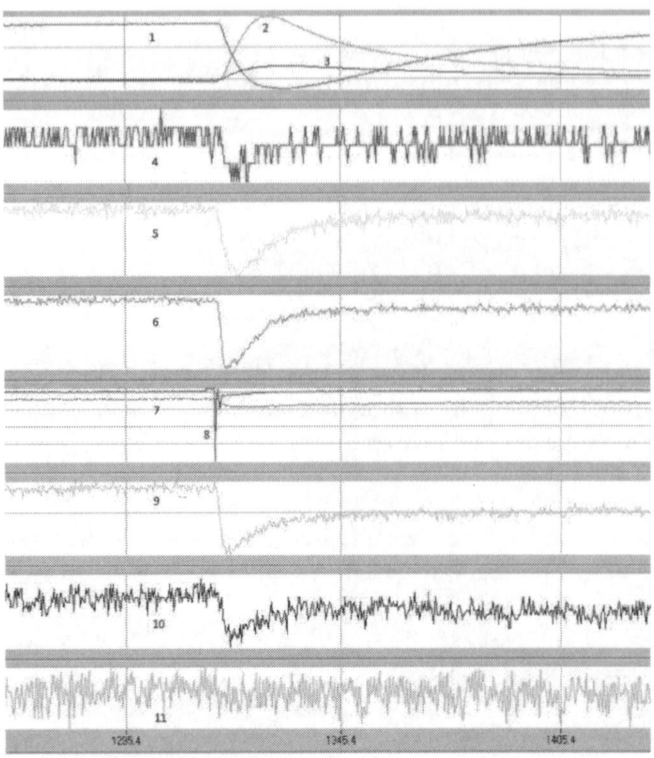

Figure 2. A sensor array response to a 5 ml air-vapor sample under 437 *ppm* n-Hexane concentration.

These data show that a satisfactory sensitivity of sensors can be achieved by the increase of mass share of carbon nanotubes up to 20–25 %. However, the necessary concentration of nanotubes exceeds significantly the required amount of carbon black (10 %). Thus, sensor № 11 (10 % MWCNT), for example, does not show the sensitivity to any of the used analytes.

The comparison of sensors' responses with carbon black (for instance, № 5 CB 10 % wt) and sensors with carbon nanotubes (for example, № 6 MWCNT 25 % wt) shows that in case of n-hexane, petroleum ether and acetone, the modification of a polymer by carbon nanotubes does not lead to a noticeable change in sensors' characteristics. The responses of such sensors are almost identical taking into account the noise level in their signals. However, in case of ethanol, the sensor № 6 has shown a minor improvement of sensitivity compared to the sensor № 5.

6. Conclusions

Carbon nanotubes DEALTOM demonstrate a good possibility of their use as a conductive filler for composite sensors based on polysiloxanes, showing expected results in sensitivity and selectivity. It should be noted that DEALTOM MWCNTs in chemical, physical and mechanical characteristics are suitable for modifying of any composite materials since they are characterized by very high hardness and corrosion resistance (they are insoluble neither in «aqua regia», nor in strong alkaline solutions). It has allowed us to use them as a suitable material to work with rather aggressive reagents. Moreover, a big surface size of nanotubes provides their high adsorbability.

References

[1] James D, Scott S M, Ali Z and O'Hare W T 2005 *Microchim. Acta* **149** p 1
[2] Scott S M, James D and Ali Z 2007 *Microchim. Acta* **156** p 183
[3] Spitalsky Z, Tasis D, Papagelis K and Galiotis C 2010 *Prog. Polym. Sci.* **35** 357
[4] Llobet E 2013 *Sens. Actuators B Chem.* **179** p 32
[5] Herrero-Carrón F, Yánez D J, Rodríguez F B and Varona P 2015 *Sens. Actuators B Chem.* **206** p 555
[6] Skutin E D, Budanova E M and Leshchinskii S S 2004 *J. Anal. Chem.* **59** p 1142
[7] Barton A F M *CRC Handbook of Solubility Parameters and Other Cohesion Parameters*; CRC Press: Boca Raton, Florida, 1983; Chapter p 13
[8] LaFratta C N and Walt D R 2008 *Chem. Rev.* **108** p 614
[9] Soumen D and Jayaraman V 2014 *Prog. Mater. Sci.* **66** p 112
[10] Skutin E D, Podgorniy S O, Zemtsov A E and Gaberkorn O V 2016 *Procedia Engineering* **152** p 493

Vacuum boilers developed heating surfaces technic and economic efficiency evaluation

E N Slobodina[1], A G Mikhailov[1], B A Semenov[2]

[1]Omsk State Technical University, 11, Mira Pr., Omsk 644050, Russian Federation
[2]Yu. A. Gagarin Saratov State Technical University,77, Politekhnicheskay, Saratov 410054, Russian Federation

Abstract. The vacuum boilers as manufacturing proto types application analysis was carried out, the possible directions for the heating surfaces development are identified with a view to improving the energy efficiency. Economic characteristics to evaluate the vacuum boilers application efficiency (Net Discounted Income (NDI), Internal Rate of Return (IRR), Profitability Index (PI) and Payback Period) are represented. The given type boilers application technic and economic efficiency criteria were established. NDI changing curves depending on the finning coefficient and operating pressure were obtained as a result of the conducted calculation studies.

1. Introduction

As part of Russia's Energy Strategy implementation for the period until 2020, there is a need for the energy-saving technologies development and high efficiency equipment manufacturing in order to reduce the unit production costs and energy resources usage.

The application of the vacuum fire tube boiler with the developed heat exchange surfaces in a volume including the vacuum heat carrier as a manufacturing prototype for the independent heat supply systems is a heat power engineering development promising area.

The chosen method of the fire tube boiler operation efficiency improvement by means of the heat exchange intensification at boiling and condensing in vacuum by using the finning is not well known, there is no published information on the practical application. It demonstrates the relevance of the scientific research chosen direction.

2. The Problem Statement

One of the main principles of the investment projects efficiency evaluation requires the comparison of the related to the project results and costs throughout its duration.

The project is considered to be effective if the return of the original investments and required return are provided for the investors furnished the capital [1].

The decisions on the investment project implementation expediency, on choosing the optimal investment alternative cannot be adopted intuitively. For minimizing risks of malinvestment, the following complex of key evaluation parameters discussed below is applied:

- Net Discounted Income (NDI)
- Internal Rate of Return (IRR),
- Profitability Index (PI),
- Payback Period.

3. Theory

The main investment project efficiency parameter is the net discounted income.
The net discounted income(integral effect, net present value) is the accumulated discounted income over the whole accounting period is defined according to the following formula:

$$NDI = \sum_{t=0}^{T}(R_t - Ex_t)\alpha_t - I$$

where NDI is the net discounted income,
R_t is the annual revenue from heat energy sales,
Ex_t is the costs of goods sold
I is the investments,
α_t is the discount coefficient defined by the formula:

$$\alpha_t = (1 + E)^{-t},$$

where E is the discount rate.
The investment project is effective if the net discounted income is not negative ($NDI \geq 0$), i.e. if the discounted income value (at the defined discount rate value) is not less than the discounted costs value, it is reasonable to prefer the investment project with NDI maximum values [1-3].
The internal rate of return (IRR) is the discount rate (E) value at which the net discounted income equals zero. Therefore, IRR is often referred to as a verification discount as it makes possible to define the discount rate boundary value dividing the investments into acceptable and unprofitable. If the discount rate E is positive and less than IRR, then the project is effective. If the discount rate E is greater than IRR, then the project is in effective. The comparison of IRR with the discount rate makes possible to evaluate the project safety margin. A great difference between these values testifies the project sustainability.
The best project is selected upon NDI criterion and the resolution on participation in such a project is made on the basis of IRR.
The profitability index is the accumulated discounted inflow and real money outflow ratio.
The profitability index is calculated by the formula:

$$PI = \frac{\sum_{t=0}^{T}(R_t - Ex_t)\alpha_t}{I}.$$

One of the advantages of this parameter is the possibility of using as a project static stability. The higher the profitability index, the significantly greater the project safety margin. The profitability index is associated with the net discounted income, if the net discounted income is positive (NDI>0), then PI>1, the project is effective. For unprofitable projects, NDI<0, then the profitability index is less than 1 (PI<1) [3, 4].
The payback period is the duration of the shortest period after which the net discounted income becomes positive and subsequently remains non-negative.

$$\sum_{t=0}^{T_{pp}} K_t/(1 + E)^t \leq \sum_{t=0}^{T_{pp}} (R_t - C_t)/(1 + E)^t,$$

where T_{pp} is the payback period parameter for the branch.
Taking into account the above, it is possible to conclude that the efficiency parameters shared use optimality conditions are required to be fulfilled for the maximum economic efficiency of investment projects.
In this regard, in the first stage, it is necessary to calculate the costs emerged during the project implementation.
The annual costs are defined by the formula:

$$C_a = C_f + C_{ee} + C_w + C_d + C_m + C_{wg} + C_s$$

Where C_f are the fuel costs, C_{ee} are the electric energy costs, C_w are the water costs, C_d are the depreciation costs, C_m are the equipment maintenance costs, C_{wg} are the wage costs, C_s are the social needs costs.

The annual fuel costs:

$$C_f = B \cdot P_{gas}$$

Where P_{gas} is the gas price, B is the fire tube boiler fuel consumption.
For the annual electric energy costs defining it is necessary to calculate the annual energy consumption by the formula:

$$N_{annual} = \sum N_{con.} n h_{max},$$

where $N_{con.}$ is the fire tube boiler capacity, n is the number of consumers, pcs – 1 ea, h_{max} is the number of active load claimed maximum using hours.
The annual electric energy costs:

$$C_{ee} = N_{annual} \cdot T_e,$$

where T_e is the electricity tariff (according to the enterprise's data).
The annual water costs:

$$C_w = G_w^b \cdot P_w$$

where $G_в^к$ is the fire tube boiler annual water consumption, P_w is the chemically purified water price.
The annual depreciation costs:

$$C_d = \alpha_d \cdot I$$

where α_d is the depreciation deductions rate for the stationary hot water boiler.
The annual maintenance costs:

$$C_m = \alpha_m \cdot I_a$$

where α_m is the maintenance deductions rate.
The annual wage costs:

$$C_{wg} = \left(\sum C_{wg}^{p} n_p + И_{wg}^{e} n_e\right)$$

Where n is the number of operating personnel, persons.
The annual social needs costs:

$$C_s = \alpha_s \cdot C_{wg}$$

where α_s is the social needs deductions rate.
The production cost of heat energy generated by the boiler room is calculated by the formula:

$$S = \frac{C_a}{Q_a^b}$$

where Q_a^b is the annual heat generation by the boiler room.
The project profit is formed by the fact that fire tube boiler modernization will make possible to reduce gas consumption and water treatment costs:

$$PP = C_{f/b} - C_{f/a.r.} - C_w .$$

The net discounted income (NDI), profitability index (PI) and investments payback period is calculated according to the formula as the integral efficiency parameters:

$$n = {}^{I}/_{NP} \leq T_s$$

4. Experiment results
The cost of the acquired and upgraded equipment, installation work is represented in tables 1 and 2:

Table 1. The fire tube boiler modernizing capital costs estimate.

The name of the acquire dequipment, construction and installation works	No.	The cost per item, RUB	The total cost, RUB
Fire tube boiler having the capacity of 500 kW	1	382600	382600
Equipment costs for the fire tube boiler furnace modernization	1	133910	133910
Installation work cost			46870
Total estimated cost(I)			563380

Table 2. The fire tube boiler operating costs.

The parameter name	UM	The parameter value
The installed capacity	kW	400
The construction project investments	RUB	563380
The heat energy generation full costs, including:	Rubles per year	2286550
Fuel costs	Rubles per year	1897345
Electric energy costs	Rubles per year	4112
Water costs	Rubles per year	236856
Maintenance costs	Rubles per year	901
Labor costs	Rubles per year	96000
Social needs deductions	Rubles per year	28800
Heat energy production unit cost	RUB/GJ	115
P_{gas} = RUB5000 thousand per 1000 m^3ofgas,VATincluded; C_{ee} = 3.56 RUB/kW his the electricity tariff (according to the enterprise's data); P_w = 27.15 RUB/$m^3$is the chemically purified water price; α_d = 4 % is the depreciation deductions rate for the stationary hot water boiler;α_m = 4 % is the maintenance deductions rate; α_s = 30 % is thesocial needs deductions rate.		

5. Results discussion

The net discounted income (NDI) graph of the boilermodernization possible variations at the different finning coefficients (Ψ) in the pressure range from 60.80 to 101.33 kPa is represented in figure 1.

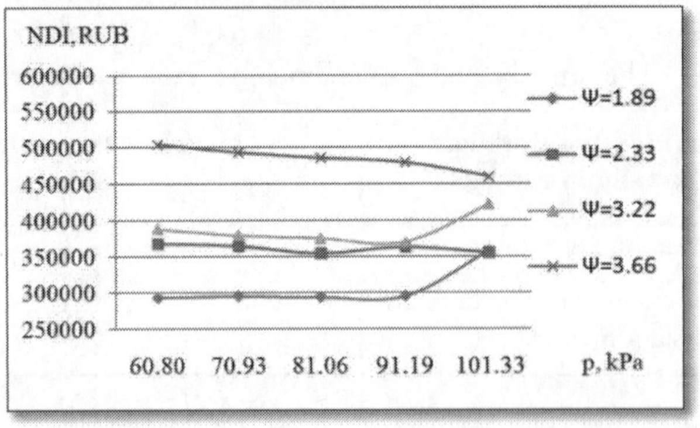

Figure 1. The net discounted income of the boiler modernization possible variations.

The nature of the NDI change is complicated and consequently a large scale optimization is required. Integral efficiency parameters calculations made possible to determine boiler operation areas not only technologically, but also economically profitable. Figure 2 shows three-dimensional graphical dependence of the net discounted income on the finning coefficient and boiler operating pressure.

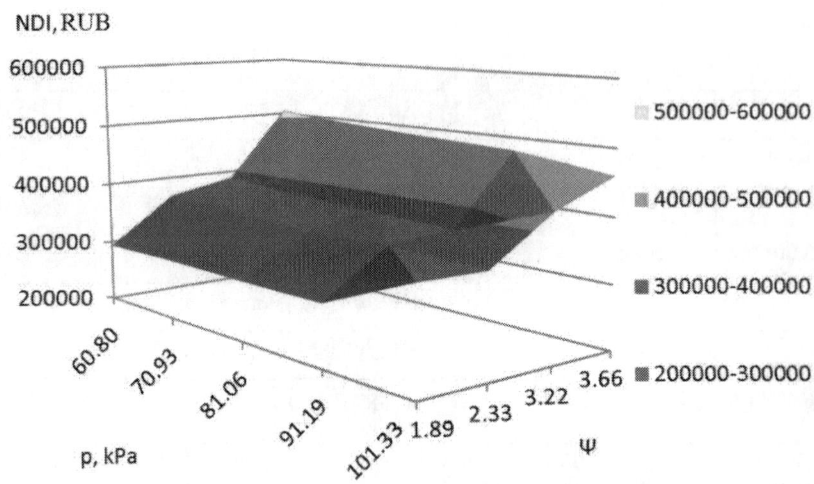

Figure 2. The net discounted income dependence on the finning coefficient and saturated vapors pressure.

6. Conclusions

Analyzing the received data it is possible to define the optimization parameters rational values at which the maximum technical and economic effect can beachieved. The highest NDI is observed at the saturated vapors pressure of p = 60.80 kPa, finning coefficient of Ψ = 3.66 and is defined by the graphical method.

References

[1] Vilenskij P L,Livshic V N and Smoljak S A 2002 *The investment projects efficiency evaluation. Theory and Practice* (Moscow: Delo)

IOP Conf. Series: Journal of Physics: Conf. Series **944** (2017) 012111 doi:10.1088/1742-6596/944/1/012111

[2] Annaratone D 2008 Steam generators (Springer-Verlag Berlin Heidelberg)

[3] Fajndshmit E 2012 *The investment projects efficiency evaluation*(Moscow)
[4] Savchuk V P 2005 Company finance management (Moscow: BINOMLaboratorijaznanij)
[5] Slobodina E N 2016 The peculiarities of the boiling process at subatmospheric pressure *Proc. Int. Conf. on Oil and Gas Engineering (Omsk)* vol 152 (Amsterdam: North-Holland/American Elsevier) pp 400-405

Acknowledgment
The research work was conducted as part of the scientific project No 17087B «Young researcher of OmSTU»

Composition characteristics and regularities of dissolving of hydroxyapatite materials obtained in water solutions with varied content of silicate ions

A P Solonenko[1,2]

[1] Omsk State Medical University, Lenina str., 12, Omsk 644099, Russia
[2] Omsk State Technical University, Mira ave., 11, Omsk 644050, Russia

E-mail: annsolonenko@gmail.com

Abstract. Research aimed at developing new bioactive materials for the repair of defects in bone tissues, do not lose relevance due to the strengthening of the regenerative approach in medicine. From this point of view, materials based on calcium phosphates, including silicate ions, consider as one of the most promising group of substances. Methods of synthesis and properties of hydroxyapatite doped with various amounts of SiO_4^{4-} ions are described in literature. In the present work synthesis of a solid phase in the systems $Ca(NO_3)_2$ - $(NH_4)_2HPO_4$ - Na_2SiO_3 - NH_4OH - H_2O (C_{Ca}/C_P = 1.70) performed with a wide range of sodium silicate additive concentration ($y = C_{Si}/C_P = 0 \div 5$). It is established that under the studied conditions at $y \geq 0.3$ highly dispersed poorly crystallized apatite containing isomorphic impurities of CO_3^{2-} and SiO_4^{4-} precipitates in a mixture with calcium hydrosilicate and SiO_2. It is shown that the resulting composites can gradually dissolve in physiological solution and initiate passive formation of the mineral component of hard tissues.

1. Introduction

Investigation of human bone matrix have shown that the basic mineral matter of hard tissues in composition and structure is close to hydroxyapatite (HA, $Ca_{10}(PO_4)_6(OH)_2$) [1]. The similarity of biogenic and synthetic salts has caused widespread use of synthetic salts in medicine as a biocompatible material for the reconstruction of bone defects. Stoichiometric HA form and it various ion-substituted modifications found the application. The possibility of introducing metal ions (Na^+, K^+, Mg^{2+}, Sr^{2+}, etc.) and anions (Cl^-, F^-, CO_3^{2-}, SiO_4^{4-}, etc.) into apatite structure is being studied [1]. Special attention is paid to silicate ions because silicon plays an important role in the early stage of cartilage and bone growth, it has a positive effect on the synthesis of collagen type I by osteoblast cells, promotes an increase in the density of the mineral constituent of bone tissue by stimulating the precipitation of basic calcium phosphate [2]. In this connection, the synthesis of HA containing various amounts of SiO_4^{4-} ions is being actively investigated. It is also promising to create composites that include silicate-substituted apatite and silicon phase that initiates the active incorporation of synthetic material into the bone components formation process.

2. Problem formulation

The main method for the synthesis of ion-substituted HA is precipitation of this sparingly soluble salt from an aqueous solution. As a rule, the amount of modifying additives is small in comparison with the concentrations of the main precipitation-forming ions.

The task of this work is the synthesis and physicochemical investigation of solid phases formed in alkaline aqueous calcium-phosphate systems containing varying amounts of SiO_4^{4-}, as well as studying the regularities of dissolution of the resulting powder materials in physiological solution.

3. Materials and methods

Synthesis was carried out in aqueous solutions $Ca(NO_3)_2$-$(NH_4)_2HPO_4$-Na_2SiO_3-NH_4OH-H_2O at room temperature (t = 20-22 °C). $Ca(NO_3)_2 \cdot 4H_2O$ (chemically pure), $(NH_4)_2HPO_4$ (pure for analysis), $Na_2SiO_3 \cdot 5H_2O$ (pure), NH_4OH (pure for analysis), HNO_3 (pure for analysis) and distilled water were used as the reagents for the systems preparation. The ratio of the concentrations of the initial components was set equal to $\boldsymbol{x} = C_{Ca}/C_P = 1.70$, and the fraction of the silicon agent varied in the range $\boldsymbol{y} = C_{Si}/C_P = 0 \div 5$. For comparison, the solid phase was synthesized in the absence of phosphate ions, at $C_{Ca}/C_{Si} = 1$. The ionic strength of solutions in each series of experiments did not exceed 0.6 mol/l. Reaction media were prepared by a dropwise addition of a solution containing predetermined amounts $(NH_4)_2HPO_4$ and / or Na_2SiO_3 to a solution of $Ca(NO_3)_2$. After mixing the components, aqueous suspension pH was adjusted to 10.00±0.05 with NH_4OH (concentrated) and / or HNO_3 (1:1) and allowed to crystallize in a plastic vessel. Each synthesis was carried out in triplicate.

The formed solid phases were separated from the supernatant by filtration. Part of the solution was taken for chemical analysis. Calcium was determined photometrically by reaction with Arsenazo III in an alkaline medium, phosphates were determined photometrically by the color of phosphomolybdic complex which form in an acid medium (UV method), the silicates were fixed in yellow form of molybdosilicic acid [Guidance document 52.24.433-2005, Russia]. Solid phase's chemical composition was determined like a difference of initial and residual precipitating ions concentrations in the systems.

The resulting precipitates were dried at 120 °C to constant weight and then at 300 °C for 2 hours. The solid phases were weighed on analytical scales and ground in a porcelain mortar. Powder-like samples were examined using X-ray diffraction analysis (XRD, diffractometers «D8 Advance» (Bruker) and «XRD-7000» (Shimadzu)), Fourier transform spectroscopy (FTIR-spectrometer «FT-801» (SIMEX); powders were pressed in tablets with KBr).

Regularities of samples dissolution in 0.9% NaCl, isotonic with human blood plasma, were studied. For kinetic measurements 0.2000 ± 0.0002 g of samples were placed in a plastic vessel with tight-fitting lid and 50.0 ml of 0.9% NaCl solution were added. The measurements were carried out at room temperature (t = 20÷22 °C). During the experiment, system's pH in time changes were monitored (pH-meter «SevenEasy S20» with electrode for pH measurement «InLab®413» (Mettler Toledo)) and concentrations of calcium ions, phosphates and silicates passing into the liquid phase were controled. To determine the concentration of Ca^{2+}, PO_4^{3-} and SiO_4^{4-}, aliquots of solutions were withdrawn at predetermined intervals. A corresponding volume of fresh 0.9% NaCl was added to the system to maintain the constancy of its volume during the measurements.

4. Results

XRD analysis showed that precipitates, obtained from the systems $Ca(NO_3)_2$-$(NH_4)_2HPO_4$-Na_2SiO_3-NH_4OH-H_2O in all range of variation of the silicon agent concentrations and treated at 300 °C, contain only HA crystalline phase (figure 1). A calculation using the Selyakov-Scherrer formula [3] showed that the effective diameter of the apatite particles in the samples is about 17 nm (table 1). For $\boldsymbol{y} \geq 1$, the crystallite size decreases to 8 nm. The parameters of the unit cell of the obtained HA samples were calculated according to the diffraction data (table 1). An increase of the parameter c relative to the reference value characteristic to $Ca_{10}(PO_4)_6(OH)_2$ was noted, the variation of a value in the interval 9.417-9.429 and increase of the cell volume with increasing $\boldsymbol{y}$ were observed. These changes may be related with isomorphous substitutions in the anion apatite sublattice.

IOP Conf. Series: Journal of Physics: Conf. Series **944** (2017) 012112 doi:10.1088/1742-6596/944/1/012112

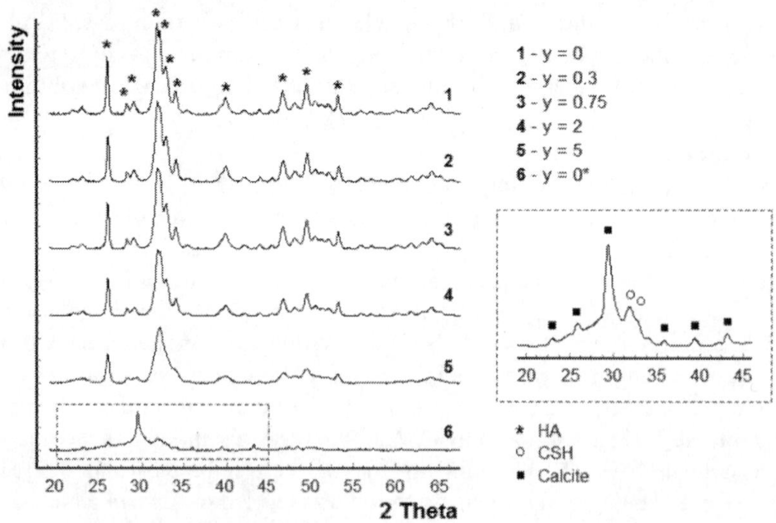

Figure 1. Diffractograms of samples obtained in systems with a variable amount of silicate ions.

Table 1. Characteristics of solid phases (XRD data)

Sample	*y*	Size of HA crystall, nm	Parameters of HA unit cell	
			a, Å[a]	*c*, Å[a]
N1	0	17.9 ± 0.3	9.427 ± 0.002	6.879 ± 0.002
N2	0.1	19.1 ± 0.3	9.428 ± 0.002	6.883 ± 0.002
N3	0.3	17.2 ± 0.4	9.427 ± 0.002	6.883 ± 0.002
N4	0.5	17.0 ± 0.4	9.428 ± 0.002	6.885 ± 0.002
N5	0.75	17.6 ± 0.4	9.429 ± 0.002	6.888 ± 0.002
N6	1	14.9 ± 0.3	9.425 ± 0.002	6.888 ± 0.002
N7	2	13.3 ± 0.3	9.424 ± 0.002	6.889 ± 0.002
N8	5	8.6 ± 0.2	9.417 ± 0.004	6.898 ± 0.004

[a] The reference values for the stoichiometric HA: a = 9.419 Å, c = 6.883 Å [card № 55-0592 of the powder diffraction database ICDD PDF_2, version 2006].

Thus, FTIR results show the presence of CO_3^{2-} and SiO_4^{4-} ions in the precipitated apatite (figure 2). Absorption bands that appear in the IR spectra of the samples at 875 cm^{-1} (ν_2), 1420 and 1460 cm^{-1} (ν_3) indicate the partial replacement of PO_4-tetrahedra in HA structure by the groups CO_3^{2-} which appear in the reaction media as a result of absorption CO_2 from air. The decrease of intensity of O-H absorption bands (630 and 3570 cm^{-1}, figure 2) could be the evidence of silicates incorporate into HA composition, which is observed as a result of the following mechanism of isomorphic substitution:

$$PO_4^{3-} + OH^- \Leftarrow SiO_4^{4-} + \square_{OH} \quad (1)$$

Molar coefficients Ca/P and Ca/(P+Si) for obtained solid phases were calculated from results of chemical analysis (table 2). It is established that for all powders Ca/P value corresponds to non-stoichiometric apatite (nHA) with phosphorus deficiency, which arises as a result of isomorphic anion substitutions. The values of Ca/(P+Si) coefficients significantly decrease with the ***y*** growth, which is probably due to the appearance of a non-apatite component in the composition of the powders.

This assumption was confirmed by FTIR data. In addition to the absorption bands of the groups included in HA structure, peaks of the silicate phase are fixed in the spectra of the solid phases deposited in systems $Ca(NO_3)_2$-$(NH_4)_2HPO_4$-Na_2SiO_3-NH_4OH-H_2O at $\boldsymbol{y} \geq 0.3$. At 790-800 cm^{-1} the mode of stretching vibrations of siloxane bridges (Si-O-Si) in the composition of silicon oxide [1, 4] and calcium silicate hydrate (CSH, $rCaO \cdot mSiO_2 \cdot nH_2O$) is recorded in the spectra of the samples [5]. The shoulder at 1200 cm^{-1} is due to stretching vibrations of Si-O bonds in SiO_2 [4] and CSH for which the molar ratio Ca/Si ≤ 0.7 [6]. These reflexes become more pronounced as the percentage of silicate ions in the initial reaction system (*y*) increases, indicating an increase of the content of these components in solid phase.

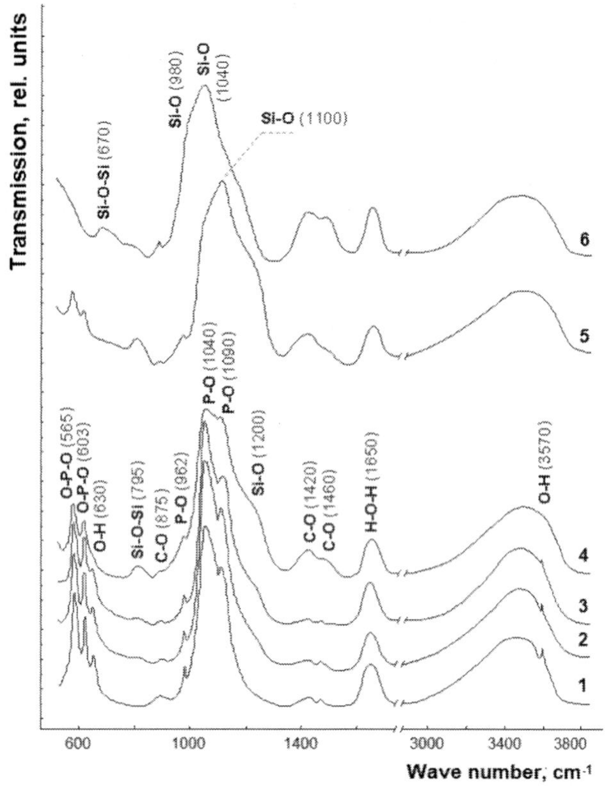

Figure 2. FTIR spectra of the samples obtained at $y = 0$ (1), $y = 0.3$ (2), $y = 0.75$ (3), $y = 2$ (4), $y = 5$ (5), $y = 0$* Ca/Si = 1 (6).

Table 2. Samples composition

Sample	*y*	$\frac{Ca}{P}$	$\frac{Ca}{P+Si}$	Stoichiometric amounts of ions in samples		
				PO_4^{3-} (*n*)	SiO_4^{4-}	Δ[a]
N1	0	1.71	1.71	5.06	0	0.94
N2	0.1	1.73	1.67	5.04	0.16	0.96
N3	0.3	1.74	1.45	5.00	1.01	1.00
N4	0.5	1.74	1.25	4.99	1.97	1.01
N5	0.75	1.74	1.05	4.99	3.24	1.01
N6	1	1.75	0.92	4.98	4.48	1.02
N7	2	1.78	0.62	4.89	9.26	1.11
N8	5	1.88	0.29	4.61	25.3	1.39
N9	0[b]	-	0.49	-	-	-

[a] Δ - the difference between the number of PO_4^{3-} ions in stoichiometric HA and HA obtained in this work (Δ=6-n).
[b] Ca/Si = 1.

To study the bioactivity of synthetic materials, the solubility of powders in a weakly acid medium (pH = 6.20) was studied in a 0.9% solution of NaCl isotonic with human plasma.
According to figure 3a, at the initial stage of dissolving of each material the pH rapidly grow from the initial acidity of the 0.9% NaCl solution (pH ≈ 6.20). The most intensive grow of pH happen during

the first day, when the solid particles interact with the unsaturated solution. The maximum pH jump in a given period is fixed for powders with high silicon content (N7-N9). For them pH reaches values of the order of 9÷10.5 units. When N1-N6 samples dissolve, a gradual increase in the alkalinity of the solutions is observed with increasing ***y***. However, the pH of the liquid phase does not exceed 7.1 ± 0.1. According to the current understanding of HA and calcium silicates degradation chemistry [7, 8], a sharp increase in pH at the initial stage of dissolution of the material occurs as a result of the absorption of protons and their binding to water molecules, dihydrogen phosphate ions and silanol groups. Obviously, the jump in pH is the stronger the more SiO_4^{4-}, PO_4^{3-} and OH^- ions are contained in the substance. The maximum appearance on a curves pH = f(τ), which characterize the resorption of samples rich in silicon (N7-N9), is probably caused by secondary precipitation processes.

In the course of the experiments it was established that, depending on the materials composition, their behavior in the physiological solution is different. Thus, when the control samples N1 and N9 are resorbed, the Ca^{2+} concentration in the liquid phase increases. In this case, the solubility of CSH powder (N9) exceeds threefold given characteristic of the sample composed of apatite (N1). The amounts of PO_4^{3-} and SiO_4^{4-}, passing into the solution, also increase in time. Calculation of the molar coefficients Ca/P and Ca/Si showed that in case of N1 degradation calcium and phosphate ions emerge into the solution in approximately equal amounts, whereas in the case of CSH (N9) resorption, the Ca/Si ratio is close to 2/3. As a result, Ca^{2+}, PO_4^{3-} and SiO_4^{4-} gradually accumulate in solution. It indicates the absence of parallel processes in which this ions could be consumed. Thus, powders based only on HA or CSH are able to dissolve in physiological solution, but they do not stimulate the subsequent precipitation of calcium phosphates.

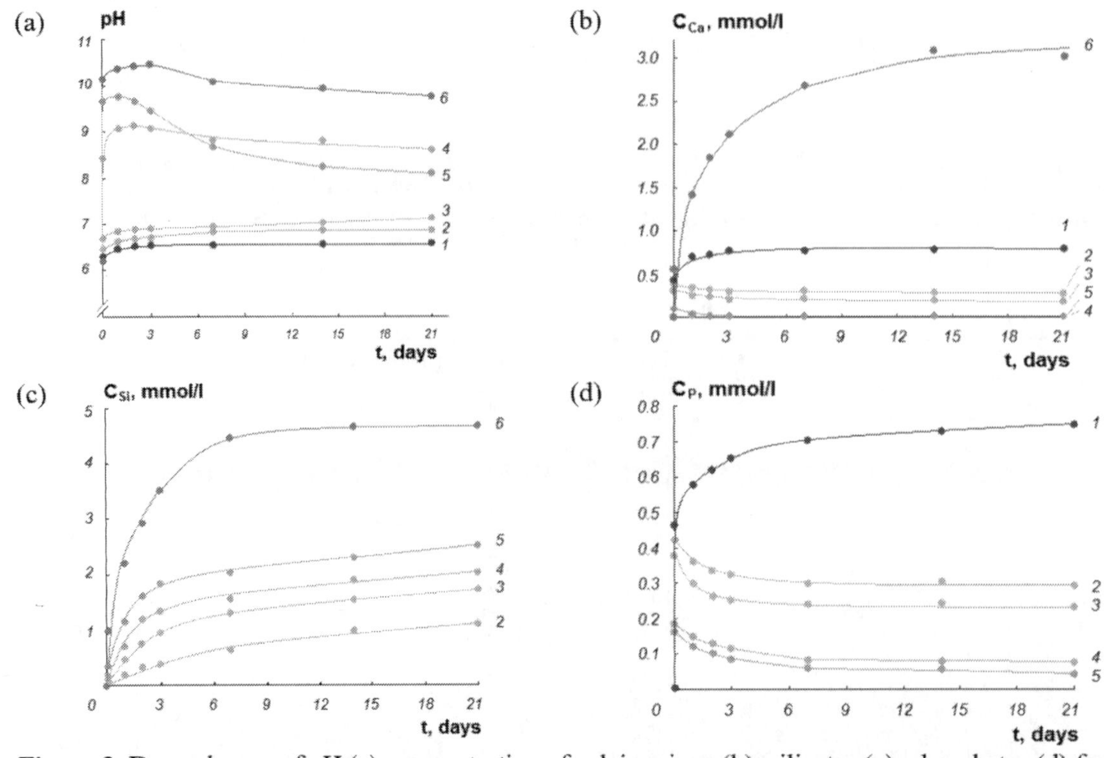

Figure 3. Dependences of pH (a), concentration of calcium ions (b), silicates (c), phosphates (d) from the time of dissolution samples N1 (1), N3 (2), N5 (3), N7 (4), N8 (5), N9 (6) in 0.9% NaCl.

During degradation in the weakly acid medium of N2-N8 samples, a regular increase in the concentration of SiO_4^{4-} in solutions is recorded. There is a positive correlation between the number of ions of this type released into the solution and the content of silicon in the solid phase. However, the expected accumulation of calcium as a result of the dissolution of solid phases (N2-N8) does not occur. The curves $C(Ca^{2+})$ = f(τ) have the same descending character as $C(PO_4^{3-})$ = f(τ). The amounts of metal ions and phosphates decrease synchronously with time (figure 3b-d). In this case, the higher the silicon content in the sample the lesser amounts of Ca^{2+} and PO_4^{3-} are found in the solution. Obviously, the uptake of Ca^{2+} and PO_4^{3-} occurs as a result of the HA precipitation reaction:

$$10\ Ca^{2+} + 6\ PO_4^{3-} + 2\ OH^- \rightarrow Ca_{10}(PO_4)_6(OH)_2 \quad (2)$$

A decrease of heterogeneous systems pH under 1 to 3 days study of samples N7, N8 interaction with physiological solution (figure 3a) may also be associated with the intensive progress of this process. The described processes cause the fact that over time and for silicon rich samples significant amounts of silicates are found in the solution. In general, ion concentrations in the systems (C_i) decrease in the following sequence: $SiO_4^{4-} > PO_4^{3-} > Ca^{2+}$. This indicates that silicates take active participation in the process of materials resorption.

5. Discussion

According to results of investigation in studied systems $Ca(NO_3)_2$ - $(NH_4)_2HPO_4$ - Na_2SiO_3 - NH_4OH - H_2O in all range of variation of the silicon agent concentrations highly dispersed poorly crystallized silicon-containing carbonatehydroxyapatite precipitates. The excess of silicate ions introduced into the system release as amorphous SiO_2 and CSH.
Based on the results of experiments with 0.9% NaCl, it was found that silicate phase (CSH) predominantly dissolves in weakly acidic media. As a result concentrations of Ca^{2+} and SiO_4^{4-} ions in the liquid phase increased. The reaction of HA sedimentation passing under these conditions, recorded by the synchronous decrease in the amounts of calcium and phosphate ions in solutions, characterizes the synthetic samples as bioactive. The materials are capable to initiate a passive (without the participation of specialized cells) formation of hard tissues mineral component. The described patterns suggest that materials based on the developed compositions can be effective in restoring damaged areas of hard tissues in dentistry and bone tissue engineering.

6. Conclusion

Highly dispersed poorly crystallized apatite containing CO_3^{2-} and SiO_4^{4-} isomorphic impurities forms in systems $Ca(NO_3)_2$ - $(NH_4)_2HPO_4$ - Na_2SiO_3 - NH_4OH - H_2O. At $y \geq 0.3$ silicate phase (CSH) and SiO_2 additionally deposit. *In vitro* experiments showed that material's solubility increases as the content of silicate phase in the samples composition increases and exceeds this characteristic of apatite. The $Ca_{10}(PO_4)_6(OH)_2$ precipitation, recorded by the synchronous decrease of calcium and phosphate ions amounts in 0.9% NaCl solution, characterizes synthetic mixtures as bioactive. The materials are capable to initiate a passive (without the participation of specialized cells) formation of hard tissues mineral component.
Materials based on the developed compositions can be effective in the restoration of damaged areas of hard tissues in dentistry and bone tissue engineering.

Acknowledgments
The reported study was funded by Omsk State Technical University according to the research project № 17045B.
The author is grateful for the help in performing physicochemical studies by PhD A.I. Blesman and PhD D.A. Polonyankin.

References

[1] Barinov S M and Komlev V S 2005 *Calcium phosphate based bioceramics* (Moscow: Science) p 204
[2] Marchat D, Zymelka M, Coelho C, Gremillard L, Joly-pottuz L, Babonneau F, Esnouf C, Chevalier J and Bernache-assollant D 2013 *Acta Biomater.* **9** 6992
[3] Danilchenko S N 2007 *SumSU herald. Phys., math., mechan.* **2** pp 33–59
[4] Chyukin G D 2008 *Dispersed silica surface chemistry and structure* (Moscow: Printing house Paladin, LLC «Printa») p 172
[5] Meiszteries A, Rosta L, Peterlik H, Rohonczy J, Kubuki S, Henits P, Sinko K 2010 *J. Phys. Chem.* A **114** 10403–11

IOP Conf. Series: Journal of Physics: Conf. Series **944** (2017) 012112 doi:10.1088/1742-6596/944/1/012112

[6] Yu P, Kirkpatrick R J, Poe B, McMillan P F and Cong X 1999 *J. Am. Ceram. Soc.* **82** pp 8–742
[7] Ni S, Lin K, Chang J and Chou L 2008 *J. Biomed. Res.* A **85** pp 72–82
[8] Dorozhkin S V 2012 *World J. Methodol.* **2** pp 1–17

IOP Conf. Series: Journal of Physics: Conf. Series **944** (2017) 012113 doi:10.1088/1742-6596/944/1/012113

Investigation of the influence of the open cell foam models geometry on hydrodynamic calculation

O V Soloveva[1,2], S A Solovev[2,3], R R Khusainov[1], O S Popkova[1] and D O Panenko[1]

[1]Institute of Heat Power Engineering, Kazan State Power Engineering University, Krasnoselskaja st. 51, Kazan, 420066, Russia
[2]Institute of Chemistry, Kazan Federal University, Kremlevskaja st. 18 Kazan, 420008, Russia
[3]Institute of Mathematics and Mechanics, Kazan Federal University, Kremlevskaja st. 18, Kazan, 420008, Russia

Abstract. A geometrical model of the open cell foam was created as an ordered set of intersecting spheres. The proposed model closely describes a real porous cellular structure. The hydrodynamics flow was calculated on the basis of a simple model in the ANSYS Fluent software package. A pressure drop was determined, the value of which was compared with the experimental data of other authors. As a result of the conducted studies, we found that a porous structure with smoothed faces provides the smallest pressure drop with the same porosity of the package. Analysis of the calculated data demonstrated that the approximation of an elementary porous cell substantially distorts the flow field. This is undesirable in detailed modeling of the open cell foam.

1. Introduction

In recent years, the international scientific community has a great interest in the study of open cell foam materials. This is due to the fact that metallic, ceramic and polyurethane porous materials are the basis for carrying out a lot of heat and mass transfer processes when an advanced surface area and a low coefficient of aerodynamic resistance are necessary even at high flow rates. As a clear example of the use of cellular porous media, we can consider chemical catalysts based on the open cell foam and filters for air purification. A good review of the porous media application is given in the article [1].

Analysis of literature data shows that models of a three-dimensional porous medium can be obtained either with the help of tomography, or mathematically using a geometrical model. However, computed tomography requires laborious processing in graphic editors before performing numerical simulation, removing excess volumes and smoothing surfaces that can significantly change the hydrodynamic processes in a porous medium. Also 3D modeling based on computer tomography for obtaining geometry and direct numerical simulation of hydrodynamics inside a porous medium require enormous computational resources and are possible to a greater extent on server computing stations.

With the growth of computational capabilities there are papers devoted to a detailed computer modeling of porous media based on various simplified models. Computer modeling of the porous structure as a model of ceramic open cell foam is presented in paper by Mitrichev I.I. et.al. [2]. Authors took a sphere with twelve segments obtained by clipping the spheres as a geometric cell of the material. Thus, they formed a structure of 13 elementary cells, which corresponds to the face-centered cubic packing of spheres. The open cell foam model consisted of 21 layers of such cells.

In the work of Hellmann [3] direct numerical simulation (DNS) using the computer tomography and a physical experiment for the flow of dusty gas in a porous medium were conducted, semi-empirical dependences for a pressure drop were obtained, a comparison was made with the experimental data of the authors Wake and Brown [4], whose data were far compared to a Hellmann's experiment. Helmann and coauthors obtained their own semi-empirical formula, but the data of DNS do not correlate well with the proposed semi-empirical expression.

In the work of authors T. Horneber, C.Rauh, A.Delgado [5] modeling of hydrodynamics presented for various three-dimensional structures of a porous medium: the tetrakaidecahedron geometry, that is a Kelvin cell geometry, the cubic lattice and the diamond structure. Calculations were carried out at different Reynolds and Euler numbers and were contributed to the understanding of optimization techniques of the porous medium.

The authors of [6] conducted research of the heat transfer process, and calculated the value of the pressure drop in a periodic unit cell of the porous medium, which is a structure of the sphere-centered open-cell tetrakaidecahedron.

A single cell model was proposed in the work [7] of authors Enrico Bianchi and others, this model they named as a macroscopic approach, which neglects small-scale details of open cell foams. Also in this work, the pressure drop is estimated on the basis of Forchheimer equation, which is the law of flow motion in a porous medium.

Maxime Lacroix and colleagues [8] tested a model of a porous cellular medium that is a dodecahedron (a dodecahedron (12 pentagonal faces, 20 vertices and 30 edges or struts)). The results of the calculations correlated well with experimental studies for the pressure drop depending on the gas velocity for different values of the porosity of the medium. Based on the recent studies, a new model for the pressure drop was introduced, as amended.

In large paper by T. P. de Carvalho et. al. [9], numerical modeling based on X-ray computer tomography and an experimental study of hydrodynamics for five different commercial metallic foam samples on a wide range of velocities, mainly on a turbulent regime, were performed. A change in the pressure drop was also analyzed depending on the thickness of the samples. The permeability and the Forchheimer coefficient were obtained from the calculations and the experimental study. Also article [10] demonstrates microtomography-based CFD simulation, that was carried out to determine effective thermal conductivity and permeability.

In the article [11] by Xin Hu et. al. for the numerical simulation of heat transfer with phase change material, a model of the micro-foam structure was used, which is an array of solid cube cells with nine spherical pores arranged in a BCC lattice. Direct numerical simulations performed to determine the effective thermal conductivity as a function of porosity and it is defined that the effective thermal conductivity is a non-linear function of the micro-foam porosity. Also temperature distributions in the cells and the temperature profiles at different time steps were shown.

Article of Saurish Das et.al. [12] is devoted to a random open cell foam. The hydrodynamics and heat transfer parameters where calculated with the help of developed Immersed Boundary Method for complex geometries, that is described in details in the paper. The article presents a graph of the change of non-dimensional drag force with Re which is also obtained by numerical modeling of gas flow in a porous medium based on computed tomography in comparison with the expression within the correlation by Dietrich using various experimental data. We can see their good agreement, at the same time the Ergun correlation is far from the pressure drop for foams. In addition there are graphs of the non-dimensional heat transfer coefficient (Nu) with Re.

In the work [13] by Sergey Lopatin and coauthors different types of catalytic cartridges made as a glass-fiber: spiral cartridge, prismatic cartridge, radial block from prismatic axial cartridges, catalytic elements of various shapes. Studies to determine the differential pressure for different types of clusters were carried out.

Also geometrically different cartridges for catalysts (ceramic foams, beads and a honeycomb monolith) were presented in paper [14], which are necessary for chemical reaction. The work is devoted to determining the pressure drop and parameters of heat and mass exchange processes for comparing different types of packages.

2. Problem formulation and solution method

The detailed model of the porous structure is characterized by three parameters: cell diameter (pore), porosity and fiber diameter (medium size of rigid bridge in the considered structure). In the literature, these parameters are connected by a single equation, for example, [1, 15, 16]. In modeling, the main task is to match these parameters to a real high-porosity structure.

A model of the open cell foam is constructed by an ordered set of intersecting spheres (figure 1,a). The diameter of the sphere, the distance between their centers and the relative positioning determine the parameters of the model.

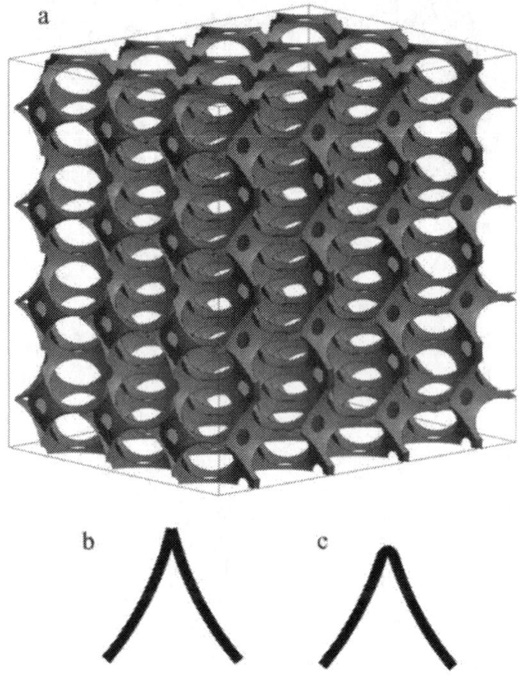

Figure 1. Model of the open cell foam

The same model is often used, for example, [2, 11] because of the simplicity of its construction. However, the main problem of the model is a large number of thin sharp edges at the intersection of spheres (Fig. 1, b). One of simple ways to get around this problem is smoothing out simple curves (Fig. 1, c).

The purpose of this work is to determine how smoothing in the open cell foam affects the resistance of the medium. As a characteristic of the resistance, the pressure drop is considered for a given gas flow rate.

2.1. Numerical model of gas motion in the open cell foam

There are models of gas flow in a porous medium, such as the Darcy and Forchheimer model. However, they can not be used for arbitrary structure because of the presence of an unknown permeability parameter.

In this work, the constructed three-dimensional model was calculated in the CFD-solver ANSYS Fluent by the finite volume method. To correctly account for the gas movement in the cells of the structure, the number of elements is 100,000 per 1 mm^3. Example of the mesh is shown in figure 2.

IOP Conf. Series: Journal of Physics: Conf. Series **944** (2017) 012113 doi:10.1088/1742-6596/944/1/012113

Figure 2. Mesh example

The gas flow was considered in the approximation of the Navier-Stokes equations:

$$\nabla \cdot \vec{U} = 0, \tag{1}$$

$$\left(\vec{U} \cdot \nabla\right)\vec{U} = -\frac{1}{\rho}\nabla P + \nu\Delta\vec{U}, \tag{2}$$

where $\vec{U}$ is vector of gas velocity, ρ is the gas density, P is the pressure, ν is the kinematic coefficient of gas viscosity.

We determined the pressure fields after the calculation convergence.

2.2. Semi-empirical model of pressure drop in the open cell foam

Modern technologies make it possible to do experimental studies of gas flow through open cell foam materials. A lot of experimental works are devoted to pressure drop calculations. Some researchers try to construct semi-empirical models of the pressure drop dependence on the filtration rate based on the results of the performed experiments.

For example, in work [4] the expression is obtained

$$\Delta P = l \cdot (A \cdot Po^2 \cdot U + B \cdot U^3 + O(U^5)), \tag{3}$$

where l – porous structure size, Po – porosity of the structure, in ppi, A, B and $O(U^5)$ – parameters obtained from experiments.

In work [3] an adapted formula is presented

$$\Delta P = l \cdot (A \cdot Po^2 \cdot U + B \cdot U^3), \tag{4}$$

where A and B – parameters obtained from new experiments of the work [3].

We notice that there is a comparison with the results of work [4] in the latest paper [3]. There are significant discrepancies in the results, and, consequently, discrepancies in the constructed semi-empirical formulas.

3. Results and discussions

Figure 3 shows the results of calculations of the pressure drop in Pa per 1 mm of the layer thickness which depends on the filtration rate in m/s.

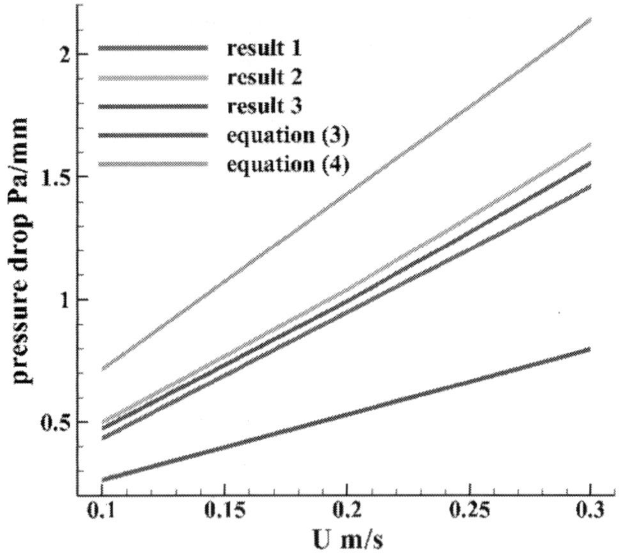

Figure 3. The curves of the pressure drop dependence on the filtration rate for different models of porous structure

The first calculation was made for a porous medium model constructed by a set of spheres without smoothing with a cell diameter $d_c = 506\,\mu\text{m}$ and porosity $\varepsilon = 0.9555$. The second calculation was made after smoothing the faces. At the same time, with the similar cell diameter $d_c = 506\,\mu\text{m}$, the porosity was $\varepsilon = 0.9583$. The third calculation was made for a structure without smoothing for porosity $\varepsilon = 0.9555$, the resulting cell diameter is $d_c = 50706\,\mu\text{m}$.

The results of the calculations were between the two semi-empirical formulas of [3] and [4]. A porous structure with smoothed faces provides the smallest pressure drop in the performed calculations. The differences between calculations 2 and 3 are obvious, because the porosity and the cell size in case 3 are larger. In case 1, the porosity is the same as in case 3, however, the cell size is smaller. But smoothing made it possible to reduce the size of bridge between cells, and consequently to increase the transparency and so to reduce the pressure drop. The results of calculations with the approximation of a porous structure without smoothing faces show a higher pressure drop in the selection of porosity and cell diameter.

Thus, a simple approximation of a highly porous cellular structure can make a significant error in the drag force calculating.

4. References

[1] Garrido G I, Patcas F C, Lang S and Kraushaar-Czarnetzki B 2008 *Chem. Eng. Sci.* **63** 5202
[2] Mitrichev I I, Koltsova E M, Zhensa A V 2012 *Fundamental Res.* **11** 440
[3] Hellmann A, Pitz M, Schmidt K, Haller F and Ripperger S 2015 *Aerosol Sci. Tech.* **49** 16
[4] Wake D and Brown R C 1991 *J. Aersosol Sci.* **22(6)** 693
[5] Horneber T, Rauh C and Delgado A 2014 *Chem. Eng. Sci.* **117** 229
[6] Bai M and Chung J N 2011 *Int. J. Therm. Sci.* **50** 869
[7] Bianchi E, Groppi G, Schwieger W, Tronconi E and Freund H 2015 *Chem. Eng. J.* **264** 268
[8] Lacroix M, Nguyen P, Schweich D, Huu C Ph, Savin-Poncet S and Edouard D 2007 *Chem. Eng. Sci.* **62** 3259

[9] De Carvalho T P, Morvan H P, Hargreaves D M, Oun H and Kennedy A 2017 *Transp. Porous Med.* **117** 311
[10] Zafari M, Panjepour M, Emami M D and Meratian M 2015 *Appl. Therm. Eng.* **80** 347
[11] Hu X, Wan H and Patnaik S S 2015 *Int. J. Heat and Mass Transfer* **88** 617
[12] Das S, Deen N G and Kuipers J A M 2016 *Cat. Today* **273** 140
[13] Lopatin S, Mikenin P, Pisarev D, Baranov D, Zazhigalov S and Zagoruiko A 2015 *Chem. Eng. J.* **282** 58
[14] Patcas F C, Garrido G I and Kraushaar-Czarnetzki B 2007 *Chem. Eng. Sci.* **62** 3984
[15] Kenny L C, Aitken R J, Beaumont G and Görner P 2001 *J. Aerosol Sci.* **32** 271
[16] Vincent J H, Aitken R J and Mark D 1993 *J. Aerosol Sci.* **24** 929

Investigation of internal elements impaction on particles circulation in a fluidized bed reactor

S A Solovev[1,2], O V Soloveva[2,3], A V Antipin[1] and E V Shamsutdinov[3]

[1]Institute of Mathematics and Mechanics, Kazan Federal University, Kremlevskaja st. 18, Kazan, 420008, Russia
[2]Institute of Chemistry, Kazan Federal University, Kremlevskaja st. 18 Kazan, 420008, Russia
[3]Institute of Heat Power Engineering, Kazan State Power Engineering University, Krasnoselskaja st. 51, Kazan, 420066, Russia

Abstract. A numerical study of the fluidized bed apparatus in the presence of various internal elements is carried out. A chemical reaction for temperature-dependent processes with heat absorption is considered. The task of incoming heated catalyst granules to the reactor is investigated. The main emphasis is focused on the circulation flows of the catalyst particles, heating of the reactor, and the efficiency of the chemical reaction. The analysis of the impact of various design elements on the efficiency of the reactor is carried out. The influence of feeding heated catalyst device design on the effectiveness of whole reactor heating is educed. The influence of the presence of fine particles on the efficiency of the reaction for different reactor design features is also educed.

1. Introduction

Fluidized bed apparatuses are widely used in the chemical industry [1-3]. Their advantage is the high speed of heat and mass transfer between components of the reaction which are resided in different aggregation states. Numerical calculations of fluidization are usually based on Eulerian-Eulerian approach when the carrier (gas, liquid) and the discrete (solid) phase are considered as continuous. To account for the interaction of particles in the fluidized bed, by analogy with the kinetic theory of gases, we added an equation describing the change in the kinetic energy of particles due to their collision (e.g., [4]). Solving these equations numerically allows to carry out calculations of different possible mechanisms of the fluidization process, both in terms of foundational research and practical application.

Each element of the chemical fluidized bed reactor can have a significant impact on hydrodynamics, processes of heat and mass transfer, and the overall efficiency of the apparatus. For example, obviously the shape of the gas feeder will affect the particle motion. Modelling and comparative analysis of two large fluidized bed units with various feed gas supply devices are carried out in [5]. A downward movement of the solid particles along the walls with a high particle concentration is observed in fluidization in cylindrical columns. This effect ensures continuous circulation of the particles in the apparatus. The location of internal elements on the way of the particles main flow stream may affect the operation of the apparatus as a whole. This may improve the chemical reactions efficiency by optimum distribution of the catalyst in the gas flow in the chemical reactors catalytic processes. The changes in the character of the movement and particles concentration along the wall with deflectors are shown in [6]. The works [7, 8] concentrate on the experimental and numerical study of the ring baffles located on the walls of cylindrical apparatus effect on the hydrodynamics of the fluidized bed.

In this paper we consider the fluidized bed apparatus of cylindrical shape with a uniform gas feeding. The ring baffles, deflectors and the grids were investigated as the internal components. The aim of the present paper was to determine the influence of the internal elements on the circulation flow of gas and particles in the apparatus. The pictures of the main circulating catalyst stream were created and the results of the catalyst influence on the concentration were determined. The effect of internal elements on the behavior of the catalyst in reactor and the efficiency of chemical were investigated.

2. Problem formulation and mathematical model

In the present work we study the fluidized bed cylindrical apparatus, designed for reactions with the heat absorption in the presence of a catalyst. In the lower part there is a gas feeder at the temperature of 550 ^{O}C. There are microspherical catalyst particles with the diameter of 20-200 μm in the apparatus. Along the axis is located a vertical pipe for the feeding of the heated catalyst regenerated at 650 ^{O}C. In the middle of the reactor, in the working area, are located angular falling through type grids with the free-sectional area approximately of 30% (figure 1, a).

A vertical pipe for the heated catalyst feeding is located in the center of the feeder, as will be shown below, can have a significant impact on the catalyst circulation. Another version of the feeder layout is therefore considered. There is no a vertical pipe in the center. A catalyst feeder simulates a pipe, embedded on the side of the apparatus, and also provides the catalyst flow in the center of the apparatus (figure 1, b).

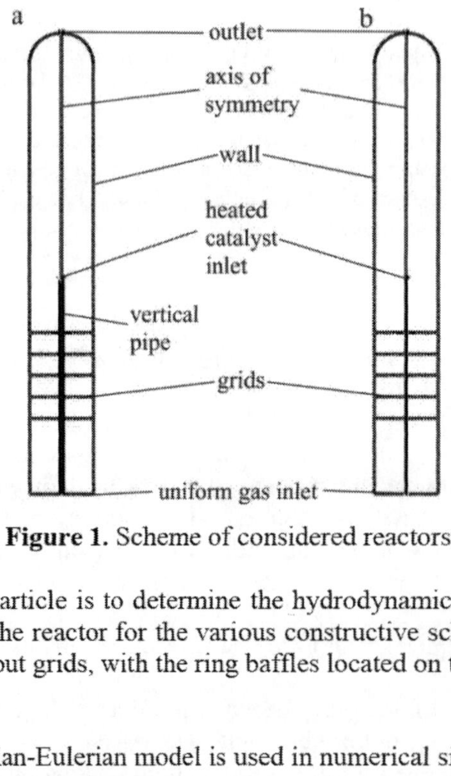

Figure 1. Scheme of considered reactors

The main objective of the article is to determine the hydrodynamic and thermal characteristics of the processes taking place in the reactor for the various constructive scheme of internal elements. We also consider the options without grids, with the ring baffles located on the walls and the other ones.

2.1. Fluidized bed model

Continuous multi-phase Eulerian-Eulerian model is used in numerical simulations of the fluidized bed. For each of the phases the equations of mass, momentum and energy conservation are implemented.

Conservation of mass

$$\frac{\partial \alpha_i \rho_i}{\partial t} + \nabla \cdot (\alpha_i \rho_i \vec{v}_i) = 0, \quad (1)$$

where α_i is the volume fraction of the i-th phase, ρ_i is the real density, $\vec{v}_i$ is the velocity.

Conservation of momentum

$$\frac{\partial \alpha_i \rho_i \vec{v}_i}{\partial t} + \nabla \cdot (\alpha_i \rho_i \vec{v}_i \vec{v}_i) = -\alpha_i \nabla p + \nabla \cdot \bar{\bar{\tau}}_i + \alpha_i \rho_i \vec{g} + \sum_j \vec{R}_{ij}, \tag{2}$$

where $\bar{\bar{\tau}}_i = \alpha_i \mu_i (\nabla \vec{v}_i + \nabla \vec{v}_i^T) + \alpha_i (\lambda_i - (2/3)\mu_i) \nabla \cdot \vec{v}_i \bar{\bar{I}}$ is the stress tensor, p is the pressure, μ_i, λ_i are shear and bulk viscosity, $\bar{\bar{I}}$ is a unity tensor, $\vec{R}_{ij} = K_{ij}(\vec{v}_i - \vec{v}_j)$ is the force of interphase interaction.

Conservation of energy

$$\frac{\partial \alpha_i \rho_i h_i}{\partial t} + \nabla \cdot (\alpha_i \rho_i \vec{v}_i h_i) = \alpha_i \frac{\partial p_i}{\partial t} + \bar{\bar{\tau}}_i : \vec{v}_i + \sum_j Q_{ij}, \tag{3}$$

where h_i is the enthalpy, Q_{ij} is the heat exchange intensity, p_i is the solid phase granule pressure.

To account for the interaction of solid phases particles is used an equation that considers the change of energy through a change in temperature of the particles [9]:

$$\frac{3}{2}\left[\frac{\partial}{\partial t}(\alpha_i \rho_i \Theta_i) + \nabla \cdot (\alpha_i \rho_i \vec{v}_i \Theta_i)\right] = \left(-p_i \bar{\bar{I}} + \bar{\bar{\tau}}_i\right) : \nabla \vec{v}_i + \nabla \cdot \left(k_{\Theta_i} \nabla \Theta_i\right) - \gamma_{\Theta_i} + \phi_{ij}, \tag{4}$$

where k_{Θ_i} is the diffusion coefficient of the granules, Θ_i is the temperature of the solid phase granules, γ_{Θ_i} is the energy dissipation due to the particles collision, ϕ_{ij} is the energy exchange between *i*-th solid phase and *j*-th gaseous phase.

To close the system use the relationships obtained from the experimental studies of the fluidized bed. The interphases interaction coefficient is determined experimentally and depends on the type of the interacting phases. For two types of phase: liquid (gaseous) phase (*i*-th) – solid granular phase (*j*-th) model used in [4], in the case of two solid granular phases is the model of [10]. The energy dissipation γ_{Θ_i} due to the particles collision is written by the model obtained in [11]. Coefficient of the heat exchange intensity is determined for the interaction of two phases by the relation, given in [12]. The motion considered in the reactor at the accepted loads of the raw gas feeding and parameters of the used catalyst is turbulent. In the model accepted for the calculations a dispersion $\kappa - \varepsilon$ model of turbulence was used in which the motion of "secondary" solid granular phases is generated on the background of the turbulent motion of "primary" gas phase.

2.2. CFD solver

In this paper differential equations that describe the hydrodynamic and thermal processes in the computational model were solved in CFD ANSYS Fluent for unsteady flow regime. The whole computational domain was divided into elements of a triangular or rectangular shape in different subregions, the dimensions of which are sufficient to determine the characteristic factors of the investigated phenomenon. For a given initial distribution of the catalyst in the reactor, after a while the solution reached quasi-stationary regime, at which characteristic hydrodynamic and thermal pictures were calculated.

Boundary conditions were set on all elements of each constructed model of apparatus in accordance with the mechanism of work and the solver used. Since the reactor model assumes rotationally symmetric, then on the axis of the constructed model conditions of axial symmetry «axis» were set. In all non-permeable surfaces «wall» conditions were accepted. In models of raw gas feeder nozzle and in the outlet of the centrally located pipe of the catalyst feed were set «velocity-inlet» (the value of the flow rate) conditions. In the area at the top of the apparatus model there were «outflow» (free flow yield) conditions.

2.3. Chemical reaction efficiency

To estimate the characteristics of the reaction kinetics and apparatus efficiency the simplified chemical reaction efficiency function for control volume was introduced as

$$F_{ef}(x,y,z) = V_{cat}(x,y,z)T_{gas}(x,y,z), \tag{5}$$

where $V_{cat}(x,y,z)$ is catalyst volume of fraction, $T_{gas}(x,y,z)$ is gas temperature at a point of reactor. The higher gas temperature and the catalyst concentration in the reactor point is more preferable for the chemical reaction at this point. To determine the integral value for total reactor unit volume V characterizing the efficiency function, we consider the coefficient

$$C_{ef} = \frac{\left(\int_V F_{ef} dv\right)}{V_{cat,max} T_{gas,max} V} * 100. \tag{6}$$

3. Results and discussions

Figure 2 demonstrates the pictures of temperature fields which are shown on a scale from gas feed flow temperature (blue areas) to the temperature of the catalyst feed (yellow areas). The basic case of the reactor in the presence of a central pipe for the heated regenerated catalyst and grids in the cross sections of the block (figure 2, a). Large concentration is observed in the upper part of the apparatus not occupied by the coarse particles. When considering the monodisperse catalyst, the heated particles received through a pipe roll along the pipe wall to the bottom part of the block. The presence of the grids contributes to the appearance of longitudinal fluctuations due to the additional obstacles they create on the way of ascending granules. Thus, the catalyst gives heat off to the incoming gas and the particles in the space between grids and hitting the bottom of the reactor rises up slowly heating the gas and the particles that are in the block. The presence of fine particles changes the picture of the temperature field in the block. The reactor is warmed up slower than in the case of coarse particles only. The fine particles form their own fast circulating flow in the upper part of the block, so that there are large particles too. The heated particles partially roll down along the pipe and are partially entrained by the circulating flow of fine particles. There is a small amount of particles at the top of the block and they quickly transfer the heat to upward feed-gas, which in turn rapidly leaves the reactor. Thus, a substantial portion of the heat is transferred to the gas in the area with a low content of catalyst particles.

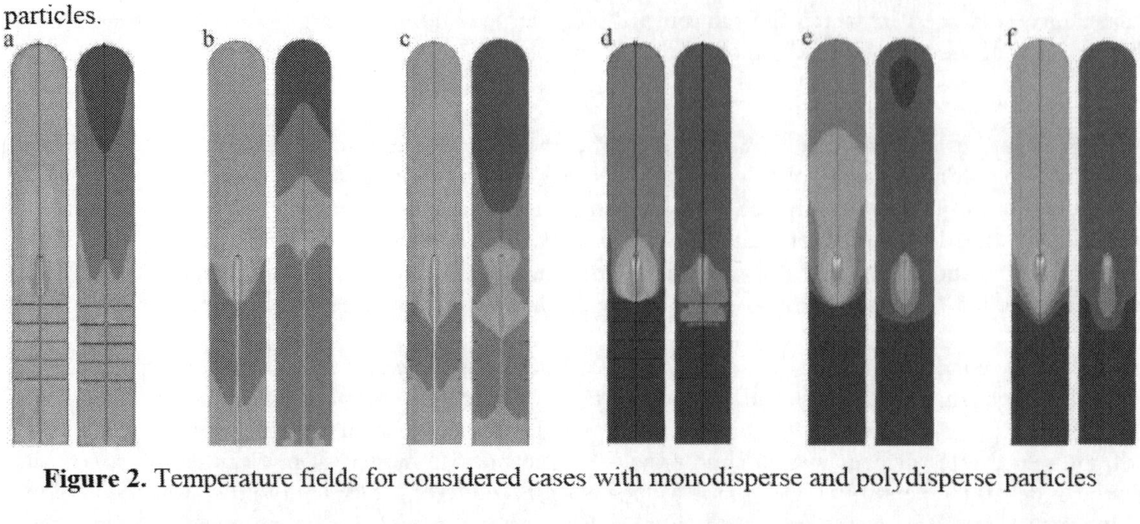

Figure 2. Temperature fields for considered cases with monodisperse and polydisperse particles

To evaluate the effect of the grids on the block the heating calculations were also performed in the absence of the grids (case 2, figure 2, b). The reactor temperature was lower compared to a case 1 both for the monodisperse catalyst and in the presence of the fine particles. The reason for it is the coarse particles form a fast circulating flow in the lower part of the apparatus. After getting down the heated particles rise up faster and give off the heat to the gas at the boundary of large particles. The same happens in the presence of fine particles. Two fast circulating flows of the fine particles in the upper part and the large particles at the bottom of the reactor can be observed.

In the third case (figure 2, c) the location of ring baffles on the walls of the reactor in the absence of grids in the central part of the block is considered. This is done in an attempt to eliminate the fast circulating flow of the case 2 while maintaining a uniform concentration of catalyst. The analysis of the calculation results of case 3 shows that catalyst concentration fields, temperature fields, and circulation flows are in the intermediate state between the cases 1 and 2. Despite the presence of large deflectors, there is upstream large catalyst particles flow around them. The total flow circulation rate slows down, which is beneficial to heat exchange with the gas-feed compared to the case 2. The presence of fine particles has the same effect as in the above cases 1 and 2. However, in this case there is a wide and complex zone of large and small particles circulating streams conjugation due to the presence of deflectors on the walls.

Another option of the construction may consist in a removal of the vertical feed pipe of the regenerated catalyst (figure 2, d). Let's assume an option (case 4) when the pipe enters the block from the side and in our model is simulated by the insulated feed channel above the top grid. As in case 1, the coarse particles move slowly in the bottom part of the apparatus, and the fine particles form a fast circulating flow in the upper part of the block. However, unlike the case 1, the heated particles get to the upper grids and begin to engage in circulating currents of coarse or fine particles. The pictures of temperature fields show that the lower part of the reactor is hardly warmed. Thus, the particles transfer heat to feed-gas in the middle of the block where the particle concentration is low. Also the calculations without grids (case 5, figure 2, e) but with annular baffles (case 6, figure 2, f) as in the cases 2 and 3 were carried out.

To analyze the efficiency of the chemical reaction we are interested in the previously described coefficients were counted based on catalyst concentration and gas temperature. The results are shown in table 1.

Table 1. Chemical reaction efficiency coefficient values for calculated cases

Case	Efficiency	
	1 fraction of catalyst	2 fractions of catalyst
Case 1: central pipe, grids	2.622	1.696
Case 2: central pipe	1.695	1.139
Case 3: central pipe, ring baffles	1.735	1.376
Case 4: without central pipe, grids	0.066	0.097
Case 5:without central pipe	0.032	0.089
Case 6: without central pipe, ring baffles	0.017	0.065

It is seen that in the presence of the central vertical pipe the efficiency is much higher. This result is achieved by a fast downflow of the heated particles entering the reactor. This mechanism ensures the feed of regenerated particles into the bottom of the block where they are involved in heat transfer and circulation of the dense bed of large particles. In turn, the presence of small particles for cases 1, 2 and 3 results in the decrease of efficiency since partially regenerated particles are captured by the fast stream and appear at the top of the apparatus. However, for cases 4, 5 and 6 the presence of the fine particles increases the efficiency of the chemical reaction. In the absence of a vertical pipe as the primary mechanism for the initial movement of the regenerated particles (cases 4, 5, 6) heated granules received in the reactor are entrained in the main circulating flows of the apparatus. Since the fine particles move faster along all the reactor height they contribute to the more rapid penetration of incoming particles into the bottom of the block where a dense bed of coarse particles is located. In these cases, the presence of fine particles allows to obtain the greater efficiency in comparison with monodisperse catalyst.

IOP Conf. Series: Journal of Physics: Conf. Series **944** (2017) 012114 doi:10.1088/1742-6596/944/1/012114

4. Conclusion

Each reactor has its design features which can have a significant effect on hydrodynamics and heat transfer in a fluidized bed. The reactor central pipe has a significant impact on the heating, providing a mechanism for the initial movement of the heated catalyst particles received into the bottom of the unit. The presence of fine particles provides a fast circulating flow in the upper part of the reactor which can have both positive and negative effect on unit performance, depending on the design features of the apparatus. The most effective design features are those that provide a slow movement of coarse particles, a dense bed located in the lower part of the apparatus. Results of this research may be useful in the calculation and design of the fluidized bed apparatus. It should be noted though that each apparatus has its own individual features that can change the representation of the discontinuous phase circulation and heat transfer significantly.

5. References

[1] Kunii D and Levenspiel O 1991 *Fluidization Engineering* (Boston: Butterworth-Heinemann)
[2] Yang W C 2003 *Handbook of Fluidization and Fluid-Particle System* (New York: Taylor and Francis Group)
[3] Basu P 2006 *Combustion and Gasification in Fluidized Beds* (Boca Raton: CRC press)
[4] Gidaspow D 1994 *Multiphase Flow and Fluidization* (Boston: Academic Press)
[5] Solov'ev S A, Egorov A G, Lamberov A A, Egorova S R and Kataev A N 2016 *Catal. Ind.* **8** 48
[6] Golriz M R and Grace J R 2002 *Int. J. Heat Mass Transfer* **45** 1149
[7] Samruamphianskun T, Piumsomboon P and Chalermsinsuwan B 2012 *Chem. Eng. J.* **210** 237
[8] Chalermsinsuwan B, Samruamphianskun T and Piumsomboon 2014 Chem. Eng. Res. Des. **92** 2479
[9] Ding J and Gidaspow D 1990 *AIChE J.* **36** 523
[10] Syamlal M 1987 *The Particle-Particle Drag Term in a Multiparticle Model of Fluidization* (Morgantown WV: EG & G Washington Analytical Service Center)
[11] Lun C K K, Savage S B and Jeffrey D J 1984 *J. Fluid Mech.* **140** 223
[12] Gunn D 1978 *Int. J. Heat Mass Transfer* **21** 467

IOP Conf. Series: Journal of Physics: Conf. Series **944** (2017) 012115 doi:10.1088/1742-6596/944/1/012115

Complex research of vanadium dioxide in the homoginuity field

Vad I Surikov[1], N A Semenyuk[1], Val I Surikov[1], Y V Kuznetsova[2] and O Y Pavlovskaya[1]

[1]Omsk State Technical University, 11 Mira Pr., Omsk 644050, Russia
[2]LLC "Surgutneftegas", Surgut, Russia

e-mail: sem-natal@mail.ru

Abstract. The complex research results of vanadium dioxide $VO_{2\pm X}$ obtained with X-ray analysis method, a method of vacuum adiabatic calorimeter and Faraday method are discussed. The samples temperature dependences of electrical resistivity, heat capacity, and also magnetic susceptibility are presented and they prove that these characteristics jump abruptly in the phase transition area. Thermophysical, electrical, magnetic properties of vanadium dioxide are established to vary with the oxygen content in the samples. Values of Debye characteristic temperatures θ_D for samples $VO_{2\pm X}$ determined at temperature 150 K are presented. Values of entropy change as an amount of electronic and phonon components at phase transition were calculated. The results obtained allow to consider the phase transition of a metal semiconductor as a Mott-Peierls transition, and also to use vanadium dioxide in temperature sensors manufacture.

1. Introduction

Vanadium dioxide ($VO_{2\pm X}$) and solid solutions based on it were tested in about 340 K structural phase transition [1, 2] (without changing the aggregation state) accompanied with a change of electrical (electrical resistivity jump may be up to 10 orders), optical, magnetic and other properties. It is widely used while manufacturing temperature sensors in automatic control systems of various technological processes.

2. Statement of the problem

The substances $VO_{2\pm X}$ within the homogeneity field have been synthesized and electrical resistivity, heat capacity and magnetic susceptibility as well as their temperature dependence in about temperatures of the phase transition metal-semiconductor (PTMS) have been studied on these samples (~340 K).

3. Theory

Samples for the study were prepared by dissociation of vanadium pentoxide at temperature 1300 K in vacuum. The obtained samples, and X-ray diffraction analysis shows it, are close in composition to stoichiometric analysis. Compounds VO_{2-X} were obtained with initial sample dissociation in high vacuum, and VO_{2+X} samples were obtained with oxidation of the sample in air at 600 K.

The samples obtained were subjected to x-ray analysis on ray diffractometer Shimadzu Maxima X XRD-7000, it showed that samples $VO_{1.990}$, $VO_{1.995}$,$VO_{1.998}$,$VO_{2.00}$ $VO_{2.010}$, $VO_{2.030}$, selected for the studies are single-phase fine powders of black color. The analysis performed on an electron, scanning microscope JEOLJCM-5700 made possible to clarify the elemental composition of each sample.

The samples electrical resistivity was measured on the compressed tablets using a standard two-contact method with a relative error less than 10%. The heat capacity was studied with a vacuum adiabatic calorimeter method of Strelkov type. The sample temperature was controlled with a platinum resistance thermometer, the error in determining the heat capacity did not exceed 1%. The magnetic susceptibility was determined with the Faraday method. The error in determining the susceptibility was about 5% and is mainly related to the sample weight determination.

4. The results of the experiment

The heat capacity temperature dependence of vanadium dioxide near the phase transition is presented in Fig.1. The obtained results show that a sharp increase in the heat capacity is observed at 338 K, and it confirms the presence of a phase transition at given temperature. Analogous temperature dependences are also observed for other samples studied.

The electrical resistivity temperature dependence of the sample $VO_{2,00}$, is shown at Fig. 2 obtained in the sample heating and cooling mode. A jump in the electrical resistivity is clearly observed in PTMS, as well as hysteresis in the phase transition.

Similar results were obtained for other samples. PTMS temperatures determined from the data on the heat capacity (temperatures corresponding to the peak C_p), and determined from data on research ρ(T) while heating the sample correspond with each other.

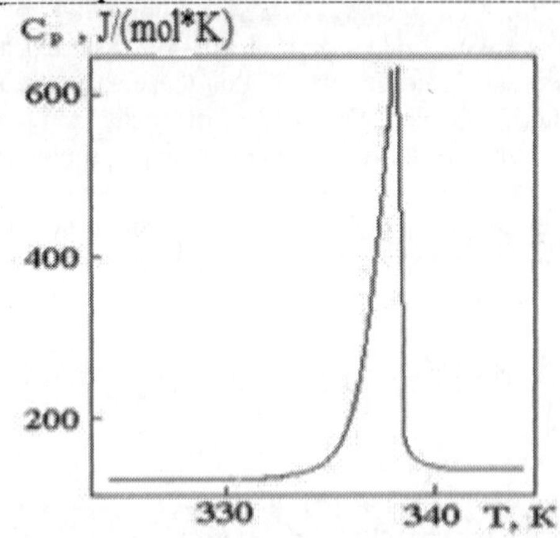

Figure 1. Temperature dependence of the heat capacity $VO_{2.00}$.

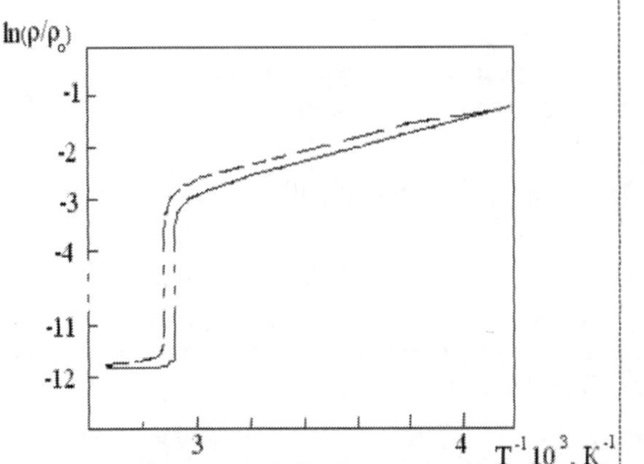

Figure 2. Temperature dependence of electrical resistivity $VO_{2.00}$.

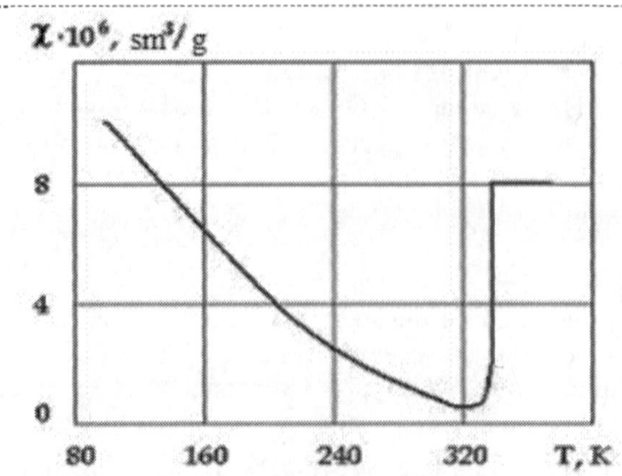

Figure 3. Temperature dependence of the magnetic susceptibility $VO_{2,00}$.

Vanadium dioxide is well known to be observed with the temperature hysteresis of PTMS phase transition [3]. For the samples, it can be established from the results of electrical resistivity temperature dependences study.

Hysteresis for the samples under study does not exceed 8 K, it has a minimum value for the sample $VO_{2.00}$ (4.0 K), and it changes little and irregularly with changes in the oxygen composition. It should be noted that the hysteresis appreciably depends on the sample heating temperature: the higher (near T_{MP}) heated sample, the lower the hysteresis. In this article all the samples, when determining the hysteresis of the phase transition, were heated to temperature 370 K. The values of the phase transition temperature determined according to electrical resistivity ($T_{MP\rho}$) and heat capacity (T_{MPc}) as well as temperature hysteresis transition (ΔT_{MD}) are shown in Table 1.

Estimates of the entropy changes during metal-insulator phase transitions are presented for all the samples studied in the same Table 1 [4]. The entropy change at the phase transition was determined from the experimental values $C_p(T)$:

$$\Delta S = \Sigma\,(\Delta C \cdot \Delta T/T) \quad (1)$$

where ΔC is difference between the set value of the heat capacity and the heat capacity value determined with extrapolation of regular branch C_p from the area below the phase transition temperature for a given T, step width ΔT was assumed equal to 1K. Temperature T corresponds to the middle of the value ΔT, the values of the Debye characteristic temperatures θ_D are given for these samples determined at 150 K.

Table 1. Values of the phase transition temperature $T_{MD,\rho}$, $T_{MD,c}$, hysteresis ΔT_{MD}, entropy phase transition ΔS and debue temperature characteristic θ_D.

Compounds	$T_{MD,\rho}$, K	$T_{MD,c}$	ΔT_{MD}, K	ΔS,J/K	θ_D, K
$VO_{1.990}$	338	338.0	4	25.2	1040
$VO_{1.995}$	339	338.4	8	25.2	950
$VO_{1.998}$	339	339.6	6	25.8	890
$VO_{2.00}$	340	340.0	4	16.5	700
$VO_{2.010}$	342	340.5	6	16.0	1090
$VO_{2.030}$	343	341.0	4	20.2	1190

From the magnetic susceptibility measurement completed the density of electronic states near the Fermi level $g(E_F)$ [1] can be estimated, if to assume that the discontinuity of the magnetic susceptibility ($\Delta\chi$) at PTMS relates primarily to the contribution to susceptibility appeared in the 3d-zone of vanadium conduction electrons $g(E_F)$

$$N(E_F) = \Delta\chi/2\mu_B^2 \quad (2)$$

Where μ_B is Bohr magneton.

Electronic heat capacity coefficient (γ) also allows to evaluate $g(E_F)$:

$$\gamma = (2/3)(\pi\kappa)^2 N(E_F) \quad (3)$$

Value γ can be estimated from the change in the entropy electronic component in the phase transition (ΔS_{El}):

$$\Delta S_{El} = \gamma\, T_{MD}. \quad (4)$$

where κ is the Boltzmann constant in equation (3).

The joint solution of equations (2) – (4) allowed determining the entropy electronic component of the phase transition, ΔS_{El}.

The experimentally determined values $\Delta\chi$ and the calculated values γ are presented in Table 2. The values ΔS_{El} and lattice contribution ΔS_{lat} to the entropy are changed when PTMS.

Table 2. Values of experimentally determined $\Delta\chi$, calculated g and values ΔS_{el} and ΔS_{let} at PTMS.

Compounds	$\Delta\chi \cdot 10^6$, g/sm^3	$\gamma \cdot 10^4$, J/(mol*K^2)	ΔS_{EL}, J/K	ΔS_{let}, J/K
$VO_{1.990}$	5.3	51	3.4	21.8
$VO_{1.995}$	9.2	88	6.0	19.5
$VO_{1.998}$	9.2	88	6.0	19.5
$VO_{2.00}$	7.0	67	4.6	11.9
$VO_{2.010}$	5.9	56	3.8	12.2
$VO_{2.030}$	6.1	58	4.0	16.2

The lattice contribution was estimated under the assumption of entropy additivity in the phase transition.

$$\Delta S = \Delta S_{El} + \Delta S_{lat}. \quad (5)$$

The dependences ΔS_{El} and ΔS_{lat} on phase transition temperatures T_{PTMS} (phase transition temperature have been determined from the data on the heat capacity) and they are presented in Fig. 4 and 5.

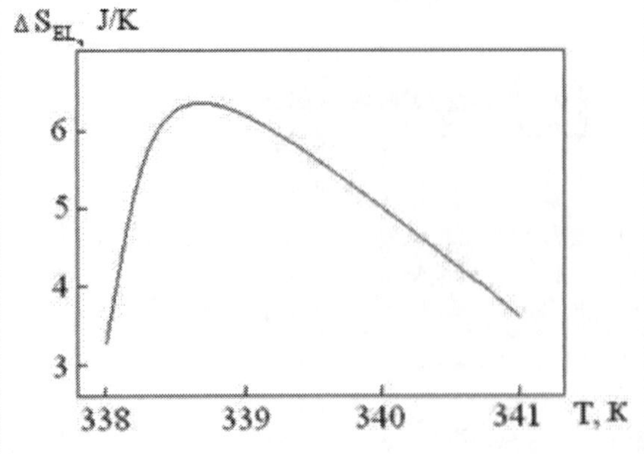

Figure 4. The entropy electronic component PTMS

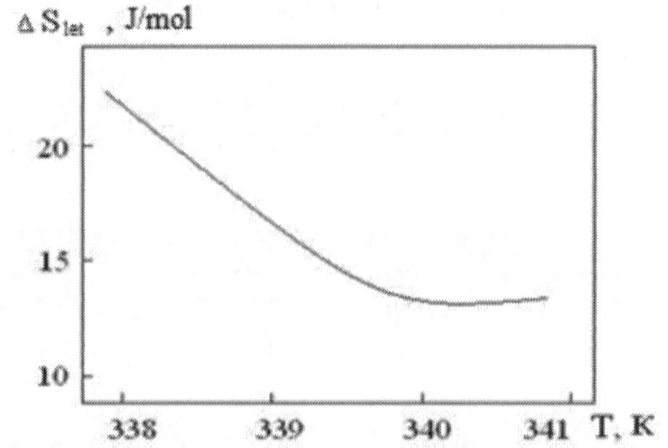

Figure 5. The entropy lattice component PTMS.

5. The discussion of results

Analyzing the results obtained, first of all it should be noted that the samples obtained are powdery. Due to our method of synthesis, each particle is inhomogeneous in volume. It leads to a non-simultaneous temperature transition in different micro-regions of each vanadium dioxide separate particle. Depending on the oxygen content in the sample, the peripheral layer of a single powder is either depleted (in the reduction reaction case) or enriched (in the oxidation reaction case) with oxygen. When moving from the periphery to the center of the powder, the oxygen concentration most likely varies nonlinearly, but in any case it changes. It leads to the fact that when studying the heat capacity and magnetic susceptibility, such heterogeneity of the composition over each sample individual powder volume changes the temperature width PTMS: it is the smallest for the initial sample and increases for the remaining samples, regardless of the oxygen composition. The electric current, with the method of study, flows along the surfaces of the powders and in this case it is not an integral characteristic. In determining the characteristic temperatures as in [3], C_p is assumed to be approximately equal to C_v.

From the results obtained it is evident that ΔS_{El} and ΔS_{lat} well correlated with T_{MD}. In our opinion, it suggests that both the Mott electronic mechanism and the Peierls mechanism determine PTMS in vanadium dioxide.

Entropy change of phase transitions ΔS at PTMS for samples VO_{2-x} does not differ and significantly exceeds ΔS (more than 20 %) for samples VO_{2+X}.

As the results show, the Debye temperatures within the homogeneity field extremely changed having a minimum value for $VO_{2.00}$. One can assume that in the samples of oxygen as deficient as excess the crystal lattice undergoes distortion. These distortions are due to crystal lattice defects such as Schottky defects that can cause a change in the boundary Debye frequencies, and, consequently, the Debye temperatures.

6. Conclusions

From the obtained experimental results it can be concluded that the reduced samples are more inhomogeneous in their composition (they have more temperature range of the phase transition) than the oxidized ones. So, the dissociation VO_2 proceeds at a much slower rate than the oxidation in an oxygen stream.

Measurement results processing allowed to estimate the electronic ΔS_{El} and phonon ΔS_{lat} entropy components PTMS. Correlations ΔS_{El} and ΔS_{lat} with T_{MD} allow to consider PTMS and, accordingly, the mechanism of the transition as Mott-Peierls transition.

References

[1] Bugaev A, Zacharchenya B and Chudnovsky F 1979 Metal-semiconductor phase transition and its application (Leningrad: Nauka) p 183

[2] Mott N 1974 Metal-insulator transitions Taylor & Francis Ltd (London) p 278

[3] Surikov V, Lyakh O, Prokudina N and Danilov S 2014 *Russian Physics Journal* T 57 8 pp 1111–1115

[4] Surikov Vadim; Danilov S and Surikov Valeriy 1981 *Izvestiya vysshikh uchebnykh zavedenii Fizika* 4 Tomsk pp 112–114

IOP Conf. Series: Journal of Physics: Conf. Series **944** (2017) 012116 doi:10.1088/1742-6596/944/1/012116

Method of sections in analytical calculations of pneumatic tires

V N Tarasov and I V Boyarkina

Siberian State Automobile and Highway University (SibADI), 5 Prospect Mira St., Omsk, 644050, Russia

E-mail: tarasov_vladimir07@mail.ru

Abstract. Analytical calculations in the pneumatic tire theory are more preferable in comparison with experimental methods. The method of section of a pneumatic tire shell allows to obtain equations of intensities of internal forces in carcass elements and bead rings. Analytical dependencies of intensity of distributed forces have been obtained in tire equator points, on side walls (poles) and pneumatic tire bead rings. Along with planes in the capacity of secant surfaces cylindrical surfaces are used for the first time together with secant planes. The tire capacity equation has been obtained using the method of section, by means of which a contact body is cut off from the tire carcass along the contact perimeter by the surface which is normal to the bearing surface. It has been established that the Laplace equation for the solution of tasks of this class of pneumatic tires contains two unknown values that requires the generation of additional equations. The developed computational schemes of pneumatic tire sections and new equations allow to accelerate the pneumatic tire structure improvement process during engineering.

1. Introduction

Pneumatic-tire wheels have been widely adopted as movers and bearing elements of land transport vehicles. A pneumatic-tire wheel consists of a pneumatic tire, performed in the form of a torus, mounted on a metal rim. The internal airproof cavity of a tire is filled with compressed air. The structure and manufacturing technology of pneumatic tires characterize the development level of the modern engineering. However, the theory of pneumatic tires doesn't reflect to the fullest extent the pneumatic and physical essence of pneumatic tires and is need of further development.

The relationship of environmental factors with processes inside a tire is established in mechanics by means of the method of sections. The theory of pneumatic tires is based on works of V.L. Biderman [1] during the 60s of the past century, in which the methods of sections were not properly adopted in calculations of pneumatic tires. The usage of the Laplace equation is a positive start, where non-traditional concentrated forces are used, but intensities of distributed forces per unit length of the tire section.

Insufficient attention is given in works of V.L. Biderman [1]. B.L. Bukhin [2], V.I. Knoroz [3] to the research of the interaction of the tire contract area with bearing surface from the point of view of its influence on capacity, tire wear, tire drag, road stability and other parameters.

In the work of A. E. Belkin and others [4] the mechanics of tires is divided into external and internal one. The external mechanics bears relationship to the car dynamics and characteristics, and the internal one – to processes occurred inside a tire.

From the moment of the computer appearance a great number of researchers use the finite element method for the research of internal processes inside a tire.

In work [5] S.V. Sheshenin notes that models taking into account huge tire deflections, viscoelastic properties of rubber are very complicated. That's why the matrix of resulting linear equation systems is turned out to be ill-conditioned.

In work [6] M.A. Salaani investigates the problems of the tire contact area with the bearing surface using Hermitian cubic functions and Hooke's law. The distribution model of normal and tangential forces of tire traction gives reliable results.

In work [7] J. Wallaschek and B. Wies notes that the modeling of tires has reached a sufficiently high level, allowing to take into account the deformation rates of elastic elements, friction force of a tire tread with road and many other factors.

In work [8] P.E. Stelzig considers a pneumatic tire like an object subjected to huge deformations within specified severe limitations. He considers the operation of cords and strips, forming the supporting carcass of a tire. A complex process modeling unit is used, taking into account pre-distortions for ensuring the approximation of special functions of the limitation of variations, maintaining natural kinematic functions.

The fulfilled review shows that the natural complication of research methods has occurred. The method of sections is insufficiently used, allowing to establish the relation of external factors of tires with internal process in tire elements.

2. Task definition

Computational schemes of tire carcass sections were developed and equations were derived for the determination of the tire capacity, as well as equations of internal force intensities in the most dangerous points and elements of the carcass: in tire equator points, on bead rings etc. The obtained results are based on equilibrium equations using linear intensities of distributed forces along the section perimeters.

3. Theory

In the work of V.N. Tarasov [9] for the first time it is suggested to use the method of section for the research of pneumatic tire parameters. The cutoff of the contact body from the tire carcass was performed along the contact perimeter by the closed secant surface which is normal to the bearing surface.

As a result of the application of the method of sections a pneumatic-tire wheel is considered as a mechanism, where the cutoff contact body interacts with the carcass, at the same time the compressed air inside the tire is the main physical factor, determining the pneumatic-tire wheel capacity and operability.

Thanks to the method of sections in the theory of pneumatic tire the internal forces transfer into the category of external forces, which ensure the carcass equilibrium without the cutoff contact body and equilibrium of the free cutoff contact body. The tire is considered as a thin-wall shell, loaded with the internal pressure. From the equilibrium condition the intensity of forces in the carcass walls is determined.

In engineering for the calculation of thin-wall shells the Laplace equation is used in the works of V.L. Biderman [1], S.P. Timoshenko [10]

$$\frac{T_m}{\rho_m}+\frac{T_t}{\rho_t}=p_w, \tag{1}$$

where T_m, T_t – correspondingly the intensity of distributed meridional and circumferential forces per unit length of the tire section (Figure 1); ρ_m, ρ_t – shell radiuses, tangential to meridional and circumferential directions; p_w – tire air pressure.

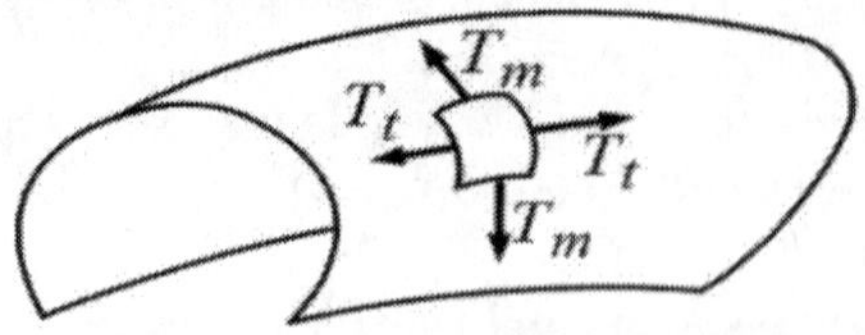

Figure 1. Computational scheme to the Laplace equation for the tire carcass infinitely small element.

Equation (1) contains two unknown values T_m and T_t. In these cases it's necessary to write down an additional equilibrium equation the Laplace equation.

In work [11] V.N. Tarasov and I.V. Boyarkina proposed the method of section of the tire carcass by cylindrical surface with r_o radius along the main parallels of a pneumatic tire. At the same time the tire carcass is sectioned by the vertical surface *a-a* along the tire equator (Figure 2).

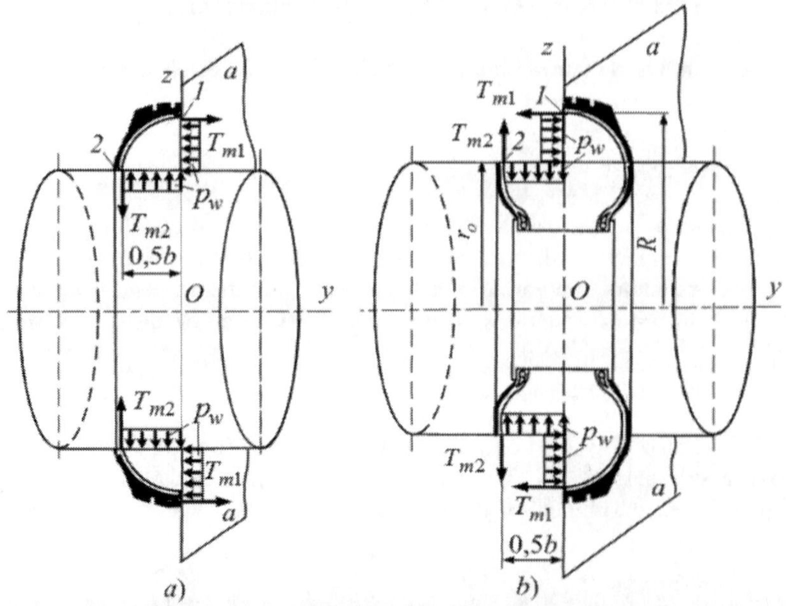

a) *b*)

Figure 2. Equilibrium of sectioned tire elements:
a) equilibrium of one fourth part of the torus *b*) equilibrium of three fourth part of the torus.

Figure 2,*a* shows the equilibrium of the cut off one fourth part of the tire torus, and Figure 2,*b* shows the carcass equilibrium without the cut off element of the tirep. Point *1* on the tire equator and point *2* on the tire side walls (poles) are considered as the most important for the tire carcass strength, and they determine the tire section geometry on Figure 2,*b*.

On Figure 2,*b* we have the system of parallel forces relative to the axis *Oy*, being in equilibrium, and the system of radial concurrent forces to the axis *Oy*, which is also in equilibrium. That's why the indicated systems of forces cannot rotate the considered and they are in equilibrium. Conditional vectors T_{m1}, T_{m2} represent intensities of specific forces per unit length of the section perimeter (N/m).

For these forces it is possible to derive the equilibrium equation in the form of the equality of air pressure active forces and carcass reaction forces for relative lengths of the section perimeters $2\pi R$, $2\pi r_o$

$$\sum F_{ky} = 0; \quad -T_{m1} \cdot 2\pi R + p_w \pi \left(R^2 - r_o^2\right) = 0; \tag{2}$$

$$T_{m2} \cdot 2\pi r_o - p_w \cdot 2\pi r_o 0{,}5b = 0\,, \tag{3}$$

where R – radius of the torus internal surface for the tire equator; r_o – average radius of the side walls of the torus internal surface; b – width of the internal surface of the tire profile.

From equations (2), (3) expressions for the determination of intensities of specific meridional forces T_{m1} and T_{m2} in the tire section perimeters were obtained

$$T_{m1} = \frac{p_w(R^2 - r_o^2)}{2R}, \text{ (N/m)}; \tag{4}$$

$$T_{m2} = p_w\, 0{,}5b, \text{ (N/m)}. \tag{5}$$

Formulas (4), (5) in the pneumatic tire theory allow to determine the values of the meridional specific forces in points *1* and *2* on the equator and poles of the pneumatic tire.

Figure 3 shows the computational scheme for the determination of the radial intensity of the distributed forces T_{m1}.

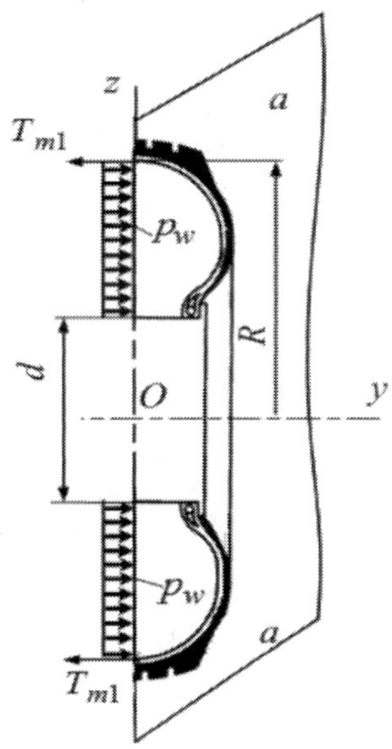

Figure 3. Equilibrium of the tire sectioned elements by the plane *a-a*.

The equilibrium equation for Figure 3 is as follows:

$$T_{m1} \cdot 2\pi R = \pi\left(R^2 - (0{,}5d)^2\right)p_w\,. \tag{6}$$

From where

$$T_{m1} = \frac{p_w\left(R^2 - (0{,}5d)^2\right)p_w}{2R}\,. \tag{7}$$

Formula (7) doesn't coincide with formula (4). That's why formula (7) is the correct one, which gives the greater absolute result and it's recommended for the determination of the intensity T_{m1}.

In work [1] formula (4) is known, but more exact formula (7) was derived for the first time in this work.

Figure 4 shows the section for the analytic definition of the intensity of carcass radial distributed forces for pole points on the tire side wall.

The equilibrium equation of intensities of carcass reaction forces and tire air pressure is as follows:

$$2T_{m2} \cdot 2\pi r_o - 2\pi r_o b p_w = 0\,. \tag{8}$$

From where

$$T_{m2} = 0{,}5bp_w\,. \tag{9}$$

Equation (9) coincides with equation (2), which is absent in technical literature and represents a new result in the pneumatic tire theory.

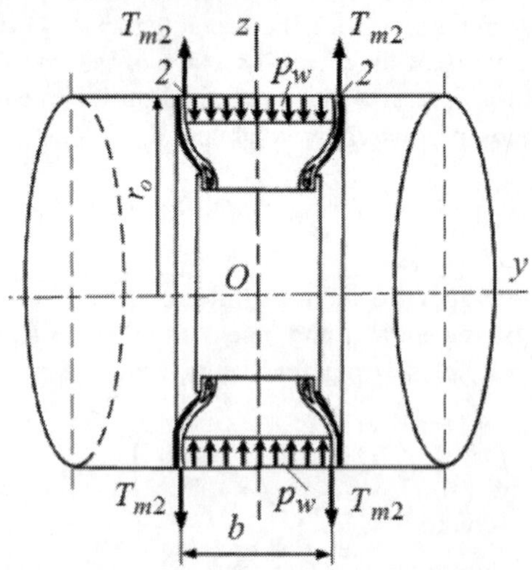

Figure 4. Equilibrium of tire elements, cut off by cylindrical secant surface with r_o radius.

Figure 5 represents the computational scheme in order to write down the analytical equation of forces in bead rings. Cylindrical secant surface with $r_{B.K}$ radius cuts off the seat rim from the tire. The ring radius $r_{B.K}$ is greater than the radius of the seat rim disk.

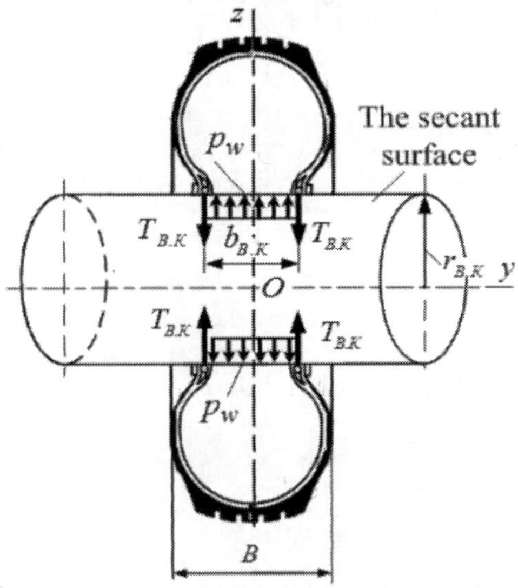

Figure 5. Equilibrium of tire elements, cut off by cylindrical secant surface with $r_{B.K}$ radius.

The equilibrium equation of he distributed intensities $T_{B.K}$ and air pressure forces is as follows:

$$2T_{B.K}2\pi r_{B.K} = 2\pi r_{B.K}b_{B.K}p_w. \quad (10)$$

$$T_{B.K} = 0{,}5b_{B.K}p_w. \quad (11)$$

Let's consider the possibilities of Laplace equation (1) usage for the pneumatic tire calculation.
For point *1* on the tire equator (see Figure 2,*b*) the Laplace equation is as follows:

$$\frac{T_{m1}}{0{,}5H} + \frac{T_{t1}}{R} = p_w, \quad (12)$$

where H – tire profile height.

Laplace equation (12) contains two unknown values.

For point *2* of the tire (see Figure 2,*b*) the Laplace equation is as follows:

$$\frac{T_{m2}}{0,5b}+\frac{T_{t1}}{\infty}=p_w. \tag{13}$$

In point *2* the tire carcass has only one curvature, then from equation (13) we'll find

$$T_{m2}=0,5b_{B.K}p_w. \tag{14}$$

Obtained formula (14) coincides with solution (11).

The Laplace equation for pneumatic tires allows to solve a limited class of tasks for objects with one wall curvature. Such objects include, for example, a thin-wall cylindrical pipe, the Laplace equation for which is as follows:

$$\frac{T_m}{r}+\frac{T_t}{\infty}=p_w, \tag{15}$$

where r – pipe section radius.

Then the intensity for the pipe wall is equal to $T_m = rp_w$. If the force ($T_m \cdot 1$) effects on one running meter of the pipe wall, then the pipe wall stress will be equal to

$$\sigma=\frac{T_m}{\delta}=\frac{rp_w}{\delta}, \tag{16}$$

where δ – pipe wall thickness.

Formula (15) coincides with the known formula for the determination of the thin-wall pipe strength.

Figure 6 shows the computational schemes for the analytical determination of the pneumatic tire capacity.

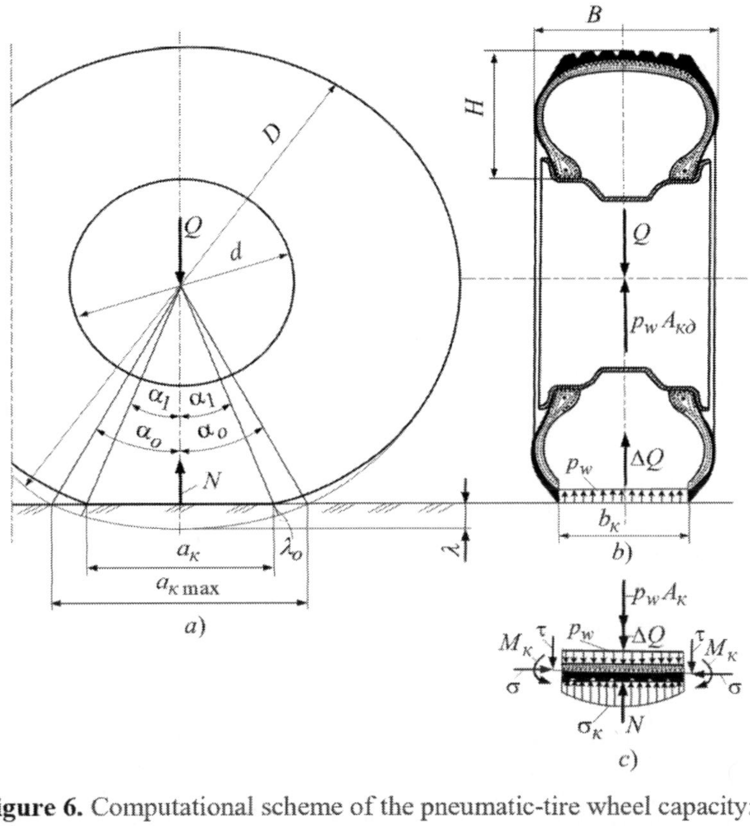

Figure 6. Computational scheme of the pneumatic-tire wheel capacity: *a*) general scheme of loading and deflection; *b*) tire shell equilibrium without cutoff contact; *c*) cutoff contact body equilibrium.

Figure 6,*a* shows the general scheme of the tire loading and deflection; Figure 6,*b* shows the cutoff of the contact body from the tire carcass along the contact perimeter by the closed surface, which is normal to the bearing surface; Figure 6,*c* shows the cutoff contact body equilibrium.

For the shell equilibrium without the cutoff contact, according to the Newton's third law, lifting force is applied to the shell, which is equal to the product of the tire air pressure p_w and actual contour area of the contact $A_{кд}$, force Q on the wheel axle and vertical force ΔQ of the carcass.

For the non-cutoff tire on Figure 6,*a* the equilibrium equation is true:

$$\sum F_{zк} = 0; \quad -Q + N = 0; \quad Q = N = \sigma_к A_{кд}, \tag{17}$$

where $\sigma_к$ – average specific pressures in the contour area of the tire contact.

The contact body equilibrium on Figure 6,*c* is ensured by the wheel lifting force $p_w A_{кд}$, acting downwards on the contact body according to the Newton's third law, by the support reaction N and vertical force of the carcass ΔQ.

In closed surface, cutting off contact from the tire shell, normal stresses σ and elementary moments $M_к$ appear, which have no the resultant force, as they are mutually balanced according to the symmetry condition. Tangential vertical specific forces have the resultant vertical force of the carcass:

$$\Delta Q = \int_S \tau dS, \tag{18}$$

where S – vertical surface area, cutting off the contact body from the tire shell; τ – tangential stresses in the secant surface.

The equilibrium equation of the cutoff contact body shown on Figure 6,*c* is as follows:

$$\sum F_{zк} = 0; \quad N - \Delta Q - p_w A_{кд} = 0. \tag{19}$$

From which, taking into account (17) we'll obtain

$$Q = p_w A_{кд} + \Delta Q, \tag{20}$$

where Q – wheel capacity, vertical force on the wheel axle, which is normal to the bearing surface; $p_w A_{кд}$ – pneumatic tire lifting force, equal to the product of the tire air pressure p_w and actual contour area of the contact $A_{кд}$; ΔQ – vertical force of the tire carcass.

4. Research results

The determination of the vertical force ΔQ according to formula (18) is a complex task, that's why in this article we'll determine the force ΔQ from equation (20) according to formula:

$$\Delta Q = Q - p_w A_{кд}. \tag{21}$$

For the solution of equation (21) the new definition is introduced – theoretical contour area of the area contact, determined according to the Pascal formula:

$$A_к = Q / p_w, \tag{22}$$

where Q – capacity, normal force on the axle; p_w – tire air pressure.

Using formulas (22) and (21) we can obtain the carcass vertical force formula:

$$\Delta Q = (A_к - A_{кд}) p_w. \tag{23}$$

From formula (23) it follows, that the force ΔQ is an ambiguous value.

If $A_к = A_{кд}$, then $\Delta Q = 0$ – it means that the load Q on the wheel axle is taken by compressed air in the tire.

If $A_к > A_{кд}$ – it means that the carcass takes a part of the vertical force ΔQ, applied to the wheel axle.

If $A_к < A_{кд}$, then $\Delta Q < 0$ – in this case the carcass vertical force ΔQ possesses a negative value, the wheel lifting force appears to be an excessive one, i.e. it not only takes the force Q, but also creates an additional vertical tension force during the interaction of the contact body and shell carcass.

For the determination of the carcass vertical force ΔQ it's necessary to determine the actual contour area of the contact $A_{\kappa\partial}$ by the experimental or analytical method. Let's consider the analytical solution of this task.

It's established that the tire elements at the inlet to the contact radially deflect long before the approach to the bearing surface (see Figure 6,*a*). At the moment of the tread elements coming into contact, the tire possesses the radial initial deflection λ_o, which is connected with the normal tire deflection λ by the expression:

$$\kappa = \lambda_o / \lambda, \tag{24}$$

where κ – coefficient of the tire initial deflection.

For truck pneumatic tires, using Figure 6,*a*, the angle α_1 of the tire contact may be determined according to the formula from work [12]

$$\alpha_1 = \arccos \frac{\frac{0,5D}{\lambda} - 1}{\frac{0,5D}{\lambda} - \kappa}. \tag{25}$$

The tire contact area length (see Figure 6,*a*) is determined according to the formula:

$$a_\kappa = 2(0,5D - \kappa\lambda)\sin\alpha_1, \tag{26}$$

where a_κ – contact area length; $\kappa\lambda = \lambda_o$ – tire initial deflection.

In formulas (24)-(26) the coefficient κ is the characteristic of the tire kinematic possibilities.

If κ=0, then the tire shell will possess perfect elasticity, in the result of which the tread elements come into contact without a preliminary deflection λ_o=0, and the contact area has the maximum length $a_\kappa = a_{\kappa\max}$. Such imaginary tire has the greater tire drag in comparison with the tire which possesses the coefficient κ<1.

If conditionally set κ=1, then we obtain the tire, which $\lambda_o = \lambda$ and the tire contact length $a_\kappa = 0$. In this limiting case the tire has no contact area, as $a_\kappa = 0$, the load Q on the wheel axle is taken by the tire carcass. Such tire has minimum rolling resistance.

It means that according to this indicator radial tires surpass cross-biased tires, because they have a tough breaker cylindrical strip under the tire tread.

For cross-biased tire 12.00-20 according to the research results the coefficient κ has the value $\kappa = 0,2 \div 0,4$, for radial tires $\kappa \geq 0,5$.

The actual area of the tire contact is determined according to the formula:

$$A_{\kappa\partial} = (a_\kappa - b_\kappa)b_\kappa + \frac{\pi b_\kappa^2}{4}, \tag{27}$$

where a_κ, b_κ – length and width of the tire contact area.

The tire contact area according to formula (27) is considered as the area of a simple oval.

5. Discussion of results

The application of the method of sections to pneumatic tires gives new impetus to the development of the theory of pneumatic tires. The tire contact contour area, tire carcass lifting force, tire carcass vertical load represent important components of the pneumatic tire theory.

New equations have been obtained for the determination of loads in tire elements, which represent the result of the developed computational schemes of the pneumatic tire sections by cylindrical surfaces with the usage of the intensity of distributed forces in the pneumatic tire sections during the generation of equations.

Each equation and formula proposed in this article is preceded by a new computational scheme of the tire section by planes or cylindrical surfaces, where intensities of distributed forces are used along the section perimeters. Formulas are proposed for the tire contact area with the bearing surface, which are useful not only for the capacity calculation, but they also allow to determine the influence of the tire contact area form on the tire rolling resistance and wear.

6. Summary and conclusion

1. The application of the method of sections for pneumatic tires allows to deepen the theoretical knowledge relative to internal processes in pneumatic tires.

2. The contact body cutoff from the tire carcass has allowed to clarify the definition of the tire capacity, which depends on the contact contour area, geometric dimensions, air pressure and tire ply rating.

3. The developed computational schemes of the pneumatic tire sections and new equations will allow to accelerate the structure improvement process of pneumatic tires during their engineering.

References

[1] Biderman V L [et al.] 1963 *Motor car tires* (structure, calculation, testing, operation) (Edited by V L M Goskhimizdat Publishing House) p 384
[2] Bukhin B L 1988 *Introduction to the mechanics of pneumatic tires* (Moscow: Chemistry) p 224
[3] Knoroz V I [et al.] 1976 *Motor car tire operation* (Moscow: Transport) p 238
[4] Belkin A E, Bukhin B L, Muknih O N and Narskaya N L 1997 Some models and methods of pneumatic tire mechanics Vehicle System Dynamics *Int. Jornal of Vehicle Mechanics and Mobility* vol **27** pp 250–571
[5] Sheshenin S V 2007 Three-dimensional modeling of tires *Mechanics of Solids* vol **42** pp 338–345
[6] Salaani M A 2009 Analytical of tyre forces as a function of normal pressure distributions *International Jornal of Heavy Vehicle Systems* vol **16** no **1/2** pp 154–188
[7] Wallaschek J and Wies B 2013 Tyre tread-blosk friction: modeling, simulation and experimental validation *Vehicle System Dynamics Int. Jornal of Vehicle Mehanics and Mobility* vol **51** State of the art Papers of the 23rd IAVSD Symposium pp 1017–1026
[8] Sterlzig P E 2012 Homogenization of many- body structures subject to large deformations *ESAIM: Control Optimisation and Calculus of Variation* vol **18** pp 91-123
[9] Tarasov V N 1965 Tire capacity with liquid ballast *Tractors and agricultural machinery* no **8** pp 10–12
[10] Timoshenko S P and Voinovsky-Kriger S 1966 *Theory of Plates and Shells* (Moscow: Science) p 636
[11] Tarasov V N, Boyarkina I V and Degtiar V V 2015 Calculation method of pneumatic-tire wheel capacity and tire carcass strength of vehicles *Construction and road mahines* no **5** pp 47–52
[12] Tarasov V N 1965 Theoretical rolling radius of elastic wheels *Vehicle manufacturing industry* no **1** pp 5–6

IOP Conf. Series: Journal of Physics: Conf. Series **944** (2017) 012117 doi:10.1088/1742-6596/944/1/012117

Algorithms for constructing optimal paths and statistical analysis of passenger traffic

S P Trofimov, N G Druzhinina and O G Trofimova

Dept. of radio electronics for information systems, Ural Federal University, Yekaterinburg, Russia

e-mail: tsp61@mail.ru

Abstract. Several existing information systems of urban passenger transport (UPT) are considered. Author's UPT network model is presented. To a passenger a new service is offered that is the best path from one stop to another stop at a specified time. The algorithm and software implementation for finding the optimal path are presented. The algorithm uses the current UPT schedule. The article also describes the algorithm of statistical analysis of trip payments by the electronic E-cards. The algorithm allows obtaining the density of passenger traffic during the day. This density is independent of the network topology and UPT schedules. The resulting density of the traffic flow can solve a number of practical problems. In particular, the forecast for the overflow of passenger transport in the «rush» hours, the quantitative comparison of different topologies transport networks, constructing of the best UPT timetable. The efficiency of the proposed integrated approach is demonstrated by the example of the model town with arbitrary dimensions.

1. Transport network and its modeling

In a small town, a car is capable to satisfy much of the personal travel demands. Increasing size of the city leads to the growing need for other modes of transport. Traditional types are tram, trolley, bus, metro, taxi, rental cycling, and improving pedestrian comfort posts. With growth of the economic and environmental benefits of the non-car modes of transport, their deployment leads to creation of complex transport systems and, especially, their effective information support. Moreover, adequate structure and organization of transport are extremely important for the transport systems of large cities and agglomerations [1].

Existing urban public transport network components have varsions degrees of stability. Thus, for example, modernizations of the laid communications are performed extremely rare. The leader in management of the urban passenger transport-ray (UPT) in Yekaterinburg is the municipal enterprise «Tram and trolleybus management (TTM)». The enterprise also rarely modifies structure of the existing routes. The most flexible component is the schedule of routes. Determination quality of the public transport networks needs the criteria for comparing different versions of these UPT networks and schedules. Substantiation of necessity of modifying the network of routes and schedules needs new quantitative assessments (estimations) of the quality.

Ignoring of determination such estimations and arbitrary modification of the UPT components can lead to appearing the frequent traffic congestion, overloading/underloading separate lines, nodes and transport units, and increasing in the number of accidents [2]. Moreover, jams may occur inside the vehicle in the form of overloading of the machine with a corresponding reduction of the ride comfort. These strains are also to be included into modeling.

Improving the UPT management system must be accompanied by tracking changes of all its components. Unwarranted use of rather complex experiments on real systems may have a negative impact on the UPT. The UPT development is recognized as the first and most effective measure to combat traffic jams.

The difficulty of the traffic flow formalization (including public transport) has become a major cause of inconsistency of the research results and requirements of practice.

Under these conditions, the UPT improving is impossible without adequate mathematical models and numerical experiments.

The novelty of the work is to develop an algorithm for finding the optimal transport routes for a model city using the model schedules and model of the metropolis transport network.

2. Scientific approaches to the study of traffic flows

There are models designed for prediction of the transport and passenger flows in networks with known geometry and characteristics and known placing various objects on the territory [3, 4, 5, 6, 7, 8, 9, 10, 11, 12, 13, 14, 15]. Models of this type are used to support decisions in the field of urban planning, to analyze the impact of various measures on the organization of movement, to choose alternative development projects and other transport network, and so on.

Our model of passenger flows uses optimal strategies, since the user of public transport often chooses his path to the target point according to some optimality criterion. Preference is given to the criterion of the minimal time of attaining the goal. This criterion and another one (of the shortest path) can be taken with some weights as the «summary» criterion.

3. Information on the public transport

Enlargement of public transport systems requires sophisticated JavaScript passenger information about modes of public transport [16, 17, 18, 19].

Existing information systems (IS) do not provide residents of Yekaterinburg with complete information necessary for a comfortable and efficient use of the UPT. In particular, the information services do not give to a passenger the optimal path with respect to the start time, driving situation, and the current schedule.

Using of the optimal path, passenger in the IS could have social and economic benefits for all major Russian cities including Yekaterinburg.

Among the existing indicators (characterizing the transport system of the city and directly related to the quality of the UPT) are such groups of indicators as Planning, Performance traffic, and Mobility [20]. In the latter group, the most interesting for us are the following ones: the «average travel time» and the «average length of travel by public transport». However, these criteria are integral and does not contain any algorithmic component. They do not take into account the quality of services rendered to a concrete passenger.

We offer new quality criteria of the UPT activity. These criteria are passenger oriented and contain algorithmic components. Our information system implements intelligent services related to optimization of the passenger travel time depending on the schedule of vehicles. In foreign metropolitan areas [16], in contrast to the Russian cities [17, 18, 19], such services are the norm.

4. Problem of accounting the traffic situation in the UPT information systems

Effectiveness of the UPT falls significantly in the case of overloading the transport networks. Neither one system is able to precisely predict the traffic jams and their behavior even for short periods. This is due to the following properties of traffic:

- forecast of the stochastic traffic flow is possible only with a certain probability, as it directly depends on such random factors as weather conditions, accidents, etc.;
- non-stationary traffic flows and fluctuations in their characteristics occur in at least three cycles: daily, weekly, and seasonal;
- behavior of each individual driver (of a traffic vehicle) is unpredictable.

Thus, inclusion of data on the traffic jams into the information systems of public transport does not lead to diminishing the role of a random factor. Improving the accuracy of the information about the work of the UPT is in strict keeping at the schedules. So, the authority should recognize the absolute priority of public transport in the urban passenger transport. The quality of public transport is one of the components of quality of the city administration.

5. Model properties

The authors developed a model UPT network, close to the real conditions of the metropolis.

Implemented model has the following properties:

- stops and routes are shown in a graph;
- model uses the tram and trolleybus routes with possibility of adding new types of the passenger transport;
- timetables for each traffic route are implemented;
- visual representation of the city transport network;
- finding the optimal path from the start stop to the end one with respect to the time of arrival of the passenger at the starting point; at the same time, all transfers, the time spent on the trip and waiting for transport should be shown.

The model allows determining the following indicators:

- minimum number of transfers, with which the passenger with guarantee can get from one stop to another provided that the transport is moving strictly according to the schedule;
- the average waiting time for a passenger at each stop.

6. Mathematical model of the transport network, the algorithm and software implementation for finding the optimal route

Mathematical model of the transport network allows us to build the trajectories of the urban passenger transport of a model city.

Let X the set of stops numbered by natural numbers. A road section between stops i and j containing no other stops is called the stage P_{ij}. The stage is a directed section. The set of all existing stages is denoted as P. The tram routes are enumerated.

Trajectory of each route is a sequence of related stages

$$T(n) = \{P_{1,2}, ..., P_{m(n)-1,m(n)}\}, \quad (1)$$

where $m(n)$ is the number of stops in the route n, including the start stop and end one. Trajectory may contain duplicated stages.

Let a train of the route n came out from the initial stop at the instant t_{start}. The instant of its arrival at the next stops form a sequence

$$V(n, t_{\text{start}}) = \{t_1 = t_{\text{start}}, t_2(n, t_{\text{start}}), ..., t_{m(n)}(n, t_{\text{start}})\} \quad (2)$$

Let a route schedule is a pair of the route trajectory and the time sequence

$$G(n, t_{\text{start}}) = \{T(n), V(n, t_{\text{start}})\} \quad (3)$$

The route may have several schedules within a day.

Consider the algorithm for finding the optimal route based on the principle of dynamic programming. Let us a passenger be at the stop A at the instant t_{start} and needs to reach the stop B in a minimum time. His path can contain transfers.

Let the mark $M(i)$ for an arbitrary stop i is a pair of a schedule and the time of arrival at this stop for this schedule

$$M(i) = \{G(n, t_{\text{start}}), t_i(n, t_{\text{start}})\} \quad (4)$$

Mark of the stop i allows us to determine the route and a previous stop, from which we come to this stop from the i-one. A mark with the earlier arrival time overwrites the previous mark. The starting stop A has not a mark.

Marking the stops is carried out on the wave principle: the first considered stops located from A with a radius equal to one stage, then two stages, *etc*.

The algorithm terminates when viewing all the stops of the last radius and none of the marks has not changed.

If the end stop *B* is marked with the time of arrival t_{end}, the view of stages coming from other adjacent stops and marked with the time later than t_{end} is not performed.

Check by increasing the radius, we came to some marked stop *i*. Let the time of new mark be worse than the time of the previous mark *M*(*i*). Then we keep this mark *M*(*i*), and further increasing the radius through the stop *i* is not performed.

Summary screen of the simulation example is shown in figure 1. Stops coincide with crossroads and are numbered from 1 to 400. The optimal trajectory is colored. The color changing (digits) indicates a transfer.

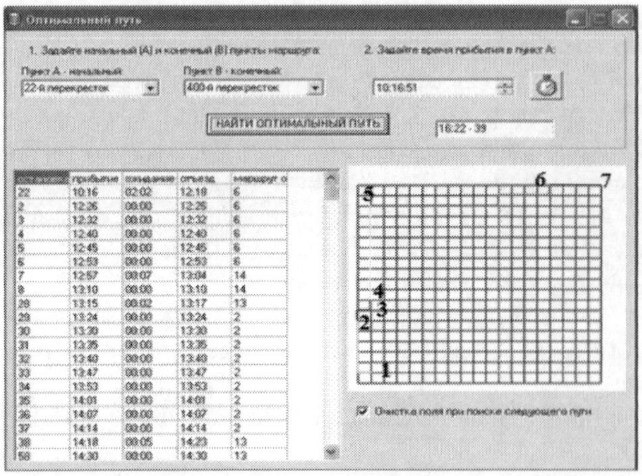

Figure 1. The result of searching for the optimal path.

In this paper, we demonstrate possibility of application of the pro-led calculations not for a single person, but for the entire population of the city. This allows one to get the experimental data and to calculate the new statistics quality UPT.

The simulation results show that the algorithm can be applied to the real transport network with real timetables.

7. Sensitivity analysis of the optimal path w. r. t. perturbation of the initial data

As a rule, passengers of the public transport do not have possibility to influence the route schedules, but rather are forced to adjust to them. For more convenient use of the UPT a passenger should know how the optimal path and minimum travel time vary when changing the initial conditions of the trip. Maybe, the passenger gets to its destination faster if he moves to the next stop with a large number of routes (sensitivity to perturbation of the starting point of the passenger path). Maybe, he should go out earlier (sensitivity to perturbation of the start time and end point).

We implemented the sensitivity analysis of the optimal solutions when the parameters of the original model vary slightly. Usually, such analysis is carried out in optimization problems in the technical, economic, and other fields.

Figure 2 shows changing in the time optimal path when changing the start time from the starting point. In the calculations, the actual tram schedule of Yekaterinburg was used.

The simulation results show: the start time being in the range of 15 h 32 min to 21 h 32 min, the minimum travel time from the stop of «Enthusiasts» to the stop «Peace» is changed from 30 min to 50 min.

High sensitivity is a disadvantage for the UPT and leads to the overloading on the optimal transport trajectories. Sensitivity coefficients should be among the quality criteria of the UPT timetables.

We investigated other important criteria for evaluating the UPT activities, such as covering the city, guaranteed travel time, number of transfers, waiting time, etc.

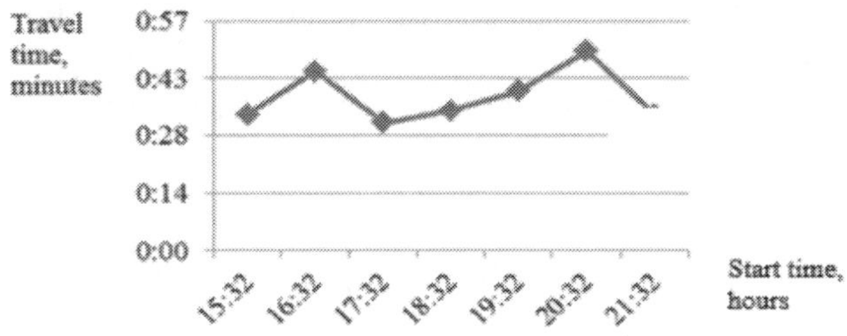

Figure 2. Analysis of dependence of the optimal path on the start time.

8. Statistical analysis of passenger flow

The UPT information system stores large statistical data related to the motion of passengers [21]. Traditional sources for this material are video cameras on the transport stops and in the interior of the passenger saloon. Processing this data allows counting the number of incoming and outgoing passengers.

We have implemented a different method. We developed analysis of system of the urban passenger electric transport on the basis of information about electronic travel payments and traffic navigation data of the mobile unit. The objective of the study is identifying the traffic flow with electronic fare payment and the navigation system of motion of the trams. We can determine the motion of passengers paying fares with an electronic card. The possibility of «gluing» of several transactions in a single itinerary for a passenger is realized; it solves the problem of the "last mile", i. e., the definition of the unknown place and arrival time of the passenger.

We have received a distribution of passenger flow using basic information about the motion of passenger: the time and place of the start and end points of his trip. Estimate of the distribution of urban the passenger electric transport carried out between the administrative areas of the city with reference to the routing scheme of the tram motion. Although the number of stops in each administrative district is different, this allowed us to determine the direction of motion of passenger traffic through several parts of the city.

The analysis is performed for the real traffic flow in the work and day time. The system works with incomplete input data (considered only one type of public transport) [22]. The grouping of stops on the administrative districts allowed building the OLAP-cube with information about the passenger path.

The statistical analysis cube allows us to define extremums of the passenger traffic, and the city administration can mitigate these extremums for modernization of the urban transport network.

9. Conclusions

An algorithm for finding the optimal route of the passenger is presented. The algorithm uses dynamic programming method, complicated by the presence of the time factor. Optimal routes form the first of three modules of UPT network analysis. The second module involves assessing correspondence matrix of passenger traffic using electronic e-cards. The passenger's transfers as he moves from the start to the end point of the path are glued into one route. This passenger route in practice is optimal. The third module will enable to calculate the quantitative assessment of the quality of the transport network topology and route schedules. The third module uses the correspondence matrix, resulting in the second module.

References

[1] Vukan R Vuchick 2011 *Transportation for Livable Cities* (Moscow: Territorija budushhego) p 412
[2] Nasakin R 2007 *Computers* **29(697)** pp 39–43
[3] Koryagin M and Katargin V 2016 *Transport* **31** pp 63–69

[4] Lagerev R Y, Lagerev S Y and Nemchinov S S 2013 *Vestnik of Irkutsk State Technical University* **12(83)** pp 185–190
[5] Fedorov V P, Bulycheva N V, Pahomova O M and Losin L A 2008 *Transport of the Russian Federation* **16–17** pp 64–67
[6] Daamen W 2004 *Modeling Passenger Flows in Public Transport Facilities* (Netherlands: Delft University Press) p 377
[7] Solecka K 2016 *Scientific Journal of Silesian University of Technology. Series Transport* **92** pp 129–139
[8] Franceries E, Liver K 2011 *Transport System Telematics* **4(4)** pp 4–10
[9] Birr K, Jamroz K and Kustra W 2014 *Transportation Research Procedia* **3** pp 359–365
[10] Habarov V I and Teselkin A A 2016 *Research Vestnik of Novosibirsk State Technical University* 1(62) pp 91–105
[11] Zvarych E B, Koryagin M E and Slobodenyuk O YU 2008 *Questions of modern science and practice. University V I Vernadsky* **2** pp 33–42
[12] Morozov A S and Chernikov A A 2017 Social and Economic Problems of City Transport Systems and Areas of Their Influence Development and Functioning *Collected articles of the XVIII International (twentieth-seconds held in Yekaterinburg) scientific and practical conference* ed S A Vaksman (Yekaterinburg: ASB Publisher) **1** pp 149–162
[13] Fedorov V P, Bulycheva N V, Pahomova O M and Losin L A 2008 *Transport of Russian Federation* 3–4(16–17) pp 64–67
[14] Pugachev I N, Burkov S M, Markelov G YA and Olejnik A V 2013 *Development strategy use of intelligent transport systems in the management of urban traffic* ed S M Burkov (Habarovsk: Tihookeanskij State University) p 141
[15] Markelov G YA, Plesovskih S V, Burkov S M and Pugachev I N 2015 *Monitoring system based on satellite navigation* ed G YA Markelov (Habarovsk: Tihookeanskij State University) p 183
[16] *Transport of London* Available at: https://tfl.gov.uk/
[17] *Tram-trolleybus management of Yekaterinburg* Available at: www.ettu.ru
[18] *Rusavtobus* Available at: http://eka.rusavtobus.ru/
[19] Postolit A V and Prushak A L 2015 COMPASS – electronic route network *The certificate of state registration of computer programs 2015610368*, **2(100)** (Moscow: The Federal service for intellectual property)
[20] Vaksman S A, Gerasimov N I and Slepuhina I A 2012 *Information technology in the management of urban public transport (tasks, experience, problems)* (Yekaterinburg: AMB)
[21] Druzhinina N G, Skhirtladze K In and Trofimova O G 2016 Statistical analysis of passenger traffic on the example of an electronic fare collection system registration *The certificate of state registration of computer programs 2016661213.* **10 (120)** (Moscow: The Federal service for intellectual property)
[22] Druzhinina N G, Trofimov S P and Trofimova O G 2014 The formation of transaction traffic at the urban electric transport *The certificate of state registration of computer programs 2014661819* **12(98)** (Moscow: The Federal service for intellectual property)

IOP Conf. Series: Journal of Physics: Conf. Series **944** (2017) 012118 doi:10.1088/1742-6596/944/1/012118

Experimental research of thermal loading of the rocket payload fairing element during the atmospheric phase of the descent trajectory

V Trushlyakov, Yu Iordan, D Davydovich and K Zharikov, M Dron'

Omsk State Technical University, 11, Mira ave., Omsk, 644050, Russia

e-mail: u1tro@mail.ru

Abstract. The thermal loading physical simulation in the experimental wind tunnel on the design element of the payload fairing made of carbon fiber was done. The experimental study is given in the speed range below 70 m/s, which corresponds to the interval of heights of the descent trajectory of the payload fairing half below 10 km. The values of heat transfer coefficient are obtained. The analysis of the results is carried out.

1. Introduction

When launching rockets in the boost phase of flight the spent rocket design elements (RDE), such as boosters, steps, fairings, interstage compartments separate.

The fall of the separated parts of the Earth's surface is a source of economic, environmental and social issues arising in the implementation of space-rocket activity. The speed of the separated parts to the atmosphere is significantly different from 1 km/s to 7 km/s, which leads to various environmental impacts: from the fall almost as much as RDE, for example, fairings to almost complete combustion of RDE (boosters the second and third steps) in the dense layers of the atmosphere.

At the present time, the problem of RDE destruction returning from space into the Earth's atmosphere having velocities of about 7 km/s, is described in the significant number of works [1–2]. The developed mathematical models include the estimation of the centre of mass and around centre of mass motion parameters, the thermal load from aerodynamic effects and assess the RDE strength. The specificity of this RDE class is that their parameters while entering the atmosphere (velocity of entrance and entrance angles) are within the range of velocities about 1.0 km/s. Because of their aerodynamic characteristics, the velocity in the dense layers of the atmosphere significantly decreases, reaching values about several tens of m/s, respectively, the aerodynamic heating of RDE is significantly lower and RDE reaches the Earth's surface without a lot of destruction. It leads to the necessity of providing large territories for their impact areas [3].

2. Statement of the research problem

As possible technology, circuit and design solutions for minimizing the creation of space debris, on the example of the RDE (payload fairing halves and adapter sections), their combustion on the atmospheric phase of the descent trajectory after separation from the launch vehicle using the installation of additional heat sources, for example, pyrotechnic composition (PC) producing the required quantity of heat, is proposed [4].

IOP Conf. Series: Journal of Physics: Conf. Series **944** (2017) 012118 doi:10.1088/1742-6596/944/1/012118

At the first stage of the research, with the purpose to estimate the amount of the required heat, providing RDE combustion or destruction into fine particles (fine dispersion) on the atmospheric phase of the descent trajectory, it is necessary to define the heat fluxes determined by the aerodynamic heating, including heat loss and heat coming from the PC burning. [5]

The article presents experimental research of the thermal state of the RDE, made of polymer composite materials: carbon fiber KMU-4L, textolite PTK. The experimental study is given in the speed range below 70 m/s, which corresponds to the interval of heights of the descent trajectory of the payload fairing half below 10 km.

3. Physical experiments review

The purpose of the experiments is to determine the RDE surface temperature variation and, based on the calculation of heat transfer coefficients at aerodynamic loading in an experimental wind tunnel (EWT). The experimental bench described in [6] allows to simultaneously simulating the aerodynamic and thermal influences on the test sample.

The experimental stand consists of system for air preparation, EWT, measuring system, valves and control equipment.

In accordance with the conditions of the RDE aerodynamic load on the atmospheric phase of the descent trajectory, experimental stand and the EWT should implement the following conditions:

- the air flow rate in the range of 1...460 m/s (Mach 1.4);
- turbulence air flow Eo = 0.05...0.3;
- the air flow temperature in the range of 260...295K;
- the flow compression ε = 0.6...0.625;
- the humidity of the air stream should not exceed 30%;
- the RDE blowing angles 0...90 °
- the size of the EWT camera: 174mm x 50 mm;
- RDE specifications: overall dimensions must not exceed 40mm x 40mm x 15mm.

Measured parameters and measurement accuracy:

The temperature is measured in at three point (1 – is the flow temperature away from the sample; 2 – is the RDE temperature from the oncoming flow; 3 – is the RDE temperature on the back side).

Figure 1 shows a fragment of EWT. The EWT is a direct-flow type with a closed working part of circular cross-section with a diameter of 174 mm and a length of 280 mm. Aerodynamic flow to the working part has a medium level of turbulence (Eo = 0.1) in the absence damping screen to provide more realistic conditions for the experiment.

The degree of flow compression is adjusted by replacing the nozzle to ensure a wide range of flow rates at the entrance to the working part with the same initial pressure.

The EWT consists of the working part (POS. 6) made of a welded cylindrical structures made of steel 41CrMO Steel, mounted on a pedestal. To the wall of the working part is welded to the diffuser main part. With the two screws (POS. 12) inside the working part includes a holder (POS. 7) for mounting the RDE (POS. 8) with screw clamps. On top of the case has 4 fitting (POS. 15) with a diameter of 6 mm for the installation of the thermocouples and measuring the temperature. To estimate flow velocity in front of the RDE envisaged the possibility of installation in the fitting (POS. 2) velocity sensor. Additionally, in horizontal section, has two fitting symmetrically relative to each other. The first is to measure the pressure in the working part, the second for supplying power from the power source to the heating element (POS. 1) in the plate form.

The counterpart is made in the form of a cover with nozzles (POS. 9), which is connected by means of bolted connections (POS. 10) with the working part. The tightness of the seals (POS. 11).

The working part is connected with a supply line and a gas receiver by means of the pipe to straighten the flow. After straightening the flow, gas for RDE blowing flows through set vortex flowmeter "EMIS Vortex 200".

Between the pipe (POS. 3) and the working part, replacement set of critical nozzles (POS. 4) made of polymer materials in the 3D printer.

Figure 2 shows photographs of the working parts and nozzles with critical cross-section of 8 and 20 mm.

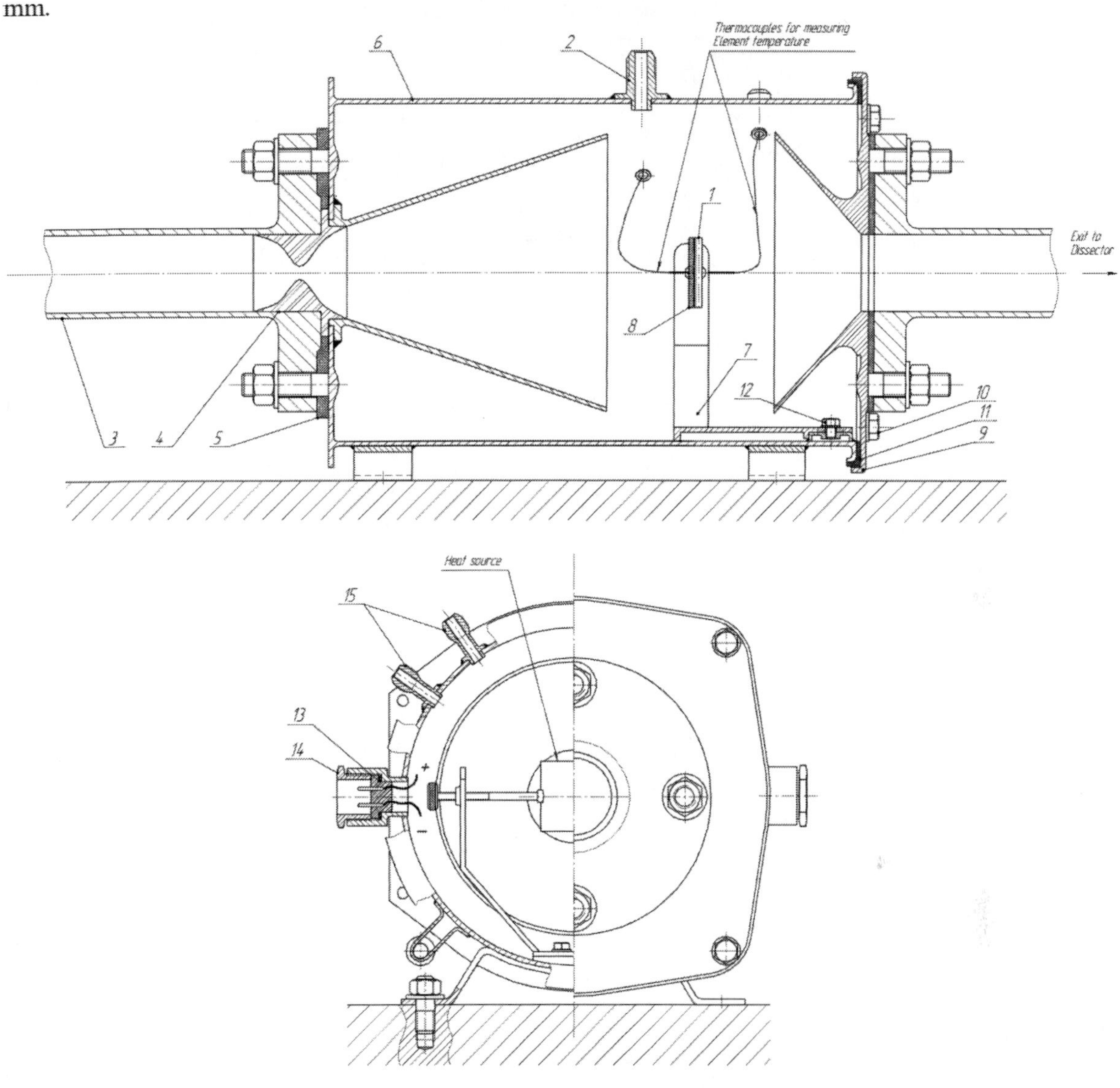

Figure 1. Scheme of the EWT working part.
1 – is the heating element; 2 – is the fitting for the velocity measuring; 3 – is the flow straightener; 4 – is the replacement of critical nozzles; 5 – is the seal; 6 – is the case of the working part; 7 – is the holder; 8 – is the RDE; 9 – is the counterpart; 10 – is the counterpart fixture; 11 – is the counterpart sealing; 12 – is the holder fixture; 13 – is the fitting; 14 – is the heating element power strip ; 15 – fitting for temperature measuring.

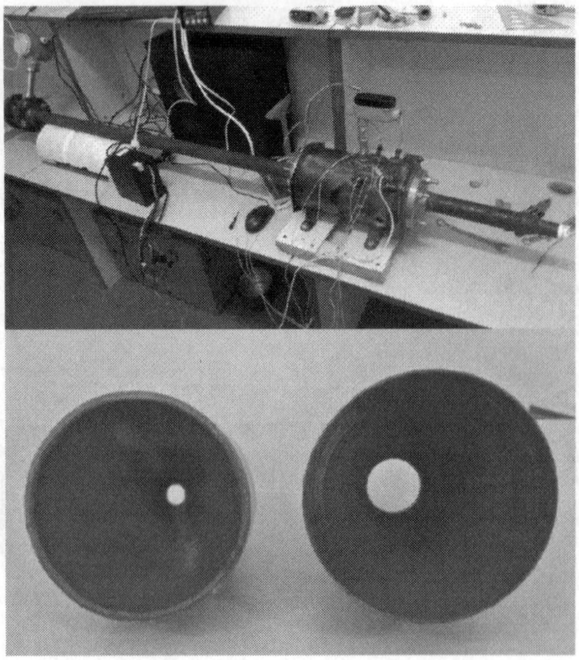

a) b)

Figure 2. The elements of EWT.
a) The EWT with a registration system of measurement; b) The nozzles.

The methodology of the experiments involves the following sequence:

- pre-purging of air receivers, the possible withdrawal of moisture from the tanks;
- injection receivers gas to a pressure up to 1.5 MPa;
- installation of RDE on the holder and connect the heating element plate;
- testing, installing, configuring, thermocouples and velocity sensors;
- check the metering data on flow, pressure, velocity and temperature measuring-computing complex;
- check the metering data on flow, pressure, velocity and temperature measuring-computing complex;
- connect the power source to the heating element;
- conducting experiments;
- saving and processing of the received data after the end of the experiment, disconnect power from the stand, removal of the RDE, purge the system and release residual pressure from the receivers.

The study was conducted series of experiments, where pre-heating of RDE to the established temperature, with subsequent input of the oncoming flow.

In all cases, RDE was a plate of rectangular cross section. As the test materials were chosen two of the most used: carbon fiber KMU-4L, textolite PTK.

In all cases RDE in the form of plate with a side length of 29 mm to 33 mm, thickness 2 mm. It was placed vertically at an angle of attack equal to 90 at a distance of 180 mm from the critical section (figure 3). With RDE clamps it is attached to the heater for maximum contact area.

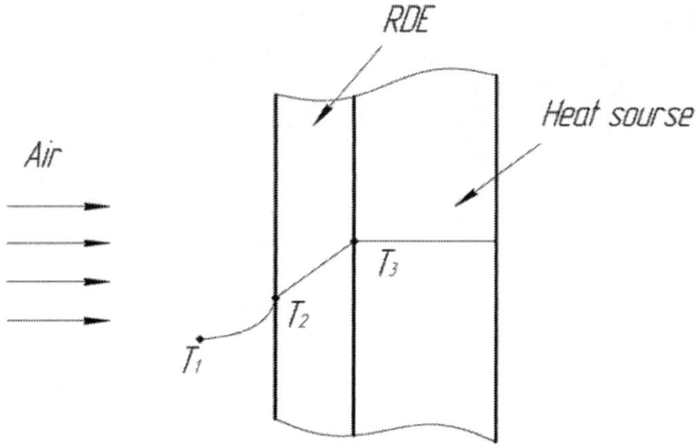

Figure 3. Sample loading scheme.

Samples of RDE are pre-heated to a temperature of 423 K on the outer surface (600K to side of heater) at a flow velocity of V= 0 m/s. After reaching the set temperature the samples were submitted by the oncoming flow at various velocities V: 0 m/s, 30 m/s, 50 m/s, 70 m/s. The temperature measurements were made at three points: in the place of the air flow on the RDE, directly from the heater surface (as occurs contact heat transfer can be considered that the temperature on the back side of the RDE is equal to the temperature of the heater) and in the area of lack of RDE. The experiment was completed after the establishment of constant temperature on RDE during blowing the air flow.

Processing of the obtained results was carried out using a number of dependencies, described in [7]:

- The temperature is recalculated in the values of the excess temperature ΔT, [K] as the temperature difference;
- Define the values of ln(ΔT) c accuracy to the third decimal place;
- Plot the ln(ΔT)=f(τ);
- Through the points of the graph are averaging line and its slope is determined by the cooling rate:

$$m = \frac{\ln(\Delta T_1) - \ln(\Delta T_2)}{\tau_2 - \tau_1}; \tag{1}$$

- Indices 1 and 2 refer to any two points lying on a straight averaging.

The total heat transfer coefficient is determined from the equation:

$$\alpha_\Sigma = \frac{M}{F} C_f m, \tag{2}$$

where: M/F=2,01 [kg/m3] is the constant of the sample (assuming that the mass and size of the sample does not change during the experiment); Cf [J/(kg·K)] – is specific heat capacity; m[c-1] – is the cooling rate specified in the graphic.

Given the constancy of the coefficients entering in (2):

$$\alpha_\Sigma = 49.245 \cdot m. \tag{3}$$

The average value of the radiation heat transfer coefficient is determined by the average temperature during the experiment. As the environmental temperature we accept temperature of a room:

$$\alpha_R = C \frac{\left(\frac{T_s}{100}\right)^4 - \left(\frac{T_f}{100}\right)^4}{T_s - T_f}, \tag{4}$$

where C=4.762 W/(m2·K4) is the emissivity coefficient of the plate surface (the degree of blackness for the material of 0.84); Tf [K] is the environmental temperature far from the plate; Ts, [K] is the mean value of temperature during the experiment.

In the end, the heat transfer coefficient can be calculated:

$$\alpha = \alpha_{\Sigma} - \alpha_{R} . \quad (5)$$

The total error of determination does not exceed ±10%.

4. The results of the experiments

In tables 1 and 2 are the values of the heat transfer coefficient is set $\alpha_{эк}$ for two types of materials (textolite PTK and carbon fiber KMU-4L).

The last column shows the comparison of the simulation results obtained in [8].

Table 1. The results of experimental studies material: textolite PTK

The velocity flow, V [m/s]	T_3[K]/T_2 [K]	The experiment time [s]	The cooling rate, m [s^{-1}]	α_{Σ}, [W/m2 ·K]	α_R, [W/m2 ·K]	α_{ex}, [W/m2 ·K]	α_{teor}, p. 3 [W/m2 ·K]
10	600/423	65	0.0277	85.097	10.968	74.129	70.362
30	600/423	65	0.0297	91.241	11.210	80.031	82.361
50	600/423	65	0.0341	104.758	11.613	93.145	98.166
70	600/423	65	0.0383	117.661	11.955	105.706	101.645

Table 2. The results of experimental studies material: KMU-4L

The velocity flow, V [m/s]	T_3[K]/T_2 [K]	The experiment time [s]	The cooling rate, m [s^{-1}]	α_{Σ}, [W/m2 ·K]	α_R, [W/m2 ·K]	α_{ex}, [W/m2 ·K]	α_{teor}, p. 3 [W/m2 ·K]
10	600/423	90	0,0280	113.475	12.913	100.562	90.231
30	600/423	90	0,0333	134.928	13.249	121.679	112.875
50	600/423	90	0,0388	157.005	13.583	143.422	135.002
70	600/423	90	0,0420	170.005	13.888	156.117	150.226

where: T2 and T3 is initial temperature in accordance with figure 3.

Figures 4a and 5a present the results of experimental studies for the above materials. For comparison, in figures 4b and 5b the results of calculations of temperatures from the time of leakage flow [8].

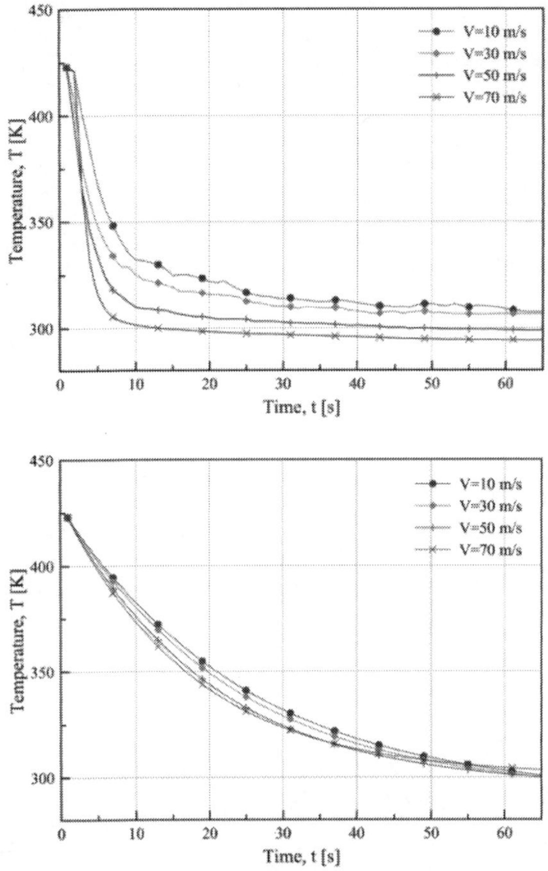

a) b)

Figure 4. Graphs of temperature change from the blowing time of the material Textolite PTK at an initial temperature of $T = 423$ K: a) experimental, b) theoretical [8].

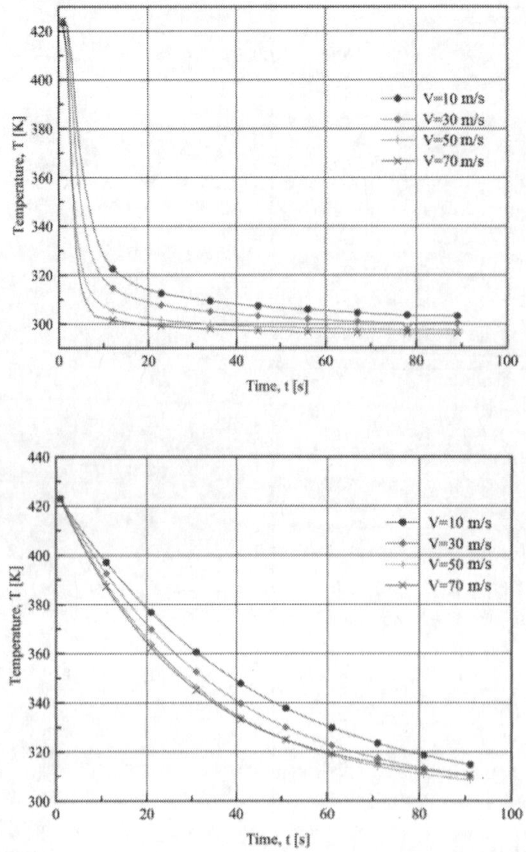

a) b)

Figure 5. Graphs of temperature change from the blowing time of the material Carbon fiber KMU-4L at an initial temperature of $T = 423$ K: a) experimental, b) theoretical [8].

5. Results and discussion

The difference of heat transfer coefficients, defined based on numerical [8] and an experimental simulation is less than 10 %.

A theoretical model [8], which compared the proposed airflow physical model to the various surfaces, was calculated under the same initial conditions as in the case of the experiment.

Initial and boundary conditions:

- RDE model dimensions – 30 x 30 mm;
- The heat capacity of the PTK material – 1470 J/kg·K;
- The heat capacity of the KMU-4L material – 1300 J/kg·K;
- Oncoming flow parameters: temperature 295 K; pressure 101325 PA; density of 1.22 kg/m3; flow velocity from 10 to 70 m/s, perpendicular to the plane of the RDE model.

In terms of reliability, as with any land-based studies (trials) in wind tunnels, the model remains motionless, and the energy expended for receiving the gas stream flowing around the model.

When modelling external flow around a jet area F_C several times more cross sectional area of the model F_0:

$$f_з = \frac{F_c}{F_0} \quad (6)$$

The flow-cluttering coefficient is $f_з = 14.056$.

According to the results of the data of physical modelling the insight into the values of heat transfer coefficients of the materials (PTK, KMU-4L) at low free-stream velocities is obtained.

It is planned to change in the angle of oncoming flow and expansion of the velocities range up to transonic velocities, the use of real PC as a source of thermal energy, the consideration of sandwich materials with honeycomb core with the addition of a real PC.

6. Conclusion

1. The formulation of experimental studies of the thermal loading RDE rocket fairing during the atmospheric phase of the descent trajectory, realizing the aerodynamic loading corresponding to the real conditions is carried out.
2. Experimental studies have shown the indicative requirements for the amount of heat flow from the PC is required to ensure the heating process to the RDE combustion temperature.
3. The experimental stand and experimental wind tunnel for studying the thermal regime of the RDE from an additional source of heat and aerodynamic inflow and entrainment of heat at free-stream velocities corresponding to the real values and experiments programme and methods of processing the results of experiments are developed. The drection for further research is obtained

Acknowledgements

The research was supported by the Russian science fund grant for the project "Development of scientific-technical fundamentals of the combustion of the separated elements of space rockets with the goal of reducing the acreage of the impact areas" the Agreement No. 16-19-10091 18.05.2016

References

[1] Shoemaker M, Van der Ha J, Abe S and Fujita K 2013 Trajectory Estimation of the Hayabusa Spacecraft during Atmospheric Disintegration vol **50** *Journal of Spacecraft and Rockets* pp 326–336
[2] Fritsche B, Klinkrad H, Kashkovsky A and Grinberg E 2000 Spacecraft disintegration during uncontrolled atmospheric Re-entry vol **47** *Acta Astronautica* pp 513–522
[3] Trushlyakov V, Zharikov K, Dron' M, Iordan Yu, Davydovich D, Pivkina A and Monogarov K 2017 Utilization of Thermite energy for Disruption of Rocket Separated Elements Made of Composite Polymeric Material *4th IAA Symposium Space Fly Safety (St-Petersburg. Russia. July 3–5)*
[4] Trushlyakov V, Shatrov Ya, Lempert D, Iordan Yu and Zarko V 2016 *Rocket payload fairing* Patent Russia no 2581636
[5] Trushlyakov V Davydovich D 2017 The use of pyrotechnic mixtures for dispergating fairings during atmospheric re-entry vol **174** *Procedia Engineering* pp 4–10
[6] Trushlyakov V, Zharikov K, Iordan Yu, Davydovich D, Dron' M, Lesnyak I and Sevoyan V 2016 *Method of modeling of processes of heat and mass transfer with the environment element of the design of the aircraft and device for its implementation* Patent Russia no 2016134556
[7] Isachenko V and Sukomel A 1981 *Heat transfer* (Moscow: Energiya) p 346
[8] Trushlyakov V, Zharikov K, Dron' M, Iordan Yu and Davydovich D 2017 *Proc. 7th European conference for aeronautics and space sciences* (Milan)

IOP Conf. Series: Journal of Physics: Conf. Series **944** (2017) 012119 doi:10.1088/1742-6596/944/1/012119

Theoretical and experimental researches of the liquid evaporation during thermal vacuum influences

V Trushlyakov[1], A Panichkin[2], O Prusova[1], K Zharikov[1] and M Dron'[1]

[1]Omsk State Technical University, 11, Mira ave., Omsk, 644050, Russia
[2]Institute of mathematics n. a. S. Sobolev SB RAS, Novosibirsk, Russia

e-mail: olga.prusova@mail.ru

Abstract. The mathematical model of the evaporation process of model liquid with the free surface boundary conditions of the "mirror" type under thermal vacuum influence and the numerical estimates of the evaporation process parameters are developed. An experimental stand, comprising a vacuum chamber, an experimental model tank with a heating element is designed; the experimental data are obtained. A comparative analysis of numerical and experimental results showed their close match.

1. Introduction

The method of thermal vacuum drying is widely used in various industries, for example, in [1, 2] a method and a device for thermal vacuum drying of wet particulate materials are described. Heating of the raw materials occurs in a reservoir in which thermally insulated space between the walls a resistive heater is placed. In [3–6], the methods of vacuum drying and a device for vacuum drying used in food, medical, microbiological and chemical industries are described. Heating of the dried product takes place on the heated shelves of the heating elements unit. In [3–5], conductive heating of the heated shelves is carried out. In [6] heating elements have an internal cavity forming a reservoir inlet and outlet of the hot heat carrier.

However, these methods are only used for the dehydration of porous and particulate materials. In [7], to dry the internal cavity of the pipeline to the required moisture content, initial vacuuming and then blowing the pipeline with pre-drained gas is suggested. In [8], for drying the pipeline, the cavity is pre-evacuated to a pressure of 0.2 to 0.005 kgf/cm^2, and then, maintaining the achieved vacuum, water vapour is pumped from the cavity of the pipeline until complete evaporation of the water film. Final drying to the required residual moisture content takes place with venting the cavity gas, which drying is carried out by its extensions inside the drained cavity due to the vacuum maintained there.

In [9, 10], the results of numerical studies of the influence of various parameters such as initial temperature, pressure, density and drop diameter, specific heat and thermal conductivity, and pressure in the vacuum chamber on the process of drops vacuum freeze are presented. The estimation of these parameters impact on the processes of drops cooling and freezing is presented. It is noted that the process of mass transfer at the droplet surface is different for the micro droplets (<100 μm) and macro droplets (>1 mm).

In [11, 12], the results of experimental and theoretical studies of the evaporation – freezing a pendent water drop in the conditions of deep vacuum are described.

IOP Conf. Series: Journal of Physics: Conf. Series **944** (2017) 012119 doi:10.1088/1742-6596/944/1/012119

As it can be seen from the above, the drying process of reservoirs and pipelines via vacuum by evaporation of the water film is widely used nowadays. However, theoretical and experimental studies were carried out only for a single drop of liquid. Below is a discussion of theoretical and experimental studies of the evaporation process from the surface layer of the model liquid (ML) on the example of water for boundary conditions of «mirror» type (water film) under the conditions of decompression and thermal loading. The efficiency of the evaporation process is proposed to evaluate the ratio of the evaporated ML mass and the energy spent on evaporation.

2. Statement of the research problem

The study of the ML evaporation process, in an experimental model tank (EMT) under reduced pressure in a vacuum chamber (VC) and temperature effects, involves the following physical-mathematical model.

At the initial moment t_0 ML, volume V_W and temperature T_W, in the EMT, is placed in the VC volume V_V and the air temperature T_V with the boundary position of the free surface of ML in the form of a "mirror" area of S_0. EMT has a form of a tub (figure 2) with mass m_S (volume V_S, density ρ_S) for the ML placement and the possibility of the ML heating. The initial pressure in the VC is atmospheric P_a with the partial pressure of water vapour P_p in the corresponding mass of dry air m_a and model liquid m_p. Pumping air from the VC is given as a continuous function of the volume dV element during the time dt.

In the process of air pumping with a pumping rate of dV/dt from time t_1 to $t_2 = t_1 + dt$ with its volume increased from V_1 to $V_2 = V_1 + dV$ simultaneous decrease in the pressure from P_1 to P_2, and temperature from T_1 to T_2 and the density from ρ_{v1} to ρ_{v2} will occur. It is assumed that the share of pumped air volume dV gets the same changes of parameters like the volume of air V_V remaining in the VC, i.e.:

$$P_2 = P_1 - dP;\ T_2 = T_1 - dT;\ \rho_{v2} = \rho_{v1} - d\rho_v.$$

The density of the mixture of dry air with the mass m_a and water vapour with the mass m_p is defined as:

$$\rho_v = (m_a + m_p)/V_V, \tag{1}$$

where VC volume V_V is constant in time, and $V_1 = V_V$.

During weak external and internal heat transfer, the process of pumping can be considered adiabatic with ratio of specific heats for the air and water vapour k=1.4, so, for this process the following formula is valid [13]

$$P_1V_1^k = P_2V_2^k = (P_1 - dP)(V_1 + dV)^k = const; \tag{2}$$

$$T_1V_1^{k-1} = T_2V_2^{k-1} = (T_1 - dT)(V_1 + dV)^k = const. \tag{3}$$

In the process of ML evaporation from t_1 to $t_2 = t_1 + dt$ with a rate of dm_p/dt on the value of dm_p the pressure in the VC reaches:

$$P_2 = \left(\frac{m_a}{\mu_a} + \frac{m_p + dm_p}{\mu_p}\right)R_0T_2/V_2, \tag{4}$$

where $R_0 = 8.314$ J/(mole·K) – is gas constant; $\mu_a = 0.02897$ kg/mole and $\mu_p = 0.01802$ kg/mole – is the molar mass of dry air and water vapour. The temperature and pressure in (4) are related by

$$T_1^k / T_2^k = P_1^{k-1} / P_2^{k-1} = const, \tag{5}$$

from which and from (3) the temperature is determined:

$$T_2 = T_1 + dT = [(\frac{m_a}{\mu_2} + \frac{m_p + dm_p}{\mu_П})R_0/(V_1 + dV_2)]^{k-1}T_1^k / P_1^{k-1} \tag{6}$$

with the subsequent substitution in (4) for determining P_2.

Density (1) at time t_1 with a mass of dry air and a vapour of ML for time t_2 will be

$$\rho_{v2}=(m_{a1}+m_{p1}+dm_p)/(V_1+dV), \quad (7)$$

from where the mass m_a of dry air and vapour ML m_p at time t_2, remaining in the volume V_V VC will follow

$$m_{a2}=m_{a1}V_V/(V_V+dV); \;\; m_{p2}=(m_{p1}+dm_p)V_V/(V_V+dV). \quad (8)$$

To simulate the process of ML evaporation at the boundary with air, the dependences on temperature and pressure, and other parameters characterizing the liquid and the air, are used. It is known, for example, that the dependence of the evaporation rate from the surface S on the liquid temperature T_W is given by [13, 14]

$$W_W(T_W)=V_W(T_0)\cdot\exp(\frac{LM(T_W-T_0)}{RT_WT_0}), \quad (9)$$

where T_0 – is the freezing temperature (K); M – is the molar mass of liquid (kg/mole); L – is the specific heat of vaporization (J/kg); R_0 – universal gas constant (J/(kg·K)).

To simulate the evaporation rate from unit area W_s (kg/(m^2·s)) the formula in the following form is used

$$W_W(P,T_W)=K_1\cdot(P_W(T_W)-P_d)(\frac{P_0}{P})^q, \quad (10)$$

where K_1 is the coefficient of condition of the ML evaporated surface (the still «mirror» type or undulating) and the movement of air (kg/(Pa·m^2·s)); P_W is the partial pressure of evaporated ML saturation at its current temperature T_W (Pa); P_d is the partial pressure of mg vapour in air (Pa); P_0 is the initial atmospheric pressure (Pa); P is the current pressure in the VC (Pa); q is the linearity for the given ML (0.66÷1.18).

For still air in the VC the K_1value, in accordance with [15], is taken equal to $1.39\cdot10^{-9}$ kg/(Pa·m^2·s).

To account for the condensation of ML the rate formula similar to (10), with partial saturation pressure P_V under the air temperature T_V, is considered

$$W_V(P,T_V)=K_1(P_V(T_V)-P_d)(\frac{P_0}{P})^q. \quad (11)$$

The VC will be considered in the following regions (figure 1): W_0 – EMT; W_1 is the region of the liquid film in the EMT; W_2 is the air environment in the VC; W_3 is the VC wall. Between these regions heat and mass transfer will be modelled with the given parameters, such as the mass m_i, the heat capacity C_{pi}, the thermal conductivity λ_i, the density ρ_i (i = 0, 1, 2, 3 for base coat, liquid, air and the case wall of VC respectively) and the initial temperatures T_S , T_W , T_V, T_K.

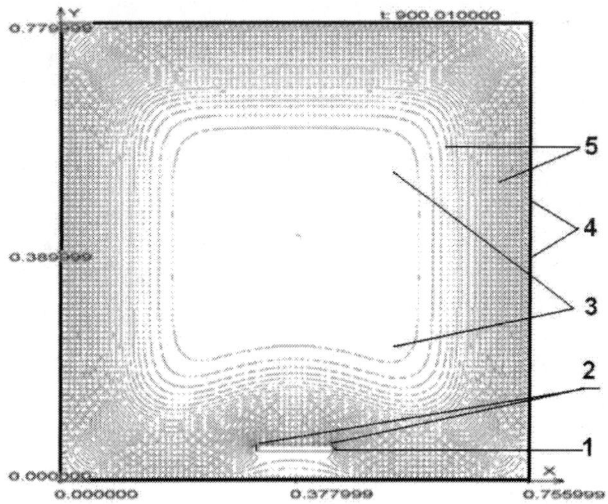

Figure 1. The region of numerical modelling in the VC with the temperature distribution at time t = 900 s: 1 – the EMT (W_0); 2 – the film (mirror) of the liquid (W_1); 3 – the air (W_2); 4 – the VC case wall (W_3); 5 – the contours of temperature *(in the preliminary estimates at this stage of the research the air velocity and liquid was not considered because of their smallness).*

We assume that at the initial moment $t_0 = 0$ in the VC and outside it the preset atmospheric pressure P_0 = 101323.2 Pa with a certain ML moisture percentage

$$C_1 = 100\cdot(P_{pn}-P_p)/\ P_{pn}, \tag{12}$$

where $P_{pn}(T_V)$ is the partial pressure of saturated air at liquid; $P_p=(m_p/\mu_p)R_0T_V/V_V$ is the partial vapour pressure of ML in the air; m_p is the mass of the liquid vapour in the air; μ_p = 0.01802 kg/mole is the molar mass of the vapour; R_0 = 8.314 J/(mole·K) is the universal gas constant.

The value of $P_{pn}(T_V)$ is tabular and at T_V= 20 °C is equal to 2338.43 Pa.

Consequently, we formulate the initial-boundary value problem of evaporation of a viscous incompressible heat-conducting fluid from the EMT and pumping of viscous heat-conducting air from the rectangular region with the wall thickness h, the length L_0 as a two-dimensional problem.

3. Equations of heat transfer

We will consider two-dimensional problem in one section of the VC, given the size of the EMT width, with defined boundary conditions under the velocity of the air ΔV_V (m^3/s) and a decrease in the internal air pressure P_V and temperature T_V according to the equations (4) – (6).

For numerical simulation of heat transfer in each of the regions W_0, W_1, W_2, W_3 the equation of heat transfer for a particular environment (air, liquid, solid) will have the following form[13]:

$$\frac{\partial T}{\partial t}+(V\cdot\nabla)T=\lambda\nabla^2T, \tag{13}$$

where $V = V$(u,v) – is the velocity vector of the moving environment (for air and liquid); T – is the environment temperature; λ – is the thermal diffusivity ($K = \lambda C_p\rho$ – is the conductivity).

For the simulation of heat and mass transfer let us define the region sizes in metres: W_0 = [0.372, 0.384]x[0.050, 0.052]; W_1 = [0.372, 0.384]x[0.052, 0.053]; W_2 = [0.018, 0.738]x[0.018, 0.762] – W_0 – W_1; W_3 = [0.000, 0.756]x[0.000, 0.780] – W_0 – W_1 – W_2.

For the heat equation (13), the conditions of heat fluxes conservation from some regions to others [13] are used.

We indicate the VC size on the x-axis: x_0=0 m, x_1=0.018 m, x_2=0.372 m, x_3=0.384 m, x_4=0.738 m, x_5=0.756 m; *the size on the y-axis:* y_0=0 m, y_1=0.018 m, y_2=0.050 m, y_3=0.052 m, y_4=0.053 m, y_5=0.762 m, y_6=0.780 m.

On the boundary W_0, W_1, W_2, having respective coefficients heat conductivity $K_0=\lambda_0 C_{p0} \rho_0$, $K_1=\lambda_1 C_{p1} \rho_1$, $K_2=\lambda_2 C_{p2} \rho_2$, between themselves, we shall take into account the conservation of heat fluxes in the form:

$$K_2 \frac{\partial T}{\partial y}(x,y_2^-) = K_0 \frac{\partial T}{\partial y}(x,y_1^+),\ K_0 \frac{\partial T}{\partial y}(x,y_3^-) = K_1 \frac{\partial T}{\partial y}(x,y_3^+),$$

$$K_1 \frac{\partial T}{\partial t}(x,y_4^-) = K_2 \frac{\partial T}{\partial y}(x,y_4^+),\ x_2 \le x \le x_3, \tag{14}$$

On the boundary of the region W_2 with the heat-conducting VC case with a coefficient of thermal conductivity $K_3=\lambda_3 C_{p3} \rho_3$ for T accepts the following conditions, similar to (14), are accepted:

$$K_3 \frac{\partial T}{\partial x}(x_1^-,y) = K_2 \frac{\partial T}{\partial x}(x_1^+,y),\ K_2 \frac{\partial T}{\partial x}(x_4^-,y) = K_3 \frac{\partial T}{\partial x}(x_4^+,y),\ y_1 \le y \le y_5,$$

$$T_x^{'}(x,0^+) = 0,\ T_x^{'}(x,y_6^-) = 0,\ x_0 \le x \le x_5,$$

$$T_x^{'}(0^+,y) = 0,\ T_x^{'}(x_5^-,y) = 0,\ y_0 \le y \le y_6. \tag{15}$$

At the initial time $t = 0$ are specified: the temperature in all the calculated regions and the air pressure from the partial vapour pressure of ML further based on the time interval $[0,t^*]$ to the critical point t^*, when the ML freezing will occur (in the mathematical model the process of ML freezing is not used).

4. Experimental studies

1. Preparatory experiments are the following:
   - Testing of experimental stand measuring complex (temperature and pressure sensors);
   - Testing of the vacuum creation system (vacuum pump, pneumatic valves, connectors and valves, pressure sensor);
   - Testing of individual elements of the experiments programme (the evaporation time of a given mass of model liquid, etc.).
2. The main experiments in the VC are the following:
   - The evaporation time measurement of a given mass of ML with heating and without heating;
   - The ML mass before and after the experiment;
   - The ML temperature measurement, in the EMT, gas-vapour mixture and pressure in the VC.

Figure 2 presents a schematic structural diagram of experimental stand for the research of ML evaporation under thermal vacuum influences.

The measured parameters of ML evaporation process are:

- The ML temperature (K);
- The pressure in the VC (Pa);
- The evaporation time (sec);
- The power used for evaporation of a given ML mass (W).

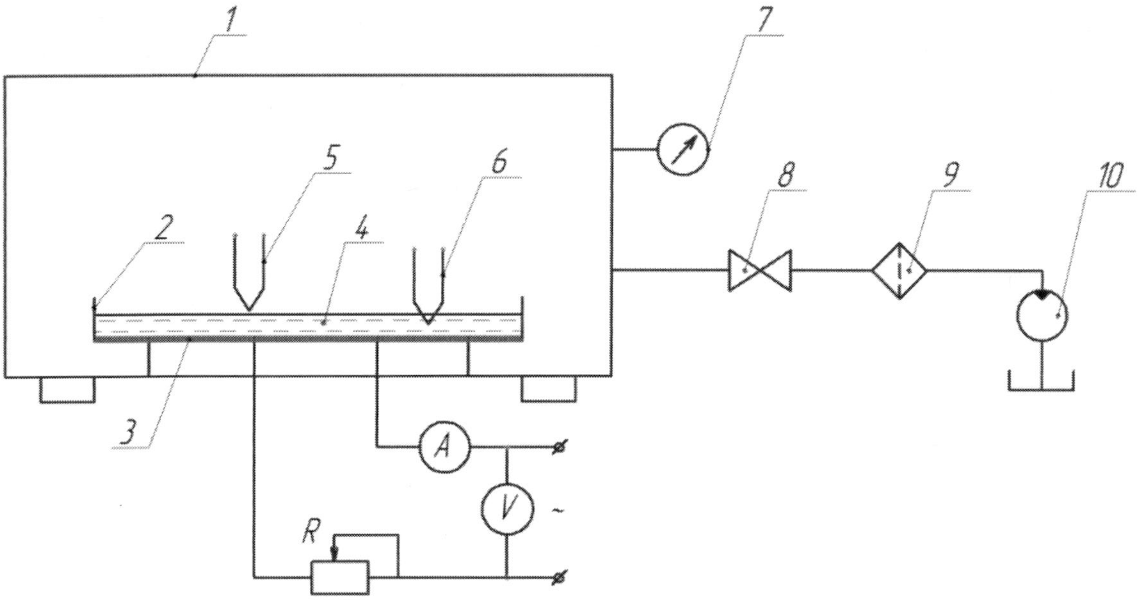

Figure 2. The experimental stand vacuum scheme: *1* – the vacuum chamber; *2* – the EMT; *3* – the heating element; *4* – the ML; *5, 6* – the thermocouples; *7* – the pressure sensor; *8* – the valve; *9* – the filter; *10* –the vacuum pump.

5. The test calculation results

Numerical calculations were carried out in accordance with (12) – (15) including evacuating air from the VC volume V_v = 0.463 m^3 with a velocity of dV/dt = 0.0065 m^3/s when using the ML evaporation model according to the formulas (10) – (11) and the model of gas-dynamic quantities in equations (1) – (8) change.

In the formula (10) for the boundary conditions of the ML evaporable surface "mirror the coefficient K_1= 1.39·10^{-8} kg/(Pa·m^2·s) and linearity q = 1.18" were used.

The calculation of heat transfer parameters in equations (13) – (15) with the use of algorithms based on economical difference schemes [16] was carried out.

Based on the developed software product [16] numerical modelling of heat and mass transfer with ML evaporation surface in the process of reducing atmospheric pressure in the VC was carried out.

When using a computational grid with N_x = 80 and N_y = 80 the test calculation of the three options for the ML evaporation in the VC with the amount of ML = 5.0 ml, 10.0 ml and 20.0 ml, which constituted the initial levels of liquid in the tub is 0.2 mm, 0.4 mm and 0.8 mm, respectively, was made.

Fig. 1 shows the picture of isolines of the temperature characteristic of the heat transfer with a volume of 20.0 ml at a pumping rate V_1 = 6.5 l/s and time of 900 s.

The calculations for these three options with the ML amount 5.0 ml, 10.0 ml and 20.0 ml without heating was carried out up to time 360 s, 600 s and 900 s, respectively. For the given time points, the level of ML (*h*) and temperature (*T*) with an initial moister content of 0% in the chamber was made up of the following values:

- *h*: 0.094 mm, 0.140 mm, 0.223 mm;
- *T*: 270.9 K, 271.2 K, 271.4 K.

Similar data were obtained for a moister content of 100% in-chamber:

- *h*: 0.097 mm, 0.143 mm, 0.226 mm;
- *T*: 271.3 K, 271.6 K, 271.6 K,

This indicates greater ML evaporation at a lower initial moisture content in VC.

6. Comparison of numerical and experimental results

In the course of the calculations and experiments the following dependence of ML temperature (figures 3, 4) and the pressure in the VC (figures 5, 6) from the time process were obtained. Figure 7 shows the results of ML evaporation at the time, which suggests that during heating, ML begins to evaporate after 200 s and corresponds to the measured values for the residual mass of the ML at the time of freezing.

Figure 8 presents the variation of the VC evacuate pump operation and the total energy of the gas and ML from the time of the process without ML heating. Table 1 shows the comparison of the consumed useful energy from the pump power of 500 W without heating and with ML heating using the rangette power of 3.57 W for pumping air and ML evaporation. The calculations showed that the chances of freezing costs amounted 167179 J and 177146 J, respectively for these two experiments.

In figures 9 and 10 the mass variation and vapour partial pressure in the tub from the time of the process without heating and with heating are shown.

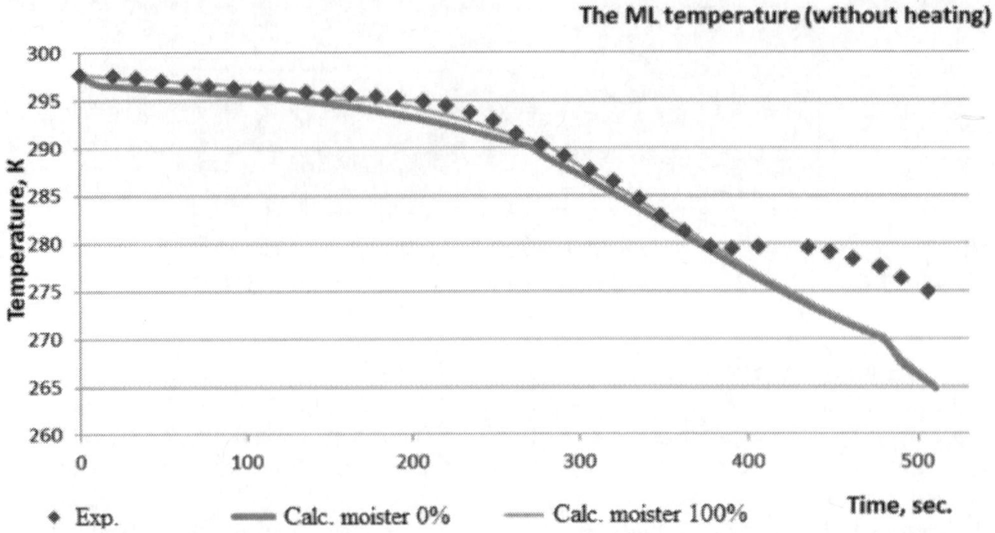

Figure 3. The ML temperature dependence on time (without heating).

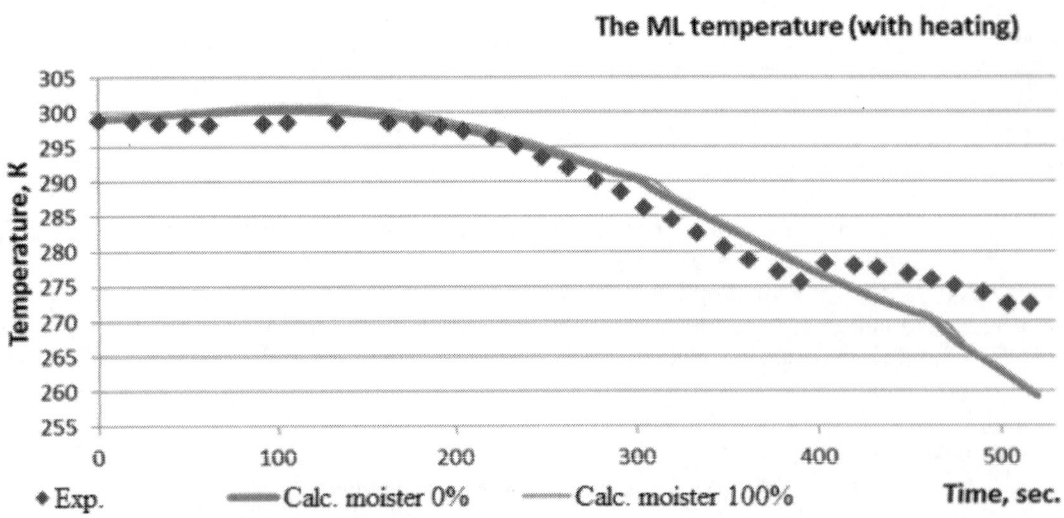

Figure 4. The ML temperature dependence on time (with heating).

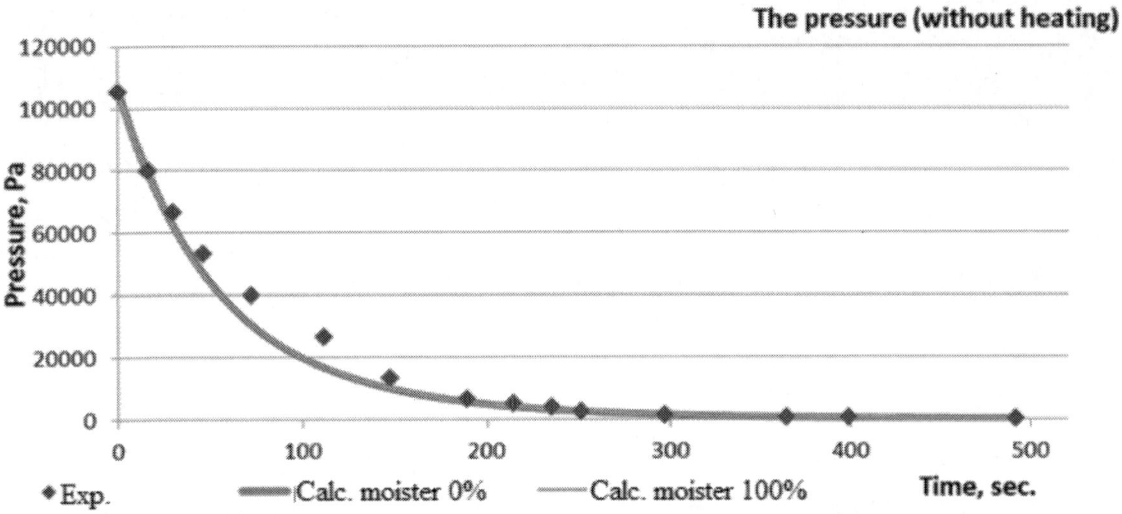

Figure 5. The ML pressure dependence on time (without heating).

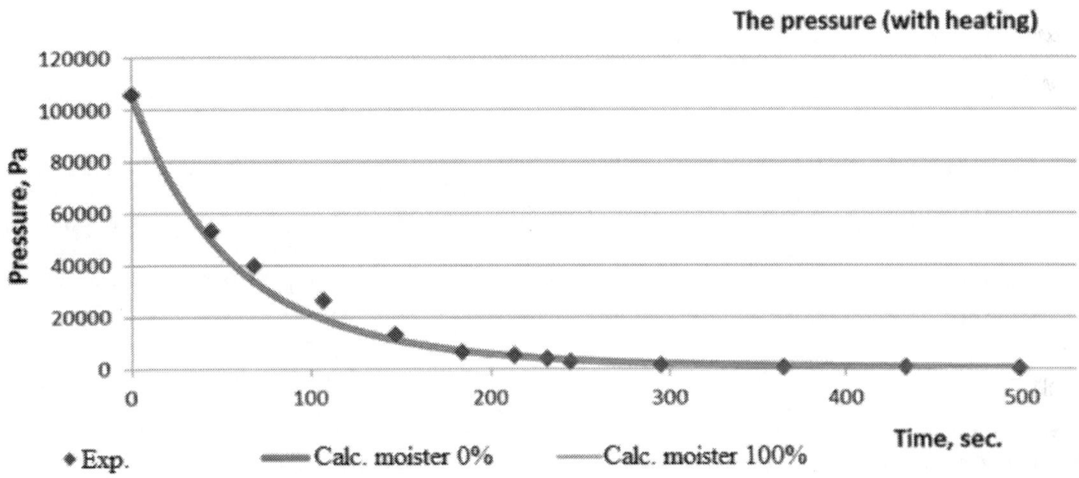

Figure 6. The ML pressure dependence on time (with heating).

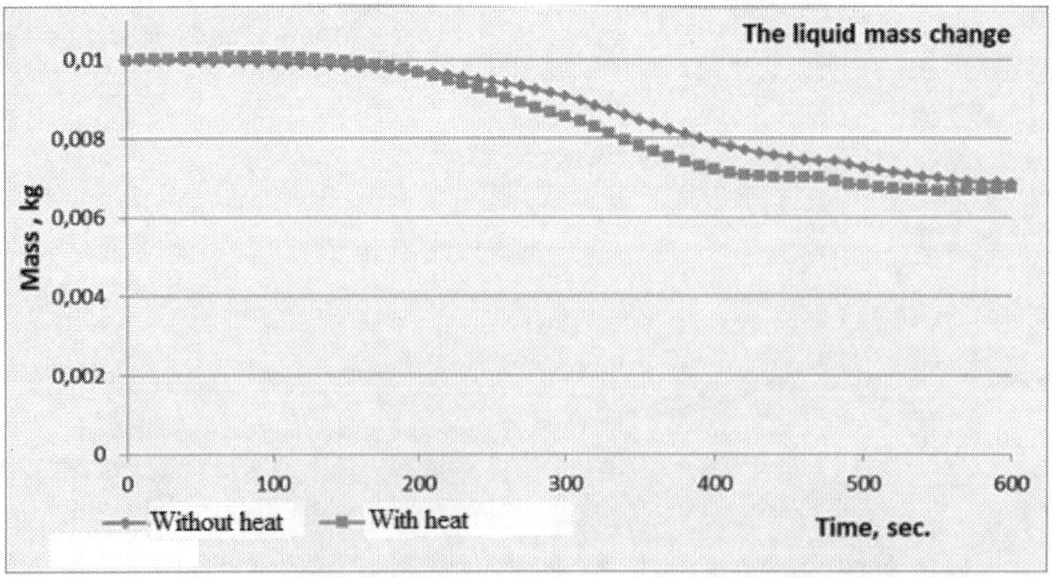

Figure 7. The ML mass dependence in EMT on time (without heating and with heating).

Table 1. The value of the useful operation of pumping and rangette with an initial moister content of 0% and 100%.

The time t, sec	The amount of pumping operation and ML heating, J.			
	Without heating		Heating – 3.57 W	
	The moister content 0%	The moister content 100%	The moister content 0%	The moister content 100%
100	31427.0	31434.0	29849.0	29854.0
200	77455.0	77457.0	74292.0	74290.0
300	122893.0	122868.0	118665.0	118617.0
400	1672565.0	167175.0	162223.0	162080.0
500	211122.0	210994.0	205009.0	204759.0
600	254880.0	254691.0	243102.0	242328.0

From the results presented in table 1 it follows that the introduction of ML heating leads to a reduction in total energy costs: adding heating energy in the amount of ~ 2000 J was a reduction in total cost of ~ 12000 J.

For the combination of effect parameters (heating and vacuumization), the effect of moister content is not essential.

Thus, there is an optimum ratio of energy consumption for the creation of a vacuum and heating.

To illustrate this phenomenon, an example with the following initial data is provided: to reduce energy consumption in order to pump the air and increase the order of the energy consumption to heat, the quantity of the evaporated liquid increased by three times while reducing the overall energy consumption by six times.

If you enter criteria $m_w / E_\Sigma = K$, for the calculation in Table 1 this ratio K_1 = 16.37 g/MJ, and for the latest example K_2 = 220.94 g/(MJ).

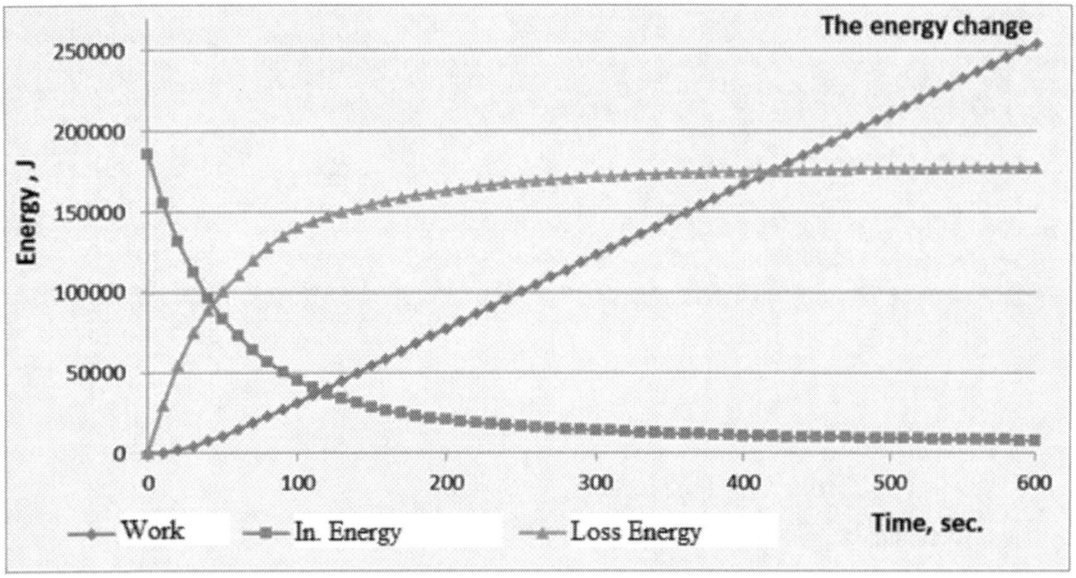

Figure 8. The dependence of the work and the total energy of the gas and ML on the time of the process without ML heating.

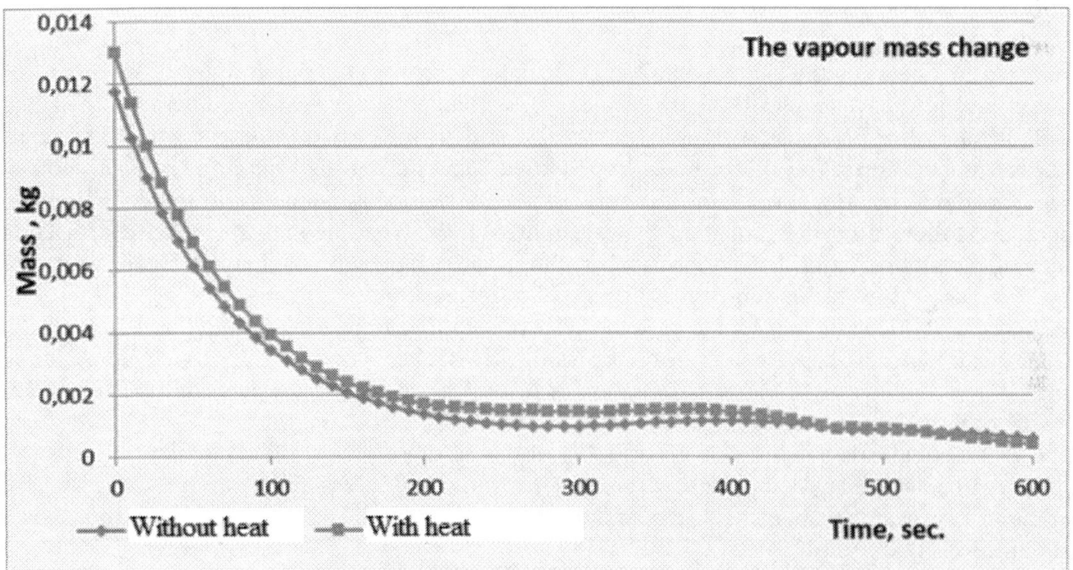

Figure 9. The vapour mass dependence in EMT on time (without heating and with heating).

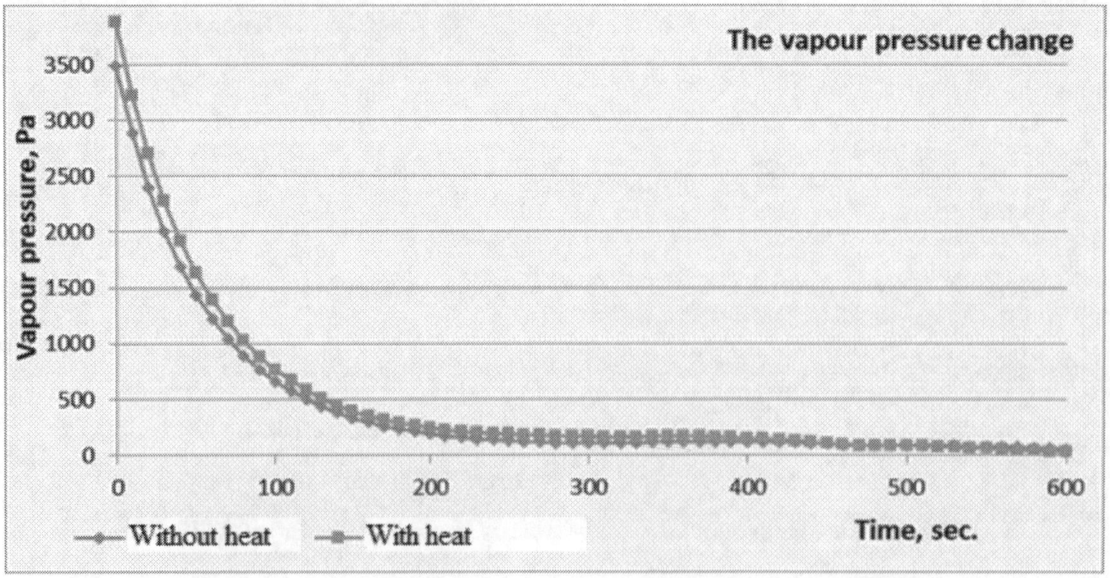

Figure 10. The vapour partial pressure dependence in EMT on time (without heating and with heating).

7. Results and discussion

The discrepancy between the experimental and calculated results on temperature by 2–3 K after 200 s and the pressure to 15% around 100 s by time can be explained as follows. In numerical simulations, the ML level in the EMT has a uniform thickness, and in the real experiment ML takes uneven thickness with the formation of separate droplets and separate boiling (figure 11, *a*). To 400–440 seconds (without heating and with heating) in the experiment, sometimes there is increased evaporation, followed by the formation of ice (figure 11, *b*). This leads to the release of a certain amount of heat when icing that is observed in the graphs: first, with a sharp decrease in the ML temperature, and then with the subsequent overshoot of increasing it.

In the field of 100 s, there is an increase in pressure resulting from evaporation that is more active. It is also possible to explain the uneven thickness of the ML layer due to the active evaporation and boiling. Where the layer is thinner, evaporation is faster, the pressure increases and the temperature decreases slightly.

In the numerical calculations, this no effect is observed due to the uniform thickness of the ML: The ML value temperature gradually decreases to zero, and pressure – up to 400 Pa. In the experiments and calculations of zero temperature and pressure of 400 Pa onset occurs almost at the same time (~520s).

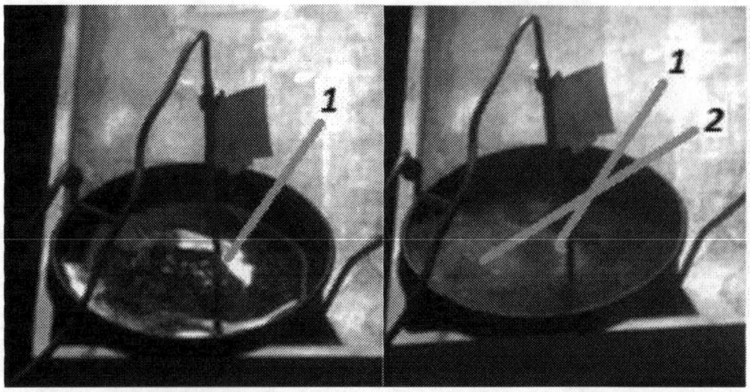

a) b)

Figure 11. Phase transitions of the ML in the EMT during the experiments:
a – evaporation; b – freezing; *1* – is the liquid; *2* – is the ice.

In the experiments without heating, the ML evaporated mass amounted to 2.15 g (21.5% of initial mass). In the experiments with heating, the ML evaporated mass amounted to 2.9 g (29 % of initial mass).

The power expended on the evaporation of a given ML mass, without heating is equal to 86.2 W, while with heating – to 89.6 W.

In both cases (without heating and with heating) at the end of the experiment ML freezing occurred: in the experiment without heating – at 613 Pa and the ML temperature 275 K, and with heating at 573 Pa and the ML temperature 273 K. This is due to the low ML temperature and the reduced pressure in the VC.

Thus, the disadvantage of using vacuum pump for vacuum drying is constant pressure decrease in VC, which leads to ML freezing. The vacuum pump does not control pressure change.

To maintain a constant reduced pressure at a certain level it is proposed to use a vortex tube, the principle of which is based on the Ranque-Hilsch Effect [17]. It is possible to use a counterflow vortex tube, the method of calculation which has been tested in [18].

8. Conclusion

1. The mathematical model of the evaporation process of model liquid with the free surface boundary conditions of the "mirror" type in thermal vacuum influences and the numerical estimates of the parameters of evaporation process is developed.
2. The experimental stand, comprising a vacuum chamber, the experimental model tank with heated is developed. Experimental researches were conducted.
3. A comparative analysis of numerical and experimental results showed a close match.
4. The downside of using vacuum pump for ML evaporation is the inability to control the pressure inside the vacuum chamber, so in the future it is suggested to consider a vortex tube for producing a vacuum.

Acknowledgements
The research was supported by the grant of the Ministry of education and science of the Russian Federation "Improvement of environmental safety and economic efficiency of launch vehicles with main liquid rocket engines", No. 9.1023.2017 PCH

References
[1] Kutovoj V 2008 *Device for thermal vacuum drying* Patent Russia no 2315928
[2] Kutovoj V, Kazarinov Yu, Lucenko A, Nikolaenko A and Koshelnikov V 2014 *Energy saving method for preparation of solid fuel for burning* vol **8** (Energosberezhenie. Energy. Energoaudit) pp 61–66
[3] Ermakov S 2007 *Device vacuum drying and method vacuum drying* Patent Russia no 2299385
[4] Shabetnik G 1998 *Method of vacuum drying of materials and device for vacuum drying of materials* Patent Russia no 2121638
[5] Kovalev L and Kovaleva N 2007 *Method and device for energy-saving dehydration and drying in vacuum* Patent Russia no 2295681
[6] Kovalev L and Kovaleva N 2005 *Method of low-temperature vacuum dehydration of materials and device for its implementation* Patent Russia no 2246079
[7] Gubanok I, Dubinsky V, Egorov I, Posin V, Usenko M and Endeka Yu 2006 *Method of drying of pipelines and device for its implementation* Patent Russia no 2272974
[8] Naumejko V, Naumejko S and Naumejko A 2007 *Method and device for drying of gas pipelines* Patent Russia no 2300062
[9] Zhang Z, Zhang Y, Zhao L, Zhang W and Zhao S 2015 Parameter Sensitivity of the Microdroplet Vacuum Freezing Process *Mathematical Problems in Engineering* vol **2** p 8

[10] Zhang Z, Gao J and Zhang S 2016 Heat and Mass Transfer of the Droplet Vacuum Freezing Process Based on the Diffusion-controlled Evaporation and Phase Transition Mechanism *Scientific Reports* vol **6** p 7
[11] Du W, Zhao J and Li K 2012 Experimental study on thermal-dynamical behaviours of liquid droplets during quick depressurization *Journal of Engineering Thermophysics* vol **33** pp 1349–1352
[12] Shin H, Lee Y and Jurng J 2000 Spherical-shaped ice particle production by spraying water in a vacuum chamber *Applied Thermal Engineering* vol **20** pp 439–454
[13] Frenkel' Ya 1975 *Kinetic theory of liquids* (Moscow: Science) p 592
[14] Chechyotkin A 1971 *High temperature heat carriers* (Moscow: Energiya) p 496
[15] Zakharevich A, Velikanov V and Vympin E 2015 The ignition fragment wood heated to a high temperature of the plate *MATEC Web of Conf. Heat and Mass Transfer in the Thermal Control System of Technical and Technological Energy Equipment* (Tomsk) vol 23 p 5
[16] Panichkin A and Varepo L 2013 Numerical calculation of the free motion of a small volume of viscous incompressible liquid between two rotating cylinders *Computational technologies* vol 2 pp 62–71
[17] Korkodinov Ya and Hurmatullin O 2012 The Use of the Ranque-Hilsch Effect *Bulletin of Perm national research Polytechnic University* vol 14 pp 42–54
[18] Kuznetsov V and Makarov V 2016 *Vortex tube: experiment and theory* (Omsk: OmSTU publishing) p 239

Finite-element analysis of elastic sound-proof coupling thermal state

V G Tsyss, I M Strokov and M Yu Sergaeva

Omsk State Technical University, 11, Mira ave., Omsk, 644050, Russia

e-mail: mjsergaewa@mail.ru

Abstract. The aim is in calculated determining of the elastic rubber-metal element thermal state of soundproof coupling ship shafting under variable influence during loads in time. Thermal coupling calculation is performed with finite element method using NX Simens software with Nastran solver. As a result of studies, the following results were obtained: - a volumetric picture of the temperature distribution over the array of the deformed coupling body is obtained; - time to reach steady-state thermal coupling mode has been determined; - dependences of maximum temperature and time to reach state on the established operation mode on rotation frequency and ambient temperature are determined. The findings prove the conclusion that usage of finite element analysis modern software can significantly speed up problem solving. Key words – soundproof coupling, thermal state, torque, finite element model.

1. Introduction

Elastic soundproof couplings are used in ship shafting of diesel generator sets, or propeller drives on ships. During operation, the couplings are subjected to repeated dynamic loading and undergo significant dissipative heating in some cases resulting to their premature failure. In this connection, the practical task is the problem of thermal coupling calculation, its feature is that energy dissipation leads to self-heating of the coupling body rubber under dynamic loading and it must be considered when assessing its efficiency.

2. The problem setting

Elastic coupling in the process of operation except for a constant load in the form of torque is also subject to time-varying dynamic loads. Dynamic loads acting on the coupling can be divided into two types depending on the causes of their occurrence. The loads connected with angular and radial displacement of half-couplings are considered the first. The nature of their variation is harmonic, and the frequency is equal to the coupling rotation frequency. The shaft rotation unevenness is considered the second, the moment inconsistency of the coupling loaded is connected. They include:

Uneven torque on the shaft of the power plant.

Uneven torque on the shaft of the actuator.

The problem is to estimate the thermal state of a coupling loaded with a harmonic moment

$$M(t) = M_1 + M_2 \cdot \sin(\omega t) \qquad (1)$$

where M_1 is permanent component of the moment; M_2 is periodic component of the moment; ω is angular frequency of the torque change.

The following initial load data was taken for calculation:

$$M_1 = 0 \text{ N}\mu;$$

$$M_2 = 1 \text{ κN}\mu;$$

$$\omega = 30\ \chi^{-1}$$

Thermal coupling calculation was performed with finite element method using the NX Siemens software with Nastran solver.

3. Theory

Heat generation in a unit volume of the rubber elastic element is proportional to the mean for the deformation cycle of the variable voltages power value [1]:

$$Q_\psi = \frac{\omega}{2\pi}\int_0^{2\pi/\omega}\left(\sigma_x\dot{\varepsilon}_x + \sigma_y\dot{\varepsilon}_y + \sigma_z\dot{\varepsilon}_z + \tau_{xy}\dot{\gamma}_{xy} + \tau_{xz}\dot{\gamma}_{xz} + \tau_{yz}\dot{\gamma}_{yz}\right)dt, \qquad (2)$$

where ω is angular frequency; $\sigma_x, \sigma_y, \sigma_z, \tau_{xy}, \tau_{xz}, \tau_{yz}$ are axial and tangential stresses; $\varepsilon_x, \varepsilon_y, \varepsilon_z, \gamma_{xy}, \gamma_{xz}, \gamma_{yz}$ are axial and shear deformations.

In the case the rubber follows a linear viscoelastic deformation law described, for example, with the generalized Maxwell model, the function Q_ψ can be expressed in terms of the energy dissipation coefficient ψ [1]:

$$Q_\psi = \frac{\omega\psi}{2\pi} G_0\left(2\varepsilon_{ax}^2 + 2\varepsilon_{ay}^2 + 2\varepsilon_{az}^2 + \gamma_{axy}^2 + \gamma_{axz}^2 + +\gamma_{ayz}^2\right). \qquad (3)$$

where $\varepsilon_{ax}, \varepsilon_{ay}, \varepsilon_{az}, \gamma_{axy}, \gamma_{axz}, \gamma_{ayz}$ are amplitude values of axial and shear deformations. Experimentally, the dissipation coefficient is determined with the decrement of free vibrations, the amplitude of the resonant oscillations or oscillatory process phase pattern observed on the oscilloscope screen (Fig. 1). In the latter case, the dissipation coefficient is numerically equal to the ratio of the hysteresis loop square area to the area of OAB triangle:

$$\psi = \frac{A_D}{A_E} \qquad (4)$$

where A_D is energy dissipated as heat for deformation cycle; A_E is potential energy of deformation (work of elastic forces).

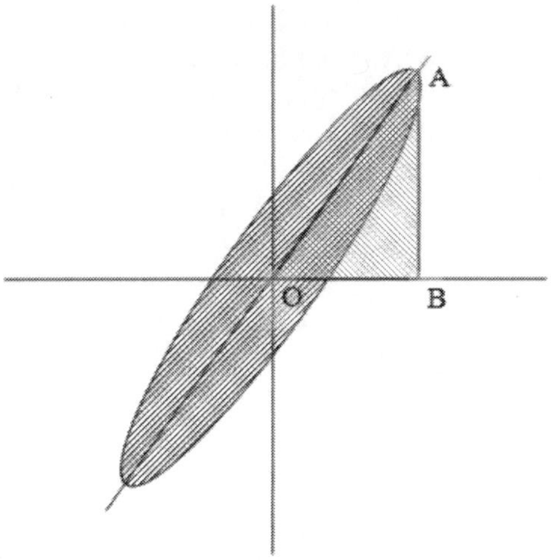

Figure 1. Hysteresis loop

It follows from equation (4), that if the material dissipation coefficient ψ is known and elastic strain energy A_E has been determined then it is possible to determine the energy dissipated in the form of heat per deformation cycle A_D:

$$A_D = \psi A_E. \quad (5)$$

Then, knowing the rotation frequency, the amount of heat generated in the material per unit of time can be calculated

$$Q = f \cdot A_D \quad (6)$$

where f is deformation frequency.

In the end, having made the heat balance equation, it is possible to determine the thermal state of the coupling [3]:

$$([L] + [C_h])\{T\} = \{Q\} \quad (7)$$

where $[L]$ is thermal conductivity matrix of the structure; $[C_h]$ is heat exchange matrix; $\{Q\}$ is heat generation vector

4. The calculation results

In modern finite element analysis software it is possible to obtain a volumetric picture of deformation energy distribution along a deformed body. Some software such as NX (Siemens), allow extracting the numerical values of the nodes (elements) depending on the nodes (elements) coordinates.

The values obtained can be recalculated using formulas (5) and (6) in the amount of heat released per unit volume of the body per time unit.

The next step is to solve the non-stationary thermal problem, where obtained values are used as the load, like heat generation per volume unit.

I stage. Simulating

To solve the problem in NX it was created finite element coupling model (Fig. 2).

To create the model 8-node hexagonal elements CHEXA (8) were used. Since this model is used both in structural and thermal analysis, then both structural and thermal characteristics of materials must be specified for it. For steel elements the material "Steel" from NX materials database was taken, and all the required characteristics are set. For elastic rubber elements it is necessary to develop a material and mechanical and thermal characteristics were indicated for it. Since the static problem is solved in a linear formulation, it is sufficient to take the value of Young's modulus E=3 MPa and the Poisson's coefficient is ν=0.49. Thermal characteristics for rubber are [2]:

thermal expansion coefficient is $\alpha=15\cdot10^{-4}$ K^{-1};

thermal conductivity is Λ=0.22 W/(m·K);

specific heat is C=1.67 kJ/(kg·K).

Figure 2. Finite element model of the rubber-metal coupling element

II stage. The solution of the static problem

To find the elastic deformation energy a structural analysis of the coupling model was performed. The solution type is "SOL 101 Linear statics".

To use the obtained result for the thermal coupling calculation, it is necessary to obtain an array of deformation energy density numerical values with elements depending on its coordinates. So the option "Determine value" is used, where a grid of elastic coupling element is chosen, and the extracted results are stored in electronic Excel table format as shown in Fig. 3.

Till to start thermal calculation it is necessary to prepare the initial data in a spreadsheet, i.e. to recalculate the energy density of deformation having been obtained in the previous stage with formulas (5) and (6) into the amount of heat released per volume unit per time unit. To do this the file saved in the previous stage is opened.

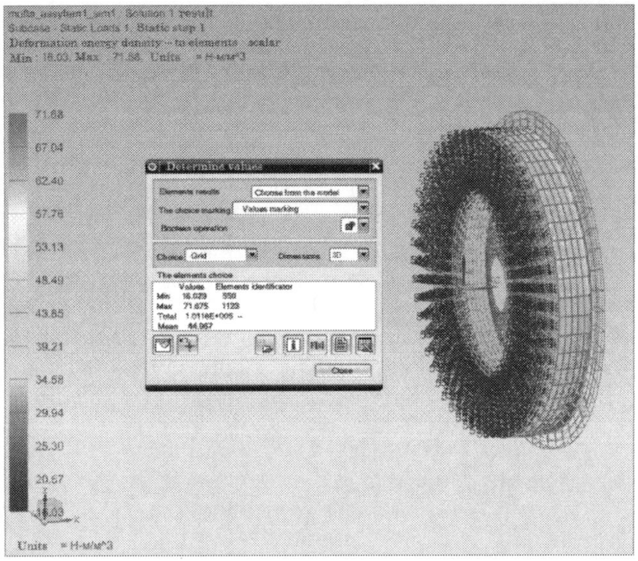

Figure 3. The values of the deformation energy density in an electronic spreadsheet format Excel

To start with, unnecessary information is removed, because the desired file must contain only four data columns with headings - three columns of coordinates (x, y, z) and one column with values of heat generated amount. Therefore, all redundant rows and columns are deleted from the file - three columns are obtained with the coordinates (A, B, C) and column with values of the deformation energy density (E). The column E is recalculated into column (F) depending on

$$\Theta = \phi \cdot \psi \cdot A_E,$$

Where $f = \omega / 2\pi = 30/6{,}24 = 4{,}8$ Гц is frequency of deformation, $\psi = 0.31$ is coefficient of dissipation, A_E is energy of elastic deformation (from the Table). It is written in the Excel editor for the second cell:

$$F_2 = E_2 \cdot 0.31 \cdot 4.8$$

and values are stretched for the entire column F. As a result, the column F will have values of the heat amount evolved per time unit. Values of column F should be moved into column D, and the columns E and F are removed. The resulting file is saved for later use.

III stage. The solution of the heat state problem

To determine the thermal coupling state a new solution based on the solver NX THERMAL/FLOW is developed, type of analysis is thermal.

Solver settings are as follows (unspecified settings remain by default):

Type of solution is transition process;

Transition process settings are:

The start time is 0 s;

The end is to perform until stationary mode;

The maximum number of intervals is 10;

To finish solution if the temperatures change between the time intervals is less than 1o C;

The units of the solution are left by default;

Ambient conditions are left by default;

Parameters of results are left by default, but if needed they can be added;

Initial conditions:

The initial temperature is uniform;

Specifying conditions for solving the problem are:

a) specification of the conditions between the contacting surfaces.

Conditions of contact between the rubber and the metal are set using simulation object "Thermal connection" (Fig. 4). As settings, the interacting surfaces of rubber and metal are indicated as "Primary area" and "Secondary area". Further the interaction type is indicated as "Heat transfer coefficient" and "Coefficient" 1150 W/(m^2·°C). Two such objects are developed for each half coupling.

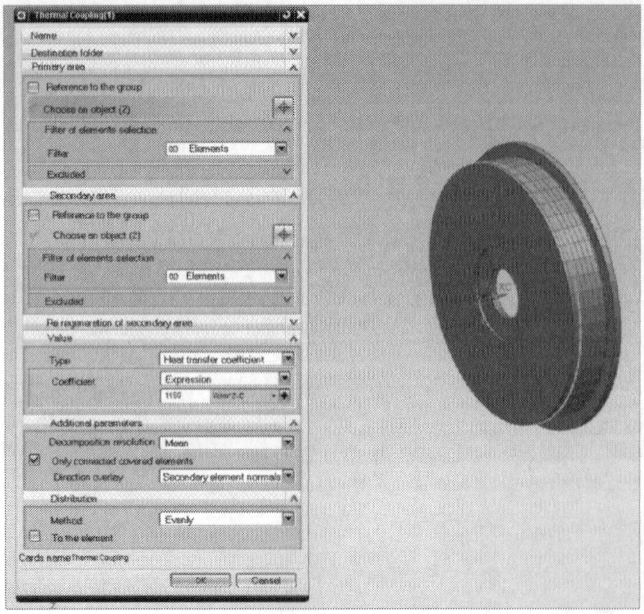

Figure 4. Setting the thermal contact between rubber and metal

To set the heat exchange conditions with the discarded parts, "Boundary resistance" simulation tool is used. The area of application will be the surface interacting with the discarded part of the structure - in this case with the mounting shaft flange (Fig. 5). Interaction type is "Heat transfer coefficient", equal to 2000 W/(m^2·°C). The same condition is developed on the second connecting surface.

IOP Conf. Series: Journal of Physics: Conf. Series **944** (2017) 012120 doi:10.1088/1742-6596/944/1/012120

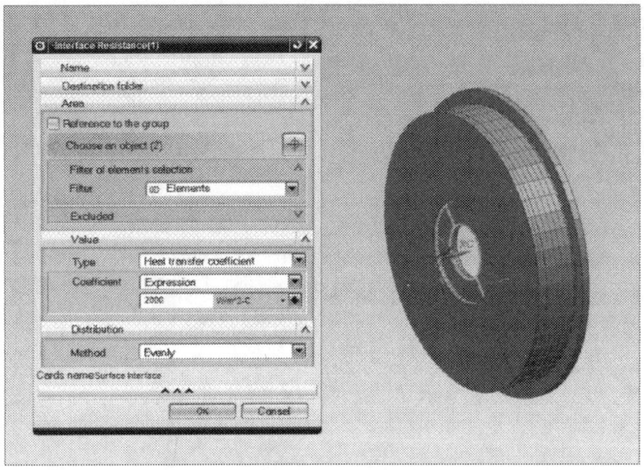

Figure 5. Setting the boundary thermal resistance

b) specification of boundary conditions.

As boundary conditions, convective heat exchange with the surrounding environment is specified. To do this "Convection to Environment" tool is used. As surfaces dissipating heat (only the outer surfaces of the metal and rubber) are indicated (Fig. 6). Convection coefficient is considered 22 W/(m$^{2}\cdot{}^{o}$C).

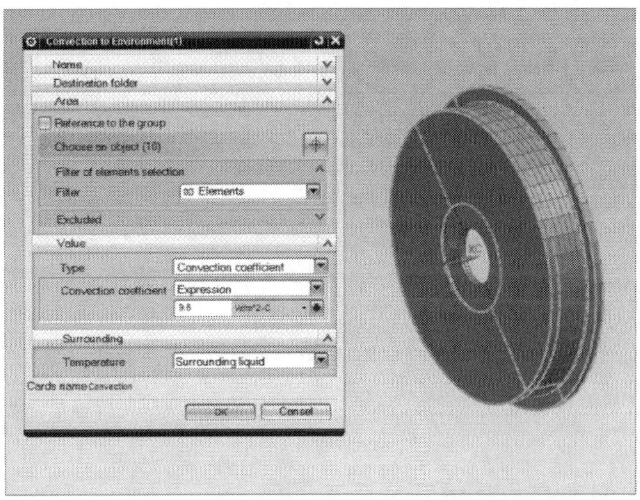

Figure 6. Setting the conditions for convective heat transfer

c) load specification.

The thermal load is set by means of "Thermal loads" tool. The distribution method is "Spatial." The load application will be the elastic coupling element. The load field is set using the table designer, and where the Table having been prepared in the previous calculation with values of heat amount released by the material volume unit is substituted.

After all the above settings, the task is started for calculation.

When completing the calculation, the solver gives a series of results, and the most interesting is the self-heating temperature of the rubber coupling array among them. The picture of thermal coupling state into the steady thermal mode is presented in Fig. 7 in a color chart form. The Figure shows that the array of coupling rubber during operation is heated to temperature of 31.1^{o}C. The most heated is the central (internal) part of the rubber mass, and it is a physically adequate result. The temperature field is shifted to the inner part of the coupling, since there is no convective heat exchange with the environment.

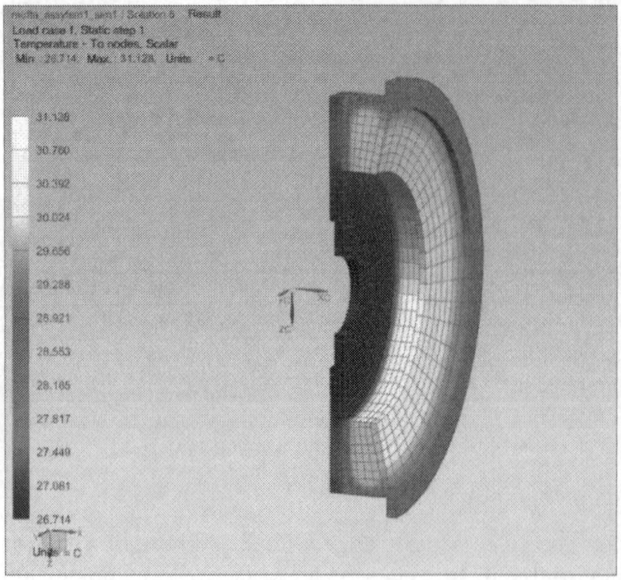

Figure 7. The color chart of the temperature distribution

The dependence of the temperature change in the most heated point of the rubber coupling mass in the process of output to the steady thermal mode is given in Fig. 8.

The Figure shows that the elastic coupling reaches the established thermal mode in 17000 s, and it is equal to 4.7 hours of operation.

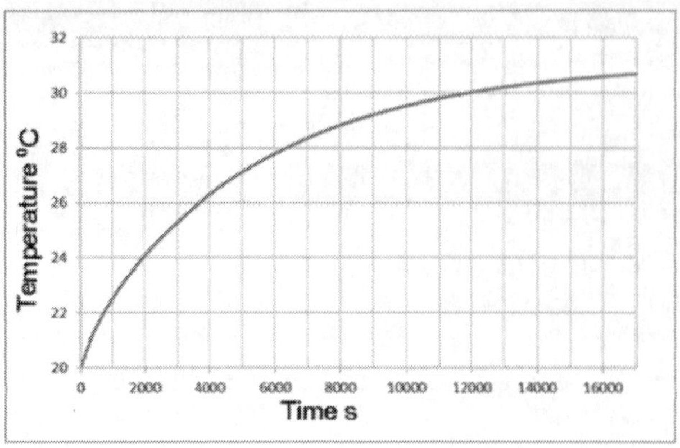

Figure 8. Graph of temperature change during the time of reaching the steady-state thermal mode.

The effect of various operational parameters on the thermal state of the soundproof coupling is determined.

Effect of the rotation speed

An analysis of the thermal coupling state for different rotation speeds is completed and the dependence of the maximum temperature inside the rubber of the coupling on the rotation speed is developed. The graph of this dependence is given in Fig. 9.

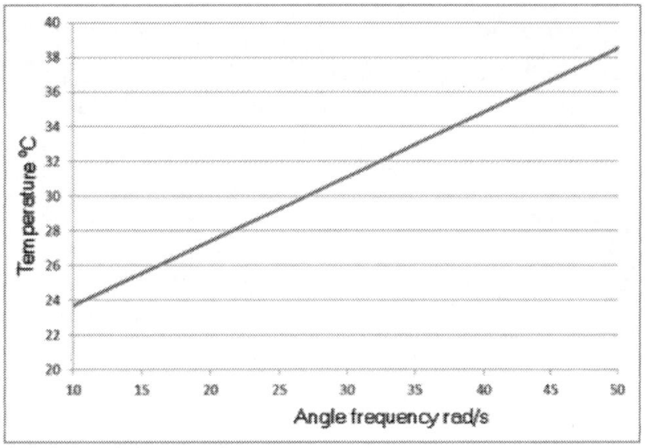

Figure 9. Change in the steady-state temperature depending on the rotation speed

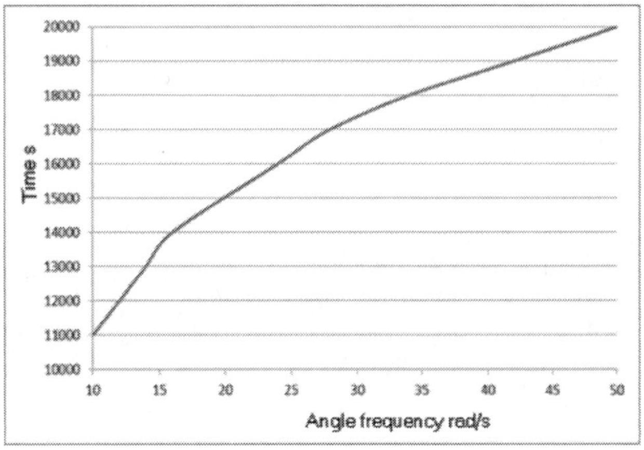

Figure 10. Time to reach steady state mode

The time change in reaching the steady-state thermal mode of the coupling operation is shown in Fig. 10.

Impact of changes in external temperature.

The thermal coupling state is defined at various ambient temperatures: from 20°C till 30°C. To do this, calculations under various environmental conditions are performed. According to the results, graphs of coupling self-heating temperature dependences are plotted (Fig. 11) and time to reach steady-state thermal mode (Fig. 12) on the ambient temperature.

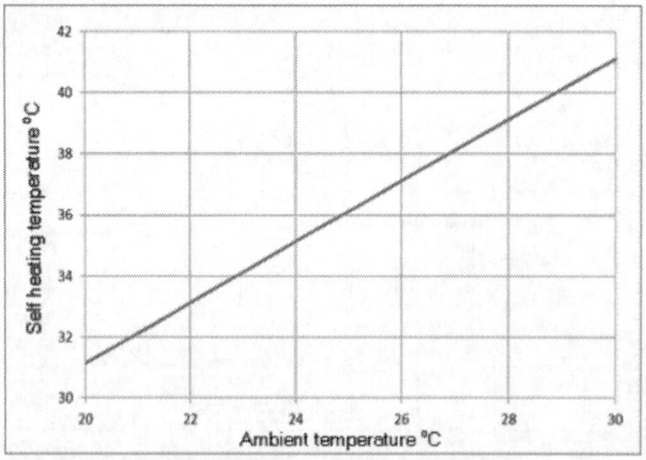

Figure 11. Effect of ambient temperature on the self-heating temperature of the coupling

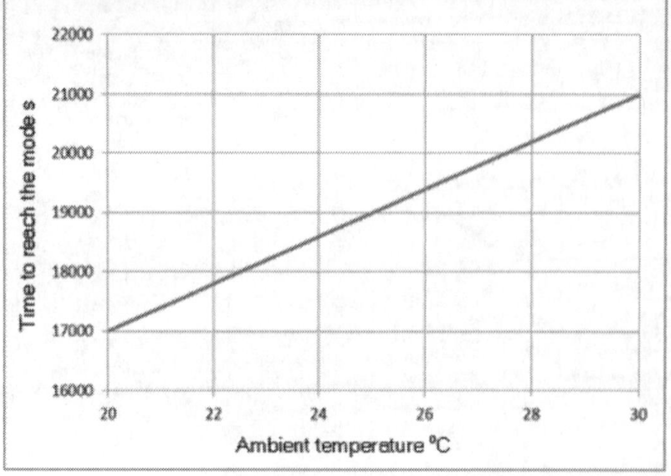

Figure 12. Effect of ambient temperature to the steady operation mode

5. Results and discussions

The studies proved that the rubber coupling body during operation is heated to temperature of 31.1°C. The most heated is central (inner) part of rubber and the temperature field is shifted to the inside part of the coupling as there is no convective heat transfer with environment from the inner side.

The picture of soundproof coupling thermal state in the steady state mode is given in Fig. 7 in a color chart form. The dependence of changes in temperature in the most heated point of the rubber coupling in the process of reaching steady state thermal mode is given in Fig. 8. The Figure shows that the soundproof coupling reaches the steady thermal mode for 17000 seconds, and it is equal to 4.7 hours of operation.

6. Conclusion

As a result of the carried out researches the thermal calculation of an elastic soundproof coupling made of rubber was made. Based on the results of the calculations, the following results were obtained:

Fields of temperature distribution over the coupling.

Time to reach the steady-state thermal mode of operation.

Dependences of maximum temperature and time of reaching to a mode on coupling rotation speed and on ambient temperature.

The obtained results allow drawing a conclusion that application of modern finite element analysis software allows to speed up the solution of the problems with a sufficient degree of accuracy.

References

[1] Mihailov Y K and Ivanov B S 1987 *Couplings with nonmetallic elastic elements: theory and calculation* (Leningrad: Mechanical Engineering) p 145
[2] Starastenko I N 2007 Calculation of heat in the pre-strained rubber cylinder with harmonic loading *KubGAU/ Scientific journal 27* p 3
[3] Goncharov P S et al 2012 *NX Advanced Simulation. Engineering analysis* (Moscow: DMK Press) p 504
[4] Mancuso Jon R 1999 *Couplings and joints. Design, selection, and application* (NewYork: Marcel Dekker) p 581
[5] Francis Alex B 2014 *Experimental, Numerical and Analytical Characterization of Torsional Disk Coupling Systems (Theses and Dissertations)* p 625
[6] Reddy J N 2004 *An Introduction to Nonlinear Finite Element Analysis* (USA: Oxford University Press)
[7] Logan D L 2012 *A First Course in the Finite Element Method, 5th ed* (USA: Cengage Learning)
[8] Tsyss V G, Strokov I M and Sergaeva M Y 2015 The issues of life extension of seismic isolation system of circular tanks for storage of liquefied petroleum gasws *Procedia Engineering* vol 113 pp 395–401
[9] Tsyss V G, Strokov I M and Sergaeva M Y 2015 Dampener resource of seismic isolation absorber system of circular tanks for liquid hydrocarbons storage *Procedia Enginering* vol 113 pp 402–407.
[10] Burkardt J, Du Q, Gunzburger M and Lee H.-C 2003 *Reduced order modeling of complex systems* (NA03 Dundee) pp 29–38

IOP Conf. Series: Journal of Physics: Conf. Series **944** (2017) 012121 doi:10.1088/1742-6596/944/1/012121

Recognition of incomplete sequences using Fisher scores and hidden Markov models

V E Uvarov, A A Popov and T A Gultyaeva

Faculty of Applied Mathematics and Computer Science, Novosibirsk State Technical University, Novosibirsk, Russia

Abstract. We propose a method for recognition of incomplete sequences which makes use of feature vectors built from fisher scores for log likelihood function of hidden Markov model. We use support vector machines method to classify the feature vectors. The proposed method was compared to a previously developed method for incomplete sequence recognition based on marginalization of missing observations which can be considered as a state of the art method. The proposed method proved to be more effective than the SOTA method in situations when the percent of missing observations in training and testing sequences is high (more than 20% in our experiment). Thus, we suggest using the proposed method in situations when big percent of data is missing but the recognition still must be done.

1. Introduction

Hidden Markov model (HMM) conception was presented in 1970-s [1]. However, problems which concern using HMMs in case of incomplete data remain are poorly investigated. These problems are quite relevant since in complex systems, e.g. when receiving signals from spacecrafts or aircrafts, one has to deal with datastreams of various sources in noisy environments when there is a high possibility of data loss or corruption. In this paper, we deal with the problem of missing observations in sequences. From now on we will refer to such sequences as incomplete. We consider a case when such missing observations are not generated by random process itself but rather occur randomly in sequences because of some external interference.

2. Hidden Markov model

2.1. Structure of hidden Markov model

Hidden Markov model (HMM) describes a random process which appears to be in one of the N hidden states at each time $t \in \{1, ..., T\}$ and transits to another or to the same state according to some transition probabilities. The states are hidden from the observer however they can be inferred from the observed sequences. In this paper, we consider HMM with continuous multivariate observation density, namely a mixture of multivariate normal distributions (often such HMM is called Gaussian HMM or GHMM).

We shall denote a hidden state of GHMM at time t as q_t, observation at time t as $\boldsymbol{o}_t$ and observation without specific time as $\boldsymbol{o}$. GHMM can be specified by initial state distribution $\Pi = \left\{\pi_i = p\left(q_1 = s_i\right),\ i = \overline{1,N}\right\}$, transition probabilities matrix $A = \left\{a_{ij} = p\left(q_{t+1} = s_j \mid q_t = s_i\right),\ i,j = \overline{1,N}\right\}$ and conditional multivariate distributions $B = \left\{b_i\left(\boldsymbol{o}\right) = f\left(\boldsymbol{o} \mid q = s_i\right), i = \overline{1,N}, \boldsymbol{o} \in R^Z\right\}$. Here conditional multivariate distributions are mixtures of multivariate normal distributions $b_i(\boldsymbol{o}) = \sum_{m=1}^{M} \tau_{im} g(\boldsymbol{o}; \mu_{im}, \Sigma_{im}),\ i = \overline{1,N}, \boldsymbol{o} \in R^Z$, where M - number of mixture components for each hidden state, $\tau_{im} \geq 0$ – weight of m-th mixture component in i-th hidden state ($\sum_{m=1}^{M} \tau_{im} = 1,\ i = \overline{1,N}$), μ_{im} – mean of normal distribution from m-th component of i-th hidden state, Σ_{im} – covariance matrix of normal distribution from m-th component of i-th hidden state and $g(\boldsymbol{o};\mu_{im},\Sigma_{im}), \boldsymbol{o}\in R^Z$ – multivariate normal probability density function, i.e. $g(\boldsymbol{o};\mu_{im},\Sigma_{im}) = \frac{1}{\sqrt{(2\pi)^Z|\Sigma_{im}|}} e^{-0.5(\boldsymbol{o}-\mu_{im})^T\Sigma_{im}^{-1}(\boldsymbol{o}-\mu_{im})}, \boldsymbol{o}\in R^Z$. Thus, some specific GHMM can be specified by a set of parameters $\lambda=\{\Pi,A,B\}$ [2].

2.2. Recognition of complete sequences using HMM

The final aim of recognition of a sequence O using HMMs is to refer this sequence to one of the classes which correspond to random processes with indexes $\overline{1,D}$ described by HMMs $\lambda_1,...,\lambda_D$. Usually a maximum likelihood (ML) criterion is used for choosing the process which generated the sequence with the highest probability. In that case sequence O is referred to a class r^* for which the loglikelihood function is maximal, i.e. $r^*=\arg\max_{r\in 1,...D}\left(\ln p(O|\lambda_r)\right)$.

To calculate the loglikelihood function of sequence O being generated by random process corresponding to model λ one usually uses forward-backward algorithm [3]. To calculate $\ln p(O|\lambda)$ one needs only the forward part of the algorithm so we will describe only it. When working with probabilities (values which usually are very close to zero) it is highly recommended to use a scaling procedure to prevent underflow. Effective scaling method is described in [2].

The scaled forward part of the forward-backward algorithm computes scaled forward variables $p(\{\boldsymbol{o}_1,\boldsymbol{o}_2,...,\boldsymbol{o}_t\},q_t=s_i|\lambda)$, $t=\overline{1,T}$, $i=\overline{1,N}$, i.e. probabilities of sequence $\{\boldsymbol{o}_1,\boldsymbol{o}_2,...,\boldsymbol{o}_t\}$ being generated by random process described by model λ and that this process was in state s_i at time t. Algorithm that computes scaled forward variables and loglikelihood function:

1) initialization:

$$\tilde{\alpha}_1(i)=\pi_i b_i(\boldsymbol{o}_1), i=\overline{1,N}; \tag{1}$$

2) induction:

$$\tilde{\alpha}_{t+1}(i)=b_i(\boldsymbol{o}_{t+1})\left[\sum_{j=1}^{N}\alpha_t'(j)a_{ji}\right],\ i=\overline{1,N},\ t=\overline{1,T-1}, \tag{2}$$

where $\alpha_t'(j)=\frac{\tilde{\alpha}_t(j)}{\sum_{n=1}^{N}\tilde{\alpha}_t(n)}$, $j=\overline{1,N}$, $t=\overline{1,T-1}$. (3)

We introduce scaling coefficients:

$$c_t=\left(\sum_{i=1}^{N}\tilde{\alpha}_t(i)\right)^{-1},\ t=\overline{1,T}, \tag{4}$$

hence:

$$\alpha_t'(i)=c_t\tilde{\alpha}_t(i),\qquad i=\overline{1,N},\quad t=\overline{1,T-1},$$

$$\alpha_t'(i)=\left(\prod_{\tau=1}^{t}c_\tau\right)\alpha_t(i),\quad i=\overline{1,N},\ t=\overline{1,T-1}.$$

Thus, loglikelihood function for a sequence O and HMM λ can be computed as follows:

$$\ln\left[p(O|\lambda)\right]=-\sum_{t=1}^{T}\ln c_t. \tag{5}$$

3. Recognition of incomplete sequences using HMM

Prior to recognition of incomplete sequences using HMMs it is necessary to estimate the parameters of HMMs corresponding to random processes, i.e. to train HMMs. However, in this article we do not consider HMM training on incomplete sequences. This problem was covered in our previous works [4,

5, 6] where marginalization of missing observations approach was applied to modify the Baum-Welch training algorithm. We shall define incomplete sequence as a sequence O which contains some undefined (missing) observations. We denote missing observation as $\varnothing$. Thus, $O=\{\boldsymbol{o}_t \in R^*, t=\overline{1,T}\}$, $R^*=R^Z \cup \{\varnothing\}$.

3.1. Recognition of incomplete sequences using marginalization of missing observations

This approach is based on maximum likelihood criterion. It is obvious that the probabilities $b_i(\boldsymbol{o}_t)$, $i=\overline{1,N}$, $t=\overline{1,T}$ in formulas (1)-(5) which are used for calculation of loglikelihood function cannot be evaluated if $\boldsymbol{o}_t=\varnothing$ since the actual value of $\boldsymbol{o}_t$ is unknown. To use these formulas for incomplete sequences one must define $b_i(\varnothing)$, $i=\overline{1,N}$ to some reasonable values.

The idea of the proposed approach is to consider that any value from R^Z could have been in place of a missing observation before the data corruption occurred [7]. Keeping this in mind we shall integrate $b_i(\varnothing)$, $i=\overline{1,N}$ over all possible values of $\varnothing$:

$$b_i(\varnothing)=\int_{R^*} b_i(\boldsymbol{x})d\boldsymbol{x}=1, \quad i=\overline{1,N}.$$

This result can be proven by the fact that only one observation $\boldsymbol{x}$ can be generated by a random process at a time and that $b_i(\boldsymbol{x})$ is the conditional probability density of observation $\boldsymbol{x}$ generation when the random process was at hidden state s_i, $i=\overline{1,N}$. Using the same idea, we shall define probability density of multivariate normal distribution from the mixtures for missing observation [7]:

$g(\varnothing,\mu_{im},\Sigma_{im})=\int_{R^*} g(\boldsymbol{x},\mu_{im},\Sigma_{im})d\boldsymbol{x}=1$, $i=\overline{1,N}$, $m=\overline{1,M}$. Now probabilities $b_i(\boldsymbol{o}_t)$, $i=\overline{1,N}$, $t=\overline{1,T}$ are defined for all the values of $\boldsymbol{o}_t \in R^*$ hence formulas (1)-(5) can be extended for the case of incomplete sequences. The modified forward variables calculation algorithm (scaled) is presented below:

1) initialization:

$$\tilde{\alpha}_1(i)=\begin{cases}\pi_i, & \boldsymbol{o}_1=\varnothing \\ \pi_i b_i(\boldsymbol{o}_1), & \text{otherwise}\end{cases}, \quad i=\overline{1,N};$$

2) induction:

$$\tilde{\alpha}_{t+1}(i)=\begin{cases}\sum_{j=1}^{N}\alpha'_t(j)a_{ji}, & \boldsymbol{o}_{t+1}=\varnothing \\ b_i(\boldsymbol{o}_{t+1})\left[\sum_{j=1}^{N}\alpha'_t(j)a_{ji}\right], & \text{otherwise}\end{cases}, \quad \begin{matrix} i=\overline{1,N}, \\ t=\overline{1,T-1}\end{matrix},$$

where $\alpha'_t(j)=\dfrac{\tilde{\alpha}_t(j)}{\sum_{n=1}^{N}\tilde{\alpha}_t(n)}$, $j=\overline{1,N}$, $t=\overline{1,T-1}$. The scaling coefficients and loglikelihood function are calculated by the original formulas (4) and (5).

We shall name the used approach as a "marginalization of missing observations" since here we marginalize out the random variable $\boldsymbol{x}$ from distribution $b_i(\boldsymbol{x})$, $i=\overline{1,N}$ for missing observations. It is easy to show that now we can perform incomplete sequence recognition using ML criterion because all the necessary formulas were extended to the case of missing observations. In our previous studies the marginalization approach outperformed all other strategies of dealing with missing values [5, 6].

3.2. Recognition of incomplete sequences using Fisher scores

The recognition of sequences using HMMs can be done not only with the maximum likelihood criterion. A method for recognition of complete sequences using Fisher scores and HMMs was previously investigated in our studies. This method showed an advantage over the ML criterion when HMMs of competing random processes were similar and when training and test sequences which were exposed to various noise conditions [8]. However, the case of completely missing observations was not included in that study. Since the missing observations can be interpreted as an effect of some noise conditions, it is advisable to investigate the applicability of this method to the recognition of incomplete sequences using HMMs which were trained on incomplete sequences.

For the sake of simplicity, we will consider the case of two-class classification with this method. Each sequence is described with a feature vector which consists of fisher scores for a sequence and each of the competing HMMs. Let $\{O^1, O^2, ..., O^K\}$ be a set of training sequences and λ_1, λ_2 - two HMMs corresponding to two competing random processes. For each training sequence O from $\{O^1, O^2, ..., O^K\}$ a feature vector $\begin{pmatrix} \frac{\partial \ln p(O|\lambda_1)}{\partial \eta}\Big|_{\lambda_1} \\ \frac{\partial \ln p(O|\lambda_2)}{\partial \eta}\Big|_{\lambda_2} \end{pmatrix}$ is built. The transposed versions of these vectors are stacked together into a training matrix X where columns correspond to features (Fisher scores) and rows – to sequences. In addition, an answers vector is formed $Y = \{y_1, ..., y_K\}$, where $y_k \in \{1,2\}$ is an index of HMM which corresponds to a random process that generated O^k, $k = \overline{1,K}$. In the next step a support vector machines (SVM) classifier is trained using training matrix X and answers vector Y. To recognize a new sequence, a similar feature vector should be built and classified using the trained support vector machines classifier [9]. The described two-class problem can be easily generalized to a multiclass problem using "one-vs-all" or "one-vs-one" strategy widely used for binary classifiers.

Below we present formulas for calculation of Fisher scores for HMM [8]. Based on formula (5) we get:

$$\frac{\partial \ln p(O|\lambda)}{\partial \eta} = -\sum_{k=1}^{K}\left(\sum_{t=1}^{T}\frac{1}{c_t^k}\frac{\partial c_t^k}{\partial \eta}\right). \tag{6}$$

To calculate the derivative of scaling coefficient with respect to some HMM parameter η one can use the following formula:

$$\frac{\partial c_t}{\partial \eta} = -c_t^2 \sum_{i=1}^{N}\frac{\partial \tilde{\alpha}_t(i)}{\partial \eta}, t = \overline{1,T}. \tag{7}$$

To calculate $\frac{\partial \tilde{\alpha}_t(i)}{\partial \eta}$, $i = \overline{1,N}$ we shall differentiate step-by-step the algorithm for calculation of scaled forward variables:

1 step:

$$\frac{\partial \tilde{\alpha}_1(i)}{\partial \eta} = \frac{\partial \alpha_1(i)}{\partial \eta}, \quad i = \overline{1,N}; \tag{8}$$

2 step:

$$\begin{aligned}\frac{\partial \tilde{\alpha}_t(i)}{\partial \eta} = &\left[\sum_{j=1}^{N}\left(\frac{\partial \alpha'_{t-1}(j)}{\partial \eta}a_{ji} + \alpha'_{t-1}(j)\frac{\partial a_{ji}}{\partial \eta}\right)\right]b_i(t) + \\ &+\sum_{j=1}^{N}\left(\alpha'_{t-1}(j)a_{ji}\right)\frac{\partial b_i(t)}{\partial \eta}\end{aligned}; \tag{9}$$

where $\frac{\partial \alpha'_{t-1}(j)}{\partial \eta} = \frac{\partial c_{t-1}}{\partial \eta}\tilde{\alpha}_{t-1}(j) + \frac{\partial \tilde{\alpha}_{t-1}(j)}{\partial \eta}c_{t-1}$, $i = \overline{1,N}$, $t = \overline{2,T}$.

IOP Conf. Series: Journal of Physics: Conf. Series **944** (2017) 012121 doi:10.1088/1742-6596/944/1/012121

Thus, to calculate $\frac{\partial \tilde{\alpha}_t(i)}{\partial \eta}, i=\overline{1,N}$ we first need to calculate derivatives $\frac{\partial \tilde{\alpha}_1(i)}{\partial \eta}, \frac{\partial b_i(t)}{\partial \eta}, \frac{\partial a_{ij}}{\partial \eta}$, $i,j=\overline{1,N}, t=\overline{1,T}$. To calculate derivative with respect to covariance matrix element in case of non-diagonal covariance matrix one must first differentiate the elements of inverse covariance matrix, hence here we will consider only diagonal covariance matrixes Σ_{im}, $i=\overline{1,N}$, $m=\overline{1,M}$.

Below we present formulas to calculate the derivatives $\frac{\partial \tilde{\alpha}_1(i)}{\partial \eta}, \frac{\partial b_i(t)}{\partial \eta}, \frac{\partial a_{ij}}{\partial \eta}$ for all the possible kinds of η parameter (which can be any of the following HMM parameters: $\pi_i, a_{ij}, \tau_{im}, \mu_{im}^z, \Sigma_{im}^{zz}$, $i,j=\overline{1,N}, m=\overline{1,M}, z=\overline{1,Z}$).

$$\frac{\partial \alpha_1(i)}{\partial \pi_j} = \begin{cases} b_i(1), & i=j \\ 0, & i \neq j \end{cases}, \quad i,j=\overline{1,N}, \tag{10}$$

$$\frac{\partial b_i(t)}{\partial \pi_j} = 0, \quad i,j=\overline{1,N}, \quad t=\overline{1,T}, \tag{11}$$

$$\frac{\partial \alpha_1(i)}{\partial a_{i_1 j_1}} = 0, \quad i,i_1,j_1=\overline{1,N}, \tag{12}$$

$$\frac{\partial b_i(t)}{\partial a_{i_1 j_1}} = 0, \quad i,i_1,j_1=\overline{1,N}, \ t=\overline{1,T}, \tag{13}$$

$$\frac{\partial a_{ij}}{\partial x} = \begin{cases} 1, & x=a_{ij} \\ 0, & x \neq a_{ij} \end{cases}, \quad i,j=\overline{1,N}, \tag{14}$$

$$\frac{\partial b_i(t)}{\partial \tau_{i_1 m}} = \begin{cases} g(o_t;\mu_{im},\Sigma_{im}), & i=i_1 \\ 0, & i \neq i_1 \end{cases}, \quad \begin{matrix} i,i_1=\overline{1,N} \\ t=\overline{1,T} \\ m=\overline{1,M} \end{matrix}, \tag{15}$$

$$\frac{\partial \alpha_1(i)}{\partial \tau_{i_1 m}} = \begin{cases} \pi_i \frac{\partial b_i(1)}{\partial \tau_{i_1 m}}, & i=i_1 \\ 0, & i \neq i_1 \end{cases}, \quad i,i_1=\overline{1,N}, \ m=\overline{1,M}, \tag{16}$$

$$\frac{\partial b_i(t)}{\partial \mu_{i_1 m}^z} = \begin{cases} 0.5\tau_{im} g(o_t;\mu_{im},\Sigma_{im}) \frac{o_t^z - \mu_{im}^z}{\Sigma_{im}^{zz}}, & i=i_1 \\ 0, & i \neq i_1 \end{cases}, \tag{17}$$
$$i,i_1=\overline{1,N}, \ t=\overline{1,T}, \ m=\overline{1,M}, \ z=\overline{1,Z}$$

$$\frac{\partial \alpha_1(i)}{\partial \mu_{i_1 m}^z} = \begin{cases} \pi_i \frac{\partial b_i(1)}{\partial \mu_{i_1 m}^z}, & i=i_1 \\ 0, & i \neq i_1 \end{cases}, \quad \begin{matrix} i,i_1=\overline{1,N} \\ m=\overline{1,M} \\ z=\overline{1,Z} \end{matrix}, \tag{18}$$

$$\frac{\partial b_i(t)}{\partial \Sigma_{i_1 m}^{zz}} = \begin{cases} 0.5\tau_{im} g(o_t;\mu_{im},\Sigma_{im}) \left(\left(\frac{o_t^z - \mu_{im}^z}{\Sigma_{im}^{zz}} \right)^2 - \frac{1}{|\Sigma_{im}|} \right), & i=i_1 \\ 0, & i \neq i_1 \end{cases}, \tag{19}$$
$$i,i_1=\overline{1,N}, \ t=\overline{1,T}, \ m=\overline{1,M}, \ z=\overline{1,Z}$$

$$\frac{\partial \alpha_1(i)}{\partial \Sigma_{i_1 m}^{zz}} = \begin{cases} \pi_i \dfrac{\partial b_i(1)}{\partial \Sigma_{im}^{zz}}, i = i_1 \\ 0, \qquad i \neq i_1 \end{cases}, i, i_1 = \overline{1,N}, m = \overline{1,M}, \quad z = \overline{1,Z}. \tag{20}$$

Formulas (6)-(20) can be extended to the case of incomplete sequences using the marginalization of missing observations approach which was described in the previous subsection. Thus in (6)-(20) we consider $b_i(\varnothing) = 1, \quad i = \overline{1,N}$ and $g(\varnothing, \mu_{im}, \Sigma_{im}) = 1, \quad i = \overline{1,N}, \quad m = \overline{1,M}$. In addition, we should make additional changes to formula (17):

$$\frac{\partial b_i(t)}{\partial \mu_{i_1 m}^z} = \begin{cases} 0.5\tau_{im} g(o_t; \mu_{im}, \Sigma_{im}) \dfrac{o_t^z - \mu_{im}^z}{\Sigma_{im}^{zz}}, & i = i_1 \text{ and } o_t \neq \varnothing \\ 0, & \text{otherwise} \end{cases},$$

$$i, i_1 = \overline{1,N}, \ t = \overline{1,T}, \ m = \overline{1,M}, \quad z = \overline{1,Z},$$

and formula (19):

$$\frac{\partial b_i(t)}{\partial \Sigma_{i_1 m}^{zz}} = \begin{cases} 0.5\tau_{im} g(o_t; \mu_{im}, \Sigma_{im}) \left(\left(\dfrac{o_t^z - \mu_{im}^z}{\Sigma_{im}^{zz}} \right)^2 - \dfrac{1}{|\Sigma_{im}|} \right), & \begin{matrix} i = i_1 \text{ and} \\ o_t \neq \varnothing \end{matrix} \\ 0, & \text{otherwise} \end{cases},$$

$$i, i_1 = \overline{1,N}, \ t = \overline{1,T}, \ m = \overline{1,M}, \quad z = \overline{1,Z}.$$

4. Evaluation

In this section, we compare methods for recognition of incomplete sequences: the proposed one that uses Fisher scores and HMM and state-of-the-art method based on maximum likelihood criterion.

For the experiment we used two HMMs λ_1 and λ_2 with the following parameters: the number of hidden states $N = 3$, number of mixture components $M = 3$, dimensionality of observations $Z = 2$. Initial distribution vector $\Pi = [1, 0, 0]$, transition probabilities matrix $A = \begin{pmatrix} 0.1 + \Delta A & 0.7 - \Delta A & 0.2 \\ 0.2 & 0.2 + \Delta A & 0.6 - \Delta A \\ 0.8 - \Delta A & 0.1 & 0.1 + \Delta A \end{pmatrix}$, mixture components weights $\left\{\tau_{im}, i = \overline{1,N}, m = \overline{1,M}\right\} = \begin{pmatrix} 0.3 + \Delta\tau & 0.4 - \Delta\tau & 0.3 \\ 0.3 & 0.4 + \Delta\tau & 0.3 - \Delta\tau \\ 0.3 - \Delta\tau & 0.4 & 0.3 + \Delta\tau \end{pmatrix}$ (here rows correspond to hidden states and columns – to mixture components), mean vectors of mixture components $\left\{\mu_{im}, i = \overline{1,N}, m = \overline{1,M}\right\} = \begin{pmatrix} (0 - \Delta\mu \quad 0 + \Delta\mu)^T & (1 - \Delta\mu \quad 1 + \Delta\mu)^T & (2 - \Delta\mu \quad 2 + \Delta\mu)^T \\ (3 - \Delta\mu \quad 3 + \Delta\mu)^T & (4 - \Delta\mu \quad 4 + \Delta\mu)^T & (5 - \Delta\mu \quad 5 + \Delta\mu)^T \\ (6 - \Delta\mu \quad 6 + \Delta\mu)^T & (7 - \Delta\mu \quad 7 + \Delta\mu)^T & (8 - \Delta\mu \quad 8 + \Delta\mu)^T \end{pmatrix}$ (here rows correspond to hidden states and columns – to mixture components), all the covariance matrixes of mixture components $\left\{\Sigma_{im}, i = \overline{1,N}, m = \overline{1,M}\right\}$ were diagonal with the same value on diagonal: $0.1 + \Delta\sigma$. The first model had $\Delta A = 0$, $\Delta\tau = 0$, $\Delta\mu = 0$, $\Delta\sigma = 0$ and the second model had $\Delta A = 0.05$, $\Delta\tau = 0.05$, $\Delta\mu = 0.01$, $\Delta\sigma = 0.01$. Such choice of parameters makes the recognition task extremely difficult since the competing random processes corresponding to the two HMMs have very similar properties and observations generated by them are difficult to distinguish. Using each of the λ_1 and λ_2 HMMs we generated $K = 100$ training and test sequences of length $T = 100$ each of which contained G missing observations (G was changed in range from 0 to 90 during the experiment) in random places. Using incomplete training sequences, we estimated parameters of the models λ_1 and λ_2 using the modified Baum-Welch algorithm and marginalization approach [4, 5, 6]. In addition, we estimated parameters of support vector machines classifier with the approach described in subsection II.B and selected the best SVM hyperparameters using cross-validation on 4 folds on the training set. Then we used λ_1 and λ_2 to recognize the test incomplete sequences using the proposed method that uses Fisher scores and HMM (dashed line) and method based on maximum likelihood criterion (solid line). The recognition accuracy on test set is displayed on the y-axis and the percent of missing observations in training and test sequences is displayed on the x-axis in figure 1. The presented results are the average of 50 launches made with different initial seeds for random numbers generator.

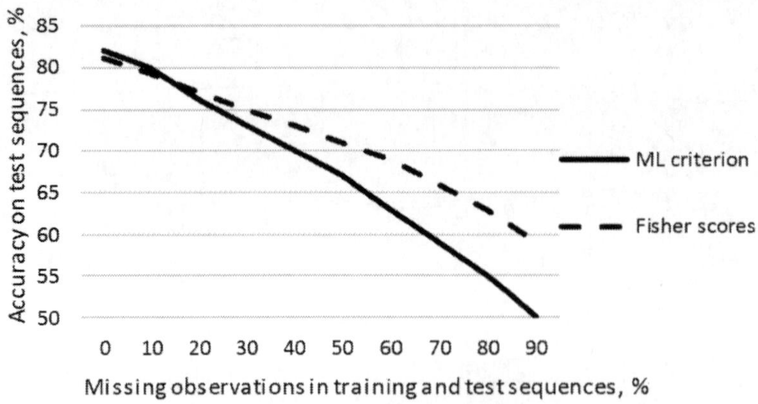

Figure 1. Recognition accuracy on the test set of the Fisher scores and ML criterion methods.

The proposed Fisher scores method starts to outperform the ML criterion method roughly after 20% of missing observations in training and test sequences. The advantage of the proposed method raises proportionally to the percent of missing observations: it is up to 10% more accurate when 90% of data is missing.

5. Conclusion

We introduced a novel method to recognize incomplete sequences which is based on Fisher kernels and hidden Markov models assisted by support vector machines classifier. The proposed method was compared to the state-of-the-art method of incomplete sequence recognition which utilizes maximum likelihood criterion. The comparison showed that the proposed method provides more recognition accuracy than the SOTA method when the percent of missing data in sequences is large (more than 20% in our experiment) and when competing hidden Markov models' parameters are close. Thus, the proposed method is recommended to be used in situations when there are a lot of missing data but the recognition of incomplete sequences still must be done to some extent. However, in normal situation when HMMs are not very close and the percent of missing data is not high we advise against using the proposed method since it gives additional computational overhead comparing to the SOTA method.

References

[1] Baum L and Petrie T 1966 Statistical inference for probabilistic functions of finite state Markov chains *The Annals of Mathematical Statistics* **37** pp 1554-1563

[2] Rabiner L 1989 A Tutorial on Hidden Markov Models and Selected Applications in Speech Recognition *Proceedings of the IEEE* **77** pp 257-285

[3] Baum L and Egon J 1967 An inequality with applications to statistical estimation for probabilistic functions of a Markov process and to a model for ecology *Bulletin of the American Meteorological Society* **73** pp 360-363

[4] Popov A, Gultyaeva T and Uvarov V 2016 Training hidden Markov models on incomplete sequences *Proceedings of Russian scientific conference Information processing and mathematical modelling* pp 125-139
[5] Popov A, Gultyaeva T and Uvarov V 2016 A Comparison of Some Methods for Training Hidden Markov Models on Sequences with Missing Observations *Proceedings of 11th International Forum on Strategic Technology IFOST-2016* **1** pp 431-435
[6] Popov A, Gultyaeva T and Uvarov V 2016 Training hidden Markov models on sequences with missing observations *Proceedings of 13th International Conference on Actual Problems of Electronic Instrument Engineering (APEIE 2016)* **1** pp 317-320
[7] Cooke M, Green P, Josifovski L and Vizinh A 2001 Robust automatic speech recognition with missing and unreliable acoustic data *Speech Communication* **34** pp 267-285
[8] Gultyaeva T, Popov A, Kokoreva V and Uvarov V 2015 Classification of observation sequences described by Hidden Markov Models *Proceedings of the International Workshop Applied Methods of Statistical Analysis Nonparametric approach AMSA-2015* pp 136–143
[9] Boser B, Guyon I and Vapnik V 1992 A training algorithm for optimal margin classifiers *Proceedings of the fifth annual workshop on Computational learning theory* p 144

IOP Conf. Series: Journal of Physics: Conf. Series **944** (2017) 012122 doi:10.1088/1742-6596/944/1/012122

Portable data collection terminal in the automated power consumption measurement system

S V Vologdin[1], I D Shushkov[1] and E K Bysygin[2]

[1] Kalashnikov Izhevsk State Technical University, 7, Studencheskaya str., Izhevsk, 426069, Russia
[2] Udmurtenergo Branch of IDGC of Center and Volga Region, Electric Energy Billing Department Deputy, 30, Sovetskaja str., Izhevsk, 426004, Russia

e-mail: vologdin_sv@mal.ru

Abstract. Aim of efficiency increasing, automation process of electric energy data collection and processing is very important at present time. High cost of classic electric energy billing systems prevent from its mass application. Udmurtenergo Branch of IDGC of Center and Volga Region developed electronic automated system called "Mobile Energy Billing" based on data collection terminals. System joins electronic components based on service-oriented architecture, WCF services. At present time all parts of Udmurtenergo Branch electric network are connected to "Mobile Energy Billing" project. System capabilities are expanded due to flexible architecture.

1. Introduction

Control over electric energy consumption is very important at present time. Application of different Automated intellectual measuring systems of automated informative electric power accounting system solves a number of tasks, existing in case of manual data collection of electric energy consumption.

Automation of data collection eliminates "human factor" from billing process, allowing getting exact information on electric energy consumption. Automation of data collection and processing reduces operation costs [1].

Prior to using Automated informative electric power accounting system it is necessary to complete a number of tasks: installation of measuring device bundled with PLC-modem, switches and GSM-gates installation, setting software to collect data and process it. Completion of all tasks to apply Automated informative electric power accounting system requires high financial expenses. High cost of such systems is a great issue for its application [2].

Application of Data Collection Terminals (DCT) used to automate the process of data collection for electric energy billing is alternative to Automated informative electric power accounting system. Development and application of electronic information analytic system of electric energy billing is performed by Udmurtenergo Branch of IDGC of Center and Volga Region (referred to as Udmurtenergo) in cooperation with Kalashnikov Izhevsk State Technical University.

The purpose of development of an information system is to Optimize the business processes of the formation of productive electricity supply by reducing operating costs labor production personnel, increasing data quality and speed of their processing, with the use of geo information systems and DCT.

2. Task description

Mobile system for electric energy billing based on Data Collection Terminals solves a number of tasks:

- Paper workflow reduction.
- Elimination of human Inspector errors of data collection and its registration in database.
- Reduction of data collection time.
- Data collection and transfer in real time.
- Data receiving from DCT.
- DCT workflow control.
- Control over data access from remote location.
- Estimation of staff personal productivity based on operation data of the system.

"Mobile Energy Billing" is the software installed in DCT should perform:

- Collection and storage of data from measurement devices.
- Data transfer to Regional Information Center (RIC) based on GSM Technologies.
- Location of Measurement device based on GPS.
- Service control over energy consumption.
- Mobile reports printing based on Wi-Fi technology.

Information system of Regional Information Center should comply with:

- DCT operation.
- Analytic processing of data transferred from DCT.
- Monitoring of Electric energy consumption based on GIS module (geo information system).
- Management solution development to increase efficiency of electric energy consumption.

Equipment and technologies of automated informative electric power accounting system is shown on Figure 1.

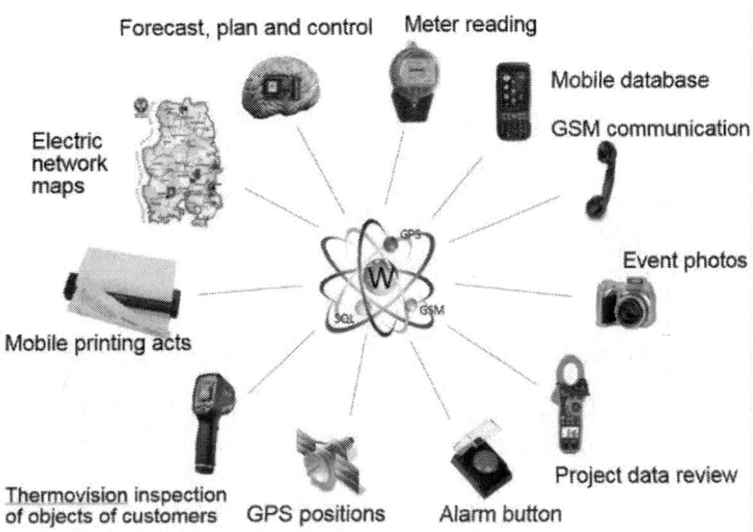

Figure 1. Equipment and technologies of automated informative electric power accounting system.

3. Theory

Prior to the introduction of DCT, the business process is a "tool check" began with the fact that the controller from the group of electricity metering produces the search for the necessary user and relevant information about metering device (date of verification, the seals, the coefficients of the transformation, the names and types of instruments) in the database of the branch, and then prints the act of conducting instrumental inspections. Further advances to the customer and carries out the relevant work, which involves the process.

For data taken with transformers and metering device manually calculated and taken into account the consumed power. Calculated instrument error. If the error is within acceptable limits, then the conclusion of the suitability of the metering device.

All records are maintained manually, which implies that mistakes and errors (the human factor). The process takes about 20-30 minutes per user, if there are errors or unexplained results (too large values), then this time can take about 50-60 minutes. At the end of the day all statements are logged and dealt technician RES. Replication to a Central database every 10 days. Further sending data in OAO Energosbyt Plus occurs 2-3 times per month.

With the exception of the intermediate step — the registration act — a test reference on the performed inspection/replacement/installation PU, decreases the probability of an erroneous entry operator data in the database. This increases the speed of all operations

Also DCT is a basic photo/video camera for the implementation of fixation of readings, revealed the facts of non-contracted/non-contracted consumption of electric power; a GPS receiver to monitor the movements of the controllers. Based on the data obtained with the GPS receiver, you can schedule crawls of consumers for the efficient operation of the controllers.

After receipt of the information in the database, with GIS you can display all the information according to the data obtained from DCT and place them on the map. For each indication you can see (value, time of removal, the data consumer controller who directed the reading, etc.). The continued collection coordinates with a GPS receiver gives you the ability to overlay routes and closely monitor the work of the controller. All data remains in the database, which gives the opportunity to use the route history for any date and display on the map.

Data can be collected from any type of measurement device, so there is no additional expenses for high cost equipment. Mobile system concept is shown on Figure 2.

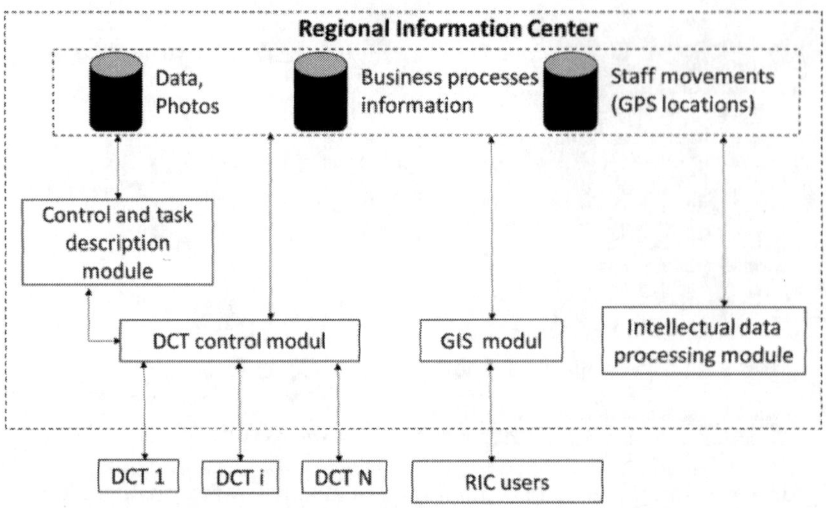

Figure 2. Mobile system concept.

DCT – high technology electronic device, mobile computer made in dust and moisture protected case (protection class IP54 and higher). Terminal includes:

- Built in barcode scanner.
- Wireless interfaces (Bluetooth, Wi-Fi).
- GPRS/GSM modules.
- GPS receiver.
- Base terminal software.

Data collection takes place on consumer's side. While first visit data collection department inspector finds consumer in DCT Database (referred to as DB) by input consumer data such as account number, address, measurement device code, barcode etc. Terminal DB contains information over the measurement device type, last inspection date and previous measurement data. Next, inspector checks all data, inputs measurement data for present date, sets unique barcode, puts barcode on measurement device case.

In case consumer is not found in DB or in case of non–authorized independent wire connection beyond the balance border (consumption without contract) takes place inspector should draw up a report. In case of electric energy theft inspector should draw up a non-registered consumption report using data from DB and mobile printer. Changes of DCT DB data are transferred to DB of a Branch through GPRS channel or through specially designed interface cable in the end of inspector working day. While next visit inspector reads barcode set on measurement device using barcode scanner of DCT first, then checks all data within DB and inputs changes if necessary.

First terminal reads measuring device barcode, then measurement device data are put into DCT. DCT estimates data are correct comparing it with previous data input. Data are transferred to Udmurtenergo RIC by GSM connection. Besides, terminal transfers to RIC its location data which is then used by GIS module (Geo Information system) to show it on the map.

Data send by terminal are received by DCT control module on the RIC side which commands and tasks are accepted by terminals. Here system and operational data are processed. Data processed are stored in database (DB). Further processing is executed by other modules of system, such as GIS-module, intellectual data processing modules (image recognition, prediction). Each module executes its own task.

Image recognition task is carried out by intellectual processing module. Facing part image of measuring device is the object to recognize which contains following data:

- Measuring device readings.
- Measuring device model (type).
- Factory number.
- Barcode.
- Seals of Electric Energy Supplier.

Image recognition allows to compare data put by inspector with recognition module results, thus verify data in integrated processing center [3].

Data collected can be shown on the map by GIS module. Value of electric energy consumption, date and time of readings, consumer data, inspector data can be restored for each reading. Frequent collection of GPS location data allows making routes and control inspector job in detail. All the data is stored in DB allowing to use routes history at any period of time and show it on the map.

Modules of system are connected directly to database. Access to the data stored in database is carried out by procedures and functions specially designed in each module. Over time, databases size increases, their architecture changes which require existing methods change of each module and decrease the efficiency of developers.

Realization of single way to operate with database server significantly simplifies developer task. This way includes system modules universal rules development allowing interaction with databases. These rules can be developed in additionally designed module based on services. Thus system has service-oriented architecture (SOA) [4].

Applying SOA gives following advantages to software developers:

- Elimination of the same software program code usage in different system modules.
- Applying WCF services (configuration allowing access to WCF service functions) for modules, developed in different platforms.
- Scalability increase of software developed by combining existing services in new solutions.
- Applying of functionality already been designed on the side of a service.

Software service development is based on WCF technology designed to create distributed information systems. WCF technology is the program framework being a part of the .NET platform. WCF is used for data exchange between applications. Configuration of technology allows creation of flexible operation between system modules [5].

Clients, system modules apply different WCF service functions by multiple endpoints created. Client sending message in definite format to endpoints is able to use service functions. Contract between client and service defines format of the message send by client. After message has been delivered to endpoint, service starts operation sending its result to client (Figure 3).

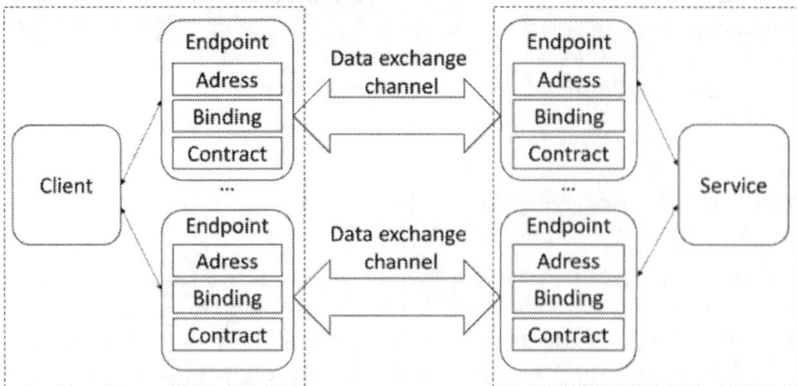

Figure 3. Client and WCF service interaction.

4. Results

Mobile Energy Billing Project has become a winner of two Russian contests [6]. At present time all 27 regions of Udmurtenergo brunch electric network are connected to Mobile Energy Billing Project. Developed Information system allows to process analyze and store data transferred from terminals with minimum investments [2].

Application of WCF services differentiates rights and responsibilities for system modules. Demand for system functionality increases over time and new services can be developed to solve a definite task. This application can help to increase system functionality with minimum time lost.

Implemented and planned tasks:

- Data Collection and transmission in real time (subject to availability of communication channels) is successfully implemented.
- The Daily formation of the productive supply of electric power was successfully implemented.
- Dynamic management of production process – in the process of implementation.
- Automated data validation is successfully implemented.
- Reducing the time of receipt of the data object to an operator from several days to several minutes – is successfully implemented.
- Geographic information the binding of objects of customers and network organization, data access control on the ground – has been successfully implemented.
- Installation of the rules execution of works with an automated control performance – in the process of implementation.

- Execution of pending queries and processes – planned implementation.
- Performance Evaluation of the personnel according to operational data of the system – is successfully implemented.
- A system of prompt notification of emergency situations – in the process of implementation.
- Energy Prediction using artificial neural networks – planned implementation.
- Establishment of a monitoring and management process being – implemented at the moment.

The expected results from the implementation of the program complex are:

- A significant reduction in operating costs for the readings of meters.
- Automation of process of removal of indications of conducting instrumental inspections and other actions with the electricity metering devices, as well as with the data consumer.
- Reducing the number of operator errors when introducing data.
- A significant decrease in the volume of paperwork, or complete failure of clearance of works on paper.
- Visualization of spatial data.
- Improving the reliability of the accounting system as a whole.
- Reduced influence of human factor on the functioning of the system.

5. Conclusions

Programs developers are confident that complete implementation of system allows to eliminate divisions with electric energy supplier. Objective and operation data transfer to database is the main functions of the project. In case of success startup of Mobile Energy Billing project it will be replicated in other regional branches of IDGC of Center and Volga Region.

References

[1] Kopylov S I, Lipa O A and Lipa D A 2016 Electric Power Losses are Prerequisites for Implementation of Automated Information and Measuring System of Commercial Electricity Metering (AIMS CEM) *Machinery and Equipment for Rural Area* **4** pp 29–33

[2] Vologdin S V, Bysygin E K, Shushkov I D, Bricyn K V and Ryabov P I 2016 Automated electricity metering in the domestic sector *Proc. Int. Conf. Fundamental and Applied scientific research: up-to-date aspects, achievements and innovations* (Penza: Nauka i Prosveshhenie) pp 84–88

[3] Vologdin S V, Bysygin E K, Shushkov I D, Bricyn K V and Ryabov P I 2016 The allocation of the readings of the energy metering method of viola-jones using the OPENCVSHARP library *Proc. Int. Conf. Science and innovation in XXI century: up-to-date aspects, achievements and ways of development* (Penza: Nauka i Prosveshhenie) pp 21–24

[4] Videnin S A, Kostuk A V and Vasiliev E V 2016 Service-oriented architecture in modern information systems *Modern scientific research and innovations* **7** pp 121–23

[5] Samoilova T A 2015 Development of service-oriented information systems technology analysis *Proc. Int. Conf. Modern information technologies and IT-education* (Moscow: Fund League Internet Media) pp 134–39

[6] Konstantinova O Two victories in Russian contests: «Udmurtenergo» presented innovation «Mobile Energy Billing». [Online]. Available: http://udmurt.media/news/biznes/11016/

IOP Conf. Series: Journal of Physics: Conf. Series **944** (2017) 012123 doi:10.1088/1742-6596/944/1/012123

Technical and economic assessment of design efficiency of information and measuring systems

V M Yurov[1], E N Eremin[2], Ya Zh Baisagov[1] and V V Arkhipov[1]

[1]Karaganda state industrial university, 30, Republic str., Temirtau 101400, Republic of Kazakhstan
[2]Omsk State Technical University, 11, Mira Ave., Omsk 644050, Russia

e-mail: weld_techn@mail.ru

Abstract. A thermodynamic approach to the analysis of information-measuring systems (IMS) is developed in the work. Expressions for efficiency of IMS are obtained. The connection between the amount of processor memory and the amount of incoming information and the accuracy of the IMS is obtained. It is shown that the probability of information loss in IMS decreases with the increase in the amount of information from the object. Using the analogy method, economic aspects of IMS design are considered. The innate ability of IMS and Moore's law are considered. The proposed approach and the resulting formulas will be useful in the design of new IMS.

1. Introduction

Information and measuring systems (IMS) have become an indispensable attribute of modern technological processes and industries, modern communication systems, transport and various communications, navigation systems and space stations and many other [1–3].

The rapid development of microelectronics, nanoelectronics and devices based on them has given rise to the urgent need to constantly update the IMS on a new element base and new physical principles that meet modern requirements of accuracy of measurement processes [4–6].

In this connection, problems arose in the design of complex IMS, requiring the use of modern mathematical methods, computer modeling and non-trivial circuitry solutions [7–9].

To the purely academic aspect of the above-mentioned problems, economic requirements are also being added: not too much cost of the design work of the IMS; Not too high cost of technology for their manufacture In this connection, there arises the need to evaluate the direction in the design of IIS both in terms of the physical principles underlying their functioning and in terms of their technical and economic feasibility [10].

The fundamentals of the thermodynamics of information processes were laid by L. Brillouin [11] and continued by P.P. Poplavsky [12]. In our paper [13], on which we shall rely on in the further presentation, IMS are considered from the point of view of nonequilibrium statistical thermodynamics. For small systems, nonequilibrium thermodynamics in the context of information theory was applied in [14].

2. Response function of IMS

The table 1 presents an analogy between thermodynamics and IMS parameters [13].

Every measurement process is connected with the interaction of certain objects (or field and objects) and every process of interaction is accompanied by dissipative processes leading to the loss of some information during the measurement.

Table 1. Thermodynamics and IMS

Thermodynamics	IMS	Thermodynamics	IMS
Free energy, work A	Memory capacity W		
Amount of substance ν	Number of sensors (communication channels) n	Production of entropy σ	Production of information σ
Entropy S	Amount of information I	Coefficient of efficiency η	Efficiency η
Temperature T(t)	Precision Δ	Internal energy U	Energy intensity E

The most important parameter of IMS (and for any system) is its efficiency η. In thermodynamics, it corresponds to the efficiency of a thermal machine:

$$\eta = 1 - T_c / T_h, \tag{1}$$

where T_h and T_c are the temperatures of the hot and cold sources respectively.
For IMS, equation (1) will look like this:

$$\eta = 1 - \Delta_{in} / \Delta_{out}, \tag{2}$$

where, according to the Table 1, Δ_{in} and Δ_{out} are input and output accuracies of IMS respectively. The first is determined by the sensitivity of the sensor, and the second is determined by the sensitivity of the sensor and the parameters of the unifying converter. It follows from the formula (2) that the efficiency of IMS is largely determined by its structure.
If, as the response function Φ [13], to take the efficiency of the IMS η, then we get:

$$\eta = \frac{kT}{C} \cdot \frac{A}{G^0} \cdot m, \tag{3}$$

where A is work (energy); T is the temperature;G^0 is the Gibbs potential; ν is the amount of the substance; k is the Boltzmann constant; C is a constant.
Using Table 1, we obtain the following expression for the effeciency of IMS:

$$\eta = C_1 \frac{nW\Delta}{W - I\Delta}. \tag{4}$$

Here C_1 is a constant. The limit value η is 1 and from formula (4) it follows:

$$W = \frac{I \cdot \Delta}{1 - C_1 n\Delta}. \tag{5}$$

The formula (5) defines a rule of a choice of the processor at designing IMS. It follows that the amount of processor memory is determined, basically, by the product of the amount of information coming from the object under investigation, and the accuracy of the IMS. The latter, as a rule, is inversely proportional to the signal-to-noise ratio and tends to the optimum value when the noise level decreases. Note that the correct choice of the processor determines to a greater extent the cost of the IMS being developed. The choice of a processor with a large memory is not always justified.
Let's consider one more example. The probability of dissipative processes in a thermodynamic system is in most cases determined by the Arrhenius law:

$$p = \nu \exp(-Q / kT), \tag{6}$$

IOP Conf. Series: Journal of Physics: Conf. Series **944** (2017) 012123 doi:10.1088/1742-6596/944/1/012123

where τ is the relaxation time; Q is the activation energy.
For IMS, τ is the response time of the system (IMS speed), and Q=A-W. Taking (5) into account, we get:

$$p = \frac{1}{\tau}\exp\left(-\frac{\alpha I}{1 - C_1 n\Delta}\right), \tag{7}$$

where α is the coefficient of dimension.
It follows from formula (7) that the loss of information in the IMS is less, the larger the measurement time τ and the greater the amount of information I from the object. It is necessary to make the following remark. As follows from (7), the deceleration of the transient process (i.e., with increasing τ), the probability of dissipative processes decreases. However, in practice, such a path is unacceptable and, conversely, modern and future IMS should have a high speed to transmit a large amount of information. This is achieved by moving from a scalar representation to a vector representation (digital IMS).
Asymptotically the best (in the sense of the product $\Delta S\cdot\tau$) is Shannon redundant coding, using a bit representation with additional (check) bits. Here, increasing the dimensionality of the vector space allows to detect and correct errors of some multiplicity, which reduces the requirements for the probability of distortion (and the signal-to-noise ratio) in single bit. At the same time, $\Delta S \sim \ln(1/\Delta)$ is attained with a simultaneous increase in time only by a factor of $\ln(1/\Delta)$ [12].
The probability of loss of information in the IMS decreases with increasing amount of information from the object (the formula (7)). Such increasing is possible due to the increase in the sensitivity of the sensor, the creation of a new type of sensors, new methods of measurement, the improvement of the systems for receiving and processing information. The key importance is the development of new types of sensors and receiving and processing systems based on nanotechnology information [4–6].

3. Economic aspects of IMS designing

Table 2 presents an analogy between IMS and economic indicators [13].

Table 2. IMS and microeconomics

IMS	Microeconomics	IMS	Microeconomics
Memory capacity W	Basic resource M	Production of information σ	Dissipation of capital σ
Number of sensors (communication channels) n	Stock of resource N	Efficiency of IMS η	Profitability of the system P
The amount of information I	Related capital F	Energy intensity of IMS E	Total capital U=M+F
Accuracy of IMS Δ	Price c(t)		

According to the formula (4) and table 2, for the economic efficiency of designing and manufacturing of IMS, we have the following expression:

$$\eta = C_1 \frac{N \cdot M \cdot c(t)}{M - Fc(t)}. \tag{8}$$

Formula (8) shows that the efficiency of design and manufacturing of IMS is determined, basically, by the underlying resource M and its stock N, as well as its price with c(t). This practically corresponds to the standard approach in the economics of enterprises of the electronic industry.
With the maximum efficiency $\eta=1$, the value (cost) of the design and manufacture of IMS will be determined by the ratio:

$$c(t) = \beta \frac{M}{F + C_1 NM}. \tag{9}$$

Here β is the coefficient of dimension. The associated capital F is defined as all the money invested by the firm or investor in the production of the products they are going to sell.

However, in modern pricing, there is a process of reducing the share of "simple", material resources with a parallel increase in the share of various compensation - for software, project development, financial markets, trademarks, etc. There are changes in the content and arrangement of accents in the components of the price. So, the main purpose of advertising is not to inform about the properties of the goods, but to ensure attention (distribution of image advertising, elements of provocation, shocking). All this has a significant impact on the change in the nature of competitive relations.

We now introduce into the formula (9) the main parameter of the IMS-the amount of memory W, which is an analog of the basic resource M:

$$c(t) = \gamma \frac{W}{F + C_1 NM}. \quad (10)$$

Here γ is the coefficient of dimension.

The formula (10) shows the increase in the cost of IMS with an increase in its memory capacity and a decrease in the cost of IMS with an increase in investment volume F.

4. «Innate» abilities IMS and moore's law

Expanding the exponent in the denominator Φ [13] in a series and neglecting small terms, it is not difficult to obtain in the linear approximation, setting Φ equal to the IMS efficiency η:

$$\eta_0 = \varepsilon \ln W, \quad (11)$$

where ε is the parameter of the model; W characterizes the amount of IMS resources, which is proportional to the amount of IMS memory, sensitivity of sensors and a number of other parameters, which will be discussed below.

At the initial moment of the formation of the system W=ε so that

$$\eta_0 = \varepsilon \ln \varepsilon. \quad (12)$$

The resulting expression is an innate ability of IMS.

Now let's turn to expression (11) and make a few remarks. If the innate ability of IMS (~ε) is small, then the increase in W resources due to the modernization of IMS will slightly change its efficiency. This is due to the logarithmic dependence of η on W. For example, an increase in IMS resources by a factor of 100 leads to a change in η only by a factor of ~ 5. Such IMS should be either substantially reconstructed or eliminated.

The obtained equation allows to determine experimentally the innate ability of IMS. If the ratio of the output signal / input signal is taken as the efficiency of the IMS, then it is possible to determine η_1, η_2, ... from the given W_1, W_2, ... and, thus, the inherent ability of the IMS. Thus, it is possible to conduct an analysis of IMS from the point of their technical solvency and economic prospects.

Efficiency of IMS is defined as the ratio of the time of its development t to the period of its existence T, then from (11) for the time dependence of W we get:

$$W = W_0 \exp\left(\frac{t}{\varepsilon T}\right), \quad (13)$$

where $W_0 = \varepsilon \ln \varepsilon$.

In the 1960s, at the very beginning of the information revolution, Gordon Moore, later one of the founders of Intel Corporation, drew attention to an interesting pattern in the development of computers. He noticed that the amount of computer memory is doubling about every two years. This pattern became a kind of empirical rule in the computer industry, and soon it turned out that not only memory, but every indicator of computer performance - chip size, processor speed, etc. - obeys this rule [15].

Moore's law is an empirical observation originally made by Gordon Moore, according to which (in the

IOP Conf. Series: Journal of Physics: Conf. Series **944** (2017) 012123 doi:10.1088/1742-6596/944/1/012123

modern formulation) the number of transistors placed on an integrated circuit chip doubles every 24 months.
In addition to predicting the exponential growth of the density of the transistors, Moore made another important and seemingly paradoxical conclusion. Reducing the size of transistors should inevitably lead to the fact that integrated circuits based on them will be cheaper, more powerful and more affordable. From this it followed that the electronic industry as a whole would change.
The main characteristic of IMS is the amount of processor memory that is proportional to its resources, so equation (13) is a mathematical expression of Moore's law. However, unlike the usual interpretations of Moore's law, our equation contains an innate ability, which is an essential fact. The fact is that an exponential dependence of the type (13) is characteristic for many processes in nature and society, far from microelectronics, but the innate ability of the system is always present.
In 2007, Moore stated that the law would obviously soon cease to operate because of the atomic nature of the substance and the speed limit of light. One of the physical limitations on the miniaturization of electronic circuits is also the Landauer principle, according to which logic circuits that are not reversible must produce heat in an amount proportional to the amount of erasable (irretrievably lost) data. The possibilities for removing heat are physically limited [16].
The boundedness of Moore's law follows naturally from (13). At t = T, the exponential dependence becomes W = const. Graphically it looks like it is shown in Fig. 1.

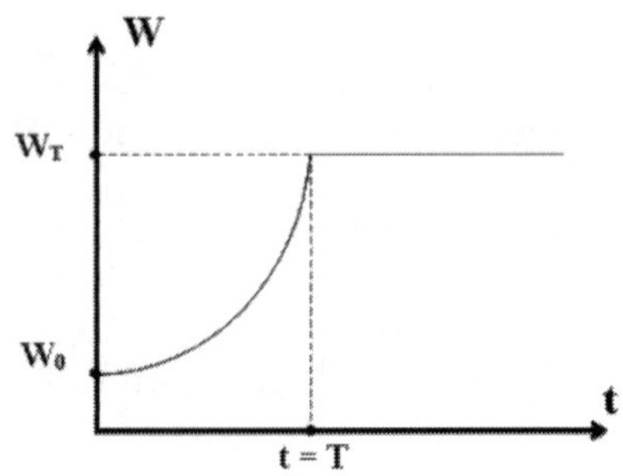

Figure 1. The time dependence of the memory capacity of the IMS processor

It should be noted that Moore's law (in the interpretation of Gordon Moore) is not being fulfilled with such precision to call it a law or even an empirical regularity. It is possible that the hype surrounding Moore's Law is Intel's clever marketing move. Nevertheless, Moore's law, like the similar "exponential laws", reflects some general trends in the development of science, technology, human society, etc.

5. Forecast resources of IMS

Now it is generally accepted that information and information-measuring systems belong to the class of communication systems.
The generalized concept of the "resource" of the communication system was first introduced by L.I. Rozonoer [17]. In this work, the exchange and distribution of the resource in the system were considered as occurring according to laws analogous to the law of energy distribution in a closed system of mechanical particles. Later, the notion of the "resource" of the communication system began to be associated with the presence of some set of communications connecting the elements of the system, and with the characteristics of these communications.
After the linearization of the expression obtained by us, the response function Φ of the system has the form:

$$\Phi = \beta \frac{E}{\Delta G^0} \cdot \overline{N}, \qquad (14)$$

where E is the "capacity" of the communication channel in the system; $\overline{N}$ is the average number of channels in the system; ΔG^0 is the Gibbs energy of a thermostat (external environment); β is some constant of the theory, the magnitude of which is calculated for each particular system according to the procedure set forth in [13].

For ideal processes $\Delta G^0 = \Delta F$ and, taking (14) into account, we obtain:

$$W_x = \frac{\beta E \overline{N}}{\Phi}. \qquad (15)$$

If the "volume" of IMS is denoted by V, then the total resources of the system will be equal to

$$W = \frac{\beta E \overline{N}}{\Phi} \cdot V. \qquad (16)$$

It was intuitively expected that IMS resources would increase with the increase in the number of communication channels and the channel capacity of the system.

For the IMS response function from (16) we have:

$$\Phi = \beta \varepsilon \ln \varepsilon \cdot E \cdot \overline{N} \cdot V \cdot \exp(-t / \varepsilon T). \qquad (17)$$

Thus, the IMS response function contains its main parameters (here the "volume" of the system can be taken to be equal to the amount of processor memory) and the inherent ability of IMS.

In the simplest case, as a response function, we can take the ratio of the output signal to the input signal and use the autocorrelation functions of the processes at the input and output of the system.

We note that most of the parameters in (17) can be determined experimentally, which has an important applied value.

6. Conclusions

Currently, in many countries, particularly in the United States, the cost of works on the automation of IMS design is more than 1/3 of the cost of developing large projects, which indicates the complexity, cost and urgency of automated design of IMS.

The thermodynamic approach and formulas obtained in the present paper can significantly reduce the cost of automation design work for IMS.

References

[1] Badiru A 2005 *Handbook of Industrial and Systems Engineering* (CRC Press, New York)

[2] Morris A S and Langari R 2012 *Measurement and Instrumentation: Theory and Application* (Elsevier/AP)

[3] Webster J G 2014 *Measurement, Instrumentation, and Sensors Handbook* (CRC Press, Taylor & Francis Group)

[4] Kaul A B et al. 2012 *Microelectronics to Nanoelectronics: Materials, Devices & Manufacturability* (CRC Press)

[5] Ismail R, Ahmadi M T and Anwar S 2012 *Advanced Nanoelectronics* (CRC Press)

[6] Khanna V K 2016 *Integrated Nanoelectronics. Nanoscale CMOS, Post-CMOS and Allied Nanotechnologies* (Springer India)

[7] Meyers R A et al. 2009 *Encyclopedia of Complexity and Systems Science* (Springer Science + Buisiness Media)

[8] Duffy V G et al. 2008 *Handbook of Digital Human Modeling. Research for Applied Ergonomics and Human Factors Engineering* (CRC Press)

[9] Demin A and Dmitrieva S 2016 *Applying math modelling methods for forecasting the engineering system states* (ITMO University, St-Pb.)

[10] Kanaracus C 2008 *Gartner: global IT spending growth stable* (InfoWorld)
[11] Brillouin L 1959 *La science et la theorie de information* (Masson)
[12] Poplavskiy R P 1981 *Thermodynamics of information processes* (Science, Moscow)
[13] Yurov V M, Kolesnikov V A, Ismailov Zh T and Baisagov Ya Zh 2013 *Thermodynamics of information-measuring systems* (Karaganda)
[14] Sagawa T 2013 *Thermodynamics of Information Processing in Small Systems* (Springer, Kyoto, Japan)
[15] Pakhomov S A 2003 *Computer press* no 1 pp 16–22
[16] Dmitriyev A S 2012 *Thermal processes in nanostructures* (Fizmatlit, Moscow)
[17] Rozonoer L I 1973 *Automation and telemechanics* no 5 pp 115–133; no 6 pp 65–80; no 8 pp82–104

The investigation of social networks based on multi-component random graphs

V N Zadorozhnyi[1], **E B Yudin**[2]

[1]Omsk state technical university, pr. Mira 11, 644050, Omsk, Russia
[2]Sobolev Institute of Mathematics, Siberian Branch of the Russian Academy of Sciences, 4 Acad. Koptyug avenue, 630090, Novosibirsk, Russia

E-mail:zwn2015@yandex.ru, udinev@asoiu.com

Abstract. The methods of non-homogeneous random graphs calibration are developed for social networks simulation. The graphs are calibrated by the degree distributions of the vertices and the edges. The mathematical foundation of the methods is formed by the theory of random graphs with the nonlinear preferential attachment rule and the theory of Erdős–Rényi random graphs. In fact, well-calibrated network graph models and computer experiments with these models would help developers (owners) of the networks to predict their development correctly and to choose effective strategies for controlling network projects.

1. Introduction

For the last two decades, the mathematical foundation of Network Science – random graphs theory – has been enriched by a developed mathematical apparatus providing an opportunity to construct and investigate various graph models of real growing networks. The theory [1, 2] of random graphs with the nonlinear preferential attachment rule (the NPA rule) is notable for wide opportunities of the graph models calibration by statistical data on the simulated networks. Thus, the NPA graph theory allows one to form adequate models for a wide class of networks and apply these models to predict the course of the studied networks development and/or to design effective strategies for the network project control.

The NPA graph is grown from a small 'seed'-graph by adding another graph increment – a new vertex with random number x of outgoing arcs at moments $t = t_1, t_2, \ldots$. The ends of the graph increment arcs attach to any randomly selected graph vertices. Probability p_i of the arc attachment to vertex i is in proportion with the vertex weight f, determined by its degree k_i:

$$p_i = \frac{f(k_i)}{\Sigma_j f(k_j)}, \quad i, j = 1, \ldots, N, \tag{1}$$

where N is the number of the graph vertices.

At infinite addition of the graph increments, an infinite graph is formed. Therefore, the NPA random graph is set by two parameters – probabilities distribution $\{r_k\} = \{r_g, \ldots, r_h\}$ for a random variable x (the number of the increment arcs) and weight function $f(k) \geq 0$. Moreover, $f(k) > 0$ then and only then, when $g \leq k \leq M$ (where $g \geq 0$, $M \leq \infty$). Function $f(k)$ can be specified as subscripted variable f_k or as weighting sequence $\{f_k\}$.

The article reveals non-uniformity of social networks structure, and the methods accounting for this non-uniformity are developed for calibrating the NPA graphs: the graphs are calibrated simultaneously by degree distributions of vertices and edges.

2. Theory

2.1. Vertex degree distribution.
At the given distribution $\{r_k\}$ of the graph arcs number x and the given weights $\{f_k\}$, stationary (at $t \to \infty$) graph vertex degree distribution (VDD) is determined by the recurrence formula found in [1]:

$$Q_k = \frac{r_k\langle f\rangle + mf_{k-1}Q_{k-1}}{\langle f\rangle + mf_k}, \quad k = g, g+1, g+2, \ldots, \tag{2}$$

whereQ_k is probability of a randomly selected vertex of an infinite graph having degree$k$; $m = \Sigma_k(kr_k)$ is the mean number of the arcs in the increment, and value $\langle f\rangle$ of the vertex mean weight is calculated by solving equation system (2) with additional equation

$$\langle f\rangle = \sum_{k=g}^{M} f_k Q_k\,. \tag{3}$$

The equation resulting from the graph growth is used as a control one. This is an equation of medium degree $\langle k\rangle$ of the doubled mean increment degree:

$$\langle k\rangle = \sum_{k=g}^{M+1} kQ_k = 2m\,. \tag{4}$$

A simple procedure for a quick numerical solution of system (3), (4) in spreadsheet tables (for example, in Excel) is described in [1].

2.2. Arcs/edges degree distribution.
The problem of acrs/edges degree distribution for the NPA graphs was solved in general terms for the first time in [3]. At given $\{r_k\}$ and $\{f_k\}$, the graph is grown with stationary distribution of arc probabilities, determined by the recurrence formula

$$Q_{l,k} = \frac{f_{k-1}(lr_lQ_{k-1} + m^2Q_{l,k-1}) + f_{l-1}m^2Q_{l-1,k}}{m(\langle f\rangle + mf_k + mf_l)}, \quad l,k = g, g+1, g+2, \ldots, \tag{5}$$

where $Q_{l,k}$ is the probability of a randomly selected arc outgoing the vertex with degree l and ingoing the vertex with degree k. Matrix $\mathbf{Q} = \|Q_{l,k}\|$ is calculated by formula (5) line by line. Probabilities Q_k are determined above.
If the NPA graph is used for simulating networks with undirected links, the arcs are substituted with edges. Denoting by $\Theta_{l,k}$ a stationary probability of a randomly selected end of a randomly selected edge being incident to the vertex with degree l, and the other end of the edge to the vertex with degree k, we find $\Theta_{l,k} = (Q_{l,k} + Q_{k,l})/2$. Therefore, matrix $\mathbf{\Theta} = \|\Theta_{l,k}\|$ of the graph edges degree distribution (EDD) can be calculated by matrix $\mathbf{Q} = \|Q_{l,k}\|$ of arcs degree distribution through transformation:

$$\mathbf{\Theta} = \frac{1}{2}(\mathbf{Q} + \mathbf{Q}^{\mathrm{T}}), \tag{6}$$

where T is the matrix transposition symbol.

2.3. The problem of complex calibration for the NPA graphs.
Problem statement: Given the specified VDD $\{Q_k\}$ of the simulated network (possibly smoothed one) and the known empirical EDD $\tilde{\mathbf{\Theta}} = \|\tilde{\Theta}_{l,k}\|$, one needs to find probabilities $\{r_k\}$ and weights $\{f_k\}$,

inducing the graph with the assigned VDD $\{Q_k\}$ and causing minimal difference in the graph EDD $\mathbf{\Theta}$ from the given degree distribution $\widetilde{\mathbf{\Theta}}$:

$$r(\mathbf{\Theta},\widetilde{\mathbf{\Theta}}) = \left(\sum_{l,k=g}^{u} (\Theta_{l,k} - \widetilde{\Theta}_{l,k})^2 \right)^{1/2} \rightarrow \min . \quad (7)$$

Parameter $u > g$ is selected by the EDD diagram for the simulated network so that two-dimensional interval $[g, u]^2$ covers the area of typical values $\Theta_{l,k}$. Varied parameters in problem (7) are probabilities $\{r_k\} = r_g, \ldots, r_h$ and weighting sequence $\{f_k\}$. Unlike the solution proposed for this problem in [4], here it is unnecessary to use weights $\{f_k\}$ providing accurate realization of the given VDD $\{Q_k\}$. If the required degree distributions $\{Q_k\}$ and $\widetilde{\mathbf{\Theta}}$ can be realized approximately at natural weights $f_k = k$, such weights are considered as the problem solution. And if in this case one may find an appropriate simple definition of probabilities $\{r_k\}$, such $\{r_k\}$ are preferred to optimal ones in terms of problem (7).

3. Characteristic features of social networks structure

Publication [4] demonstrates complex calibration of a growing random graph, simulating autonomous systems (AS) network, including 22,963 nodes and 48,436 links between them. The quality of the complex graph calibration carried out due to the data on AS network is shown by the diagrams in figure 1. The horizontal axes in the EDD diagrams include values l and k, while the vertical axes have probabilities $\Theta_{l,k}$ of a randomly selected graph edge being incident to the vertices with degrees l and k ($\Theta_{l,k} = \Theta_{k,l}$ at any l, k). When speaking of the simulated network, instead of the words *vertex* and *edge* one uses the words *node* and *link*. The comparison of the last two diagrams in figure 1 shows that in AS network simulation the algorithm of a complex calibration proposed in [4] led to a high-quality result. This confirms that the mechanism of AS network growth is well-described by rule (1).

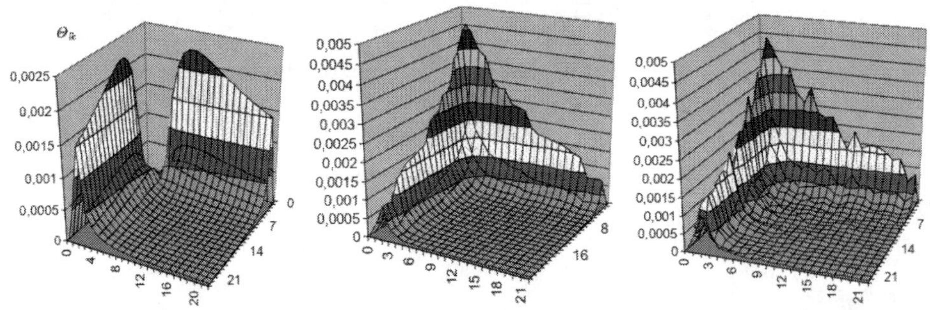

Figure 1. The edge degree distribution of the NPA random graph (after the calibration by the vertex degree distribution – on the left, and after the complex calibration – in the centre) and the degree distribution of the links between AS network nodes (on the right).

When simulating *social networks*, however, it becomes obvious that applying this algorithm of the complex calibration does not lead to as good results as in engineering AS network simulation.

Let us examine, for instance, the result (figure 2) of applying the algorithm when simulating social network Brightkite by the data [5] on its topology. At the time of the data collection, Brightkite network included 58,228 nodes and 214,078 links.The dissimilarity in two diagrams' forms in figure 2 can still be called tolerable. But the range of probabilities $\Theta_{l,k}$ (plotted vertically) for the graph EDD is twice as narrow as that of link degree distribution for Brightkite network. Therefore, the resulting calibration cannot be considered appropriate.

As a result of the further investigation, a hypothesis arose that Brightkite network is composed of two networks with differing structures and therefore it should be simulated by composing (combining) two NPA graphs (two components), isolated or loosely connected ones. Obviously, such components can be called the components with an autonomous structure. Various suppositions on two main components of Brightkite network were studied. Significant resemblance of the graph EDD and network link degree distribution is achieved when combining the following two components.

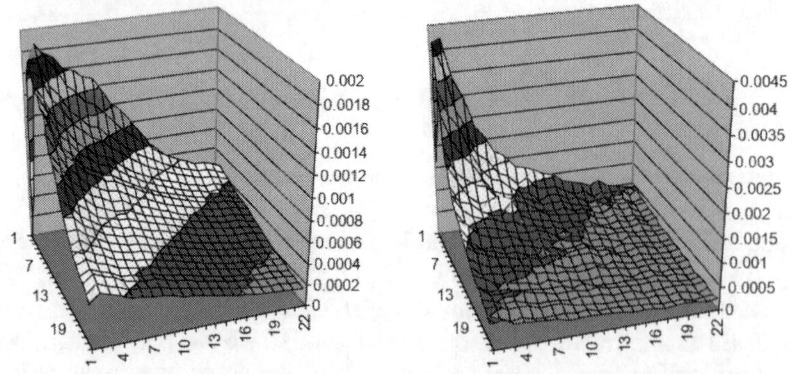

Figure 2. The edge degree distribution for the graph calibrated by the methods [4] (on the left) and links degree distribution in Brightkite network.

The first component is fixed, it is the Barabasi-Albert tree representing a special case of the NPA graph, specified by parameters $m = g = 1$ (i.e. $r_1 = 1$) and $f_k = k$ ($k = 1, 2, \ldots$). The second component is a complement NPA graph, with its composition minimizing the target function (7). Let the composition include N vertices, and the included tree and the complement graph have $N_1 = \rho N$ and $N_2 = (1 - \rho)N$ vertices, correspondingly ($0 < \rho < 1$). It is easy to notice that VDD $\{Q_k\}$ of the composition graph is a combination of the components' VDD:

$$Q_k = \rho Q'_k + (1-\rho)Q''_k, \quad k = 1, 2, \ldots, \tag{8}$$

where Q'_k, Q''_k are the probabilities that the vertices of the first and second components have degree k, correspondingly. In (8) probabilities Q'_k for the tree are estimated by formula (3) calculation, and probabilities Q_k are given (it is a smoothed vertex degree distribution for Brightkite network). Therefore, at known ρ, formula (8) determines univalently probabilities Q''_k of degrees k for the complement graph vertices. Consequently, parameters m, m' and m'' of these graphs are connected by the formula $m = \rho m' + (1 - \rho)m''$, where $m = 3.6765$ (as in a real network), $m' = 1$ (for the BA tree). This implies that at given ρ we get $m'' = (m - \rho m')/(1 - \rho) = (m - \rho)/(1 - \rho)$.

In a similar way, EDD $\{\Theta_{l,k}\}$ for the two-component graph is a combination of its components' EDD:

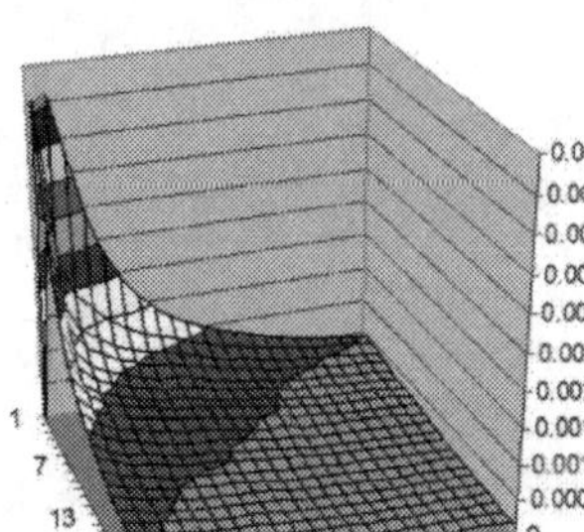

Figure 3. The calculated edge degree distribution of the calibrated two-component graph.

$$\Theta_{l,k} = \gamma\Theta'_{l,k} + (1-\gamma)\Theta''_{l,k}, \quad l, k = 1, 2, \ldots, \tag{9}$$

where γ is the share of the tree edges in the number of the whole graph edges:

$$\gamma = m'\rho N / mN = m'\rho / m = \rho / m. \tag{10}$$

At specified ρ, problem (7) is solved by varying the parameters only of the complement graph. After that ρ is refined, and problem (7) is solved again. In two such iterations, value ρ is determined as equal to about 0.225. The resulting distribution $\{r_k\}$ of the second component includes probabilities $\{r_1, \ldots, r_{40}\}$, providing the minimum of the target function (7). The weight function of the second component is linear: $f_k = k$ ($k = 1, 2, \ldots$); this makes the model more natural as a whole figure 3 shows the diagram for EDD of the graph combining two described components with an autonomous structure.

The EDD diagram has the form almost similar to that of the diagram for the links degree distribution (see figure 2 on the right) in Brightkite network (the diagram has "noises" caused by statistical errors). More accurate range coincidence of the compared diagrams can be easily achieved by a small decrease of parameter ρ, also increasing the resemblance of the diagram forms.

In all considered examples, VDD for the calibrated graphs coincides with the nodes degree distribution for the simulated networks, and the calculations validity is verified by the graph simulation modeling (SM).

The leap in the graph model accuracy as a result of combining two differing NPA graphs suggests that simulated network Brightkite is also a composition of networks with autonomous structures.

Further, the article deals with geosocial ("geocontact', "geolocational") network Gowalla. During the complex graph calibration to simulate this network, one reveals a number of characteristic structural features that cannot be explained when considering the structure as being uniform.

4. The hypothesis for non-uniformity of Gowalla

The data on the nodes of Gowalla network and the links between them, gathered in February, 2009 - October, 2010, are published in [6]. These data were used to calculate the degree distribution $\{Q_k\}$ of the network nodes and the degree distribution $\widetilde{\boldsymbol{\Theta}}$ of the links between them (figure 4). This figure (on the right) also displays EDD of the NPA graph, calibrated by the method proposed in [4], that provides the accurate realization of the specified VDD and uses non-linear weights. This is the best result of the complex calibration achieved by applying the method.

Just like in case of Brightkite network simulation, the approach to Gowalla network as to a uniform structure led to no satisfactory calibration. Consequently, supposition arises that Gowalla network is a composition of networks with autonomous structures, as well.

5. Unexpectedproblemsandunexpectedsolutions

The attempts to obtain the graph EDD corresponding to the links degree distribution in Gowalla network by combiningtwo NPA graphs with an autonomous structure (by analogy with Brighkite network simulation) led to no success. The result of such calibration is always the EDD diagrams, having flat or concave slopes of the diagram surface and/or inappropriate ranges of values $\Theta_{l,k}$ on the vertical axis (as in the last diagram in figure 4). The problem (*the first one*) proved to be rather unexpected.

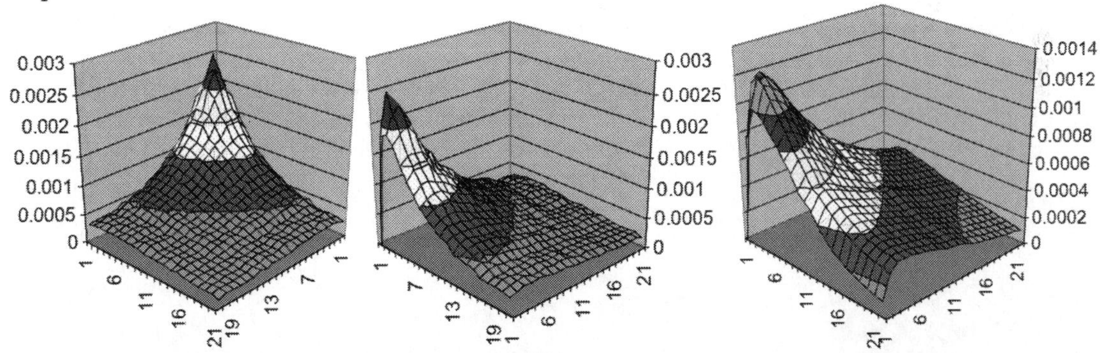

Figure 4. The links degree distribution in Gowalla network is on the left and in the center (two perspectives of the three-dimensional diagram), the best result of the uniform NPA graph calibration is on the right.

The convex slope of the diagram surface for the graph EDD (see the first two images in figure 4) can result if one of the graph components is theErdős–Rényi graph (the ER graph) [7], i.e. "classical" random graph. But the component representing the NPA graph with linear weights f_k must be present as well. It ensures necessary slow (as a power function) decrease in probabilities $Q_{l,k}$ with increase in l and k, observed in Gowalla network.

The other two problems arose when searching for such components (the ER graph and the NPA graph with linear weights), their composition would allow one to obtain EDD close to the links degree distribution in Gowalla network. One of these problems is as follows: even though the ER component allows one to achieve rather good resemblance of the graph forms for the compared degree distributions, it makes a too steep slope of the diagram surface for EDD (much steeper than necessary). Moreover, it cannot be compensated by selecting an appropriate second component (the NPA graph). Another problem is that it is impossible to obtain a sharp peak on the diagram surface for EDD (see the last diagram in figure 4) when calibrating.

When comparing Poisson degree distribution of the ER graph vertices with the standard one, it is obvious that one needs to increase VDD dispersion to obtain more gentle slope on the surface of EDD diagram. It should be made, however, without changing the mathematical expectation of the degree distribution. But it is impossible since the mathematical expectation of Poisson distribution equals the dispersion. To solve this problem, the concept "autocorrelated ER graph" (the AER graph) was adopted. When constructing this graph between random attachments of vertex pairs, positive correlation was introduced. Using the AER graph as a component solved all three problems: the surface of the EDD diagram was convex one at calibration, its slope became more gentle, and the surface peak turned out sharp.

6. The description of the calibrated graph

The graph on N vertices, calibrated by the data on Gowalla network consists of two components. The first component is the AER graph on $N_1 = \rho N$ vertices, where $\rho \approx 0.35$. The second component is the NPA graph on $N_2 = (1-\rho)N$ vertices. If taking N=100 000, we get N_1 = 35000, N_2 = 65 000.

The AER component is constructed in the following way. N_1 isolated vertices are taken. The edge is lead from vertex $i = 1$ to vertex $j = i + 1$ (then to vertex $j = i + 2, j = i + 3, \ldots, j = N_1$) with probability $p = (p_a + z_{j-1})/2$, where $p_a = a/(N_1 - 1)$; $a = 2.75$ is the mean vertex degree. Random variable $z_{j-1} = 0$, if the edge is not traced to vertex $j - 1$, and $z_{j-1} = 1$, if the edge is led to vertex $j - 1$. Then, the edges connecting vertex $i = 2$ (vertex $i = 3, 4, \ldots, N_1 - 1$) to vertices $j = i + 1, i + 2, \ldots, N_1$ are distributed in a similar way. After the graph having been grown, all isolated vertices are deleted from it, as well as isolated vertex pairs connected by a single edge.

The second component, the NPA graph, is grown at linear weights $f_k = k$ and simple limited degree distribution $\{r_k\}$, determined by formula $r_k \approx 0.3004(k-0.1259)^{-1.2562}$, $k = 1, \ldots, 50$.

The diagram of the calculated EDD for the described two-component graph is shown in figure 5 on the left (deleting isolated pairs of the connected vertices was not taken into account in calculation). In the center there is EDD for the graph, grown by simulation modeling. Comparing these diagrams of the graph EDD with the diagram of the links degree distribution inGowalla network (see figure 4 in the center) and considering the manual selection of the components parameters for the calibrated graph, the conclusion can be made that this graph reproduces the structure of Gowalla network satisfactorily. The last diagram on figure 5 demonstrates that VDD $\{Q_k\}$ of the calibrated graph describes the network nodes degree distribution adequately.

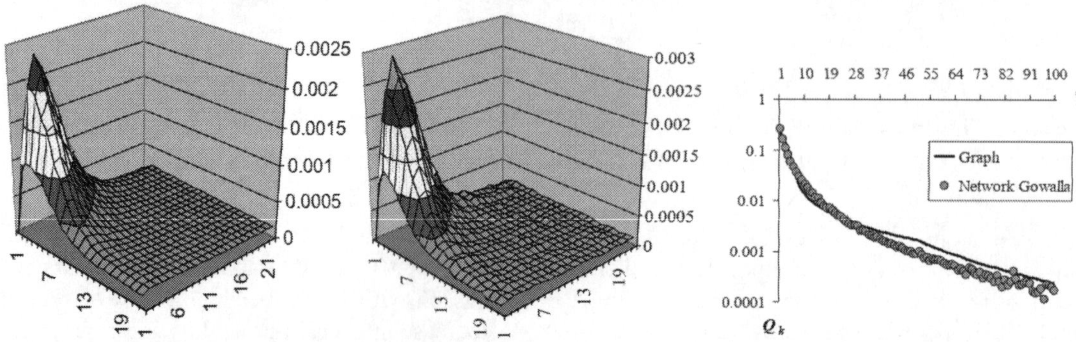

Figure 5. The calculated edge degree distribution of the calibrated two-component graph is on the left, the edge degree distribution of the calibrated graph obtained through SM is in the centre, the graph vertices and network nodes degree distributions are compared on the right.

The quality of the calibration result obtained manually indicates the possibility of its significant refinement by formulating and solving the resulting extreme problem in an appropriate way. There are no «artificial» technical adjustments (complex linear weights and «optimized» sets $\{r_k\}$) and no real mechanisms of the networks formation unexplained logically, which allows us to suppose that the simulated network Gowalla does consist of the sub-networks with an autonomous structure.
It should also be mentioned that the supposition on the success correlation at the edges distribution in the AER component of the calibrated graph seems natural since it correlates with the methods applied in Gowalla network to stimulate the users' activity [8, 9].

7. Conclusion

The research findings demonstrate that the structure of some social networks is non-uniform. Social network Brighkite can be represented as a composition (combination) of two large components with autonomous structures. Both components are properly described by the preferential attachment graphs. Geosocial networkGowalla can be treated as the composition of an autocorrelated random Erdős–Rényi network and a network described by the preferential attachment graph. The appropriate modification of the Erdős–Rényi graph, an autocorrelated ER graph introduced in the present work, takes into account anautocorrelatedbehaviour of the network users making connections, the behavior affected by the memory of previous successes or failures.
In fact, well-calibrated network graph models and computer experiments with these models would help developers (owners) of the networks to predict their development correctly and to choose effective strategies for controlling network projects. The story [9] of Gowalla network creation, its unexpected loss to competing networks and the subsequent forced sale to Facebook owners, demonstrates the significance of such networks simulation.

Acknowledgments

The reported study was funded by RFBR, according to the research project No 16-31-60023 mol_a_dk.

References

[1] Zadorozhnyi V N 2010 Stochastic graphs and nonlinear preferential attachment rule *Control Sciences* Vol 6 pp 2–11
[2] Zadorozhnyi V N and Yudin E B Growing network: models following nonlinear preferential attachment rule 2015 *Physica A: Statistical Mechanics and its Applications*Vol 428 pp 111–132 DOI: 10.1016/j.physa.2015.01.052
[3] Zadorozhnyi V N 2016 Growing network: the dynamic of degree distribution related nodes *Omsk Scientific Bulletin*Vol2 (146) pp 81–86
[4] Zadorozhnyi V N and Yudin E B 2017 Calibration of random graphs of preferential attachment by the degree distributions of vertexes and edges endpoints *The Journal Omsk Scientific Bulletin* Vol 151 pp 114–119
[5] Leskovec J 2011 Brightkite. *Dataset information* URL: http://snap.stanford.edu/data/loc-brightkite.html
[6] Leskovec JGowalla. *Dataset information* 2010 URL: http://snap.stanford.edu/data/loc-gowalla.html
[7] Erdos P and Renyi A1960 On the evolution of random graphs *Publications of the Mathematical Institute of the Hungarian Academy of Sciences* 5 pp 17–61
[8] Cho E, Myers S A and Leskovec J 2011 Friendship and Mobility: Friendship and Mobility: User Movement in Location-Based Social Networks *KDD '11 Proceedings of the 17th ACM SIGKDD international conference on Knowledge discovery and data mining* pp 1082–1090 DOI:10.1145/2020408.2020579
[9] Campagna M, Floris R, Massa P and Mura R 2017 Social Media Geographic Information: The Community Perspective in Planning Knowledge *Enriching Urban Spaces with Ambient Computing, the Internet of Things, and Smart City Design* Eds S Konomiand G Roussos pp184–208 DOI:10.4018/978-1-5225-0827-4.ch010

Building an authorization model for external means of protection of APCS based on the Internet of things

A A Zaharov, O V Nissenbaum, K Y Ponomaryov, E S Nesgovorov

Tyumen State University, 6, Volodarskogo, Tyumen, 625000, Russia

e-mail: o.v.nissenbaum@utmn.ru

Abstract. In this paper we study application of Internet of Thing concept and devices to secure automated process control systems. We review different approaches in IoT (Internet of Things) architecture and design and propose them for several applications in security of automated process control systems. We consider an Attribute-based encryption in context of access control mechanism implementation and promote a secret key distribution scheme between attribute authorities and end devices. Keywords – Internet of things, automated process control, information security, access control.

1. Introduction

Contemporary oil and gas development area has several specific features namely:

- sparse control objects;
- hard environmental and climatic conditions;
- specific social infrastructure in hydrocarbon production areas;
- variety of the automated process control systems (APCS);
- high system reliability requirements;
- high standards of health care and requirements for safety of service staff.

All these features require new approaches to management of production profitability by implementation of technical solutions and special control systems focused on energy-saving technologies, minimization of human labor involved to the production process.

The concept of the Internet of Things (IoT) evolves over time. The key idea is following: it is an information system consisting of various integrated physical entities. In this concept entity is a device that has an ability to transmit data to computer networks (e.g. sensor, smartphone, computer, video camera, etc.). All these devices are connected in the same network; therefore, they can interact with each other and an external environment. There are different options of IoT implementation [1]: production control systems (noise, vibration, pipe deformation, corrosion, soil movement, temperature and pressure measurement can be identified by displacement, pressure sensors, etc.), intelligent energy monitoring system, industrial processes automation, etc.

Distributed networks collecting information from sensors can also be used to control physical secure perimeter. In this case, the Internet of Things concept is applied not to the automated control systems, but to the physical environment in which they are located to build secure perimeter. However, there are many problems and limitations intrinsic to IoT itself, not to several application area. Therefore, we have several problems to consider with IoT itself in light of adaptation to the automated process control system.

It is necessary to keep in mind following IoT requirements, described by E. Leloglu [2]:

- resource constraints: In IoT architecture, most of nodes lack of storage capacity, power and CPU, generally using low-bandwidth communication channels. It is challenge to apply some security techniques;
- data volumes: although some IoT applications use brief and infrequent communication channels, there are considerable number of IoT system such as sensor-based, logistics and large scale system that have potentials to entail huge volume of data on central network or servers;
- privacy protection: since a great number of RFID systems are short of suitable authentication mechanism, anyone can tracks tags and find the identity of the objects carrying them. Intruders can not only read the data, but can also modify or even delete data as well;
- scalability: the IoT network consists of a large number of nodes. The proposed security mechanism on IoT should be scalable;
- autonomic control: traditional computers need users to configure and adapt them to different application domains and different communication environments. However, objects in IoT network should establish connections spontaneously, and organize/configure themselves for adapting to the platform they are operating in. This kind of control also involves some techniques and mechanisms such as self-configuring, self-optimizing, self-management, self-healing and self-protecting.

Several scientific sourced focuses a significant attention on issues of access control in IoT networks. There is also a need for detailed and granular information control and an ability to take into account the location coordinates with limited computing resources.

An IoT-system often has a distributed architecture: all objects of the network can extract, process, combine and provide information to other objects. All sensors are able to interact with each other within the wire or wireless network. A distributed network connecting heterogeneous devices is a platform to create complex software products. In order to construct such structure, it is necessary to answer the following questions:

- How to authenticate many devices and sensors in dynamic environment?
- How to implement the access control mechanism?
- How should data providers and processors learn about each other's true identifiers, properties and network addresses?
- How to achieve connectivity in conditions of heterogeneous network elements with different interaction methods?

2. Main part

Access control is an important point of information systems security, such as computer networks that include data collection sensors. Due to the growing trend of sensor intellectualization, there is a need to distribute access at the application level of the interaction model, which provides support for processes and management of network objects.

It is possible to build an access control system with Attribute-Based Encryption (ABE) cryptographic scheme in distributed networks. ABE schemes were introduced by Sahai and Waters [3]. Multiple attributes, that regulate access to specific information, have to be defined when building systems with Attribute-Based Encryption. Each message in this system has a certain set of attributes. Every user secret key contains access tree with certain attributes. The message data can be decrypted only when message attributes satisfy access tree requirements. This concept is called Key Policy (KP-ABE). Trusted centers issue keys to users and also verify attribute values. Ciphertext Policy (CP-ABE) is an example of another technique in which the access tree is encrypted into a data packet and the user key includes the validation attributes.

Another access control model is based on registration of devices on the authentication server (SA), generating a common secret key. Attribute Authority conduct the same operations. Thus, each node of the system and each trusted center have a common secret key with SA. Access control is achieved by applying attribute encryption, since all messages between devices are encrypted with secret keys based on the access structure. Attributes are different characteristics of the end devices or application area. It is necessary to consider the method to distribute keys from trusted centers to sensors, because attributes can change in a dynamic environment and in response to new changes there is a need to obtain a new secret key. We will describe this process using the Otway-Rees protocol [4]. We assume that device initiates the transfer of a new set of attributes and AA has a mechanism for authenticating it.

Table 1. Notifications

Symbol	Value
N	Node that requests new key when attribute value has been changed
AA	Identifier of Attribute Authority: trusted center that distributes keys
SA	Identifier of Service Authority: trusted center that shares keys between AA and N
I	Session number identifier
E_N	Symmetric-key encryption with key common for N and SA
E_A	Symmetric-key encryption with key common for AA and SA
K_{ABE}	Key of ABE algorithm based on new attributes of N
$attr_{new}$	New set of attributes
R_N, R_A	Random numbers created by N and AA respectively
K	Session key

1. The device N sends to AA the session number, the random number and the set of attributes that are encrypted on the common with SA key.

$$\mathrm{N} \rightarrow \mathrm{AA}: I, N, AA, E_N(R_N, attr_{new}, I, N, AA).$$

2. The Attribute Authority transmits the received encrypted SA message, and also encrypts its pseudo-random number with their common key.

$$AA \rightarrow SA: I, N, AA, E_N(R_N, attr_{new}, I, N, AA), E_A(R_A, I, N, AA).$$

3. SA decrypts received messages, extracts parameters I, N, AA and checks if they match with parameters transmitted in the clear text. If values do not match, SA must abort the protocol session or send a request for resubmission. It also generates a shared session key for N and AA, extracts the attributes and a signature from the message from N and passes them to AA.

$$SA \rightarrow AA: I, E_N(K, R_N, attr_{new}), E_A(K, R_A, attr_{new}).$$

4. At this stage, AA checks an equality of the received random number generated earlier and checks the session number. After that, it generates a new encryption key based on the received attributes and sends key to N encrypted with a symmetric cipher using the session key obtained from SA. At that moment, AA can approve SA authenticity, otherwise SA would not be able to decrypt the message and return the same pseudo-random number.

$$AA \rightarrow N: I, E_N(K, R_N, attr_{new}), E_K(K_{ABE})$$

5. The device N decrypts messages. It retrieves the key containing the access tree (KP-ABE) or the attributes (CP-ABE) using a session key. It is also necessary to check if the set of attributes match the previous values. N verifies SA by checking a pseudo-random number. There may be additional step with AA returning delivery message.

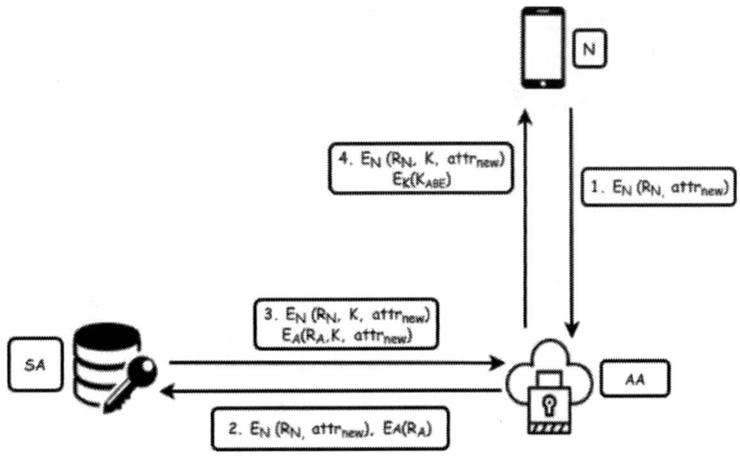

Figure 1. Otway-Rees protocol

Despite a broad presence of information on the capabilities and models of attribute encryption in the scientific literature, many issues are still open and have to be resolved.

- How to verify the authenticity of attributes during a private key generation?
- How to transfer secret keys in a secure way in a dynamic and open environment?
- What scheme does not have disadvantages of the attribute encryption? For example, how to exclude the possibility of the key sharing by user and transferring its parts to other participants?
- How to hide the attributes of participants and messages in CP-ABE schemes?

There is a large number of documented attribute encryption types [5-9], including papers studying the application of attribute encryption in IoT [10, 11] and most of them implements different approaches to the transformation of data. However, there is no survey evaluating their advantages and disadvantages.

The Internet of things is rapidly developing area, and the number of new devices connected is growing exponentially. Hence, there is an obvious problem with the fact that the network ontology is unknown and changes dynamically. Thus, there is a chance that the services become unavailable and applications cannot find the required devices due to their unavailability (e.g. when changing the geo position) [12, 13].

To solve this problem, it was proposed to use a special software that stores the ontology and handles all requests in the system. Object's data can be supplemented with special fields to simplify the work of the service. For example, store its attributes (if they do not contain confidential information). When registering a new sensor, all information about it (a unique URI, the subnet segment in which it is located, etc.) is stored in databases. Further, when a request is received from the business logic application, it is redirected to the software service where additional authorization is possible, for example, verifying the access rights to perform the required operation, and in case of authorized operation, the service returns information about the logical or physical location of the requested object.

In the context of a distributed system, each subnet can have its own instance of the entity for a flexibility and scalability. In this case, there is a difficulty with the distribution of information between all nodes of the system and maintaining its relevance. As one of options, routing protocols can be implemented by grouping information about hosts on a set of defined parameters and forwarding it to related entities. In this scheme, the problem of maintaining the relevance of data is critical and require a separate analysis.

The issue of actualization, that is, an authenticated update of device attributes, is put in one way or another in all applied research of ABE. There are various approaches to solving this problem, starting with fairly simple ones, for example, centralized lists of attributes, up to very complex schemes that allow to confirm the updating of attributes by various supervisors within their authority to change one or another attribute [14]. In the IoT with a decentralized structure, it seems appropriate to apply the latter approach.

3. Conclusion

Modern network concepts, such as the Internet of things, can be used as a system of physical protection for production facilities. In this case, the system is a distributed network for collecting and analyzing data from sensors.

This article describes the problems of a distributed approach of building the Internet of things, which is the most effective in terms of capabilities and the most difficult to implement at the same time. The main problems of this approach are authorization and access control models, registration and authentication mechanisms and ontology schemes for service discovery. The large number of devices and the dynamic nature of the interaction between them are properties of the Internet of things which give rise to additional questions of fault tolerance, control and scalability. In order to build an access control system, it was suggested to use attribute encryption. Introduced a model for distributing secret keys in conditions of a large number of devices and trusted centers.

The merits of attribute encryption are flexible settings of access structures. Attribute based encryption allows encryption of a message between all devices in the network. At the same time if knowledge of the attributes is not secret, it increases privacy risks for specific application areas. The management of cryptographic keys is the bottleneck of ABE in a distributed network. The solving of this problem is the Otway-Riis protocol with a single trusted center for all nodes. Each entity needs to store only one key for communication with this trusted center. However, this scheme generates failover risks due to a single communication node. The shared secret key is generated during the registration process on the trusted the authentication server. It is planned to implement a software prototype in a laboratory, which allows to identify other possible shortcomings.

References

[1] Kupriyanovskiy V P, Namiot D Ye, Drozhzhinov V I, Kupriyanovskaya Yu V and Ivanov M O 2016 Internet of Things in industrial plants *Int. Journal of Open Information Technologies* **4 (12)** pp 69-78

[2] Leloglu E 2017 A Review of Security Concerns in Internet of Things *Journal of Computer and Communications* **5** pp 121–136

[3] Sahai A and Waters B 2005 Fuzzy Identity-Based Encryption *Advances in Cryptology* V (Eurocrypt) pp 457–73

[4] Otway D and Rees O 1987 Efficient and Timely Mutual Authentication *Operating Systems Review* **24 (1)** pp 8–10

[5] Waters B 2011 Ciphertext-policy attribute-based encryption: An expressive, efficient, and provably secure realization *Int. Workshop on Public Key Cryptography* **6571** pp 53–70

[6] Ostrovsky R, Sahai A and Waters B 2007 Attribute-based encryption with non-monotonic access structures *Proc. of the 14th ACM conf. on Computer and communications security* (ACM) pp 195–203

[7] Lewko A 2010 Fully secure functional encryption: Attribute-based encryption and (hierarchical) inner product encryption Annual *Int. Conf. on the Theory and Applications of Cryptographic Techniques* **6110** (Springer Berlin Heidelberg) pp 62–91

[8] Lewko A and Waters B 2011 Decentralizing attribute-based encryption *Annual Int. Conf. on the Theory and Applications of Cryptographic Techniques* (Springer Berlin Heidelberg) pp 568–88

[9] Wang G, Liu Q and Wu J 2010 Hierarchical attribute-based encryption for fine-grained access control in cloud storage services *Proc. of the 17th ACM conf. on Computer and communications security* (ACM) pp 37–735

[10] Wang X, Zhang J, Schooler E M and Ion M 2014 Performance evaluation of attribute-based encryption: Toward data privacy in the IoT *Communications (ICC) 2014 IEEE Int. Conf. on* (IEEE) pp 30–725

IOP Conf. Series: Journal of Physics: Conf. Series **944** (2017) 012125 doi:10.1088/1742-6596/944/1/012125

[11] Yao X, Chen Z and Tian Y 2015 A lightweight attribute-based encryption scheme for the Internet of Things *Future Generation Computer Systems* **49** pp 12–104

[12] Hachem S, Teixeira T and Issarny V 2011 Ontologies for the internet of things *Proc. of the 8th Middleware Doctoral Symp.* (ACM) p 3

[13] Wang W, De S, Toenjes R, Reetz E and Messner K 2012 A comprehensive ontology for knowledge representation in the internet of things *Trust, Security and Privacy in Computing and Communications (TrustCom) 2012 IEEE 11th Int. Conf. on* (IEEE) pp 98–1793

[14] Chase M 2007 Multi-authority attribute based encryption *Theory of Cryptography Conf.* (Springer Berlin Heidelberg) pp 34–515

IOP Conf. Series: Journal of Physics: Conf. Series **944** (2017) 012126 doi:10.1088/1742-6596/944/1/012126

Fiber-optical method of pyrometric measurement of melts temperature

V A Zakharenko, Ya R Veprikova

Omsk State Technical University, 11, Mira ave., Omsk, 644050, Russia

e-mail: yana.veprikova.1996@mail.ru

Abstract. There is a scientific problem of non-contact measurement of the temperature of metal melts now. The problem is related to the need to achieve the specified measurement errors in conditions of uncertainty of the blackness coefficients of the radiating surfaces. The aim of this work is to substantiate the new method of measurement in which the influence of the blackness coefficient is eliminated. The task consisted in calculating the design and material of special crucible placed in the molten metal, which is an emitter in the form of blackbody (BB). The methods are based on the classical concepts of thermal radiation and calculations based on the Planck function. To solve the problem, the geometry of the crucible was calculated on the basis of the Goofy method which forms the emitter of a blackbody at the immersed in the melt. The paper describes the pyrometric device based on fiber optic pyrometer for temperature measurement of melts, which implements the proposed method of measurement using a special crucible. The emitter is formed by the melt in this crucible, the temperature within which is measured by means of fiber optic pyrometer. Based on the results of experimental studies, the radiation coefficient $\varepsilon^{*} > 0.999$, which confirms the theoretical and computational justification is given in the article.

1. Introduction

There are most technological processes are driven by temperature in modern metallurgical industry. Increased the importance of accurate temperature measurements in the most sought after processes for the smelting of iron, steel, non-ferrous metals. For solving problems of temperature measurements of melts at different times have proposed various kinds of methods. For example, direct pyrometric measurement of the radiation exposed surface of the melt, the method of the sighting tube, the method being blown pipe, dip of the contact of the thermocouples in the melt. At present, there are two methods for solving this problem most widespread: contact and pyrometric methods.

In the factory practice, in general, the method of contact measurements is used with the use of one-time thermoelectric transient transducers for short-term immersion. The advantage of the contact method measurements is sufficient accuracy, and the disadvantages are one-time use of measuring instruments and their greater inertness [1, 2]. An example of a wide application of contact measurements in industrial thermometry for measuring the temperature of melts is a thermocouple with replaceable measuring inserts. Such converters are widely used in our country and abroad in platinum and tungsten-rhenium thermocouples for measuring the temperature of metal melts [3, 4, 5]. The converters consist of two parts: a body (rod) and a replaceable thermocouple measuring package of one-time use.

The advantages of the pyrometric method of temperature measurement include non-contactness and high speed of measurement, and the disadvantages are a high error associated primarily with the uncertainties of the blackness coefficients [6, 7]. A large number of publications [2, 13] have been devoted to solutions to reducing the uncertainties associated with the uncertainties in the blackness coefficients in pyrometric measurements.

2. Problem definition

It should be noted that the processes of high-speed heating have become widespread in recent years in metallurgy and problems have arisen to automatically maintain the preset metal temperatures [5], when the control of the quality of heating is of great importance for increasing productivity and reducing specific fuel consumption. In this regard, the use of contact means, and especially one-time, for automated management of heating processes is difficult. The use of pyrometers for the solution of these tasks in the conditions of operation of the furnaces is faced with the difficulty of considering the influence of the following factors: the presence of the absorbing gas medium, the uncertainty degrees of blackness melts (unknown dependence on temperature, slag, mill scale), the effect of multipath radiation. There is a pyrometer for measuring the temperature of the melt in this work, where these factors are overcome by forming in the melt of the emitter in the form of a model of a blackbody and application of a fiber-optical pyrometer is offered. In this pyrometer the detuning from the effects of acquisitions gases, accompanying the processes of smelting, is due to the use of fiber. The elimination of the uncertainties associated with the unknown ratio of black will occur due to the formation of the measuring surface in the form of emitter type of model a blackbody. In such a pyrometer, the temperature of the melt is measured by recording the radiation of this cavity. For this, it is necessary to immerse the hollow body in the form of a crucible in a melt and to visor a fiber optic pyrometer (similar to that used in replaceable thermocouples) to its bottom [8]. As material of a crucible it is necessary to apply the brand of glasses with high melting point and the greatest transparency in the spectral sensitivity of the fiber optic pyrometer, then the melt, forming a radiating cavity after placement of the crucible and will be a radiator type of model a blackbody.

3. Theory

For meaningful calculations of the emitter was chosen the cylindrical shape of a blackbody, as this form cavities suitable for practical manufacture. For such a model have been carried out below the calculations of geometrical parameters of the emitter in the form of a blackbody model. The calculations were performed based on methods described in [9].

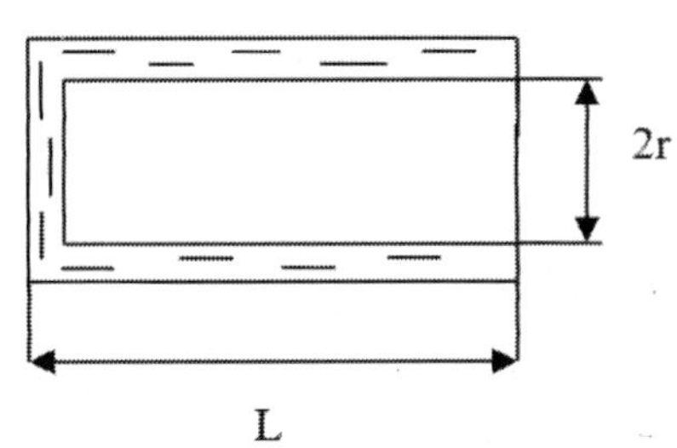

Figure 1. Sketch of the blackbody cavity.
L – cylinder length, r – cylinder radius.

The calculation was carried out according to the formula Guffe (1):

$$\varepsilon' = \frac{\varepsilon[1+(1-\varepsilon)\left(\frac{\sigma}{S}-\frac{\sigma}{S_0}\right)]}{\varepsilon\left(1-\frac{\sigma}{S}\right)+\frac{\sigma}{S}} \quad (1)$$

Then, when the emissivity of the material of the crucible ε = 0,9 (glass type KI) and the ratio $\frac{L}{r} = 10$ estimated relative emissivity of a cavity is equal to $\varepsilon' = 0.999$, where ε – is the emissivity of the cavity walls; σ – is square holes in the cavity; S – is total surface area of the cavity; S_0 is the distance between the plane of the holes and the farthest point of the cavity.

With diaphragm to ensure more isothermal cavity emitter design will be like (Fig. 2).

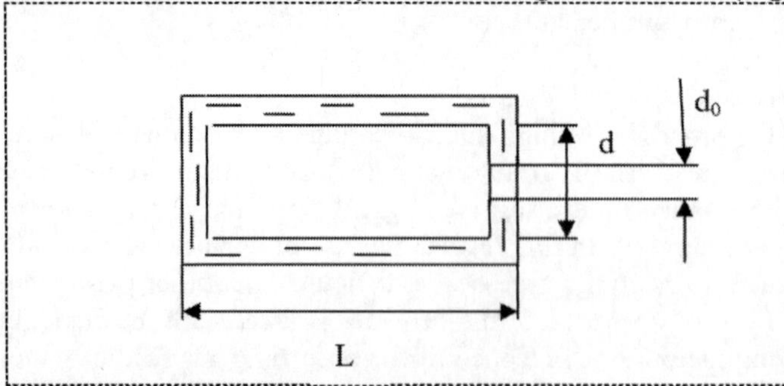

Figure 2. Diaphragm cylindrical cavity.
d_0 – is the diameter of the aperture, d – is the diameter of the cylinder.

Emissivity of the crucible material ε = 0,9 and the relation $\frac{d_0}{d} = 0.1$, and preservation conditions $\frac{L}{r} = 10$, the settlement coefficient of radiation will be $\varepsilon' > 0.999$.
Simplified formula for diaphragm cylindrical cavity

$$\varepsilon' = 1 - \frac{\rho * r_{\partial}^2}{(L/r)^2} \qquad (2)$$

where, $r_д$ – diaphragm opening radius; ρ – cavity wall material reflection coefficient, the settlement coefficient of radiation will be $\varepsilon' > 0.999$.
By results of calculations the following geometrical sizes of a design of a radiating crucible have been received: L = 100 mm; r = 10 mm; d_0 = 2 mm.
Quartz glass brand KI was chosen as the material for the manufacture of the cavity of a blackbody on the basis of its spectral characteristics, which is shown in Fig. 3 [10].

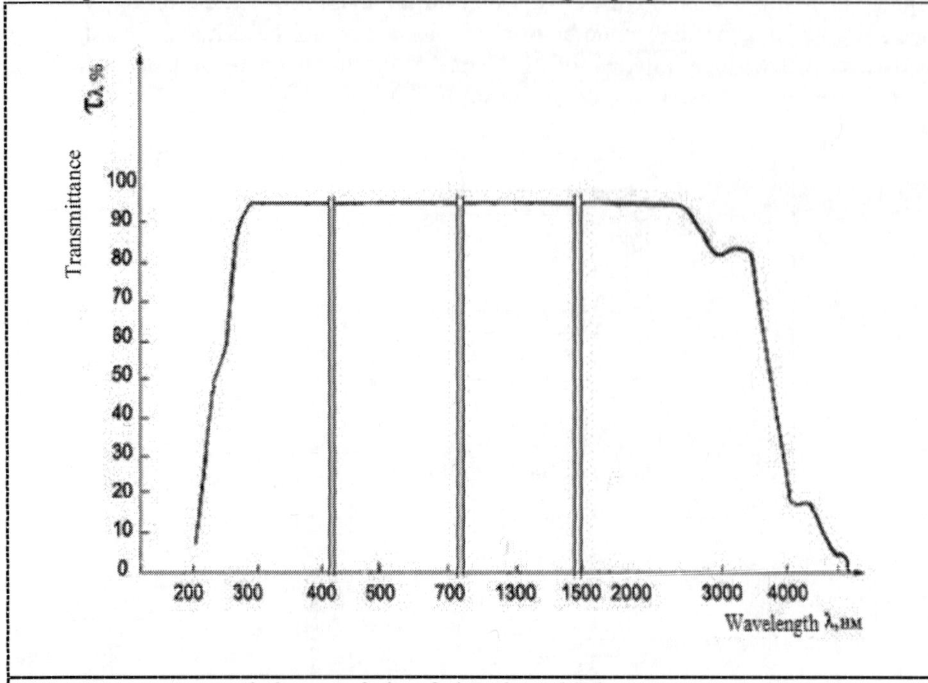

Figure 3. Spectral characteristic of glass brand KI.

Calculations carried out using the above Planck function (3) [10, 12, 13] have shown that about 90% of the radiation lies in the region of 200-3000 nanometers at melting temperatures in the range 1100-2500. It allows to claim that losses on absorption in crucible walls will be less than 10% of glass KI brand. These losses exceed (35-40) % for brands of other glasses, for example, KV-2, KV-1 brands.

$$r(\lambda) := \left(e^{\frac{h \cdot c}{k \cdot T1 \cdot \lambda}} - 1 \right)^{-1} \cdot \frac{2 \cdot \pi \cdot h \cdot c^2}{\lambda^5} \quad (3),$$

where, r(λ) is the spectral density of irradiance; C – light speed; λ – wavelength; T1 is the temperature in Kelvin; h – Planck constant; k – Boltzmann constant.

According to the calculations' results, changes in the spectral densities of the energy luminosities of surfaces were obtained at temperatures of 1100 ° C and 2500 ° C, shown in Fig.4.

As a fiber optic, the pyrometer type PD-10-02 is proposed to be used, serially produced by JSC "SPE" Etalon ".

It should be noted that in the proposed measurement method, crucibles can be used in the pyrometer design from any materials whose melting point exceeds the range of measured temperatures (for example, silicon carbide). Then a radiating cavity is formed by the molten metal, and isothermal surface of the crucible is heated to a temperature of molten metal. In this case, the measurement time constant will increase until the time of thermodynamic equilibrium establishment between the crucible and the molten metal.

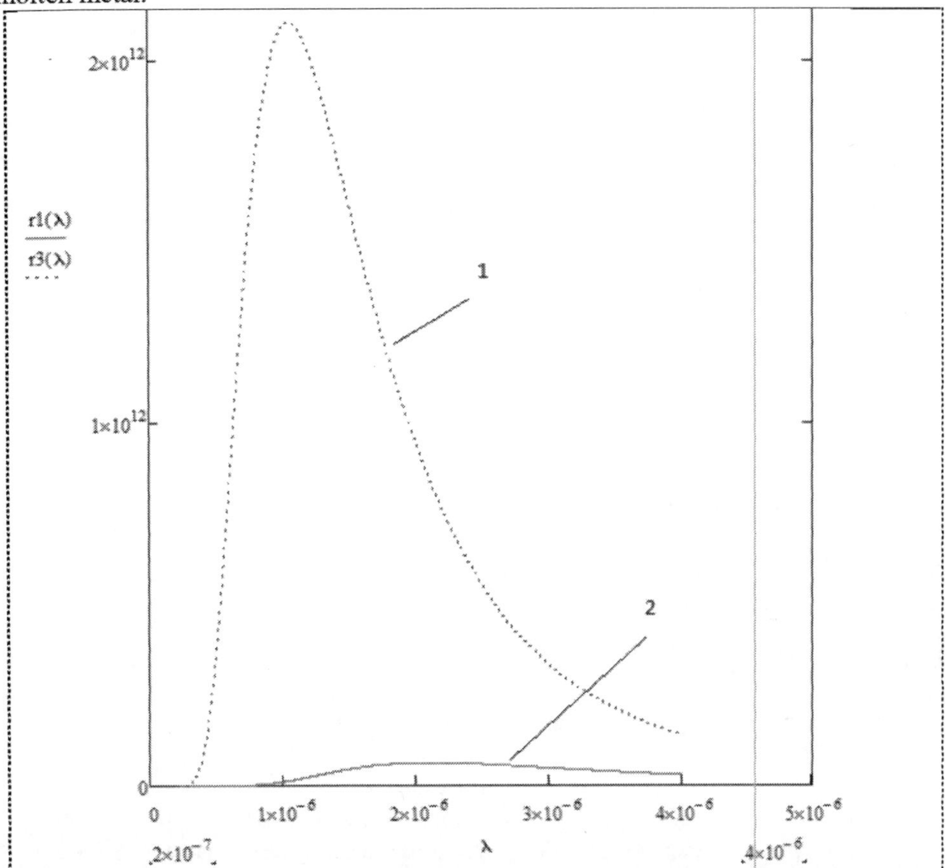

Figure 4. The dependence of irradiance of the wavelength at the temperature 1100°C and 2500°C.
1 – dependence at a temperature 2500°C, 2 – dependence at a temperature of 1100°.

IOP Conf. Series: Journal of Physics: Conf. Series **944** (2017) 012126 doi:10.1088/1742-6596/944/1/012126

3. Experimental results

The measuring device realizing the offered method has been installed in trial operation of the Omsk enterprise JSC "Omsktransmash" on furnaces for isothermal training of the cutting tool in the environment of fusion chlorine - barium salt.

As a result of experimental observations after the room of a crucible in the furnace and vising on the bottom of a fiber-optical pyrometer increase of indications of a pyrometer was observed too long (within 30 minutes). The initial emissivity of the material at the steady state value of indications of the thermometer did not match the calculated value of $\varepsilon^{*} > 0.999$ and was 0.85. During the hour of operation, the readings of the pyrometer rose by 30 degrees, which corresponded to the radiation coefficient $\varepsilon^{*} > 0.999$. At shutdown of the furnace the regular thermocouple has been rechecked in the quick-response tubular furnace with using PPO converter 2 categories.

EMF (electromotive force) of the working thermocouple at a temperature of 1100 degrees corresponded to the nominal statistical characteristics according to GOST (governmental standard). The operating characteristics of the furnace are given in Table 1.

The operating characteristics of the furnace are given in Table. 1.

Table 1. General characteristics of the studied hardening furnace.

Characteristic	Interpretation
Type	Furnace hardening tool (FHT)
Bathtub sizes	Bath diameter 500 mm, depth 1000 mm
Environment hardening	Chlorine-barium salt
Heating type	Electric arc
Denomination	Furnace for isothermal annealing of the cutting tool
Type standard tools of measurement for the control and temperature regulation	Potentiometer recording KSP-3 with a class of accuracy 0,5 together thermocouple of the type TPR(B) 001-2-1250-3-0,5/0,5 with corner cover of own manufacture.
Operating temperature	1240°C.

Conclusions

Such pyrometers can be used to measure the temperatures of different melts in the range of 1100-2500 ° C instead of the currently widely used expensive platinum-rhodium and tungsten-ferric thermocouples type TPR-2085 and TBP 2085. The advantage of the presented tools temperature measuring of melts in front of the variety of thermometers available on the market today, is the detuning of the measurement results from the acquisitions of gas environments, uncertainties associated with unknown coefficients blackness melts, influences of the rereflected radiations

References

[1] Ivanov G A, Pervadchuk V P 2011 *Production Technology and properties of quartz optical fibres* (Perm: Publishing house Perm. nat. Polytechnic. University) p 171

[2] Fukuyama H, Waseda Y 2011 *High-Temperature measurements of materials* (Springer Berlin Heidelberg)
[3] Dvurechenskii A V, Petrov V A, Rezni V Yu 1977 *Measurement of emissivity of partially transparent materials at high temperatures by high-speed spectrometry method* (Translated from Izmeritel'naya Tekhnika) pp 55–57
[4] Kovalev I I, Muchenik P G, Gordon A R, Krylov V M 1970 *Thermophysical measurementsents* (Translated from lzmeritel 'naya Tekhnika) no 8 pp 56–57
[5] Dahl A I 2010 *The stability of base-metal thermocouples in air from 800 to 2200°F* (National Bureau of Standards, Washington, USA: Proceedings «Ther-mometric metals and alloys») p 1238–1266
[6] Taymanov R, SapozhnikovaK, Druzhinin I 2011 Sensor Devices with Metrological Self-Check *Sensors & Transducers* vol 10 pp 30–45
[7] Taymanov R, SapozhnikovaK 2010 Metrological Self-Check and Evolution of Metrology *Measurement* № 43 pp 869–877
[8] Wurtman, a, Samarin, A M, Moscow: Nauka 1969 *Method for studying the properties of metal rasplava* (Moscow: Nauka) p 197
[9] Ivliev A D, Morilov V V, Kurichenko A A, Meshkov V V, Goi1 S A 2014 Methods of measuring the thermal Diffusivity of molten ferrous and nonferrous metals *Measurement Techniques* vol 57 no 3
[10] Kriksunov L Z 1978 *A guide to the basics of infrared technology* (Moscow Sov Radio) p 400
[11] Debasish Das, Dilip Kumar Pratihar, Gour Gopal Roy 2016 *Electron Beam Melting of Steel Plates* (Springer India: Temperature Measurement Using Thermocouples and Prediction Through Finite Element Analysis) DOI 10.1007/978-81-322-2740-3_57
[12] Tsvetkov V B, Seregin V F, Veber A A, Pyrkov Yu N, Rusanov S Ya 2014 Method for Measuring Optical Characteristics of Opaque and Translucent Solids at Temperatures to 1600◦C *Physics of Wave Phenomena* vol 22, no 4 pp 255–261
[13] Zakharenko V A, Klikushin Yu N, Ponomarev D B *A Compensation – Type Pyrometer* (Measurement Techniqes.- NY, Springe) vol 57, no 6 pp 679–683

IOP Conf. Series: Journal of Physics: Conf. Series **944** (2017) 012127 doi:10.1088/1742-6596/944/1/012127

Visual modeling in an analysis of multidimensional data

A A Zakharova, E V Vekhter, A V Shklyar and A J Pak

Tomsk Polytechnic University, avenue of Lenin, 30, 634050, Tomsk, Russia

Abstract. The article proposes an approach to solve visualization problems and the subsequent analysis of multidimensional data. Requirements to the properties of visual models, which were created to solve analysis problems, are described. As a perspective direction for the development of visual analysis tools for multidimensional and voluminous data, there was suggested an active use of factors of subjective perception and dynamic visualization. Practical results of solving the problem of multidimensional data analysis are shown using the example of a visual model of empirical data on the current state of studying processes of obtaining silicon carbide by an electric arc method. There are several results of solving this problem. At first, an idea of possibilities of determining the strategy for the development of the domain, secondly, the reliability of the published data on this subject, and changes in the areas of attention of researchers over time.

1. Tasks of modeling

Computer visualization is a convenient and common way of presenting information. A consequence of this was the desire of researchers to use this method in solving different scientific and practical. As a result of the systematic and active participation of visual objects in procedures for obtaining the results of scientific research, there is a need in transition from an exclusively auxiliary role of visualization to its justified use as an independent scientific tool [1].

Definition of the visual data model as a visually perceptible image correlates to this data according to some rule. As a model, this image must meet a number of requirements [2]. First of all, visual models are a realized representation of the information defining them, they become an example of a sensible perception of knowledge, which should allow them to analyze, store and process.

The artificial nature of the object, obtained as a result of the system of principles of comparison, implies the preservation of the possibility of subsequent transformations or manipulation of properties of the object. Visual transformability determines the functionality of models as a tool for research [3]. For problems of visual analysis, this property is embodied in the form of the possibility of studying the visible representation of the quasi-reality of model space with the involvement of a natural and well-understood perceptual apparatus.

2. The problem of modeling

The necessity of comprehension and the ability of the researcher to manipulate the space of the visual model for the fulfillment of its purpose face the problem of multidimensionality of the studied data. In this case, authors of the visual model should overcome the contradiction between the need to reduce the dimension of the data and to preserve visual interpretability of the model, which is determined by capabilities of users [4]. Possible solutions of this problem in visual analysis are the search for a variant of the decomposition of the original problem using data of lower dimensionality or the choice of a visual representation that creates a higher dimensionality of the model's space due to the attraction of additional visual (color, movement) or mental (memory, perception) informative components.

The created visual model always corresponds to the original only in the part of properties. The investigated object can be associated with a set of models, each of which is determined not only by the properties of the original, but also by the task of the research. In addition, the model, being an artificial analogy, is always subjective. This means the existence of an authorial factor that manifests itself in influencing the way of representation a real or imaginary object [5].

The subjectivity of visual modeling and perception are reserves, the use of which can have a positive effect on achieving the intended result of modeling. In overcoming the problem of modeling multidimensional data, this circumstance can be an effective factor, but requires justification and a preliminary study of the problem, the data properties and the target audience.

3. Requirements for a visual solution

Solving the problem of visual analysis of data leads to obtaining new information that can have a reverse effect on the user's opinion, and therefore there are a number of requirements for the procedure for obtaining an answer of the research question. For example, the features of using visual models are associated with the concentration of attention on the object of studies. This leads to the need for simultaneous visualization of both the background information and the intermediate information needed to answer the research question.

Changing the visual model during the analysis of the data creates the need to verify the correspondence of the new shape to the conditions of the initial analysis task. The phase of such assay also becomes a part of the visual image and changes its perception. The output from an infinite cycle of verification of the received solutions is achieved by means of the visual model itself. For this purpose, the following characteristics have been distinguished:

- Predictability of the consequences of decision making. A significant influence on the speed and quality of the solution is provided by the addition of the visual model by means which demonstrate expected results of various solutions. The most important is the form of representation of the solution that fulfills the function of emotional persuasion.
- Compliance with additional criteria. The question of the research can be formulated in such a way that there are a sufficiently large number of answers. Just one would be chosen as the correct solution which meets the requirements coming from external criteria, in relation to the task under study. Probability of such progress of analysis for a specific task should be taken into account while creating a model, because there is a requirement to present external information using the linguistic and expressive means of the existing model.
- The uniqueness or originality of the solution. The condition for obtaining a solution which is characterized as a new one, is unusual, different from others in some property, but is not inferior to all others in terms of effectiveness. This condition is typical for tasks in which the user's opinion becomes the main criterion for verification. In this case, the model should have means of simultaneously presenting information about the set of possible solutions to the user in a not complicate for the analysis process form.
- Dimensionless of the representation. In the visual model there might be several shapes of data or possible variants of solutions obtained. It allows the researcher to form his opinion, relying on the relative rather than the absolute values in the data. This corresponds to the peculiarities of visual perception, for which a direct comparison of the properties of the observed objects is characteristic. In addition, this makes it possible to exclude the attribute values of the data from the researcher's attention, increasing concentration on solving the main problem and reduction the analysis time.

4. Visual model of multidimensional data

Basing on formulated requirements, authors have developed a methodology for constructing and software tools for creating a visual model designed to solve operational analysis problems of a significant amount of experimental data. In general, this problem is characterized by the following features:

- The objective of the analysis is to identify data of low reliability.
- The data being examined is a set of descriptions of objects, which belong to a given domain. The number of objects can be more than 10, i.e. it is known to be greater than the number of objects simultaneously available for comparative observation.
- The description of each object in the source data is a set of characteristics (multidimensional data). Sets of features of individual objects may differ, i.e. intersect only partially.
- Types of variables that are separated attributes can be either numeric or categorical.
- The time for obtaining the solution to the analysis problem is limited.

The obtained visual model of multidimensional data corresponds to the conditions of the problem and the requirements to the solution:

- Computer implementation of the visual model (Figure 1) has the ability to represent data of different dimensions and types, creating a visual space for their comparative analysis [6]. Data is imported in text format, which simplifies the exchange with third-party applications.
- The desire to shorten the time of visual analysis leads to the appearance in the structure of the model of the means for automatically constructing the visual image using predefined expressive means. This makes the formalization of the analysis process stronger.
- Using the built visual model is regulated by the user interface, which allows changing both the amount of data involved in the analysis and the metaphor of the presentation. The combination of analysis tools in the space of the visual model, reduces the working time in 3-4 times.
- In solving the problem of representing multidimensional data, decomposition of data into individual components selected by the user and dimensionless display of variables take part simultaneously. This allows to use similar images for 5 types of variables, creating comfortable conditions for their simultaneous study.
- The principle of visual grouping, supported by graphic elements, combines individual variables into a single observable object. This solution preserves a holistic perception of the amount of data pertaining to a particular research object. In the example above, the data dimension is bigger than 10.
- In the comparative analysis, conditions to achieve the modeling goal while simultaneously studying a large number of objects were created (134 in a test study, limited only by computational capabilities).
- The study time is determined by the speed of building the model. In a test task the average time is 6 minutes.
- The model allows evaluating data features ("focus points", "excessive deviations") without determining their formal characteristics at first.
- Model management allows making quick cluster analysis, formulating hypotheses describing internal patterns.

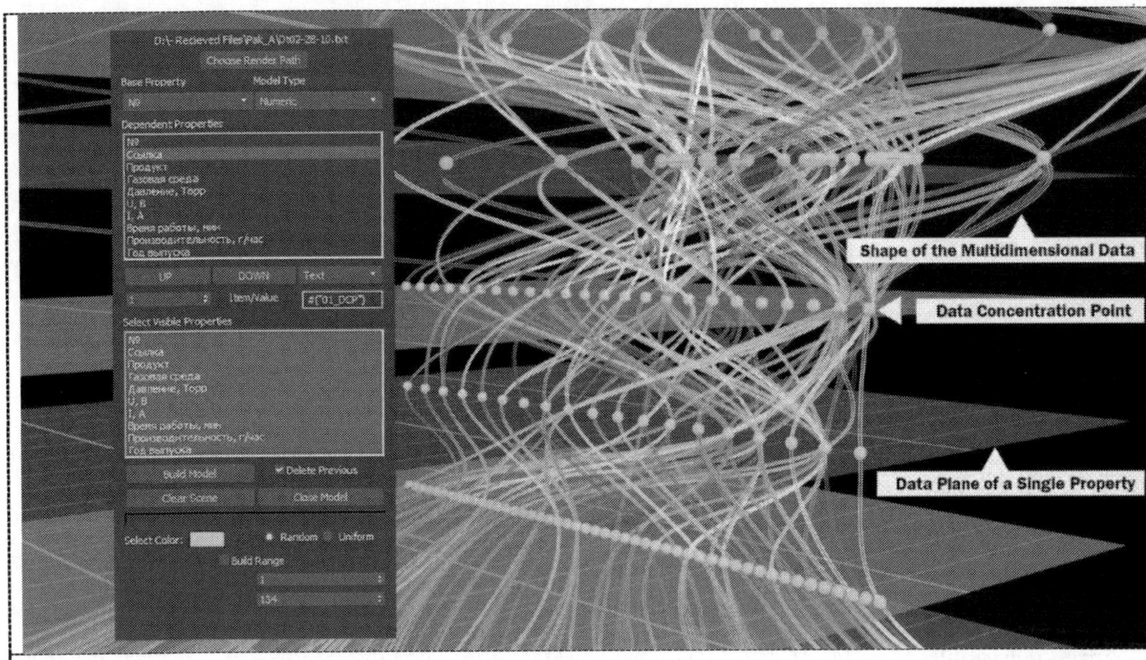

Figure 1. The example of a visual model of multidimensional data.

5. External requirements to the visual model

There are many external conditions which are not relevant to the essence of the analysis, but it could affect on its results [7]. Such factors are significant for the activity of the researcher, for the use of his cognitive potential or to control the user's attitude to the task. It can be an essential part of the visual model and has an impact on its functionality. The developed technique of construction of visual models is focused on improving the effectiveness of the analysis under controlled use of the following factors:

- Limitation of analysis time. It is not a mandatory requirement to have a high decision-making speed for any model and therefore it becomes a resource for using the model. In case of urgent obtaining the result, the type of model and the ways of interaction with it can significantly change. Reducing the time of analysis is achieved through the use of simplified presentation metaphors and the ability to exclude non-informative objects from the perceived image.
- Variability of the data. In some issues, a constant change of the state of the initial data causes a change in the type of their study. For example, in addition to the speed of analysis, it may be necessary to take into account not only the current values of the studied parameters, but also the entire history of changes. The ability to represent variable data is formed when you enter time into a set of object's features and build groups of object's images.
- Attraction of subjectivity. If the characteristics of the research need to be addressed to the opinion of the user, including personal experience, emotional evaluation or associative thinking, this leads to an additional personification of the model aimed at activating the respective abilities of the user. The use of subjective experience is achieved by including in the initial data objects that correspond to previously achieved successful solutions.

6. Conclusion

The search for a common approach to solve the problem of visual research of multidimensional data allows using visual data models as an effective research tool. The basis of this decision is a methodical approach to constructing a visual model and using the patterns of visual perception. Additional prospects for the use of visual models are associated with the full and valid attraction of such computer modeling capabilities as simulation of movements, formation of abstract images and the high speed of their obtaining.

Involvement of the individual user's thinking, his personal knowledge and experience is necessary at all stages of visual analysis, including for the formation of a hypothesis of the solution and for making a decision about its reliability. The use of resources of subjective perception and dynamic controllability of the state of the visual model can provide an opportunity for an operative study of multidimensional data and the construction of hypotheses, which is necessary for finding solutions to analysis problems.

The project has been completed as a governmental task № 2.1642.2017/ПЧ –a project about "Cognitive methods of visualization and analysis of multidimensional data during modeling of nonlinear dynamic systems".

References

[1] Chen C 2013 *Mapping Scientific Frontiers* (London: Springer)
[2] Bondarev A E and Galaktionov V *A* 2015 *Multidimensional data analysis and visualization for time-dependent CFD problems* (Programming and Computer Software) 41(5) pp 247–252. DOI: 10.1134/S0361768815050023
[3] Zakharova A A and Shklyar A V 2016 Visual presentation of different types of data by dynamic sign structures *Scientific Visualization* 8(4) pp 28–37.
[4] Chen C 2005 Top 10 unsolved information visualization problems *IEEE Computer Graphics and Applications* 25(4) pp 12–16.
[5] Averbukh V 2013 Semiotic Approach to Forming the Theory of Computer Visualization [Semioticheskij podhod k formirovaniju teorii komp'juternoj vizualizacii] *Scientific Visualization* 5(1) pp 1–25
[6] Zavyalov D A, Zakharova A A, Shklyar A V and Bagutdinov R A 2017 An integrated approach to modeling by an example of a landfill of disposal of liquid oil waste (*Software systems and computational methods [Programmnye sistemy i vychislitel'nye metody*]) pp 22–30
[7] Eppler M and Burkhard R.A 2007 Visual representations in knowledge management: Framework and cases *Journal of Knowledge Management* vol 11(4) pp 112–122

IOP Conf. Series: Journal of Physics: Conf. Series **944** (2017) 012128 doi:10.1088/1742-6596/944/1/012128

Update of the equations of the limit state of the structural material with the realization of their deformation

E V Zenkov

Department of Strength of Materials and Structural Mechanics, Irkutsk National Research Technical University, 83, Lermontov, Irkutsk, Russian Federation

Abstract. Two methods are given in the article by considering the type of stressed-Deformed state (SDS) based on equations limit condition and analyzing the results of laboratory tests of special specimens for mechanical testing, focus having destruction thereof in the same view of SDS as in focus possible destruction of the structural member. The considered limited use of these methods in terms of considering physically consistent strength criterion type Pisarenko-Lebedev. A revised design-experimental procedure for determining the strength of the material of the structure, combining therein the elements of these two methods, consisting in determining the strength parameters of construction material, entering criterion equation Pisarenko-Lebedev, considering the actual appearance of the region-of-interest SDS structure. The implementation of the procedure is performed on the basis of the selection of the respective experimental laboratory specimens for mechanical testing, plan SDS in working zone coinciding with a SDS: structure whose strength is evaluated. The refinement process limit state equations demonstrated in determining 50CrV4 steel strength parameters, being in a state of biaxial stretching. Design -experimentally determined by, that steel for a given voltage limit value is almost a quarter of its value is reduced compared to the conventional tensile strength. value is reduced compared to the conventional tensile strength.

1. Introduction

Evaluation of the structural strength of parts of machines and mechanisms must in some cases be carried out taking into account the kind stressed-deformed state (SDS) in a possible fire destruction [1-5]. Such account is important, for example, when evaluating the static strength of choke assemblies of pressure vessels, elements railway car-side frame, the whole rolled wheel strength analysis of various parts in the form of plates and shells, bendable in two different directions by the action of temperature differences and in other cases. A number of techniques for strength calculation [6-10] relies on the results of laboratory tests of specimens, focus having destruction thereof in the same view of SDS in the working area. For this purpose, the testing equipment typically used with several actuators creating multidirectional action on the test sample.

Accounting type of SDS can be carried out in two ways. The first of these is represented by the so-called combined strength criteria (hereinafter, the limit state equations) of the type Pisarenko-Lebedev, Yagna-Buzhinsky, Drucker-Prager, etc. [3], and is based on a preliminary determination of the strength characteristics of the material of the design in question, determined by quasistatic failure laboratory samples under the conditions of typical types of loading of this material - uniaxial stretching, compression and shearing (respectively determining the values of σ_t, σ_c and τ_s). These criteria are combined strength criteria, since they combine with the different weight factors two terms corresponding to the destruction by a cut (the first term) and separation. From this point of view, the

factor limiting the accuracy of this method is the difference between the real type of SDS of the design and the type of SDS of the samples tested before failure in determining the values of σ_t, σ_c and τ_s. In addition, the need for a preliminary determination of these quantities and a corresponding variety of laboratory testing techniques complicates the implementation of this method, but does not exclude it
The second way of accounting for the type of SDS relies on the results of laboratory tests of special samples having the same type of SDS in the center of their destruction as in the center of possible destruction of the structural element in question. When implementing the second method, a special testing technique with several power drives is used, creating multidirectional effects on the test sample, which also makes it difficult to habitually use this method and restrains its use. In this paper, we consider a more precise calculation and experimental calculation technique that combines the elements of these two methods, allowing for the use of standard single-drive test machines to take into account the actual type of SDS of the structural element in the source of destruction. At the same time, refinement of the calculation is achieved through the use of experimental data on the destruction of a laboratory sample, whose stressed state simulates in the working zone the stress state of a real design.
These guidelines, written in the style of a submission to *J. Phys.: Conf. Ser.*, show the best layout for your paper using Microsoft Word. If you don't wish to use the Word template provided, please use the following page setup measurements.

2. A technique for solving the problems

Consider the equation of the limiting state of the Pisarenko-Lebedev type [1, 3], used in evaluating the static strength of structures made of isotropic materials. The corresponding condition at which the quasistatic material destruction occurs, accompanied by the appearance of cracks, has the form

$$\alpha\sigma_i^{lim} + (1-\alpha)\sigma_1^{lim} A^{1-P} = \sigma_t, \tag{1}$$

where σ_i^{lim} is the intensity of stresses in a possible failure point

$$\sigma_i^{lim} = \frac{1}{\sqrt{2}}\sqrt{\left(\sigma_1^{lim} - \sigma_2^{lim}\right)^2 + \left(\sigma_2^{lim} - \sigma_3^{lim}\right)^2 + \left(\sigma_3^{lim} - \sigma_1^{lim}\right)^2}, \tag{2}$$

$\sigma_1{}^{lim}$, $\sigma_2{}^{lim}$, $\sigma_3{}^{lim}$ – the main stresses arising in this source; P is the Smirnov-Alyaev coefficient [4, 5], which characterizes the form of SDS at the point under consideration, defined by

$$P = \frac{\sigma_1^{lim} + \sigma_2^{lim} + \sigma_3^{lim}}{\sigma_i^{lim}}, \tag{3}$$

(In the case of biaxial stretching, the value of P reaches the value $P = 2$, in the case of a simple uniaxial tension $P = 1$, a simple uniaxial contraction $P = -1$); α and A are empirical constants characterizing the strength of the material and independent of the type and level of SDS of the material of the component in the center of its destruction (hereinafter, the strength parameters of the material) and calculated by formulas

$$\alpha = \frac{\sigma_t}{\sigma_c}, \quad A = \frac{\varphi - \sqrt{3}\alpha}{1-\alpha}, \quad \varphi = \frac{\sigma_t}{\tau_s}. \tag{4}$$

As already noted earlier, the accuracy of the equation of the limiting state (1) in a number of cases is insufficient for engineering practice due to the difference in the real type of SDS in the possible source of destruction of the structure and the type of SDS of the samples tested before failure in determining the quantities σ_t, σ_c, τ_s and entering into equation (1) by means of the constants α and A, which do not depend on the coefficient P and the level of the material's SDS in the source of its destruction. In [11-12], the value of the quantity Π determined by the equality (3) significantly affects the location of the source of destruction: with an increase in P (according to Smirnov-Alyaev – with an increase in the "hardness" of the type of SDS), the limiting values of the stress intensity and the first principal voltage decrease .

In order to refine the structure material limit state equations and, accordingly, increase the reliability of its strength calculation, a calculation-experimental approach is proposed that consists in

IOP Conf. Series: Journal of Physics: Conf. Series **944** (2017) 012128 doi:10.1088/1742-6596/944/1/012128

determining the strength parameters of the material of the structure under study, entering into equation (1), taking into account the real type of SDS of the most loaded area of the structure.

The essence of the proposed approach is as follows. It can be seen from equation (1) that α and A can be regarded as the coefficients of the empirical formula characterizing the structural strength of the material of the design in question with certain calculated values of the quantities σ_1^{lim}, σ_i^{lim} and P^{lim}. The parameters α and A can be calculated directly from equation (1), knowing the remaining parameters – σ_1^{lim}, σ_i^{lim} and P^{lim}. From this it follows that α and A do not in this case have a traditional physical meaning, originally inherent in the Pisarenko-Lebedev test, calculated from (4), but are constant or unimportant parameters from the kind of SDS of the material from which the design is made. Taking into account the foregoing, the strength parameters α and A can be modified, which allows them to be considered as the coefficients of the empirical formula, determined from the test results before the destruction of the samples simulating the estimated construction's SDS in its area of interest. The parameters α and A can be determined in a more complex way and, at least weakly, depend on the value of the coefficient P determined by (3).

In the paper, as in the Pisarenko-Lebedev equation, the assumption of the independence of the parameters α and A entering into equation (1) remains on the intensity values σ_i^{lim} and the first principal stress σ_1^{lim} in a possible failure point, but the dependence of these parameters on the quantity P Equation (1) with this approach is an approximation of the true limit state equation constructed for a specific (or relatively narrow range of variation) the value of the coefficient P that characterizes one or another type of SDS. This type is determined by the design features of the samples tested before destruction, and the source of their destruction should be characterized by the value of P, which coincides (or is close in value) with the value of P for the area of the structure calculated for strength.

When implementing the proposed approach, after finding the calculated value of the coefficient P to determine the values of the parameters α and A entering into equation (1), it is necessary to select appropriate experimental laboratory samples so that their SDS type coincides (or was close) with the type of construction SDS, strength Which is evaluated. In this case, to determine the two parameters α and A, it is sufficient to test before the destruction of two laboratory samples differing from each other in geometric dimensions and the design characteristics of the SDS at the time of destruction of their material. In accordance with this, the determination of the strength parameters α and A is carried out on the basis of the following algorithm:

- selection of geometric parameters of two different design variants of special-shaped specimens with values of P, close to P for the area of the design being evaluated – $P^{1\,lim}$ and $P^{2\,lim}$;
- testing before destruction of design variants of samples;
- calculation of σ_1^{lim} and σ_i^{lim} – characteristics of the SDS level entering into equation (1) in the working area of the destroyed selected samples;
- determination of the values of α and A from the resolving system of equations.

$$\alpha\sigma_i^{1lim} + (1-\alpha)\sigma_1^{1lim} A^{1-P^{1\,lim}} = \sigma_t$$

$$\alpha\sigma_i^{2lim} + (1-\alpha)\sigma_1^{2lim} A^{1-P^{2\,lim}} = \sigma_t, \qquad (5)$$

where the values of the values σ_1^{1lim}, σ_i^{1lim} and P^{1lim} correspond to the experimentally established moment of destruction of the first of the selected samples, σ_1^{2lim}, σ_i^{2lim} and P^{2lim} – the second of the destroyed samples. The refinement of the calculation is achieved in the proposed technique due to the use in it of experimental data on the destruction of laboratory samples obtained under real-design SDS conditions. The circumstance that causes difficulties in the implementation of the approach to the refinement of the equation of the limiting state of the construction material is the need to have strength characteristics, taking into account their dependence on the ratio of the principal stresses. The authors of the paper proposed a prismatic specimen for evaluating the strength of a material under a complex stress state [9], which makes it possible to obtain information on the strength of a material, depending on the required coefficient P, by loading it on a standard testing machine (see Figure 1).

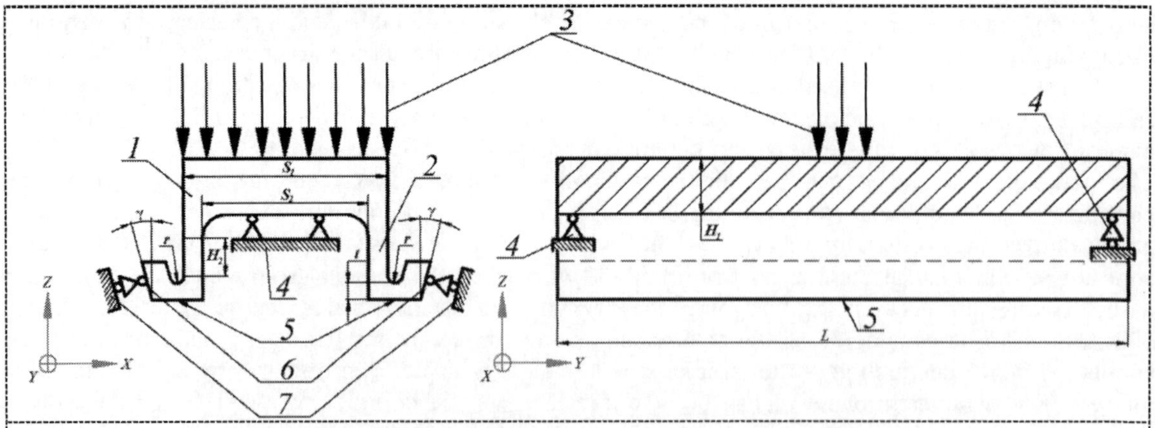

Figure 1. Geometric parameters and loading scheme of the prismatic sample
1 - laboratory sample, 2 - lateral protrusions of the sample, 3 - force generated by the pusher of the testing machine, 4 - end supports, 5 - working area, 6 - inclined support surfaces of the sample, 7 - inclined surfaces of the prismatic support

3. Results of computational and experimental studies

With the aim of realizing the process of refinement of the equation of the limiting state of the material of the structure, taking into account the peculiarities of its deformation, the strength parameters α and *A* for 50CrV4 steel under the conditions of biaxial stretching ($1<P<2$) were determined. For the study, a SDS was chosen, characterized by a value of $P = 1.85$. This condition corresponds to the type of SDS at the time of destruction of the choke unit of the pressure vessel at its outer surface in the joint area of the branch pipe with the housing [11].

A tensile test of standard circular samples of 50CrV4 steel was preliminarily carried out. The results showed that the selected steel is characterized by yield stress 1050 MPa, tensile strength 1300 MPa and a relative elongation $\delta = 7.5\%$. [14, 15, 19, 20].

According to the method in order to determine the values of the two parameters α and A entering into equation (1), it is first necessary to destroy two design variants of the samples, differing from each other in size and corresponding values of the values σ_1^{1lim}, σ_i^{1lim} and σ_1^{2lim}, σ_i^{2lim} at the time of destruction, but coinciding (Or close) by the value of the coefficient *P* in their working zones. In this case, equations of the form (1) corresponding to the two chosen constructive variants form a system of two nonlinear algebraic equations with respect to the unknown quantities α and *A*, which can be solved by the method of successive approximations. The possibility and order of selecting the necessary samples is given, for example, in [15, 19, 20]. In accordance with the algorithm, two series of samples were produced - No. 1 and No. 2 - three samples in each series. Their dimensions (see Table 1) were chosen so that in the work area of the samples their SDS was characterized by a value of $P = 1.85$ [19-20].

Table 1. the main dimensions of the experimental prismatic samples, mm

H	H_1	H_2	H_3	S	L	t No. 1 No. 2	r	γ, Degrees
33	12	11	21	44	220	4 6	2	15

The experimental destruction of prismatic samples whose geometric dimensions for the series No. 1 and No. 2 are shown in Table 1 was carried out on a typical single-wheeled testing machine Instron 5989. It showed that for the forces F_1 and F_2 (in Figure 1 – item 3) at the moment destruction (according to the results of averaging the experimental data) the following equalities were satisfied:

$$F_1 = 205 \text{ kN}; F_2 = 235 \text{ kN}. \quad (6)$$

The obtained values of the forces F_1 and F_2 were used as initial values for the numerical analysis of the design values of the design variants at the moment of their destruction. In this case, pre-designed computational deformation models, supported by computational tests and full-scale experiment, were used. Numerical analysis was carried out using the finite element method, taking into account the possible appearance of elastoplastic deformations in the sample material. In this case, the equations of plastic flow with isotropic hardening (the Prandtl-Reuss equation [17]) are used as a mathematical model of the process of development of elastoplastic deformations. Thus, for the first of the selected prismatic samples ($P^{1\ lim}$ =1,9), the calculated values σ_1^{1lim}, σ_i^{1lim} were determined, and for the second ($P^{2\ lim}$ =1,8) of the destroyed specimens – σ_1^{2lim}, σ_i^{2lim}. The results of numerical analysis of prismatic samples of the series No. 1 ($P^{1\ lim}$ =1,9) at the moment of their destruction (the moment of occurrence of the crack) are shown in Fig. 2a is the stress intensity distribution, in Fig. 2b is the distribution of the component of the first principal voltage. The results of a numerical analysis of the deformation of prismatic samples of series No. 2 ($P^{2\ lim}$ =1,8) at the moment of their destruction are shown in Fig. 3a is the stress intensity distribution, in Fig. 2b is the distribution of the component of the first principal voltage.

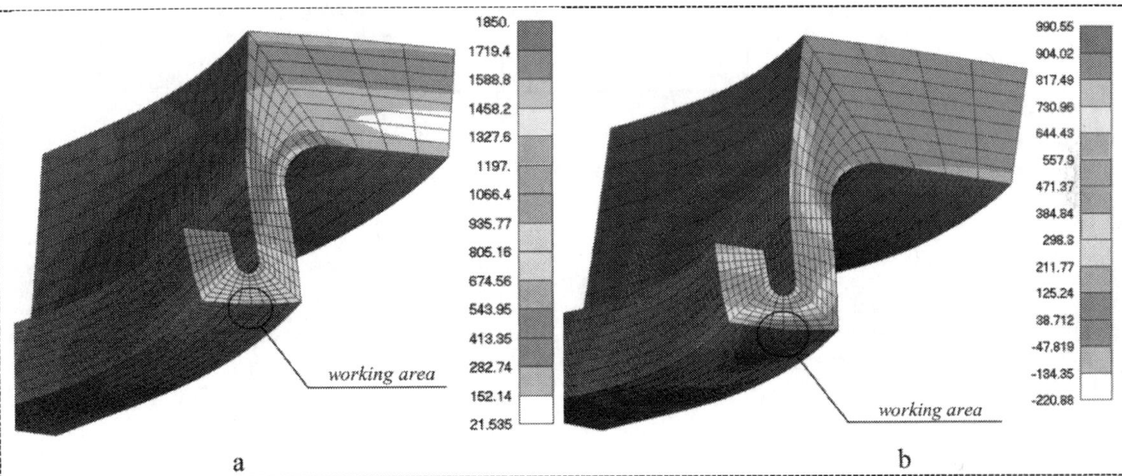

Figure 2. Estimated stress distribution at the moment of destruction of samples of series No. 1, MPa (type of quarter of the sample carved from it by two planes of symmetry)
a - stress intensities σ_i^{1lim}, b - first principal σ_1^{1lim}

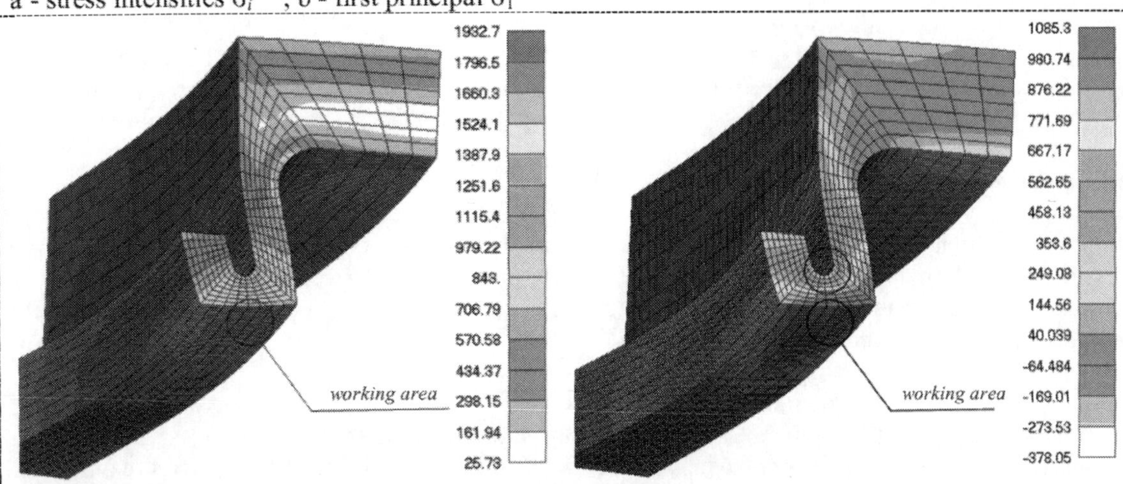

Figure 3. Estimated stress distribution at the moment of destruction of samples of series No. 2, MPa (type of quarter of the sample carved from it by two planes of symmetry):
a - intensity of stress σ_i^{2lim}, b - first principal σ_1^{2lim}

In aggregate, the calculated and experimental data obtained are presented in Table 2. The obtained experimental results show that the effect of biaxial stretching in the source of destruction of the structure can be significant - for the tested samples of the series No. 1 the limiting stress intensity σ_1^{Ilim} the moment of their destruction is 895 MPa, which is almost a quarter of the value of the tensile strength in the steel 50CrV4 under study $\sigma_i^{lim} = \sigma_t = 1270$ MPa, determined under conditions of uniaxial tension ($P = 1$). The result obtained qualitatively coincides with the result of the experimental studies of Y.A. Wilimok, K.A. Nazarova, A.K. Evdokimova [20], who established a similar effect of a significant reduction (by more than a quarter) of the value σ_i^{lim} under conditions of biaxial stretching of steel X10CrNiTi18-10 with the SDS type factor $P \approx 2$. obtained on special testing equipment.

Table 2. Calculated characteristics of SDS samples at the time of their destruction

No	working area			
	σ_1^{lim}, MPa	σ_2^{lim}, MPa	σ_i^{lim}, MPa	P
1	985	743	892	1,9
2	1050	615	917	1,8

To determine the strength parameters of equation (1), substituting the calculated values of the SDS characteristics into the system of equations (5), obtained from the experiment (see Table 2), we obtain

$$\begin{cases} \alpha 892 + (1-\alpha) 985 A^{1-1,9} = 1270 \\ \alpha 917 + (1-\alpha) 1050 A^{1-1,8} = 1270 \end{cases}. \tag{7}$$

Solving the system of equations (7) with respect to the quantities α and A by the method of successive approximations, we obtain the following equalities:

$$\alpha = 0.73; A = 0.40. \tag{8}$$

These values allow us to use equations (1) for a more accurate calculation of the strength of structural elements made of steel 50CrV4 and characterized by the equality $P \approx 1.85$. If we take into account the fact that the parameters α and A entering into equation (1) have a traditional physical meaning inherent in (1), i.e. Independent of P, we can verify that for steel 50CrV4, $\alpha = 0.6$, $A = 0.75$ for any P. This circumstance contradicts the well-known assertion of the dependence of the strength of a material on the type of SDS, and the resulting equation (8) confirms it. Taking into account the obtained values of $\alpha = 0.73$ and $A = 0.4$, it can be seen that the influence of the SDS on the strength parameters can be significant.

4. Conclusions

The proposed calculation-experimental approach to the refinement of the equation of the limiting state of the construction material, based on a preliminary calculation of the type of SDS in a possible source of structural failure and subsequent quasistatic testing prior to the destruction of the corresponding prismatic samples, made it possible to determine the numerical values of the strength parameters entering into the equation of the limiting state of the Pisarenko-Lebedev, characterized by a value of $P \approx 1.85$ and made of the steel in question.

It has been calculated experimentally that for steel 50CrV4, from which the experimental prismatic samples were made for laboratory tests, the limiting value of the first principal stress under biaxial tension ($P \geq 1.8$) is reduced by almost a quarter in comparison with the value of its strength limit determined Under conditions of uniaxial tension ($P = 1$).

5. References

[1] Pisarenko G S, Yakovlev A P, Matveev V V 2008 *Reference book on the resistance of materials* (Kiev: Publish. house Delta) p 816

[2] Lebedev A A, Kovalchuk B I, et al 2003 *The mechanical properties of structural materials under complex stress state* (Kiev: Publish. house Dom «In Ure») p 540

[3] Kogaev V P, Makhutov N A, Gusenkov A P 1985 *The calculations of machine parts and structures for strength and durability* (Moscow: Mechanical engineering) p 224

[4] Tsvik L B, Zenkov E V 2015 Determining the strength characteristics of materials prismatic experimental samples of the biaxial stretching *Bulletin of Machine Building* No 1 pp 42-46

[5] Esiev T S, Basiev K D, Steklov O I 1997 A sample for testing a metal under a biaxial stress state *Patent RF No 2073842 publ 20.02.1997 Bulletin of Inventions RF* No.7

[6] Gagarin Yu A, Pichkov S N 2000 Investigation of the behavior of defects in the fields of tensile and compressive stresses *Problems of strength and plasticity* No 1 pp 11-116

[7] Vansovich K A, Jadrov V I 2013 Experimental study of the rate of growth of surface cracks in an aluminum alloy AK6 and in steel 20 under biaxial loading *Izvestiya of the Samara Scientific Center of the Russian Academy of Sciences* No 4 pp 436-438

[8] Gagarin Y A, et al 1978 On the influence of the nature of the stress state of plasticity and fracture of structural steel *Strength problems* No 6 pp 70-75

[9] Vazhentsev O G 1993 Prismatic sample for evaluation of mechanical properties of the material *Description of the invention to the author's certificate of the USSR No 1793320 A1 kl G 01 N3/28 publ 07.02.93* No 5

[10] Tsvik L B, et al 1993 Strengthening holes and static strength of axially symmetric choke assemblies *Problems of machine building and machine reliability* No 1 pp 58-65

[11] Zenkov EV, Tsvik L B, Zapolskiy D V 2014 Prismatic sample to estimate the strength of the material *Patent RF No 2516599 publ 27.03.2014 Bulletin of Inventions RF* No 9

[12] Targ S M 1986 *Short course of theoretical mechanics* (Moscow: Higher education school) p 416

[13] Galin L A 1980 *Contact problems of the theory of elasticity and viscoelasticity* (Moscow: Science) p 304

[14] Zenkov E V, Andreev A A 2013 Methods of experimental study of deformation fields through the use of digital optical system *Problems of transport of Eastern Siberia: Proceedings of the IV All-Russian scientific-practical conference* (Irkutsk, IrGUPS) chapter 1 pp. 95-99

[15] Zenkov E V, Tsvik L B 2013 Settlement and experimental evaluation of the stress-strain state of the laboratory sample to fillet *Bulletin of IrSTU* (Irkutsk, IrSTU) No 9 pp 70-78

[16] Sutton M A, Orteu J-J, Schreier H 2009 *Image Correlation for Shape, Motion and Deformation Measurements* (South Carolina: University of South Carolina) p 364

[17] Malinin N N 1975 *Applied theory of plasticity and creep* (Moscow: Mashinostroenie) p 398

[18] Lohr R D, Ellison E G 1980 Biaxial high strain fatigue testing of Cr-Mo-V steel *Fatigue Eng. Mater, and Structure* No 1 pp 19-37

[19] Zenkov E V, Tsvik L B 2013 Stress-Strain State of Prismatic Samples with Hollow Chamfers *Russian Engineering Research* Vol 33 No 10 pp 562-565

[20] Zenkov E V, Tsvik L B 2015 Formation of divergent testing efforts and experimental evaluation of material strength under biaxial stretching *PNRPU Mechanics Bulletin* No 4 pp 110-120

Dynamics of inductors for heating of the metal under deformation

L S Zimin, A S Yeghiazaryan and A N Protsenko

Samara State Technical University, Samara, Russia

e-mail:epp@samgtu.ru

Abstract. Current issues of creating powerful systems for hot sheet rolling with induction heating application in mechanical engineering and metallurgy were discussed. Electrodynamical and vibroacoustic problems occurring due to the induction heating of objects with complex shapes, particularly the slabs heating prior to rolling, were analysed. The numerical mathematical model using the method of related contours and the principle of virtual displacements is recommended for electrodynamical calculations. For the numerical solution of the vibrational problem, it is reasonable to use the finite element method (FEM). In general, for calculating the distribution forces, the law of Biot-Savart-Laplace method providing the determination of the current density of the skin layer in slab was used. The form of the optimal design of the inductor based on maximum hardness was synthesized while researching the vibrodynamic model of the system "inductor-metal" which provided allowable sound level meeting all established sanitary standards.

1. Introduction

In the case of induction heating of bodies with rapidly changing surface curvature, which primarily include slabs[1, 2, 3], a problem occurs that is determined by the fact that electromagnetic processes in the system "inductor-metal" are characterized by not only the release of heat energy in the work piece but also in the inductor [4, 5]. In this case, it is also necessary to consider the density of the electromagnetic field and its associated electrodynamics efforts. Electrodynamics forces in the melting furnaces can improve the mixing of the liquid metal, but during induction heating for the deformation, the mechanical manifestation of electromagnetic energy plays a very negative role, due to the occurrence the problem of resistance of inductor against vibration arising under the electrodynamics forces [6,7, 8]. This problem is especially actual for the induction heating of rectangular billets, particularly made of non-magnetic metals. This phenomenon can be used in the production for obtaining of artificial vibrations with a view to their beneficial use, for ex-ample, when exposed to the molding sand [9]. In design of inductors for heating of cylindrical billets there are no conditions for the occurrence of significant vibrations (circular cross section has a high natural stiffness) and in the inductors of rectangular shape it is necessary to take into account the low stability of the straight sections of the windings of the inductor. Moreover, the problem is not the mechanical strength of the copper pipe of the inductor, since the resulting bending stresses are much less than acceptable for copper, but the vibration and accompanying noise, which, if not to accept special measures greatly exceed sanitary norms for industrial buildings [10, 11,12, 13].

2. Statement of the problem

In most cases, there is a problem related to dynamic properties of inductors designed for heating of large billets of rectangular shape, for example, slabs, whose weight is measured in tons. The resolution of this problem is possible by creating a vibration-resistant induction heating systems (ELIS), on the basis of the study of electrodynamics and related vibroacoustic processes. It is advisable carry out Electrodynamics calculations with the numerical mathematical model built using the related contours and displacements principle. Thus, the task of optimal design [14,15] of inductors, robust against the vibration and noise, is based on the study of electrodynamics and associated vibroacoustic processes According to the criterion of reduced costs, construct of the inductor, while ensuring acceptable level of vibration and of noise must have a minimum cost. It is equivalent to minimizing its weight of the given composite materials for the vibration-resistant sheath of the inductor. Attempts intuitive approach to the selection of the best forms of shell of minimum volume because of a non-obviousness optimal design lead to contradictory and sometimes erroneous results. It is therefore appropriate to formulate and solve the optimization problem for the shape of the shell of the inductor. For this purpose it is necessary to choose the decisive optimization criterion, to build a simplified, idealized model of the inductor, and with the consideration real structural constraints formalize the conditions for its optimality, and develop an efficient optimization algorithm that allows getting the concrete results. The criterion of optimization it is advisable to take the rigidity of the inductor that is the equivalent of frequency of own oscillations.The accuracy of the theoretical research is confirmed by the patents [16, 17, 18].

3. Theory

3.1. Electromagnetic problem

If to base on the plane-parallel magnetic field model, then for the given case it is better to solve internal electrodynamics problem in 2D space for body's crosssection:

$$\frac{d^2\dot{H}(x,y)}{dx^2}+\frac{d^2\dot{H}(x,y)}{dy^2}=\dot{k}^2\dot{H}(x,y) \tag{1}$$

$$\dot{H}(x,y)\big|_{д}=H_0, \tag{2}$$

$$x\in[0,b],\quad y\in[0,d],\quad \dot{k}=(j+1)\frac{1}{\Delta_э} \tag{3}$$

$$P(x,y,\frac{b}{\Delta_э},\beta)=\frac{1}{2}\gamma(\dot{E}_x\ddot{E}_x+\dot{E}_y\ddot{E}_y) \tag{4}$$

$$\dot{E}_x=\frac{1}{\gamma}\frac{d\dot{H}(x,y)}{dy},\quad \dot{E}_y=\frac{1}{\gamma}\frac{d\dot{H}(x,y)}{dx} \tag{5}$$

$$\beta=\frac{b}{d};\ \xi=\frac{d}{\Delta_э} \tag{6}$$

where *E, H* - are electric and magnetic field strengths; μ, μ_0 - are relative and absolute permeability; γ - is the specific electrical conductivity; *d, b* - the thickness and width of the heated rectangular billet (slab); $\Delta_э$ - the current penetration depth in the slab.

3.2. Heat problem.

Using (4), internal sources of the electromagnetic power distribution were explored, as well as the influence of crosssection size and penetrating depth.

The results are shown on Figure 1, where the power density on the crosssection's wide side center, on the surface, is equal to 1.

The variants, shown on Figure1a, has the same parameters as the aluminum slabs heating on 50 Hz frequency with the thickness of 0.28-0.35 m. Figure 1b - the same, but in the case of titanium slabs.

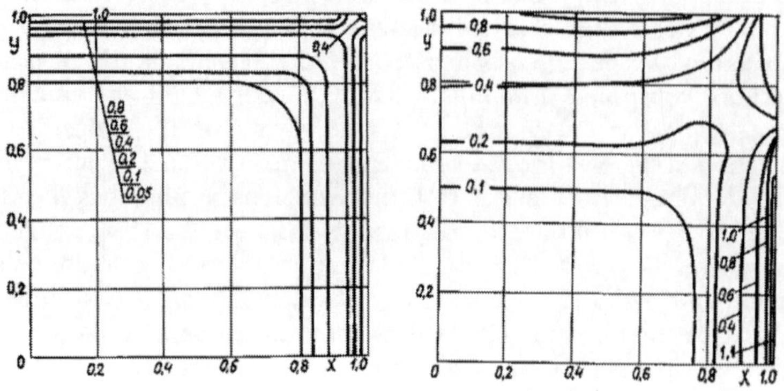

a) b)

Figure 1. Relative power P(x,y)/P(1,0): a) ξ=7.5; b) ξ=1.

To make a long story short, the term, penetrating depth, isn't suitable for halfbordered body (Shtainmetz formula).The power density changing is determined by ξ as well as byβ.

The Furies' equation solving for exothermic rectangular prism (see (4)), made isotherm creation possible (Figure 2) for the same ξ and β as on Figure1. The isotherm analysis shows that the temperature gradients kind is the same, to the electromagnetic sources distribution, and they are also determined by $\xi$and β; with ξ and β lowing-negative.

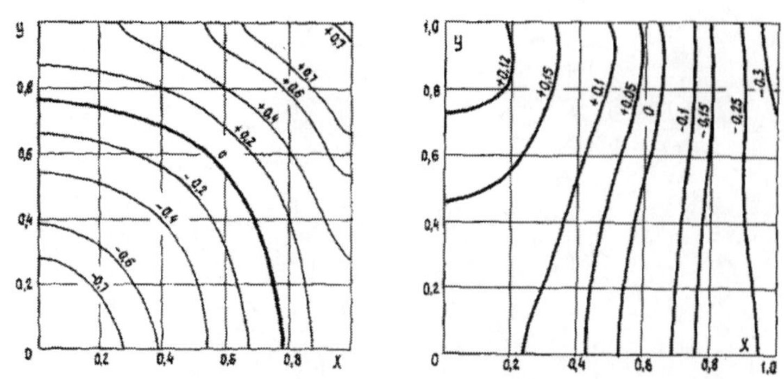

a) b)

Figure 2. Isotherms: a) ξ=7.5; b) ξ=1.

The heat efficiency (η_T) also plays big role. So, at the rectangular slabs induction heating, the process is characterized by regional effect-temperature gradients appearance on the crosssection. The strength of this regional effect is influenced by the electrical and heat properties of the heated metal, by its crosssection's sizes, by the current frequency and by the amount of the heat loss.

3.3. Electrodynamics forces in the "inductor - metal" system

Electromagnetic processes in the "inductor - metal" system are characterized by thermal power liberation as well as by electrodynamics forces affecting the resistance of an inductor to vibration and

noise in the process of induction heating for metal forming. Rectangular inductors are particularly sensitive to this problem, while inductors for heating of cylinders have a greater natural rigidity.
The task of optimization - based design of inductors, resistant to acoustic vibration, is set on the basis of electrodynamics and acoustic vibration processes analysis.
Force interaction in the "inductor - metal" system is expressed as

$$\bar{f} = \mu\mu_0\gamma\left[\bar{E}\cdot\bar{H}\right] = -\frac{1}{2}\mu\mu_0 grad\bar{H}^2 + \mu\mu_0\left(\bar{H}\nabla\right)\bar{H} \tag{7}$$

where μ, μ_0 - are relative and absolute permeability's.
When the field is a plane - parallel one (H=H_0), and the angular frequency of the harmonic current is ω=$2\pi f$ the centrifugal pressure, uniformly distributed over the inner surface of the inductor, will be

$$Q(\tau) = \frac{\mu_0}{4} H_0^2 \left(1 + \cos 2\omega\tau\right) \tag{8}$$

The value of the mean pressure, equal to the amplitude in this case, with the specific surface power P_s (W/m^2) absorbed by a rectangular billet being taken into consideration, is derived from the formula

$$Q_m = \frac{\mu_0}{4} H_0^2 = 3{,}15\cdot 10^{-4}\frac{P_s}{\sqrt{\rho\mu f}\cdot F(\beta,\xi)} \tag{9}$$

where ρ - is the specific electrical resistivity; F(β,ξ) -is the function, dependent on the ratio of the heated billet dimensions to the depth of its penetration, and tending to unit with the increment of the arguments.
Figure 3 shows the distribution of electrodynamics pressure, acting along x- and y- axes, with different depths of slab penetration in the inductor (9).
The model and the program devised allow determining forces in multiphase inductors. The design was executed for an inductor with the dimensions of 1,7*0,37*2,7 m and for a slab, with the cross section of 1,24*0,28 m and with various lengths. The pressure in the site of joint figure 4 is of great interest.

3.4. Optimization - based design of an inductor resistant to vibration

An intuitive approach to the choice of the best possible form of a cover, protecting a small inductor against vibration, often results in errors.

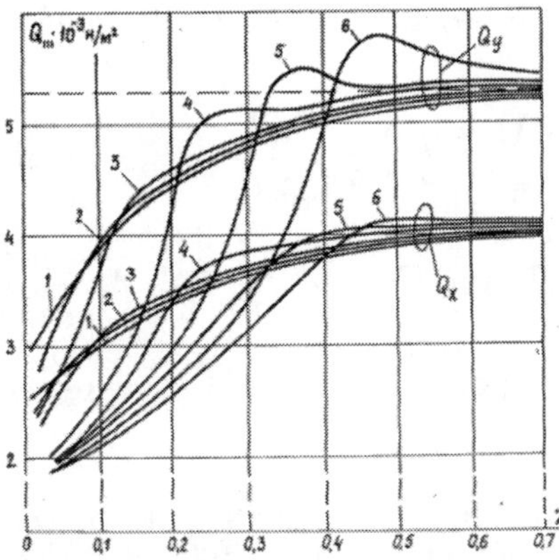

Figure 3. Electrodynamics pressure at penetration (m):
1 - 0.015; 2 - 0.035; 3 - 0.085; 4 - 0.185; 5 - 0.285; 6 - 0.385

Thus the task to optimize the form of inductors cover by rigidity index or by its equivalent, i.e. internal vibration frequency is set.

For the mathematical formalization of the task, inductor (its broad side) is replaced by an ideal vibrator system in the form of a prismatic beam with stiffly fixed ends, its length being equal to the

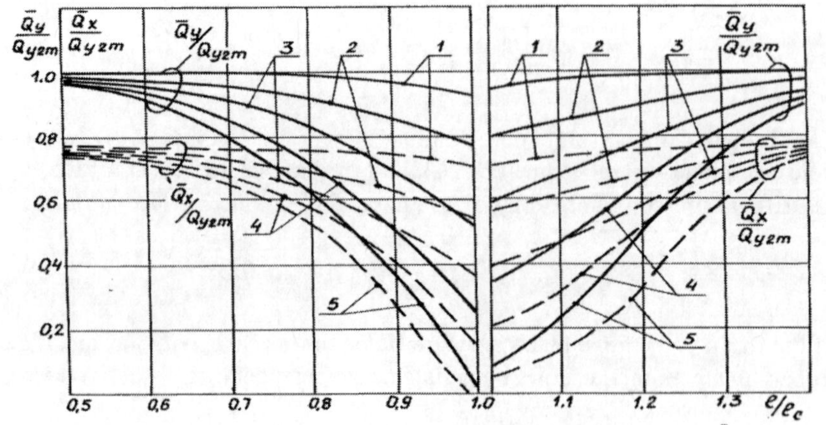

Figure 4. Electrodynamics pressure in the site of joint at φ:
1 - 30°; 2 - 60°; 3 - 90°; 4 - 120°; 5 - 180°

inductor's width (*L*), and its width being equal to the inductor's length (*B*). The above mentioned simplification is legitimate for the more 3 relationship of inductor's window dimensions, when the difference between the bending moments at the edges of the inductor's broad side and similar beam moments at the sites of fixation is as large as 14% or less, which, as a first approximation, allows to neglect the influence of the inductor's narrow sides.

The thickness of the cover, protecting inductors against vibration "*h*", and, if the inductor's length "*B*" is set, then the area of its cross section are considered to be design parameters in the task of optimization. The equation of free cross section vibrations in the ideal construction is

$$\frac{d^2}{dx^2}\left[EI(x)\frac{d^2W(x)}{dx^2}\right]-p^2\gamma F(x)W(x)=0,\quad x\in[0,L] \tag{10}$$

where *E, I* - is the modulus of elasticity and the moment of inertia of the inductor's broad side crosssection; *p* - is the frequency of internal vibrations; *γ* - is the tightness of material.

Stretch values are introduced

$$x=\frac{X}{L},\quad \alpha(x)=\frac{F(X)L}{V},\quad F(X)=Bh(X),$$
$$I(X)=aF^3(X),\quad a=\frac{1}{12}B^2,\quad \Lambda=\frac{p^2\gamma L^6}{aEV^2}, \tag{11}$$

where: *V* - is the volume of the cover; *Λ* - is the value, characterizing the internal vibration frequency.

In a stretch form the equation (10) becomes

$$\left(\alpha^3(x)W''(x)\right)''-\Lambda\alpha(x)W(x)=0,\quad x\in[0,1] \tag{12}$$

with boundary conditions

$$W(0)=W(1)=W'(0)=W'(1)=0 \tag{13}$$

Internal vibration frequency of an inductor is determined as follows

$$\Delta = \int_0^1 a^3(x)(W''(x))^2 dx \left(\int_0^1 \alpha(x) W^2(x) dx \right)^{-1} \tag{14}$$

It is required to maximize (14) by a real limitation of geometrical dimensions and the volume of the cover

$$\int_0^1 \alpha(x)dx = 1; \quad \alpha(x) \geq \overline{\alpha} \tag{15}$$

where α - is a given minimum value.
The task is executed using the maximal principle device. The form of the optimal cover is shown in figure5.

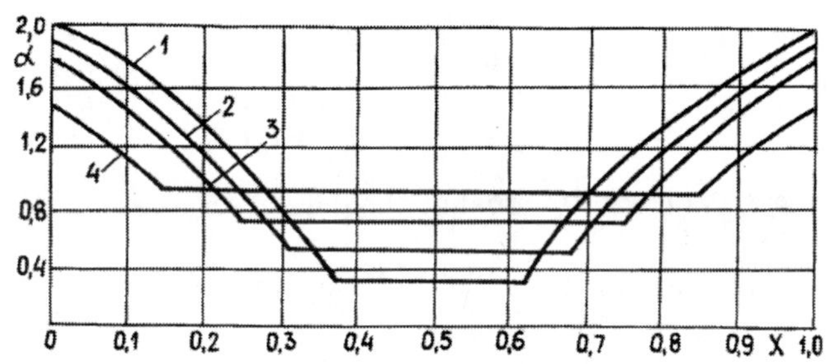

Figure 5. The form of optional cover at α:
1 - 0.3; 2 - 0.5; 3 - 0.7; 4 - 0.9

Optimal constructions resistant to vibration are 1.5 – 2.4 times as rigid versus similar conventional constructions and they provide noise abatement (4 - 6 dB) decreasing the specific consumption of materials to 15 - 20%.

3.5. Field tests

Field tests were conducted with inductors having commercial of 50 Hz while heating aluminum, titanium and steel slabs. Inductors' capacity was up to 3000 kW. The coil was enclosed either in an epoxy or in a glass - reinforced plastic compound [19, 20]. In the process of titanium slabs heating, the level of noise was as high as 85 dB or less, while for steel slabs - 65 dB or less and for aluminum slab - 130 dB. When the last inductor was enclosed in a concrete cover of an optimal form (figure6) the level of noise was reduced to 80 dB with the capacity of 2500 kW.
The field tests have proved the maximum error to be as 8% or less, which meets the requirements.

4. Conclusions

Algorithm and software are offered for the automatic design of induction installations to heat metal for pressurized processing. The optimal solutions are found from the condition of reaching extremum of a summated economic index of the work of the corresponding technological complex "heating - deformation"; they represent optimal regime, constructive, power and technological parameters of the induction heating system.
The developed system of automatic design fulfills the needed complex of electromagnetic, thermal, electrodynamics and vibro-acoustic calculations, supplies the selection of optimal constructive parameters.

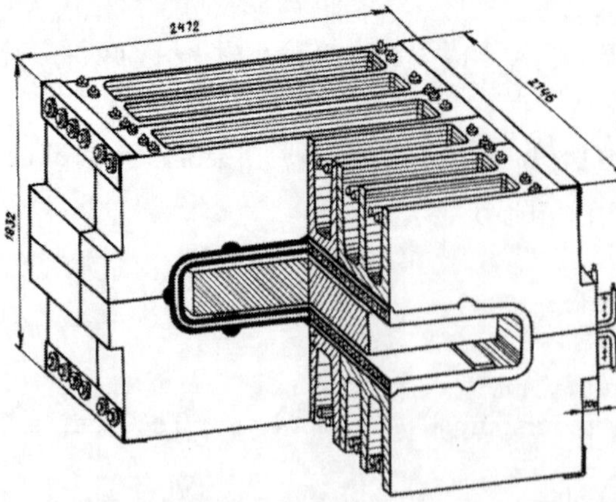

Figure 6. Inductor enclosed in a concrete cover

References

[1] Ross N V 1977 Megawatt inductor heating for rolling, forging and extrusion *World Electrotech. Congress*4A 65

[2] Howell CN 1972Slab heating at the Philadelphia Mint *IEEE Conference Rec. 7th Annu. Meet. IAEEE Ind. Appl. Soc.*pp 60–355

[3] 1982 SvensktStal to install ASEA induction heater for slabs and blooms *ASEA***5**p 135

[4] Zimin L S 1969 Induction heating bodies of rectangular shape *Electroheat***86**pp 5–24

[5] Yeghiazaryan A S andZimin L S 2010 Transverse edge effect during induction heating *Vestnik of Samara state technical University***7** pp33–231

[6] Lupi S 1979 The numerical calculation of forces in induction heating Systems*Int. Appl. Soc. IEEE: 14th Annu. Meet.*pp 31–1226

[7] Di Pieri C andLupi S 1980 Electromagnetic forces in medium frequency induction heating *UIE9: 9th Int. Congress*12

[8] Lupi S andOrefice C 1981Azonyelectrodinamichenelriscaldamentoadinduzione di piastreferromagnetiche*L'EnergiaElettrica***10** pp 44–437

[9] Zimin L S, Tikhonov I N and Lukin K A10.08.72 Shaking Table 347100 A. S. USSR №1630865 p 24

[10] GOST 12.1.003-83 Permissible levels of noise in industrial premises Introduced 01.07 84

[11] NcQueen Douglas H 1983 Noise Control of Induction Furnace*Noise Contr. Eng.* **3** pp 25–118

[12] Diehl GM 1974 *Sound and Vibrating***8**p 92

[13] Junger M andFeit D 1972 Sound, structures and their interaction *Mass. MIT Press*p 214

[14] Niordson FI 1965 On the optimal design of vibrating beam*Quart. Appl. Math.***23**pp 47–53

[15] Karihaloo BL andNiordson FI 1973 Optimum design of vibrating cantilevers*Optimization Theory and Appl.***11** pp 54–638

[16] Ross N V Apparatus for induction heating of slabs Youngstown, Ohio, assignor to Ajax Magnethermic Corporation, Warren, Ohio, a Corporation of Ohio (USA). Pat. 3485985 USA.

[17] Ross NV Induction heating Youngstown, Ohio, assignor to Ajax Magnethermic Corporation, Warren, Ohio, a Corporation of Ohio (USA). Pat. 3424886 USA.

[18] Ross N V Method and apparatus for induction heating of slabs. Youngstown, Ohio, assignor to Ajax Magnethermic Corporation, Warren, Ohio, a Corporation of Ohio (USA). Pat. 3505492. USA.

[19] Zimin L S, Getselev Z N, Astanin S D, Rapoport E Yaand Saburov V V Induction furnace. 154555 A. S. USSR №773323 Stated. 10.04.62. pub. 14.03.63. B. I. No. 10

[20] Zimin, L. S., Sutyagin A. F., hasayev L. I., Yaitskov S. A. Device for induction heating of slabs. 1669085 A. S. USSR №4425222.Publ. 07.08.91. B. I. No. 29.

Model for Sucker-Rod Pumping Unit Operating Modes Analysis Based on SimMechanics Library

A M Zyuzev[1] and M V Bubnov[1]

[1] Electric drive and automation of industrial plants department, Ural Federal University, Yekaterinburg, Russia

E-mail: a.m.zyuzev@urfu.ru and m.v.bubnov@urfu.ru

Abstract. The article provides basic information about the process of a sucker-rod pumping unit (SRPU) model developing by means of SimMechanics library in the MATLAB Simulink environment. The model is designed for the development of a pump productivity optimal management algorithms, sensorless diagnostics of the plunger pump and pumpjack, acquisition of the dynamometer card and determination of a dynamic fluid level in the well, normalization of the faulty unit operation before troubleshooting is performed by staff as well as equilibrium ratio determining by energy indicators and outputting of manual balancing recommendations to achieve optimal power consumption efficiency. Particular attention is given to the application of various blocks from SimMechanics library to take into account the pumpjack construction principal characteristic and to obtain an adequate model. The article explains in depth the developed tools features for collecting and analysis of simulated mechanism data. The conclusions were drawn about practical implementation possibility of the SRPU modelling results and areas for further development of investigation.

Keywords: SimMechanics, sucker-rod pumping unit, dynamometer card, wattmeter card, diagnostics

1. Introduction

While oil extraction by means of sucker rod pumps has been going on for a long time, and this extraction method is widely found across the world, however, research of sucker-rod pumping unit (SRPU) operating modes continues to be a big challenge. According to the existing rules of SRPU operation, the main tool of proper operational technical conditions control is dynamometry. Specialist-technologist is able to distinguish one SRPU fault or another by look of a dynamometer card. Accuracy of such diagnostics depended solely on the experience and attentiveness of the operating staff. Furthermore, another challenge is the necessity to install attached equipment (dynamograph) on the pumpjack, which is a labour-consuming procedure itself [1]. In addition, reliability of cables routing to the attached apparatus which placed at the pumpjack moving parts is principally cannot be high. Hence, the need for sensorless SRPU diagnostics tools. A correct model, which would be able to reflect the key SRPU working fault types, is needed to develop such tools.

The design of pumpjack mechanism is an oscillating crank drive of sucker-rod pumping unit. It comprises V-belt transmission, reducer and dual planar four-bar linkage. A connecting rod of the machine makes a non-linear movement, which complicates a calculating procedure of coordinates for all mechanism points at each time instant.

To facilitate the calculation of such mechanism it is better to use special application packages for mechanics modelling, in particular Simulink SimMechanics library from MATLAB software. Aside from a possibility to build of mechanic system of any complexity, further advantages for choosing such software are calculation accuracy improvement by greater account of bodies moments of inertia, possibility of choice between joints with any set of degrees of freedom, assignment different types of interaction of bodies, visualization of mechanism movements.

2. Model building

This article describes creation of the model of rear-mounted sucker-rod pumping unit SRPD8-3-5500 manufactured by the Ural Transport Machinery Plant. The overall scheme of its design is shown in figure 1.

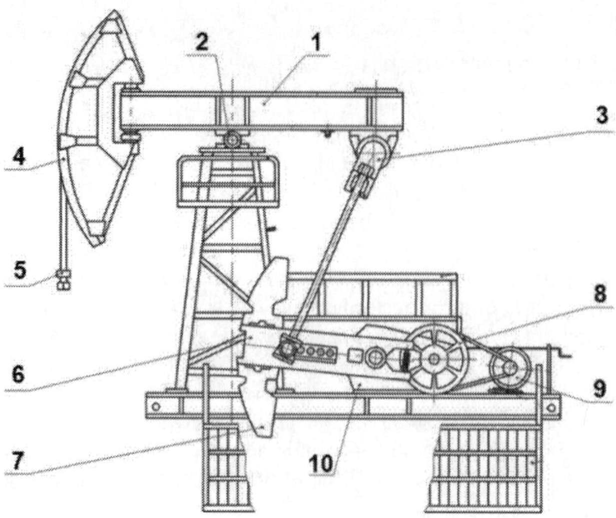

Figure 1. Overall scheme of sucker-rod pumping unit SRPD8-3-5500 design.

2.1. Frame

To simplify the model and to reduce calculation time it is sufficient to consider mechanism in only one plane. Machine model construction is performed using standard blocks from Simulink SimMechanics library. “Body” block is used to build the mechanism segments. These segments concern crank (6, figure 1), connecting rod (3, figure 1) and balance beam (1, figure 1). Section ends coordinates define the segment dimensions. Besides, coordinates of masses center, which may lie on both within and outside of the segment, should be specified. Two basic mass are considered – horse head (4, figure 1) and counterbalance (7, figure 1). The coordinate system is chosen according to conventional principle for this mechanism so the vertical axis Y is oriented through pressure and rotation point of balance beam (2, figure 1) and the horizontal axis X is oriented through crank (6, figure 1) rotation point. The pumpjack structure placed at the mentioned axes is shown graphically on figure 2.

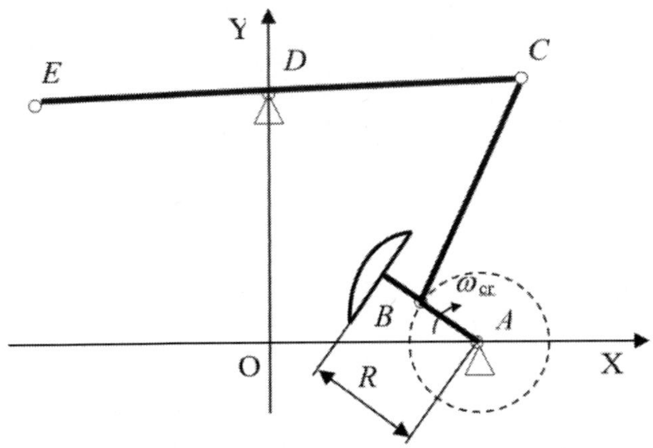

Figure 2. Design of pumpjack structure at calculation axes.

A SimMechanics model assembling begins with a "Ground" block, which is a stationary mount. "Machine Environment" block, which set out modelled environment parameters (one of which is the gravitational acceleration), is connected to the first used "Ground" block.

"Body" blocks are connected to "Ground" block and with each other by elements from "Joints" group, each with different degrees of freedom from absence in "Weld" block (strictly fixed connection) to six in "Six-DoF" block (rotational and translation for each of the three spatial axes). To set machine in motion, you should either attach "Body Actuator" block to a segment or attach "Joint Actuator" block a joint. The second option is more efficient because it ensures rotary motion around assigned axis. Blocks of pivot assembly type "Revolute" with one rotational degree of freedom were used to serve that purpose. These joints include pivot point of balance (2, figure 1), clamps of conrod (3, figure 1) and output shaft of gear reducer (10, figure 1). Z-axis, which is perpendicular to the mechanism plane, is chosen as rotational axis.

2.2. Reducer and V-belt transmission
The next model elements of no small importance are reducer and V-belt transmission (8, figure 1). They are simulated by "Gear Constraint" block which is imitation of two coupled cogwheels and its parameters is defined a desired gear ratio. At the same time, cogwheels are modelled by "Body" blocks as one point, which showcases a rotation center. In determining of their coordinates, it is essential that the distance between the rotation centers enable them to contact based on the configured "Gear Constraint" ratio. To simplify model at the present stage reducer and V-belt transmission are modelled only by common transmission ratio without giving consideration to backlash of gear and stretch of elastic transmission. The electric engine 4A200L6UZ drives the whole structure (6, figure 1). The engine model is defined based on induction motor equations written in common variables in a two-phase coordinate system, which stay fixed with respect to stator.

SimMechanics enables to display a graphic representation of a simulated object after a calculation is started which shows the visual picture of mutual location and motion of machine parts. figure 3, 4 and 5 demonstrate three mechanism conditions, which reflect all its key features: horizontal, upper dead center and bottom dead center. All other features will be considered below.

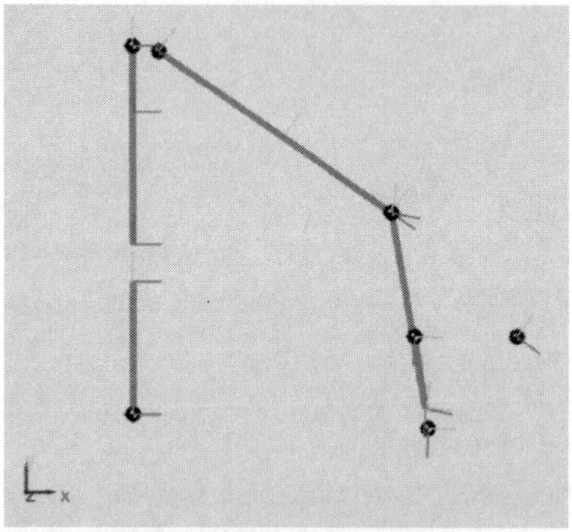

Figure 3. Horizontal state of balance beam.

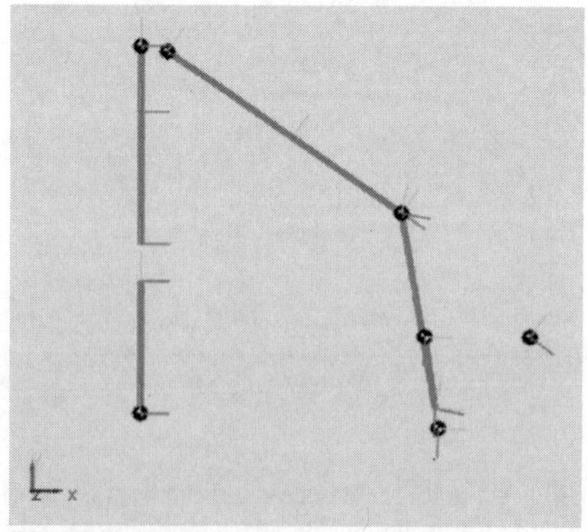

Figure 4. Upper dead center.

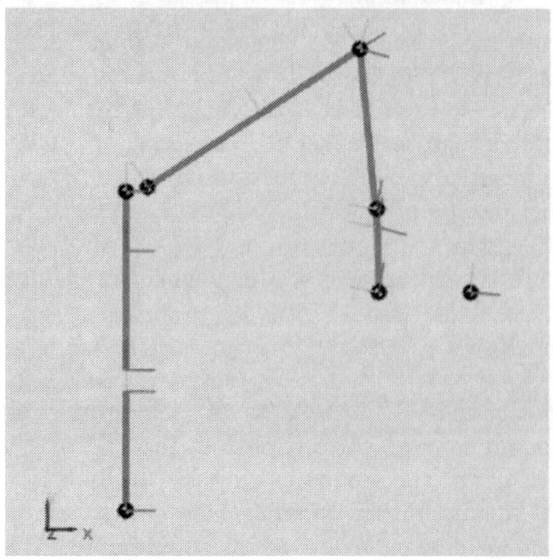

Figure 5. Bottom dead center.

2.3. Horse head

Now, let us move on to the horse head (4, figure 1) modeling that is integral part of any pumpjack construction [2]. A special form of horse head makes the polished rod hanging point (5, figure 1) moves with mechanism motion along the vertical line above well mouth, thereby avoiding skew and rubbing of the polished rod. The cambered horse head shape is circle segment with center in pivot point of balance and radius equal to the front side of balance. Therefore, balance horse head could be seen as a part of barrel, on which polished rod hanging ropes wrapped around during balance lifting and lowering. From this point of view, it is required to create flexible hanging using "Body" blocks of SimMechanics connected in chain to obtain circular arc while wrapped around a barrel. But this approach, firstly, is complicating the model, as far as a significant increase in number of elements is needed, secondly, has limited validity, because accuracy of circular arc approximation by sections depended on the sections amount. To avoid these problems, but at the same time save reliable information about a polished rod displacement another approach was embraced. Obviously, if one imagines horse head as cogwheel and polished rod hanging as pinion rack, then polished rod displacement would not be affected (if negligible elastic properties of hanging would not be taken into account) both vertically and horizontally (horizontal displacement is absent). However, standard facilities of SimMechanics does not permit to fulfil this interaction principle for blocks.

Therefore, it was decided to adopt the equivalent solution and embody pinion rack as part of gear wheel with infinite radius. Of course, wheel has finite radius in the model, but its radius much bigger then balance front arm, in our case it is thousand times bigger. One issue with this approach is that polished rod hanging point will divert from vertical line above well mouth, though slightly but it gives rise to mistakes of model compilation. Therefore, hanging is connected with the point on the big wheel and polished rod by "Revolute" joint. To ensure the strictly vertical upward and downward movement of polished rod and sucker rod, there are connected by "Custom Joint" block, where the one progressive degree of freedom along Y-axis is defined. Although horse head itself is not shown on graphic representation of mechanism, however figure 4 and 5 indicates that in extreme balance positions polished rod and sucker rod keep their vertical position.

2.4. Sucker rod and plunger pump

The next step in model preparing is the imitation of sucker rod movement and plunger pump operation. Those processes are visually characterized by dynamometer card arranged as relation between force in polished rod hanging point and position of this hanging point [3]. Representative dynamometer cards obtained from model are presented in figure 6.

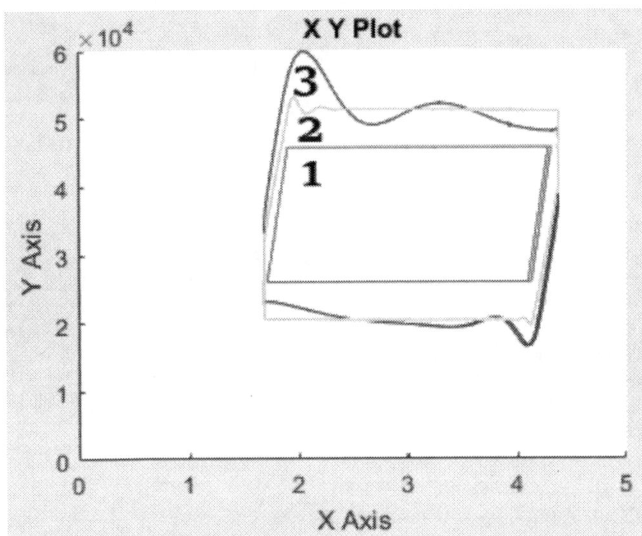

Figure 6. Representative dynamometer cards of plunger pump operation.

The submersible plunger pump operation is that the suction valve is open during upward stroke and pump receive fluid from the well, the discharge valve is open during downward stroke thus portion of fluid pushed up into the tubing string. Hence, polished rod force is equal to the sucker rod weight in fluid at the downward stroke and taken on additional weight of the liquid column above fluid surface in the well so-called dynamic fluid level at the upward stroke. Thus, polished rod force comprise sucker rod weight in fluid depending on the pump depths and liquid column weight depending on the dynamic fluid level.

Without taking into account features of valve group operation such as a sealing, incomplete closing and so on such load can be represented in model by variable mass which value is changed at the time of dead center passage and determined by specified pump depths and dynamic fluid level. That is why deep dynamometer card near pump plunger during normal operation under the assumptions is represented by rectangle whose height is determined by dynamic fluid level and width is determined by polished rod stroke length. From another observation point on the polished rod hanging, we see a rather different picture through expansion and contraction of the sucker rod during upward and downward stroke force in this point steadily increasing and decreasing, which is reflected in the inclined sections of dynamometer card (1, figure 6). A dynamometer card of static pump operation referred to as theoretical, because it does not take into account the dynamic and friction forces but it is a very important tool in dynamometry, which is able to evaluate a real dynamometer card.

To simulate action of elastic forces in sucker rod "Joint Spring & Damper" block is used, which is connected with "Custom Joint" block. Block parameters includes three positions: elastic modulus, damping coefficient and initial distance setpoint in not deformed condition (which is equal to zero if parts coupled).

Additionally, it is necessary to model the friction force arising between gland packing and polished rod. It is static friction in nature, which changes its sign during its passage through dead center. Herewith polished rod do not start a movement until tractive force is less than static friction that cause vertical section on the dynamometer card. Thereby specified conditions allow getting shape of dynamometer card shown on figure 6: on the full motor speed (3) and on decreased four times speed (2). Overall look of the obtained pumpjack model in SimMechanics is shown on figure 7.

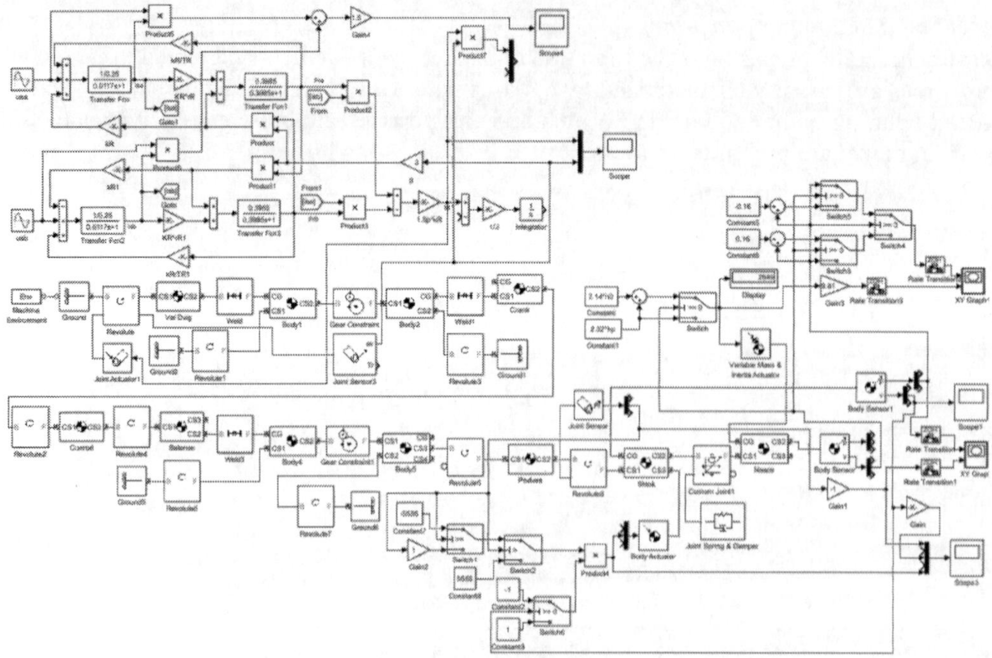

Figure 7. Overall look of pumpjack model in SimMechanics.

3. Simulation results

Simulation results has already partly been represented above since they were necessary for explanation of model structure. The most important and essential, of course, is dynamometry (figure 6), because it is general method of sucker-rod pumping unit diagnostics. Another important diagram of SRPU operation is wattmeter card (figure 8) of drive motor (relation between consuming electric power). Its advantage in respect to dynamometer card for diagnostics purposes is that wattmeter card might be obtained by standard tools without additional attached equipment and sensors [4]. However, it is more complicated for analytic processing and it is required development of algorithmic base. Such parameters as torque and speed of drive motor also available in the model (figure 9).

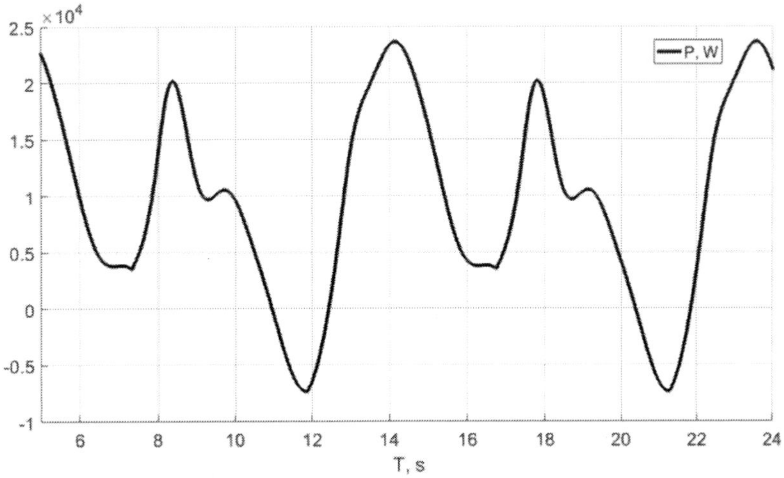

Figure 8. Wattmeter card of SRPU motor.

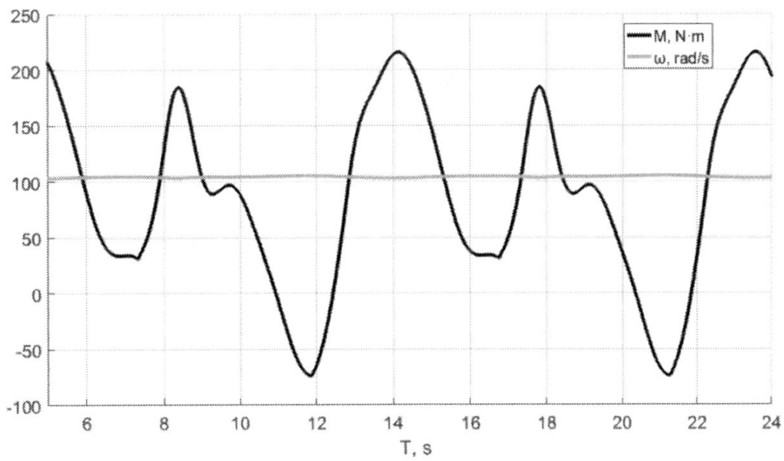

Figure 9. Torque and speed of SRPU motor.

4. Conclusion

Hence rather detailed model of sucker-rod pumping unit, which is satisfied to equipment diagnostics purposes, is developed. It served to simulate different typical faults of equipment to develop means and algorithms of revealing such faults without using of special sensors and external dynamometer systems. Several objectives of sensorless data acquisition such as dynamometer card, dynamic fluid level and downhole pressure are still promising for diagnostics and automated control purposes. Further efforts will be focused on design and improvement of model along these lines.

References

[1] Lindh T, Montonen J-H, Grachev M, and Niemelä M 2015 Generating surface dynamometer cards for a sucker-rod pump by using frequency converter estimates and a process identification run *POWERENG* (Riga: IEEE) pp 416-420

[2] Fozao K F, M. Lissouck M, Lontsi F, Ngasa A and Mbanda N 2015 Hands-on model of sucker rod pumping facility for oil well production *Journal of Petroleum and Gas Engineering* vol 6(4) pp 45-53

[3] Khakimyanov M I and Khusainov F F 2016 The information processing of dynamometer cards at controllers of automation power drives sucker rod pumps *ICPDS* (Perm: IEEE) pp 1-6

[4] Khakimyanov M I, Shafikov I N and Khusainov F F 2015 Control of sucker rod pumps energy consumption *SIBCON* (Omsk: IEEE) pp 1-4

AUTHOR INDEX

AUTHOR INDEX

AUTHOR INDEX

AUTHOR INDEX